PLANT DESIGN AND ECONOMICS FOR CHEMICAL ENGINEERS

McGraw-Hill Chemical Engineering Series

Editorial Advisory Board

BUILDING THE LITERATURE OF A PROFESSION

Fifteen prominent chemical engineers first met in New York more than 50 years ago to plan a continuing literature for their rapidly growing profession. From industry came such pioneer practitioners as Leo H. Baekeland, Arthur D. Little, Charles L. Reese, John V. N. Dorr, M. C. Whitaker, and R. S. McBride. From the universities came such eminent educators as William H. Walker, Alfred H. White, D. D. Jackson, J. H. James, Warren K. Lewis, and Harry A. Curtis. H. C. Parmelee, then editor of *Chemical and Metallurgical Engineering*, served as chairman and was joined subsequently by S. D. Kirkpatrick as consulting editor.

After several meetings, this committee submitted its report to the McGraw-Hill Book Company in September 1925. In the report were detailed specifications for a correlated series of more than a dozen texts and reference books which have since become the McGraw-Hill Series in Chemical Engineering and which became the cornerstone of the chemical engineering curriculum.

From this beginning there has evolved a series of texts surpassing by far the scope and longevity envisioned by the founding Editorial Board. The McGraw-Hill Series in Chemical Engineering stands as a unique historical record of the development of chemical engineering education and practice. In the series one finds the milestones of the subject's evolution: industrial chemistry, stoichiometry, unit operations and processes, thermodynamics, kinetics, and transfer operations.

Chemical engineering is a dynamic profession, and its literature continues to evolve. McGraw-Hill and its consulting editors remain committed to a publishing policy that will serve, and indeed lead, the needs of the chemical engineering profession during the years to come.

The Series
Bailey and Ollis: *Biochemical Engineering Fundamentals*
Bennett and Myers: *Momentum, Heat, and Mass Transfer*
Beveridge and Schechter: *Optimization: Theory and Practice*
Carberry: *Chemical and Catalytic Reaction Engineering*
Churchill: *The Interpretation and Use of Rate Data—The Rate Concept*
Clarke and Davidson: *Manual for Process Engineering Calculations*
Coughanowr and Koppel: *Process Systems Analysis and Control*
Danckwerts: *Gas Liquid Reactions*
Gates, Katzer, and Schuit: *Chemistry of Catalytic Processes*
Harriott: *Process Control*
Johnson: *Automatic Process Control*
Johnstone and Thring: *Pilot Plants, Models, and Scale-up Methods in Chemical Engineering*
Katz, Cornell, Kobayashi, Poettmann, Vary, Elenbaas, and Weinaug: *Handbook of Natural Gas Engineering*
King: *Separation Processes*
Knudsen and Katz: *Fluid Dynamics and Heat Transfer*
Lapidus: *Digital Computation for Chemical Engineers*
Luyben: *Process Modeling, Simulation, and Control for Chemical Engineers*
McCabe and Smith, J. C.: *Unit Operations of Chemical Engineering*
Mickley, Sherwood, and Reed: *Applied Mathematics in Chemical Engineering*
Nelson: *Petroleum Refinery Engineering*
Perry and Chilton (Editors): *Chemical Engineers' Handbook*
Peters: *Elementary Chemical Engineering*
Peters and Timmerhaus: *Plant Design and Economics for Chemical Engineers*
Reed and Gubbins: *Applied Statistical Mechanics*
Reid, Prausnitz, and Sherwood: *The Properties of Gases and Liquids*
Satterfield: *Heterogeneous Catalysis in Practice*
Sherwood, Pigford, and Wilke: *Mass Transfer*
Slattery: *Momentum, Energy, and Mass Transfer in Continua*
Smith, B. D.: *Design of Equilibrium Stage Processes*
Smith, J. M.: *Chemical Engineering Kinetics*
Smith, J. M., and Van Ness: *Introduction to Chemical Engineering Thermodynamics*
Thompson and Ceckler: *Introduction to Chemical Engineering*
Treybal: *Liquid Extraction*
Treybal: *Mass Transfer Operations*
Van Winkle: *Distillation*
Volk: *Applied Statistics for Engineers*
Walas: *Reaction Kinetics for Chemical Engineers*
Wei, Russell, and Swartzlander: *The Structure of the Chemical Processing Industries*
Whitwell and Toner: *Conservation of Mass and Energy*

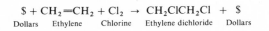

$$\$ + CH_2{=}CH_2 + Cl_2 \rightarrow CH_2ClCH_2Cl + \$$$

Dollars   Ethylene   Chlorine   Ethylene dichloride   Dollars

The complete plant—the complete economic process. Here is the design engineer's goal. (*C. F. Braun and Co.*)

PLANT DESIGN AND ECONOMICS FOR CHEMICAL ENGINEERS

Third Edition

Max S. Peters
Klaus D. Timmerhaus

Professors of Chemical Engineering
University of Colorado

McGraw-Hill Book Company

New York St. Louis San Francisco Auckland Bogotá Düsseldorf
Johannesburg London Madrid Mexico Montreal New Delhi
Panama Paris São Paulo Singapore Sydney Tokyo Toronto

Library of Congress Cataloging in Publication Data

Peters, Max Stone, date
    Plant design and economics for chemical engineers.

    (McGraw-Hill chemical engineering series)
    Bibliography: p.
    Includes indexes.
    1. Chemical plants—Design and construction.
I. Timmerhaus, Klaus D., joint author. II. Title.
TP155.5.P4 1980 660'.28   79-12841
ISBN 0-07-049582-3

PLANT DESIGN AND ECONOMICS FOR CHEMICAL ENGINEERS

1 2 3 4 5 6 7 8 9 0 D O D O 8 9 8 7 6 5 4 3 2 1 0

This book was set in Times Roman. The editor was Julienne V. Brown; the cover was
designed by Rafael Hernandez; the production supervisor was Donna Piligra.
R. R. Donnelley & Sons Company was printer and binder.

CONTENTS

PREFACE

Advances in the level of understanding of chemical engineering principles, combined with the availability of new tools and new techniques, have led to an increased degree of sophistication which can now be applied to the design of industrial chemical operations. This third edition takes advantage of the widened spectrum of chemical engineering knowledge by the inclusion of considerable material on profitability evaluation, optimum design methods, continuous interest compounding, statistical analyses, cost estimation, and methods for problem solution including use of computers. Special emphasis is placed on the economic and engineering principles involved in the design of chemical plants and equipment. An understanding of these principles is a prerequisite for any successful chemical engineer, no matter whether the final position is in direct design work or in production, administration, sales, research, development, or any other related field.

The expression *plant design* immediately connotes industrial applications; consequently, the dollar sign must always be kept in mind when carrying out the design of a plant. The theoretical and practical aspects are important, of course; but, in the final analysis, the answer to the question "Will we realize a profit from this venture?" almost always determines the true value of the design. The chemical engineer, therefore, should consider plant design and applied economics as one combined subject.

The purpose of this book is to present economic and design principles as applied in chemical engineering processes and operations. No attempt is made to train the reader as a skilled economist, and, obviously, it would be impossible to present all the possible ramifications involved in the multitude of different plant designs. Instead, the goal has been to give a clear concept of the important principles and general methods. The subject matter and manner of presentation are such that the book should be of value to advanced chemical engineering undergraduates, graduate students, and practicing engineers. The information should also be of interest to administrators, operation supervisors, and research or development workers in the process industries.

The first part of the text presents an overall analysis of the major factors involved in process design, with particular emphasis on economics in the process industries and in design work. The various costs involved in industrial processes, capital investments and investment returns, cost estimation, cost accounting, optimum economic design methods, and other subjects dealing with economics are covered both qualitatively and quantitatively. The remainder of the text deals with methods and important factors in the design of plants and equipment. Generalized subjects, such as waste disposal, structural design, and equipment fabrication, are included along with design methods for different types of process equipment. Basic cost data and cost correlations are also presented for use in making cost estimates.

Illustrative examples and sample problems are used extensively in the text to illustrate the applications of the principles to practical situations. Problems are included at the ends of most of the chapters to give the reader a chance to test the understanding of the material. Practice-session problems, as well as longer design problems of varying degrees of complexity, are included in Appendix C.

Suggested additional references are presented at the end of most of the chapters to show the reader where added information can be obtained. It should be noted that most of these references are taken from the chemical engineering literature, and these references should be considered merely as suggestions and not as an all-inclusive listing. In general, only references from 1967 or later are included at the ends of the chapters. Earlier references are listed in the first and second editions of this book.

A large amount of cost data is presented in tabular and graphical form. The table of contents for the book includes a condensed listing of the cost data presented in the various figures, and additional cost information on specific items of equipment or operating factors can be located by reference to the subject index. To simplify use of the extensive cost data given in this book, all cost figures are updated by means of the all-industry Marshall and Swift cost index of 561 applicable for January 1, 1979. Because exact prices can be obtained only by direct quotations from manufacturers, caution should be exercised in the use of the data for other than approximate cost-estimation purposes.

The book would be suitable for use in a one- or two-semester course for advanced undergraduate or graduate chemical engineers. It is assumed that the reader has a background in stoichiometry, thermodynamics, and chemical engineering principles as taught in normal first-degree programs in chemical engineering. Detailed explanations of the development of various design equations and methods are presented. The book provides a background of design and economic information with a large amount of quantitative interpretation so that it can serve as a basis for further study to develop complete understanding of the general strategy of process engineering design.

Although nomographs, simplified equations, and shortcut methods are included, every effort has been made to indicate the theoretical background and assumptions for these relationships. The true value of plant design and economics for the chemical engineer is not found merely in the ability to put numbers in an equation and solve for a final answer. The true value is found in obtaining an understanding of the reasons *why* a given calculation method gives a satisfactory result. This understanding gives the engineer the confidence and ability necessary to proceed when new problems are encountered for which there are no predetermined methods of solution. Thus, throughout the study of plant design and economics, the engineer should always attempt to understand the assumptions and theoretical factors involved in the various calculation procedures and never fall into the habit of robot-like number plugging.

Because applied economics and plant design deal with practical applications of chemical engineering principles, a study of these subjects offers an ideal way for tying together the entire field of chemical engineering. The final result of a plant design may be expressed in dollars and cents, but this result can only be achieved through the application of various theoretical principles combined with industrial and practical knowledge. Both theory and practice are emphasized in this book, and aspects of all phases of chemical engineering are included.

The authors are indebted to the many industrial firms and individuals who have supplied information and comments on the material presented in this edition. The authors also express their appreciation to the many persons who have, on the basis of use of earlier editions, supplied constructive criticism and helpful suggestions on the presentation for this edition.

Max S. Peters
Klaus D. Timmerhaus

PROLOGUE

THE INTERNATIONAL SYSTEM OF UNITS (SI)

As the United States moves toward acceptance of the International System of Units, or the so-called SI units, it is particularly important for the design engineer to be able to think in both the SI units and the U.S. customary units. From an international viewpoint, the United States is the last major country to accept SI, but it will be many years before the U.S. conversion will be sufficiently complete for the design engineer, who must deal with the general public, to think and write solely in SI units. For this reason, a mixture of SI and U.S. customary units will be found in this text.

For those readers who are not familiar with all the rules and conversions for SI units, Appendix A of this text presents the necessary information. This appendix gives descriptive and background information for the SI units along with a detailed set of rules for SI usage and lists of conversion factors presented in various forms which should be of special value for chemical engineering usage.

Chemical engineers in design must be totally familiar with SI and its rules. Reading of Appendix A is recommended for those readers who have not worked closely and extensively with SI.

PLANT DESIGN AND ECONOMICS
FOR CHEMICAL ENGINEERS

INTRODUCTION

In this modern age of industrial competition, a successful chemical engineer needs more than a knowledge and understanding of the fundamental sciences and the related engineering subjects such as thermodynamics, reaction kinetics, and computer technology. The engineer must also have the ability to apply this knowledge to practical situations for the purpose of accomplishing something that will be beneficial to society. However, in making these applications, the chemical engineer must recognize the economic implications which are involved and proceed accordingly.

Chemical engineering design of new chemical plants and the expansion or revision of existing ones require the use of engineering principles and theories combined with a practical realization of the limits imposed by industrial conditions. Development of a new plant or process from concept evaluation to profitable reality is often an enormously complex problem. A plant-design project moves to completion through a series of stages such as is shown in the following:

1. Inception
2. Preliminary evaluation of economics and market
3. Development of data necessary for final design
4. Final economic evaluation
5. Detailed engineering design
6. Procurement
7. Erection
8. Startup and trial runs
9. Production

This brief outline suggests that the plant-design project involves a wide variety of skills. Among these are research, market analysis, design of individual pieces of equipment, cost estimation, computer programming, and plant-location surveys. In fact, the services of a chemical engineer are needed in each step of the outline, either in a central creative role, or as a key advisor.

CHEMICAL ENGINEERING PLANT DESIGN

As used in this text, the general term *plant design* includes all engineering aspects involved in the development of either a new, modified, or expanded industrial plant. In this development, the chemical engineer will be making economic evaluations of new processes, designing individual pieces of equipment for the proposed new venture, or developing a plant layout for coordination of the overall operation. Because of these many design duties, the chemical engineer is many times referred to here as a *design engineer*. On the other hand, a chemical engineer specializing in the economic aspects of the design is often referred to as a *cost engineer*. In many instances, the term *process engineering* is used in connection with economic evaluation and general economic analyses of industrial processes, while *process design* refers to the actual design of the equipment and facilities necessary for carrying out the process. Similarly, the meaning of plant design is limited by some engineers to items related directly to the complete plant, such as plant layout, general service facilities, and plant location.

The purpose of this book is to present the major aspects of plant design as related to the overall design project. Although one person cannot be an expert in *all* the phases involved in plant design, it is necessary to be acquainted with the general problems and approach in each of the phases. The process engineer may not be connected directly with the final detailed design of the equipment, and the designer of the equipment may have little influence on a decision by management as to whether or not a given return on an investment is adequate to justify construction of a complete plant. Nevertheless, if the overall design project is to be successful, close teamwork is necessary among the various groups of engineers working on the different phases of the project. The most effective teamwork and coordination of efforts are obtained when each of the engineers in the specialized groups is aware of the many functions in the *overall* design project.

PROCESS DESIGN DEVELOPMENT

The development of a process design, as outlined in Chap. 2, involves many different steps. The first, of course, must be the inception of the basic idea. This idea may originate in the sales department, as a result of a customer request, or to meet a competing product. It may occur spontaneously to someone who is acquainted with the aims and needs of a particular company, or it may be the result of an orderly research program or an offshoot of such a program. The operating division of the company may develop a new or modified chemical, generally as an intermediate in the final product. The engineering department of the company may originate a new process or modify an existing process to create new products. In all these possibilities, if the initial analysis indicates that the idea may have possibilities of developing into a worthwhile project, a preliminary research or investigation program is initiated. Here, a general survey of the possibilities for a successful process is made considering the physical and chemical operations involved as well as the economic aspects. Next comes the process-research phase including preliminary market surveys, laboratory-scale experiments, and production of research samples of the final product. When the potentialities of the process are fairly well established, the project is ready for the development phase. At this point, a pilot plant or a commercial-development plant may be constructed. A pilot plant is a small-scale replica of the full-scale final plant, while a commercial-development plant is usually made from odd pieces of equipment which are already available and is not meant to duplicate the exact setup to be used in the full-scale plant.

Design data and other process information are obtained during the development stage. This information is used as the basis for carrying out the additional phases of the design project. A complete market analysis is made, and samples of the final product are sent to prospective customers to determine if the product is satisfactory and if there is a reasonable sales potential. Capital-cost estimates for the proposed plant are made. Probable returns on the required investment are determined, and a complete cost-and-profit analysis of the process is developed.

Before the final process design starts, company management normally becomes involved to decide if significant capital funds will be committed to the project. It is at this point that the engineers' preliminary design work along with the oral and written reports which are presented become particularly important because they will provide the primary basis on which management will decide if further funds should be provided for the project. When management has made a firm decision to proceed with provision of significant capital funds for a project, the engineering then involved in further work on the project is known as *capitalized engineering* while that which has gone on before while the consideration of the project was in the development stage is often referred to as *expensed engineering*. This distinction is used for tax purposes to allow capitalized engineering costs to be amortized over a period of several years.

If the economic picture is still satisfactory, the final process-design phase is ready to begin. All the design details are worked out in this phase including controls, services, piping layouts, firm price quotations, specifications and designs for individual pieces of equipment, and all the other design information necessary for the construction of the final plant. A complete construction design is then made with elevation drawings, plant-layout arrangements, and other information required for the actual construction of the plant. The final stage consists of procurement of the equipment, construction of the plant, startup of the plant, overall improvements in the operation, and development of standard operating procedures to give the best possible results.

The development of a design project proceeds in a logical, organized sequence requiring more and more time, effort, and expenditure as one phase leads into the next. It is extremely important, therefore, to stop and analyze the situation carefully before proceeding with each subsequent phase. Many projects are discarded as soon as the preliminary investigation or research on the original idea is completed. The engineer working on the project must maintain a realistic and practical attitude in advancing through the various stages of a design project and not be swayed by personal interests and desires when deciding if further work on a particular project is justifiable. Remember, if the engineer's work is continued on through the various phases of a design project, it will eventually end up in a proposal that money be invested in the process. If no tangible return can be realized from the investment, the proposal will be turned down. Therefore, the engineer should have the ability to eliminate unprofitable ventures before the design project approaches a final-proposal stage.

GENERAL OVERALL DESIGN CONSIDERATIONS

The development of the overall design project involves many different design considerations. Failure to include these considerations in the overall design project may, in many instances, alter the entire economic situation so drastically as to make the venture unprofitable. Some of the factors involved in the development of a complete plant design include plant location, plant layout, materials of construction, structural design, utilities, buildings, storage, materials handling, safety, waste disposal, federal, state, and local laws or codes, and patents. Because of their importance, these general overall design considerations are considered in detail in Chap. 3.

Record keeping and accounting procedures are also important factors in general design considerations, and it is necessary that the design engineer be familiar with the general terminology and approach used by accountants for cost and asset accounting. This subject is covered in Chap. 4.

COST ESTIMATION

As soon as the final process-design stage is completed, it becomes possible to make accurate cost estimations because detailed equipment specifications and definite plant-facility information are available. Direct price quotations based on detailed specifications can then be obtained from various manufacturers. However, as mentioned earlier, no design project should proceed to the final stages before costs are considered, and cost estimates should be made throughout all the early stages of the design when complete specifications are not available. Evaluation of costs in the preliminary design phases is sometimes called "guesstimation" but the appropriate designation is *predesign cost estimation*. Such estimates should be capable of providing a basis for company management to decide if further capital should be invested in the project.

The chemical engineer (or cost engineer) must be certain to consider all possible factors when making a cost analysis. Fixed costs, direct production costs for raw materials, labor, maintenance, power, and utilities must all be included along with costs for plant and administrative overhead, distribution of the final products, and other miscellaneous items.

Chapter 5 presents many of the special techniques that have been developed for making predesign cost estimations. Labor and material indexes, standard cost ratios, and special multiplication factors are examples of information used when making design estimates of costs. The final test as to the validity of any cost estimation can come only when the completed plant has been put into operation. However, if the design engineer is well acquainted with the various estimation methods and their accuracy, it is possible to make remarkably close cost estimations even before the final process design gives detailed specifications.

FACTORS AFFECTING PROFITABILITY OF INVESTMENTS

A major function of the directors of a manufacturing firm is to maximize the long-term profit to the owners or the stockholders. A decision to invest in fixed facilities carries with it the burden of continuing interest, insurance, taxes, depreciation, manufacturing costs, etc., and also reduces the fluidity of the company's future actions. Capital-investment decisions, therefore, must be made with great care. Chapters 6 and 9 present guidelines for making these capital-investment decisions.

Money, or any other negotiable type of capital, has a time value. When a manufacturing enterprise invests money, it expects to receive a return during the time the money is being used. The amount of return demanded usually depends on the degree of risk that is assumed. Risks differ between projects which might otherwise seem equal on the basis of the best estimates of an overall plant design. The risk may depend upon the process used, whether it is well established or a complete innovation; on the product to be made, whether it is a staple item or a completely new product; on the sales forecasts, whether all sales will be outside the company or whether a significant fraction is internal, etc. Since means for incorporating different levels of risk into *profitability* forecasts are not too well established, the most common methods are to raise the minimum acceptable *rate of return* for the riskier projects.

Time value of money has been integrated into investment-evaluation systems by means of *compound-interest* relationships. Dollars, at different times, are given different degrees of importance by means of compounding or discounting at some preselected compound-interest rate. For any assumed interest value of money, a known amount at any one time can be converted to an equivalent but different amount at a different time. As time passes, money can be invested to increase at the interest rate. If the time when money is needed for investment is in the future, the present value of that investment can be calculated by discounting from the time of investment back to the present at the assumed interest rate.

Expenses, as outlined in Chap. 7, for various types of taxes and insurance can materially affect the economic situation for any industrial process. Because modern taxes may amount to a major portion of a manufacturing firm's net earnings, it is essential that the chemical engineer be conversant with the fundamentals of taxation. For example, income taxes apply differently to projects with different proportions of fixed and working capital. Profitability, therefore, should be based on income after taxes. Insurance costs, on the other hand, are normally only a small part of the total operational expenditure of an industrial enterprise; however, before any operation can be carried out on a sound economic basis, it is necessary to determine the insurance requirements to provide adequate coverage against unpredictable emergencies or developments.

Since all physical assets of an industrial facility decrease in value with age, it is normal practice to make periodic charges against earnings so as to distribute the first cost of the facility over its expected service life. This *depreciation* expense as detailed in Chap. 8, unlike most other expenses, entails no current outlay of cash. Thus in a given accounting period, a firm has available, in addition to the net profit, additional funds corresponding to the depreciation expense. This cash is *capital recovery*, a partial regeneration of the first cost of the physical assets.

Income-tax laws permit recovery of funds by two accelerated depreciation schedules as well as by straight-line methods. Since cash-flow timing is affected, choice of depreciation method affects profitability significantly. Depending on the ratio of depreciable to nondepreciable assets involved, two projects which look equivalent before taxes, or rank in one order, may rank entirely differently when considered after taxes. Though cash costs and sales values may be equal on two projects, their reported net incomes for tax purposes may be different, and one will show a greater net profit than the other.

OPTIMUM DESIGN

In almost every case encountered by a chemical engineer, there are several alternative methods which can be used for any given process or operation. For example, formaldehyde can be produced by catalytic dehydrogenation of methanol, by controlled oxidation of natural gas, or by direct reaction between CO and H_2 under special conditions of catalyst, temperature, and pressure. Each of these processes contains many possible alternatives involving variables such as gas-mixture composition, temperature, pressure, and choice of catalyst. It is the responsibility of the chemical engineer, in this case, to choose the best process and to incorporate into the design the equipment and methods which will give the best results. To meet this need, various aspects of chemical engineering plant-design optimization are described in Chap. 10 including presentation of design strategies which can be used to establish the desired results in the most efficient manner.

Optimum Economic Design

If there are two or more methods for obtaining exactly equivalent final results, the preferred method would be the one involving the least total cost. This is the basis of an *optimum economic design*. One typical example of an optimum economic design is determining the pipe diameter to use when pumping a given amount of fluid from one point to another. Here the same final result (i.e., a set amount of fluid pumped between two given points) can be accomplished by using an infinite number of different pipe diameters. However, an economic balance will show that one particular pipe diameter gives the least total cost. The total cost includes the cost for pumping the liquid and the cost (i.e., fixed charges) for the installed piping system.

A graphical representation showing the meaning of an optimum economic pipe diameter is presented in Fig. 1-1. As shown in this figure, the pumping cost increases with decreased size of pipe diameter because of frictional effects, while the fixed charges for the pipeline become lower when smaller pipe diameters are used because of the reduced capital investment. The optimum economic diameter is located where the sum of the pumping costs and fixed costs for the pipeline becomes a minimum, since this represents the point of least total cost. In Fig. 1-1, this point is represented by *E*.

The chemical engineer often selects a final design on the basis of conditions

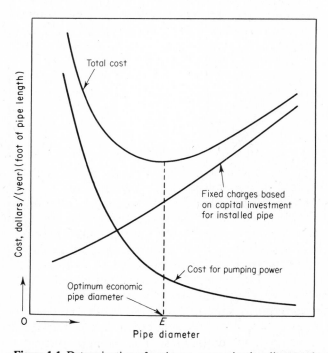

Figure 1-1 Determination of optimum economic pipe diameter for constant mass-throughput rate.

giving the least total cost. In many cases, however, alternative designs do not give final products or results that are exactly equivalent. It then becomes necessary to consider the quality of the product or the operation as well as the total cost. When the engineer speaks of an optimum economic design, it ordinarily means the cheapest one selected from a number of equivalent designs. Cost data, to assist in making these decisions, are presented in Chaps. 13 through 15.

Various types of optimum economic requirements may be encountered in design work. For example, it may be desirable to choose a design which gives the maximum profit per unit of time or the minimum total cost per unit of production.

Optimum Operation Design

Many processes require definite conditions of temperature, pressure, contact time, or other variables if the best results are to be obtained. It is often possible to make a partial separation of these optimum conditions from direct economic considerations. In cases of this type, the best design is designated as the *optimum operation design*. The chemical engineer should remember, however, that economic considerations ultimately determine most quantitative decisions. Thus, the optimum operation design is usually merely a tool or step in the development of an optimum economic design.

An excellent example of an optimum operation design is the determination of operating conditions for the catalytic oxidation of sulfur dioxide to sulfur

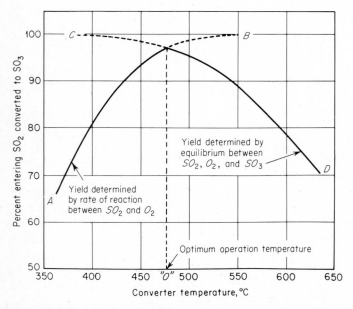

Figure 1-2 Determination of optimum operation temperature in sulfur dioxide converter.

trioxide. Suppose that all the variables, such as converter size, gas rate, catalyst activity, and entering-gas concentration, are fixed and the only possible variable is the temperature at which the oxidation occurs. If the temperature is too high, the yield of SO_3 will be low because the equilibrium between SO_3, SO_2, and O_2 is shifted in the direction of SO_2 and O_2. On the other hand, if the temperature is too low, the yield will be poor because the reaction rate between SO_2 and O_2 will be low. Thus, there must be one temperature where the amount of sulfur trioxide formed will be a maximum. This particular temperature would give the optimum operation design. Figure 1-2 presents a graphical method for determining the optimum operation temperature for the sulfur dioxide converter in this example. Line *AB* represents the maximum yields obtainable when the reaction rate is controlling, while line *CD* indicates the maximum yields on the basis of equilibrium conditions controlling. Point *O* represents the optimum operation temperature where the maximum yield is obtained.

The preceding example is a simplified case of what an engineer might encounter in a design. In reality, it would usually be necessary to consider various converter sizes and operation with a series of different temperatures in order to arrive at the optimum operation design. Under these conditions, several equivalent designs would apply, and the final decision would be based on the optimum economic conditions for the equivalent designs.

PRACTICAL CONSIDERATIONS IN DESIGN

The chemical engineer must never lose sight of the practical limitations involved in a design. It may be possible to determine an exact pipe diameter for an optimum economic design, but this does not mean that this exact size must be used in the final design. Suppose the optimum diameter were 3.43 in. (8.71 cm). It would be impractical to have a special pipe fabricated with an inside diameter of 3.43 in. Instead, the engineer would choose a standard pipe size which could be purchased at regular market prices. In this case, the recommended pipe size would probably be a *standard* $3\frac{1}{2}$-in.-diameter pipe having an inside diameter of 3.55 in. (9.02 cm).

If the engineer happened to be very conscientious about getting an adequate return on all investments, he or she might say, "A standard 3-in.-diameter pipe would require less investment and would probably only increase the total cost slightly; therefore, I think we should compare the costs with a 3-in. pipe to the costs with the $3\frac{1}{2}$-in. pipe before making a final decision." Theoretically, the conscientious engineer is correct in this case. Suppose the total cost of the installed $3\frac{1}{2}$-in. pipe is $5000 and the total cost of the installed 3-in. pipe is $4500. If the total yearly savings on power and fixed charges, using the $3\frac{1}{2}$-in. pipe instead of the 3-in. pipe, were $25, the yearly percent return on the extra $500 investment would be only 5 percent. Since it should be possible to invest the extra $500 elsewhere to give more than a 5 percent return, it would appear that the 3-in.-diameter pipe would be preferred over the $3\frac{1}{2}$-in.-diameter pipe.

The logic presented in the preceding example is perfectly sound. It is a typical example of investment comparison and should be understood by all chemical engineers. Even though the optimum economic diameter was 3.43 in., the good engineer knows that this diameter is only an exact mathematical number and may vary from month to month as prices or operating conditions change. Therefore, all one expects to obtain from this particular optimum economic calculation is a good estimation as to the best diameter, and investment comparisons may not be necessary.

The practical engineer understands the physical problems which are involved in the final operation and maintenance of the designed equipment. In developing the plant layout, crucial control valves must be placed where they are easily accessible to the operators. Sufficient space must be available for maintenance personnel to check, take apart, and repair equipment. The engineer should realize that cleaning operations are simplified if a scale-forming fluid is passed through the inside of the tubes rather than on the shell side of a tube-and-shell heat exchanger. Obviously, then, sufficient plant-layout space should be made available so that the maintenance workers can remove the head of the installed exchanger and force cleaning worms or brushes through the inside of the tubes or remove the entire tube bundle when necessary.

The theoretical design of a distillation unit may indicate that the feed should be introduced on one particular tray in the tower. Instead of specifying a tower with only one feed inlet on the calculated tray, the practical engineer will include inlets on several trays above and below the calculated feed point since the actual operating conditions for the tower will vary and the assumptions included in the calculations make it impossible to guarantee absolute accuracy.

The preceding examples typify the type of practical problems the chemical engineer encounters. In design work, theoretical and economic principles must be combined with an understanding of the common practical problems that will arise when the process finally comes to life in the form of a complete plant or a complete unit.

THE DESIGN APPROACH

The chemical engineer has many tools to choose from in the development of a profitable plant design. None, when properly utilized, will probably contribute as much to the optimization of the design as the use of high-speed computers. Many problems encountered in the process development and design can be solved rapidly with a higher degree of completeness with high-speed computers and at less cost than with ordinary hand or desk calculators. Generally overdesign and safety factors can be reduced with a substantial savings in capital investment.

At no time, however, should the engineer be led to believe that plants are designed around computers. They are used to determine design data and are used as models for optimization once a design is established. They are also used to maintain operating plants on the desired operating conditions. The latter function is a part of design and supplements and follows process design.

The general approach in any plant design involves a carefully balanced combination of theory, practice, originality, and plain common sense. In original design work, the engineer must deal with many different types of experimental and empirical data. The engineer may be able to obtain accurate values of heat capacity, density, vapor-liquid equilibrium data, or other information on physical properties from the literature. In many cases, however, exact values for necessary physical properties are not available, and the engineer is forced to make approximate estimates of these values. Many approximations also must be made in carrying out theoretical design calculations. For example, even though the engineer knows that the ideal-gas law applies exactly only to simple gases at very low pressures, this law is used in many of the calculations when the gas pressure is as high as 5 or more atmospheres (507 kPa). With common gases, such as air or simple hydrocarbons, the error introduced by using the ideal gas law at ordinary pressures and temperatures is usually negligible in comparison with other uncertainties involved in design calculations. The engineer prefers to accept this error rather than to spend time determining virial coefficients or other factors to correct for ideal gas deviations.

In the engineer's approach to any design problem, it is necessary to be prepared to make many assumptions. Sometimes these assumptions are made because no absolutely accurate values or methods of calculation are available. At other times, methods involving close approximations are used because exact treatments would require long and laborious calculations giving little gain in accuracy. The good chemical engineer recognizes the need for making certain assumptions but also knows that this type of approach introduces some uncertainties into the final results. Therefore, assumptions are made only when they are necessary and essentially correct.

Another important factor in the approach to any design problem involves economic conditions and limitations. The engineer must consider costs and probable profits constantly throughout all the work. It is almost always better to sell many units of a product at a low profit per unit than a few units at a high profit per unit. Consequently, the engineer must take into account the volume of production when determining costs and total profits for various types of designs. This obviously leads to considerations of customer needs and demands. These factors may appear to be distantly removed from the development of a plant design, but they are extremely important in determining its ultimate success.

TWO

PROCESS DESIGN DEVELOPMENT

A principal responsibility of the chemical engineer is the design, construction, and operation of chemical plants. In this responsibility, the engineer must continuously search for additional information to assist in these functions. Such information is available from numerous sources, including recent publications, operation of existing process plants, and laboratory and pilot-plant data. This collection and analysis of all pertinent information is of such importance that chemical engineers are often members, consultants, or advisers of even the basic research team which is developing a new process or improving and revising an existing one. In this capacity, the chemical engineer can frequently advise the research group on how to provide considerable amounts of valuable design data.

Subjective decisions are and must be made many times during the design of any process. What are the best methods of securing sufficient and usable data? What is sufficient and what is reliable? Can better correlations of the data be devised, particularly ones that permit more valid extrapolation?

The chemical engineer should always be willing to consider completely new designs. An attempt to understand the controlling factors of the process, whether chemical or physical, helps to suggest new or improved techniques. For example, consider the commercial processes of aromatic nitration and alkylation of isobutane with olefins to produce high-octane gasolines. Both reactions involve two immiscible liquid phases and the mass-transfer steps are essentially rate controlling. Nitro-aromatics are often produced in high yields (up to 99 percent); however, the alkylation of isobutane involves numerous side reactions and highly complex chemistry that is less well understood. Several types of reactors have been used for each reaction. Then radically new and simplified reactors were developed, based apparently on a better understanding of the chemical and physical steps involved.

DESIGN-PROJECT PROCEDURE

The development of a design project always starts with an initial idea or plan. This initial idea must be stated as clearly and concisely as possible in order to define the scope of the project. General specifications and pertinent laboratory or chemical engineering data should be presented along with the initial idea.

Types of Designs

The methods for carrying out a design project may be divided into the following classifications, depending on the accuracy and detail required:

1. Preliminary or quick-estimate designs
2. Detailed-estimate designs
3. Firm process designs or detailed designs

Preliminary designs are ordinarily used as a basis for determining whether further work should be done on the proposed process. The design is based on approximate process methods, and rough cost estimates are prepared. Few details are included, and the time spent on calculations is kept at a minimum.

If the results of the preliminary design show that further work is justified, a *detailed-estimate design* may be developed. In this type of design, the cost-and-profit potential of an established process is determined by detailed analyses and calculations. However, exact specifications are not given for the equipment, and drafting-room work is minimized.

When the detailed-estimate design indicates that the proposed project should be a commercial success, the final step before developing construction plans for the plant is the preparation of a *firm process design*. Complete specifications are presented for all components of the plant, and accurate costs based on quoted prices are obtained. The firm process design includes blueprints and sufficient information to permit immediate development of the final plans for constructing the plant.

Feasibility Survey

Before any detailed work is done on the design, the technical and economic factors of the proposed process should be examined. The various reactions and physical processes involved must be considered, along with the existing and potential market conditions for the particular product. A preliminary feasibility survey of this type gives an indication of the probable success of the project and also shows what additional information is necessary to make a complete evaluation. Following is a list of items that should be considered in making a feasibility survey:

1. Raw materials (availability, quantity, quality, cost)
2. Thermodynamics and kinetics of chemical reactions involved (equilibrium, yields, rates, optimum conditions)

3. Facilities and equipment available at present
4. Facilities and equipment which must be purchased
5. Estimation of production costs and total investment
6. Profits (probable and optimum, per pound of product and per year, return on investment)
7. Materials of construction
8. Safety considerations
9. Markets (present and future supply and demand, present uses, new uses, present buying habits, price range for products and by-products, character, location, and number of possible customers)
10. Competition (overall production statistics, comparison of various manufacturing processes, product specifications of competitors)
11. Properties of products (chemical and physical properties, specifications, impurities, effects of storage)
12. Sales and sales service (method of selling and distributing, advertising required, technical services required)
13. Shipping restrictions and containers
14. Plant location
15. Patent situation and legal restrictions

When detailed data on the process and firm product specifications are available, a complete market analysis combined with a consideration of all sales factors should be made. This analysis can be based on a breakdown of items 9 through 15 as indicated in the preceding list.

Process Development

In many cases, the preliminary feasibility survey indicates that additional research, laboratory, or pilot-plant data are necessary, and a program to obtain this information may be initiated. Process development on a pilot-plant or semi-works scale is usually desirable in order to obtain accurate design data. Valuable information on material and energy balances can be obtained, and process conditions can be examined to supply data on temperature and pressure variations, yields, rates, grades of raw materials and products, batch versus continuous operation, materials of construction, operating characteristics, and other pertinent design variables.

Design

If sufficient information is available, a preliminary design may be developed in conjunction with the preliminary feasibility survey. In developing the preliminary design the chemical engineer must first establish a workable manufacturing process for producing the desired product. Quite often a number of alternative processes or methods may be available to manufacture the same product. Except for those processes obviously undesirable, each method should be given consideration.

The first step in preparing the preliminary design is to establish the *bases for design*. In addition to the known specifications for the product and availability of raw materials, the design will be controlled by such items as the expected annual operating factor (fraction of the year that the plant will be in operation), temperature of the cooling water, available steam pressures, fuel used, value of by-products, etc. The next step consists of preparing a simplified flow diagram showing the processes that are involved and deciding upon the unit operations which will be required. A preliminary material balance at this point may very quickly eliminate some of the alternative cases. Flow rates and stream conditions for the remaining cases are now evaluated by complete material balances, energy balances, and a knowledge of raw-material and product specifications, yields, reaction rates, and time cycles. The temperature, pressure, and composition of every process stream is determined. Stream enthalpies, percent vapor, liquid, and solid, heat duties, etc., are included where pertinent to the process.

Unit process principles are used in the design of specific pieces of equipment. (Assistance with the design and selection of various types of process equipment is given in Chaps. 13 through 15.) Equipment specifications are generally summarized in the form of tables and included with the final design report. These tables usually include the following:

1. *Columns (distillation)*. In addition to the number of plates and operating conditions it is also necessary to specify the column diameter, materials of construction, plate layout, etc.
2. *Vessels*. In addition to size, which is often dictated by the holdup time desired, materials of construction and any packing or baffling should be specified.
3. *Reactors*. Catalyst type and size, bed diameter and thickness, heat-interchange facilities, cycle and regeneration arrangements, materials of construction, etc., must be specified.
4. *Heat exchangers and furnaces*. Manufacturers are usually supplied with the duty, corrected log mean-temperature difference, percent vaporized, pressure drop desired, and materials of construction.
5. *Pumps and compressors*. Specify type, power requirement, pressure difference, gravities, viscosities, and working pressures.
6. *Instruments*. Designate the function and any particular requirement.
7. *Special equipment*. Specifications for mechanical separators, mixers, driers, etc.

The foregoing is not intended as a complete checklist, but rather as an illustration of the type of summary that is required. (The headings used are particularly suited for the petrochemical industry; others may be desirable for different industries.) As noted in the summary, the selection of materials is intimately connected with the design and selection of the proper equipment.

As soon as the equipment needs have been firmed up, the utilities and labor requirements can be determined and tabulated. Estimates of the capital investment and the total product cost (as outlined in Chap. 5) complete the preliminary-design calculations. Economic evaluation plays an important part in any process design. This is particularly true not only in the selection of a specific process, choice of raw materials used, operating conditions chosen, but also in the specification of equipment. No design of a piece of equipment or a process is complete without an economical evaluation. In fact, as mentioned in Chap. 1, no design project should ever proceed beyond the preliminary stages without a consideration of costs. Evaluation of costs in the preliminary-design phases greatly assists the engineer in further eliminating many of the alternative cases.

The final step, and an important one in preparing a typical process design, involves writing the report which will present the results of the design work. Unfortunately this phase of the design work quite often receives very little attention by the chemical engineer. As a consequence, untold quantities of excellent engineering calculations and ideas are sometimes discarded because of poor communications between the engineer and management.†

Finally, it is important that the preliminary design be carried out as soon as sufficient data are available from the feasibility survey or the process-development step. In this way, the preliminary design can serve its main function of eliminating an undesirable project before large amounts of money and time are expended.

The preliminary design and the process-development work give the results necessary for a detailed-estimate design. The following factors should be established within narrow limits before a detailed-estimate design is developed:

1. Manufacturing process
2. Material and energy balances
3. Temperature and pressure ranges
4. Raw-material and product specifications
5. Yields, reaction rates, and time cycles
6. Materials of construction
7. Utilities requirements
8. Plant site

When the preceding information is included in the design, the result permits accurate estimation of required capital investment, manufacturing costs, and potential profits. Consideration should be given to the types of buildings, heating, ventilating, lighting, power, drainage, waste disposal, safety facilities, instrumentation, etc.

Firm process designs (or detailed designs) can be prepared for purchasing and construction from a detailed-estimate design. Detailed drawings are made for the fabrication of special equipment, and specifications are prepared for purchasing standard types of equipment and materials. A complete plant layout is prepared, and blueprints and instructions for construction are developed. Piping diagrams and other construction details are included. Specifications are given for

---

† See Chap. 12 for assistance in preparing more concise and clearer design reports.

warehouses, laboratories, guard-houses, fencing, change houses, transportation facilities, and similar items. The final firm process design must be developed with the assistance of persons skilled in various engineering fields, such as architectural, ventilating, electrical, and civil.

Construction and Operation

When a definite decision to proceed with the construction of a plant is made, there is usually an immediate demand for a quick plant startup. Timing, therefore, is particularly important in plant construction. Long delays may be encountered in the fabrication of major pieces of equipment, and deliveries often lag far behind the date of ordering. These factors must be taken into consideration when developing the final plans and may warrant the use of the Project Evaluation and Review Technique (PERT) or the Critical Path Method (CPM).† The chemical engineer should work closely with construction personnel during the final stages of construction and purchasing designs. In this way, the design sequence can be arranged to make certain the important factors that might delay construction are given first consideration. Construction of the plant may be started long before the final design is 100 percent complete. Correct design sequence is then essential in order to avoid construction delays.

During construction of the plant, the chemical engineer should visit the plant site to assist in interpretation of the plans and learn methods for improving future designs. The engineer should also be available during the initial startup of the plant and the early phases of operation. Thus, by close teamwork between design, construction, and operations personnel, the final plant can develop from the drawing-board stage to an operating unit that can function both efficiently and effectively.

DESIGN INFORMATION FROM THE LITERATURE

A survey of the literature will often reveal general information and specific data pertinent to the development of a design project. One good method for starting a literature survey is to obtain a recent publication dealing with the subject under investigation. This publication will give additional references, and each of these references will, in turn, indicate other sources of information. This approach permits a rapid survey of the important literature.

Chemical Abstracts, published semimonthly by the American Chemical Society, can be used for comprehensive literature surveys on chemical processes and operations.‡ This publication presents a brief outline and the original reference of published articles dealing with chemistry and related fields. Yearly and decennial indexes of subjects and authors permit rapid location of articles concerning specific topics.

---

† For further discussion of these methods consult Chap. 10.
‡ Abstracts of general engineering articles are available in the *Engineering Index*.

A primary source of information on all aspects of chemical engineering principles, design, costs, and applications is "The Chemical Engineers' Handbook" published by McGraw-Hill Book Company with J. H. Perry and C. H. Chilton as editors for the 5th edition as published in 1973. This reference should be in the personal library of all chemical engineers involved in the field.

Regular features on design-related aspects of equipment, costs, materials of construction, and unit processes are published in *Chemical Engineering*. In addition to this publication, there are many other periodicals that publish articles of direct interest to the design engineer. The following periodicals are suggested as valuable sources of information for the chemical engineer who wishes to keep abreast of the latest developments in the field: *American Institute of Chemical Engineers' Journal, Chemical Engineering Progress, Chemical and Engineering News, Chemical Week, Chemical Engineering Science, Industrial and Engineering Chemistry Fundamentals, Industrial and Engineering Chemistry Process Design and Development, Journal of the American Chemical Society, Journal of Physical Chemistry, Hydrocarbon Processing, Engineering News-Record, Oil and Gas Journal, Canadian Journal of Chemical Engineering, Transactions of the Institution of Chemical Engineers* (London), *The Chemical Engineer* (London), *Transactions of the Faraday Society*, and *The Hydrocarbon Processing Catalog*.

A large number of textbooks covering the various aspects of chemical engineering principles and design are available.† In addition, many handbooks have been published giving physical properties and other basic data which are very useful to the design engineer. Examples of handbooks and other data sources of particular value to the chemical engineer are presented in the reference section at the end of this chapter.

Trade bulletins are published regularly by most manufacturing concerns, and these bulletins give much information of direct interest to the chemical engineer preparing a design. Some of the trade-bulletin information is condensed in an excellent reference book on chemical engineering equipment, products, and manufacturers. This book is known as the "Chemical Engineering Catalog,"‡ and contains a large amount of valuable descriptive material.

New information is constantly becoming available through publication in periodicals, books, trade bulletins, government reports, university bulletins, and many other sources. Many of the publications are devoted to shortcut methods for estimating physical properties or making design calculations, while others present convenient compilations of essential data in the form of nomographs or tables.

The effective design engineer must make every attempt to keep an up-to-date knowledge of the advances in the field. Personal experience and contacts, attendance at meetings of technical societies and industrial expositions, and reference to the published literature are very helpful in giving the engineer the background information necessary for a successful design.

---

† For example, see the *Chemical Engineering Series* listing at the front of this text.
‡ Published annually by Reinhold Publishing Corporation, New York.

FLOW DIAGRAMS

The chemical engineer uses flow diagrams to show the sequence of equipment and unit operations in the overall process, to simplify visualization of the manufacturing procedures, and to indicate the quantities of materials and energy transfer. These diagrams may be divided into three general types: (1) qualitative, (2) quantitative, and (3) combined-detail.

A qualitative flow diagram indicates the flow of materials, unit operations involved, equipment necessary, and special information on operating temperatures and pressures. A quantitative flow diagram shows the quantities of materials required for the process operation. An example of a qualitative flow diagram for the production of nitric acid is shown in Fig. 2-1. Figure 2-2 presents a quantitative flow diagram for the same process.

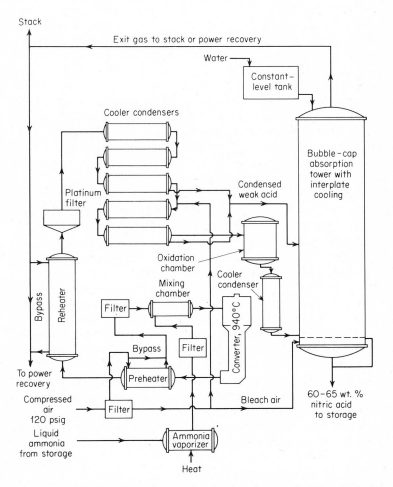

Figure 2-1 Qualitative flow diagram for the manufacture of nitric acid by the ammonia-oxidation process.

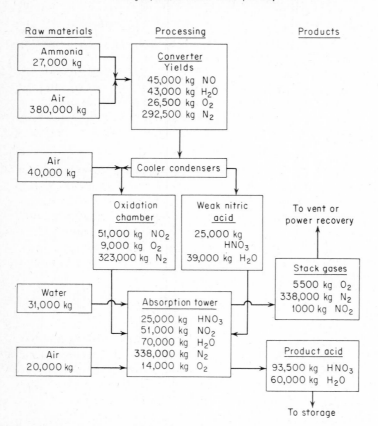

Basis: One operating day

Unit designed to produce 153,500 kilograms of
61 weight percent nitric acid per day

Figure 2-2 Quantitative flow diagram for the manufacture of nitric acid by the ammonia-oxidation process.

Preliminary flow diagrams are made during the early stages of a design project. As the design proceeds toward completion, detailed information on flow quantities and equipment specifications become available, and combined-detail flow diagrams can be prepared. This type of diagram shows the qualitative flow pattern and serves as a base reference for giving equipment specifications, quantitative data, and sample calculations. Tables presenting pertinent data on the process and the equipment are cross-referenced to the drawing. In this way, qualitative information and quantitative data are combined on the basis of one flow diagram. The drawing does not lose its effectiveness by presenting too much information; yet the necessary data are readily available by direct reference to the accompanying tables.

A typical combined-detail flow diagram shows the location of temperature and pressure regulators and indicators, as well as the location of critical control valves and special instruments. Each piece of equipment is shown and is designated by a defined code number. For each piece of equipment, accompanying tables give essential information, such as specifications for purchasing, specifications for construction, type of fabrication, quantities and types of chemicals involved, and sample calculations.

Equipment symbols and flow-sheet symbols, particularly for detailed equipment flow sheets, are given in the Appendix.

THE PRELIMINARY DESIGN

In order to amplify the remarks made earlier in this chapter concerning the design-project procedure, it is appropriate at this time to look more closely at a specific preliminary design. Because of space limitations, only a brief presentation of the design will be attempted at this point.† However, sufficient detail will be given to outline the important steps which are necessary to prepare such a preliminary design. The problem presented here is a practical one of a type frequently encountered in the chemical industry; it involves both process design and economic considerations.

Problem Statement

A conservative petroleum company has recently been reorganized and the new management has decided that the company must diversify its operations into the petrochemical field if it wishes to remain competitive. The research division of the company has suggested that a very promising area in the petrochemical field would be in the development and manufacture of biodegradable synthetic detergents using some of the hydrocarbon intermediates presently available in the refinery. A survey by the market division has indicated that the company could hope to attain 2.5 percent of the detergent market if a plant with an annual production of 15 million pounds were to be built. To provide management with an investment comparison, the design group has been instructed to proceed first with a preliminary design and an updated cost estimate for a nonbiodegradable detergent producing facility similar to ones supplanted by recent biodegradable facilities.

Literature Survey

A survey of the literature reveals that the majority of the nonbiodegradable detergents are alkylbenzene sulfonates (ABS). Theoretically, there are over

† Completion of the design is left as an exercise for the reader.

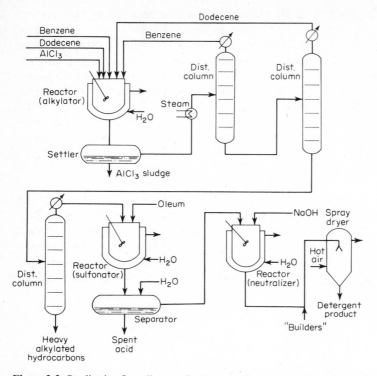

Figure 2-3 Qualitative flow diagram for the manufacture of sodium dodecylbenzene sulfonate.

80,000 isomeric alkylbenzenes in the range of C_{10} to C_{15} for the alkyl side chain. Costs, however, generally favor the use of dodecene (propylene tetramer) as the starting material for ABS.

There are many different schemes in the manufacture of ABS. Most of the schemes are variations of the one shown in Fig. 2-3 for the production of sodium dodecylbenzene sulfonate. A brief description of the process is as follows:

This process involves reaction of dodecene with benzene in the presence of aluminum chloride catalyst; fractionation of the resulting crude mixture to recover the desired boiling range of dodecylbenzene; sulfonation of the dodecylbenzene and subsequent neutralization of the sulfonic acid with caustic soda; blending the resulting slurry with chemical "builders"; and drying.

Dodecene is charged into a reaction vessel containing benzene and aluminum chloride. The reaction mixture is agitated and cooled to maintain the reaction temperature at about 115°F maximum. An excess of benzene is used to suppress the formation of byproducts. Aluminum chloride requirement is 5 to 10 wt % of dodecene.

After removal of aluminum chloride sludge, the reaction mixture is fractionated to recover excess benzene (which is recycled to the reaction vessel), a light alkylaryl hydrocarbon, dodecylbenzene, and a heavy alkylaryl hydrocarbon.

Sulfonation of the dodecylbenzene may be carried out continuously or batch-wise under a variety of operating conditions using sulfuric acid (100 percent), oleum (usually 20 percent SO_3), or anhydrous sulfur trioxide. The optimum sulfonation temperature is usually in the range of 100 to 140°F depending on the strength of acid employed, mechanical design of the equipment, etc. Removal of spent sulfuric acid from the sulfonic acid is facilitated by adding water to reduce the sulfuric acid strength to about 78 percent. This dilution prior to neutralization results in a final neutralized slurry having approximately 85 percent active agent based on the solids. The inert material in the final product is essentially Na_2SO_4.

The sulfonic acid is neutralized with 20 to 50 percent caustic soda solution to a pH of 8 at a temperature of about 125°F. Chemical "builders" such as trisodium phosphate, tetrasodium pyrophosphate, sodium silicate, sodium chloride, sodium sulfate, carboxymethyl cellulose, etc., are added to enhance the detersive, wetting, or other desired properties in the finished product. A flaked, dried product is obtained by drum drying or a bead product is obtained by spray drying.

The basic reactions which occur in the process are the following.

Alkylation:

$$C_6H_6 + C_{12}H_{24} \xrightarrow{AlCl_3} C_6H_5 \cdot C_{12}H_{25}$$

Sulfonation:

$$C_6H_5 \cdot C_{12}H_{25} + H_2SO_4 \longrightarrow C_{12}H_{25} \cdot C_6H_4 \cdot SO_3H + H_2O$$

Neutralization:

$$C_{12}H_{25} \cdot C_6H_4 \cdot SO_3H + NaOH \longrightarrow C_{12}H_{25} \cdot C_6H_4 \cdot SO_3Na + H_2O$$

A literature search indicates that yields of 85 to 95 percent have been obtained in the alkylation step, while yields for the sulfonation process are substantially 100 percent, and yields for the neutralization step are always 95 percent or greater. All three steps are exothermic and require some form of jacketed cooling around the stirred reactor to maintain isothermal reaction temperatures.

Laboratory data for the sulfonation of dodecylbenzene, described in the literature, provide additional information useful for a rapid material balance. This is summarized as follows:

1. Sulfonation is essentially complete if the ratio of 20 percent oleum to dodecylbenzene is maintained at 1.25.
2. Spent sulfuric acid removal is optimized with the addition of 0.244 lb of water to the settler for each 1.25 lb of 20 percent oleum added in the sulfonation step.
3. A 25 percent excess of 20 percent NaOH is suggested for the neutralization step.

Operating conditions for this process, as reported in the literature, vary somewhat depending upon the particular processing procedure chosen.

Material and Energy Balance

The process selected for the manufacture of the nonbiodegradable detergent is essentially continuous even though the alkylation, sulfonation, and neutralization steps are semicontinuous steps. Provisions for possible shutdowns for repairs and maintenance are incorporated into the design of the process by specifying plant operation for 300 calendar days per year. Assuming 90 percent yield in the alkylator and a sodium dodecylbenzene sulfonate product to be 85 percent active with 15 percent sodium sulfate as inert, the overall material balance is as follows:

Input components:

$$\text{Product (85\% active)} = \frac{(15 \times 10^6)(0.85)}{(300)(348.5)} = 122 \text{ lb mol/day}$$

$$C_6H_6 \text{ feed} = (122)\left(\frac{1}{0.95}\right)\left(\frac{1}{0.90}\right) = 142.7 \text{ lb mol/day}$$

$$= (142.7)(78.1) = 11,145 \text{ lb/day}$$

$C_{12}H_{24}$ feed = 142.7 lb mol/day

$$= (142.7)(168.3) = 24,016 \text{ lb/day}$$

20% oleum in = (1.25)(11,145 + 24,016) = 43,951 lb/day

Dilution H_2O in = (0.244/1.25)(43,951) = 8579 lb/day

20% NaOH in = (1.25)(43,951) = 55,085 lb/day

$AlCl_3$ catalyst in = (0.05)(11,145 + 24,016) = 1758 lb/day

Alkylation process:

Alkylate yield = (0.9)(142.7)(246.4) = 31,645 lb/day

Unreacted C_6H_6 = (0.1)(11,145) = 1114 lb/day

Unreacted $C_{12}H_{24}$ = (0.1)(24,016) = 2402 lb/day

Sulfur balance:

Sulfur in = (43,951)(1.045)(32.1/98.1) = 15,029 lb/day

Sulfur out = sulfur in detergent + sulfur in spent acid

$$\text{Sulfur in detergent} = \frac{(50,000)(0.85)(32.1)}{(348.5)} + \frac{(50,000)(0.15)(32.1)}{(142)}$$

$$= 3915 + 1695 = 5610 \text{ lb/day}$$

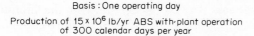

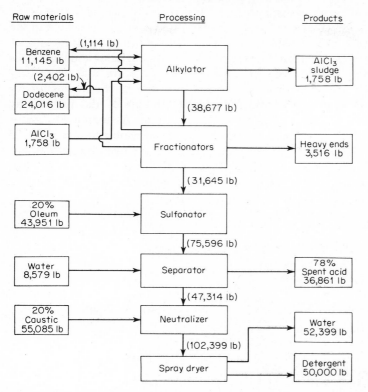

Figure 2-4 Quantitative flow diagram for the manufacture of sodium dodecylbenzene sulfonate.

Sulfur out in acid $= 15,029 - 5610 = 9419$ lb/day

Weight of 78% $H_2SO_4 = (9419)\left(\dfrac{98.1}{32.1}\right)\left(\dfrac{1}{0.78}\right) = 36,861$ lb/day

The weight of the heavy alkylaryl hydrocarbon is obtained by difference as 3516 lb/day.

The material balance summary made by the design group for the process shown in Fig. 2-3 is given on a daily basis in Fig. 2-4. After a complete material balance is made, the mass quantities are used to compute energy balances around each piece of equipment. Temperature and pressure levels at various key points in the process, particularly at the reactors, serve as guides in making these heat balances. The complete calculations for the material and energy balances for each piece of equipment, because of their length, are not presented in this discussion.

Equipment Design and Selection

Equipment design for this preliminary process evaluation involves determining the size of the equipment in terms of volume, flow per unit time, or surface area. Some of the calculations associated with the alkylation unit are presented in the following to indicate the extent of the calculations which are sometimes adequate for a preliminary design.

ALKYLATION UNIT EQUIPMENT DESIGN AND SELECTION

Reactor Volume

Assume a 4-h cycle and operation of the alkylator at constant temperature and pressure of 115°F and 1 atm, respectively. The volume of reactants per day is

$$V_r = \frac{(12,259)}{(8.34)(0.88)} + \frac{(26,418)}{(8.34)(0.762)} + \frac{(1758)(7.48)}{(2.44)(62.4)}$$

$$= 1670 + 4160 + 86 = 5918 \text{ gal/day}$$

$$= \frac{5918}{6} = 985 \text{ gal/cycle}$$

If the reactor is 75 percent full on each cycle, the volume of reactor needed is

$$V_R = \frac{985}{0.75} = 1313 \text{ gal}$$

Select a 1300-gal, glass-lined, stirred reactor.

Heat of Reaction Calculation

$$C_6H_{6(l)} + C_{12}H_{24(l)} \rightarrow C_6H_5 \cdot C_{12}H_{25(l)}$$

$$\Delta H_r = \Delta H_{f(C_6H_5 \cdot C_{12}H_{25})l} - \Delta H_{f(C_6H_6)l} - \Delta H_{f(C_{12}H_{24})l}$$

The heats of formation ΔH_f of dodecylbenzene and dodecene are evaluated using standard thermochemistry techniques outlined in most chemical engineering thermodynamic texts. The heat of formation of benzene is available in the literature.

$$\Delta H_{f(C_6H_5 \cdot C_{12}H_{25})l} = -54,348 \text{ cal/g mol}$$

$$\Delta H_{f(C_{12}H_{24})l} = -51,239 \text{ cal/g mol}$$

$$\Delta H_{f(C_6H_6)l} = 11,717 \text{ cal/g mol}$$

Thus,

$$\Delta H_r = -54,348 - 11,717 + 51,239 = -14,826 \text{ cal/g mol}$$

$$= -26,687 \text{ Btu/lb mol}$$

Assume heat of reaction is liberated in 3 h of the 4-h cycle ($\frac{1}{6}$ of an operating day):

$$Q_r = \left(\frac{11,145}{78.1}\right)\left(\frac{1}{6}\right)\left(\frac{1}{3}\right)(-26,687) = -211,500 \text{ Btu/h}$$

Use a 10°F temperature difference for the cooling water to find the mass of cooling water required to remove the heat of reaction.

$$m_{H_2O} = \frac{Q_r}{C_p \Delta T} = \frac{211,500}{(1)(10)} = 21,150 \text{ lb/h}$$

$$q_{f(H_2O)} = \frac{21,150}{(60)(8.33)} = 42.3 \text{ gpm}$$

The volumetric flow rate is, therefore, 42.3 gpm. Select a 45-gpm centrifugal pump, carbon steel construction.

Heat Transfer Area Needed to Cool Reactor

Assume water inlet of 80°F with a 10°F temperature rise. A reasonable overall heat transfer coefficient for this type of heat transfer may be calculated as 45 Btu/(h)(ft^2)(°F).

$$\Delta T_{lm} = \frac{(115 - 80) - (115 - 90)}{2.303 \log \frac{35}{25}} = 29.7°F$$

$$A = \frac{Q}{U \, \Delta T_{lm}} = \frac{211,500}{(45)(29.7)} = 158 \text{ ft}^2$$

A 1300-gal stirred reactor has approximately 160 ft^2 of jacket area. Therefore, the surface area available is sufficient to maintain isothermal conditions in the reactor.

Sizing of Storage Tanks

Provide benzene and dodecene storage for six days:

$$V_{\text{benzene}} = (1670)(6) = 10,020 \text{ gal}$$

$$V_{\text{dodecene}} = (4160)(6) = 24,960 \text{ gal}$$

Select a 10,000-gal carbon steel tank for benzene storage and a 25,000-gal carbon steel tank for dodecene storage.

Provide holding tank storage for one day:

$$V_{\text{holding}} = 5918 \text{ gal}$$

Select a 6000-gal carbon steel tank for holding tank.

Table 1 Equipment specifications for alkylation unit†

No. req'd.	Item and description	Size	Mat'l. const.
1	T-1, storage tank for benzene	10,000 gal	Carbon steel
1	T-2, storage tank for dodecene	25,000 gal	Carbon steel
1	T-3, holding tank for alkylate	6,000 gal	Carbon steel
1	P-1, pump (centrifugal) for benzene transfer from T-1 to R-1	30 gpm	Carbon steel
1	P-2, pump (centrifugal) for dodecene transfer from T-2 to R-1	70 gpm	Carbon steel
1	P-3, pump (centrifugal) for pumping cooling water to jacket of R-1	45 gpm	Carbon steel
1	P-4, pump (positive displacement) for alkylate transfer from T-3 to C-1	10 gpm	Cast iron
1	R-1, reactor (stirred) alkylator	1,300 gal	Glass-lined

† See Fig. 2-5.

Sizing Other Pumps

Provide benzene and dodecene filling of reactor in 10 min:

$$q_{f(\text{benzene})} = \frac{1670}{(6)(10)} = 27.8 \text{ gpm}$$

Select a 30-gpm centrifugal pump, carbon steel construction.

$$q_{f(\text{dodecene})} = \frac{4160}{(6)(10)} = 69.3 \text{ gpm}$$

Select a 70-gpm centrifugal pump, carbon steel construction. The alkylate pump used to transfer alkylate from the holding tank to the benzene fractionator must operate continuously. Thus,

$$q_{f(\text{alkylate})} = \frac{(1670 + 4160)}{(24)(60)} = 4 \text{ gpm}$$

Select a 10-gpm positive displacement pump, carbon steel construction. A summary of the equipment needs for the alkylation unit in this preliminary process design is presented in Table 1. The preparation of similar equipment lists for the other process units completes the equipment selection and design phase of the preliminary design. Figure 2-5 shows a simplified equipment diagram for the proposed process and includes the specified size or capacity of each piece of process equipment.

Economics

The purchased cost of each piece of process equipment may now be estimated from published cost data or from appropriate manufacturers' bulletins. Re-

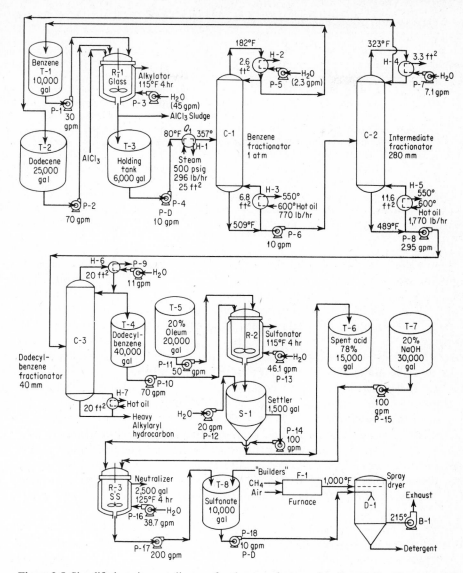

Figure 2-5 Simplified equipment diagram for the manufacture of sodium dodecylbenzene sulfonate.

gardless of the source, the published purchased-cost data must always be corrected to the current cost index. This procedure is described in detail in Chap. 5.

For the alkylation unit, purchased-equipment costs may be estimated using the equipment-specification information of Table 1 and the cost data presented in Chaps. 13 through 15 of this text. Table 2 presents these costs updated to January 1, 1979. The required fixed-capital investment for the nonbiodegradable

Table 2 Estimated purchased-equipment cost for alkylation unit†

Designation	Item	Purchased cost
T-1	Storage tank	$13,400
T-2	Storage tank	22,600
T-3	Holding tank	10,000
P-1	Centrifugal pump (with motor)	910
P-2	Centrifugal pump (with motor)	980
P-3	Centrifugal pump (with motor)	910
P-4	Positive-displacement pump	4,000
R-1	Jacketed (stirred) reactor	36,000
	Total	$88,800

† January 1, 1979 costs. See Fig. 2-5.

detergent manufacturing process may be estimated from the total purchased-equipment cost using the equipment-cost ratio method outlined in Table 17 of Chap. 5. The total purchased-equipment cost is presented in Table 3 and is the basis for the estimated fixed-capital cost tabulation given in Table 4. The probable error in this method of estimating the fixed-capital investment is as much as ±30 percent.

An evaluation of the operating labor and utilities requirements of the process must be made before the total product cost can be estimated. Details for evaluating these direct production costs are given in Chap. 5 and Appendix B. The estimate of the total product cost for the manufacture of 15 million lb per year of detergent, based on methods outlined in Chap. 5, is presented in Table 5.

Once the total product cost has been estimated, the design group is in a position to evaluate for management the attractiveness of the proposed process using such measures of profitability as rate of return, payout time, or present worth. These methods are fully outlined in Chap. 9. The design report, as mentioned previously, completes the preliminary design.

Table 3 Summary of purchased-equipment cost for complete process unit

Process unit	Purchased cost
Alkylation	$ 88,800
Fractionators	108,400
Sulfonation	151,100
Neutralization	100,900
Spray dryer	242,600
Auxiliary units	26,400
Total	$718,200

Table 4 Fixed-capital investment estimate†

Items	Cost
Purchased equipment	$ 718,200
Purchased-equipment installation	337,500
Instrumentation and controls	129,300
Piping (installed)	474,000
Electrical (installed)	79,000
Buildings (including services)	129,300
Yard improvements	71,800
Service facilities (installed)	502,700
Land (purchase not required)	
Engineering and supervision	237,000
Construction expenses	294,500
Contractor's fee	150,800
Contingency	301,600
Fixed-capital investment	$3,425,700
Working capital	617,300
Total capital investment	$4,043,000

   † Equipment-cost ratio percentages used in Table 4 are factors applicable to a fluid-processing plant as outlined in Chap. 5.

Table 5   Total product cost estimate

Items	Costs
Direct production costs	
Raw materials	$1,548,800
Operating labor	594,000
Direct supervisory and clerical labor	118,800
Utilities	350,000
Maintenance and repairs	68,500
Operating supplies	10,300
Fixed charges	
Depreciation	342,600
Local taxes	68,500
Insurance	34,300
Plant-overhead costs	468,800
General expenses	
Administration	117,200
Distribution and selling	475,900
Research and development	237,900
Financing (interest)	323,400
Annual total product cost	$4,759,000
Total product cost per pound	$0.317

Summary

The preliminary design presented in this section was developed to show the logi-cal step-by-step approach which is quite often followed for each new process design. The exact procedure may vary from company to company and from one design engineer to another. Likewise, the assumptions and rule-of-thumb factors used may vary from one company to the next depending to a large extent on design experience and company policy. Nevertheless, the basic steps for a process design are those as outlined in this preliminary design covering the manu-facture of a common household item.

No attempt has been made to present a complete design. In fact, to minimize the length, many assumptions were made which would have been verified or justified in a normal process design. Neither were any alternative solutions con-sidered even though some were suggested by the literature survey. The investiga-tion of these various alternatives is left to the reader.

COMPARISON OF DIFFERENT PROCESSES

In the course of a design project it is necessary to determine the most suitable process for obtaining a desired product. Several different manufacturing methods may be available for making the same material, and the various processes must be compared in order to select the one best suited to the existing conditions.

The comparison can be accomplished through the development of complete designs. In many cases, however, all but one or two of the possible processes can be eliminated by a weighted comparison of the essential variable items, and detailed design calculations for each process may not be required. The following items should be considered in a comparison of this type:

1. Technical factors
   a. Process flexibility
   b. Continuous operation
   c. Special controls involved
   d. Commercial yields
   e. Technical difficulties involved
   f. Energy requirements
   g. Special auxiliaries required
   h. Possibility of future developments
   i. Health and safety hazards involved
2. Raw materials
   a. Present and future availability
   b. Processing required
   c. Storage requirements
   d. Materials handling problems
3. Waste products and by-products
   a. Amount produced

  b. Value
  c. Potential markets and uses
  d. Manner of discard
  e. Environmental aspects
 4. Equipment
  a. Availability
  b. Materials of construction
  c. Initial costs
  d. Maintenance and installation costs
  e. Replacement requirements
  f. Special designs
 5. Plant location
  a. Amount of land required
  b. Transportation facilities
  c. Proximity to markets and raw-material sources
  d. Availability of service and power facilities
  e. Availability of labor
  f. Climate
  g. Legal restrictions and taxes
 6. Costs
  a. Raw materials
  b. Energy
  c. Depreciation
  d. Other fixed charges
  e. Processing and overhead
  f. Special labor requirements
  g. Real estate
  h. Patent rights
  i. Environmental controls
 7. Time factor
  a. Project completion deadline
  b. Process development required
  c. Market timeliness
  d. Value of money
 8. Process considerations
  a. Technology availability
  b. Raw materials common with other processes
  c. Consistency of product within company
  d. General company objectives

Batch Versus Continuous Operation

When comparing different processes, consideration should always be given to the advantages of continuous operation over batch operation. In many cases, costs can be reduced by using continuous instead of batch processes. Less labor is required, and control of the equipment and grade of final product is simplified. Whereas batch operation was common in the early days of the chemical industry, most processes have been switched completely or partially to continuous operation. The advent of many new types of control instruments has made this transition possible, and the design engineer should be aware of the advantages inherent in any type of continuous operation.

EQUIPMENT DESIGN AND SPECIFICATIONS

The goal of a "plant design" is to develop and present a complete plant that can operate on an effective industrial basis. To achieve this goal, the chemical engineer must be able to combine many separate units or pieces of equipment into one smoothly operating plant. If the final plant is to be successful, each piece of equipment must be capable of performing its necessary function. The design of equipment, therefore, is an essential part of a plant design.

The engineer developing a process design must accept the responsibility of preparing the specifications for individual pieces of equipment and should be acquainted with methods for fabricating different types of equipment. The importance of choosing appropriate materials of construction in this fabrication must be recognized. Design data must be developed, giving sizes, operating conditions, number and location of openings, types of flanges and heads, codes, variation allowances, and other information. Many of the machine-design details are handled by the fabricators, but the chemical engineer must supply the basic design information.

SCALE-UP IN DESIGN

When accurate data are not available in the literature or when past experience does not give an adequate design basis, pilot-plant tests may be necessary in order to design effective plant equipment. The results of these tests must be scaled up to the plant capacity. A chemical engineer, therefore, should be acquainted with the limitations of scale-up methods and should know how to select the essential design variables.

Pilot-plant data are almost always required for the design of filters unless specific information is already available for the type of materials and conditions involved. Heat exchangers, distillation columns, pumps, and many other types of conventional equipment can usually be designed adequately without using pilot-plant data.

Table 6 presents an analysis of important factors in the design of different types of equipment.† This table shows the major variables that characterize the

---

† Adapted from Johnstone, R. E., and M. W. Thring, "Pilot Plants, Models, and Scale-up Methods," McGraw-Hill Book Company, New York, 1957.

Table 6 Factors in equipment scale-up and design

Type of equipment	Is pilot plant usually necessary?	Major variables for operational design (other than flow rate)	Major variables characterizing size or capacity	Maximum scale-up ratio based on indicated characterizing variable	Approximate recommended safety or over-design factor, %
Agitated batch crystallizers	Yes	Solubility-temperature relationship	Flow rate Heat transfer area	>100:1	20
Batch reactors	Yes	Reaction rate Equilibrium state	Volume Residence time	>100:1	20
Centrifugal pumps	No	Discharge head	Flow rate Power input Impeller diameter	>100:1 >100:1 10:1	10
Continuous reactors	Yes	Reaction rate Equilibrium state	Flow rate Residence time	>100:1	20
Cooling towers	No	Air humidity Temperature decrease	Flow rate Volume	>100:1 10:1	15
Cyclones	No	Particle size	Flow rate Diameter of body	10:1 3:1	10
Evaporators	No	Latent heat of vaporization Temperatures	Flow rate Heat-transfer area	>100:1 >100:1	15
Hammer mills	Yes	Size reduction	Flow rate Power input	60:1 60:1	20
Mixers	No	Mechanism of operation System geometry	Flow rate Power input	>100:1 20:1	20
Nozzle-discharge centrifuges	Yes	Discharge method	Flow rate Power input	10:1 10:1	20 20

(continued)

Table 6 Factors in equipment scale-up and design (Continued)

Type of equipment	Is pilot plant usually necessary?	Major variables for operational design (other than flow rate)	Major variables characterizing size or capacity	Maximum scale-up ratio based on indicated characterizing variable	Approximate recommended safety or over-design factor, %
Packed columns	No	Equilibrium data Superficial vapor velocity	Flow rate Diameter Height to diameter ratio	>100:1 10:1	15
Plate columns	No	Equilibrium data Superficial vapor velocity	Flow rate Diameter	>100:1 10:1	15
Plate-and-frame filters	Yes	Cake resistance or permeability	Flow rate Filtration area	>100:1 >100:1	20
Reboilers	No	Temperatures Viscosities	Flow rate Heat-transfer area	>100:1 >100:1	15
Reciprocating compressors	No	Compression ratio	Flow rate Power input Piston displacement	>100:1 >100:1 >100:1	10
Rotary filters	Yes	Cake resistance or permeability	Flow rate Filtration area	>100:1 25:1	20
Screw conveyors	No	Bulk density	Flow rate Diameter Drive horsepower	90:1 8:1	20
Screw extruders	No	Shear rate	Flow rate Power input	100:1 100:1	20 10
Sedimentation centrifuges	No	Discharge method	Flow rate Power input	10:1 10:1	20 20
Settlers	No	Settling velocity	Volume Residence time	>100:1	15
Spray columns	No	Gas solubilities	Flow rate Power input	10:1	20

(continued)

Table 6 Factors in equipment scale-up and design (Continued)

Type of equipment	Is pilot plant usually necessary?	Major variables for operational design (other than flow rate)	Major variables characterizing size or capacity	Maximum scale-up ratio based on indicated characterizing variable	Approximate recommended safety or overdesign factor, %
Spray condensers	No	Latent heat of vaporization Temperatures	Flow rate Height to diameter ratio	70:1 12:1	20
Tube-and-shell heat exchangers	No	Temperatures Viscosities Thermal conductivities	Flow rate Heat-transfer area	>100:1 >100:1	15

size or capacity of the equipment and the maximum scale-up ratios for these variables. Information on the need for pilot-plant data, safety factors, and essential operational data for the design is included in Table 6.

SAFETY FACTORS

Some examples of recommended safety factors for equipment design are shown in Table 6. These factors represent the amount of overdesign that would be used to account for changes in the operating performance with time.

The indiscriminate application of safety factors can be very detrimental to a design. Each piece of equipment should be designed to carry out its necessary function. Then, if uncertainties are involved, a reasonable safety factor can be applied. The role of the particular piece of equipment in the overall operation must be considered along with the consequences of underdesign. Fouling, which may occur during operation, should never be overlooked when a design safety factor is determined. Potential increases in capacity requirements are sometimes used as an excuse for applying large safety factors. This practice, however, can result in so much overdesign that the process or equipment never has an opportunity to prove its economic value.

In general design work, the magnitudes of safety factors are dictated by economic or market considerations, the accuracy of the design data and calculations, potential changes in the operating performance, background information available on the overall process, and the amount of conservatism used in developing the individual components of the design. Each safety factor must be chosen on the basis of existing conditions, and the chemical engineer should not hesitate to use a safety factor of zero if the situation warrants it.

SPECIFICATIONS

A generalization for equipment design is that standard equipment should be selected whenever possible. If the equipment is standard, the manufacturer may have the desired size in stock. In any case, the manufacturer can usually quote a lower price and give better guarantees for standard equipment than for special equipment.

The chemical engineer cannot be an expert on all the types of equipment used in industrial plants and, therefore, should make good use of the experience of others. Much valuable information can be obtained from equipment manufacturers who specialize in particular types of equipment.

Before a manufacurer is contacted, the engineer should evaluate the design needs and prepare a preliminary specification sheet for the equipment. This preliminary specification sheet can be used by the engineer as a basis for the preparation of the final specifications, or it can be sent to a manufacturer with a request for suggestions and fabrication information. Preliminary specifications for equipment should show the following:

1. Identification
2. Function
3. Operation
4. Materials handled
5. Basic design data
6. Essential controls
7. Insulation requirements
8. Allowable tolerances
9. Special information and details pertinent to the particular equipment, such as materials of construction including gaskets, installation, necessary delivery date, supports, and special design details or comments

Final specifications can be prepared by the engineer; however, care must be exercised to avoid unnecessary restrictions. The engineer should allow the potential manufacturers or fabricators to make suggestions before preparing detailed specifications. In this way, the final design can include small changes that reduce the first cost with no decrease in the effectiveness of the equipment. For example, the tubes in standard heat exchangers are usually 8, 12, 16, or 20 ft long, and these lengths are ordinarily kept in stock by manufacturers and maintenance departments. If a design specification called for tubes 15 ft long, the manufacturer would probably use 16-ft tubes cut off to the specified length. Thus, an increase from 15 to 16 ft for the specified tube length could cause a reduction in the total cost for the unit, because the labor charge for cutting the standard-length tubes would be eliminated. In addition, replacement of tubes might become necessary after the heat exchanger has been in use, and the replacement costs with 16-ft tubes would probably be less than with 15-ft tubes.

HEAT EXCHANGER

Identification: Item *Condenser*
Item No. *H-5*
No. required *1*

Date *1-1-79*

By *JRL*

Function: *Condense overhead vapors from methanol fractionation column*

Operation: *Continuous*

Type: *Horizontal*
Fixed tube sheet
Expansion ring in shell
Duty *3,400,00* Btu/h   Outside area *470* sq ft

Tube side:
Fluid handled *Cooling water*
Flow rate *380 gpm*
Pressure *20* psig
Temperature *15°C* to *25°C*
Head material Carbon steel

Tubes: *1* in. diam. *14* BWG
1.25" Centers △ Pattern
225 Tubes each *8* ft long
2 Passes
Tube material *Carbon steel*

Shell side:
Fluid handled *Methanol vapor*
Flow rate *7000 lb/h*
Pressure *0* psig
Temperature *65°C* to *(constant temp.)*

Shell: *22* in. diam. *1* Passes
(Transverse baffles *Tube*
support Req'd)
(Longitudinal baffles *0* Req'd)
Shell material *Carbon steel*

Utilities: *Untreated cooling water*
Controls: *Cooling-water rate controlled by vapor temperature in vent line*
Insulation: *2-in. rock cork or equivalent; weatherproofed*
Tolerances: *Tubular Exchangers Manufacturers Association (TEMA) standards*
Comments and drawings: *Location and sizes of inlets and outlets are shown*
on drawing

Figure 2-6 Specification sheet for heat exchangers using U.S. Customary units.

Figures 2-6 and 2-7 show typical types of specification sheets for equipment. These sheets apply for the normal type of equipment encountered by a chemical engineer in design work. The details of mechanical design, such as shell or head thicknesses, are not included, since they do not have a direct effect on the performance of the equipment. However, for certain types of equipment involving unusual or extreme operating conditions, the engineer may need to extend the specifications to include additional details of the mechanical design. Locations and sizes of outlets, supports, and other essential fabrication information can be presented with the specifications in the form of comments or drawings.

SIEVE-TRAY COLUMN

Identification: Item _____   Date _____
Item No. _____
No. required _____   By _____

Function:

Operation:

Materials handled:	*Feed*	*Overhead*	*Reflux*	*Bottoms*
Quantity	_____	_____	_____	_____
Composition	_____	_____	_____	_____
Temperature	_____	_____	_____	_____

Design data: No. of trays _____   Reflux ratio _____
Pressure_____   Tray spacing _____
Functional height _____   Skirt height _____
Material of construction _____
Diameter: Liquid density _____ lb/ft^3 (____ kg/m^3)
Vapor density _____ lb/ft^3 (____ kg/m^3)
Maximum allowable vapor velocity (superficial) _____ ft/s (____ m/s)
Maximum vapor flow rate _____ ft^3/s (____ m^3/s)
Recommended inside diameter _____
Hole size and arrangement _____
Tray thickness _____

Utilities:
Controls:
Insulation:
Tolerances:
Comments and drawings:

Figure 2-7 Specification sheet for sieve-tray distillation column.

MATERIALS OF CONSTRUCTION

The effects of corrosion and erosion must be considered in the design of chemical plants and equipment. Chemical resistance and physical properties of constructional materials, therefore, are important factors in the choice and design of equipment. The materials of construction must be resistant to the corrosive action of any chemicals that may contact the exposed surfaces. Possible erosion caused by flowing fluids or other types of moving substances must be considered, even though the materials of construction may have adequate chemical resistance. Structural strength, resistance to physical or thermal shock, cost, ease of fabrication, necessary maintenance, and general type of service required, including operating temperatures and pressures, are additional factors that influence the final choice of constructional materials.

If there is any doubt concerning suitable materials for the construction of equipment, reference should be made to the literature,† or laboratory tests should be carried out under conditions similar to the final operating conditions. The results from laboratory tests indicate the corrosion resistance of the material and also the effects on the product caused by contact with the particular material. Further tests on a pilot-plant scale may be desirable in order to determine the amount of erosion resistance or the effects of other operational factors.

COMPUTER-AIDED DESIGN

In addition to understanding the basic engineering and economic principles and their applications, the design engineer needs special tools to assist in making the large number of calculations required in a design project. Many of these calculations are repetitive in nature and therefore are readily adaptable to computer solution using data-file programs. In recent years, effective systems have been developed which allow input of commands and data to a language interpreter for the system which then calls for appropriate calculational subsystems from data files, makes the calculations needed, and reports the results. This approach is generally referred to as *computer-aided design* and is a major development in industry to allow more efficient and comprehensive design analyses.‡

The "first generation" of the programs for computer-aided design developed in the 1950's and early 1960's and had limited capabilities even though the potential value was very apparent. This included programs such as the Chevron Heat and Material Balancing Program, the Kellogg Flexible Flow-Sheet, PACER, and CHESS. By the early 1970's a "second generation" of programs was well under way allowing much more sophisticated and effective use. Included here were Monsanto's FLOWTRAN, Exxon's COPE, Union Carbide's IPES, Du Pont's CPES, and Chiyoda's CAPES programs. Because of the large expenditure the individual industries put into the development of the programs, many of the programs were highly proprietary.

In 1972, in order to investigate the possibility of getting some of the proprietary programs out into the general educational market, a committee of the Commission on Education of the National Academy of Engineering carried out an evaluation of industrial, computer-aided, steady-state, process design and simulation programs that might be available for use at colleges and universities. This committee had the name of CACHE (Computer Aids for Chemical Engineering Education) and has made significant progress in these activities. CACHE

---

† Detailed information on materials of construction is presented in Chap. 11 and in the references for that chapter.

‡ Evans, L. B., and W. D. Seider, The Requirements of an Advanced Computing System, *Chem. Eng. Progr.*, **72**(6):80 (1976); A. I. Llewelyn *et al.*, Computer-Aided Design (CAD) Hardware and Technology Feature, *Computer-Aided Design*, **9**(4):222 (Oct., 1977); See also the list of suggested additional references at the end of this chapter.

concluded that Monsanto's FLOWTRAN was the most suitable for use at colleges and universities and has published numerous articles and reports on the use of this program.†‡

While the overall development of computer-aided design is rapidly moving into a "third generation" of programs which are very sophisticated, there are many programs currently available which are extremely useful for design calculations. A summary of many of these programs particularly appropriate for use in the chemical process industries has been developed through a grant to the Massachusetts Institute of Technology by the U.S. Department of Energy. The purpose of the project was to develop an advanced-software computing system to meet the needs of the chemical-process engineers in the 1980's with the system being named ASPEN (Advanced System for Process Engineering). A software survey to obtain sources of computer programs that would fall in the categories of commercial, university, proprietary industrial, specialty, and public was carried out, and the results have been published in a series in *Chemical Engineering.*§

---

† FLOWTRAN was first conceived in 1961 by the Applied Mathematics Department of Monsanto Company for process design and simulation and has undergone continuous development since then. In 1966, the system was put into general use at Monsanto and, from 1969 to 1973, outside companies could use FLOWTRAN through Monsanto Enviro-Chem Systems, Inc. by commercial computer networks. After 1973, FORTRAN was licensed and special arrangements were made by Monsanto to allow use in colleges and universities.

‡ Seader, J. D., W. D. Seider, and A. C. Pauls, "FLOWTRAN Simulation—An Introduction," 2d ed., CACHE Corp., 77 Massachusetts Ave., Cambridge, Massachusetts, 1977; Hughes, R. R., "CACHE Use of FLOWTRAN on United Computing Systems," Univ. of Wisconsin, Madison, Wisconsin, 1975; Clark, J. P., "Exercises in Process Simulation Using FLOWTRAN," CACHE Corp., 77 Massachusetts Ave., Cambridge, Massachusetts, 1977; Rosen, E. M. and A. C. Pauls, Computer-Aided Chemical Process Design—The FLOWTRAN System, *Comput. Chem. Eng.*, 1(1):11 (1977).

§ Peterson, J. N., C.-C. Chen, and L. B. Evans, Computer Programs for Chemical Engineers: 1978, Part 1 (Programs on 1. Steady-state flow sheet simulator: Power plant, 2. Dynamic flow sheet simulator, 3. Special process package: Coal conversion processing—Solid handling—Waste treatment), *Chem. Eng.*, 85(13):145 (June 5, 1978); Part 2 (Programs on unit operations: Absorption—Crystallization—Distillation—Evaporation—Extraction—Flash—Furnace—Heat exchange—Leaching—Pressure Change—Reactor design), *Chem. Eng.*, 85(15):69 (July 3, 1978); Part 3 (Programs on 1. Thermodynamics: Chemical equilibrium—Vapor liquid equilibrium—Thermodynamic properties—Physical properties—Transport properties—Metallurgical thermochemical data, 2. Cost estimation: Heat exchange—Vessel—Wastewater treatment plant, 3. Economic evaluation), *Chem. Eng.*, 85(17):79 (July 31, 1978); Part 4 (Programs on 1. Mathematics: Data regression—Equation solving—Optimization—Partitioning and tearing, 2. Piping, 3. Tanks and vessels, 4. Dynamics and process control, 4. Instrument data analysis, 5. Miscellaneous: Plant management—Coefficient deck generation—Water quality simulation—Stream assay analysis—Hybrid simulation compiler—Flowchart drawing—Matrix assembly—Hybrid computer simulation—Cosited ventures analysis—Interactive system—Digital analog simulator—Ventures analysis—Plotting—Unit conversion—Manpower utilization), *Chem. Eng.*, 85(19):107 (Aug. 28, 1978). See also Chen, C.-C., and L. B. Evans, More Computer Programs for Chemical Engineers, *Chem. Eng.* 86(11):167 (May 21, 1979).

A total of 391 programs are listed in the four-part series referred to in the preceding paragraph. Examples of four of these programs are given in the following:†

Category	Program name and description	Person to contact for further information
Unit operation, reactor (204)	**Name: One and Two-Dimensional Model of Catalytic Reactors for Hydrazine Decomposition (United Aircraft Corporation)**—The program is based upon a one-dimensional model of the reactor system, which describes the behavior of reactors having radially uniform injection profiles and catalyst bed configurations, while the other program is based on a two-dimensional model that permits consideration of nonuniform radial injection and of catalyst bed configurations exhibiting both radial and axial nonuniformities. The programs consider a typical reactor where liquid hydrazine, injected into the reaction chamber, decomposes to ammonia, nitrogen and hydrogen; the ammonia can in turn decompose to nitrogen and hydrogen. The program is completely self-contained, and input parameters allow for consideration of buried injector schemes, various distributions of catalyst pellets throughout the chamber, as well as variations in inlet temperature, pressure, flowrate and reactor length. The program is intended as a tool for use in engineering design of reactor hardware, and is written so that implementation for use on various large scale digital computers should require only a minor effort.	Mrs. Shirley Parten COSMIC Suite 112, Barrow U. of Georgia Athens, GA 30602
Unit operation, distillation (CACHE) (94)	**Name: Design and Cost of a Distillation Tower and Auxiliary Equipment**—Performs process and mechanical design of multicomponent distillation tower, reflux drum, condenser and reboiler and also roughly sizes pumps and motors. Process calculations use the Fenske-Underwood-Gilliland shortcut method. Mechanical design based on Souders-Brown correlation for maximum vapor velocity. Column and reflux drum wall thickness calculated according to ASME Pressure Vessel Code. Cost of equipment ob-	Robert S. Kirk U. of Massachusetts Amherst, MA 01002

(continued)

† Quoted with permission.

Category	Program name and description	Person to contact for further information
	tained from equations fitted to published data. The number of components handled is ten, but this may be readily increased.	
Cost estimation, heat exchanger (269)	**Name: Shell and Tube Heat Exchange Cost Estimation, Program 5066 (Phillips Petroleum Co. Program Package 1)**—Estimates cost of a shell-and-tube heat exchanger by summing the cost of the individual component costs. User may specify almost any standard shell-and-tube arrangement. Required input data have been kept to a minimum by having tables of cost, materials, etc. stored internally. An estimate based on current material current costs may be obtained by using cost index factors to update the cost data stored in the tables. Phillips compared estimated and purchase cost of 49 exchangers. Results were that estimated costs should lie between +0.45 and 20% of actual cost.	William R. Vickroy V. P., Marketing McDonnell-Douglas Automation Co. St. Louis, MO 63166
Thermodynamics, transport properties (CACHE) (263)	**Name: Evaluation of Gas Phase Diffusion Coefficients from the Chapman-Enskog Equation**—Computes binary mass and molar diffusivities in multicomponent gas mixtures. The binary diffusivities and specified compositions are then used to compute effective diffusivities for each component in the mixture (assuming the remaining ones are stagnant). The computations of binary diffusivities using the Chapman-Enskog method will be executed in spite of a total or partial lack of exact data concerning the required Lennard-Jones parameters. In such situations where data are lacking, the Lennard-Jones parameters are estimated from the critical constants for the gas. The collision function which appears in the Chapman-Enskog equation is evaluated by means of an empirical equation.	E. N. Bart or E. C. Roche, Jr. Dept of Chem. Eng. & Chem. Newark College of Engineering 323 High St. Newark, NJ 07102

Computer-aided design has become an important component of the overall design process and will become increasingly important for simplifying the calculational effort as more programs of greater sophistication become available. Many industrial concerns will continue to develop their own proprietary programs, but useful programs and methods will more and more become available to the public sector, and the design engineer should be ready to take advantage of them. Problem 21 at the end of this chapter represents a simple illustration of the type of time saving that can be accomplished in chemical reaction analysis by using computer-aided design.

SUGGESTED ADDITIONAL REFERENCES FOR PROCESS DESIGN DEVELOPMENT

Handbooks

"American Institute of Physics Handbook," D. E. Gray, McGraw-Hill Book Company, New York, 1972.

"Chemical Engineers' Handbook," 5th ed., R. H. Perry and C. H. Chilton, McGraw-Hill Book Company, New York, 1974.

"Compressed Air and Gas Handbook," 4th ed., J. P. Rollins, Compressed Air and Gas Institute, New York, 1973.

"Energy Technology Handbook," D. M. Considine, McGraw-Hill Book Company, New York, 1977.

"Environmental Engineers' Handbook," B. G. Lipták, Chilton Book Company, Philadelphia, Pennsylvania, 1974.

"Facilities and Plant Engineering Handbook," B. T. Lewis and J. P. Marron, McGraw-Hill Book Company, New York, 1973.

"Fire Protection Handbook," 14th ed., G. P. McKinnon and K. Tower, National Fire Protection Association, Boston, Massachusetts, 1976.

"Handbook of Advanced Wastewater Treatment," 2d ed., R. L. Culp, G. M. Wesner, and G. L. Culp, Van Nostrand Reinhold, New York, 1978.

"Handbook of Air Pollution," J. P. Sheehy, W. C. Achinger, and R. A. Simon, HEW, R. A. Taft Sanitary Engineering Center, Cincinnati, Ohio, 1968.

"Handbook of Chemistry and Physics," 58th ed., R. C. Weast, CRC Press, Cleveland, Ohio, 1977–1978.

"Handbook of Environmental Control," R. G. Bond and C. P. Straub, CRC Press, West Palm Beach, Florida, 1978.

"Handbook of Environmental Management," C. J. Hilado, Technomic Publishing Company, Westport, Connecticut, 1972.

"Handbook of Heat Transfer," W. M. Rohsenow and J. P. Hartnett, McGraw-Hill Book Company, New York, 1972.

"Handbook of Heating, Ventilating, and Air Conditioning," 6th ed., J. Porges, Newnes-Butterworth, London, 1971.

"Handbook of Industrial Chemistry," J. A. Kent, Van Nostrand Reinhold, New York, 1974.

"Handbook of Industrial Control Computers," T. J. Harrison, J. Wiley and Sons, New York, 1972.

"Handbook of Industrial Noise Control," L. L. Faulkner, Industrial Press, New York, 1976.

"Handbook of Measurement and Control," E. E. Herceg, Schaevitz Engineering, Pennsuaken, New Jersey, 1972.

"Handbook of Process Stream Analysis," K. J. Clevett, Halsted Press, New York, 1974.

"Handbook of Solid Waste Disposal," J. L. Pavoni, J. E. Heer, Jr., and D. J. Hagerty, Van Nostrand Reinhold, New York, 1975.

"Handbook of System and Product Safety," W. Hammer, Prentice-Hall, Englewood Cliffs, New Jersey, 1972.

"Handbook of Tables for Applied Engineering Sciences," R. E. Bolz and G. L. Tuve, CRC Press, Inc., West Palm Beach, Florida, 1973.

"Handbook of Tables for Probability and Statistics, 2d ed., W. H. Beyer, CRC Press, Cleveland, Ohio, 1968.

"Handbook on Air Pollution Control," F. L. Cross, Jr., Technomic Publishing Company, Westport, Connecticut, 1973.
"Industrial Pollution Control Handbook," H. F. Lund, McGraw-Hill Book Company, New York, 1970.
"Industrial Safety Handbook," W. Handley, McGraw-Hill Book Company, New York, 1977.
"Instrument Engineers' Handbook," B. G. Lipták, Chilton Book Company, Philadelphia, Pennsylvania, 1969–1970.
"Maintenance Engineering Handbook, L. C. Morrow, McGraw-Hill Book Company, New York, 1966.
"Materials Handbook," G. S. Brady, McGraw-Hill Book Company, New York, 1971.
"Mechanical Engineers' Handbook," A. Parrish, Butterworths, London, 1973.
"Petroleum Processing Handbook," W. F. Bland and R. L. Davidson, McGraw-Hill Book Company, New York, 1967.
"Piping Handbook," 5th ed., R. C. King and S. Crocker, McGraw-Hill Book Company, New York, 1967.
"Practical Petroleum Engineers' Handbook," J. Zaba and W. T. Doherty, Gulf Publishing Company, Houston, Texas, 1970.
"Pollution Engineering Practice Handbook," P. N. Cheremisnoff and R. A. Young, Ann Arbor Science Publishers, Ann Arbor, Michigan, 1975.
"Process Instruments and Controls Handbook," 2d ed., D. M. Considine, McGraw-Hill Book Company, New York, 1974.
"Standard Handbook of Engineering Calculations," T. G. Hicks, McGraw-Hill Book Company, New York, 1972.
"Tribology Handbook," M. J. Neale, Butterworths, London, 1973.

Computer-aided design

Adams, J. A., and D. F. Rogers, "Computer-Aided Heat Transfer Analysis," McGraw-Hill Book Company, New York, 1973.
Allen, D. H., How to Use Mixed-Integer Programming, *Chem. Eng.*, **83**(7):114 (1976).
Aristovich, V. Y., and A. I. Levin, Method for Calculating Rectification of Multi-component Mixtures on Digital Computers, *Intern. Chem. Eng.*, **8**(1):1 (1968).
Baltzell, H. J., and R. G. Jones, Computerized Piping Cost Estimating System (PCES), *1970 Trans. Am. Assoc. Cost Eng.* (1970), p. 120.
Benenati, R. F., Solving Engineering Problems on Programmable Pocket Calculators, *Chem. Eng.*, **84**(5):201 (1977); **84**(6):129 (1977).
Beychok, M. R., Program Calculators for Design Study, *Hydrocarbon Process.*, **55**(9):261 (1976).
Birdwell, J. R., and W. W. Shull, Computers Write Our Piping Specs., *Hydrocarbon Process.*, **49**(8):103 (1970).
Blecker, H. G., Computer Simulation of Definitive Estimates, *1973 Trans. Am. Assoc. Cost Eng.* (1973), p. 143.
Bluck, D., and A. J. Sheppard, Computer Programs in Process Plant Design, *Chem. Eng. (London)*, 751 (December 1975).
Boddy, D. E., "Engineering Design Computation Manual," Holt, Rinehart and Winston, New York, 1969.
Bresler, S. A., and M. T. Kuo, Cost Estimating by Computer, *Chem. Eng.*, **79**(12):84 (1972).
—— and ——, More Programs for Cost Estimating by Computer, *Chem. Eng.*, **79**(14):130 (1972).
Briddell, E. T., Process Design by Computer, *Chem. Eng.*, **81**(3):60 (1974); **81**(5):113 (1974): **81**(7):77 (1974).
Brown, I. D., Computer-Aided Pipe Sketching, *Chem. Eng. Progr.*, **67**(10):41 (1971).
Buckley, P. S., R. K. Cox, and W. L. Luyben, How to Use a Small Calculator in Distillation Column Design, *Chem. Eng. Progr.*, **74**(6):49 (1978).
Burningham, D. W., and F. D. Otto, Which Computer Design for Absorbers? *Hydrocarbon Process.*, **46**(10):163 (1967).

Butterworth, D., and L. B. Cousins, Use of Computer Programs in Heat-Exchanger Design, *Chem. Eng.*, **83**(14):72 (1976).

Campbell, J. R., and J. R. F. Alonso, To Get Curves from Data Points, *Hydrocarbon Process.*, **57**(1):123 (1978).

Canfield, F. B., Estimate K-Values with the Computer, *Hydrocarbon Process.*, **50**(4):137 (1971).

Carter, A. G., Computers and Chemical Plant Engineering, *British Chem. Eng.*, **15**(11):1427 (1970).

Chai, D. K., A Practical Approach to Using the C.O.M.E. Project on Complex Items, *1973 Trans. Am. Assoc. Cost Eng.* (1973), p. 110.

Crowe, C. M., A. E. Hamielec, T. W. Hoffman, A. I. Johnson, D. R. Woods, and P. T. Shannon, "Chemical Plant Simulation," Prentice-Hall, Englewood Cliffs, New Jersey, 1971.

DeCicco, R. W., Economic Evaluation of Research Projects—By Computer, *Chem. Eng.*, **75**(12):84 (1968).

deLesdernier, D. L., and J. T. Sommerfeld, Computer Program Sizes Pipe, *Hydrocarbon Process.*, **51**(3):112 (1972).

Distefano, G. P., and W. Richards, II, Hybrid Computers in the CPI, *Chem. Eng.*, **75**(10):195 (1968).

Economopoulos, A. P., A Fast Computer Method for Distillation Calculations, *Chem. Eng.*, **85**(10):91 (1978).

——, Computer Design of Sieve Trays and Tray Columns, *Chem. Eng.*, **85**(27):109 (1978).

Ewell, R. B., and G. Gadmer, Design Cat Crackers by Computer, *Hydrocarbon Process.*, **57**(4):125 (1978).

Evans, L. B., D. G. Steward, and C. R. Sprague, Computer-Aided Chemical Process Design, *Chem. Eng. Progr.*, **64**(4):39 (1968).

—— and W. D. Seider, The Requirements of an Advanced Computing System, *Chem. Eng. Progr.*, **72**(6):80 (1976).

Frith, K. M., Your Computer Can Help You Estimate Physical-Property Data, *Chem. Eng.*, **79**(4):72 (1972).

Goffin, J. W., Computerization Saves Time in Detailed Estimates, *1973 Trans. Am. Assoc. Cost Eng.* (1973), p. 125.

Hariu, O. H., and R. C. Sage, Crude Split Figures by Computer, *Hydrocarbon Process.*, **48**(4):143 (1969).

Henley, E. J., *et al.*, "Computer Programs for Chemical Engineering," Vols. 1–7, Aztec Publishing Company, Austin, Texas, 1972–1974.

Hodge, B., and J. P. Mantey, Applying FORTRAN to Engineering Problems, *Chem. Eng.*, **75**(16):151 (1968).

—— and ——, Elements of Digital Computers, *Chem. Eng.*, **74**(20):180 (1967); **74**(22):167 (1967).

—— and ——, FORTRAN—Subprograms and Specification Statements, *Chem. Eng.*, **75**(13):271 (1968).

Hughson, R. V., and E. H. Steymann, Computer Programs for Chemical Engineers, *Chem. Eng.*, **78**(16):66 (1971).

—— and ——, More Computer Programs for Chemical Engineers, *Chem. Eng.*, **78**(29):63 (1971).

Hyman, A., "The Computer in Design," Studio Vista, London, 1973.

Hyman, M. H., Fundamentals of Engineering/HPI Plants with the Digital Computer, *Petro/Chem. Eng.*, **40**(5):53 (1968); **40**(6):43 (1968); **40**(9):33 (1968); **40**(10):37 (1968); **40**(11):33 (1968); **40**(12):36 (1968); **41**(2):30 (1969); **41**(4):29 (1969); **41**(6):46 (1969); and **41**(9):31 (1969).

Jones, P. R., and S. Katell, Computer Usage for Evaluation of Design Parameters and Cost of Heat Exchangers, *1968 Trans. Am. Assoc. Cost Eng.* (1968).

Johnson, D. W., and C. P. Colver, Mixture Properties by Computer, *Hydrocarbon Process.*, **47**(12):79 (1968); **48**(1):127 (1969); **48**(3):113 (1969).

Kinsler, M. R., and T. F. Tindel, Computer Cost Analysis Tools for Advanced System Cost Effectiveness Evaluation, *1970 Trans. Am. Assoc. Cost Eng.*, (1970), p. 132.

Klumpar, I. V., Process Predesign by Computer, *Chem. Eng.*, **76**(20):114 (1969).

——, Project Evaluation by Computer, *Chem. Eng.*, **77**(14):76 (1970).

Korelitz, T. H., Integrating Plant Design With Process Simulation, *Chem. Eng.*, **78**(13):98 (1971).

Lederman, P. B., Process Design With Computers, *Chem. Eng.*, **75**(20):221 (1968).

———, Equipment Design by Computers, *Chem. Eng.*, **75**(22):151 (1968).

———, Flowsheet Simulation and Beyond, *Chem. Eng.*, **75**(25):127 (1968).

———, Computers in Research and Development, *Chem. Eng.*, **75**(27):107 (1968).

Lee, W., and M. T. Tayyabkhan, Optimize Process Runs . . . Get More Profit, *Hydrocarbon Process.*, **49**(9):286 (1970).

Lin, T. D., and R. S. H. Mah, A Sparse Computation System for Process Design and Simulation: Part 1. Data Structures and Processing Techniques, *AIChE J.*, **24**(5):830 (1978).

Lion, A. R., and W. C. Edmister, Make Equilibrium Calculations by Computer, *Hydrocarbon Process.*, **54**(8):119 (1975).

Loux, P. C., How to Evaluate Computer Programs, *Chem. Eng.*, **78**(16):66 (1971).

Ludwig, W. R., and R. P. Peterson, Computer Design Helps Small Chemical Company, *Chem. Eng.*, **76**(5):98 (1969).

Mah, R. S. H., and T. D. Lin, A Performance Evaluation Based on the Simulation of a Natural Gas Liquefaction Process: Part II, *AIChE J.*, **24**(5):839 (1978).

Martin, J. P., A Proposed Approach to Estimating Vessel Costs by Computer, *1968 Trans. Am. Assoc. Cost Eng.* (1968).

Modell, D. J., Minimum Cost Formulations by Computer, *1970 Trans. Am. Assoc. Cost Eng.* (1970), p. 64.

Nguyen, H.-X., Computer Program Expedites Packed-Tower Design, *Chem. Eng.*, **85**(26):181 (1978).

Niccoli, L. G., R. T. Jaske, and P. A. Witt, System Costs Say Optimize Cooling, *Hydrocarbon Process.*, **49**(10):97 (1970).

Null, H. R., "Phase Equilibrium in Process Design," J. Wiley and Sons, New York, 1970.

Orr, H. D., and R. A. Stutz, A Computerized Estimating Package, *1973 Trans. Am. Assoc. Cost Eng.* (1973), p. 206.

Paskusz, G. F., R. L. Motard, G. S. Dawkins, and D. Wills, "Syllabus for Computer Aided Design," Univ. of Houston, Houston, Texas (1969).

Peterson, J. N., C.-C. Chen, and L. B. Evans, Computer Programs for Chemical Engineering: 1978, *Chem. Eng.*, **85**(13):145 (1978); **85**(15):69 (1978); **85**(17):79 (1978); **85**(19):107 (1978).

Richter, S. H., Control Valve Sizing by Computer, *Petro/Chem. Eng.*, **40**(13):23 (1968).

Rissler, K., Heat Exchanger Design by Computer, *Chem. Proc. Eng.*, **52**(10):61 (1971).

Ritchey, K. R., F. B. Canfield, and T. B. Calland, Heavy-Oil Distillation via Computer Simulation, *Chem. Eng.*, **83**(16):79 (1976).

Roberts, J. A., Computer-Aided Pipe Sketching, *Innovation*, **3**(1):9 (1972).

Robins, D. L., and M. M. Mattia, Computer Program Helps Design Stacks for Curbing Air Pollution, *Chem. Eng.*, **75**(3):119 (1968).

Rosen, E. M., and A. C. Pauls, Computer Aided Chemical Process Design—The Flowtran System, *Comput. Chem. Eng.*, **1**(1):11 (1977).

Russell, R. A., Improve Your Efficiency in Writing Computer Programs, *Chem. Eng.*, **84**(9):111 (1977).

Sarma, N. V. L. S., P. J. Reddy, and P. S. Murti, A Computer Design Method for Vertical Thermosyphon Reboilers, *I&EC Process Des. & Dev.*, **12**(3):278 (1973).

Shah, Y. T., Computer Simulation of Transport Processes, *I&EC Prod. Res. & Develop.*, **11**(3):269 (1972).

Silver, L., and S. Bacher, Computer Estimation in Batch Processes, *Chem. Eng. Progr.*, **65**(5):56 (1969).

Simon, H., and T. H. Whelan, Use Computer to Select Optimum Control Valve, *Hydrocarbon Process.*, **49**(7):103 (1970).

Sommer, M., E. F. Cooke, III, J. E. Goehring, and J. J. Haydel, Process Control in a Petrochemical Complex, *Chem. Eng. Progr.*, **67**(10):54 (1971).

Sommerfeld, J. T., and G. L. Perry, Compressibility Factors by Computer, *Hydrocarbon Process.*, **47**(10):109 (1968).

Spencer, R. A., Jr., Predicting Heat-Exchanger Performance by Successive Summation, *Chem. Eng.*, **85**(27):121 (1978).

Spitzer, H., Pipework Detailing by Computer, *Chem. Proc. Eng.*, **50**(11):86 (1969).

———, Computer Approach to Pipe Detailing, *Chem. Eng. (London)*, **252**:305 (August 1971).

Tarrer, A. R., H. C. Lim, and L. B. Koppel, Finding the Economically Optimum Heat Exchanger, *Chem. Eng.*, **78**(22):79 (1971).

Thorngren, J. T., Reboiler Computer Evaluation, *I&EC Process Des. & Dev.*, **11**(1):39 (1972).

Tsubaki, M., Computer-Aided Process Analysis and Design, *Chem. Eng. Progr.*, **69**(9):78 (1973).

Usami, H., Computer Checks Steam Cracker, *Hydrocarbon Process.*, **51**(1):103 (1972).

van Eenennaam, J., and A. C. A. v. Kesteren, Isometric Pipework Drawing by Computer, *Chem. Proc. Eng.*, **52**(9):53 (1971).

Westerberg, A. W., and F. C. Edie, Computer-Aided Design, *Chem. Eng. J.*, **2**(1):9 (1971); **2**(1):17 (1971); **2**(2):114 (1971).

Wild, N. H., Program for Discounted-Gas-Flow Return on Investment, *Chem. Eng.*, **84**(10):137 (1977).

Wills, J. S., Size Vapor Piping by Computer, *Hydrocarbon Process.*, **49**(5):149 (1970).

Winter, P., and R. G. Newell, Data Bases in Design Management, *Chem. Eng. Progr.*, **73**(6):97 (1977).

Woosley, R. D., G. K. Baker, and D. J. Stubblefield, Use of the Computer to Select Design Parameters for Solid Adsorbent Dehydration of Gas Streams, *Chem. Eng. Symp. Ser.*, **67**(117):98 (1971).

Computer Programs for Chemical Engineers—1973, *Chem. Eng.*, **80**(19):121 (1973); **80**(21):127 (1973).

How to Computerize Plant Costs Estimations, *Petro/Chem. Eng.*, **42**(5):39 (1970).

Survey Report: Computer Applications in Chemical Engineering, *Chem. Eng. (London)*, **222**:345 (1968).

Engineering design

Avriel, M., and D. J. Wilde, Engineering Design under Uncertainty, *I&EC Process Des. & Dev.*, **8**(1):124 (1969).

Baasel, W. D., "Preliminary Chemical Engineering Plant Design," American Elsevier, New York, 1976.

Backhurst, J. R., and J. H. Harker, "Process Plant Design," American Elsevier Publishing Company, New York, 1973.

Leathers, J. M., Future Changes in Chemical Engineering, *Chem. Eng.*, **84**(12):119 (1977).

Ludwig, E. E., Changes in Process Equipment Caused by the Energy Crisis, *Chem. Eng.*, **82**(6):61 (1975).

Equipment specification

Bresler, S. A., A Hard Look at Equipment Warranties, *Chem. Eng.*, **75**(7):86 (1968).

———, Compressors, *Chem. Eng.*, **77**(12):161 (1970).

Brocato, L. J., Writing Specifications for a Laboratory Automation System, *Am. Lab.*, **5**(9):47 (1973).

Elliott, I. G., Improve Your Specification Writing, *Chem. Proc. Eng.*, **49**(7):76 (1968).

Evers, W. J., and E. E. Ludwig, How to Specify Centrifugal Pumps, *Petro/Chem. Eng.*, **40**(5):50 (1968).

Franzel, H. L., Guide to Trouble-Free Quality Control of Equipment Engineering, *Chem. Eng.*, **77**(12):135 (1970).

Hertz, M. J., Does Your Equipment Warranty Really Protect You? *Chem. Eng.*, **75**(8):137 (1968).

Lord, R. C., P. E. Minton, and R. P. Slusser, Heat Exchangers, *Chem. Eng.*, **77**(12):153 (1970).

McLaren, D. B., and J. C. Upchurch, Distillation, *Chem. Eng.*, **77**(12):139 (1970).

Nailen, R. L., Electric-Motor Drives, *Chem. Eng.*, **77**(12):181 (1970).

Penney, W. R., Mixers, *Chem. Eng.*, **77**(12):171 (1970).

Rubin, F. L., How to Specify Heat Exchangers, *Chem. Eng.*, **75**(8):130 (1968).

Wetherhorn, D., Evaporators, *Chem. Eng.*, **77**(12):187 (1970).

Flow diagrams

Forder, G. J., and H. P. Hutchison, A Flowsheet Drawing Program for a Visual Display Computer, *Can. J. Chem. Eng.*, **48**(1):79 (1970).

Hill, R. G., Drawing Effective Flowsheet Symbols, *Chem. Eng.*, **75**(1):84 (1968).
Roberts, J. A., Computer-Aided Pipe Sketching, *Innovation*, **3**(1):9 (1972).

Material and energy balances

Agarwal, J. C., I. V. Klumpar, and F. D. Zybert, A Simple Material Balance Model, *Chem. Eng. Progr.*, **74**(6):68 (1978).
Boston, J. F., and S. L. Sullivan, Jr., An Improved Algorithm for Solving the Mass Balance Equations in Multistage Separation Processes, *Can. J. Chem. Eng.*, **50**(5):663 (1972).
Brinsko, J. A., How to Make a Steam Balance, *Hydrocarbon Process.*, **57**(11):227 (1978).
Miller, R., Jr., Process Energy Systems, *Chem. Eng.*, **75**(11):130 (1968).
Murthy, A. K. S., A Least-Squares Solution to Mass Balance Around a Chemical Reactor, *I&EC Process Des. & Dev.*, **12**(3):246 (1973).
———, Material Balance Around a Chemical Reactor, *I&EC Process & Des. & Dev.*, **13**(4):347 (1974).
Sisson, W., Calculate Heat Energy Absorbed by Boiler Fluid, *Power Eng.*, **75**(2):33 (1971).
Slack, J. B., Steam Balance: A New Exact Method, *Hydrocarbon Process.*, **48**(3):154 (1969).

Materials of construction

See references listed in Chap. 11.

Pilot plants

Drew, J. W., and A. F. Ginder, How to Estimate the Cost of Pilot-Plant Equipment, *Chem. Eng.*, **77**(3):100 (1970).
Ellis, C. E., Unattended Pilot Plants, *Chem. Eng. Progr.*, **64**(10):50 (1968).
Hudson, W. G., Pilot Plant Cost Control, *Chem. Eng. Progr.*, **64**(10):39 (1968).
Katzen, R., When is the Pilot Plant Necessary? *Chem. Eng.*, **75**(7):95 (1968).
Katell, S., Justifying Pilot Plant Operations, *Chem. Eng. Progr.*, **69**(4):55 (1973).
Marinak, M. J., Safety Checklist, *Chem. Eng. Progr.*, **63**(11):59 (1967).
May, V. T., Process Development Costs and Experience, *Chem. Eng. Progr.*, **69**(2):71 (1973).
Ohsol, E. O., Do We Really Need Pilot Plants? *Research/Development*, **23**(3):38 (1972).
———, What Does It Cost to Pilot a Process? *Chem. Eng. Progr.*, **69**(4):17 (1973).

Process Design

Baasel, W. D., "Preliminary Chemical Engineering Plant Design," American Elsevier, New York, 1976.
Backhurst, J. R., and J. H. Harker, "Process Plant Design," American Elsevier Publishing Company, New York, 1973.
Bluck, D., and A. J. Sheppard, Computer Programs in Process Plant Design, *Chem. Eng. (London)*, 751 (December 1975).
Bodman, S. W., "The Industrial Practice of Chemical Process Engineering," The M.I.T. Press, Cambridge, Massachusetts, 1968.
Broughton, D. B., Molex: Case History of a Process, *Chem. Eng. Progr.*, **64**(8):60 (1968).
Burkhardt, D. B., Increasing Conversion Efficiency, *Chem. Eng. Progr.*, **64**(11):66 (1968).
Canova, F., Matching Turbomachinery to a Process, *Chem. Eng.*, **76**(12):178 (1969).
Clark, J. P., How to Design Chemical Plants on the Back of an Envelope, Facts and Their Interrelation., *Chemtech*, **6**(1):23 (1976).
Dosher, J. R., Trends in Petroleum Refining, *Chem. Eng.*, **77**(17):96 (1970).
Drew, J. W., Design for Solvent Recovery, *Chem. Eng. Progr.*, **71**(2):92 (1975).
Fair, J. R., Process Design: Past, Present and Future, *Chem. Eng.*, **80**(1):98 (1973).
Glass, J. A., Process Engineering, *Chem. Eng.*, **75**(1):56 (1968).
Grau, J. C., Project Engineering, *Chem. Eng.*, **75**(1):66 (1968).
Happel, J., and D. G. Jordan, "Chemical Process Economics," 2d ed., Marcel Dekker, New York, 1975.

Hendry, J. E., D. F. Rudd, and J. D. Seader, Synthesis in the Design of Chemical Processes, *AIChE J.*, **19**(1), (1973).

Jordan, D. G., "Chemical Process Development," Vols. 1 and 2, J. Wiley and Sons, New York, 1968.

Kerns, G. D., How to Check Process Designs, *Hydrocarbon Process.*, **51**(1):100 (1972).

King, C. J., Understanding and Conceiving Chemical Processes, *AIChE Monograph Ser.*, **70**(8):1 (1974).

Klumpar, I. V., Process Predesign by Computer, *Chem. Eng.*, **76**(20):114 (1969).

Lee, H. H., L. B. Koppel, and H. C. Lim, Integrated Approach to Design and Control of a Class of Countercurrent Processes, *I&EC Process Des. & Dev.*, **11**(3):376 (1972).

Ligi, J. J., Processing with Cryogenics, *Hydrocarbon Process.*, **48**(6):118 (1969).

Lin, T. D., and R. S. H. Mah, A Sparse Computation System for Process Design and Simulation: Part 1. Data Structures and Processing Techniques, *AIChE J.*, **24**(5):830 (1978).

Llovet, J. E., H. J. Klooster, and D. G. Chapel, Refinery Design (circa 1984), *Chem. Eng. Progr.*, **71**(6):85 (1975).

Lobstein, R., "Guide to Chemical Plant Planning," Noyes Development Corporation, Park Ridge, New Jersey, 1969.

Lowe, E. I., and A. E. Hidden, "Computer Control in Process Industries," Peter Peregrinus, London, 1971.

Ludwig, E. E., "Applied Process Design for Chemical and Petrochemical Plants," Vol. 1, 2, and 3, Gulf Publishing Company, Houston, Texas, 1964.

——, Changes in Process Equipment Caused by the Energy Crisis, *Chem. Eng.*, **82**(6):61 (1975).

Mah, R. S. H., and T. D. Lin, Part II, A Performance Evaluation Based on the Simulation of a Natural Gas Liquefaction Process, *AIChE J.*, **24**(5):839 (1978).

Nahas, R. S., How Feedstocks Affect Plant Design, *Hydrocarbon Process.*, **54**(7):97 (1975).

Null, H. R., "Phase Equilibrium in Process Design," J. Wiley and Sons, New York, 1970.

Pickert, P. E., A. P. Bolton, and M. A. Lanewala, Process Design with Molecular Sieve Catalysts, *Chem. Eng.*, **75**(16):139 (1968).

Powers, G. J., Heuristic Synthesis in Process Development, *Chem. Eng. Progr.*, **68**(8):88 (1972).

Rabb, A., Use Entropy for Quick Evaluations, *Hydrocarbon Process.*, **48**(6):133 (1969).

Robbins, J., New Approaches to Process Design, *Chem. Eng.* (*London*), 298 (October 1970).

Rudd, D. F., G. J. Powers, and J. J. Siirola, "Process Synthesis," Prentice-Hall, Englewood Cliffs, New Jersey, 1973.

Rudkin, J., From Bright Idea to Plant Production, *Chem. Eng.*, **82**(3):69 (1975).

Siirola, J. J., G. J. Powers, and D. F. Rudd, Synthesis of System Designs: III. Toward a Process Concept Generator, *AIChE J.*, **17**(3):677 (1971).

Smith, B. D. (ed.), "Design Case Studies," Vol. 1-22, Washington University, St. Louis, Missouri, 1966–1976.

Wells, G. L., and P. M. Robson, "Computation for Process Engineers," J. Wiley and Sons, New York, 1973.

Williams, V. C., Cryogenics, *Chem. Eng.*, **77**(25):92 (1970).

AIChE Student Contest Problems, 1966–72, American Institute of Chemical Engineers, New York, 1966–72.

1973 Student Contest Problem and First Prize-Winning Solution, *AIChE Student Chap. Bull.*, **14**(2):13, (1973).

1974 Student Contest Problem, *AIChE Student Chap. Bull.*, **15**(2):30 (1974).

1975 Student Contest Problem and First Prize-Winning Solution, *AIChE Student Chap. Bull.*, **16**(2):23 (1975).

1976 Student Contest Problem and First Prize-Winning Solution, *AIChE Student Chap. Bull.*, **17**(2):39 (1976).

1977 Student Contest Problem and First Prize-Winning Solution, *AIChE Student Chap. Bull.*, **18**(2):55 (1977).

Process simulation

Barnes, F. J., and C. J. King, Synthesis of Cascade Refrigeration and Liquefaction Systems, *I&EC Process Des. & Dev.*, **13**(4):421 (1974).

Chen, J. W., F. L. Cunningham, and J. A. Buege, Computer Simulation of Plant-Scale Multi-column Adsorption Processes under Periodic Countercurrent Operation, *I&EC Process Des. & Dev.*, **11**(3):430 (1972).

Chou, A., A. M. Fayon, and B. L. Bauman, Simulations Provide Blueprint for Distillation Operation, *Chem. Eng.*, **83**(12):131 (1976).

Crowe, C. M., A. E. Hamielec, T. W. Hoffman, A. I. Johnson, D. R. Woods, and P. T. Shannon, "Chemical Plant Simulation," Prentice-Hall, Englewood Cliffs, New Jersey, 1971.

Franks, R. G. E., "Modeling and Simulation in Chemical Engineering," J. Wiley and Sons, New York, 1972.

Friut, W. M., G. V. Reklaitis, and J. M. Woods, Simulation of Multiproduct Batch Chemical Processes, *Chem. Eng. J.*, **8**(3):199 (1974).

Gerdes, F. O., P. J. Hoftyzer, J. F. Kemkes, M. Van Loon, and C. Schweigman, Mathematical Models for Chemical Process Plant, *Chem. Eng. (London)*, 267 (September 1970).

Glueck, A. R., Simulation of Chemical Processes, *Chem. Eng. Progr.*, **69**(10):95 (1973).

Goldman, M. R., Simulating Multi-Component Batch Distillation, *Brit. Chem. Eng.*, **15**(11):1450 (1970).

―――― and E. R. Robinson, The Computer Simulation of Batch Distillation Processes, *Brit. Chem. Eng.*, **13**(12):602 (1968).

Harvey, D. J., and J. R. Fowler, Putting Evaporators to Work: Dynamic Process Modeling of a Quadruple Effect Evaporation System, *Chem. Eng. Progr.*, **72**(4):47 (1976).

Hendry, J. E., D. F. Rudd, and J. D. Seader, Synthesis in the Design of Chemical Processes, *AIChE J.*, **19**(1):1 (1973).

Hyman, M. H., Simulate Methane Reformer Reactions, *Hydrocarbon Process.*, **47**(7):131 (1968).

Jackson, C. A., and R. A. Troupe, Cut-Set Simulation, *Chem. Eng. Progr.*, **69**(10):98 (1973).

King, C. J., D. W. Gantz, and F. J. Barnes, Systematic Evolutionary Process Synthesis, *I&EC Process Des. & Dev.*, **11**(2):271 (1972).

Korchinsky, W. J., Modelling of Liquid-Liquid Extraction Columns: Use of Published Model Correlations in Design, *Can. J. Chem. Eng.*, **52**(4):468 (1974).

Lin, T. D., and R. S. H. Mah, A Sparse Computation System for Process Design and Simulation: Part I. Data Structures and Processing Techniques, *AIChE J.*, **24**(5):830 (1978).

Lubyen, W. L., "Process Modeling, Simulation, and Control for Chemical Engineers," McGraw-Hill Book Company, New York, 1973.

Lunde, P. J., Modeling, Simulation, and Operation of a Sabatier Reactor, *I&EC Process Des. & Dev.*, **13**(3):226 (1974).

Mah, R. S. H., and T. D. Lin, Part II. A Performance Evaluation Based on the Simulation of a Natural Gas Liquefaction Process, *AIChE J.*, **24**(5):839 (1978).

McGalliard, R. L., and A. W. Westerberg, Structural Sensitivity Analysis in Design Synthesis, *Chem. Eng. J.*, **4**(2):127 (1972).

Nishida, N., and A. Ichikawa, Synthesis of Optimal Dynamic Process Systems by a Gradient Method, *I&EC Process Des. & Dev.*, **14**(3):236 (1975).

Peters, N., and P. E. Barker, An Appraisal of the use of Pacer, Gemcs, and Concept for Chemical Plant Simulation and Design, *Chem. Eng. (London)*, 149 (March 1974.)

Powers, G. J., Heuristic Synthesis in Process Development, *Chem. Eng. Progr.*, **68**(8):88 (1972).

Rathore, R. N. S., and G. J. Powers, A Forward Branching Scheme for the Synthesis of Energy Recovery Systems, *I&EC Process Des. & Dev.*, **14**(2):175 (1975).

Ripps, D. L., and B. H. Wood, Jr., A Program-Oriented General Process Simulator, *I&EC Process Des. & Dev.*, **11**(2):179 (1972).

Robinson, E. R., and M. R. Goldman, The Simulation of Multi-Component Batch Distillation Processes on a Small Digital Computer, *Brit. Chem. Eng.*, **14**(6):318 (1969).

Rudd, D. F., G. J. Powers, and J. J. Siirola, "Process Synthesis," Prentice Hall, Englewood Cliffs, New Jersey, 1973.

Sass, A., Simulation of the Heat-Transfer Phenomena in a Rotary Kiln, *I&EC Process Des. & Dev.*, **6**(4):532 (1967).

Schmidt, J. R., and D. R. Clark, Analog Simulation Techniques for Modelling Parallel Flow Heat Exchangers, *Simulation*, **12**(1):15 (1969).

Shah, Y. T., Computer Simulation of Transport Processes, *I&EC Prod. Res. & Dev.*, **11**(3):269 (1972).

Smith, T. G., and T. W. Cadman, Learn About Computers, Process Simulation, *Hydrocarbon Process.*, **47**(5):171 (1968).

Siirola, J. J., and D. F. Rudd, Computer-Aided Synthesis of Chemical Process Designs, *I&EC Fund.*, **10**:353 (1971).

Sparrow, R. E., G. J. Forder, and D. W. T. Rippin, The Choice of Equipment Sizes for Multiproduct Batch Plants. Heuristics vs. Branch and Bound, *I&EC Process Des. & Dev.*, **14**(3):197 (1975).

Umeda, T., and M. Nishio, Comparison Between Sequential and Simultaneous Approaches in Process Simulation, *I&EC Process Des. & Dev.*, **11**(2):153 (1972).

Scale-up

Backhurst, J. R., and J. H. Harker, " Process Plant Design," American Elsevier Publishing Company, New York, 1973.

Barona, N., and H. W. Prengle, Jr., Design Reactors This Way for Liquid-Phase Processes, *Hydrocarbon Process.*, **52**(3):63 (1973).

Cappello, V. F., Simplifying Scale-up Cost Estimation, *Chem Eng.*, **79**(17):99 (1972).

Chase, J. D., Plant Cost vs. Capacity: New Way to use Exponents, *Chem. Eng.*, **77**(7):113 (1970).

Chen, J. W., J. A. Buege, F. L. Cunningham, and J. I. Northam, Scale-up of a Column Adsorption Process by Computer Simulation, *I&EC Process Des. & Dev.*, **7**(1):26 (1968).

Connolly, J. B., and R. L. Winter, Approaches to Mixing Operation Scale-up, *Chem. Eng. Progr.*, **65**(8):70 (1969).

Davies, G. A., G. V. Jeffreys, and F. Ali, Design and Scale-up of Gravity Settlers, *Chem. Eng. (London)*, 378 (November 1970).

Devia, N., and W. L. Luyben, Reactors: Size Versus Stability, *Hydrocarbon Process.*, **57**(6):119 (1978).

Fuchs, R., D. D. Y. Ryu, and A. E. Humphrey, Effect of Surface Aeration on Scale-up Procedures for Fermentation Processes, *I&EC Process Des. & Dev.*, **10**(2):190 (1971).

Henry, H. C., and J. B. Gilbert, Scale-up of Pilot Plant Data for Catalytic Hydroprocessing, *I&EC Process Des. & Dev.*, **12**(3):328 (1973).

Jope, J. A., Variable-Unit Scale-up Methods Applied to Thermal Systems, *Chem. Eng. (London)*, 291 (August 1972).

————, Improved Scale-up Design Using Variable-unit Analysis, *Chem. Eng. (London)*, 382 (July-August 1973).

Jordan, D. G., "Chemical Process Development," Vols. 1 and 2, J. Wiley and Sons, New York, 1968.

Karr, A. E., and T. C. Lo, Scale-up of Large Diameter Reciprocating-plate Extraction Columns, *Chem. Eng. Progr.*, **72**(11):68 (1976).

Keairns, D. L., Scale-up of Continuous Flow Stirred Tank Reactors, *Can. J. Chem. Eng.*, **47**(4):395 (1969).

Kern, D. Q., Converting Research to Design Use, *Chem. Eng. Progr.*, **65**(7):77 (1969).

Leamy, G. H., Scale-up of Gas-Dispersion Mixers, *Chem. Eng.*, **80**(24):115 (1973).

May, V. T., Process Development Costs and Experience, *Chem. Eng. Progr.*, **69**(2):71 (1973).

Miller, D. N., Scale-up of Agitated Vessels, *I&EC Process Des. & Dev.*, **10**(3):365 (1971).

————, Scale-up of Agitated Vessels Gas-Liquid Mass Transfer, *AIChE J.*, **20**(3):445 (1974).

Ohsol, E. O., Do We Really Need Pilot Plants? *Research/Development*, **23**(3):38 (1972).

Rautzen, R. R., R. R. Corpstein, and D. S. Dickey, How to use Scale-up Methods for Turbine Agitators, *Chem. Eng.*, **83**(23):119 (1976).

Rosen, A. M., and V. S. Krylov, Theory of Scaling-up and Hydrodynamic Modelling of Industrial Mass Transfer Equipment, *Chem. Eng. J.*, **7**(2):85 (1974).

Small, W. M., Scale-up Problems in Reactor Design, *Chem. Eng. Progr.*, **65**(7):81 (1969).

Smith, J. M., Scale-down to Research, *Chem. Eng. Progr.*, **64**(8):78 (1968).

Soper, W. G., Scale Modeling, *Intern. Sci. and Tech.*, **62**(2):60 (1967).

Tanaka, T., Scale-up Theory of Jet Mills on Basis of Comminution Kinetics, *I&EC Process Des. & Dev.*, **11**(2):238 (1972).

van Klaveren, N., Nomograms for Scale-up of Agitated Vessels, *Chem. Proc. Eng.*, **50**(9):127 (1969).

Wang, R. H., and L. T. Fan, Methods for Scaling-up Tumbling Mixers, *Chem. Eng.*, **81**(11):88 (1974).
Is There an Economy of Size for Tomorrow's Refinery? *Hydrocarbon Process.*, **54**(5):111 (1975).

PROBLEMS

1 Using *Chemical Abstracts* as a basis, list the original source, title, author, and brief abstract of three articles published since 1964 dealing with three different processes for producing formaldehyde.

2 Prepare, in the form of a flow sheet, an outline showing the sequence of steps in the complete development of a plant for producing formaldehyde. A detailed analysis of the points to be considered at each step should be included. The outline should take the project from the initial idea to the stage where the plant is in efficient operation.

3 A process for making a single product involves reacting two liquids in a continuously agitated reactor and distilling the resulting mixture. Unused reactants are recovered as overhead and are recycled. The product is obtained in sufficiently pure form as bottoms from the distillation tower.

(*a*) Prepare a qualitative flow sheet for the process, showing all pieces of equipment.

(*b*) With cross reference to the qualitative flow sheet, list each piece of equipment and tabulate for each the information needed concerning chemicals and the process in order to design the equipment.

4 Figure 2-1 presents a qualitative flow diagram for the manufacture of nitric acid by the ammonia-oxidation process. Figure 2-2 presents a quantitative flow diagram for the same process. With the information from these two figures, prepare a quantitative energy balance for the process and size the equipment in sufficient detail for a preliminary cost estimate.

5 A search of the literature reveals many different processes for the production of acetylene. Select four different processes, prepare qualitative flow sheets for each, and discuss the essential differences between each process. When would one process be more desirable than the others? What are the main design problems which would require additional information? What approximations would be necessary if data are not available to resolve these questions?

6 Ethylene is produced commercially in a variety of different processes. Feed stocks for these various processes range from refinery gas, ethane, propane, butane, pentane, natural gasoline, light and heavy naphthas to gas and oil and heavier fractions. Prepare three different qualitative flow sheets to handle a majority of these feed stocks. What are the advantages and disadvantages of each selected process?

7 Gather all the available information on one of the ethylene processes for which a flow sheet was prepared in the preceding problem and make a preliminary material balance for the production of 50 million lb/yr of ethylene. Assume an operating factor of 90 percent.

8 One method of preparing acetaldehyde is by the direct oxidation of ethylene. The process employs a catalytic solution of copper chloride containing small quantities of palladium chloride. The reactions may be summarized as follows:

$$C_2H_4 + 2CuCl_2 + H_2O \xrightarrow{\text{PdCl}_2} CH_3CHO + 2HCl + 2CuCl$$

$$2CuCl + 2HCl + \tfrac{1}{2}O_2 \longrightarrow 2CuCl_2 + H_2O$$

In the reaction, $PdCl_2$ is reduced to elemental palladium and HCl, and is reoxidized by $CuCl_2$. During catalyst regeneration the CuCl is reoxidized with oxygen. The reaction and regeneration steps can be conducted separately or together.

In the process, 99.8 percent ethylene, 99.5 percent oxygen, and recycle gas are directed to a vertical reactor and are contacted with the catalyst solution under slight pressure. The water evaporated during the reaction absorbs the exothermic heat evolved, and make-up water is fed as necessary to maintain the catalytic solution concentration. The reacted gases are water-scrubbed and the resulting acetaldehyde solution is fed to a distillation column. The tail gas from the scrubber is recycled to the reactor. Inerts are eliminated from the recycle gas in a bleed stream which flows to an auxiliary reactor for additional ethylene conversion.

Prepare, in the form of a flow sheet, the sequence of steps in the development of a plant to produce acetaldehyde by this process. An analysis of the points to be considered at each step should be included. List the additional information that will be needed to complete the preliminary design evaluation.

9 Prepare a simplified equipment flow sheet for the acetaldehyde process outlined in Prob. 8. Identify temperature, pressure, and composition, wherever possible, at each piece of equipment.

10 Prepare a material balance and a quantitative flow sheet for the production of 7800 kg/h of acetaldehyde using the process described in the previous problem. Assume an operating factor of 90 percent and a 95 percent yield based on the ethylene feed. Both ethylene and oxygen enter the process at 930 kPa.

11 Using the information developed in Prob. 10, make a basic energy balance around each piece of equipment and for the entire process. Prepare a quantitative flow sheet to outline the results of the basic energy balance.

12 Prepare a material balance for the production of 7800 kg/h of acetaldehyde using the process described in Prob. 8. However, because 99.5 percent oxygen is unavailable, it will be necessary to use 830 kPa air as one of the raw materials. What steps of the process will be affected by this substitution in feed stocks? Assume an operating factor of 90 percent and a 95 percent yield based on the ethylene feed.

13 Synthesis gas may be prepared by a continuous, noncatalytic conversion of any hydrocarbon by means of controlled partial combustion in a fire-brick lined reactor. In the basic form of this process, the hydrocarbon and oxidant (oxygen or air) are separately preheated and charged to the reactor. Before entering the reaction zone, the two feed stocks are intimately mixed in a combustion chamber. The heat produced by combustion of part of the hydrocarbon pyrolyzes the remaining hydrocarbons into gas and a small amount of carbon in the reaction zone. The reactor effluent then passes through a waste-heat boiler, a water-wash carbon-removal unit, and a water cooler-scrubber. Carbon is recovered in equipment of simple design in a form which can be used as fuel or in ordinary carbon products.

Prepare a simplified equipment flow sheet for the process, with temperatures and pressure conditions at each piece of equipment.

14 Make a material balance and a qualitative flow sheet for the synthesis gas process described in Prob. 13. Assume an operating factor of 95 percent and a feed stock with an analysis of 84.6 percent C, 11.3 percent H_2, 3.5 percent S, 0.13 percent O_2, 0.4 percent N_2 and 0.07 percent ash (all on a weight basis). The oxidant in this process will be oxygen having a purity of 95 percent. Production is to be 8.2 m^3/s.

15 Prepare an energy balance and a suitable flow sheet for the synthesis gas production requested in Prob. 14.

16 Size the equipment that is necessary for the synthesis gas production outlined in Probs. 13 and 14.

17 Estimate the required utilities for the synthesis gas plant described in the previous four problems. Compare these results with the published results of Singer and ter Haar in *Chem. Eng. Progr.,* **57**(7):68 (1961).

18 Repeat the calculations of Probs. 14 to 17 by substituting air as the oxidant in place of the 95 percent purity oxygen.

19 In the face of world food shortages accompanying an exploding world population, many engineers have suggested that the world look to crude oil as a new source of food. Explore this possibility and prepare a flow sheet which utilizes the conversion of petroleum to food by organic microorganisms. What are the problems that must be overcome to make this possibility an economic reality?

20 A chemical engineering consultant for a large refinery complex has been asked to investigate the feasibility of manufacturing 1.44×10^{-2} kg/s of thiophane, an odorant made from a combination of tetrahydrofuran (THF) and hydrogen sulfide. The essential reaction is given below:

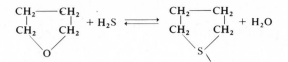

The process consists essentially of the following steps:

a. THF is vaporized and mixed with H_2S in a ratio of 1.5 moles H_2S to one mole of THF and reacted over an alumina catalyst at an average temperature of 672 K and 207 kPa.

b. Reactor vapors are cooled to 300 K and phase separated.

c. The noncondensable gases are removed and burned in a fume furnace while the crude thiophane is caustic washed in a batch operation.

d. The caustic treated thiophane is then batch distilled in a packed tower and sent to storage before eventual shipment to commercial use.

e. Recoverable THF is recycled back to the reactors from the batch column.

f. The aqueous bottoms stream is stored for further processing in the plant.

g. Carbon deposition on the catalyst is heavy (4 percent of THF feed) and therefore provision for regeneration of the catalyst must be made.

Assist the consultant in analyzing this process with a complete flow sheet and material balance, assuming 85 percent operating factor, 80 percent conversion in the reactor and 90 percent recovery after the reactor. Outline the types of equipment necessary for the process. Determine approximate duties of heat exchangers and list overall heat balances on the plant. It is known that the heat of formation of THF is -59.4 kcal/g mol, H_2S is -4.77 kcal/g mol, and thiophane -17.1 kcal/g mol.

What additional information would be required in order to complete the project analysis?

Physical properties:

THF   MW = 72   sp gr = 0.887   Boiling pt. = 65°C

Vap. press. at 25°C = 176 mm Hg

Thiophane, MW = 88   Boiling pt. = 121°C

21 Toluene is converted to benzene catalytically according to the reaction:

$$C_7H_8 + H_2 \rightarrow C_6H_6 + CH_4 + 20{,}513 \text{ Btu/lb mol}$$

For the following reactor feed, use the AREAC block of the CACHE FLOWTRAN simulation to determine the exit temperature for adiabatic operation at 90 percent conversion of toluene.

Species	lb mol/h
H_2	2000
CH_4	1000
C_6H_6	50
C_7H_8	400

Temperature = 800°F.   Pressure = 500 psia.

For solution, see Exercise 12 on page 81 of "Exercises in Process Simulation Using FLOW-TRAN" by J. Peter Clark published by CACHE Corp., 77 Massachusetts Ave., Cambridge, Massachusetts in 1977. For information on the AREAC block, see pages A-146, A-147, and 15 of "FLOWTRAN Simulation—An Introduction" by J. D. Seader, W. D. Seider, and A. C. Pauls, 2d ed., published by CACHE Corp., in 1977.

THREE

GENERAL DESIGN CONSIDERATIONS

The development of a complete plant design requires consideration of many different subjects. As indicated previously, the role of costs and profits is very important. The application of engineering principles in the design of individual pieces of equipment is equally important. In addition, many other factors must be considered, such as plant location, plant layout, plant operation and control, utilities, structural design, storage, materials handling, waste disposal, health and safety, patents, and legal restrictions.

Before proceeding any further with the actual process design development and its associated economics, it will be desirable to consider an overall view of the various functions involved in a complete plant design. Particular emphasis in this view will be placed on the general considerations which serve to integrate the design and yield a successful final plant.

PLANT LOCATION†

The geographical location of the final plant can have a strong influence on the success of an industrial venture. Much care must be exercised in choosing the plant site, and many different factors must be considered. Primarily, the plant should be located where the minimum cost of production and distribution can be obtained, but other factors, such as room for expansion and general living conditions, are also important.

---

† An excellent reference for updated information on plant location is provided by the periodic site surveys published in *Chemical Week*.

An approximate idea as to the plant location should be obtained before a design project reaches the detailed estimate stage, and a firm location should be established upon completion of the detailed-estimate design. The choice of the final site should first be based on a complete survey of the advantages and disadvantages of various geographical areas and, ultimately, on the advantages and disadvantages of available real estate. The following factors should be considered in choosing a plant site:

1. Raw materials
2. Markets
3. Energy availability
4. Climate
5. Transportation facilities
6. Water supply
7. Waste disposal
8. Labor supply
9. Taxation and legal restrictions
10. Site characteristics
11. Flood and fire protection
12. Community factors

The factors that must be evaluated in a plant-location study indicate the need for a vast amount of information, both quantitative (statistical) and qualitative. Fortunately, a large number of agencies, public and private, publish useful information of this type greatly reducing the actual original gathering of the data.

Raw materials The source of raw materials is one of the most important factors influencing the selection of a plant site. This is particularly true if large volumes of raw materials are consumed, because location near the raw-materials source permits considerable reduction in transportation and storage charges. Attention should be given to the purchased price of the raw materials, distance from the source of supply, freight or transportation expenses, availability and reliability of supply, purity of the raw materials, and storage requirements.

Markets The location of markets or intermediate distribution centers affects the cost of product distribution and the time required for shipping. Proximity to the major markets is an important consideration in the selection of a plant site, because the buyer usually finds it advantageous to purchase from nearby sources. It should be noted that markets are needed for byproducts as well as for major final products.

Energy availability Power and steam requirements are high in most industrial plants, and fuel is ordinarily required to supply these utilities. Consequently, power and fuel can be combined as one major factor in the choice of a plant site.

Electrolytic processes require a cheap source of electricity, and plants using electrolytic processes are often located near large hydroelectric installations. If the plant requires large quantities of coal or oil, location near a source of fuel supply may be essential for economic operation. The local cost of power can help determine whether power should be purchased or self-generated.

Climate If the plant is located in a cold climate, costs may be increased by the necessity for construction of protective shelters around the process equipment, and special cooling towers or air-conditioning equipment may be required if the prevailing temperatures are high. Excessive humidity or extremes of hot or cold weather can have a serious effect on the economic operation of a plant, and these factors should be examined when selecting a plant site.

Transportation facilities Water, railroads, and highways are the common means of transportation used by major industrial concerns. The kind and amount of products and raw materials determine the most suitable type of transportation facilities. In any case, careful attention should be given to local freight rates and existing railroad lines. The proximity to railroad centers and the possibility of canal, river, lake, or ocean transport must be considered. Motor trucking facilities are widely used and can serve as a useful supplement to rail and water facilities. If possible, the plant site should have access to all three types of transportation, and, certainly, at least two types should be available. There is usually need for convenient air and rail transportation between the plant and the main company headquarters, and effective transportation facilities for the plant personnel are necessary.

Water supply The process industries use large quantities of water for cooling, washing, steam generation, and as a raw material. The plant, therefore, must be located where a dependable supply of water is available. A large river or lake is preferable, although deep wells or artesian wells may be satisfactory if the amount of water required is not too great. The level of the existing water table can be checked by consulting the state geological survey, and information on the constancy of the water table and the year-round capacity of local rivers or lakes should be obtained. If the water supply shows seasonal fluctuations, it may be desirable to construct a reservoir or to drill several standby wells. The temperature, mineral content, silt or sand content, bacteriological content, and cost for supply and purification treatment must also be considered when choosing a water supply.

Waste disposal In recent years, many legal restrictions have been placed on the methods for disposing of waste materials from the process industries. The site selected for a plant should have adequate capacity and facilities for correct waste disposal. Even though a given area has minimal restrictions on pollution, it should not be assumed that this condition will continue to exist. In choosing a plant site, the permissible tolerance levels for various methods of waste disposal

should be considered carefully, and attention should be given to potential requirements for additional waste-treatment facilities.

Labor supply The type and supply of labor available in the vicinity of a
proposed plant site must be examined. Consideration should be given to prevailing pay rates, restrictions on number of hours worked per week, competing
industries that can cause dissatisfaction or high turnover rates among the workers, racial problems, and variations in the skill and intelligence of the workers.

Taxation and legal restrictions State and local tax rates on property, income,
unemployment insurance, and similar items vary from one location to another.
Similarly, local regulations on zoning, building codes, nuisance aspects, and
transportation facilities can have a major influence on the final choice of a plant
site. In fact, zoning difficulties and obtaining the many required permits can
often be much more important in terms of cost and time delays than many of the
factors discussed in the preceding sections.

Site characteristics The characteristics of the land at a proposed plant site
should be examined carefully. The topography of the tract of land and the soil
structure must be considered, since either or both may have a pronounced effect
on construction costs. The cost of the land is important, as well as local building
costs and living conditions. Future changes may make it desirable or necessary
to expand the plant facilities. Therefore, even though no immediate expansion is
planned, a new plant should be constructed at a location where additional space
is available.

Flood and fire protection Many industrial plants are located along rivers or near
large bodies of water, and there are risks of flood or hurricane damage. Before
choosing a plant site, the regional history of natural events of this type should be
examined and the consequences of such occurrences considered. Protection from
losses by fire is another important factor in selecting a plant location. In case of
a major fire, assistance from outside fire departments should be available. Fire
hazards in the immediate area surrounding the plant site must not be
overlooked.

Community factors The character and facilities of a community can have quite
an effect on the location of the plant. If a certain minimum number of facilities
for satisfactory living of plant personnel do not exist, it often becomes a burden
for the plant to subsidize such facilities. Cultural facilities of the community are
important to sound growth. Churches, libraries, schools, civic theaters, concert
associations, and other similar groups, if active and dynamic, do much to make a
community progressive. The problem of recreation deserves special consideration. The efficiency, character, and history of both state and local government
should be evaluated. The existence of low taxes is not in itself a favorable situation unless the community is already well developed and relatively free of debt.

Selection of the Plant Site

The major factors in the selection of most plant sites are (1) raw materials, (2) markets, (3) energy supply, (4) climate, (5) transportation facilities, and (6) water supply. For a preliminary survey, the first four factors should be considered. Thus, on the basis of raw materials, markets, energy supply, and climate, acceptable locations can usually be reduced to one or two general geographical regions. For example, a preliminary survey might indicate that the best location for a particular plant would be in the south-central or south-eastern part of the United States.

In the next step, the effects of transportation facilities and water supply are taken into account. This permits reduction of the possible plant location to several general target areas. These areas can then be reduced further by considering all the factors that have an influence on plant location.

As a final step, a detailed analysis of the remaining sites can be made. Exact data on items such as freight rates, labor conditions, tax rates, price of land, and general local conditions can be obtained. The various sites can be inspected and appraised on the basis of all the factors influencing the final decision. Many times, the advantages of locating a new plant on land or near other facilities already owned by the concern that is building the new plant outweigh the disadvantages of the particular location. In any case, however, the final decision on choosing the plant site should take into consideration all the factors that can affect the ultimate success of the overall operation.

PLANT LAYOUT

After the process flow diagrams are completed and before detailed piping, structural, and electrical design can begin, the layout of process units in a plant and the equipment within these process units must be planned. This layout can play an important part in determining construction and manufacturing costs, and thus must be planned carefully with attention being given to future problems that may arise. Since each plant differs in many ways and no two plant sites are exactly alike, there is no one ideal plant layout. However, proper layout in each case will include arrangement of processing areas, storage areas, and handling areas in efficient coordination and with regard to such factors as:

1. New site development or addition to previously developed site
2. Type and quantity of products to be produced
3. Type of process and product control
4. Operational convenience and accessibility
5. Economic distribution of utilities and services
6. Type of buildings and building-code requirements
7. Health and safety considerations
8. Waste-disposal problems

Figure 3-1 Scale model showing details of plant layout. *(Courtesy of the M. W. Kellogg Company.)*

9. Auxiliary equipment
10. Space available and space required
11. Roads and railroads
12. Possible future expansion

Preparation of the Layout

Scale drawings, complete with elevation indications can be used for determining the best location for equipment and facilities. Elementary layouts are developed first. These show the fundamental relationships between storage space and operating equipment. The next step requires consideration of the operational sequence and gives a primary layout based on flow of materials, unit operations, storage, and future expansion. Finally, by analyzing all the factors that are involved in plant layout, a detailed recommendation can be presented, and drawings and elevations, including isometric drawings of the piping systems, can be prepared.

Templates, or small cutouts constructed to a selected scale, are useful for making rapid and accurate layouts, and three-dimensional models are often made. The use of such models for making certain a proposed plant layout is correct has found increasing favor in recent years.

Figure 3-1 shows a view of a typical model for an industrial plant. Errors in a plant layout are easily located when three-dimensional models are used, since the operations and construction engineers can immediately see errors which might have escaped notice on two-dimensional templates or blueprints. In addition to increasing the efficiency of a plant layout, models are very useful during plant construction and for instruction and orientation purposes after the plant is completed.

PLANT OPERATION AND CONTROL

In the design of an industrial plant, the methods which will be used for plant operation and control help determine many of the design variables. For example, the extent of instrumentation can be a factor in choosing the type of process and setting the labor requirements. It should be remembered that maintenance work will be necessary to keep the installed equipment and facilities in good operating condition. The engineer must recognize the importance of such factors which are directly related to plant operation and control and must take them into proper account during the development of a design project.

INSTRUMENTATION

Instruments are used in the chemical industry to measure process variables, such as temperature, pressure, density, viscosity, specific heat, conductivity, pH, humidity, dew point, liquid level, flow rate, chemical composition, and moisture content. By use of instruments having varying degrees of complexity, the values of these variables can be recorded continuously and controlled within narrow limits.

Automatic control has been accepted generally throughout the chemical industry, and the resultant savings in labor combined with improved ease and efficiency of operations has more than offset the added expense for instrumentation. (In many instances, control is now being achieved through the use of high-speed computers. In this capacity, the computer serves as a vital tool in the operation of the plant.) Effective utilization of the many instruments employed in a chemical process is achieved through centralized control, whereby one centrally located control room is used for the indication, recording, and regulation of the process variables. Panel boards have been developed which present a graphical representation of the process and have the instrument controls and indicators mounted at the appropriate locations in the overall process. This helps a new operator to quickly become familiar with the significance of the instrument readings, and rapid location of any operational variance is possible. An example of a graphic panel in a modern industrial plant is shown in Fig. 3-2.

Because of the many variables found in processing and the wide range over

Figure 3-2 Example of a graphic panel for a modern industrial plant with a computer-controlled system. (*Courtesy of C. F. Braun & Company.*)

which these variables must be determined and controlled, the assistance of a skilled instrumentation engineer is essential in setting up a control system. Instrumentation problems caused by transmission lags, cycling due to slow or uncompensated response, radiation errors, or similar factors are commonly encountered in plant operation, but most of these problems can be eliminated if the control system is correctly designed.

MAINTENANCE

Many of the problems involved in maintenance are caused by the original design and layout of plant and equipment. Sufficient space for maintenance work on equipment and facilities must be provided in the plant layout, and the engineer needs to consider maintenance requirements when making decisions on equipment.

Too often, the design engineer is conscious only of first costs and fails to recognize that maintenance costs can easily nullify the advantages of a cheap initial installation. For example, a close-coupled motor pump utilizing a high-speed motor may require less space and lower initial cost than a standard motor combined with a coupled pump. However, if replacement of the impeller and shaft becomes necessary, the repair cost with a close-coupled motor pump is much greater than with a regular coupled pump. The use of a high-speed motor reduces the life of the impeller and shaft, particularly if corrosive liquids are involved. If the engineer fails to consider the excessive maintenance costs that may result, an error in recommending the cheaper and smaller unit can be made.

Similarly, a compact system of piping, valves, and equipment may be cheap and convenient for the operators' use, but maintenance of the system may require costly and time-consuming dismantling operations.

UTILITIES

The primary sources of raw energy for the supply of power are found in the heat of combustion of fuels and in elevated water supplies. Fuel-burning plants are of greater industrial significance than hydroelectric installations because the physical location of fuel-burning plants is not restricted. At the present time, the most common sources of energy are oil, gas, coal, and nuclear energy. The decreasing availability of the first two sources of energy will necessitate the use of alternate forms of energy in the not-too-distant future.

In the chemical industries, power is supplied primarily in the form of electrical energy. Agitators, pumps, hoists, blowers, compressors, and similar equipment are usually operated by electric motors, although other prime movers, such as steam engines, internal-combustion engines, and hydraulic turbines are sometimes employed.

When a design engineer is setting up the specifications for a new plant, a decision must be made on whether to use purchased power or have the plant set up its own power unit. It may be possible to obtain steam for processing and heating as a by-product from the self-generation of electricity, and this factor may influence the final decision. In some cases, it may be justified to provide power to the plant from two independent sources to permit continued operation of the plant facilities if one of the power sources fails.

Power can be transmitted in various forms, such as mechanical energy, electrical energy, heat energy, and pressure energy. The engineer should recognize the different methods for transmitting power and must choose the ones best suited to the particular process under development.

Steam is generated from whatever fuel is cheapest, usually at pressures of 450 psig (3100 kPa) or more, expanded through turbines or other prime movers to generate the necessary plant power, and the exhaust steam is used in the process as heat. The quantity of steam used in a process depends upon the thermal requirements, plus the mechanical power needs, if such power is generated in the plant.

Water for industrial purposes can be obtained from one of two general sources: the plant's own source or a municipal supply. If the demands for water are large, it is more economical for the plant to provide its own water source. Such a supply may be obtained from drilled wells, rivers, lakes, dammed streams, or other impounded supplies. Before a company agrees to go ahead with any new project, it must ensure itself of a sufficient supply of water for all industrial, sanitary, and safety demands, both present and future.

The value of an abundance of good water supplies is reflected in the selling price of plant locations that have such supplies. Any engineering techniques which are required to procure, conserve, and treat water significantly increase the operational cost for a plant or process. Increased costs of water processing have made maximum use of the processed water essential. In fact, the high costs of constructing and operating a waste treatment plant have led to concentration of industrial wastes with the smallest amount of water, except where treatment processes require dilution.

STRUCTURAL DESIGN

One of the most important aspects in structural design for the process industries is a correct foundation design with allowances for heavy equipment and vibrating machinery used. The purpose of the foundation is to distribute the load so that excessive or damaging settling will not occur. The type of foundation depends on the load involved and the material on which the foundation acts. It is necessary, therefore, to know the characteristics of the soil at a given plant site before the structural design can be started.

The allowable bearing pressure varies for different types of soils, and the soil should be checked at the surface and at various depths to determine the bearing characteristics. The allowable bearing pressure for rock is 30 or more ton/ft^2 (30 × 10^4 kg/m^2), while that for soft clay may be as low as 1 ton/ft^2 (1 × 10^4 kg/m^2). Intermediate values of 4 to 10 ton/ft^2 (4 × 10^4 to 10 × 10^4 kg/m^2) apply for mixtures of gravel with sand, hard clay, and hardpan.

A foundation may simply be a wall founded on rock or hardpan, or it may be necessary to increase the bearing area by the addition of a footing. Plain concrete is usually employed for making footings, while reinforced concrete, containing steel rods or bars, is commonly used for foundation walls. If possible, a foundation should extend below the frost line, and it should always be designed to handle the maximum load. Pilings are commonly used for supporting heavy equipment or for other special loads.

Maintenance difficulties encountered with floors and roofs should be given particular attention in a structural design. Concrete floors are used extensively in the process industries, and special cements and coatings are available which make the floors resistant to heat or chemical attack. Flat roofs are often specified for industrial structures. Felt saturated with coal-tar pitch combined with a coal-tar pitch–gravel finish is satisfactory for roofs of this type. Asphalt-saturated felt may be used if the roof has a slope of more than $\frac{1}{2}$ in./ft (4.17 cm/m).

Corrosive effects of the process, cost of construction, and climatic effects must be considered when choosing structural materials. Steel and concrete are the materials of construction most commonly used, although wood, aluminum, glass blocks, cinder blocks, glazed tile, bricks, and other materials are also of importance. Allowances must be made for the type of lighting and drainage, and sufficient structural strength must be provided to resist normal loads as well as extreme loads due to high winds or other natural causes.

In any type of structural design for the process industries, the function of the structure is more important than the form. The style of architecture should be subordinated to the need for supplying a structure which is adapted to the proposed process and has sufficient flexibility to permit changes in the future. Although cost is certainly important, the engineer preparing the design should never forget the fact that the quality of a structure remains apparent long after the initial cost is forgotten.

STORAGE

Adequate storage facilities for raw materials, intermediate products, final products, recycle materials, off-grade materials, and fuels are essential to the operation of a process plant. A supply of raw materials permits operation of the process plant regardless of temporary procurement or delivery difficulties. Storage of intermediate products may be necessary during plant shutdown for emergency repairs while storage of final products makes it possible to supply the customer even during a plant difficulty or unforeseen shutdown. An additional need for adequate storage is often encountered when it is necessary to meet seasonal demands from steady production.

Bulk storage of liquids is generally handled by closed spherical or cylindrical tanks to prevent the escape of volatiles and minimize contamination. Since safety is an important consideration in storage-tank design, the American Petroleum Institute† and the National Fire Protection Association‡ publish rules for safe design and operation. Floating roof tanks are used to conserve valuable products with vapor pressures which are below atmospheric pressure at the storage temperature. Liquids with vapor pressures above atmospheric must be stored in vapor-tight tanks capable of withstanding internal pressure. If flammable liquids are stored in vented tanks, flame arresters must be installed in all openings except connections made below the liquid level.

Gases are stored at atmospheric pressure in wet- or dry-seal gas holders. The wet-gas holder maintains a liquid seal of water or oil between the top movable inside tank and the stationary outside tank. In the dry-seal holder the seal between the two tanks is made by means of a flexible rubber or plastic curtain. Recent developments in bulk natural gas or gas-product storage show that pumping the gas into underground strata is the cheapest method available. High-pressure gas is stored in spherical or horizontal cylindrical pressure vessels.

Solid products and raw materials are either stored in weather-tight tanks with sloping floors or in outdoor bins and mounds. Solid products are often packed directly in bags, sacks, or drums.

† American Petroleum Institute, 50 W. 50th St., New York.
‡ National Fire Protection Association, 60 Batterymarch St., Boston, Mass.

MATERIALS HANDLING

Materials-handling equipment is logically divided into continuous and batch types, and into classes for the handling of liquids, solids, and gases. Liquids and gases are handled by means of pumps and blowers; in pipes, flumes, and ducts; and in containers such as drums, cylinders, and tank cars. Solids may be handled by conveyors, bucket elevators, chutes, lift trucks, and pneumatic systems. The selection of materials-handling equipment depends upon the cost and the work to be done. Factors that must be considered in selecting such equipment include:

1. Chemical and physical nature of material being handled
2. Type and distance of movement of material
3. Quantity of material moved per unit time
4. Nature of feed and discharge from materials-handling equipment
5. Continuous or intermittent nature of materials handling

The major movement of liquid and gaseous raw materials and products within a plant to and from the point of shipment is done by pipeline. Many petroleum plants also transport raw materials and products by pipeline. When this is done, local and federal regulations must be strictly followed in the design and specification of the pipeline.

Movement of raw materials and products outside of the plant is usually handled either by rail, ship, truck, or air transportation. Some type of receiving or shipping facilities, depending on the nature of the raw materials and products, must be provided in the design of the plant. Information for the preparation of such specifications can usually be obtained from the transportation companies serving the area.

In general, the materials-handling problems in the chemical engineering industries do not differ widely from those in other industries except that the existence of special hazards, including corrosion, fire, heat damage, explosion, pollution, and toxicity, together with special service requirements, will frequently influence the design. The most difficult of these hazards often is corrosion. This is generally overcome by the use of a high-first-cost, corrosion-resistant material in the best type of handling equipment or by the use of containers which adequately protect the equipment.

WASTE DISPOSAL

Increasingly, chemical engineers find it necessary to be versed in the latest federal and state regulations involving environmental protection, worker safety, and health. This need is especially great for engineers in design-related functions, such as capital-cost estimating, process and equipment design, and plant layout. It is particularly important to learn what is legally required by the Environmental Protection Agency (EPA), the Occupational Safety and Health Administra-

Table 1 Federal repositories of federal regulations

1. *Federal Register* (FR)—Published daily, Monday through Friday, excepting federal holidays. Provides regulations and legal notices issued by federal agencies. The *Federal Register* is arranged in the same manner as the CFR (see below), as follows:
   a. Title—Each title represents a broad area that is subject to federal regulations. There are a total of 50 titles. For example, Title 29 involves labor, and Title 40 is about protection of the environment.
   b. Chapter—Each chapter is usually assigned to a single issuing agency. For example, Title 29, Chapter XVII, covers the Occupational Safety and Health Administration; Title 40, Chapter I, covers the Environmental Protection Agency.
   c. Part—Chapters or subchapters are divided into parts, each consisting of a unified body of regulations devoted to a specific subject. For example, Title 40, Chapter I, Subchapter C, Part 50, is National Primary and Secondary Ambient Air Quality Standards. Title 29, Chapter XVII, Part 1910, is Occupational Safety and Health Standards. Parts can further be divided into subparts, relating sections within a part.
   d. Section—The section is the basic unit of the CFR (see below), and ideally consists of a short, simple presentation of one proposition.
   e. Paragraph—When internal division of a section is necessary, sections are divided into paragraphs (which may even be further subdivided).
2. *FR Index*—Published monthly, quarterly, and annually. The index is based on a consolidation of contents entries appearing in the month's issues of the *Federal Register* together with broad subject references. The quarterly and annual index consolidates the previous three months' and 12 months' issues, respectively.
3. *Code of Federal Regulations* (CFR)—Published quarterly and revised annually. A codification in book form of the general and permanent rules published in the *Federal Register* by the executive departments and agencies of the federal government.
4. *CFR General Index*—Revised annually. July 1. Contains broad subject and title references.
5. *Cumulative List of CFR Sections Affected*—Published monthly and revised annually according to the following schedule: Titles 1–16 as of Jan. 1; 17–27 as of April 1; 28–41 as of July 1; 42–50 as of Oct. 1. The CFR is also revised according to these dates. Provides users of the CFR with amendatory actions published in the *Federal Register*.

tion (OSHA), and corresponding regulatory groups at the state and local level. As a minimum, every design engineer should understand how the federal regulatory system issues and updates its standards.

Every design engineer must be certain that a standard being used has not been revised or deleted. To be sure that a regulation is up to date, it must first be located in the most recent edition of the *Code of Federal Regulations* (CFR). Next, the *Cumulative List of CFR Sections Affected* must be checked to see if actions have been taken since the CFR was published. If action has been taken, the *Cumulative List* will indicate where the changes can be found in the *Federal Register*. The latter provides the latest regulations and legal notices issued by various federal agencies. To aid the design engineer, Table 1 presents a listing of federal repositories for environmental and safety regulations.

It should be noted that maintaining an awareness of federal regulations is not an end in itself, but a necessary component for legally acceptable plant design.

FEDERAL ENVIRONMENTAL REGULATIONS

Environmental Regulations

Several key aspects of the U.S. Federal environmental regulation as spelled out in legislation entitled *Protection of the Environment* (Title 40, Chapter 1 of the CFR) are listed in Table 2. This checklist must also consider applicable state and local codes. Often these may be more stringent than the federal codes or may single out and regulate specific industries.

Note that Part 6 of Title 40, Chapter 1, in Table 2 requires the preparation of an Environmental Impact Statement (EIS). The National Environment Policy Act (NEPA), signed into law January 1, 1970, requires that federal agencies prepare such a statement in advance of any major "action" that may significantly alter the quality of the environment. To prepare the EIS, the federal

Table 2 Key aspects of U.S. federal environmental regulation
Based on Title 40 of the CFR

	Title 40—Protection of Environment
Chapter I—Environmental Protection Agency	
Part	Subchapter A—General
6	Preparation of Environmental Impact Statements (Information for the designer in preparing an EIA.)
	Subchapter C—Air Programs
50	National primary and secondary ambient air quality standards
53	Ambient air monitoring reference and equivalent methods
60	Standards of performance for new stationary sources
61	National emission standards for hazardous air pollutants
81	Air quality control regions, criteria, and control techniques
	Subchapter D—Water Programs
112	Oil pollution prevention
120	Water quality standards
122	Thermal discharges
128	Pretreatment standards
129	Toxic pollutant effluent standards
133	Secondary treatment information
	Subchapter E—Pesticide Programs
	Subchapter H—Ocean Dumping
	Subchapter N—Effluent Guidelines and Standards
Chapter IV—Low-Emission Vehicle Certification Board	
Chapter V—Council on Environmental Quality	
Part	
1500	Preparation of Environmental Impact Statements: guidelines

agencies require the preparation of an Environmental Impact Assessment (EIA). The latter is required to be a full-disclosure statement. This includes project parameters that will have a positive environmental effect, negative impact, or no impact whatsoever. Generally, design engineers will only be involved with a small portion of the EIA preparation, in accordance with their expertise. However, each individual should be aware of the total scope of work necessary to prepare the EIA, as well as the division of work. This will minimize costly duplication, as well as provide the opportunity for developing feasible design alternatives.

Even though this legislation has been in effect for a number of years, there is still considerable uncertainty over the meaning of various parts of the law as well as over the procedures and requirements needed to obtain mandatory permits for construction. As noted by Buehner,[†] numerous steps are required when the environment is taken into consideration during a design project. These, as a minimum, include developing a timetable for completing required environmental statements, incorporating this timetable into the critical-path planning for the project, determining what environmental standards require compliance by the project, obtaining baseline data, examining existing data to determine environmental safety of the project, preparing effluent and emission summary with possible alternatives to meet acceptable standards, and finally preparing the environmental statement or report. Since it may require a full year to obtain baseline data such as air quality, water quality, ambient noise levels, ecological studies, and social surveys, emissions and effluents studies should take place concurrently to avoid delay in preparing the EIA. The emissions and effluents studies must include all "significant" sources of pollution. Omission of data could cause inconsistencies that could result in further time delays when negotiating with the regulatory agencies issuing the many required construction permits.

It becomes clear that environmental considerations not only can play a major factor in the choice of selecting a plant site but can also be quite costly. The American Petroleum Institute[‡] has estimated that the preparation of an EIA for each site considered may range from $50,000 for small projects to $1.5 million for a large petroleum refinery. On the other hand, a detailed environmental assessment may quickly eliminate possible sites because of their highly restrictive standards.

Development of a Pollution Control System

Developing a pollution control system involves an engineering evaluation of several factors which encompass a complete system. These include investigation of the pollution source, determining the properties of the pollution emissions,

---

[†] F. W. Buehner, *Chem. Eng.*, **84**(12):161 (June 6, 1977).

[‡] The Economic Impact of Environmental Regulations on the Petroleum Industry—Phase II Study, American Petroleum Institute, June 11, 1976.

Table 3 Air pollution control equipment characteristics

Control equipment	Optimum† size particle, microns	Optimum concentration, grains/ft³	Temperature limitations, °F	Pressure drop, in. H_2O	Efficiency	Space requirements‡	Collected pollutant	Remarks
					Particulates pollutant			
Mechanical collectors								
Settling chamber	> 50	> 5	700	< 0.1	< 50	L	Dry dust	⎫ Good as precleaner
Cyclone	5–25	> 1	700	1–5	50–90	M	Dry dust	⎬ Low initial cost
Dynamic precipitator	> 10	> 1	700	Fan	< 80	M	Dry dust	
Impingement separator	> 10	> 1	700	< 4	< 80	S	Dry dust	
Bag filter	< 1	> 0.1	500	> 4	> 99	L	Dry dust	Bags sensitive to humidity, filter velocity, and temperature
Wet collector								
Spray tower	25	> 1	40–700	0.5	< 80	L	Liquid	1. Waste treatment required
Cyclonic	> 5	> 1	40–700	> 2	< 80	L	Liquid	2. Visible plume possible
Impingement	> 5	> 1	40–700	> 2	< 80	L	Liquid	3. Corrosion
Venturi	< 1	> 0.1	40–700	1–60	< 99	S	Liquid	4. High temperature operation possible
Electrostatic precipitator	< 1	> 0.1	850	< 1	95–99	L	Dry or wet dust	Sensitive to varying condition and particle properties
					Gaseous pollutant			
Gas scrubber		> 1%	40–100	< 10	> 90	M–L	Liquid	Same as wet collector
Gas adsorber		§	40–100	< 10	> 90	L	Solid or liquid	Adsorbent life critical ⎱ High initial and operating cost
Direct incinerator		Combustible vapors	2000	< 1	< 95	M	None	High operating costs
Catalytic combustion		Combustible vapors	1000	> 1	< 95	L	None	Contaminants could poison catalyst

† Minimum particle size (collected at approximately 90% efficiency under usual operation conditions)

‡ Space requirements: S = small, M = moderate, L = large

§ Adsorber (concentrations less than 2 ppm non-regenerative system; greater than 2 ppm regenerative system)

design of the collection and transfer systems, selection of the control device, and dispersion of the exhaust to meet applicable regulations.

A key responsibility of the design engineer is to investigate the pollutants and the total volume dispersed. It is axiomatic that the size of equipment is directly related to the volume being treated and thus equipment costs can be reduced by decreasing the exhaust volume. Similarly, stages of treatment are related to the quantity of pollutants that must be removed. Any process change that favorably alters the concentrations will result in savings. Additionally, consideration should be given to changing raw materials used and even process operations if a significant reduction in pollution source can be attained. The extent to which source correction is justified depends on the cost of the proposed treatment plant.

For example, the characteristics of equipment for air pollution control, as specified in Table 3, often limit the temperature and humidity of inlet streams to these devices. Three methods generally considered for cooling gases below 500°F are dilution with cool air, quenching with a water spray, and the use of cooling columns. Each approach has advantages and disadvantages. The method selected will be dependent on cost and limitation imposed by the control device.

Selection of the most appropriate control device requires consideration of the pollutant being handled and the features of the control device. Often, poor system performance can be attributed to the selection of a control device that is not suited to the pollutant characteristics. An understanding of the equipment operating principles will enable the design engineer to avoid this problem.

AIR POLLUTION ABATEMENT

The latest changes (1977) in the U.S. Clean Air Act Amendments have changed the regulatory ground rules so that almost any air-pollutant-emitting new facility or modification is subject to the provisions of the law. For most situations, a New Source Review (NSR) application will have to be filed before construction is allowed. Source categories covered at this time include petroleum refineries, sulfur recovery plants, carbon black plants, fuel conversion plants, chemical process plants, fossil-fuel boilers (greater than 250 MM Btu/h heat input), and petroleum storage and transfer facilities (greater than 300,000-barrel capacity).

To obtain a construction permit, a new or modified source governed by the Clean Air Act must meet certain requirements. These include a demonstration that "best available control technology" (BACT) is to be used at the source. In addition, an air quality review must demonstrate that the source will not cause or contribute to a violation of the Ambient Air Quality Standard (AAQS) or maximum allowable increase over the baseline concentration of sulfur dioxide and particulates in any area. (Three different clean air areas have been designated, with class I the most pristine encompassing national parks and forests). Only when these steps indicate that the ambient air will not be significantly impacted by the source may a construction permit be issued.

Air pollution control equipment can essentially be classified into two major categories, those suitable for removing particulates and those associated with removing gaseous pollutants. Particulates are generally removed by mechanical forces, while gaseous pollutants are removed by chemical and physical means.

Particulate Removal

The separating forces in a cyclone are the centrifugal and impact forces impacted on the particulate matter. Similar forces account for the particulate capture in mechanical collectors such as impingement and dynamic separators. In settling chambers, the separation is primarily the result of gravitational forces on the particulates. The mechanism in a wet collector involves contact between a water spray and the gaseous pollutant stream. Separation results primarily from a collision between the particulates and the water droplets. Separation also occurs because of gravitational forces on the large particles, or electrostatic and thermal forces on the small particles. The main separating forces in a bag filter are similar to those described in the wet collector, i.e., collision or attraction between the particle and the filter of the bag. Finally, the principal components in an electrostatic precipitator are a discharge plate and a collecting surface. The separation is affected by charging the particles with a high voltage and allowing the charged particles to be attracted to the oppositely charged collection plates.

To obtain the greatest efficiency in particulate removal, particular attention must be given to particle diameter and the air velocity. The particle size determines the separating force required, while the effectiveness of the control equipment is related to the stream velocity. Generally, the greater the relative velocity between the air stream and the collision obstacle for the particulates, the more effective the separating mechanism. The electrostatic precipitator is an exception to this generality, since here the particle diameter influences the migratory velocity and the power required to maintain the electrical field influences the equipment performance. Figure 3-3 illustrates the characteristics of various pollution particulates and the range of application for several control devices as related to particle size.

A review of Table 3 and Fig. 3-3 indicates that large-diameter particles can be removed with low-energy devices such as settling chambers, cyclones, and spray chambers. Submicron particles must be removed with high-energy units such as bag filters, electrostatic precipitators, and venturi scrubbers. Intermediate particles can be removed with impingement separators or low-energy wet collectors. Obviously, other equipment performance characteristics as noted in Table 3 will also have their influence on the final equipment selection. Costs for much of the equipment considered in this section are given in Chap. 13 (Materials Transfer, Handling, and Treatment Equipment).

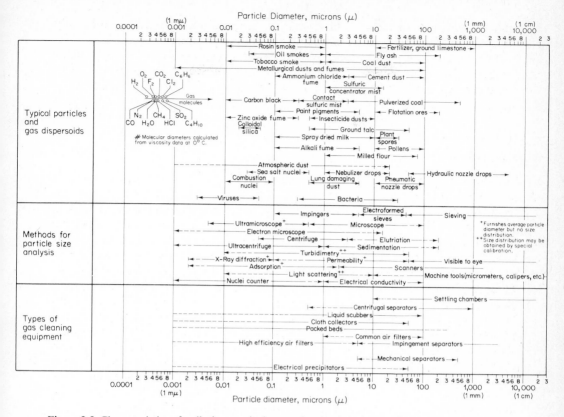

Figure 3-3 Characteristics of pollution particulates and control equipment for removal.

Noxious Gas Removal

Gaseous pollutants can be removed from air streams either by absorption, adsorption, condensation, or incineration. A list of typical gaseous pollutants that can be treated with these four methods is given in Table 4. Generally, condensation is not utilized as a method for removing a solvent vapor from air or other carrier gas unless the concentration of the solvent in the gas is high and the solvent is worth recovery. Since condensation cannot remove all of the solvent, it can only be used to reduce the solvent concentration in the carrier gas.

Gas-liquid absorption processes are normally carried out in vertical, countercurrent flow through packed, plate, or spray towers. For absorption of gaseous streams, good liquid-gas contact is essential and is partly a function of proper equipment selection. Optimization of absorbers or scrubbers (as applied to noxious gas removal) is also important. The power consumption of a modern, high-energy scrubber at its peak can be considerable because of the high pres-

Table 4 Typical gaseous pollutants and their sources

Key element	Pollutant	Source
S	SO_2	Boiler, flue gas
	SO_3	Sulfuric acid manufacture
	H_2S	Natural gas processing, sewage treatment, paper and pulp industry
	R-SH (mercaptans)	Petroleum refining, pulp and paper
N	NO, NO_2	Nitric acid manufacturing, high-temperature oxidation processes, nitration processes
	NH_3	Ammonia manufacturing
	Other basic N compounds, pyridines, amines	Sewage, rendering, pyridine base, solvent processes
Halogen:		
F	HF	Phosphate fertilizer, aluminium
	SiF_4	Ceramics, fertilizers
Cl	HCl	HCl mfg., PVC combustion, organic chlorination processes
	Cl_2	Chlorine manufacturing
C	Inorganic	
	CO	Incomplete combustion processes
	CO_2	Combustion processes (not generally considered a pollutant)
	Organic	
	Hydrocarbons—paraffins, olefins, and aromatics	Solvent operations, gasoline, petrochemical operations, solvents
	Oxygenated hydrocarbons—aldehydes ketones, alcohols, phenols, and oxides	Partial oxidation processes, surface coating operations, petroleum processing, plastics, ethylene oxide
	Chlorinated solvents	Dry-cleaning, degreasing operations

sure drop involved. The latter difficulty has been alleviated in the spray scrubber with its low pressure drop even when handling large volumes of flue gases. In addition, sealing and plugging which can be a problem in certain scrubbing processes (e.g., using a limestone-slurry removal of sulfur dioxide from a flue gas) do not present difficulties when a spray is used in a chemically balanced system.

The use of dry adsorbents like activated carbon and molecular sieves has received considerable attention in removing final traces of objectionable gaseous pollutants. Adsorption is generally carried out in large, horizontal fixed beds often equipped with blowers, condensers, separators, and controls. A typical installation usually consists of two beds; one is onstream while the other is being regenerated.

For those processes producing contaminated gas streams that have no recovery value, incineration may be the simplest route when the gas streams are combustible. There are presently two methods in common use; direct flame and catalytic oxidation. The former usually has lower capital-cost requirements, but higher operating costs, particularly if an auxiliary fuel is required. Either method provides a clean, odorless effluent if the exit-gas temperature is sufficiently high.

Each technique for removing pollutants from process gas streams is economically feasible under certain conditions. Each specific instance must be carefully analyzed before a commitment is made to any type of approach.

WATER POLLUTION ABATEMENT

Better removal of pollutants from wastewater effluents was mandated by the federal government in the Water Pollution Control Act of 1970 (P.L. 92-500). Since then the performance requirements for the various treatment technologies have been raised to new and higher standards with additional legislation aimed at regulating the amounts of toxic and hazardous substances discharged as effluents. Recent legal and enforcement efforts by various governmental agencies to define toxic and hazardous substances give evidence of the demands that will be placed on pollution-technology in the future. The trend in effluent standards is definitely away from the broad, nonspecific parameters (such as chemical oxygen demand or biochemical oxygen demand), and towards limits on specific chemical compounds.

The problems of handling a liquid waste effluent are considerably more complex than those of handling a waste gas effluent. The waste liquid may contain dissolved gases or dissolved solids, or it may be a slurry in either concentrated or dilute form. Because of this complexity, priority should first be given to the possibility of recovering part or all of the waste products for reuse or sale. Frequently, money can be saved by installing recovery facilities rather than more expensive waste-treatment equipment. If product recovery is not capable of solving a given waste-disposal problem, waste treatment must be used. One of the functions of the design engineer then is to decide which treatment process, or combination of processes, will best perform the necessary task of cleaning up the wastewater effluent involved. This treatment can be either physical, chemical, or biological in nature, depending upon the type of waste involved and the amount of removal necessary.

Physical Treatment

The first step in any wastewater treatment process is to remove large floating or suspended particles by sources. This is usually followed by sedimentation or gravity settling. Where sufficient land area is available, earthen basins are sometimes used to remove settleable solids from dilute wastewater. Otherwise, circular clarifiers with rotating sludge scrapers or rectangular clarifiers with continuous chain sludge scrapers are used. These units permit removal of settled sludge from the floor of the clarifiers and scum removal from the surface.

Numerous options for improving the operation of clarifiers are presently available.

Sludge from primary or secondary treatment that has been initially concentrated in a clarifier or thickener can be further concentrated by vacuum filtration or centrifugation. The importance of first concentrating a thin slurry by clarifier or thickener action needs to be recognized. For example, concentrating the sludge from 5 to 10 percent solids before centrifuging can result in a 250 percent increase in solids recovery for the same power input to the centrifuge.

Solid-liquid separation by flotation may be achieved by gravity alone or induced by dissolved-air or vacuum techniques. The mechanisms and driving forces are similar to those found in sedimentation, but the separation rate and solids concentration can be greater in some cases.

Adsorption processes, and in particular those using activated carbon, are also finding increased use in wastewater treatment for removal of refractory organics, toxic substances, and color. The primary driving forces for adsorption are a combination of (1) the hydrophobic nature of the dissolved organics in the wastewater and (2) the affinity of the organics to the solid adsorbent. The latter is due to a combination of electrostatic attraction, physical adsorption, and chemical adsorption. Operational arrangements of the adsorption beds are similar to those described for gaseous adsorption.

Three different membrane processes, ultrafiltration, reverse osmosis, and electrodialysis are receiving increased interest in pollution-control applications as end-of-pipe treatment and for inplant recovery systems. There is no sharp distinction between ultrafiltration and reverse osmosis. In the former, the separation is based primarily on the size of the solute molecule which, depending upon the particular membrane porosity, can range from about 2 to 10,000 millimicrons. In the reverse osmosis process, the size of the solute molecule is not the sole basis for the degree of removal, since other characteristics of the solute such as hydrogen-bonding and valency affect the membrane selectivity. In contrast to these two membrane processes, electrodialysis employs the removal of the solute (with some small amount of accompanying water) from solution rather than the removal of the solvent. The other major distinction is that only ionic species are removed. The advantages due to electrodialysis are primarily due to these distinctions.

Chemical Treatment

In wastewater treatment, chemical methods are generally used to remove colloidal matter, color, odor, acids, alkalis, heavy metals, and oil. Such treatment is considered as a means of stream upgrading by coagulation, emulsion breaking, precipitation, or neutralization.

Coagulation is a process that removes colloids from water by the addition of chemicals. The chemicals upset the stability of the system by neutralizing the colloid charge. Additives commonly used introduce a large multivalent cation such as Al^{+3} (from alum) or Fe^{+3} (from ferric chloride). Emulsion breaking is similar to coagulation. The emulsions are generally broken with a combination of acidic reagents and polyelectrolytes. The common ion effect can also be useful in wastewater treatment. In this case an unwanted salt is removed from solution by adding a second soluble salt to increase one of the ion concentrations. Coagulant aids may then be needed to remove the precipitate.

One method for treating acid and alkaline waste products is by neutralization with lime or sulfuric acid (other available materials may also be suitable). Even though this treatment method may change the pH of the waste stream to the desired level, it does not remove the sulfate, chloride, or other ions. Therefore, the possibility of recovering the acid or alkali by distillation, concentration, or in the form of a useful salt should always be considered before neutralization or dilution methods are adopted.

Chemical oxidation is frequently another tool in wastewater treatment. Chemical oxidants in wide use today are chlorine, ozone, and hydrogen peroxide. The historical use of chlorine and ozone has been in the disinfection of water and wastewater. All three oxidizers are, however, receiving increased attention for removing organic materials from wastewaters that are resistant to biological or other treatment processes. The destruction of cyanide and phenols by chlorine oxidation is well known in waste-treatment technology. However, the use of chlorine for such applications has come under intense scrutiny because of the uncertainty in establishing and predicting the products of the chlorine oxidation reactions and their relative toxicity. Ozone, on the other hand, with only a short half-life is found to be effective in many applications for color removal, disinfection, taste and odor removal, iron and manganese removal, and in the oxidation of many complex inorganics, including lindane, aldrin, surfactants, cyanides, phenols, and organo-metal complexes. With the latter, the metal ion is released and can be removed by precipitation.

The need for chemical reduction of wastewaters occurs less often. The most common reducing agents are ferrous chloride or sulfate which may be obtained from a variety of sources.

Biological Treatment

In the presence of the ordinary bacteria found in water, many organic materials will oxidize to form carbon dioxide, water, sulfates, and similar materials. This process consumes the oxygen dissolved in the water and may cause a depletion of dissolved oxygen. A measure of the ability of a waste component to consume the oxygen dissolved in water is known as the *biochemical oxygen demand*. The biochemical oxygen demand (BOD) of a waste stream is often the primary factor that determines its importance as a pollutant. The biochemical oxygen demand of sewage, sewage effluents, polluted waters, or industrial wastes is the oxygen, reported as parts per million, consumed during a set period of time by bacterial action on the decomposable organic matter.

One of the more common biological wastewater treatment procedures today involves the use of concentrated masses of microorganisms to break down organic matter, resulting in the stabilization of organic wastes. These organisms are broadly classified as aerobic, anaerobic, and aerobic-anaerobic facultative. Aerobic organisms require molecular oxygen for metabolism, anaerobic organisms derive energy from organic compounds and function in the absence of oxygen, while facultative organisms may function in either an aerobic or anaerobic environment.

Basically, the aerobic biological processes involve either the activated sludge process or the fixed-film process. The activated sludge process is a continuous system in which aerobic biological growths are mixed with wastewater and the resulting excess flocculated suspension separated by gravity clarification or air flotation. The predominant fixed-film process has, in the past, been the conventional trickling filter. In this process, wastewater trickles over a biological film fixed to an inert medium. Bacterial action in the presence of oxygen breaks down the organic pollutants in the wastewater. (An attempt to improve on the biological efficiency of the fixed-film process has resulted in the recent development of a rotating disk biological contactor.) A comparison of biological loading and area requirements for various aerobic biological processes is shown in Table 5.

Many organic industrial wastes (including those from food processing, meat packing, pulp and paper, refining, leather tanning, textiles, organic chemicals, and petrochemicals) are amenable to biological treatment; however, a fair number may prove to be refractory, i.e., nonbiodegradable. Thus, although the BOD removal may be excellent, the removal of the *chemical oxygen demand* (COD) may be low. Process evaluation prior to system design should center on characterization of the waste stream, particularly to determine the presence of inhibitory or toxic components relative to biological treatment, and the establishment of pollutant removal rates, oxygen requirements, nutrient requirements (nitrogen and phosphorus), sludge production, and solids settleability.

Table 5 Comparative requirements for processing 1700 lb BOD/day by various aerobic systems

System	Area, acres	Biological loading lb BOD/1000 ft^3	BOD removal, %
Stabilization pond	57†	0.09–0.23	70–90
Aerated lagoon	5.75‡	1.15–1.60	80–90
Activated sludge			
Extended	0.23	11.0–30.0	95‡
Conventional	0.08	33.0–400	90
High rate	0.046	57.0–150	70
Trickling filter			
Rock	0.2–0.5	0.7–50	40–70
Plastic disks	0.02–0.05	20–200	50–70

† 5 ft deep
‡ 10 ft deep

Anaerobic treatment is important in the disposal of municipal wastes but has not been widely used on industrial wastes. It has found some use in reducing highly concentrated BOD wastes, particularly in the food and beverage industries as a pretreatment or roughing technique. However, compared to aerobic systems, anaerobic treatment is more sensitive to toxic materials and is more difficult to control.

SOLID WASTE DISPOSAL

Solid wastes differ from air and water pollutants since these wastes remain at the point of origin until a decision is made to collect and dispose of them. There are several means of disposal available including recycling, chemical conversion, incineration, pyrolysis, and landfill. Federal regulations, local conditions, and overall economics generally determine which method is the most acceptable.

Recycling and Chemical Conversion

Resource recovery is a factor often overlooked in waste disposal. For example, specific chemicals may often be recovered by stripping, distillation, leaching, or extraction. Valuable solids such as metals and plastics can be recovered by magnets, electrical conductivity, jigging, flotation, or hand picking. Process wastes may at times also be converted to saleable products or innocuous materials that can be disposed of safely. The former would include hydrogenation of organics to produce fuels, acetylation of waste cellulose to form cellulose acetate, or nitrogen and phosphorus enrichment of wastes to produce fertilizer.

Incineration

The controlled oxidation of solid, liquid, or gaseous combustible wastes to final products of carbon dioxide, water, and ash is known as incineration. Since sulfur and nitrogen-containing waste materials will produce their corresponding oxides, they should not be incinerated without considering their effect on air quality. A variety of incinerator designs are available. Multiple-chamber incinerators, rotary kilns, and multiple-hearth furnaces are most widely used in industrial waste disposal.

Incineration has in the past provided certain advantages, particularly where land disposal sites are not available or are too remote for economic hauling. A properly designed and carefully operated incinerator can be located adjacent to a process plant and can be adjusted to handle a variety and quantity of wastes. Not only can heat recovery through steam generation reduce operating costs, but it can also save on pollution control equipment. Additionally, the residue is a small fraction of the original weight and volume of the waste and is suitable for landfill.

Pyrolysis

The most acceptable route to recycling wastes in the future may be through pyrolytic techniques in which wastes are heated in an airfree chamber at temperatures as high as 3000°F (1650°C). Pyrolysis seems to provide several advantages over incineration. These systems encounter far fewer air-pollution problems, handle larger throughputs resulting in lower capital costs, provide their own fuel, degrade marginally burnable materials, and have the added potential for recovering chemicals or synthesis gas.

Landfill

Sanitary landfill is basically a simple technique that involves spreading and compacting solid wastes into cells that are covered each day with soil. Care needs to be exercised that wastes disposed of in this fashion are either inert to begin with or are capable of being degraded by microbial attack to harmless compounds. The principal problems encountered in landfill operation are the production of leachates that may contaminate the surrounding groundwater and the potential hazards associated with the accumulation of flammable gases produced during the degradation of the waste material. A number of methods are available to prevent these hazards.†

THERMAL POLLUTION CONTROL

Temperature affects nearly every physical property of concern in water quality management including density, viscosity, vapor pressure, surface tension, gas solubility, and gas diffusion. The solubility of oxygen is probably the most important of these parameters, inasmuch as dissolved oxygen in water is necessary to maintain many forms of aquatic life. This potential damage to the aquatic environment by changes in temperature, the reduction in the assimilative capacity of organic wastes due to increased temperature, and the federal enactment of more stringent water-temperature standards has led design engineers to investigate various offstream cooling systems to handle thermal discharges from processes and plants. Cooling towers are most often considered for this service, followed by cooling ponds and spray ponds in that order.

Cooling towers may be classified on the basis of the fluid used for heat transfer and on the basis of the power supplied to the unit. In wet cooling towers, the condenser cooling water and ambient air are intimately mixed. Cooling results from the evaporation of a portion of the water and to a lesser extent from the loss of sensible heat to the air. In dry cooling towers, the temperature

---

† Mead, B. E., and W. G. Wilkie, Leachate Prevention and Control from Sanitary Landfills, Paper 35b, 68th National AIChE Meeting, Houston, Texas, March 1971.

reduction of the condenser water depends upon conduction and convection for the transfer of heat from the water to the air.

Mechanical draft cooling towers either force or induce the air which serves as the heat-transfer medium through the tower. For their driving force, natural draft cooling towers depend upon the density difference between the air leaving the tower and the air entering the tower.

Cooling ponds are generally only considered for heat removal when suitable land is available at a reasonable price, since such systems are simple, cheap, and frequently less water-intensive. It is normally assumed that all heat discharged to a cooling pond is lost through the air-water interface. With low heat-transfer rates, large surface areas are required.

When land costs are too high, spray ponds often provide a viable alternative to cooling ponds. It is estimated that a spray pond requires only about 5 to 10 percent of the area of a cooling pond due to the more intimate air-water contact. In addition, drift losses and corrosion problems are less severe than in cooling towers.

HEALTH AND SAFETY

Toxic and corrosive chemicals, fires, explosions, plant personnel falling accidents, and mechanical equipment are the major health and safety hazards encountered in the operation of plants in the process industries. As a consequence, when potentials for hazardous exposures of any kind exist, interrelationships of activities must be considered by the design engineer if safe and healthy working conditions are to be provided.

Safety Regulations

The expressed intent of the Occupational Safety and Health Act (OSHA) of 1970 is "to assure so far as possible every working man and woman in the Nation safe and healthful working conditions and to preserve our human resources . . .". The act of 1970 presently affects more than five million workplaces and sixty million employees. Nearly 500 amendments to the Act have been introduced since the original legislation. A recent printing of the OSHA standards can be found in Title 29, Chapter XVII, Part 1910 of the CFR.

Two of the standards directly related to worker health and important in design work are *Toxic Hazardous Substances* and *Occupational Noise Exposure*. The first of these two concerns the normal release of toxic and carcinogenic substances, carried via vapors, fumes, dust fibers, or other media. Compliance with the Act requires the designer to make calculations of concentrations and exposure time of plant personnel to toxic substances during normal operation of a process or plant. These releases could emanate from various types of seals and from control-valve packings or other similar sources. Normally, the designer can meet the limits set for exposure to toxic substances by specifying special valves, seals, vapor-recovery systems, and appropriate ventilation systems.

The list of materials declared hazardous is being updated at a rapid rate. Acceptable material exposure times and concentrations, likewise, are undergoing continuous revision. Thus, it is important that the *Federal Register* be examined closely before beginning the detailed design of a project. A new publication, *Chemical Regulation Reporter*†, detailing these proposed and new regulations is now available to the design engineer. This weekly information service includes information concerning the *Toxic Substances Control Act* (a law administered by EPA rather than OSHA).

The *Occupational Noise Exposure* standard requires a well-planned, timely execution of steps to conform to the 90-dBA rule in the design stages of a project. Since many cities have adopted EPA's recommended noise-level criteria, or have stringent regulations of their own, design-stage noise control must also consider noise leaving the plant. It is a good idea, during plant design, to prepare two noise specifications: one to define the designer's own scope of work and the other to set vendor noise-level requirements for various pieces of equipment.

Other standards in the safety area that are most often cited by OSHA and which must be considered in detailed designs are the *National Electric Code* and *Machinery and Machinery Guarding*. A cursory investigation by a designer of these and other OSHA standards quickly points out several problems, particularly in interpretation. The standards frequently do not allow for alternate designs that provide equivalent protection. Some sections are very specific, while others are rather vague. Additionally, some sections refer to other sets of codes such as ASME and ANSI. As a result, when a designer cannot obtain a satisfactory interpretation of a regulation from the standards, the regional or area OSHA should be contacted and an interpretation requested. Since many states also have approved plans comparable to that of the federal government, the designer must also be aware of these regulations.

Chemical Hazards

Many chemicals can cause severe burns if permitted to come into contact with living tissue. Dehydration by strong dehydrating agents, digestion by strong acids and bases, and oxidation by strong oxidizing agents can destroy living tissue. Eyes and the mucous membranes of the nose and throat are particularly susceptible to the effects of corrosive dusts, mists, and gases. In addition to having the ability to destroy living tissue, many chemicals are very toxic, flammable, or detonable. Tolerance levels for toxic chemicals have been set by federal regulations. Flammability and detonability limits are available in most good handbooks.

---

† *Chemical Regulation Reporter*, Bureau of National Affairs, Inc., 1231 25th Street, N.W., Washington, D.C. 20037.

Fire and Explosion Hazards

The chances that a single fire or explosion will spread to adjoining units can be reduced by careful plant layout and judicious choice of constructional materials. Hazardous operations should be isolated by location in separate buildings or by the use of brick fire walls. Brick or reinforced concrete walls can serve to limit the effects of an explosion, particularly if the roof is designed to lift easily under an explosive force.

Equipment should be designed to meet the specifications and codes of recognized authorities, such as the American Standards Association, American Petroleum Institute, American Society for Testing Materials, Factory Mutual Laboratories, National Fire Protection Association, and Underwriters' Laboratories. The design and construction of pressure vessels and storage tanks should follow API and ASME Codes, and the vessels should be tested at 1.5 to 2 or more times the design pressure. Adequate venting is necessary, and it is advisable to provide protection by using both spring-loaded valves and rupture disks.

Possible sources of fire are reduced by eliminating all unnecessary ignition sources, such as flames, sparks, or heated materials. Matches, smoking, welding and cutting, static electricity, spontaneous combustion, and non-explosion-proof electrical equipment are all potential ignition sources. The installation of sufficient fire alarms, temperature alarms, fire-fighting equipment, and sprinkler systems must be specified in the design.

Personnel Safety

Every attempt should be made to incorporate facilities for health and safety protection of plant personnel in the original design. This includes, but is not limited to, protected walkways, platforms, stairs, and work areas. Physical hazards, if unavoidable, must be clearly defined. In such areas, means for egress must be unmistakable. All machinery must be guarded with protective devices.†
In all cases, medical services and first-aid must be readily available for all workers.

Noise Abatement

The design engineer should include noise studies in the design stage of any industrial facility. Generally, acoustical problems left for field resolution cost roughly twice as much. Unnecessary costs incurred in post-construction noise work may include the replacement of insulation, redesign of piping configuration to accommodate silencers, modification of equipment, additional labor costs, and possible downtime to make necessary changes. Considerable judgment, therefore, must be exercised by the designer to establish final design-stage noise

---

† A general requirement for safeguarding all machinery is provided in Section 212 of the Occupational Safety Standard for General Industry (OSHA Standards, 29 CFR 1910).

Table 6 Equipment noise sources, levels, and potential control solutions

Equipment	Sound level in dBA† at 3 ft	Possible noise control treatments
Air coolers	87–94	Aerodynamic fan blades, decrease rev./min and increase pitch, tip and hub seals, decrease pressure drop.
Compressors	90–120	Install mufflers on intake and exhaust, enclosure of machine casing, vibration isolation and lagging of piping systems.
Electric motors	90–110	Acoustically lined fan covers, enclosures, and motor mutes.
Heaters and furnaces	95–110	Acoustic plenums, intake mufflers, ducts lined and damped.
Valves	< 80 to 108	Avoid sonic velocities, limit pressure drop, and mass flow, replace with special low noise valves, vibration isolation and lagging.
Piping	90–105	Inline silencers, vibration isolation and lagging.

† Defined as the sound intensity measured in units equal to ten times the logarithm of the square of the relative pressure associated with the sound wave.

recommendations. These should not only consider the results of the equipment data-analysis procedure, but should also recognize additional factors such as administrative controls, feasibility of redesign, economic alternatives, intrusion of noise into the community, and the basic limitations of the equations employed in the applicable computer programs.

To attain efficient, effective, and practical noise control, it is necessary to understand the individual equipment or process noise sources, their acoustic properties and characteristics, and how they interact to create the overall noise situation. Table 6 presents typical process design equipment providing high noise levels and potential solutions to this problem.

PATENTS

A patent is essentially a contract between an inventor and the public. In consideration of full disclosure of the invention to the public, the patentee is given exclusive rights to control the use and practice of the invention. A patent gives the holder the power to prevent others from using or practicing the invention for a period of 17 years from the date of granting. In contrast, trade secrets and certain types of confidential disclosures can receive protection under common-law rights only as long as the secret information is not public knowledge.

A new design should be examined to make certain no patent infringements are involved. If the investigation can uncover even one legally expired patent covering the details of the proposed process, the method can be used with no fear of patent difficulties. Although most large corporations have patent attorneys to handle investigations of this type, the design engineer can be of considerable assistance in determining if infringements are involved. An engineer, therefore, should have a working knowledge of the basic practices and principles of patent law.†

Patentable Inventions

A patent may be obtained on any new and useful process, machine, method of manufacture, composition of matter, or plant, provided it has not been known or used by others before the person applying for the patent made his invention or discovery. The invention must not have been described in a printed publication or been in public use or sale for more than 1 year prior to the patent application. A patentable item must result from the use of creative ability above and beyond that which would be expected in the regular work of a person skilled in the particular field. Thus, a patentable item cannot be something requiring merely mechanical skill, and a patent will not be granted for a change in a previously known item or process unless the change involves something entirely new.

Patent Applications

A patent application consists of (1) a petition, directed to the Commissioner of Patents and requesting the grant of a patent, (2) an oath, sworn to before a notary public or other designated officer, (3) specifications and claims, in which the claims to be patented are indicated along with detailed specifications including drawings and other pertinent information, and (4) the filing fee. When the application is filed, the patent office gives it a serial number and informs the applicant of the number and date of filing. The application is then examined, and, after a period of time, the official action on the claim is sent to the applicant or his attorney. If the official action requires a reply, the applicant must respond within 6 months or the application is considered to be abandoned.

In giving the specifications and claims, every effort must be made to present a full disclosure. A patent can be declared invalid if it is shown that the patentee purposely held back some essential part of the invention.

A patent application must be signed by the person or persons who made the invention. It is seldom advisable, however, for the inventor to handle the application without the assistance of an experienced patent attorney. There are so many legal phases and complex proceedings involved in obtaining full patent coverage that competent counsel is practically a necessity.

---

† Statutes and general rules applying to United States patents are presented in the two pamphlets "Patent Laws" and "Rules of Practice of the United States Patent Office in Patent Cases," U.S. Government Printing Office.

Foreign Patents

A United States patent is enforceable only in the United States and its territories. Similarly, patents granted in a foreign country apply only to that particular country. A patent must be obtained in each foreign country where protection of the invention is desired.

Under the regulations set up by the International Convention for the Protection of Industrial Property, an inventor can obtain the benefit of a previous foreign filing date provided the application is filed in this country within 12 months of the foreign filing date. A statutory requirement for the granting of a patent is that it cannot have been patented by the inventor or the legal representatives in a foreign country on an application filed more than 12 months before being filed in this country.

Interferences

The situation often arises in which two or more independent patent applications covering essentially the same invention are on file in the U.S Patent Office. When this occurs, a proceeding called an *interference* is instituted to determine who is entitled to the patent. An interference may also be instituted between a pending application and a granted patent. It is obvious that patents on identical inventions cannot be granted to separate parties, even though each may be an original inventor. Interferences are decided on the basis of priority. The patent is granted to the applicant who was first to conceive the idea for the invention, providing reasonable diligence was exercised in reducing it to practice.

Because of the role of priority in any type of interference proceeding, it is very important for an inventor to maintain complete records. A written description and sketches should be prepared by an inventor as soon as possible after the conception of an idea that might eventually be patentable. This material should be disclosed to one or more witnesses who should indicate in writing that they understand the purpose, method, and structure of the invention. The disclosure should be signed and dated by the inventor and the witnesses. Similarly, all subsequent work, modifications, and new ideas on the process should be recorded in chronological order in a permanent file, such as a bound notebook. Each page should be dated and signed by the person who makes the entries, and a competent witness should periodically examine, sign, and date the notes.

Infringement

The infringement of a patent may consist of making, using, or selling the invention covered by the patent without express permission of the holders. A *contributory infringement* involves the assistance or cooperation with another in the unauthorized making, using, or selling of a patented invention.

If the infringement is deliberate, the court may award the plaintiff as much as three times the actual damages caused the patentee plus three times the earned profits. In case the infringement is unintentional or innocent, the award to the plaintiff is no more than the actual loss proved.

Assignment of Patent Rights

A patentee may treat the patent as personal property with the right to sell it outright or license it in any form desired. Thus, the patent rights can be assigned exclusively to one concern, or the assignment of partial rights can be made. The inventor can be paid by a lump sum, a periodic payment, a percent of the profit, a set amount for each unit produced, or in any other way that is agreeable to the parties involved. An assignment on a patent should be recorded with the U.S. Patent Office.

Contracts are frequently made between employer and employee in which the employee agrees to assign the employer all rights to patents developed through the use of the concern's facilities or information. If no definite agreement is made with respect to patent rights, an employer has a free and nonexclusive right to use any patented invention or discovery made by an employee at the expense of the employer's time and facilities. This is known as *shop rights.* Legally, the patent belongs to the employee, who may sell the invention to others but cannot control in any way the shop right of the employer.

SUGGESTED ADDITIONAL REFERENCES FOR GENERAL DESIGN CONSIDERATIONS

Air Pollution

Abbott, J. H., and D. C. Drehmel, Control of Fine Particulate Emissions, *Chem. Eng. Progr.,* **72**(12):47 (1976).

Adlhart, O. J., S. G. Hindin, and R. E. Kenson, Processing Nitric Acid Tail Gas, *Chem. Eng. Progr.,* **67**(2):73 (1971).

Akitsune, K., and T. Takae, Abatement of Prilling Tower Effluent, *Chem. Eng. Progr.,* **69**(6):72 (1973).

Alonso, J. R. F., Estimating the Costs of Gas-Cleaning Plants, *Chem. Eng.,* **78**(28):86 (1971).

Bailey, T., and F. M. Wall, Heat Transfer—Ethylene Furnace Design, *Chem. Eng. Progr.,* **74**(7):45 (1978).

Ball, F. J., G. N. Brown, J. E. Davis, A. J. Repik, and S. L. Torrence, Take Sulfur Out of Waste Gases, *Hydrocarbon Process.,* **51**(10):125 (1972).

Barber, J. C., Energy Requirements for Pollution Abatement, *Chem. Eng. Progr.,* **71**(12):42 (1976).

Barrett, D. F., and J. R. Small, Emission Monitoring of SO_2 and NO_x, *Chem. Eng. Progr.,* **69**(12):35 (1973).

Barry, C. B., Reduce Claus Sulfur Emission, *Hydrocarbon Process.,* **51**(4):102 (1972).

Barthel, Y., Y. Bistri, A. Deschamps, R. Renault, J. C. Simadoux, and R. Dutriau, Treat Claus Tail Gas, *Hydrocarbon Process.,* **50**(5):89 (1971).

Bartok, W., A. R. Crawford, and A. Skopp, Control of NO_x Emissions from Stationary Sources, *Chem. Eng. Progr.,* **67**(2):64 (1971).

Beavon, D. K., Add-On Process Slashes Claus Tailgas Pollution, *Chem. Eng.,* **78**(28):71 (1971).

Becker, J. H., and J. N. Driscoll, Stationary Source Monitoring I: Nonmethane Hydrocarbons, *Am. Lab.,* **6**(12):31 (1974).

Beltran, M., Engineering/Economic Aspects of Odor Control, *Chem. Eng. Progr.,* **70**(5):57 (1974).

Bernardin, F. E. Jr., Selecting and Specifying Activated-carbon-adsorption Systems, *Chem. Eng.*, **83**(22):77 (1976).

Beychok, M. R., Coping with SO_2, *Chem. Eng.*, **81**(22):79 (1974).

———, Environmental Factors in Producing Supplemental Fuels, *Hydrocarbon Process.*, **54**(10):78 (1975).

Bibbero, R. J., and I. G. Young, "Systems Approach to Air Pollution Control," J. Wiley and Sons, New York, 1974.

Bodurtha, F. T., P. A. Palmer, and W. H. Walsh, Discharge of Heavy Gases from Relief Valves, *Chem. Eng. Progr.*, **69**(4):37 (1973).

Bonnifay, P., R. Dutriau, S. Frankowiak, and A. Deschamps, Pollution Abatement: Partial and Total Sulfur Recovery, *Chem. Eng. Progr.*, **68**(8):51 (1972).

Brady, J. D., D. W. Cooper, and M. T. Rei, A Wet Collector of Fine Particles, *Chem. Eng. Progr.*, **73**(8):45 (1977).

Browder, T. J., Modern Sulfuric Acid Technology, *Chem. Eng. Progr.*, **67**(5):45 (1971).

Brown, G. N., S. L. Torrence, A. J. Repik, J. L. Stryker, and F. J. Ball, Pollution Abatement: SO_2 Recovery via Activated Carbon, *Chem. Eng. Progr.*, **68**(8):55 (1972).

Buckingham, P. A., and H. R. Homan, Sulfur and the Energy Industry, *Hydrocarbon Process.*, **50**(8):121 (1971).

Bump, R. L., Electrostatic Precipitators in Industry, *Chem. Eng.*, **84**(2):129 (1977).

Byers, R. L., Test for Gas Stream Particulates, *Hydrocarbon Process.*, **52**(10):92 (1973).

Calvert, S., How to Choose a Particulate Scrubber, *Chem. Eng.*, **84**(18):54 (1977).

———, Get Better Performance from Particulate Scrubbers, *Chem. Eng.*, **84**(23):133 (1977).

———, Guidelines for Selecting Mist Eliminators, *Chem. Eng.*, **85**(5):109 (1978).

Caplan, F., Plant Notebook, Sizing Electrostatic Precipitators, *Chem. Eng.*, **85**(9):153 (1978).

Celenza, G. J., Designing Air Pollution Control Systems, *Chem. Eng. Progr.*, **66**(11):31 (1970).

Charlton, J., R. Sarteur, and J. M. Sharkey, Identify Refinery Odors, *Hydrocarbon Process.*, **54**(5):97 (1975).

Cheng, L., Collection of Airborne Dust by Water Sprays, *I&EC Proc. Design & Develop.*, **12**(3):221 (1973).

Chilton, T. H., Reducing SO_2 Emission from Stationary Sources, *Chem. Eng. Progr.*, **67**(5):69 (1971).

Chopey, N. P., Desulfurization ... Taking Coal's Sulfur Out, *Chem. Eng.*, **79**(16):86 (1972).

Cines, M. R., D. M. Haskell, and C. G. Houser, Molecular Sieves for Removing HS from Natural Gas, *Chem. Eng. Progr.*, **72**(8):89 (1976).

Collins, J. J., L. L. Fornoff, K. D. Manchanda, W. C. Miller, and D. C. Lovell, The PuraSiv S Process for Removing Acid Plant Tail Gas, *Chem. Eng. Progr.*, **70**(6):58 (1974).

Coloff, S. G., M. Cooke, R. J. Drago, and S. F. Sleva, Ambient Air Monitoring of Gaseous Pollutants, *Am. Lab.*, **5**(7):10 (1973).

Constance, J. D., Calculate Effective Stack Height Quickly, *Chem. Eng.*, **79**(19):81 (1972).

Cooper, H. B. H., Jr., Can You Measure Odor? *Hydrocarbon Process.*, **52**(10):97 (1973).

Cornell, C. G., and D. A. Dahlstrom, Sulfur Dioxide Removal in a Double-Alakli Plant, *Chem. Eng. Progr.*, **69**(12):47 (1973).

Cortelyou, C. G., Commercial Processes for SO_2 Removal, *Chem. Eng. Progr.*, **65**(9):69 (1969).

Crocker, B. B., Minimizing Air Pollution Control Costs, *Chem. Eng. Progr.*, **64**(4):79 (1968).

———, Water Vapor in Effluent Gases: What To Do about Opacity Problems, *Chem. Eng.*, **75**(15):109 (1968).

———, Monitoring Plant Air Pollution, *Chem. Eng. Progr.*, **70**(1):41 (1974).

Cross, F. L., Jr., and H. F. Schiff, Continuous Source Monitoring, *Chem. Eng.*, **80**(14):125 (1973).

Crow, J. H., and J. C. Baumann, Versatile Process Uses Selective Absorption, *Hydrocarbon Process.*, **53**(10):131 (1974).

Davies, D. S., R. A. Jiminez, and P. A. Lemke, Economics of Neutralizing Concentrated Sulfuric Acid, *Chem. Eng. Progr.*, **70**(6):68 (1974).

Davis, J. C., Desulfurization—Add-on Processes Stem H_2S, *Chem. Eng.*, **79**(11):66 (1972).

———, Desulfurization—SO_2 Removal Still Prototype, *Chem. Eng.*, **79**(13):52 (1972).

———, Desulfurization—More SO_2 from Resid Options, *Chem. Eng.*, **79**(15):36 (1972).

————, Pulpers Apply Odor Control, *Chem. Eng.*, **78**(13):52 (1971).

————, Taking Malodors' Measure, *Chem. Eng.*, **80**(11):86 (1973).

————, SO$_x$ Control Held Feasible, *Chem. Eng.*, **80**(25):76 (1973).

Day, R. W., The Hydrocyclone in Process and Pollution Control, *Chem. Eng. Progr.*, **69**(9):67 (1973).

Djololian, C., and D. Billaud, Absorbing Fluorine Compounds from Waste Gases, *Chem. Eng. Progr.*, **74**(11):46 (1978).

Doerschlag, C., and G. Miczek, How to Choose a Cyclone Dust Collector, *Chem. Eng.*, **84**(4):64 (1977).

Dorsey, J. A., and J. O. Burckle, Particulate Emissions and Process Monitors, *Chem. Eng. Progr.*, **67**(8):92 (1971).

Dravnieks, A., Measuring Industrial Odors, *Chem. Eng.*, **81**(22):91 (1974).

Dunlap, R. W., and M. R. Deland, Latest Clean Air Requirements, *Hydrocarbon Process.*, **57**(10):91 (1978).

Duprey, R. L., Sulfur Oxide Regulation: The Status of SO$_x$ Emission Limitations, *Chem. Eng. Progr.*, **68**(2):70 (1972).

Edwards, W. M., and P. Huang, The Kellogg-Weir Air Quality Control System, *Chem. Eng. Progr.*, **73**(8):64 (1977).

Eisenfelder, D. J., and J. W. Dolen, Jr., Air Pollution Control Can Save Money, *Chem. Eng. Progr.*, **70**(5):48 (1974).

Elkin, H. F., and R. A. Constable, Source/Control of Air Emissions, *Hydrocarbon Process.*, **51**(10):113 (1972).

Epstein, M., L. Sybert, S. C. Wang, C. C. Leivo, and F. T. Princiotta, Limestone Wet Scrubbing of SO$_2$, *Chem. Eng. Progr.*, **70**(6):53 (1974).

Ermenc, E. D., Controlling Nitric Oxide Emission, *Chem. Eng.*, **77**(12):193 (1970).

Fair, J. R., B. B. Crocker, and H. R. Null, Trace-Quantity Engineering, *Chem. Eng.*, **79**(17):60 (1972).

———— ———— and ————, Sampling and Analyzing Trace Qualities, *Chem. Eng.*, **79**(21):146 (1972).

Falkenberry, H. L., and A. V. Slack, SO$_2$ Removal by Limestone Injection, *Chem. Eng Progr.*, **65**(12):61 (1969).

Felder, R. M., G. W. Miller, and J. K. Ferrell, Continuous Stack Monitoring Using Polymer Interfaces, *Chem. Eng. Progr.*, **74**(6):86 (1978).

Fischer, H., Burner/Fire Box Design Improves Sulfur Recovery, *Hydrocarbon Process.*, **53**(10):125 (1974).

Frost, A. C., Slurry Bed Adsorption with Molecular Sieves, *Chem. Eng. Progr.*, **70**(5):70 (1974).

Furkert, H., Two Routes to Sulfuric Prove Tough on Pollution, *Chem. Eng.*, **76**(1):70 (1969).

Gall, R. L., and E. J. Piasecki, SO$_2$ Processing: The Double Alkali Wet Scrubbing System, *Chem. Eng. Progr.*, **71**(5):72 (1975).

Gammell, D. M., Blanket Tanks for Gas Control, *Hydrocarbon Process.*, **55**(12):101 (1976).

Gamse, R. N., and J. Speyer, Economic Impact of Sulfur Dioxide Pollution Controls, *Chem. Eng. Progr.*, **70**(6):45 (1974).

Gifford, D. C., Operation of a Limestone Wet Scrubber, *Chem. Eng. Progr.*, **69**(6):86 (1973).

Goar, B., Impure Feeds Cause Claus Plant Problems, *Hydrocarbon Process.*, **53**(7):129 (1974).

Goddin, C. S., E. B. Hunt, and J. W. Palm, CBA Process Ups Claus Recovery, *Hydrocarbon Process.*, **53**(10):122 (1974).

Grimm, C., J. Z. Abrams, W. W. Leffman, I. A. Raben, and C. LaMantia, Flue Gas Desulfurization—The Colstrip Flue Gas Cleaning System, *Chem. Eng. Progr.*, **74**(2):51 (1978).

Haas, L. A., and S. E. Khalafalla, Removing Sulfur Dioxide by Carbon Monoxide Reduction, *Instr. & Control Systems*, **45**(3):101 (1972).

Haase, D. J., and D. G. Walker, The COSORB Process, *Chem. Eng. Progr.*, **70**(5):74 (1974).

Hanf, E. W., and J. W. MacDonald, Economic Evaluation of Wet Scrubbers, *Chem. Eng. Progr.*, **71**(3):48 (1975).

Hardison, L. C., and E. J. Dowd, Emission Control via Fluidized Bed Oxidation, *Chem. Eng. Progr.*, **73**(8):31 (1977).

Harwood, C. F., P. C. Siebert, and D. K. Oestreich, Optimizing Baghouse Performance . . . To Control Asbestos Emissions, *Chem. Eng. Progr.*, **73**(1):54 (1977).

Hayford, J. S., Process Cleans Tail Gases, *Hydrocarbon Process.*, **52**(10):95 (1973).

—— L. P. VanBrocklin, and M. A. Kuck, Stauffer's Aquaclaus Process, *Chem. Eng Progr.*, **69**(12):54 (1973).

Helfritch, D. J., Performance of an Electrostatically Aided Fabric Filter, *Chem. Eng. Progr.*, **73**(8):54 (1977).

Hellman, T. M., and F. H. Small, Characterization of Petrochemical Odors, *Chem. Eng. Progr.*, **69**(9):75 (1973).

Herr, G. A., Odor Destruction: A Case History, *Chem. Eng. Progr.*, **70**(5):65 (1974).

Holland, W. D., and R. E. Conway, Three Multi-Stage Stack Samplers, *Chem. Eng. Progr.*, **69**(6):93 (1973).

Horlacher, W. R., R. E. Barnard, R. K. Teague, and P. L. Hayden, Pollution Abatement: Four SO_2 Removal Systems, *Chem. Eng. Progr.*, **68**(8):42 (1972).

Horzella, T. I., Selecting, Installing and Maintaining Cyclone Dust Collectors, *Chem. Eng.*, **85**(3):85 (1978).

Humphries, J. J., S. B. Zdonik, and E. J. Parsi, An SO_2 Removal and Recovery Process, *Chem. Eng. Progr.*, **67**(5):64 (1971).

Hunter, S. C., and W. A. Carter, Combustion Technology for NO_x Emission Reduction, *Chem. Eng. Progr.*, **73**(8):66 (1977).

Iammartino, N. R., Technology Gears Up to Control Fine Particles, *Chem. Eng.*, **79**(18):50 (1972).

——, Cleaning Up Claus Offgas, *Chem. Eng.*, **81**(20):56 (1974).

Iya, K., Reduce NO_x in Stack Gases, *Hydrocarbon Process.*, **51**(11):163 (1972).

Jahnig, C. E., and R. R. Bertrand, Environmental Aspects of Coal Gasification, *Chem. Eng. Progr.*, **72**(8):51 (1976).

James, G. R., Stripping Ammonium Nitrate from Vapors, *Chem. Eng. Progr.*, **69**(6):79 (1973).

Jimeson, R. M., and R. R. Maddocks, Sulfur Compound Cleanup: Trade-offs in Selecting SO_x Emission Controls, *Chem. Eng. Progr.*, **72**(8):80 (1976).

Jones, H. R., "Air and Gas Cleanup Equipment," Noyes Data Corporation, Park Ridge, New Jersey, 1972.

Jonker, P. E., W. J. Porter, and C. B. Scott, Control Floating Roof Tank Emissions, *Hydrocarbon Process.*, **56**(5):151 (1977).

Jones, J. W., Disposal of Flue-Gas-Cleaning Wastes, *Chem. Eng.*, **84**(4):79 (1977).

Kaplan, N., and M. A. Maxwell, Removal of SO_2 from Industrial Waste Gases, *Chem. Eng.*, **84**(22):127 (1977).

Kent, G. R., Make Profit from Flares, *Hydrocarbon Process.*, **51**(10):121 (1972).

Kiang, Y.-H., Controlling Vinyl Chloride Emissions, *Chem. Eng. Progr.*, **72**(12):37 (1976).

Kiovsky, J. R., P. B. Koradia, and D. S. Hook, Molecular Sieves for SO_2 Removal, *Chem. Eng. Progr.*, **72**(8):98 (1976).

Klasens, H. A., Analyze Stack Gases via Sampling or Optically, in Place, *Chem. Eng.*, **84**(25):201 (1977).

Klooster, H. J., G. A. Vogt, and G. F. Braun, Optimizing the Design of Relief and Flare Systems, *Chem. Eng. Progr.*, **71**(1):39 (1975).

—— —— and D. G. Bernhart, Refinery Odor Control, *Hydrocarbon Process.*, **55**(4):121 (1976).

Knoepke, J., Tracer-Gas System Determines Flow Volume of Flue Gases, *Chem. Eng.*, **84**(3):91 (1977).

Koch, W. H., and W. Licht, New Design Approach Boosts Cyclone Efficiency, *Chem. Eng.*, **84**(24):80 (1977).

Koehler, G. R., Alkaline Scrubbing Removes Sulfur Dioxide, *Chem. Eng. Progr.*, **70**(6):63 (1974).

Krill, H., and K. Storp, H_2S Adsorbed from Tailgas, *Chem. Eng.*, **80**(17):84 (1973).

Kusik, C. L., J. I. Stevens, R. M. Nadkarni, P. A. Huska, and D. W. Lee, Energy Use and Air Pollution Control in New Process Technology, *Chem. Eng. Progr.*, **73**(8):36 (1977).

LaMantia, C. R., R. R. Lunt, and I. S. Shah, Dual Alkali Process for Sulfur Dioxide Removal, *Chem. Eng. Progr.*, **70**(6):66 (1974).

Lasater, R. C., and J. H. Hopkins, Removing Particulates from Stack Gases, *Chem. Eng.*, **84**(22):111 (1977).

Lord, H. C., CO_2 Measurements Can Correct for Stack-Gas Dilution, *Chem. Eng.*, **84**(3):95 (1977).

Lovett, W. D., and F. T. Cunniff, Air Pollution Control by Activated Carbon, *Chem. Eng. Progr.*, **70**(5):43 (1974).

Lucas, R. L., Gas-Solids Separations: State-of-the-Art, *Chem. Eng. Progr.*, **70**(12):52 (1974).

Ludwig, J. H., and P. W. Spaite, Control of Sulfur Oxide Pollution, *Chem. Eng. Progr.*, **63**(6):82 (1967).

Mace, G. R., and D. Casaburi, Lime vs. Caustic for Neutralizing Power Plant Effluents, *Chem. Eng. Progr.*, **73**(8):86 (1977).

Mahajan, K. K. Tall Stack Design Simplified, *Hydrocarbon Process.*, **54**(9):217 (1975).

Martin, R. R., F. S. Manning, and E. D. Reed, Watch for Elevated Dew Points in SO_3-Bearing Stack Gases, *Hydrocarbon Process.*, **53**(6):143 (1974).

Maurin, P. G., and J. Jonakin, Removing Sulfur Oxides from Stacks, *Chem. Eng.*, **77**(9):173 (1970).

McCarthy, J. E., Choosing a Flue-Gas Desulfurization System, *Chem. Eng.*, **85**(6):79 (1978).

McCutchen, G. D., NO_x Emission Trends and Federal Regulation, *Chem. Eng. Progr.*, **73**(8):58 (1977).

McFarlin, W. A., Design Considerations for Mobile Air Quality Monitoring Laboratories, *Am. Lab.*, **5**(7):75 (1973).

McKee, H. C., Air-Pollution Instrumentation, *Chem. Eng.*, **83**(13):155 (1976).

McKinney, C. M., Sulfur in Products vs. Crude Oil, *Hydrocarbon Process.*, **51**(10):117 (1972).

Meisen, A., and H. A. Bennett, Consider All Claus Reactions, *Hydrocarbon Process.*, **53**(11):171 (1974).

Miller, W. E., The Cat-Ox Process at Illinois Power, *Chem. Eng. Progr.*, **70**(6):49 (1974).

Miller, W. C., Adsorption Cuts SO_2, NO_x, Hg, *Chem. Eng.*, **80**(18):62 (1973).

Moores, C. W., Control PA Emissions, *Hydrocarbon Process.*, **54**(10):100 (1975).

Morrow, N. L., R. S. Brief, and R. R. Bertrand, Air Sampling and Analysis, *Chem. Eng.*, **79**(10):125 (1972).

Moulton, F. H., Environmental Restrictions and Chlorinated Solvents, *Chem. Eng. Progr.*, **69**(10):85 (1973).

Naber, J. E., J. A. Wesselingh, and W. Groenendaal, New Shell Process Treats Claus Off-gas, *Chem. Eng. Progr.*, **69**(12):29 (1973).

Nichols, R. A., Hydrocarbon-Vapor Recovery, *Chem. Eng.*, **80**(6):85 (1973).

Noll, K., and J. Duncan, "Industrial Air Pollution Control," Ann Arbor Science Publishers, Ann Arbor, Michigan, 1973.

O'Connell, W. L., How to Attack Air-Pollution Control Problems, *Chem. Eng.*, **83**(22):97 (1976).

Olds, F. C., and L. DuBridge, Views on Air Pollution, *Power Eng.*, **74**(7):30 (1970).

Osag, T. R., J. A. Smith, F. L. Bunyard, and G. B. Crane, Fluoride Emission Control Costs, *Chem. Eng. Progr.*, **72**(12):33 (1976).

Ouwerkerk, C., Design for Selective H_2S Absorption, *Hydrocarbon Process.*, **57**(4):89 (1978).

Owen, L. T., Selecting In-Plant Dust-Control Systems, *Chem. Eng.*, **81**(21):120 (1974).

Painter, D. E., "Air Pollution Technology," Reston Publishing Company, Reston, Virginia, 1974.

Parekh, R., Equipment for Controlling Gaseous Pollutants, *Chem. Eng.*, **82**(21):129 (1975).

Pearson, M. J., Developments in Claus Catalysts, *Hydrocarbon Process.*, **52**(2):81 (1973).

Peters, J. M., Predicting Efficiency of Fine-Particle Collectors, *Chem. Eng.*, **80**(9):99 (1973).

Pfoutz, B. D., and L. L. Stewart, Electrostatic Precipitators—Materials of Construction, *Chem. Eng. Progr.*, **71**(3):53 (1975).

Pierce, R. R., Estimating Acid Dewpoints in Stack Gases, *Chem. Eng.*, **84**(8):125 (1977).

Potter, B. H., and T. L. Craig, Pollution Abatement: Commercial Experience with an SO_2 Recovery Process, *Chem. Eng. Progr.*, **68**(8):53 (1972).

Princiotta, F. T., Flue Gas Desulfurization—Advances in SO_2 Stack Gas Scrubbing, *Chem. Eng. Progr.*, **74**(2):58 (1978).

Pruessner, R. D., and L. D. Broz, Hydrocarbon Emission Reduction Systems, *Chem. Eng. Progr.*, **73**(8):69 (1977).

Quane, D. E., Reducing Air Pollution at Pharmaceutical Plants, *Chem. Eng. Progr.*, **70**(5):51 (1974).

Rashmi, P., Equipment for Controlling Gaseous Pollutants, *Chem. Eng.*, **82**(21):129 (1975).

Raymond, W. J., and A. G. Sliger, Flue Gas Desulfurization—The Kellogg/Weir Scrubbing System, *Chem. Eng. Progr.*, **74**(2):75 (1978).

Reed, R. D., Nitrogen Oxides Problems in Industry, *Chem. Eng.*, **84**(22):153 (1977).

Ring, T. A., and J. M. Fox, Stack Gas Cleanup Progress, *Hydrocarbon Process.*, **53**(10):119 (1974).

Robins, D. L., and M. M. Mattia, Computer Program Helps Design Stacks for Curbing Air Pollution, *Chem. Eng.*, **75**(3):119 (1968).

Rochelle, G. T., and C. J. King, Flue Gas Desulfurization—Alternatives for Stack Gas Desulfurization by Throwaway Scrubbing, *Chem. Eng. Progr.*, **74**(2):65 (1978).

Rosebrook, D. D., Fugitive Hydrocarbon Emissions, *Chem. Eng.*, **84**(22):143 (1977).

Rosenberg, H. S., R. B. Engdahl, J. H. Oxley, and J. M. Genco, SO_2 Processing: The Status of SO_2 Control Systems, *Chem. Eng. Progr.*, **71**(5):66 (1975).

———, How Good is Flue Gas Desulfurization?, *Hydrocarbon Process.*, **57**(5):132 (1978).

Ross, R. D., Pollution Abatement: Incineration of Solvent-Air Mixtures, *Chem. Eng. Progr.*, **68**(8):59 (1972).

———, "Air Pollution and Industry," Van Nostrand Reinhold, Cincinnati, Ohio, 1972.

Ross, S. S., Designing Your Plant for Easier Emission Testing, *Chem. Eng.*, **79**(14):112 (1972).

——— and L. J. White, International Pollution Control, *Chem. Eng.*, **79**(10):137 (1972).

———, The Revolutionary Field of Air Pollution Instrumentation, *Chem. Eng.*, **79**(20):37 (1972).

Rossano, A. T., and H. B. H. Cooper, Sampling and Analysis—Air Pollution Control, *Chem. Eng.*, **75**(22):142 (1968).

Rushton, J. D., Handling Sulfur Compounds from Pulp Mills, *Chem. Eng. Progr.*, **69**(12):39 (1973).

Rymarz, T. M., How to Specify Pulse-Jet Filters, *Chem. Eng.*, **82**(7):97 (1975).

——— and D. H. Klipstein, Removing Particulates from Gases, *Chem. Eng.*, **82**(21):113 (1975).

Sargent, G. D., Gas/Solid Separations, *Chem. Eng.*, **78**(4):11 (1971).

Schmidt, T. R., Ground-Level Detector Tames Flare-stack Flames, *Chem. Eng.*, **84**(8):121 (1977).

Schneider, G. G., T. I. Horzella, J. Cooper, and P. J. Striegl, Electrostatic Precipitators: How They are Used in the CPI, *Chem. Eng.*, **82**(17):97 (1975).

——— ——— ——— and ———, Selecting and Specifying Electrostatic Precipitators, *Chem. Eng.*, **82**(11):94 (1975).

Selmeczi, J. G., and D. A. Stewart, Flue Gas Desulfurization—The Thiosorbic Flue Gas Desulfurization Process, *Chem. Eng Progr.*, **74**(2):41 (1978).

Semrau, K. T., Practical Process Design of Particulate Scrubbers, *Chem. Eng.*, **84**(20):87 (1977).

Shah, I. S., Pulp Plant Pollution Control, *Chem. Eng. Progr.*, **64**(9):66 (1968).

———, Removing SO_2 and Acid Mist with Venturi Scrubbers, *Chem. Eng. Progr.*, **67**(5):51 (1971).

———, MgO Absorbs Stackgas SO_2, *Chem. Eng.*, **79**(14):80 (1972).

Shen, T. T., Online Instruments Expedite Emission Tests, *Chem. Eng.*, **82**(11):109 (1975).

Siddiqi, A. A., and J. W. Tenini, FGD—a Viable Alternative, *Hydrocarbon Process.*, **56**(10):104 (1977).

Sisson, B., Calculating the Fly Ash Discharged to a Stack, *Chem. Eng.*, **83**(8):154 (1976).

Slack, A. V., Air Pollution: The Control of SO_2 from Power Stacks, *Chem. Eng.*, **74**(25):188 (1967).

———, Flue Gas Desulfurization: An Overview, *Chem. Eng. Progr.*, **72**(8):94 (1976).

———, Flue Gas Desulfurization—Lime-Limestone Scrubbing: Design Considerations, *Chem. Eng. Progr.*, **74**(2):71 (1978).

Smith, R. S., Control Stack Gas Pollution, *Hydrocarbon Process.*, **51**(9):223 (1972).

Spaite, P. W., and R. E. Harrington, Abatement Goes Global, *Power Eng.*, **75**(2):42 (1971).

Squires, A. M., Air Pollution: The Control of SO_2 from Power Stacks, *Chem. Eng.*, **72**(26):101 (1967); **74**(23):260 (1967); **74**(24):133 (1967).

———, Keeping Sulfur Out of the Stacks, *Chem. Eng.*, **77**(9):181 (1970).

Stone, L. K., Emissions from Coal Conversion Processes, *Chem. Eng. Progr.*, **71**(12):52 (1976).

Straitz, J. F., III, Make the Flare Protect the Environment, *Hydrocarbon Process.*, **56**(10):131 (1977).

Strom, S. S., and W. Downs, A Systematic Approach to Limestone Scrubbing, *Chem. Eng. Progr.*, **70**(6):55 (1974).

Svarovsky, L., Gas Cyclone Selection Procedure, *Chem. Eng.* (*London*), 133 (March 1975).

Tamaki, A., SO_2 Processing: The Thoroughbred 101 Desulfurization Process, *Chem. Eng. Progr.*, **71**(5):55 (1975).

Teale, J. M., Fast Payout from In-Plant Recovery of Spent Solvents, *Chem. Eng.*, **84**(3):98 (1977).
Teller, A. J., Air Pollution Control, *Chem. Eng.*, **79**(10):93 (1972).
Thomson, S. J., and R. H. Crow, Energy Cost of NO_x Control, *Hydrocarbon Process.*, **55**(5):95 (1976).
Tucker, W. G., and J. R. Burleigh, SO_2 Emission Control from Acid Plants, *Chem. Eng. Progr.*, **67**(5):57 (1971).
Turk, A., Industrial Odor Control, *Chem. Eng.*, **77**(9):199 (1970).
Uno, T., S. Fukui, M. Atsukawa, M. Higashi, H. Yamada, and K. Kamei, Scale-up of a SO_2 Control Process, *Chem. Eng. Progr.*, **66**(1):61 (1970).
Vandegrift, A. E., L. J. Shannon, and P. G. Gorman, Controlling Fine Particles, *Chem. Eng.*, **80**(14):107 (1973).
Vanderlinde, L. G., Smokeless Flares, *Hydrocarbon Process.*, **53**(10):99 (1974).
Vasan, S., SO_2 Processing: The Citrex Process for SO_2 Removal, *Chem. Eng. Progr.*, **71**(5):61 (1975).
Verschueren, K., "Handbook of Environmental Data on Organic Chemicals," Van Nostrand Reinhold Company, New York (1978).
Waid, D. E., Pollution Abatement: Controlling Pollutants via Thermal Incineration, *Chem. Eng. Progr.*, **68**(8):57 (1972).
Walling, J. C., Ins and Outs of Gas Filter Bags, *Chem. Eng.*, **77**(25):162 (1970).
Walther, J. E., and H. R. Amberg, Odor Control in the Kraft Pulp Industry, *Chem. Eng. Progr.*, **66**(3):73 (1970).
——— ——— and H. Hamby, III, Meeting New Pollution Requirements at a Paper Mill, *Chem. Eng. Progr.*, **69**(6):100 (1973).
Welty, A. B., Jr., Flue Gas Desulfurization Technology, *Hydrocarbon Process.*, **50**(10):104 (1971).
Wiedersum, G. C., Control of Power Plant Emissions, *Chem. Eng. Progr.*, **66**(11):49 (1970).
Wiley, S. K., Controlling Vapor Losses, *Chem. Eng.*, **80**(21):116 (1973).
Wilkins, P. E., The Clean Air Act: Implications for Industry, *Instr. & Control Systems*, **45**(1):27 (1972).
Wilks, P. A., Jr., Infrared Analysis of Process Emissions, *Chem. Eng.*, **78**(28):120 (1971).
Yamagucki, M., K. Matsushita, and K. Takami, Remove NO_x from HNO_3 Tail Gas, *Hydrocarbon Process.*, **55**(8):101 (1976).
Yocom, J. E., and R. A. Duffee, Controlling Industrial Odors, *Chem. Eng.*, **77**(13):160 (1970).
York, O. H., and E. W. Poppele, Two-Stage Mist Eliminators for Sulfuric Acid Plants, *Chem. Eng. Progr.*, **66**(11):67 (1970).
Yulish, J., Sulfur-Recovery Processes Compete for Leading Role, *Chem. Eng.*, **78**(13):58 (1971).
Zabolotny, E. R., and R. W. Kuhr, Waste Treatment Advances: Hot Gas Purification, *Chem. Eng. Progr.*, **72**(10):69 (1976).
Zanker, A., Estimate Tank Breathing Loss, *Hydrocarbon Process.*, **56**(1):117 (1977).
———, Plant Notebook, Estimating the Dew Points of Stack Gases, *Chem. Eng.*, **85**(9):154 (1978).
Basic Technology, *Chem. Eng.*, **77**(9):165 (1970).
Air Pollution Control, *Chem. Eng.*, **78**(14):131 (1971).
Air Pollution Codes, *Chem. Eng.*, **80**(14):11 (1973).

Construction

Armstrong, R., Better Ways to Build Process Plants, *Chem. Eng.*, **79**(8):86 (1972).
Bookout, A. R., The Ch.E.'s Impact on Construction, *Chem. Eng. Progr.*, **68**(2):37 (1972).
Coppen, J. L., Managing Small Design-Construction Projects, *Chem. Eng.*, **81**(25):85 (1974).
Ferguson, C. C., Role of the Ch.E. in Engineering/Construction Firms, *Chem. Eng.*, **80**(1):100 (1973).
Foster, N., "Construction Estimates from Take-Off to Bid," 2d ed., McGraw-Hill Book Company, New York, 1972.
Friedrich, E. R., Jr., Floors for Process Areas, *Chem. Eng.*, **81**(13):157 (1974).
King, R. A., How to Achieve Effective Project Control, *Chem. Eng.*, **84**(14):117 (1977).

Contracts

Bergtraun, E. M., Contracting for New Construction, *Chem. Eng.*, **84**(13):133 (1977).
Geiger, G. H., "Supplementary Readings in Engineering Design," McGraw-Hill Book Company, New York, 1975.

Loring, R. J., The Proposal Manager's Work, *Chem. Eng.*, **77**(18):98 (1970).
Moon, G. B., A Law Primer for the Chemical Engineer, *Chem. Eng.*, **85**(9):114 (1978).

Design practice

Albright, L. F., F. H. Van Munster, and J. C. Forman, High Pressure, Principles and Process Trends, *Chem. Eng.*, **75**(20):194 (1968).
Kern, R., Obtain Minimum Cost, Plant Layout, *Chem. Eng.*, **84**(11):131 (1977).
Landrum, R. J., Equipment, *Chem. Eng.*, **77**(22): 75 (1970).
Norden, R. B., Modern Engineering Design Practices, *Chem. Eng.*, **72**(16): 91 (1965).
Zudkevitch, D., Imprecise Data Impacts Plant Design and Operation, *Hydrocarbon Process.*, **54**(3):97 (1975).

Energy conservation

Abadie, V. H., Turboexpanders Recover Energy, *Hydrocarbon Process.*, **52**(7):93 (1973).
Balfoort, J. P., Improved Hot-Gas Expanders for Cat Cracker Flue Gas, *Hydrocarbon Process.*, **55**(3):141 (1976).
Bannon, R. P., and S. Marple, Jr., Heat Transfer—Heat Recovery in Hydrocarbon Distillation, *Chem. Eng Progr.*, **74**(7):41 (1978).
Barber, R. E., Rankine-Cycle Systems for Waste Heat Recovery, *Chem. Eng.*, **81**(25):101 (1974).
Barbier, J.-C., Save Energy When Making Gasoline, *Hydrocarbon Process.*, **56**(9):85 (1977).
Barlow, J. A., Energy Recovery in a Petrochemical Plant, *Chem. Eng.*, **82**(14):93 (1975).
Barnhart, J. H., Energy Analysis of a Claus Plant, *Chem. Eng. Progr.*, **74**(5):58 (1978).
Barrows, G. L., Save Energy with Ceramic Fiber Insulation, *Hydrocarbon Process.*, **56**(10):187 (1977).
Beatty, R. E., and R. G. Krueger, How the Steam Trap Saves Plant Energy, *Chem. Eng. Progr.*, **74**(9):94 (1978).
Bennett, R. C., Heat Transfer—Recompression Evaporation, *Chem. Eng. Progr.*, **74**(7):67 (1978).
Berman, H. L., Fired Heaters—IV, How to Reduce Your Fuel Bill, *Chem. Eng.*, **85**(20):165 (1978).
Bogart, M. J. P., Save Energy in Ammonia Plants, *Hydrocarbon Process.*, **57**(4):141 (1978).
Bojnowski, J. H., J. W. Crandall, and R. M. Hoffman, Modernized Separation System Saves More Than Energy, *Chem. Eng. Progr.*, **71**(10):50 (1975).
Brant, A. J., What is Your E-Q?, *Hydrocarbon Process.*, **54**(7):88 (1975).
Braun, S. S., Power Recovery Cuts Energy Costs, *Hydrocarbon Process.*, **52**(5):81 (1973).
Briley, G. C., Conserve Energy . . . Refrigerate with Waste Heat, *Hydrocarbon Process.*, **55**(5):173 (1976).
Brown, C. L., and D. Figenscher, Preheat Process Combustion Air, *Hydrocarbon Process.*, **52**(7):115 (1973).
Campbell, P. R., and P. S. Chappelear, Power Cycle Recovers Lost Energy, *Hydrocarbon Process.*, **57**(4):104 (1978).
Cantrell, C. J., Jr., Waste-Heat Recovery Cuts Plant Fuel Costs, *Oil and Gas. J.*, **74**(45):183 (1976).
Cantwell, E. N., Exhaust Controls Use Energy, *Hydrocarbon Process.*, **53**(7):94 (1974).
Casten, J. W., Heat Transfer—Mechanical Recompression Evaporators, *Chem. Eng. Progr.*, **74**(7):61 (1978).
Congram, G. E., Refiners Whittle Dow Fuel Costs, *Oil & Gas J.*, **73**(15):57 (1975).
Connolly, J. R., Energy Conservation in Fluid Mixing, *Chem. Eng. Progr.*, **72**(5):52 (1976).
Cordero, R., The Cost of Missing Pipe Insulation, *Chem. Eng.*, **84**(4):77 (1977).
Cummings, W. P., Save Energy in Adsorption, *Hydrocarbon Process.*, **54**(2):97 (1975).
Curt, R. P., Economic Insulation Thickness, *Hydrocarbon Process.*, **55**(3):137 (1976).
Danekind, W. E., Stream Management in a Refinery, *Hydrocarbon Process.*, **55**(12):71 (1976).
Dockendorff, J. D., and P. J. Cheng, Energy-Conscious Evaporators, *Chem. Eng. Progr.*, **72**(5):56 (1976).
Doolin, J. H., Select Pumps to Cut Energy Cost, *Chem. Eng.*, **84**(2):137 (1977).
Drennan, J. N., Btu's vs. Buck—Who Wins? **74**(5):81 (1978).

Duckham, H., and J. Fleming, Better Plant Design Saves Energy, *Hydrocarbon Process.*, **55**(7):78 (1976).

Dugan, C. F., W. L. Van Nostrand, and J. L. Haluska, How Antifoulants Reduce the Energy Consumption of Refineries. *Chem. Eng. Progr.*, **74**(5):53 (1978).

Fair, J. R., Advances in Distillation System Design, *Chem. Eng. Progr.*, **73**(11):78 (1977).

Farin, W. G., Low-Cost Evaporation Method Saves Energy by Reusing Heat, *Chem. Eng.*, **83**(5):100 (1976).

Fleming, J., H. Duckham, and J. Styslinger, Recover Energy with Exchangers, *Hydrocarbon Process.*, **55**(7):101 (1976).

Fleming, J. B., J. R. Lambrix, and M. R. Smith, Energy Conservation in New-Plant Design, *Chem. Eng.*, **81**(2):112 (1974).

Flynn, B. L., Jr., Wet Air Oxidation for Black Liquor Recovery, *Chem. Eng. Progr.*, **72**(5):66 (1976).

Franzke, A., Save Energy with Hydraulic Power Recovery Turbines, *Hydrocarbon Process.*, **54**(3):107 (1975).

Freshwater, D. C., and E. Ziogon, Reducing Energy Requirements in Unit Operations, *Chem. Eng. J.*, **11**(3):215 (1976).

Fuchs, W., G. R. James, and K. J. Stokes, Economics of Flue Gas Heat Recovery, *Chem. Eng. Progr.*, **73**(11):65 (1977).

Funk, G. L., B. F. Houston, and G. D. Stacy, Making Energy Conservation Pay Through Automation, *Chem. Eng. Progr.*, **74**(5):66 (1978).

Gasparini, L., Fuel Saving in Chemical-Plant Power Stations, *Chem. Eng.*, **81**(20):123 (1974).

Geyer, G. R., and P. E. Kline, Energy Conservation Schemes for Distillation Processes, *Chem. Eng. Progr.*, **72**(5):49 (1976).

Gilbert, L. F., Precise Combustion-Control Saves Fuel and Power, *Chem. Eng.*, **83**(13):145 (1976).

Goossens, G., Principles Govern Energy Recovery, *Hydrocarbon Process.*, **57**(8):133 (1978).

Grace, E. C., and H. L. Fidoe, U.S. Refineries Conserve Energy, *Hydrocarbon Process.*, **55**(5):120 (1976).

Grace, J. A., and K. C. Khurana, Investing Capital for Energy Savings, *Oil & Gas J.*, **73**(35):105 (1975).

Gyger, R. F., and E. L. Doerflein, Save Energy—Use O_2 on Waste, *Hydrocarbon Process.*, **55**(7):96 (1976).

Hallman, D. W., Save Energy by Cogeneration, *Hydrocarbon Process.*, **57**(5):151 (1978).

Hanna, W. T., and W. J. Frederick, Jr., Energy Conservation in The Pulp and Paper Industries, *Chem. Eng. Progr.*, **74**(5):71 (1978).

Harbert, W. D., Preflash Saves Heat in Crude Unit, *Hydrocarbon Process.*, **57**(7):123 (1978).

Harrison, M. R., and C. M. Pelanne, Cost-Effective Thermal Insulation, *Chem. Eng.*, **84**(27):63 (1977).

Haslam, A., P. Brook, and H. Isalski, Recycle H_2 in NH_3 Purge Gas, *Hydrocarbon Process.*, **55**(1):103 (1976).

Hayden, R., Save Energy: Better Column Control, *Hydrocarbon Process.*, **55**(5):122 (1976).

Hayden, J. E., and W. H. Levers, Design Plants to Save Energy, *Hydrocarbon Process.*, **52**(7):72 (1973).

Heaton, L. S., and E. J. Smet, New Approaches Help to Cut Energy Use, *Oil & Gas J.*, **74**(33):134 (1976).

Hendler, H. E., Efficiency Control in Combustion Processes, *Chem. Eng. Progr.*, **71**(10):39 (1975).

Hess, M., Ammonia: Coal versus Gas, *Hydrocarbon Process.*, **55**(11):97 (1976).

Hickok, H. N., Save Electrical Energy, *Hydrocarbon Process.*, **57**(7):131 (1978); **57**(8):127 (1978).

Holden, G. F., Strategic Energy Planning, *Chem. Eng.*, **82**(2):94 (1975).

Hopkins, P. S., and F. L. Greve, Refinery Energy Saving—Energy-Savings Projects Can Yield Some Big Returns, *Oil & Gas J.*, **73**(29):73 (1975).

———— and ————, Process Energy Savings: Complex but Rewarding, *Oil & Gas J.*, **73**(35):77 (1975).

Houghton, J., and R. K. Bailes, Economics of Multiple Effect Evaporators, *Chem. Eng. (London)*, 82 (February 1975).

Hsiang, T. C. H., and D. R. Woods, The Influence of Design and Operating Variables on Energy

Consumption and Separation Efficiency of a Hydrocyclonic Concentrator, *Can. J. Chem. Eng.*, **50**(5):607 (1972).

Huang, F., and R. Elshout, Optimizing the Heat Recovery of Crude Units, *Chem. Eng. Progr.*, **72**(7):68 (1976).

Huckins, H. A., Conserving Energy in Chemical Plants, *Chem. Eng. Progr.*, **74**(5):85 (1978).

Hughes, R., and V. Deumaga, Insulation Saves Energy, *Chem. Eng.*, **81**(11):95 (1974).

Jenkins, B. A., More-Careful Control of Ventilation for Enclosed Operations Saves Energy, *Chem. Eng.*, **81**(18):60 (1974).

Jones, F. A., Build and Run Plants to Save Energy, *Hydrocarbon Process.*, **53**(7):89 (1974).

Judd, D. K., Selexol Unit Saves Energy, *Hydrocarbon Process.*, **57**(4):122 (1978).

Kennedy, J. P., Changing Economy: Its Impact on Equipment, *Chem. Eng.*, **82**(6):54 (1975).

Kline, P. E., Technical Task Force Approach to Energy Conservation, *Chem. Eng. Progr.*, **70**(2):23 (1974).

Knight, W. P., Evaluate Waste Heat Steam Generators, *Hydrocarbon Process.*, **57**(7):126 (1978).

Krueding, A. P., Cat Cracker Power Recovery Techniques, *Chem. Eng. Progr.*, **71**(10):56 (1975).

Kydd, P. H., Integrated Gasification Gas Turbine Cycles, *Chem. Eng. Progr.*, **71**(10):62 (1975).

Lambert, L. C., Versatile Oil Burner Offers High Efficiency and Reduced Emissions, *Chem. Eng.*, **81**(18):51 (1974).

Latour, P. R., Energy Conservation via Process Computer Control, *Chem. Eng. Progr.*, **72**(4):76 (1976).

Lieberman, N. P., Change Controls to Save Energy, *Hydrocarbon Process.*, **57**(2):93 (1978).

Llovet, J. E., H. J. Klooster, and D. G. Chapel, Refinery Design (circa 1984), *Chem. Eng. Progr.*, **71**(6):85 (1975).

Lott, J. L., and C. M. Sliepcevich, Use Design Innovation to Save Energy, *Hydrocarbon Process.*, **54**(7):81 (1975).

Ludwig, E. E., Changes in Process Equipment Caused by the Energy Crisis, *Chem. Eng.*, **82**(6):61 (1975).

Lyche, D. W., and D. M. Heffelfinger, Engineered Standards: A Route to Energy Savings, *Chem. Eng. Progr.*, **74**(5):60 (1978).

Maloney, K. F., Economic Potential of Steam Turbines in the HPI, *Hydrocarbon Process.*, **54**(11):261 (1975).

Martin, R. B., Guide to Better Insulation, *Chem. Eng.*, **82**(10):98 (1975).

Mathur, P. K., and T. W. F. Russell, Chemical Recycle Can Save Energy, *Hydrocarbon Process.*, **57**(7):89 (1978).

May, D. L., Cutting Boiler Fuel Costs with Combustion Controls, *Chem. Eng.*, **82**(7):53 (1975).

McClaskey, B. M., and J. A. Lundquist, Can You Justify Hydraulic Turbines?, *Hydrocarbon Process.*, **55**(10):163 (1976).

McCracken, C. K., Time for Forced Draft Burners?, *Hydrocarbon Process.*, **56**(2):89 (1977).

Mergens, E. H., How Shell Oil Conserves Energy, *Hydrocarbon Process.*, **56**(7):120 (1977).

Millar, G., Use of Turboexpanders with FCC, *Hydrocarbon Process.*, **57**(7):111 (1978).

Mix, T. J., J. S. Dweck, M. Weinberg, and R. C. Armstrong, Energy Conservation in Distillation, *Chem. Eng. Progr.*, **74**(4):49 (1978).

Mol, A., Which Heat Recovery System?, *Hydrocarbon Process.*, **52**(7):109 (1973).

Monroe, E. S., Energy Conservation and Vacuum Pumps, *Chem. Eng. Progr.*, **71**(10):69 (1975).

Monroe, E. S., Jr., How to Test Steam Traps, *Chem. Eng.*, **82**(18):99 (1975).

———, Select the Right Steam Trap, *Chem. Eng.*, **83**(1):129 (1976).

Moore, G. F., Save Energy in Plant Operations, *Hydrocarbon Process.*, **52**(7):67 (1973).

Nailen, R. L., Recover Energy with Induction Generators, *Hydrocarbon Process.*, **57**(7):103 (1978).

Newkirk, R. W., Machinery Costs vs. Energy Conservation, *Chem. Eng.*, **81**(13):138 (1974).

Nishio, M., and A. I. Johnson, Strategy for Energy System Expansion, *Chem. Eng. Progr.*, **73**(1):73 (1977).

Null, H. R., Heat Pumps in Distillation, *Chem. Eng Progr.*, **72**(7):58 (1976).

O'Brien, N. G., Reducing Column Steam Consumption, *Chem. Eng. Progr.*, **72**(7):65 (1976).

O'Sullivan, T. F., H. R. McChesney and W. H. Pollock, Coal-Fired Process Heaters?, *Hydrocarbon Process.*, **57**(7):95 (1978).

Panesar, K. S., Select Pumps to Save Energy, *Hydrocarbon Process.*, **57**(10):127 (1978).

Perreault, E. A., and P. J. Prutzman, Strategies for Curtailing Electric Power, *Chem. Eng.*, **84**(11):153 (1977).

Petterson, W. C., and T. A. Wells, Energy-Saving Schemes in Distillation, *Chem. Eng.*, **84**(20):78 (1977).

Pettman, M. J., and G. C. Humphreys, Improve Designs to Save Energy, *Hydrocarbon Process.*, **54**(1):77 (1975).

Pinto, A., and P. L. Rogerson, Impact of High Fuel Cost on Plant Design, *Chem. Eng. Progr.*, **73**(7):95 (1977).

Prather, B. V., and E. P. Young, Energy for Wastewater Treatment, *Hydrocarbon Process.*, **55**(5):88 (1976).

Rathore, R. N. S., and G. J. Powers, A Forward Branching Scheme for the Synthesis of Energy Recovery Systems, *I&EC Proc. Design & Develop.*, **14**(2):175 (1975).

Reay, D., Energy Conservation in Industrial Drying, *Chem. Eng. (London)*, 507 (July–August 1976).

Reed, R. D., Save Energy at Your Heater, *Hydrocarbon Process.*, **52**(7):119 (1973).

———, Recover Energy From Furnace Stacks; Both Old and New Installations Need Study for Improvement, *Hydrocarbon Process.*, **55**(1):127 (1976).

Rex, M. J., Choosing Equipment for Process Energy Recovery, *Chem. Eng.*, **82**(16):98 (1975).

Reynolds, J. A., Saving Energy and Costs in Pumping Systems, *Chem. Eng.*, **83**(1):135 (1976).

Ring, T. A., and J. M. Fox, Session: Advances in Processing, *Hydrocarbon Process.*, **55**(5):119 (1976).

Robertson, J. C., Energy Conservation in Existing Plants, *Chem. Eng.*, **81**(2):104 (1974).

Rollwage, W. A., Energy Conservation in Chemical Plants, *Chem. Eng. Progr.*, **71**(10):44 (1975).

Rosenblad, A. E., Putting Evaporators to Work: Evaporator Systems for Black Liquor Concentration, *Chem. Eng. Progr.*, **72**(4):53 (1976).

Rozycki, J., Energy Conservation via Recompression Evaporation, *Chem. Eng. Progr.*, **72**(5):69 (1976).

Ryskamp, C. J., H. L. Wade, and R. B. Britton, Improve Crude Unit Operation, *Hydrocarbon Process.*, **55**(5):81 (1976).

Sander, U., Waste Heat Recovery in Sulfuric Acid Plants, *Chem. Eng. Progr.*, **73**(3):61 (1977).

Schumacher, C. E., and B. Y. Girgis, Conserving Utilities' Energy in New Construction, *Chem. Eng.*, **81**(4):133 (1974).

Seifert, W., Heat-Transfer Fluid Conserves 30 Billion Btu in Tracing System, *Chem. Eng.*, **81**(18):61 (1974).

Shaner, R. L., Energy Scarcity: A Process Design Incentive, *Chem. Eng. Progr.*, **74**(5):47 (1978).

Shepherd, D. G., Pick Up Energy from Low Heat Sources, *Hydrocarbon Process.*, **56**(12):141 (1977).

Shinskey, F. G., Energy-Conserving Control Systems for Distillation Units, *Chem. Eng. Progr.*, **72**(5):73 (1976).

———, Control Systems Can Save Energy, *Chem. Eng. Progr.*, **74**(5):43 (1978).

Simo, F. E., Which Flow Control: Valve or Pump?, *Hydrocarbon Process.*, **52**(7):103 (1973).

Smith, D. E., W. S. Stewart, and D. E. Griffin, Distill with Composition Control, *Hydrocarbon Process.*, **57**(2):99 (1978).

Smith, R. D., and R. B. Scollon, Balancing Boilers Against Plant Loads, *Chem. Eng.*, **83**(7):125 (1976).

Snyder, N. W., Energy Recovery and Resource Recycling, *Chem. Eng.*, **81**(22):65 (1974).

Soderlind, C., Tank Insulation Can Pay Off, *Hydrocarbon Process.*, **52**(7):122 (1973).

Starr, C., Conserve Energy Resources, *Hydrocarbon Process.*, **53**(7):87 (1974).

Steenberg, L. R., Fuel Recovery from Flare Systems, *Chem. Eng. Progr.*, **70**(7):74 (1974).

Steen-Johnsen, H., Turbines Using Too Much Steam? *Hydrocarbon Process.*, **52**(7):99 (1973).

Summerell, H., Recovering Energy from Stacks, *Chem. Eng.*, **83**(7):147 (1976).

Sutton, G. P., Save Energy with Pumps, *Hydrocarbon Process.*, **57**(6):103 (1978).

Swartz, A., A Guide for Troubleshooting Multiple-Effect Evaporators, *Chem. Eng.*, **85**(11):175 (1978).

Szablya, J. F., Units Aid Energy Work, *Hydrocarbon Process.*, **57**(7):142 (1978).

Taylor, R. I., How Exxon Conserves Energy, *Chem. Eng. Progr.*, **71**(10):35 (1975).

———, Refiners Can Save Energy, Too, *Hydrocarbon Process.*, **55**(7):91 (1976).

Teller, W. M., W. Diskant, and L. Malfitani, Conserving Fuel by Heating with Hot Water Instead of Steam, *Chem. Eng.*, **83**(13):185 (1976).

Troyan, J. E., Energy Conservation Programs Require Accurate Records, *Chem. Eng.*, **85**(26):189 (1978).

Tunnah, B. G., and B. Appelbaum, Use of Fuel Cells in Refining, *Hydrocarbon Process.*, **57**(7):119 (1978).

Tyreus, B. D., and W. L. Luyben, Two Towers Cheaper Than One? *Hydrocarbon Process.*, **54**(7):93 (1975).

Valentine, A. C., and S. V. Wildman, Energy Conservation from Steam Turbines, *Chem. Eng. (London)*, 516 (September 1975).

Vogt, G. A., and M. J. Wolters, Steam Balance As a Working Tool, *Chem. Eng. Progr.*, **72**(5):62 (1976).

Waid, D. E., Energy from Waste Gases, *Chem. Eng. Progr.*, **74**(5):78 (1978).

Whitcomb, M. G., Jr., and F. M. Orr, Plan Plant Energy Conservation, *Hydrocarbon Process.*, **52**(7):65 (1973).

Wickl, R., and R. Sparmann, Can Steam Turbines Save Energy? *Hydrocarbon Process.*, **53**(7):109 (1974).

Williams, C. C., and M. A. Albright, Better Data Saves Energy, *Hydrocarbon Process.*, **55**(5):115 (1976).

Wolf, W., High Flux Tubing Conserves Energy, *Chem. Eng. Progr.*, **72**(7):53 (1976).

Womack, J. W., Improved Waste Heat Boiler Operation, *Chem. Eng. Progr.*, **72**(7):56 (1976).

Woodard, A. M., Reduce Process Heater Fuel, *Hydrocarbon Process.*, **53**(7):106 (1974).

———, Upgrading Process Heater Efficiency, *Chem. Eng. Progr.*, **71**(10):53 (1975).

Wylie, R., Energy-Saving Turbine Uses Wastewater to Raise Output, *Chem. Eng.*, **81**(18):54 (1974).

———, Air Preheaters Can Save Fuel in Many Furnace Installations, *Chem. Eng.*, **81**(18):56 (1974).

Yates, W., Better Steam Trapping Cuts Energy Waste, *Hydrocarbon Process.*, **54**(11):267 (1975).

Zdonik, S. B., Techniques for Saving Energy in Processes and Equipment, *Chem. Eng.*, **84**(14):99 (1977).

Distillation, Prime Target for Energy Conservation, *Oil & Gas J.*, **76**(15):92 (1978).

Energy costs

Balzhiser, R. E., Energy Options to the Year 2000, *Chem. Eng.*, **84**(1):73 (1977).

Barber, J. C., Energy Requirements for Pollution Abatement, *Chem. Eng. Progr.*, **71**(12):42 (1976).

Beychok, M. R., Environmental Factors in Producing Supplemental Files, *Hydrocarbon Process.*, **54**(10):78 (1975).

Booth, H. R., R. D. Ehrlich, F. J. Keneshea, P. D. Knecht, H. Lawroski, R. L. Haymark, A. G. Silvester, and W. R. Thompson, Nuclear Power Today, *Chem. Eng.*, **82**(22): 102 (1975).

Clair, D. R., How Energy Costs Affect Petrochemicals, *Hydrocarbon Process.*, **52**(5):137 (1973).

Danekind, W. E., Stream Management in a Refinery, *Hydrocarbon Process.*, **55**(12):71 (1976).

Dickinson, W. C., Economics of Process Heat from Solar Energy, *Chem. Eng.*, **84**(3):101 (1977).

Honea, F. I., Energy Requirements for Environmental Control Equipment, *Chem. Eng.*, **81**(22):55 (1974).

Jelen, F. C., and C. L. Yaws, How Energy Affects Life Cycle Costs, *Hydrocarbon Process.*, **56**(7):89 (1977).

Kusik, C. L., J. I. Stevens, R. M. Nadkarni, P. A. Huska, and D. W. Lee, Energy Use and Air Pollution Control .in New Process Technology, *Chem. Eng. Progr.*, **73**(8):36 (1977).

Mai, K. L., Energy and Petrochemical Raw Materials Through 1990, *Chem. Eng.*, **84**(12):122 (1977).

May, D. L., First Steps in Cutting Steam Costs, *Chem. Eng.*, **80**(26):228 (1973).

Nelson, W. L., Energy Requirements, *Oil & Gas J.*, **72**(52):155 (1974).

Nichols, M. W., Balancing Requirements for World Oil and Energy, *Chem. Eng. Progr.*, **70**(10):36 (1974).

Punwani, D. V., and A. M. Rader, Gas from Peat—A Good Source of Heat, *Hydrocarbon Process.*, **57**(4):107 (1978).

Saxton, J. C., M. P. Kramer, D. L. Robertson, M. A. Fortune, N. E. Leggett, and R. G. Capell, Federal Findings on Energy for Industrial Chemicals, *Chem. Eng.*, **81**(18):71 (1974).

Spitz, P. H., and G. N. Ross, What Is Feedstock Worth? *Hydrocarbon Process.*, **55**(4):143 (1976).

Steinmeyer, D., Energy Price Impacts Designs, *Hydrocarbon Process.*, **55**(11):205 (1976).

Stevens, J. I., and C. L. Kusik, The Effects of Water-Pollution Control on Energy Consumption, *Chem. Eng.*, **84**(17):139 (1977).

Swearingen, J. S., Power from Hot Geothermal Brines, *Chem. Eng. Progr.*, **73**(7):83 (1977).

Thompson, R. G., and R. J. Lievano, What Does a "Clean" Environment Cost? *Hydrocarbon Process.*, **54**(10):73 (1975).

Wilson, W. B., and J. M. Kovacik, Electricity: Generate or Buy? *Hydrocarbon Process.*, **55**(12):75 (1976).

Winton, J. W., Plant Sites 1978, *Chem. Week.*, **121**(24):49 (1977).

Woodard, A. M., Control Flue Gas to . . . Improve Heater Efficiency, *Hydrocarbon Process.*, **54**(5):165 (1975).

Fire protection

Babbidge, L. G., C. C. Partridge, and J. A. D'Angelo, Fire-Test Valves for Reliability, *Hydrocarbon Process.*, **56**(12):179 (1977).

Bennett, T. C., Automate for Plant Emergencies, *Hydrocarbon Process.*, **51**(12):57 (1972).

Biggers, E. W., and T. C. Smith, Train Operators for Safety, *Hydrocarbon Process.*, **51**(12):54 (1972).

Brown, L. E., H. R. Wesson, and J. R. Welker, Predict LNG Fire Radiation, *Hydrocarbon Process.*, **53**(5):141 (1974).

Browning, R. L., Calculating Loss Exposures, *Chem. Eng.*, **76**(25):239 (1969).

Deuschle, R., and F. Tiffany, Barrier Intrinsic Safety, *Hydrocarbon Process.*, **52**(12):111 (1973).

Gillespie, P. J., and L. R. DiMaio, How Foam Can Protect HPI Plants, *Hydrocarbon Process.*, **56**(8):111 (1977).

Guise, A. B., How to Fight Natural Gas Fires, *Hydrocarbon Process.*, **54**(8):76 (1975).

Haessler, W. M., Fire Extinguishing Chemicals, *Chem. Eng.*, **80**(5):95 (1973).

Kletz, T., Protect Pressure Vessels from Fire, *Hydrocarbon Process.*, **56**(8):98 (1977).

LeRoy, N. L., and D. M. Johnson, The Story of a Reservoir Fire, *Hydrocarbon Process.*, **48**(5):129 (1969).

Ludwig, E. E., Designing Process Plants to Meet OSHA Standards, *Chem. Eng.*, **80**(20):88 (1973).

Mahley, H. S., Fight Tank Fires, Subsurface, *Hydrocarbon Process.*, **54**(8):72 (1975).

Methner, J. C., Hard Sell for Fire Fighters, *Chem. Eng.*, **75**(4):112 (1968).

Monroy, A. D., and V. M. Gonzales Majul, Stop Fires in EO Plants, *Hydrocarbon Process.*, **56**(9):175 (1977).

Nelson, R. W., Know Your Insurer's Expectation, *Hydrocarbon Process.*, **56**(8):103 (1977).

Pearson, L., When It's Time for Startup, *Hydrocarbon Process.*, **56**(8):116 (1977).

Perry, M., Five-Squad Fire Brigade Organized for Over-all Plant Protection, *Plant Eng.*, **21**(5):143 (1967).

Rains, W. A., Fire Resistance—How to Test for It, *Chem. Eng.*, **84**(27):97 (1977).

Saia, S. A., Vapor Clouds and Fires in a Light . . . Hydrocarbon Plant, *Chem. Eng. Progr.*, **72**(11):56 (1976).

Schuster, R., Fire Insurance Can Work for You, *Plant Eng.*, **21**(11):149 (1967).

Searson, A. H., Fire in a Catalytic Reforming Unit, *Chem. Eng. Progr.*, **68**(5):65 (1972).

Slama, W. R., and R. E. McMahon, Testing Plastics for Fire Snuff-Out, *Chem. Eng.*, **77**(24):120 (1970).

Underwood, H. C., Jr., R. E. Sourwine, and C. D. Johnson, Organize for Plant Emergencies, *Chem. Eng.*, **83**(21):118 (1976).

Vervalin, C. H., Fire and Safety Information: A Description of Key Resources, Part I, *Hydrocarbon Process.*, **51**(4):163 (1972).

———, Learn from HPI Plant Fires, *Hydrocarbon Process.*, **51**(12):49 (1972).

———, Who's Publishing on Fire and Safety? *Hydrocarbon Process.*, **52**(1):128 (1973).

———, Learn from HPI Tank Fires, *Hydrocarbon Process.*, **52**(8):81 (1973).

———, What's New in Fire and Safety Training? *Hydrocarbon Process.*, **52**(4):185 (1973); **53**(3):163 (1974).

———, Fire/Safety Information Resources, *Hydrocarbon Process.*, **54**(8):65 (1975).

Wahl, G. F., Methods of Controlling Fires in Flammable Liquids, *Plant Eng.*, **21**(2):122 (1967).

Waldman, S., Fireproofing in Chemical Plants, *Chem. Eng. Progr.*, **63**(8):71 (1967).

Welker, J. R., H. R. Wesson, and L. E. Brown, Use Foam to Disperse LNG Vapors? *Hydrocarbon Process.*, **53**(2):119 (1974).

H. R. Wesson, J. R. Welker, and L. E. Brown, Control LNG-Spill Fires, *Hydrocarbon Process.*, **51**(12):61 (1972).

——— ——— ———and C. M. Sliepvevich, Fight LNG Fires with Foam? *Hydrocarbon Process.*, **52**(10):165 (1973).

——— ——— ——— and ———, Fight LNG Spill Fires with Dry Chemicals, *Hydrocarbon Process.*, **52**(11):234 (1973).

West, H., and L. E. Brown, Analyze Fire Protection Systems, *Hydrocarbon Process.*, **56**(8):89 (1977).

Whitehorn, V. J., and H. W. Brown, How to Handle a Safety Inspection, *Hydrocarbon Process.*, **46**(4):125 (1967); **46**(5):227 (1967).

Woodard, A. M., How to Design a Plant Firewater System, *Hydrocarbon Process.*, **52**(10):103 (1973).

Zielinski, R. M., Case Study of a Reactor Fire, *Chem. Eng. Progr.*, **63**(8):59 (1967).

Fire Prevention Success Formula: Planning, Teamwork, Attention to Details, *Plant Eng.*, **21**(9):140 (1967).

Foundation design

Arya, S. C., R. P. Drewyer, and G. Pincus, Foundation Design for Reciprocating Compressors, *Hydrocarbon Process.*, **56**(5):223 (1977).

——— and ———, Foundation Design for Vibrating Machines, *Hydrocarbon Process.*, **54**(11):273 (1975).

Brown, A. A., New Look at Tower Foundation Design, *Hydrocarbon Process.*, **47**(4):174 (1968).

———, How to Design Pile-Supported Foundations for Flare Stacks, *Hydrocarbon Process.*, **48**(12):139 (1969).

———, Tank Foundation Design, *Hydrocarbon Process.*, **53**(10):153 (1974).

Char, C. V., New Approach to . . . Exchanger Foundation Design, *Hydrocarbon Process.*, **54**(3):121 (1975).

Czerniak, E., Foundation Design Guide for Stacks and Towers, *Hydrocarbon Process.*, **48**(6):95 (1969).

Lee, J. P., and N. C. Chokshi, Foundation Design for Rotating Fans, *Hydrocarbon Process.*, **57**(10):131 (1978).

Molnar, I. A., How to Design Foundations for Elevated Towers, *Hydrocarbon Process.*, **50**(4):129 (1971).

Sing, K. P., Design of Skirt Mounted Supports, *Hydrocarbon Process.*, **55**(4):199 (1976).

Instrumentation

Andrew, W. G., "Applied Instrumentation in the Process Industries," Gulf Publishing Company, Houston, Texas, 1974.

Angerhofer, A. W., Instrumenting an Air Separation Process, *Instr. Tech.*, **15**(3):41 (1968).

Bartz, A. M., and H. D. Ruhl, Jr., Process Control via Infrared Analyzers, *Chem. Eng. Progr.*, **64**(8):45 (1968).

Brown, C. W., Electric Pipe Tracing, *Chem. Eng.*, **82**(13):172 (1975).

Brown, J. E., Onstream Process Analyzers, *Chem. Eng.*, **75**(10):164 (1968).

Byrne, E. J., Measuring and Controlling, *Chem. Eng.*, **76**(8):189 (1969).

Considine, D. M., Liquid Level Measurement Systems: Their Evaluation and Selection, *Chem. Eng.*, **75**(4):137 (1968).

Corcoran, W. S., and J. Honeywell, Practical Methods for Measuring Flows, *Chem. Eng.*, **82**(14):86 (1975).

Crocker, B. B., Monitoring Particulate Emissions, *Chem. Eng. Progr.*, **71**(3):74 (1975).

Davies, R., Rapid Response Instrumentation for Particle Size Analyses, A Review, *Am. Lab.*, **5**(12):1 (1973); **6**(1):73 (1974); **6**(2):47 (1974).

Fair, J. R., B. B. Crocker, and H. R. Null, Sampling and Analyzing Trace Qualities, *Chem. Eng.*, **79**(21):146 (1972).

Felton, G. L., Low Flow Measurement with the Integral Orifice, *Chem. Eng. Progr.*, **68**(1):43 (1972).

Forman, E. R., Instrumentation, *Chem. Eng.*, **75**(1):74 (1968).

Foster, R. A., Guidelines for Selecting Online Process Analyzers, *Chem. Eng.*, **82**(6):65 (1975).

Gassett, L. D., Instruments . . . Pneumatic or Electronic? *Chem. Eng.*, **76**(12):136 (1969).

Harshe, B. L., Estimating Instrument Accuracy, *Chem. Eng.*, **81**(6):93 (1974).

Herrick, L. K., Jr., Instrumentation for Monitoring Toxic and Flammable Work Areas, *Chem. Eng.*, **83**(22):147 (1976).

Jackson, C., A Practical Vibration Primer, Part 7—Instrumentation for Analysis, *Hydrocarbon Process.*, **57**(3):119 (1978).

Jolls, K. R., and R. L. Riedinger, Measuring Methods: Volt-Ohm-Millimeter, *Chem. Eng.*, **79**(22):67 (1972).

——— and ———, Recorders and Oscilloscopes: Applied Electronics, *Chem. Eng.*, **79**(24):117 (1972).

Kardos, P. W., Response of Temperature-Measuring Elements, *Chem. Eng.*, **84**(18):79 (1977).

Kehoe, T. J., Online Process Analyzers, *Chem. Eng.*, **76**(12):117 (1969).

Kern, R., Measuring Flow in Pipes with Orifices and Nozzles, *Chem. Eng.*, **82**(3):72 (1975).

———, How to Size Flowmeters, *Chem. Eng.*, **82**(5):161 (1975).

Kern, R., Plant Layout: Instrument Arrangements for Ease of Maintenance and Convenient Operation, *Chem. Eng.*, **85**(9):127 (1978).

Kinsman, S., Instrumentation for Filtration Tests, *Chem. Eng. Progr.*, **70**(12):48 (1974).

Lapidot, H., Weight of Contents in Cylindrical Storage Tanks, *Chem. Eng.*, **81**(2):123 (1974).

Lawford, V. N., How to Select Liquid-Level Instruments, *Chem. Eng.*, **80**(24):109 (1973).

Lieberman, A., Optical Instruments Monitor Liquid-borne Solids, *Chem. Eng.*, **85**(28):105 (1978).

Lieberman, N. P., Instrumenting a Plant to Run Smoothly, *Chem. Eng.*, **84**(19):140 (1977).

———, Change Controls to Save Energy, *Hydrocarbon Process*, **57**(2):93 (1978).

Lipták, B. G., Process Instrumentation for Slurries and Viscous Materials, *Chem. Eng.*, **74**(3):133 (1967).

———, Instruments to Measure and Control Slurries and Viscous Materials, *Chem. Eng.*, **74**(4):151 (1967).

———,"Instrument Engineer's Handbook," Chilton Book Company, Philadelphia, Pennsylvania, 1969–70.

———, Costs of Process Instruments, *Chem. Eng.*, **77**(19):60 (1970).

———, Costs of Viscosity, Weight, Analytical Instruments, *Chem. Eng.*, **77**(20):175 (1970).

———, Safety Instruments and Control-Valves Costs, *Chem. Eng.*, **77**(24):94 (1970).

———, "Instrumentation in the Processing Industries," Chilton Book Company, Philadelphia, Pennsylvania, 1973.

———, Higher Profits via Advanced Instrumentation, *Chem. Eng.*, **82**(13):152 (1975).

Lommatsch, E. A., Pneumatic versus Electronic Instrumentation, *Chem. Eng.*, **83**(13):159 (1976).

Lovett, O. P., Jr., Control Valves, *Chem. Eng.*, **78**(23):129 (1971).

Masek, J. A., Where to Locate Thermowells, *Hydrocarbon Process.*, **51**(4):147 (1972).

McAllister, D. G., Jr., How to Select a System for Drying Instrument Air, *Chem. Eng.*, **80**(5):39 (1973).

McCoy, J. N., Liquid Level Measurement, *Petrol. Eng.*, **47**(7):62 (1975).

McDonald, D. P., Process Control: Choosing an Oxygen Analyser, *Chem. Proc. Eng.*, **53**(5):38 (1972).

McDonough, R., Selecting Sight Flow Indicators, *Chem. Eng.*, **84**(14):113 (1977).

McKee, H. C., Air-Pollution Instrumentation, *Chem. Eng.*, **83**(13):155 (1976).

Morrow, N. L., R. S. Brief, and R. R. Bertrand, Sampling and Analyzing Air Pollution Sources, *Chem. Eng.*, **79**(4):85 (1972).

Mulvany, H., M. Togneri, and E. Snyder, Computer Updates Instrument List, *Hydrocarbon Process.*, **51**(8):93 (1972).

Nisenfeld, A. E., and C. H. Cho, Parallel Compressor Control . . . What Should be Considered, *Hydrocarbon Process.*, **57**(2):147 (1978).

Parker, D. T., To Measure Moisture in a Vapor, *Hydrocarbon Process.*, **52**(11):215 (1973).

Paulis, N. J., and D. Silvermetz, Instrumentation for Slurry Systems, *Chem. Eng.*, **84**(9):107 (1977).

Ross, S. S., The Revolutionary Field of Air Pollution Instrumentation, *Chem. Eng.*, **79**(20):37 (1972).

Rowton, E. E., Digital Meters Gain Favor as Process Instruments, *Chem. Eng.*, **77**(3):111 (1970).

Ryan, J. B., Pressure Control, *Chem. Eng.*, **82**(3):63 (1975).

Ryan, J. M., R. S. Timmins, and J. F. O'Donnell, Production Scale Chromatography, *Chem. Eng. Progr.*, **64**(8):53 (1968).

Scales, J. W., "Water Quality Instrumentation," Instrument Society of America, Pittsburgh, Pennsylvania, 1972.

Schieber, J. R., Continuous Monitoring, *Chem. Eng.*, **77**(9):111 (1970).

Scull, W. L., Selecting Temperature Controls for Heaters, *Chem. Eng.*, **82**(11):128 (1975).

Smith, C. L., Liquid-Measurement Technology, *Chem. Eng.*, **85**(8):155 (1978).

Smith, D. E., W. S. Stewart, and D. E. Griffin, Distill with Composition Control, *Hydrocarbon Process.*, **57**(2):99 (1978).

Solomon, P., Monitor Claus Emissions, *Hydrocarbon Process.*, **52**(5):120 (1973).

Spolidoro, E. F., Comparing Positive-Displacement Meters, *Chem. Eng.*, **75**(12):91 (1968).

Strobel, H. A., "Chemical Instrumentation: A Systematic Approach," 2d ed., Addison-Wesley Publishing Company, Reading, Massachusetts, 1973.

Thompson, R. G., Water-Pollution Instrumentation, *Chem. Eng.*, **83**(13):151 (1976).

Thorsen, T., and R. Oen, How to Measure Industrial Wastewater Flow, *Chem. Eng.*, **82**(4):95 (1975).

Utterback, V. C., Online Process Analyzers, *Chem. Eng.*, **83**(13):141 (1976).

van Eijk, F. P., Instrument Trip System Maintenance and Improvement Program, *Chem. Eng. Progr.*, **71**(1):48 (1975).

Wallace, L. M., Sighting in On Level Instruments, *Chem. Engr.*, **83**(4):95 (1976).

Warren, C. W., How to Read Instrument Flow Sheets, *Hydrocarbon Process.*, **54**(7):163 (1975); **54**(9):191 (1975).

Waters, A. W., Design and Application of a Low Cost Pneumatic Composition Transmitter, *Chem. Eng. Progr.*, **74**(6):80 (1978).

Weston, F. C., Improve Solids Handling Instrumentation Systems, *Hydrocarbon Process.*, **53**(3):87 (1974).

Whitaker, N. R., Guide to Trouble-Free Plant Operation . . . Instrumentation, *Chem. Eng.*, **79**(14):96 (1972).

Wightman, E. J., "Instrumentation in Process Control," Butterworths, London, 1972.

Yard, J., Low-Flow Measurement, *Chem. Eng.*, **81**(8):74 (1974).

Zacharias, E. M., Jr., and D. W. Franz, Sound Velocimeters Monitor Process Streams, *Chem. Eng.*, **80**(2):101 (1973).

Zientara, D. E., Measuring Process Variables, *Chem. Eng.*, **79**(20):18 (1972).

Chemical Engineering 1969 Guide to Process Instrument Elements, *Chem. Eng.*, **76**(12):137 (1969).

A Guide to Selecting Circular Chart Recorders, *Instr. & Control Sys.*, **45**(7):69 (1972).

Guide to Selecting Dial Pressure Gages, *Instr. & Control Sys.*, **45**(8):63 (1972).

"Instrumentation for Environmental Monitoring: Air, Water, Radiation," Lawrence Berkeley Laboratory, University of California, Berkeley, California, 1972.

Insulation

Abramowitz, J. L., and R. Cordero, How to Select Insulation for Hot Pipes, *Chem. Eng.*, **82**(15):88 (1975).

——, Economic Pipe Insulation for Cold Systems, *Chem. Eng.*, **83**(23):105 (1976).

Barnhart, J. M., Economic Thickness of Thermal Insulation, *Chem. Eng. Progr.*, **70**(8):50 (1974).

Barrows, G. L., Save Energy with Ceramic Fiber Insulation, *Hydrocarbon Process.*, **56**(10):187 (1977).

Bertram, C. G., V. J. Desai, and E. Interess, Designing Steam Tracing, *Chem. Eng.*, **79**(7):74 (1972).

Briggs, M. A., Estimation of Economic Lagging Thickness, *Chem. Eng.* (*London*), 513 (September 1975).

Cordero, R., The Cost of Missing Pipe Insulation, *Chem. Eng.*, **84**(4):77 (1977).

Cross, T. A., Design Key to Good Insulation Systems, *Oil & Gas J.*, **73**(11):126 (1975).

———, Economics Dictate Tank Insulation, *Oil & Gas J.*, **73**(35):73 (1975).

Curt, R. P., A New Approach to Economic Insulation Thickness, *Hydrocarbon Process.*, **55**(3):137 (1976).

Foster, C. S., Foam Insulation for Tanks and Vessels, *Chem. Eng. Progr.*, **70**(8):55 (1974).

Harrison, M. R., and C. M. Pelanne, Cost-Effective Thermal Insulation, *Chem. Eng.*, **84**(27):63 (1977).

House, F. K., Pipe Tracing and Insulation, *Chem. Eng.*, **75**(13):243 (1968).

Hughes, R., and V. Deumaga, Insulation Saves Energy, *Chem. Eng.*, **81**(11):95 (1974).

Isaacs, M., Selecting Efficient, Economical Insulation, *Chem. Eng.*, **76**(6):143 (1969).

Kabbani, A. S., E1, Estimate Insulation Thickness, *Hydrocarbon Process.*, **49**(3):145 (1970).

Ligi, J. J., Processing with Cryogenics, *Hydrocarbon Process.*, **48**(4):93 (1969).

Marks, J. B., and K. D. Holton, Protection of Thermal Insulation, *Chem. Eng. Progr.*, **70**(8):46 (1974).

Martin, R. B., Guide to Better Insulation, *Chem. Eng.*, **82**(10):98 (1975).

Menicatti, S., Check Tank Insulation Economics, *Hydrocarbon Process.*, **48**(4):133 (1969).

Paros, S. V., Choosing Cold Insulation for Your Piping and Storage? *Hydrocarbon Process.*, **55**(11): 257 (1976).

Polastri, F., Check Pipe Insulation Thickness, *Hydrocarbon Process.*, **50**(11):231 (1971).

Smith, J. C., and J. S. Kummins, Protective Coatings for Foam Insulation, *Chem. Eng. Progr.*, **70**(8):57 (1974).

Soderlind, C., Tank Insulation Can Pay Off, *Hydrocarbon Process.*, **52**(7):122 (1973).

Turner, W. C., Criteria for Installing Insulation Systems in Petrochemical Plants, *Chem. Eng. Progr.*, **70**(8):41 (1974).

Wells, G. L., Insulation Thickness Schedules, *Chem. Proc. Eng.*, **53**(4):35 (1972).

Winner, G. R., New INSULO System Estimates Piping Insulation by Computer, *Petro/Chem. Eng.*, **43**(6):48 (1971).

Process Piping, *Chem. Eng.*, **76**(8):95 (1969).

Maintenance

Balaam, E., Dynamic Predictive Maintenance for Refinery Equipment, *Hydrocarbon Process.*, **56**(5):131 (1977).

Bergtraun, E. M., In-Plant vs. Contract Maintenance, *Chem. Eng.*, **84**(7):131 (1977).

Bloch, H. P., Improve Safety and Reliability of Pumps and Drivers, *Hydrocarbon Process.*, **56**(3):133 (1977); **56**(4):181 (1977).

Boyce, M. P., How to Achieve Online Availability of Centrifugal Compressors, *Chem. Eng.*, **85**(13):115 (1978).

Brooke, J. M., Inhibitors: New Demands for Corrosion Control, *Hydrocarbon Process.*, **49**(1):121 (1970).

Brown, P. J., Compressor and Engine Maintenance, *Chem. Eng. Progr.*, **68**(6):77 (1972).

Cason, R. L., Estimate the Downtime Your Improvements Will Save, *Hydrocarbon Process.*, **51**(1):73 (1972).

Cotz, V. J., "Plant Engineer's Manual and Guide," Prentice-Hall, Englewood Cliffs, New Jersey, 1973.

David, H. M., How to Improve Compressor Operation and Maintenance, *Hydrocarbon Process.*, **53**(1):93 (1974).

Engle, J. P., Cleaning Boiler Tubes Chemically, *Chem. Eng.*, **78**(24):154 (1971).

Finley, H. F., How Cost-Effective is Your Maintenance Organization? *Hydrocarbon Process.*, **51**(1):81 (1972).

———, Maintenance Management for Today's High Technology Plants, *Hydrocarbon Process.*, **57**(1):101 (1978).

Forman, E. R., How to Improve Online Control-Valve Performance, *Chem. Eng.*, **85**(13):128 (1978).

Goyal, S. K., Improve Turnaround Manpower Planning, *Hydrocarbon Process.*, **54**(1):71 (1975).

Grannell, T. E., An Approach to Cost Reduction in Maintenance Engineering, *AACE Bull.*, **16**(1):20 (1974).

Haluska, J. L., What to Follow if Your Goal Is Effective Fouling Control, *Hydrocarbon Process.*, **55**(7):153 (1976).

Hancock, W. P., How to Control Pump Vibration, *Hydrocarbon Process.*, **53**(3):107 (1974).

Helzner, A. E., Operating Performance of Steam-Heated Reboilers, *Chem. Eng.*, **84**(4):73 (1977).

Hinchley, P., Waste Heat Boilers: Problems & Solutions, *Chem. Eng. Progr.*, **73**(3):90 (1977).

Hopkins, C. D., Preventive Maintenance of Gas Plant Equipment, *Chem. Eng. Progr.*, **66**(6):60 (1970).

Isles, D. L., Maintaining Steam Traps for Best Efficiency, *Hydrocarbon Process.*, **56**(1):103 (1977).

Jackson, C., A Practical Vibration Primer, *Hydrocarbon Process.*, **54**(4):161 (1975).

James, R., Jr., Maintenance of Rotating Equipment; Pump Maintenance, *Chem. Eng. Progr.*, **72**(2):35 (1976).

Jumper, O., What Can be Done About Maintenance Inventories? *Hydrocarbon Process.*, **49**(1):129 (1970).

Kelly, W. J., Maintaining Venturi-Tray Scrubbers, *Chem. Eng.*, **85**(27):133 (1978).

Kobrin, G., Evalute Equipment Condition by Field Inspection and Tests, *Hydrocarbon Process.*, **49**(1):115 (1970).

Lee, R. P., How Poor Design Causes Equipment Failures, *Chem. Eng.*, **84**(3):129 (1977).

———, Some Unusual Failure Modes, *Chem. Eng.*, **84**(1):107 (1977).

Loucks, C. M., Boosting Capacities with Chemicals, *Chem. Eng.*, **80**(5):79 (1973).

Love, F. S., Pumps: A Trouble-Shooting Guide, *Hydrocarbon Process.*, **51**(1):91 (1972).

McClelland, G. D., Maintenance and Safe Operation of High-Pressure Equipment, *Chem. Eng.*, **75**(20):202 (1968).

McGill, J. C., and E. C. McGill, Save Lost Hydrocarbons, *Hydrocarbon Process.*, **57**(5):158 (1978).

McWilliam, J. D., Protecting Demineralizers from Organic Fouling, *Chem. Eng.*, **85**(12):80 (1978).

Meyer, R. J., Solve Vertical Pump Vibration Problems, *Hydrocarbon Process.*, **56**(8):45 (1977).

Mistrot, D. J., Downtime, Maintenance Challenge HPI, *Petro/Chem. Eng.*, **40**(12):14 (1968).

Mruk, G. K., J. D. Halloran, and R. M. Kolodziej, New Method Predicts Startup Torque, Part 2: Field Measurements and Abnormal Motor Conditions, *Hydrocarbon Process.*, **57**(5):229 (1978).

Mutty, P. C., Re-examine Procedures to Reduce Maintenance Costs, *Plant Eng.*, **21**(5):130 (1967).

Niño, L. E., Predict Failures and Overhaul Interval to Minimize Downtime, *Hydrocarbon Process.*, **53**(1):108 (1974).

Norden, R. B., Maintenance Painting, *Chem. Eng.*, **80**(5):85 (1973).

Pattison, D. A., Lubrication for Process Plants, *Chem. Eng.*, **82**(9):99 (1975).

Perkins, R. L., Maintenance Costs and Equipment Reliability, *Chem. Eng.*, **82**(7):126 (1975).

Perugini, J. J., More About Antifoulants, *Hydrocarbon Process.*, **55**(7):161 (1976).

Pippitt, R. R., Maintenance of Rotating Equipment; Maintenance of Kilns, Calciners, and Dryers, *Chem. Eng. Progr.*, **72**(2):41 (1976).

Radway, J. E., Measurement Technique Evaluates Boiler Fuel Additives, *Hydrocarbon Process.*, **56**(2):89 (1977).

Raynesford, J. D., Use Dynamic Absorbers to Reduce Vibration, *Hydrocarbon Process.*, **54**(4):167 (1975).

Reynolds, J. A., Pump Installation and Maintenance, *Chem. Eng.*, **78**(23):67 (1971).

Roebuck, A. H., Safe Chemical Cleaning—The Organic Way, *Chem. Eng.*, **85**(17):107 (1978).

Ryman, D. J., and J. E. Steenbergen, Maintenance of Rotating Equipment; Field Balancing of Rotary Equipment, *Chem. Eng. Progr.*, **72**(2):46 (1976).

Santini, F., "MBO" for Maintenance Operations, *Hydrocarbon Process.*, **55**(1):159 (1976).

Sarappo, J. W., Contract Maintenance: Its Place in Chemical Plants, *Chem. Eng.*, **76**(26):264 (1969).

Shah, G. C., Troubleshooting Distillation Columns, *Chem. Eng.*, **85**(17):70 (1978).

Sharp, E. C., Jr., Operating and Maintenance Records for Heating Equipment, *Chem. Eng.*, **84**(9):141 (1977).

Shaw, C. P., Jr., Better Performance from Antifriction Bearings, *Hydrocarbon Process.*, **56**(4):185 (1977).

Simon, E. L., and H. S. Gonzalez, Maintenance Control for Medium-Size Plants, *Chem. Eng.*, **78**(19):102 (1971).

Skabo, R. R. Internal Maintenance of Vessels, *Chem. Eng. Progr.*, **66**(10):33 (1970).

Sohre, J. S., You Can Predict Reliability of Turbomachinery, *Hydrocarbon Process.*, **49**(1):100 (1970).

Templeton, H. C., Valve Installation, Operation, and Maintenance, *Chem. Eng.*, **78**(23):141 (1971).

Trotter, J. A., New Tools for Modern Maintenance, *Chem. Eng.*, **80**(5):87 (1973).

————, Techniques of Predictive Maintenance, *Chem. Eng.*, **77**(18):66 (1970).

Weaver, F. L., Reliable Overspeed Protection for Steam Turbines, *Hydrocarbon Process.*, **56**(4):173 (1977).

Zielinski, M., Microcomputer Systems: Where Do You Begin Trouble-Shooting? *Hydrocarbon Process.*, **57**(1):109 (1978).

Market research

Jackson, J. W., and J. H. Black, Predicting Markets for Chemical Products, *1969 Trans. Am. Assoc. Cost Eng.* (1969).

———— and ————, Predict Petrochemical Production, *Hydrocarbon Process.*, **48**(7):143 (1969).

Laurent, P. A., Future Demand for Petroleum Products, *Chem. Eng. Progr.*, **65**(12):29 (1969).

Leibson, I., and C. A. Trischman, Jr., Avoiding Pitfalls in Developing a Major Capital Project, *Chem. Eng.*, **78**(18):103 (1971).

Mapstone, G. E., Forecasting for Sales and Production, *Chem. Eng.*, **80**(11):126 (1973).

Martino, J. P., Technological Forecasting for the Chemical Process Industries, *Chem. Eng.*, **78**(29):54 (1971).

Massey, D. J., and J. H. Black, Predicting Selling Prices for Chemical Products, *1969 Trans. Am. Assoc. Cost Eng.* (1969).

Miller, R. M., and H. F. Tureff, Use of Growth Curves in Forecasting, *1969 Trans. Am. Assoc. Cost Eng.* (1969).

Nelles, M., Accelerated Manufacturing Planning, *ASTME Vectors*, **4**(6):35 (1969).

O'Hara, J. B., Forecast Capital Spending Better, *Hydrocarbon Process.*, **52**(1):151 (1973).

Ohsol, E. O., Evaluation of Research Projects—Marketing Cost Estimation, *Chem. Eng. Progr.*, **67**(4):19 (1971).

————, Estimating Marketing Costs, *Chem. Eng.*, **78**(10):116 (1971).

Rothermel, T. W., Market Simulation Makes a Science out of Forecasting, *Chem. Eng.*, **76**(6):157 (1969).

Sawyer, F. G., Petrochemicals in '70: They Still Look Good, Knock on Wood, *Hydrocarbon Process.*, **49**(1):129 (1970).

Spitz, P. H., 1972 and Beyond (Raw-Material and Energy Challenges), *Chem. Eng.*, **79**(1):77 (1972).

Starczewski, J., How to Seek New Ventures, *Hydrocarbon Process.*, **48**(12):163 (1969).

Strelzoff, S. Z., Ethylene and Propylene: Booming Building Blocks, *Chem. Eng.*, **77**(18):75 (1970).

Swager, W. L., Technological Forecasting in R & D, *Chem. Eng. Progr.*, **65**(12):39 (1969).

Forecast: The CPI and You in the 1970's, *Chem. Eng.*, **77**(1):80 (1970).

Noise control

Allen, E. E., How to Combat Control Valve Noise, *Chem. Eng. Progr.*, **71**(8):43 (1975).

————, Valves Can Be Quiet, *Hydrocarbon Process.*, **51**(10):137 (1972).

Arcuri, K., Quiet-Steamline Design to Comply with OSHA Rules, *Chem. Eng.*, **80**(1):134 (1973).

Barrington, E. A., Acoustic Vibrations in Tubular Exchangers, *Chem. Eng. Progr.*, **69**(7):62 (1973).

Baumann, H. D., Control Valve Sound Pressure Level Prediction, *Instr. Control Sys.*, **44**(4):93 (1971).
————, Control-Valve Noise: Cause and Cure, *Chem. Eng.*, **78**(11):120 (1971).
Bell, L. H., "Fundamentals of Industrial Noise Control," 2d ed., Harmony Publications Company, Trumbull, Connecticut, 1974.
Bell, L. H., "Fundamentals of Noise Control," Harmony Publications Company, Trumbull, Connecticut, 1977.
Beranek, L. L., Industrial Noise Control, *Chem. Eng.*, **77**(9):227 (1970).
Beychok, M. R., Environmental Factors in Producing Supplemental Fuels, *Hydrocarbon Process.*, **54**(10):78 (1975).
Bruce, R. D., and R. E. Werchan, Noise Control in the Petroleum and Chemical Industries, *Chem. Eng. Progr.*, **71**(8):57 (1975).
Bruggink, R. H., and J. R. Shadley, A Workable Furnace Noise Control Program, *Chem. Eng. Progr.*, **69**(10):56 (1973).
Cahill, L. B., Defining Plant Noise Problems, *Chem. Eng. Progr.*, **69**(10):65 (1973).
Connor, W. K., Noise Control, *Chem. Eng.*, **80**(14):165 (1973).
Dear, T., Large Gas Handling Plants in Noise Control, *Chem. Eng. Progr.*, **70**(2):65 (1974).
Everett, W. S., Proper Installation of Exhaust-Vent Silencers, *Chem. Eng.*, **83**(5):116 (1976).
Graham, J. B., Fan Selection and Installation, *Chem. Eng. Progr.*, **71**(10): 74 (1975).
Harris, R. W., "Introduction to Noise Analysis," T. J. Ledwidge, London, 1974.
Judd, S. H., Noise Abatement in Process Plants, *Chem. Eng.*, **78**(1):139 (1971).
————, Noise Abatement in Existing Refineries, *Chem. Eng. Progr.*, **71**(8):31 (1975).
Kugler, B. A., Noise Control Design for New Plants, *Chem. Eng. Progr.*, **71**(8):48 (1975).
Lacey, B. G., Noise and Its Control in Process Plants, *Chem. Eng.*, **76**(13):74 (1969).
Lou, S. C., Noise-Control Design for Process Plants, *Chem. Eng.*, **80**(27):77 (1973).
Ludwig, E. E., Designing Process Plants to Meet OSHA Standards, *Chem. Eng.*, **80**(20):88 (1973).
Magrab, E. B., "Environmental Noise Control," J. Wiley and Sons, New York, 1975.
May, D. N., ed., "Handbook of Noise Assessment," Van Nostrand Reinhold, New York, 1978.
McLarty, T. E., Selecting Silencers to Suppress Plant Noise, *Chem. Eng.*, **83**(8):104 (1976).
Norman, R. S., Predict Plant Noise Problems, *Hydrocarbon Process.*, **52**(10):89 (1973).
Middleton, A. H., Noise from Chemical Plant, *Chem. Eng.* (*London*), **115** (February 1976).
Paddock, S. G., OHSA and Machinery Noise Control, *Petrol. Eng.*, **46**(12):20 (1974).
Richings, W. V., Noise Control in Chemical Processing, *Chem. Proc. Eng.*, **48**(10):77 (1967); **48**(11):66 (1967); **49**(1):86 (1968).
Schweisheimer, W., Noise: Another Pollutant, *ASTME Vectors*, **4**(4):21 (1969).
Seebold, J. G., Furnace Roar Control, *Chem. Eng. Progr.*, **71**(8):53 (1975).
————, How to Reduce Control Valve and Furnace Combustion Noise, *Hydrocarbon Process.*, **51**(3):97 (1972).
————, Flare Noise: Causes and Cures, *Hydrocarbon Process.*, **51**(10):143 (1972).
————, Smooth Piping Reduces Noise—Fact or Fiction? *Hydrocarbon Process.*, **52**(9):189 (1973).
————, Control Plant Noise This Way, *Hydrocarbon Process.*, **54**(8):80 (1975).
————, Reduce Noise from Pulsating Combustion in Elevated Flares, *Hydrocarbon Process.*, **54**(9):225 (1975).
Shore, D., Toward Quieter Flaring, *Chem. Eng. Progr.*, **69**(10):60 (1973).
Spanaus, D. D., and K. N. Allan, Controlling Noise in Coal Preparation Plants, *Allis-Chalmers Eng. Rev.*, **38**(2):23 (1973).
Thomas, D. E., Jr., R. James, Jr., and C. R. Sparks, Some Plant Noise Problems and Solutions, *Hydrocarbon Process.*, **51**(10):149 (1972).
Thumann, A. Interdisciplinary Plant-Noise Control, *Chem. Eng.*, **81**(17):120 (1974).
Welker, R. H., Minimize Noise—Turbulence in Gas Piping, *Petrol. Eng.*, **45**(2):73 (1973).
Werchan, R. E., and R. D. Bruce, Process Plant Noise Can Be Controlled, *Chem. Eng. Progr.*, **69**(10):51 (1973).
Wick, C. H., Controlling Industrial Noise, Part 1: Analysis and Measurement, *Mfg. Eng. & Mgt.*, **70**(3):21 (1973).

Winnerling, H. A., Quieting of Process Machinery, *Chem. Eng. Progr.*, **69**(6):96 (1973).

Patents

Brenner, E. J., Developments in the Patent Field, *Chem. Eng. Progr.*, **64**(4):13 (1968).

Geiger, G. H., "Supplementary Readings in Engineering Design," McGraw-Hill Book Company, New York, 1975.

Maynard, J. T., "Understanding Chemical Patents," The American Chemical Society, Washington, D.C., 1978.

Moon, G. B., A Law Primer for the Chemical Engineer, *Chem. Eng.*, **85**(9):114 (1978).

Schaal, E. A., Hints on Reading Patents, *Chem. Eng.*, **84**(24):89 (1977).

———, You and Your Patent Lawyer, *Chem. Eng.*, **85**(13): 143 (1978).

Smith, D. B., D. M. Young, and R. L. Miller, Process Technology for License or Sale, *Chem. Eng.*, **77**(8):115 (1970).

Whale, A. R., and B. W. Sandt, The Engineer's Guide to Patent Infringement, *Chem. Eng.*, **79**(6):107 (1972).

———, Patents for Profits and Progress, *Chem. Eng.*, **78**(3):68 (1971).

Plant layout

Bush, M. J., and G. L. Wells, Unit Plot Plans for Plant Layout, *Brit. Chem. Eng.*, **16**(4/5):325 (1971); **16**(6):514 (1971).

House, F. F., An Engineer's Guide to Process-Plant Layout, *Chem. Eng.*, **76**(16):120 (1969).

Kaess, D., Jr., Guide to Trouble-Free Plant Layout, *Chem. Eng.*, **77**(12):122 (1970).

Kern, R., Obtain Minimum Cost, Plant Layout, *Chem. Eng.*, **84**(11):131 (1977).

———, Specifications are the Key to Successful Plant Design, *Chem. Eng.*, **84**(14):123 (1977).

———, Arrangements of Process and Storage Vessels, *Chem. Eng.*, **84**(24):93 (1977).

———, How to Get the Best Process-Plant Layouts for Pumps and Compressors, *Chem. Eng.*, **84**(26):131 (1977).

———, Piperack Design for Process Plants, *Chem. Eng.*, **85**(3):105 (1978).

———, Space Requirements and Layout for Process Furnaces, *Chem. Eng.*, **85**(5):117 (1978).

———, Instrument Arrangements for Ease of Maintenance and Convenient Operation, *Chem. Eng.*, **85**(9):127 (1978).

———, How to Arrange the Plot Plan for Process Plants, *Chem. Eng.*, **85**(11):191 (1978).

———, Arranging the Housed Chemical Process Plant, *Chem. Eng.*, **85**(16):123 (1978).

———, Controlling the Cost Factors in Plant Design, *Chem. Eng.*, **85**(18):141 (1978).

Ludwig, E. E., Designing Process Plants to Meet OSHA Standards, *Chem. Eng.*, **80**(20):88 (1973).

Maclean, W. D., Construction Site Selection—A U.S. Viewpoint, *Hydrocarbon Process.*, **56**(6):111 (1977).

Mixon, G. M., Chemical Plant Lighting, *Chem. Eng.*, **74**(12):113 (1967).

Patterson, G. C., Fundamentals of Engineering Off-Sites & Utilities for the HPI; Product Shipping, *Petro/Chem. Eng.*, **39**(1):34 (1967).

———, Fundamentals of Engineering Off-sites and Utilities for the HPI; Refinery Sewer Systems and Waste Treatment Facilities, *Petro/Chem. Eng.*, **39**(2):50 (1967).

Sachs, G., Economic and Technical Factors in Chemical Plant Lay-out, *Chem. Eng.* (*London*), 304 (October 1970).

Spitzgo, C. R., Guidelines for Overall Chemical-Plant Layout, *Chem. Eng.*, **83**(20):103 (1976).

Whitmer, J. S., Plant Engineer's Experiences When Planning, Constructing New Plant, *Plant Eng.*, **21**(10):136 (1967).

Evaluate Plant Facilities for Tomorrow's Railroad Equipment, *Plant Eng.*, **21**(6):166 (1967).

Plant location

Biggert, E. C., State and City Information Sources, *Chem. Eng.*, **79**(18):94 (1972).

DeHarpporte, D. R., Cooling Tower Site Considerations, *Power Eng.*, **74**(8):49 (1970).

Ludwig, E. E., Designing Process Plants to Meet OSHA Standards, *Chem. Eng.*, **80**(20):88 (1973).

Martin, R., How to Get Site Selection Data, *Petro/Chem. Eng.*, **39**(4):40 (1967).
———, Prime Factors in Petrochemical Plant Site Selection, *Petro/Chem. Eng.*, **40**(4):54 (1968).
Mendel, O., How Location Affects U.S. Plant-Construction Costs, *Chem. Eng.*, **79**(28):120 (1972).
Moores, C. W., Before the Environmental Assessment, *Hydrocarbon Process.*, **56**(3):173 (1977).
Oppel, E. I., Why and How to Evaluate Customer-Oriented Plant Sites, *Petro/Chem. Eng.*, **39**(4):35 (1967).
Roskill, O. W., The Location of Chemical Plants, *Chem. Eng.* (*London*), 29 (January 1976).
Scharpf, C. A., Siting a Thermal Multi-purpose Energy Center, *Chem. Eng. Progr.*, **68**(5):26 (1972).
Speir, W. B., Choosing and Planning Industrial Sites, *Chem. Eng.*, **77**(26):69 (1970).
Weismantel, G. E., Plant-Siting Barriers Grow, *Chem. Eng.*, **84**(13):69 (1977).
Winton, J. W., Plant Sites '67, *Chem. Week*, **101**(18):71 (1967).
———, Plant Sites 1969, *Chem. Week*, **105**(17):59 (1969).
———, Plant Sites 1971, *Chem. Week*, **109**(15):35 (1971).
———, Plant Sites 1972, *Chem. Week*, **111**(15):35 (1972).
———, Plant Sites 1974, *Chem. Week*, **113**(16):29 (1973).
———, Plant Sites 1975, *Chem. Week*, **115**(17):33 (1974).
———, Plant Sites 1976, *Chem. Week*, **117**(17):27 (1975).
———, Plant Sites 1977, *Chem. Week*, **119**(19):35 (1976).
———, Plant Sites 1978, *Chem. Week*, **121**(24):49 (1977).
Yocom, J. E., G. F. Collins, and N. E. Bowne, Plant Site Selection, *Chem. Eng.*, **78**(14):164 (1971).

Pollution regulation and control

Barber, J. C., The Cost of Pollution Control, *Chem. Eng.*, **64**(9):78 (1978).
——— and T. D. Farr, Fluoride Recovery from Phosphorus Production, *Chem. Eng. Progr.*, **66**(11):56 (1970).
Bennett, G. F., Information Sources, *Chem. Eng.*, **80**(14):39 (1973).
Berkau, E. E., D. A. Denny, G. S. Thompson, and D. L. Becker, Multimedia Assessment: An Integrated Approach to Pollution, *Chem. Eng.*, **84**(17):148 (1977).
Beychok, M. R., Environmental Factors in Producing Supplemental Fuels, *Hydrocarbon Process.*, **54**(10):78 (1975).
Boothe, J. N., "Cleaning Up—The Cost of Refinery Pollution," Council on Economic Priorities, New York, 1975.
Buehner, F. W., OSHA, EPA, and Plant Design, *Chem. Eng.*, **84**(12):161 (1977).
Byrd, J. F., How to Organize a Control Program, *Chem. Eng.*, **75**(22):50 (1968).
Cecil, L. K., Manage the Environment with Imagination, *Hydrocarbon Process.*, **51**(10):79 (1972).
Cochran, L. G., J. P. Gay, H. N. Hill, Jr., and R. C. Lawrence, Pollution Abatement: Pollution Instrumentation Techniques, *Chem. Eng. Progr.*, **68**(8):76 (1972).
Crocker, B. B., Preventing Hazardous Pollution During Plant Catastrophes, *Chem. Eng.*, **77**(10):97 (1970).
Crowley, J. B., Good Sampling Saves Money, *Hydrocarbon Process.*, **51**(10):164 (1972).
D'Ambra, F. K., and Z. C. Dobrowolski, Pollution Control for Vacuum Systems, *Chem. Eng.*, **80**(15):95 (1973).
Dosher, J. R., Trends in Petroleum Refining, *Chem. Eng.*, **77**(17):96 (1970).
Fair, J. R., Advances in Distillation System Design, *Chem. Eng. Progr.*, **73**(11):78 (1977).
Fisher, T. F., M. L. Kasboh, and J. R. Rivero, The Purox System, *Chem. Eng. Progr.*, **72**(10):75 (1976).
Fogiel, M., "Pollution Control Technology," Research & Education Association, New York, 1973.
Fox, R. D., Pollution Control at the Source, *Chem. Eng.*, **80**(18):72 (1973).
Gardiner, W. C., and F. Munoz, Mercury Removed From Waste Effluent via Ion Exchange, *Chem. Eng.*, **78**(19):57 (1971).
Gross, R. W., Jr., Waste Treatment Advances: The UNOX Process, *Chem. Eng. Progr.*, **72**(10):51 (1976).
Harrison, E. B., What the New Clean Water Act Means to HPI Plant Managers, *Hydrocarbon Process.*, **57**(2):165 (1978).

Heath, D. P., How Water Regs Impact Refining, *Hydrocarbon Process.*, **57**(5): 135 (1978).

Himmelstein, K. J., R. D. Fox, and T. H. Winter, In-Place Regeneration of Activated Carbon, *Chem. Eng. Progr.*, **69**(11):65 (1973).

Iammartino, N. R., Mercury Cleanup Routes—II, *Chem. Eng.*, **82**(3):36 (1975).

Iverstine, J. C., How Pollution Control Affects Plant Operations, *Chem. Eng.*, **81**(22):103 (1974).

Jenkins, D. M., and W. J. Sheppard, What Refinery Pollution Abatement Costs, *Hydrocarbon Process.*, **56**(5):154 (1977).

Jones, H. R., "Pollution Control in the Petroleum Industry," Noyes Data Corporation, Park Ridge, New Jersey, 1973.

Kalen, B., and F. A. Zenz, Filtering Effluent from a Cat Cracker, *Chem. Eng. Progr.*, **69**(6):67 (1973).

Kilburn, P. D., and M. W. Legatski, Environmental Planning Starts with Process Development, *Hydrocarbon Process.*, **53**(10):95 (1974).

Koches, C. F., and S. B. Smith, Reactivate Powdered Carbon, *Chem. Eng.*, **79**(9):46 (1972).

Lanouette, K. H., Treatment of Phenolic Wastes, *Chem. Eng.*, **84**(22):99 (1977).

Lederman, P. B., H. S. Skovronek, and P. E. Des Rosiers, Pollution Abatement in the Phamaceutical Industry, *Chem. Eng. Progr.*, **74**(4):93 (1975).

Lewis, C. R., R. E. Edwards, and M. A. Santoro, Incineration of Industrial Wastes, *Chem. Eng.*, **83**(22):115 (1976).

Liptàk, B. G., "Environmental Engineer's Handbook," Chilton Book Company, Philadelphia, Pennsylvania, 1974.

Loven, A. W., Perspectives on Carbon Regeneration, *Chem. Eng. Progr.*, **69**(11):56 (1973).

Marshall, J. R., Current Status, Future Outlook, *Chem. Eng.*, **75**(22):14 (1968).

————, Current Legislation, Section I, Pollution Control Laws, *Chem. Eng.*, **77**(9):17 (1970).

Mattia, M. M., Process for Solvent Pollution Control, *Chem. Eng. Progr.*, **66**(12):74 (1970).

McBride, J. A., Regulatory Controls on Radioactive Effluents, *Chem. Eng. Progr.*, **66**(4):74 (1970).

McMullen, A. I., J. F. Monk, and M. J. Stuart, Detection and Measurement of Pollutants of Water Surfaces, *Am. Lab.*, **7**(2):87 (1975).

Mencher, S. K., Change Your Process to Alleviate Your Pollution Problem, *Petro/Chem. Eng.*, **39**(5):21 (1967).

Meredith, H. H., Jr., and W. L. Lewis, Desulfurization and the Petroleum Industry, *Chem. Eng. Progr.*, **64**(9):57 (1968).

Mills, J. L., Continuous Monitoring, *Chem. Eng.*, **77**(9):217 (1970).

Moores, C. W., FMC's Glycerine Plant Utilizes Centralized Pollution Control, *Petro/Chem. Eng.*, **42**(2):19 (1970).

————, Before the Environmental Assessment, *Hydrocarbon Process.*, **56**(3):173 (1977).

Mueller, J., How to Prepare an EIR, *Chem. Eng.*, **81**(22):39 (1974).

Newman, D. J., Nitric Acid Plant Pollutants, *Chem. Eng. Progr.*, **67**(2):79 (1971).

Nilsen, J., Cleanup: What's it Worth? *Chem. Eng.*, **79**(13):48 (1972).

Oeben, R. W., O. Muniz, and R. A. Payne, Pollution Control at Carbide Ponce, *Chem. Eng.*, **80**(14):145 (1973).

Olds, F. C., Thermal Effects: A Report on Utility Action, *Power Eng.*, **74**(4):26 (1970).

Passow, N. R., U.S. Environmental Regulations Affecting the Chemical Process Industries, *Chem. Eng.*, **85**(26):173 (1978).

Paulson, E. G., How to Get Rid of Toxic Organics, *Chem. Eng.*, **84**(22):21 (1977).

Paulus, J. J., Environmental Monitoring Helps Resolve Pollution Problems, *Westinghouse Eng.*, **33**(5):150 (1973).

Perrotti, A. E., and C. A. Rodman, Enhancement of Biological Waste Treatment by Activated Carbon, *Chem. Eng. Progr.*, **69**(11):63 (1973).

Putnam, B., and M. Manderson, Iron Pyrites From High Sulfur Coals, *Chem. Eng. Progr.*, **64**(9):60 (1968).

Racine, W. J., Plant Designed to Protect Environment, *Hydrocarbon Process.*, **51**(3):115 (1972).

Reed, D. T., E. F. Klen, and D. A. Johnson, Use Side Stream Softening to Reduce Pollution, *Hydrocarbon Process.*, **56**(11):339 (1977).

Rickles, R. N., Desulfurization's Impact on Sulfur Markets, *Chem. Eng. Progr.*, **64**(9):53 (1968).

Ripley, K. D., Monitoring Industrial Effluents, *Chem. Eng.*, **79**(10):119 (1972).

Ross, L. W., Predict Pollutants Dispersion, *Hydrocarbon Process.*, **47**(8):144 (1968).

Ross, R. D., "Industrial Waste Dispersal," Van Nostrand Reinhold, Cincinnati, Ohio, 1968.

Ross, S. S., Current Legislation (Laws), *Chem. Eng.*, **78**(14):9 (1971).

———, Federal Laws and Regulations, *Chem. Eng.*, **79**(10):9 (1972).

Sax, N. I., "Dangerous Properties of Industrial Materials," 4th ed. Van Nostrand Reinhold Company, New York, 1978.

Schieber, J. R. Continuous Monitoring, *Chem. Eng.*, **77**(9):111 (1970).

Sittig, M., "Pollutant Removal Handbook," Noyes Data Corporation, Park Ridge, New Jersey, 1973.

Sittenfield, M., Toxic Substances Control Act, *Am. Lab.*, **10**(9):37 (1978).

Snowden, F. C., Defining the Problems of Air and Water Pollution, *Instr. Tech.*, **15**(5):37 (1968).

Snyder, N. W., Energy Recovery and Resource Recycling, *Chem. Eng.*, **81**(22):65 (1974).

Tearle, K., "Industrial Pollution Control: Its Practical Implications," Business Books, London, 1973.

Verschueren, K., "Handbook of Environmental Data on Organic Chemicals," Van Nostrand Reinhold Company, New York, 1978.

Vervalin, C. H., What Governments Expect, *Hydrocarbon Process.*, **51**(10):85 (1972).

———, Environmental-Management Resources: How to Keep up with What's Happening, *Hydrocarbon Process.*, **52**(2):113 (1973).

———, Contact These Sources for Environmental Information, *Hydrocarbon Process.*, **52**(10):71 (1973).

Wallace, M. J., Basic Concepts of Toxicology, *Chem. Eng.*, **85**(10):72 (1978).

White, L. J., State Laws and Enforcement, *Chem. Eng.*, **79**(10):13 (1972).

Williams, D. L., Microprocessors Enhance Computer Control of Plants, *Chem. Eng.*, **84**(15):95 (1977).

Yocum, J. E., G. F. Collins, and N. E. Bowne, Plant Site Selection, *Chem. Eng.*, **78**(14):164 (1971).

Zarytkiewicz, E. D., Federal Environmental Laws and Regulations, *Chem. Eng.*, **82**(21):9 (1975).

The Challenge of Pollution Control—State Regulations, *Chem. Eng.*, **75**(22):25 (1968).

State Regulations, *Chem. Eng.*, **77**(9):23 (1970).

Basic Techniques, Water Pollution Control, *Chem. Eng.*, **77**(9):63 (1970).

Ocean Pollution and Marine Waste Disposal, *Chem. Eng.*, **78**(3):60 (1971).

State Regulations, *Chem. Eng.*, **78**(14):19 (1971).

"Modern Pollution Control Technology," Research and Education Association, New York, 1978.

Process control

Allen, P., Control of Chemical Plant, *Chem. Proc. Eng.*, **49**(10):93 (1968).

Arant, J. B., Special Control Valves Reduce Noise and Vibration, *Chem. Eng.*, **79**(5):92 (1972).

———, Applying Ratio Control to Chemical Processing, *Chem. Eng.*, **79**(21):155 (1972).

Baker, W. J., Direct Digital Control of Batch Processes Pays Off, *Chem. Eng.*, **76**(27):121 (1969).

Ball, J., C. Brez, and J. Cassiday, Batch-Control Improvement through Computers, *Chem. Eng.*, **81**(16):93 (1974).

Bibbero, R. J., "Microprocessors in Instruments and Control," Wiley-Interscience, Somerset, New Jersey, 1978.

Biles, W. R., and H. L. Sellars, A Computer Control System, *Chem. Eng. Progr.*, **65**(8):33 (1969).

Bizarro, L. A., Networking Computers for Process Control, *Chem. Eng.*, **83**(26):151 (1976).

Bloch, M. G., and P. P. Lifland, Catalytic Reforming Improved by Moisture Metering, *Chem. Eng. Progr.*, **69**(9):49 (1973).

Block, B., Control of Batch Distillations, *Chem. Eng.*, **74**(2):147 (1967).

Bojnowski, J. J., R. M. Groghan, Jr., and R. M. Hoffman, Direct and Indirect Material Balance Control, *Chem. Eng. Progr.*, **72**(9):54 (1976).

Borut, R., Trends in Control Room Design, *Chem. Eng.*, **79**(20):75 (1972).

Boyd, D. M., Fractionation Column Control, *Chem. Eng. Progr.*, **71**(6):55 (1975).

Brodmann, M. T., and C. L. Smith, Computer Control of Batch Processors, *Chem. Eng.*, **83**(19):191 (1976).

Buckley, P. S., Controls for Sidestream Drawoff Columns, *Chem. Eng. Progr.*, **65**(5):45 (1969).

———, Protective Controls for Chemical Reactor, *Chem. Eng.*, **77**(8):145 (1970).

———, R. K. Cox, and D. L. Rollins, Inverse Response in a Distillation Column, *Chem. Eng. Progr.*, **71**(6):83 (1975).

Burman, L. K., and R. N. Maddox, Dynamic Control of Distillation Columns, *I & EC Proc. Design & Develop.*, **8**(4):433 (1969).

Cusset, B. F., and D. A. Mellichamp, On-Line Identification of Process Dynamics. A Multifrequency Response Method, *I & EC Proc. Design & Develop.*, **14**(4):359 (1975).

Callisen, F. I., Control Rooms of the Future, *Chem. Eng.*, **76**(12):107 (1969).

Castellano, E. N., C. A. McCain, and F. W. Nobles, Digital Control of a Distillation System, *Chem. Eng. Progr.*, **74**(4):56 (1978).

Corripio, A. B., and C. L. Smith, Computer Simulation to Evaluate Control Strategies, *Instr. Control Sys.*, **44**(1):87 (1971).

Coughanowr, D. R., New Algorithms for Control, *Chem. Eng.*, **76**(12):130 (1969).

Cox, R. K., and J. P. Shunta, Tracking Action Improves Continuous Control, *Chem. Eng. Progr.*, **69**(9): 56 (1973).

Cox, W. E., Computer Control . . . How Will it Pay? *Hydrocarbon Process.*, **57**(4):169 (1978).

Daigre, L. C., III, and G. R. Nieman, Computer Control of Ammonia Plants, *Chem. Eng. Progr.*, **70**(2):50 (1974).

Dann, J. N., and T. L. Henson, Computer Simulation of a Chemical Plant, *Chem. Eng. Progr.*, **74**(6):64 (1978).

del Valle, J. L., Use of pH Value, Conductivity and Redox Potential Measuring Equipment for Monitoring and Controlling Processes in Chemical Plants, *Siemens Rev.*, **41**(9):411 (1974).

Ewing, R. W., G. L. Glahn, R. P. Larkins, and W. N. Zartman, Generalized Process Control Programming System, *Chem. Eng. Progr.*, **63**(1):104 (1967).

Driskell, L. R., New Approach to Control Valve Sizing, *Hydrocarbon Process.*, **48**(7):131 (1969).

Engstrom, F. G., and D. A. Hambleton, Automatic Batch Control, *Chem. Eng.*, **79**(20):81 (1972).

Fauth, G. F., and F. G. Shinskey, Advanced Control of Distillation Columns, *Chem. Eng. Progr.*, **71**(6):49 (1975).

Feitler, H., and C. R. Townsend, Automatic Cooling Tower Control, *Chem. Eng. Progr.*, **65**(5):63 (1969).

Freedman, B. G., Model Reference Control for Batch Digesters, *Chem. Eng. Progr.*, **72**(4):82 (1976).

Friedly, J. C., "Dynamic Behavior of Processes," Prentice-Hall, Englewood Cliffs, New Jersey, 1972.

Friedmann, P. G., and J. A. Moore, For Process Control . . . Select the Key Variable, *Chem. Eng.*, **79**(13):85 (1972).

Funk, G. L., B. F. Houston, and G. D. Stacy, Making Energy Conservation Pay Through Automation, *Chem. Eng. Progr.*, **74**(5):66 (1978).

Gallier, P. W., Operating Compressors, Via Computer, *Chem. Eng. Progr.*, **64**(6):71 (1968).

——— and L. C. McCune, Simple Internal Reflux Control, *Chem. Eng. Progr.*, **70**(9):71 (1974).

Garibian, S. K., and E. C. Koeninger, Computers and Batch Systemization, *Chem. Eng. Progr.*, **68**(3):73 (1972).

Gassett, L. D., Instruments . . . Pneumatic or Electronic? *Chem. Eng.*, **76**(12):136 (1969).

Gervasio, V., and G. Rozzi, Computer Control Cuts Reflux Ratios, *Petro/Chem. Eng.*, **41**(5):18 (1969).

Gould, L. A., "Chemical Process Control: Theory and Applications," Addison-Wesley, Reading, Massachusetts, 1969.

Graham, G. E., Analyzing Process-Control Loops, *Chem. Eng.*, **83**(16):72 (1976).

Hammett, J. L., Jr., and L. A. Lindsay, Advanced Computer Control of Ethylene Plants Pays Off, *Chem. Eng.*, **83**(24):115 (1976).

Hanson, N. M., B. Y. Maughan, and J. D. Cornell, Scanning and Processing Plant Variables, *Chem. Eng. Progr.*, **65**(8):36 (1969).

Harbold, H. S., How to Control Biological-Waste-Treatment Processes, *Chem. Eng.*, **83**(26):157 (1976).

Hatfield, J. M., Process Control by Computer—A Review, *Chem. Eng. (London)*, 171 (March 1976).

Hayden, R., Save Energy: Better Column Control, *Hydrocarbon Process.*, **55**(5):122 (1976).

Hileman, J. R., Justifying a Minicomputer for Process Control, *Chem. Eng.*, **79**(12):61 (1972).

Hougen, J. O., Boiler Control System Design, *Chem. Eng. Progr.*, **74**(6):83 (1978).

Hovorka, R. B., C. R. Cutler, K. C. Young, and A. Chou, The Control System in Action, *Chem. Eng. Progr.*, **65**(8):46 (1969).

Johnson, E. F., "Automatic Process Control Principles," McGraw-Hill Book Company, New York, 1967.

Jones, C., Near-Infrared Analyzers Refine Process Control, *Chem. Eng.*, **85**(22):111 (1978).

Jordan, L., Gas Chromatographs Serviced by a Minicomputer System, *Chem. Eng. Progr.*, **69**(9):53 (1973).

Joseph, B., C. B. Brosilow, J. C. Howell, and W. R. D. Kerr, Multi-temps Give Better Control, *Hydrocarbon Process.*, **55**(3):127 (1976).

Jud, H. G., Basics of Control System Theory, *Plant Eng.*, **21**(10):156 (1967).

Kehoe, T. J., Online Process Analyzers, *Chem. Eng.*, **76**(12):117 (1969); **79**(20):33 (1972).

Kennedy, J. P., Changing Economy: Its Impact on Equipment, *Chem. Eng.*, **82**(6):54 (1975).

Kemp, D. W., and D. G. Ellis, Computer Control of Fractionation Plants, *Chem. Eng.*, **82**(26):115 (1975).

Kestenbaum, A., R. Shinnar, and F. E. Than, Design Concepts for Process Control, *I & EC Fund.*, **15**(1):2 (1976).

Kortlandt, D., and R. L. Zwart, Find Optimum Timing for Sampling Product Quality, *Chem. Eng.*, **78**(25):66 (1971).

Lane, J. W., Four Examples Where Process Computers Pay Off, *Instr. Tech.*, **15**(7):46 (1968).

Lapidus, L., The Control of Nonlinear Systems via Second-Order Approximations, *Chem. Eng. Progr.*, **63**(12):64 (1967).

Latour, P. R., Energy Conservation Via Process Computer Control, *Chem. Eng. Progr.*, **72**(4):76 (1976).

Lawley, H. G., and T. A. Kletz, High-Pressure-Trip Systems for Vessel Protection, *Chem. Eng.*, **82**(10):81 (1975).

Lawrence, J. A., and A. A. Buster, Guide to Trouble-Free Plant Operation . . . Computer-Process Interface, *Chem. Eng.*, **79**(14):102 (1972).

Lee, H. H., L. B. Koppel, and H. C. Lim, Integrated Approach to Design and Control of a Class of Counter-current Processes, *I & EC Proc. Design & Develop.*, **11**(3):376 (1972).

———— ———— and ————, Optimal Sensor Locations and Controller Settings for Class of Counter-current Processes, *I & EC Proc. Design & Develop.*, **12**(1):37 (1973).

Lee, W., and V. W. Weekman, Jr., Advanced Control Practice in the Chemical Process Industry: A View from Industry, *AIChE J.*, **22**(1):27 (1976).

Lovett, O. P., Jr., Control Valves, *Chem. Eng.*, **78**(23):129 (1971).

Luyben, W. L., "Process Modeling, Simulation, and Control for Chemical Engineers," McGraw-Hill Book Company, New York, 1973.

———— and M. Melcic, Consider Reactor Control Lags, *Hydrocarbon Process.*, **57**(3):115 (1978).

Martin, R. L., Simple Solutions to Control-Problems, *Chem. Eng.*, **85**(12):103 (1978).

———— and W. O. Webber, Combination Control: Better Response, *Hydrocarbon Process.*, **49**(3):149 (1970).

Masek, J. A., How to Design Control Panels, *Chem. Eng.*, **83**(13):135 (1976).

Maselli, S. A., and O. Miller, Automate Controls for Profit, *Hydrocarbon Process.*, **51**(4):107 (1972).

Mayfield, R., Backup for Batches, *Chem. Eng.*, **81**(12):79 (1974).

Miller, A. R., The Automation Dilemma: Choosing a Control Technology, *Instr. Control Sys.*, **45**(6):87 (1972).

Miller, J. A., P. W. Murrill, and C. L. Smith, How to Apply Feedforward Control, *Hydrocarbon Process.*, **48**(7):165 (1969).

McCoy, R. E., Comparing Flow-Ratio Control Systems, *Chem. Eng.*, **76**(13):92 (1969).

McLeod, D. P., Computer Economics, *Chem. Proc. Eng.*, **49**(10):79 (1968).

McNeill, G. A., and J. D. Sacks, High Performance Column Control, *Chem. Eng. Progr.*, **65**(3):33 (1969).

Moore, J. F., and N. F. Gardner, Process Control in the 1970's, *Chem. Eng.*, **76**(12):94 (1969).

Murray, J. J., Combined Analog and Digital Control, *Chem. Eng.*, **83**(13):165 (1976).

Murrill, P. W., R. W. Pike, and C. L. Smith, Transient Response in Dynamic-Systems Analyses, *Chem. Eng.*, **75**(27):103 (1968).

Nisenfeld, A. E., Feedforward Control for Azeotropic Distillations, *Chem. Eng.*, **75**(20):227 (1968).

————, Shutdown Features of In-line Process Control, *Chem. Eng. Progr.*, **68**(4):54 (1972).

————, Applying Control Computers to an Integrated Plant, *Chem. Eng. Progr.*, **69**(9): 45 (1973).

————, Cost Comparisons of Analog-Control and Computer-Control Systems, *Chem. Eng.*, **82**(17):104 (1975).

———— and C. H. Cho, Parallel Compressor Control . . . What Should be Considered, *Hydrocarbon Process.*, **57**(2):147 (1978).

———— and J. Harbison, Benefits of Better Distillation Control, *Chem. Eng. Progr.*, **74**(7):88 (1978).

Noonan, R. P., What Kind of Computer for Your Plant? *Chem. Eng.*, **76**(12):112 (1969).

Payne, W. R., Toxicology and Process Design, *Chem. Eng.*, **85**(10):83 (1978).

Pennington, E. N., and L. V. Wilson, Computer Control Economics, *Chem. Eng.*, **76**(12):123 (1969).

Pike, D. H., and M. E. Thomas, Optimal Control of a Continuous Distillation Column, *I & EC Proc. Design & Develop.*, **13**(2):97 (1974).

Podio, A. L., Control Elements, *Petrol Eng.*, **4**(46):70a (1974).

Rasmussen, E. J., Alarm and Shutdown Devices Protect Process Equipment, *Chem. Eng.*, **82**(10):74 (1975).

Rijnsdorp, J. E., Chemical Process Systems and Automatic Control, *Chem. Eng. Progr.*, **63**(7):97 (1967).

Ritter, R. A., and H. Andre, Evaporator Control System Design, *Can. J. Chem. Eng.*, **48**(6):696 (1970).

Schagrin, E. F., How Much Do Minicomputer Control Systems Cost? *Chem. Eng.*, **78**(7):103 (1971).

Scull, W. L., Selecting Temperature Controls for Heaters, *Chem. Eng.*, **82**(11):128 (1975).

Sellars, H. L., F. H. Lyons, N. M. Hanson, and J. D. Cornell, Program Scheduling and Data Handling, *Chem. Eng. Progr.*, **65**(8):42 (1969).

Shah, M. J., Computer Control of Ethylene Production, *I & EC*, **59**(5):71 (1967).

————, Automation of Remote Instruments with an On-line Computer, *Chem. Eng. Progr.*, **68**(4):50 (1972).

Shinksey, F. G., Energy Conserving Control Systems for Distillation Units, *Chem. Eng. Progr.*, **72**(5):73 (1976).

————, Control Systems Can Save Energy, *Chem. Eng. Progr.*, **74**(5):43 (1978).

Simon, H., What's New In . . . Automatic Process Control, *Chem. Eng.*, **79**(20):49 (1972).

Skrokov, M. R., The Benefits of Microprocessor Control, *Chem. Eng.*, **83**(21):133 (1976).

Smith, C. L., and P. W. Murrill, The Dynamics of Spot Samples, *Hydrocarbon Process.*, **46**(12):109 (1967); **47**(1):155(1968).

———— and ————, Process Dynamics—Try Solving This Way, *Hydrocarbon Process.*, **46**(2):105 (1967).

———— and M. T. Brodmann, Process Control: Trends for the Future, *Chem. Eng.*, **83**(13):129 (1976).

Sommer, M., E. F. Cooke, III, J. E. Goehring, and J. J. Haydel, Process Control in a Petrochemical Complex, *Chem. Eng. Progr.*, **67**(10):54 (1971).

Sorell, G., Process Control (Designing for Corrosion Resistance), *Chem. Eng.*, **77**(22):83 (1970).

Soule, L. M., Basic Concepts of Industrial Process Control, *Chem. Eng.*, **76**(20):133 (1969).

————, Basic Control Modes, *Chem. Eng.*, **76**(23):115 (1969).

————, Tuning Process Controllers, *Chem. Eng.*, **76**(26):101 (1969).

————, Single-Loop Circuits May Need New Control Functions, *Chem. Eng.*, **77**(1):103 (1970).

————, Multiple Variables Require Special Control Techniques, *Chem. Eng.*, **77**(2):130 (1970).

————, Feedforward Control Improves System Response, *Chem. Eng.*, **77**(4):113 (1970).

————, Auxiliary Stations for Control Systems, *Chem. Eng.*, **77**(5):142 (1970).

Spence, L. A., Methods of Obtaining Stable Control, *Chem. Eng. Progr.*, **68**(3):50 (1972).

Stainthorp, F. P., and R. S. Benson, Computer Aided Design of Process Control Systems, *Chem. Eng. (London)*, 531 (1974).

Steve, E. H., Logic Diagram Boosts Process-Control Efficiency, *Chem. Eng.*, **78**(6):96 (1971).

Steymann, E. H., Justifying Process Computer Control, *Chem. Eng.*, **75**(4):124 (1968).

Stout, T. M., Justifying Process Control Computers: Selection and Costs, *Chem. Eng.*, **79**(20):89 (1972).

Strader, N. R., II, CRT Consoles: New Look in Control Rooms, *Chem. Eng.*, **80**(10):83 (1973).

Svrek, W. Y., and H. W. Wilson, Case History of A Column Control Scheme, *Chem. Eng. Progr.*, **67**(2):45 (1971).

Tyreus, B., and W. L. Luyben, Control of a Binary Distillation Column with Sidestream Drawoff, *I & EC Proc. Design & Develop.*, **14**(4):391 (1975).

——— and ———, Controlled Heat Integrated Distillation Columns, *Chem. Eng. Progr.*, **72**(9):59 (1976).

van Eijk, F. P., Instrument Trip System Maintenance and Improvement Program, *Chem. Eng. Progr.*, **71**(1):48 (1975).

van Horn, L. D., Computer Control—How to Get Started, *Hydrocarbon Process.*, **57**(9):243 (1978).

Wade, H. L., C. H. Jones, T. B. Rooney, and L. B. Evans, Cyclic Distillation Control, *Chem. Eng. Progr.*, **65**(3):40 (1969).

Wareham, R. L., Using Microprocessors for Process Control, *Chem. Eng.*, **83**(13):163 (1976).

Waters, A. W., Design & Application of a Low Cost Pneumatic Composition Transmitter, *Chem. Eng. Progr.*, **74**(6):80 (1978).

Wherry, T. C., and J. R. Parsons, Guide to Profitable Computer Control, *Hydrocarbon Process.*, **46**(4):179 (1967).

Wightman, E. J., "Instrumentation in Process Control," Butterworths, London, 1972.

Williams, T. J., Computers and Process Control, *I & EC*, **59**(12):53 (1967).

Zielinski, M., Microcomputer Systems: Where Do You Begin Trouble-Shooting? *Hydrocarbon Process.*, **57**(1):109 (1978).

Zientara, D. E., Measuring Process Variables, *Chem. Eng.*, **79**(20):19 (1972).

Process licensing

Gillespie, T. G., Jr., and G. S. Schaffel, Transfer of Licensed Information, *Chem. Eng. Progr.*, **68**(2):44 (1972).

Pilat, H. L., Evaluating Outside Technology, *Chem. Eng. Progr.*, **68**(2):40 (1972).

Schaffel, G. S., Transferring Licensed Process Technology, *Chem. Eng.*, **77**(8):118 (1970).

Smith, D. B., D. M. Young, and R. L. Miller, Process Technology for License or Sale, *Chem. Eng.*, **77**(8):115 (1970).

Safety

Adcock, C. T., and J. D. Weldon, Vapor Release and Explosion in a Light Hydrocarbon Plant, *Chem. Eng. Progr.*, **63**(8):54 (1967).

Alba, C., Size Rupture Discs by Nomograph, *Hydrocarbon Process.*, **49**(9):297 (1970).

Allen, O. M., and O. I. Hanna, Arrow Diagrams Improve Operational Safety, *Chem. Eng.*, **78**(21):166 (1971).

Atallah, S., Security for CPI Plants, *Chem. Eng.*, **84**(21):139 (1977).

Baker, G. F., T. A. Kletz, and H. A. Knight, Olefin Plant Safety During the Last 15 Years, *Chem. Eng. Progr.*, **73**(9):64 (1977).

Barnwell, J., Designing Safety into an Ethylene Plant, *Chem. Eng. Progr.*, **74**(10):66 (1978).

Bartknecht, W., Explosion Pressure Relief, *Chem. Eng. Progr.*, **73**(9):45 (1977).

Benjaminsen, J. M., and P. H. van Wiechen, Mean Time to Electrical Explosions, *Hydrocarbon Process.*, **47**(8):121 (1968).

Bennett, T. C., Automate for Plant Emergencies, *Hydrocarbon Process.*, **51**(12):57 (1972).

Bently, D. E., Monitor Machinery Condition for Safe Operation, *Hydrocarbon Process.*, **53**(11):205 (1974).

Bergtraun, E. M., Safety Planning for Construction and Alteration Projects, *Chem. Eng.*, **85**(5):133 (1978).

Biggers, E. W., and T. C. Smith, Train Operators for Safety, *Hydrocarbon Process.*, **51**(12):54 (1972).

Blair, D. A., Operate Your Refinery Safely, *Hydrocarbon Process.*, **56**(4):241 (1977).

Bonilla, J. A., Estimate Safe Flare-Headers Quickly, *Chem. Eng.*, **85**(9):135 (1978).

Bott, E. C. B., Chemical Process Hazards, *Chem. Eng.* (*London*), 705 (November 1974).

Bouilloud, P., Calculation of Maximum Flow Rate Through Safety Valves, *Brit. Chem. Eng.*, **15**(11):1447 (1970).

Boyle, W. J., Jr., Sizing Relief Area for Polymerization Reactors, *Chem. Eng. Progr.*, **63**(8):61 (1967).

Browning, R. L., Calculating Loss Exposures, *Chem. Eng.*, **76**(25):239 (1969).

———, Estimating Loss Probabilities, *Chem. Eng.*, **76**(27):135 (1969).

Buchanan, J. R., How to Obtain Nuclear Safety Data, *Power Eng.*, **74**(12):46 (1970).

Buehler, J. H., R. H. Freeman, R. G. Keister, M. P. McCready, B. I. Pesetsky, and D. T. Watters, Report on Explosion at Union Carbide's Texas City Butadiene Refining Unit, *Chem. Eng.*, **77**(19):77 (1970).

Buehner, F. W., OSHA, EPA and Plant Design, *Chem. Eng.*, **84**(12):161 (1977).

Burklin, C. R., Safety Standards, Codes and Practices for Plant Design, *Chem. Eng.*, **79**(22):56 (1972); **79**(23):113 (1972); **79**(25):143 (1972).

Burns, B. W., Design Control Centers to Resist Explosions, *Hydrocarbon Process.*, **46**(11):257 (1967).

Calabrese, A. M., and L. D. Krejci, Safety Instrumentation for Ammonia Plants, *Chem. Eng. Progr.*, **70**(2):54 (1974).

Clark, C. R. N., Bursting Discs in Process Plant Protection, *Chem. Proc. Eng.*, **51**(3):431 (1970).

Cocks, R. E., and J. E. Rogerson, Organizing a Process Safety Program, *Chem. Eng.*, **85**(23):138 (1978).

Coffee, R. D., Design Your Plant for Survival, *Hydrocarbon Process.*, **55**(5):301 (1976).

Constance, J. D., Control of Explosive or Toxic Air-Gas Mixtures, *Chem. Eng.*, **78**(9):121 (1971).

———, Pressure Ventilate for Explosion Protection, *Chem. Eng.*, **78**(21):156 (1971).

Cordes, R. J., Use Compressors Safely, *Hydrocarbon Process.*, **56**(10):227 (1977); **56**(11):469 (1977).

Crocker, B. B., Preventing Hazardous Pollution During Plant Catastrophes, *Chem. Eng.*, **77**(10):97 (1970).

Crom, R. C. W., Safeguarding Against Shock Hazards, *Chem. Eng.*, **84**(7):90 (1977).

Dartnell, R. C., Jr., and T. A. Ventrone, Explosion of a Para-Nitro-Meta-Cresol Unit, *Chem. Eng. Progr.*, **67**(6): 58 (1971).

Davenport, J. A., Prevent Vapor Cloud Explosions, *Hydrocarbon Process.*, **56**(3):205 (1977).

———, A Survey of Vapor Cloud Incidents, *Chem. Eng. Progr.*, **73**(9):54 (1977).

de Heer, H. J., Calculating How Much Safety Is Enough, *Chem. Eng.*, **80**(4):121 (1973).

———, Choosing an Economical Automatic Protective System, *Chem. Eng.*, **82**(6):73 (1975).

Deloney, H. C., an Evaluation of Intrinsically Safe Instrumentation, *Chem. Eng.*, **79**(12):67 (1972).

DeRenzo, D. J., Unit Operations for Treatment of Hazardous Industrial Wastes, Noyes Data Corporation, Park Ridge, New Jersey, 1978.

Deuschle, R., and F. Tiffany, Barrier Intrinsic Safety, *Hydrocarbon Process.*, **52**(12):111 (1973).

Dorsey, J. S., Static Sparks—How to Exorcize "Go Devils," *Chem. Eng.*, **83**(19):203 (1976).

Ecker, H. W., B. A. James, and R. H. Toensing, Electrical Safety: Designing Purged Enclosures, *Chem. Eng.*, **81**(10):93 (1974).

Eckstrom, C. M., Inplant Health Hazards, *Chem. Eng.*, **80**(14):173 (1973).

Eichel, F. G., Electrostatics, *Chem. Eng.*, **74**(6):153 (1967).

Ford, H., K. E. Bett, J. Rogan, and H. F. Gardner, Design Criteria for Pressure Vessels, *Chem. Eng. Progr.*, **68**(11):77 (1972).

Frankland, P. B., Relief Valves . . . What Needs Protection? *Hydrocarbon Process.*, **57**(4):189 (1978).

Freeman, R. H., and M. P. McCready, Butadiene Explosion at Texas City, *Chem. Eng. Progr.*, **67**(6):45 (1971).

Fried, J., Meeting Safety Standards in the Laboratory, *Am. Lab.*, **9**(13):79 (1977).

Gans, M., The A to Z of Plant Startup, *Chem. Eng.*, **83**(6):72 (1976).

Geiger, G. H., "Supplementary Readings in Engineering Design," McGraw-Hill Book Company, New York, 1975.

Gelburd, R. M., A Comprehensive Approach to Occupational Safety and Health, *Chem. Eng.*, **84**(8):114 (1977).

Gibson, N., and G. F. P. Harris, The Calculation of Dust Explosion Vents, *Chem. Eng. Progr.*, **72**(11):62 (1976).

Gibson, S. B., Risk Criteria in Hazard Analysis, *Chem. Eng. Progr.*, **72**(2):59 (1976).

———, Reliability Engineering Applied to the Safety of New Projects, *Chem. Eng. (London)*, 105 (February 1976).

Graf, F. A., Jr., Safe Handling of Hazardous Materials, *Chem. Eng. Progr.*, **63**(11):67 (1967).

Graham, J. J., The Fluidized Bed Phthalic Anhydride Process, *Chem. Eng. Progr.*, **66**(9):54 (1970).

Guill, A. W., Protective Design for Reactor, Compressor, and High Pressure Piping, *Chem. Eng. Progr.*, **68**(11):61 (1972).

Haessler, W. M., Fire Extinguishing Chemicals, *Chem. Eng.*, **80**(5):95 (1973).

Halley, P. D., OSHA Standards Impact Refining, *Hydrocarbon Process.*, **56**(5):137 (1977).

Hatfield, P. E., Safety Around Equipment, *Chem. Eng. Progr.*, **68**(11):65 (1972).

Hearfield, F., The Philosophy of Loss Prevention, *Chem. Eng. (London)*, 257 (April 1976).

Heron, P. M., Accidents Caused by Plant Modifications, *Chem. Eng. Progr.*, **72**(11):45 (1976).

Herrick, L. K., Jr., Instrumentation for Monitoring Toxic and Flammable Work Areas, *Chem. Eng.*, **83**(22):147 (1976).

Hettig, S. B., Jr., Designing for Safety, *Chem. Eng.*, **75**(6):170 (1968).

Hickes, W. F., Fundamentals of Intrinsic Safety, *Chem. Eng.*, **76**(14):139 (1969).

———, Intrinsic Safety, *Chem. Eng.*, **79**(9):64 (1972).

Hilado, C. J., How to Predict If Materials Will Burn, *Chem. Eng.*, **77**(27):174 (1970).

Hix, A. H., Safety and Instrumentation Systems, *Chem. Eng. Progr.*, **68**(5):43 (1972).

Hodgson, J. D., The New Occupational Safety and Health Act, *Chem. Eng.*, **78**(15):108 (1971).

House, F. F., An Engineer's Guide to Process-Plant Layout, *Chem. Eng.*, **76**(16):120 (1969).

Howard, W. B., Hazards with Flammable Mixtures, *Chem. Eng. Progr.*, **66**(9):59 (1970).

Isaacs, M., Pressure-Relief Systems, *Chem. Eng.*, **78**(5):113 (1971).

Ito, T., and N. Sawada, Ground Flares Aid Safety, *Hydrocarbon Process.*, **55**(6):175 (1976).

Jarvis, H. C., Butadiene Explosion at Texas City, *Chem. Eng. Progr.*, **67**(6):41 (1971).

Jenkins, J. H., P. E. Kelly, and C. B. Cobb, Design for Better Safety Relief, *Hydrocarbon Process.*, **56**(8):93 (1977).

Johnson, D. M., The Billion Gallon Tank Farm, *Chem. Eng. Progr.*, **65**(4):41 (1969).

Jolls, K. R., and R. L. Riedinger, Impedance: An Electrical Effect in A.C. Circuits, *Chem. Eng.*, **79**(21):165 (1972).

Jones, D. W., Use a Safety Audit Program, *Hydrocarbon Process.*, **54**(8):67 (1975).

Kauffmann, W. M., Safe Operation of Oxygen Compressors, *Chem. Eng.*, **76**(23):140 (1969).

Kern, R., Pressure-Relief Valves for Process Plants, *Chem. Eng.*, **84**(5):187 (1977).

Kilby, J. L., Flare System Explosions, *Chem. Eng. Progr.*, **64**(6):49 (1968).

Kinsley, G. R., Jr., Controlling Cross-Connections Keeps Plant Drinking-Water Pure, *Chem. Eng.*, **85**(9):121 (1978).

Kirven, J. B., and D. P. Handke, Plan Safety for New Refining Facilities, *Hydrocarbon Process.*, **52**(6):121 (1973).

Kiyoura, R., and K. Urano, Mechanism, Kinetics, and Equilibrium of Thermal Decomposition of Ammonium Sulfate, *I & EC Proc. Design & Develop.*, **9**(4):489 (1970).

Kletz, T. A., Emergency Isolation Valves for Chemical Plants, *Chem. Eng. Progr.*, **71**(9):63 (1975).

———, Accidents Caused by Reverse Flow, *Hydrocarbon Process.*, **55**(3):187 (1976).

———, Preventing Catastrophic Accidents, *Chem. Eng.*, **83**(8):124 (1976).

———, A Three-Pronged Approach To Plant Modification, *Chem. Eng. Progr.*, **72**(11):48 (1976).

———, Evaluate Risk in Plant Design, *Hydrocarbon Process.*, **56**(5):297 (1977).

———, Practical Applications of Hazard Analysis, *Chem. Eng. Progr.*, **74**(10):47 (1978).

Klooster, H. J., G. A. Vogt, and G. F. Braun, Optimizing the Design of Relief and Flare Systems, *Chem. Eng. Progr.*, **71**(1):39 (1975).

Kolodner, H. J., Use a Fault-Tree Approach, *Hydrocarbon Process.*, **56**(9):303 (1977).

Krigman, A., and R. Redding, Test for Intrinsic Safety, *Hydrocarbon Process.*, **53**(10):198 (1974).

Kurz, G. R., Inert-Gas Protection, *Chem. Eng.*, **80**(7):112 (1973).

Lambert, H. E., Fault Trees for Location Sensors in Process Systems, *Chem. Eng. Progr.*, **73**(8):81 (1977).

Lambertin, W. J., and F. H. Vaughan, Will Existing Equipment Fail by Catastrophic Brittle Fracture? *Hydrocarbon Process.*, **53**(9):217 (1974).

Lawley, H. G., and T. A. Kletz, High-Pressure-Trip Systems for Vessel Protection, *Chem. Eng.*, **82**(10):81 (1975).

———, Size Up Plant Hazards This Way, *Hydrocarbon Process.*, **55**(4):247 (1976).

Lawrence, W. E., and E. E. Johnson, Design for Limiting Explosion Damage, *Chem. Eng.*, **81**(1):96 (1974).

Lee, R. H., Electrical Grounding: Safe or Hazardous? *Chem. Eng.*, **76**(16):158 (1969).

Leeah, C. J., The Changing Face of HPI Safety, *Hydrocarbon Process.*, **47**(10):157 (1968).

———, Poor Plant Design Courts Disaster, *Hydrocarbon Process.*, **47**(11):248 (1968).

———, Good Procedures Save Lives, Money, *Hydrocarbon Process.*, **48**(1):149 (1969).

LeVine, R. Y., Electrical Safety in Process Plants, *Chem. Eng.*, **79**(9):51 (1972).

Liptàk, B. G., Safety Instruments and Control-Valves Costs, *Chem. Eng.*, **77**(24):94 (1970).

Ludwig, E. E., Project Managers Should Know OSHA, *Hydrocarbon Process.*, **52**(6):135 (1973).

———, Designing Process Plants to Meet OSHA Standards, *Chem. Eng.*, **80**(20):88 (1973).

Lynch, M. E., How to Investigate a Plant Disaster, *Hydrocarbon Process.*, **46**(8):161 (1967).

Magison, E. C., Engineering Instruments for Safety, *Instr. Tech.*, **16**(2):41 (1969).

Marinak, M. J., Safety Checklist, *Chem. Eng. Progr.*, **63**(11):59 (1967).

Marshall, V. C., Process-Plant Safety—A Strategic Approach, *Chem. Eng.*, **82**(27):58 (1975).

———, The Strategic Approach to Safety, *Chem. Eng. (London)*, 260. (April 1976).

Mitsui, R., and T. Tanaka, Simple Models of Dust Explosion. Predicting Ignition Temperature and Minimum Explosive Limit in Terms of Particle Size, *I & EC Process. Design & Develop.*, **12**(3):384 (1973).

May, W. G., W. McQueen, and R. H. Whipp, Dispersion of LNG Spills, *Hydrocarbon Process.*, **52**(5):105 (1973).

McBride, J. A., Regulatory Controls on Radioactive Effluents, *Chem. Eng. Progr.*, **66**(4):74 (1970).

McClelland, G. D., Maintenance and Safe Operation of High-Pressure Equipment, *Chem. Eng.*, **75**(20):202 (1968).

McConnaughey, W. E., M. E. Welsh, R. J. Lakey, and R. M. Goldman, Hazardous Materials Transportation, *Chem. Eng. Progr.*, **66**(2):57 (1970).

McFarland, I., Safety in Pressure Vessel Design, *Chem. Eng. Progr.*, **66**(6):56 (1970).

McKay, F. F., G. R. Worrell, B. C. Thornton, and H. L. Lewis, If an Ethylene Pipe Line Ruptures, *Hydrocarbon Process.*, **56**(11):487 (1977).

Miller, E. G., and T. E. Peterson, How HPI Designs, Operates, Trains for Safety, *Petro/Chem. Eng.*, **40**(9):16 (1968).

Monroy, A. D., and V. M. Gonzales Majul, Stop Fires in EO Plants, *Hydrocarbon Process.*, **56**(9):175 (1977).

Moon, G. B., A Law Primer for the Chemical Engineer, *Chem. Eng.*, **85**(9):114 (1978).

Morton, W. I., Safety Techniques for Workers Handling Hazardous Materials, *Chem. Eng.*, **83**(22):127 (1976).

Nailen, R. L., How to Apply Electric Motors in Explosive Atmospheres, *Hydrocarbon Process.*, **54**(2):101 (1975).

———, If Fire Strikes Your Plant, Can Firemen Get to It? *Plant. Eng.*, **21**(1):145 (1967).

Naughton, D. A., Turbine Accidents Which Should Not Have Happened, *Chem. Eng. Progr.*, **64**(6):63 (1968).

Nelson, R. W., Know Your Insurer's Expectations, *Hydrocarbon Process.*, **56**(8):103 (1977).

Nilsen, J. M., OSHA: Acronym for Trouble, *Chem. Eng.*, **79**(6):58 (1972).

O'Neal, L. R., C. M. Elwonger, and W. T. Hughes, Fatigue-Vibration Problems in Piping Systems, *Chem. Eng. Progr.*, **68**(11):68 (1972).

Ostroot, G., Jr., Case History: A Two Burner Boiler Explosion, *Hydrocarbon Process.*, **55**(12):85 (1976).

Owen, L. T., Selecting In-Plant Dust-Control Systems, *Chem. Eng.*, **81**(21):120 (1974).

Parker, J. U., Anatomy of a Plant Safety Inspection, *Hydrocarbon Process.*, **46**(1):197 (1967).

Pence. T. G., Transportation, *Chem. Eng.*, **76**(8):23 (1969).

Perciful, J. C., and R. S. Edwards, Transfer of Information from the Chemist, *Chem. Eng. Progr.*, **63**(11):69 (1967).

Peterson, P., Explosions in Flare Stacks, *Chem. Eng. Progr.*, **63**(8):67 (1967).

——— and H. R. Cutler, Explosion Protection for Centrifuges, *Chem. Eng. Progr.*, **69**(4):42 (1973).

Picciotti, M., Design for Ethylene Plant Safety, *Hydrocarbon Process.*, **57**(3):191 (1978); **57**(4):261 (1978).

Powers, G. J., and S. A. Lapp, Computer-Aided Fault Tree Synthesis, *Chem. Eng. Progr.*, **72**(4):89 (1976).

Preddy, D. L., Guidelines for Safety and Loss Prevention, *Chem. Eng.*, **76**(8):94 (1969).

Prentice, J. S., S. E. Smith, and L. S. Virtue, Safety in Polyethylene Plant Compressor Areas, *Chem. Eng. Progr.*, **70**(9):49 (1974).

Price, F. C., Are Long-Distance Pipelines Getting Safer? *Chem. Eng.*, **81**(21):94 (1974).

Prugh, R. W., Preliminary Evaluation of Safety Hazards, *Chem. Eng. Progr.*, **63**(11):49 (1967).

Rasmussen, E. J., Alarm and Shutdown Devices Protect Process Equipment, *Chem. Eng.*, **82**(10):74 (1975).

Rearick, J. S., How to Design Pressure Relief Systems, *Hydrocarbon- Process.*, **48**(8):104 (1969); **48**(9):161 (1969).

Redding, R. J., Electrical Safety in a Chemical Plant, *Chem. Proc. Eng.*, **50**(5):98 (1969).

———, Intrinsic Safety: A European View, *Chem. Eng.*, **76**(14):137 (1969).

———, Safety Developments in Instrumentation, *Chem. Proc. Eng.*, **51**(3):47 (1970).

Reed, R. D., Design and Operation of Flare Systems, *Chem. Eng. Progr.*, **64**(6):53 (1968).

Richter, S. H., Size Relief Systems for Two-Phase Flow, *Hydrocarbon Process.*, **57**(7):145 (1978).

Robey, N. T., Apathy Can be Dangerous, *Hydrocarbon Process.*, **50**(7):129 (1971).

Royalty, C. A., and B. D. Woosley, Design Criteria for High Pressure Piping, *Chem. Eng. Progr.*, **68**(11):72 (1972).

Saia, S. A., Vapor Clouds and Fires in a Light . . . Hydrocarbon Plant, *Chem. Eng Progr.*, **72**(11):56 (1976).

Sax, N. I., "Dangerous Properties of Industrial Materials," 4th ed., Van Nostrand Reinhold, Cincinnati, Ohio, 1978.

Scherberger, R. F., Industrial Hygiene Control Methods, *Chem. Eng.*, **84**(8):118 (1977).

Schmit, K. H., T. E. Lee, J. H. Mitchen, and E. E. Spiteri, How to Cope with Aluminum Alkyls, *Hydrocarbon Process.*, **57**(11):341 (1978).

Schwartz, R., and M. Keller, Environmental Factors vs. Flare Application, *Chem. Eng. Progr.*, **73**(9):41 (1977).

Schwab, R. F., and W. H. Doyle, Hazards in Phthalic Anhydride Plants, *Chem. Eng. Progr.*, **66**(9):49 (1970).

Sengupta, M., and F. Y. Staats, A New Approach to Relief Valve Load Calculations, *Hydrocarbon Process.*, **57**(5):160 (1978).

Smith, D., Pneumatic Transport and Its Hazards, *Chem. Eng. Progr.*, **66**(9):41 (1970).

Shelly, P. G., and E. J. Sills, Monomer Storage and Protection, *Chem. Eng. Progr.*, **65**(4):29 (1969).

Shield, B. H., Battery Failure Causes Major Failure of Gas Turbine, *Chem. Eng. Progr.*, **63**(8):58 (1978).

Short, W. A., Electrical Equipment for Hazardous Locations, *Chem. Eng.*, **79**(9):59 (1972).

Showalter, D. R., "How to Make the OSHA-1970 Work for You," Ann Arbor Science Publishers, Ann Arbor, Michigan, 1972.

Silver, L., Screening for Chemical Process Hazards, *Chem. Eng. Progr.*, **63**(8):43 (1967).

Smith, C. W., Toxic Substances Control is Here, *Hydrocarbon Process.*, **56**(1):213 (1977).

Snyder, A. I., Safe Design, Construction and Installation of Pressure Vessels, *Brit. Chem. Eng.*, **16**(6):499 (1971).

Snyder, I. G., Jr., Implementing a Good Safety Program, *Chem. Eng.*, **81**(11):112 (1974).

Statesir, W. A., Explosive Reactivity of Organics and Chlorine, *Chem. Eng. Progr.*, **69**(4):52 (1972).

Stender, J. H., What the HPI Can Expect from OSHA, *Hydrocarbon Process.*, **53**(7):190 (1974).

Stephens, T. J. R., and C. B. Livingston, Explosion of a Chlorine Distillate Receiver, *Chem. Eng. Progr.*, **69**(4):45 (1973).

Stuart, L. R., You Can Have Safer Construction, *Hydrocarbon Process.*, **53**(8):106 (1974).

Sutherland, M. E., and H. W. Wegert, An Acetylene Decomposition Incident, *Chem. Eng. Progr.*, **69**(4):48 (1973).

Telesmanic, M. J., Axial Flow Compressor Accidents, *Chem. Eng. Progr.*, **64**(6):68 (1968).

Townsend, D. I., Hazard Evaluation of Self-Accelerating Reactions, *Chem. Eng. Progr.*, **73**(9):80 (1977).

Underwood, H. C., Jr., R. E. Sourwine, and C. D. Johnson, Organize for Plant Emergencies, *Chem. Eng.*, **83**(21):118 (1976).

van Eijnatten, A. L. M., Explosion in a Naphtha Cracking Unit, *Chem. Eng. Progr.*, **73**(9):69 (1977).

Vervalin, C. H., Keep up with OSHA Developments, *Hydrocarbon Process.*, **50**(12):118 (1971); **51**(2):107 (1972); **51**(7):117 (1972).

————, Fire and Safety Information: A Description of Key Resources, *Hydrocarbon Process.*, **51**(4):163 (1972).

————, Know Where OSHA is Taking You, *Hydrocarbon Process.*, **51**(12):65 (1972).

————, Learn From HPI Plant Fires, *Hydrocarbon Process.*, **51**(12):49 (1972).

————, Who's Publishing on Fire and Safety? *Hydrocarbon Process.*, **52**(1):128 (1973).

————, Environmental-Management Resources: How to Keep up with What's Happening, *Hydrocarbon Process.*, **52**(2):113 (1973).

————, What's New in Fire and Safety Training? *Hydrocarbon Process.*, **52**(4):185 (1973).

————, Learn from HPI Tank Fires, *Hydrocarbon Process.*, **52**(8):81 (1973).

————, What's New in Fire and Safety Training? *Hydrocarbon Process.*, **53**(3):163 (1974).

————, HPI Looks at Toxicity, *Hydrocarbon Process.*, **54**(4):225 (1975).

————, Fire/Safety Information Resources, *Hydrocarbon Process.*, **54**(8):65 (1975).

————, Will Your Plant Burn Tomorrow? *Hydrocarbon Process.*, **54**(12):145 (1975).

————, Loss Prevention in the HPI, *Hydrocarbon Process.*, **55**(8):182 (1976).

————, HPI Loss-Incident Case Histories, *Hydrocarbon Process.*, **57**(2):183 (1978).

Vincent, G. C., Rupture of a Nitroaniline Reactor, *Chem. Eng Progr.*, **67**(6):51 (1971).

Wafelman, H. E., and H. Buhrmann, Maintain Safer Tank Storage, *Hydrocarbon Process.*, **56**(1):229 (1977).

Wahrmund, R. C., Musts for Emergency Relief System Design, *Petro/Chem. Eng.*, **40**(12):21 (1968).

Waldman, S., Fireproofing in Chemical Plants, *Chem. Eng. Progr.*, **63**(8):71 (1967).

Walker, J. J., Sizing Relief Areas for Distillation Columns, *Chem. Eng. Progr.*, **66**(9):38 (1970).

Walls, W. L., Fire Protection for LNG Plants, *Hydrocarbon Process.*, **50**(9):205 (1971).

Warren, J. H., R. F. Lange, and D. L. Rhynard, Conductivity Additives Are Best, *Hydrocarbon Process.*, **53**(12):111 (1974).

———— and A. A. Corona, This Method Tests Fire Protective Coatings, *Hydrocarbon Process.*, **54**(1):121 (1975).

Webb, H. E., Jr., What to do When Disaster Strikes, *Chem. Eng.*, **84**(16):47 (1977).

Weiby, P., and K. R. Dickinson, Monitoring Work Areas for Explosive and Toxic Hazards, *Chem. Eng.*, **83**(22):139 (1976).

Weismantel, G. E., A Fresh Look at Intrinsic Safety, *Chem. Eng.*, **76**(14):132 (1969).

Welker, J. R., H. R. Wesson, and L. E. Brown, Use Foam to Disperse LNG Vapors? *Hydrocarbon Process.*, **53**(2):119 (1974).

Wesson, H. R., J. R. Welker, and L. E. Brown, Control LNG-spill Fires, *Hydrocarbon Process.*, **51**(12):61 (1972).

Whelan, T. W., and S. J. Thomson, Reduce Relief System Costs, *Hydrocarbon Process.*, **54**(8):83 (1975).

White, L. J., Occupational Safety and Health Act, *Chem. Eng.*, **80**(5):13 (1973).

Whitehorn, V. J., and H. W. Brown, How to Handle a Safety Inspection, *Hydrocarbon Process.*, **46**(4):125 (1967); **46**(5):227 (1967).

Wilks, P. A., Jr., OSHA Compliance Testing of Toxic Vapors, *Am. Lab.*, **5**(12):67 (1973).

Wirth, G. F., Preventing and Dealing with In-Plant Hazardous Spills, *Chem. Eng.*, **82**(17): 82 (1975).

Wissmiller, I. L., New Relief Valve Gas Equations, *Hydrocarbon Process.*, **49**(5): 123 (1970).

Wittig, S. L. K., High Volume Submillisecond Pressure Relief Emergency System, *I & EC Proc. Design & Develop.*, **9**(4): 605 (1970).

Woinsky, S. G., Predicting Flammable-Material Classifications, *Chem. Eng.*, **79**(26): 81 (1972).

Wolnex, G. A., Ten Important Steps to Safety, *Plant Eng.*, **21**(10): 131 (1967).

Wood, W. S., Transporting, Loading and Unloading Hazardous Materials, *Chem. Eng.*, **80**(15): 72 (1973).

Yowell, R. L., Bisphenol-A Dust Explosions, *Chem. Eng. Progr.*, **64**(6): 58 (1968).

Yurkanin, R. M., and E. O. Claussen, Safety Aspects of Electrical Heating Systems, *Chem. Eng.*, **77**(27): 164 (1970).

Zanker, A., Nomograph Helps Determine Static Electrical Charge from Fluid Flow, *Hydrocarbon Process.*, **55**(3): 133 (1976).

Ziefle, R. G., Designing Safe High Pressure Polyethylene Plants, *Chem. Eng. Progr.*, **68**(11): 51 (1972).

Zielinski, R. M., Case Study of a Reactor Fire, *Chem. Eng. Progr.*, **63**(8): 59 (1967).

"Chemical Process Hazards III," C. F. Hodgson and Son Limited, London, 1967.

Occupational Safety and Health Act, *Hydrocarbon Process.*, **50**(4): 147 (1971).

Occupational Safety and Health Act, *Chem. Eng.*, **80**(14): 157 (1973).

Scale-models

Babcock, J. A., How to Get the Most out of Engineering Models, *Chem. Eng.*, **80**(4): 112 (1973).

Birkhoff, J., One Step Isometrics from Models, *Hydrocarbon Process.*, **51**(7): 95 (1972).

Horowitz, J., and W. C. Hullender, Engineering, Design and Construction via Models: Computer Graphics and Scale Models, *Chem. Eng. Progr.*, **68**(6): 45 (1972).

Miller, R. E., Jr., Scale Modeling of Large and Small Plant Projects, *Chem. Eng.*, **78**(27): 69 (1971).

Rowland, J. R., Engineering, Design and Construction via Models: The Concept, Principles, and Function of Models, *Chem. Eng. Progr.*, **68**(6): 41 (1972).

Steele, L. W., and R. E. Miller, Engineering, Design and Construction via Models: More Ways to use Engineering Models, *Chem. Eng. Progr.*, **68**(6): 58 (1972).

Utley, C. O., Engineering, Design and Construction via Models: Using Models on Foreign Projects, *Chem. Eng. Progr.*, **68**(6): 63 (1972).

Design Models for Chemical Plant, *Chem. Proc. Eng.*, **50**(11): 95 (1969).

Shut-down problems

Fisher, J. T., Designing Emergency Shutdown Systems, *Chem. Eng.*, **80**(28): 138 (1973).

Williams, G. P., Causes of Ammonia Plant Shutdowns, *Chem. Eng. Progr.*, **74**(9): 88 (1978).

Start-up problems

Arndt, J. H., and D. B. Kiddoo, Special Purpose Gearing—How to Avoid Startup Problems, *Hydrocarbon Process.*, **54**(3): 113 (1975).

Butzert, H. E., Ammonia/Methanol Plant Startup, *Chem. Eng. Progr.*, **72**(1): 56 (1976).

Clark, M. E., E. M. DeForest, and L. R. Steckley, Aches and Pains of Plant Startup, *Chem. Eng. Progr.*, **67**(12): 25 (1971).

Dolle, J., and D. Gilbourne, LNG: Startup of the Skikda LNG Plant, *Chem. Eng. Progr.*, **72**(1): 39 (1976).

Dyess, C. E., Putting a 70,000 bbl/day Crude Still Onstream, *Chem. Eng Progr.*, **69**(8): 98 (1973).

Feldman, R. P., Economics of Plant Startups, *Chem. Eng.*, **76**(24): 87 (1969).

Gans, M., The A to Z of Plant Startup, *Chem. Eng.*, **83**(6): 72 (1976).

Grieve, P., Plant Startup as a Career, *Chem. Eng.*, **76**(19): 148 (1969).

Love, F. S., Troubleshooting Distillation Problems, *Chem. Eng. Progr.*, **71**(6): 61 (1975).

Lowe, R. E., Starting FCC Power Recovery, *Hydrocarbon Process.*, **57**(8): 103 (1978).

Matley, J., Keys to Successful Plant Startups, *Chem. Eng.*, **76**(19): 110 (1969).

Parnell, D. C., Claus Sulfur Recovery Unit Startups, *Chem. Eng. Progr.*, **69**(8): 89 (1973).

Parsons, R. H., Guidelines for Plant Startup, *Chem. Eng. Progr.*, **67**(12):25 (1971).

Pearson, L., When It's Time for Startup, *Hydrocarbon Process.*, **56**(8):116 (1977).

Puckorius, P. R., Proper Startup Projects, Cooling-Tower Systems, *Chem. Eng.*, **85**(1):101 (1978).

Richards, J. W., Empirical Nomograms in Plant Startup, *Chem. Proc. Eng.*, **51**(8):50 (1970).

Robinson, J. R., Startup of a Bender® Process Unit, *Chem. Eng. Progr.*, **69**(8):101 (1973).

Ryan, G. T., Managing the Project Startup, *Chem. Eng. Progr.*, **68**(12):65 (1972).

Severa, J. E., Startup of a Crude/Vacuum Distillation Unit, *Chem. Eng. Progr.*, **69**(8):85 (1973).

Stainthorp, F. P., and B. West, Computer Controlled Plant Startup, *Chem. Eng.* (*London*), 526 (September 1974).

Swain, R. T., and B. J. Hopper, Contractor Problems During Startup, *Chem. Eng. Progr.*, **67**(12):32 (1971).

Thermal pollution

Dickey, B. R., E. S. Grimmett, and D. C. Kilian, Waste Heat Disposal via Fluidized Beds, *Chem. Eng. Progr.*, **70**(1):60 (1974).

Hall, W. A., Cooling Tower Plume Abatement, *Chem. Eng. Progr.*, **67**(7):52 (1971).

Jimeson, R. M., and G. G. Adkins, Waste Heat Disposal in Power Plants, *Chem. Eng. Progr.*, **67**(7):64 (1971).

Nester, D. M., Salt Water Cooling Tower, *Chem. Eng. Progr.*, **67**(7):49 (1971).

Remirez, R., Thermal Pollution: Hot Issue for Industry, *Chem. Eng.*, **75**(7):48 (1968).

Zarić, Z. P., "Thermal Effluent Disposal from Power Generation," Hemisphere Publishing Corporation, Washington, D.C., 1978.

Utilities

Arnstein, R., and L. O'Connell, What's the Optimum Heat Cycle for Process Utilities? *Hydrocarbon Process.*, **47**(6):88 (1968).

Brooke, M., Water, *Chem. Eng.*, **77**(27):135 (1970).

Donohue, J. M., and W. W. Hales, Improve Cooling Water Treatment, *Hydrocarbon Process.*, **47**(6):101 (1968).

Gambhir, S. P., T. J. Heil, and T. F. Schuelke, Steam Use and Distribution, *Chem. Eng.*, **85**(28):91 (1978).

Gambs, G. C., and A. A. Rauth, The Energy Crisis, *Chem. Eng.*, **78**(12):57 (1971).

Horn, B. C., On-Line Optimization of Plant Utilities, *Chem. Eng. Progr.*, **74**(6):76 (1978).

Imhof, H., Protecting Process Plants from Power Failures, *Chem. Eng.*, **80**(8):56 (1973).

Jones, F. A., Utilities Consumption in Natural Gas Plants, *Chem. Eng. Progr.*, **68**(10):73 (1972).

Margolis, I., Plant Electrical Systems, *Chem. Eng.*, **85**(28):81 (1978).

McAllister, D. G., Jr., Air, *Chem. Eng.*, **77**(27):138 (1970).

Monroe, L. R., Process Plant Utilities—Steam, *Chem. Eng.*, **77**(27):130 (1970).

Price, H. A., D. G. McAllister, Jr., Inert Gas, *Chem. Eng.*, **77**(27):142 (1970).

Swengel, F. M., A New Era of Power Supply Economics, *Power Eng.*, **74**(3):30 (1970).

Wilson, W. B., and J. M. Kovacik, Electricity: Generate or Buy? *Hydrocarbon Process.*, **55**(12):75 (1976).

Yuen, M. H., Electricity, *Chem. Eng.*, **77**(27):144 (1970).

Industrial On-site Generation, *Power Eng.*, **72**(5):54 (1968).

Waste disposal

Alsentzer, H. A., Pollution Abatement: Treatment of Industrial Wastes at Regional Facilities, *Chem. Eng. Progr.*, **68**(8):73 (1972).

Anderson, C. R., Nuclear Fuel Recovery, *Power Eng.*, **73**(1):50 (1969).

Annessen, R. J., and G. D. Gould, Sour-Water Processing Turns Problem into Payout, *Chem. Eng.*, **78**(7):67 (1971).

Barber, J. C., Waste Effluent: Treatment and Reuse, *Chem. Eng. Progr.*, **65**(6):70 (1969).

Barker, W. G., and D. Schwarz, Engineering Processes for Waste Control, *Chem. Eng. Progr.*, **65**(1):58 (1969).

Becker, K. P., and C. J. Wall, Fluid Bed Incineration of Wastes, *Chem. Eng. Progr.*, **72**(10):61 (1976).

Beckner, J. L., Consider Ash Disposal, *Hydrocarbon Process.*, **55**(2)1:107 (1976).

Bennett, G. F., and L. Lash, Industrial Waste Disposal Made Profitable, *Chem. Eng. Progr.*, **70**(2):75 (1974).

Beychok, M. R., Wastewater Treatment, *Hydrocarbon Process.*, **50**(12):109 (1971).

Blanco, R. E., H. W. Godbee and E. J. Frederick, Incorporating Industrial Wastes in Insoluble Media, *Chem. Eng. Progr.*, **66**(2):51 (1970).

Boback, M. W., J. O. Davis, K. N. Ross, and J. B. Stevenson, Disposal of Low-Level Radioactive Wastes from Pilot Plants, *Chem. Eng. Progr.*, **67**(4):81 (1971).

Bonniaud, R., C. Sombret, and A. Barbe, Continuous Vitrification of Radioactive Wastes, *Chem. Eng. Progr.*, **72**(3):47 (1976).

Bowerman, F. R., Solid Waste Disposal, *Chem. Eng.*, **77**(9):147 (1970).

Brenton, R. W., Treatment of Sugarbeet Wastes by Recycling, *Indus. Wastes*, Jan./Feb. (1972), p. 14.

Browning, J. E., Garbage-Pipelines' Progress, *Chem. Eng.*, **78**(17):60 (1971).

Chapman, C. C., H. T. Blair, and W. F. Bonner, Waste Vitrification at Battelle Northwest, *Chem. Eng. Progr.*, **72**(3):58 (1976).

Coleman, L. W., and L. F. Cheek, Liquid Waste Incineration, *Chem. Eng. Progr.*, **64**(9):83 (1968).

Dahlstrom, D. A., L. D. Lash, J. L. Boyd, Biological and Chemical Treatment of Industrial Wastes, *Chem. Eng. Progr.*, **66**(11):41 (1970).

DeRenzo, D. J., "Unit Operations for Treatment of Hazardous Industrial Wastes," Noyes Data Corporation, Park Ridge, New Jersey, 1978.

Deviny, W. M., Disposal of Hazardous Chemicals, *Chem. Eng. Progr.*, **63**(11):56 (1967).

Dlouhy, P. E., and D. A. Dahlstrom, Food and Fermentation Waste Disposal, *Chem. Eng. Progr.*, **65**(1):52 (1969).

Eckenfelder, W. W., Jr., and J. L. Barnard, Treatment-Cost Relationship for Industrial Wastes, *Chem. Eng. Progr.*, **67**(9):76 (1971).

Eldredge, R. W., Solid Waste Management, *Chem. Eng.*, **77**(9):143 (1970).

Gantz, R. G., L. W. Cresswell, and J. F. Gauen, Pontoon System Automated for Slop Recovery, *Hydrocarbon Process.*, **54**(3):93 (1975).

Goodlett, C. B., Putting Evaporators to Work: Concentration of Aqueous Radioactive Waste, *Chem. Eng. Progr.*, **72**(4):63 (1976).

Grove, G. W., Use Land Farming for Oily Waste Disposal, *Hydrocarbon Process.*, **57**(5):138 (1978).

Grover, P., A Waste Stream Management System, *Chem. Eng. Progr.*, **73**(12):71 (1977).

Gurnham, C. F., Material Conservation Through Waste Salvage, *Chem. Eng. Progr.*, **67**(1):67 (1971).

Halzel, G. C., Flyash Disposal, *Power Eng.*, **73**(6):44 (1969).

Hardenbergh, W. A., and E. R. Rodie, "Water Supply and Waste Disposal," International Textbook Company, Scranton, Pennsylvania, 1967.

Hartz, K. E., Shredding Speeds Solid Waste Reduction, *Allis-Chalmers Eng. Rev.*, **38**(2):4 (1973).

Heaney, F. L., and C. V. Keane, Solid Waste Disposal, *1970 Trans. Am. Assoc. Cost Eng.*, 1970, p. 1.

Herbert, W., Recycling Municipal Waste, *Chem. Eng.*, **79**(1):66 (1972).

Hoffman, F., How to Select a pH Control System for Neutralizing Waste Acids, *Chem. Eng.*, **79**(24):105 (1972).

Johnson, C. D., Modern Radwaste Systems: An Overview, *Chem. Eng. Progr.*, **72**(3):43 (1976).

Jones, J., Converting Solid Wastes and Residues to Fuel, *Chem. Eng.*, **85**(1):87 (1978).

Kearney, M. S., and R. D. Walton, Jr., Long-Term Management of High Level Wastes, *Chem. Eng. Progr.*, **72**(3):61 (1976).

Kiang, Y.-H., Liquid Waste Disposal System, *Chem. Eng. Progr.*, **72**(1):71 (1976).

Klass, D. L., Make SNG From Waste and Biomass, *Hydrocarbon Process.*, **55**(4):76 (1976).

Krikau, F. G., Neutralization is Key to Acid-Liquor Waste Disposal, *Chem. Eng.*, **75**(24):124 (1968).

Kumar, J., and J. A. Jedlicka, Selecting and Installing Synthetic Pond-Linings, *Chem. Eng.*, **80**(3):67 (1973).

LaRiviers, J. R., Packaging and Storing Radioactive Wastes, *Chem. Eng. Progr.*, **66**(2):42 (1970).

Lash, L. D., and E. G. Kominek, Primary-Waste Treatment Methods, *Chem. Eng.*, **82**(21):49 (1975).

——— and G. L. Shell, Treating Polymer Wastes, *Chem. Eng. Progr.*, **65**(6):63 (1969).

Legler, B. M., and G. R. Bray, Concentration and Storage of High Level Wastes, *Chem. Eng. Progr.*, **72**(3):52 (1976).

Lerch, R. E., C. R. Cooley, and J. M. Atwood, Acid Digestion—A New Method for Treatment of Nuclear Waste, *Westinghouse Eng.*, **33**(5):146 (1973).

Lewis, C. R., R. E. Edwards, and M. A. Santoro, Incineration of Industrial Wastes, *Chem. Eng.*, **83**(22):115 (1976).

Lindsey, A. W., Ultimate Disposal of Spilled Hazardous Materials, *Chem. Eng.*, **82**(23):107 (1975).

McHarg, W. H., Designing the Optimum System for Biological-Waste-Treatment, *Chem. Eng.*, **80**(29):46 (1973).

Medwed, P. M., Water Wastes Carefully Treated for Stream Discharge, *Plant Eng.*, **21**(9):114 (1967).

Mencher, S. K., Minimizing Waste in the Petrochemical Industry, *Chem. Eng. Progr.*, **63**(10):80 (1967).

Monroe, E. S., Jr., Burning Waste Waters, *Chem. Eng.*, **75**(20):215 (1968).

Nilsen, J. M., Weighing the Options for Industrial Sludge Disposal, *Chem. Eng.*, **79**(19):28 (1972).

Novak, R. G., Eliminating or Disposing of Industrial Solid Wastes, *Chem. Eng.*, **77**(21):78 (1970).

———, J. J. Cudahy, M. D. Denove, R. L. Standifer, and W. E. Wass, How Sludge Characteristics Affect Incinerator Design, *Chem. Eng.*, **84**(10):131 (1977).

Patterson, G. C., Refinery Sewer Systems and Waste Treatment Facilities, *Petro/Chem. Eng.*, **39**(2):50 (1967).

Perry, R. A., Mercury Recovery from Process Sludges, *Chem. Eng. Progr.*, **70**(3):73 (1974).

Petrie, J. C., R. I. Donovan, R. E. Van der Cook, and W. R. Christensen, Putting Evaporators to Work: Vacuum Evaporator-Crystalizer Handles Radioactive Waste, *Chem. Eng. Progr.*, **72**(4):65 (1976).

Ramalho, R. S., Principles of Activated Sludge Treatment, *Hydrocarbon Process.*, **57**(11):275 (1978).

Reiter, W. M., and R. Sobel, Waste Control Management, *Chem. Eng.*, **80**(14):59 (1973).

——— and W. F. Stocker, In-Plant Waste Abatement, *Chem. Eng. Progr.*, **70**(1):55 (1974).

Ross, R. D., Problems with Liquid Waste? Why Not Burn It? *Plant Eng.*, **21**(8):120 (1967).

Santoleri, J. J., Chlorinated Hydrocarbon Waste Disposal and Recovery Systems, *Chem. Eng. Progr.*, **69**(1):68 (1973).

Saxton, J. C., and M. Narkus-Kramer, EPA Findings on Solid Wastes from Industrial Chemicals, *Chem. Eng.*, **82**(9):107 (1975).

Scher, J. A., Solid Wastes Characterization Techniques, *Chem. Eng. Progr.*, **67**(3):81 (1971).

Schneider, K. J., Solidification of Radioactive Wastes, *Chem. Eng. Progr.*, **66**(2):35 (1970).

Sebastian, F. P., and P. J. Cardinal, Jr., Solid Waste Disposal, *Chem. Eng.*, **75**(22):112 (1968).

Stubblefield, F. E., and E. B. Jackson, Improved Control of Radioactive Wastes, *Chem. Eng. Progr.*, **70**(3):87 (1974).

Sueyoshi, H., and Y. Kitaoka, Make Fuel from Plastic Wastes, *Hydrocarbon Process.*, **51**(10):161 (1972).

Talbot, J. S., Deep Wells, *Chem. Eng.*, **75**(22):108 (1968).

Templeton, W. L., Disposal of Liquid Wastes into Coastal Waters, *Chem. Eng. Progr.*, **42**(2):45 (1970).

Thomson, S. J., How to Design Activated Sludge Units, *Hydrocarbon Process.*, **54**(8):99 (1975).

van Geel, J. N. C., H. Eschrich, and E. J. Detilleux, Conditioning High Level Radioactive Wastes, *Chem. Eng. Progr.*, **72**(3):49 (1976).

Wall, C. J., J. T. Graves, and E. J. Roberts, How to Burn Salty Sludges, *Chem. Eng.*, **82**(2):77 (1975).

Wathen, P. R., M. M. Clemmens, and W. Zabban, Three-Way Cooperation Develops Complete Waste Treatment System, *Plant Eng.*, **21**(6):113 (1967).

Weismantel, G. E., ed., Chemical Engineering Application in Solid Waste Management, *Chem. Eng. Progr. Symp. Series*, **68**(122), 1972.

Welty, R. K., Solar Evaporation of Fluoride Wastes, *Chem. Eng. Progr.*, **72**(3):54 (1976).

Werner, K., Catalytic Oxidation of Industrial Waste Gases, *Chem. Eng.*, **75**(23):179 (1968).

Wilson, D. G., ed., "Handbook of Solid Waste Management," Van Nostrand Reinhold Company, New York, 1978.

White, W. F., A Centrifugal for Industrial Wastes, *Chem. Eng. Progr.*, **65**(6):74 (1969).

Tables Show How to Treat Petrochemical Wastes, *Petro/Chem. Eng.*, **39**(5):25 (1967).

Solid Waste Disposal, *Chem. Eng.*, **78**(14):155 (1971).

"Solid Waste Management, Technology Assessment," Van Nostrand Reinhold Company, New York, 1978.

Water and water treatment

Anderson, D., The Economics of Industrial Waste Treatment, *Chem. Eng. (London)*, 422 (November 1972).

Askew, T., Selecting Economic Boiler-Water Pretreatment Equipment, *Chem. Eng.*, **80**(9):114 (1973).

Baker, C. D., E. W. Clark, W. V. Jesernig, and C. H. Huether, Recovering para-Cresol from Process Effluent, *Chem. Eng. Progr.*, **69**(8):77 (1973).

Barber, J. C., Waste Effluent: Treatment and Reuse, *Chem. Eng. Progr.*, **65**(6):70 (1969).

Bellew, E. F., Comparing Chemical Precipitation Methods for Water Treatment, *Chem. Eng.*, **85**(6):85 (1978).

Bennett, J. E., Non-Diaphragm Electrolytic Hypochlorite Generators, *Chem. Eng. Progr.*, **70**(12):60 (1974).

Bernardin, F. E., Jr., Selecting and Specifying Activated-Carbon-Adsorption Systems, *Chem. Eng.*, **83**(22):77 (1976).

Beychok, M. R., Wastewater Treatment, *Hydrocarbon Process.*, **50**(12):109 (1971).

Blackburn, J. W., Removal of Salts from Process Wastewaters, *Chem. Eng.*, **84**(22):33 (1977).

Bowen, L. B., J. H. Mallinson, and J. H. Cosgrove, Zinc Recovery from Rayon Plant Sludge, *Chem. Eng. Progr.*, **73**(5):50 (1977).

Braunscheidel, D. E., and R. G. Gyger, UNOX System for Water Waste Treatment, *Chem. Eng. Progr.*, **72**(11):71 (1976).

Brooke, J. M., Corrosion Inhibitor Economics, *Hydrocarbon Process.*, **49**(9):299 (1970).

Brooke, M., Water, *Chem. Eng.*, **77**(27):135 (1970).

Browning, J. E., Activated Carbon Bids for Wastewater Treatment Jobs, *Chem. Eng.*, **77**(19):32 (1970).

―――, New Water-Cleanup Roles for Powdered Activated Carbon, *Chem. Eng.*, **79**(4):36 (1972).

Brunet, M. J., and R. H. Parsons, How Mobil Reduced Sour Water Stripping Problems, *Hydrocarbon Process.*, **51**(10):107 (1972).

Brunotts, V. A., R. T. Lynch, and G. R. Van Stone, Granular Carbon Handles Concentrated Waste, *Chem. Eng. Progr.*, **69**(8):81 (1973).

Bush, K. E., Refinery Wastewater Treatment and Reuse, *Chem. Eng.*, **83**(8):113 (1976).

Cadman, T. W., and R. W. Dellinger, Techniques for Removing Metals from Process Wastewater, *Chem. Eng.*, **81**(8):79 (1974).

Carnes, B. A., Water-Pollution Control, *Chem. Eng.*, **79**(28):97 (1972).

Cecil, L. K., Water Reuse and Disposal, *Chem. Eng.*, **76**(10):92 (1969).

Chambers, D. B., and W. R. T. Cottrell, Flotation: Two Fresh Ways to Treat Effluents, *Chem. Eng.*, **83**(16):95 (1976).

Characklis, W. G., and A. W. Busch, Industrial Wastewater Treatment, *Chem. Eng.*, **79**(10):61 (1972).

Ciaccio, L. L., Instrumentation in Water Reuse and Pollution Control, *Am. Lab.*, **3**(12):21 (1971).

Comeaux, R. V., Basic Cooling Water Inhibitor Guide, *Hydrocarbon Process.*, **46**(12):129 (1967).

Crame, L. W., Cut Wastewater Treatment Costs, *Hydrocarbon Process.*, **55**(5):92 (1976).

Cross, E. F., Sampling and Analysis—Water Pollution Control, *Chem. Eng.*, **75**(22):75 (1968).

Crossley, T. J., Lime Treatment and Ammonia Recovery of Liquid Waste, *Chem. Eng. Progr.*, **72**(10):81 (1976).

Culp, R. L., G. M. Wesner, and G. L. Culp, "Handbook of Advanced Wastewater Treatment," 2d ed., Van Nostrand Reinhold Company, New York (1978).

Cummings-Saxton, J., Chemical-Industry Costs of Water Pollution Abatement, *Chem. Eng.*, **83**(24):106 (1976).

Curtin, S. D., and R. M. Silverstein, Corrosion and Fouling Control of Cooling Waters, *Chem. Eng. Progr.*, **67**(7):39 (1971).

Dahlstrom, D., Sludge Dewatering, *Chem. Eng.*, **75**(22):103 (1968).

Daly, K. F., The Luminescence Biometer in the Assessment of Water Quality and Wastewater Analysis, *Am. Lab.*, **6**(12):38 (1974).

Dammer, R. H., Rubber-Lined Reservoirs Solve Waste Water Storage Problems, *Plant Eng.*, **21**(12):119 (1967).

DeAryan, D. L., Ion Exchange Due for Computerization, *Petro/Chem. Eng.*, **40**(13):32 (1968).

Dev, L., and P. W. Kelso, Steam Stripping of Kraft Mill Condensates, *Chem. Eng. Progr.*, **74**(1):72 (1978).

Dierks, R. D., and W. F. Bonner, Putting Evaporators to Work: Wiped Film Evaporator for High Level Wastes, *Chem. Eng. Progr.*, **72**(4):61 (1976).

DeJohn, P. B., and A. D. Adams, Activated Carbon Improves Wastewater Treatment, *Hydrocarbon Process.*, **54**(10):104 (1975).

Donohue, J. M., and W. W. Hales, Improve Cooling Water Treatment, *Hydrocarbon Process.*, **47**(6):101 (1968).

———, Making Cooling Water Safe for Steel and Fish, Too, *Chem. Eng.*, **78**(22):98 (1971).

——— and C. C. Nathan, Unusual Problems in Cooling Water Treatment, *Chem. Eng. Progr.*, **71**(7):88 (1975).

Dresner, L., Preliminary Economic Analysis of Donnan Softening, *I & EC Proc. Design & Develop.*, **12**(2):148 (1973).

Earhart, J. P., K. W. Won, H. Y. Wong, J. M. Prausnitz, and C. J. King, Recovery of Organic Pollutants via Solvent Extraction, *Chem. Eng. Progr.*, **73**(5):67 (1977).

Eckenfelder, W. W., Jr., Economics of Wastewater Treatment, *Chem. Eng.*, **76**(18):109 (1969).

———, "Applications of New Concepts of Physical-Chemical Wastewater Treatment," Pergamon Press, New York, 1972.

Eliassen, R., and G. Tchobanoglous, Advanced Treatment Processes—Water Pollution Control, *Chem. Eng.*, **75**(22):95 (1968).

Erskine, D. B., and W. G. Schuliger, Activated Carbon Processes for Liquids, *Chem. Eng. Progr.*, **67**(11):41 (1971).

Evans, R. R., and R. S. Millward, Equipment for Dewatering Waste Streams, *Chem. Eng.*, **82**(21):83 (1975).

Finelt, S., and J. R. Crump, Predict Wastewater Generation, *Hydrocarbon Process.*, **56**(8):159 (1977).

——— and ———, Pick the Right Water Reuse System, *Hydrocarbon Process.*, **56**(10):111 (1977).

Fitzgerald, C. L., M. M. Clemens, and P. B. Reilly, Jr., Coagulants for Waste Water Treatment, *Chem. Eng. Progr.*, **66**(1):36 (1970).

Ford, D. L., and L. F. Tischler, Guide to Wastewater Treatment: Biological-System Developments, *Chem. Eng.*, **84**(17):131 (1977).

——— and R. L. Elton, Removal of Oil and Grease from Industrial Wastewaters, *Chem. Eng.*, **84**(22):49 (1977).

Foster, W. B., Automation in the Monitoring of Water Resources, *Am. Lab.*, **6**(12):25 (1974).

Franzen, A. E., V. G. Skogan, and J. F. Grutsch, Pollution Abatement: Tertiary Treatment of Process Water, *Chem. Eng. Progr.*, **68**(8):65 (1972).

Gantz, R. G., Sour Water Stripper Operations, *Hydrocarbon Process.*, **54**(5):85 (1975).

Gasper, K. E., Process Corrosion Control: Non-chromate Methods of Cooling Water Treatment, *Chem. Eng. Progr.*, **74**(3):52 (1978).

Gehm, H. W., and J. I. Bregman, ed., "Handbook of Water Resources and Pollution Control," Van Nostrand Reinhold Company, New York, 1978.

Geinopolos, A., and W. J. Katz, Primary Treatment—Water Pollution Control, *Chem. Eng.*, **75**(22):78 (1968).

Goodman, B. L., and K. A. Mikkelson, Advanced Wastewater Treatment (Removing Phosphorus and Suspended Solids), *Chem. Eng.*, **77**(9):75 (1970).

Grier, J. C., and R. J. Christensen, Biocides Give Flexibility in Water Treatment, *Hydrocarbon Process.*, **54**(11): 283 (1975).

Grieves, C. G., M. K. Stenstrom, J. D. Walk, and J. F. Grutsch, Powdered Carbon Improves Activated Sludge Treatment, *Hydrocarbon Process.*, **56**(10):125 (1977).

Grutsch, J. F., and R. C. Mallatt, Optimize the Effluent System; Activated Sludge Process, *Hydrocarbon Process.*, **55**(3):105 (1976).

———— and ————, Optimize the Effluent System; Electrochemistry of Destabilization, *Hydrocarbon Process.*, **55**(5):221 (1976).

———— and ————, Optimize the Effluent System; Approach to Chemical Treatment, *Hydrocarbon Process.*, **55**(6):115 (1976).

———— and ————, Optimize the Effluent System; Multi-Media Filters, *Hydrocarbon Process.*, **55**(7):113 (1976).

———— and ————, Optimize the Effluent System; Biochemistry of Activated Sludge Process, *Hydrocarbon Process.*, **55**(8):137 (1976).

Hager, D. G., Waste Water Treatment Via Activated Carbon, *Chem. Eng. Progr.*, **72**(10):57 (1976).

Hamilton, C. E., ed., " Manual on Water," ASTM Publication, Philadelphia, Pennsylvania, 1978.

Hammer, M. J., "Water and Waste-Water Technology; SI Version," J. Wiley and Sons, New York, 1977.

Harbold, H. S., How to Control Biological-Waste-Treatment Processes, *Chem. Eng.*, **83**(26):157 (1976).

Hardenbergh, W. A., and E. R. Rodie, "Water Supply and Waste Disposal," International Textbook Company, Scranton, Pennsylvania, 1967.

Harland, J. R., J. A. Oberteuffer, and D. J. Goldstein, High Gradient Magnetic Filtration, *Chem. Eng. Progr.*, **72**(10):79 (1976).

Hayes, R. C., Advanced Water Treatment Via Filter, *Chem. Eng. Progr.*, **65**(6):81 (1969).

Helsel, R. W., Removing Carboxylic Acids from Aqueous Wastes, *Chem. Eng. Progr.*, **73**(5):55 (1977).

Henshaw, T. B., Adsorption/Filtration Plant Cuts Phenols from Effluent, *Chem. Eng.*, **78**(12):47 (1971).

Hiser, L. L., Selecting a Wastewater Treatment Process., *Chem. Eng.*, **77**(26):76 (1970).

Higgins, I. R., Continuous Ion Exchange of Process Water, *Chem. Eng. Progr.*, **65**(6):59 (1969).

Hutton, D. G., Improved Biological Waste Water Treatment, *Innovation*, **3**(1):6 (1972).

Jackson, G. S., and H. G. Blecker, Economic Evaluation of Wastewater Management Systems by Computer, *1973 Trans. Am. Assoc. Cost Eng.*, (1973), p. 38.

Jaeschke, L., and K. Trobisch, Treat HPI Wastes Biologically, *Hydrocarbon Process.*, **46**(7):111 (1967).

James, E. W., W. F. Maguire, and W. L. Harpel, Using Wastewater as Cooling-system Makeup Water, *Chem. Eng.*, **83**(18):95 (1976).

Jones, H. R., " Wastewater Cleanup Equipment 1973," 2d ed., Noyes Data Corporation, Park Ridge, New Jersey, 1973.

Kaup, E. C., Design Factors in Reverse Osmosis, *Chem. Eng.*, **80**(8):46 (1973).

Kemmer, F. N., and K. Odland, Chemical Treatment—Water Pollution Control, *Chem. Eng.*, **75**(22):83 (1968).

Kinsley, G. R., Jr., Controlling Cross-Connections Keeps Plant Drinking-Water Pure, *Chem. Eng.*, **85**(9):121 (1978).

Kirk, J. C., How Water Laws Affect the HPI, *Hydrocarbon Process.*, **46**(7):107 (1967).

Klett, R. J., Treat Sour Water for Profit, *Hydrocarbon Process.*, **51**(10):97 (1972).

Knowles, C. L., Jr., Improving Biological Processes, *Chem. Eng.*, **77**(9):103 (1970).

Knowlton, H. E., Why Not Use a Rotating Disk? *Hydrocarbon Process.*, **56**(9):227 (1977).

Kohn, P. M., Water Treatment System Cuts Organics, *Chem. Eng.*, **84**(17):108 (1977).

Kovalcik, R. N., Single Waste-Treatment Vessel Both Flocculates and Clarifies, *Chem. Eng.*, **85**(14):117 (1978).

Krisher, A. S., Low-Toxicity Cooling-Water Inhibitors—How They Stack Up, *Chem. Eng.*, **85**(4):115 (1978).

————, Raw Water Treatment in the CPI, *Chem. Eng.*, **85**(19):79 (1978).

Kumar, J., and J. P. Fairfax, Rating Alternatives to Chromates in Cooling-Water Treatment, *Chem. Eng.*, **83**(9):111 (1976).

Kunin, R., and D. G. Downing, Ion-Exchange System Boasts More Pulling Power, *Chem. Eng.*, **78**(15):67 (1971).

Kunz, R. G., A. F. Yen, and T. C. Hess, Cooling-Water Calculations, *Chem. Eng.*, **84**(16):61 (1977).

Lanouette, K. H., Heavy Metals Removal, *Chem. Eng.*, **84**(22):73 (1977).

————, Treatment of Phenolic Wastes, *Chem. Eng.*, **84**(22):99 (1977).

Larkman, D., Physical/Chemical Treatment, *Chem. Eng.*, **80**(14):87 (1973).

Lederman, P. B., Physical and Chemical Methods, *Chem. Eng.*, **84**(17):135 (1977).

Leitner, G. F., Reverse Osmosis for Water Recovery and Reuse, *Chem. Eng. Progr.*, **69**(6):83 (1973).

Lesperance, T. W., Biological Treatment—Water Pollution Control, *Chem. Eng.*, **75**(22):89 (1968).

Levi, E. J., New Developments in Basics of Cooling-Water Treatment, *Chem. Eng.*, **81**(12):88 (1974).

Leyden, D. E., X-ray Emission Spectrometry and Environmental Water Analysis, *Am. Lab.*, **6**(11):24 (1974).

Liptàk, B. G., Higher Profits via Advanced Instrumentation, *Chem. Eng.*, **82**(13):152 (1975).

Matsch, L. C., and W. C. Dedeke, Using Pure Oxygen for Secondary Treatment, *Chem. Eng. Progr.*, **69**(8):75 (1973).

Mattson, R. J., and V. J. Tomsic, Improved Water Quality, *Chem. Eng. Progr.*, **65**(1):62 (1969).

McCoy, J. W., "The Chemical Treatment of Cooling Water," Chemical Publishing Company, New York, 1974.

McGovern, J. G., Inplant Wastewater Control, *Chem. Eng.*, **80**(11):137 (1973).

McIlhenny, W. F., Recovery of Additional Water from Industrial Waste Water, *Chem. Eng. Progr.*, **63**(6):76 (1967).

Melin, G. A., J. L. Niedzwiecki, and A. M. Goldstein, Optimum Design of Sour Water Strippers, *Chem. Eng. Progr.*, **71**(6):78 (1975).

Michalson, A. W., High Quality Water via Ion Exchange, *Chem. Eng. Progr.*, **64**(10):67 (1968).

Milios, P., Water Reuse at a Coal Gasification Plant, *Chem. Eng. Progr.*, **71**(6):99 (1975).

Miller, R. L., Desalting Shapes Up, *Chem. Eng.*, **79**(19):24 (1972).

Minott, J. D., A Water Treating System for a VCM Plant, *Chem. Eng. Progr.*, **69**(8):71 (1973).

Mohler, E. F., Jr., and L. T. Clere, Bio-oxidation Process Saves H_2O, *Hydrocarbon Process.*, **52**(10):84 (1973).

Moore, J. R., Jr., Wastewater Requirements Multiply Solids Problem, *Hydrocarbon Process.*, **55**(10):98 (1976).

Moores, C. W., Wastewater Biotreatment: What It Can and Cannot Do, *Chem. Eng.*, **79**(29):63 (1972).

Mulligan, T. J., and R. D. Fox, Treatment of Industrial Wastewaters, *Chem. Eng.*, **83**(22):49 (1976).

Nathan, M. F., Choosing a Process for Chloride Removal, *Chem. Eng.*, **85**(3):93 (1978).

Newkirk, R. W., and P. J. Schroeder, R.O. Can Help with Wastes, *Hydrocarbon Process.*, **51**(10):103 (1972).

Newton, W. E., Plant Water Pollution: A Calculating Solution, *Instr. Tech.*, **15**(12):55 (1968).

Nogaj, R. J., Selecting Wastewater Aeration Equipment, *Chem. Eng.*, **79**(8):95 (1972).

Oldshue, J. Y., The Case for Deep Aeration Tanks, *Chem. Eng. Progr.*, **66**(11):73 (1970).

Othmer, D. F., Oxygenation of Aqueous Wastes: the PROST System, *Chem. Eng.*, **84**(13):117 (1977).

Pallanich, P. J., Pure Oxygen Treatment of Pesticide Plant Waste Water, *Chem. Eng. Progr.*, **74**(4):79 (1978).

Parker, C. L., and C. V. Fong, Estimation of Operating Costs for Industrial Waste Water Treatment Facilities, *AACE Bull.*, **18**(6):207 (1976).

Partridge, E. P., and E. G. Paulson, Water: Its Economic Reuse Via the Closed Cycle, *Chem. Eng.*, **74**(21):244 (1967).

Paulson, E. G., Reducing Fluoride in Industrial Wastewater, *Chem. Eng.*, **84**(22):89 (1977).

Pavoni, J. L., ed., "Handbook of Water Quality Management Planning," Van Nostrand Reinhold Company, New York, 1978.

Pollio, F. X., R. Kunin, and J. W. Petralia, Treat Sour Water by Ion Exchange, *Hydrocarbon Process.*, **48**(5):124 (1969).

Prather, B. V., and E. P. Young, Energy for Wastewater Treatment, *Hydrocarbon Process.*, **55**(5):88 (1976).

Prengle, H. W., Jr., J. R. Crump, S. Curtice, T. P. John, A. P. Gutierrez, and T. T. Tseng, Recycle Waste Water by Ion Exchange, *Hydrocarbon Process.*, **54**(4):173 (1975).

———— C. E. Mauk, R. W. Legan, and C. G. Hewes, III, Ozone UV Process Effective Wastewater Treatment, *Hydrocarbon Process.*, **54**(10):82 (1975).

Quartulli, O. J., Stop Wastes: Reuse Process Condensate, *Hydrocarbon Process.*, **54**(10):94 (1975).

Rabosky, J. G., and D. L. Koraido, Gaging and Sampling Industrial Wastewaters, *Chem. Eng.*, **80**(1):111 (1973).

Ramalho, R. S., "Introduction to Wastewater Treatment," Academic Press, New York, 1977.

Rickles, R. N., Joint Industrial/Municipal Waste Treatment, *Chem. Eng.*, **77**(9):129 (1970).

Rizzo, J. L., and A. R. Shepherd, Treating Industrial Wastewater with Activated Carbon, *Chem. Eng.*, **84**(1):95 (1977).

Robinson, W. E., Waste Water Treatment in a Kraft Mill, *Chem. Eng. Progr.*, **65**(6):70 (1969).

Robitaille, D. R., and J. G. Bilek, Molybdate Cooling-Water Treatments, *Chem. Eng.*, **83**(27):77 (1976).

Roffman, H. K., and A. Roffman, Water that Cools but Does Not Pollute, *Chem. Eng.*, **83**(13):167 (1976).

Rosenblad, A. E., Putting Evaporators to Work: Evaporator Systems for Black Liquor Concentration, *Chem. Eng. Progr.*, **72**(4):53 (1976).

Rosenzweig, M. D., Mercury Cleanup Routes—I, *Chem. Eng.*, **82**(2):60 (1975).

Rucker, J. E., and R. W. Oeben, Waste Water Control Facilities in a Petrochemical Plant, *Chem. Eng. Progr.*, **66**(11):63 (1970).

Ryan, L. F., and R. M. Brown, Water Treatment for Reactors, *Power Eng.*, **73**(1):36 (1969).

Sack, W. A., Factors Affecting Wastewater Reclamation and Recycle, *1970 Trans. Am. Assoc. Cost Eng.*, (1970), p. 7.

Sawyer, G. A., New Trends in Wastewater Treatment and Recycle, *Chem. Eng.*, **79**(16):120 (1972).

Seels, F. H., Industrial Water Pretreatment, *Chem. Eng.*, **80**(5):27 (1973).

Shade, H. I., Waste Water Nitrification, *Chem. Eng. Progr.*, **73**(5):45 (1977).

Shell, G. L., J. L., Boyd, and D. A. Dahlstrom, Upgrading Waste Treatment Plants, *Chem. Eng.*, **78**(14):97 (1971).

Silverstein, R. M., and S. D. Curtis, Cooling Water, *Chem. Eng.*, **78**(18):85 (1971).

Skogen, D. B., Treat HPI Wastes with Bugs, *Hydrocarbon Process.*, **46**(7):105 (1967).

Smith, C. V., and D. Di Gregorio, Advance Wastewater Treatment, *Chem. Eng.*, **77**(9):71 (1970).

Stevens, B. W., and J. W. Kerner, Recovering Organic Materials from Wastewater, *Chem. Eng.*, **82**(3):84 (1975).

Stevens, J. I., and C. L. Kusik, The Effects of Water-Pollution Control on Energy Consumption, *Chem. Eng.*, **84**(17):139 (1977).

Stickney, W. W., and T. M. Fosberg, Putting Evaporators to Work: Treating Chemical Wastes by Evaporation, *Chem. Eng. Progr.*, **72**(4):41 (1976).

Sussman, S., Facts on Water Use in Cooling Towers, *Hydrocarbon Process.*, **54**(7):147 (1975).

Sutphin, E. M., Trailer-Mounted Pilot Plants for Water Conservation, *Chem. Eng. Progr.*, **69**(8):79 (1973).

Thompson, A. R., Cooling Towers, *Chem. Eng.*, **75**(22):100 (1968).

Thomson, S. J., Data Improves Separator Design, *Hydrocarbon Process.*, **52**(10):81 (1973).

Thorsen, T., and R. Oen, How to Measure Industrial Wastewater Flow, *Chem. Eng.*, **82**(4):95 (1975).

Tyer, B. R., and C. H. Hayes, Oxygenation Equipment—Aerators and Pure Oxygen Systems, *Chem. Eng.*, **82**(21):75 (1975).

Troscinski, E. S., and R. G. Watson, Controlling Deposits in Cooling-Water Systems, *Chem. Eng.*, **77**(5):125 (1970).

Weil, R. V., Water Conservation in the Petroleum Industry, *Chem. Eng. Progr.*, **65**(11):69 (1969).

Walko, J. F., Controlling Biological Fouling in Cooling Systems—Part 1, *Chem. Eng.*, **79**(24):128 (1972).

Watkins, J. P., Controlling Sulfur Compounds in Wastewaters, *Chem. Eng.*, **84**(22):61 (1977).

Watson, I. C., W. R. T. Oakes, Jr., and E. F. Miller, Desalting Inland Brackish Water by Ion Exchange, *1970 Trans. Am. Assoc. Cost Eng.*, (1970), p. 14.

Welder, B. Q., Water Must Be Managed, *Hydrocarbon Process.*, **46**(7):103 (1967).

Wirth, F., Recover MA from PA Scrubber Water, *Hydrocarbon Process.*, **54**(8):103 (1975).

Wiseman, S., and B. T. Bawden, Removing Refractory Compounds from Waste Water, *Chem. Eng. Progr.*, **73**(5):60 (1977).

Zanitsch, R. H., and R. T. Lynch, Selecting a Thermal Regeneration System for Activated Carbon, *Chem. Eng.*, **85**(1):95 (1978).

Zogorski, J. S., and S. D. Faust, Removing Phenols via Activated Carbon, *Chem. Eng. Progr.*, **73**(5):65 (1977).

How Much and What's in HPI Waste Water Streams, *Hydrocarbon Process.*, **46**(7):109 (1967).

A Plant Engineering Roundtable, Water Pollution, *Plant Eng.*, **21**(10):120 (1967).

Curing Chromate Pollution, *Chem. Eng.*, **77**(4):124 (1970).

Basic Techniques, Section II Water Pollution Control, *Chem. Eng.*, **77**(9):63 (1970).

Water Pollution Control, *Chem. Eng.*, **78**(14):65 (1971).

Direct Oxygenation of Wastewater, *Chem. Eng.*, **78**(27):66 (1971).

Get Zero Discharge with Brine Concentration, *Hydrocarbon Process.*, **52**(10):77 (1973).

PROBLEMS

1 A process for preparing acetaldehyde is by direct oxidation of ethylene. (This process is described in Prob. 8 of Chap. 2.) Completely analyze the various factors which should be considered in choosing a plant site for this process. With this information, outline possible geographical locations for the plant noting the advantages and disadvantages of each site.

2 The trend in the fertilizer industry during the past few years has been to go to larger and larger fertilizer plants. In terms of plant location, what are the more important factors that should be considered and which factors become even more important as the size of the plant is increased? Are these factors of equal importance regardless of the type of fertilizer produced? Analyze this situation for ammonia, urea, and phosphate fertilizer process plants.

3 Make an analysis of various means of transportation available at Chicago, Ill., Houston, Tex., and Denver, Colo. Use this analysis to determine what form of transportation should be recommended from these three locations for the shipment of the following finished chemical products:

Quantity	Frequency	Distance of shipment
100 lb solid	once a month	100 miles
50,000 lb solid	once a month	100 miles
50,000 gal liquid	once a week	100 miles
50,000 gal liquid	once a week	500 miles
50,000 gal liquid	daily	500-mile radius

Since destination of the chemical product can greatly affect the mode of transportation recommended, consider various destinations in this analysis, e.g., coastal location, river location, mountain location, etc. What approximate cost would have to be added to each shipment of finished chemical products as shipping charges in the above analysis?

4 Prepare a plant layout complete with elevation drawings for the nitric acid unit shown in Figs. 2-1 and 2-2. The unit operates at a pressure of 120 psig. Use the following sizes: absorption tower, 40 ft high by 5 ft in diameter; cooler condensers, each 12 ft long by 2 ft in diameter; converter, 3 ft long by 2 ft in diameter; mixing chamber, 5 ft long by 1 ft in diameter; preoxidation chamber, 8 ft long by 3 ft in diameter; preheater, 10 ft long by 2 ft in diameter; reheater, 16 ft long by 2 ft in diameter; all other pieces of equipment except storage tanks are approximately 3 by 3 by 3 ft. Indicate location of panel boards, instruments, and control valves. Preheat the entering air by using some of the heat from the gases leaving the converter. Indicate how the gases leaving the absorption tower might be used for power recovery.

5 Develop a complete plant layout for the sodium dodecylbenzene sulfonate plant that is described in Chap. 2. Indicate location of instruments, control valves, and panel boards. Provide possibilities for future expansion or revision of the plant.

6 A synthesis gas process is described in Probs. 13 through 18 of Chap. 2. Prepare a plant layout for a production of 25 MM scf/day which can use either air or 95 percent purity oxygen as the oxidant in this process.

7 Completely instrument the synthesis gas plant considered in the previous problem.

8 Determine the utilities requirement for the nitric acid unit shown in Figs. 2-1 and 2-2. Note that the exit gas from the absorption tower is to be used for power recovery.

9 Outline and present solutions to the materials handling, waste disposal, and safety problems that are encountered in the nitric acid unit shown in Figs. 2-1 and 2-2.

10 Figure 2-5 shows a simplified equipment diagram for the manufacture of sodium dodecylbenzene sulfonate. Develop acceptable schemes of handling the alkylaryl hydrocarbon effluent from the bottom of the dodecylbenzene fractionator and the $AlCl_3$ sludge from the alkylator.

11 Provide a design for handling the spent acid which is removed from the settler after the sulfonation process in the sodium dodecylbenzene sulfonate process. The final waste effluent must meet the waste-disposal standards of your area.

12 The following information has been obtained during a test for the BOD of a given industrial waste: Fifteen ml of the waste sample was added to a 500-ml BOD bottle, and the bottle was then filled with standard dilution water seeded with sewage organisms. A second 500-ml BOD bottle was filled with the standard dilution water. After 5 days at 20°C, the blank and the diluted waste sample were analyzed for dissolved oxygen content by the sodium azide modification of the Winkler method. The blank results indicated a dissolved-oxygen content of 9.0 ppm. Results for the diluted sample showed a dissolved-oxygen content of 4.0 ppm. On the basis of the following assumptions, determine the BOD for the waste: The specific gravities of the liquids are 1.0; the waste sample contains no dissolved oxygen.

13 List all the areas in the sodium dodecylbenzene sulfonate process that could be potential danger areas with respect to health and safety of operating personnel. What precautions in each instance should be taken to minimize the hazard?

14 List the major fields which should be considered for plant safety, and discuss the responsibilities of the design engineer in each one.

15 Obtain a copy of a patent dealing with a chemical process, and outline the claims presented in this patent. Indicate the method used for presenting the specifications, and explain how the inventor presented proof of his claims.

FOUR

COST AND ASSET ACCOUNTING

The design engineer, by analyses of costs and profits, attempts to predict whether capital should be invested in a particular project. After the investment is made, records must be maintained to check on the actual financial results. These records are kept and interpreted by accountants. The design engineer, of course, hopes that the original predictions will agree with the facts reported by the accountant. There is little chance for agreement, however, if both parties do not consider the same cost factors, and comparison of the results is simplified if the same terminology is used by the engineer and the accountant.

The purpose of accounting is to record and analyze any financial transactions that have an influence on the utility of capital. Accounts of expenses, income, assets, liabilities, and similar items are maintained. These records can be of considerable value to the engineer, since they indicate where errors were made in past estimates and give information that can be used in future evaluations. Thus, the reason why the design engineer should be acquainted with accounting procedures is obvious. Although it is not necessary to know all the details involved in accounting, a knowledge of the basic principles as applied in economic evaluations is an invaluable aid to the engineer.

This chapter presents a survey of the accounting procedures usually encountered in industrial operations. Its purpose is to give an understanding of the terminology, basic methods, and manner of recording and presenting information as employed by industrial accountants.

OUTLINE OF ACCOUNTING PROCEDURE

The diagram in Fig. 4-1 shows the standard accounting procedure, starting with the recording of the original business transactions and proceeding to the final preparation of summarizing balance sheets and income statements. As the day-by-day business transactions occur, they are recorded in the *journal*. A single journal may be used for all entries in small businesses, but large concerns ordinarily use several types of journals, such as cash, sales, purchase, and general journals.

The next step is to assemble the journal entries under appropriate account headings in the *ledger*. The process of transferring the daily journal entries to the ledger is called *posting*.

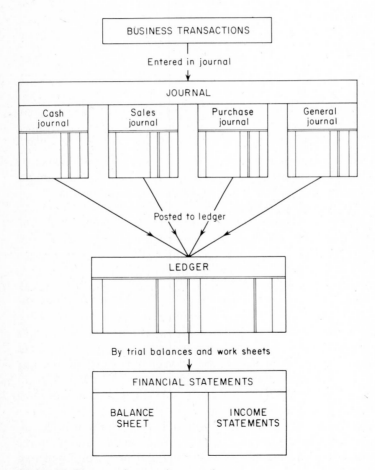

Figure 4-1 Diagram of accounting procedure.

Statements showing the financial condition of the business concern are prepared periodically from the ledger accounts. These statements are presented in the form of *balance sheets* and *income statements*. The balance sheet shows the financial condition of the business at a particular time, while the income statement is a record of the financial gain or loss of the organization over a given period of time.

BASIC RELATIONSHIPS IN ACCOUNTING

In the broadest sense, an *asset* may be defined as anything of value, such as cash, land, equipment, raw materials, finished products, or any type of property. At any given instant, a business concern has a certain monetary value because of its *assets*. At the same instant, many different persons may have a just claim, or *equity*, to ownership of the concern's assets. Certainly, any creditors would have a just claim to partial ownership, and the owners of the business should have some claim to ownership. Under these conditions, a fundamental relationship in accounting can be written as

$$\text{Assets} = \text{equities} \tag{1}$$

Equities can be divided into two general classes as follows: (1) *Proprietorship*—the claims of the concern or person who owns the asset; and (2) *liabilities*—the claims of anyone other than the owner. The term *proprietorship* is often referred to as *net worth* or simply as *ownership* or *capital*. Thus, Eq. (1) can be written as†

$$\text{Assets} = \text{liabilities} + \text{proprietorship} \tag{2}$$

The meaning of this basic equation can be illustrated by the following simple example. Five students have gone together and purchased a secondhand automobile worth $1000. Because they did not have the necessary $1000 they borrowed $400 from one of their parents. Therefore, as far as the students are concerned, the value of their asset is $1000, their proprietorship is $600, and their liability is $400.

Equation (2) is the basis for balancing assets against equities at any given instant. A similar equation can be presented for balancing costs and profits over any given time period. The total income must be equal to the sum of all costs and profits, or

$$\text{Total income} = \text{costs} + \text{profits} \tag{3}$$

Any engineering accounting study can ultimately be reduced to one of the forms represented by Eq. (2) or (3).

---

† Equation (2) is sometimes expressed simply as "Assets = liabilities," where "liabilities" is synonymous with "equities."

THE BALANCE SHEET

A balance sheet for an industrial concern is based on Eq. (1) or (2) and shows the financial condition at any given date. The amount of detail included varies depending on the purpose. Consolidated balance sheets based on the last day of

ABC PETROLEUM COMPANY
CONSOLIDATED BALANCE SHEET, DECEMBER 31, 1978

Assets		*Liabilities and stockholders' equity*	
Current assets		**Current liabilities**	
Cash	$ 93,879,000	Accounts payable	$160,307,000
Notes and accounts		Long-term debt (due	
receivable	219,374,000	within one year)	3,514,000
Inventories:		Accrued taxes	58,938,000
Crude oil, petroleum		Other accruals	20,796,000
products, and		Total current	
merchandise	152,021,000	liabilities	$243,555,000
Materials and		**Long-term debt**	$333,738,000
supplies (at cost and		**Deferred credits**	
condition value)	25,524,000	Federal income taxes	$ 86,845,000
Total current		Other	45,579,000
assets	$490,798,000	Total deferred	
Investments and long-		credits	$132,424,000
term receivables		**Reserve for**	
(at cost)	$309,249,000	**contingencies**	$ 24,197,000
Properties, plants, and		**Stockholders' equity**	
equipment (at cost)		Common stock, $5 par	
Production	$1,390,688,000	value	
Manufacturing	669,915,000	Shares authorized,	
Transportation	195,255,000	50,000,000	
Marketing	188,405,000	Shares issued,	
Other	79,823,000	34,465,956	$ 172,330,000
	$2,524,086,000	Capital in excess of	
Less reserves for		par value of	
depreciation,		common stock	264,238,000
depletion, and		Earnings employed	
amortization	1,315,416,000	in the business	918,165,000
Total	$1,208,670,000		$1,354,733,000
Prepaid and deferred		Less treasury stock	
charges	$ 20,347,000	(at cost), 1,096,627	
TOTAL ASSETS	$2,029,064,000	shares	59,583,000
		Total stock-	
		holders' equity	$1,295,150,000
		TOTAL	
		LIABILITIES	
		AND STOCK-	
		HOLDERS'	
		EQUITY	$2,029,064,000

Figure 4-2 A consolidated balance sheet.

the fiscal year are included in the annual report of a corporation. These reports are intended for distribution to stockholders, and the balance sheets present the pertinent information without listing each individual asset and equity in detail.

Assets are commonly divided into the classifications of current, fixed, and miscellaneous. *Current assets*, in principle, represent capital which can readily be converted into cash. Examples would be accounts receivable, inventories, cash, and marketable securities. These are *liquid assets*. On the other hand, *fixed assets*, such as land, buildings, and equipment, cannot be converted into immediate cash. Deferred charges, other investments, notes and accounts due after 1 year, and similar items are ordinarily listed as miscellaneous assets under separate headings.

Modern balance sheets often use the general term *liabilities* in place of *equities*. *Current liabilities* are grouped together and include all liabilities such as accounts payable, debts, and tax accruals due within 12 months of the balance-sheet date. The *net working capital* of a company can be obtained directly from the balance sheet as the difference between current assets and current liabilities. Other liabilities, such as long-term debts, deferred credits, and reserves are listed under separate headings. Proprietorship, stockholders' equity, or capital stock and surplus complete the record on the equity (or liability) side of the balance sheet.

Consolidated balance sheets are ordinarily presented with assets listed on the left and liabilities, including proprietorship, listed on the right. As indicated in Eq. (1), the total value of the assets must equal the total value of the equities. A typical balance sheet of this type is presented in Fig. 4-2.

The value of property items, such as land, buildings, and equipment, is usually reported as the value of the asset at the time of purchase. Depreciation reserves are also indicated, and the difference between the original property cost and the depreciation reserve represents the book value of the property. Thus, in depreciation accounting, separate records showing accumulation in the depreciation reserve must be maintained. In the customary account, *reserve for depreciation* is not actually a separate fund but is merely a bookkeeping method for recording the decline in property value.

The ratio of total current assets to total current liabilities is called the *current ratio*. The ratio of immediately available cash (i.e., cash plus U.S. Government and other marketable securities) to total current liabilities is known as the *cash ratio*. The current and cash ratios are valuable for determining the ability to meet financial obligations, and these ratios are examined carefully by banks or other loan concerns before credit is extended. From the data presented in Fig. 4-2, the current ratio for the company on December 31, 1978, was

$$\frac{\$490,798,000}{\$243,555,000} = 2.02$$

and the cash ratio was $93,879,000/$243,555,000 = 0.386.

THE INCOME STATEMENT

A balance sheet applies only at one specific time, and any additional transactions cause it to become obsolete. Most of the changes that occur in the balance sheet are due to revenue received from the sale of goods or services and costs incurred in the production and sale of the goods or services. Income-sheet accounts of all income and expense items, such as sales, purchases, depreciation, wages, salaries, taxes, and insurance, are maintained, and these accounts are summarized periodically in *income statements.*

A consolidated income statement is based on a given time period. It indicates surplus capital and shows the relationship among total income, costs, and profits over the time interval. The transactions presented in income-sheet accounts and income statements, therefore, are of particular interest to the

---

XYZ CHEMICAL COMPANY AND CONSOLIDATED SUBSIDIARIES
CONSOLIDATED INCOME STATEMENT FOR THE YEAR
ENDING DECEMBER 31, 1978

Income	
Net sales	$341,822,557
Dividends from subsidiary and associated companies	798,483
Other	2,534,202
Total (or gross) income	$345,155,242
Deductions	
Cost of goods sold	$243,057,056
Selling and administrative expenses	42,167,634
Research expenses	10,651,217
Provision for employees' bonus	649,319
Interest expenses	3,323,372
Net income applicable to minority interests	143,440
Other	2,608,694
Total deductions	$302,600,732
Income before provision for income taxes	$ 42,554,510
Less provision for income taxes	18,854,000
Net income (net profit)	$ 23,700,510
Earned surplus at beginning of year	90,436,909
Total surplus	$114,137,419
Surplus deductions	
Preferred dividends ($3.85 per share)	$ 721,875
Common dividends ($2.50 per share)	13,148,300
Total surplus deductions	$ 13,870,175
EARNED SURPLUS AT END OF YEAR	$100,267,244

Figure 4-3 A consolidated income statement.

engineer, since they represent the facts which were originally predicted through cost and profit analyses.

The terms *gross income* or *gross revenue* used by accountants refer to the total amount of capital received as a result of the sale of goods or service. *Net income* or *net revenue* is the total profit remaining after deducting all costs, including taxes.

Figure 4-3 is a typical example of a consolidated income statement based on a time period of 1 year. As indicated by Eq. (3), the total income ($345,155,242) equals the total cost ($302,600,732 + $18,854,000) plus the net income or profit ($23,700,510).

The role of interest on borrowed capital is clearly indicated in Fig. 4-3. Since the accountant considers interest as an expense arising from the particular method of financing, the cost due to interest is listed as a separate expense.

MAINTAINING ACCOUNTING RECORDS

Balance sheets and income statements are summarizing records showing the important relationships among assets, liabilities, income, and costs at one particular time or over a period of time. Some method must be used for recording the day-to-day events. This is accomplished by the use of *journals* and *ledgers*.

A journal is a book, group of vouchers, or some other convenient substitute in which the original record of a transaction is listed, while a ledger is a group of accounts giving condensed and classified information from the journal.

Debits and Credits

When recording business transactions, a *debit* entry represents an addition to an account, while a *credit* entry represents a deduction from an account. In more precise terms, a debit entry is one which increases the assets or decreases the equities, and a credit entry is one which decreases the assets or increases the equities.

Since bookkeeping accounts must always show a balance between assets and equities, any single transaction must affect both assets and equities. Each debit entry, therefore, requires an equal and offsetting credit entry. For example, if a company purchased a piece of equipment by a cash payment, the assets of the company would be increased by the value of the equipment. This represents an addition to the account, and would therefore, be listed as a debit. However, the company had to pay out cash to obtain the equipment. This payment must be recorded as a credit entry, since it represents a deduction from the account. At least one debit entry and one credit entry must be made for each business transaction in order to maintain the correct balance between assets and equities. This is known as *double-entry bookkeeping*.

JOURNAL

Date 19__		Analysis	F	Debit	Credit
					Page 1
April	3	Salaries	403	$ 758|00	$
		Cash	112		758|00
		Payment of salaries for week ending April 3			
	4	Rent	314	300|00	
		Cash	112		300|00
		Building rental for month of April			
	5	Cash	112	1041|00	
		Sales	201		1041|00
		Product *A* to X Company as per invoice No. 6839			
	8	Equipment	104	1800|00	
		Notes payable	521		1800|00
		Equipment for plant—6%, 90-day note to Y Company			
	9	Purchases	608	88|00	
		Z Company	842		88|00
		Tools on open account—Terms—30 days			

Figure 4-4 A typical journal page.

The Journal

A typical example of a journal page is shown in Fig. 4-4. The date is indicated in the first two columns. An analysis of the account affected by the particular transaction is listed in the third column, with debits listed first and credits listed below and offset to the right. A brief description of the item is included if necessary. The amounts of the individual debit and credit entries are indicated in the last two columns, with debits always shown on the left. When the journal entry is posted in the ledger, the number of the ledger account or the page number is entered in the fourth column of the journal page. This column is usually designated by "F" for "Folio."

The Ledger

Separate ledger accounts may be kept for various items, such as cash, equipment, accounts receivable, inventory, accounts payable, and manufacturing expense. A typical ledger sheet is shown in Fig. 4-5. The ledger sheets serve as a secondary record of business transactions and are used as the intermediates between journal records and balance sheets, income statements, and general cost records.

LEDGER

Cash   112

Date 19—		Analysis	F	Debit		Date 19—		Analysis	F	Credit	
April	1	Balance forward	J-1	$ 945	00	April	3	Salaries	J-1	$ 758	00
	5	Sales	J-1	1041	00		4	Rent	J-1	300	00
	10	Sales	J-2	86	00		10	Purchases	J-2	175	00
	11	R Company	J-2	700	00		12	Insurance	J-2	455	00
	11	Sales	J-2	550	00		13	Taxes	J-2	875	00
	12	Sales	J-2	94	00		17	Salaries	J-3	821	00
	18	S Company	J-3	1200	00		21	Purchases	J-3	985	00
	22	Sales	J-3	175	00		29	Office	J-3	158	00
	28	Sales	J-3	548	00			supplies			
	29	Sales	J-3	630	00		30	Purchases	J-3	154	00
	30	Sales	J-3	74	00						
		Total		$6043	00			Balance		1362	00
								Total		$6043	00
		Balance forward		$1362	00						

Figure 4-5 Typical ledger sheet that has been closed and balanced.

COST ACCOUNTING METHODS

In the simplest form, cost accounting is the determination and analysis of the cost of producing a product or rendering a service. This is exactly what the design engineer does when estimating costs for a particular plant or process, and cost estimation is one type of cost accounting.

Accountants in industrial plants maintain records on actual expenditures for labor, materials, power, etc., and the maintenance and interpretation of these records is known as *actual* or *post-mortem* cost accounting. From these data, it is possible to make accurate predictions of the future cost of the particular plant or process. These predictions are very valuable for determining future capital requirements and income, and represent an important type of cost accounting known as *standard* cost accounting. Deviations of standard costs from actual costs are designated as *variances*.

There are many different types of systems used for reporting costs, but all the systems employ some method for classifying the various expenses. One common type of classification corresponds to that presented in Chap. 5 (Cost Estimation). The total cost is divided into the basic groups of manufacturing costs and general expenses. These are further subdivided, with administrative, distribution, selling, financing, and research and development costs included under general expenses. Manufacturing costs include direct production costs, fixed charges, and plant overhead.

Each of the subdivided groups can be classified further as indicated in Chap. 5. For example, direct production costs can be broken down into costs for raw materials, labor, supervision, maintenance, supplies, power, utilities, laboratory charges, and royalties.

Each concern has its own method for distributing the costs on its accounts. In any case, all costs are entered in the appropriate journal account, posted in the ledger, and ultimately reported in a final cost sheet or cost statement.

Accumulation, Inventory, and Cost-of-Sales Accounts

In general, basic cost-accounting methods require posting of all costs in so-called *accumulation accounts*. There may be a series of such accounts to handle the various costs for each product. At the end of a given time period, such as one month, the accumulated costs are transferred to *inventory* accounts, which give a summary of all expenditures during the particular time interval. The amounts of all materials produced or consumed are also shown in the inventory accounts. The information in the inventory account is combined with data on the amount of product sales and transferred to the *cost-of-sales* account. The cost-of-sales accounts give the information necessary for determining the profit or loss for each product sold during the given time interval. One type of inventory account is shown in Fig. 4-6, and a sample cost-of-sales account is presented in Fig. 4-7.

When several products or by-products are produced by the same plant, allocation of the cost to each product must be made on some predetermined basis.

MANUFACTURING COST WORKS INVENTORY

Refining of crude product D For month of January, 19___

Cost element Name	Unit	Units on hand Start of month	End of month	Units used per unit pro-duced	Total units used or pro-duced	Price per unit	Total cost	Cost per unit pro-duced
Crude product D	Gallons	13,000	11,000	1.5000	150,000	$0.2787	$41,800	$0.4180
Operating wages	Hours			0.0150	1,500	1.2500	1,975	0.0198
Operating supplies							890	0.0089
Maintenance wages	Hours			0.0250	2,500	2.0000	5,000	0.0500
Maintenance materials							10,500	0.1050
Utilities							8,000	0.0800
Depreciation	$ investment			5.0000	500,000	0.0100	5,000	0.0500
Overhead							3,800	0.0380
Total cost and production	Gallons	5,000	4,000		100,000	$0.7697	$76,965	$0.7697

Figure 4-6. Example of one type of inventory account.

COST-OF-SALES ACCOUNT

Product E   *For month of June, 19___*

Item	This month	Last month	Year to date
Sales, lb	475,000	590,000	3,220,000
	$/unit	$/unit	$/unit
Selling price	$0.200	$0.200	$0.200
Cost of sales:			
Manufacturing cost	0.120	0.100	0.105
Freight and delivery	0.007	0.008	0.007
Selling expense	0.018	0.020	0.016
Administrative expense	0.025	0.020	0.022
Research expense	0.010	0.008	0.008
Total cost of sales	$0.180	$0.156	$0.158
Profits before taxes	$0.020	$0.044	$0.042

Figure 4-7 Example of cost-of-sales account.

Although the allocation of raw-material and direct labor costs can be determined directly, the exact distribution of overhead costs may become quite complex, and the final method depends on the policies of the particular concern involved.

Materials Costs

The variation in costs due to price fluctuations can cause considerable difficulty in making the transfer from accumulation accounts to inventory and cost-of-sales accounts. For example, suppose an accumulation account showed the following;

ACCUMULATION ACCOUNT
Item: Chemical A for use in producing product X

Date 19___	Received	Cost	Balance on hand	Delivered for use in process
May 2	5,000 lb	$0.0360/lb	5,000 lb	
May 15	10,000 lb	$0.0390/lb	15,000 lb	
May 17			9,000 lb	6,000 lb

In transferring the cost of chemical A to the inventory and cost-of-sales accounts, there is a question as to what price applies for chemical A. There are three basic methods for handling problems of this type.

1. *The current-average method.* The average price of all the inventory on hand at the time of delivery or use is employed in this method. In the preceding example, the current-average price for chemical *A* would be $0.0380 per pound.
2. *The first-in first-out* (or *fifo*) *method.* This method assumes the oldest material is always used first. The price for the 6000 lb of chemical *A* would be $0.0360 per pound for the first 5000 lb and $0.0390 for the remaining 1000 lb.
3. *The last-in first-out* (or *lifo*) *method.* With this method, the most recent prices are always used. The price for the 6000 lb of chemical *A* would be transferred as $0.0390 per pound.

Any of these methods can be used. The current-average method presents the best picture of the true cost during the given time interval, but it may be misleading if used for predicting future costs.

The information presented in this chapter shows the general principles and fundamentals of accounting which are of direct interest to the engineer. However, the many aspects of accounting make it impossible to present a complete coverage of all details and systems in one chapter or even in one book.† The exact methods used in different businesses may vary widely depending on the purpose and the policies of the organization, but the basic principles are the same in all cases.

SUGGESTED ADDITIONAL REFERENCES FOR COST AND ASSET ACCOUNTING

Cost accounting

Abbott, J. T., R. R. Janssen, and C. M. Merz, Finance for Engineers, *Chem. Eng.*, **80**(15): 108 (1973).

Allison, J. B., G. D. Rucker, and R. W. Dorsey, Project Cash Flow—Refinancing Existing Plants, *Hydrocarbon Process.*, **57**(3):81 (1978).

Barish, N. N., and S. Kaplan, "Economic Analysis for Engineering and Managerial Decision Making," 2d ed., McGraw-Hill Book Company, New York, 1978.

Cason, R. L., Go Broke While Showing a Profit? *Hydrocarbon Process.*, **54**(7): 179 (1975).

Clark, F. D., and A. B. Lorenzoni, "Applied Cost Engineering," Marcel Dekker, Inc., New York (1978).

Clark, F. D., How to Assess Inflation of Plant Costs, *Chem. Eng.*, **82**(14): 70 (1975).

Clark, W. G., Project Cash Flow, Concept to Completion, *Hydrocarbon Process.*, **57**(3):87 (1978).

Dixon, S., Short-term Planning—The Accountant's Contribution, *Chem. Eng.* (*London*), **258**:153 (April 1971).

Eastman, N. S., Cost-Improvement Techniques Spur Widespread Savings, *Chem. Eng.*, **80**(28):102 (1973).

Forbes, M. C., Cost Accounting for Pollution Control, *Hydrocarbon Process.*, **48**(10):145 (1969).

Hanna, J. H., Control the Cost of New Plants, *Hydrocarbon Process.*, **53**(7):183 (1974).

† C. T. Horngren, "Cost Accounting—A Managerial Emphasis," 4th ed., Prentice-Hall, Inc., Englewood Cliffs, N.J., 1977; W. E. Thomas, "Readings in Cost Accounting, Budgeting, and Control," 5th ed., South-Western Publishing Company, Cincinnati, Ohio, 1978.

Holland, F. A., F. A. Watson, and J. K. Wilkinson, How to Allocate Overhead Cost and Appraise Inventory, *Chem. Eng.*, **81**(12):83 (1974).

—— —— and ——, How to Assess Your Company's Progress, *Chem. Eng.*, **81**(19):119 (1974).

—— —— and ——, Principles of Accounting, *Chem. Eng.*, **81**(14):93 (1974).

Jelen, F. C., "Cost and Optimization Engineering," McGraw-Hill Book Company, New York, 1970.

Jelen, F. C., and C. L. Yaws, Project Cash Flow—Description and Interpretation, *Hydrocarbon Process.*, **57**(3):77 (1978).

Krause, W. A., Cost Control: A Contractor's Viewpoint, *Chem. Eng. Progr.*, **64**(12):15 (1968).

Newsome, R. C., Gas Pipeline Cost Accounting Practices, presented at 12th National Meeting American Association Cost Engineers, Houston, Texas, June 17–19, 1968.

Nitchie, E. B., Accounting Data and Methods Help Control Costs and Evaluate Profits, *Chem. Eng.*, **74**(1):87 (1967).

——, Modern Accounting Methods Supply Vital Cost Information, *Chem. Eng.*, **74**(2):165 (1967). (1967).

Pekar, P. P., Jr., Project Control: A Case for a New Evolutionary Time Series Prediction, *AACE Bull.*, **18**(6):195 (1976).

Poulton, C., and P. Zakaib, Construction Administration in a Medium-Sized Industrial Company, *1973 Trans. Am. Assoc. Cost Eng.*, (1973), p. 93.

Riggs, J. L., "Engineering Economics," McGraw-Hill Book Company, New York, 1977.

Seeney, D., Project Planning and Control, *AACE Bull.*, **16**(5):135 (1974).

Tarquin, A. J., and L. Blank, "Engineering Economy: A Behavioral Approach," McGraw-Hill Book Company, New York, 1976.

Wei, J., T. W. F. Russell, and M. W. Schwartzlander, "The Structure of the Chemical Processing Industries: Function and Economics," McGraw-Hill Book Company, New York, 1978.

Capital Budgets Keep Expanding, *Petro/Chem. Eng.*, **40**(5):PM7 (1968).

PROBLEMS

1. Prepare a balance sheet applicable at the date when the X Corporation had the following assets and equities:

Cash	$20,000	Common stock sold	$50,000
Accounts payable:		Machinery and equipment	
B Company	2,000	(at present value)	18,000
C Corporation	8,000	Furniture and fixtures (at	
Accounts receivable	6,000	present value)	5,000
Inventories	15,000	Government bonds	3,000
Mortgage payable	5,000	Surplus	2,000

2. During the month of October, the following information was obtained in an antifreeze retailing concern:

Salaries	$ 3,000
Delivery expenses	700
Rent	400
Sales	15,100
Antifreeze available for sale during October (at cost)	20,200
Antifreeze inventory on Oct. 31 (at cost)	11,600
Other expenses	1,200
Earned surplus before income taxes as of Sept. 30	800

Prepare an income statement for the month of October giving as much detail as possible.

3. The following information applies to E Company on a given date:

Long-term debts	$ 1,600
Debts due within 1 year	1,000
Accounts payable	2,300
Machinery and equipment (at cost)	10,000
Cash in bank	3,100
Prepaid rent	300
Government bonds	3,000
Social security taxes payable	240
Reserve for depreciation	600
Reserve for expansion	1,200
Inventory	1,600
Accounts receivable	1,700

Determine the current ratio, cash ratio, and working capital for Company E at the given date.

4. On Aug. 1, a concern had 10,000 lb of raw material on hand which was purchased at a cost of $0.030 per pound. In order to build up the reserve, 8000 lb of additional raw material was purchased on Aug. 15 at a cost of $0.028 per pound, and 2 days later 6000 lb was purchased from another supplier at $0.031 per pound. If none of the raw material was used until after the last purchase, determine the total cost of 12,000 lb of the raw material on an inventory or cost-of-sales account for the month of August by (*a*) the current-average method, (*b*) the "fifo" method, and (*c*) the "lifo" method.

5. Prepare a complete list, with sample form sheets, of all cost-accounting records which should be maintained in a large plant producing ammonia, nitric acid, and ammonium nitrate. Explain how these records are used in recording, summarizing, and interpreting costs and profits.

CHAPTER FIVE

COST ESTIMATION

An acceptable plant design must present a process that is capable of operating under conditions which will yield a profit. Since net profit equals total income minus *all* expenses, it is essential that the chemical engineer be aware of the many different types of costs involved in manufacturing processes. Capital must be allocated for direct plant expenses, such as those for raw materials, labor, and equipment. Besides direct expenses, many other indirect expenses are incurred, and these must be included if a complete analysis of the total cost is to be obtained. Some examples of these indirect expenses are administrative salaries, product-distribution costs, and costs for interplant communications.

A capital investment is required for any industrial process, and determination of the necessary investment is an important part of a plant-design project. The total investment for any process consists of fixed-capital investment for physical equipment and facilities in the plant plus working capital which must be available to pay salaries, keep raw materials and products on hand, and handle other special items requiring a direct cash outlay. Thus, in an analysis of costs in industrial processes, capital-investment costs, manufacturing costs, and general expenses including income taxes must be taken into consideration.

CASH FLOW FOR INDUSTRIAL OPERATIONS

Figure 5-1 shows the concept of cash flow for an overall industrial operation based on a support system serving as the source of capital or the sink for capital receipts. Input to the capital sink can be in the form of loans, stock issues, bond

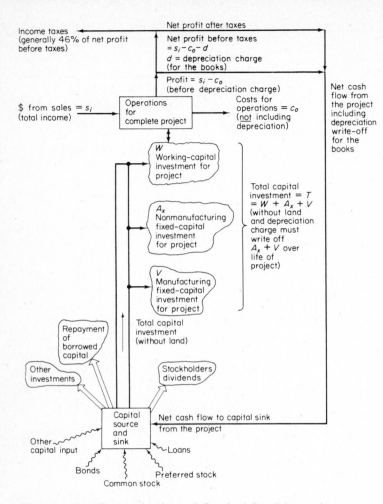

Figure 5-1 Tree diagram showing cash flow for industrial operations.

releases, and other funding sources including the net cash flow returned to the capital sink from each project. Output from the capital source is in the form of total capital investments for each of the company's industrial operations, dividends to stockholders, repayment of debts, and other investments.

The tree-growth concept, as shown in Fig. 5-1, depicts a trunk output to start the particular industrial operation designated as the total capital investment. This total capital investment includes all the funds necessary to get the project underway. This encompasses the regular manufacturing fixed-capital investment and the working-capital investment along with the investment required for all necessary auxiliaries and nonmanufacturing facilities. The cash flow for the capital investments can usually be considered as in a lump sum or in-an-instant such as for the purchase of land with a lump-sum payment or the provision of working capital as one lump sum at the start of the operation of the completed plant. Fixed capital for equipment ideally can be considered as in-an-instant for each piece of equipment although the payments, of course, can be spread over the entire construction period when considering the fixed-capital investment for a complete plant. Because income from sales and necessary operating costs can occur on an irregular time basis, a constant reservoir of working capital must be kept on hand continuously to draw from or add to as needed.

The rectangular box in Fig. 5-1 represents the overall operations for the complete project with working-capital funds moving in and out as needed but maintaining a constant fund as available working capital. Cash flows into the operations box as total dollars of income (s_i) from all sales while actual costs for the operations, such as for raw materials and labor, are shown as outflow costs (c_o). These cash flows for income and operating expenses can be considered as continuous and represent rates of flow at a given point in time using the same time basis, such as dollars per day or dollars per year. Because depreciation charges to allow eventual replacement of the equipment are in effect costs which are paid back to the company capital sink, these charges are not included in the costs for operations shown in Fig. 5-1. The difference between the income (s_i) and operating costs (c_o) represents gross profits before depreciation or income-tax charges ($s_i - c_o$) and is represented by the vertical line rising out of the operations box.

Depreciation, of course, must be recognized as a cost before income-tax charges are made and before net profits are reported to the stockholders. Consequently, removal of depreciation in the cash-flow diagram as a charge against profit is accomplished at the top of the tree diagram in Fig. 5-1 with the depreciation charge (d) entering the cash-flow stream for return to the capital sink. The resulting new profit of $s_i - c_o - d$ is taxable, and the income-tax charge is shown as a cash-flow stream deducted at the top of the diagram. The remainder, or net profit after taxes, is now clear profit which can be returned to the capital sink, along with the depreciation charge, to be used for new investments, dividends, or repayment of present investment as indicated by the various trunks emanating from the capital source in Fig. 5-1.

Cumulative Cash Position

The cash-flow diagram shown in Fig. 5-1 represents the steady-state situation for cash flow with s_i, c_o, and d all based on the same time increment. Figure 5-2 is for the same type of cash flow for an industrial operation except that it depicts the situation over a given period of time as the *cumulative cash position*. The time period chosen is the estimated life period of the project, and the time value of money is neglected.

In the situation depicted in Fig. 5-2, land value is included as part of the total capital investment to show clearly the complete sequence of steps in the full life cycle for an industrial process. The zero point on the abscissa represents that

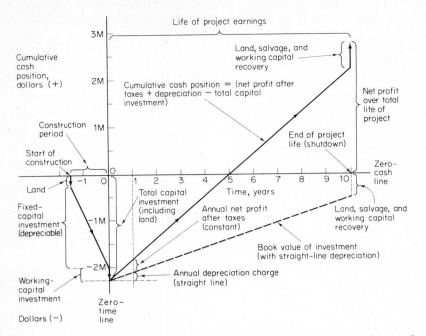

Figure 5-2 Graph of cumulative cash position showing effects of cash flow with time for an industrial operation neglecting time value of money.

time at which the plant has been completely constructed and is ready for operation. The total capital investment at the zero point in time includes land value, fixed-capital and auxiliaries investment, and working capital. The cash position is negative by an amount equivalent to the total capital investment at zero time, but profits in the ideal situation come in from the operation as soon as time is positive. Cash flow to the company, in the form of net profits after taxes and depreciation charges, starts to accumulate and gradually pays off the full capital investment. For the conditions shown in Fig. 5-2, the full capital investment is paid off in five years. After that time, profits accumulate on the positive side of the cumulative cash position until the end of the project life at which time the project theoretically is shut down and the operation ceases. At that time, the working capital is still available, and it is assumed that the land can still be sold at its original value. Thus, the final result of the cumulative cash position is a net profit over the total life of the project, or a cash flow into the company capital sink (in addition to the depreciation cash flow for investment payoff) over the ten-year period, as shown in the upper right-hand bracket in Fig. 5-2.

The relationships presented in Fig. 5-2 are very important for the understanding of the factors to be considered in cost estimation. To put emphasis on the basic nature of the role of cash flow, including depreciation charges, Fig. 5-2 has been simplified considerably by neglecting the time value of money and using straight-line relationships of constant annual profit and constant annual depreciation. In the chapters to follow, more complex cases will be considered in detail.

FACTORS AFFECTING INVESTMENT AND PRODUCTION COSTS

When a chemical engineer determines costs for any type of commercial process, these costs should be of sufficient accuracy to provide reliable decisions. To accomplish this, the engineer must have a complete understanding of the many factors that can affect costs. For example, many companies have reciprocal arrangements with other concerns whereby certain raw materials or types of equipment may be purchased at prices lower than the prevailing market prices. Therefore, if the chemical engineer bases the cost of the raw materials for the process on regular market prices, the result may be that the process is uneconomical. If the engineer had based the estimate on the actual prices the company would have to pay for the raw materials, the economic picture might have been altered completely. Thus the engineer must keep up to date on price fluctuations, company policies, governmental regulations, and other factors affecting costs.

Sources of Equipment

One of the major costs involved in any chemical process is for the equipment. In many cases, standard types of tanks, reactors, or other equipment are used, and a substantial reduction in cost can be made by employing idle equipment or by purchasing second-hand equipment. If new equipment must be bought, several independent quotations should be obtained from different manufacturers. When the specifications are given to the manufacturers, the chances for a low cost estimate are increased if the engineer does not place overly strict limitations on the design.

Price Fluctuations

In our modern economic society, prices may vary widely from one period to another, and this factor must be considered when the costs for an industrial process are determined. It would obviously be ridiculous to assume that plant operators or supervisors could be hired today at the same wage rate as in 1970. The same statement applies to comparing prices of equipment purchased at different times. The chemical engineer, therefore, must keep up-to-date on price and wage fluctuations. One of the most complete sources of information on existing price conditions is the *Monthly Labor Review* published by the U.S. Bureau of Labor Statistics. This publication gives up-to-date information on present prices and wages for different types of industries.

Company Policies

Policies of individual companies have a direct effect on costs. For example, some concerns have particularly strict safety regulations and these must be met in every detail. Accounting procedures and methods for determining depreciation costs vary among different companies. The company policies with reference to labor unions should be considered, because these will affect overtime labor charges and the type of work the operators or other employees can do. Labor-union policies may even dictate the amount of wiring and piping that can be done on a piece of equipment before it is brought into the plant, and, thus, have a direct effect on the total cost of installed equipment.

Operating Time and Rate of Production

One of the factors that has an important effect on costs is the fraction of the total available time during which the process is in operation. When equipment stands idle for an extended period of time, the labor costs are usually low; however, other costs, such as those for maintenance, protection, and depreciation, continue even though the equipment is not in active use.

Operating time, rate of production, and sales demand are closely inter-related. The ideal plant should operate under a time schedule which gives the maximum production rate while maintaining economic operating methods. In this way, the total cost per unit of production is kept near a minimum because the fixed costs are utilized to the fullest extent. This ideal method of operation is based on the assumption that the sales demand is sufficient to absorb all the material produced. If the production capacity of the process is greater than the sales demand, the operation can be carried on at reduced capacity or periodically at full capacity.

Figure 5-3 gives a graphical analysis of the effect on costs and profits when the rate of production varies. As indicated in this figure, the fixed costs remain constant and the total product cost increases as the rate of production increases. The point where the total product cost equals the total income is known as the *break-even point*. Under the conditions shown in Fig. 5-3, an ideal production rate for this chemical processing plant would be approximately 450,000 kg/month, because this represents the point of maximum net earnings.

The effects of production rate and operating time on costs should be recognized. By considering sales demand along with the capacity and operating characteristics of the equipment, the engineer can recommend the production rates and operating schedules that will give the best economic results.

Governmental Policies

The national government has many regulations and restrictions which have a direct effect on industrial costs. Some examples of these are import and export tariff regulations, restrictions on permissible depreciation rates, income-tax rules, and environmental regulations.

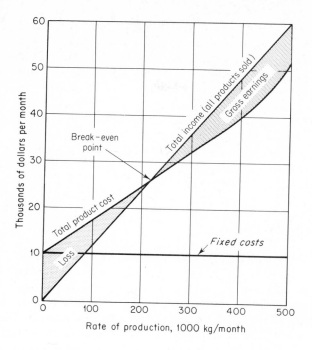

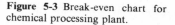

Figure 5-3 Break-even chart for chemical processing plant.

Prior to 1951, there were strict governmental regulations against rapid depreciation write-offs for industrial equipment. These restrictions increased the income-tax load for new companies during their first few years of existence and tended to discourage new enterprise. Therefore, during the Korean War, a fast amortization policy for certain defense installations was authorized. This policy permitted at least part of the value of the installation to be written off in 5 years as compared to an average of 10 to 15 years under the old laws. In 1954, a new law was passed permitting approximately two-thirds of the total investment for any process to be written off as depreciation during the first half of the useful life. A rapid write-off of this type is very desirable for some concerns because it reduces income taxes during the early years of the plant life.

Governmental policies with reference to capital gains and gross-earnings taxes should be understood when costs are determined. Suppose a concern decides to sell some valuable equipment before its useful life is over. The equipment has a certain asset or unamortized value, but the offered price may be more than the unamortized value. This profit over the unamortized value would have been taxable as long-term capital gain at 25 percent if it had been held for more than six months in the tax years of 1968 and earlier. However, changes in the tax laws were made in 1969 and again in 1976 so that long-term capital gains require a longer time (nine months in 1977 and one year after 1977). The rules were also changed relative to the simple 25 percent tax on long-term capital gains. Nevertheless, even after the 1977 rules went into effect, net long-term capital gains generally continued to result in less percentage tax than ordinary gains. Therefore, in the example referred to where a long-term capital gain would be realized by selling equipment, if a fast depreciation method had been used by the company, the capital gain would have been fairly large; however, this gain would probably be taxed at a low rate (perhaps as low as 28 percent), while the amount saved through the fast-depreciation allowances may have been at an income-tax rate of about 50 percent.†

The preceding examples illustrate why the chemical engineer should understand the effects of governmental regulations on costs. Each company has its own methods for meeting these regulations, but changes in the laws and alterations in the national and company economic situation require constant surveillance if optimum cost conditions are to be maintained.

CAPITAL INVESTMENTS

Before an industrial plant can be put into operation, a large sum of money must be supplied to purchase and install the necessary machinery and equipment. Land and service facilities must be obtained, and the plant must be erected complete with all piping, controls, and services. In addition, it is necessary to have money available for the payment of expenses involved in the plant operation.

The capital needed to supply the necessary manufacturing and plant facilities is called the *fixed-capital investment*, while that necessary for the operation of the plant is termed the *working capital*. The sum of the fixed-capital investment and the working capital is known as the *total capital investment*. The fixed-capital portion may be further subdivided into *manufacturing fixed-capital investment* and *nonmanufacturing fixed-capital investment*.

Fixed-Capital Investment

Manufacturing fixed-capital investment represents the capital necessary for the installed process equipment with all auxiliaries that are needed for complete process operation. Expenses for piping, instruments, insulation, foundations, and site preparation are typical examples of costs included in the manufacturing fixed-capital investment.

Fixed capital required for construction overhead and for all plant components that are not directly related to the process operation is designated as the nonmanufacturing fixed-capital investment. These plant components include the land, processing buildings, administrative and other offices, warehouses, laboratories, transportation, shipping, and receiving facilities, utility and waste-disposal facilities, shops, and other permanent parts of the plant. The construction overhead cost consists of field-office and supervision expenses, home-office expenses, engineering expenses, miscellaneous construction costs, contractor's fees, and

---

† For a discussion of income-tax rates, see Chap. 7 (Taxes and Insurance).

contingencies. In some cases, construction overhead is proportioned between manufacturing and nonmanufacturing fixed-capital investment.

Working Capital

The working capital for an industrial plant consists of the total amount of money invested in (1) raw materials and supplies carried in stock, (2) finished products in stock and semifinished products in the process of being manufactured, (3) accounts receivable, (4) cash kept on hand for monthly payment of operating expenses, such as salaries, wages, and raw-material purchases, (5) accounts payable, and (6) taxes payable.

The raw-materials inventory included in working capital usually amounts to a 1-month supply of the raw materials valued at delivered prices. Finished products in stock and semifinished products have a value approximately equal to the total manufacturing cost for 1 month's production. Because credit terms extended to customers are usually based on an allowable 30-day payment period, the working capital required for accounts receivable ordinarily amounts to the production cost for 1 month of operation.

The ratio of working capital to total capital investment varies with different companies, but most chemical plants use an initial working capital amounting to 10 to 20 percent of the total capital investment. This percentage may increase to as much as 50 percent or more for companies producing products of seasonal demand because of the large inventories which must be maintained for appreciable periods of time.

ESTIMATION OF CAPITAL INVESTMENT

Of the many factors which contribute to poor estimates of capital investments, the most significant one is usually traceable to sizable omissions of equipment, services, or auxiliary facilities rather than to gross errors in costing. A check list of items covering a new facility is an invaluable aid in making a complete estimation of the fixed-capital investment. Table 1 gives a typical list of these items.

Types of Capital Cost Estimates

An estimate of the capital investment for a process may vary from a predesign estimate based on little information except the size of the proposed project to a detailed estimate prepared from complete drawings and specifications. Between these two extremes of capital-investment estimates, there can be numerous other estimates which vary in accuracy depending upon the stage of development of the project. These estimates are called by a variety of names, but the following five categories represent the accuracy range and designation normally used for design purposes:

Table 1 Breakdown of fixed-capital investment items for a chemical process

Direct Costs

1. *Purchased equipment*
   All equipment listed on a complete flow sheet
   Spare parts and noninstalled equipment spares
   Surplus equipment, supplies, and equipment allowance
   Inflation cost allowance
   Freight charges
   Taxes, insurance, duties
   Allowance for modifications during startup

2. *Purchased-equipment installation*
   Installation of all equipment listed on complete flow sheet
   Structural supports, insulation, paint

3. *Instrumentation and controls*
   Purchase, installation, calibration, computer tie-in

4. *Piping*
   Process piping—carbon steel, alloy, cast iron, lead, lined, aluminum, copper, ceramic, plastic,
      rubber, reinforced concrete
   Pipe hangers, fittings, valves
   Insulation—piping, equipment

5. *Electrical equipment and materials*
   Electrical equipment—switches, motors, conduit, wire, fittings, feeders, grounding, instrument
      and control wiring, lighting, panels
   Electrical materials and labor

6. *Buildings (including services)*
   Process buildings—substructures, superstructures, platforms, supports, stairways, ladders, access
      ways, cranes, monorails, hoists, elevators
   Auxiliary buildings—administration and office, medical or dispensary, cafeteria, garage, product
      warehouse, parts warehouse, guard and safety, fire station, change house, personnel building,
      shipping office and platform, research laboratory, control laboratory
   Maintenance shops—electric, piping, sheet metal, machine, welding, carpentry, instrument
   Building services—plumbing, heating, ventilation, dust collection, air conditioning, building
      lighting, elevators, escalators, telephones, intercommunication systems, painting, sprinkler
      systems, fire alarm

7. *Yard improvements*
   Site development—site clearing, grading, roads, walkways, railroads, fences, parking areas,
      wharves and piers, recreational facilities, landscaping

8. *Service facilities*
   Utilities—steam, water, power, refrigeration, compressed air, fuel, waste disposal
   Facilities—boiler plant incinerator, wells, river intake, water treatment, cooling towers, water
      storage, electric substation, refrigeration plant, air plant, fuel storage, waste disposal plant,
      environmental controls, fire protection
   Nonprocess equipment—office furniture and equipment, cafeteria equipment, safety and
      medical equipment, shop equipment, automotive equipment, yard material-handling equip-
      ment, laboratory equipment, locker-room equipment, garage equipment, shelves, bins, pallets,
      hand trucks, housekeeping equipment, fire extinguishers, hoses, fire engines, loading stations
   Distribution and packaging—raw-material and product storage and handling equipment, product
      packaging equipment, blending facilities, loading stations

Table 1 Breakdown of fixed-capital investment items for a chemical process (*Continued*)

Direct Costs

9. *Land*
   Surveys and fees
   Property cost

Indirect costs

1. *Engineering and supervision*
   Engineering costs–administrative, process, design and general engineering, drafting, cost engineering, procuring, expediting, reproduction, communications, scale models, consultant fees, travel
   Engineering supervision and inspection

2. *Construction expenses*
   Construction, operation and maintenance of temporary facilities, offices, roads, parking lots, railroads, electrical, piping, communications, fencing
   Construction tools and equipment
   Construction supervision, accounting, timekeeping, purchasing, expediting
   Warehouse personnel and expense, guards
   Safety, medical, fringe benefits
   Permits, field tests, special licenses
   Taxes, insurance, interest

3. *Contractor's fee*

4. *Contingency*

---

1. Order-of-magnitude estimate (ratio estimate) based on similar previous cost data; probable accuracy of estimate over ±30 percent.
2. Study estimate (factored estimate) based on knowledge of major items of equipment; probable accuracy of estimate up to ±30 percent.
3. Preliminary estimate (budget authorization estimate; scope estimate) based on sufficient data to permit the estimate to be budgeted; probable accuracy of estimate within ±20 percent.
4. Definitive estimate (project control estimate) based on almost complete data but before completion of drawings and specifications; probable accuracy of estimate within ±10 percent.
5. Detailed estimate (contractor's estimate) based on complete engineering drawings, specifications, and site surveys; probable accuracy of estimate within ±5 percent.

Figure 5-4 shows the relationship between probable accuracy and quantity and quality of information available for the preparation of these five levels of estimates.† The approximate limits of error in this listing are plotted and show

† Adapted from a method presented by W. T. Nichols, *Ind. Eng. Chem.*, **43**(10):2295 (1951).

Error range %

40
30
20
10
0
10
20
30
40

Most probable cost

		Detailed estimate ±5% range	Definitive estimate ±10% range	Preliminary estimate ±20% range	Study estimate ±30% range	Order-of-magnitude estimate >±30% range
	Required information					
Site	Location	•	•	•	•	
	General description	•	•	•		
	Soil bearing	•	•			
	Location & dimensions R.R., roads, impounds, fences	•	•	•		
	Well-developed site plot plan & topographical map	•	•	•		
	Well-developed site facilities	•				
Process flow sheet	Rough sketches				•	
	Preliminary			•		
	Engineered	•	•			
Equipment list	Preliminary sizing & material specifications	•			•	•
	Engineered specifications	•	•			
	Vessel sheets	•	•			
	General arrangement					
	(a) Preliminary				•	
	(b) Engineered	•	•			
Building and structures	Approximate sizes & type of construction				•	•
	Foundation sketches			•	•	
	Architectural & construction	•	•	•		
	Preliminary structural design			•		
	General arrangements & elevations	•	•			
	Detailed drawings	•				
Utility requirements	Rough quantities (steam, water, electricity, etc.)					•
	Preliminary heat balance			•		
	Preliminary flow sheets			•		
	Engineered heat balance	•	•			
	Engineered flow sheets	•	•			
	Well-developed drawings	•				
Piping	Preliminary flow sheet & specifications				•	•
	Engineered flow sheet	•	•			
	Piping layouts & schedules	•				
Insulation	Rough specifications				•	
	Preliminary list of equipment & piping to be insulated		•			
	Insulation specifications & schedules	•	•			
	Well-developed drawings or specifications	•				
Instrumentation	Preliminary instrument list				•	
	Engineered list & flow sheet	•	•			
	Well-developed drawings	•				
Electrical	Preliminary motor list – approximate sizes				•	•
	Engineered list & sizes	•	•			
	Substations, number & sizes, specifications	•	•			
	Distribution specifications	•				
	Preliminary lighting specifications			•		
	Preliminary interlock, control, & instrument wiring specs.		•			
	Engineered single-line diagrams (power & light)	•	•			
	Well-developed drawings	•				
Man-hours	Engineering & drafting	•	•	•	•	
	Labor by craft	•				
	Supervision	•				
Project scope standard processes	Product, capacity, location & site requirements. Utility & service requirements. Building & auxiliary requirements. Row materials & finished product handling & storage requirements					•

Figure 5-4 Cost-estimating information guide.

Table 2 Typical average costs for making estimates (1979)†

Cost of project	Less than $2,000,000	$2,000,000 to $10,000,000	$10,000,000 to $100,000,000
Order-of-magnitude estimate	$ 2,000	$ 4,000	$ 8,000
Study estimate	12,000	25,000	35,000
Preliminary estimate	30,000	50,000	80,000
Definitive estimate	50,000	100,000	200,000
Detailed estimate	130,000	320,000	630,000

† Adapted from A. Pikulik and H.E. Diaz, Cost Estimating for Major Process Equipment, *Chem. Eng.*, 84(21): 106 (Oct. 10, 1977).

an envelope of variability. There is a large probability that the actual cost will be more than the estimated cost where information is incomplete or in time of rising-cost trends. For such estimates, the positive spread is likely to be wider than the negative, e.g., +40 and −20 percent for a study estimate. Table 2 illustrates the wide variation that can occur in the cost of making a capital-investment estimate depending on the type of estimate.

Predesign cost estimates (defined here as order-of-magnitude, study, and preliminary estimates) require much less detail than firm estimates such as the definitive or detailed estimate. However, the predesign estimates are extremely important for determining if a proposed project should be given further consideration and to compare alternative designs. For this reason, most of the information presented in this chapter is devoted to predesign estimates, although it should be understood that the distinction between predesign and firm estimates gradually disappears as more and more detail is included.

It should be noted that the predesign estimates may be used to provide a basis for requesting and obtaining a capital appropriation from company management. Later estimates, made during the progress of the job, may indicate that the project will cost more or less than the amount appropriated. Management is then asked to approve a *variance* which may be positive or negative.

COST INDEXES

Most cost data which are available for immediate use in a preliminary or predesign estimate are based on conditions at some time in the past. Because prices may change considerably with time due to changes in economic conditions, some method must be used for updating cost data applicable at a past date to costs that are representative of conditions at a later time.† This can be done by the use of cost indexes.

---

† See Chap. 10 for a discussion of the strategy to use in design estimates to consider the effects of inflation or deflation on costs and profits in the future.

A cost index is merely an index value for a given point in time showing the cost at that time relative to a certain base time. If the cost at some time in the past is known, the equivalent cost at the present time can be determined by multiplying the original cost by the ratio of the present index value to the index value applicable when the original cost was obtained.

$$\text{Present cost} = \text{original cost} \left(\frac{\text{index value at present time}}{\text{index value at time original cost was obtained}} \right)$$

Cost indexes can be used to give a general estimate, but no index can take into account all factors, such as special technological advancements or local conditions. The common indexes permit fairly accurate estimates if the time period involved is less than 10 years.

Many different types of cost indexes are published regularly.† Some of these can be used for estimating equipment costs; others apply specifically to labor, construction, materials, or other specialized fields. The most common of these indexes are the *Marshall and Swift all-industry and process-industry equipment indexes*, the *Engineering News-Record construction index*, the *Nelson refinery construction index*, and the *Chemical Engineering plant cost index*. Table 3 presents a list of values for various types of indexes over the past 15 years.

Marshall and Swift Equipment Cost Indexes‡

The Marshall and Swift (formerly known as Marshall and Stevens) equipment indexes are normally divided into two categories. The all-industry equipment index is simply the arithmetic average of the individual indexes for 47 different types of industrial, commercial, and housing equipment. The process-industry equipment index is a weighted average of eight of these, with the weighting based on the total product value of the various process industries. The percentages used for the weighting in a typical year are as follows: cement, 2; chemicals, 48; clay products, 2; glass, 3; paint, 5; paper, 10; petroleum, 22; and rubber, 8.

The Marshall and Swift indexes are based on an index value of 100 for the year 1926. These indexes take into consideration the cost of machinery and major equipment plus costs for installation, fixtures, tools, office furniture, and

---

† For a detailed summary of various cost indexes, see *Eng. News-Record*, **178**(11):87 (1967); and *Chem. Eng.*, **70**(4):143 (Feb. 18, 1963); **73**(9)184 (April 25, 1966); **76**(10):134 (May 5, 1969); **79**(25):168 (Nov. 13, 1972); **82**(9):117 (April 28, 1975). See also the list of suggested references at the end of this chapter.

‡ Values for the Marshall and Swift equipment-cost indexes are published in each issue of *Chemical Engineering*. For a complete description of these indexes, see R. W. Stevens, *Chem. Eng.*, **54**(11):124 (Nov., 1947). See also *Chem. Eng.*, **82**(9):117 (April 28, 1975) and **85**(11):189 (May 8, 1978).

Table 3 Cost indexes as annual averages

Year	Marshall and Swift installed-equipment indexes, 1926 = 100		Eng. News-Record construction index			Nelson refinery construction index, 1946 = 100	Chemical engineering plant cost index 1957–1959 = 100
	All-industry	Process-industry	1913 = 100	1949 = 100	1967 = 100		
1964	242	241	936	196	87	252	103
1965	245	244	971	204	91	261	104
1966	253	252	1019	214	95	273	107
1967	263	260	1070	224	100	287	110
1968	273	268	1155	242	108	304	114
1969	285	283	1269	266	119	329	119
1970	303	301	1385	290	129	365	126
1971	321	321	1581	331	148	406	132
1972	332	332	1753	368	164	439	137
1973	344	344	1895	397	177	468	144
1974	398	403	2020	423	189	523	165
1975	444	452	2212	464	207	576	182
1976	472	479	2401	503	224	616	192
1977	505	514	2577	540	241	653	204
1978	545	554	2776	578	258	701	219
1979 (Jan.)	561†	569	2872	598	267	729	230

† All costs presented in this text are based on this value of the Marshall and Swift index unless otherwise indicated.

Table 3a Labor and material indexes as annual averages

(Basis: 1967 = 100. Construction Materials Producer Price Index and Hourly Earnings Index for Construction Workers. Adapted from *Monthly Labor Review*)

	Year							
	1964	1965	1966	1967	1968	1969	1970	1971
Labor index	86	90	95	100	107	116	128	139
Materials index	95	96	99	100	106	112	113	120

	Year							
	1972	1973	1974	1975	1976	1977	1978	Jan. 1979
Labor index	147	155	164	176	187	197	210	218
Materials index	127	139	161	174	188	205	228	241

other minor equipment. All costs reported in this text are based on a Marshall and Swift all-industry index of 561 as reported for January 1, 1979 unless indicated otherwise.

Engineering News-Record Construction Cost Index†

Relative construction costs at various dates can be estimated by use of the *Engineering News-Record* construction index. This index shows the variation in labor rates and materials costs for industrial construction. It employs a composite cost for 2500 lb of structural steel, 1088 fbm of lumber, 6 bbl of cement, and 200 h of common labor. The index is usually reported on one of three bases: an index value of 100 in 1913, 100 in 1949, or 100 in 1967.

Nelson Refinery Construction Cost Index‡

Construction costs in the petroleum industry are the basis of the Nelson construction index. The total index percentages are weighted as follows: skilled labor, 30; common labor, 30; iron and steel, 24; building materials, 8; and miscellaneous equipment, 8. An index value of 100 is used for the base year of 1946.

Chemical Engineering Plant Cost Index§

Construction costs for chemical plants form the basis of the *Chemical Engineering* plant cost index. The four major components of this index are weighted by percentage in the following manner: equipment, machinery, and supports, 61; erection and installation labor, 22; buildings, materials, and labor, 7; and engineering and supervision, 10. The major component, equipment, is further subdivided and weighted as follows: fabricated equipment, 37; process machinery, 14; pipe, valves, and fittings, 20; process instruments and controls, 7; pumps and compressors, 7; electrical equipment and materials, 5; and structural supports, insulation, and paint, 10. All index components are based on 1957–1959 = 100.

---

† The *Engineering News-Record* construction index appears weekly in the *Engineering News-Record*. For a complete description of this index and the revised basis, see *Eng. News-Record*, **143**(9):398 (1949); **178**(11):87 (1967). History is in March issue each year; for example, see *Eng. News-Record*, **200**(12):69 (March 23, 1978).

‡ The |Nelson refinery construction index is published the first week of each month in the *Oil and Gas Journal*. For a complete description of this index, see *Oil Gas J.*, **63**(14):185 (1965); **63**(27):117 (1965); **65**(20):97 (1967); **74**(48):68 (1976).

§ The *Chemical Engineering* plant cost index is published every other week in *Chemical Engineering*. A complete description of this index is in *Chem. Eng.*, **70**(4):143 (Feb. 18, 1963) with recapping and updating in issues of **73**(9):184 (April 25, 1966); **76**(10):134 (May 5, 1969); **79**(25):168 (Nov. 13, 1972); and **82**(9):117 (April 28, 1975).

Other Indexes and Analysis

There are numerous other indexes presented in the literature which can be used for specialized purposes. For example, cost indexes for materials and labor for various types of industries are published monthly by the U.S. Bureau of Labor Statistics in the *Monthly Labor Review*. These indexes can be useful for special kinds of estimates involving particular materials or unusual labor conditions. Another example of a cost index which is useful for world-wide comparison of cost changes with time is the *EPE Plant Cost Indices International* (1970 = 100) published periodically in *Engineering and Process Economics*. This presents cost indexes for plant costs for various countries in the world including Australia, Austria, Belgium, Canada, Denmark, France, Germany, Ireland, Italy, Netherlands, Norway, South Africa, Spain, Sweden, the United Kingdom, and the United States.

Unfortunately, all cost indexes are rather artificial; two indexes covering the same types of projects may give results that differ considerably. The most that any index can hope to do is to reflect average changes. The latter may at times have little meaning when applied to a specific case. For example, a contractor may, during a slack period, accept a construction job with little profit just to keep his construction crew together. On the other hand, if there are current local labor shortages, a project may cost considerably more than a similar project in another geographical location.

For use with process-equipment estimates and chemical-plant investment estimates, the *Marshall and Swift* equipment cost indexes and the *Chemical Engineering* plant cost indexes are recommended. These two cost indexes give very similar results, while the *Engineering News-Record* construction cost index, relative with time, has increased much more rapidly than the other two because it does not include a productivity improvement factor. Similarly, the Nelson refinery construction index has shown a very large increase with time and should be used with caution and only for refinery construction.

COST FACTORS IN CAPITAL INVESTMENT

Capital investment, as defined earlier, is the total amount of money needed to supply the necessary plant and manufacturing facilities plus the amount of money required as working capital for operation of the facilities. Let us now consider the proportional costs of each major component of fixed-capital investment as outlined previously in Table 1 of this chapter. The cost factors presented here are based on a careful study by Bauman and associates[†] plus additional

---

† H. C. Bauman, "Fundamentals of Cost Engineering in the Chemical Industry," Reinhold Publishing Corporation, New York, 1964.

Table 4 Typical percentages of fixed-capital investment values for direct and indirect cost segments for multipurpose plants or large additions to existing facilities

Component	Range, %
Direct costs	
Purchased equipment	15–40
Purchased-equipment installation	6–14
Instrumentation and controls (installed)	2–8
Piping (installed)	3–20
Electrical (installed)	2–10
Buildings (including services)	3–18
Yard improvements	2–5
Service facilities (installed)	8–20
Land	1–2
Total direct costs	
Indirect costs	
Engineering and supervision	4–21
Construction expense	4–16
Contractor's fee	2–6
Contingency	5–15
Total fixed-capital investment	

data and interpretations from other more recent sources† with input based on modern industrial experience.

Table 4 summarizes the typical variation in component costs as percentages of fixed-capital investment for multiprocess *grass-roots* plants or large *battery-limit* additions. A *grass-roots* plant is defined as a complete plant erected on a new site. Investment includes all costs of land, site development, battery-limit facilities, and auxiliary facilities. A geographical boundary defining the coverage of a specific project is a *battery limit*. Usually this encompasses the manufacturing area of a proposed plant or addition, including all process equipment but excluding provision of storage, utilities, administrative buildings, or auxiliary facilities unless so specified. Normally this excludes site preparation and therefore may be applied to the extension of an existing plant.

Example 1. Estimation of fixed-capital investment using ranges of process-plant component costs Make a study estimate of the fixed-capital investment

†|J. E. Haselbarth, Updated Investment Costs for 60 Types of Chemical Plants, *Chem. Eng.*, **74**(25): 214 (Dec. 4, 1967); K. M. Guthrie, Capital Cost Estimating, *Chem. Eng.*, **76**(6): 114 (Mar. 24, 1969); K. M. Guthrie, Capital and Operating Costs for 54 Chemical Processes, *Chem. Eng.*, **77**(13): 140 (June 15, 1970); R. H. Perry and C. H. Chilton, "Chemical Engineers' Handbook," 5th ed., McGraw-Hill Book Company, Inc., New York, 1973; K. M. Guthrie, "Process Plant Estimating, Evaluation, and Control," Craftsman Book Company of America, Solana Beach, California, 1974; D. H. Allen and R. C. Page, Revised Techniques for Predesign Cost Estimating, *Chem. Eng.*, **82**(5): 142 (Mar. 3, 1975); W. D. Baasel, "Preliminary Chemical Engineering Plant Design," American Elsevier Publishing Company, Inc., New York, 1976.

for a process plant if the purchased-equipment cost is $100,000. Use the ranges of process-plant component cost outlined in Table 4 for a process plant handling both solids and fluids with a high degree of automatic controls and essentially outdoor operation.

SOLUTION:

Components	Assumed % of total	Cost	Ratioed % of total
Purchased equipment	25	$100,000	23.0
Purchased-equipment installation	9	36,000	8.3
Instrumentation (installed)	7	28,000	6.4
Piping (installed)	8	32,000	7.3
Electrical (installed)	5	20,000	4.6
Buildings (including services)	5	20,000	4.6
Yard improvements	2	8,000	1.8
Service facilities (installed)	15	60,000	13.8
Land	1	4,000	0.9
Engineering and supervision	10	40,000	9.2
Construction expense	12	48,000	11.0
Contractor's fee	2	8,000	1.8
Contingency	8	32,000	7.3
		$436,000	100.0

Range will vary from $371,000 to $501,000 for normal conditions; if economy is inflationary, it may vary from $436,000–$566,000.

Purchased Equipment

The cost of purchased equipment is the basis of several predesign methods for estimating capital investment. Sources of equipment prices, methods of adjusting equipment prices for capacity, and methods of estimating auxiliary process equipment are therefore essential to the estimator in making reliable cost estimates.

The various types of equipment can often be divided conveniently into (1) processing equipment, (2) raw-materials handling and storage equipment, and (3) finished-products handling and storage equipment. The cost of auxiliary equipment and materials, such as insulation and ducts, should also be included.

The most accurate method of determining process equipment costs is to obtain firm bids from fabricators or suppliers. Often, fabricators can supply quick estimates which will be very close to the bid price but will not involve too much time. Second best in reliability are cost values from the file of past purchase orders. When used for pricing new equipment, purchase-order prices must be corrected to the current cost index. Limited information on process-equipment costs has also been published in various engineering journals. Costs, based on January 1, 1979 prices, for a large number of different types and capacities of equipment are presented in Chaps. 13 through 15. A convenient reference to these various cost figures is given in the Table of Contents and in the subject index.

Estimating Equipment Costs by Scaling

It is often necessary to estimate the cost of a piece of equipment when no cost data are available for the particular size of operational capacity involved. Good results can be obtained by using the logarithmic relationship known as the *six-tenths-factor rule*, if the new piece of equipment is similar to one of another capacity for which cost data are available. According to this rule, if the cost of a given unit at one capacity is known, the cost of a similar unit with X times the capacity of the first is approximately $(X)^{0.6}$ times the cost of the initial unit.

$$\text{Cost of equip. } a = \text{cost of equip. } b \left(\frac{\text{capac. equip. } a}{\text{capac. equip. } b}\right)^{0.6} \tag{1}$$

The preceding equation indicates that a log-log plot of capacity versus equipment cost for a given type of equipment should be a straight line with a slope equal to 0.6. Figure 5-5 presents a plot of this sort for shell-and-tube heat exchangers. However, the application of the 0.6 rule of thumb for most purchased equipment is an oversimplification of a valuable cost concept since the actual values of the cost capacity factor vary from less than 0.2 to greater than 1.0 as shown in Table 5. Because of this, the 0.6 factor should only be used

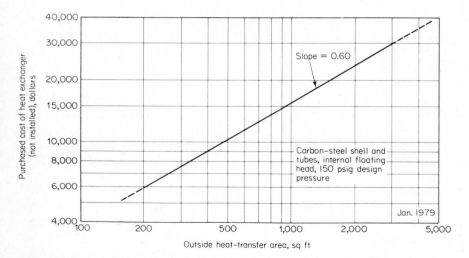

Figure 5-5 Application of "six-tenth-factor" rule to costs for shell-and-tube heat exchangers.

Table 5 Typical exponents for equipment cost vs. capacity

Equipment	Size range	Exponent
Blender, double cone rotary, c.s.	50–250 ft^3	0.49
Blower, centrifugal	10^3–10^4 ft^3/min	0.59
Centrifuge, solid bowl, c.s.	10–10^2 hp drive	0.67
Crystallizer, vacuum batch, c.s.	500–7000 ft^3	0.37
Compressor, reciprocating, air-cooled, two-stage, 150 psi discharge	10–400 ft^3/min	0.69
Compressor, rotary, single-stage, sliding vane, 150 psi discharge	10^2–10^3 ft^3/min	0.79
Dryer, drum, single vacuum	10–10^2 ft^2	0.76
Dryer, drum, single atmospheric	10–10^2 ft^2	0.40
Evaporator (installed), horizontal tank	10^2–10^4 ft^2	0.54
Fan, centrifugal	10^3–10^4 ft^3/min	0.44
Fan, centrifugal	2 × 10^4–7 × 10^4 ft^3/min	1.17
Heat exchanger, shell and tube, floating head, c.s.	100–400 ft^2	0.60
Heat exchanger, shell and tube, fixed sheet, c.s.	100–400 ft^2	0.44
Kettle, cast iron, jacketed	250–800 gal	0.27
Kettle, glass lined, jacketed	200–800 gal	0.31
Motor, squirrel cage, induction, 440 volts, explosion proof	5–20 hp	0.69
Motor, squirrel cage, induction, 440 volts, explosion proof	20–200 hp	0.99
Pump, reciprocating, horizontal, cast iron (includes motor)	2–100 gpm	0.34
Pump, centrifugal, horizontal, cast steel (includes motor)	10^4–10^5 gpm × psi	0.33
Reactor, glass lined, jacketed (without drive)	50–600 gal	0.54
Reactor, S.S., 300 psi	10^2–10^3 gal	0.56
Separator, centrifugal, c.s.	50–250 ft^3	0.49
Tank, flat head, c.s.	10^2–10^4 gal	0.57
Tank, c.s., glass lined	10^2–10^3 gal	0.49
Tower, c.s.	10^3–2 × 10^6 lb	0.62
Tray, bubble cup, c.s.	3–10 ft diameter	1.20
Tray, sieve, c.s.	3–10 ft diameter	0.86

in the absence of other information.† In general, the cost-capacity concept should not be used beyond a tenfold range of capacity, and care must be taken to make certain the two pieces of equipment are similar with regard to type of construction, materials of construction, temperature and pressure operating range, and other pertinent variables.

Example 2. Estimating cost of equipment using scaling factors and cost index The purchased cost of a 50-gal glass-lined, jacketed reactor (without

† See C. A. Miller, Capital Cost Estimation—A Science Rather than an Art, *Cost Engineers' Notebook*, ASCE, A-1.000 (June, 1968) and C. A. Miller, Current Concepts in Capital Cost Forecasting, *Chem. Eng. Progr.*, **69**(5):77 (1973). See also the suggested references at the end of this chapter.

drive) was $3000 on January 1, 1971. Estimate the purchased cost of a similar 300-gal, glass-lined, jacketed reactor (without drive) on January 1, 1976. Use the Marshall and Swift equipment-cost index (all industry) to update the purchase cost of the reactor.

SOLUTION Marshall and Swift equipment-cost index (all industry)†

$$\text{January 1, 1971} \quad 311$$

$$\text{January 1, 1976} \quad 451$$

From Table 5, the equipment vs. capacity exponent is given as 0.41:

$$\text{Cost of reactor} = (\$3000)\left(\frac{451}{311}\right)\left(\frac{300}{50}\right)^{0.41}$$

(January 1, 1976)
$$\text{Cost} = \$9070$$

Purchased-equipment costs for vessels, tanks, and process- and materials-handling equipment can often be estimated on the basis of weight. The fact that a wide variety of types of equipment have about the same cost per unit weight is quite useful, particularly when other cost data are not available. Generally, the cost data generated by this method are sufficiently reliable to permit order-of-magnitude estimates.

Purchased-Equipment Installation

The installation of equipment involves costs for labor, foundations, supports, platforms, construction expenses, and other factors directly related to the erection of purchased equipment. Table 6 presents the general range of installation cost as a percentage of the purchased-equipment cost for various types of equipment.

Installation labor cost as a function of equipment size shows wide variations when scaled from previous installation estimates. Table 7 shows exponents varying from 0.0 to 1.56 for a few selected pieces of equipment.

Tubular heat exchangers appear to have zero exponents, implying that direct labor cost is independent of size. This reflects the fact that such equipment is set with cranes and hoists, which, when adequately sized for the task, recognize no appreciable difference in size or weight of the equipment. The higher labor exponent for installing carbon-steel towers indicates the increasing complexity of tower internals (trays, downcomers, etc.) as tower diameter increases.

Analyses of the total installed costs of equipment in a number of typical chemical plants indicate that the cost of the purchased equipment varies from 65 to 80 percent of the installed cost depending upon the complexity of the equip-

---

† The values from Table 3 were not used here since those listed are annual average cost-index values rather than first-of-the-year values.

Table 6 Installation cost for equipment as a percentage of the purchased-equipment cost†

Type of equipment	Installation cost, %
Centrifugal separators	20–60
Compressors	30–60
Dryers	25–60
Evaporators	25–90
Filters	65–80
Heat exchangers	30–60
Mechanical crystallizers	30–60
Metal tanks	30–60
Mixers	20–40
Pumps	25–60
Towers	60–90
Vacuum crystallizers	40–70
Wood tanks	30–60

† Adapted from K. M. Guthrie, "Process Plant Estimating, Evaluation, and Control," Craftsman Book Company of America, Solana Beach, California, 1974.

ment and the type of plant in which the equipment is installed. Installation costs for equipment, therefore, are estimated to vary from 25 to 55 percent of the purchased-equipment cost.

Insulation Costs

When very high or very low temperatures are involved, insulation factors can become important, and it may be necessary to estimate insulation costs with a

Table 7 Typical exponents for equipment installation labor vs. size

Equipment	Size range	Exponent
Conduit, aluminum	0.5–2-in. diam.	0.49
Conduit, aluminum	2–4-in. diam.	1.11
Motor, squirrel cage, induction, 440 volts	1–10 hp	0.19
Motor, squirrel cage, induction, 440 volts	10–50 hp	0.50
Pump, centrifugal, horizontal	0.5–1.5 hp	0.63
Pump, centrifugal, horizontal	1.5–40 hp	0.09
Tower, c.s.	Constant diam.	·0.88
Tower, c.s.	Constant height	1.56
Transformer, single phase, dry	9–225 kva	0.58
Transformer, single phase, oil, class A	15–225 kva	0.34
Tubular heat exchanger	Any size	0.00

great deal of care. Expenses for equipment insulation and piping insulation are often included under the respective headings of equipment-installation costs and piping costs.

The total cost for the labor and materials required for insulating equipment and piping in ordinary chemical plants is approximately 8 to 9 percent of the purchased-equipment cost. This is equivalent to approximately 2 percent of the total capital investment.

Instrumentation and Controls

Instrument costs, installation-labor costs, and expenses for auxiliary equipment and materials constitute the major portion of the capital investment required for instrumentation. This part of the capital investment is sometimes combined with the general equipment groups. Total instrumentation cost depends on the amount of control required and may amount to 6 to 30 percent of the purchased cost for all equipment. Computers are commonly used with controls and have the effect of increasing the cost associated with controls.

For the normal solid-fluid chemical processing plant, a value of 13 percent of the purchased equipment is normally used to estimate the total instrumentation cost. This cost represents approximately 3 percent of the total capital investment. Depending on the complexity of the instruments and the service, additional charges for installation and accessories may amount to 50 to 70 percent of the purchased cost, with the installation charges being approximately equal to the cost for accessories.

Piping

The cost for piping covers labor, valves, fittings, pipe, supports, and other items involved in the complete erection of all piping used directly in the process. This includes raw-material, intermediate-product, finished-product, steam, water, air, sewer, and other process piping. Since process-plant piping can run as high as 80 percent of purchased-equipment cost or 20 percent of fixed-capital investment, it is understandable that accuracy of the entire estimate can be seriously affected by the improper application of estimation techniques to this one component.

Piping estimation methods involve either some degree of piping take-off from detailed drawings and flow sheets or using a factor technique when neither detailed drawings nor flow sheets are available. Factoring by percent of purchased-equipment cost and percent of fixed-capital investment is based strictly on experience gained from piping costs for similar previously installed chemical-process plants. Table 8 presents a rough estimate of the piping costs for various types of chemical processes. Additional information for estimating piping costs is presented in Chap. 13. Labor for installation is estimated as approximately 40–50 percent of the total installed cost of piping. Material and labor for pipe insulation is estimated to vary from 15 to 25 percent of the total installed cost of the piping and is influenced greatly by the extremes in temperature which are encountered in the process streams.

Table 8 Estimated cost of piping

Type of process plant	Percent of purchased-equipment			Percent of fixed-capital investment
	Material	Labor	Total	Total
Solid †	9	7	16	4
Solid-fluid ‡	17	14	31	7
Fluid §	36	30	66	13

† A coal briquetting plant would be a typical solid-processing plant.
‡ A shale oil plant with crushing, grinding, retorting, and extraction would be a typical solid-fluid processing plant.
§ A distillation unit would be a typical fluid-processing plant.

Electrical Installations

The cost for electrical installations consists primarily of installation labor and materials for power and lighting, with building-service lighting usually included under the heading of building-and-services costs. In ordinary chemical plants, electrical-installations cost amounts to 10 to 15 percent of the value of all purchased equipment. However, this may range to as high as 40 percent of purchased-equipment cost for a specific process plant. There appears to be little relationship between percent of total cost and percent of equipment cost, but there is a better relationship to fixed-capital investment. Thus, the electrical installation cost is generally estimated between 3 to 10 percent of the fixed-capital investment.

The electrical installation consists of four major components, namely, power wiring, lighting, transformation and service, and instrument and control wiring. Table 9 shows these component costs as ratios of the total electrical cost.

Table 9 Component electrical costs as percent of total electrical cost

Component	Range, %	Typical value, %
Power wiring	25–50	40
Lighting	7–25	12
Transformation and service	9–65	40
Instrument control wiring	3–8	5

The lower ratio is generally applicable to grass-roots single-product plants; the higher percentages apply to complex chemical plants and expansions to major chemical plants.

Table 10 Cost of buildings including services based on purchased-equipment cost

Type of process plant†	Percentage of purchased-equipment cost		
	New plant at new site (Grass roots)	New unit at existing site (Battery limit)	Expansion at an existing site
Solid	68	25	15
Solid-fluid	47	29	7
Fluid	45	5–18‡	6

   † See Table 8 for definition of types of process plants.
   ‡ The lower figure is applicable to petroleum refining and related industries.

Buildings Including Services

The cost for buildings including services consists of expenses for labor, materials, and supplies involved in the erection of all buildings connected with the plant. Costs for plumbing, heating, lighting, ventilation, and similar building services are included. The cost of buildings, including services for different types of process plants, is shown in Tables 10 and 11 as a percentage of purchased-equipment cost and fixed-capital investment.

Yard Improvements

Costs for fencing, grading, roads, sidewalks, railroad sidings, landscaping, and similar items constitute the portion of the capital investment included in yard improvements. Yard-improvements cost for chemical plants approximates 10 to 20 percent of the purchased-equipment cost. This is equivalent to approximately 2 to 5 percent of the fixed-capital investment. Table 12 shows the range in variation for various components of yard improvements in terms of the fixed-capital investment.

Table 11 Cost of buildings and services as a percentage of fixed-capital investment for various types of process plants

Type of process plant†	New plant at new site	New unit at existing site	Expansion at an existing site
Solid	18	7	4
Solid-fluid	12	7	2
Fluid	10	2–4‡	2

   † See Table 8 for definition of types of process plants.
   ‡ The lower figure is applicable to petroleum refining and related industries.

Table 12 Typical variation in percent of fixed-capital investment for yard improvements

Yard improvement	Range, %	Typical value, %
Site clearing	0.4–1.2	0.8
Roads and walks	0.2–1.2	0.6
Railroads	0.3–0.9	0.6
Fences	0.1–0.3	0.2
Yard and fence lighting	0.1–0.3	0.2
Parking areas	0.1–0.3	0.2
Landscaping	0.1–0.2	0.1
Other improvements	0.2–0.6	0.3

Service Facilities

Utilities for supplying steam, water, power, compressed air, and fuel are part of the service facilities of an industrial plant. Waste disposal, fire protection, and miscellaneous service items, such as shop, first aid, and cafeteria equipment and facilities, require capital investments which are included under the general heading of service-facilities cost.

The total cost for service facilities in chemical plants generally ranges from 30 to 80 percent of the purchased-equipment cost with 55 percent representing an average for a normal solid-fluid processing plant. For a single-product, small, continuous-process plant, the cost is likely to be in the lower part of the range. For a large, new, multiprocess plant at a new location, the costs are apt to be near the upper limit of the range. The cost of service facilities, in terms of capital investment, generally ranges from 8 to 20 percent with 13 percent considered as an average value. Table 13 lists the typical variations in percentages of fixed-capital investment that can be encountered for various components of service facilities. Except for entirely new facilities, it is unlikely that all service facilities will be required in all process plants. This accounts to a large degree for the wide variation range assigned to each component in Table 13. The range also reflects the degree to which utilities which depend on heat balance are used in the process. Service facilities largely are functions of plant physical size and will be present to some degree in most plants. However, not always will there be a need for each service-facility component. The omission of these utilities would tend to increase the relative percentages of the other service facilities actually used in the plant. Recognition of this fact, coupled with a careful appraisal as to the extent that service facilities are used in the plant, should result in selecting from Table 13 a reasonable cost ratio applicable to a specific process design.

Land

The cost for land and the accompanying surveys and fees depends on the location of the property and may vary by a cost factor per acre as high as thirty to

Table 13 Typical variation in percent of fixed-capital investment for service facilities

Service facilities	Range, %	Typical value, %
Steam generation	2 .6–6.0	3.0
Steam distribution	0.2–2.0	1.0
Water supply, cooling, and pumping	0.4–3.7	1.8
Water treatment	0.5–2.1	1.3
Water distribution	0.1–2.0	0.8
Electric substation	0.9–2.6	1.3
Electric distribution	0.4–2.1	1.0
Gas supply and distribution	0.2–0.4	0.3
Air compression and distribution	0.2–3.0	1.0
Refrigeration including distribution	1.0–3.0	2.0
Process waste disposal	0.6–2.4	1.5
Sanitary waste disposal	0.2–0.6	0.4
Communications	0.1–0.3	0.2
Raw-material storage	0.3–3.2	0.5
Finished-product storage	0.7–2 .4	1.5
Fire-protection system	0.3–1.0	0.5
Safety installations	0.2–0.6	0.4

fifty between a rural district and a highly industrialized area. As a rough average, land costs for industrial plants amount to 4 to 8 percent of the purchased-equipment cost or 1 to 2 percent of the total capital investment. Because the value of land usually does not decrease with time, this cost should not be included in the fixed-capital investment when estimating certain annual operating costs, such as depreciation.

Engineering and Supervision

The costs for construction design and engineering, drafting, purchasing, accounting, construction and cost engineering, travel, reproductions, communications, and home office expense including overhead constitute the capital investment for engineering and supervision. This cost, since it cannot be directly charged to equipment, materials, or labor, is normally considered an indirect cost in fixed-capital investment and is approximately 30 percent of the purchased-equipment cost or 8 percent of the total direct costs of the process plant. Typical percentage variations of fixed-capital investment for various components of engineering and supervision are given in Table 14.

Construction Expense

Another expense which is included under indirect plant cost is the item of construction or field expense and includes temporary construction and operation,

Table 14 Typical variation in percent of fixed-capital investment for engineering and services

Component	Range, %	Typical value, %
Engineering	1.5–6.0	2.2
Drafting	2.0–12.0	4.8
Purchasing	0.2–0.5	0.3
Accounting, construction, and cost engineering	0.2–1.0	0.3
Travel and living	0.1–1.0	0.3
Reproductions and communications	0.2–0.5	0.2
Total engineering and supervision (including overhead)	4.0–21.0	8.1

construction tools and rentals, home office personnel located at the construction site, construction payroll, travel and living, taxes and insurance, and other construction overhead. This expense item is occasionally included under equipment installation, or more often under engineering, supervision, and construction. If construction or field expenses are to be estimated separately, then Table 15 will be useful in establishing the variation in percent of fixed-capital investment for this indirect cost. For ordinary chemical-process plants the construction expenses average roughly 10 percent of the total direct costs for the plant.

Contractor's Fee

The contractor's fee varies for different situations, but it can be estimated to be about 2 to 8 percent of the direct plant cost or 1.5 to 6 percent of the fixed-capital investment.

Table 15 Typical variation in percent of fixed-capital investment for construction expenses

Component	Range, %	Typical value, %
Temporary construction and operations	1.0–3.0	1.7
Construction tools and rental	1.0–3.0	1.5
Home office personnel in field	0.2–2.0	0.4
Field payroll	0.4–4.0	1.0
Travel and living	0.1–0.8	0.3
Taxes and insurance	1.0–2.0	1.2
Startup materials and labor	0.2–1.0	0.4
Overhead	0.3–0.8	0.5
Total construction expenses	4.2–16.6	7.0

Contingencies

A contingency factor is usually included in an estimate of capital investment to compensate for unpredictable events, such as storms, floods, strikes, price changes, small design changes, errors in estimation, and other unforeseen expenses, which previous estimates have shown statistically to be likely to occur. This factor may or may not include allowance for escalation. Contingency factors ranging from 5 to 15 percent of the direct and indirect plant costs are commonly used, with 8 percent being considered a fair average value.

Startup Expense

After plant construction has been completed, there are quite frequently changes that have to be made before the plant can operate at maximum design conditions. These changes involve expenditures for materials and equipment and result in loss of income while the plant is shut down or is operating at only partial capacity. Capital for these startup changes should be part of any capital appropriation because they are essential to the success of the venture. These expenses may be as high as 12 percent of the fixed-capital investment. In general, however, an allowance of 8 to 10 percent of the fixed-capital investment for this item is satisfactory.

Startup expense is not necessarily included as part of the required investment; so it is not presented as a component in the summarizing Table 26 for capital investment at the end of this chapter. In the overall cost analysis, startup expense may be presented as a one-time-only expenditure in the first year of the plant operation or as part of the total capital investment depending on the company policies.

METHODS FOR ESTIMATING CAPITAL INVESTMENT

Various methods can be employed for estimating capital investment. The choice of any one method depends upon the amount of detailed information available and the accuracy desired. Seven methods are outlined in this chapter, with each method requiring progressively less detailed information and less preparation time. Consequently, the degree of accuracy decreases with each succeeding method. A maximum accuracy within approximately ± 5 percent of the actual capital investment can be obtained with method A.

Method A. Detailed-item estimate A detailed-item estimate requires careful determination of each individual item shown in Table 1. Equipment and material needs are determined from completed drawings and specifications and are priced either from current cost data or preferably from firm delivered quotations. Estimates of installation costs are determined from accurate labor rates, efficiencies, and employee-hour calculations. Accurate estimates of engineering, drafting, field supervision employee-hours, and field expenses must be detailed in the same manner. Complete site surveys and soil data must be available to minimize errors in site development and construction cost estimates. In fact, in this type of estimate, an attempt is made to firm up as much of the estimate as possible by obtaining quotations from vendors and suppliers. Because of the extensive data necessary and the large amounts of engineering time required to prepare such a detailed-item estimate, this type of estimate is almost exclusively only prepared by contractors bidding on lump-sum work from finished drawings and specifications.

Method B. Unit-cost estimate The unit-cost method results in good estimating accuracies for fixed-capital investment provided accurate records have been kept of previous cost experience. This method, which is frequently used for preparing definitive and preliminary estimates, also requires detailed estimates of purchased price obtained either from quotations or index-corrected cost records and published data. Equipment installation labor is evaluated as a fraction of the delivered-equipment cost. Costs for concrete, steel, pipe, electricals, instrumentation, insulation, etc., are obtained by take-offs from the drawings and applying unit costs to the material and labor needs. A unit cost is also applied to engineering employee-hours, number of drawings, and specifications. A factor for construction expense, contractor's fee, and contingency is estimated from previously completed projects and is used to complete this type of estimate. A cost equation summarizing this method can be given as†

$$C_n[\Sigma(E + E_L) + \Sigma(f_x M_x + f_y M'_L) + \Sigma f_e H_e + \Sigma f_d d_n](f_F) \tag{2}$$

where C_n = new capital investment
$\quad E$ = purchased-equipment cost
$\quad E_L$ = purchased-equipment labor cost
$\quad f_x$ = specific material unit cost, e.g., f_p = unit cost of pipe
$\quad M_x$ = specific material quantity in compatible units
$\quad f_y$ = specific material labor unit cost per employee-hour
$\quad M'_L$ = labor employee-hours for specific material
$\quad f_e$ = unit cost for engineering
$\quad H_e$ = engineering employee-hours
$\quad f_d$ = unit cost per drawing or specification
$\quad d_n$ = number of drawings or specifications
$\quad f_F$ = construction or field expense factor always greater than 1

Approximate corrections to the base equipment cost of complete, main-plant items for specific materials of construction or extremes of operating pressure and temperature can be applied in the form of factors as shown in Table 16.

Method C. Percentage of delivered-equipment cost This method for estimating the fixed or total-capital investment requires determination of the delivered-equipment cost. The other items included in the total direct plant cost are then estimated as percentages of the delivered-equipment cost. The additional com-

† H. C. Bauman, "Fundamentals of Cost Engineering in the Chemical Industry," Reinhold Publishing Corporation, New York, 1964.

Table 16 Correction factors for operating pressure, operating temperature, and material of construction to apply for fixed-capital investment for major plant items†‡

Operating pressure, psia (atm)	Correction factor
0.08 (0.005)	1.3
0.2 (0.014)	1.2
0.7 (0.048)	1.1
8 (0.54) to 100 (6.8)	1.0 (base)
700 (48)	1.1
3000 (204)	1.2
6000 (408)	1.3

Operating temperature, °C	Correction factor
−80	1.3
0	1.0 (base)
100	1.05
600	1.1
5,000	1.2
10,000	1.4

Material of construction	Correction factor
Carbon steel–mild	1.0 (base)
Bronze	1.05
Carbon/molybdenum steel	1.065
Aluminum	1.075
Cast steel	1.11
Stainless steel	1.28 to 1.5
Worthite alloy	1.41
Hastelloy C alloy	1.54
Monel alloy	1.65
Nickel/inconel alloy	1.71
Titanium	2.0

† Adapted from D. H. Allen and R. C. Page, Revised Techniques for Predesign Cost Estimating, *Chem. Eng.*, **82**(5): 142 (March 3, 1975).

‡ It should be noted that these factors are to be used *only* for complete, main-plant items and serve to correct from the base case to the indicated conditions based on pressure or temperature extremes that may be involved or special materials of construction that may be required. For the case of small or single pieces of equipment which are completely dedicated to the extreme conditions, the factors given in this table may be far too low and factors or methods given in other parts of this book must be used.

ponents of the capital investment are based on average percentages of the total direct plant cost, total direct and indirect plant costs, or total capital investment. This is summarized in the following cost equation:

$$C_n = [\Sigma E + \Sigma(f_1 E + f_2 E + f_3 E + \cdots)](f_I) \tag{3}$$

where $f_1, f_2 \ldots$ = multiplying factors for piping, electrical, instrumentation, etc.
f_I = indirect cost factor always greater than 1.

The percentages used in making an estimation of this type should be determined on the basis of the type of process involved, design complexity, required materials of construction, location of the plant, past experience, and other items dependent on the particular unit under consideration. Average values of the various percentages have been determined for typical chemical plants, and these values are presented in Table 17.

Estimating by percentage of delivered-equipment cost is commonly used for preliminary and study estimates. It yields most accurate results when applied to projects similar in configuration to recently constructed plants. For comparable plants of different capacity, this method has sometimes been reported to yield definitive estimate accuracies.

Example 3. Estimation of fixed-capital investment by percentage of delivered-equipment cost Prepare a study estimate of the fixed-capital investment for the process plant described in Example 1 if the delivered-equipment cost is $100,000.

SOLUTION Use the ratio factors outlined in Table 17 with modifications for instrumentation and outdoor operation

Components	Cost
Purchased equipment (delivered), E	$100,000
Purchased equipment installation, 39% E	39,000
Instrumentation (installed), 28% E	28,000
Piping (installed), 31% E	31,000
Electrical (installed), 10% E	10,000
Buildings (including services), 22% E	22,000
Yard improvements, 10% E	10,000
Service facilities (installed), 55% E	55,000
Land, 6% E	6,000
Total direct plant cost D	301,000
Engineering and supervision, 32% E	32,000
Construction expenses, 34% E	34,000
Total direct and indirect cost $(D + I)$	367,000
Contractor's fee, 5% $(D + I)$	18,000
Contingency, 10% $(D + I)$	37,000
Fixed-capital investment	$422,000

Table 17 Ratio factors for estimating capital-investment items based on delivered-equipment cost

Values presented are applicable for major process plant additions to an existing site where the necessary land is available through purchase or present ownership.† The values are based on fixed-capital investments ranging from under $1 million to over $10 million.

Item	Percent of delivered-equipment cost for		
	Solid-processing plant‡	Solid-fluid-processing plant‡	Fluid-processing plant‡
Direct costs			
Purchased equipment-delivered (including fabricated equipment and process machinery) §	100	100	100
Purchased-equipment installation	45	39	47
Instrumentation and controls (installed)	9	13	18
Piping (installed)	16	31	66
Electrical (installed)	10	10	11
Buildings (including services)	25	29	18
Yard improvements	13	10	10
Service facilities (installed)	40	55	70
Land (if purchase is required)	6	6	6
Total direct plant cost	264	293	346
Indirect costs			
Engineering and supervision	33	32	33
Construction expenses	39	34	41
Total direct and indirect plant costs	336	359	420
Contractor's fee (about 5% of direct and indirect plant costs)	17	18	21
Contingency (about 10% of direct and indirect plant costs)	34	36	42
Fixed-capital investment	387	413	483
Working capital (about 15% of total capital investment)	68	74	86
Total capital investment	455	487	569

† Because of the extra expense involved in supplying service facilities, storage facilities, loading terminals, transportation facilities, and other necessary utilities at a completely undeveloped site, the fixed-capital investment for a new plant located at an undeveloped site may be as much as 100 percent greater than for an equivalent plant constructed as an addition to an existing plant.

‡ See Table 8 for definition of types of process plants.

§ Includes pumps and compressors.

Table 18 Lang multiplication factors for estimation of fixed-capital investment or total capital investment

Factor × delivered-equipment cost = fixed-capital investment or total capital investment for major additions to an existing plant.

Type of plant	Factor for	
	Fixed-capital investment	Total capital investment
Solid-processing plant	3.9	4.6
Solid-fluid-processing plant	4.1	4.9
Fluid-processing plant	4.8	5.7

Method D. "Lang" factors for approximation of capital investment This technique, proposed originally by Lang† and used quite frequently to obtain order-of-magnitude cost estimates, recognizes that the cost of a process plant may be obtained by multiplying the basic equipment cost by some factor to approximate the capital investment. These factors vary depending upon the type of process plant being considered. The percentages given in Table 17 are rough approximations which hold for the types of process plants indicated. These values, therefore, may be combined to give Lang multiplication factors that can be used for estimating the total direct plant cost, the fixed-capital investment, or the total capital investment. Factors for estimating the fixed-capital investment or the total capital investment are given in Table 18. It should be noted that these factors include costs for land and contractor's fees.

Greater accuracy of capital investment estimates can be achieved in this method by using not one but a number of factors. One approach is to use different factors for different types of equipment. To obtain battery-limit costs, Hand‡ recommended the following factors: 4.0 for fractionating columns, pressure vessels, pumps, and instruments; 3.5 for heat exchangers; 2.5 for compressors; and 2.0 for fired heaters.

Another approach is to use separate factors for erection of equipment, foundations, utilities, piping, etc., or even to break up each item of cost into material and labor factors.§ With this approach, each factor has a range of values and the

---

† H. J. Lang, *Chem. Eng.*, **54**(10):117 (1947); H. J. Lang, *Chem. Eng.*, **55**(6):112 (1948).

‡ W. E. Hand, *Petrol. Refiner*, **37**(9):331 (1958).

§ Further discussions on these methods may be found in. W. D. Baasel, " Preliminary Chemical Engineering Plant Design," American Elsevier Publishing Company, Inc., New York, 1976; S. G. Kirkham, Preparation and Application of Refined Lang Factor Costing Techniques, *AACE Bul.*, **15**(5):137 (Oct., 1972); C. A. Miller, Capital Cost Estimating—A Science Rather Than an Art, *Cost Engineers' Notebook*, ASCE A-1000 (June, 1978). See also the suggested references at the end of this chapter.

chemical engineer must rely on past experience to decide, in each case, whether to use a high, average, or low figure.

Since tables are not convenient for computer calculations it is better to combine the separate factors into an equation similar to the one proposed by Hirsch and Glazier[†]

$$C_n = f_I[E(1 + f_F + f_p + f_m) + E_i + A] \tag{4}$$

where the three installation-cost factors are, in turn, defined by the following three equations:

$$\log f_F = 0.635 - 0.154 \log 0.001E - 0.992\frac{e}{E} + 0.506\frac{f_v}{E} \tag{5}$$

$$\log f_p = -0.266 - 0.014 \log 0.001E - 0.156\frac{e}{E} + 0.556\frac{p}{E} \tag{6}$$

$$\log f_m = 0.344 + 0.033 \log 0.001E + 1.194\frac{t}{E} \tag{7}$$

and the various parameters are defined accordingly:

E = purchased-equipment on an f.o.b. basis
f_I = indirect cost factor always greater than 1 (normally taken as 1.4)
f_F = cost factor for field labor
f_p = cost factor for piping materials
f_m = cost factor for miscellaneous items, including the materials cost for insulation, instruments, foundations, structural steel, building, wiring, painting, and the cost of freight and field supervision
E_i = cost of equipment already installed at site
A = incremental cost of corrosion-resistant alloy materials
e = total heat exchanger cost (less incremental cost of alloy)
f_v = total cost of field-fabricated vessels (less incremental cost of alloy)
p = total pump plus driver cost (less incremental cost of alloy)
t = total cost of tower shells (less incremental cost of alloy)

Note that Eq. (4) is designed to handle both purchased equipment on an f.o.b. basis and completely installed equipment. To simplify the work involved in the use of Eq. (4), Walas has combined the equations and prepared an easy-to-use nomograph.[‡]

Method E. Power factor applied to plant-capacity ratio This method for study or order-of-magnitude estimates relates the fixed-capital investment of a new process plant to the fixed-capital investment of similar previously constructed

† J. H. Hirsch and E. M. Glazier, *Chem. Eng. Progr.*, **56**(12): 37 (1960).
‡ S. M. Walas, *Chem. Eng. Progr.*, **57**(6):68 (1961).

plants by an exponential power ratio. That is, for certain similar process plant configurations, the fixed-capital investment of the new facility is equal to the fixed-capital investment of the constructed facility C multiplied by the ratio R, defined as the capacity of the new facility divided by the capacity of the old, raised to a power x. This power has been found to average between 0.6 and 0.7 for many process facilities. Table 19 gives the capacity power factor (x) for various kinds of processing plants.

$$C_n = C(R)^x \qquad (8)$$

A closer approximation for this relationship which involves the direct and indirect plant costs has been proposed as

$$C_n = f[D(R)^x + I] \qquad (9)$$

where f is a lumped cost-index factor relative to the original installation cost. D is the direct cost and I is the total indirect cost for the previously installed facility of a similar unit on an equivalent site. The value of x approaches unity when the capacity of a process facility is increased by adding identical process units instead of increasing the size of the process equipment. The lumped cost-index factor f is the product of a geographical labor cost index, the corresponding area labor productivity index, and a material and equipment cost index. Table 20 presents the relative median labor rate and productivity factor for various geographical areas in the United States.

Example 4. Estimating relative costs of construction labor as a function of geographical area If a given chemical process plant is erected near Dallas (Southwest area) with a construction labor cost of $100,000 what would be the construction labor cost of an identical plant if it were to be erected at the same time near Los Angeles (Pacific Coast area) for the time when the factors given in Table 20 apply?

SOLUTION:
Relative median labor rate—Southwest   0.90 from Table 20
Relative median labor rate—Pacific Coast 1.04 from Table 20

$$\text{Relative labor rate ratio} = \frac{1.04}{0.90} = 1.155$$

Relative productivity factor—Southwest   1.06 from Table 20
Relative productivity factor—Pacific Coast 0.81 from Table 20

$$\text{Relative productivity factor ratio} = \frac{0.81}{1.06} = 0.764$$

Construction labor cost of Southwest to Pacific Coast = $(1.155)/(0.764) =$ 1.512
Construction labor cost at Los Angeles = $(1.512)(\$100{,}000) = \$151{,}200$

Table 19 Capital-cost data for processing plants (1979)†

Product or process	Process remarks	Typical plant size, 1000 tons/yr	Fixed-capital investment, million $	$ of fixed-capital investment per annual ton of product	Power factor (x)‡ for plant-capacity ratio
Chemical plants					
Acetic acid	CH₃OH and CO—catalytic	10	4	400	0.68
Acetone	Propylene—copper chloride catalyst	100	20	200	0.45
Ammonia	Steam reforming	100	15	150	0.53
Ammonium nitrate	Ammonia and nitric acid	100	3	30	0.65
Butanol	Propylene, CO, and H₂O—catalytic	50	25	500	0.40
Chlorine	Electrolysis of NaCl	50	17	340	0.45
Ethylene	Refinery gases	50	8	160	0.83
Ethylene oxide	Ethylene—catalytic	50	31	620	0.78
Formaldehyde (37%)	Methanol—catalytic	10	10	1000	0.55
Glycol	Ethylene and chlorine	5	9	1800	0.75
Hydrofluoric acid	Hydrogen fluoride and H₂O	10	5	500	0.68
Methanol	CO₂, natural gas, and steam	60	8	130	0.60
Nitric acid (high strength)	Ammonia—catalytic	100	4	40	0.60
Phosphoric acid	Calcium phosphate and H₂SO₄	5	2	400	0.60
Polyethylene (high density)	Ethylene—catalytic	5	10	2000	0.65
Propylene	Refinery gases	10	2	200	0.70
Sulfuric acid	Sulfur—catalytic	100	2	20	0.65
Urea	Ammonia and CO₂	60	5	80	0.70

Table 19 Capital-cost data for processing plants (1979) (*Continued*)

Product or process	Process remarks	Typical plant size, 1000 bbl/day	Fixed-capital investment, million $	$ of fixed-capital investment per bbl/day	Power factor (x)‡ for plant-capacity ratio
Refinery units					
Alkylation (H$_2$SO$_4$)	Catalytic	10	12	1200	0.60
Coking (delayed)	Thermal	10	16	1600	0.38
Coking (fluid)	Thermal	10	10	1000	0.42
Cracking (fluid)	Catalytic	10	10	1000	0.70
Cracking	Thermal	10	3	300	0.70
Distillation (atm.)	65% vaporized	100	20	200	0.90
Distillation (vac.)	65% vaporized	100	12	120	0.70
Hydrotreating	Catalytic desulfurization	10	2	200	0.65
Reforming	Catalytic	10	18	1800	0.60
Polymerization	Catalytic	10	3	300	0.58

† Adapted from K. M. Guthrie, Capital and Operating Costs for 54 Chemical Processes, *Chem. Eng.*, 77 (13): 140 (June 15, 1970) and K. M. Guthrie, "Process Plant Estimating, Evaluation, and Control," Craftsman Book Company of America, Solana Beach, California, 1974. See also J. E. Haselbarth, Updated Investment Costs for 60 Chemical Plants, *Chem. Eng.*, 74 (25): 214 (Dec. 4, 1967) and D. Drayer, How to Estimate Plant Cost-Capacity Relationship, *Petro/Chem Engr.*, 42(5): 10 (1970).

‡ These power factors apply within roughly a three-fold ratio extending either way from the plant size as given.

Table 20 Relative labor rate and productivity indexes in the chemical and allied products industries for the United States (1975–1976)†

Geographical area	Relative labor rate	Relative productivity factor
New England	0.90	0.85
Middle Atlantic	0.96	0.90
South Atlantic	0.90	0.86
Midwest	0.99	1.08
Gulf	1.16	1.32
Southwest	0.90	1.06
Mountain	0.93	0.91
Pacific Coast	1.04	0.81

† Adapted from J. M. Winton, Plant Sites 1978, *Chem. Week,* **121** (24): 49 (Dec. 14, 1977). Productivity, as considered here, is an economic term that gives the value added (products minus raw materials) per dollar of total payroll cost. Relative values were determined by taking the average of Winton's weighted state values in each region divided by the weighted average value of all the regions. See also Tables 23 and 24 of this chapter; H. Popper and G. E. Weismantel, Costs and Productivity in the Inflationary 1970's, *Chem Eng.,* **77**(1): 132 (Jan. 12, 1970); and C. H. Edmondson, *Hydrocarbon Process.,* **53**(7): 167 (1974).

To determine the fixed-capital investment required for a new similar single-process plant at a new location with a different capacity and with the same number of process units, the following relationship has given good results:

$$C_n = R^x[f_E E + f_M M + f_L f_F e_L(E_L + f_y M'_L)](f_I) \frac{C}{C - I} \qquad (10)$$

where f_E = current equipment cost index relative to cost of the purchased equipment

f_M = current material cost index relative to cost of material

M = material cost

f_L = current labor cost index in new location relative to E_L and M'_L at old location

e_L = labor efficiency index in new location relative to E_L and M'_L at old location

E_L = purchased-equipment labor cost

M'_L = labor employee-hours for specific material

f_y = specific material labor cost per employee-hour

C = original capital investment

In those situations where estimates of fixed-capital investment are desired for a similar plant at a new location and with a different capacity, but with multiples of the original process units, Eq. (11) often gives results with somewhat better than study-estimate accuracy.

$$C_n = [R f_E E + R^x f_M M + R^x f_L f_F e_L (E_L + f_y M_L)](f_I) \frac{C}{C - I} \qquad (11)$$

More accurate estimates by this method are obtained by subdividing the process plant into various process units, such as crude distillation units, reformers, alkylation units, etc., and applying the best available data from similar previously installed process units separately to each subdivision. Table 19 lists some typical process unit capacity-cost data and exponents useful for making this type of estimate.

Example 5. Estimation of fixed-capital investment with power factor applied to plant-capacity ratio If the process plant, described in Example 1, was erected in the Dallas area for a fixed-capital investment of $436,000 in 1970, determine what the estimated fixed-capital investment would have been in 1975 for a similar process plant located near Los Angeles with twice the process capacity but with an equal number of process units? Use the power-factor method to evaluate the new fixed-capital investment and assume the factors given in Table 20 apply.

SOLUTION If Eq. (8) is used with a 0.6 power factor and the Marshall and Swift process-industry index (Table 3), the fixed-capital investment is

$$C_n = C f_E (R)^x$$

$$C_n = (436{,}000) \left(\frac{444}{303} \right) (2)^{0.6} = \$968{,}000$$

If Eq. (8) is used with a 0.7 power factor and the Marshall and Swift process-industry index (Table 3), the fixed-capital investment is

$$C_n = (436{,}000) \left(\frac{444}{303} \right) (2)^{0.7} = \$1{,}038{,}000$$

If Eq. (9) is used with a 0.6 power factor, the Marshall and Swift process-industry index (Table 3), and the relative labor and productivity indexes (Table 20), the fixed-capital investment is

$$C_n = f[D(R)^x + I]$$

where $f = f_E f_L e_L$, and D and I are obtained from Example 1,

$$C_n = \left(\frac{444}{303} \right) \left(\frac{1.04}{0.90} \right) \left(\frac{1.06}{0.81} \right) [(308{,}000)(2)^{0.6} + 128{,}000]$$

$$C_n = (1.465)(1.512)(467{,}000 + 128{,}000)$$

$$C_n = \$1{,}318{,}000$$

If Eq. (9) is used with a 0.7 power factor, the Marshall and Swift process-industry index (Table 3), and the relative labor and productivity indexes (Table 20), the fixed-capital investment is

$$C_n = \$1,392,000$$

Results obtained using this procedure have shown high correlation with fixed-capital investment estimates that have been obtained with more detailed techniques. Properly used, these factoring methods can yield quick fixed-capital investment requirements with accuracies sufficient for most economic-evaluation purposes.

Method F. Investment cost per unit of capacity Many data have been published giving the fixed-capital investment required for various processes per unit of annual production capacity such as those shown in Table 19. Although these values depend to some extent on the capacity of the individual plants, it is possible to determine the unit investment costs which apply for average conditions. An order-of-magnitude estimate of the fixed-capital investment for a given process can then be obtained by multiplying the appropriate investment cost per unit of capacity by the annual production capacity of the proposed plant. The necessary correction for change of costs with time can be made with the use of cost indexes.

Method G. Turnover ratios A rapid evaluation method suitable for order-of-magnitude estimates is known as the "turnover ratio" method. Turnover ratio is defined as the ratio of gross annual sales to the fixed-capital investment,

$$\text{Turnover ratio} = \frac{\text{gross annual sales}}{\text{fixed-capital investment}} \tag{12}$$

where the product of the annual production rate and the average selling price of the commodities produced is the gross annual sales figure. The reciprocal of the turnover ratio is sometimes defined as the *capital ratio* or the *investment ratio*.† Turnover ratios of up to 5 are common for some business establishments and some are as low as 0.2. For the chemical industry, as a very rough rule of thumb, the ratio can be approximated as 1.

ORGANIZATION FOR PRESENTING CAPITAL INVESTMENT ESTIMATES BY COMPARTMENTALIZATION

The methods for estimating capital investment presented in the preceding sections represent the fundamental approaches that can be used. However, the direct application of these methods can often be accomplished with considerable

---

† When the term *investment ratio* is used, the investment is usually considered to be the total capital investment which includes working capital as well as other capitalized costs.

improvement by considering the fixed-capital investment requirement by parts. With this approach, each identified part is treated as a separate unit to obtain the total investment cost directly related to it. Various forms of compartmentalization for this type of treatment have been proposed. Included in these are (1) the *modular estimate*,† (2) the *unit-operations estimate*,‡ (3) the *functional-unit estimate*,§ and (4) the *average-unit-cost estimate*.¶

The same principle of breakdown into individual components is used for each of the four approaches. For the *modular estimate*, the basis is to consider individual modules in the total system with each module consisting of a group of similar items. For example, all heat exchangers might be included in one module, all furnaces in another, all vertical process vessels in another, etc. The total cost estimate is considered under six general groupings including chemical processing, solids handling, site development, industrial buildings, offsite facilities, and project indirects. As an example of an equipment cost module for heat exchangers, the module would include the basic delivered cost of the piece of equipment with factors similar to Lang factors being presented for supplemental items needed to get the equipment ready for use such as piping, insulation, paint, labor, auxiliaries, indirect costs, and contingencies.

In presenting the basic data for the module factors, the three critical variables are size or capacity of the equipment, materials of construction, and operating pressure with temperature often being given as a fourth critical variable. It is convenient to establish the base cost of all equipment as that constructed of carbon steel and operated at atmospheric pressure. Factors, such as are presented in Table 16, are then used to change the estimated costs of the equipment to account for variation in the preceding critical variables. Once the equipment cost for the module is determined, various factors are applied to obtain the final fixed-capital investment estimate for the item completely installed and ready for operation. Figure 5-6 shows two typical module approaches with Fig. 5-6a representing a module that applies to a "normal" chemical process where the overall Lang factor for application to the f.o.b. cost of the original equipment is 3.482 and Fig. 5-6b representing a "normal" module for a piece of mechanical equipment where the Lang factor has been determined to be 2.456.

The modules referred to in the preceding can be based on combinations of

† W. J. Dodge *et al.*, Metropolitan New York Section of AACE, The Module Estimating Technique as an Aid in Developing Plant Capital Costs, *Trans. AACE* (1962); K. M. Guthrie, Capital Cost Estimating, *Chem. Eng.*, **76**(6):114 (March 24, 1969); K. M. Guthrie, "Process Plant Estimating, Evaluation, and Control," Craftsman Book Company of America, Solana Beach, California, 1974.

‡ E. F. Hensley, "The Unit-Operations Approach," American Association of Cost Engineers, Paper presented at Annual Meeting, 1967.

§ A. V. Bridgewater, The Functional-Unit Approach to Rapid Cost Estimation, *AACE Bull.* **18**(5):153 (1976).

¶ C. A. Miller, New Cost Factors Give Quick Accurate Estimates, *Chem. Eng.*, **72**(19):226 (Sept. 13, 1965); C. A. Miller, Current Concepts in Capital Cost Forecasting, *Chem. Eng. Progr.*, **69**(5):77 (1973).

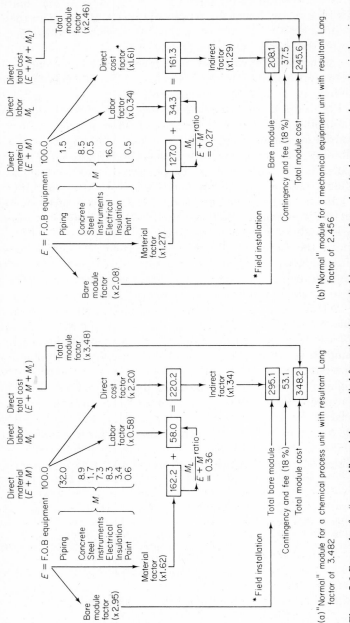
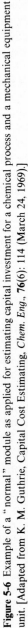

Figure 5-6 Example of a "normal" module as applied for estimating capital investment for a chemical process and a mechanical equipment unit. [Adapted from K. M. Guthrie, Capital Cost Estimating, *Chem. Eng.,* **76**(6): 114 (March 24, 1969).]

equipment that involve similar types of operations requiring related types of auxiliaries. An example would be a distillation operation requiring the distillation column with the necessary auxiliaries of reboiler, condenser, pumps, holdup tanks, and structural supports. This type of compartmentalization for estimating purposes can be considered as resulting in a so-called *unit-operations estimate*. Similarly, the *functional-unit estimate* is based on the grouping of equipment by function such as distillation or filtration and including the fundamental pieces of equipment as the initial basis with factors applied to give the final estimate of the capital investment.

The *average-unit-cost* method puts special emphasis on the three variables of size of equipment, materials of construction, and operating pressure as well as on the type of process involved. In its simplest form, all of these variables and the type of process can be accounted for by one number so that a given factor to convert the process equipment cost to total fixed-capital investment can apply for each "average unit cost." "Average unit cost" is defined as the total cost of the process equipment divided by the number of equipment items in that particular process. As the "average unit cost" increases, the size of the factor for converting equipment cost to total fixed-capital investment decreases with a range of factor values applicable for each "average unit cost" depending on the particular type of process, operating conditions, and materials of construction.

ESTIMATION OF TOTAL PRODUCT COST

Methods for estimating the total capital investment required for a given plant are presented in the first part of this chapter. Determination of the necessary capital investment is only one part of a complete cost estimate. Another equally important part is the estimation of costs for operating the plant and selling the products. These costs can be grouped under the general heading of *total product cost*. The latter, in turn, is generally divided into the categories of *manufacturing costs* and *general expenses*. Manufacturing costs are also known as *operating* or *production costs*. Further subdivision of the manufacturing costs is somewhat dependent upon the interpretation of direct and indirect costs.

Accuracy is as important in estimating total product cost as it is in estimating capital investment costs. The largest sources of error in total-product-cost estimation are overlooking elements of cost. A tabular form is very useful for estimating total product cost and constitutes a valuable checklist to preclude omissions. Figure 5-7 provides a suggested checklist which is typical of the costs involved in chemical processing operations.

Total product costs are commonly calculated on one of three bases: namely, daily basis, unit-of-product basis, or annual basis. The annual cost basis is probably the best choice for estimation of total product cost because (1) the effect of seasonal variations is smoothed out, (2) plant on-stream time or equipment-operating factor is considered, (3) it permits more-rapid calculation of operating costs at less than full capacity, and (4) it provides a convenient way of

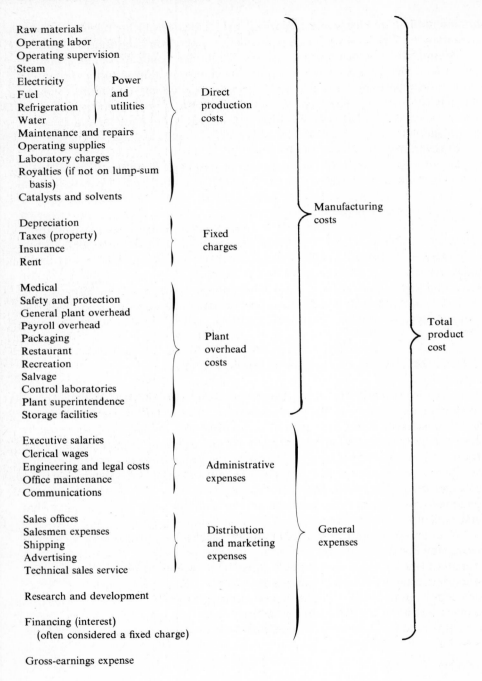

Raw materials
Operating labor
Operating supervision
Steam
Electricity
Fuel
Refrigeration
Water
Maintenance and repairs
Operating supplies
Laboratory charges
Royalties (if not on lump-sum basis)
Catalysts and solvents

Power and utilities

Direct production costs

Depreciation
Taxes (property)
Insurance
Rent

Fixed charges

Manufacturing costs

Medical
Safety and protection
General plant overhead
Payroll overhead
Packaging
Restaurant
Recreation
Salvage
Control laboratories
Plant superintendence
Storage facilities

Plant overhead costs

Total product cost

Executive salaries
Clerical wages
Engineering and legal costs
Office maintenance
Communications

Administrative expenses

Sales offices
Salesmen expenses
Shipping
Advertising
Technical sales service

Distribution and marketing expenses

General expenses

Research and development

Financing (interest)
  (often considered a fixed charge)

Gross-earnings expense

Figure 5-7 Costs involved in total product cost for a typical chemical process plant.

considering infrequently occurring but large expenses such as annual turn-around costs in a refinery.

The best source of information for use in total-product-cost estimates is data from similar or identical projects. Most companies have extensive records of their operations, so that quick, reliable estimates of manufacturing costs and general expenses can be obtained from existing records. Adjustments for increased costs as a result of inflation must be made, and differences in plant site and geographical location must be considered.

Methods for estimating total product cost in the absence of specific information are discussed in the following paragraphs. The various cost elements are presented in the order shown in Fig. 5-7.

Manufacturing Costs

All expenses directly connected with the manufacturing operation or the physical equipment of a process plant itself are included in the manufacturing costs. These expenses, as considered here, are divided into three classifications as follows: (1) direct production costs, (2) fixed charges, and (3) plant-overhead costs.

Direct production costs include expenses directly associated with the manufacturing operation. This type of cost involves expenditures for raw materials (including transportation, unloading, etc.); direct operating labor; supervisory and clerical labor directly connected with the manufacturing operation; plant maintenance and repairs; operating supplies; power; utilities; royalties; and catalysts.

It should be recognized that some of the variable costs listed here as part of the direct production costs have an element of fixed cost in them. For instance, maintenance and repair decreases, but not directly, with production level because a maintenance and repair cost still occurs when the process plant is shut down.

Fixed charges are expenses which remain practically constant from year to year and do not vary widely with changes in production rate. Depreciation, property taxes, insurance, and rent require expenditures that can be classified as fixed charges.

Plant-overhead costs are for hospital and medical services; general plant maintenance and overhead; safety services; payroll overhead including pensions, vacation allowances, social security, and life insurance; packaging, restaurant and recreation facilities, salvage services, control laboratories, property protection, plant superintendence, warehouse and storage facilities, and special employee benefits. These costs are similar to the basic fixed charges in that they do not vary widely with changes in production rate.

General Expenses

In addition to the manufacturing costs, other general expenses are involved in any company's operations. These general expenses may be classified as

(1) administrative expenses, (2) distribution and marketing expenses, (3) research and development expenses, (4) financing expenses, and (5) gross-earnings expenses.

Administrative expenses include costs for executive and clerical wages, office supplies, engineering and legal expenses, upkeep on office buildings, and general communications.

Distribution and marketing expenses are costs incurred in the process of selling and distributing the various products. These costs include expenditures for materials handling, containers, shipping, sales offices, salesmen, technical sales service, and advertising.

Research and development expenses are incurred by any progressive concern which wishes to remain in a competitive industrial position. These costs are for salaries, wages, special equipment, research facilities, and consultant fees related to developing new ideas or improved processes.

Financing expenses include the extra costs involved in procuring the money necessary for the capital investment. Financing expense is usually limited to interest on borrowed money, and this expense is sometimes listed as a fixed charge.

Gross-earnings expenses are based on income-tax laws. These expenses are a direct function of the gross earnings made by all the various interests held by the particular company. Because these costs depend on the company-wide picture, they are often not included in predesign or preliminary cost-estimation figures for a single plant, and the probable returns are reported as the gross earnings obtainable with the given plant design. However, when considering net profits, the expenses due to income taxes are extremely important, and this cost must be included as a special type of general expense.

DIRECT PRODUCTION COSTS

Raw Materials

In the chemical industry, one of the major costs in a production operation is for the raw materials involved in the process. The amount of the raw materials which must be supplied per unit of time or per unit of product can be determined from process material balances. In many cases, certain materials act only as an agent of production and may be recoverable to some extent. Therefore, the cost should be based only on the amount of raw materials actually consumed as determined from the overall material balances.

Direct price quotations from prospective suppliers are preferable to published market prices. For preliminary cost analyses, market prices are often used for estimating raw-material costs. These values are published regularly in journals such as the *Chemical Marketing Reporter* (formerly the *Oil, Paint and Drug Reporter*).

Freight or transportation charges should be included in the raw-material costs, and these charges should be based on the form in which the raw materials are to be purchased for use in the final plant. Although bulk shipments are cheaper than smaller-container shipments, they require greater storage facilities and inventory. Consequently, the demands to be met in the final plant should be considered when deciding on the cost of raw materials.

The ratio of the cost of raw materials to total plant cost obviously will vary considerably for different types of plants. In chemical plants, raw-material costs are usually in the range of 10 to 50 percent of the total product cost.

Operating Labor

In general, operating labor may be divided into skilled and unskilled labor. Hourly wage rates for operating labor in different industries at various locations can be obtained from the U.S. Bureau of Labor *Monthly Labor Review.* For chemical processes, operating labor usually amounts to about 15 percent of the total product cost.

In preliminary costs analyses, the quantity of operating labor can often be estimated either from company experience with similar processes or from published information on similar processes. Because the relationship between labor requirements and production rate is not always a linear one, a 0.2 to 0.25 power of the capacity ratio when plant capacities are scaled up or down is often used.†

If a flow sheet and drawings of the process are available, the operating labor may be estimated from an analysis of the work to be done. Consideration must be given to such items as the type and arrangement of equipment, multiplicity of units, amount of instrumentation and control for the process, and company policy in establishing labor requirements. Table 21 indicates some typical labor requirements for various types of process equipment.

† F. P. O'Connell, *Chem. Eng.,* **69**(4): 150 (1962).

Table 21 Typical labor requirements for process equipment

Type of equipment	Workers/unit/shift
Dryer, rotary	$\frac{1}{2}$
Dryer, spray	1
Dryer, tray	$\frac{1}{2}$
Centrifugal separator	$\frac{1}{4} - \frac{1}{2}$
Crystallizer, mechanical	$\frac{1}{6}$
Filter, vacuum	$\frac{1}{8} - \frac{1}{4}$
Evaporator	$\frac{1}{4}$
Reactor, batch	1
Reactor, continuous	$\frac{1}{2}$
Steam plant (100,000 lb/h)	3

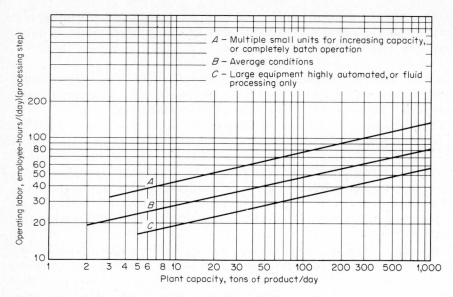

Figure 5-8 Operating labor requirements for chemical process industries.

Another method of estimating labor requirements as a function of plant capacity is based on adding up the various principal processing steps on the flow sheets.† In this method, a process step is defined as any unit operation, unit process, or combination thereof, which takes place in one or more units of integrated equipment on a repetitive cycle or continuously, e.g., reaction, distillation, evaporation, drying, filtration, etc. Once the plant capacity is fixed, the number of employee-hours per ton of product per step is obtained from Fig. 5-8 and multiplied by the number of process steps to give the total employee-hours per ton of production. Variations in labor requirements from highly automated processing steps to batch operations are provided by selection of the appropriate curve on Fig. 5-8.

Example 6. Estimation of labor requirements Consider a highly automated processing plant having a capacity of 100 tons/day of product and requiring principal processing steps of heat transfer, reaction, and distillation. What are the average operating labor requirements for an annual operation of 300 days?

SOLUTION The process plant is considered to require three process steps. From Fig. 5-8, for a capacity of 100 tons product/day, the highly automated

† H. E. Wessel, New Graph Correlates Operating Labor for Chemical Processes, *Chem. Eng.*, **59**(7):209 (July, 1952).

process plant requires 33 employee-hours/day/processing step. Thus, for 300 days annual operation, operating labor required = (3)(33)(300) = 29,700 employee-hours/year.

Because of new technological developments including computerized controls and long-distance control arrangements, the practice of relating employee-hour requirements directly to production quantities for a given product can give inaccurate results unless very recent data are used. As a general rule of thumb,† the labor requirements for a fluids-processing plant, such as an ethylene oxide plant or others as shown in Table 22, would be in the low range of $\frac{1}{3}$ to 2 employee-hours per ton of product; for a solid-fluids plant, such as a polyethylene plant, the labor requirement would be in the intermediate range of 2 to 4 employee-hours per ton of product; for plants primarily engaged in solids processing, such as a coal briquetting plant, the large amount of materials handling would make the labor requirements considerably higher than for other types of plants with a range of 4 to 8 employee-hours per ton of product being reasonable. The data shown in Fig. 5-8 and Table 22, where plant capacity and specific type of process are taken into account, are much more accurate than the preceding rule of thumb if the added necessary information is available.

In determining costs for labor, account must be taken of the type of worker required, the geographical location of the plant, the prevailing wage rates, and worker productivity. Table 20 presents data that can be used as a guide for relative median labor rates and productivity factors for workers in various geographical areas of the United States. Tables 23 and 24 provide data on labor rates in dollars per hour for the U.S. Gulf Coast region and average labor indexes to permit estimation of prevailing wage rates.

Direct Supervisory and Clerical Labor

A certain amount of direct supervisory and clerical labor is always required for a manufacturing operation. The necessary amount of this type of labor is closely related to the total amount of operating labor, complexity of the operation, and product quality standards. The cost for direct supervisory and clerical labor averages about 15 percent of the cost for operating labor. For reduced capacities, supervision usually remains fixed at the 100-percent-capacity rate.

Utilities

The cost for utilities, such as steam, electricity, process and cooling water, compressed air, natural gas, and fuel oil, varies widely depending on the amount of consumption, plant location, and source. For example, costs for a few selected

---

† J. E. Haselbarth, Updated Investment Costs for 60 Chemical Plants, *Chem. Eng.*, **74**(25):214 (Dec. 4, 1967).

Table 22 Operating labor, fuel, steam, power, and water requirements for various processes†

	Capacity thousand ton/yr	Operating labor and supervision workhours/ ton	Maintenance labor and supervision workhours/ ton	Power and utilities, per ton or bbl			
				Fuel MM Btu/h	Steam lb/h	Power kWh	Water gpm
			Chemical plants				
Acetone	100	0.518	0.315	.....	1.73	310	5.18
Acetic acid	10	1.483	0.984	.....	......	180	0.58
Butadiene	100	0.345	0.285	.....	0.012	130	0.73
Ethylene oxide	100	0.232	0.104	.....	4.88	140	0.148
Formaldehyde	100	0.259	0.328	.....	34.6	200	0.029
Hydrogen peroxide	100	0.288	0.352	.....	2.62	160	0.186
Isoprene	100	0.230	0.325	.....	0.81	710	0.001
Phosphoric acid	10	1.85	0.442	.....	0.18	40	0.03
Polyethylene	100	0.259	0.295	.....	0.23	450	0.0004
Urea	100	0.238	0.215	.....	0.33	135	0.0002
Vinyl acetate	100	0.432	0.528	.....	1.34	275	0.27
			Refinery units				
	Thousand bbl/day	Workhours/ bbl	Workhours/ bbl				
Alkylation	10	0.007	0.0895	.....	10.83	0.07	1.48
Coking (delayed)	10	0.011‡	0.0096	0.007	1.85	0.07	....
Coking (fluid)	10	0.0096	0.0058	0.012	2.55	0.06	0.64
Cracking (fluid)	10	0.0122	0.0115	.....	(4.73)§	0.02	0.33
Cracking (thermal)	10	0.0096	0.0025	0.012	(2.55)§	0.06	0.64
Distillation (atm)	10	0.0048	0.0042	0.004	0.25	0.03	0.16
Distillation (vac)	10	0.0024	0.0154	0.003	0.95	0.04	0.18
Hydrotreating	10	0.0048	0.0028	0.006	0.92	0.01	0.14
Reforming, catalyt.	10	0.0048	0.0078	0.002	1.38	0.23	0.28
Polymerization	10	0.0024	0.0158	.....	4.85	0.07	0.43

† Based on information from K. M. Guthrie, Capital and Operating Costs for 54 Chemical Processes, *Chem. Eng.*, **77**(13): 140 (June 15, 1970).
‡ Includes two coke cutters (1 shift/day).
§ Net steam generated.

utilities in the U.S. Gulf Coast region are given in Table 23. A more detailed list of average rates for various utilities is presented in Appendix B. The required utilities can sometimes be estimated in preliminary cost analyses from available information about similar operations as shown in Table 22. If such information is unavailable, the utilities must be estimated from a preliminary design. The utility may be purchased at predetermined rates from an outside source, or the service may be available from within the company. If the company supplied its own service and this is utilized for just one process, the entire cost of the service

Table 23 Cost tabulation for selected utilities and labor†‡

1979 costs based on U.S. Gulf Coast location

	Cost
Fuel costs	
Gas at well head including gathering-system costs:	
Existing contracts, $/million Btu	1.10
New contracts, $/million Btu	2.00
Fuel oil in $/million Btu with 6.25 million Btu/bbl	2.30
Gas transmission costs in ¢/100 miles	4.50
Plant fuel gas in $/million Btu	2.30
Purchased power for midcontinent USA in ¢/kWh	3.70
Water costs	
Process water (treated) in ¢/1000 gal	50
Cooling water in ¢/1000 gal	6
Labor rates (ENR Skilled Labor Index = 260. See Table 24)	
Supervisor, $/h	10.50
Operators, $/h	9.00
Helpers, $/h	8.00
Chemists, $/h	8.50
Labor burden as % of direct labor§	25
Plant general overhead as % of total labor + burden	40

† Based on information from C. C. Johnnie and D. K. Aggarwal, Calculating Plant Utility Costs, *Chem. Eng. Progr.,* **73**(11): 84 (1977).

‡ See Appendix B for a more detailed listing of utility and related costs.

§ Labor burden refers to costs the company must pay associated with and above the base labor rate, such as for Social Security, insurance, and other benefits.

installation is usually charged to the manufacturing process. If the service is utilized for the production of several different products, the service cost is apportioned among the different products at a rate based on the amount of individual consumption.

Steam requirements include the amount consumed in the manufacturing process plus that necessary for auxiliary needs. An allowance for radiation and line losses must also be made.

Electrical power must be supplied for lighting, motors, and various process-equipment demands. These direct-power requirements should be increased by a factor of 1.1 to 1.25 to allow for line losses and contingencies. As a rough approximation, utility costs for ordinary chemical processes amount to 10 to 20 percent of the total product cost.

Maintenance and Repairs

A considerable amount of expense is necessary for maintenance and repairs if a plant is to be kept in efficient operating condition. These expenses include the cost for labor, materials, and supervision.

Table 24 Engineering News-Record labor indexes to permit estimation of prevailing wage rates by location†

(See Table 23 for values of labor rates as $/h)

Location	ENR Skilled Labor Index (December values). (Based on 1967 = 100)									
	1967	1970	1971	1972	1973	1974	1975	1976	1977	1978
Atlanta	103	136	155	190	177	186	192	205	214	224
Baltimore	103	144	168	177	186	191	208	224	228	237
Birmingham	104	123	149	167	173	193	205	217	229	251
Boston	104	138	152	169	178	191	205	215	227	236
Chicago	102	141	159	172	181	193	203	215	229	239
Cincinnati	106	152	182	188	195	210	225	234	251	268
Cleveland	108	150	166	175	182	194	208	223	237	252
Dallas	101	137	148	166	178	187	197	217	226	245
Denver	98	139	159	170	180	193	207	222	243	252
Detroit	104	153	169	178	186	200	215	228	241	264
Kansas City	101	148	171	186	197	213	224	239	254	271
Los Angeles	103	132	154	170	178	194	213	242	262	280
Minneapolis	102	146	165	170	175	190	203	217	230	245
New Orleans	104	134	147	160	174	192	203	225	239	255
New York	101	130	153	161	170	179	192	207	217	216
Philadelphia	102	146	163	176	185	199	213	227	238	248
Pittsburgh	109	136	160	168	174	184	194	207	225	236
St Louis	101	131	143	155	161	174	189	202	213	228
San Francisco	103	134	147	163	168	188	210	228	235	259
Seattle	103	135	146	156	164	184	203	219	238	264
Montreal	107	135	140	149	167	185	218	239	279	304
Toronto	105	152	173	194	211	226	253	290	316	336

† Engineering News-Record, **202**(12): 74 (March 22, 1979). Published monthly in the second ENR issue of the month.

Annual costs for equipment maintenance and repairs may range from as low as 2 percent of the equipment cost if service demands are light to 20 percent for cases in which there are severe operating demands. Charges of this type for buildings average 3 to 4 percent of the building cost. In the process industries, the total plant cost per year for maintenance and repairs is roughly equal to an average of 6 percent of the fixed-capital investment. Table 25 provides a guide for estimation of maintenance and repair costs as a function of process conditions.

For operating rates less than plant capacity, the maintenance and repair cost is generally estimated as 85 percent of that at 100 percent capacity for a 75 percent operating rate, and 75 percent of that at 100 percent capacity for a 50 percent operating rate.

Operating Supplies

In any manufacturing operation, many miscellaneous supplies are needed to keep the process functioning efficiently. Items such as charts, lubricants, test

Table 25 Estimation of costs for maintenance and repairs

	Maintenance cost as percentage of fixed-capital investment (on annual basis)		
Type of operation	Wages	Materials	Total
Simple chemical processes	1–3	1–3	2–6
Average processes with normal operating conditions	2–4	3–5	5–9
Complicated processes, severe corrosion operating conditions, or extensive instrumentation	3–5	4–6	7–11

chemicals, custodial supplies, and similar supplies cannot be considered as raw materials or maintenance and repair materials, and are classified as operating supplies. The annual cost for this type of supplies is about 15 percent of the total cost for maintenance and repairs.

Laboratory Charges

The cost of laboratory tests for control of operations and for product-quality control is covered in this manufacturing cost. This expense is generally calculated by estimating the employee-hours involved and multiplying this by the appropriate rate. For quick estimates, this cost may be taken as 10 to 20 percent of the operating labor.

Patents and Royalties

Many manufacturing processes are covered by patents, and it may be necessary to pay a set amount for patent rights or a royalty based on the amount of material produced. Even though the company involved in the operation obtained the original patent, a certain amount of the total expense involved in the development and procurement of the patent rights should be borne by the plant as an operating expense. In cases of this type, these costs are usually amortized over the legally protected life of the patent. Although a rough approximation of patent and royalty costs for patented processes is 0 to 6 percent of the total product cost, the engineer must use judgment because royalties vary with such factors as the type of product and the industry.

Catalysts and Solvents

Costs for catalysts and solvents can be significant and depend upon the specific manufacturing processes chosen.

FIXED CHARGES

Certain expenses are always present in an industrial plant whether or not the manufacturing process is in operation. Costs that are invariant with the amount of production are designated as *fixed costs* or *fixed charges*. These include costs for depreciation, local property taxes, insurance, and rent. Expenses of this type are a direct function of the capital investment. As a rough approximation, these charges amount to about 10 to 20 percent of the total product cost.

Depreciation

Equipment, buildings, and other material objects comprising a manufacturing plant require an initial investment which must be written off as a manufacturing expense. In order to write off this cost, a decrease in value is assumed to occur throughout the usual life of the material possessions. This decrease in value is designated as *depreciation.*

Since depreciation rates are very important in determining the amount of income tax, the U.S. Bureau of Internal Revenue has established allowable depreciation rates based on the probable useful life of various types of equipment and other fixed items involved in manufacturing operations. While several alternative methods may be used for determining the rate of depreciation, a straight-line method is usually assumed for engineering projects. In applying this method, a useful-life period and a salvage value at the end of the useful life are assumed, with due consideration being given to possibilities of obsolescence and economic changes. The difference between initial cost and the salvage value divided by the total years of useful life gives the annual cost due to depreciation.

The annual depreciation rate for machinery and equipment ordinarily is about 10 percent of the fixed-capital investment, while buildings are usually depreciated at an annual rate of about 3 percent of the initial cost.

Local Taxes

The magnitude of local property taxes depends on the particular locality of the plant and the regional laws. Annual property taxes for plants in highly populated areas are ordinarily in the range of 2 to 4 percent of the fixed-capital investment. In less populated areas, local property taxes are about 1 to 2 percent of the fixed-capital investment.

Insurance

Insurance rates depend on the type of process being carried out in the manufacturing operation and on the extent of available protection facilities. On an annual basis, these rates amount to about 1 percent of the fixed-capital investment.

Rent

Annual costs for rented land and buildings amount to about 8 to 12 percent of the value of the rented property.

PLANT OVERHEAD COSTS

The costs considered in the preceding sections are directly related with the production operation. In addition, however, many other expenses are always involved if the complete plant is to function as an efficient unit. The expenditures required for routine plant services are included in *plant-overhead costs*. Nonmanufacturing machinery, equipment, and buildings are necessary for many of the general plant services, and the fixed charges and direct costs for these items are part of the plant-overhead costs.

Expenses connected with the following comprise the bulk of the charges for plant overhead:

Hospital and medical services
General engineering
Safety services
Cafeteria and recreation facilities
General plant maintenance and overhead
Payroll overhead including employee benefits
Control laboratories
Packaging
Plant protection
Janitor and similar services
Employment offices
Distribution of utilities
Shops
Lighting
Interplant communications and transportation
Warehouses
Shipping and receiving facilities

These charges are closely related to the costs for all labor directly connected with the production operation. The plant-overhead cost for chemical plants is about 50 to 70 percent of the total expense for operating labor, supervision, and maintenance.

ADMINISTRATIVE COSTS

The expenses connected with top-management or administrative activities cannot be charged directly to manufacturing costs; however, it is necessary to include the *administrative costs* if the economic analysis is to be complete. Salaries and wages for administrators, secretaries, accountants, stenographers, typists, and similar workers are part of the administrative expenses, along with costs for office supplies and equipment, outside communications, administrative buildings, and other overhead items related with administrative activities. These costs may vary markedly from plant to plant and depend somewhat on whether the plant under consideration is a new one or an addition to an old plant. In the absence of more-accurate cost figures from company records, or for a quick estimate, the administrative costs may be approximated as 20 to 30 percent of the operating labor.

DISTRIBUTION AND MARKETING COSTS

From a practical viewpoint, no manufacturing operation can be considered a success until the products have been sold or put to some profitable use. It is necessary, therefore, to consider the expenses involved in selling the products. Included in this category are salaries, wages, supplies, and other expenses for sales offices; salaries, commissions, and traveling expenses for salesmen; shipping expenses; cost of containers; advertising expenses; and technical sales service.

Distribution and marketing costs vary widely for different types of plants depending on the particular material being produced, other products sold by the company, plant location, and company policies. These costs for most chemical plants are in the range of 2 to 20 percent of the total product cost. The higher figure usually applies to a new product or to one sold in small quantities to a large number of customers. The lower figure applies to large-volume products, such as bulk chemicals.

RESEARCH AND DEVELOPMENT COSTS

New methods and products are constantly being developed in the chemical industries. These accomplishments are brought about by emphasis on research and development. *Research and development costs* include salaries and wages for all personnel directly connected with this type of work, fixed and operating expenses for all machinery and equipment involved, costs for materials and supplies, direct overhead expenses, and miscellaneous costs. In the chemical industry, these costs amount to about 2 to 5 percent of every sales dollar.

FINANCING

Interest

Interest is considered to be the compensation paid for the use of borrowed capital. A fixed rate of interest is established at the time the capital is borrowed; therefore, interest is a definite cost if it is necessary to borrow the capital used to make the investment for a plant. Although interest on borrowed capital is a fixed charge, there are many persons who claim that interest should not be considered as a manufacturing cost. It is preferable to separate interest from the other fixed charges and list it as a separate expense under the general heading of management or financing cost. Annual interest rates amount to 5 to 10 percent of the total value of the borrowed capital.

When the capital investment is supplied directly from the existing funds of a company, it is a debatable point whether interest should be charged as a cost. For income-tax calculations, interest on owned money cannot be charged as a cost. In design calculations, however, interest can be included as a cost unless there is assurance that the total capital investment will be supplied from the company's funds and the company policies permit exclusion of interest as a cost.

GROSS-EARNINGS COSTS

The total income minus the total production cost gives the *gross earnings* made by the particular production operation, which can then be treated mathematically by any of several methods to measure the profitability of the proposed venture or project. These methods will be discussed later in Chaps. 6 and 9.

Because of income-tax demands, the final *net profit* is often much less than the gross earnings. Income-tax rates are based on the gross earnings received from all the company interests. Consequently, the magnitude of these costs varies widely from one company to another.

On an annual basis, the corporate income-tax laws for the United States in 1979 required payment of a 17, 20, 30, and 40 percent normal tax on the 1st, 2nd, 3rd, and 4th $25,000, respectively, of the annual gross earnings of a corporation plus 46 percent of all annual gross earnings above $100,000. In addition, if other levies, such as state income taxes, were included, the overall tax rate could have been even higher. Tax rates vary from year to year depending on Federal and state regulations as is shown in the following example where 1974 Federal-tax rates are considered.

Example 7. Break-even point, gross earnings, and net profit for a process plant The annual direct production costs for a plant operating at 70 percent capacity are $280,000 while the sum of the annual fixed charges, overhead costs, and general expenses is $200,000. What is the break-even point in units of production per year if total annual sales are $560,000 and the product sells at $40 per unit? What were the annual gross earnings and net profit for this plant at 100 percent capacity in the year 1974 when corporate income taxes required a 22 percent normal tax on the total gross earnings plus a 26 percent surtax on gross earnings above $25,000?

SOLUTION The break-even point (Fig. 5-3) occurs when the total annual product cost equals the total annual sales. The total annual product cost is the sum of the fixed costs (includes fixed charges, overhead, and general expenses) and the direct production costs for n units per year. The total annual sales is the product of the number of units and the selling price per unit. Thus

$$\text{Direct production cost/unit} = \frac{280,000}{(560,000/40)} = \$20/\text{unit}$$

and the number of units needed for a break-even point is given by

$$200,000 + 20n = 40n$$

$$n = \frac{200,000}{20} = 10,000 \text{ units/year}$$

This is $[(10,000)/(14,000/0.7)]100 = 50\%$ of the present plant operating capacity.

$$\text{Gross annual earnings} = \text{total annual sales} - \text{total annual product cost}$$

$$= \frac{14,000}{0.7} \text{ units } (\$40/\text{unit})$$

$$- \left[200,000 + \frac{14,000}{0.7} \text{ units } (\$20/\text{unit}) \right]$$

$$= 800,000 - 600,000$$

$$= \$200,000$$

$$\text{Net annual earnings} = \text{gross annual earnings} - \text{income taxes}$$

$$= 200,000 - [(0.22)(200,000) + (0.26)(175,000)]$$

$$= 200,000 - 89,500$$

$$= \$110,500$$

CONTINGENCIES

Unforeseen events, such as strikes, storms, floods, price variations, and other *contingencies*, may have an effect on the costs for a manufacturing operation. When the chemical engineer predicts total costs, it is advisable to take these factors into account. This can be accomplished by including a contingency factor equivalent to 1 to 5 percent of the total product cost.

SUMMARY

This chapter has outlined the economic considerations which are necessary when a chemical engineer prepares estimates of capital investment cost or total product cost for a new venture or project. Methods for obtaining predesign cost estimates have purposely been emphasized because the latter are extremely important for determining the feasibility of a proposed investment and to compare alternative designs. It should be remembered, however, that predesign estimates are often based partially on approximate percentages or factors that are applicable to a particular plant or process under consideration. Tables 26 and 27 summarize the predesign estimates for capital investment costs and total product costs, respectively. The percentages indicated in both tables give the ranges

Table 26 Estimation of capital investment cost (showing individual components)

The percentages indicated in the following summary of the various costs constituting the capital investment are approximations applicable to ordinary chemical processing plants. It should be realized that the values given can vary depending on many factors, such as plant location, type of process, complexity of instrumentation, etc.

  I. **Direct costs** = material and labor involved in actual installation of complete facility (70–85% of fixed-capital investment)
    A. Equipment + installation + instrumentation + piping + electrical + insulation + painting (50–60% of fixed-capital investment)
        1. Purchased equipment (15–40% of fixed-capital investment)
        2. Installation, including insulation and painting (25–55% of purchased-equipment cost)
        3. Instrumentation and controls, installed (6–30% of purchased-equipment cost)
        4. Piping, installed (10–80% of purchased-equipment cost)
        5. Electrical, installed (10–40% of purchased-equipment cost)
    B. Buildings, process and auxiliary (10–70% of purchased-equipment cost)
    C. Service facilities and yard improvements (40–100% of purchased-equipment cost)
    D. Land (1–2% of fixed-capital investment or 4–8% of purchased-equipment cost)
 II. **Indirect costs** = expenses which are not directly involved with material and labor of actual installation of complete facility (15–30% of fixed-capital investment)
    A. Engineering and supervision (5–30% of direct costs)
    B. Construction expense and contractor's fee (6–30% of direct costs)
    C. Contingency (5–15% of fixed-capital investment)
III. **Fixed-capital investment** = direct costs + indirect costs
 IV. **Working capital** (10–20% of total capital investment)
  V. **Total capital investment** = fixed-capital investment + working capital

---

Table 27 Estimation of total product cost (showing individual components)

The percentages indicated in the following summary of the various costs involved in the complete operation of manufacturing plants are approximations applicable to ordinary chemical processing plants. It should be realized that the values given can vary depending on many factors, such as plant location, type of process, and company policies.

Percentages are expressed on an annual basis.

  I. **Manufacturing cost** = direct production costs + fixed charges + plant overhead costs
    A. Direct production costs (about 60% of total product cost)
        1. Raw materials (10–50% of total product cost)
        2. Operating labor (10–20% of total product cost)
        3. Direct supervisory and clerical labor (10–25% of operating labor)
        4. Utilities (10–20% of total product cost)
        5. Maintenance and repairs (2–10% of fixed-capital investment)
        6. Operating supplies (10–20% of cost for maintenance and repairs, or 0.5–1% of fixed-capital investment)
        7. Laboratory charges (10–20% of operating labor)
        8. Patents and royalties (0–6% of total product cost)
    B. Fixed charges (10–20% of total product cost)
        1. Depreciation (depends on life period, salvage value, and method of calculation—about 10% of fixed-capital investment for machinery and equipment and 2–3% of building value for buildings)
        2. Local taxes (1–4% of fixed-capital investment)
        3. Insurance (0.4–1% of fixed-capital investment)
        4. Rent (8–12% of value of rented land and buildings)

(continued)

Table 27 Estimation of total product cost (showing individual components) (*Continued*)

      C. Plant-overhead costs (50–70% of cost for operating labor, supervision, and maintenance, or 5–15% of total product cost); includes costs for the following: general plant upkeep and overhead, payroll overhead, packaging, medical services, safety and protection, restaurants, recreation, salvage, laboratories, and storage facilities.

  II. **General expenses** = administrative costs + distribution and selling costs + research and development costs

      A. Administrative costs (about 15% of costs for operating labor, supervision, and maintenance, or 2–6% of total product cost); includes costs for executive salaries, clerical wages, legal fees, office supplies, and communications

      B. Distribution and selling costs (2–20% of total product cost); includes costs for sales offices, salesmen, shipping, and advertising

      C. Research and development costs (2–5% of every sales dollar or about 5% of total product cost)

      D. Financing (interest)† (0–10% of total capital investment)

 III. **Total product cost**‡ = manufacturing cost + general expenses

 IV. **Gross-earnings cost** (gross earnings = total income − total product cost; amount of gross-earnings cost depends on amount of gross earnings for entire company and income-tax regulations; a general range for gross-earnings cost is 30–60% of gross earnings)

---

  † Interest on borrowed money is often considered as a fixed charge.

  ‡ If desired, a contingency factor can be included by increasing the total product cost by 1–5%.

encountered in typical chemical plants. Because of the wide variations in different types of plants, the factors presented should be used *only* when more accurate data are not available.

NOMENCLATURE FOR CHAPTER 5

A = incremental cost of corrosion-resistant alloy materials

A_x = nonmanufacturing fixed-capital investment

c_o = costs for operations (*not* including depreciation)

C = original capital investment

C_n = new capital investment

d = depreciation charge

d_n = number of drawings and specifications

D = total direct cost of plant

e = total heat exchanger cost (less incremental cost of alloy)

e_L = labor efficiency index in new location relative to cost of E_L and M'_L

E = purchased-equipment cost (installation cost not included) on f.o.b. basis

E_i = installed-equipment cost (purchased and installation cost included)

E_L = purchased-equipment labor cost (base)

f = lumped cost index relative to original installation cost

f_1, f_2 = multiplying factors for piping, electrical, instrumentation, etc.

f_d = unit cost per drawing and specification
f_e = unit cost for engineering
f_E = current equipment cost index relative to cost of E
f_F = construction or field-labor expense factor always greater than 1
f_I = indirect cost factor always greater than 1
f_L = current labor cost index in new location relative to cost of E_L and M'_L
f_M = current material cost index relative to cost of M
f_m = cost factor for miscellaneous items
f_p = cost factor for piping materials
f_v = total cost of field-fabricated vessels (less incremental cost of alloy)
f_x = specific material unit cost, e.g., f_p = unit cost of pipe
f_y = specific material labor unit cost per employee-hour
H_e = engineering employee-hours
I = total indirect cost of plant
M = material cost
M'_L = labor employee-hours for specific material
M_L = direct labor cost for equipment installation and material handling
M_x = specific material quantity in compatible units
p = total pump plus driver cost (less incremental cost of alloy)
R = ratio of new to original capacity
s_i = total income from sales
t = total cost of tower shells (less incremental cost of alloy)
T = total capital investment
V = manufacturing fixed-capital investment
W = working-capital investment
x = exponential power for cost-capacity relationships

SUGGESTED ADDITIONAL REFERENCES FOR COST ESTIMATION

Building costs

Cherry, J. R., Cost of Concrete Industrial Buildings, *Civil Eng.*, **38**(1):42 (1968).
Godfrey, R. S., *et al*, "Building Construction Cost Data," 34th ed., Robert Snow Means Company, Inc., Duxbury, Massachusetts, 1975.
Guthrie, K. M., Capital Cost Estimating, *Chem. Eng.*, **76**(6):114 (1969).
Jones, L. R., Building-Cost Escalation, *Chem. Eng.*, **77**(16):170 (1970).
Knox, W. G., Estimating the Cost of Process Buildings via Volumetric Ratios, *Chem. Eng.*, **75**(13):292 (1968).
McKee-Berger-Mansueto, Inc., "Building Cost File," Western Edition, 4th ed., Construction Publishing Company, Inc., New York, 1975.
Moselle, G., "National Construction Estimator," 25th ed., Craftsman Book Company of America, Solana Beach, California, 1977.
Wallace, D. M., Saudi Arabia Building Costs, *Hydrocarbon Process.*, **55**(11):189 (1976).

Capital investment costs

Buehler, J. D., and G. J. Figge, Operating vs. Capital Costs: Evaluating Trade-off Benefits, *Chem. Eng.*, **78**(3):96 (1971).

de la Mare, R. F., Parameters Affecting Capital Investment, *Chem. Eng. (London)*, 227 (April 1975).

Drayer, D. E., How to Estimate Plant Cost-Capacity Relationship, *Petro/Chem. Eng.*, **42**(5):10 (1970).

Garcia-Borras, T., Research-Project Evaluations, Part 1, *Hydrocarbon Process.*, **55**(12):137(1976).

Grigsby, E. K., E. W. Mills, and D. C. Collins, What Will Future Refineries Cost? *Hydrocarbon Process.*, **52**(5):133 (1973).

Guthrie, K. M., Capital Cost Estimating, *Chem. Eng.*, **76**(6):114 (1969).

———, Capital and Operating Costs for 54 Chemical Processes, *Chem. Eng.*, **77**(13):140 (1970).

———, "Process Plant Estimating Evaluation and Control," Craftsman Book Company of America, Solana Beach, California, 1974.

Henderson, D. H., Maximizing the Capital Dollar, *Chem. Eng. Progr.*, **66**(8):11 (1970).

Holland, F. A., F. A. Watson, and J. K. Wilkinson, Capital Costs and Depreciation *Chem. Eng.*, **80**(17):118 (1973).

——— ——— and ———, How to Estimate Capital Costs, *Chem. Eng.*, **81**(7):71 (1974).

Jelen, F. C., "Cost and Optimization Engineering," McGraw-Hill Book Company, New York, 1970.

Johnson, P. W., and F. A. Peters, A Computer Program for Calculating Capital and Operating Costs, United States Department of the Interior, Bureau of Mines Information Circular 8426 (1969).

Klumpar, I. V., Project Evaluation by Computer, *Chem. Eng.*, **77**(14):76 (1970).

Miller, C. A., Capital Cost Estimating—A Science Rather than an Art, *AACE Cost Eng. Notebook*, A-1.000, June 1978.

Taylor, M. P., Why a Capital Investment Estimate?, presented at American Association Cost Engineers 12th National Meeting, Houston, Texas, June 17–19, 1968.

Williama, L. F., Capital Cost Estimating from the Viewpoint of Process Plant Contractor—Part II, *AACE Bull.*, **15**(1):5 (1973).

Winter, O., Preliminary Economic Evaluation of Chemical Processes at the Research Level, *I&EC*, **61**(4):45 (1969).

Yen, Y.-C., Estimating Plant Costs in the Developing Countries, *Chem. Eng.*, **79**(15):89 (1972).

Zimmerman, O. T., Elements of Capital Cost Estimation, *Cost Eng.*, **13**(4):4 (1968).

Zudkevitch, D., Imprecise Data Impacts Plant Design and Operation, *Hydrocarbon Process.*, **54**(3):97 (1975).

Developments in Capital Cost Estimating, *Chem. Eng. (London)*, 39 (January 1976).

Construction costs

Armstrong, R., Better Ways to Build Process Plants, *Chem. Eng.*, **79**(8):86 (1972).

Blough, R. M., Effect of Scheduled Overtime on Construction Projects, *AACE Bull.*, **15**(5):155 (1973).

Borneman, W. F., Estimating Electrical Construction Costs for Commercial and Industrial Projects, presented at American Association of Cost Engineers, 12th National Meeting, Houston, Texas, June 17–19, 1968.

Brown, B. E., Estimating Heavy Rigging, presented at American Association of Cost Engineers 12th National Meeting, Houston, Texas, June 17–19, 1968.

Foster, N., "Construction Estimates from Take-Off to Bid," 2d ed., McGraw-Hill Book Company, New York, 1972.

Gallagher, J. T., Analyzing Field Construction Costs, *Chem. Eng.*, **75**(11):182 (1968).

Galopin, F. E., Electrical Equipment and Materials Erection, *Hydrocarbon Process.*, **47**(7):112 (1968).

Godfrey, R. S., *et al.*, "Building Construction Cost Data," 34th ed., Robert Snow Means Company, Inc., Duxbury, Massachusetts, 1975.

Greek, B. F., Inflation of Plant Construction Costs Eases, *Chem. Eng. News*, **55**(15):12 (1977).

Guthrie, K. M., Field-Labor Predictions for Conceptual Projects, *Chem. Eng.*, **76**(7):170 (1969).

Hirschmann, W. B., Has the Cost of Building New Refineries Really Gone Up? *Chem. Eng. Progr.*, **67**(8):39 (1971).

Hudson, W. G., Pilot Plant Cost Control, *Chem. Eng. Progr.*, **64**(10):39 (1968).

Jacobs, J. J., Cost-Plus vs. Fixed-Fee Construction Contracts. *Chem. Eng.*, **80**(2):109 (1973).
Jarrold, R., and J. DeDevitis, Minimizing Construction Costs, *Chem. Eng. Progr.*, **66**(8):17 (1970).
Jones, L. R., Overtime-Pay Problems, *Chem. Eng.*, **78**(15):114 (1971).
Kennedy, D. E., Estimating Conceptual Costs for Construction Projects, *Plant Eng.*, **27**(16):72 (1973); **27**(19):140 (1973); **27**(8):128 (1973); **27**(13):88 (1973); **28**(8):120 (1974); **26**(26):50 (1972).
Mendel, O., How Location Affects U.S. Plant-Construction Costs, *Chem. Eng.*, **79**(28):120 (1972).
McKee-Berger-Mansueto, Inc., " Building Cost File, Western Edition," 4th ed., Construction Publishing Company, Inc., New York, 1976.
Moselle, G., " National Construction Estimator," 26th ed., Craftsman Book Company of America, Solano Beach, California, 1978.
Myles, K. W., Construction Cost Estimating—Concrete Slab on Grade, *Plant Eng.*, **21**(10):176 (1967).
————, Construction Cost Estimating—Brick and Concrete Block, *Plant Eng.*, **22**(11):56 (1969).
————, Construction Cost Estimating—Siding and Windows, *Plant Eng.*, **23**(4):61 (1969).
Nelson, W. L., What is Cost of Deeper Pipeline Ditching, *Oil & Gas J.*, **65**(3):97 (1967).
Pekar, P. P., Jr., Project Control: A Case for a New Evolutionary Time Series Prediction, *AACE Bull.*, **18**(6):195 (1976).
Weinheimer, W. R., Percent Your Indirect Field Costs, *1969 Trans. Am. Assoc. Cost Eng.* (1969).

Cost estimation

Abraham, C. T., and R. D. Prasad, Stochastic Model for Manufacturing Cost Estimation, *IBM J. Research and Development*, **13**(4):343 (1969).
Adams, L., Relative Metal Economy of Pressure-Vessel Steels, *Chem. Eng.*, **76**(27):150 (1969).
Agarwal, J. C., W. E. Sonstelie, and I. V. Klumpar, Preliminary Economic Analysis, *AACE Bull.*, **14**(1):7 (1972).
Allen, D. H., and R. C. Page, Revised Technique for Predesign Cost Estimating, *Chem. Eng.*, **82**(5):142 (1975).
Baltzell, H. J., Estimate Refinery Cost Better, *Hydrocarbon Process.*, **57**(1):129 (1978).
Blecker, H. G., Computer Simulation of Definitive Estimates, *1973 Trans. Am. Assoc. Cost Eng.* (1973), p. 143.
Bresler, S. A., and M. T. Kuo, Cost Estimating by Computer, *Chem. Eng.*, **79**(12):84 (1972).
———— and ————, More Programs for Cost Estimating by Computer, *Chem. Eng.*, **79**(14):130 (1972).
Bridgwater, A. V., The Functional Unit Approach to Rapid Cost Estimation, *AACE Bull.*, **18**(5):153 (1976).
Chase, J., Plant Cost vs. Capacity: New Way to Use Exponents, *Chem. Eng.*, **77**(7):113 (1970).
Clark, F. D., and A. B. Lorenzoni, "Applied Cost Engineering," Marcel Dekker, Inc., New York, (1978).
Drayer, D. E., How to Estimate Plant Cost-Capacity Relationship, *Petro/Chem. Eng.*, **42**(5):10 (1970).
Drew, J. W., and A. F. Ginder, How to Estimate the Cost of Pilot-Plant Equipment, *Chem. Eng.*, **77**(3):100 (1970).
Enyedy, G., Jr., Cost Engineering, *Chem. Eng.*, **75**(1):77 (1968).
————, The COME Computer Program Correlates Cost Data, *AACE Bull.*, **15**(3):93 (1973).
————, The Systems Approach to Estimating Equipment Costs, presented at 12th National Meeting of American Association of Cost Engineers, Houston, Texas, June 17–19, 1968.
Gallagher, J. T., Efficient Estimating of Worldwide Plant Costs, *Chem. Eng.*, **78**(12):196 (1969).
Garcia-Borras, T., Research-Project Evaluations, *Hydrocarbon Process.*, **55**(12):137 (1976); **56**(1):171 (1977).
Garnett, E. W., Firm Bid from Preliminary Drawings, presented at 12th National Meeting American Association of Cost Engineers, Houston, Texas, June 17–19, 1968.
Gary, J. H., and G. E. Handwerk, " Petroleum Refining: Technology and Economics," Marcel Dekker, Inc., New York, 1975.

Gladstone, J., Analyzing Low Visibility Items in the Mechanical Estimate, *AACE Bull.*, **18**(6):201 (1976).

Grigsby, E. K., E. W. Mills, and D. C. Collins, What Will Future Refineries Cost? *Hydrocarbon Process.*, **52**(5):133 (1973).

Guthrie, K. M., Capital Cost Estimating, *Chem. Eng.*, **76**(6):114 (1969).

———, Rapid Calc Charts, *Chem. Eng.*, **76**(1):138 (1969).

———, "Process Plant Estimating Evaluation and Control," Craftsman Book Company, Solana Beach, California, 1974.

Hemphill, R. B., A Method for Predicting the Accuracy of a Construction Cost Estimate, presented at 12th National Meeting American Association of Cost Engineers, Houston, Texas, June 17–19, 1968.

Hoerner, G. M., Jr., Nomograph Updates Process Equipment Costs, *Chem. Eng.*, **83**(11):141 (1976).

Holland, F. A., F. A. Watson, and J. K. Wilkinson, How to Estimate Capital Costs, *Chem. Eng.*, **81**(7): 71 (1974).

——— ——— and ———, "Introduction to Process Economics," J. Wiley and Sons, New York, 1974.

Humphreys, K. K., Preliminary Capital and Operating Cost Estimation—A Short Course, *1970 Trans. Am. Assoc. Cost Eng.* (1970), p. 203.

Jelen, F. C., "Cost and Optimization Engineering," McGraw-Hill Book Company, New York, 1970.

Jenckes, L. C., How to Estimate Operating Costs and Depreciation, *Chem. Eng.*, **77**(27):168 (1970).

———, Developing and Evaluating a Manufacturing-Cost Estimate, *Chem. Eng.*, **78**(1):168 (1971).

Johnson, R. J., Costs of Overseas Plants, *Chem. Eng.*, **76**(5):146 (1969).

Keim, C. R., Meaningful Production Costs, *Chem. Eng.*, **78**(26):184 (1971).

Kirkham, S. G., Preparation and Application of Refined Lang Factor Costing Techniques, *AACE Bull.*, **14**(5):137 (1972).

Landrum, R. J., Equipment, *Chem. Eng.*, **77**(22):75 (1970).

Lipták, G. C., Costs of Process Instruments, *Chem. Eng.*, **77**(19):60 (1970).

———, Costs of Viscosity, Weight, Analytical Instruments, *Chem. Eng.*, **77**(20):175 (1970).

Long, W. P., and E. D. Montrone, Economics, *Chem. Eng.*, **77**(22):41 (1970).

Mapstone, G. E., Find Exponents for Cost Estimates, *Hydrocarbon Process.*, **48**(5):165 (1969).

Massey, D. J., and J. H. Black, Predicting Chemical Prices, *Chem. Eng.*, **76**(23):150 (1969).

Matthews, R. L., and J. F. Adams, Predicting Manpower Needs in Engineering Departments, *Chem. Eng.*, **76**(14):152 (1969).

Mendel, O., Types of Estimates, *AACE Bull.*, **16**(4):107 (1974).

Miller, C. A., Current Concepts in Capital Cost Forecasting, *Chem. Eng. Progr.*, **69**(5):77 (1973).

Nelson, W. L., How to Find Current Cost Data, *Oil & Gas J.*, **74**(10):83 (1976).

Ohsol, E. O., Evaluation of Research Projects—Marketing Cost Estimation, *Chem. Eng. Progr.*, **67**(4):19 (1971).

———, Estimating Marketing Costs, *Chem. Eng.*, **78**(10):116 (1971).

Ostwald, P. F., "Cost Estimating for Engineers and Management," Prentice-Hall, Englewood Cliffs, New Jersey, 1974.

Pikulik, A., and H. E. Diaz, Cost Estimating for Major Process Equipment, *Chem. Eng.*, **84**(21):107 (1977).

Popper, H., "Modern Cost-Engineering Techniques," McGraw-Hill Book Company, New York, 1970.

——— and G. E. Weismantel, Costs and Productivity in the Inflationary 1970's, *Chem. Eng.*, **77**(1):132 (1970).

Schewe, M. F., Practical Aspects of the Computerized Estimating, presented at 12th National Meeting American Association of Cost Engineers, Houston Texas, June 17–19, 1968.

Silver, L., S. Bacher, and A. Sipos, Estimating Batch Process Production and Capital Costs, *1973 Trans. Am. Assoc. of Cost Eng.* (1973), p. 164.

Sommerville, R. F., Estimating Mill Costs at Low Production Rates, *Chem. Eng.*, **77**(7):148 (1970).

Swaney, J. B., Preliminary Cost Estimating, *Hydrocarbon Process.*, **52**(4):167 (1973).

Winter, O., Preliminary Economic Evaluation of Chemical Processes at the Research Level, *I&EC*, **61**(4):45 (1969).

Yen, Y.-C., Estimating Plant Costs in the Developing Countries, *Chem. Eng.*, **79**(15):89 (1972).
Is There an Economy of Size for Tomorrow's Refinery? *Hydrocarbon Process.* **54**(5):111 (1975).
How to Computerize Plant Cost Estimations, *Petro/Chem Eng.*, **42**(5):39 (1970).
Developments in Capital Cost Estimating, *Chem. Eng.* (*London*), 39 (January 1976).

Cost indexes

Caldwell, D. W., and J. H. Ortego, A Method for Forecasting the CE Plant Cost Index, *Chem. Eng.*, **82**(14):83 (1975).
Cason, R. L., Go Broke While Showing a Profit? *Hydrocarbon Process.*, **54**(7):179 (1975).
Clune, J. L., A Quantitative Discussion of Capital Cost Indexes Used in the Chemical Process Industry, *1973 Trans. Am. Assoc. Cost Eng.* (1973), p. 1.
Holland, F. A., F. A. Watson, and J. K. Wilkinson, How to Estimate Capital Costs, *Chem. Eng.*, **81**(7):71 (1974).
Kohn, P. M., C.E. Cost Indexes Maintain 13-Year Ascent, *Chem. Eng.*, **85**(11):189 (1978).
Miller, C. A., Current Concepts in Capital Cost Forecasting, *Chem. Eng. Progr.*, **59**(5):77 (1973).
————, Selection of a Cost Index, *AACE Cost Eng. Notebook*, E-1.000, June 1978.
Nelson, W. L., Cost Indexes Without Productivity Mean Little, presented at 12th National Meeting American Association of Cost Engineers, Houston, Texas, June 17–19, 1968.
————, Refinery Construction Costs Keep on Climbing, *Oil & Gas J.*, **73**(4):146 (1975).
————, Nelson Indexes—1: How the Nelson Refinery Construction-Cost Indexes Evolved, *Oil & Gas J.*, **74**(48):68 (1976).
————, Nelson Indexes—2: Index Price Bases Explained, *Oil & Gas J.*, **74**(50):59 (1976).
————, Nelson Indexes—3: Sources Cited for Quarterly Itemized Cost Indexes, *Oil & Gas J.*, **74**(51):73 (1976).
————, Nelson Indexes—4: Where to Find Base Refinery Costs, *Oil & Gas J.*, **74**(52):199 (1976).
Norden, R. B., Development of a Cost Index, presented at 12th National Meeting American Association of Cost Engineers, Houston, Texas, June 17–19, 1968.
————, C.E. Cost Indexes: A Sharp Rise Since 1965, *Chem. Eng.*, **76**(10):134 (1969).
Patterson, W. H., Preparing and Maintaining a Construction Cost Index, *1969 Trans. Am. Assoc. Cost Eng.* (1969).
Pikulik, A., and H. E. Diaz, Cost Estimating for Major Process Equipment, *Chem. Eng.*, **84**(21):107 (1977).
Popper, H., and G. E. Weismantel, Costs and Productivity in the Inflationary 1970s, *Chem. Eng.*, **77**(1):132 (1970).
Ricci, L. J., C.E. Cost Indexes Accelerate 10-Year Climb, *Chem. Eng.*, **82**(9):117(1975).
Thorsen, D. R., The Seven-Year Surge in the C.E. Cost Indexes, *Chem. Eng.*, **79**(25):168 (1972).
Walton, P. R., Obscure Error Sources in Operating Cost Estimating, presented at 12th National Meeting of American Association of Cost Engineers, Houston, Texas, June 17–19, 1968.
Webber, J., Cost Indexes—What They Can and Can't Do, presented at 12th National Meeting American Association of Cost Engineers, Houston, Texas, June 17–19, 1968.

Design costs

Loring, R. J., Cost of Preparing Proposals, Chem. Eng., **77**(25):126 (1970).

Engineering costs

Gallagher, J. T., Rapid Estimating of Engineering Cost, *Chem. Eng.*, **74**(13):250 (1967).
————, Analyzing "Cost Plus" Engineering Bids, *Chem. Eng.*, **75**(3):140 (1968).
Gladstone, J., Analyzing Low Visibility Items in the Mechanical Estimate, *AACE Bull.*, **18**(6):201 (1976).
Loring, R. J., Cost of Preparing Proposals, *Chem. Eng.*, **77**(25):126 (1970).
Matthews, R. L., and J. F. Adams, Predicting Manpower Needs in Engineering Departments, *Chem. Eng.*, **76**(14):152 (1969).
Zimmerman, O. T., Elements of Capital Cost Estimation, *Cost Eng.*, **13**(4):4 (1968).
————, Indirect Capital Costs, *Cost Eng.*, **14**(1):4 (1969).

Equipment costs

Bonano, D. R., Cost Escalation: Its Impact on Purchased Equipment, *Chem. Eng.*, **82**(14):81 (1975).

Chase, J. D., Plant Cost vs. Capacity: New Way to Use Exponents, *Chem. Eng.*, **77**(7):113 (1970).

Drew, J. W., and A. F. Ginder, How to Estimate the Cost of Pilot-Plant Equipment, *Chem. Eng.*, **77**(3):100 (1970).

Enyedy, G., Jr., Cost Data for Major Equipment, *Chem. Eng.* Progr., **67**(5):73 (1971).

———, Generate Your Own Cost Curves with the COME Computer Program, *AACE Bull.* **14**(6):171 (1972).

Garcia-Borras, T., Research-Project Evaluation, *Hydrocarbon Process.*, **56**(1):171 (1977).

Greek, B. F., Inflation of Plant Construction Costs Eases, *Chem. Eng. News*, **55**(15):12 (1977).

Guthrie, K. M., Capital Cost Estimating, *Chem. Eng.*, **76**(6):114 (1969).

———, "Process Plant Estimating Evaluation and Control," Craftsman Book Company, Solana Beach, California, 1974.

Hoerner, G. J., Jr., Nomograph Updates Process Equipment Costs, *Chem. Eng.*, **83**(11):141 (1976).

Holland, F. A., F. A. Watson, and J. K. Wilkinson, How to Estimate Capital Costs, *Chem. Eng.*, **81**(7):71 (1974).

Pikulik, A., and H. E. Diaz, Cost Estimating for Major Process Equipment, *Chem. Eng.*, **84**(21):107 (1977).

Popper, H., "Modern Cost-Engineering Techniques," McGraw-Hill Book Company, New York, 1970.

Stout, T. M., Justifying Process Control Computers: Selection and Costs, *Chem. Eng.*, **79**(20):89 (1972).

Williams, V. C., Cyrogenics, *Chem. Eng.*, **77**(25):92 (1970).

Zimmerman, O. T., Process Equipment Cost Data, *Cost. Eng.*, **18**(2):12 (1973).

How to Computerize Plant Cost Estimations, *Petro/Chem. Eng.*, **42**(5):39 (1970).

Inflation

Bonano, D. R., Cost Escalation: Its Impact on Purchased Equipment, *Chem. Eng.*, **82**(14):81 (1975).

Caldwell, D. W., and J. H. Ortego, A Method for Forecasting the C.E. Plant Cost Index, *Chem. Eng.*, **82**(14):83 (1975).

Cason, R. L., Go Broke While Showing a Profit, *Hydrocarbon Process.*, **54**(7):179 (1975).

Clark, F. D., How to Assess Inflation of Plant Costs—Cost Control for Process Plants from the Owner's View, *Chem. Eng.*, **82**(14):70 (1975).

Davidson, L. B., Investment Evaluation Under Conditions of Inflation, *J. Petrol. Technol.*, **27**(10):1183 (1975).

Elliott, D. P., What is This Thing Called Escalation? *AACE Bull.*, **19**(3):85 (1977).

Greek, B. F., Inflation of Plant Construction Costs Eases, *Chem. Eng. News*, **55**(15):12 (1977).

Holland, F. A., F. A. Watson, and J. K. Wilkinson, How to Allocate Overhead Cost and Appraise Inventory, *Chem. Eng.*, **81**(12):83 (1974).

——— ——— and ———, Inflation and Its Impact on Costs and Prices, *Chem. Eng.*, **81**(23): 107 (1974).

Holland, F. A., and F. A. Watson, Putting Inflation into Profitability Studies, *Chem. Eng.*, **84**(4):87 (1977).

——— and ———, Project Risk, Inflation, and Profitability, *Chem. Eng.*, **84**(6):133 (1977).

Jelen, F. C., Pitfalls in Profitability Analysis, *Hydrocarbon Process.*, **55**(1):111 (1976).

——— and M. S. Cole, Methods for Economic Analysis, *Hydrocarbon Process.*, **53**(7):133 (1974); **53**(9):227 (1974); **53**(10):161 (1974).

Jones, L. R., Estimating Cost Escalation, *1970 Trans. Am. Assoc. Cost Eng.* (1970), p. 58.

Maristany, B. A., Inflation and Economic Evaluations, *Chem. Eng. Progr.*, **67**(4):25 (1971).

Nelson, W. L., Appalling Inflation, *Oil & Gas J.*, **75**(42): 156 (1977).

Pikulik, A., and H. E. Diaz, Cost Estimating for Major Process Equipment, *Chem. Eng.*, **84**(21):107 (1977).

Savay, A. C., Effects of Inflation and Escalation on Plant Costs, *Chem. Eng.*, **82**(14):78 (1975).

Stuart, D. O., and R. D. Meckna, Accounting for Inflation in Present-Worth Studies, *Power Eng.*, **79**(2):45 (1975).

Tacke, D. M., and H. C. Thorne, Predict Prices With Inflation, *Hydrocarbon Process.*, **55**(4):151 (1976).

Thorne, H. C., Impact of Inflation on International Ventures, *AACE Bull.*, **20**(1):5 (1978).

Williams, L. F., The Effect of Inflation on the U.K. Process Plant and Engineering Industry, *AACE Bull.*, **19**(1):5 (1977).

Installation costs

Bloch, L. J., Estimating Machinery and Equipment Erection Costs, *1973 Trans. Am. Assoc. Cost Eng.* (1973), p. 76.

Culver, P. E., and E. H. Marsh, Estimates for Electrical Process Plant, *Chem. & Proc. Eng.*, **48**(5):89 (1967).

Drew, J. W., and A. F. Ginder, How to Estimate the Cost of Pilot-Plant Equipment, *Chem. Eng.*, **77**(3):100 (1970).

Guthrie, K. M., Capital Cost Estimating, *Chem. Eng.*, **76**(6):114 (1969).

———, Costs, *Chem. Eng.*, **76**(8):201 (1969).

Marshall, S. P., and K. L. Brandt, Installed Cost of Corrosion-Resistant Piping, *Chem. Eng.*, **78**(19):68 (1971).

Process Piping, *Chem. Eng.*, **76**(8):95 (1969).

Instrumentation costs

Brown, J. E., Onstream Process Analyzers, *Chem. Eng.*, **75**(10):164 (1968).

Deuschle, R., and A. Hoos, The Trend in Instrumentation Costs, *Plant Eng.*, **31**(9):263 (1977).

Liptàk, B. G., Costs of Process Instruments, *Chem. Eng.*, **77**(19):60 (1970).

———, Costs of Viscosity, Weight, Analytical Instruments, *Chem. Eng.*, **77**(20):75 (1970).

———, Control-Panel Costs, Process Instruments, *Chem. Eng.*, **77**(21):83 (1970).

———, Safety Instruments and Control-Valves Costs, Chem. Eng., **77**(24):94 (1970).

McMullen, A. I., J. F. Monk, and M. J. Stuart, Detection and Measurement of Pollutants of Water Surfaces, *Am. Lab.*, **7**(2):87 (1975).

Podio, A. L., Gas Flow Computers, *Pet. Eng.*, **45**(7):56 (1973).

Roberson, C., Instrument Estimating for the Cost Engineer, *AACE Bull.*, **19**(3): 107 (1976).

Ryan, J. M., R. S. Timmins, and J. F. O'Donnell, Production Scale Chromatography, *Chem. Eng. Prog.*, **64**(8):53 (1968).

Taylor, W. S., and L. T. Macgill, Instrumentation Costs from 1962 to 1968, presented at 12th National Meeting American Association of Cost Engineers, Houston, Texas, June 17–19, 1968.

Whelan, T., Instrumentation Costs in the HPI, *Hydrocarbon Process.*, **54**(9):185 (1975).

Zacharias, E. M., Jr., and D. W. Franz, Sound Velocimeters Monitor Process Streams, *Chem. Eng.*, **80**(2):101 (1973).

Zimmerman, O. T., Process and Analytical Instruments, *Cost Eng.*, **18**(1):4 (1973).

Insulation costs

Abramovitz, J. L., Economic Pipe Insulation for Cold Systems, *Chem. Eng.*, **83**(23):105 (1976).

Barnhart, J. M., Economic Thickness of Thermal Insulation, *Chem. Eng. Progr.*, **70**(8):50 (1974).

Cross, T. A., Economics Dictate Tank Insulation, *Oil & Gas J.*, **73**(35):73 (1975).

Davis, H. N., Preformed Pipe Insulation, *Air Cond. Heat & Vent.*, **66**(6):49 (1969).

House, F. K., Pipe Tracing and Insulation, *Chem. Eng.*, **75**(13):243 (1968).

Orefice, J. D., The Expanding Insulations in Construction, *Air Cond. Heat & Vent.*, **66**(6):45 (1969).

Cost Engineers' Notebook, *AACE Bull.*, **14**(6):A-4.280 (1972).

Labor costs

Drew, J. W., and A. F. Ginder, How to Estimate the Cost of Pilot-Plant Equipment, *Chem. Eng.*, **77**(3):100 (1970).

Edmondson, C. H., You *Can* Predict Construction-Labor Productivity, *Hydrocarbon Process.*, **53**(7):167 (1974).

Eldredge, W. J., and L. C. Jung, People Factor in Investments, *Hydrocarbon Process.*, **54**(4):197 (1975).

Guthrie, K. M., Field-Labor Predictions for Conceptual Projects, *Chem. Eng.*, **76**(7):170 (1969).

Holland, F. A., F. A. Watson, and J. K. Wilkinson, How to Budget and Control Manufacturing Costs, *Chem. Eng.*, **81**(10):105 (1974).

Johnnie, C. C., and D. K. Aggarwal, Calculating Plant Utility Costs, *Chem. Eng. Progr.*, **73**(11):84 (1977).

Nelson, W. L., Wages in U.S., Abroad Vary, *Oil & Gas J.*, **74**(12):66 (1976).

———, Labor Costs Key Factor in Determining Inflation Index, *Oil & Gas J.*, **69**(5):74 (1971).

———, Inflation in Refinery Wages: Effect of Location, *Oil & Gas J.*, **70** (18):116 (1972).

———, Manpower and Labor Costs, *Oil & Gas J.*, **77**(50):70 (1974).

Swaney, J. B., Preliminary Cost Estimating, Hydrocarbon Process., **52**(4):167 (1973).

Winton, J. W., Plant Sites 1971, *Chem. Week*, **109**(15): 35 (1971).

———, Plant Sites 1972, *Chem. Week*, **111**(15): 35 (1972).

———, Plant Sites 1974, *Chem. Week*, **113**(16): 29 (1973).

———, Plant Sites 1975, *Chem. Week*, **115**(17): 33 (1974).

———, Plant Sites 1976, *Chem. Week*, **117**(17): 27 (1975).

———, Plant Sites 1977, *Chem. Week*, **119**(19): 35 (1976).

———, Plant Sites 1978, *Chem. Week*, **121**(24): 49 (1977).

Maintenance costs

Buttery, L. M., New Survey of U.S. Maintenance Costs, *Hydrocarbon Process.*, **57**(1):85 (1978).

Fair, E. W., Before You Buy, What About Maintenance Costs? *Chem. Eng.*, **81**(5):132 (1974).

Goyal, S. K., How Much Manpower for Turnarounds, *Hydrocarbon Process.*, **48**(5):155 (1969).

Guthrie, K. M., Capital and Operating Costs for 54 Chemical Processes, *Chem. Eng.*, **77**(13):140 (1970).

Jordan, J. H., How to Evaluate the Advantages of Contract Maintenance, *Chem. Eng.*, **75**(7):124 (1968).

Nelson, W. L., Refinery Maintenance Costs, *Oil & Gas J.*, **69**(17):84 (1971).

———, Cost of Tank Cleaning, Repair, *Oil & Gas J.*, **70**(11):83 (1972).

———, Maintenance Material and Labor, *Oil & Gas J.*, **73**(2):57 (1975).

———, Average Process—Unit Maintenance, *Oil & Gas J.*, **73**(13):77 (1975).

Pekrul, P. J., Vibration Monitoring Increases Equipment Availability, *Chem. Eng.*, **87**(17):109 (1975).

Perkins, R. L., Maintenance Costs and Equipment Reliability, *Chem. Eng.*, **82**(7):126 (1975).

Weiss, W. H., Plant Designers Can Cut Maintenance Costs, *Petro/Chem. Eng.*, **41**(11):12 (1969).

Maintenance is Costing More, *Chem. Eng.*, **81**(21):60 (1974).

Operating costs

Abbott, J. T., R. R. Janssen, and C. M. Merz, Finance for Engineers, *Chem. Eng.*, **80**(15):108 (1973).

Buehler, J. D., and G. J. Figge, Operating vs. Capital Costs: Evaluating Tradeoff Benefits, *Chem. Eng.*, **78**(3):96 (1971).

Garcia-Borras, T., Research-Project Evaluations, *Hydrocarbon Process.*, **55**(12):137 (1976).

Grigsby, E. K., E. W. Mills, and D. C. Collins, What Will Future Refineries Cost? *Hydrocarbon Process.*, **52**(5):133 (1973).

Guthrie, K. M., Capital and Operating Costs for 54 Chemical Processes, *Chem. Eng.*, **77**(13):140 (1970).

Holland, F. A., F. A. Watson, and J. K. Wilkinson, How to Allocate Overhead Cost and Appraise Inventory, *Chem. Eng.*, **81**(12):83 (1974).

——— ——— and ———, How to Budget and Control Manufacturing Costs, *Chem. Eng.*, **81**(10):105 (1974).

Jelen, F. C., Pitfalls in Profitability Analysis, *Hydrocarbon Process.*, **55**(1):111 (1976).

Jenckes, L. C., Developing and Evaluating a Manufacturing-Cost Estimate, *Chem. Eng.*, **78**(1):168 (1971).

———, How to Estimate Operating Costs and Depreciation, *Chem. Eng.*, **77**(27):168 (1970).

———, Estimation and Optimization of Operating Costs, *1970 Trans. Am. Assoc. Cost Eng.* (1970), p. 48.

Johnnie, C. C., and D. K. Aggarwal, Calculating Plant Utility Costs, *Chem. Eng. Progr.*, **73**(11):84 (1977).

Johnson, P. W., and F. A. Peters, A Computer Program for Calculating Capital and Operating Costs, U.S. Department of the Interior, Bureau of Mines Information Circular 8426 (1969).

Leibson, I., and C. A. Trischman, How to Cut Operating Costs: Evaluate Your Feedstocks, *Chem. Eng.*, **78**(13):92 (1971).

——— and ———, Spotlight on Operating Cost, *Chem. Eng.*, **78**(12):69 (1971).

Lipták, B. G., Costs of Process Instruments, *Chem. Eng.*, **77**(19):60 (1970).

Nelson, W. L., Operating Costs Indexed for Refineries, *Oil & Gas J.*, **74**(11):129 (1976).

———, Process Costimating Series Reviewed, *Oil & Gas J.*, **74**(18):244 (1976).

Parker, C. L., and C. V. Fong, Estimation of Operating Costs for Industrial Waste Water Treatment Facilities, *AACE Bull.*, **18**(6):207 (1976).

Sommerville, R. F., New Method Gives Quick, Accurate Estimate of Distillation Costs, *Chem. Eng.*, **79**(9):71 (1972).

Walton, P. R., Sources of Error in Operating-Cost Estimates, *Chem. Eng.*, **75**(15):150 (1968).

Yen, Y.-C., Estimating Plant Costs in the Developing Countries, *Chem. Eng.*, **79**(15):89 (1972).

Zudkevitch, D., Imprecise Data Impacts Plant Design and Operation, *Hydrocarbon Process.*, **54**(3):97 (1975).

Plant and process costs

Alagy, J., L. Asselineau, C. Busson, B. Cha, and H. Sandler, Selectively Oxidize Cyclohexane, *Hydrocarbon Process.*, **47**(12):131 (1968).

Axelrod, L., R. E. Daze, and H. P. Wickham, The Large Plant Concept, *Chem. Eng. Progr.*, **64**(7):17 (1968).

Baltzell, H. J., Estimate Refinery Cost Better, *Hydrocarbon Process.*, **57**(1):129 (1978).

Barthel, Y., Y. Bistri, A. Deschamps, R. Renault, J. C. Simadoux, and R. Dutriau, Treat Claus Tail Gas, *Hydrocarbon Process.*, **50**(5):89 (1971).

Boll, C. H., Recovering Solvents by Adsorption, *Plant Eng.*, **30**(1):69 (1976).

Bolton, D. H., and D. Hanson, Economics of Low Pressures in Methanol Plants, *Chem. Eng.*, **76**(20):154 (1969).

Bourguet, J. M., Economics of Today's Plants, *Hydrocarbon Process.*, **49**(4):93 (1970).

Bowerman, F. R., Solid Waste Disposal, *Chem. Eng.*, **77**(9):147 (1970).

Bruzzone, M., W. Marconi, and S. Noe, SNAM's New Polydiene Process, *Hydrocarbon Process.*, **47**(11):179 (1968).

Bryant, H. S., C. A. Duval, L. E. McMakin, and J. I. Savoca, Mobil's Process for TPA, *Chem. Eng. Progr.*, **67**(9): 69 (1971).

Connor, J. M., Economics of Sulfuric Acid Manufacture, *Chem. Eng. Progr.*, **64**(11):59 (1968).

Cook, T. P. and R. N. Tennyson, Improved Economics in Synthesis Gas Plants, *Chem. Eng. Progr.*, **65**(11):61 (1969).

Deschner, W. W., J. Gieck, and P. Potts, Can Small Ammonia Plants Compete? *Hydrocarbon Process.*, **47**(9):261 (1968).

Di Napoli, R. N., Estimating Costs for Base-Loaded LNG Plants, *Oil & Gas. J.*, **73**(46):58 (1975).

Dosher, J. R., Trends in Petroleum Refining, *Chem. Eng.*, **77**(17):96 (1970).

Drayer, D. E., How to Estimate Plant Cost-Capacity Relationship, *Petro/Chem. Eng.*, **42**(5):10 (1970).

Eddinger, R. T., J. F. Jones, and F. E. Blanc, Development of the COED Process, *Chem. Eng. Progr.*, **64**(10):33 (1968).

Evan, D. N., J. B. Lawrence, C. L. Rambo, and R. R. Tonne, Curves Analyze Cyrogenic Process Economics, *Oil & Gas J.*, **72**(32):119 (1974).

Falkenberry, H. L., and A. V. Slack, SO$_2$ Removal by Limestone Injection, *Chem. Eng. Progr.*, **65**(12):61 (1969).

Goodman, B. L., and K. A. Mikkelson, Advanced Waste-water Treatment (Removing Phosphorus and Suspended Solids), *Chem. Eng.*, **77**(9):75 (1970).

Grigsby, E. K., E. W. Mills, and D. C. Collins, What Will Future Refineries Cost? *Hydrocarbon Process.*, **52**(5):133 (1973).

Guthrie, K. M., Capital and Operating Costs for 54 Chemical Processes, *Chem. Eng.*, **77**(13):140 (1970).

Haaga, J. C., Petrochemical Feedstocks, *Chem. Eng.*, **79**(5):130 (1972).

Haselbarth, J. E., Updated Investment Costs for 60 Types of Chemical Plants, *Chem. Eng.*, **74**(25):214 (1967).

Hedley, B., W. Powers, and R. B. Stobaugh, Methanol: How, Where, Who—Future, *Hydrocarbon Process.*, **49**(9):275 (1970).

Hendrickson, D. L., S. D. Lawson, A. R. Cunningham, J. Byrne, and E. L. Winkler, Effect of Changing Gas Volatility on Refining Costs, *Chem. Eng. Progr.*, **65**(2):51 (1969).

Hiller, H., and F. Marschner, Lürgi Makes Low-Pressure Methanol, *Hydrocarbon Process.*, **49**(9):281 (1970).

Ito, S., No Pollution or Catalyst in Continuous Oil Gasification, *Chem. Eng.*, **77**(27):113 (1970).

Keane, D. P., R. B. Stobaugh, and P. L. Townsend, Vinyl Chloride: How, Where, Who—Future, *Hydrocarbon Process.*, **52**(2):99 (1973).

Knowles, C. L., Jr., Improving Biological Processes, *Chem. Eng.*, **77**(9):103 (1970).

Kronseder, J. G., Cost of Reducing Sulfur Dioxide Emissions, *Chem. Eng. Progr.*, **64**(11):71 (1968).

——, Economics of Phosphoric Acid Processes, *Chem. Eng. Progr.*, **64**(9):97 (1968).

Logan, R. S., and R. L. Banks, Disproportionate Propylene to Make More and Better Alkylate, *Hydrocarbon Process.*, **47**(6):135 (1968).

Martin, R., Dollars and Sense of Cryogenics, *Petro/Chem. Eng.*, **39**(7):PM 12 (1967).

——, Reexamine Fluid Coking, *Petro/Chem. Eng.*, **40**(11):27 (1968).

Masamune, S., and T. Kawatani, First Commercial MHC Unit Onstream, *Hydrocarbon Process.*, **47**(12):111 (1968).

Mehta, D. D., Optimize Methanol Synthesis Gas, *Hydrocarbon Process.*, **47**(12):127 (1968).

—— and D. E. Ross, Optimize ICI Methanol Process, *Hydrocarbon Process.*, **49**(11):183 (1970).

Meredith, H. H., Jr., and W. L. Lewis, Desulfurization and the Petroleum Industry, *Chem. Eng. Progr.*, **64**(9):57 (1968).

Miller, E. F., Lowering the Cost of Reverse-Osmosis Desalting, *Chem. Eng.*, **75**(24):153 (1968).

Nelson, W. L., More on Costs of LNG Operations, *Oil & Gas J.*, **66**(16):92 (1968).

——, Cost of Plants for Recovering H$_2$S, *Oil & Gas J.*, **66**(17):199 (1968).

——, Current Cost of Sulfur Recovery Plants in U.S., *Oil & Gas J.*, **66**(22):111 (1968).

——, Investment Costs Hydrocracking; H$_2$ Manufacture, *Oil & Gas J.*, **66**(29):141 (1968).

——, Oxygen Plants are Still Expensive, *Oil & Gas J.*, **66**(40):116 (1968).

——, Ammonia Plants, *Oil & Gas J.*, **67**(23):101 (1969).

——, How to Estimate Cost of a Specific Refinery, *Oil & Gas J.*, **68**(37):99 (1970).

——, Carbon-Black Plant Costs, *Oil & Gas J.*, **69**(28):72 (1971).

——, Gas-Process Plant Costs, *Oil & Gas J.*, **70**(20):163 (1972).

——, Cost of 2-Component Fractionation Plants, *Oil & Gas J.*, **70**(24):60 (1972).

——, Control Splitter and Stripping Plants, *Oil & Gas J.*, **70**(33):96 (1972).

——, Cost of Aromatics Plants, *Oil & Gas J.*, **71**(18):188 (1973).

——, Crude Distillation Costs, *Oil & Gas J.*, **72**(8):70 (1974).

——, Vacuum Distillation Costs, *Oil & Gas J.*, **72**(9):100 (1974).

——, Plant Costs for Processing Hydrogen, *Oil & Gas J.*, **72**(10):111 (1974).

——, Sulfur Recovery Costs, *Oil & Gas J.*, **72**(11):120 (1974).

——, Hydrocracking: Hydrogen Manufacture, *Oil & Gas. J.*, **72**(12):120 (1974).

——, Coking Plant Costs, *Oil & Gas J.*, **72**(13):118 (1974).

——, Alkylation Plants: Viscosity Breaking Plants, *Oil & Gas J.*, **72**(14):74 (1974).

——, Cost of Cat-Cracking Plants, *Oil & Gas J.*, **72**(15):66 (1974).

————, Catalytic Reforming, *Oil & Gas J.*, **72**(16):132 (1974).

————, Cost of Refineries—Process Unit Costs, *Oil & Gas J.*, **72**(28):87 (1974).

————, Process Costimating Series Reviewed, *Oil & Gas J.*, **74**(19):244 (1976).

————, Where to Find Base Refinery Costs, *Oil & Gas J.*, **74**(52):200 (1976).

————, What Do Calciners Cost? *Oil & Gas J.*, **75**(52):65 (1977).

————, What Do Lube-Plants Cost? *Oil & Gas J.*, **76**(9):152 (1978).

————, Grease Compounding Facilities Are Costly, *Oil & Gas J.*, **76**(15):71 (1978).

Putnam, B., and M. Manderson, Iron Pyrites from High Sulfur Coals, *Chem. Eng. Progr.*, **64**(9):60 (1968).

Reese, K. M., Economics in the Chemical Industry, *J. Chem. Educ.*, **46**(11):725 (1969).

Reis, T., Compare Butadiene Recovery Methods, Processes, Solvents, Economics, *Petro/Chem. Eng.*, **8**(41):12 (1969).

Rickles, R. N., Desulfurization's Impact on Sulfur Markets, *Chem. Eng. Progr.*, **64**(9):53 (1968).

Royal, M. J., and N. M. Nimmo, Why LP Methanol Costs Less, *Hydrocarbon Process.*, **48**(3):147 (1969).

Sommerville, R. F., Estimating Mill Costs at Low Production Rates, *Chem. Eng.*, **77**(7):148 (1970).

Spencer, P., Petrochemical Feedstocks: Economic Potential of LNG, *Chem. Eng.*, **77**(5):160 (1970).

Spitz, P. H., Phthalic Anhydride Revisited, *Hydrocarbon Process.*, **47**(11):162 (1968).

Starmer, R., and F. Lowes, Nuclear Desalting: Future Trends, and Today's Costs, *Chem. Eng.*, **75**(19):127 (1968).

Stobaugh, R. B., G. C. Ray, and R. A. Spinek, Ethylene Oxide: How, Where, Who—Future, *Hydrocarbon Process.*, **49**(10):105 (1970).

———— S. G. McH. Clark, and G. D. Camirand, Acrylonitrile: How, Where, Who—Future, *Hydrocarbon Process.*, **50**(1):109 (1971).

Strelzoff, S. Z., Ethylene and Propylene: Booming Building Blocks, *Chem. Eng.*, **77**(18):75 (1970).

Sugai, K., Low Cost, High Yield for Forming Ethylene, *Chem. Eng.*, **77**(13):126 (1970).

Sweeney, N., Here's What Users Pay for Ammonia, *Hydrocarbon Process.*, **47**(9):265 (1968).

Wall, J. D., HPI Grows With Clean Fuels, *Hydrocarbon Process.*, **51**(5):133 (1972).

Wright, J. B., Successful Large-Scale Desalting, *Power Eng.*, **73**(7):46 (1969).

Is There an Economy of Size for Tomorrow's Refinery? *Hydrocarbon Process.*, **54**(5):111 (1975).

Productivity

Edmondson, C. H., You Can Predict Construction-Labor Productivity, *Hydrocarbon Process.*, **53**(7):167 (1974).

Nelson, W. L., Where to Find Base Refinery Costs, *Oil & Gas J.*, **74**(52):200 (1976).

————, Here's a Look at Productivity in Design and Erection of Refineries—1973 and the Future. Productivity Costimating—76, *Oil & Gas J.*, **73**(19):100 (1975).

Winton, J. W., Plant Sites 1974, *Chem. Week*, **113**(16):29 (1973).

————, Plant Sites 1978, *Chem. Week*, **121**(24):49 (1977).

Research and research costs

Bobis, A. H., and A. Atkinson, Analyzing Potential Research Projects, *Chem. Eng.*, **77**(4):95 (1970).

Corrigan, T. E., and W. O. Beavers, Research and Development, *Chem. Eng.*, **75**(1):56 (1968).

DeCicco, R. W., Economic Evaluation of Research Projects—By Computer, *Chem. Eng.*, **75**(12):84 (1968).

Garcia-Borras, T., Research-Project Evaluations, *Hydrocarbon Process.*, **55**(12):137 (1976); **56**(1): 171 (1977).

West, A. S., How to Budget R & D Expenses, *Chem. Eng. Progr.*, **66**(1):13 (1970).

Royalties

Maux, R., How to Estimate Royalties, *Hydrocarbon Process.*, **55**(5):233 (1976).

Smith, D. B., D. M. Young, and R. L. Miller, Process Technology for License or Sale, *Chem. Eng.*, **77**(8):115 (1970).

Selling costs

Dosher, J. R., Trends in Petroleum Refining, *Chem. Eng.*, **77**(17):96 (1970).
Garcia-Borras, T., Research-Project Evaluations, *Hydrocarbon Process.*, **55**(12):137 (1976): **56**(1):171 (1977).
Malloy, J. B., Projecting Chemical Product Prices, *Chem. Eng. Progr.*, **70**(9):77 (1974).
Nathanson, D. M., Forecasting Petrochemical Prices, *Chem. Eng. Progr.*, **68**(11):89 (1972).

Site improvement costs

Guthrie, K. M., Capital Cost Estimating, *Chem. Eng.*, **76**(6):114 (1969).
Nelson, W. L., Cost of Refineries—Off Site Facilities, *Oil & Gas J.*, **72**(27):114 (1974).
———, Cost of Refineries—Off Sites Breakup, *Oil & Gas J.*, **72**(29):60 (1974).

Startup costs

Derrick, G. C., and W. L. Sutor, Estimation of Industrial Chemical Plant Startup Costs, *AACE Bull.*, **19**(3):C-2.300 (1977).
Gans, M., The A to Z of Plant Startup, *Chem. Eng.*, **83**(6):72 (1976).
McCallister, R. A., Here's a Study of Sources and Cost of Startup Problems, *Oil & Gas J.*, **69**(26):72 (1971).

Total product costs

Dosher, J. R., Trends in Petroleum Refining, *Chem. Eng.*, **77**(17):96 (1970).
Grumer, E. L., Selling Price vs. Raw-Material Cost., *Chem. Eng.*, **74**(9):190 (1967).
Haas, C. M., Product Cost Programs for R & D Areas, *1973 Trans. Am. Assoc. Cost Eng.* (1973), p. 28.
Holland, F. A., F. A. Watson, and J. K. Wilkinson, Manufacturing Costs and How to Estimate Them., *Chem. Eng.*, **81**(8):91 (1974).
Nelson, W. L., Joint Product Costing, *Oil & Gas J.*, **73**(18):242 (1975); **73**(22):123 (1975); **73**(25):94 (1975).
Sommerfeld, J. T., and C. T. Lenk, Thermodynamics Helps You Predict Selling Price, *Chem. Eng.*, **77**(10):136 (1970).
Stroup, R., Jr., Breakeven Analysis, *Chem. Eng.*, **79**(1):122 (1972).

Transportation costs

Browning, J. E., and E. H. Steymann, Distribution is on the Move, *Chem. Eng.*, **77**(1):88 (1970).
Denton, W. M., Pipelines Offer Least Costly Overland Transportation Means, *Oil & Gas J.*, **66**(31):138 (1968).
Friedman, W. F., Stretch Distribution Dollars, *Chem. Eng.*, **77**(20):169 (1970).
Mapstone, G. E., Effect of Transport Costs on Optimum Plant Size, *Chem. Proc. Eng.*, **52**(12):63 (1971).
Nelson, W. L., Tanker Transportation Costs, *Oil & Gas J.*, **66**(26):100 (1968).
———, Economics of Ammonia Transportation, *Oil & Gas J.*, **67**(31):162 (1969).
———, Here are Latest Tanker Rates, *Oil & Gas J.*, **68**(25):87 (1970).
———, World-Scale Tanker Rates, *Oil & Gas J.*, **68**(39):84 (1970).
———, Latest Average Tanker Transportation Costs, *Oil & Gas J.*, **70**(46):145 (1972).
———, What are Costs of Tanker Transportation? *Oil & Gas J.*, **73**(17):92 (1975).
Pence, T. G., Transportation, *Chem. Eng.*, **76**(8):23 (1969).
Raymus, G. J., Evaluating the Options for Packaging Chemical Products, *Chem. Eng.*, **80**(23):67 (1973).
Sherwood, D. W., Economics of the Maxi-Plants, *Petro/Chem. Eng.*, **41**(9):12 (1969).
Winton, J. W., Plant Sites 1971, *Chem. Week*, **109**(15):35 (1971).
———, Plant Sites 1978, *Chem. Week.*, **121**(24):49 (1977).

Utilities costs

Arnstein, R., and L. O'Connell, What's the Optimum Heat Cycle for Process Utilities? *Hydrocarbon Process.*, **47**(6):88 (1968).

Brooke, M., Water, *Chem. Eng.*, **77**(27):135 (1970).

Fleming, J. B., J. R. Lambrix, and M. R. Smith, Energy Conservation in New-Plant Design, *Chem. Eng.*, **81**(2):112 (1974).

Guthrie, K. M., Capital and Operating Costs for 54 Chemical Processes, *Chem. Eng.*, **77**(13):140 (1970).

Johnnie, C. C., and D. K. Aggarwal, Calculating Plant Utility Costs, *Chem. Eng. Progr.*, **73**(11):84 (1977).

Le Cerda, D. J., Better Instrument and Plant Air Systems, *Hydrocarbon Process.*, **47**(6):107 (1968).

McAllister, D. G., Air, *Chem. Eng.*, **77**(27):138 (1970).

Miller, C. A., Current Concepts in Capital Cost Forecasting, *Chem. Eng. Progr.*, **59**(5):77 (1973).

Miller, R., Jr., Process Energy Systems, *Chem. Eng.*, **75**(11):130 (1968).

Nelson, W. L., Fuel, Investment Govern Refinery-Steam Cost, *Oil & Gas J.*, **73**(50):87 (1975).

———, Utility Requirements, *Oil & Gas J.*, **74**(24):73 (1976); **74**(25):122 (1976).

———, What are Utility Needs? *Oil & Gas J.*, **75**(45):118 (1977).

Price, H. A., and D. G. McAllister, Jr., Inert Gas, *Chem. Eng.*, **77**(27):142 (1970).

van Loosen, J., What You Need to Know About Boilers, *Hydrocarbon Process.*, **47**(6):95 (1968).

Zimmerman, O. T., Power Production, *Cost Eng.*, **14**(4):7 (1969).

———, Standby Electric Generating Sets, *Cost Eng.*, **18**(2):8 (1973).

Working capital costs

Holland, F. A., F. A. Watson, and J. K. Wilkinson, Capital Costs and Depreciation, *Chem. Eng.*, **80**(17):118 (1973).

——— ——— and ———, How to Evaluate Working Capital for a Company, *Chem. Eng.*, **81**(16):101 (1974).

Lyda, T. B., How Much Working Capital Will the New Project Need? *Chem. Eng.*, **79**(21):182 (1972).

PROBLEMS

1 The purchased cost of a shell-and-tube heat exchanger (floating head and carbon-steel tubes) with 100 ft^2 of heating surface was $1400 in 1970. What will be the purchased cost of a similar heat exchanger with 200 ft^2 of heating surface in 1970 if the purchased-cost-capacity exponent is 0.60 for surface area ranging from 100 to 400 ft^2? If the purchased-cost-capacity exponent for this type of exchanger is 0.81 for surface areas ranging from 400 to 2000 ft^2, what will be the purchased cost of a heat exchanger with 1000 ft^2 of heating surface in 1970?

2 Plot the 1975 purchased cost of the shell-and-tube heat exchanger outlined in the previous problem as a function of the surface area from 100 to 2000 ft^2. Note that the purchased-cost-capacity exponent is not constant over the range of surface area requested.

3 The purchased and installation costs of some pieces of equipment are given as a function of weight rather than capacity. An example of this is the installed costs of large tanks. The 1970 cost for an installed aluminum tank weighing 100,000 lb was $188,000. For a size range from 200,000 to 1,000,000 lb, the installed cost-weight exponent for aluminum tanks is 0.93. If an aluminum tank weighing 700,000 lb is required, what is the present capital investment needed?

4 What weight of installed stainless-steel tank could have been obtained for the same capital investment as in the previous problem? The 1970 cost for an installed stainless-steel tank weighing 300,000 lb was $410,000. The installed cost-weight exponent for stainless tanks is 0.88 for a size range from 300,000 to 700,000 lb.

5 The purchased cost of a 1400-gal stainless-steel tank in 1970 was $4300. The tank is cylindrical with flat top and bottom, and the diameter is 6 ft. If the entire outer surface of the tank is to be covered with 2 in. thickness of magnesia block, estimate the present total cost for the installed and insulated tank. The Jan. 1, 1975 cost for the 2-in. magnesia block was $1.40 per ft^2 while the labor for installing the insulation was $3.00 per ft^2.

6 A one-story warehouse 120 by 60 ft is to be added to an existing plant. An asphalt-pavement service area 60 by 30 ft will be added adjacent to the warehouse. It will also be necessary to put in 500 lin ft of railroad siding to service the warehouse. Utility service lines are already available at the warehouse site. The proposed warehouse has a concrete floor and steel frame, walls, and roof. No heat is necessary, but lighting and sprinklers must be installed. Estimate the total cost of the proposed addition. Consult App. B for necessary cost data.

7 The purchased cost of equipment for a solid-processing plant is $500,000. The plant is to be constructed as an addition to an existing plant. Estimate the total capital investment and the fixed-capital investment for the plant. What percentage and amount of the fixed-capital investment is due to cost for land and contractor's fee?

8 The purchased-equipment cost for a plant which produces pentaerythritol (solid-fluid-processing plant) is $300,000. The plant is to be an addition to an existing formaldehyde plant. The major part of the building cost will be for indoor construction, and the contractor's fee will be 7 percent of the direct plant cost. All other costs are close to the average values found for typical chemical plants. On the basis of this information, estimate the following:

  (*a*) The total direct plant cost.
  (*b*) The fixed-capital investment.
  (*c*) The total capital investment.

9 Estimate by the turnover-ratio method the fixed-capital investment required for a proposed sulfuric acid plant (battery limit) which has a capacity of 140,000 tons of 100 percent sulfuric acid per year (contact-catalytic process) using the data from Table 19 for 1979 with sulfuric acid cost at $53 per ton. The plant may be considered as operating full time. Repeat using the cost-capacity-exponent method with data from Table 19. Repeat using the data presented by Haselbarth in the reference listed in the footnote to Table 19.

10 The total capital investment for a chemical plant is $1 million, and the working capital is $100,000. If the plant can produce an average of 8000 kg of final product per day during a 365-day year, what selling price in dollars per kilogram of product would be necessary to give a turnover ratio of 1.0?

11 A process plant was constructed in the Philadelphia area (Middle Atlantic) at a labor cost of $200,000 in 1970. What would the average labor costs for the same plant be in the Miami, Florida area (South Atlantic) if it were constructed in late 1978? Assume, for simplicity, that the relative labor rate and relative productivity factor have remained essentially constant during this time period.

12 A cost estimate is being prepared for a new plant, and one portion of the project requires estimation of the costs for the installations necessary to supply steam, cooling-tower water, and electricity. The process requires 14,000 lb/h of 100-psig steam, 40,000 gph of cooling-tower water, and 500 kVA of electricity. Estimate the installed cost for these utilities as of January 1, 1979.

13 A company has been selling a soap containing 30 percent by weight water at a price of $10 per 100 lb f.o.b. (i.e., freight on board, which means the laundry pays the freight charges). The company offers an equally effective soap containing only 5 percent water. The water content is of no importance to the laundry, and it is willing to accept the soap containing 5 percent water if the delivered costs are equivalent. If the freight rate is 70 cents per 100 lb, how much should the company charge the laundry per 100 lb f.o.b. for the soap containing 5 percent water?

14 Estimate the total operating cost per day for labor, power, steam, and water in a plant producing 100 tons of acetone per day from the data given in Table 22. Use average costs for utilities. Consider all water as city water. The steam pressure may be assumed to be 100 psig. Labor rates average $8.00 per employee-hour. Electricity must be purchased.

15 The total capital investment for a conventional chemical plant is $1,500,000, and the plant produces 3 million kg of product annually. The selling price of the product is $0.82/kg. Working capital amounts to 15 percent of the total capital investment. The investment is from company funds, and no interest is charged. Raw-materials costs for the product are $0.09/kg, labor $0.08/kg, utilities

$0.05/kg, and packaging $0.008/kg. Distribution costs are 5 percent of the total product cost. Estimate the following:

(a) Manufacturing cost per kilogram of product.
(b) Total product cost per year.
(c) Profit per kilogram of product before taxes.
(d) Profit per kilogram of product after taxes (use current rate).

16 Estimate the manufacturing cost per 100 lb of product under the following conditions:
  Fixed-capital investment = $2 million
  Annual production output = 10 million lb of product
  Raw-materials cost = $0.12/lb of product
  Utilities
    100 psig steam = 50 lb/lb of product
    Purchased electrical power = 0.4 kWh/lb of product
    Filtered and softened water = 10 gal/lb of product
  Operating labor = 20 men per shift at $8.00 per employee-hour
  Plant operates three hundred 24-h days per year
  Corrosive liquids are involved
  Shipments are in bulk carload lots
  A large amount of direct supervision is required
  There are no patent, royalty, interest, or rent charges
  Plant-overhead costs amount to 50 percent of the cost for operating labor, supervision, and maintenance

17 The manufacture of 1 ton of liquid ammonia requires the following raw materials and labor: coke 1.78 tons, water 31,000 gal, coal 2.3 tons, and direct labor 21.8 employee-hours. Estimate material and labor costs from these data. Using these direct production costs as constituting 90 percent of all the direct production costs, estimate the total product cost using average values in Table 27 for other costs. How does this estimated total product cost compare with the current selling price for liquid ammonia? This problem shows the effect of technological advances on product cost.

18 A company has direct production costs equal to 50 percent of total annual sales and fixed charges, overhead, and general expenses equal to $200,000. If management proposes to increase present annual sales of $800,000 by 30 percent with a 20 percent increase in fixed charges, overhead, and general expenses, what annual sales dollar is required to provide the same gross earnings as the present plant operation? What would be the net profit if the expanded plant were operated at full capacity with an income tax on gross earnings fixed at 48 percent? What would be the net profit for the enlarged plant if total annual sales remained the same as at present? What would be the net profit for the enlarged plant if the total annual sales actually decreased to $700,000?

19 A chemical plant producing phenol is reported to require 2200 lb benzol, 2800 lb 66°Be sulfuric acid, 3400 lb caustic soda (solid), 4000 lb steam, 80 kWh electricity, and 40 employee-hours of labor per ton of phenol. What are the direct production costs per ton of product? Estimate the total product cost using average values from Table 27 for other costs and compare this with the current price of phenol.

20 A process plant making 2000 tons per year of a product selling for $0.80 per lb has annual direct production costs of $2 million at 100 percent capacity and other fixed costs of $700,000. What is the fixed cost per pound at the break-even point? If the selling price of the product is increased by 10 percent, what is the dollar increase in net profit at full capacity if the income tax rate is 48 percent of gross earnings?

21 A rough rule of thumb for the chemical industry is that $1 of annual sales requires $1 of fixed-capital investment. In a chemical processing plant where this rule applies, the total capital investment is $2,500,000 and the working capital is 20 percent of the total capital investment. The annual total product cost amounts to $1,500,000. If the national and regional income-tax rates on gross earnings total 50 percent, determine the following:

(a) Percent of total capital investment returned annually as gross earnings.
(b) Percent of total capital investment returned annually as net profit.

22 The total capital investment for a proposed chemical plant which will produce $1,500,000 worth of goods per year is estimated to be $1 million. It will be necessary to do a considerable amount of research and development work on the project before the final plant can be constructed, and management wishes to estimate the permissible research and development costs. It has been decided that the net profits from the plant should be sufficient to pay off the total capital investment plus all research and development costs in 7 years. A return after taxes of at least 12 percent of sales must be obtained, and 50 percent of the research and development cost is tax-free (i.e., income-tax rate for the company is 50 percent of the gross earnings). Under these conditions, what is the total amount the company can afford to pay for research and development?

23 A chemical processing unit has a capacity for producing 1 million kg of a product per year. After the unit has been put into operation, it is found that only 500,000 kg of the product can be disposed of per year. An analysis of the existing situation shows that all fixed and other invariant charges, which must be paid whether or not the unit is operating, amount to 35 percent of the total product cost when operating at full capacity. Raw-material costs and other production costs that are directly proportional to the quantity of production (i.e., constant per kilogram of product at any production rate) amount to 40 percent of the total product cost at full capacity. The remaining 25 percent of the total product cost is for variable overhead and miscellaneous expenses, and the analysis indicates that these costs are directly proportional to the production rate during operation raised to the 1.5 power. What will be the percent change in total cost per kilogram of product if the unit is switched from the 1-million-kg-per-year rate to a time and rate schedule which will produce 500,000 kg of product per year at the least total cost? All costs referred to above are on a per-kilogram basis.

CHAPTER SIX

INTEREST AND INVESTMENT COSTS

A considerable amount of confusion exists among engineers over the role of interest in determining costs for a manufacturing operation. The confusion is caused by the attempt to apply the classical economist's definition of interest. According to the classical definition, interest is the money returned to the owners of capital for use of their capital. This would mean that any profit obtained through the use of capital could be considered as interest. Modern economists seldom adhere to the classical definition. Instead, they prefer to substitute the term *return on capital* or *return on investment* for the classical *interest*.

Engineers define interest as the *compensation paid for the use of borrowed capital*. This definition permits distinction between profit and interest. The rate at which interest will be paid is usually fixed at the time the capital is borrowed, and a guarantee is made to return the capital at some set time in the future or on an agreed-upon pay-off schedule.

TYPES OF INTEREST

Simple Interest

In economic terminology, the amount of capital on which interest is paid is designated as the *principal*, and *rate of interest* is defined as the amount of interest earned by a unit of principal in a unit of time. The time unit is usually taken as one year. For example, if $100 were the compensation demanded for giving someone the use of $1000 for a period of one year, the principal would be $1000, and the rate of interest would be $100/1000 = 0.1$ or 10 percent/year.

The simplest form of interest requires compensation payment at a constant interest rate based only on the original principal. Thus, if $1000 were loaned for a total time of 4 years at a constant interest rate of 10 percent/year, the simple interest earned would be

$$\$1000 \times 0.1 \times 4 = \$400$$

If P represents the principal, n the number of time units or interest periods, and i the interest rate based on the length of one interest period, the amount of simple interest I during n interest periods is

$$I = Pin \tag{1}$$

The principal must be repaid eventually; therefore, the entire amount S of principal plus simple interest due after n interest periods is

$$S = P + I = P(1 + in) \tag{2}$$

Ordinary and Exact Simple Interest

The time unit used to determine the number of interest periods is usually 1 year, and the interest rate is expressed on a yearly basis. When an interest period of less than 1 year is involved, the *ordinary* way to determine simple interest is to assume the year consists of twelve 30-day months, or 360 days. The *exact* method accounts for the fact that there are 365 days in a normal year. Thus, if the interest rate is expressed on the regular yearly basis and d represents the number of days in an interest period, the following relationships apply:

$$\text{Ordinary simple interest} = Pi\,\frac{d}{360} \tag{3}$$

$$\text{Exact simple interest} = Pi\,\frac{d}{365} \tag{4}$$

Ordinary interest is commonly accepted in business practices unless there is a particular reason to use the exact value.

Compound Interest

In the payment of simple interest, it makes no difference whether the interest is paid at the end of each time unit or after any number of time units. The same total amount of money is paid during a given length of time, no matter which method is used. Under these conditions, there is no incentive to pay the interest until the end of the total loan period.

Interest, like all negotiable capital, has a time value. If the interest were paid at the end of each time unit, the receiver could put this money to use for earning additional returns. *Compound interest* takes this factor into account by stipulating that interest is due regularly at the end of each interest period. If payment is not made, the amount due is added to the principal, and interest is charged on this converted principal during the following time unit. Thus, an initial loan of $1000 at an annual interest rate of 10 percent would require payment of $100 as interest at the end of the first year. If this payment were not made, the interest for the second year would be $($1000 + $100)(0.10) = 110, and the total *compound amount* due after 2 years would be

$$\$1000 + \$100 + \$110 = \$1210$$

The compound amount due after any discrete number of interest periods can be determined as follows:

Period	Principal at start of period	Interest earned during period (i = interest rate based on length of one period)	Compound amount S at end of period
1	P	Pi	$P + Pi = P(1 + i)$
2	$P(1 + i)$	$P(1 + i)(i)$	$P(1 + i) + P(1 + i)(i) = P(1 + i)^2$
3	$P(1 + i)^2$	$P(1 + i)^2(i)$	$P(1 + i)^2 + P(1 + i)^2(i) = P(1 + i)^3$
n	$P(1 + i)^{n-1}$	$P(1 + i)^{n-1}(i)$	$P(1 + i)^n$

Therefore, the total amount of principal plus compounded interest due after n interest periods and designated as S is†

$$S = P(1 + i)^n \tag{5}$$

The term $(1 + i)^n$ is commonly referred to as the *discrete single-payment compound-amount factor*. Values for this factor at various interest rates and numbers of interest periods are given in Table 1.

Figure 6-1 shows a comparison among the total amounts due at different times for the cases where simple interest, discrete compound interest, and continuous interest are used.

NOMINAL AND EFFECTIVE INTEREST RATES

In common industrial practice, the length of the discrete interest period is assumed to be 1 year and the fixed interest rate i is based on 1 year. However, there are cases where other time units are employed. Even though the actual interest period is not 1 year, the interest rate is often expressed on an annual basis. Consider an example in which the interest rate is 3 percent per period and the interest is compounded at half-year periods. A rate of this type would be

---

† For the analogous equation for continuous interest compounding, see Eq. (12).

Table 1 Discrete compound-interest factor $(1 + i)^n$ at various values of i and n†

Value of $(1 + i)^n$ at indicated percent interest

Number of interest periods, n	1%	2%	3%	4%	5%	6%	7%	8%	10%	12%	14%	16%	18%	20%
1	1.0100	1.0200	1.0300	1.0400	1.0500	1.0600	1.0700	1.0800	1.1000	1.1200	1.1400	1.1600	1.1800	1.2000
2	1.0201	1.0404	1.0609	1.0816	1.1025	1.1236	1.1449	1.1664	1.2100	1.2544	1.2996	1.3456	1.3924	1.4400
3	1.0303	1.0612	1.0927	1.1249	1.1576	1.1910	1.2250	1.2597	1.3310	1.4049	1.4815	1.5609	1.6430	1.7280
4	1.0406	1.0824	1.1255	1.1699	1.2155	1.2625	1.3108	1.3605	1.4641	1.5735	1.6890	1.8106	1.9388	2.0736
5	1.0510	1.1041	1.1593	1.2167	1.2763	1.3382	1.4026	1.4693	1.6105	1.7623	1.9254	2.1003	2.2878	2.4883
6	1.0615	1.1262	1.1941	1.2653	1.3401	1.4185	1.5007	1.5869	1.7716	1.9738	2.1950	2.4364	2.6996	2.9860
7	1.0721	1.1487	1.2299	1.3159	1.4071	1.5036	1.6058	1.7138	1.9487	2.2107	2.5023	2.8262	3.1855	3.5832
8	1.0829	1.1717	1.2668	1.3686	1.4775	1.5938	1.7182	1.8509	2.1436	2.4760	2.8592	3.2784	3.7589	4.2998
9	1.0937	1.1951	1.3048	1.4233	1.5513	1.6895	1.8385	1.9990	2.3579	2.7731	3.2520	3.8030	4.4355	5.1598
10	1.1046	1.2190	1.3439	1.4802	1.6289	1.7908	1.9672	2.1589	2.5937	3.1058	3.7072	4.4114	5.2338	6.1917
11	1.1157	1.2434	1.3842	1.5395	1.7103	1.8983	2.1049	2.3316	2.8531	3.4785	4.2262	5.1173	6.1759	7.4301
12	1.1268	1.2682	1.4258	1.6010	1.7959	2.0122	2.2522	2.5182	3.1384	3.8960	4.8179	5.9360	7.2876	8.9161
13	1.1381	1.2936	1.4685	1.6651	1.8856	2.1329	2.4098	2.7196	3.4523	4.3635	5.4924	6.8858	8.5994	10.699
14	1.1495	1.3195	1.5126	1.7317	1.9799	2.2609	2.5785	2.9372	3.7975	4.8871	6.2614	7.9875	10.147	12.839
15	1.1610	1.3459	1.5580	1.8009	2.0789	2.3966	2.7590	3.1722	4.1772	5.4736	7.1380	9.2655	11.974	15.407
16	1.1726	1.3728	1.6047	1.8730	2.1829	2.5404	2.9522	3.4259	4.5950	6.1304	8.1373	10.748	14.129	18.488
17	1.1843	1.4002	1.6528	1.9479	2.2920	2.6928	3.1588	3.7000	5.0545	6.8660	9.2765	12.468	16.672	22.186
18	1.1961	1.4282	1.7024	2.0258	2.4066	2.8543	3.3799	3.9960	5.5599	7.6900	10.575	14.462	19.673	26.623
19	1.2081	1.4568	1.7535	2.1068	2.5270	3.0256	3.6165	4.3157	6.1159	8.6128	12.056	16.777	23.214	31.948
20	1.2202	1.4859	1.8061	2.1911	2.6533	3.2071	3.8697	4.6610	6.7275	9.6463	13.744	19.461	27.393	38.338
21	1.2324	1.5157	1.8603	2.2788	2.7860	3.3996	4.1406	5.0338	7.4003	10.804	15.668	22.574	32.324	46.005
22	1.2447	1.5460	1.9161	2.3699	2.9253	3.6035	4.4304	5.4365	8.1403	12.100	17.861	26.186	38.142	55.206
23	1.2572	1.5769	1.9736	2.4647	3.0715	3.8197	4.7405	5.8715	8.9543	13.552	20.362	30.376	45.008	66.247
24	1.2697	1.6084	2.0328	2.5633	3.2251	4.0489	5.0724	6.3412	9.8497	15.179	23.212	35.236	53.109	79.497
25	1.2824	1.6406	2.0938	2.6658	3.3864	4.2919	5.4274	6.8485	10.835	17.000	26.462	40.874	62.669	95.396
30	1.3478	1.8114	2.4273	3.2434	4.3219	5.7435	7.6123	10.063	17.449	29.960	50.950	87.044	143.37	237.38
35	1.4166	1.9999	2.8139	3.9461	5.5160	7.6861	10.677	14.785	28.102	52.800	98.101	180.31	328.00	590.67
40	1.4889	2.2080	3.2620	4.8010	7.0400	10.286	14.974	21.725	45.259	93.051	188.88	378.72	750.38	1469.8
45	1.5648	2.4379	3.7816	5.8412	8.9850	13.765	21.002	31.920	72.891	163.99	363.68	795.44	1716.7	3657.2
50	1.6446	2.6916	4.3839	7.1067	11.467	18.420	29.457	46.902	117.39	289.00	700.24	1670.7	3927.3	9100.4

† Percent interest = $(i)(100)$.

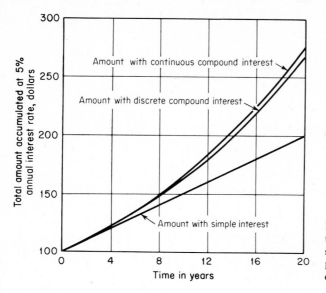

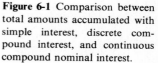

Figure 6-1 Comparison between total amounts accumulated with simple interest, discrete compound interest, and continuous compound nominal interest.

referred to as " 6 percent compounded semiannually." Interest rates stated in this form are known as *nominal interest rates*. The actual annual return on the principal would not be exactly 6 percent but would be somewhat larger because of the compounding effect at the end of the semiannual period.

It is desirable to express the exact interest rate based on the original principal and the convenient time unit of 1 year. A rate of this type is known as the *effective interest rate*. In common engineering practice, it is usually preferable to deal with effective interest rates rather than with nominal interest rates. The only time that nominal and effective interest rates are equal is when the interest is compounded annually.

Nominal interest rates should always include a qualifying statement indicating the compounding period. For example, using the common annual basis, $100 invested at a nominal interest rate of 20 percent compounded annually would amount to $120.00 after 1 year; if compounded semiannually, the amount would be $121.00; and, if compounded continuously, the amount would be $122.14. The corresponding effective interest rates are 20.00 percent, 21.00 percent, and 22.14 percent, respectively.

If nominal interest rates are quoted, it is possible to determine the effective interest rate by proceeding from Eq. (5).

$$S = P(1 + i)^n \tag{5}$$

In this equation, S represents the total amount of principal plus interest due after n periods at the periodic interest rate i. Let r be the nominal interest rate under conditions where there are m conversions or interest periods per year. Then the interest rate based on the length of one interest period is r/m, and the amount S after 1 year is

$$S_{\text{after 1 year}} = P\left(1 + \frac{r}{m}\right)^m \tag{6}$$

Designating the effective interest rate as i_{eff}, the amount S after 1 year can be expressed in an alternate form as

$$S_{\text{after 1 year}} = P(1 + i_{\text{eff}}) \tag{7}$$

By equating Eqs. (6) and (7), the following equation can be obtained for the effective interest rate in terms of the nominal interest rate and the number of periods per year:

$$\text{Effective annual interest rate} = i_{\text{eff}} = \left(1 + \frac{r}{m}\right)^m - 1 \tag{8}$$

Similarly, by definition,

$$\text{Nominal annual interest rate} = m\left(\frac{r}{m}\right) = r \tag{9}$$

Example 1. Applications of different types of interest It is desired to borrow $1000 to meet a financial obligation. This money can be borrowed from a loan agency at a monthly interest rate of 2 percent. Determine the following:

(a) The total amount of principal plus simple interest due after 2 years if no intermediate payments are made.

(b) The total amount of principal plus compounded interest due after 2 years if no intermediate payments are made.

(c) The nominal interest rate when the interest is compounded monthly.

(d) The effective interest rate when the interest is compounded monthly.

SOLUTION:

(a) Length of one interest period = 1 month

Number of interest periods in 2 years = 24

For simple interest, the total amount due after n periods at periodic interest rate i is

$$S = P(1 + in) \tag{2}$$

P = initial principal = $1000
$i = 0.02$ on a monthly basis
$n = 24$ interest periods in 2 years

$$S = \$1000(1 + 0.02 \times 24) = \$1480$$

(b) For compound interest, the total amount due after n periods at periodic interest rate i is

$$S = P(1 + i)^n \tag{5}$$

$$S = \$1000(1 + 0.02)^{24} = \$1608$$

(c) Nominal interest rate $= 2 \times 12 = 24\%$ per year compounded monthly
(d) Number of interest periods per year $= m = 12$
Nominal interest rate $= r = 0.24$

$$\text{Effective interest rate} = \left(1 + \frac{r}{m}\right)^m - 1 \tag{8}$$

$$\text{Effective interest rate} = \left(1 + \frac{0.24}{12}\right)^{12} - 1 = 0.268 = 26.8\%$$

CONTINUOUS INTEREST

The preceding discussion of types of interest has considered only the common form of interest in which the payments are charged at periodic and discrete intervals, where the intervals represent a finite length of time with interest accumulating in a discrete amount at the end of each interest period. Although in practice the basic time interval for interest accumulation is usually taken as one year, shorter time periods can be used as, for example, one month, one day, one hour, or one second. The extreme case, of course, is when the time interval becomes infinitesimally small so that the *interest is compounded continuously.*

The concept of continuous interest is that the cost or income due to interest flows regularly, and this is just as reasonable an assumption for most cases as the concept of interest accumulating only at discrete intervals. The reason why continuous interest has not been used widely is that most industrial and financial practices are based on methods which executives and the public are used to and can understand. Because normal interest comprehension is based on the discrete-interval approach, little attention has been paid to the concept of continuous interest even though this may represent a more realistic and idealized situation.

The Basic Equations for Continuous Interest Compounding

Equations (6), (7), and (8) represent the basic expressions from which continuous-interest relationships can be developed. The symbol r represents the nominal interest rate with m interest periods per year. If the interest is compounded continuously, m approaches infinity, and Eq. (6) can be written as

$$S_{\text{after } n \text{ years}} = P \lim_{m \to \infty} \left(1 + \frac{r}{m}\right)^{mn} = P \lim_{m \to \infty} \left(1 + \frac{r}{m}\right)^{(m/r)(rn)} \tag{10}$$

The fundamental definition for the base of the natural system of logarithms ($e = 2.71828$) is†

$$\lim_{m \to \infty} \left(1 + \frac{r}{m}\right)^{m/r} = e = 2.71828\ldots \tag{11}$$

Thus, with continuous interest compounding at a nominal annual interest rate of r, the *amount* S an initial *principal* P will compound to in n years is‡§

$$S = Pe^{rn} \tag{12}$$

Similarly, from Eq. (8), the effective annual interest rate i_{eff}, which is the conventional interest rate that most executives comprehend, is expressed in terms of the nominal interest rate r compounded continuously as

$$i_{\text{eff}} = e^r - 1 \tag{13}$$

$$r = \ln\left(i_{\text{eff}} + 1\right) \tag{14}$$

Therefore,

$$e^{rn} = \left(1 + i_{\text{eff}}\right)^n \tag{15}$$

---

† See any book on advanced calculus. For example, W. Fulks, "Advanced Calculus," 3d ed., pp. 55–56, John Wiley & Sons, Inc., New York, 1978.

‡ The same result can be obtained from calculus by noting that, for the case of continuous compounding, the differential change of S with time must equal the nominal continuous interest rate times S, or $dS/dn = rS$. This expression can be integrated as follows to give Eq. (12):

$$\int_P^S \frac{dS}{S} = r \int_o^n dn$$

$$\ln \frac{S}{P} = rn \qquad \text{or} \qquad S = Pe^{rn} \tag{12}$$

§ A generalized way to express both Eq. (12) and Eq. (5), with direct relationship to the other interest equations in this chapter, is as follows:

Future worth = present worth × compound interest factor

$$S = PC$$

or

Future worth × discount factor = present worth

$$SF = P$$

$$\text{Discount factor} = F = \frac{1}{\text{compound interest factor}} = \frac{1}{C}$$

Although the various factors for different forms of interest expressions are derived in terms of interest rate in this chapter, the overall concept of interest evaluations is simplified by the use of the less-complicated nomenclature where designated factors are applied. Thus, expressing both Eqs. (12) and (5) as $SF = P$ would mean that F is e^{-rn} for the continuous interest case of Eq. (12) and $(1 + i)^{-n}$ for the discrete interest case of Eq. (5). See Table 4 for further information on this subject.

and

$$S = Pe^{rn} = P(1 + i_{\text{eff}})^n \qquad (16)$$

As is illustrated in the following example, a conventional interest rate (i.e., effective annual interest rate) of 22.14 percent is equivalent to a 20.00 percent nominal interest rate compounded continuously. Note, also, that a nominal interest rate compounded daily gives results very close to those obtained with continuous compounding.

Example 2. Calculations with continuous interest compounding For the case of a nominal annual interest rate of 20.00 percent, determine
(a) The total amount to which one dollar of initial principal would accumulate after one 365-day year with daily compounding.
(b) The total amount to which one dollar of initial principal would accumulate after one year with continuous compounding.
(c) The effective annual interest rate if compounding is continuous.

SOLUTION:
(a) Using Eq. (6), $P = \$1.0$, $r = 0.20$, $m = 365$

$$S_{\text{after 1 year}} = P\left(1 + \frac{r}{m}\right)^m = (1.0)\left(1 + \frac{0.20}{365}\right)^{365} = \$1.2213$$

(b) Using Eq. (12),

$$S = Pe^{rn} = (1.0)(e)^{(0.20)(1)} = \$1.2214$$

(c) Using Eq. (13),

$$i_{\text{eff}} = e^r - 1 = 1.2214 - 1 = 0.2214 \text{ or } 22.14\%$$

Tabulated values of i_{eff} and the corresponding r with continuous interest compounding are shown in Table 2.

Example 3. Use of analog computer to give graphical representation of compounded amounts Present the analog computer diagram and the resulting plot from the analog computer for the total amount accumulated with time when $100 is invested initially at a nominal interest rate of 20 percent with continuous interest compounding. Cover a time period of 0 to 20 years. For comparison, show on the plot the results with a simple interest rate of 20 percent and an effective annual interest rate of 20 percent.

SOLUTION The equation to be solved on the analog computer is

$$S = Pe^{rn}$$

where S will be plotted versus n

$$P = \$100$$

$$r = 0.20$$

Table 2 Effective annual interest rates compared to equivalent nominal interest rates with continuous interest

Effective annual rate of return, %	Nominal continuous rate of return, %	Effective annual rate of return, %	Nominal continuous rate of return, %
1	0.99504	35	30.010
2	1.9803	40	33.647
3	2.9559	45	37.156
4	3.9221	50	40.547
5	4.8790	60	47.000
6	5.8269	70	53.063
7	6.7659	80	58.779
8	7.6961	90	64.185
9	8.6178	100	69.315
10	9.5310	110	74.194
15	13.976	120	78.846
20	18.232	130	83.291
25	22.314	140	87.547
30	26.236	150	91.629

Setting up in differential form for the analog computer,

$$\frac{dS}{dn} = Pre^{rn} = Sr$$

The limits for S are \$100 to \$$S$ corresponding to n limits of 0 to n years. The analog computer diagram is:

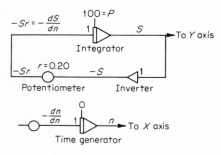

The resulting scaled plot from the analog computer is shown in Fig. 6-2.

Example 4. Use of digital computer to give tabulated values of amount accumulated with continuous interest compounding Present the digital computer program and the tabulated printout to six significant figures giving the amount to which an initial principal of \$100 will accumulate year by year from 1 to 20 years with continuous interest compounding based on a nom-

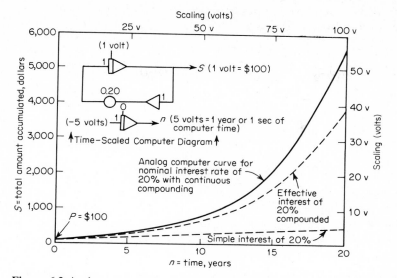

Figure 6-2 Analog computer plot showing effect of continuous interest compounding. (Example 3).

inal interest rate of 20 percent. (NOTE: *The result should be the same as shown in Fig. 6-2 except, in this case, the exact numbers to six significant figures are given for each year.*)

SOLUTION The equation to be solved on the digital computer is

$$S = Pe^{rn}$$

where S will be evaluated to six significant figures for

$$n = 1, 2, 3, \ldots, 20$$

$$P = \$100$$

$$r = 0.20$$

The Fortran IV program and the computer print-out follow:

```
            $IBJOB      MAP
            $IBFTC DECK 1
1               DO 1 N = 1,20
2               AN = N
3                 S = 100.*EXP(.20*AN)
4       1       WRITE(6,2)N,S
5       2       FORMAT(I4,F12.3)
6               END
            $ENTRY
```

Printout

1	122.140	11	902.501
2	149.182	12	1102.318
3	182.212	13	1346.374
4	222.554	14	1644.465
5	271.828	15	2008.554
6	332.012	16	2453.253
7	405.520	17	2996.410
8	495.303	18	3659.823
9	604.965	19	4470.118
10	738.906	20	5459.815

(NOTE: *The preceding illustrates the method used to prepare tabulated results of factors and emphasizes the simplicity of the procedure with a digital computer. This set of results represents a standard exponential function available in tabulated form in standard mathematical tables. See Prob. 9 at the end of this chapter for a requested computer solution for a more complicated continuous-interest case.*)

PRESENT WORTH AND DISCOUNT

It is often necessary to determine the amount of money which must be available at the present time in order to have a certain amount accumulated at some definite time in the future. Because the element of time is involved, interest must be taken into consideration. The *present worth* (or *present value*) of a future amount is the present principal which must be deposited at a given interest rate to yield the desired amount at some future date.†

In Eq. (5), S represents the amount available after n interest periods if the initial principal is P and the discrete compound-interest rate is i. Therefore, the present worth can be determined by merely rearranging Eq. (5).

$$\text{Present worth} = P = S\frac{1}{(1+i)^n} \tag{17}$$

The factor $1/(1+i)^n$ is commonly referred to as the *discrete single-payment present-worth factor*.

Similarly, for the case of continuous interest compounding, Eq. (12) gives

$$\text{Present worth} = P = S\frac{1}{e^{rn}} \tag{18}$$

Some types of capital are in the form of bonds having an indicated value at a future date. In business terminology, the difference between the indicated future value and the present worth (or present value) is known as the *discount*.

---

† In the analyses presented in this chapter, effects of inflation or deflation on future worth are not considered. See Chap. 10 (Optimum Design and Design Strategy) for information on the strategy for dealing with inflation or deflation in design economic evaluations.

Example 5. Determination of present worth and discount A bond has a maturity value of $1000 and is paying discrete compound interest at an effective annual rate of 3 percent. Determine the following at a time four years before the bond reaches maturity value:

(a) Present worth.
(b) Discount.
(c) Discrete compound rate of effective interest which will be received by a purchaser if the bond were obtained for $700.
(d) Repeat part (a) for the case where the nominal bond interest is 3 percent compounded continuously.

SOLUTION:

(a) By Eq. (17), present worth $= S/(1 + i)^n = \$1000/(1 + 0.03)^4 = \$888$
(b) Discount $=$ future value $-$ present worth $= \$1000 - \$888 = \$112$
(c) Principal $= \$700 = S/(1 + i)^n = \$1000/(1 + i)^4$

$$i = \left(\frac{1000}{700}\right)^{1/4} - 1 = 0.0935 \qquad \text{or} \qquad 9.35\%$$

(d) By Eq. (18), present worth $= S/e^{rn} = \$1000/e^{(0.03)(4)} = \$869$

ANNUITIES

An *annuity* is a series of equal payments occurring at equal time intervals. Payments of this type can be used to pay off a debt, accumulate a desired amount of capital, or receive a lump sum of capital that is due in periodic installments as in some life-insurance plans. Engineers often encounter annuities in depreciation calculations, where the decrease in value of equipment with time is accounted for by an annuity plan.

The common type of annuity involves payments which occur *at the end of each interest period*. This is known as an *ordinary annuity*. Interest is paid on all accumulated amounts, and the interest is compounded each payment period. An *annuity term* is the time from the beginning of the first payment period to the end of the last payment period. The *amount of an annuity* is the sum of all the payments plus interest if allowed to accumulate at a definite rate of interest from the time of initial payment to the end of the annuity term.

Relation between Amount of Ordinary Annuity and the Periodic Payments

Let R represent the uniform periodic payment made during n discrete periods in an ordinary annuity. The interest rate based on the payment period is i, and S is the amount of the annuity. The first payment of R is made at the end of the first period and will bear interest for $n - 1$ periods. Thus, at the end of the annuity term, this first payment will have accumulated to an amount of $R(1 + i)^{n-1}$. The second payment of R is made at the end of the second period and will bear interest for $n - 2$ periods giving an accumulated amount of $R(1 + i)^{n-2}$. Similarly, each periodic payment will give an additional accumulated amount until the last payment of R is made at the end of the annuity term.

By definition, the amount of the annuity is the sum of all the accumulated amounts from each payment; therefore,

$$S = R(1 + i)^{n-1} + R(1 + i)^{n-2} + R(1 + i)^{n-3} + \cdots + R(1 + i) + R \quad (19)$$

To simplify Eq. (19), multiply each side by $(1 + i)$ and subtract Eq. (19) from the result. This gives

$$Si = R(1 + i)^n - R \quad (20)$$

or

$$S = R\frac{(1 + i)^n - 1}{i} \quad (21)$$

The term $[(1 + i)^n - 1]/i$ is commonly designated as the *discrete uniform-series compound-amount factor* or the *series compound-amount factor*.

Continuous Cash Flow and Interest Compounding

The expression for the case of continuous cash flow and interest compounding, equivalent to Eq. (21) for discrete cash flow and interest compounding, is developed as follows:

As before, let r represent the nominal interest rate with m conversions or interest periods per year so that $i = r/m$ and the total number of interest periods in n years is mn. With m annuity payments per year, let $\bar{R}$ represent the total of all ordinary annuity payments occurring regularly and uniformly throughout the year so that $\bar{R}/m$ is the uniform annuity payment at the end of each period. Under these conditions, Eq. (21) becomes

$$S = \frac{\bar{R}}{m}\frac{[1 + (r/m)]^{(m/r)(rn)} - 1}{r/m} \quad (22)$$

For the case of continuous cash flow and interest compounding, m approaches infinity, and Eq. (22), by use of Eq. (11), becomes†

$$S = \bar{R}\left(\frac{e^{rn} - 1}{r}\right) \quad (23)$$

---

† The same result is obtained from the calculus by noting that the definition of $\bar{R}$ is such that the differential charge in S with n is equal to $\bar{R}$, which is the constant gradient during the year, plus the contribution due to interest, or $dS/dn = \bar{R} + rS$. This expression can be integrated as follows to give Eq. (23):

$$\int_0^S \frac{dS}{\bar{R} + rS} = \int_0^n dn$$

$$\ln\frac{\bar{R} + rS}{\bar{R}} = rn \quad \text{or} \quad S = \bar{R}\left(\frac{e^{rn} - 1}{r}\right) \quad (23)$$

Present Worth of an Annuity

The *present worth of an annuity* is defined as the principal which would have to be invested at the present time at compound interest rate i to yield a total amount at the end of the annuity term equal to the amount of the annuity. Let P represent the present worth of an ordinary annuity. Combining Eq. (5) with Eq. (21) gives, for the case of discrete interest compounding,

$$P = R\frac{(1 + i)^n - 1}{i(1 + i)^n} \tag{24}$$

The expression $[(1 + i)^n - 1]/[i(1 + i)^n]$ is referred to as the *discrete uniform-series present-worth factor* or the *series present-worth factor*, while the reciprocal $[i(1 + i)^n]/[(1 + i)^n - 1]$ is often called the *capital-recovery factor*.

For the case of continuous cash flow and interest compounding, combination of Eqs. (12) and (23) gives the following equation which is analogous to Eq. (24):

$$P = \bar{R}\frac{e^{rn} - 1}{re^{rn}} \tag{25}$$

Example 6. Application of annuities in determining amount of depreciation with discrete interest compounding A piece of equipment has an initial installed value of $12,000. It is estimated that its useful life period will be 10 years and its scrap value at the end of the useful life will be $2000. The depreciation will be charged as a cost by making equal charges each year, the first payment being made at the end of the first year. The depreciation fund will be accumulated at an annual interest rate of 6 percent. At the end of the life period, enough money must have been accumulated to account for the decrease in equipment value. Determine the yearly cost due to depreciation under these conditions.

(NOTE: *This method for determining depreciation is based on an ordinary annuity and is known as the sinking-fund method.*)

SOLUTION This problem is a typical case of an ordinary annuity. Over a period of 10 years, equal payments must be made each year at an interest rate of 6 percent. After 10 years, the amount of the annuity must be equal to the total amount of depreciation.

Amount of annuity $= S$
Total amount of depreciation $= \$12,000 - \$2000 = \$10,000 = S$
Equal payments per year $= R =$ yearly cost due to depreciation
Number of payments $= n = 10$
Annual interest rate $= i = 0.06$

From Eq. (21),

$$R = S\frac{i}{(1 + i)^n - 1} = \$10,000\frac{0.06}{(1.06)^{10} - 1} = \$759/\text{year}$$

Yearly cost due to depreciation $= \$759$

Example 7. Application of annuities in determining amount of depreciation with continuous cash flow and interest compounding Repeat Example 6 with continuous cash flow and nominal annual interest of 6 percent compounded continuously.

SOLUTION This problem is solved in exactly the same manner as Example 6, except the appropriate Eq. (23) for the continuous-interest case is used in place of the discrete-interest equation.

Amount of annuity = S
Total amount of depreciation = \$12,000 − \$2000 = S
Equal payments per year based on continuous cash flow and interest compounding = $\bar{R}$ = yearly cost due to depreciation
Number of years = n = 10
Nominal interest rate with continuous compounding = r = 0.06

From Eq. (23),

$$\bar{R} = S\frac{r}{e^{rn} - 1} = \$10,000\,\frac{0.06}{e^{(0.06)(10)} - 1} = \$730/\text{year}$$

Yearly cost due to depreciation = \$730

Special Types of Annuities

One special form of an annuity requires that payments be made at the beginning of each period instead of at the end of each period. This is known as an *annuity due*. An annuity in which the first payment is due after a definite number of years is called a *deferred annuity*. Determination of the periodic payments, amount of annuity, or present value for these two types of annuities can be accomplished by methods analogous to those used in the case of ordinary annuities.†

PERPETUITIES AND CAPITALIZED COSTS

A *perpetuity* is an annuity in which the periodic payments continue indefinitely. This type of annuity is of particular interest to engineers, for in some cases they may desire to determine a total cost for a piece of equipment or other asset under conditions which permit the asset to be replaced perpetually without considering inflation or deflation.

Consider the example in which the original cost of a certain piece of equipment is \$12,000. The useful-life period is 10 years, and the scrap value at the end of the useful life is \$2000. The engineer reasons that this piece of equipment, or its replacement, will be in use for an indefinitely long period of time, and it will be necessary to supply \$10,000 every 10 years in order to replace the equipment. He therefore wishes to provide a fund of sufficient size so that it will earn enough interest to pay for the periodic replacement. If the discrete annual interest rate is

† For further information, see E. P. DeGarmo, "Engineering Economy," 4th ed., The Macmillan Company, New York, 1967.

6 percent, this fund would need to be $12,650. At 6 percent interest compounded annually, the fund would amount to $($12,650$)(1 + 0.06)^{10} = $22,650$ after 10 years. Thus, at the end of 10 years, the equipment can be replaced for $10,000 and $12,650 will remain in the fund. This cycle could now be repeated indefinitely. If the equipment is to perpetuate itself, the theoretical amount of total capital necessary at the start would be $12,000 for the equipment plus $12,650 for the replacement fund. The total capital determined in this manner is called the *capitalized cost*. Engineers use capitalized costs principally for comparing alternative choices.†

In a perpetuity, such as in the preceding example, the amount required for the replacement must be earned as compounded interest over a given length of time. Let P be the amount of present principal (i.e., the present worth) which can accumulate to an amount S during n interest periods at periodic interest rate i. Then, by Eq. (5),

$$S = P(1 + i)^n \qquad (5)$$

If perpetuation is to occur, the amount S accumulated after n periods minus the cost for the replacement must equal the present worth P. Therefore, letting C_R represent the replacement cost,

$$P = S - C_R \qquad (26)$$

Combining Eqs. (5) and (26),

$$P = \frac{C_R}{(1 + i)^n - 1} \qquad (27)$$

The capitalized cost is defined as the original cost of the equipment plus the present value of the renewable perpetuity. Designating K as the capitalized cost and C_V as the original cost of the equipment,‡

$$K = C_V + \frac{C_R}{(1 + i)^n - 1} \qquad (28)$$

Example 8. Determination of capitalized cost A new piece of completely installed equipment costs $12,000 and will have a scrap value of $2000 at the end of its useful life. If the useful-life period is 10 years and the interest is compounded at 6 percent per year, what is the capitalized cost of the equipment?

SOLUTION The cost for replacement of the equipment at the end of its useful life (assuming costs unchanged) = $12,000 − $2000 = $10,000.

---

† For further discussion of capitalized costs used in engineering, see Chap. 9 and F. C. Jelen and M. S. Cole, Methods for Economic Analysis, Part I, *Hydrocarbon Proc.*, **53**(7):133 (1974); Part II, *Hydrocarbon Proc.*, **53**(9):227 (1974).

‡ For the continuous-interest-compounding expression equivalent to the discrete-interest-compounding case given in Eq. (28), see Prob. 14 at the end of the chapter.

By Eq. (28)

$$\text{Capitalized cost} = C_V + \frac{C_R}{(1 + i)^n - 1}$$

where $C_V = \$12{,}000$
$\quad C_R = \$10{,}000$
$\quad\quad i = 0.06$
$\quad\quad n = 10$

$$\text{Capitalized cost} = \$12{,}000 + \frac{\$10{,}000}{(1 + 0.06)^{10} - 1}$$

$$= \$12{,}000 + \$12{,}650 = \$24{,}650$$

Example 9. Comparison of alternative investments using capitalized costs
A reactor, which will contain corrosive liquids, has been designed. If the reactor is made of mild steel, the initial installed cost will be $5000, and the useful-life period will be 3 years. Since stainless steel is highly resistant to the corrosive action of the liquids, stainless steel, as the material of construction, has been proposed as an alternative to mild steel. The stainless-steel reactor would have an initial installed cost of $15,000. The scrap value at the end of the useful life would be zero for either type of reactor, and both could be replaced at a cost equal to the original price. On the basis of equal capitalized costs for both types of reactors, what should be the useful-life period for the stainless-steel reactor if money is worth 4 percent compounded annually?

SOLUTION By Eq. (28), the capitalized cost for the mild-steel reactor is

$$K = C_V + \frac{C_R}{(1 + i)^n - 1} = \$5000 + \frac{\$5000}{(1 + 0.04)^3 - 1}$$

$$K = \$5000 + \$40{,}030 = \$45{,}030$$

Therefore, the capitalized cost for the stainless-steel reactor must also be $45,030.
For the stainless-steel reactor,

$$\$45{,}030 = C_V + \frac{C_R}{(1 + i)^n - 1} = \$15{,}000 + \frac{\$15{,}000}{(1 + 0.04)^n - 1}$$

Solving algebraically for n,

$$n = 10.4 \text{ years}$$

Thus, the useful-life period of the stainless-steel reactor should be 10.4 years for the two types of reactors to have equal capitalized costs. If the stainless-steel reactor would have a useful life of more than 10.4 years, it would be the recommended choice, while the mild-steel reactor would be recommended if the useful life using stainless steel were less than 10.4 years.

RELATIONSHIPS FOR CONTINUOUS CASH FLOW AND CONTINUOUS INTEREST OF IMPORTANCE FOR PROFITABILITY ANALYSES

The fundamental relationships dealing with continuous interest compounding can be divided into two general categories: (1) those that involve instantaneous or lump-sum payments, such as a required initial investment or a future payment that must be made at a given time, and (2) those that involve continuous payments or continuous cash flow, such as construction costs distributed evenly over a construction period or regular income that flows constantly into an overall operation. Equation (12) is a typical example of a lump-sum formula, while Eqs. (23) and (25) are typical of continuous-cash-flow formulas.

The symbols S, P, and R represent discrete lump-sum payments as future worth, present principal (or present worth), and end-of-period (or end-of-year) payments, respectively. A bar above the symbol, such as $\bar{S}$, $\bar{P}$, or $\bar{R}$, means that the payments are made *continuously* throughout the time period under consideration.[†] For example, consider the case where construction of a plant requires a continuous flow of cash to the project for one year, with the plant ready for operation at the end of the year of construction. The symbol $\bar{P}$ represents the total amount of cash put into the project on the basis of one year with a continuous flow of cash. At the end of the year, the compound amount of this $\bar{P}$ is

$$S_{\text{at end of one individual year}} = \bar{P}\frac{e^r - 1}{r} = P_{\text{at startup}} \tag{29}$$

If the plant is ready for operation after one year of construction time and the startup of the plant is designated as zero time, the future worth of the plant construction cost after n years with continuous interest compounding is

$$S_{\text{after } n \text{ years in operation}} = (P_{\text{at startup}})e^{rn} = \bar{P}\frac{e^r - 1}{r}e^{rn} \tag{30}$$

For profitability analyses, certain discounting or compounding factors based on continuous interest compounding are of sufficient importance that tables have been prepared which give values of the factors for various interest rates and time periods. Table 3 gives examples of tabulated factors for the following cases:[‡][§]

[†] It should be noted that $\bar{S}$, $\bar{P}$, and $\bar{R}$ represent sums accumulated by continuous payment over an indicated time period without any interest accumulation. $\bar{R}$ represents a periodic accumulation normally based on one year, while $\bar{S}$ and $\bar{P}$ represent accumulations during a given period of time. Thus, $\bar{S}$ and $\bar{P}$ are interchangeable depending on the basic form of equation being used, and $\bar{S}$, $\bar{P}$, and $\bar{R}$ are interchangeable if the time period under consideration is limited to one year or one basic interest period.

[‡] See Table 4 for a summary of the significance and meaning of the factors presented in Table 3. Extended values of the factors for parts (a) to (d) of Table 3 are given in Tables 5 to 8.

[§] For illustrations of the applications of continuous interest compounding and continuous cash flow to cases of profitability evaluation, see Examples 2 and 3 in Chap. 9.

Table 3 Discount and compounding factors for continuous interest and cash flows[†]

r as percent

Discount factors to give present worths for cash flows which

	1%	5%	10%	15%	20%	25%	30%	40%	50%	60%	80%	100%
(a) Occur in an instant at a point in time after the reference point												
$n = 1$	0.990	0.951	0.905	0.861	0.819	0.779	0.741	0.670	0.606	0.549	0.449	0.368
2	0.980	0.905	0.819	0.741	0.670	0.606	0.549	0.449	0.368	0.301	0.202	0.135
3	0.970	0.861	0.741	0.638	0.549	0.472	0.407	0.301	0.223	0.165	0.091	0.050
4	0.961	0.819	0.670	0.549	0.449	0.368	0.301	0.202	0.135	0.091	0.041	0.018
5	0.951	0.779	0.606	0.472	0.368	0.286	0.223	0.135	0.082	0.050	0.018	0.007
10	0.905	0.606	0.368	0.223	0.135	0.082	0.050	0.018	0.007	0.002		
15	0.861	0.472	0.223	0.105	0.050	0.024	0.011	0.002	0.001			
20	0.819	0.368	0.135	0.050	0.018	0.007	0.002					
25	0.779	0.286	0.082	0.024	0.007	0.002	0.001					
$1.0 \left(\dfrac{1}{e^{rn}} \right) = F_a$												
(b) Occur uniformly over one-year periods after the reference point												
$n = 1$. 1st year	0.995	0.975	0.952	0.929	0.906	0.885	0.864	0.824	0.787	0.752	0.688	0.632
2. 2nd year	0.985	0.928	0.861	0.799	0.742	0.689	0.640	0.552	0.477	0.413	0.309	0.232
3. 3rd year	0.975	0.883	0.779	0.688	0.608	0.537	0.474	0.370	0.290	0.226	0.139	0.086
4. 4th year	0.966	0.840	0.705	0.592	0.497	0.418	0.351	0.248	0.176	0.124	0.062	0.032
5. 5th year	0.956	0.799	0.638	0.510	0.407	0.326	0.260	0.166	0.106	0.068	0.028	0.012
$1.0 \left(\dfrac{e^r - 1}{r} \right) e^{-rn} = F_b$												
(c) Occur uniformly over a period of years—For period of years = T = 5 years												
$n = 5$. 1st 5 years	0.975	0.885	0.787	0.704	0.632	0.571	0.518	0.432	0.367	0.317	0.245	0.199
10. 2nd 5 years	0.928	0.689	0.477	0.332	0.232	0.164	0.116	0.058	0.030	0.016	0.004	0.001
15. 3rd 5 years	0.883	0.537	0.290	0.157	0.086	0.047	0.026	0.008	0.002	0.001		
20. 4th 5 years	0.840	0.418	0.176	0.074	0.032	0.013	0.006	0.001				
25. 5th 5 years	0.799	0.326	0.106	0.035	0.012	0.004	0.001					
$\dfrac{1}{T} \left(\dfrac{e^{rT} - 1}{r} \right) e^{-rn} = F_c$												

244

Table 3 Discount and compounding factors for continuous interest and cash flows† (Continued)

r as percent	1%	5%	10%	15%	20%	25%	30%	40%	50%	60%	80%	100%
(d) Decline to zero at a constant rate over a period of years starting with the reference point												
1st 5 years	0.983	0.922	0.852	0.791	0.736	0.687	0.643	0.568	0.506	0.456	0.377	0.320
1st 10 years	0.968	0.852	0.736	0.643	0.568	0.506	0.456	0.377	0.320	0.278	0.219	0.180
1st 15 years	0.952	0.791	0.643	0.536	0.456	0.394	0.347	0.278	0.231	0.198	0.153	0.124
1st 20 years	0.936	0.736	0.568	0.456	0.377	0.320	0.278	0.219	0.180	0.153	0.117	0.095
1st 25 years	0.922	0.687	0.506	0.394	0.320	0.269	0.231	0.180	0.147	0.124	0.095	0.077

$$\frac{2}{rn_T}[1-(1-e^{-rn_T})/rn_T] = F_d$$

Compounding factors to give future worths for cash flows which

	1%	5%	10%	15%	20%	25%	30%	40%	50%	60%	80%	100%
(e) Occur in an instant at a point in time before the reference point												
½ year before	1.005	1.025	1.051	1.078	1.105	1.133	1.162	1.221	1.284	1.350	1.492	1.649
1 year before	1.010	1.051	1.105	1.162	1.221	1.284	1.350	1.492	1.649	1.822	2.226	2.718
1½ years before	1.015	1.078	1.162	1.252	1.350	1.455	1.568	1.822	2.117	2.460	3.320	4.482
2 years before	1.020	1.105	1.221	1.350	1.492	1.649	1.822	2.226	2.718	3.320	4.953	7.389
3 years before	1.030	1.162	1.350	1.568	1.822	2.117	2.460	3.320	4.482	6.050	11.023	20.086

$$1.0(e^{rn}) = C_e$$

	1%	5%	10%	15%	20%	25%	30%	40%	50%	60%	80%	100%
(f) Occur uniformly before the reference point												
½ year before	1.002	1.013	1.025	1.038	1.052	1.065	1.079	1.107	1.136	1.166	1.230	1.297
1 year before	1.005	1.025	1.052	1.079	1.107	1.136	1.166	1.230	1.297	1.370	1.532	1.718
1½ years before	1.008	1.038	1.079	1.121	1.166	1.213	1.263	1.370	1.489	1.622	1.933	2.321
2 years before	1.010	1.052	1.107	1.166	1.230	1.297	1.370	1.532	1.718	1.933	2.471	3.194
3 years before	1.015	1.079	1.166	1.263	1.370	1.489	1.622	1.933	2.321	2.805	4.176	6.362

$$\frac{1.0}{T}\left(\frac{e^T-1}{r}\right) = C_f$$

† r = nominal interest compounded continuously, percent/100; n = number of years; T and n_T = number of years in a time period. Extended values of factors for parts (a), (b), and (d) are given See Table 4 for significance and meaning of compounding factors. in Tables 5, 6, and 8, and Table 7 gives extended values for part (c) with n = T.

Table 4 Summary of significance and meaning of discount and compounding factors presented in Tables 3, 5, 6, 7, and 8†

As indicated in the footnote for Eq. (12), the common interest expressions can be written in simplified form by using discount-factor and compound-interest-factor notation. Following is a summary showing the significance and meaning of the compounding factors presented in Table 3. Derivations of the factors are presented in the text.

For part (a) in Table 3 and for Table 5

F_a = Discount factor to give present worth for cash flows which occur in an instant at a point in time after the reference point.

$$P = F_a S \qquad F_a = e^{-rn}$$

For part (b) in Table 3 and for Table 6

F_b = Discount factor to give present worth for cash flows which occur uniformly over one-year periods after the reference point. ($\bar{S}$ is the total cash flow for the nth year.)

$$P = F_b \bar{S}_{(\text{nth year})} \qquad F_b = \left(\frac{e^r - 1}{r}\right) e^{-rn}$$

For part (c) in Table 3 and for Table 7 with $n = T$

F_c = Discount factor to give present worth for cash flows which occur uniformly over a period of years T. ($\bar{S}$ is the total cash flow for the T-year period.)

$$P = F_c \bar{S}_{(T\text{-year period})} \qquad F_c = \frac{1}{T}\left(\frac{e^{rT} - 1}{r}\right) e^{-rn}$$

NOTE: *For the case when the period of years T is based on the period immediately after the reference point, $n = T$, and $F_c = (1 - e^{-rT})/rT$. This is the factor presented in Table 7.*

For part (d) in Table 3 and for Table 8

F_d = Discount factor to give present worth for cash flows which decline to zero at a constant rate over a period of years n_T starting with the reference point. ($\bar{S}$ is the total cash flow for the n_T-year period.)

$$P = F_d \bar{S}_{(\text{declining to zero at constant rate in } n_T \text{ years})}$$

$$F_d = \frac{2}{rn_T}\left(1 - \frac{1 - e^{-rn_T}}{rn_T}\right)$$

For part (e) in Table 3

C_e = Compounding factor to give future worth for cash flows which occur in an instant at a point in time before the reference point.

$$S = C_e P \qquad C_e = e^{rn} = \frac{1}{F_a}$$

NOTE: *Table 5 gives reciprocal values of C_e.*

Table 4 Summary of significance and meaning of discount and compounding factors presented in Tables 3, 5, 6, 7, and 8† (*Continued*)

For part (f) in Table 3

C_f = Compounding factor to give future worth for cash flows which occur uniformly over a period of years T before the reference point. ($\bar{P}$ is the total cash flow for the T-year period.)

$$S = C_f \bar{P} \qquad C_f = \frac{e^{rT} - 1}{rT}$$

For $n = T$, $C_f = F_c/F_a = F_c C_e$.

NOTE: *On basis of above relationship, C_f can be generated from F_a and F_c values given in Tables 5 and 7.*

---

† r = nominal interest compounded continuously, percent/100; n = number of years; T and n_T = number of years in a time period.

(a) *Discount factors to give present worths for cash flows which occur in an instant at a point in time after the reference point.* These factors are used to convert one dollar of money, which must be available in an instant after time n (such as scrap value, working capital, or land value), to the present worth of this one dollar with continuous interest compounding. The appropriate equation for calculating the factor, therefore, is based on Eq. (12), and

$$\text{Factor} = 1.0 \frac{1}{e^{rn}} = F_a \tag{31}$$

For example, if the nominal continuous interest rate is 20 percent and the time period is 5 years, the appropriate factor, as shown in Table 3, is

$$\text{Factor} = \frac{1}{e^{(0.2)(5)}} = \frac{1}{2.7183} = 0.368$$

(b) *Discount factors to give present worths for cash flows which occur uniformly over one-year periods after the reference point.* For this situation, the factor would convert one dollar of money, as the total yearly amount flowing continuously and uniformly during the year (such as cash receipts for one year), to the present worth of this one dollar at zero time with continuous interest compounding. Thus, $\bar{R}$ (or $\bar{S}$) for the year in question is 1.0, and the appropriate equation for calculating the factor, based on Eqs. (23) and (12), is

$$\text{Factor} = 1.0 \frac{e^r - 1}{r} e^{-rn} = F_b \tag{32}$$

As an example, if r represents 20 percent and n is the fifth year, the appropriate factor, as shown in Table 3, is

$$\text{Factor} = \frac{e^{(0.2)} - 1}{0.2}\frac{1}{e^{(0.2)(5)}} = \frac{1.2214 - 1}{0.2}\frac{1}{2.7183} = 0.407$$

(c) *Discount factors to give present worths for cash flows which occur uniformly over a period of years.* For this situation, a total amount of one dollar over a given time period is used as the basis. The cash flows continuously and uniformly during the entire period, and the factor converts the total of one dollar put in over the given time period to the present worth at zero time. This condition would be applicable to a case where cash receipts are steady over a given period of time, such as for five years. Designating T as the time period involved, the total amount put in each year is $\$1/T$, and the factor, based on Eqs. (23) and (12), is

$$\text{Factor} = \frac{1}{T}\frac{e^{rT} - 1}{r}e^{-rn} = F_c \tag{33}$$

For example, if the time period involved is the second five years (i.e., the 6th through the 10th years) and r represents 20 percent, the appropriate factor, as shown in Table 3, is

$$\text{Factor} = \frac{1}{5}\left(\frac{e^{(0.2)(5)} - 1}{0.2}\right)\left(\frac{1}{e^{(0.2)(10)}}\right) = \frac{1}{5}\left(\frac{2.7183 - 1}{0.2}\right)\frac{1}{7.3891} = 0.232$$

(d) *Discount factors to give present worths for cash flows declining to zero at a constant rate over a period of years starting with the reference point.* For this case, the assumption is made that the continuous cash flow declines linearly with time from the initial flow at time zero to zero flow at time n_T. A situation similar to this exists when the sum-of-the-years-digits method is used for calculating depreciation in that depreciation allowances decline linearly with time from a set value in the first year to zero at the end of the life.†‡ For the case of continuous cash flow declining to zero at a constant rate over a time period of n_T, the linear equation for $\bar{R}$ is

$$\bar{R} = a - gn \tag{34}$$

where g = the constant declining rate or the gradient
$\bar{R}$ = instantaneous value of the cash flow
a = a constant

---

† See Chap. 8 (Depreciation) for information on the sum-of-the-years-digits method for calculating depreciation.
‡ Equation (35) does not represent a true sum-of-the-years-digits factor. Normally, the constant declining rate or gradient for the sum-of-the-years-digits method of depreciation is $1/\sum_1^{n_T} n = 2/n_T(n_T + 1)$. For the true case of continuous cash flow declining to zero at a constant rate, n_T is replaced by $n_T m$ as $m \to \infty$, and the constant gradient becomes $2/(n_T)^2$.

The discount factor is based on a total amount of one dollar of cash flow over the time period n_T and converts this total of one dollar to the present worth at time zero. Under these conditions, g equals $2/(n_T)^2$†, and the factor is‡

$$\text{Factor} = \frac{2}{rn_T}\left(1 - \frac{1 - e^{-rn_T}}{rn_T}\right) = F_d \qquad (35)$$

As an example, if the cash flow declines at a constant rate to zero in 5 years and r is equivalent to 20 percent, the appropriate factor, as shown in Table 3, is

$$\text{Factor} = \frac{2}{(0.2)(5)}\left[1 - \frac{1}{(0.2)(5)}\left(1 - \frac{1}{e^{(0.2)(5)}}\right)\right]$$

$$= 2\left(1 - 1 + \frac{1}{2.7183}\right) = 0.736$$

(e) *Compounding factors to give future worths for cash flows which occur in an instant at a point in time before the reference point.* These factors merely show the future worth to which one dollar of principal, such as that for land purchase, will compound at continuous interest. Based on Eq. (12), the factor is

$$\text{Factor} = 1.0e^{rn} = C_e \qquad (36)$$

For example, with r equivalent to 20 percent and a purchase made $1\frac{1}{2}$ years before the reference point, the appropriate factor, as shown in Table 3, is

$$\text{Factor} = e^{(0.2)(1.5)} = 1.350$$

(f) *Compounding factors to give future worths for cash flows which occur uniformly before the reference point.* The basis for these factors is a uniform and

---

† By definition of terms and conditions, $\bar{R}$ is zero when $n = n_T$ and $\bar{R}$ is a when $n = 0$. Also, if a total of one dollar is the cash flow during n_T

$$\int_0^{n_T} \bar{R}\, dn = an_T - \frac{g(n_T)^2}{2} = \$1.0$$

Because $\bar{R}$ is zero when $n = n_T$, $a = gn_T$.
Therefore,

$$\$1.0 = g(n_T)^2 - \frac{g(n_T)^2}{2} = \frac{g(n_T)^2}{2} \qquad \text{and} \qquad g = \frac{2}{(n_T)^2}$$

‡ This can be derived by assuming an ordinary annuity with $\$gn_T$ for the first year, $\$g(n_T - 1)$ for the second year, etc., to $\$g$ for year n_T. The result with discrete interest compounding is

$$\text{Factor} = g\left[\frac{in_T - 1 + (1 + i)^{-n_T}}{i^2}\right]$$

Letting $g = 2/(n_T)^2$ and replacing i by r/m and n_T by mn_T gives Eq. (35) for $m \to \infty$.

continuous flow of cash amounting to a total of one dollar during the given time period of T years, such as for construction of a plant. The factor converts this one dollar to the future worth at the reference time and is based on Eq. (23).

$$\text{Factor} = \frac{1.0}{T} \frac{e^{rT} - 1}{r} = C_f \tag{37}$$

As an example, for the case of continuous compounding at r equivalent to 20 percent for a period from 3 years before the reference time, the appropriate factor, as shown in Table 3, is

$$\text{Factor} = \frac{1}{3} \frac{e^{(0.2)(3)} - 1}{0.2} = \frac{1.8221 - 1}{(3)(0.2)} = 1.370$$

TABLES FOR INTEREST AND CASH-FLOW FACTORS

Tables of interest and cash-flow factors, such as are illustrated in Tables 1, 5, 6, 7, and 8 of this chapter, are presented in all standard interest handbooks and textbooks on the mathematics of finance as well as in appendices of most textbooks on engineering economy.[†] Exponential functions for continuous compounding are available in the standard mathematical tables. The development of tables for any of the specialized factors is a relatively simple matter with the ready availability of digital computers, as is illustrated in Example 4 of this chapter.

The end-of-year convention is normally adopted for discrete interest factors (or for lump-sum payments) wherein the time unit of one interest period is assumed to be one year with interest compounding (or with lump-sum payments being made) at the end of each period. Thus, the *effective interest rate* is the form of interest most commonly understood and used by management and business executives.

In the tabulation of factors for continuous interest compounding and continuous cash flow, the nominal interest rate r is used for calculating the factors, but the tables are sometimes based on the effective interest rate. To avoid confusion between effective and nominal interest rates, the tables should always present a clear statement in the heading as to the type of interest basis used if there is any possibility for misunderstanding. In case such a necessary statement is not included with the continuous-interest table, the interest figures quoted are probably nominal, but it is advisable to check several of the factors by use of exponential tables to make certain that nominal, rather than effective, interest rates are quoted.

---

† For example, see the appendices of E. L. Grant, W. G. Ireson, and R. C. Leavenworth, "Principles of Engineering Economy," 6th ed., Ronald Press Company, New York, 1976; or G. A. Taylor, "Managerial and Engineering Economy," 2d ed., D. Van Nostrand Company, Inc., New York, 1975.

Table 5 Discount factors (F_a) with continuous interest to give present worths for cash flows which occur in an instant at a point in time after the reference point††‡

100rn	0	1	2	3	4	5	6	7	8	9
0	1.0000	0.9901	0.9802	0.9704	0.9608	0.9512	0.9418	0.9324	0.9231	0.9139
10	0.9048	0.8958	0.8869	0.8781	0.8694	0.8607	0.8521	0.8437	0.8353	0.8270
20	0.8187	0.8106	0.8025	0.7945	0.7866	0.7788	0.7711	0.7634	0.7558	0.7483
30	0.7408	0.7334	0.7261	0.7189	0.7118	0.7047	0.6977	0.6907	0.6839	0.6771
40	0.6703	0.6637	0.6570	0.6505	0.6440	0.6376	0.6313	0.6250	0.6188	0.6126
50	0.6065	0.6005	0.5945	0.5886	0.5827	0.5770	0.5712	0.5655	0.5599	0.5543
60	0.5488	0.5434	0.5379	0.5326	0.5273	0.5220	0.5169	0.5117	0.5066	0.5016
70	0.4966	0.4916	0.4868	0.4819	0.4771	0.4724	0.4677	0.4630	0.4584	0.4538
80	0.4493	0.4449	0.4404	0.4360	0.4317	0.4274	0.4232	0.4190	0.4148	0.4107
90	0.4066	0.4025	0.3985	0.3946	0.3906	0.3867	0.3829	0.3791	0.3753	0.3716
100	0.3679	0.3642	0.3606	0.3570	0.3535	0.3499	0.3465	0.3430	0.3396	0.3362
110	0.3329	0.3296	0.3263	0.3230	0.3198	0.3166	0.3135	0.3104	0.3073	0.3042
120	0.3012	0.2982	0.2952	0.2923	0.2894	0.2865	0.2837	0.2808	0.2780	0.2753
130	0.2725	0.2698	0.2671	0.2645	0.2618	0.2592	0.2567	0.2541	0.2516	0.2491
140	0.2466	0.2441	0.2417	0.2393	0.2369	0.2346	0.2322	0.2299	0.2276	0.2254
150	0.2231	0.2209	0.2187	0.2165	0.2144	0.2122	0.2101	0.2080	0.2060	0.2039
160	0.2019	0.1999	0.1979	0.1959	0.1940	0.1921	0.1901	0.1882	0.1864	0.1845
170	0.1827	0.1809	0.1791	0.1773	0.1755	0.1738	0.1720	0.1703	0.1686	0.1670
180	0.1653	0.1637	0.1620	0.1604	0.1588	0.1572	0.1557	0.1541	0.1526	0.1511
190	0.1496	0.1481	0.1466	0.1451	0.1437	0.1423	0.1409	0.1395	0.1381	0.1367
200	0.1353	0.1340	0.1327	0.1313	0.1300	0.1287	0.1275	0.1262	0.1249	0.1237
210	0.1225	0.1212	0.1200	0.1188	0.1177	0.1165	0.1153	0.1142	0.1130	0.1119
220	0.1108	0.1097	0.1086	0.1075	0.1065	0.1054	0.1044	0.1033	0.1023	0.1013
230	0.1003	0.0993	0.0983	0.0973	0.0963	0.0954	0.0944	0.0935	0.0926	0.0916
240	0.0907	0.0898	0.0889	0.0880	0.0872	0.0863	0.0854	0.0846	0.0837	0.0829
250	0.0821	0.0813	0.0805	0.0797	0.0789	0.0781	0.0773	0.0765	0.0758	0.0750
260	0.0743	0.0735	0.0728	0.0721	0.0714	0.0707	0.0699	0.0693	0.0686	0.0679
270	0.0672	0.0665	0.0659	0.0652	0.0646	0.0639	0.0633	0.0627	0.0620	0.0614
280	0.0608	0.0602	0.0596	0.0590	0.0584	0.0578	0.0573	0.0567	0.0561	0.0556
290	0.0550	0.0545	0.0539	0.0534	0.0529	0.0523	0.0518	0.0513	0.0508	0.0503
300	0.0498	0.0493	0.0488	0.0483	0.0478	0.0474	0.0469	0.0464	0.0460	0.0455
310	0.0450	0.0446	0.0442	0.0437	0.0433	0.0429	0.0424	0.0420	0.0416	0.0412
320	0.0408	0.0404	0.0400	0.0396	0.0392	0.0388	0.0384	0.0380	0.0376	0.0373
330	0.0369	0.0365	0.0362	0.0358	0.0354	0.0351	0.0347	0.0344	0.0340	0.0337
340	0.0334	0.0330	0.0327	0.0324	0.0321	0.0317	0.0314	0.0311	0.0308	0.0305
350	0.0302	0.0299	0.0296	0.0293	0.0290	0.0287	0.0284	0.0282	0.0279	0.0276
360	0.0273	0.0271	0.0268	0.0265	0.0263	0.0260	0.0257	0.0255	0.0252	0.0250
370	0.0247	0.0245	0.0242	0.0240	0.0238	0.0235	0.0233	0.0231	0.0228	0.0226
380	0.0224	0.0221	0.0219	0.0217	0.0215	0.0213	0.0211	0.0209	0.0207	0.0204
390	0.0202	0.0200	0.0198	0.0196	0.0194	0.0193	0.0191	0.0189	0.0187	0.0185

Continued

Table 5 Discount factors (F_a) with continuous interest to give present worths for cash flows which occur in an instant at a point in time after the reference point†‡ *(Continued)*

100rn	0	1	2	3	4	5	6	7	8	9
400	0.0183	0.0181	0.0180	0.0178	0.0176	0.0174	0.0172	0.0171	0.0169	0.0167
410	0.0166	0.0164	0.0162	0.0161	0.0159	0.0158	0.0156	0.0155	0.0153	0.0151
420	0.0150	0.0148	0.0147	0.0146	0.0144	0.0143	0.0141	0.0140	0.0138	0.0137
430	0.0136	0.0134	0.0133	0.0132	0.0130	0.0129	0.0128	0.0127	0.0125	0.0124
440	0.0123	0.0122	0.0120	0.0119	0.0118	0.0117	0.0116	0.0114	0.0113	0.0112
450	0.0111	0.0110	0.0109	0.0108	0.0107	0.0106	0.0105	0.0104	0.0103	0.0102
460	0.0101	0.0100	0.0099	0.0098	0.0097	0.0096	0.0095	0.0094	0.0093	0.0092
470	0.0091	0.0090	0.0089	0.0088	0.0087	0.0087	0.0086	0.0085	0.0084	0.0083
480	0.0082	0.0081	0.0081	0.0080	0.0079	0.0078	0.0078	0.0077	0.0076	0.0075
490	0.0074	0.0074	0.0073	0.0072	0.0072	0.0071	0.0070	0.0069	0.0069	0.0068

100rn	0	10	20	30	40	50	60	70	80	90
500	0.0067	0.0061	0.0055	0.0050	0.0045	0.0041	0.0037	0.0033	0.0030	0.0027
600	0.0025	0.0022	0.0020	0.0018	0.0017	0.0015	0.0014	0.0012	0.0011	0.0010
700	0.0009	0.0008	0.0007	0.0007	0.0006	0.0006	0.0005	0.0005	0.0004	0.0004
800	0.0003	0.0003	0.0003	0.0002	0.0002	0.0002	0.0002	0.0002	0.0002	0.0001
900	0.0001	0.0001	0.0001	0.0001	0.0001	0.0001	0.0001	0.0001	0.0001	0.0001
1000	0.0000									

† r = nominal interest compounded continuously, percent/100; n = number of years. See Tables 3 and 4 for information on F_a.

‡ The columns represent the unit increments from 1 to 9 for the 10-digit intervals of $100rn$ shown in the left column.

Table 6 Discount factors (F_b) with continuous interest to give present worths for cash flows which occur uniformly over one-year periods after the reference point†

Year	1%	2%	3%	4%	5%	6%	7%	8%	9%	10%	11%	12%	13%	14%	15%
0–1	0.9950	0.9901	0.9851	0.9803	0.9754	0.9706	0.9658	0.9610	0.9563	0.9516	0.9470	0.9423	0.9377	0.9332	0.9286
1–2	0.9851	0.9705	0.9560	0.9418	0.9278	0.9141	0.9005	0.8872	0.8740	0.8611	0.8483	0.8358	0.8234	0.8112	0.7993
2–3	0.9753	0.9512	0.9278	0.9049	0.8826	0.8608	0.8396	0.8189	0.7988	0.7791	0.7600	0.7413	0.7230	0.7053	0.6879
3–4	0.9656	0.9324	0.9004	0.8694	0.8395	0.8107	0.7829	0.7560	0.7300	0.7050	0.6808	0.6574	0.6349	0.6131	0.5921
4–5	0.9560	0.9140	0.8737	0.8353	0.7986	0.7635	0.7299	0.6979	0.6672	0.6379	0.6099	0.5831	0.5575	0.5330	0.5096
5–6	0.9465	0.8959	0.8479	0.8026	0.7596	0.7190	0.6806	0.6442	0.6098	0.5772	0.5463	0.5172	0.4895	0.4634	0.4386
6–7	0.9371	0.8781	0.8229	0.7711	0.7226	0.6772	0.6346	0.5947	0.5573	0.5223	0.4894	0.4588	0.4299	0.4029	0.3775
7–8	0.9278	0.8607	0.7985	0.7409	0.6874	0.6377	0.5917	0.5490	0.5093	0.4726	0.4385	0.4069	0.3775	0.3502	0.3250
8–9	0.9185	0.8437	0.7749	0.7118	0.6538	0.6006	0.5517	0.5068	0.4655	0.4276	0.3928	0.3609	0.3314	0.3045	0.2797
9–10	0.9094	0.8270	0.7520	0.6839	0.6219	0.5656	0.5144	0.4678	0.4254	0.3869	0.3519	0.3201	0.2910	0.2647	0.2407
10–11	0.9003	0.8106	0.7298	0.6571	0.5916	0.5327	0.4796	0.4318	0.3888	0.3501	0.3152	0.2839	0.2556	0.2301	0.2072
11–12	0.8914	0.7946	0.7082	0.6312	0.5628	0.5016	0.4472	0.3986	0.3553	0.3168	0.2824	0.2518	0.2244	0.2000	0.1783
12–13	0.8825	0.7788	0.6873	0.6065	0.5353	0.4724	0.4169	0.3680	0.3248	0.2866	0.2530	0.2233	0.1970	0.1739	0.1535
13–14	0.8737	0.7634	0.6670	0.5827	0.5092	0.4449	0.3888	0.3397	0.2968	0.2593	0.2266	0.1981	0.1730	0.1512	0.1321
14–15	0.8650	0.7483	0.6473	0.5599	0.4844	0.4190	0.3625	0.3136	0.2713	0.2347	0.2030	0.1757	0.1519	0.1314	0.1137
15–16	0.8564	0.7335	0.6282	0.5380	0.4608	0.3946	0.3380	0.2895	0.2479	0.2123	0.1819	0.1558	0.1334	0.1143	0.0979
16–17	0.8479	0.7189	0.6096	0.5169	0.4383	0.3716	0.3151	0.2672	0.2266	0.1921	0.1629	0.1382	0.1172	0.0993	0.0842
17–18	0.8395	0.7047	0.5916	0.4966	0.4169	0.3500	0.2938	0.2467	0.2071	0.1739	0.1460	0.1225	0.1029	0.0864	0.0725
18–19	0.8311	0.6908	0.5741	0.4772	0.3966	0.3296	0.2740	0.2277	0.1893	0.1573	0.1308	0.1087	0.0903	0.0751	0.0624
19–20	0.8228	0.6771	0.5571	0.4584	0.3772	0.3104	0.2554	0.2102	0.1730	0.1423	0.1171	0.0964	0.0793	0.0653	0.0537
20–21	0.8147	0.6637	0.5407	0.4405	0.3588	0.2923	0.2382	0.1940	0.1581	0.1288	0.1049	0.0855	0.0697	0.0568	0.0462
21–22	0.8065	0.6505	0.5247	0.4232	0.3413	0.2753	0.2221	0.1791	0.1445	0.1165	0.0940	0.0758	0.0612	0.0493	0.0398
22–23	0.7985	0.6376	0.5092	0.4066	0.3247	0.2593	0.2071	0.1653	0.1320	0.1054	0.0842	0.0673	0.0537	0.0429	0.0343
23–24	0.7906	0.6250	0.4941	0.3907	0.3089	0.2442	0.1931	0.1526	0.1207	0.0954	0.0754	0.0596	0.0472	0.0373	0.0295
24–25	0.7827	0.6126	0.4795	0.3753	0.2938	0.2300	0.1800	0.1409	0.1103	0.0863	0.0676	0.0529	0.0414	0.0324	0.0254
25–30	0.7596	0.5772	0.4386	0.3334	0.2535	0.1928	0.1466	0.1115	0.0849	0.0646	0.0492	0.0374	0.0285	0.0217	0.0165
30–35	0.7226	0.5223	0.3775	0.2730	0.1974	0.1428	0.1033	0.0748	0.0541	0.0392	0.0284	0.0205	0.0149	0.0108	0.0078
35–40	0.6874	0.4726	0.3250	0.2235	0.1538	0.1058	0.0728	0.0501	0.0345	0.0238	0.0164	0.0113	0.0078	0.0054	0.0037
40–45	0.6538	0.4276	0.2797	0.1830	0.1197	0.0784	0.0513	0.0336	0.0220	0.0144	0.0094	0.0062	0.0041	0.0027	0.0017
45–50	0.6219	0.3869	0.2407	0.1498	0.0933	0.0581	0.0362	0.0225	0.0140	0.0087	0.0054	0.0034	0.0021	0.0013	0.0008

Year	16%	17%	18%	19%	20%	21%	22%	23%	24%	25%	26%	27%	28%	29%	30%
0–1	0.9241	0.9196	0.9152	0.9107	0.9063	0.9020	0.8976	0.8933	0.8890	0.8848	0.8806	0.8764	0.8722	0.8681	0.8640
1–2	0.7875	0.7759	0.7644	0.7531	0.7421	0.7311	0.7204	0.7098	0.6993	0.6891	0.6790	0.6690	0.6592	0.6495	0.6400
2–3	0.6710	0.6546	0.6385	0.6228	0.6075	0.5926	0.5781	0.5639	0.5501	0.5367	0.5235	0.5107	0.4982	0.4860	0.4741
3–4	0.5718	0.5522	0.5333	0.5150	0.4974	0.4804	0.4639	0.4481	0.4327	0.4179	0.4037	0.3899	0.3765	0.3637	0.3513
4–5	0.4873	0.4659	0.4455	0.4259	0.4072	0.3894	0.3723	0.3560	0.3404	0.3255	0.3112	0.2976	0.2846	0.2721	0.2602
5–6	0.4152	0.3931	0.3721	0.3522	0.3334	0.3156	0.2988	0.2829	0.2678	0.2535	0.2400	0.2272	0.2151	0.2036	0.1928
6–7	0.3538	0.3316	0.3108	0.2913	0.2730	0.2558	0.2398	0.2247	0.2106	0.1974	0.1850	0.1734	0.1626	0.1524	0.1428
7–8	0.3015	0.2798	0.2596	0.2409	0.2235	0.2074	0.1924	0.1786	0.1657	0.1538	0.1427	0.1324	0.1229	0.1140	0.1058
8–9	0.2569	0.2360	0.2168	0.1992	0.1830	0.1681	0.1544	0.1419	0.1303	0.1197	0.1100	0.1011	0.0929	0.0853	0.0784
9–10	0.2189	0.1991	0.1811	0.1647	0.1498	0.1363	0.1239	0.1127	0.1025	0.0933	0.0848	0.0772	0.0702	0.0638	0.0581
10–11	0.1866	0.1680	0.1513	0.1362	0.1227	0.1105	0.0995	0.0896	0.0807	0.0726	0.0654	0.0589	0.0530	0.0478	0.0430
11–12	0.1590	0.1417	0.1264	0.1126	0.1004	0.0895	0.0798	0.0711	0.0634	0.0566	0.0504	0.0450	0.0401	0.0357	0.0319
12–13	0.1355	0.1196	0.1055	0.0932	0.0822	0.0726	0.0641	0.0565	0.0499	0.0441	0.0389	0.0343	0.0303	0.0267	0.0236
13–14	0.1154	0.1009	0.0882	0.0770	0.0673	0.0588	0.0514	0.0449	0.0393	0.0343	0.0300	0.0262	0.0229	0.0200	0.0175
14–15	0.0984	0.0851	0.0736	0.0637	0.0551	0.0477	0.0413	0.0357	0.0309	0.0267	0.0231	0.0200	0.0173	0.0150	0.0130
15–16	0.0838	0.0718	0.0615	0.0527	0.0451	0.0387	0.0331	0.0284	0.0243	0.0208	0.0178	0.0153	0.0131	0.0112	0.0096
16–17	0.0714	0.0606	0.0514	0.0436	0.0369	0.0313	0.0266	0.0225	0.0191	0.0162	0.0137	0.0117	0.0099	0.0084	0.0071
17–18	0.0609	0.0511	0.0429	0.0360	0.0303	0.0254	0.0213	0.0179	0.0150	0.0126	0.0106	0.0089	0.0075	0.0063	0.0053
18–19	0.0519	0.0431	0.0358	0.0298	0.0248	0.0206	0.0171	0.0142	0.0118	0.0098	0.0082	0.0068	0.0057	0.0047	0.0039
19–20	0.0442	0.0364	0.0299	0.0246	0.0203	0.0167	0.0137	0.0113	0.0093	0.0077	0.0063	0.0052	0.0043	0.0035	0.0029
20–21	0.0377	0.0307	0.0250	0.0204	0.0166										
21–22	0.0321	0.0259	0.0209	0.0169	0.0136										
22–23	0.0274	0.0218	0.0175	0.0139	0.0111										
23–24	0.0233	0.0184	0.0146	0.0115	0.0091										
24–25	0.0199	0.0156	0.0122	0.0095	0.0075										
25–30	0.0126	0.0096	0.0073	0.0056	0.0043										
30–35	0.0057	0.0041	0.0030	0.0022	0.0016										
35–40	0.0025	0.0018	0.0012	0.0008	0.0006										
40–45	0.0011	0.0008	0.0005	0.0003	0.0002										
45–50	0.0005	0.0003	0.0002	0.0001	0.0001										

† Percent is nominal interest compounded continuously = 100r. Year indicates one-year period in which cash flow occurs. See Tables 3 and 4 for information on F_b.

Table 7 Discount factors (F_c) with continuous interest to give present worths for cash flows which occur uniformly over a period of T years after the reference point††‡

$100rT$	0	1	2	3	4	5	6	7	8	9
0	1.0000	0.9950	0.9901	0.9851	0.9803	0.9754	0.9706	0.9658	0.9610	0.9563
10	0.9516	0.9470	0.9423	0.9377	0.9332	0.9286	0.9241	0.9196	0.9152	0.9107
20	0.9063	0.9020	0.8976	0.8933	0.8891	0.8848	0.8806	0.8764	0.8722	0.8681
30	0.8639	0.8598	0.8558	0.8517	0.8477	0.8438	0.8398	0.8359	0.8319	0.8281
40	0.8242	0.8204	0.8166	0.8128	0.8090	0.8053	0.8016	0.7979	0.7942	0.7906
50	0.7869	0.7833	0.7798	0.7762	0.7727	0.7692	0.7657	0.7622	0.7588	0.7554
60	0.7520	0.7486	0.7452	0.7419	0.7386	0.7353	0.7320	0.7288	0.7256	0.7224
70	0.7192	0.7160	0.7128	0.7097	0.7066	0.7035	0.7004	0.6974	0.6944	0.6913
80	0.6883	0.6854	0.6824	0.6795	0.6765	0.6736	0.6707	0.6679	0.6650	0.6622
90	0.6594	0.6566	0.6537	0.6510	0.6483	0.6455	0.6428	0.6401	0.6374	0.6348
100	0.6321	0.6295	0.6269	0.6243	0.6217	0.6191	0.6166	0.6140	0.6115	0.6090
110	0.6065	0.6040	0.6016	0.5991	0.5967	0.5942	0.5918	0.5894	0.5871	0.5847
120	0.5823	0.5800	0.5777	0.5754	0.5731	0.5708	0.5685	0.5663	0.5641	0.5618
130	0.5596	0.5574	0.5552	0.5530	0.5509	0.5487	0.5466	0.5444	0.5424	0.5402
140	0.5381	0.5361	0.5340	0.5320	0.5299	0.5279	0.5259	0.5239	0.5219	0.5199
150	0.5179	0.5160	0.5140	0.5121	0.5102	0.5082	0.5064	0.5044	0.5026	0.5007
160	0.4988	0.4970	0.4952	0.4933	0.4915	0.4897	0.4879	0.4861	0.4843	0.4825
170	0.4808	0.4790	0.4773	0.4756	0.4739	0.4721	0.4704	0.4687	0.4671	0.4654
180	0.4637	0.4621	0.4605	0.4588	0.4571	0.4555	0.4540	0.4523	0.4508	0.4491
190	0.4476	0.4460	0.4445	0.4429	0.4414	0.4399	0.4383	0.4368	0.4354	0.4338
200	0.4323	0.4308	0.4294	0.4279	0.4265	0.4250	0.4236	0.4221	0.4207	0.4193
210	0.4179	0.4165	0.4151	0.4137	0.4123	0.4109	0.4096	0.4082	0.4069	0.4055
220	0.4042	0.4029	0.4015	0.4002	0.3989	0.3976	0.3963	0.3950	0.3937	0.3925
230	0.3912	0.3899	0.3887	0.3874	0.3862	0.3849	0.3837	0.3825	0.3813	0.3801
240	0.3789	0.3777	0.3765	0.3753	0.3741	0.3729	0.3718	0.3706	0.3695	0.3683
250	0.3672	0.3660	0.3649	0.3638	0.3627	0.3615	0.3604	0.3593	0.3582	0.3571
260	0.3560	0.3550	0.3539	0.3528	0.3517	0.3507	0.3496	0.3486	0.3476	0.3465
270	0.3455	0.3445	0.3434	0.3424	0.3414	0.3404	0.3393	0.3384	0.3374	0.3364
280	0.3354	0.3344	0.3335	0.3325	0.3315	0.3306	0.3296	0.3287	0.3277	0.3268
290	0.3259	0.3249	0.3240	0.3231	0.3221	0.3212	0.3203	0.3194	0.3185	0.3176
300	0.3167	0.3158	0.3150	0.3141	0.3132	0.3123	0.3115	0.3106	0.3098	0.3089
310	0.3080	0.3072	0.3064	0.3055	0.3047	0.3039	0.3030	0.3022	0.3014	0.3006
320	0.2998	0.2990	0.2982	0.2974	0.2966	0.2958	0.2950	0.2942	0.2934	0.2926
330	0.2919	0.2911	0.2903	0.2896	0.2888	0.2880	0.2873	0.2865	0.2858	0.2850
340	0.2843	0.2836	0.2828	0.2821	0.2814	0.2807	0.2799	0.2792	0.2785	0.2778
350	0.2771	0.2764	0.2757	0.2750	0.2743	0.2736	0.2729	0.2722	0.2715	0.2709
360	0.2702	0.2695	0.2688	0.2682	0.2675	0.2669	0.2662	0.2655	0.2649	0.2642
370	0.2636	0.2629	0.2623	0.2617	0.2610	0.2604	0.2598	0.2591	0.2585	0.2579
380	0.2573	0.2567	0.2560	0.2554	0.2548	0.2542	0.2536	0.2530	0.2524	0.2518
390	0.2512	0.2506	0.2500	0.2495	0.2489	0.2483	0.2477	0.2471	0.2466	0.2460

Table 7 Discount factors (F_c) **with continuous interest to give present worths for cash flows which occur uniformly over a periof of** T **years after the reference point**††‡ (*Continued*)

$100rT$	0	1	2	3	4	5	6	7	8	9
400	0.2454	0.2449	0.2443	0.2437	0.2432	0.2426	0.2421	0.2415	0.2410	0.2404
410	0.2399	0.2393	0.2388	0.2382	0.2377	0.2372	0.2366	0.2361	0.2356	0.2350
420	0.2345	0.2340	0.2335	0.2330	0.2325	0.2319	0.2314	0.2309	0.2304	0.2299
430	0.2294	0.2289	0.2284	0.2279	0.2274	0.2269	0.2264	0.2259	0.2255	0.2250
440	0.2245	0.2240	0.2235	0.2230	0.2226	0.2221	0.2216	0.2212	0.2207	0.2202
450	0.2198	0.2193	0.2188	0.2184	0.2179	0.2175	0.2170	0.2166	0.2161	0.2157
460	0.2152	0.2148	0.2143	0.2139	0.2134	0.2130	0.2126	0.2121	0.2117	0.2113
470	0.2108	0.2104	0.2100	0.2096	0.2091	0.2087	0.2083	0.2079	0.2074	0.2070
480	0.2066	0.2062	0.2058	0.2054	0.2050	0.2046	0.2042	0.2038	0.2034	0.2030
490	0.2026	0.2022	0.2018	0.2014	0.2010	0.2006	0.2002	0.1998	0.1994	0.1990

$100rT$	0	10	20	30	40	50	60	70	80	90
500	0.1987	0.1949	0.1912	0.1877	0.1843	0.1811	0.1779	0.1749	0.1719	0.1690
600	0.1663	0.1636	0.1610	0.1584	0.1560	0.1536	0.1513	0.1491	0.1469	0.1448
700	0.1427	0.1407	0.1388	0.1369	0.1351	0.1333	0.1315	0.1298	0.1282	0.1265
800	0.1250	0.1234	0.1219	0.1206	0.1190	0.1176	0.1163	0.1149	0.1136	0.1123
900	0.1111	0.1099	0.1087	0.1075	0.1064	0.1053	0.1043	0.1031	0.1020	0.1010
1000	0.1000	0.0990	0.0980	0.0971	0.0962	0.0952	0.0943	0.0935	0.0926	0.0917
1100	0.0909	0.0901	0.0893	0.0885	0.0877	0.0869	0.0862	0.0855	0.0847	0.0840
1200	0.0833	0.0826	0.0820	0.0813	0.0806	0.0800	0.0794	0.0787	0.0781	0.0775
1300	0.0769	0.0763	0.0758	0.0752	0.0746	0.0741	0.0735	0.0730	0.0725	0.0719
1400	0.0714	0.0709	0.0704	0.0699	0.0694	0.0690	0.0685	0.0680	0.0676	0.0671
1500	0.0667	0.0662	0.0658	0.0654	0.0649	0.0645	0.0641	0.0637	0.0633	0.0629
1600	0.0625	0.0621	0.0617	0.0613	0.0610	0.0606	0.0602	0.0599	0.0595	0.0592
1700	0.0588	0.0585	0.0581	0.0578	0.0575	0.0571	0.0568	0.0565	0.0562	0.0559
1800	0.0556	0.0552	0.0549	0.0546	0.0543	0.0541	0.0538	0.0535	0.0532	0.0529
1900	0.0526	0.0524	0.0521	0.0518	0.0515	0.0513	0.0510	0.0508	0.0505	0.0502
2000	0.0500	0.0497	0.0495	0.0492	0.0490	0.0487	0.0485	0.0483	0.0481	0.0478

† r = nominal interest compounded continuously, percent/100; $T = n$ = number of years in time period.   See Tables 3 and 4 for information on F_c.
‡ The columns represent the unit increments from 1 to 9 or 10 to 90 for the intervals of $100rT$ shown in the left column.

Table 8 Discount factors (F_d) **with continuous interest to give present worths for cash flows which decline to zero at a constant rate over a period of years** n_T **starting with the reference point**†‡

$100rn_T$	0	1	2	3	4	5	6	7	8	9
0	1.0000	0.9967	0.9934	0.9901	0.9868	0.9835	0.9803	0.9771	0.9739	0.9707
10	0.9675	0.9643	0.9612	0.9580	0.9549	0.9518	0.9487	0.9457	0.9426	0.9396
20	0.9365	0.9335	0.9305	0.9275	0.9246	0.9216	0.9187	0.9158	0.9129	0.9100
30	0.9071	0.9042	0.9013	0.8985	0.8957	0.8929	0.8901	0.8873	0.8845	0.8818
40	0.8790	0.8763	0.8736	0.8708	0.8682	0.8655	0.8628	0.8602	0.8575	0.8549
50	0.8523	0.8497	0.8471	0.8445	0.8419	0.8394	0.8368	0.8343	0.8317	0.8292
60	0.8267	0.8242	0.8218	0.8193	0.8169	0.8144	0.8120	0.8096	0.8072	0.8048
70	0.8024	0.8000	0.7976	0.7953	0.7930	0.7906	0.7883	0.7860	0.7837	0.7814
80	0.7791	0.7769	0.7746	0.7724	0.7701	0.7679	0.7657	0.7635	0.7613	0.7591
90	0.7570	0.7548	0.7527	0.7505	0.7484	0.7462	0.7441	0.7420	0.7399	0.7378
100	0.7358	0.7337	0.7316	0.7295	0.7275	0.7255	0.7235	0.7215	0.7195	0.7175
110	0.7155	0.7135	0.7115	0.7095	0.7076	0.7057	0.7037	0.7018	0.6999	0.6980
120	0.6961	0.6942	0.6923	0.6904	0.6885	0.6867	0.6848	0.6830	0.6812	0.6794
130	0.6776	0.6758	0.6740	0.6722	0.6704	0.6686	0.6668	0.6650	0.6632	0.6615
140	0.6598	0.6580	0.6563	0.6546	0.6529	0.6512	0.6495	0.6478	0.6461	0.6444
150	0.6428	0.6411	0.6394	0.6377	0.6361	0.6345	0.6329	0.6313	0.6297	0.6281
160	0.6265	0.6249	0.6233	0.6217	0.6201	0.6186	0.6170	0.6154	0.6139	0.6124
170	0.6109	0.6093	0.6078	0.6063	0.6048	0.6033	0.6018	0.6003	0.5988	0.5973
180	0.5959	0.5944	0.5929	0.5914	0.5900	0.5886	0.5871	0.5856	0.5842	0.5828
190	0.5814	0.5800	0.5786	0.5772	0.5758	0.5745	0.5731	0.5717	0.5703	0.5690
200	0.5677	0.5663	0.5649	0.5636	0.5623	0.5610	0.5596	0.5583	0.5570	0.5557
210	0.5544	0.5531	0.5518	0.5505	0.5492	0.5480	0.5467	0.5454	0.5441	0.5429
220	0.5417	0.5404	0.5391	0.5379	0.5367	0.5355	0.5342	0.5330	0.5318	0.5306
230	0.5294	0.5282	0.5270	0.5258	0.5246	0.5234	0.5222	0.5210	0.5198	0.5187
240	0.5176	0.5164	0.5152	0.5141	0.5130	0.5119	0.5107	0.5096	0.5085	0.5074
250	0.5063	0.5052	0.5041	0.5030	0.5019	0.5008	0.4997	0.4986	0.4975	0.4964
260	0.4953	0.4942	0.4931	0.4920	0.4910	0.4900	0.4889	0.4878	0.4868	0.4858
270	0.4848	0.4837	0.4827	0.4817	0.4807	0.4797	0.4787	0.4777	0.4767	0.4757
280	0.4747	0.4737	0.4727	0.4717	0.4707	0.4698	0.4688	0.4678	0.4668	0.4658
290	0.4649	0.4639	0.4629	0.4620	0.4611	0.4602	0.4592	0.4582	0.4573	0.4564
300	0.4555	0.4545	0.4536	0.4527	0.4518	0.4509	0.4500	0.4491	0.4482	0.4473
310	0.4464	0.4455	0.4446	0.4437	0.4428	0.4419	0.4410	0.4401	0.4392	0.4384
320	0.4376	0.4367	0.4358	0.4349	0.4340	0.4332	0.4324	0.4316	0.4308	0.4300
330	0.4292	0.4283	0.4274	0.4266	0.4258	0.4250	0.4242	0.4234	0.4226	0.4218
340	0.4210	0.4202	0.4194	0.4186	0.4178	0.4170	0.4162	0.4154	0.4146	0.4138
350	0.4131	0.4123	0.4115	0.4107	0.4099	0.4091	0.4083	0.4076	0.4069	0.4062
360	0.4055	0.4047	0.4039	0.4031	0.4023	0.4016	0.4009	0.4002	0.3995	0.3988
370	0.3981	0.3973	0.3965	0.3958	0.3951	0.3944	0.3937	0.3930	0.3923	0.3916
380	0.3909	0.3902	0.3895	0.3888	0.3881	0.3874	0.3867	0.3860	0.3853	0.3846
390	0.3840	0.3833	0.3826	0.3819	0.3812	0.3805	0.3798	0.3791	0.3785	0.3779

Table 8 Discount factors (F_d) with continuous interest to give present worths for cash flows which decline to zero at a constant rate over a period of years n_T starting with the reference point†‡ (*Continued*)

$100rn_T$	0	1	2	3	4	5	6	7	8	9
400	0.3773	0.3766	0.3759	0.3752	0.3745	0.3738	0.3732	0.3726	0.3720	0.3714
410	0.3708	0.3701	0.3694	0.3687	0.3681	0.3675	0.3669	0.3663	0.3657	0.3651
420	0.3645	0.3638	0.3632	0.3626	0.3620	0.3614	0.3608	0.3602	0.3596	0.3590
430	0.3584	0.3578	0.3572	0.3566	0.3560	0.3554	0.3548	0.3542	0.3536	0.3530
440	0.3525	0.3519	0.3513	0.3507	0.3501	0.3495	0.3489	0.3483	0.3478	0.3473
450	0.3468	0.3462	0.3456	0.3450	0.3444	0.3438	0.3432	0.3427	0.3422	0.3417
460	0.3412	0.3406	0.3400	0.3394	0.3388	0.3383	0.3378	0.3373	0.3368	0.3363
470	0.3358	0.3352	0.3346	0.3341	0.3336	0.3331	0.3326	0.3321	0.3316	0.3311
480	0.3306	0.3300	0.3294	0.3289	0.3284	0.3279	0.3274	0.3269	0.3264	0.3259
490	0.3254	0.3249	0.3244	0.3239	0.3234	0.3229	0.3224	0.3219	0.3214	0.3209

$100rn_T$	0	10	20	30	40	50	60	70	80	90
500	0.3205	0.3157	0.3111	0.3065	0.3020	0.2978	0.2936	0.2895	0.2856	0.2817
600	0.2779	0.2742	0.2707	0.2672	0.2637	0.2604	0.2572	0.2540	0.2509	0.2479
700	0.2449	0.2420	0.2392	0.2364	0.2337	0.2311	0.2285	0.2260	0.2235	0.2212
800	0.2188	0.2165	0.2143	0.2121	0.2098	0.2076	0.2056	0.2036	0.2016	0.1995
900	0.1975	0.1957	0.1939	0.1920	0.1902	0.1884	0.1867	0.1850	0.1834	0.1817
1000	0.1800	0.1784	0.1769	0.1753	0.1738	0.1723	0.1709	0.1694	0.1680	0.1667
1100	0.1653	0.1639	0.1626	0.1613	0.1601	0.1588	0.1576	0.1563	0.1551	0.1539
1200	0.1528	0.1516	0.1505	0.1494	0.1483	0.1472	0.1462	0.1451	0.1441	0.1430
1300	0.1420	0.1410	0.1401	0.1390	0.1382	0.1372	0.1363	0.1354	0.1345	0.1336
1400	0.1327	0.1318	0.1309	0.1301	0.1292	0.1284	0.1276	0.1268	0.1260	0.1252
1500	0.1244	0.1236	0.1229	0.1221	0.1214	0.1207	0.1200	0.1193	0.1186	0.1179
1600	0.1172	0.1165	0.1158	0.1151	0.1145	0.1139	0.1133	0.1126	0.1120	0.1113
1700	0.1107	0.1101	0.1095	0.1089	0.1084	0.1078	0.1072	0.1066	0.1060	0.1054
1800	0.1049	0.1044	0.1039	0.1033	0.1028	0.1023	0.1018	0.1012	0.1007	0.1002
1900	0.0997	0.0992	0.0987	0.0982	0.0978	0.0973	0.0968	0.0963	0.0959	0.0955
2000	0.0950	0.0946	0.0942	0.0938	0.0933	0.0928	0.0923	0.0919	0 0915	0.0911

† r = nominal interest compounded continuously, percent/100; n_T = number of years in time period for cash flow to decrease to zero. See Tables 3 and 4 for information on F_d.

‡ The columns represent the unit increments from 1 to 9 or 10 to 90 for the intervals of $100rn_T$ shown in the left column.

COSTS DUE TO INTEREST ON INVESTMENT

Money, or any other negotiable type of capital, has a time value. When a business concern invests money, it expects to receive a return during the time the money is tied up in the investment. The amount of return demanded usually is related to the degree of risk that the entire investment might be lost.

One of the duties of a design engineer is to determine the net return or profit which can be obtained by making an investment. It is necessary, therefore, to find the total cost involved. Too often, the engineer fails to recognize the time value of money and neglects the effects of interest on cost. According to the modern definition of interest, the cost due to time value of an investment should be included as interest for that portion of the total capital investment which comes from outside sources.

Borrowed Capital versus Owned Capital

The question sometimes arises as to whether interest on owned capital can be charged as a true cost. The modern definition of interest permits a definite answer of "no" to this question. Court decisions and income-tax regulations verify this answer.

Interest Effects in a Small Business

In small business establishments, it is usually quite easy to determine the exact source of all capital. Therefore, the interest costs can be obtained with little difficulty. For example, suppose that a young chemical engineer has $20,000 and decides to set up a small plant for producing antifreeze from available raw materials. For a working-capital plus fixed-capital investment of $20,000, the chemical engineer determines that the proposed plant can provide a total yearly profit of $8000 before income taxes. Since the investment was a personal one, interest obviously could not be included as a cost. If it had been necessary to borrow the $20,000 at an annual interest rate of 10 percent, interest would have been a cost, and the total profit would have been $8000 − (0.10) × ($20,000) = $6000 per year.

Interest Effects in a Large Business

In large business establishments, new capital may come from issue of stocks and bonds, borrowing from banks or insurance companies, funds set aside for replacement of worn-out or obsolete equipment, profits received but not distributed to the stockholders, and other sources. Therefore, it is often difficult to designate the exact source of new capital, and the particular basis used for determining interest costs should be indicated when the results of a cost analysis are reported.

An approximate breakdown showing the various sources of new capital for large corporations typical of the chemical industry is presented in Table 9.

Table 9 Source of new capital for corporations

Source of capital	Approximate amount of total new capital, %
External financing (loans from banks or other concerns, issue of stocks and bonds)	25
Profits earned but not distributed to stockholders as dividends	30
Depreciation funds set aside	25
Miscellaneous	20

SOURCE OF CAPITAL

One source of new capital is outside loans. Interest on such loans is usually at a fixed rate, and the annual cost can be determined directly.

New capital may also be obtained from the issue of bonds, preferred stock, or common stock. Interest on bonds and preferred-stock dividends must be paid at fixed rates. A relatively low interest rate is paid on bonds because the bondholder has first claim on earnings, while higher rates are paid on preferred stock because the holder has a greater chance to lose the entire investment. The holder of common stock accepts all the risks involved in owning a business. The return on common stock, therefore, is not at a fixed rate but varies depending on the success of the company which issued the stock. To compensate for this greater risk, the return on common stock may be much higher than that on bonds or preferred stock.

Income-Tax Effects

The effect of high income-tax rates on the cost of capital is very important. In determining income taxes, interest on loans and bonds can be considered as a cost, while the return on both preferred and common stock cannot be included as a cost. Since corporate income taxes can amount to more than half of the gross earnings, the source of new capital may have a considerable influence on the net profits.

If the annual income-tax rate for a company is 48 percent, every dollar spent for interest on loans or bonds would have a true cost after taxes of only 52 cents. Thus, after income taxes are taken into consideration, a bond issued at an annual interest rate of 6 percent would actually have an interest rate of only $6 \times \frac{52}{100} = 3.1$ percent. On the other hand, the dividends on preferred stock must be paid from net profits after taxes. If preferred stock has an annual dividend rate of 7 percent, the equivalent rate before taxes would be $7 \times \frac{100}{52} = 13.5$ percent.

Despite the fact that it may be cheaper to use borrowed capital in place of

Table 10 Typical costs for externally financed capital
Income tax rate = 48% of (total income−total pretax cost)

Source of capital	Indicated interest or dividend rate, %/year	Actual interest or dividend rate before taxes, %/year	Actual interest or dividend rate after taxes, %/year
Bonds	6	6	3.1
Bank or other loans	7	7	3.6
Preferred stock	7	13.5	7
Common stock	0	17.3	9

other types of capital, it is unrealistic to finance each new venture by using borrowed capital. Every corporation needs to maintain a balanced capital structure and is therefore hesitant about placing itself under a heavy burden of debt.

A comparison of interest or dividend rates for different types of externally financed capital is presented in Table 10.

METHODS FOR INCLUDING COST OF CAPITAL IN ECONOMIC ANALYSES

The cost of new capital obtained from bonds, loans, or preferred stock can be determined directly from the stated interest or dividend rate, adjusted for income taxes. However, the cost of new capital obtained from the issue of common stock is not so obvious, and some basis must be set for determining this cost. Probably the fairest basis is to consider the viewpoint of existing holders of common stock. If new common stock is issued, its percent return should be at least as much as that obtained from the old common stock; otherwise, the existing stockholders would receive a lower return after the issue of the new stock. Therefore, from the viewpoint of the existing stockholders, the cost of new common stock is the present rate of common-stock earnings.

A major source of new capital is from internal capital, including, primarily, undistributed profits and depreciation funds. Since this definitely is owned capital, it is not necessary to consider interest as a cost. However, some concerns prefer to assign a cost to this type of capital, particularly if comparisons of alternative investments are to be made. The reasoning here is that the owned capital could be loaned out or put into other ventures to give a definite return.

Two methods are commonly used for determining the cost of owned capital. In the first method, the capital is charged at a low interest rate on the assumption that it could be used to pay off funded debts or invest in risk-free loans. The second method requires interest to be paid on the owned capital at a rate equal to the present return on all the company's capital.

Design-Engineering Practice for Interest and Investment Costs

Many alternative methods are used by engineers when determining interest costs in an economic analysis of a design project. In preliminary designs, one of the following two methods is usually employed:

1. No interest costs are included. This assumes that all the necessary capital comes from owned capital, and any comparisons to alternative investments must be on the same basis.
2. Interest is charged on the total capital investment at a set interest rate. Rates equivalent to those charged for bank loans or bonds are usually employed. Under these conditions, the total profit represents the increase over the return that would be obtained if the company could invest the same amount of money in an outside loan at the given interest rate.

As the design proceeds to the final stages, the actual source of the new capital should be considered in detail, and more-refined methods for determining interest costs can be used.

When interest is included as a cost, there is some question as to whether the interest costs should be based on the initial investment or on the average investment over the life of the project. Although this is a debatable point, the accepted design practice is to base the interest costs on the initial investment.

Because of the different methods used for treating interest as a cost, a definite statement should be made concerning the particular method employed in any given economic analysis. Interest costs become especially important when making comparisons among alternative investments. These comparisons, as well as the overall cost picture, are simplified if the role of interest in the economic analysis is clearly defined.

NOMENCLATURE FOR CHAPTER 6

a = a constant
C = compound interest factor
C_R = cost for replacement or other asset, dollars
C_V = original cost of equipment or other asset, dollars
d = number of days in an interest period, days, or derivative
e = base of the natural logarithm = 2.71828 ...
F = discount factor
g = constant declining rate or gradient
i = interest rate based on the length of one interest period, percent/100
i_{eff} = effective interest rate—exact interest rate based on an interest period of one year, percent/100
I = total amount of interest during n interest periods, dollars

K = capitalized cost, dollars

m = number of interest periods per year

n = number of time units or interest periods

n_T = number of time units necessary for cash flow to decrease to zero

P = principal or present worth of capital on which interest is paid, dollars

$\bar{P}$ = principal or present worth considered as occurring regularly throughout the time period, dollars

r = nominal interest rate—approximate interest rate based on an interest period of one year or continuous interest rate, percent/100

R = uniform periodic payments made during n periods in an ordinary annuity, dollars/period

$\bar{R}$ = total of all ordinary annuity payments occurring regularly throughout the time period, dollars/period

S = future worth—amount of principal or present worth plus interest due after n interest periods, dollars

$\bar{S}$ = future worth considered as occurring continuously throughout the time period, dollars

T = time period, years

SUGGESTED ADDITIONAL REFERENCES FOR INTEREST AND INVESTMENT COSTS

Cost of capital

Childs, J. F., Should Your Pet Project be Built? What Should the Profit Test be? *Chem. Eng.*, **75**(5): 188 (1968).

Douglas, F. R., R. A. Bellin, B. J. Blewitt, W. H. Kapfer, F. J. Marsik, A. P. Narins, and L. W. Pullen, Benchmark for Profitability, *Chem. Eng.*, **76**(25): 274 (1969).

Holland, F. A., F. A. Watson, and J. K. Wilkinson, Inflation and Its Impact on Costs and Prices, *Chem. Eng.*, **81**(23): 107 (1974).

Murphy, L. A., What is Capital? *Petro/Chem. Eng.*, **10**(41): PM5 (1969).

Neely, C., and J. E. Browning, Sources of Capital for Growth of Process Plants, *Chem. Eng.*, **84**(12): 142 (1977).

Discounted cash flow

Agarwal, J. C., and I. V. Klumpar, Profitability, Sensitivity, and Risk Analysis for Project Economics, *Chem. Eng.*, **82**(20): 66 (1975).

Allen, D. H., Investment Decisions—Evaluation Techniques in Perspective, *Chem. Eng. (London)*, 42 (January 1975).

Cason, R. L., Go Broke While Showing a Profit? *Hydrocarbon Process.*, **54**(7): 179 (1975).

de la Mare, R. F., Parameters Affecting Capital Investment, *Chem. Eng. (London)*, 227 (April 1975).

Drayer, D. E., How to Estimate Plant Cost-Capacity Relationship, *Petro/Chem. Eng.*, **42**(5): 10 (1970).

Ebly, R. W., Comparison of Methods for Evaluating Capital Equipment Replacement, *1973 Trans. Am. Assoc. Cost Eng.*, (1973) p. 23.

Foord, A., Investment Appraisal for Chemical Engineers, *Chem. Proc. Eng.*, **48**(7): 71 (1967); **48**(10): 95 (1967); and **49**(2): 87 (1968).

Gambro, A. J., K. Muenz, and M. Abrahams, Optimize Ethylene Complex, *Hydrocarbon Proc.*, **51**(3): 73 (1972).

Holland, F. A., F. A. Watson, and J. K. Wilkinson, Methods of Estimating Project Profitability, *Chem. Eng.*, **80**(22):80 (1973).

—— —— and ——, Sensitivity Analysis of Project Profitabilities, *Chem. Eng.*, **80**(25):115 (1973).

—— —— and ——, Time, Capital, and Interest Affect Choice of Project, *Chem. Eng.*, **80**(27):83 (1973).

—— —— and ——, Probability Techniques for Estimates of Profitability, *Chem. Eng.*, **81**(1):105 (1974).

—— —— and ——, Inflation and Its Impact on Costs and Prices, *Chem. Eng.*, **81**(23):107 (1974).

Holland, F. A., and F. A. Watson, Economic Penalties of Operating a Process at Reduced Capacity, *Chem. Eng.*, **84**(1):91 (1977).

—— and ——, Putting Inflation into Profitability Studies, *Chem. Eng.*, **84**(4):87 (1977).

Jelen, F. C., and M. S. Cole, Methods for Economic Analysis, *Hydrocarbon Process.*, **53**(10):161 (1974).

—— and C. L. Yaws, Project Cash Flow—Description and Interpretation, *Hydrocarbon Process.*, **57**(3):77 (1978).

Kapier, W. H., Appraising Rate of Return Methods, *Chem. Eng. Progr.*, **65**(11):55 (1969).

Klumpar, I. V., Determining Economic and Process Variables of Ventures, *Chem. Eng. Progr.*, **67**(4):74 (1971).

Leibson, I., and C. A. Trischman, Jr., Should You Make or Buy Your Major Raw Materials? *Chem. Eng.*, **79**(4):76 (1972).

—— and ——, When and How to Apply Discounted Cash Flow and Present Worth, *Chem. Eng.*, **78**(28):97 (1971).

Leung, T. K. Y., New Nomograph: Quick Route to Discounted Cash Flow, *Chem. Eng.*, **77**(12):208 (1970).

Malloy, J. B., Instant Economic Evaluation, *Chem. Eng. Progr.*, **65**(11):47 (1969).

Mapstone, G. E., Figure Discount Cash Flow Quickly, *Hydrocarbon Process.*, **54**(12):99 (1975).

Merckx, L. J., Compare Large HPI Plants by Index of Technical Success, *Hydrocarbon Process.*, **50**(8):103 (1971).

Park, W. R., Graphical Short-Cuts in Discounted Cash Flow Analysis, presented at 12th National American Association of Cost Engineers Meeting, Houston, Texas, June 17–19, 1968.

Petty, W. J., The Capital Expenditure Decision-Making Process of Large Corporations, *Eng. Econ.*, **20**(3):159 (1975).

Piekarski, J. A., Fundamentals of Manufacturing Economics, *1973 Trans. Am. Assoc. Cost Eng.* (1973), p. 221.

Polk, H. K., Economic Evaluation of Prime Movers, *Pet. Eng.*, **46**(12):34 (1974).

Tarquin, A. J., and L. Blank, "Engineering Economy: A Behavioral Approach," McGraw-Hill Book Company, New York, 1976.

Watson, F. A., Simplified Calculation of Discounted Cash Flow Rate of Return, *Chem. Eng. (London)*, 437 (August 1975).

Wild, N. H., Return on Investment Made Easy, *Chem. Eng.*, **83**(8):153 (1976).

Interest

Barish, N. N., and S. Kaplan, "Economic Analysis for Engineering and Managerial Decision Making," 2d ed., McGraw-Hill Book Company, New York, 1978.

Holland, F. A., F. A. Watson, and J. K. Wilkinson, Methods of Estimating Project Profitability, *Chem. Eng.*, **80**(22):80 (1973).

—— —— and ——, Time, Capital, and Interest Affect Choice of Project, *Chem. Eng.*, **80**(27):83 (1973).

—— —— and ——, Time Value of Money, *Chem. Eng.*, **80**(21):123 (1973).

Piekarski, J. A., Fundamentals of Manufacturing Economics, *1973 Trans. Am. Assoc. Cost Eng.* (1973), p. 221.

Present worth

Allen, D. H., Investment Decisions—Evaluation Techniques in Perspective, *Chem. Eng. (London)*, 42 (January 1975).

Barish, N. N., and S. Kaplan, "Economic Analysis for Engineering and Managerial Decision Making," 2d ed., McGraw-Hill Book Company, New York, 1978.

Beenhaker, H. L., Sensitivity Analysis of the Present Value of a Project, *Eng. Econ.*, **20**(2):123 (1975).

Congelliere, R. H., Correcting Economic Analyses, *Chem. Eng.*, **77**(25):109 (1970).

Curt, R. P., A New Approach to Economic Insulation Thickness, *Hydrocarbon Process.*, **55**(3):137 (1976).

Davidson, L. B., Investment Evaluation Under Conditions of Inflation, *J. Petrol. Technol.*, **27**:1183 (1975).

Davis, G. O., How to Make the Correct Economic Decision on Spare Equipment, *Chem. Eng.*, **84**(25):187 (1977).

de la Mare, R. F., Parameters Affecting Capital Investment, *Chem. Eng. (London)*, 227 (April 1975).

Drayer, D. E., How to Estimate Plant Cost-Capacity Relationship, *Petro/Chem. Eng.*, **42**(5):10 (1970).

Foord, A., Investment Appraisal for Chemical Engineers, *Chem. Proc. Eng.*, **48**(10):95 (1967); **49**(2):87 (1968).

Generoso, E., Jr., and L. B. Hitchcock, Optimizing Plant Expansion—Two Cases, *I&EC*, **60**(11):12 (1968).

Harrison, M. R., and C. M. Pelanne, Cost-Effective Thermal Insulation, *Chem. Eng.*, **84**(27):63 (1977).

Haskett, C. E., Evaluation by Rate of Return or Present Value? *Pet. Eng.*, **8**(45):48 (1972).

Holland, F. A., F. A. Watson, J. K. Wilkinson, Probability Techniques for Estimates of Profitability, *Chem. Eng.*, **81**(1):105 (1974).

—— —— and ——, Sensitivity Analysis of Project Profitabilities, *Chem. Eng.*, **80**(25):115 (1973).

—— —— and ——, Time, Capital, and Interest Affect Choice of Project, *Chem. Eng.*, **80**(27):83 (1973).

Holland, F. A., and F. A. Watson, Putting Inflation into Profitability Studies, *Chem. Eng.*, **84**(4):87 (1977).

Jelen, F. C., and M. S. Cole, Methods for Economic Analysis, *Hydrocarbon Process.*, **53**(7):133 (1974); **53**(9):227 (1974).

Kapier, W. H., Appraising Rate of Return Methods, *Chem. Eng. Progr.*, **65**(11):55 (1969).

Leibson, I., and C. A. Trischman, Jr., Decision Trees: A Rapid Evaluation of Investment Risk, *Chem. Eng.*, **79**(4):99 (1972).

—— and ——, Should You Make or Buy Your Major Raw Materials? *Chem. Eng.*, **79**(4):76 (1972).

—— and ——, When and How to Apply Discounted Cash Flow and Present Worth, *Chem. Eng.*, **78**(28):97 (1971).

Malloy, J. B., Instant Economic Evaluation, *Chem. Eng. Progr.*, **65**(11):47 (1969).

——, Risk Analysis of Chemical Plants, *Chem. Eng. Progr.*, **67**(10):68 (1971).

Merckx, L. J., Compare Large HPI Plants by Index of Technical Success, *Hydrocarbon Process.*, **50**(8):103 (1971).

Piekarski, J. A., Fundamentals of Manufacturing Economics, *1973 Trans. Am. Assoc. Cost Eng.* (1973), p. 221.

Price, J., Computerized Models—the Scientific Way to Pick a Winner, *Chem. Week*, **106**(15):37 (1970).

Reul, R. I., Which Investment Appraisal Techniques Should You Use? *Chem. Eng.*, **75**(10):212 (1968).

Riggs, J. L., "Engineering Economics," McGraw-Hill Book Company, New York, 1977.

Stuart, D. O., and R. D. Meckna, Accounting for Inflation in Present-Worth Studies, *Power Eng.*, **79**(2):45 (1975).

Tarquin, A. J., and L. Blank, "Engineering Economy: A Behavioral Approach," McGraw-Hill Book Company, New York, 1976.

Young, D., and L. E. Contreras, Expected Present Worths of Cash Flows Under Uncertain Timing, *Eng. Econ.*, **20**(4):257 (1975).

Wild, N. H., Return on Investment Made Easy, *Chem. Eng.*, **83**(8):153 (1976).

————, Program for Discounted-Cash-Flow Return on Investment, *Chem. Eng.*, **84**(10):137 (1977).

PROBLEMS

1 It is desired to have $9000 available 12 years from now. If $5000 is available for investment at the present time, what discrete annual rate of compound interest on the investment would be necessary to give the desired amount?

2 What will be the total amount available 10 years from now if $2000 is deposited at the present time with nominal interest at the rate of 6 percent compounded semiannually?

3 An original loan of $2000 was made at 6 percent simple interest per year for 4 years. At the end of this time, no interest had been paid and the loan was extended for 6 more years at a new, effective, compound-interest rate of 8 percent per year. What is the total amount owed at the end of the 10 years if no intermediate payments are made?

4 A concern borrows $50,000 at an annual, effective, compound-interest rate of 10 percent. The concern wishes to pay off the debt in 5 years by making equal payments at the end of each year. How much will each payment have to be?

5 The original cost for a distillation tower is $24,000 and the useful life of the tower is estimated to be 8 years. The sinking-fund method for determining the rate of depreciation is used (see Example 6), and the effective annual interest rate for the depreciation fund is 6 percent. If the scrap value of the distillation tower is $4000, determine the asset value (i.e., total book value of equipment) at the end of 5 years.

6 An annuity due is being used to accumulate money. Interest is compounded at an effective annual rate of 8 percent, and $1000 is deposited at the beginning of each year. What will the total amount of the annuity due be after 5 years?

7 By use of an analog computer, make a plot showing the continuous single-payment compound amount factor (that is, e^{rn}) versus n for effective interest rates of 5, 10, and 20 percent over a range of n from 0 to 20.

8 By use of a digital computer, develop and present a printout of the data of effective interest versus nominal interest compounded continuously as given in Table 2.

9 By use of a digital computer, develop and present a printout of the first five lines of Table 7.

10 For total yearly payments of $5000 for 10 years, compare the compound amount accumulated at the end of 10 years if the payments are (*a*) end-of-year, (*b*) weekly, and (*c*) continuous. The effective (annual) interest is 20 percent and payments are uniform.

11 For the conditions of Prob. 10, determine the present worth at time zero for each of the three types of payments.

12 A heat exchanger has been designed for use in a chemical process. A standard type of heat exchanger with a negligible scrap value costs $4000 and will have a useful life of 6 years. Another proposed heat exchanger of equivalent design capacity costs $6800 but will have a useful life of 10 years and a scrap value of $800. Assuming an effective compound interest rate of 8 percent per year, determine which heat exchanger is cheaper by comparing the capitalized costs.

13 A new storage tank can be purchased and installed for $10,000. This tank would last for 10 years. A worn-out storage tank of capacity equivalent to the new tank is available, and it has been proposed to repair the old tank instead of buying the new tank. If the tank were repaired, it would have a useful life of 3 years before the same type of repairs would be needed again. Neither tank has any scrap value. Money is worth 9 percent compounded annually. On the basis of equal capitalized costs for the two tanks, how much can be spent for repairing the existing tank?

14 Equation (28) is the expression for capitalized cost based on discrete interest compounding. For continuous interest compounding, the expression becomes

$$K = C_V + \frac{C_R}{e^{rn} - 1}$$

Present a detailed derivation of this continuous-interest relationship going through each of the equivalent steps used in deriving Eq. (28).

15 The total investment required for a new chemical plant is estimated at $2 million. Fifty percent of the investment will be supplied from the company's own capital. Of the remaining investment, half will come from a loan at an effective interest rate of 8 percent and the other half will come from an issue of preferred stock paying dividends at a stated effective rate of 8 percent. The income-tax rate for the company is 47 percent of pre-tax earnings. Under these conditions, how many dollars per year does the company actually lose (i.e., after taxes) by issuing preferred stock at 8 percent dividends instead of bonds at an effective interest rate of 6 percent?

16 It has been proposed that a company invest $1 million in a venture which will yield a gross income of $1 million per year. The total annual costs will be $800,000 per year including interest on the total investment at an annual rate of 8 percent. In an alternate proposal, the company can invest a total of $600,000 and receive annual net earnings (before income taxes) of $220,000 from the venture. In this case, the net earnings were determined on the basis of no interest costs. The company has $1 million of its own which it wishes to invest, and it can always obtain an effective 6 percent annual interest rate by loaning out the money. What would be the most profitable way for the company to invest its $1 million?

SEVEN

TAXES AND INSURANCE

Expenses for taxes and insurance play an important part in determining the economic situation for any industrial process. Because modern taxes may amount to a major portion of a concern's earnings, it is essential for the chemical engineer to understand the basic principles and factors underlying taxation. Insurance costs ordinarily are only a small part of the total expenditure involved in an industrial operation; however, adequate insurance coverage is necessary before any operation can be carried out on a sound economic basis.

Taxes are levied to supply funds to meet the public needs of a government, while insurance is required for protection against certain types of emergencies or catastrophic occurrences. Insurance rates and tax rates can vary considerably for business concerns as compared to the rates for individual persons. The information presented in this chapter applies generally to large business establishments.

TYPES OF TAXES

Taxes may be classified into three types: (1) property taxes, (2) excise taxes, and (3) income taxes. These taxes may be levied by the Federal government, state governments, or local governments.

Property Taxes

Local governments usually have jurisdiction over *property taxes*, which are commonly charged on a county basis. In addition to these, individual cities and towns may have special property taxes for industrial concerns located within the city limits.

Property taxes vary widely from one locality to another, but the average annual amount of these charges is 1 to 4 percent of the assessed valuation. Taxes of this type are referred to as *direct* since they must be paid directly by the particular concern and cannot be passed on as such to the consumer.

Excise Taxes

Excise taxes are levied by Federal and state governments. Federal excise taxes include charges for import customs duties, transfer of stocks and bonds, and a large number of other similar items. Manufacturers' and retailers' excise taxes are levied by Federal and state governments on the sale of many products such as gasoline and alcoholic beverages. Taxes of this type are often referred to as *indirect* since they can be passed on to the consumer. Many business concerns must also pay excise taxes for the privilege of carrying on a business or manufacturing enterprise in their particular localities.

Income Taxes

In general, *income taxes* are based on gross earnings, which are defined as the difference between total income and total product cost. Revenue from income taxes is an important source of capital for both Federal and state governments. National and state laws are the basis for these levies, and the laws change from year to year. State income taxes vary from one state to another and are a function of the total income for individual concerns. Depending on the particular state and the existing laws, state income taxes may range from 0 to 5 percent or more of gross earnings.

FEDERAL INCOME TAXES

The Federal government has set up an extremely complex system for determining income taxes for business establishments. New laws are added and old laws are changed each year, and it would be impossible to present all the rules and interpretations in a few pages.† Accordingly, this section will deal only with the basic pattern of Federal income-tax regulations and give the methods generally used for determining Federal income taxes. It should be emphasized strongly that the final determination of income-tax payments should be made with the aid of legal and accounting tax experts.

---

† Complete details are available in "Income Tax Regulations" and periodic "Income Tax Bulletins" issued by the U.S. Treasury Department, Superintendent of Documents, Internal Revenue Service and in services published by private concerns, such as the multivolume guide entitled "Prentice-Hall Federal Taxes," giving the latest tax laws with explanations and examples, which is published annually by Prentice-Hall, Inc., Englewood Cliffs, New Jersey.

The corporate income-tax rate in the United States has varied widely during the past 50 years. During the period from 1913 to 1935 the tax rate based on gross earnings increased from 1 to 13.75 percent. In 1938, the rate was increased to 19 percent, and, during the Second World War, it was 40 percent plus an excess-profits tax. In 1946, the standard income-tax rate for corporations was reduced to 38 percent, but the rate was increased to 42 percent in 1950 plus an excess-profits tax. During the Korean War, the rate was 52 percent plus an excess-profits tax which could make an overall tax rate of 70 percent on gross earnings. From 1954 through 1963, the corporation income-tax rate was 52 percent with reductions to 50 percent in 1964, 48 percent in 1965, and 46 percent in 1979. Table 1 summarizes the standard tax rates for corporations during the period of 1929 to 1979, and Table 2 presents a summary of Federal income taxes on corporations as applicable from 1965 to 1979.

The figures in Table 1 indicate the wide variations in income-tax rate caused by national emergencies, the prevailing economic situation, and the desires of lawmakers in office at any particular time. Figure 7-1 presents a graphical representation showing the changes in income-tax rates for a typical chemical company from 1943 to 1977.

Many industries have special tax exemptions because of the type of product,

Table 1 United States national taxes on corporation profits

Year	Regular tax rate, %	Effective limit with wartime excess-profits tax, %
United States:		
1929	11	
1930–1931	12	
1932–1935	13.75	
1936–1937	15	
1938–1939	19	
1940	24	
1941	31	
1942–1943	40	80
1944–1945	40	72
1946–1949	38	
1950	42	52
1951	50.75	68
1952–1953	52	70
1954–1963	52	
1964	50	
1965–1978	48	
1979	46	

Table 2 Federal income taxes on corporations (1965 to 1979)

Year	Taxes	Limitations	Percent of gross earnings
1965–1974	Normal tax	On gross earnings	22
	Surtax	On gross earnings above $25,000	26
	Combined rate	On gross earnings above $25,000	48
	Capital-gains tax	Varies depending on accounting methods	25–30
1975–1978	Normal tax	On first $25,000 of gross earnings	20
		On gross earnings over $25,000	22
	Surtax	On gross earnings over $50,000	26
	Combined rate	On gross earnings over $50,000	48
	Capital-gains tax	Varies depending on accounting methods	30
1979	Normal tax	On first $25,000 of gross earnings	17
		On second $25,000 of gross earnings	20
		On third $25,000 of gross earnings	30
		On fourth $25,000 of gross earnings	40
	Surtax	Graduated as shown to reach a combined rate of 46 percent on gross earnings above $100,000	
	Combined rate	On gross earnings above $100,000	46
	Capital-gains tax	Varies depending on accounting methods	28

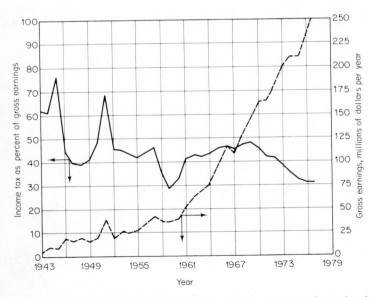

Figure 7-1 Example of variation in income-tax rate with time for a chemical company. (*Based on annual reports by Chas. Pfizer & Co. Inc.*)

market, or service involved in their business, or because the government wishes to offer particular support and inducement to concerns producing essential materials. Even for large concerns with high profits, the overall tax rate can vary widely from year to year depending on the size of available tax deductions or by carry-back or carry-forward provisions. These possible variations in income-tax effects can have an important influence on the optimum timing for expenditures or other financial transactions.

Normal Tax

A so-called *normal tax* has been levied by the Federal government on the earnings of corporations. This tax was at a rate set by the national lawmakers. For taxable years of 1965 to 1974, the normal tax in the United States was 22 percent of gross earnings. For the years 1975 through 1978, the normal tax was 20 percent of the first $25,000 of gross earnings and 22 percent of gross earnings above $25,000. For 1979, the normal tax was 17 percent of the first $25,000 of gross earnings, 20 percent of the second $25,000, 30 percent of the third $25,000, and 40 percent of the fourth $25,000.

Surtax

In addition to the normal tax, corporations have had to pay a second Federal income tax on gross earnings above a certain base limit. This additional tax is known as a *surtax*. The base limit was $25,000 per year for taxable years of 1965 to 1974 and $50,000 per year for taxable years of 1975 through 1978. The surtax on gross earnings above this limit was at a rate of 26 percent. For the year 1979, the base limit was $100,000 with a tax rate of 46 percent applying to gross earnings above this figure.

Capital-Gains Taxes

A *capital-gains tax* is levied on profits made from the sale of capital assets, such as land, buildings, or equipment. When the capital asset is sold after being in possession for more than one year as of 1978 (nine months in 1977 and six months before 1977), the profit is known as long-term capital gain. Prior to 1969, the tax rate on long-term capital gains was 25 percent. However, the tax laws have been changed since then so that they are now relatively complicated for long-term capital gains and are different for individuals and corporations.†
Basically, the rule for corporations in effect for 1979 is that any net long-term capital gain minus any net short-term capital loss is considered as net capital gain and can be taxed at a 28 percent rate, or the net capital gain can be added to the other corporation income and be taxed at the regular tax rate of that corporation if that produces a better result.

---

† For a complete and up-to-date description, see the most recent annual issue of "Prentice-Hall Federal Taxes," Prentice-Hall, Inc., Englewood Cliffs, New Jersey.

Tax Exemption for Dividends Received

Corporations are given a partial tax exemption for dividends received. In general, only 15 percent of such dividends are considered as taxable income for corporations, with the remaining 85 percent being tax exempt.

Contributions

Corporate contributions to appropriate organizations, as defined by the income-tax laws, can be deducted as an expense up to 5 percent of the taxable income. Thus, for a corporation which is paying income tax at a 46 percent rate, a contribution of $10,000 would represent an actual cost to the corporation after taxes of only $5400.

Carry-back and Carry-forward of Losses

The preceding analyses of taxes have been based on the assumption that the corporations involved were operating at a profit. In case the situation was one in which a loss resulted, some method of tax accounting needs to be available for this case of negative taxable income. To handle this possible situation, tax regulations permit the corporation to use the loss to offset profits in other years by *carry-back* or *carry-forward* of losses. Tax laws put into effect in 1958 permit a corporation to carry its losses back as charges against profits for as many as three years before the loss or, if necessary, to carry the losses forward as charges against profits for as many as five years after the loss.

Investment Credit

The 1971 Revenue Act of the United States provided for a special first-year tax deduction on new investments for machinery, equipment, and certain other assets used in production processes, in the form of a 7 percent "*investment credit*" for the first year of the life for assets with over 7 years of estimated service life. The investment credit rate was increased to 10 percent for the years 1975 and later, with the possibility of a higher rate. The investment credit amount is limited to the first $25,000 of the corporation's tax liability for the year plus 50 to 90 percent (depending on the year) of the corporation's tax liability above $25,000.

Tax revisions, such as those referred to in the preceding sections, are often made for the primary purpose of stimulating or controlling investments and the national economy. Accordingly, one can expect regular changes in the tax regulations, and the assistance of responsible tax experts who keep up with the latest developments is recommended for final evaluation of economic effects.

Taxes and Depreciation

Because Federal income taxes are based on gross earnings, which means that all costs have been deducted, the U.S. Treasury Department has devoted considerable effort to controlling one of the major costs in industrial operations, i.e., the cost for depreciation. The subject of depreciation costs is considered in Chap. 8, where some of the tax regulations by the U.S. Treasury Department are discussed.

In determining the influence of depreciation costs on income taxes, it should be clear that depreciation costs represent a deduction from taxable gross earnings. Thus, if d is the depreciation cost for the year and ϕ is the fractional tax rate,

$$\text{Tax "credit" for depreciation} = \phi d \tag{1}$$

Funds set aside for depreciation, although they represent a cost, normally go directly into the corporation treasury. Therefore, if S represents the total annual income or revenue and C represents the total annual costs with the exceptions of depreciation and taxes.

Net annual cash flow to company after taxes

$$= (S - C - d)(1 - \phi) + d = (S - C)(1 - \phi) + \phi d \tag{2}$$

The preceding equation is applied to various situations of cash flow in Table 3 of Chap. 9.

Excess-Profits Tax

During times of national emergency, certain types of business concerns can realize extremely high income and profit. This is true in particular for concerns producing military necessities during wartime. An *excess-profits tax* may be levied, therefore, to supply the national government with part of these profits.

The system for determining excess-profits taxes is extremely complex. In general, the amount of the tax is based on the normal past earnings of a concern or on the total capital investment. Special provisions are made for new corporations or for concerns which do not have a normal past history to use as a basis. The excess-profits taxes are very unpopular with businessmen, and there is always considerable opposition to the levying of these taxes.

Tax Returns

Income-tax returns may be reported on a cash basis or on an accrual basis. When the cash basis is used, only money actually received or paid out during the year is reported. With the accrual method, income and expenses are included as of the time they were incurred, even though final payment has not yet been made.

Returns may be based on a standard calendar year or on a fiscal year. Any date may be chosen as the end of the fiscal year, and it is usually advisable to choose a time when the work of assembly and determination of the tax will be the most convenient. The tax payment itself is usually made in installments.

OTHER TAXES

The Federal Insurance Contribution Act levies a social security tax on most employers and also requires a certain percentage of employees' wages to be withheld. Special local assessments for tax purposes are often encountered, and concerns doing business in foreign countries must pay taxes based on the laws of the foreign countries involved.

The question sometimes arises in cost accounting whether certain service charges and license fees can be considered as taxes. If the charge can be regarded as part of a public duty to support government, it is legally correct to designate the charge as a tax. When the exaction is for a service and the amount charged is a reasonable fee for the service actually received, the cost cannot be considered as a tax. Fees for building permits, government inspections, formation of corporations, bridge and road tolls, and certain types of licenses cannot be charged as taxes because the primary purpose of these fees is to serve for control and regulation rather than to support government.

If a corporation is organized by an individual for the purpose of avoiding high personal-income taxes, the organization is classed as a personal-holding company, and special tax rates apply. The amount of income tax which must be paid by a private business exceeds that required of an equivalent corporation at surprisingly small gross earnings. Consequently, a private business should make a periodic analysis of the advantages and disadvantages of becoming incorporated.

INSURANCE

The annual insurance cost for ordinary industrial concerns is approximately 1 percent of the capital investment. Despite the fact that insurance costs may represent only a small fraction of total costs, it is necessary to consider insurance requirements carefully to make certain the economic operation of a plant is protected against emergencies or unforeseen developments.

The design engineer can aid in reducing insurance requirements if he or she understands the factors which must be considered in obtaining adequate insurance. In particular, the engineer should be aware of the different types of insurance available and the legal responsibilities of a concern with regard to accidents or other unpredictable emergencies.

LEGAL RESPONSIBILITY

A concern can obtain insurance to protect itself against loss of property owing to any of a number of different causes. In case a property loss occurs and the loss is covered by insurance, payment will be made for the damage even though the loss was caused by the owner's negligence.

Protection against unforeseen emergencies, other than direct property loss, can also be obtained through insurance. For example, injuries to employees or persons near the danger area may occur due to a fire or explosion, and the concern involved should have insurance adequate to handle claims made in these cases. It is, of course, impossible to insure against every possible emergency, but it is necessary to consider the results of a potential occurrence, and the legal responsibility for various types of events should be understood. The payments required for settling a case in which legal responsibility has been proved may be much greater than any costs due to direct property damage.

An *assumed liability* is one which the concern accepts in the form of a written contract or statement, while a *legal liability* is always in effect whether or not it is stated in writing. Legal liabilities include civil responsibility for events occurring because of damage or injuries due to negligence. A stronger type of legal liability is known as *criminal liability*. This is involved in cases where gross negligence or reckless disregard for the life and property of others is claimed.

The design engineer should be familiar with the legal aspects of any laws or regulations governing the type of plant or process involved in a design. In case of an accident, failure to comply with the definite laws involved is a major factor in fixing legal responsibility. Compliance with all existing laws, however, is not a sufficient basis for disallowance of legal liability. Every known safety feature should be included and extraordinary care in the complete operation must be proved before a good case can be presented for disallowing legal liability.

Many contracts include *hold-harmless agreements* wherein the legal responsibility for an accident or other type of event is indicated as part of a written agreement. Any new lease or contract should be examined by an expert to make certain all hold-harmless agreements are clearly stated and understood by both parties.

Any concern producing a product which may be dangerous to life or property has a legal responsibility to indicate the potential hazard by use of warning labels or other protective methods. The manufacturer must supply safe shipping containers and make certain that any hazards involved in their handling or use are clearly indicated. Legal liability also holds for defective or misrepresented products.

A manufacturing establishment may have on its property some object which would be highly attractive as a place for children to play. Two examples would be a quarry pit and a sand pile. An object of this type is known as an *attractive nuisance*, and the concern may be liable for injuries to children if the injuries are a result of their playing around or in the object. The liability would apply even though the children were obviously trespassing. High fences or some other effective safety measure should be used to keep children from gaining admittance to an attractive nuisance.

An industrial concern has a legal responsibility for property belonging to others as long as the property is on the concern's premises. This responsibility is known as a *bailee's liability*. The property may be stored equipment or materials, finished products, or products in process. If the property is damaged or destroyed, the bailee's liability is roughly a function of the degree of care used in safeguarding the property. In case the damaged or destroyed property is insured by the owner, the insurance company will pay the claim; however, the insurance company can then exercise its *subrogation rights* and attempt to force the bailee to pay the full amount received by the owner.

TYPES OF INSURANCE

Many different types of insurance are available for protection against property loss or charges based on legal liability. Despite every precaution, there is always the possibility of an unforeseen event causing a sudden drain on a company's finances, and an efficient management protects itself against such potential emergencies by taking out insurance. In order to make an intelligent analysis of insurance requirements for any kind of operation, it is necessary to understand the physical factors involved in carrying out the process and to be aware of the types of insurance available.

The major insurance requirements for manufacturing concerns can be classified as follows:

1. Fire insurance and similar emergency coverage on buildings, equipment, and all other owned, used, or stored property. Included in this category would be losses caused by lightning, wind- or hailstorms, floods, automobile accidents, explosions, earthquakes, and similar occurrences.
2. Public-liability insurance, including bodily injury and property loss or damage, on all operations such as those involving automobiles, elevators, attractive nuisances, bailee's charges, aviation products, or any company function carried on at a location away from the plant premises.
3. Business-interruption insurance. The loss of income due to a business interruption caused by a fire or other emergency may far exceed any loss in property. Consequently, insurance against a business interruption of this type should be given careful consideration.
4. Power-plant, machinery, and special-operations hazards.
5. Workmen's-compensation insurance.
6. Marine and transportation insurance on all property in transit.
7. Comprehensive crime coverage.
8. Employee-benefit insurance, including life, hospitalization, accident, health, personal property, and pension plans.

Self-Insurance

On an average basis, insurance companies pay out loss claims amounting to 55 to 60 cents for each dollar received. The balance is used for income taxes, salaries, commissions, administrative costs, inspection costs, and various overhead costs. Theoretically, a saving of 40 to 45 cents per dollar paid for insurance could be achieved by self-insurance. If insurance requirements are great, this saving could amount to a very large sum, and it would be worthwhile to consider the possibilities of self-insurance.

A careful analysis of all risks involved is necessary when considering self-insurance on possible losses or emergencies. The final decision should not be based on whether or not the insurable event will occur, because this is impossible to predict. Instead, the decision should be based on the total loss involved if the event or a series of such events were to occur. If an industrial concern has a number of widespread interests and sufficient funds available to handle simultaneous major losses in several of these interests, it might be reasonable to consider self-insurance on some of the potential hazards. On the other hand, if a single potential event could ruin the economic standing of the company, it would be very inadvisable to assume the risk involved in self-insurance.

There are several different ways of applying self-insurance. One method involves depositing money equivalent to an insurance premium into a special company fund. This fund can then be used to handle any losses or emergencies which may occur. At the outset, this fund would be small and would be inadequate to handle any major losses. Consequently, if this method is used, it may be necessary to supply an original base fund or else assume a disproportionate amount of risk until the fund has built up to a practical value. Under ordinary conditions, the premiums paid into a self-insurance reserve are not tax-deductible.

A second method may be used in which the company assumes all the risk and no payments are made into a reserve fund. This method is designated as "self-assumption of risk." Partial self-insurance may be obtained through the purchase of deductible insurance from regular agencies. The purchaser assumes the risk up to a certain amount and the insurance company agrees to pay for any additional losses.

The effects of income taxes should be considered when making a final decision regarding insurance. Because the premiums for standard insurance are tax-deductible, the actual cost after taxes for the protection may be much less than the direct premium charge. Another advantage of standard insurance is the inspection services supplied by the insurance companies. These companies require periodic inspections by specialists to make certain that the insurance rates are adequate, and the reports of these inspectors often indicate new ideas or methods for increasing the safety of the operation.

The overall policies of the particular manufacturing concern dictate the type and amount of insurance which will be held. It should be realized, however, that a well-designed insurance plan requires a great deal of skilled and informed investigation by persons who understand all the aspects of insurance as well as the problems involved in the manufacturing operation.

SUGGESTED ADDITIONAL REFERENCES FOR TAXES AND INSURANCE

Income taxes

Jelen, F. C., and M. S. Cole, Methods for Economic Analysis, *Hydrocarbon Process.*, **53**(9):227 (1974); **53**(10):161 (1974).

Jelen, F. C., and C. L. Yaws, Project Cash Flow—Description and Interpretation, *Hydrocarbon Process.*, **57**(3):77 (1978).

Matley, J., Will New Tax Laws Save or Cost You Money? *Chem. Eng.*, **84**(7):125 (1977).

Piekarski, J. A., Fundamentals of Manufacturing Economics, *1973 Trans. Am. Assoc. Cost Eng.* (1973), p. 221.
Weismantel, G. E., Can U.S. Firms Still Compete Abroad? *Chem. Eng.*, **84**(18):25 (1977).

Taxes

Winton, J. W., Plant Sites 1971, *Chem. Week*, **109**(15):35 (1971).
———, Plant Sites 1972, *Chem. Week*, **111**(15):35 (1972).
———, Plant Sites 1974, *Chem. Week*, **113**(16):29 (1973).
———, Plant Sites 1975, *Chem. Week*, **115**(17):33 (1974).
———, Plant Sites 1976, *Chem. Week*, **117**(17):27 (1975).
———, Plant Sites 1977, *Chem. Week*, **119**(19):35 (1976).
———, Plant Sites 1978, *Chem. Week*, **121**(24):49 (1977).

PROBLEMS

1 The fixed-capital investment for an existing chemical plant is $20 million. Annual property taxes amount to 1 percent of the fixed-capital investment, and state income taxes are 5 percent of the gross earnings. The net income per year after all taxes is $2 million, and the Federal income taxes amount to 48 percent of gross earnings. If the same plant had been constructed at a location where property taxes were 4 percent of the fixed-capital investment and state income taxes were 2 percent of the gross earnings, what would be the net income per year after taxes, assuming all other cost factors were unchanged?

2 The gross earnings for a small corporation were $54,000 in 1960. What would have been the percent reduction in Federal income taxes paid by the company if the tax rates in effect in 1973 had been in effect in 1960? (In 1960, surtax was 22 percent of all gross earnings above $25,000 and normal tax was 30 percent.)

3 During the period of one taxable year at a manufacturing plant, the total income on a cash basis was $21 million. Five million dollars of immediate debts due the company was unpaid at the end of the year. The company paid out $15 million on a cash basis during the year, and all of this amount was tax-deductible as a product cost. The company still owed $3 million of tax-deductible bills at the end of the year. If the total Federal income tax for the company amounts to 48 percent of the gross earnings, determine the amount of Federal income tax due for the year on a cash basis and also on an accrual basis.

4 Complete fire and allied-coverage insurance for one unit of a plant requires an annual payment of $700 based on an investment value of $100,000. If income taxes over a 10-year period average 40 percent of gross earnings, by how much is the net income, after taxes, reduced during this 10-year period owing to the cost of the insurance?

5 Self-insurance is being considered for one portion of a chemical company. The fixed-capital investment involved is $50,000, and insurance costs for complete protection would amount to $400 per year. If self-insurance is used, a reserve fund will be set up under the company's jurisdiction, and annual insurance premiums of $300 will be deposited in this fund under an ordinary annuity plan. All money in the fund can be assumed to earn interest at a compound annual rate of 5 percent. Neglecting any charges connected with administration of the fund, how much money should be deposited in the fund at the beginning of the program in order to have enough money accumulated to replace a complete $50,000 loss after 10 years?

6 A corporation shows a gross earnings or net profit before Federal income taxes of $200,000 in the taxable years of 1973, 1978, and 1979. The taxable income for the corporation in all 3 years is the $200,000 given, and there are no special tax exemptions for the corporation so that the situation described in this chapter under the headings of *normal tax* and *surtax* applies (see Table 2). Determine the Federal income tax paid by the corporation in each of the 3 years and find the amount of tax saved in 1979 because of the tax regulations in effect that year as compared to 1978 and 1973. Repeat for the case of a gross earnings of $50,000 instead of $200,000.

EIGHT

DEPRECIATION

An analysis of costs and profits for any business operation requires recognition of the fact that physical assets decrease in value with age. This decrease in value may be due to physical deterioration, technological advances, economic changes, or other factors which ultimately will cause retirement of the property. The reduction in value due to any of these causes is a measure of the *depreciation*.† The economic function of depreciation, therefore, can be employed as a means of distributing the original expense for a physical asset over the period during which the asset is in use.

Because the engineer thinks of depreciation as a measure of the decrease in value of property with time, depreciation can immediately be considered from a cost viewpoint. For example, suppose a piece of equipment had been put into use 10 years ago at a total cost of $31,000. The equipment is now worn out and is worth only $1000 as scrap material. The decrease in value during the 10-year period is $30,000; however, the engineer recognizes that this $30,000 is in reality

---

† According to the Bureau of Internal Revenue Code, depreciation is defined as "A reasonable allowance for the exhaustion, wear, and tear of property used in the trade or business including a reasonable allowance for obsolescence." The terms *amortization* and *depreciation* are often used interchangeably. Amortization is usually associated with a definite period of cost distribution, while depreciation usually deals with an unknown or estimated period over which the asset costs are distributed. Depreciation and amortization are of particular significance as an accounting concept which serves to reduce taxes.

a cost incurred for the use of the equipment. This depreciation cost was spread over a period of 10 years, and sound economic procedure would require part of this cost to be charged during each of the years. The application of depreciation in engineering design, accounting, and tax studies is almost always based on costs prorated throughout the life of the property.

Meaning of Value

From the viewpoint of the design engineer, the total cost due to depreciation is the original or new value of a property minus the value of the same property at the end of the depreciation period. The original value is usually taken as the total cost of the property at the time it is ready for initial use. In engineering design practice, the total depreciation period is ordinarily assumed to be the length of the property's useful life, and the value at the end of the useful life is assumed to be the probable scrap or salvage value of the components making up the particular property.

It should be noted here that the engineer cannot wait until the end of the depreciation period to determine the depreciation costs. These costs must be prorated throughout the entire life of the property, and they must be included as an operating charge incurred during each year. The property value at the end of the depreciation period and the total length of the depreciation period cannot be known with certainty when the initial yearly costs are determined. Consequently, it is necessary to estimate the final value of the property as well as its useful life. In estimating property life, the various factors which may affect the useful-life period, such as wear and tear, economic changes, or possible technological advances, should be taken into consideration.

When depreciation is not used in a prorated-cost sense, various meanings can be attached to the word *value*. One of these meanings involves appraisal of both initial and final values on the basis of conditions at a certain time. The difference between the estimated cost of new equivalent property and the appraised value of the present asset is known as the *appraised depreciation*. This concept involves determination of the values of two assets at one date as compared with the engineering-cost concept, which requires determination of the value of one asset at two different times.

Purpose of Depreciation as a Cost

Consideration of depreciation as a cost permits realistic evaluation of profits earned by a company and, therefore, provides a basis for determination of Federal income taxes. Simultaneously, the consideration of depreciation as a cost provides a means whereby funds are set aside regularly to provide recovery of the invested capital. When accountants deal with depreciation, they must follow certain rules which are established by the U.S. Bureau of Internal Revenue for determination of income taxes. These rules deal with allowable life for the depreciable equipment and acceptable mathematical procedures for allocating the depreciation cost over the life of the asset.

Although any procedure for depreciation accounting can be adopted for internal company evaluations, it is highly desirable to keep away from the necessity of maintaining two sets of accounting books. Therefore, the engineer should be familiar with Federal regulations relative to depreciation and should follow these regulations as closely as possible in evaluating depreciation as a cost.

TYPES OF DEPRECIATION

The causes of depreciation may be physical or functional. *Physical depreciation* is the term given to the measure of the decrease in value due to changes in the physical aspects of the property. Wear and tear, corrosion, accidents, and deterioration due to age or the elements are all causes of physical depreciation. With this type of depreciation, the serviceability of the property is reduced because of physical changes. Depreciation due to all other causes is known as *functional depreciation.*

One common type of functional depreciation is *obsolescence*. This is caused by technological advances or developments which make an existing property obsolete. Even though the property has suffered no physical change, its economic serviceability is reduced because it is inferior to improved types of similar assets that have been made available through advancements in technology.

Other causes of functional depreciation could be (1) change in demand for the service rendered by the property, such as a decrease in the demand for the product involved because of saturation of the market, (2) shift of population center, (3) changes in requirements of public authority, (4) inadequacy or insufficient capacity for the service required, (5) termination of the need for the type of service rendered, and (6) abandonment of the enterprise. Although some of these situations may be completely unrelated to the property itself, it is convenient to group them all under the heading of functional depreciation.

Because depreciation is measured by decrease in value, it is necessary to consider all possible causes when determining depreciation. Physical losses are easier to evaluate than functional losses, but both of these must be taken into account in order to make fair allowances for depreciation.

Depletion

Capacity loss due to materials actually consumed is measured as *depletion*. Depletion cost equals the initial cost times the ratio of amount of material used to original amount of material purchased. This type of depreciation is particularly applicable to natural resources, such as stands of timber or mineral and oil deposits.

Costs for Maintenance and Repairs

The term *maintenance* conveys the idea of constantly keeping a property in good condition; *repairs* connotes the replacing or mending of broken or worn parts of a property. The costs for maintenance and repairs are direct operating expenses which must be paid from income, and these costs should not be confused with depreciation costs.

The extent of maintenance and repairs may have an effect on depreciation cost, because the useful life of any property ought to be increased if it is kept in good condition. However, a definite distinction should always be made between costs for depreciation and costs for maintenance and repairs.

SERVICE LIFE

The period during which the use of a property is economically feasible is known as the *service life* of the property. Both physical and functional depreciation are taken into consideration in determining service life, and, as used in this book, the term is synonymous with *economic* or *useful life*. In estimating the probable service life, it is assumed that a reasonable amount of maintenance and repairs will be carried out at the expense of the property owner.

Many data are available concerning the probable life of various types of property. Manufacturing concerns, engineers, and the U.S. Bureau of Internal Revenue have compiled much information of this sort. All of these data are based on past records, and there is no certainty that future conditions will be unchanged. Nevertheless, by statistical analysis of the various data, it is possible to make fairly reliable estimates of service lives.

The U.S. Bureau of Internal Revenue recognizes the importance of depreciation as a legitimate expense, and the Bureau has issued formal statements which list recommended service lives for many types of properties.† Prior to July 12, 1962, Federal regulations for service lives and depreciation rates were based on the so-called *Bulletin "F"*, *"Income Tax Depreciation and Obsolescence— Estimated Useful Lives and Depreciation Rates"* as originally published by the U.S. Bureau of Internal Revenue in 1942. In July, 1962, the *Bulletin "F"* regulations were replaced by a set of new guidelines based on four groups of depreciable assets. The 1971 Revenue Act of the United States provided more flexibility in choosing depreciation life by allowing a choice of depreciation life of 20 percent longer or shorter than the guideline lives called for by earlier tax laws for machinery, equipment, or other assets put in service after December 31, 1970. This is known as the *Class Life Asset Depreciation Range System (ADR)*.

Table 1 presents estimated service lives for equipment based on the four group guidelines as recommended by the Bureau of Internal Revenue in the 1962 Federal regulations. These values, along with similar values as presented in *Bulletin "F"*, can serve as an indication of acceptable and useful lives to those not using other procedures.

---

† For an up-to-date presentation of Federal income-tax regulations as related to depreciation including estimation of service lives, see the latest annual issue of "Prentice-Hall Federal Taxes," Prentice-Hall, Inc., Englewood Cliffs, New Jersey.

Table 1. Estimated life of equipment

The following tabulation for estimating the life of equipment in years is an abridgement of information from "Depreciation—Guidelines and Rules" (Rev. Proc. 62-21) issued by the Internal Revenue Service of the U.S. Treasury Department as Publication No. 456 (7-62) in July, 1962. The original publication should be consulted for exact accounting details. See Table 2 for an extended and more flexible interpretation including repair allowance as approved by the Federal regulations in 1971.

	Life, years
Group I: General business assets	
1. Office furniture, fixtures, machines, equipment	10
2. Transportation	
a. Aircraft	6
b. Automobile	3
c. Buses	9
d. General-purpose trucks	4–6
e. Railroad cars (except for railroad companies)	15
f. Tractor units	4
g. Trailers	6
h. Water transportation equipment	18
3. Land and site improvements (not otherwise covered)	20
4. Buildings (apartments, banks, factories, hotels, stores, warehouses)	40–60
Group II: Nonmanufacturing activities (excluding transportation, communications, and public utilities)	
1. Agriculture	
a. Machinery and equipment	10
b. Animals	3–10
c. Trees and vines	variable
d. Farm buildings	25
2. Contract construction	
a. General	5
b. Marine	12
3. Fishing	variable
4. Logging and sawmilling	6–10
5. Mining (excluding petroleum refining and smelting and refining of minerals)	10
6. Recreation and amusement	10
7. Services to general public	10
8. Wholesale and retail trade	10
Group III: Manufacturing	
1. Aerospace industry	8
2. Apparel and textile products	9
3. Cement (excluding concrete products)	20
4. Chemicals and allied products	11
5. Electrical equipment	
a. Electrical equipment in general	12
b. Electronic equipment	8
6. Fabricated metal products	12
7. Food products, except grains, sugar and vegetable oil products	12
8. Glass products	14
9. Grain and grain-mill products	17

(*Continued*)

Table 1 Estimated life of equipment (*Continued*)

	Life, years
Group III: Manufacturing (continued)	
10. Knitwear and knit products	9
11. Leather products	11
12. Lumber, wood products, and furniture	10
13. Machinery unless otherwise listed	12
14. Metalworking machinery	12
15. Motor vehicles and parts	12
16. Paper and allied products	
a. Pulp and paper	16
b. Paper conversion	12
17. Petroleum and natural gas	
a. Contract drilling and field service	6
b. Company exploration, drilling, and production	14
c. Petroleum refining	16
d. Marketing	16
18. Plastic products	11
19. Primary metals	
a. Ferrous metals	18
b. Nonferrous metals	14
20. Printing and publishing	11
21. Scientific instruments, optical and clock manufacturing	12
22. Railroad transportation equipment	12
23. Rubber products	14
24. Ship and boat building	12
25. Stone and clay products	15
26. Sugar products	18
27. Textile mill products	12–14
28. Tobacco products	15
29. Vegetable oil products	18
30. Other manufacturing in general	12
Group IV: Transportation, communications, and public utilities	
1. Air transport	6
2. Central steam production and distribution	28
3. Electric utilities	
a. Hydraulic	50
b. Nuclear	20
c. Steam	28
d. Transmission and distribution	30
4. Gas utilities	
a. Distribution	35
b. Manufacture	30
c. Natural-gas production	14
d. Trunk pipelines and storage	22
5. Motor transport (freight)	8
6. Motor transport (passengers)	8
7. Pipeline transportation	22
8. Radio and television broadcasting	6

Table 1 Estimated life of equipment (*Continued*)

	Life, years
Group IV: Transportation, communications, and public utilities (continued)	
9. Railroads	
a. Machinery and equipment	14
b. Structures and similar improvements	30
c. Grading and other right of way improvements	variable
d. Wharves and docks	20
10. Telephone and telegraph communications	variable
11. Water transportation	20
12. Water utilities	50

Table 2 gives a partial listing of the *Class Life Asset Depreciation Range System* as recommended for use by Federal regulations in 1971. The table shows the basic guideline life period as recommended in the earlier regulations along with the 20 percent variation allowed plus recommended guidelines for repair and maintenance allowance. Although these values are recommended by the Bureau of Internal Revenue, the Bureau does not require taxpayers to use the indicated lives. However, if other life periods are used, the taxpayer must be prepared to support the claim.

There has been considerable demand for a wider choice of service lives for properties, and the widespread revision and reinterpretation of the national income-tax laws in 1954, 1962, and 1971 met part of this demand. During times of national emergencies, the United States Congress may approve rapid-amortization policies to make it more attractive for conerns to invest in additional plants and equipment needed for the national welfare. Certificates of necessity can be obtained for certain types of industries, and these certificates permit writing off various percentages of the value of new equipment and facilities over a period of five years.

SALVAGE VALUE

Salvage value is the net amount of money obtainable from the sale of used property over and above any charges involved in removal and sale. If a property is capable of further service, its salvage value may be high. This is not necessarily true, however, because other factors, such as location of the property, existing price levels, market supply and demand, and difficulty of dismantling, may have an effect. The term *salvage value* implies that the asset can give some type of further service and is worth more than merely its scrap or junk value.

If the property cannot be disposed of as a useful unit, it can often be dismantled and sold as junk to be used again as a manufacturing raw material. The profit obtainable from this type of disposal is known as the *scrap*, or *junk, value*.

Table 2 Class life asset depreciation range†

Description of class life asset	Asset depreciation range (in years)			Annual asset guide-line repair allowance, percentage of cost
	Lower limit	Asset guide-line period	Upper limit	
Assets used in business activities:				
Office furniture, fixtures, and equipment	8	10	12	2
Information systems, computers, peripheral equipment	5	6	7	7.5
Data handling equipment, except computers	5	6	7	15
Airplanes, except commercial	5	6	7	14
Automobiles, taxis	2.5	3	3.5	16.5
Buses	7	9	11	11.5
Light general-purpose trucks	3	4	5	16.5
Heavy general-purpose trucks	5	6	7	10
Railroad cars and locomotives, except owned by railroad transportation companies	12	15	18	8
Tractor units for use over-the-road	3	4	5	16.5
Trailers and trailer-mounted containers	5	6	7	10
Vessels, barges, tugs and similar water-transportation equipment	14.5	18	21.5	6
Land improvements		20		
Industrial steam and electricity genera-tion and/or distribution systems	22.5	33.5	33.5	2.5
Assets used in agriculture	8	10	12	11
Assets used in mining	8	10	12	6.5
Assets used in drilling of oil and gas wells	5	6	7	10
Assets used in exploration for and production of petroleum and natural gas deposits	5	6	7	10
Assets used in petroleum refining	13	16	19	7
Assets used in marketing of petroleum and petroleum products	13	16	19	4
Assets used in contract construction other than marine	4	5	6	12.5
Assets used in marine contract construction	9.5	12	14.5	5

† Values were excerpted from the listing given with full description of each category in the 1978 "Prentice-Hall Federal Taxes" guide as updated from the original 1971 Federal regulation to the March 21, 1977 Revenue Procedure in the Internal Revenue Bulletin. The official documents originally setting up the ADR system were U.S. Treasury Decision 7128 in 1971 and Revenue Procedure 71-25 in 1971.

Table 2 Class life asset depreciation range† (*Continued*)

Description of class life asset	Asset depreciation range (in years)			Annual asset guide-line repair allowance, percentage of cost
	Lower limit	Asset guide-line period	Upper limit	
Assets used in the manufacture of:				
Grain and grain-mill products	9.5	12	14.5	5
Sugar and sugar products	13.5	17	20.5	6
Vegetable oils and vegetable-oil products	14.5	18	21.5	4.5
Other food and kindred products	14.5	18	21.5	3.5
Tobacco and tobacco products	12	15	18	5
Knitted goods	6	7.5	9	7
Nonwoven fabrics	8	10	12	15
Wood products and furniture	5	6	7	10
Pulp and paper	8	10	12	6.5
Chemicals and allied products	9	11	13	5.5
Rubber products	11	14	17	5
Finished plastic products	9	11	13	5.5
Glass products	11	14	17	12
Cement	16	20	24	3
Machinery	8	10	12	11
Electrical equipment	9.5	12	14.5	5.5
Motor vehicles	9.5	12	14.5	9.5
Assets used in electric, gas, water, and steam utility services:				
Electric utility nuclear production plant	16	20	24	3
Electric utility steam production plant	22.5	28	33.5	5
Electric utility transmission and distribution plant	24	30	36	4.5
Gas utility distribution facilities	28	35	42	2
Gas utility manufactured gas production plant	24	30	36	2
Substitute natural gas—coal gasification (Lurgi process with advanced methanation)	14.5	18	21.5	15
Natural gas production plant	11	14	17	4.5
Liquefied natural gas plant	17.5	22	26.5	4.5
Water utilities	40	50	60	1.5
Central steam utility production and distribution	22.5	28	33.5	2.5

† Values were excerpted from the listing given with full description of each category in the 1978 "Prentice-Hall Federal Taxes" guide as updated from the original 1971 Federal regulation to the March 21, 1977 Revenue Procedure in the Internal Revenue Bulletin. The official documents originally setting up the ADR system were U.S. Treasury Decision 7128 in 1971 and Revenue Procedure 71–25 in 1971.

Salvage value, scrap value, and service life are usually estimated on the basis of conditions at the time the property is put into use. These factors cannot be predicted with absolute accuracy, but improved estimates can be made as the property increases in age. It is advisable, therefore, to make new estimates from time to time during the service life and make any necessary adjustments of the depreciation costs. Because of the difficulties involved in making reliable, future estimates of salvage and scrap values, engineers often neglect the small error involved and designate these values as zero.

PRESENT VALUE

The *present value* of an asset may be defined as the value of the asset in its condition at the time of valuation. There are several different types of present values, and the standard meanings of the various types should be distinguished.

Book Value, or Unamortized Cost

The difference between the original cost of a property, and all the depreciation charges made to date is defined as the *book value* (sometimes called *unamortized cost*). It represents the worth of the property as shown on the owner's accounting records.

Market Value

The price which could be obtained for an asset if it were placed on sale in the open market is designated as the *market value*. The use of this term conveys the idea that the asset is in good condition and that a buyer is readily available.

Replacement Value

The cost necessary to replace an existing property at any given time with one at least equally capable of rendering the same service is known as the *replacement value*.

It is difficult to predict future market values or replacement values with a high degree of accuracy because of fluctuations in market demand and price conditions.† On the other hand, a future book value can be predicted with absolute accuracy as long as a constant method for determining depreciation costs is used. It is quite possible for the market value, replacement value, and book value of a property to be widely different from one another because of unrealistic depreciation allowances or changes in economic and technological factors.

---

† See Chap. 10 (Optimum Design and Design Strategy) for a discussion on inflation and the strategy for considering it.

METHODS FOR DETERMINING DEPRECIATION

Depreciation costs can be determined by a number of different methods, and the design engineer should understand the bases for the various methods. The Federal government has definite rules and regulations concerning the manner in which depreciation costs may be determined. These regulations must be followed for income-tax purposes as well as to obtain most types of governmental support. Since the methods approved by the government are based on sound economic procedures, most industrial concerns use one of the government-sanctioned methods for determining depreciation costs, both for income-tax calculations and for reporting the concern's costs and profits.‡ It is necessary, therefore, that the design engineer keep abreast of current changes in governmental regulations regarding depreciation allowances.

In general, depreciation accounting methods may be divided into two classes: (1) arbitrary methods giving no consideration to interest costs, and (2) methods taking into account interest on the investment. Straight-line, declining-balance, and sum-of-the-years-digits methods are included in the first class, while the second class includes the sinking-fund and the present-worth methods.

Straight-Line Method

In the *straight-line method* for determining depreciation, it is assumed that the value of the property decreases linearly with time. Equal amounts are charged for depreciation each year throughout the entire service life of the property. The annual depreciation cost may be expressed in equation form as follows:

$$d = \frac{V - V_s}{n} \tag{1}$$

where d = annual depreciation, \$/year
V = original value of the property at start of the service-life period, completely installed and ready for use, dollars
V_s = salvage value of property at end of service life, dollars
n = service life, years

The asset value (or book value) of the equipment at any time during the service life may be determined from the following equation:

$$V_a = V - ad \tag{2}$$

where V_a = asset or book value, dollars, and a = the number of years in actual use.

---

‡ An alternate procedure often used by industrial concerns is to use straight-line depreciation for reporting profits and one of the accelerated-depreciation methods for income-tax calculations.

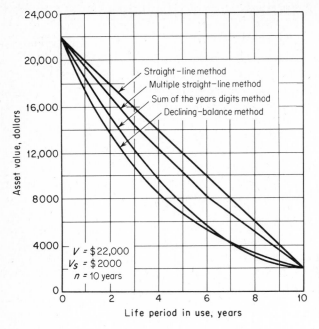

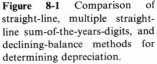

Figure 8-1 Comparison of straight-line, multiple straight-line sum-of-the-years-digits, and declining-balance methods for determining depreciation.

Because of its simplicity, the straight-line method is widely used for determining depreciation costs. In general, design engineers report economic evaluations on the basis of straight-line depreciation unless there is some specific reason for using one of the other methods.

Because it is impossible to estimate exact service lives and salvage values when a property is first put into use, it is sometimes desirable to reestimate these factors from time to time during the life period of the property. If this is done, straight-line depreciation can be assumed during each of the periods, and the overall method is known as *multiple straight-line depreciation.* Figure 8-1 shows how the asset value of a property varies with time using the straight-line and the multiple straight-line methods for determining depreciation.

The straight-line method may be applied on the basis of units of production or predicted amount of service output, instead of life years. The depreciation may be based on miles, gallons, tons, number of unit pieces produced, or other measures of service output. This so-called *unit-of-production* or *service-output* method is particularly applicable when depletion occurs, as in the exploitation of natural resources. It should also be considered for properties having useful lives that are more dependent on the number of operations performed than on calendar time.

Declining-Balance (or Fixed Percentage) Method

When the declining-balance method is used, the annual depreciation cost is a fixed percentage of the property value at the beginning of the particular year.

The fixed-percentage (or declining-balance) factor remains constant throughout the entire service life of the property, while the annual cost for depreciation is different each year. Under these conditions, the depreciation cost for the first year of the property's life is $V\mathfrak{f}$, where $\mathfrak{f}$ represents the fixed-percentage factor.

At the end of the first year

$$\text{Asset value} = V_a = V(1 - \mathfrak{f}) \tag{3}$$

At the end of the second year

$$V_a = V(1 - \mathfrak{f})^2 \tag{4}$$

At the end of a years

$$V_a = V(1 - \mathfrak{f})^a \tag{5}$$

At the end of n years (i.e., at the end of service life)

$$V_a = V(1 - \mathfrak{f})^n = V_s \tag{6}$$

Therefore,

$$\mathfrak{f} = 1 - \left(\frac{V_s}{V}\right)^{1/n} \tag{7}$$

Equation (7) represents the textbook method for determining the fixed-percentage factor, and the equation is sometimes designated as the *Matheson formula*. A plot showing the change of asset value with time using the declining-balance depreciation method is presented in Fig. 8-1. Comparison with the straight-line method shows that declining-balance depreciation permits the investment to be paid off more rapidly during the early years of life. The increased depreciation costs in the early years are very attractive to concerns just starting in business, because the income-tax load is reduced at the time when it is most necessary to keep all pay-out costs at a minimum.

The textbook relationship presented in Eq. (7) is seldom used in actual practice, because it places too much emphasis on the salvage value of the property and is certainly not applicable if the salvage value is zero. To overcome this disadvantage, the value of the fixed-percentage factor is often chosen arbitrarily using a sound economic basis.

Prior to 1954, the United States government would not accept any depreciation method which permitted depreciation rates more than 50 percent greater than those involved in the straight-line method. In 1954, the laws were changed to allow rates up to twice those for the straight-line method. Under these conditions, one arbitrary method for choosing the value of $\mathfrak{f}$ is to fix it at two times the reciprocal of the service life n.† This permits approximately two-thirds of the depreciable value to be written off in the first half of the useful life.‡

† The salvage value is considered to be zero, and the fixed-percentage factor is based on the straight-line rate of depreciation during the first year.

‡ Based on the 1954 tax revision for depreciation accounting, any method can be used if the depreciation for the first two-thirds of the useful life of the property does not exceed the total of such allowances if they had been computed by the double declining-balance method.

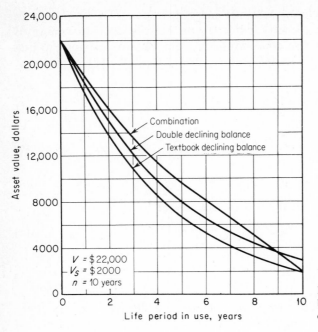

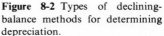

Figure 8-2 Types of declining-balance methods for determining depreciation.

Figure 8-2 shows the effect of time on asset value when the declining-balance method of depreciation is used with an arbitrarily chosen value of f. It should be noted that the value of the asset cannot decrease to zero at the end of the service life and may possibly be greater than the salvage or scrap value. To handle this difficulty, it is sometimes desirable to switch from the declining-balance to the straight-line method after a portion of the service life has expired. This is known as the *combination method*. It permits the property to be fully depreciated during the service life, yet also gives the advantage of faster early-life write-offs. A curve showing this type of depreciation is presented in Fig. 8-2.

The main advantage of the declining-balance and the combination methods is that they permit greater depreciation allowances in the early life of the property than in the later life. They are particularly applicable for units in which the greater proportion of the production occurs in the early part of the useful life or when operating costs increase markedly with age.

Example 1. Determination of depreciation by straight-line and declining-balance methods The original value of a piece of equipment is $22,000, completely installed and ready for use. Its salvage value is estimated to be $2000 at the end of a service life estimated to be 10 years. Determine the asset (or book) value of the equipment at the end of 5 years using:
(*a*) Straight-line method.
(*b*) Textbook declining-balance method.

(c) Double declining-balance method (i.e., the declining-balance method using a fixed-percentage factor giving a depreciation rate equivalent to twice the minimum rate with the straight-line method).

SOLUTION:
(a) Straight-line method:

$$V = \$22,000$$

$$V_s = \$2000$$

$$n = 10 \text{ years}$$

$$d = \frac{V - V_s}{n} = \frac{20,000}{10} = \$2000/\text{year}$$

Asset value after 5 years $= V_a$, where $a = 5$, or

$$V_a = V - ad = 22,000 - (5)(2000) = \$12,000$$

(b) Textbook declining-balance method:

$$\hat{f} = 1 - \left(\frac{V_s}{V}\right)^{1/n} = 1 - \left(\frac{2000}{22,000}\right)^{1/10} = 0.2131$$

Asset value after 5 years is

$$V_a = V(1 - \hat{f})^a = (22,000)(1 - 0.2131)^5 = \$6650$$

(c) Double declining-balance method:

Using the straight-line method, the minimum depreciation rate occurs in the first year when $V = \$22,000$ and the depreciation $= \$2000$. This depreciation rate is $2000/22,000$, and the double declining-balance (or double fixed-percentage) factor is $(2)(2000/22,000) = 0.1818 = \hat{f}$. (It should be noted that the double declining-balance method is often applied to cases where the salvage value is considered to be zero. Under this condition, the double fixed-percentage factor for this example would be 0.2000.)

Asset value after 5 years is

$$V_a = V(1 - \hat{f})^a = (22,000)(1 - 0.1818)^5 = \$8060$$

Sum-of-the-Years-Digits Method

The *sum-of-the-years-digits method* is an arbitrary process for determining depreciation which gives results similar to those obtained by the declining-balance method. Larger costs for depreciation are allotted during the early-life years than during the later years. This method has the advantage of permitting the asset value to decrease to zero or a given salvage value at the end of the service life.

In the application of the sum-of-the-years-digits method, the annual depreciation is based on the number of service-life years remaining and the sum of the arithmetic series of numbers from 1 to n, where n represents the total service life. The yearly depreciation factor is the number of useful service-life years remaining divided by the sum of the arithmetic series. This factor times the total depreciable value at the start of the service life gives the annual depreciation cost.

As an example, consider the case of a piece of equipment costing \$20,000 when new. The service life is estimated to be 5 years and the scrap value \$2000. The sum of the arithmetic series of numbers from 1 to n is $1 + 2 + 3 + 4 + 5 = 15$. The total depreciable value at the start of the service life is \$20,000 − \$2000 = \$18,000. Therefore, the depreciation cost for the first year is $(\$18,000)(\frac{5}{15}) = \$6000$, and the asset value at the end of the first year is \$14,000. The depreciation cost for the second year is $(\$18,000)(\frac{4}{15}) = \$4800$. Similarly, the depreciation costs for the third, fourth, and fifth years, respectively, would be \$3600, \$2400, and \$1200.† Figure 8-1 presents a curve showing the change with time in asset value when the sum-of-the-years-digits method is used for determining depreciation.

Sinking-Fund Method

The use of compound interest is involved in the *sinking-fund method*. It is assumed that the basic purpose of depreciation allowances is to accumulate a sufficient fund to provide for the recovery of the original capital invested in the property. An ordinary annuity plan is set up wherein a constant amount of money should theoretically be set aside each year. At the end of the service life, the sum of all the deposits plus accrued interest must equal the total amount of depreciation.

Derivation of the formulas for the sinking-fund method can be accomplished by use of the following notations in addition to those already given:

$i = $ annual interest rate expressed as a fraction

$R = $ uniform annual payments made at end of each year (this is the annual depreciation cost), dollars

$V - V_s = $ total amount of the annuity accumulated in an estimated service life of n years (original value of property minus salvage value at end of service life), dollars

---

† Equations which apply for determining annual depreciation by the sum-of-the-years-digits method are

$$d_a = \text{depreciation for year } a = \frac{(n - a + 1)}{\sum\limits_{1}^{n} a}(V - V_s)$$

$$= \frac{2(n - a + 1)}{n(n + 1)}(V - V_s)$$

According to the equations developed for an ordinary annuity in Chap. 6 (Interest and Investment Costs),

$$R = (V - V_s) \frac{i}{(1 + i)^n - 1} \tag{8}$$

The amount accumulated in the fund after a years of useful life must be equal to the total amount of depreciation up to that time. This is the same as the difference between the original value of the property V at the start of the service life and the asset value V_a at the end of a years. Therefore,

$$\text{Total amount of depreciation after } a \text{ years} = V - V_a \tag{9}$$

$$V - V_a = R \frac{(1 + i)^a - 1}{i} \tag{10}$$

Combining Eqs. (8) and (10),

$$V - V_a = (V - V_s) \frac{(1 + i)^a - 1}{(1 + i)^n - 1} \tag{11}$$

Asset (or book) value after a years $= V_a$

$$V_a = V - (V - V_s) \frac{(1 + i)^a - 1}{(1 + i)^n - 1} \tag{12}†$$

Since the value of R represents the annual depreciation cost, the yearly cost for depreciation is constant when the sinking-fund method is used. As shown in Fig. 8-3, this method results in book values which are always greater than those obtained with the straight-line method. Because of the effects of interest in the sinking-fund method, the annual decrease in asset value of the property is less in the early-life years than in the later years.

Although the sinking-fund viewpoint assumes the existence of a fund into which regular deposits are made, an actual fund is seldom maintained. Instead, the money accumulated from the depreciation charges is put to work in other interests, and the existence of the hypothetical fund merely serves as a basis for this method of depreciation accounting.

The sinking-fund theory of cost accounting is now used by only a few concerns, although it has seen considerable service in the public-utilities field. Theoretically, the method would be applicable for depreciating any property that did not undergo heavy service demands during its early life and stood little chance of becoming obsolete or losing service value due to other functional causes.

The same approach used in the sinking-fund method may be applied by

---

† Exactly the same result for asset value after a years is obtained if an annuity due (i.e., equal periodic payments at beginning of each year) is used in place of an ordinary annuity. The periodic payment with an annuity due would be $R/(1 + i)$. In accepted engineering practice, the sinking-fund method is based on an ordinary annuity plan.

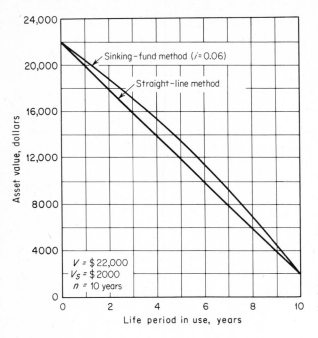

Figure 8-3 Asset values of property when depreciated by interest (sinking fund) and no-interest (straight-line) methods.

analyzing depreciation on the basis of reduction with time of future profits obtainable with a property. When this is done, it is necessary to use an interest rate equivalent to the annual rate of return expected from the use of the property. This method is known as the *present-worth method* and gives results similar to those obtained with the conventional sinking-fund approach. The sinking-fund and the present-worth methods are seldom used for depreciation cost accounting but are occasionally applied for purposes of comparing alternative investments.

ADDITIONAL FIRST-YEAR ALLOWANCE

For property assets purchased after 1957 with at least 6 years of estimated life, the Federal tax laws provide for an "*additional first-year allowance*" as an extra depreciation charge, generally in the year of purchase. This additional depreciation write-off for tax purposes can amount to as much as 20 percent of the investment. However, the total write-off for all assets cannot exceed $10,000 in any one year so that this tax saving would probably only be of any significance to small businesses or individuals where an increased tax write-off based on an additional $10,000 deduction might be of some importance. If the tax write-off as additional first-year allowance is taken, the regular depreciation can then be applied to the asset using as an investment basis the original property value minus the extra allowance.

SINGLE-UNIT AND GROUP DEPRECIATION

In depreciation accounting procedures, assets may be depreciated on the basis of individual units or on the basis of various types of property groups or classifications. The *single-unit method* requires keeping records on each individual asset. Although the application of this method is simple, the large number of detailed records required makes the accounting expenses very high.

To simplify the accounting procedures, many concerns combine their various assets into groups for depreciation purposes. There are several types of group accounts employed, and the most common among these are composite accounts, classified accounts, and vintage-group accounts.

A *composite account* includes all depreciable assets in one single group, and an overall depreciation rate is applied to the entire account. With this method, the composite depreciation rate must be redetermined when important changes occur in the relative distribution of the service lives of the individual assets.

Instead of including all assets in a single depreciation account, it is possible to classify properties into general types, such as machinery and equipment, office furniture and fixtures, buildings, and transportation equipment. The records for these groups are known as *classified accounts*. A classified account is similar to a composite account because many items are included in the same group, regardless of life characteristics.

Another approach to group depreciation is to include in each account all similar assets having approximately the same service lives. These accounts are known as *vintage-group accounts*. A separate record is kept for each group and the same depreciation rate is applied to all the items included in each account. With this method, the advantages of single-unit depreciation are obtained since life characteristics serve as the basis. If a large number of items are contained in a vintage-group account, the overall depreciation results can be quite accurate because the law of averages will apply to the true service lives as compared with the estimated service lives.

ADJUSTMENT OF DEPRECIATION ACCOUNTS

The estimated service life and salvage value of a property are seldom exactly equal to the actual service life and salvage value. It is, therefore, advisable to adjust depreciation accounts by making periodic reestimations of the important variables. When a property is retired under conditions which do not permit exact agreement between estimated and actual values, the difference between the book depreciation and the actual depreciation may be handled in one of the following ways: (1) The gain or loss may be credited or charged on the financial record for the current period; (2) the difference may be credited or charged to a special depreciation reserve; or (3) the difference may be carried on the books for amortization during a reasonable future period.

According to the Federal income-tax laws, any gain on the retirement of a property is taxed as a capital gain. However, losses cannot be subtracted from the taxable income unless the maximum expected life was used. Because of the losses involved when a property must be retired before the end of its estimated service life, some concerns prefer to use a combination of methods 2 and 3 indicated in the preceding paragraph. A special depreciation reserve is built up by continuing the book depreciation of properties whose actual service lives exceed the estimated service lives. This fund is then used to handle losses due to early retirement of assets. The final choice of method for adjusting depreciation accounts depends on the accounting policies of the individual concern and on income-tax regulations.

EVALUATION OF DEPRECIATION METHODS

Comparison of the various depreciation methods shows that the declining-balance and the sum-of-the-years-digits methods give similar results. In both cases, the depreciation costs are greater in the early-life years of the property than in the later years. Annual depreciation costs are constant when the straight-line, sinking-fund, or present-worth method is used. Because interest effects are included in the sinking-fund and present-worth methods, the annual decrease in asset value with these two methods is lower in the early-life years than in the later years. The straight-line method is widely used for depreciation cost accounting because it is very simple to apply, both to groups and single units, and it is acceptable for cost-accounting purposes and for income-tax determinations.

From the viewpoint of financial protection, it is desirable to make a greater charge for property depreciation during early life than during later life. This can be accomplished by use of the declining-balance or sum-of-the-years-digits method. The difficulties of accurate application to group accounts and income-tax restrictions have served to suppress the widespread usage of these methods. However, in recent years, a large number of industrial concerns have started using declining-balance and sum-of-the-years-digits depreciation, with many companies finding it desirable to use the combination method.

The liberalized tax laws passed in 1954 permitted use of double declining-balance depreciation as well as sum-of-the-years-digits depreciation for income-tax calculations. In general, these laws gave approval to any depreciation method which did not give faster write-offs during the first two-thirds of an asset's useful life than the double declining-balance method. These regulations are not applicable to assets with service lives of less than 3 years.

The final choice of the best depreciation method depends on a number of different factors. The type and function of the property involved is, of course, one important factor. Also, it is desirable to use a simple formula giving results as accurate as the estimated values involved. The advisability of keeping two separate sets of books, one for income-tax purposes and one for company purposes, should be considered. The final decision involves application of good judgment and an analysis of the existing circumstances.

NOMENCLATURE FOR CHAPTER 8

a = length of time in actual use, years
d = annual depreciation, \$/year
f = fixed-percentage or declining-balance factor
i = annual interest rate expressed as a fraction, percent/100
n = service life, years
R = uniform annual payments made in an ordinary annuity, dollars
V = original value of a property at start of service-life period, completely installed and ready for use, dollars
V_a = asset or book value, dollars
V_s = salvage value of property at end of service life, dollars

SUGGESTED ADDITIONAL REFERENCES FOR DEPRECIATION

Depreciation

Barish, N. N., and S. Kaplan, "Economic Analysis for Engineering and Managerial Decision Making," 2d ed., McGraw-Hill Book Company, New York, 1978.
Cason, R. L., Go Broke While Showing a Profit, *Hydrocarbon Process.*, **54**(7):179 (1975)
Holland, F. A., F. A. Watson, and J. K. Wilkinson, Capital Costs and Depreciation, *Chem. Eng.*, **80**(17):118 (1973).
——— ——— and ———, Profitability of Invested Capital, *Chem. Eng.*, **80**(19):139 (1973).
Jelen, F. C., Pitfalls in Profitability Analysis, *Hydrocarbon Process.*, **55**(1):111 (1976).
——— and M. S. Cole, Methods for Economic Analysis, *Hydrocarbon Process.*, **53**(7):133 (1974); **53**(9):227 (1974); **53**(10):161 (1974).
——— and C. L. Yaws, Project Cash Flow—Description and Interpretation, *Hydrocarbon Process.*, **57**(3):77 (1978).
Jenckes, L. C., How to Estimate Operating Costs and Depreciation, *Chem. Eng.*, **77**(27):168 (1970).
Nelson, W. L., Literature Depreciation Rates, *Oil & Gas J.*, **64**(23):66 (1971).
———, What Depreciation Means in the Estimation of Process Costs, *Oil & Gas J.*, **69**(33):90 (1971).
Mapstone, G. E., The Present Value of Plant Depreciation Allowances, *Chem. Proc. Eng.*, **52**(4):66 (1971).
Pfeiffer, R. R., Planned Equipment Replacement, *AACE Bull.*, **12**(2):35 (1970).
Piekarski, J. A., Fundamentals of Manufacturing Economics, *1973 Trans. Am. Assoc. Cost Eng.* (1973), p. 221.

PROBLEMS†

1 A reactor of special design is the major item of equipment in a small chemical plant. The initial cost of the completely installed reactor is \$60,000, and the salvage value at the end of the useful life is estimated to be \$10,000. Excluding depreciation costs for the reactor, the total annual expenses for the plant are \$100,000. How many years of useful life should be estimated for the reactor if 12 percent of the total annual expenses for the plant are due to the cost for reactor depreciation? The straight-line method for determining depreciation should be used.

† To simplify the application of the basic principles of depreciation in these problems, the so-called "additional first-year allowance" permitted for income-tax purposes should *not* be included.

2 The initial installed cost for a new piece of equipment is $10,000, and its scrap value at the end of its useful life is estimated to be $2000. The useful life is estimated to be 10 years. After the equipment has been in use for 4 years, it is sold for $7000. The company which originally owned the equipment employs the straight-line method for determining depreciation costs. If the company had used an alternative method for determining depreciation costs, the asset (or book) value for the piece of equipment at the end of 4 years would have been $5240. The total income-tax rate for the company is 48 percent of all gross earnings. Capital-gains taxes amount to 25 percent of the gain. How much net saving after taxes would the company have achieved by using the alternative (in this case, reducing-balance) depreciation method instead of the straight-line depreciation method?

3 A piece of equipment originally costing $40,000 was put into use 12 years ago. At the time the equipment was put into use, the service life was estimated to be 20 years and the salvage and scrap value at the end of the service life were assumed to be zero. On this basis, a straight-line depreciation fund was set up. The equipment can now be sold for $10,000, and a more advanced model can be installed for $55,000. Assuming the depreciation fund is available for use, how much new capital must be supplied to make the purchase?

4 The original investment for an asset was $10,000, and the asset was assumed to have a service life of 12 years with $2000 salvage value at the end of the service life. After the asset has been in use for 5 years, the remaining service life and final salvage value are reestimated at 10 years and $1000, respectively. Under these conditions, what is the depreciation cost during the sixth year of the total life if straight-line depreciation is used?

5 A property has an initial value of $50,000, service life of 20 years, and final salvage value of $4000. It has been proposed to depreciate the property by the text-book declining-balance method. Would this method be acceptable for income-tax purposes if the income-tax laws do not permit annual depreciation rates greater than twice the minimum annual rate with the straight-line method?

6 A piece of equipment having a negligible salvage and scrap value is estimated to have a service life of 10 years. The original cost of the equipment was $40,000. Determine the following:

   (*a*) The depreciation charge for the fifth year if double declining-balance depreciation is used.

   (*b*) The depreciation charge for the fifth year if sum-of-the-years-digits depreciation is used.

   (*c*) The percent of the original investment paid off in the first half of the service life using the double declining-balance method.

   (*d*) The percent of the original investment paid off in the first half of the service life using the sum-of-the-years-digits method.

7 The original cost of a property is $30,000, and it is depreciated by a 6 percent sinking-fund method. What is the annual depreciation charge if the book value of the property after 10 years is the same as if it had been depreciated at $2500/year by the straight-line method?

8 A concern has a total income of $1 million/year, and all expenses except depreciation amount to $600,000/year. At the start of the first year of the concern's operation, a composite account of all depreciable items shows a value of $850,000, and the overall service life is estimated to be 20 years. The total salvage value at the end of the service life is estimated to be $50,000. Forty-five percent of all profits before taxes must be paid out as income taxes. What would be the reduction in income-tax charges for the first year of operation if the sum-of-the-years-digits method were used for depreciation accounting instead of the straight-line method?

9 The total value of a new plant is $2 million. A certificate of necessity has been obtained permitting a write-off of 60 percent of the initial value in 5 years. The balance of the plant requires a write-off period of 15 years. Using the straight-line method and assuming negligible salvage and scrap value, determine the total depreciation cost during the first year.

10 A profit-producing property has an initial value of $50,000, a service life of 10 years, and zero salvage and scrap value. By how much would annual profits before taxes be increased if a 5 percent sinking-fund method were used to determine depreciation costs instead of the straight-line method?

11 In order to make it worthwhile to purchase a new piece of equipment, the annual depreciation costs for the equipment cannot exceed $3000 at any time. The original cost of the equipment is $30,000, and it has zero salvage and scrap value. Determine the length of service life necessary if the equipment is depreciated (*a*) by the sum-of-the-years-digits method, and (*b*) by the straight-line method.

12 The owner of a property is using the unit-of-production method for determining depreciation costs. The original value of the property is $55,000. It is estimated that this property can produce 5500 units before its value is reduced to zero; i.e., the depreciation cost per unit produced is $10. The property produces 100 units during the first year, and the production rate is doubled each year for the first 4 years. The production rate obtained in the fourth year is then held constant until the value of the property is paid off. What would have been the annual depreciation cost if the straight-line method based on time had been used?

NINE

PROFITABILITY, ALTERNATIVE INVESTMENTS, AND REPLACEMENTS

The word *profitability* is used as the general term for the measure of the amount of profit that can be obtained from a given situation. Profitability, therefore, is the common denominator for all business activities.

Before capital is invested in a project or enterprise, it is necessary to know how much profit can be obtained and whether or not it might be more advantageous to invest the capital in another form of enterprise. Thus, the determination and analysis of profits obtainable from the investment of capital and the choice of the best investment among various alternatives are major goals of an economic analysis.

There are many reasons why capital investments are made. Sometimes, the purpose is merely to supply a service which cannot possibly yield a monetary profit, such as the provision of recreation facilities for free use of employees. The profitability for this type of venture cannot be expressed on a positive numerical basis. The design engineer, however, usually deals with investments which are expected to yield a tangible profit.

Because profits and costs are considered which will occur in the future, the possibilities of inflation or deflation affecting future profits and costs must be recognized. The strategy for handling effects of inflation or deflation is discussed in Chap. 10.

Investments may be made for replacing or improving an existing property, for developing a completely new enterprise, or for other purposes wherein a profit is expected from the outlay of capital. For cases of this sort, it is extremely important to make a careful analysis of the capital utilization.

PROFITABILITY STANDARDS

In the process of making an investment decision, the profits anticipated from the investment of funds should be considered in terms of a minimum profitability standard.† This profitability standard, which can normally be expressed on a direct numerical basis, must be weighed against the overall judgment evaluation for the project in making the final decision as to whether or not the project should be undertaken.

The judgment evaluation must be based on the recognition that a quantified profitability standard can serve only as a guide. Thus, it must be recognized that the profit evaluation is based on a prediction of future results so that assumptions are necessarily included. Many intangible factors, such as future changes in demand or prices, possibility of operational failure, or premature obsolescence, cannot be quantitized. It is in areas of this type that judgment becomes critical in making a final investment decision.

A primary factor in the judgment decision is the consideration of possible alternatives. For example, the alternatives to continuing the operation of an existing plant may be to replace it with a more efficient plant, to discontinue the operation entirely, or to make modifications in the existing plant. In reaching the final decision, the alternatives should be considered two at a time on a mutually exclusive basis.

An obvious set of alternatives involves either making the capital investment in a project or investing the capital in a safe venture for which there is essentially no risk and a guaranteed return. In other words, the second alternative involves the company's decision as to the cost of capital.

Cost of Capital

Methods for including the cost of capital in economic analyses have been discussed in Chap. 6. Although the management and stockholders of each company must establish the company's characteristic cost of capital, the simplest approach is to assume that investment of capital is made at a hypothetical cost or rate of return equivalent to the total profit or rate of return over the full expected life of the particular project. This method has the advantage of putting the profitability analysis of all alternative investments on an equal basis, thereby permitting a clear comparison of risk factors. This method is particularly useful for preliminary estimates, but it may need to be refined further to take care of income-tax effects for final evaluation.

BASES FOR EVALUATING PROJECT PROFITABILITY

Total profit alone cannot be used as the deciding profitability factor in determining if an investment should be made. The profit goal of a company is to maxi-

---

† One often-used basis for the minimum profitability standard is the value of money to the company, expressed as a rate, based on earnings after taxes.

mize income above the cost of the capital which must be invested to generate the income. If the goal were merely to maximize profits, any investment would be accepted which would give a profit, no matter how low the return or how great the cost. For example, suppose that two equally sound investments can be made. One of these requires $100,000 of capital and will yield a profit of $10,000/year, and the second requires $1 million of capital and will yield $25,000/year. The second investment gives a greater yearly profit than the first, but the annual *rate of return* on the second investment is only

$$(\$25,000/\$1,000,000) \times (100) = 2.5 \text{ percent}$$

while the annual rate of return on the $100,000 investment is 10 percent. Because reliable bonds and other conservative investments will yield annual rates of return in the range of 6 to 9 percent, the $1 million investment in this example would not be very attractive; however, the 10 percent return on the $100,000 capital would make this investment worthy of careful consideration. Thus, for this example, the rate of return, rather than the total amount of profit, is the important profitability factor in determining if the investment should be made.

The basic aim of a profitability analysis is to give a measure of the attractiveness of the project for comparison to other possible courses of action. It is, therefore, very important to consider the exact purpose of a profitability analysis before the standard reference or base case is chosen. If the purpose is merely to present the total profitability of a given project, a simple statement of total profit per year or annual rate of return may be satisfactory. On the other hand, if the purpose is to permit comparison of several different projects in which capital might be invested, the method of analysis should be such that all cases are on the same basis so that direct comparison can be made among the appropriate alternatives.

Mathematical Methods for Profitability Evaluation

The most commonly used methods for profitability evaluation, as illustrated in Fig. 9-1, fall under the following headings:

1. Rate of return on investment
2. Discounted cash flow based on full-life performance
3. Net present worth
4. Capitalized costs
5. Payout period

Each of these methods has its advantages and disadvantages, and much has been written on the virtues of the various methods. Because no single method is best for all cases, the engineer should understand the basic ideas involved in each method and be able to choose the one best suited to the needs of the particular situation.

Rate of return on investment In engineering economic studies, rate of return on investment is ordinarily expressed on an annual percentage basis. The yearly

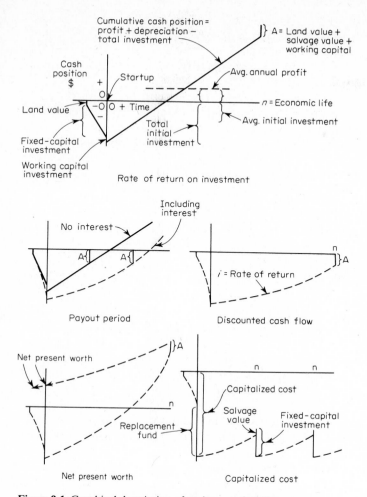

Rate of return on investment

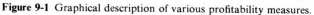

Figure 9-1 Graphical description of various profitability measures.

profit divided by the total initial investment necessary represents the fractional return, and this fraction times 100 is the standard *percent return on investment*.††

† The normal procedure is to base the percent return on investment on the total initial investment. However, because equipment depreciates during its useful life, it is sometimes convenient to base the rate of return on the average estimated investment during the life of the project. With this method, the rate of return is determined by dividing the average annual profit or saving by one-half the initial fixed-capital investment (or initial fixed-capital investment minus the estimated salvage value at the end of the useful life) plus the working-capital investment.

‡ An article by J. Linsley, Return on Investment: Discounted and Undiscounted, *Chem. Eng.*, **86**(11): 201 (May 21, 1979), suggests that "return on investment" can be defined as net, after-tax profit plus depreciation divided by capital investment. This definition of return on investment where depreciation cash flow is included as part of the return is not used in this book. Instead, this method of handling cash flow is included in the profitability methods reported for discounted-cash-flow Profitability Index and Net Present Worth.

Profit is defined as the difference between income and expense. Therefore, profit is a function of the quantity of goods or services produced and the selling price. The amount of profit is also affected by the economic efficiency of the operation, and increased profits can be obtained by use of effective methods which reduce operating expenses.

To obtain reliable estimates of investment returns, it is necessary to make accurate predictions of profits and the required investment. To determine the profit, estimates must be made of direct production costs, fixed charges including depreciation, plant overhead costs, and general expenses. Profits may be expressed on a before-tax or after-tax basis, but the conditions should be indicated. Both working capital and fixed capital should be considered in determining the total investment.†

Returns Incorporating Minimum Profits as an Expense

The standard method for reporting rate of return on investment has been outlined in the preceding paragraphs. Another method which is sometimes used for reporting rate of return is based on the assumption that it must be possible to obtain a certain minimum profit or return from an investment before the necessary capital outlay will be desirable. This minimum profit is included as a fictitious expense along with the other standard expenses. When return on investment is determined in this manner, the result shows the *risk earning rate*, and it represents the return over and above that necessary to make the capital expenditure advisable. If the return is zero or larger, the investment will be attractive. This method is sometimes designated as *return based on capital recovery with minimum profit*.

The inclusion of minimum profit as an expense is rather unrealistic, especially when it is impossible to designate the exact return which would make a given investment worthwhile. One difficulty with this method is the tendency to use a minimum rate of return equal to that obtained from present investments. This, of course, gives no consideration to the element of risk involved in a new venture. Despite these objections, the use of returns incorporating minimum profits as an expense is acceptable providing the base, or minimum, return and the general method employed are clearly indicated.

Example 1. Determination of rate of return on investment—consideration of income-tax effects A proposed manufacturing plant requires an initial fixed-capital investment of $900,000 and $100,000 of working capital. It is estimated that the annual income will be $800,000 and the annual expenses including depreciation will be $520,000 before income taxes. A minimum

---

† Under some conditions, such as a profitability analysis based on a small component of an overall operation, the return on investment can be based on the fixed-capital investment instead of the total investment.

annual return of 15 percent before income taxes is required before the investment will be worthwhile. Income taxes amount to 48 percent of all pre-tax
profits.

Determine the following:

(a) The annual percent return on the total initial investment before income
taxes.
(b) The annual percent return on the total initial investment after income
taxes.
(c) The annual percent return on the total initial investment before income
taxes based on capital recovery with minimum profit.
(d) The annual percent return on the average investment before income
taxes assuming straight-line depreciation and zero salvage value.

SOLUTION

(a) Annual profit before income taxes = $800,000 − $520,000 = $280,000.
Annual percent return on the total initial investment before income
taxes = [280,000/(900,000 + 100,000)](100) = 28 percent.
(b) Annual profit after income taxes = ($280,000)(0.52) = $145,600.
Annual percent return on the total initial investment after income
taxes = [145,600/(900,000 + 100,000)](100) = 14.6 percent.
(c) Minimum profit required per year before income taxes =
($900,000 + $100,000)(0.15) = $150,000.
Fictitious expenses based on capital recovery with minimum profit =
$520,000 + $150,000 = $670,000/year. Annual percent return on the total
investment based on capital recovery with minimum annual rate of
return of 15 percent before income taxes = [(800,000 − 670,000)/(900,000
+ 100,000)](100) = 13 percent.
(d) Average investment assuming straight-line depreciation and zero salvage
value = $900,000/2 + $100,000 = $550,000.
Annual percent return on average investment before income taxes =
(280,000/550,000)(100) = 51 percent.

The methods for determining rate of return, as presented in the preceding
sections, give "point values" which are either applicable for one particular year
or for some sort of "average" year. They do not consider the time value of
money, and they do not account for the fact that profits and costs may vary
significantly over the life of the project.

One example of a cost that can vary during the life of a project is depreciation cost. If straight-line depreciation is used, this cost will remain constant;
however, it may be advantageous to employ a declining-balance or sum-of-the-
years-digits method to determine depreciation costs, which will immediately
result in variations in costs and profits from one year to another. Other predict-
able factors, such as increasing maintenance costs or changing sales volume, may
also make it necessary to estimate year-by-year profits with variation during the
life of the project. For these situations, analyses of project profitability cannot be

made on the basis of one point on a flat time-versus-earning curve, and profitability analyses based on discounted cash flow may be appropriate. Similarly, time-value-of-money considerations may make the discounted-cash-flow approach desirable when annual profits are constant.

DISCOUNTED CASH FLOW

Rate of Return Based on Discounted Cash Flow†

The method of approach for a profitability evaluation by discounted cash flow takes into account the time value of money and is based on the amount of the investment that is unreturned at the end of each year during the estimated life of the project. A trial-and-error procedure is used to establish a rate of return which can be applied to yearly cash flow so that the original investment is reduced to zero (or to salvage and land value plus working-capital investment) during the project life. Thus, the rate of return by this method is equivalent to the maximum interest rate (normally, after taxes) at which money could be borrowed to finance the project under conditions where the net cash flow to the project over its life would be just sufficient to pay all principal and interest accumulated on the outstanding principal.

To illustrate the basic principles involved in discounted-cash-flow calculations and the meaning of rate of return based on discounted cash flow, consider the case of a proposed project for which the following data apply:

Initial fixed-capital investment = $100,000
Working-capital investment = $10,000
Service life = 5 years
Salvage value at end of service life = $10,000

Year	Predicted after-tax cash flow to project based on total income minus all costs except depreciation, $ (expressed as end-of-year situation)
0	(110,000)
1	30,000
2	31,000
3	36,000
4	40,000
5	43,000

† Common names of methods of return calculations related to the discounted-cash-flow approach are *profitability index, interest rate of return, true rate of return,* and *investor's rate of return.*

Designate the discounted-cash-flow rate of return as i. This rate of return represents the after-tax interest rate at which the investment is repaid by proceeds from the project. It is also the maximum after-tax interest rate at which funds could be borrowed for the investment and just break even at the end of the service life.

At the end of five years, the cash flow to the project, compounded on the basis of end-of-year income, will be

$$(\$30{,}000)(1 + i)^4 + (\$31{,}000)(1 + i)^3 + (\$36{,}000)(1 + i)^2$$
$$+ (\$40{,}000)(1 + i) + \$43{,}000 = S \qquad (1)$$

The symbol S represents the future worth of the proceeds to the project and must just equal the future worth of the initial investment compounded at an interest rate i corrected for salvage value and working capital. Thus,

$$S = (\$110{,}000)(1 + i)^5 - \$10{,}000 - \$10{,}000 \qquad (2)$$

Setting Eq. (1) equal to Eq. (2) and solving by trial and error for i gives $i = 0.207$, or the discounted-cash-flow rate of return is 20.7 percent.

Table 1 Computation of discounted-cash-flow rate of return

Year (n')	Estimated cash flow to project, $	Trial for $i = 0.15$		Trial for $i = 0.20$		Trial for $i = 0.25$		Trial for $i = 0.207$†	
		Discount factor, $\dfrac{1}{(1 + i)^{n'}}$	Present value, $	Discount factor, $\dfrac{1}{(1 + i)^{n'}}$	Present value, $	Discount factor, $\dfrac{1}{(1 + i)^{n'}}$	Present value, $	Discount factor, $\dfrac{1}{(1 + i)^{n'}}$	Present value, $
0	(110,000)								
1	30,000	0.8696	26,100	0.8333	25,000	0.8000	24,000	0.829	24,900
2	31,000	0.7561	23,400	0.6944	21,500	0.6400	19,800	0.687	21,200
3	36,000	0.6575	23,300	0.5787	20,700	0.5120	18,400	0.570	20,500
4	40,000	0.5718	22,900	0.4823	19,300	0.4096	16,400	0.472	18,800
5	{ 43,000 { +20,000	0.4971	31,300	0.4019	25,300	0.3277	20,600	0.391	24,600
	Total		127,000		111,800		99,200		110,000
Ratio = $\dfrac{\text{total present value}}{\text{initial investment}}$			1.155		1.016		0.902		1.000
									Trial is satisfactory

† As illustrated in Fig. 9-2, interpolation to determine the correct rate of return can be accomplished by plotting the ratio (total present value/initial investment) versus the trial interest rate for three bracketing values and reading the correct rate from the curve where the ratio = 1.0.

NOTE: *In this example, interest was compounded annually on an end-of-year basis and continuous interest compounding was ignored. Also, construction period and land value were not considered. The preceding effects could have been included in the analysis for a more sophisticated treatment using the methods presented in Examples 2 and 3 of this chapter.*

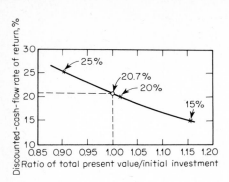

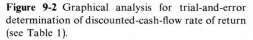

Figure 9-2 Graphical analysis for trial-and-error determination of discounted-cash-flow rate of return (see Table 1).

Some of the tedious and time-consuming calculations can be eliminated by applying a *discount factor* to the annual cash flows and summing to get a present value equal to the required investment. The discount factor for end-of-year payments and annual compounding is

$$d_{n'} = \frac{1}{(1 + i)^{n'}} = \text{discount factor} \qquad (3)$$

where i = rate of return

n' = year of project life to which cash flow applies

This discount factor, $d_{n'}$, is the amount that would yield one dollar after n' years if invested at an interest rate of i. The discounted-cash-flow rate of return can be determined by the trial-and-error method illustrated in Table 1, where the annual cash flows are discounted by the appropriate discount factor to a total present value equal to the necessary initial investment.†

Example 2. Discounted-cash-flow calculations based on continuous interest compounding and continuous cash flow

Using the discount factors for continuous interest and continuous cash flow presented in Tables 5 to 8 of Chapter 6, determine the continuous discounted-cash-flow rate of return r for the example presented in the preceding section where yearly cash flow is continuous. The data follow.

Initial fixed-capital investment = $100,000
Working-capital investment = $10,000
Service life = 5 years
Salvage value at end of service life = $10,000

---

† The significance of the use of discount factors, as illustrated in Table 1 and Example 2, can be seen by dividing both sides of Eq. (1) and Eq. (2) by $(1 + i)^5$, or by $(1 + i)^n$ for the general case where n is the estimated service life in years.

Year	Predicted after-tax cash flow to project based on total income minus all costs except depreciation with cash flow occurring continuously, $ (total of continuous cash flow for year indicated)
1	30,000
2	31,000
3	36,000
4	40,000
5	43,000

SOLUTION The following tabulation shows the final result of the trial-and-error solution using the factors F_a and F_b from Tables 5 and 6 in Chap. 6:

Year	Estimated continuous cash flow to project, $	Trial for $r = 0.225$		Present value, $
		Discount factor		
		F_b (from Table 6, Chap. 6)	F_a (from Table 5, Chap. 6)	
0	(110,000) In an instant			
0–1	30,000	0.8954		26,850
1–2	31,000	0.7151		22,200
2–3	36,000	0.5710		20,550
3–4	40,000	0.4560		18,250
4–5	43,000	0.3648		15,650
5	+20,000 In an instant		0.3246	6,500
				Total 110,000

Trial is satisfactory

Because the assumed trial value of $r = 0.225$ discounted all the cash flows to the present worth of $110,000, the continuous interest rate of 22.5 percent represents the discounted-cash-flow rate of return for this example which can be compared to the value of 20.7 percent shown in Table 1 for the case of discrete interest compounding and instantaneous cash flow.

NET PRESENT WORTH

In the preceding treatment of discounted cash flow, the procedure has involved the determination of an index or interest rate which discounts the annual cash flows to a zero present value when properly compared to the initial investment. This index gives the rate of return which includes the profit on the project, payoff of the investment, and normal interest on the investment. A related approach, known as the method of *net present worth* (or *net present value* or *venture worth*), substitutes the cost of capital at an interest rate i for the discounted-cash-flow rate of return. The cost of capital can be taken as the average rate of return the company earns on its capital, or it can be designated as the minimum acceptable return for the project. The net present worth of the project is then the difference between the present value of the annual cash flows and the initial required investment.

To illustrate the method for determining net present worth, consider the example presented in Table 1 for the case where the value of capital to the company is at an interest rate of 15 percent. Under these conditions, the present value of the cash flows is $127,000 and the initial investment is $110,000. Thus, the net present worth of the project is

$$\$127,000 - \$110,000 = \$17,000$$

Work Sheet for Calculating Present Value and Net Present Worth

An example of a work sheet that can be used for handling discounted-cash-flow presentations to determine present value and net present worth is given in Table 2. The definitions as given in lines 17 and 18 of this table clearly show the preferred distinction between the terms *net present worth* and *present value* as used in this text. The table is particularly useful because it makes certain the user handles depreciation cash flow correctly by subtracting depreciation costs to determine tax costs (see lines 10 and 11) and including depreciation cash flow to determine the annual cash income (see lines 9 and 12). Line 15 shows four values of discount factors for 15 percent interest based on (a) continuous uniform cash flow and continuous interest compounding, (b) continuous uniform cash flow and finite (year-end) interest compounding, (c) finite (year-end) cash flow and continuous interest compounding, and (d) finite (year-end) cash flow and finite (year-end) interest compounding.

Lines 1, 2, and 3 (investments) in Table 2 would normally only be filled in for the first column (discount factor of 1.000) which is designated as the zero year for the operation, with the unit actually going into operation at the start of the so-called first year. It is assumed that working capital and salvage value will be recovered in a lump sum at the end of the estimated service life, so these values are listed on lines 1, 2, and 14 as positive (incoming funds) numbers in the end-of-life column. Since these are lump-sum instantaneous values, the discount factor to apply to them is the finite (year-end) cash flow factor as shown in line 15 in the end-of-life column.

The investment-tax-credit value shown in line 13 of Table 2 would ordinarily apply for the first year of operation only. Line 14 gives the annual cash flows for each of the operating years with the zero-year column giving only the total capital investment. In line 17, the present value of the annual cash flows to the project is obtained by summing the individual present values for each year of operation including the present value of the working-capital and salvage-value recovery at the end of the service life. Line 18 merely applies the definition of net present worth as used in this text as the difference between the total present value of the annual cash flows to the project and the initial required investment.

Notes should be included with the table to explain the basis for special factors used, such as escalation factors, startup costs, and depreciation method. The notes can also be used to explain the methods used for estimating the various items as, for example, note 5 in Table 2 to show the methods used for estimating lines 5, 6, and 7.

The format shown in Table 2 is intended as an example, and a real case would undoubtedly include more columns to represent a life of more than five years. Similarly, capital is normally spent during the period of one or two years before operations begin and sales are made. Thus, the factors in the zero-year column could be changed to values other than 1.000 using methods presented in Chap. 6 (Interest and Investment Costs) as illustrated in Example 3 of this chapter.

CAPITALIZED COSTS†

The *capitalized-cost* profitability concept is useful for comparing alternatives which exist as possible investment choices within a single overall project. For example, if a decision based on profitability analysis were to be made as to whether stainless steel or mild steel should be used in a chemical reactor as one part of a chemical plant, capitalized-cost comparison would be a useful and appropriate approach. This particular case is illustrated in Example 9 of Chap. 6.

Capitalized cost related to investment represents the amount of money that must be available initially to purchase the equipment and simultaneously provide sufficient funds for interest accumulation to permit perpetual replacement of the equipment. If only one portion of an overall process to accomplish a set objective is involved and operating costs do not vary, then the alternative giving the least capitalized cost would be the desirable economic choice.

The basic equation for capitalized cost for equipment was developed in Chap. 6 as Eq. (28), which can be written as follows:

$$K = C_V + \frac{C_R}{(1 + i)^n - 1} = \frac{C_R(1 + i)^n}{(1 + i)^n - 1} + V_s \tag{4}$$

---

† For an analysis of the meaning of capitalized costs, development of related equations, and references, see Chap. 6 (Interest and Investment Costs).

Table 2 Work sheet for presenting discounted-cash flow, present-value, and net-present-worth determinations

Project Title: _____

Notes:
1. Dollar values can be in thousands of dollars and rounded to the nearest $1,000.
2. For line 13, a credit of 10% of line 1 is allowed for the first year by Federal income-tax regulations in effect in 1979.
3. For lines 11 and 15, company policies will dictate which tax rate, interest, and discount factors to use.
4. The estimated service life for this example is taken as 5 years.
5. For lines 5, 6, and 7, see Table 27 of Chapter 5 for estimating information and basis.

Line	Item Numbers in ( ) designate line	1978 0	1979 1st	1980 2nd	1981 3rd	1982 4th	1983 5th	End-of-life working capital and salvage value
1.	Fixed-capital investment							
2.	Working capital							
3.	Total capital investment (1 + 2)							
4.	Annual income (sales)							
5.	Annual manufacturing cost							
	(a) Raw materials							
	(b) Labor							
	(c) Utilities							
	(d) Maintenance and repairs							
	(e) Operating supplies							
	(f) Laboratory charges							
	(g) Patents and royalties							
	(h) Local taxes and insurance							
	(i) Plant overhead							
	(j) Other (explain in Notes)							
5-T.	Total of line 5							

314

No.	Item							(see heading above)
6.	Annual general expenses							
	(a) Administrative							
	(b) Distribution and selling							
	(c) Research and development							
	(d) Interest							
	(e) Other (explain in Notes)							
6-T.	Total of line 6							
7.	Total product cost (5-T + 6-T)							
8.	Annual operating income (4 − 7)							
9.	Annual depreciation							
10.	Income before tax (8 − 9)							
11.	Income after 46% tax (0.54 × 10)							
12.	Annual cash income (9 + 11)							
13.	Investment tax credit (0.1 × 1) (first year only)							
14.	Annual cash flow (3 + 12 + 13)							
15.	Discount factors for 15% interest							
	(a) See footnote †	1.000	0.929	0.799	0.688	0.592	0.510	0.472
	(b) See footnote ‡	1.000	0.933	0.812	0.706	0.614	0.534	0.497
	(c) See footnote §	1.000	0.861	0.741	0.638	0.549	0.472	0.472
	(d) See footnote ¶	1.000	0.870	0.756	0.658	0.572	0.497	0.497
16.	Annual present value (14 × 15)							

17. TOTAL present value of annual cash flows (sum of line 16 is *not* including 0 year)
= _____ in dollars or thousands of dollars

18. Net present worth = total present value of annual cash flows − total capital investment = line 17 − line 3 = _____ in dollars or thousands of dollars

† Continuous uniform cash flow and continuous nominal interest (r) of 15%.
‡ Continuous uniform cash flow and finite effective interest (i) of 15%.
§ Finite (year-end) cash flow and continuous nominal interest (r) of 15%.
¶ Finite (year-end) cash flow and finite effective interest (i) of 15%.

where $\quad K$ = capitalized cost
$\quad\quad C_V$ = original cost of equipment
$\quad\quad C_R$ = replacement cost
$\quad\quad V_s$ = salvage value at end of estimated useful life
$\quad\quad n$ = estimated useful life of equipment
$\quad\quad i$ = interest rate

$\dfrac{(1 + i)^n}{(1 + i)^n - 1}$ = capitalized-cost factor

Inclusion of Operating Costs in Capitalized-Costs Profitability Evaluation

The capitalized-costs concept can be extended to include operating costs by adding an additional capitalized cost to cover operating costs during the life of the project. Each annual operating cost is considered as equivalent to a necessary piece of equipment that will last one year.†

The procedure is to determine the present or discounted value of each year's cost by the method illustrated in Table 1. The sum of these present values is then capitalized by multiplying by the capitalized-cost factor given with Eq. (4). The total capitalized cost is the sum of the capitalized cost for the initial investment and that for the operating costs plus the working capital. This procedure is illustrated as part of Example 5 in this chapter.

PAYOUT PERIOD

Payout period, or *payout time*,‡ is defined as the minimum length of time theoretically necessary to recover the original capital investment in the form of cash flow to the project based on total income minus all costs except depreciation. Generally, for this method, original capital investment means only the original, depreciable, fixed-capital investment, and interest effects are neglected. Thus,

$$\frac{\text{Payout period in years}}{\text{(no interest charge)}} = \frac{\text{depreciable fixed-capital investment}}{\text{avg profit/yr} + \text{avg depreciation/yr}} \quad (5)$$

Another approach to payout period takes the time value of money into consideration and is designated as *payout period including interest*. With this method, an appropriate interest rate is chosen representing the minimum acceptable rate of return. The annual cash flows to the project during the estimated life

---

† If annual operating cost is constant and the cost is considered as an end-of-year cost, the capitalized cost for operation is equal to the annual operating cost divided by i. Continuous interest compounding can be used to resolve the problem of whether an operating cost is an end-of-year or start-of-year cost. The effects of depreciation methods and taxes may be very important when capitalized costs are used to compare design alternatives involving operating costs.

‡ Other equivalent names are *payback period, payback time, payoff period, payoff time,* and *cash-recovery period*

are discounted at the designated interest rate to permit computation of an average annual figure for profit plus depreciation which reflects the time value of money.† The time to recover the fixed-capital investment plus compounded interest on the total capital investment during the estimated life by means of the average annual cash flow is the payout period including interest, or

Payout period including interest =

$$\frac{\begin{array}{c}\text{depreciable fixed-} \\ \text{capital investment}\end{array} + \begin{array}{c}\text{interest on total capital} \\ \text{investment during} \\ \text{estimated service life}\end{array}}{(\text{avg profit/yr} + \text{avg depreciation/yr})_{\text{as constant annuity}}} \qquad (6)$$

This method tends to increase the payout period above that found with no interest charge and reflects advantages for projects that earn most of their profits during the early years of the service life.

USE OF CONTINUOUS INTEREST COMPOUNDING

In the preceding presentation of methods for profitability evaluation, where interest was considered, it was generally treated as finite-period interest compounded annually. By use of the relationships developed in Chap. 6 (Interest and Investment Costs), it is a simple matter to convert to the case of continuous interest compounding in place of finite interest compounding.

For example, the discount factor $d_{n'} = 1/(1 + i)^{n'}$, given as Eq. (3), becomes

$$d_{n'} = \frac{1}{e^{rn'}} \qquad (7)$$

for the case of continuous interest compounding with r representing the nominal continuous interest. The preceding equation follows directly from Eq. (18) of Chap. 6.

The application of continuous interest compounding, along with a method of profitability evaluation which includes construction costs and other prestart-up costs, is illustrated in the following example.

Example 3. Determination of profitability index with continuous interest compounding and prestartup costs Determine the discounted-cash-flow rate of return (i.e., the *profitability index*) for the overall plant project described in the following, and present a plot of cash position versus time to illustrate the solution.

One year prior to startup of the plant, the necessary land is purchased at a cost of $200,000.

---

† This discounting procedure is similar to that illustrated in the footnote to part (*a*) of Example 5 in this chapter. Continuous interest tables, such as Tables 5 to 8 in Chap. 6, can also be used.

During the year prior to startup, the plant is under construction with money for the construction and related activities flowing out uniformly during the entire year starting at zero dollars and totaling $600,000 for the year.

A working-capital investment of $200,000 is needed at the time the plant starts operation and must be retained indefinitely.

Salvage value for the plant at the end of the estimated useful life is $100,000.

The estimated useful life is 10 years.

Estimations of operating costs, income, and taxes indicate that the annual cash flow to the project (i.e., net profit plus depreciation per year) will be $310,000 flowing uniformly throughout the estimated life. This is an after-tax figure.

The concept of continuous interest compounding and continuous cash flow will be used. Neglect any effects due to inflation or deflation.

SOLUTION The procedure for this problem is similar to that illustrated in Table 1 in that a trial-and-error method is used with various interest rates until that rate is found which decreases the net cash position to zero at the end of the useful life. Let r represent the profitability index or discounted-cash-flow rate of return with continuous cash flow and continuous interest compounding.

1. Determination of cash position at zero time (i.e., at time of plant startup) in terms of unknown profitability index r

Land value: The in-an-instant value of the land is $200,000 one year before the zero reference point of plant startup time. The land value at zero time, therefore, is the future worth of this $200,000 after one year with continuous interest compounding. Thus, by Eq. (36) of Chap. 6 or part (e) of Table 3 in Chap. 6,

$$\text{Compounded land value at zero time} = \$200,000(e^r)$$

Construction cost: The total construction cost of the plant during the one year prior to startup is $600,000 occurring uniformly during the year. The compounded construction cost at zero time, therefore, is the future worth of this $600,000 after one year flowing uniformly throughout the year with continuous compounding. Thus, by Eq. (37) of Chap. 6 or part (f) of Table 3 in Chap. 6,

$$\text{Compounded construction cost at zero time} = \$600,000 \frac{e^r - 1}{r}$$

Working-capital investment: The working-capital investment of $200,000 must be supplied at the time of plant startup or at the reference point of zero time.

Summary of cash position at zero time:

Total cash position at zero time

$$= CP_{\text{zero time}} = \$200,\!000(e^r) + \$600,\!000\,\frac{e^r - 1}{r} + \$200,\!000$$

2. Determination of cash position at end of estimated useful life (i.e., ten years from zero time) in terms of profitability index r

At the end of the useful life with the correct value of r, the total cash position, taking into account the working-capital investment, the salvage value, and the land value, must be zero.

After plant startup, the annual cash flow to the project (i.e., net profit plus depreciation) is \$310,000 flowing continuously and uniformly, and this annual figure is constant throughout the estimated useful life.

The following procedure for evaluating the total cash position at the end of the estimated useful life is analogous to the procedure used in establishing Eqs. (1), (2), (3), and (7) of this chapter.

At the end of each year, the compounded cash flow to the project, with continuous uniform flow and continuous compounding, gives, by Eq. (37) of Chap. 6 or part (f) of Table 3 in Chap. 6, a future worth ($S_{\text{each year}}$) of†

$$S_{\text{each year}} = \$310,\!000\,\frac{e^r - 1}{r} \tag{A}$$

At the end of 10 years, the total future worth (S) of the cash flow to the project, by Eq. (36) of Chap. 6 or part (e) of Table 3 in Chap. 6, becomes

$$S = (\$310,\!000)\frac{e^r - 1}{r}\,(e^{9r} + e^{8r} + e^{7r} + \cdots + e^r + 1) \tag{B}$$

The future worth of the total cash flow to the project after 10 years must be equal to the future worth of the total cash position at zero time ($CP_{\text{zero time}}$) compounded continuously for 10 years minus salvage value, land value, and working-capital investment. Therefore, by Eq. (36) of Chap. 6 or part (e) of Table 3 in Chap. 6,

$$S = (CP_{\text{zero time}})(e^{10r}) - \$100,\!000 - \$200,\!000 - \$200,\!000 \tag{C}$$

---

† The concept of continuous and uniform cash flow with continuous interest compounding is obviously an assumption which is reasonable for some cash flow, such as costs for raw material and labor. However, it is clear that some major portions of the cash flow may not approximate continuous flow. For this reason, the annual cash flow is often estimated as an end-of-year figure, and the interest factor in Eq. (A) is eliminated.

3. Determination of profitability index r

Equating Eq. (B) to Eq. (C) gives the following result with r as the only unknown, and a trial-and-error solution will give the profitability index r.

$$(\$310,000)\frac{e^r - 1}{r}(e^{9r} + e^{8r} + e^{7r} + \cdots + e^r + 1) - (CP_{\text{zero time}})(e^{10r})$$
$$+ \$100,000 + \$200,000 + \$200,000 = 0 \qquad (D)$$

The trial-and-error approach can be simplified by dividing Eq. (D) by e^{10r} and substituting the expression for $CP_{\text{zero time}}$ to give the present-value or discounted-cash-flow equation as follows:

$$(\$310,000)\frac{e^r - 1}{r}\sum_{n'=1}^{n'=10}\frac{1}{e^{rn'}} - \$200,000(e^r) - \$600,000\frac{e^r - 1}{r}$$
$$- \$200,000 + (\$100,000 + \$200,000 + \$200,000)\frac{1}{e^{10r}} = 0 \qquad (E)$$

where $1/e^{rn'}$ represents the discount factor for continuous cash flow and continuous compounding as given in Eq. (7) of this chapter.

Because the compounded annual flow to the project is constant for each year at $(\$310,000)[(e^r - 1)/r]$, the year-by-year use of the discount factor, as illustrated in Table 1 of this chapter, can be replaced by a one-step process wherein the equivalent present value is determined from Eq. (25) of Chap. 6, $P = \bar{R}[(e^{rn} - 1)/re^{rn}]$, with $\bar{R} = \$310,000$, so that

$$\$310,000\frac{e^r - 1}{r}\sum_{n'=1}^{n'=10}\frac{1}{e^{rn'}} = 310,000\frac{e^{rn} - 1}{re^{rn}}$$

Similarly, in Eq. (D), the future-worth expression for the cash flow, $\$310,000[(e^r - 1)/r](e^{9r} + e^{8r} + e^{7r} + \cdots + e^r + 1)$, can be replaced by $\$310,000[(e^{rn} - 1)/r]$, as shown by Eq. (23) of Chap. 6.

With these simplifications, either Eq. (D) or (E) can be used for the trial-and-error solution for r. Table 3 shows the method of solution using the present-value Eq. (E) to give the correct value of $r = 0.26$.

Thus, the profitability index or discounted-cash-flow rate of return for this example is 26%.

4. Graphical representation of problem solution

Equation (D) can be generalized, with the simplifications indicated in the preceding section of this problem, to give the following (note that direct land value, salvage value, and working-capital investment are now included in the cash position):

Table 3 Computation of profitability index for example 3†

$n = 10$ years

Basis: Eq. (E) and zero (present-value) time at plant startup

Trial for	$r = 0.20$	$r = 0.25$	$r = 0.40$	$r = 0.26$‡
a. Present value of cash flow to project	$\dfrac{e^r - 1}{r} = 1.107$	1.136	1.230	1.142
($310,000) $\dfrac{e^r - 1}{r} \sum\limits_{n'=1}^{n'=10} \dfrac{1}{e^{rn'}}$	$\dfrac{e^{rn} - 1}{re^{rn}} = 4.31$	3.68	2.45	3.565
or				
($310,000 $\dfrac{e^{rn} - 1}{re^{rn}}$	$1,335,000	$1,140,000	$760,000	$1,107,000
b. Present value of land ($200,000) e^r	$e^r = 1.221$ $244,000	1.284 $257,000	1.492 $298,000	1.297 $259,000
c. Present value of construction cost	$\dfrac{e^r - 1}{r} = 1.107$	1.136	1.230	1.142
($600,000) $\dfrac{e^r - 1}{r}$	$664,000	$682,000	$738,000	$685,000
d. Present value of working-capital investment	$200,000	$200,000	$200,000	$200,000
e. Present value of terminal land, working capital, and salvage value based on interest compounded continuously for n years	$\dfrac{1}{e^{nr}} = 0.135$	0.0822	0.0183	0.744
($500,000) $\dfrac{1}{e^{nr}}$	$68,000	$41,000	$9,000	$37,000
f. Total of all present values with interest of r. Should be zero at correct value of r. $f = a - b - c - d + e$	$295,000 Rate too low	$42,000 Rate too low	-$467,000 Rate too high	$0 Trial is satisfactory

† Graphical methods, special tables for particular cases, MAPI worksheets and terminology, computer solutions, and rules of thumb are available to simplify the type of calculations illustrated in this table. For example, see G. A. Taylor, "Managerial and Engineering Economy: Economic Decision-Making, 2d ed., D. Van Nostrand Company, Inc., Princeton, New Jersey, 1975.

‡ See Fig. 9–2 for example of graphical interpolation procedure.

Cash position at time n

$$= \text{(annual constant cash flow to project)} \ \frac{e^{rn} - 1}{r}$$

$$- \text{(land value)} e^{r(n+Z)} - \text{(construction cost)} \frac{e^{rY} - 1}{Yr} e^{nr}$$

$$- \text{(working-capital investment)}(e^{nr}) \hspace{2cm} (F)$$

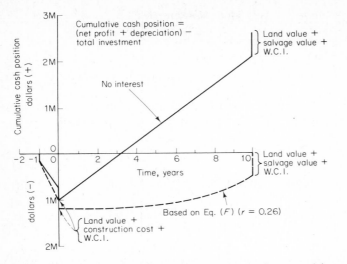

Figure 9-3 Illustrative plot showing cash position vs. time to explain graphically the solution to Example 3. Dashed line is with interest or profitability index of 26 percent. Solid line is with no interest charge. (Note that method for calculating depreciation is not important except for income taxes.)

where Z is the time period in years the land is owned before startup and Y is the time period in years required for construction. In this example, Z and Y are both 1.0.

Figure 9-3 is the requested plot of cash position versus time for the case of $r = 0.26$ based on Eq. (F) and illustrates the concepts involved in the solution of this problem showing the cases with interest (or profitability index) and without interest.

DETERMINING ACCEPTABLE RETURNS

It is often possible to make a profit by the investment of capital, but it is not always easy to determine if a given return is sufficient to justify an investment. Many factors must be considered when deciding if a return is acceptable, and it is not possible to give one figure which will apply for all cases.

When dealing with ordinary industrial operations, profits cannot be predicted with absolute accuracy. Risk factors, therefore, must be given careful consideration, and the degree of uncertainty involved in predicted returns on investments plays an important role in determining what returns are acceptable.

A certain amount of risk is involved in any type of investment, but the degree of risk varies widely for different types of enterprises. For example, there is very little uncertainty involved in predicting returns on capital invested in government bonds, and the chances of losing the original capital are very small.

However, money invested in a wildcat mining enterprise would stand a good chance of being lost completely with no return whatsoever.

If capital is available for investment in a proposed enterprise, it would also be available for use in other ventures. Therefore, a good basis for determining an acceptable return is to compare the predicted return and the risks involved with returns and risks in other types of investments.

Very conservative investments, such as government bonds, pay low returns in the range of 5 to 7 percent, but the risk involved is practically negligible. Preferred stocks yield returns of about 7 to 9 percent. There is some risk involved in preferred-stock investments since a business depression or catastrophe could cause reduction in returns or even a loss of the major portion of the capital investment. Common stocks may yield very high returns; however, the returns fluctuate considerably with varying economic conditions, and there is always the possibility of losing much or all of the original investment.

It can be stated that moderate risks are involved in common-stock investments. Certainly, at least moderate risks are involved in most industrial projects. In general, a 20 percent return before income taxes would be the minimum acceptable return for any type of business proposition, even if the economics appeared to be completely sound and reliable. Many industrial concerns demand a predicted pretax return of at least 35 percent based on reliable economic estimates before they will consider investing capital in projects that are known to be well engineered and well designed.

The final decision as to an acceptable return depends on the probable accuracy of the predicted return and on the amount of risk the investor wishes to take. Availability of capital, personal opinions, and intangible factors, such as the response of the public to changes or appearances, may also have an important effect on the final decision.

ALTERNATIVE INVESTMENTS

In industrial operations, it is often possible to produce equivalent products in different ways. Although the physical results may be approximately the same, the capital required and the expenses involved can vary considerably depending on the particular method chosen. Similarly, alternative methods involving varying capital and expenses can often be used to carry out other types of business ventures. It may be necessary, therefore, not only to decide if a given business venture would be profitable, but also to decide which of several possible methods would be the most desirable.

The final decision as to the best among alternative investments is simplified if it is recognized that each dollar of additional investment should yield an adequate rate of return. In practical situations, there are usually a limited number of choices, and the alternatives must be compared on the basis of incremental increases in the necessary capital investment.

The following simple example illustrates the principle of investment comparison. A chemical company is considering adding a new production unit which will require a total investment of $1,200,000 and will yield an annual profit of $240,000. An alternative addition has been proposed requiring an investment of $2 million and yielding an annual profit of $300,000. Although both of these proposals are based on reliable estimates, the company executives feel that other equally sound investments can be made with at least a 14 percent annual rate of return. Therefore, the minimum rate of return required for the new investment is 14 percent.

The rate of return on the $1,200,000 unit is 20 percent, and that for the alternative addition is 15 percent. Both of these returns exceed the minimum required value, and it might appear that the $2 million investment should be recommended because it yields the greater amount of profit per year. However, a comparison of the incremental investment between the two proposals shows that the extra investment of $800,000 gives a profit of only $60,000, or an incremental return of 7.5 percent. Therefore, if the company had $2 million to invest, it would be more profitable to accept the $1,200,000 proposal and put the other $800,000 in another investment at the indicated 14 percent return.

A general rule for making comparisons of alternative investments can be stated as follows: *The minimum investment which will give the necessary functional results and the required rate of return should always be accepted unless there is a specific reason for accepting an alternative investment requiring more initial capital.* When alternatives are available, therefore, the base plan would be that requiring the minimum acceptable investment. The alternatives should be compared with the base plan, and additional capital would not be invested unless an acceptable incremental return or some other distinct advantage could be shown.

Alternatives When an Investment Must Be Made

The design engineer often encounters situations where it is absolutely necessary to make an investment and the only choice available is among various alternatives. An example of this might be found in the design of a plant requiring an evaporation operation. The evaporator unit must have a given capacity based on the plant requirements, but there are several alternative methods for carrying out the operation. A single-effect evaporator would be satisfactory. However, the operating expenses would be lower if a multiple-effect evaporator were used, because of the reduction in steam consumption. Under these conditions, the best number of effects could be determined by comparing the increased savings with the investment required for each additional effect. A graphical representation showing this kind of investment comparison is presented in Fig. 9-4.

The base plan for an alternative comparison of the type discussed in the preceding paragraph would be the minimum investment which gives the necessary functional results. The alternatives should then be compared with the base plan, and an additional investment would be recommended only if it would give a definite advantage.

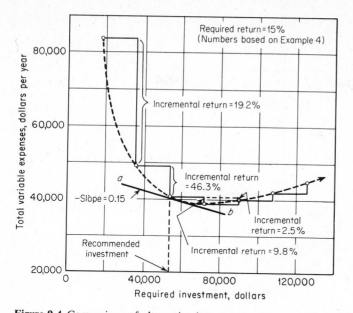

Figure 9-4 Comparison of alternative investments when one investment must be made for a given service and there are a limited number of choices.

When investment comparisons are made, alternatives requiring more initial capital are compared only with lower investments *which have been found to be acceptable.* Consider an example in which an investment of $50,000 will give a desired result, while alternative investments of $60,000 and $70,000 will give the same result with less annual expense. Suppose that comparison between the $60,000 and the $50,000 cases shows the $60,000 investment to be unacceptable. Certainly, there would be no reason to give further consideration to the $60,000 investment, and the next comparison should be between the $70,000 and the $50,000 cases. This type of reasoning, in which alternatives are compared in pairs on a mutually exclusive basis, is illustrated in the following simplified example.

An example to illustrate principles of alternative investment analysis In making a choice among various alternative investments, it is necessary to recognize the need to compare one investment to another on a mutually exclusive basis in such a manner that the *return on the incremental investment* is satisfactory. The following example illustrates this principle.

An existing plant has been operating in such a way that a large amount of heat is being lost in the waste gases. It has been proposed to save money by recovering the heat that is now being lost. Four different heat exchangers have been designed to recover the heat, and all prices, costs, and savings have been calculated for each of the designs. The results of these calculations are presented in the following:

Design	No. 1	No. 2	No. 3	No. 4
Total initial installed cost, $	10,000	16,000	20,000	26,000
Operating costs, $/yr	100	100	100	100
Fixed charges, % of initial cost/yr	20	20	20	20
Value of heat saved, $/yr	4,100	6,000	6,900	8,850

The company in charge of the plant demands at least a 10 percent annual return based on the initial investment for any unnecessary investment. Only one of the four designs can be accepted. Neglecting effects due to income taxes and the time value of money, which (if any) of the four designs should be recommended?

The first step in the solution of this example problem is to determine the amount of money saved per year for each design, from which the annual percent return on the initial investment can be determined. The net annual savings equals the value of heat saved minus the sum of the operating costs and fixed charges; thus,

For design No. 1,

$$\text{Annual savings} = 4100 - (0.2)(10{,}000) - 100 = \$2000$$

$$\text{Annual percent return} = \frac{2000}{10{,}000}(100) = 20\%$$

For design No. 2,

$$\text{Annual savings} = 6000 - (0.2)(16{,}000) - 100 = \$2700$$

$$\text{Annual percent return} = \frac{2700}{16{,}000}(100) = 16.9\%$$

For design No. 3,

$$\text{Annual savings} = 6900 - (0.2)(20{,}000) - 100 = \$2800$$

$$\text{Annual percent return} = \frac{2800}{20{,}000}(100) = 14\%$$

For design No. 4,

Annual savings $= 8850 - (0.2)(26,000) - 100 = \$3550$

Annual percent return $= \dfrac{3550}{26,000}(100) = 13.6\%$

Because the indicated percent return for each of the four designs is above the minimum of ten percent required by the company, any one of the four designs would be acceptable, and it now becomes necessary to choose one of the four alternatives.

Alternative analysis by method of return on incremental investment Analysis by means of return on incremental investment is accomplished by a logical step-by-step comparison of an acceptable investment to another which might be better. If design No. 1 is taken as the starting basis, comparison of design No. 2 to design No. 1 shows that an annual saving of $\$2700 - \$2000 = \$700$ results by making an additional investment of $\$16,000 - \$10,000 = \$6,000$. Thus, the percent return on the incremental investment is $\frac{700}{6000}(100) = 11.7$ percent, and design No. 2 is acceptable by company policy in preference to design No. 1. This logical procedure results in the following tabulation and the choice of design No. 2 as the final recommendation:

Design No. 1 is acceptable.

Comparing design No. 1 to design No. 2, annual percent return $= \frac{700}{6000}(100) = 11.7$ percent. Thus, design No. 2 is acceptable and is preferred over design No. 1.

Comparing design No. 2 to design No. 3, annual percent return $= \frac{100}{4000}(100) = 2.5$ percent. Thus, design No. 3 compared to design No. 2 shows that the return is unacceptable and design No. 2 is preferred.

Comparing design No. 2 to design No. 4, annual percent return $= \frac{850}{10000}(100) = 8.5$ percent. Thus, design No. 4 is not acceptable when compared to design No. 2, and design No. 2 is the alternative that should be recommended.

Alternative analysis incorporating minimum return as a cost Identical results to the preceding are obtained by choosing the alternative giving the greatest annual profit or saving if the required return is included as a cost for each case. For the heat-exchanger example, the annual cost for the required return would be 10 percent of the total initial investment; thus,

For design No. 1, annual savings above required return $= 2000 - (0.1) \times (10,000) = \$1000$.

For design No. 2, annual savings above required return $= 2700 - (0.1) \times (16,000) = \$1100$.

For design No. 3, annual savings above required return = $2800 - (0.1) \times (20,000) = \$800$.

For design No. 4, annual savings above required return = $3550 - (0.1) \times (26,000) = \$950$.

Because annual saving is greatest for design No. 2, this would be the recommended alternative which is the same result as was obtained by the direct analysis based on return on incremental investment.

This simplified example has been used to illustrate the basic concepts involved in making comparisons of alternative investments. The approach was based on using the simple return on initial investment in which time value of money is neglected. Although this method may be satisfactory for preliminary and rough estimations, for final evaluations a more sophisticated approach is needed in which the time value of money is considered along with other practical factors to assure the best possible chance for future success. Typical more advanced approaches of this type are presented in the following sections.

ANALYSIS WITH SMALL INVESTMENT INCREMENTS

The design engineer often encounters the situation in which the addition of small investment increments is possible. For example, in the design of a heat exchanger for recovering waste heat, each square foot of additional heat-transfer area can cause a reduction in the amount of heat lost, but the amount of heat recovered per square foot of heat-transfer area decreases as the area is increased. Since the investment for the heat exchanger is a function of the heat-transfer area, a plot of net savings (or net profit due to the heat exchanger) versus total investment can be made. A smooth curve of the type shown in Fig. 9-5 results.

The point of maximum net savings, as indicated by O in Fig. 9-5, represents a classical optimum condition. However, the last incremental investment before this maximum point is attained is at essentially a zero percent return. On the basis of alternative investment comparisons, therefore, some investment less than that for maximum net savings should be recommended.

The exact investment where the incremental return is a given value occurs at the point where the slope of the curve of net savings versus investment equals the required return. Thus, a straight line with a slope equal to the necessary return is tangent to the net-savings-versus-investment curve at the point representing the recommended investment. Such a line for an annual return on incremental investment of 20 percent is shown in Fig. 9-5, and the recommended investment for this case is indicated by RI. If an analytical expression relating net savings and investment is available, it is obvious that the recommended investment can be determined directly by merely setting the derivative of the net savings with respect to the investment equal to the required incremental return.

The method described in the preceding paragraph can also be used for continuous curves of the type represented by the dashed curve in Fig. 9-4. Thus, the line ab in Fig. 9-4 is tangent to the dashed curve at the point representing the recommended investment for the case of a 15 percent incremental return.

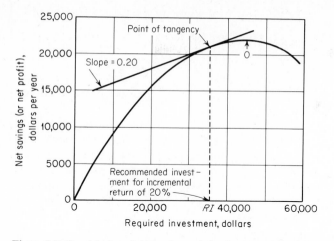

Figure 9-5 Graphical method for determining investment for a given incremental return when investment increments can approach zero.

Example 4. Investment comparison for required operation with limited number of choices

A plant is being designed in which 450,000 lb per 24-h day of a water–caustic soda liquor containing 5 percent by weight caustic soda must be concentrated to 40 percent by weight. A single-effect or multiple-effect evaporator will be used, and a single-effect evaporator of the required capacity requires an initial investment of $18,000. This same investment is required for each additional effect. The service life is estimated to be 10 years, and the salvage value of each effect at the end of the service life is estimated to be $6000. Fixed charges minus depreciation amount to 20 percent yearly, based on the initial investment. Steam costs $0.60 per 1000 lb, and administration, labor, and miscellaneous costs are $40 per day, no matter how many evaporator effects are used.

Where X is the number of evaporator effects, $0.9X$ equals the number of pounds of water evaporated per pound of steam. There are 300 operating days per year. If the minimum acceptable return on any investment is 15 percent, how many effects should be used?

SOLUTION
Basis: 1 operating day
X = total number of evaporator effects

$$\text{Depreciation per operating day (straight-line method)} = \frac{X(18,000 - 6000)}{(10)(300)}$$

$$= \$4.00X/\text{day}$$

$$\text{Fixed charges} - \text{depreciation} = \frac{X(18,000)(0.20)}{300} = \$12.00X/\text{day}$$

Pounds of water evaporated per day $= (450,000)(0.05)(\frac{95}{5}) - (450,000)(0.05)$

$$(\tfrac{60}{40}) = 393,800 \text{ lb/day}$$

$$\text{Steam costs} = \frac{(393,800)(0.60)}{X(0.9)(1000)} = \frac{\$262.5}{X} \text{ per day}$$

$X =$ no. of effects	Steam costs per day	Fixed charges minus depreciation per day	Depreciation per day	Labor, etc., per day	Total cost per day
1	$262.5	$12	$ 4	$40	$318.5
2	131.3	24	8	40	203.3
3	87.5	36	12	40	175.5
4	65.6	48	16	40	169.6
5	52.5	60	20	40	172.5
6	43.8	72	24	40	179.8

Comparing two effects with one effect,

$$\text{Percent return} = \frac{(318.5 - 203.3)(300)(100)}{36,000 - 18,000} = 192\%$$

Therefore, two effects are better than one effect.

Comparing three effects with two effects,

$$\text{Percent return} = \frac{(203.3 - 175.5)(300)(100)}{54,000 - 36,000} = 46.3\%$$

Therefore, three effects are better than two effects.

Comparing four effects with three effects,

$$\text{Percent return} = \frac{(175.5 - 169.6)(300)(100)}{72,000 - 54,000} = 9.8\%$$

Since a return of at least 15 percent is required on any investment, three effects are better than four effects, and the four-effect evaporator should receive no further consideration.

Comparing five effects with three effects,

$$\text{Percent return} = \frac{(175.5 - 172.5)(300)(100)}{90,000 - 54,000} = 2.5\%$$

Therefore, three effects are better than five effects.

Since the total daily costs continue to increase as the number of effects is increased above five, no further comparisons need to be made.

A three-effect evaporator should be used.

ANALYSIS OF ADVANTAGES AND DISADVANTAGES OF VARIOUS PROFITABILITY MEASURES FOR COMPARING ALTERNATIVES

Of the methods presented for profitability evaluation and the economic comparison of alternatives, net present worth and discounted cash flow are the most generally acceptable, and these methods are recommended. Capitalized costs have limited utility but can serve to give useful and correct results when applied to appropriate situations. Payout period does not adequately consider the later years of the project life, does not consider working capital, and is generally useful only for rough and preliminary analyses. Rates of return on original investment and average investment do not include the time value of money, require approximations for estimating average income, and can give distorted results because of methods used for depreciation allowances.

It is quite possible to compare a series of alternative investments by each of the profitability measures outlined in the early part of this chapter and find that different alternatives would be recommended depending on the evaluation technique used.† If there is any question as to which method should be used for a final determination, *net present worth* should be chosen, as this will be the most likely to maximize the future worth of the company.

Investment costs due to land can be accounted for in all the methods except payout period. Costs incurred during the construction period prior to startup can be considered correctly in both the net-present-worth and the discounted-cash-flow methods, while they are ignored in the return-on-investment methods and are seldom taken into account in determining payout period. None of the methods gives a direct indication of the magnitude of the project, although net present worth does give a general idea of the magnitude if interpreted correctly. In certain cases, such as for alternatives of different economic lives, the discounted-cash-flow rate-of-return method is very difficult to use for comparing investments correctly. The discounted-cash-flow rate-of-return method may give multiple or impossible answers for unusual cash-flow situations, such as a case where no cash investment is needed at the start and for certain replacement situations.

Consideration of Income Taxes

Income-tax effects can be included properly in all the profitability methods discussed in this chapter by using appropriate definitions of terms, such as those presented in Table 4. The methods of discounted-cash-flow rate of return and present worth are limited to consideration of cash income and cash outgo over the life of the project. Thus, depreciation, as a cost, does not enter directly into the calculations except as it may affect income taxes.

---

† This situation is illustrated in Example 5 of this chapter.

Table 4 Definitions to clarify income-tax situation for profitability evaluation

Revenue = total income (or total savings)
Net profits = revenue − all expenses − income tax
All expenses = cash expenses + depreciation
Income tax = (revenue − all expenses)(tax rate)
Cash flow = net profits + depreciation
Cash flow = (revenue)(1 − tax rate) − (cash expenses)(1 − tax rate) + (depreciation)(tax rate)
Cash flow = (revenue)(1 − tax rate) − (all expenses)(1 − tax rate) + depreciation

For the case of a 46% tax rate

$1.00 of revenue (either as sales income or savings) yields a cash inflow of $0.54.
$1.00 of cash expenses (as raw materials, labor, etc.) yields a cash outflow of $0.54.
$1.00 of depreciation yields a cash inflow of $0.46.

Net cash flow represents the difference between all cash revenues and all cash expenses with taxes included as a cash expense. Thus, discounted-cash-flow rate of return and present worth should be calculated on an after-tax basis, unless there is some particular reason for a pretax basis, such as comparison to a special alternate which is presented on a pre-tax basis.

Example 5. Comparison of alternative investments by different profitability methods

A company has three alternative investments which are being considered. Because all three investments are for the same type of unit and yield the same service, only one of the investments can be accepted. The risk factors are the same for all three cases. Company policies, based on the current economic situation, dictate that a minimum annual return on the original investment of 15 percent after taxes must be predicted for any unnecessary investment with interest on investment not included as a cost. (This may be assumed to mean that other equally sound investments yielding a 15 percent return after taxes are available.) Company policies also dictate that, where applicable, straight-line depreciation is used and, for time-value of money interpretations, end-of-year cost and profit analysis is used. Land value and prestartup costs can be ignored.

Given the following data, determine which investment, if any, should be made by alternative-analysis profitability-evaluation methods of

(*a*) Rate of return on initial investment
(*b*) Minimum payout period with no interest charge
(*c*) Discounted cash flow
(*d*) Net present worth
(*e*) Capitalized costs

Investment number	Total initial fixed-capital investment, $	Working-capital investment, $	Salvage value at end of service life, $	Service life, years	Annual cash flow to project after taxes,† $	Annual cash expenses‡ (constant for each year), $
1	100,000	10,000	10,000	5	See yearly tabulation§	44,000
2	170,000	10,000	15,000	7	52,000 (constant)	28,000
3	210,000	15,000	20,000	8	59,000 (constant)	21,000

† This is total annual income or revenue minus all costs except depreciation and interest cost for investment.

‡ This is annual cost for operation, maintenance, taxes, and insurance.   Equals total annual income minus annual cash flow.

§ For investment number 1, variable annual cash flow to project is: year 1 = $30,000, year 2 = $31,000, year 3 = $36,000, year 4 = $40,000, year 5 = $43,000.

SOLUTION

(a) Method of rate of return on initial investment

Average annual profit = annual cash flow − annual depreciation cost

The average annual profits for investment No. 1, using straight-line depreciation are as follows:

Year	Average annual profit, dollars
1	$30,000 - \dfrac{(100,000 - 10,000)}{5} = 30,000 - 18,000 = 12,000$
2	$31,000 - 18,000 = 13,000$
3	$36,000 - 18,000 = 18,000$
4	$40,000 - 18,000 = 22,000$
5	$43,000 - 18,000 = 25,000$
Total	90,000

For investment No. 1, the arithmetic average of the annual profits is 90,000/5 = $18,000.†

---

† An alternate method to obtain the average of the annual profits would be to determine the amount of the annuity R based on end-of-year payments that would compound to the same future worth as the individual profits using an interest rate i of 0.15. With this approach, the average of the annual profits for investment No. 1 would be $17,100.

The method for determining this $17,100 is to apply the series compound-amount factor $[(1 + i)^n - 1]/i$ [see Eq. (21) in Chap. 6] to the annuity to give the future worth S of the annual incomes. The expression is $(12,000)(1 + i)^4 + (13,000)(1 + i)^3 + (18,000)(1 + i)^2 + (22,000)(1 + i) + 25,000 = R[(1 + i)^5 - 1]/i$. Solving for the case of $i = 0.15$ gives $R = \$17,100$.

The annual average rate of return on the first investment is

$$\frac{18,000}{100,000 + 10,000}(100) = 16.4\% \text{ after taxes}$$

Because this return is greater than 15 percent, one of the three investments will be recommended, and it is only necessary to compare the three investments.

For investment number	Total initial investment	Average annual profit, dollars		
1	$110,000			$18,000
2	$180,000	$52,000 - \dfrac{170,000 - 15,000}{7}$	$= 52,000 - 22,100$	$= \$29,900$
3	$225,000	$59,000 - \dfrac{210,000 - 20,000}{8}$	$= 59,000 - 23,800$	$= \$35,200$

Comparing investment No. 2 with investment No. 1,

$$\text{Percent return} = \frac{29,900 - 18,000}{180,000 - 110,000}(100) = 17.0\%$$

Therefore, investment No. 2 is preferred over investment No. 1. Comparing investment No. 3 with investment No. 2,

$$\text{Percent return} = \frac{35,200 - 29,900}{225,000 - 180,000}(100) = 11.8\%$$

This return is not acceptable, and *investment No. 2 should be recommended.*

The same result would have been obtained if a minimum return of 15 percent had been incorporated as an expense.

(*b*) **Method of minimum payout period with no interest charge**

Payout period (with no interest charge)

$$= \frac{\text{depreciable fixed-capital investment}}{\text{avg profit/yr} + \text{avg depreciation/yr}}$$

For investment number   *Payout period, years*

1     $\dfrac{90,000}{18,000 + 18,000} = 2.50$

2     $\dfrac{155,000}{29,900 + 22,100} = 2.98$

3     $\dfrac{190,000}{35,200 + 23,800} = 3.22$

The payout period for investment No. 1 is least; therefore, by this method, *investment No. 1 should be recommended.*

(c) Method of discounted cash flow

For investment No. 1, as illustrated in Table 1, the rate of return based on discounted cash flow is 20.7 percent.

For investment No. 2, the discounted-cash-flow equation is

$$(52,000) \left[\frac{1}{1+i} + \frac{1}{(1+i)^2} + \cdots + \frac{1}{(1+i)^7} \right]$$

$$+ (10,000 + 15,000) \frac{1}{(1+i)^7} = \$180,000$$

By trial-and-error solution, the discounted-cash-flow rate of return is 22.9 percent.

Similarly, for investment No. 3,

$$(59,000) \left[\frac{1}{1+i} + \frac{1}{(1+i)^2} + \cdots + \frac{1}{(1+i)^8} \right]$$

$$+ (15,000 + 20,000) \frac{1}{(1+i)^8} = \$225,000$$

By trial-and-error solution, the discounted-cash-flow rate of return is 21.5 percent.

To make a choice among the three alternatives, it is necessary to make a comparison among the three possible choices. This comparison can be made in a relatively straightforward manner using discounted-cash-flow rates of return by comparing pairs of investments on a mutually exclusive basis if the various alternatives have the same economic service lives. When different lengths of service life are involved, as in this problem, the best approach is to avoid the calculated rates of return and make the investment comparison by the net present-worth method as shown in part (*d*) of this problem. It would be possible to use discounted-cash-flow rates of return for comparison between investments with different service lives by assuming that each investment could be repeated at the end of its service life until a common end point was obtained for the investments being compared; however, this method becomes very involved mathematically and is not realistic.

If the service lives of the investments being compared are not widely different, the following *approximate* method using discounted-cash-flow rate of return can be employed for the comparison.†

In comparing a pair of alternatives, the base time is chosen as the longer

---

† The method is shown to illustrate the use of discounted-cash-flow rates of return for investment comparisons. It is correct only for comparisons involving equal service lives. If service lives are different, this method tends to favor the investment with the longest service life.

of the two service lives. For the case of the investment with the shorter life, it is assumed that the accumulated amount at the end of its life can be invested at the minimum acceptable rate for the remaining time to equalize the two lives. The rate of return on the incremental investment can then be determined.

Comparison of investment No. 2 to investment No. 1. At the end of its 7-year service life, the net value of investment No. 2 is

$$(180,000)(1 + 0.229)^7 + 10,000 + 15,000 = \$785,000$$

With investment No. 1, the net value after 7 years is the amount accumulated in 5 years times the factor to let this accumulated amount be invested at 15 percent for 2 more years, or

$$[(110,000)(1 + 0.207)^5 + 10,000 + 10,000](1 + 0.15)^2 = \$398,000$$

Therefore, a gain of $785,000 - $398,000 = $387,000 is made in 7 years by an added investment of $70,000 if investment No. 2 is made instead of investment No. 1. The discounted-cash-flow rate of return for this incremental investment is found by

$$(70,000)(1 + i)^7 = 387,000$$

$$i = 0.277 \qquad \text{or} \qquad 27.7\%$$

This return is greater than 15 percent; so investment No. 2 is preferred over investment No. 1.

Comparison of investment No. 3 to investment No. 2. At the end of its 8-year service life, the net value of investment No. 3 is

$$(225,000)(1 + 0.215)^8 + 15,000 + 20,000 = \$1,105,000$$

For comparison, $180,000 invested in investment No. 2 would, with the last year at a 15 percent return, accumulate in 8 years to

$$[(180,000)(1 + 0.229)^7 + 10,000 + 15,000](1 + 0.15) = \$903,000$$

Therefore, a gain of $1,105,000 - $903,000 = $202,000 is made in 8 years by an added investment of $45,000 by making investment No. 3 instead of investment No. 2. The discounted-cash-flow rate of return for this incremental investment is found by

$$(45,000)(1 + i)^8 = 202,000$$

$$i = 0.208 \qquad \text{or} \qquad 20.8\%$$

This return is greater than 15 percent; so investment No. 3 is preferred over investment No. 2. Therefore, *investment No. 3 should be recommended.*

(*d*) **Method of net present worth**

For investment No. 1, as illustrated in Table 1, the present value of the cash flow to the project, discounted at an interest rate of 15 percent, is $127,000. Therefore, the net present worth for investment No. 1 is $127,000 − $110,000 = $17,000.

For investments 2 and 3, the present values of the cash flows to the projects are determined from the first two equations under part (*c*) of this problem, with $i = 0.15$. The resulting net present worths are:

For investment No. 2, net present worth = $226,000 − $180,000 = $46,000

For investment No. 3, net present worth = $278,000 − $225,000 = $53,000

The greatest net present worth is found for investment No. 3; therefore, *investment No. 3 should be recommended.*

(*e*) **Method of capitalized costs**

Capitalized costs for each investment situation must include the capitalized cost for the original investment to permit an indefinite number of replacements plus the capitalized present value of the cash expenses plus working capital.

Capitalized present value of cash expenses is determined as follows:

Let $C_{n'}$ be the annual cash expense in year n' of the project life. The present value of the annual cash expenses is then

$$\sum_{n'=1}^{n'=n} C_{n'} \frac{1}{(1+i)^{n'}}$$

and the capitalized present value is

$$\frac{(1+i)^n}{(1+i)^n - 1} \sum_{n'=1}^{n'=n} C_{n'} \frac{1}{(1+i)^{n'}}$$

If $C_{n'}$ is constant, as is the case for this example, the capitalized present value becomes (annual cash expenses)/i. Therefore,

Capitalized cost

$$= \frac{C_R(1+i)^n}{(1+i)^n - 1} + V_s + \frac{\text{annual cash expenses}}{i} + \text{working capital}$$

where n = service life
i = annual rate of return
C_R = replacement cost
V_s = salvage value

For investment No. 1,

$$\text{Capitalized cost} = \frac{(90,000)(1 + 0.15)^5}{(1 + 0.15)^5 - 1} + 10,000 + \frac{44,000}{0.15} + 10,000$$

$$= \$492,000$$

For investment No. 2,

$$\text{Capitalized cost} = \frac{(155,000)(1 + 0.15)^7}{(1 + 0.15)^7 - 1} + 15,000 + \frac{28,000}{0.15} + 10,000$$

$$= \$460,000$$

For investment No. 3,

$$\text{Capitalized cost} = \frac{(190,000)(1 + 0.15)^8}{(1 + 0.15)^8 - 1} + 20,000 + \frac{21,000}{0.15} + 15,000$$

$$= \$457,000$$

The capitalized cost based on a minimum rate of return of 15 percent is least for investment No. 3; therefore, *investment No. 3 should be recommended.*

Note: Methods (*a*) and (*b*) in this problem give incorrect results because the time value of money has not been included. Although investment No. 3 is recommended by methods (*c*), (*d*), and (*e*), it is a relatively narrow choice over investment No. 2. Consequently, for a more accurate evaluation, it would appear that the company management should be informed that certain of their policies relative to profitability evaluation are somewhat old fashioned and do not permit the presentation of a totally realistic situation. For example, the straight-line depreciation method may not be the best choice, and a more realistic depreciation method may be appropriate. The policy of basing time-value-of-money interpretations on end-of-year costs and profits is a simplification, and it may be better to permit the use of continuous interest compounding and continuous cash flow where appropriate. For a final detailed analysis involving a complete plant, variations in prestartup costs among alternatives may be important, and this factor should not be ignored.

REPLACEMENTS[†]

The term "replacement," as used in this chapter, refers to a special type of alternative in which facilities are currently in existence and it may be desirable to replace these facilities with different ones. Although intangible factors may have a strong influence on decisions relative to replacements, the design engineer must understand the tangible economic implications when a recommendation is made as to whether or not existing equipment or facilities should be replaced.

The reasons for making replacements can be divided into two general classes, as follows:

[†] For additional discussion and analysis of replacements, see G. A. Taylor, "Managerial and Engineering Economy: Economic Decision-Making," 2d ed., D. Van Nostrand Company, Inc., Princeton, New Jersey, 1975; E. L. Grant, W. G. Ireson, and R. S. Leavenworth, "Principles of Engineering Economy," 6th ed., Ronald Press Co., New York, 1976; and D. G. Newnan, "Engineering Economic Analysis," Engineering Press, San Jose, California, 1976.

1. An existing property *must* be replaced or changed in order to continue operation and meet the required demands for service or production. Some examples of reasons for this type of necessary replacement are:

   a. The property is worn out and can give no further useful service.

   b. The property does not have sufficient capacity to meet the demand placed upon it.

   c. Operation of the property is no longer economically feasible because changes in design or product requirements have caused the property to become obsolete.

2. An existing property is capable of yielding the necessary product or service, but more efficient equipment or property is available which can operate with lower expenses.

When the reason for a replacement falls in the first general type, the only alternatives are to make the necessary changes or else go out of business. Under these conditions, the final economic analysis is usually reduced to a comparison of alternative investments.

The correct decision as to the desirability of replacing an existing property which is capable of yielding the necessary product or service depends on a clear understanding of theoretical replacement policies plus a careful consideration of many practical factors. In order to determine whether or not a change is advisable, the operating expenses with the present equipment must be compared with those that would exist if the change were made. Practical considerations, such as amount of capital available or the benefits to be gained by meeting a competitor's standards, may also have an important effect on the final decision.

Methods of Profitability Evaluation for Replacements

The same methods that were explained and applied earlier in this chapter are applicable for replacement analyses. Net-present-worth and discounted-cash-flow methods give the soundest results for maximizing the overall future worth of a concern. However, for the purpose of explaining the *basic principles* of replacement economic analyses, the simple rate-of-return-on-investment method of analysis is just as effective as those methods involving the time value of money. Thus, to permit the use of direct illustrations which will not detract from the principles under consideration, the following analysis of methods for economic decisions on replacements uses the annual rate of return on initial investment as the profitability measure. The identical principles can be treated by more complex rate-of-return and net-present-worth solutions by merely applying the methods described earlier in this chapter.

Typical Example of Replacement Policy

The following example illustrates the type of economic analysis involved in determining if a replacement should be made: A company is using a piece of equipment which originally cost $30,000. The equipment has been in use for 5 years, and it now has a net realizable value of $6000. At the time of installation, the service life was estimated to be 10 years and the salvage value at the end of the service life was estimated to be zero. Operating costs amount to $22,000/year. At the present time, the remaining service life of the equipment is estimated to be 3 years.

A proposal has been made to replace the present piece of property by one of more advanced design. The proposed equipment would cost $40,000, and the operating costs would be $15,000/year. The service life is estimated to be 10 years with a zero salvage value. Each piece of equipment will perform the same service, and all costs other than those for operation and depreciation will remain constant. Depreciation costs are determined by the straight-line method.† The company will not make any unnecessary investments in equipment unless it can obtain an annual return on the necessary capital of at least 10 percent.

The two alternatives in this example are to continue the use of the present equipment or to make the suggested replacement. In order to choose the better alternative, it is necessary to consider both the reduction in expenses obtainable by making the change and the amount of new capital necessary. The only variable expenses are those for operation and depreciation. Therefore, the annual variable expenses for the proposed equipment would be $15,000 + $40,000/10 = $19,000.

The net realizable value of the existing equipment is $6000. In order to make a fair comparison between the two alternatives, all costs must be based on conditions at the present time. Therefore, the annual depreciation cost for the existing equipment during the remaining three years of service life would be $6000/3 = $2000. For purposes of comparison, the annual variable expenses if the equipment is retained in service would be $22,000 + $2000 = $24,000.

An annual saving of $24,000 − $19,000 = $5000 can be realized by making the replacement. The cost of the new equipment is $40,000, but the sale of the existing property would provide $6000; therefore, it would be necessary to invest only $34,000 to bring about an annual saving of $5000. Since this represents a return greater than 10 percent, the existing equipment should be replaced.

Book Values and Unamortized Values

In the preceding example, the book value of the existing property was $15,000 at the time of the proposed replacement. However, this fact was given no consideration in determining if the replacement should be made. The book value is based on past conditions, and the correct decision as to the desirability of making a replacement must be based on present conditions.

The difference between the book value and the net realizable value at any time is commonly designated as the *unamortized value*. In the example, the un-

---

† The use of the sinking-fund method for determining depreciation is sometimes advocated for replacement studies. However, for most practical situations, the straight-line method is satisfactory.

amortized value was $15,000 − $6000 = $9000. This means that a $9000 loss was incurred because of incorrect estimation of depreciation allowances.†

Much of the confusion existing in replacement studies is caused by un-amortized values. Some individuals feel that a positive unamortized value represents a loss which would be caused by making a replacement. This is not correct because the loss is a result of past mistakes, and the fact that the error was not apparent until a replacement was considered can have no bearing on the conditions existing at the present time. *When making theoretical replacement studies, unamortized values must be considered as due to past errors, and these values are of no significance in the present decision as to whether or not a replacement should be made.*

Although unamortized values have no part in a replacement study, they must be accounted for in some manner. One method for handling these capital losses or gains is to charge them directly to the profit-and-loss account for the current operating period. When a considerable loss is involved, this method may have an unfavorable and unbalanced effect on the profits for the current period. Many concerns protect themselves against such unfavorable effects by building up a surplus reserve fund to handle unamortized values. This fund is accumulated by setting aside a certain sum each accounting period. When losses due to unamortized values occur, they are charged against the accumulated fund. In this manner, unamortized values can be handled with no serious effects on the periodic profits.

Investment on which Replacement Comparison is Based

As indicated in the preceding section, the unamortized value of an existing property is based on past conditions and plays no part in a replacement study. The advisability of making a replacement is usually determined by the rate of return which can be realized from the necessary investment. It is, therefore, important to consider the amount of the investment. *The difference between the total cost of the replacement property and the net realizable value of the existing property equals the necessary investment.* Thus, the correct determination of the investment involves only consideration of the present capital outlay required if the replacement is made.

Net Realizable Value

In replacement studies, the *net realizable value* of an existing property should be assumed to be the market value. Although this may be less than the actual value of the property as far as the owner is concerned, it still represents the amount of

---

† As explained in Chap. 7 (Taxes and Insurance) and Chap. 8 (Depreciation) and as illustrated in Example 6 of this Chapter, part of this loss may be recovered by income-tax write-offs if the Internal Revenue Service will agree that the maximum expected life was used.

capital which can be obtained from the old equipment if the replacement is made. Any attempt to assign an existing property a value greater than the net realizable value tends to favor replacements which are uneconomical.

Analysis of Common Errors Made in Replacement Studies

Most of the errors in replacement studies are caused by failure to realize that a replacement analysis must be based on conditions existing at the present time. Some persons insist on trying to compensate for past mistakes by forcing new ventures to pay off losses incurred in the past. Instead, these losses should be accepted, and the new venture should be considered on its own merit.

Including unamortized value as an addition to the replacement investment This is one of the most common errors. It increases the apparent cost for the replacement and tends to prevent replacements which are really economical. Some persons incorporate this error into the determination of depreciation cost for the replacement equipment, while others include it only in finding the investment on which the rate of return is based. In any case, this method of attempting to account for unamortized values is incorrect. The unamortized value must be considered as a dead loss (or gain) due to incorrect depreciation accounting in the past.

Use of book value for old equipment in replacement studies This error is caused by refusal to admit that the depreciation accounting methods used in the past were wrong. Persons who make this error attempt to justify their actions by claiming that continued operation of the present equipment would eventually permit complete depreciation. This viewpoint is completely unrealistic because it gives no consideration to the competitive situation existing in modern business. The concern which can operate with good profit and still offer a given product or service at the lowest price can remain in business and force competitors either to reduce their profits or to cease operation.

When book values are used in replacement studies, the apparent costs for the existing equipment are usually greater than they should be, while the apparent capital outlay for the replacement is reduced. Therefore, this error tends to favor replacements which are really uneconomical.

Example 6. An extreme situation to illustrate result of replacement economic analysis A new manufacturing unit has just been constructed and put into operation by your company. The basis of the manufacturing process is a special computer for control (designated as OVT computer) as developed by your research department. The plant has now been in operation for less than one week and is performing according to expectations. A new computer (designated as NTR computer) has just become available on the market. This new computer can easily be installed at once in place of your present computer and will do the identical job at far less annual cash expense because of reduced maintenance and personnel costs. However, if the new computer is installed, your present computer is essentially worthless because you have no other use for it.

Following is pertinent economic information relative to the two computers:

	OVT computer	NTR computer
Capital investment	$2,000,000	$1,000,000
Estimated economic life	10 years	10 years
Salvage value at end of economic life	0	0
Annual cash expenses	$250,000	$50,000

What recommendation would you make relative to replacing the present $2,000,000 computer with the new computer?

SOLUTION Assuming straight-line depreciation, the annual total expenses with the NTR computer = $50,000 + $1,000,000/10 = $150,000.

For replacement economic comparison, the OVT computer is worth nothing at the present time; therefore, annual total expenses with the OVT computer = $250,000 (i.e., no depreciation charge).

The $2,000,000 investment for the OVT computer is completely lost if the NTR computer is installed; so the total necessary investment to make an annual saving of $250,000 − $150,000 is $1,000,000. Therefore, the return on the investment would be (100,000/1,000,000)(100) = 10 percent.

If your company is willing to accept a return on investment of 10 percent before taxes [or (0.52)(10) = 5.2 percent after taxes, assuming a 48 percent tax rate], the replacement should be made.

If income taxes are taken completely into consideration, the result will tend to favor the replacement. For example, if your company can write off the $2,000,000 loss in one lump sum against other capital gains which would normally be taxed at 30 percent, the net saving will be 30 percent of $2,000,000, or $600,000. Assuming a 48 percent tax rate on annual profits and dividing the $600,000 capital-gains tax saving over the ten years of the new computer life, the percent return after taxes =

$$\frac{600,000/10 + (0.52)(100,000)}{1,000,000} (100) = 11.2 \text{ percent}$$

Because of the reduced costs for the NTR computer, profitability evaluation including time value of money will tend to favor the replacement more than does the method of rate of return on investment as used for the solution of this example.

PRACTICAL FACTORS IN ALTERNATIVE-INVESTMENT AND REPLACEMENT STUDIES

The previous discussion has presented the theoretical viewpoint of alternative-investment and replacement studies; however, certain practical considerations also influence the final decision. In many cases, the amount of available capital is limited. From a practical viewpoint, therefore, it may be desirable to accept the smallest investment which will give the necessary service and permit the required return. Although a greater investment might be better on a theoretical basis, the additional return would not be worth the extra risks involved when capital must be borrowed or obtained from some other outside source.

A second practical factor which should be considered is the accuracy of the estimations used in determining the rates of return. A theoretically sound investment might not be accepted because the service life used in determining depreciation costs appears to be too long. All risk factors should be given careful consideration before making any investment, and the risk factors should receive particular attention before accepting an investment greater than that absolutely necessary.

The economic conditions existing at the particular time may have an important practical effect on the final decision. In depression periods or in times when economic conditions are very uncertain, it may be advisable to refrain from investing any more capital than is absolutely necessary. The tax situation for the corporation can also have an effect on the decision.

Many intangible factors enter into the final decision on a proposed investment. Sometimes it may be desirable to impress the general public by the external appearance of a property or by some unnecessary treatment of the final product. These advertising benefits would probably receive no consideration in a theoretical economic analysis, but they certainly would influence management's final decision in choosing the best investment.

Personal whims or prejudices, desire to better a competitor's rate of production or standards, the availability of excess capital, and the urge to expand an existing plant are other practical factors which may be involved in determining whether or not a particular investment will be made.

Theoretical analyses of alternative investments and replacements can be used to obtain a dollars-and-cents indication of what should be done about a proposed investment. The final decision depends on these theoretical results plus practical factors determined by the existing circumstances.

NOMENCLATURE FOR CHAPTER 9

$C_{n'}$ = annual cash expenses in year n', dollars
$CP_{\text{zero time}}$ = total cash position at zero time, dollars
C_R = replacement cost, dollars
C_V = original cost, dollars

$d_{n'}$ = discount factor for determining present value
e = base of the natural logarithm = 2.71828 . . .
i = annual interest rate of return, percent/100
K = capitalized cost, dollars
n = estimated service life, years
n' = year of project life to which cash flow applies
P = present value, dollars
R = end-of-year (or ordinary) annuity amount, dollars/year
$\bar{R}$ = total of all ordinary annuity payments occurring regularly through-out the time period of one year, dollars/year
r = nominal continuous interest rate, percent/100
S = future worth, dollars
V_s = estimated salvage value at end of service life, dollars
X = number of evaporator units
Y = time period for construction, years
Z = time period land is owned before plant startup, years

SUGGESTED ADDITIONAL REFERENCES FOR PROFITABILITY, ALTERNATIVE INVESTMENTS, AND REPLACEMENTS

Alternative investments

Allen, D. H., Two New Tools for Project Evaluation, *Chem. Eng.*, **74**(14):75 (1967).

Barish, N. N., and S. Kaplan, "Economic Analysis for Engineering and Managerial Decision Making," 2d ed., McGraw-Hill Book Company, New York, 1978.

Benkly, G. J., Fit Cooling Systems to the Job, *Hydrocarbon Process.*, **49**(11):219 (1970).

Brown, T. R., Economic Evaluation of Future Equipment Needs, *Chem. Eng.*, **84**(2):125 (1977).

Carpenter, D. B., The Uses of Custom Processing, *Chem. Eng.*, **84**(21):129 (1977).

Champlety, J. A., A "Business" Approach to Capital Budgeting Decisions, *Chem. Eng.*, **76**(20):127 (1969).

Chilton, T. H., Investment Return Via the Engineer's Method, *Chem. Eng. Progr.*, **65**(7):29 (1969).

Davis, G. O., How to Make the Correct Economic Decision on Spare Equipment, *Chem. Eng.*, **84**(25):187 (1977).

Ebly, R. W., Comparison of Methods for Evaluating Capital Equipment Replacement, *1973 Trans. Am. Assoc. Cost Eng.* (1973), p. 23.

Fair, E. W., Shall We Buy That New Equipment? *Chem. Eng.*, **80**(24):124 (1973).

Gazzi, L., and R. Pasero, Selection—Process Cooling Systems, *Hydrocarbon Process*, **49**(10):83 (1970).

Grange, F., A Cost Engineer's Tutorial on Resource Allocation, *AACE Bull.*, **16**(6):177 (1974).

Grumer, E. L., Ranking Plant Projects, *1970 Trans. Am. Assoc. Cost Eng.* (1970), p. 87.

Hegarty, W. P., Evaluating the Incremental Project: An Illustrative Example, *Chem. Eng.*, **75**(19):158 (1968).

Leibson, I., and C. A. Trischman, Jr., Decision Trees: A Rapid Evaluation of Investment Risk, *Chem. Eng.*, **79**(4):99 (1972).

Maristany, B. A., Equipment Decision: Repair or Replace? *Chem. Eng.*, **75**(23):210 (1968).

———, Repair and Replace Economics, presented at American Association of Cost Engineers 12th National Meeting, Houston, Texas, June 17–19, 1968.

Martinez, S., Equipment Buying Decisions, *Chem. Eng.*, **78**(8):145 (1971).

Mol, A., Which Heat Recovery System? *Hydrocarbon Process.*, **52**(7):109 (1973).

Pfeiffer, R. R., Planned Equipment Replacement, *AACE Bull.*, **12**(2):35 (1970).

Polk, H. K., Economic Evaluation of Prime Movers, *Petrol. Eng.*, **46**(12):34 (1974).

Price, J., Computerized Models—the Scientific Way to Pick a Winner, *Chem. Week*, **106**(15):37 (1970).

Ralph, C. A., Resource Allocation Logic, *Chem. Eng. Progr.*, **66**(5):31 (1970).

Shore, B., Replacement Decisions Under Capital Budgeting Constraints, *Eng. Econ.*, **20**(4):243 (1975).

Smith, E. O., Fitting Nuclear Plants into System Planning, *Power Eng.*, **73**(1):52 (1969).

Tarquin, A. J., and L. Blank, " Engineering Economy: A Behavioral Approach," McGraw-Hill Book Company, New York, 1976.

Uchiyama, T., Cooling Tower Estimates Made Easy, *Hydrocarbon Process.*, **55**(12):93 (1976).

Capitalized cost

Holland, F. A., F. A. Watson, and J. K. Wilkinson, Time Value of Money, *Chem. Eng.*, **80**(21):123 (1973).

Jelen, F. C., and M. S. Cole, Methods for Economic Analysis, *Hydrocarbon Process.*, **53**(7):133 (1974); **53**(9):227 (1974).

Mapstone, G. E., Comparison of Amortization Rates and Capitalized Costs, *Chem. Proc. Eng.*, **50**(4):141 (1969).

Reul, R. I., Which Investment Appraisal Techniques Should You Use? *Chem. Eng.*, **75**(10):212 (1968).

Discounted cash flow

See references listed in Chap. 6.

Engineering economics

Baasel, W. D., " Preliminary Chemical Engineering Plant Design," American Elsevier, New York, 1976.

Brooke, J. M., Corrosion Inhibitor Economics, *Hydrocarbon Process.*, **49**(9):299 (1970).

Brown, T. R., Economic Evaluation of Future Equipment Needs, *Chem. Eng.*, **84**(2):125 (1977).

DeGarmo, E. P., " Engineering Economy," The Macmillan Company, New York, 1967.

Hegarty, W. P., Evaluating the Incremental Project: An Illustrative Example, *Chem. Eng.*, **75**(19):158 (1968).

Holland, F. A., F. A. Watson, and J. K. Wilkinson, Engineering Economics for Chemical Engineers, *Chem. Eng.*, **80**(15):103 (1973).

—— —— and ——, " Introduction to Process Economics," J. Wiley and Sons, New York, 1974.

Jelen, F. C., " Cost and Optimization Engineering," McGraw-Hill Book Company, New York, 1970.

—— and M. S. Cole, Methods for Economic Analysis, *Hydrocarbon Process.*, **53**(9):227 (1974).

Jeynes, P. H., " Profitability and Economic Choice, " Iowa State University Press, Ames, Iowa, 1969.

Leibson, I., and C. A. Trischman, Jr., Avoiding Pitfalls in Developing a Major Capital Project, *Chem. Eng.*, **78**(18):103 (1971).

Morgan, F. L., Jr., Gas Plant Economic Risks, *Hydrocarbon Process.*, **47**(6):116 (1968).

Newnan, D. G., " Engineering Economic Analysis," Engineering Press, Inc., San Jose, California, 1977.

Park, W. R., " Cost Engineering Analysis: A Guide to the Economic Evaluation of Engineering Projects," J. Wiley and Sons, New York, 1973.

Popper, H., " Modern Cost-Engineering Techniques," McGraw-Hill Book Company, New York, 1970.

Robertson, J. M., Design for Expansion, Part 1: Economics, *Chem. Eng.*, **75**(10):179 (1968).

Smith, G. W., " Engineering Economy," 2d ed., Iowa State University Press, Ames, Iowa, 1973.

Taylor, G. A., " Managerial and Engineering Economy," 2d ed., D. Van Nostrand Company, New York, 1975.

Thompson, R. G., and R. J. Lievano, What Does a Clean Environment Cost? *Hydrocarbon Process.*, **54**(10):73 (1975).

Wells, G. L., and P. M. Robson, "Computation for Process Engineers," J. Wiley and Sons, New York, 1973.

Woods, D. R., "Financial Decision Making in the Process Industry," Prentice-Hall, Englewood Cliffs, New Jersey, 1975.

Payout time (period)

Brown, T. R., Economic Evaluation of Future Equipment Needs, *Chem. Eng.*, **84**(2):125 (1977).

Garcia- Borras, T., Part 2, Research-Project Evaluation, *Hydrocarbon Process.*, **56**(1):171 (1977).

Jelen, F. C., and M. S. Cole, Methods for Economic Analysis (Part 2), *Hydrocarbon Process.*, **53**(9):227 (1974).

Liebson, I., and C. A. Trischman, Jr., Should You Make Your Major Raw Materials? *Chem. Eng.*, **79**(4):76 (1972).

Morrow, N. L., R. S. Brief, and R. R. Bertrand, Sampling and Analyzing Air Pollution Sources, *Chem. Eng.*, **79**(4):85 (1972).

Waligura, C. L., and R. L. Motard, Data Management on Engineering and Construction Projects, *Chem. Eng. Progr.*, **73**(12):62 (1977).

Wigham, I., Designing Optimum Cooling Systems, *Chem. Eng.*, **78**(18):95 (1971).

Present worth

See references listed in Chap. 6.

Process evaluation

Adam, D. E., S. Sack, and A. Sass, Coal Gasification by Pyrolysis, *Chem. Eng. Progr.*, **70**(6):74 (1974).

Atkins, R. S., Which Process for Xylenes? *Hydrocarbon Process.*, **49**(11):127 (1970).

Barron, R. F., LNG Plants Use These Systems, *Hydrocarbon Process.*, **49**(11):192 (1970).

Bishop, R. B., Find Polystyrene Plant Costs, *Hydrocarbon Process.*, **51**(11):137 (1972).

Bresler, S. A., and J. D. Ireland, Substitute Natural Gas: Processes, Equipment, Costs, *Chem. Eng.*, **79**(23):94 (1972).

Brown, F., Make Ammonia from Coal, *Hydrocarbon Process.*, **56**(11):361 (1977).

Brownstein, A. M., Economics of Ethylene Glycol Processes, *Chem. Eng. Progr.*, **71**(9):72 (1975).

Buividas, L. J., J. A. Finneran, and O. J. Quartulli, Alternate Ammonia Feedstocks, *Chem. Eng. Progr.*, **70**(10):21 (1974).

Chalmers, F. S., Citrate Process Ideal for Claus Tailgas Cleanup, *Hydrocarbon Process.*, **53**(4):75 (1974).

Chauvel, A. R., P. R. Courty, R. Maux, and C. Petitpas, Select Best Formaldehyde Catalyst, *Hydrocarbon Process.*, **52**(9):179 (1973).

Chopey, N. P., Gas-From-Coal: An Update, *Chem. Eng.*, **81**(5):70 (1974).

Conn, A. L., Low Btu. Gas for Power Plants, *Chem. Eng. Progr.*, **69**(12):56 (1973).

Cover, A. E., W. C. Schreiner, and G. T. Skaperdas, Kellogg's Coal Gasification Process, *Chem. Eng. Progr.*, **69**(3):31 (1973).

Crossland, S., Process Liquids to SNG, *Hydrocarbon Process.*, **51**(4):89 (1972).

Davis, J. C., SO_x Control Held Feasible, *Chem. Eng.*, **80**(25):76 (1973).

Detman, R., Economics of Six Coal-to-SNG Processes, *Hydrocarbon Process.*, **56**(3):115 (1977).

Dingman, J. C., and T. F. Moore, Compare DGA and MEA Sweetening Methods, *Hydrocarbon Process.*, **47**(7):138 (1968).

Dutkiewicz, B., and P. H. Spitz, Producing SNG from Crude Oil and Naphtha, *Chem. Eng. Prog.*, **68**(12):45 (1972).

Ennis, R., and P. F. Lesur, How Small NH_3 Plants Compete, *Hydrocarbon Process.*, **56**(12):121 (1977).

Flanagan, T. P., Choosing the Right Instruments, *Chem. Eng.* (*London*), 178 (March 1976).

Foo, K. W., and I. Shortland, Compare CO Production Methods, *Hydrocarbon Process.*, **55**(5):149 (1976).

Fortuin, J. B., and D. Haag, Total Energy by Gasification, *Hydrocarbon Process.*, **53**(8):85 (1974).

Frank, M. E., and B. K. Schmid, Design of a Coal-Oil-Gas Refinery, *Chem. Eng., Progr.*, **69**(3):62 (1973).

Gall, R. L., and E. J. Piasecki, SO_2 Processing: The Double Alkali Wet Scrubbing System, *Chem. Eng. Progr.*, **71**(5):72 (1975).

Gambro, A. J., K. Muenz, and M. Abrahams, Optimize Ethylene Complex, *Hydrocarbon Process.*, **51**(3):73 (1972).

Garcia-Borras, T., Research-Project Evaluations, Part 1, *Hydrocarbon Process.*, **55**(12):137 (1976).

Gary, J. H., and G. E. Handwerk, "Petroleum Refining: Technology and Economics," Marcel Dekker, Inc., New York, 1975.

Gitchel, W. B., J. A. Meidl, and W. Burant, Jr., Activated Carbon Regeneration: Carbon Regeneration by Wet Air Oxidation, *Chem. Eng. Progr.*, **71**(5):90 (1975).

Haase, D. J., and D. G. Walker, The COSORB Process, *Chem. Eng. Progr.*, **70**(5):74 (1974).

Harper, E. A., J. R. Rust, and L. E. Dean, Trouble Free LNG, *Chem. Eng. Progr.*, **71**(11):75 (1975).

Haslam, A., Which Cycle for H_2 Recovery? *Hydrocarbon Process.*, **51**(3):101 (1972).

Hazelton, J. P., and R. N. Tennyson, SNG Refinery Configurations, *Chem. Eng. Progr.*, **69**(7):97 (1973).

Hegarty, W. P., and B. E. Moody, Evaluating the Bi-Gas SNG Process, *Chem. Eng. Progr.*, **69**(3):37 (1973).

———, Evaluating Proposed Ventures that Tie in with Existing Facilities, *Chem. Eng.*, **75**(17): 190 (1968).

Herron, D. P., Comparing Investment Evaluation Methods, *Chem. Eng.*, **74**(3):125 (1967).

Huebler, J., J. Janka, G. Seay, and P. Tarman, Pipeline Gas from Crude Oil, *Chem. Eng. Progr.*, **59**(5):91 (1973).

Hokanson, A. E., and R. Katzen, Chemical from Wood Waste, *Chem. Eng. Progr.*, **74**(1):67 (1978).

Hosoi, T., and H. G. Keister, Ethylene from Crude Oil, *Chem. Eng. Progr.*, **71**(11):63 (1975).

Hutchins, R. A., Activated Carbon Regeneration: Thermal Regeneration Costs, *Chem. Eng. Progr.*, **71**(5):80 (1975).

Jones, J., Converting Solid Wastes and Residues to Fuel, *Chem. Eng.*, **85**(1):87 (1978).

Katell, S., P. S. Lewis, and P. Wellman, The Economics of Producer Gas at Atmospheric and Elevated Pressures, *1973 Trans. Am. Assoc. Cost Eng.* (1973), p. 238.

Keane, D. P., R. B. Stobaugh, and P. L. Townsend, Vinyl Chloride: How, Where, Who—Future, *Hydrocarbon Process.*, **52**(2):99 (1973).

Kearney, D. B., Ethylene Plant Size Limitation, *Chem. Eng. Progr.*, **71**(11):68 (1975).

Kitzen, M. R., Comparing Feedstocks for Olefins Production by Pyrolysis, *Petro/Chem. Eng.*, **42**(9):24 (1970).

Klingman, G. E., and R. P. Schaaf, Make SNG from Coal? *Hydrocarbon Process.*, **51**(4):97 (1972).

Klumpar, I. V., Process Economics by Computer, *Chem. Eng.*, **77**(1):107 (1970).

———, Project Evaluation by Computer, *Chem. Eng.*, **77**(14):76 (1970).

———, Determining Economic and Process Variables of Ventures, *Chem. Eng. Progr.*, **67**(4):74 (1971).

Kuhre, C. J., and C. J. Shearer, Syn Gas from Heavy Fuels, *Hydrocarbon Process.*, **50**(12):113 (1971).

Lacey, R. E., Membrane Separation Processes, *Chem. Eng.*, **79**(19):56 (1972).

Lambrix, J. R., and C. S. Morris, Petrochemical Feedstocks for North America, *Chem. Eng. Progr.*, **68**(8):24 (1972).

Leibson, I., and C. A. Trischman, Jr., Avoiding Pitfalls in Developing a Major Capital Project, *Chem. Eng.*, **78**(18):103 (1971).

——— and ———, A Realistic Project-Development Case Study, *Chem. Eng.*, **78**(20):86 (1971).

——— and ———, Case Study Shows Project Development in Action, *Chem. Eng.*, **78**(22):85 (1971).

——— and ———, Final Stage in the Project Development Case Study, *Chem. Eng.*, **78**(25):78 (1971).

McAuley, J. H., and A. N. Mann, Evaluation of Process Economics by Computerized Design, *1969 Trans. Am. Assoc. Cost Eng.* (1969).

McMath, H. G., R. E. Lumpkin, J. R. Longanbach, and A. Sass, A Pyrolysis Reactor for Coal Gasification, *Chem. Eng. Progr.,* **70**(6):72 (1974).

Mearns, A. M., "Chemical Engineering Process Analysis," Oliver and Boyd, Edinburgh, England, 1973.

Milios, P., Water Reuse at a Coal Gasification Plant, *Chem. Eng. Progr.,* **71**(6):99 (1975).

Miller, R. L., Desalting Shapes Up, *Chem. Eng.,* **79**(19):24 (1972).

Naidel, R. W., Hydrogen Chloride Production by Combustion in Graphite Vessels, *Chem. Eng. Progr.,* **69**(2):53 (1973).

Nobles, E. J., M. Van Sickels, and S. Crossland, the CRG Process for SNG, *Chem. Eng. Progr.,* **68**(12):39 (1972).

O'Hara, J. B., N. E. Jentz, S. N. Rippee, and E. A. Mills, Producing Clean Boiler Fuel from Coal, *Chem. Eng. Progr.,* **70**(6):70 (1974).

Perry, H., Coal Conversion Technology, *Chem. Eng.,* **81**(15):88 (1974).

Phillips, R. F., Evaluate Projects Continuously until Final Capital Commitment, *Petro/Chem. Eng.,* **39**(6):27 (1967).

Pilat, H. L., Evaluating Outside Technology, *Chem. Eng. Progr.,* **68**(2):40 (1972).

Ponder, T. C., Compare Alkane Recovery Processes, *Hydrocarbon Process.,* **48**(10):141 (1969).

Powers, G. J., Heuristic Synthesis in Process Development, *Chem. Eng., Progr.,* **68**(8):88 (1972).

Ransom, E. A., Guidelines for Evaluating New Fields and Products, *Chem. Eng.,* **74**(23):286 (1967).

Roth, P. M., M. H. Rothkopf, and P. T. Grimes, Preresearch Process Analysis, *I & EC Process Des. & Dev.,* **11**(2):292 (1972).

Sawyer, J. G., and G. P. Williams, Turndown Efficiency of a Single-Train Plant, *Chem. Eng. Progr.,* **70**(2):62 (1974).

Schowalter, K. A., and N. S. Boodman, The Clean-Coke Process for Metallurgical Coke, *Chem. Eng., Progr.,* **70**(6):76 (1974).

Shearer, H. A., The COED Process Plus Char Gasification, *Chem. Eng. Progr.,* **69**(3):43 (1973).

Sherwin, M. B., and P. H. Spitz, Potential Breakthroughs in Petrochemical Technology, *Chem. Eng. Progr.,* **68**(3):69 (1972).

Smith, S. B., Activated Carbon Regeneration: The Thermal Transport Process, *Chem. Eng. Progr.,* **71**(5):87 (1975).

Spitz, P. H., Handling Your Process Engineering, *Chem. Eng.,* **74**(22):175 (1967).

Stewart, H. A., and J. L. Heck, Pressure Swing Adsorption, *Chem. Eng. Progr.,* **65**(9):78 (1969).

Stobaugh, R. B., V. A. Calarco, R. A. Morris, and L. W. Stroud, Propylene Oxide: How, Where, Who—Future, *Hydrocarbon Process.,* **52**(1):99 (1973).

Strelzoff, S., Make Ammonia from Coal, *Hydrocarbon Process.,* **53**(10):133 (1974).

———, Partial Oxidation for Syngas and Fuel, *Hydrocarbon Process.,* **53**(12):79 (1974).

Tamaki, A., SO$_2$ Processing: The Thoroughbred 101 Desulfurization Process, *Chem. Eng. Progr.,* **71**(5):55 (1975).

Taupitz, K. C., Making Liquids from Solid Fuels, *Hydrocarbon Process.,* **56**(9):219 (1977).

Taverna, M., and M. Chiti, Compare Routes to Caprolactam, *Hydrocarbon Process.,* **49**(11):137 (1970).

Teplitzky, G., Business Potential of R & D Projects, *Chem. Eng.,* **74**(12):136 (1967).

Thirkell, H., Reforming Step Is Basis of SNG Processes, *Chem. Proc. Eng.,* **53**(1):42 (1972).

Thornton, D. P., D. J. Ward, and R. A. Erickson, MRG Process for SNG, *Hydrocarbon Process.,* **51**(8):81 (1972).

van Dalen, J. D., Problems in Planning Ethylene Projects, *Chem. Eng. Progr.,* **71**(6):91 (1975).

van Dijk, C. P., and W. C. Schreiner, Hydrogen Chloride to Chlorine via the Kel-Chlor Process, *Chem. Eng. Progr.,* **69**(4):57 (1973).

Vasan, S., SO$_2$ Processing: The Citrex Process for SO$_2$ Removal, *Chem. Eng. Progr.,* **71**(5):61 (1975).

Vernède, J., How to Compare Urea Processes, *Hydrocarbon Process.,* **47**(1):143 (1968).

Ware, C. H., Jr., Improving R & D Effectiveness, *Chem. Eng.,* **85**(5):99 (1978).

Weinstein, N. J., Re-fining Schemes Compared, *Hydrocarbon Process.*, **53**(12):74 (1974).

Weiss, A. J., The CRG-Hydrogasification Process for SNG Production, *Chem. Eng. Progr.*, **59**(5):84 (1973).

Williams, K. R., and N. Van Lookeren Campagne, Non-Fossil Fuels Raise Costs, *Hydrocarbon Process.*, **52**(7):62 (1973).

Winter, O., Preliminary Economic Evaluation of Chemical Processes at the Research Level, *I & EC*, **61**(4):45 (1969).

Woodhouse, G., D. Samols, and J. Newman, The Economics and Technology of Large Ethylene Projects, *Chem. Eng.*, **81**(6):73 (1974).

Zdonik, S. B., E. J. Bassler, and L. P. Hallee, How Feedstocks Affect Ethylene, *Hydrocarbon Process.*, **53**(2):73 (1974).

SNG: The Process Options, *Chem. Eng.*, **79**(8):64 (1972).

Profitability estimation

Agarwal, J. C., and I. V. Klumpar, Profitability, Sensitivity, and Risk Analysis for Project Economics, *Chem. Eng.*, **82**(20):66 (1975).

Allen, D. H., Two New Tools for Project Evaluation, *Chem. Eng.*, **74**(14):75 (1967).

———, Investment Decisions—Evaluation Techniques in Perspective, *Chem. Eng. (London)*, 42 (January 1975).

Collier, C. A., Rate-of-Return Method for Selecting the Most Profitable Equipment, presented at American Association of Cost Engineers 12th National Meeting, Houston, Texas, June 17–19, 1968.

Douglas, F. R., R. A. Bellin, B. J. Blewitt, W. H. Kapfer, F. V. Marsik, A. P. Narins, and L. W. Pullen, Benchmark for Profitability, presented at American Association of Cost Engineers 12th National Meeting, Houston, Texas, June 17–19, 1968.

Estrup, C., Investment Profitability and Decentralization, *Brit. Chem. Eng.*, **16**(4/5):357 (1971).

Furzer, D. G., Economic Feasibility Studies for Chemical Plant, *Chem. Eng. (London)*, 33 (January 1976).

Garcia-Borras, T., Part 2, Research-Project Evaluation, *Hydrocarbon Process.*, **56**(1):171 (1977).

Gerrard, A. M., D. K. Patel, and A. A. L. Pollock, Shortcuts to Profitability Analysis, *Cost Eng.*, **16**(11):6 (1977).

Hays, J. R., Essential Planning in Investment Analysis, presented at American Association of Cost Engineers 12th National Meeting, Houston, Texas, June 17–19, 1968.

Herron, D. P., Comparing Investment Evaluation Methods, *Chem. Eng.*, **74**(3):125 (1967).

Holland, F. A., F. A. Watson, and J. K. Wilkinson, Profitability of Invested Capital, *Chem. Eng.*, **80**(19):139 (1973).

——— ——— and ———, Methods of Estimating Project Profitability, *Chem. Eng.*, **80**(22):80 (1973).

——— ——— and ———, Sensitivity Analysis of Project Profitabilities, *Chem. Eng.*, **80**(25):115 (1973).

——— ——— and ———, Time, Capital and Interest Affect Choice of Project, *Chem. Eng.*, **80**(27):83 (1973).

——— ——— and ———, Statistical Techniques Improve Decision-Making, *Chem. Eng.*, **80**(29):61 (1973).

——— ——— and ———, Probability Techniques for Estimates of Profitability, *Chem. Eng.*, **81**(1):105 (1974).

——— ——— and ———, Estimating Profitability When Uncertainties Exist, *Chem. Eng.*, **81**(3):73 (1974).

——— ——— and ———, Numerical Measures of Risk, *Chem. Eng.*, **81**(5):119 (1974).

——— ——— and ———, How to Allocate Overhead Cost and Appraise Inventory, *Chem. Eng.*, **81**(12):83 (1974).

——— ——— and ———, "Introduction to Process Economics," J. Wiley and Sons, New York, 1974.

—— and F. A. Watson, Economic Penalties of Operating a Process at Reduced Capacity, *Chem. Eng.*, **84**(1):91 (1977).

—— and ——, Putting Inflation into Profitability Studies, *Chem. Eng.*, **84**(4):87 (1977).

—— and ——, Project Risk, Inflation, and Profitability, *Chem. Eng.*, **84**(6):133 (1977).

Jelen, F. C., Pitfalls in Profitability Analysis, *Hydrocarbon Process.*, **55**(1):111 (1976).

——, "Cost and Optimization Engineering," McGraw-Hill Book Company, New York, 1970.

Kapfer, W. H., R. A. Bellin, B. J. Blewitt, F. R. Douglas, F. V. Marsik, L. W. Pullen, and J. E. Ross, Steps to Profitability, *1970 Trans. Am. Assoc. Cost. Eng.* (1970), p. 68.

Klumpar, I. V., Process Economics by Computer, *Chem. Eng.*, **77**(1):107 (1970).

——, Determining Economic and Process Variables of Ventures, *Chem. Eng. Progr.*, **67**(4):74 (1971).

Leibson, I., and C. A. Trischman, Spotlight on Operating Cost, *Chem. Eng.*, **78**(12):69 (1971).

—— and ——, How to Cut Operating Costs: Evaluate Your Feedstocks, *Chem. Eng.*, **78**(13):92 (1971).

—— and ——, How to Get Approval of Capital Projects, *Chem. Eng.*, **78**(15):95 (1971).

—— and ——, When and How to Apply Discounted Cash Flow and Present Worth, *Chem. Eng.*, **78**(28):97 (1971).

—— and ——, Decision Trees: A Rapid Evaluation of Investment Risk, *Chem. Eng.*, **79**(2):99 (1972).

—— and ——, Should You Make or Buy Your Major Raw Materials? *Chem. Eng.*, **79**(4):76 (1972).

—— and ——, Zeroing in on "Make or Buy" Decisions, *Chem. Eng.*, **79**(6):113 (1972).

—— and ——, How to Profit from Product Improvement and Development, *Chem. Eng.*, **79**(8):103 (1972).

Malloy, J. B., Risk Analysis of Chemical Plants, *Chem. Eng. Progr.*, **67**(10):68 (1971).

Merckx, L. J., Compare Large HPI Plants by Index of Technical Success, *Hydrocarbon Process.*, **50**(8):103 (1971).

Moore, C. S., Supply and Demand Curves in Profitability Analysis, *Chem. Eng.*, **75**(21):198 (1968).

Popper, H., "Modern Cost-Engineering Techniques," McGraw-Hill Book Company, New York, 1970.

Reul, R. I., Economic Productivity of Investments, *Chem. Proc. Eng.*, **48**(5):92 (1967).

Robertson, J. M., Design for Expansion, Part 1: Economics, *Chem. Eng.*, **75**(10):179 (1968).

Schenck, G. K., Revenue Estimation in Cost Engineering, *AACE Bull.*, **16**(3):65 (1974).

Stroup, R., Jr., Breakeven Analysis, *Chem. Eng.*, **79**(1):122 (1972).

Teplitzky, G., Using the Profit-and-Loss Statement, *Chem. Eng.*, **74**(13):215 (1967).

Thorngren, J. T., Probability Technique Improves Investment Analysis, *Chem. Eng.*, **74**(17):132 (1967).

Vogely, W. A., Cost Evaluation Techniques Applied to Mineral Resource Planning, presented at American Association of Cost Engineers 12th National Meeting, Houston, Texas, June 17–19, 1968.

Wild, N. H., Return on Investment Made Easy, *Chem. Eng.*, **83**(8):153 (1976).

Profitability index

Chilton, T. H., Investment Return via the Engineer's Method, *Chem. Eng. Progr.*, **65**(7):29 (1969).

Holland, F. A., F. A. Watson, and J. K. Wilkinson, Engineering Economics for Chemical Engineers, *Chem. Eng.*, **80**(15):103 (1973).

Kaitz, M., Percentage Gain on Investment—An Investment Decision Yardstick, *J. Petrol. Tech.*, **19**(5):679 (1967).

Reul, R. I., Which Investment Appraisal Techniques Should You Use? *Chem. Eng.*, **75**(10):212 (1968).

Rate of return

Abbot, J. T., R. R. Janssen, and C. M. Merz, Finance for Engineers, *Chem. Eng.*, **80**(15):108 (1973).

Agarwal, J. C., and I. V. Klumpar, Profitability, Sensitivity, and Risk Analysis for Project Economics, *Chem. Eng.*, **82**(20):66 (1975).

Allen, D. H., Two New Tools for Project Evaluation, *Chem. Eng.*, **74**(14):75 (1967).

———, Investment Decisions—Evaluation Techniques in Perspective, *Chem. Eng. (London)*, 42 (January 1975).

Asnani, G. C., Rate of Return Calculator, *Western Electric Eng.*, **18**(2):30 (1974).

Barish, N. N., and S. Kaplan, "Economic Analysis for Engineering and Managerial Decision Making," 2d ed., McGraw-Hill Book Company, New York, 1978.

Champley, J. A., A "Business" Approach to Capital-Budgeting Decisions, *Chem. Eng.*, **76**(20):127 (1969).

Childs, J. F., Should Your Pet Project Be Built? What Should the Profit Test Be? *Chem. Eng.*, **75**(5):188 (1968).

Chilton, T. H., Investment Return via the Engineer's Method, *Chem. Eng. Progr.*, **65**(7):29 (1969).

Collier, C. A., Rate-of-Return Method for Selecting the Most Profitable Equipment, presented at American Association of Cost Engineers 12th National Meeting, Houston, Texas, June, 1968.

Drayer, D. E., How to Estimate Plant Cost-Capacity Relationship, *Petro/Chem. Eng.*, **42**(5):10 (1970).

Ebly, R. W., Comparison of Methods for Evaluating Capital Equipment Replacement, *1973 Trans. Am. Assoc. Cost Eng.* (1973), p. 23.

Foord, A., Investment Appraisal for Chemical Engineers—Part 2, *Chem. Proc. Eng.*, **48**(10):95 (1967).

Harrison, M. R., and C. M. Pelanne, Cost-Effective Thermal Insultation, *Chem. Eng.*, **84**(27):63 (1977).

Haskett, C. E., Evaluation by Rate of Return or Present Value? *Petrol. Eng.*, **8**(45):48 (1972).

Hays, J. R., Essential Planning in Investment Analysis, presented at American Association of Cost Engineers 12th National Meeting, Houston, Texas, June, 1968.

Herron, D. P., Comparing Investment Evaluation Methods, *Chem. Eng.*, **74**(3):125 (1967).

Holland, F. A., F. A. Watson, and J. K. Wilkinson, Engineering Economics for Chemical Engineers, *Chem. Eng.*, **80**(15):103 (1973).

——— ——— and ———, Capital Costs and Depreciation, *Chem. Eng.*, **80**(17):118 (1973).

——— ——— and ———, Profitability of Invested Capital, *Chem. Eng.*, **80**(19):139 (1972).

——— ——— and ———, Time, Capital, and Interest Affect Choice of Project, *Chem. Eng.*, **80**(27):83 (1973).

——— ——— and ———, How to Evaluate Working Capital for a Company, *Chem. Eng.*, **81**(16):101 (1974).

——— ——— and ———, Financing Assets by Equity and Debt, *Chem. Eng.*, **81**(18):62 (1974).

Jelen, F. C., and M. S. Cole, Methods for Economic Analysis, *Hydrocarbon Process.*, **53**(7):133 (1974); **53**(9):227 (1974); **53**(10):161 (1974).

Johnnie, C. C., and D. K. Aggarwal, Calculating Plant Utility Costs, *Chem. Eng. Progr.*, **73**(11):84 (1977).

Leibson, I., and C. A. Trischman, Jr., Avoiding Pitfalls in Developing a Major Capital Project, *Chem. Eng.*, **78**(18):103 (1971).

Lyon, J. R., Using Multiple Cutoff Rates for Capital Investment, *J. Petrol. Tech.*, **27**:822 (1975).

Malloy, J. B., Instant Economic Evaluation, *Chem. Eng. Progr.*, **65**(11):47 (1969).

———, Risk Analysis of Chemical Plants, *Chem. Eng. Progr.*, **67**(10):68 (1971).

Moore, C. S., Supply and Demand Curves in Profitability Analysis, *Chem. Eng.*, **75**(21):198 (1968).

Paier, W. H., Appraising Rate of Return Methods, *Chem. Eng. Progr.*, **65**(11):55 (1969).

Reul, R. I., Economic Productivity of Investments, *Chem. Proc. Eng.*, **48**(5):92 (1967).

———, Which Investment Appraisal Techniques Should You Use? *Chem. Eng.*, **75**(10):212 (1968).

Riggs, J. L., "Engineering Economics," McGraw-Hill Book Company, New York, 1977.

Ross, R. C., Uncertainty Analysis Helps in Making Business Decision, *Chem. Eng.*, **78**(21):149 (1971).

Santangelo, J. G., Computers in New Business Evaluation, *1973 Trans. Am. Assoc. Cost Eng.* (1973), p. 45.

Tarquin, A. J., and L. Blank, "Engineering Economy: A Behavioral Approach," McGraw-Hill Book Company, New York, 1976.

Thorngren, J. T., Probability Technique Improves Investment Analysis, *Chem. Eng.*, **74**(17):132 (1967).

Ting, A. P., and S. W. Wan, Sizing CO Shift Converters, *Chem. Eng.*, **76**(11):185 (1969).
Tuckers, W., and W. E. Cline, Large Plant Reliability, *Chem. Eng. Progr.*, **67**(1):37 (1971).

Sales and profit

Holland, F. A., F. A. Watson, and J. K. Wilkinson, How to Budget and Control Manufacturing Costs, *Chem. Eng.*, **81**(10):105 (1974).
Stroup, R., Jr., Breakeven Analysis, *Chem. Eng.*, **79**(1):122 (1972).
When Is a Profit Not a Profit? *Hydrocarbon Process.*, **53**(11):217 (1974).

Venture analysis

Bobis, A. H., and A. Atkinson, Analyzing Potential Research Projects, *Chem. Eng.*, **77**(4):95 (1970).
Chinn, J. S., and W. A. Cuddy, Project Decision and Control: A Case History, *Chem. Eng. Progr.*, **67**(6):17 (1971).
Hegarty, W. P., Evaluating Proposed Ventures That Tie in With Existing Facilities, *Chem. Eng.*, **75**(17):190 (1968).
Herbert, V. D., Jr., and A. Bisio, The Risk and the Benefit: Part 2. Venture Analysis, *Chemtech*, **6**(7):422 (1976).
Jelen, F. C., and M. S. Cole, Methods for Economic Analysis, *Hydrocarbon Process.*, **53**(9):227 (1974); **53**(10):161 (1974).
Leibson, I., and C. A. Trischman, Jr., A Realistic Project-Development Case Study, *Chem. Eng.*, **78**(20):86 (1971).
Oakes, W. R. T., Jr., and R. L. Ralston, Life Cycle Cost Techniques in the Department of Defense, presented at American Association of Cost Engineers 12th National Meeting, Houston, Texas, June 17–19, 1968.
Reilly, P. M., and H. P. Johri, Decision-Making Through Opinion Analysis, *Chem. Eng.*, **76**(7):122 (1969).
Strauss, R., The Sensitivity Chart—Giving Meaning to Shaky Estimates, *Chem. Eng.*, **75**(7):112 (1968).
Woods, L. M., How Mobil Handles Venture Analysis, *Petro/Chem. Eng.*, **42**(11):16 (1970).

PROBLEMS

1 A proposed chemical plant will require a fixed-capital investment of $10 million. It is estimated that the working capital will amount to 25 percent of the total investment, and annual depreciation costs are estimated to be 10 percent of the fixed-capital investment. If the annual profit will be $3 million, determine the standard percent return on the total investment and the minimum payout period.

2 An investigation of a proposed investment has been made. The following result has been presented to management: The minimum payout period based on capital recovery using a minimum annual return of 10 percent as a fictitious expense is 10 years; annual depreciation costs amount to 8 percent of the total investment. Using this information, determine the standard rate of return on the investment.

3 The information given in Prob. 2 applies to conditions before income taxes. If 45 percent of all profits must be paid out for income taxes, determine the standard rate of return after taxes using the figures given in Prob. 2.

4 A heat exchanger has been designed and insulation is being considered for the unit. The insulation can be obtained in thicknesses of 1, 2, 3, or 4 in. The following data have been determined for the different insulation thicknesses:

	1 in.	2 in.	3 in.	4 in.
Btu/h saved	300,000	350,000	370,000	380,000
Cost for installed insulation	$1200	$1600	$1800	$1870
Annual fixed charges	10%	10%	10%	10%

What thickness of insulation should be used? The value of heat is 30 cents/1,000,000 Btu. An annual return of 15 percent on the fixed-capital investment is required for any capital put into this type of investment. The exchanger operates 300 days per year.

5 A company must purchase one reactor to be used in an overall operation. Four reactors have been designed, all of which are equally capable of giving the required service. The following data apply to the four designs:

	Design 1	Design 2	Design 3	Design 4
Fixed-capital investment	$10,000	$12,000	$14,000	$16,000
Sum of operating and fixed costs per year (all other costs are constant)	3,000	2,800	2,350	2,100

If the company demands a 15 percent return on any unnecessary investment, which of the four designs should be accepted?

6 The capitalized cost for a piece of equipment has been found to be $55,000. This cost is based on the original cost plus the present value of an indefinite number of renewals. An annual interest rate of 12 percent was used in determining the capitalized cost. The salvage value of the equipment at the end of the service life was taken to be zero, and the service life was estimated to be 10 years. Under these conditions, what would be the original cost of the equipment?

7 An existing warehouse is worth $500,000, and the average value of the goods stored in it is $400,000. The annual insurance rate on the warehouse is 1.1 percent, and the insurance rate on the stored goods is 0.95 percent. If a proposed sprinkling system is installed in the warehouse, both insurance rates would be reduced to three-quarters of the original rate. The installed sprinkler system would cost $20,000, and the additional annual cost for maintenance, inspection, and taxes would be $300. The required write-off period for the entire investment in the sprinkler system is 20 years. The capital necessary to make the investment is available. The operation of the warehouse is now giving an 8 percent return on the original investment. Give reasons why you would or would not recommend installing the sprinkler system.

8 A power plant for generating electricity is one part of a plant-design proposal. Two alternative power plants with the necessary capacity have been suggested. One uses a boiler and steam turbine while the other uses a gas turbine. The following information applies to the two proposals:

	Boiler and steam turbine	Gas turbine
Initial investment	$600,000	$400,000
Fuel costs, per year	16,000	23,000
Maintenance and repairs, per year	12,000	15,000
Insurance and taxes, per year	18,000	12,000
Service life, years	20	10
Salvage value at end of service life	0	0

All other costs are the same for either type of power plant. A 12 percent return is required on any investment. If one of these two power plants must be accepted, which one should be recommended?

9 The facilities of an existing chemical company must be increased if the company is to continue in operation. There are two alternatives. One of the alternatives is to expand the present plant. If this is done, the expansion would cost $130,000. Additional labor costs would be $150,000 per year, while additional costs for overhead, depreciation, taxes, and insurance would be $60,000 per year.

A second alternative requires construction and operation of new facilities at a location about 50 miles from the present plant. This alternative is attractive because cheaper labor is available at this location. The new facilities would cost $200,000. Labor costs would be $120,000 per year. Overhead costs would be $70,000 per year. Annual insurance and taxes would amount to 2 percent of the initial cost. All other costs except depreciation would be the same at each location. If the minimum return on any acceptable investment is 9 percent, determine the minimum service life allowable for the facilities at the distant location for this alternative to meet the required incremental return. The salvage value should be assumed to be zero, and straight-line depreciation accounting may be used.

10 A chemical company is considering replacing a batch-wise reactor with a modernized continuous reactor. The old unit cost $40,000 when new 5 years ago, and depreciation has been charged on a straight-line basis using an estimated service life of 15 years and final salvage value of $1000. It is now estimated that the unit has a remaining service life of 10 years and a final salvage value of $1000.

The new unit would cost $70,000 and would result in an increase of $5000 in the gross annual income. It would permit a labor saving of $7000 per year. Additional costs for taxes and insurance would be $1000 per year. The service life is estimated to be 12 years with a final salvage value of $1000. All costs other than those for labor, insurance, taxes, and depreciation may be assumed to be the same for both units. The old unit can now be sold for $5000. If the minimum required return on any investment is 15 percent, should the replacement be made?

11 The owner of a small antifreeze plant has a small canning unit which cost him $5000 when he purchased it 10 years ago. The unit has been completely depreciated, but the owner estimates that it will still give good service for 5 more years. At the end of 5 years the unit will be worth a junk value of $100. The owner now has an opportunity to buy a more efficient canning unit for $6000 having an estimated service life of 10 years and zero salvage or junk value. This new unit would reduce annual labor and maintenance costs by $1000 and increase annual expenses for taxes and insurance by $100. All other expenses except depreciation would be unchanged. If the old canning unit can be sold for $600, what replacement return on his capital investment will the owner receive if he decides to make the replacement?

12 An engineer in charge of the design of a plant must choose either a batch or a continuous system. The batch system offers a lower initial outlay but, owing to high labor requirements, exhibits a higher operating cost. The cash flows relevant to this problem have been estimated as follows:

	Year		Discounted-cash-flow rate of return	Net present worth at 10%
	0	1–10		
Batch system	−$20,000	$5600	25%	$14,400
Continuous system	−$30,000	$7650	22%	$17,000

Check the values given for the discounted-cash-flow rate of return and net present worth. If the company requires a minimum rate of return of 10 percent, which system should be chosen?

13 A company is considering the purchase and installation of a pump which will deliver oil at a faster rate than the pump already in use. The purchase and installation of the larger pump will require an immediate outlay of $1600, but it will recover all the oil by the end of one year. The relevant cash flows have been established as follows:

	Year			Discounted-cash-flow rate of return	Net present worth at 10%
	0	1	2		
Install larger pump	−$1600	$20,000	0	1250%	$16,580
Operate existing pump	0	$10,000	$10,000	?	$17,355

Explain the values given for the discounted-cash-flow rate of return and net present worth. If the company requires a minimum rate of return of 10 percent, which alternative should be chosen?

14. An oil company is offered a lease of a group of oil wells on which the primary reserves are close to exhaustion. The major condition of the purchase is that the oil company must agree to undertake a water-flood project at the end of five years to make possible secondary recovery. No immediate payment by the oil company is required. The relevant cash flows have been estimated as follows:

Year				Discounted-cash-flow rate of return	Net present worth at 10%
0	1–4	5	6–20		
0	$50,000	−$650,000	$100,000	?	$227,000

Should the lease-and-flood arrangement be accepted? How should this proposal be presented to the company board of directors who understand and make it a policy to evaluate by discounted-cash-flow rate of return?

15 For Example 3 in this chapter, determine the profitability index using the simplified Eq. (*D*) in the example instead of Eq. (*E*) as was used for Table 3. As a first approximation, assume the profitability index is 30 percent.

CHAPTER TEN

OPTIMUM DESIGN AND DESIGN STRATEGY

An optimum design is based on the best or most favorable conditions. In almost every case, these optimum conditions can ultimately be reduced to a consideration of costs or profits. Thus, an optimum economic design could be based on conditions giving the least cost per unit of time or the maximum profit per unit of production. When one design variable is changed, it is often found that some costs increase and others decrease. Under these conditions, the total cost may go through a minimum at one value of the particular design variable, and this value would be considered as an optimum.

An example illustrating the principles of an optimum economic design is presented in Fig. 10-1. In this simple case, the problem is to determine the optimum thickness of insulation for a given steam-pipe installation. As the insulation thickness is increased, the annual fixed costs increase, the cost of heat loss decreases, and all other costs remain constant. Therefore, as shown in Fig. 10-1, the sum of the costs must go through a minimum at the optimum insulation thickness.

Although cost considerations and economic balances are the basis of most optimum designs, there are times when factors other than cost can determine the most favorable conditions. For example, in the operation of a catalytic reactor, an optimum operation temperature may exist for each reactor size because of equilibrium and reaction-rate limitations. This particular temperature could be based on the maximum percentage conversion or on the maximum amount of final product per unit of time. Ultimately, however, cost variables need to be considered, and the development of an optimum operation design is usually merely one step in the determination of an optimum economic design.

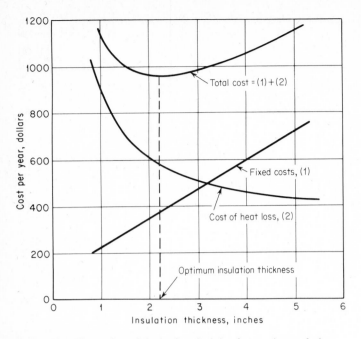

Figure 10-1 Illustration of the basic principle of an optimum design.

INCREMENTAL COSTS

The subject of incremental costs is covered in detail in Chap. 9 (Profitability, Alternative Investments, and Replacements). Consideration of incremental costs shows that a final recommended design does not need to correspond to the optimum economic design, because the incremental return on the added investment may become unacceptable before the optimum point is reached.† However, the optimum values can be used as a basis for starting the incremental-cost analyses.

This chapter deals with methods for determining optimum conditions, and it is assumed that the reader understands the role of incremental costs in establishing a final recommended design.

INTANGIBLE AND PRACTICAL CONSIDERATIONS

The various mathematical methods for determining optimum conditions, as presented in this chapter, represent on a theoretical basis the conditions that best

---

† See Fig. 9-5 and the related discussion in Chap. 9. The material presented in Chap. 10 considers optimum designs based on maximum or minimum values of a specified variable. The same type of approach could be used if the term *optimum* (referring to an investment) were defined on the basis of a stipulated incremental return.

meet the requirements. However, factors that cannot easily be quantitized or practical considerations may change the final recommendation to other than the theoretically correct optimum condition. Thus, a determination of an "optimum condition," as described in this chapter, serves as a base point for a cost or design analysis, and it can often be quantitized in specific mathematical form. From this point, the engineer must apply judgment to take into account other important practical factors, such as return on investment or the fact that commercial equipment is often available in discrete intervals of size.

As an example, consider the case where an engineer has made an estimation of the optimum pipe diameter necessary to handle a given flow stream based on minimizing the costs due to fixed charges and frictional pumping costs. The mathematical result shows that the optimum inside pipe diameter is 2.54 in. based on costs for standard (schedule 40) steel pipe. Nominal pipe diameters available commercially in this range are $2\frac{1}{2}$ in. (ID of 2.469 in.) and 3 in. (ID of 3.069 in.). The practical engineer would probably immediately recommend a nominal pipe diameter of $2\frac{1}{2}$ in. without going to the extra effort of calculating return on investment for the various sizes available. This approach would normally be acceptable because of the estimations necessarily involved in the optimum calculation and because of the fact that an investment for pipe represents only a small portion of the total investment.

Intangible factors may have an effect on the degree of faith that can be placed on calculated results for optimum conditions. Perhaps the optimum is based on an assumed selling price for the product from the process, or it might be that a preliminary evaluation is involved in which the location of the plant is not final. Obviously, for cases of this type, an analysis for optimum conditions can give only a general idea of the actual results that will be obtained in the final plant, and it is not reasonable to go to extreme limits of precision and accuracy in making recommendations. Even for the case of a detailed and firm design, intangibles, such as the final bid from various contractors for the construction,† may make it impractical to waste a large amount of effort in bringing too many refinements into the estimation of optimum conditions.

GENERAL PROCEDURE FOR DETERMINING OPTIMUM CONDITIONS

The first step in the development of an optimum design is to determine what factor is to be optimized. Typical factors would be total cost per unit of production or per unit of time, profit, amount of final product per unit of time, and percent conversion. Once the basis is determined, it is necessary to develop relationships showing how the different variables involved affect the chosen factor.

---

† As an example of the contractor's view of strategic bidding, see W. R. Park, Profit Optimization Through Strategic Bidding, *AACE Bull.*, **6**(5): 141 (1964).

Finally, these relationships are combined graphically or analytically to give the desired optimum conditions.

PROCEDURE WITH ONE VARIABLE

There are many cases in which the factor being optimized is a function of a single variable. The procedure then becomes very simple. Consider the example presented in Fig. 10-1, where it is necessary to obtain the insulation thickness which gives the least total cost. The primary variable involved is the thickness of the insulation, and relationships can be developed showing how this variable affects all costs.

Cost data for the purchase and installation of the insulation are available, and the length of service life can be estimated. Therefore, a relationship giving the effect of insulation thickness on fixed charges can be developed. Similarly, a relationship showing the cost of heat lost as a function of insulation thickness can be obtained from data on the value of steam, properties of the insulation, and heat-transfer considerations. All other costs, such as maintenance and plant expenses, can be assumed to be independent of the insulation thickness.

The two cost relationships obtained might be expressed in a simplified form similar to the following:

$$\text{Fixed charges} = \phi(x) = ax + b \tag{1}$$

$$\text{Cost of heat loss} = \phi^i(x) = \frac{c}{x} + d \tag{2}$$

$$\text{Total variable cost} = C_T = \phi(x) + \phi^i(x) = \phi^{ii}(x) = ax + b + \frac{c}{x} + d \tag{3}$$

where a, b, c, and d are constants and x is the common variable (insulation thickness).

The *graphical* method for determining the optimum insulation thickness is shown in Fig. 10-1. The optimum thickness of insulation is found at the minimum point on the curve obtained by plotting total variable cost versus insulation thickness.

The slope of the total-variable-cost curve is zero at the point of optimum insulation thickness. Therefore, if Eq. (3) applies, the optimum value can be found *analytically* by merely setting the derivative of C_T with respect to x equal to zero and solving for x.

$$\frac{dC_T}{dx} = a - \frac{c}{x^2} = 0 \tag{4}$$

$$x = \left(\frac{c}{a}\right)^{1/2} \tag{5}$$

If the factor being optimized (C_T) does not attain a usable maximum or minimum value, the solution for the dependent variable will indicate this condition by giving an impossible result, such as infinity, zero, or the square root of a negative number.

The value of x shown in Eq. (5) occurs at an optimum point or a point of inflection. The second derivative of Eq. (3), evaluated at the given point, indicates if the value occurs at a minimum (second derivative greater than zero), maximum (second derivative less than zero), or point of inflection (second derivative equal to zero). An alternative method for determining the type of point involved is to calculate values of the factor being optimized at points slightly greater and slightly smaller than the optimum value of the dependent variable.

The second derivative of Eq. (3) is

$$\frac{d^2C_T}{dx^2} = +\frac{2c}{x^3} \tag{6}$$

If x represents a variable such as insulation thickness, its value must be positive; therefore, if c is positive, the second derivative at the optimum point must be greater than zero, and $(c/a)^{1/2}$ represents the value of x at the point where the total variable cost is a minimum.

PROCEDURE WITH TWO OR MORE VARIABLES

When two or more independent variables affect the factor being optimized, the procedure for determining the optimum conditions may become rather tedious; however, the general approach is the same as when only one variable is involved.

Consider the case in which the total cost for a given operation is a function of the two independent variables x and y, or

$$C_T = \phi^{iii}(x, y) \tag{7}$$

By analyzing all the costs involved and reducing the resulting relationships to a simple form, the following function might be found for Eq. (7):

$$C_T = ax + \frac{b}{xy} + cy + d \tag{8}$$

where a, b, c, and d are positive constants.

Graphical procedure The relationship among C_T, x, and y could be shown as a curved surface in a three-dimensional plot, with a minimum value of C_T occurring at the optimum values of x and y. However, the use of a three-dimensional plot is not practical for most engineering determinations.

The optimum values of x and y in Eq. (8) can be found graphically on a two-dimensional plot by using the method indicated in Fig. 10-2. In this figure, the factor being optimized is plotted against one of the independent variables (x), with the second variable (y) held at a constant value. A series of such plots is made with each dashed curve representing a different constant value of the second variable. As shown in Fig. 10-2, each of the curves (A, B, C, D, and E)

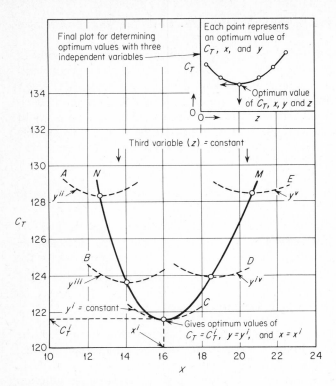

Figure 10-2 Graphical determination of optimum conditions with two or more independent variables.

gives one value of the first variable x at the point where the total cost is a minimum. The curve NM represents the locus of all these minimum points, and the optimum value of x and y occurs at the minimum point on curve NM.

Similar graphical procedures can be used when there are more than two independent variables. For example, if a third variable z were included in Eq. (8), the first step would be to make a plot similar to Fig. 10-2 at one constant value of z. Similar plots would then be made at other constant values of z. Each plot would give an optimum value of x, y, and C_T for a particular z. Finally, as shown in the insert in Fig. 10-2, the overall optimum value of x, y, z, and C_T could be obtained by plotting z versus the individual optimum values of C_T.

Analytical procedure In Fig. 10-2, the optimum value of x is found at the point where $(\partial C_T/\partial x)_{y=y^i}$ is equal to zero. Similarly, the same results would be obtained if y were used as the abscissa instead of x. If this were done, the optimum value of y (i.e., y^i) would be found at the point where $(\partial C_T/\partial y)_{x=x^i}$ is equal to zero. This immediately indicates an analytical procedure for determining optimum values.

Using Eq. (8) as a basis,

$$\frac{\partial C_T}{\partial x} = a - \frac{b}{x^2 y} \tag{9}$$

$$\frac{\partial C_T}{\partial y} = c - \frac{b}{xy^2} \tag{10}$$

At the optimum conditions, both of these partial derivatives must be equal to zero; thus, Eqs. (9) and (10) can be set equal to zero and the optimum values of $x = (cb/a^2)^{1/3}$ and $y = (ab/c^2)^{1/3}$ can be obtained by solving the two simultaneous equations. If more than two independent variables were involved, the same procedure would be followed, with the number of simultaneous equations being equal to the number of independent variables.

Example 1. Determination of optimum values with two independent variables The following equation shows the effect of the variables x and y on the total cost for a particular operation:

$$C_T = 2.33x + \frac{11,900}{xy} + 1.86y + 10$$

Determine the values of x and y which will give the least total cost.

SOLUTION *Analytical method:*

$$\frac{\partial C_T}{\partial x} = 2.33 - \frac{11,900}{x^2 y}$$

$$\frac{\partial C_T}{\partial y} = 1.86 - \frac{11,900}{xy^2}$$

At the optimum point,

$$2.33 - \frac{11,900}{x^2 y} = 0$$

$$1.86 - \frac{11,900}{xy^2} = 0$$

Solving simultaneously for the optimum values of x and y,

$$x = 16$$

$$y = 20$$

$$C_T = 121.6$$

A check should be made to make certain the preceding values represent conditions of minimum cost.

$$\frac{\partial^2 C_T}{\partial x^2} = \frac{(2)(11,900)}{x^3 y} = \frac{(2)(11,900)}{(16)^3(20)} = + \text{ at optimum point}$$

$$\frac{\partial^2 C_T}{\partial y^2} = \frac{(2)(11,900)}{xy^3} = \frac{(2)(11,900)}{(16)(20)^3} = + \text{ at optimum point}$$

Since the second derivatives are positive, the optimum conditions must occur at a point of minimum cost.

Graphical method:

The following constant values of y are chosen arbitrarily:

$$y^{ii} = 32 \qquad y^{iii} = 26 \qquad y^i = 20 \qquad y^{iv} = 15 \qquad y^v = 12$$

At each constant value of y, a plot is made of C_T versus x. These plots are presented in Fig. 10-2 as curves *A*, *B*, *C*, *D*, and *E*. A summary of the results is presented in the following table:

y	Optimum x	Optimum C_T
$y^{ii}\ = 32$	12.7	128.3
$y^{iii} = 26$	14.1	123.6
$y^i\ = 20$	16.0	121.6
$y^{iv} = 15$	18.5	123.9
$y^v\ = 12$	20.7	128.5

One curve (*NM* in Fig. 10-2) through the various optimum points shows that the overall optimum occurs at

$$x = 16$$

$$y = 20$$

$$C_T = 121.6$$

Note: In this case, a value of y was chosen which corresponded to the optimum value. Usually, it is necessary to interpolate or make further calculations in order to determine the final optimum conditions.

COMPARISON OF GRAPHICAL AND ANALYTICAL METHODS

In the determination of optimum conditions, the same final results are obtained with either graphical or analytical methods. Sometimes it is impossible to set up one analytical function for differentiation, and the graphical method must be used. If the development and simplification of the total analytical function require complicated mathematics, it may be simpler to resort to the direct graphical solution; however, each individual problem should be analyzed on the basis of the existing circumstances. For example, if numerous repeated trials are necessary, the extra time required to develop an analytical solution may be well spent.

The graphical method has one distinct advantage over the analytical method. The shape of the curve indicates the importance of operating at or very close to the optimum conditions. If the maximum or minimum occurs at a point where the curve is flat with only a gradual change in slope, there will be a considerable spread in the choice of final conditions, and incremental cost analyses may be necessary. On the other hand, if the maximum or minimum is sharp, it may be essential to operate at the exact optimum conditions.

THE BREAK-EVEN CHART FOR PRODUCTION SCHEDULE AND ITS SIGNIFICANCE FOR OPTIMUM ANALYSIS

In considering the overall costs or profits in a plant operation, one of the factors that has an important effect on the economic results is the fraction of total available time during which the plant is in operation. If the plant stands idle or operates at low capacity, certain costs, such as those for raw materials and labor, are reduced, but costs for depreciation and maintenance continue at essentially the same rate even though the plant is not in full use.

There is a close relationship among operating time, rate of production, and selling price. It is desirable to operate at a schedule which will permit maximum utilization of fixed costs while simultaneously meeting market sales demand and using the capacity of the plant production to give the best economic results. Figure 10-3 shows graphically how production rate affects costs and profits. The

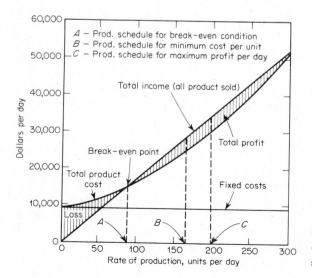

Figure 10-3 Break-even chart for operating production plant (based on situation presented in Example 2).

fixed costs remain constant while the total product cost, as well as the profit, increases with increased rate of production. The point where total product cost equals total income represents the break-even point, and the optimum production schedule must be at a production rate higher than that corresponding to the break-even point.

OPTIMUM PRODUCTION RATES IN PLANT OPERATION

The same principles used for developing an optimum design can be applied when determining the most favorable conditions in the operation of a manufacturing plant. One of the most important variables in any plant operation is the amount of product produced per unit of time. The production rate depends on many factors, such as the number of hours in operation per day, per week, or per month; the load placed on the equipment; and the sales market available. From an analysis of the costs involved under different situations and consideration of other factors affecting the particular plant, it is possible to determine an *optimum rate of production* or a so-called *economic lot size.*

When a design engineer submits a complete plant design, the study ordinarily is based on a given production capacity for the plant. After the plant is put into operation, however, some of the original design factors will have changed, and the optimum rate of production may vary considerably from the "designed capacity." For example, suppose a plant had been designed originally for the batchwise production of an organic chemical on the basis of one batch every 8 hours. After the plant has been put into operation, cost data on the actual process are obtained, and tests with various operating procedures are conducted. It is found that more total production per month can be obtained if the time per batch is reduced. However, when the shorter batch time is used, more labor is required, the percent conversion of raw materials is reduced, and steam and power costs increase. Here is an obvious case in which an economic balance can be used to find the optimum production rate. Although the design engineer may have based the original recommendations on a similar type of economic balance, price and market conditions do not remain constant, and the operations engineer now has actual results on which to base an economic balance. The following analysis indicates the general method for determining economic production rates or lot sizes.

The total product cost per unit of time may be divided into the two classifications of *operating costs* and *organization costs.* Operating costs depend on the rate of production and include expenses for direct labor, raw materials, power, heat, supplies, and similar items which are a function of the amount of material produced. Organization costs are due to expenses for directive personnel, physical equipment, and other services or facilities which must be maintained irrespective of the amount of material produced. Organization costs are independent of the rate of production.

It is convenient to consider operating costs on the basis of one unit of production. When this is done, the operating costs can be divided into two types of expenses as follows: (1) Minimum expenses for raw materials, labor, power, etc., that remain constant and must be paid for each unit of production as long as any amount of material is produced; and (2) extra expenses due to increasing the rate of production. These extra expenses are known as *superproduction costs*. They become particularly important at high rates of production. Examples of superproduction costs are extra expenses caused by overload on power facilities, additional labor requirements, or decreased efficiency of conversion. Superproduction costs can often be represented as follows:

$$\text{Superproduction costs per unit of production} = mP^n \qquad (11)$$

where P = rate of production as total units of production per unit of time
m = a constant
n = a constant

Designating h as the operating costs which remain constant per unit of production and O_c as the organization costs per unit of time, the total product cost c_T per unit of production is

$$c_T = h + mP^n + \frac{O_c}{P} \qquad (12)$$

The following equations for various types of costs or profits are based on Eq. (12):

$$C_T = c_T P = \left(h + mP^n + \frac{O_c}{P}\right)P \qquad (13)$$

$$r = s - c_T = s - h - mP^n - \frac{O_c}{P} \qquad (14)$$

$$R' = rP = \left(s - h - mP^n - \frac{O_c}{P}\right)P \qquad (15)$$

where C_T = total product cost per unit of time
r = profit per unit of production
R' = profit per unit of time
s = selling price per unit of production

OPTIMUM PRODUCTION RATE FOR MINIMUM COST PER UNIT OF PRODUCTION

It is often necessary to know the rate of production which will give the least cost on the basis of one unit of material produced. This information shows the selling price at which the company would be forced to cease operation or else operate at a loss. At this particular optimum rate, a plot of the total product cost per unit of production versus the production rate shows a minimum product cost; therefore, the optimum production rate must occur where $dc_T/dP = 0$. An analytical solution for this case may be obtained from Eq. (12), and the optimum rate P_o giving the minimum cost per unit of production is found as follows:

$$\frac{dc_T}{dP} = 0 = nmP_o^{n-1} - \frac{O_c}{P_o^2} \tag{16}$$

$$P_o = \left(\frac{O_c}{nm}\right)^{1/(n+1)} \tag{17}$$

The optimum rate shown in Eq. (17) would, of course, give the maximum profit per unit of production if the selling price remains constant.

OPTIMUM PRODUCTION RATE FOR MAXIMUM TOTAL PROFIT PER UNIT OF TIME

In most business concerns, the amount of money earned over a given time period is much more important than the amount of money earned for each unit of product sold. Therefore, it is necessary to recognize that the production rate for maximum profit per unit of time may differ considerably from the production rate for minimum cost per unit of production.

Equation (15) presents the basic relationship between costs and profits. A plot of profit per unit of time versus production rate goes through a maximum. Equation (15), therefore, can be used to find an analytical value of the optimum production rate. When the selling price remains constant, the optimum rate giving the maximum profit per unit of time is

$$P_o = \left[\frac{s - h}{(n + 1)m}\right]^{1/n} \tag{18}$$

The following example illustrates the preceding principles and shows the analytical solution for the situation presented in Fig. 10-3.

Example 2. Determination of profits at optimum production rates A plant produces refrigerators at the rate of P units per day. The variable costs per refrigerator have been found to be $\$47.73 + 0.1P^{1.2}$. The total daily fixed charges are \$1750, and all other expenses are constant at \$7325 per day. If the selling price per refrigerator is \$173, determine:

(a) The daily profit at a production schedule giving the minimum cost per refrigerator.

(b) The daily profit at a production schedule giving the maximum daily profit.

(c) The production schedule at the break-even point.

SOLUTION:

(a) Total cost per refrigerator $= c_T = 47.73 + 0.1P^{1.2} + (1750 + 7325)/P$. At production schedule for minimum cost per refrigerator,

$$\frac{dc_T}{dP} = 0 = 0.12P_o^{0.2} - \frac{9075}{P_o^2}$$

$P_o = 165$ units per day for minimum cost per unit

Daily profit at production schedule for minimum cost per refrigerator

$$= \left[173 - 47.73 - 0.1(165)^{1.2} - \frac{9075}{165} \right] 165$$

$$= \$4040$$

(b) Daily profit is

$$R' = \left(173 - 47.73 - 0.1P^{1.2} - \frac{1750 + 7325}{P} \right) P$$

At production schedule for maximum profit per day,

$$\frac{dR'}{dP} = 0 = 125.27 - 0.22P_o^{1.2}$$

$P_o = 198$ units per day for maximum daily profit

Daily profit at production schedule for maximum daily profit

$$= \left[173 - 47.73 - 0.1(198)^{1.2} - \frac{9075}{198} \right] 198$$

$$= \$4400$$

(c) Total profit per day $= \left\{ 173 - \left[47.73 + 0.1P^{1.2} + \frac{(1750 + 7325)}{P} \right] \right\} P = 0$

at break-even point.

Solving the preceding equation for P,

$$P_{\text{at break-even point}} = 88 \text{ units/day}$$

OPTIMUM CONDITIONS IN CYCLIC OPERATIONS

Many processes are carried out by the use of cyclic operations which involve periodic shutdowns for discharging, cleanout, or reactivation. This type of operation occurs when the product is produced by a batch process or when the rate of production decreases with time, as in the operation of a plate-and-frame filtration unit. In a true batch operation, no product is obtained until the unit is shut down for discharging. In semicontinuous cyclic operations, product is delivered continuously while the unit is in operation, but the rate of delivery decreases with time. Thus, in batch or semicontinuous cyclic operations, the variable of total time required per cycle must be considered when determining optimum conditions.

Analyses of cyclic operations can be carried out conveniently by using the time for one cycle as a basis. When this is done, relationships similar to the following can be developed to express overall factors, such as total annual cost or annual rate of production:

$$\text{Total annual cost} = \text{operating and shutdown costs/cycle} \\ \times \text{cycles/year} + \text{annual fixed costs} \qquad (19)$$

$$\text{Annual production} = (\text{production/cycle})(\text{cycles/year}) \qquad (20)$$

$$\text{Cycles/year} = \frac{\text{operating} + \text{shutdown time used/year}}{\text{operating} + \text{shutdown time/cycle}} \qquad (21)$$

The following example illustrates the general method for determining optimum conditions in a batch operation.

Example 3. Determination of conditions for minimum total cost in a batch operation An organic chemical is being produced by a batch operation in which no product is obtained until the batch is finished. Each cycle consists of the operating time necessary to complete the reaction plus a total time of 1.4 h for discharging and charging. The operating time per cycle is equal to $1.5P_b^{0.25}$ h, where P_b is the kilograms of product produced per batch. The operating costs during the operating period are \$20 per hour, and the costs during the discharge-charge period are \$15 per hour. The annual fixed costs for the equipment vary with the size of the batch as follows:

$$C_F = 340P_b^{0.8} \text{ dollars per batch}$$

Inventory and storage charges may be neglected. If necessary, the plant can be operated 24 h per day for 300 days per year. The annual production is 1 million kg of product. At this capacity, raw-material and miscellaneous costs, other than those already mentioned, amount to \$260,000 per year. Determine the cycle time for conditions of minimum total cost per year.

SOLUTION

$$\text{Cycles/year} = \frac{\text{annual production}}{\text{production/cycle}} = \frac{1,000,000}{P_b}$$

$$\text{Cycle time} = \text{operating} + \text{shutdown time} = 1.5P_b^{0.25} + 1.4 \text{ h}$$

$$\text{Operating} + \text{shutdown costs/cycle} = (20)(1.5P_b^{0.25}) + (15)(1.4) \text{ dollars}$$

$$\text{Annual fixed costs} = 340P_b^{0.8} + 260,000 \text{ dollars}$$

$$\text{Total annual costs} = (30P_b^{0.25} + 21)(1,000,000/P_b) + 340P_b^{0.8} \\ + 260,000 \text{ dollars}$$

The total annual cost is a minimum where $d(\text{total annual cost})/dP_b = 0$.

Performing the differentiation, setting the result equal to zero, and solving for P_b gives

$$P_{b, \text{optimum cost}} = 1630 \text{ kg per batch}$$

This same result could have been obtained by plotting total annual cost versus P_b and determining the value of P_b at the point of minimum annual cost.

For conditions of minimum annual cost and 1 million kg/year production,

$$\text{Cycle time} = (1.5)(1630)^{0.25} + 1.4 = 11 \text{ h}$$

$$\text{Total time used per year} = (11)\left(\frac{1{,}000{,}000}{1630}\right) = 6750 \text{ h}$$

$$\text{Total time available per year} = (300)(24) = 7200 \text{ h}$$

Thus, for conditions of minimum annual cost and a production of 1 million kg/year, not all the available operating and shutdown time would be used.

SEMICONTINUOUS CYCLIC OPERATIONS

Semicontinuous cyclic operations are often encountered in the chemical industry, and the design engineer should understand the methods for determining optimum cycle times in this type of operation. Although product is delivered continuously, the rate of delivery decreases with time owing to scaling, collection of side product, reduction in conversion efficiency, or similar causes. It becomes necessary, therefore, to shut down the operation periodically in order to restore the original conditions for high production rates. The optimum cycle time can be determined for conditions such as maximum amount of production per unit of time or minimum cost per unit of production.

Scale Formation in Evaporation

During the time an evaporator is in operation, solids often deposit on the heat-transfer surfaces, forming a scale. The continuous formation of the scale causes a gradual increase in the resistance to the flow of heat and, consequently, a reduction in the rate of heat transfer and rate of evaporation if the same temperature-difference driving forces are maintained. Under these conditions, the evaporation unit must be shut down and cleaned after an optimum operation time, and the cycle is then repeated.

Scale formation occurs to some extent in all types of evaporators, but it is of particular importance when the feed mixture contains a dissolved material that has an inverted solubility. The expression *inverted solubility* means the solubility decreases as the temperature of the solution is increased. For a material of this type, the solubility is least near the heat-transfer surface where the temperature is the greatest. Thus, any solid crystallizing out of the solution does so near the heat-transfer surface and is quite likely to form a scale on this surface. The most common scale-forming substances are calcium sulfate, calcium hydroxide, sodium carbonate, sodium sulfate, and calcium salts of certain organic acids.

When true scale formation occurs, the overall heat-transfer coefficient may be related to the time the evaporator has been in operation by the straight-line equation†

$$\frac{1}{U^2} = a\theta_b + d \tag{22}$$

where a and d are constants for any given operation and U is the overall heat-transfer coefficient at any operating time θ_b since the beginning of the operation.

If it is not convenient to determine the heat-transfer coefficients and the related constants as shown in Eq. (22), any quantity that is proportional to the heat-transfer coefficient may be used. Thus, if all conditions except scale formation are constant, feed rate, production rate, and evaporation rate can each be represented in a form similar to Eq. (22). Any of these equations can be used as a basis for finding the optimum conditions. The general method is illustrated by the following treatment, which employs Eq. (22) as a basis.

If Q represents the total amount of heat transferred in the operating time θ_b, and A and Δt represent, respectively, the heat-transfer area and temperature-difference driving force, the rate of heat transfer at any instant is

$$\frac{dQ}{d\theta_b} = UA\,\Delta t = \frac{A\,\Delta t}{(a\theta_b + d)^{1/2}} \tag{23}$$

The instantaneous rate of heat transfer varies during the time of operation, but the heat-transfer area and the temperature-difference driving force remain essentially constant. Therefore, the total amount of heat transferred during an operating time of θ_b can be determined by integrating Eq. (23) as follows:

$$\int_0^Q dQ = A\,\Delta t \int_0^{\theta_b} \left(\frac{1}{a\theta_b + d}\right)^{1/2} d\theta_b \tag{24}$$

$$Q = \frac{2A\,\Delta t}{a}\left[(a\theta_b + d)^{1/2} - d^{1/2}\right] \tag{25}$$

Cycle time for maximum amount of heat transfer Equation (25) can be used as a basis for finding the cycle time which will permit the maximum amount of heat transfer during a given period. Each cycle consists of an operating (or boiling) time of θ_b h. If the time per cycle for emptying, cleaning, and recharging is θ_c, the total time in hours per cycle is $\theta_t = \theta_b + \theta_c$. Therefore, designating the total time *used* for actual operation, emptying, cleaning, and refilling as H, the number of cycles during H h = $H/(\theta_b + \theta_c)$.

---

† W. McCabe and C. Robinson, *Ind. Eng. Chem.*, **16**:478 (1924).

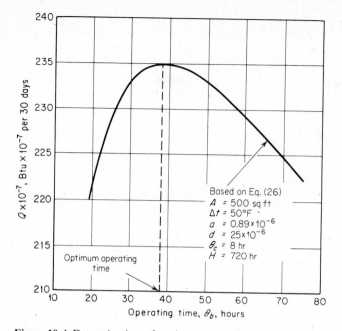

Figure 10-4 Determination of optimum operating time for maximum amount of heat transfer in evaporator with scale formation.

The total amount of heat transferred during

$$H \text{ h} = Q_H = (Q/\text{cycle}) \times (\text{cycles}/H \text{ h})$$

Therefore,

$$Q_H = \frac{2A\,\Delta t}{a}[(a\theta_b + d)^{1/2} - d^{1/2}]\frac{H}{\theta_b + \theta_c} \tag{26}$$

Under ordinary conditions, the only variable in Eq. (26) is the operating time θ_b. A plot of the total amount of heat transferred versus θ_b shows a maximum at the optimum value of θ_b. Figure 10-4 presents a plot of this type. The optimum cycle time can also be obtained by setting the derivative of Eq. (26) with respect to θ_b equal to zero and solving for θ_b. The result is

$$\theta_{b,\ \text{per cycle for maximum amount of heat transfer}} = \theta_c + \frac{2}{a}\sqrt{ad\theta_c} \tag{27}$$

The optimum boiling time given by Eq. (27) shows the operating schedule necessary to permit the maximum amount of heat transfer. All the time available for operation, emptying, cleaning, and refilling should be used. For constant operating conditions, this same schedule would also give the maximum amount of feed consumed, product obtained, and liquid evaporated.

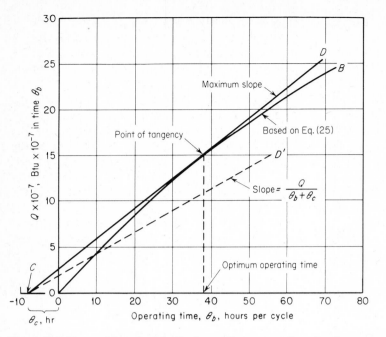

Figure 10-5 Tangential method for finding optimum operating time for maximum amount of heat transfer in evaporator with scale formation.

A third method for determining the optimum cycle time is known as the *tangential method for finding optimum conditions*, and it is applicable to many types of cyclic operations. This method is illustrated for conditions of constant cleaning time (θ_c) in Fig. 10-5, where a plot of amount of heat transferred versus boiling time is presented. Curve OB is based on Eq. (25). The *average* amount of heat transferred per unit of time during one complete cycle is $Q/(\theta_b + \theta_c)$. When the total amount of heat transferred during a number of repeated cycles is a maximum, the average amount of heat transferred per unit of time must also be a maximum. The optimum cycle time, therefore, occurs when $Q/(\theta_b + \theta_c)$ is a maximum.

The straight line CD' in Fig. 10-5 starts at a distance equivalent to θ_c on the left of the plot origin. The slope of this straight line is $Q/(\theta_b + \theta_c)$, with the values of Q and θ_b determined by the point of intersection between line CD' and curve OB. The maximum value of $Q/(\theta_b + \theta_c)$ occurs when line CD is tangent to the curve OB, and the point of tangency indicates the optimum value of the boiling time per cycle for conditions of maximum amount of heat transfer.

Cycle time for minimum cost per unit of heat transfer There are many different circumstances which may affect the minimum cost per unit of heat transferred in an evaporation operation. One simple and commonly occurring case will be considered. It may be assumed that an evaporation unit of fixed capacity is available, and a definite amount of feed and evaporation must be handled each day. The total cost for one cleaning and inventory charges is assumed to be constant no matter how much boiling time is used. The problem is to determine the cycle time which will permit operation at the least total cost.

The total cost includes (1) fixed charges on the equipment and fixed overhead expenses, (2) steam, materials, and storage costs which are proportional to the amount of feed and evaporation, (3) expenses for direct labor during the actual evaporation operation, and (4) cost of cleaning. Since the size of the equipment and the amounts of feed and evaporation are fixed, the costs included in (1) and (2) are independent of the cycle time. The optimum cycle time, therefore, can be found by minimizing the sum of the costs for cleaning and for direct labor during the evaporation.

If C_c represents the cost for one cleaning and S_b is the direct labor cost per hour during operation, the total variable costs during H h of operating and cleaning time must be

$$C_{T,\text{ for } H\text{ h}} = (C_c + S_b\theta_b)\frac{H}{\theta_b + \theta_c} \tag{28}$$

Equations (26) and (28) may be combined to give

$$C_{T,\text{ for } H\text{ h}} = \frac{aQ_H(C_c + S_b\theta_b)}{2A\,\Delta t[(a\theta_b + d)^{1/2} - d^{1/2}]} \tag{29}$$

The optimum value of θ_b for minimum total cost may be obtained by plotting C_T versus θ_b or by setting the derivative of Eq. (29) with respect to θ_b equal to zero and solving for θ_b. The result is

$$\theta_{b,\text{ per cycle for minimum total cost}} = \frac{C_c}{S_b} + \frac{2}{aS_b}\sqrt{adC_cS_b} \tag{30}$$

Equation (30) shows that the optimum cycle time is independent of the required amount of heat transfer Q_H. Therefore, a check must be made to make certain the optimum cycle time for minimum cost permits the required amount of heat transfer. This can be done easily by using the following equation, which is based on Eq. (26):

$$\theta_t = \theta_b + \theta_c = \frac{2AH'\,\Delta t}{aQ_H}[(a\theta_{b,\text{opt}} + d)^{1/2} - d^{1/2}] \tag{31}$$

where H' is the total time *available* for operation, emptying, cleaning, and recharging. If θ_t is equal to or greater than $\theta_{b,\text{opt}} + \theta_c$, the optimum boiling time indicated by Eq. (30) can be used, and the required production can be obtained at conditions of minimum cost.

The optimum cycle time determined by the preceding methods may not fit into convenient operating schedules. Fortunately, as shown in Figs. 10-4 and 10-5, the optimum points usually occur where a considerable variation in the cycle time has little effect on the factor that is being optimized. It is possible, therefore, to adjust the cycle time to fit a convenient operating schedule without causing much change in the final results.

The approach described in the preceding sections can be applied to many different types of semicontinuous cyclic operations. An illustration showing how the same reasoning is used for determining optimum cycle times in filter-press operations is presented in Example 4.

Example 4. Cycle time for maximum amount of production from a plate-and-frame filter press Tests with a plate-and-frame filter press, operated at constant pressure, have shown that the relation between the volume of filtrate delivered and the time in operation can be represented as follows:

$$P_f^2 = 2.25 \times 10^4 (\theta_f + 0.11)$$

where P_f = cubic feet of filtrate delivered in filtering time θ_f h.

The cake formed in each cycle must be washed with an amount of water equal to one-sixteenth times the volume of filtrate delivered per cycle. The washing rate remains constant and is equal to one-fourth of the filtrate delivery rate at the end of the filtration. The time required per cycle for dismantling, dumping, and reassembling is 6 h. Under the conditions where the preceding information applies, determine the total cycle time necessary to permit the maximum output of filtrate during each 24 h.

SOLUTION

Let θ_f = hours of filtering time per cycle.

Filtrate delivered per cycle = $P_{f,\text{cycle}} = 150(\theta_f + 0.11)^{1/2}$ ft^3. Rate of filtrate delivery at end of cycle is

$$\text{Washing rate} \times 4 = \frac{dP_f}{d\theta_f} = \frac{150}{2}(\theta_f + 0.11)^{-1/2} \text{ ft}^3/\text{h}$$

$$\text{Time for washing} = \frac{\text{volume of wash water}}{\text{washing rate}}$$

$$= \frac{(4)(2)(150)(\theta_f + 0.11)^{1/2}}{(16)(150)(\theta_f + 0.11)^{-1/2}} = \frac{\theta_f + 0.11}{2} \text{ h}$$

$$\text{Total time per cycle} = \theta_f + \frac{\theta_f + 0.11}{2} + 6 = 1.5\theta_f + 6.06 \text{ h}$$

$$\text{Cycles per 24 h} = \frac{24}{1.5\theta_f + 6.06}$$

Filtrate in ft^3 delivered/24 h is

$$P_{f,\text{cycle}} \text{ (cycles per 24 h)} = 150(\theta_f + 0.11)^{1/2} \frac{24}{1.5\theta_f + 6.06}$$

At the optimum cycle time giving the maximum output of filtrate per 24 h,

$$\frac{d(\text{ft}^3 \text{ filtrate delivered}/24 \text{ h})}{d\theta_f} = 0$$

Performing the differentiation and solving for θ_f,

$$\theta_{f,\text{opt}} = 3.8 \text{ h}$$

Total cycle time necessary to permit the maximum output of filtrate $= (1.5)(3.8) + 6.06 = 11.8$ h.

ACCURACY AND SENSITIVITY OF RESULTS

The purpose of the discussion and examples presented in the preceding sections of this chapter has been to give a basis for understanding the significance of *optimum conditions* plus simplified examples to illustrate the general concepts. Costs due to taxes, time value of money, capital, efficiency or inefficiency of operation, and special maintenance are examples of factors that have not been emphasized in the preceding. Such factors may have a sufficiently important influence on an optimum condition that they need to be taken into account for a final analysis. The engineer must have the practical understanding to recognize when such factors are important and when the added accuracy obtained by including them is not worth the difficulty they cause in the analysis.

A classic example showing how added refinements can come into an analysis for optimum conditions is involved in the development of methods for determining optimum economic pipe diameter for transportation of fluids. The following analysis, dealing with economic pipe diameters, gives a detailed derivation to illustrate how simplified expressions for optimum conditions can be developed. Further discussion showing the effects of other variables on the sensitivity is also presented.

FLUID DYNAMICS (OPTIMUM ECONOMIC PIPE DIAMETER)

The investment for piping and pipe fittings can amount to an important part of the total investment for a chemical plant. It is necessary, therefore, to choose pipe sizes which give close to a minimum total cost for pumping and fixed charges. For any given set of flow conditions, the use of an increased pipe diameter will cause an increase in the fixed charges for the piping system and a decrease in the pumping or blowing charges. Therefore, an optimum economic pipe diameter must exist. The value of this optimum diameter can be determined by combining the principles of fluid dynamics with cost considerations. The optimum economic pipe diameter is found at the point at which the sum of pumping or blowing costs and fixed charges based on the cost of the piping system is a minimum.

Pumping or Blowing Costs

For any given operating conditions involving the flow of a noncompressible fluid through a pipe of constant diameter, the total mechanical-energy balance can be reduced to the following form:

$$\text{Work}' = \frac{2\mathfrak{f}V^2L(1+J)}{g_cD} + B \tag{32}$$

where Work' = mechanical work added to system from an external mechanical source, ft · lbf/lbm

$\mathfrak{f}$ = Fanning friction factor, dimensionless†

V = average linear velocity of fluid, ft/s

L = length of pipe, ft

J = frictional loss due to fittings and bends, expressed as equivalent fractional loss in a straight pipe

g_c = conversion factor in Newton's law of motion, 32.17 ft · lbm/(s)(s)(lbf)

D = inside diameter of pipe, ft, subscript i means in.

B = a constant taking all other factors of the mechanical-energy balance into consideration

In the region of turbulent flow (Reynolds number greater than 2100), $\mathfrak{f}$ may be approximated for new steel pipes by the following equation:

$$\mathfrak{f} = \frac{0.04}{(N_{\text{Re}})^{0.16}} \tag{33}$$

where N_{Re} is the Reynolds number or $DV\rho/\mu$.

If the flow is viscous (Reynolds number less than 2100),

$$\mathfrak{f} = \frac{16}{N_{\text{Re}}} \tag{34}$$

By combining Eqs. (32) and (33) and applying the necessary conversion factors, the following equation can be obtained representing the annual pumping cost when the flow is turbulent:

$$C_{\text{pumping}} = \frac{0.273q_f^{2.84}\rho^{0.84}\mu_c^{0.16}K(1+J)H_y}{D_i^{4.84}E} + B' \tag{35}$$

where C_{pumping} = pumping cost as dollars per year per foot of pipe length when flow is turbulent

q_f = fluid-flow rate, ft³/s

ρ = fluid density, lb/ft³

μ_c = fluid viscosity, centipoises

---

† Based on Fanning equation written as $\Sigma(\text{friction}) = 2\mathfrak{f}V^2L/g_cD$.

K = cost of electrical energy, \$/kWh
H_y = hours of operation per year
E = efficiency of motor and pump expressed as a fraction
B' = a constant independent of D_i

Similarly, Eqs. (32) and (34) and the necessary conversion factors can be combined to give the annual pumping costs when the flow is viscous:

$$C'_{\text{pumping}} = \frac{0.024q_f^2 \mu_c K(1 + J)H_y}{D_i^4 E} + B' \tag{36}$$

where C'_{pumping} = pumping cost as dollars per year per foot of pipe length when flow is viscous.

Equations (35) and (36) apply to noncompressible fluids. In engineering calculations, these equations are also generally accepted for gases if the total pressure drop is less than 10 percent of the initial pressure.

Fixed Charges for Piping System

For most types of pipe, a plot of the logarithm of the pipe diameter versus the logarithm of the purchase cost per foot of pipe is essentially a straight line. Therefore, the purchase cost for pipe may be represented by the following equation:

$$c_{\text{pipe}} = XD_i^n \tag{37}$$

where c_{pipe} = purchase cost of new pipe per foot of pipe length, \$/ft
X = purchase cost of new pipe per foot of pipe length if pipe diameter is 1 in., \$/ft
n = a constant with value dependent on type of pipe

The annual cost for the installed piping system may be expressed as follows:

$$C_{\text{pipe}} = (1 + F)XD_i^n K_F \tag{38}$$

where C_{pipe} = cost for installed piping system as dollars per year per foot of pipe length†
F = ratio of total costs for fittings and installation to purchase cost for new pipe
K_F = annual fixed charges including maintenance, expressed as a fraction of initial cost for completely installed pipe

Optimum Economic Pipe Diameter

The total annual cost for the piping system and pumping can be obtained by adding Eqs. (35) and (38) or Eqs. (36) and (38). The only variable in the result-

---

† Pump cost could be included if desired; however, in this analysis, the cost of the pump is considered as essentially invariant with pipe diameter.

ing total-cost expressions is the pipe diameter. The optimum economic pipe diameter can be found by taking the derivative of the total annual cost with respect to pipe diameter, setting the result equal to zero, and solving for D_i. This procedure gives the following results:

For turbulent flow,

$$D_{i,\text{opt}} = \left[\frac{1.32 q_f^{2.84} \rho^{0.84} \mu_c^{0.16} K(1 + J)H_y}{n(1 + F)XEK_F} \right]^{1/(4.84 + n)} \tag{39}$$

For viscous flow,

$$D_{i,\text{opt}} = \left[\frac{0.096 q_f^2 \mu_c K(1 + J)H_y}{n(1 + F)XEK_F} \right]^{1/(4.0 + n)} \tag{40}$$

The value of n for steel pipes is approximately 1.5 if the pipe diameter is 1 in. or larger and 1.0 if the diameter is less than 1 in. Substituting these values in Eqs. (39) and (40) gives:

For turbulent flow in steel pipes,

$D_i \geq 1$ in.:

$$D_{i,\text{opt}} = q_f^{0.448} \rho^{0.132} \mu_c^{0.025} \left[\frac{0.88 K(1 + J)H_y}{(1 + F)XEK_F} \right]^{0.158} \tag{41}$$

$D_i < 1$ in.:

$$D_{i,\text{opt}} = q_f^{0.487} \rho^{0.144} \mu_c^{0.027} \left[\frac{1.32 K(1 + J)H_y}{(1 + F)XEK_F} \right]^{0.171} \tag{42}$$

For viscous flow in steel pipes,

$D_i \geq 1$ in.:

$$D_{i,\text{opt}} = q_f^{0.364} \mu_c^{0.182} \left[\frac{0.064 K(1 + J)H_y}{(1 + F)XEK_F} \right]^{0.182} \tag{43}$$

$D_i < 1$ in.:

$$D_{i,\text{opt}} = q_f^{0.40} \mu_c^{0.20} \left[\frac{0.096 K(1 + J)H_y}{(1 + F)XEK_F} \right]^{0.20} \tag{44}$$

The exponents involved in Eqs. (41) through (44) indicate that the optimum diameter is relatively insensitive to most of the terms involved. Since the exponent of the viscosity term in Eqs. (41) and (42) is very small, the value of $\mu_c^{0.025}$ and $\mu_c^{0.027}$ may be taken as unity over a viscosity range of 0.02 to 20 centipoises. It is possible to simplify the equations further by substituting average numerical values for some of the less critical terms. The following values are applicable under ordinary industrial conditions:

$$K = \$0.055/\text{kWh}$$

$$J = 0.35 \text{ or } 35 \text{ percent}$$

$$H_y = 8760 \text{ h/year}$$

$$E = 0.50 \text{ or } 50 \text{ percent}$$

$$F = 1.4$$

$$K_F = 0.20 \text{ or } 20 \text{ percent}$$

$$X = \$0.45 \text{ per foot for 1-in.-diameter steel pipe}$$

Substituting these values into Eqs. (41) through (44) gives the following simplified results:

For turbulent flow in steel pipes,

$D_i \geq 1$ in.:

$$D_{i,\text{opt}} = 3.9 q_f^{0.45} \rho^{0.13} = \frac{2.2 w_m^{0.45}}{\rho^{0.32}} \tag{45}$$

where w_m = thousands of pounds mass flowing per hour.

$D_i < 1$ in.:

$$D_{i,\text{opt}} = 4.7 q_f^{0.49} \rho^{0.14} \tag{46}$$

For viscous flow in steel pipes,

$D_i \geq 1$ in.:

$$D_{i,\text{opt}} = 3.0 q_f^{0.36} \mu_c^{0.18} \tag{47}$$

$D_i < 1$ in.:

$$D_{i,\text{opt}} = 3.6 q_f^{0.40} \mu_c^{0.20} \tag{48}$$

Depending on the accuracy desired and the type of flow, Eqs. (39) through (48) may be used to estimate optimum economic pipe diameters.† The simplified Eqs. (45) through (48) are sufficiently accurate for design estimates under ordinary plant conditions, and, as shown in Table 2, the diameter estimates obtained are usually on the safe side in that added refinements in the calculation methods generally tend to result in smaller diameters. A nomograph based on these equations is presented in Chap. 13 (Materials Transfer, Handling, and Treatment Equipment—Design and Costs).

---

† This type of approach was first proposed by R. P. Genereaux, *Ind. Eng. Chem.*, **29**:385 (1937); *Chem. Met. Eng.*, **44**(5):281 (1937) and B. R. Sarchet and A. P. Colburn, *Ind. Eng. Chem.*, **32**:1249 (1940).

ANALYSIS INCLUDING TAX EFFECTS AND COST OF CAPITAL

The preceding analysis clearly neglects a number of factors that may have an influence on the optimum economic pipe diameter, such as cost of capital or return on investment, cost of pumping equipment, taxes, and the time value of money. If the preceding development of Eq. (39) for turbulent flow is refined to include the effects of taxes and the cost of capital (or return on investment) plus a more accurate expression for the frictional loss due to fittings and bends, the result is:[†]

For turbulent flow,

$$
\frac{D_{\text{opt}}^{4.84+n}}{1 + 0.794 L_e' D_{\text{opt}}}
$$

$$
= \frac{0.000189 Y K w_s^{2.84} \mu_c^{0.16}\{[1 + (a' + b')M](1 - \phi) + ZM\}}{n(1 + F)X'[Z + (a + b)(1 - \phi)]E\rho^2} \tag{49}
$$

where D_{opt} = optimum economic inside diameter, ft

X' = purchase cost of new pipe per foot of pipe length if pipe diameter is 1 ft, based on $c_{\text{pipe}} = X' D_{\text{opt}}^n$, \$/ft

L_e' = frictional loss due to fittings and bends, expressed as equivalent pipe length in pipe diameters per unit length of pipe, $L_e' = J/D_{\text{opt}}$

w_s = pounds mass flowing per second

M = ratio of total cost for pumping installation to yearly cost of pumping power required

Y = days of operation per year

a = fraction of initial cost of installed piping system for annual depreciation

a' = fraction of initial cost of pumping installation for annual depreciation

b = fraction of initial cost of installed piping system for annual maintenance

b' = fraction of initial cost of pumping installation for annual maintenance

ϕ = fractional factor for rate of taxation

Z = fractional rate of return (or cost of capital before taxes) on incremental investment

---

[†] A similar equation is presented by J. H. Perry and C. H. Chilton, ed., "Chemical Engineers' Handbook," 5th ed., p. 5–32, McGraw-Hill Book Company, New York, 1973. See Prob. 15 at end of this chapter for derivation of Eq. (49) and comparison to form of equation given in above-mentioned reference.

Table 1 Values of variables used to obtain Eq. (50)

Turbulent flow—steel pipe—diameter 1 in. or larger

Variable	Value used	Variable	Value used
n	1.5	ϕ	0.39
L'_e	2.35	Z	0.2
Y	328	X'	2.91
K	0.0085	$a + b$	0.2
μ_c	1.0	$a' + b'$	0.4
M	0.8	E	0.5

The variable F is a function of diameter and can be approximated by $F \cong 0.75/D_{opt} + 3$.

By using the values given in Table 1, for turbulent flow in steel pipes, Eq. (49) simplifies to:

$$D_i \geq 1 \text{ in.}$$

$$D_{opt} = \frac{(1 + 1.865 D_{opt})^{0.158}(0.32)w_s^{0.45}}{(1 + F)^{0.158}\rho^{0.32}} \tag{50}$$

SENSITIVITY OF RESULTS

The simplifications made in obtaining Eqs. (45) to (48) and Eq. (50) illustrate an approach that can be used for approximate results when certain variables appear in a form where relatively large changes in them have little effect on the final results. The variables appearing in Table 1 and following Eq. (44) are relatively independent of pipe diameter, and they are raised to a small power for the final determination of diameter. Thus, the final results are not particularly sensitive to the variables listed in Table 1, and the practical engineer may decide that the simplification obtained by using the approximate equations is worth the slight loss in absolute accuracy.

Table 2 shows the extent of change in optimum economic diameter obtained by using Eq. (50) versus Eq. (45) and illustrates the effect of bringing in added refinements as well as changes in values of some of the variables.

HEAT TRANSFER (OPTIMUM FLOW RATE OF COOLING WATER IN CONDENSER)

If a condenser, with water as the cooling medium, is designed to carry out a given duty, the cooling water may be circulated at a high rate with a small change in water temperature or at a low rate with a large change in water

Table 2 Comparison of optimum economic pipe diameter estimated from Eqs. (50) and (45)

Turbulent flow—schedule 40 steel pipe—approximate 15-ft spacing of fittings

$D_{i,\, opt}$ in.		w_s,	p,
By Eq. (50)	By Eq. (45)	lb/s	lb/ft^3
10.0	11.2	450	200
5.0	6.1	4.5	2
3.0	3.9	12.5	35
1.5	2.2	0.45	2

temperature. The temperature of the water affects the temperature-difference driving force for heat transfer. Use of an increased amount of water, therefore, will cause a reduction in the necessary amount of heat-transfer area and a resultant decrease in the original investment and fixed charges. On the other hand, the cost for the water will increase if more water is used. An economic balance between conditions of high water rate–low surface area and low water rate–high surface area indicates that the optimum flow rate of cooling water occurs at the point of minimum total cost for cooling water and equipment fixed charges.

Consider the general case in which heat must be removed from a condensing vapor at a given rate designated by q Btu/h. The vapor condenses at a constant temperature of $t'°$F, and cooling water is supplied at a temperature of $t_1°$F. The following additional notation applies:

w = flow rate of cooling water, lb/h
c_p = heat capacity of cooling water, Btu/(lb)(°F)
t_2 = temperature of cooling water leaving condenser, °F
U = constant overall coefficient of heat transfer determined at optimum conditions, Btu/(h)(ft^2)(°F)
A = area of heat transfer, ft^2
Δt_{lm} = log-mean temperature-difference driving force over condenser, °F
H_y = hours the condenser is operated per year, h/year
C_w = cooling-water cost assumed as directly proportional to amount of water supplied,† $/lb
C_A = installed cost of heat exchanger per square foot of heat-transfer area, $/ft^2
K_F = annual fixed charges including maintenance, expressed as a fraction of initial cost for completely installed equipment

---

† Cooling water is assumed to be available at a pressure sufficient to handle any pressure drop in the condenser; therefore, any cost due to pumping the water is included in C_w.

The rate of heat transfer as Btu per hour can be expressed as

$$q = wc_p(t_2 - t_1) = UA \, \Delta t_{lm} = \frac{UA(t_2 - t_1)}{\ln\left[(t' - t_1)/(t' - t_2)\right]} \tag{51}$$

Solving for w,

$$w = \frac{q}{c_p(t_2 - t_1)} \tag{52}$$

The design conditions set the values of q and t_1, and the heat capacity of water may ordinarily be approximated as 1 Btu/(lb)(°F). Therefore, Eq. (52) shows that the flow rate of the cooling water is fixed if the temperature of the water leaving the condenser (t_2) is fixed. Under these conditions, the optimum flow rate of cooling water can be found directly from the optimum value of t_2.

The annual cost for cooling water is $wH_y C_w$. From Eq. (52),

$$wH_y C_w = \frac{qH_y C_w}{c_p(t_2 - t_1)} \tag{53}$$

The annual fixed charges for the condenser are $AK_F C_A$, and the total annual cost for cooling water plus fixed charges is

$$\text{Total annual variable cost} = \frac{qH_y C_w}{c_p(t_2 - t_1)} + AK_F C_A \tag{54}$$

Substituting for A from Eq. (51),

$$\text{Total annual variable cost} = \frac{qH_y C_w}{c_p(t_2 - t_1)}$$

$$+ \frac{qK_F C_A \ln\left[(t' - t_1)/(t' - t_2)\right]}{U(t_2 - t_1)} \tag{55}$$

The only variable in Eq. (55) is the temperature of the cooling water leaving the condenser. The optimum cooling-water rate occurs when the total annual cost is a minimum. Thus, the corresponding optimum exit temperature can be found by differentiating Eq. (55) with respect to t_2 (or, more simply, with respect to $t' - t_2$) and setting the result equal to zero. When this is done, the following equation is obtained:

$$\frac{t' - t_1}{t' - t_{2,\text{opt}}} - 1 + \ln \frac{t' - t_{2,\text{opt}}}{t' - t_1} = \frac{UH_y C_w}{K_F c_p C_A} \tag{56}$$

The optimum value of t_2 can be found from Eq. (56) by a trial-and-error solution, and Eq. (52) can then be used to determine the optimum flow rate of cooling water. The trial-and-error solution can be eliminated by use of Fig. 10-6, which is a plot of Eq. (56).†

---

† See Fig. 14-32 for a similar plot for counterflow coolers.

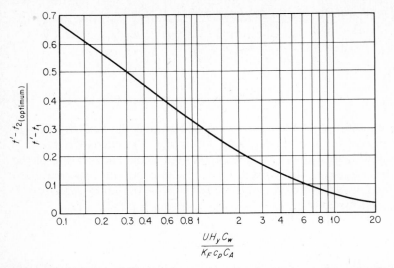

Figure 10-6 Solution of Eq. (56) for use in evaluating optimum flow rate of cooling medium in condenser.

Example 5. Optimum cooling-water flow rate in condenser A condenser for a distillation unit must be designed to condense 5000 lb (2268 kg) of vapor per hour. The effective condensation temperature for the vapor is 170°F (350 K). The heat of condensation for the vapor is 200 Btu/lb (4.65×10^5 J/kg). Cooling water is available at 70°F (294 K). The cost of the cooling water is $0.06 per 1000 gal ($5.30 per 1000 m³). The overall heat-transfer coefficient at the optimum conditions may be taken as 50 Btu/(h)(ft²)(°F) (284 J/m² · s · K). The cost for the installed heat exchanger is $21 per square foot of heat-transfer area ($226 per square meter of heat-transfer area) and annual fixed charges including maintenance are 20 percent of the initial investment. The heat capacity of the water may be assumed to be constant at 1.0 Btu/(lb)(°F) (4.2 kJ/kg · K). If the condenser is to operate 6000 h/yr, determine the cooling-water flow rate in pounds per hour and in kilograms per hour for optimum economic conditions.

SOLUTION

$$U = 50 \text{ Btu/(h)(ft}^2)(°F)$$
$$H_y = 6000 \text{ h/year}$$
$$K_F = 0.20$$
$$c_p = 1.0 \text{ Btu/(lb)(°F)}$$
$$C_A = \$21/\text{ft}^2$$

$$C_w = \frac{0.06}{(1000)(8.33)} = \$0.0000072/\text{lb}$$

$$\frac{UH_yC_w}{K_Fc_pC_A} = \frac{(50)(6000)(0.0000072)}{(0.20)(1.0)(21)} = 0.514$$

The optimum exit temperature may be obtained by a trial-and-error solution of Eq. (56) or by use of Fig. 10-6.

From Fig. 10-6, when the abscissa is 0.514,

$$\frac{t' - t_{2,\text{opt}}}{t' - t_1} = 0.42$$

where $t' = 170°F$
$t_1 = 70°F$
$t_{2,\text{opt}} = 128°F$

By Eq. (52), at the optimum economic conditions,

$$w = \frac{q}{c_p(t_2 - t_1)} = \frac{(5000)(200)}{(1.0)(128 - 70)} = 17,200 \text{ lb water/h } (7800 \text{ kg water/h})$$

MASS TRANSFER (OPTIMUM REFLUX RATIO)

The design of a distillation unit is ordinarily based on specifications giving the degree of separation required for a feed supplied to the unit at a known composition, temperature, and flow rate. The design engineer must determine the size of column and reflux ratio necessary to meet the specifications. As the reflux ratio is increased, the number of theoretical stages required for the given separation decreases. An increase in reflux ratio, therefore, may result in lower fixed charges for the distillation column and greater costs for the reboiler heat supply and condenser coolant.

As indicated in Fig. 10-7, the optimum reflux ratio occurs at the point where

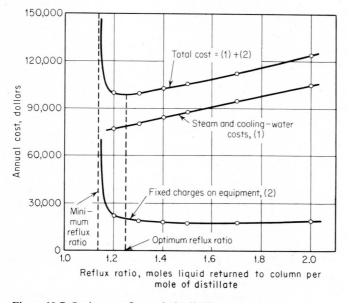

Figure 10-7 Optimum reflux ratio in distillation operation.

the sum of fixed charges and operating costs is a minimum. As a rough approximation, the optimum reflux ratio usually falls in the range of 1.1 to 1.3 times the minimum reflux ratio. The following example illustrates the general method for determining the optimum reflux ratio in distillation operations.

Example 6. Determination of optimum reflux ratio A sieve-plate distillation column is being designed to handle 700 lb mol (318 kg mol) of feed per hour. The unit is to operate continuously at a total pressure of 1 atm. The feed contains 45 mol % benzene and 55 mol % toluene, and the feed enters at its boiling temperature. The overhead product from the distillation tower must contain 92 mol % benzene, and the bottoms must contain 95 mol % toluene. Determine the following:

(a) The optimum reflux ratio as moles liquid returned to tower per mole of distillate product withdrawn.
(b) The ratio of the optimum reflux ratio to the minimum reflux ratio.
(c) The percent of the total variable cost due to steam consumption at the optimum conditions.

The following data apply:

Vapor-liquid equilibrium data for benzene-toluene mixtures at atmospheric pressure are presented in Fig. 10-8.

The molal heat capacity for liquid mixtures of benzene and toluene in all proportions may be assumed to be 40 Btu/(lb mol)(°F) (1.67 × 10⁵ J/kg mol·K).

The molal heat of vaporization of benzene and toluene may be taken as 13,700 Btu/lb mol (3.19 × 10⁷ J/kg mol).

Effects of change in temperature on heat capacity and heats of vaporiza-

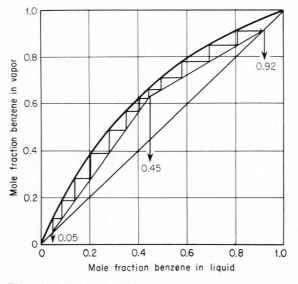

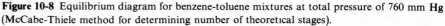

Figure 10-8 Equilibrium diagram for benzene-toluene mixtures at total pressure of 760 mm Hg (McCabe-Thiele method for determining number of theoretical stages).

tion are negligible. Heat losses from the column are negligible. Effects of pressure drop over the column may be neglected.

The overall coefficient of heat transfer is 80 Btu/(h)(ft²)(°F) (454 J/m² · s · K) in the reboiler and 100 Btu/(h)(ft²)(°F) (568 J/m² · s · K) in the condenser.

The boiling temperature is 201°F (367 K) for the feed, 179°F (367 K) for the distillate, and 227°F (381 K) for the bottoms. The temperature-difference driving force in the reflux condenser may be based on an average cooling-water temperature of 90°F (305 K), and the change in cooling-water temperature is 50°F (27.8 K) for all cases. Saturated steam at 60 psia (413.6 kPa) is used in the reboiler. At this pressure, the temperature of the condensing steam is 292.7°F (418 K) and the heat of condensation is 915.5 Btu/lb (2.13 × 10⁶ J/kg). No heat-savings devices are used.

The column diameter is to be based on a maximum allowable vapor velocity of 2.5 ft/s (0.76 m/s) at the top of the column. The overall plate efficiency may be assumed to be 70 percent. The unit is to operate 8500 h per year.

Cost data

Steam = $0.75/1000 lb ($1.65/1000 kg).
Cooling water = $0.045/1000 gal or $0.054/10,000 lb ($0.119/10,000 kg).

The sum of costs for piping, insulation, and instrumentation can be estimated to be 60 percent of the cost for the installed equipment. Annual fixed charges amount to 15 percent of the total cost for installed equipment, piping, instrumentation, and insulation.

The following costs are for the installed equipment and include delivery and erection costs:

Sieve-plate distillation column

Values may be interpolated

Diameter		$/plate
in.	(m)	
60	(1.52)	1200
70	(1.78)	1500
80	(2.03)	1850
90	(2.29)	2250
100	(2.54)	2700

Condenser—tube-and-shell heat exchanger

Values may be interpolated

Heat-transfer area		$
ft²	(m²)	
800	(74.3)	9,750
1000	(92.9)	11,250
1200	(111.5)	12,600
1400	(130.1)	13,800
1600	(148.6)	14,850

Reboiler—tube-and-shell heat exchanger
Values may be interpolated

Heat-transfer area

ft^2	(m^2)	$
1000	(92.9)	17,250
1400	(130.1)	21,150
1800	(167.2)	24,600
2200	(204.4)	27,750
2600	(241.5)	30,300

SOLUTION The variable costs involved are cost of column, cost of reboiler, cost of condenser, cost of steam, and cost of cooling water. Each of these costs is a function of the reflux ratio, and the optimum reflux ratio occurs at the point where the sum of the annual variable costs is a minimum. The total variable cost will be determined at various reflux ratios, and the optimum reflux ratio will be found by the graphical method.

Sample calculation for reflux ratio = 1.5:

Annual cost for distillation column

The McCabe-Thiele simplifying assumptions apply for this case, and the number of theoretical plates can be determined by the standard graphical method shown in Fig. 10-8. The slope of the enriching operating line is $1.5/(1.5 + 1) = 0.6$. From Fig. 10-8, the total number of theoretical stages required for the given separation is 12.1.

The actual number of plates $= (12.1 - 1)/0.70 = 16$.

The moles of distillate per hour (M_D) and the moles of bottoms per hour (M_B) may be determined by a benzene material balance as follows:

$$(700)(0.45) = (M_D)(0.92) + (700 - M_D)(0.05)$$

$$M_D = 322 \text{ moles distillate/h}$$

$$M_B = 700 - 322 = 378 \text{ moles bottoms/h}$$

Moles vapor per hour at top of column $= 322(1 + 1.5) = 805$.

Applying the perfect-gas law,

$$\text{Vapor velocity at top of tower} = 2.5 \text{ ft/s}$$

$$= \frac{(805)(359)(460 + 179)(4)}{(3600)(492)(\pi)(\text{diameter})^2}$$

$$\text{Diameter} = 7.3 \text{ ft}$$

Cost per plate for plate and vessel $= \$2145$

$$\text{Annual cost for distillation column} = (2145)(16)(1 + 0.60)(0.15)$$

$$= \$8235$$

Annual cost for condenser

Rate of heat transfer per hour in condenser = (moles vapor condensed per hour)(molal latent heat of condensation) = (805)(13,700) = 11,000,000 Btu/h.

From the basic heat-transfer-rate equation $q = UA \, \Delta t$,

$$A = \text{heat-transfer area} = \frac{(11,000,000)}{(100)(179 - 90)} = 1240 \text{ sq ft}$$

$$\text{Cost per square foot} = \frac{\$12,825}{1240}$$

$$\text{Annual cost for condenser} = \frac{12,825}{1240}(1240)(1 + 0.60)(0.15)$$

$$= \$3075$$

Annual cost for reboiler

The rate of heat transfer in the reboiler (q_r) can be determined by a total energy balance around the distillation unit.

Base energy level on liquid at 179°F.

$$\text{Heat input} = \text{heat output}$$

$$q_r + (700)(201 - 179)(40) = 11,000,000 + (378)(227 - 179)(40)$$

$$q_r = 11,110,000 \text{ Btu/h} = UA \, \Delta t$$

$$A = \text{heat-transfer area} = \frac{11,110,000}{(80)(292.7 - 227)} = 2120 \text{ ft}^2$$

$$\text{Cost per square foot} = \frac{\$27,150}{2120}$$

$$\text{Annual cost for reboiler} = \frac{27,150}{2120}(2120)(1 + 0.60)(0.15)$$

$$= \$6510$$

Annual cost for cooling water

The rate of heat transfer in the condenser = 11,000,000 Btu/h. The heat capacity of water may be taken as 1.0 Btu/(lb)(°F).

$$\text{Annual cost for cooling water} = \frac{(11,000,000)(0.054)(8500)}{(1.0)(50)(10,000)}$$

$$= \$10,110$$

Annual cost for steam

The rate of heat transfer in the reboiler = 11,110,000 Btu/h.

$$\text{Annual cost for steam} = \frac{(11,110,000)(0.75)(8500)}{(915.5)(1000)}$$

$$= \$77,550$$

Total annual variable cost at reflux ratio of 1.5

$$\$8235 + \$3075 + \$6510 + \$10,110 + \$77,550 = \$105,480$$

By repeating the preceding calculations for different reflux ratios, the following table can be prepared:

Reflux ratio	Number of actual plates required	Column diam- eter, ft	Annual cost, dollars, for					Total annual cost, dollars
			Column	Con- denser	Reboiler	Cooling water	Steam	
1.14	∞	6.7	∞	2805	5940	8,670	66,450	∞
1.2	29	6.8	13,395	2865	6060	8,910	68,250	99,480
1.3	21	7.0	9,930	2925	6195	9,300	71,250	99,600
1.4	18	7.1	8,880	3000	6360	9,705	74,400	102,345
1.5	16	7.3	8,235	3075	6510	10,110	77,550	105,480
1.7	14	7.7	7,935	3225	6810	10,935	83,550	112,455
2.0	13	8.0	7,815	3420	7200	12,150	92,700	123,285

(a) The data presented in the preceding table are plotted in Fig. 10-7. The minimum total cost per year occurs at a reflux ratio of 1.25.

$$\text{Optimum reflux ratio} = 1.25$$

(b) For conditions of minimum reflux ratio, the slope of the enriching line in Fig. 10-8 is 0.532

$$\frac{\text{Minimum reflux ratio}}{\text{Minimum reflux ratio} + 1} = 0.532$$

$$\text{Minimum reflux ratio} = 1.14$$

$$\frac{\text{Optimum reflux ratio}}{\text{Minimum reflux ratio}} = \frac{1.25}{1.14} = 1.1$$

(c) At the optimum conditions,

$$\text{Annual steam cost} = \$69,750$$

$$\text{Total annual variable cost} = \$99,000$$

$$\text{Percent of variable cost due to steam consumption} = \frac{69,750}{99,000}(100) = 70\%$$

THE STRATEGY OF LINEARIZATION FOR OPTIMIZATION ANALYSIS

In the preceding analyses for optimum conditions, the general strategy has been to establish a partial derivative of the dependent variable from which the absolute optimum conditions are determined. This procedure assumes that an absolute maximum or minimum occurs within attainable operating limits and is restricted to relatively simple conditions in which limiting constraints are not exceeded. However, practical industrial problems often involve establishing the best possible program to satisfy existing conditions under circumstances where the optimum may be at a boundary or limiting condition rather than at a true maximum or minimum point. A typical example is that of a manufacturer who must determine how to blend various raw materials into a final mix that will meet basic specifications while simultaneously giving maximum profit or least cost. In this case, the basic limitations or constraints are available raw materials, product specifications, and production schedule, while the overall objective (or *objective function*) is to maximize profit.

LINEAR PROGRAMMING FOR OBTAINING OPTIMUM CONDITIONS

One strategy for simplifying the approach to a programming problem is based on expressing the constraints and the objective in a linear mathematical form. The "straight-line" or linear expressions are stated mathematically as

$$ax_1 + bx_2 + \cdots + jx_j + \cdots + nx_n = z \tag{57}$$

where the coefficients $a \cdots n$ and z are known values and $x_1 \cdots x_n$ are unknown variables. With two variables, the result is a straight line on a two-dimensional plot, while a plane in a three-dimensional plot results for the case of three variables. Similarly, for more than three variables, the geometric result is a hyperplane.

The general procedure mentioned in the preceding paragraph is designated as *linear programming*. It is a mathematical technique for determining optimum conditions for allocation of resources and operating capabilities to attain a definite objective. It is also useful for analysis of alternative uses of resources or alternative objectives.

EXAMPLE OF APPROACH IN LINEAR PROGRAMMING

As an example to illustrate the basic methods involved in linear programming for determining optimum conditions, consider the following simplified problem. A brewery has received an order for 100 gal of beer with the special constraints that the beer must contain 4 percent alcohol by volume and it must be supplied immediately. The brewery wishes to fill the order, but no 4 percent beer is now in stock. It is decided to mix two beers now in stock to give the desired final product. One of the beers in stock (Beer A) contains 4.5 percent alcohol by volume and is valued at \$0.32 per gallon. The other beer in stock (Beer B) contains 3.7 percent alcohol by volume and is valued at \$0.25 per gallon. Water (W) can be added to the blend at no cost. What volume combination of the two beers in stock with water, including at least 10 gal of Beer A, will give the minimum ingredient cost for the 100 gal of 4 percent beer?

This example is greatly simplified because only a few constraints are involved and there are only three independent variables, i.e., amount of Beer A (V_A), amount of Beer B (V_B), and amount of water (V_W). When a large number of possible choices is involved, the optimum set of choices may be far from obvious, and a solution by linear programming may be the best way to approach the problem. A step-by-step rational approach is needed for linear programming. This general rational approach is outlined in the following with application to the blending example cited.

RATIONAL APPROACH TO PROBLEMS INVOLVING LINEAR PROGRAMMING

A systematic rationalization of a problem being solved by linear programming can be broken down into the following steps:

1. *A systematic description of the limitations or constraints.* For the brewery example, the constraints are as follows:
   a. Total volume of product is 100 gallons, or

   $$V_W + V_A + V_B = 100 \tag{58}$$

   b. Product must contain 4 percent alcohol, or

   $$0.0V_W + 4.5V_A + 3.7V_B = (4.0)(100) \tag{59}$$

   c. Volume of water and Beer B must be zero or greater than zero, while volume of Beer A must be 10 gal or greater; i.e.,

   $$V_W \geq 0 \qquad V_B \geq 0 \qquad V_A \geq 10 \quad \text{or} \quad V_A - S = 10 \tag{60}$$

   where S is the so-called "slack variable."

2. *A systematic description of the objective.* In the brewery example, the objective is to minimize the cost of the ingredients; i.e., the objective function is

   $$C = \text{cost} = \text{a minimum} = 0.0V_W + 0.32V_A + 0.25V_B \tag{61}$$

3. *Combination of the constraint conditions and the objective function to choose the best result out of many possibilities.* One way to do this would be to use an intuitive approach whereby every reasonable possibility would be considered to give ultimately, by trial and error, the best result. This approach would be effective for the brewery example because of its simplicity. However, the intuitive approach is not satisfactory for most practical situations, and linear programming can be used. The computations commonly become so involved that a computer is required for the final solution. If a solution is so simple that a

computer is not needed, linear programming would probably not be needed. To illustrate the basic principles, the brewery example is solved in the following by linear programming including intuitive solution, graphical solution, and computer solution.

From Eqs. (58) to (61), the following linearized basic equations can be written:

$$V_W + V_A + V_B = 100 \tag{58}$$

$$4.5V_A + 3.7V_B = 400 \tag{59}$$

$$0.32V_A + 0.25V_B = C = \text{minimum} \tag{61}$$

where C is designated as the objective function.
Combination of Eqs. (58) and (59) gives

$$V_A = 37.5 + 4.625V_W \tag{62}$$

Equation (62) is plotted as line OE in Fig. 10-9, and the optimum must fall on this line.
Equation (61) combined with Eq. (58) gives

$$V_A = \frac{C - 25}{0.07} + 3.57V_W \tag{63}$$

Intuitive solution It can be seen intuitively, from Eqs. (63) and (62), that the minimum value of the objective function C occurs when V_W is zero. Therefore, the optimum value of V_A, from Eq. (62), is 37.5 gal and the optimum value of V_B, from Eq. (58), is 62.5 gal.

Linear programming graphical solution Figure 10-9 is the graphical representation of this problem. Line OE represents the overall constraint placed on the problem by Eqs. (58), (59), and (60). The parallel dashed lines represent possible conditions of cost. The goal of the program is to minimize cost (that is, C) while still remaining within the constraints of the problem. The minimum value of C that still meets the constraints occurs for the line OD, and the optimum must be at point O. Thus, the recommended blend is no water, 37.5 gal of A, 62.5 gal of B, and a total cost C of $27.63 for 100 gal of blend.

Linear programming computer solution Although the simplicity of this problem makes it trivial to use a computer for solution, the following is presented to illustrate the basic type of reasoning that is involved in developing a computer program for the linearized system.

An iterative procedure must be used for the computer solution to permit the computer to make calculations for repeated possibilities until the minimum objective function C is attained. In this case, there are four variables (V_A, V_B, V_W, and S) and three nonzero constraints (total volume, final alcohol content, and

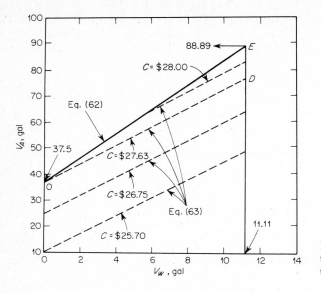

Figure 10-9 Graphical representation of linear-programming solution based on brewery example.

$V_A = 10 + S$). Because the number of real variables cannot exceed the number of nonzero constraints, one of the four variables must be zero.† Thus, one approach for a computer solution merely involves solving a four-by-three matrix with each variable alternatively being set equal to zero, followed by instruction that the desired combination is the one giving the least total cost.

The computer logic, from which the computer diagram, program, and solution can be developed directly, is presented in Tables 3 and 4.‡

GENERALIZATION OF STRATEGY FOR LINEAR PROGRAMMING

The basic problem in linear programming is to maximize or minimize a linear function of a form as shown in Eq. (57). There are various *strategies* that can be developed to simplify the methods of solution, some of which can lead to algorithms which allow rote or pure number-plugging methods of solution that are well adapted for machine solution.

In linear programming, the variables $x_1 \cdots x_n$ are usually restricted (or can be transformed) to values of zero or greater. This is known as a *nonnegativity restriction* on x_j; that is,

$$x_j \geq 0 \qquad j = 1, 2, \ldots, n$$

Consider a simple two-dimensional problem such as the following: The objective function is to maximize

$$3x_1 + 4x_2 \tag{64}$$

---

† For proof of this statement, see any book on linear programming. For example, S. I. Gass, "Linear Programming: Methods and Applications," 3rd ed., p. 54, McGraw-Hill Book Company, New York, 1969.

‡ In Prob. 18 at the end of this chapter, the student is requested to develop the full computer program and solve this problem on a computer.

Table 3 Computer logic for "linear programming" solution to brewery example

The computer must solve a linearized situation in which there are four variables and three nonzero constraints to meet a specified objective function of minimum cost. Under these conditions, one of the variables *must* be zero. Thus, one method for computer approach is to set each variable in turn equal to zero, solve the resulting three-by-three matrix, and determine the final solution to make cost a minimum.† Instead of one total computer solution with instructions for handling a three-by-four matrix, the approach will be simplified by repeating four times the solution by determinants of a standard case of three equations and three unknowns with each of the four variables alternately set equal to zero.

Basis: $E(I, J)$ and $X(J)$; where $E(I, J)$ designates the appropriate coefficient, $X(J)$ designates the appropriate variable, I designates the proper row, and J designates the proper column, Thus,

$$\text{Constraints} \begin{cases} E(1,1)X(1) + E(1,2)X(2) + E(1,3)X(3) = E(1,4) \\ E(2,1)X(1) + E(2,2)X(2) + E(2,3)X(3) = E(2,4) \\ E(3,1)X(1) + E(3,2)X(2) + E(3,3)X(3) = E(3,4) \end{cases}$$

Equivalent to
Eq. (58)
Eq. (59)
Eq. (60)

Objective function
$$\begin{cases} C(1)X(1) + C(2)X(2) + C(3)X(3) = F(C) \qquad \text{Eq. (61)} \\ C(J) \text{ designates the appropriate coefficient and } F(C) \text{ is the objective function.} \end{cases}$$

Logic and procedure based on arbitrary choice of one variable set equal to zero

Step 1: Read in to the computer on data cards the constant coefficients for the three variables being retained based on Eqs. (58), (59), and (60). Read in to computer on a data card the constant coefficients for the objective function, i.e., Eq. (61).

Step 2: Solve the resulting three equations simultaneously by determinants.

A. Evaluate the determinant of the coefficients of the system = $F(E) = D$:

B. Evaluate $DX(J) = F(W)$:

$$D \equiv \begin{vmatrix} E(1,1) & E(1,2) & E(1,3) \\ E(2,1) & E(2,2) & E(2,3) \\ E(3,1) & E(3,2) & E(3,3) \end{vmatrix} \qquad F(W) = \begin{vmatrix} W(1,1) & W(1,2) & W(1,3) \\ W(2,1) & W(2,2) & W(2,3) \\ W(3,1) & W(3,2) & W(3,3) \end{vmatrix}$$

where $W(I, J)$ designates the appropriate coefficient for the DX working matrix in which the column of equality constraints is substituted in the appropriate determinant column.

Step 3: Evaluate the objective function.

Step 4: Print out the values of the three variables and the value of the objective function.

Step 5: Repeat steps 1 to 4 three more times with each variable equal to zero.

Step 6: After all results have been printed out, choose the set of results giving the minimum value of the objective function with all variables meeting the requirements.

---

† For normal methods used for linear-programming solutions, there are various rules which can be used to determine which variable should be set equal to zero so that not all possible combinations must be tried. See later discussion of the *Simplex Algorithm.*

subject to the linear constraints of

$$2x_1 + 5x_2 \leq 10 \qquad (65)$$

$$4x_1 + 3x_2 \leq 12 \qquad (66)$$

$$x_1 \geq 0 \qquad (67)$$

$$x_2 \geq 0 \qquad (68)$$

This problem and its solution is pictured graphically in Fig. 10-10, which shows that the answer is $3x_1 + 4x_2 = 11$. From Fig. 10-10 it can be seen that the

Table 4 Computer diagram based on logic of Table 3 for linear-programmed brewery example

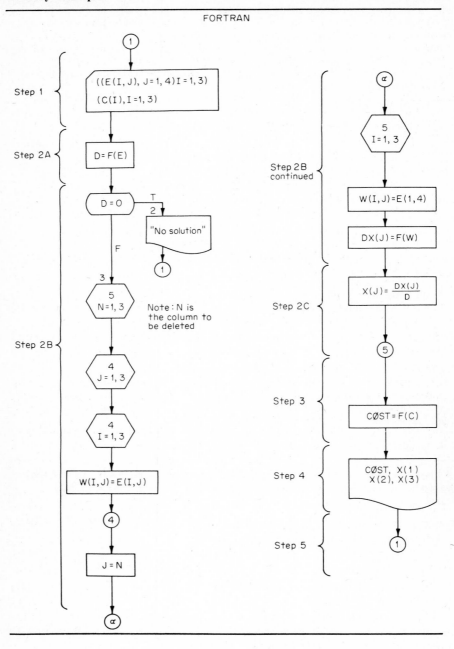

FORTRAN

Step 1

$((E(I, J), J = 1, 4) I = 1, 3)$
$(C(I), I = 1, 3)$

Step 2A

$D = F(E)$

$D = 0$ ──T──▶ "No solution" ──▶ 1

F

Step 2B

3

5
$N = 1, 3$

Note: N is the column to be deleted

4
$J = 1, 3$

4
$I = 1, 3$

$W(I, J) = E(I, J)$

4

$J = N$

α

Step 2B continued

α

5
$I = 1, 3$

$W(I, J) = E(1, 4)$

$DX(J) = F(W)$

Step 2C

$X(J) = \dfrac{DX(J)}{D}$

5

Step 3

$CØST = F(C)$

Step 4

$CØST, X(1)$
$X(2), X(3)$

Step 5

1

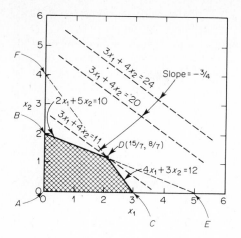

Figure 10-10 Graphical illustration of two-dimensional linear-programming solution.

linear constraints, in the form of inequalities, restrict the solution region to the cross-hatched area. This solution region is a polygon designated as *convex* because all points on the line between any two points in the cross-hatched region are in the set of points that satisfy the constraints. The set of the objective function is a family of lines with slope of $-\frac{3}{4}$. The maximum value of the objective function occurs for the line passing through the polygon vertex D. Thus, the maximum value of the objective function occurs for the case of $3x_1 + 4x_2 = 11$ at $x_1 = \frac{15}{7}$ and $x_2 = \frac{8}{7}$.

For the two-dimensional case considered in the preceding, one linear condition defines a line which divides the plane into two half-planes. For a three-dimensional case, one linear condition defines a set plane which divides the volume into two half-volumes. Similarly, for an n-dimensional case, one linear condition defines a hyperplane which divides the space into two half-spaces.

For the n-dimensional case, the region that is defined by the set of hyperplanes resulting from the linear constraints represents a *convex set* of all points which satisfy the constraints of the problem. If this is a bounded set, the enclosed space is a convex polyhedron, and, for the case of monotonically increasing or decreasing values of the objective function, the maximum or minimum value of the objective function will always be associated with a vertex or *extreme point* of the convex polyhedron. This indicates that the linear-programming solution for the model of inequality or equality constraints combined with the requested value for the objective function will involve determination of the value of the objective function at the extreme points of the set of all points that satisfy the constraints of the problem. The desired objective function can then be established by comparing the values found at the extreme points. If two extremes give the same result, then an infinite number of solutions exist as defined by all points on the line connecting the two extreme points.

SIMULTANEOUS EQUATIONS[†]

Linear programming is concerned with solutions to simultaneous linear equations where the equations are developed on the basis of restrictions on the variables. Because these restrictions are often expressed as inequalities, it is necessary to convert these inequalities to equalities. This can be accomplished by the inclusion of a new variable designated as a *slack variable*.

For a restriction of the form

$$a_1 x_1 + a_2 x_2 + a_3 x_3 \leq b \tag{70}$$

the inequality is converted to a linear equation by adding a slack variable S_4 to the left-hand side:

$$a_1 x_1 + a_2 x_2 + a_3 x_3 + S_4 = b \tag{71}$$

The slack variable takes on whatever value is necessary to satisfy the equation and normally is considered as having a nonnegativity restriction. Therefore, the slack variable would be subtracted from the left-hand side for an inequality of the form

$$a_1 x_1 + a_2 x_2 + a_3 x_3 \geq b \tag{72}$$

to give

$$a_1 x_1 + a_2 x_2 + a_3 x_3 - S_4 = b \tag{73}$$

After the inequality constraints have been converted to equalities, the complete set of restrictions becomes a set of linear equations with n unknowns. The linear-programming problem then will involve, in general, maximizing or minimizing a linear objective function for which the variables must satisfy the set of simultaneous restrictive equations with the variables constrained to be nonnegative. Because there will be more unknowns in the set of simultaneous equations than there are equations, there will be a large number of possible solutions, and the final solution must be chosen from the set of possible solutions.

If there are m independent equations and n unknowns with $m < n$, one approach is to choose arbitrarily $n - m$ variables and set them equal to zero.

[†] Cases are often encountered in design calculations where a large number of design equations and variables are involved with long and complex simultaneous solution of the equations being called for. The amount of effort involved for the simultaneous solutions can be reduced by using the so-called *structural-array algorithm* which is a purely mechanical operation involving crossing out rows for equations and columns for variables to give the most efficient order in which the equations should be solved. For details, see D. F. Rudd and C. C. Watson, "Strategy of Process Engineering," pp. 45–49, John Wiley & Sons, Inc., New York, 1968.

This gives m equations with m unknowns so that a solution for the m variables can be obtained. Various combinations of this type can be obtained so that the total possible number of solutions by this process becomes

$$\binom{n}{m} = n!/m!(n - m)!$$

representing the total number of possible combinations obtainable by taking n variables m at a time. Another approach is to let $n - m$ combinations of variables assume any zero or nonzero value which results in an infinite number of possible solutions. Linear programming deals only with the situation where the excess variables are set equal to zero.

TWO EXAMPLES TO SHOW APPROACH BY SIMULTANEOUS EQUATIONS

To illustrate the introductory ideas presented for a linear-programming problem, consider the following example which is solved by using a step-by-step simultaneous-equation approach:

A production facility is being used to produce three different products, x_1, x_2, and x_3. Each of these products requires a known number of employee-hours and machine-hours for production such that

product x_1 requires 10 employee-hours and 15 machine-hours per unit,
product x_2 requires 25 employee-hours and 10 machine-hours per unit,
product x_3 requires 20 employee-hours and 10 machine-hours per unit.

The profit per unit is \$5 for x_1, \$10 for x_2, and \$12 for x_3. Over the base production period under consideration, a total of 300 employee-hours and a total of 200 machine-hours are available. With the special restriction that all employee-hours are to be used, what mix of products will maximize profits?

For this problem, the linear constraints are:
for machine hours,

$$15x_1 + 10x_2 + 10x_3 \leq 200 \tag{74}$$

for employee-hours,

$$10x_1 + 25x_2 + 20x_3 = 300 \tag{75}$$

The objective function is to maximize profits, or

$$\text{Maximize} \quad 5x_1 + 10x_2 + 12x_3 \tag{76}$$

By including a slack variable for Eq. (74), the constraining equalities become

$$15x_1 + 10x_2 + 10x_3 + S_4 = 200 \tag{77}$$

$$10x_1 + 25x_2 + 20x_3 = 300 \tag{78}$$

Table 5 Six solutions by simultaneous equations for example problem

Solution number	x_1	x_2	x_3	S_4	Objective function $5x_1 + 10x_2 + 12x_3$
1	7.2727	9.0909	0	0	127.27
2	5	0	12.5	0	175
3	30	0	0	−250	Infeasible (negative)
4	0	−20	40	0	Infeasible (negative)
5	0	12	0	80	120
6	0	0	15	50	180

For this case, $n = 4$ and $m = 2$. Setting any two of the variables equal to zero and solving the result gives

$$\binom{n}{m} = \frac{n!}{m!\,(n-m)!} = \frac{4!}{2!\,2!} = 6$$

possible solutions. These six solutions are shown in Table 5.

Solutions 3 and 4 are infeasible because the nonnegativity restriction has been violated, while Solution 6 is a feasible solution which maximizes the objective function. Thus, Solution 6 is the desired solution for this example and represents the optimal solution.

For the example presented graphically in Fig. 10-10, the two linear constraining equations made into equalities by the slack variables S_3 and S_4 are

$$2x_1 + 5x_2 + S_3 + 0S_4 = 10 \tag{79}$$

$$4x_1 + 3x_2 + 0S_3 + S_4 = 12 \tag{80}$$

with the objective function being

$$3x_1 + 4x_2 + 0S_3 + 0S_4 = z = \text{a maximum} \tag{81}$$

Table 6 Six solutions by simultaneous equations for Fig. 10-10 example

Solution designation	x_1	x_2	S_3	S_4	Objective function $3x_1 + 4x_2$
A	0	0	10	12	0
B	0	2	0	6	8
C	3	0	4	0	9
D	15/7	8/7	0	0	11†
E	5	0	0	−8	Infeasible (negative)
F	0	4	−10	0	Infeasible (negative)

† Maximum feasible value; so the optimum solution is D.

In this case, there are two equations ($m = 2$) and four variables ($n = 4$). Thus, the approach with $n - m = 4 - 2 = 2$ of the variables being zero for each solution will involve having

$$\binom{n}{m} = \frac{n!}{m!\,(n-m)!} = \frac{4!}{2!\,2!} = 6$$

possible solutions. These solutions are shown in Fig. 10-10 as points A to F and in Table 6 as obtained by simultaneous-equation solution with the optimum result being the D solution.

The preceding examples, although they represent an approach for solving linear-programming problems, are very inefficient because of the large number of useless solutions that may be generated if many variables are involved. More efficient procedures are available, and these are discussed in the following sections.

GENERALIZATION OF LINEAR PROGRAMMING APPROACH FOR ALGORITHM SOLUTION

To permit efficient solutions for linear-programming problems, an algorithm can be developed. An algorithm, basically, is simply an objective mathematical method for solving a problem and is purely mechanical so that it can be taught to a nonprofessional or programmed for a computer. The algorithm may consist of a series of repeated steps or *iterations*. To develop this form of approach for linear-programming solutions, the set of linear inequalities which form the constraints, written in the form of "equal to or less than" equations is

$$a_{11}x_1 + a_{12}x_2 + \cdots + a_{1n}x_n \le b_1$$

$$a_{21}x_1 + a_{22}x_2 + \cdots + a_{2n}x_n \le b_2 \tag{82}$$

$$\cdots\cdots\cdots\cdots\cdots\cdots\cdots\cdots\cdots\cdots\cdots\cdots$$

$$a_{m1}x_1 + a_{m2}x_2 + \cdots + a_{mn}x_n \le b_m$$

or, in general summation form,

$$\sum_{j=1}^{n} a_{ij}x_j \le b_i \qquad i = 1, 2, \ldots, m \tag{82a}$$

for

$$x_j \ge 0 \qquad j = 1, 2, \ldots, n$$

where i refers to columns (or number of equations, m) in the set of inequalities and j refers to rows (or number of variables, n).

As has been indicated earlier, these inequalities can be changed to equalities by adding a set of slack variables, $x_{n+1} \cdots x_{n+m}$ (here x is used in place of S to simplify the generalized expressions), so that

$$a_{11}x_1 + a_{12}x_2 + \cdots + a_{1n}x_n + x_{n+1} = b_1$$
$$a_{21}x_1 + a_{22}x_2 + \cdots + a_{2n}x_n + x_{n+2} = b_2$$

$$\ldots\ldots\ldots\ldots\ldots\ldots\ldots\ldots\ldots\ldots\ldots\ldots$$

$$a_{m1}x_1 + a_{m2}x_2 + \cdots + a_{mn}x_n + x_{n+m} = b_m$$

(83)†

or, in general summation form,

$$\sum_{j=1}^{n} (a_{ij}x_j + x_{n+i}) = b_i \qquad i = 1, 2, \ldots, m \qquad (83a)$$

for

$$x_j \geq 0 \qquad j = 1, 2, \ldots, n + m$$

In addition to the constraining equations, there is an objective function for the linear program which is expressed in the form of

$$z = \text{maximum (or minimum) of } c_1x_1 + c_2x_2 + \cdots + c_jx_j + \cdots + c_nx_n \qquad (84)‡$$

where the variables x_j are subject to $x_j \geq 0$ $(j = 1, 2, \ldots, n + m)$. Note that, in this case, all variables above x_n are slack variables and make no direct contribution to the value of the objective function.

Within the constraints as indicated by Eqs. (82) and (83), a solution for values of the variables, x_j, must be found which meets the maximum or mini-

---

† This can be written in standard matrix form and notation as

$$AX = B$$

where

$$A = \begin{pmatrix} a_{11} & a_{12} & \cdots & a_{1n} & 1 & 0 & 0 & \cdots & \cdots & 0 \\ a_{21} & a_{22} & \cdots & a_{2n} & 0 & 1 & 0 & \cdots & \cdots & 0 \\ \cdots & \cdots & \cdots & \cdots & 0 & 0 & \cdots & 1 & \cdots & 0 \\ a_{m1} & a_{m2} & \cdots & a_{mn} & 0 & 0 & \cdots & \cdots & 0 & 1 \end{pmatrix}$$

$$X = \begin{pmatrix} x_1 \\ x_2 \\ \vdots \\ x_n \\ x_{n+1} \\ \vdots \\ x_{n+m} \end{pmatrix} \qquad B = \begin{pmatrix} b_1 \\ b_2 \\ \vdots \\ b_m \end{pmatrix}$$

$$x_j \geq 0 \qquad j = 1, 2, \ldots, n + m$$

and standard matrix operations of multiplying, addition, etc., can be applied.

‡ In a more compact form using matrix notation, the problem is to find the solution to $AX = B$ which maximizes or minimizes $z = cX$ where $X \geq 0$.

mum requirement of the objective function, Eq. (84). As has been demonstrated in the preceding examples, the solution to a problem of this sort must lie on an extreme point of the set of possible feasible solutions. For any given solution, the number of equations to be solved simultaneously must be set equal to the number of variables, and this is accomplished by setting n (number of variables) minus m (number of equations) equal to zero and then proceeding to obtain a solution.

While the preceding generalization is sufficient to allow for reaching a final solution ultimately, it can be very inefficient unless some sort of special method is used to permit generation of extreme-point solutions in an efficient manner to allow rapid and effective approach to the optimum condition. This is what the *simplex method* does.†

THE SIMPLEX ALGORITHM

The basis for the simplex method is the generation of extreme-point solutions by starting at any one extreme point for which a feasible solution is known and then proceeding to a neighboring extreme point. Special rules are followed which cause the generation of each new extreme point to be an improvement toward the desired objective function. When the extreme point is reached where no further improvement is possible, this will represent the desired optimum feasible solution. Thus, the simplex algorithm is an iterative process that starts at one extreme-point feasible solution, tests this point for optimality, and proceeds toward an improved solution. If an optimal solution exists, this algorithm can be shown to lead ultimately and efficiently to the optimal solution.

The stepwise procedure for the simplex algorithm is as follows (based on the optimum being a maximum):

1. State the linear-programming problem in standard equality form.
2. Establish the initial feasible solution from which further iterations can proceed. A common method to establish this initial solution is to base it on the values of the slack variables where all other variables are assumed to be zero. With this assumption, the initial matrix for the simplex algorithm can be set up with a column showing those variables which will be involved in the first solution. The coefficient for these variables appearing in the matrix table should be 1 with the rest of the column being 0.
3. Test the initial feasible solution for optimality. The optimality test is accomplished by the addition of rows to the matrix which give (a) a value of z_j for each column where z_j is defined as the sum of the objective-function coefficient for each solution variable (c_i corresponding to solution x_i in that

† The simplex method and algorithm were first made generally available when published by G. B. Dantzig in "Activity Analysis of Production and Allocations," edited by T. C. Koopmans, Chap. XXI, John Wiley & Sons, Inc., New York, 1951.

row) times the coefficient of the constraining-equation variable for that column [a_{ij} in Eq. (83a)]: (that is, $z_j = \sum_{i=1}^{m} c_i a_{ij}$ ($j = 1, 2, \ldots, n$)), (b) c_j [see Eq. (84)], and (c) $c_j - z_j$. If $c_j - z_j$ is positive for at least one column, then a better program is possible.

4. Iteration toward the optimal program is accomplished as follows: Assuming that the optimality test indicates that the optimal program has not been found, the following iteration procedure can be used:

a. Find the column in the matrix with the maximum value of $c_j - z_j$ and designate this column as k. The incoming variable for the new test will be the variable at the head of this column.

b. For the matrix applying to the initial feasible solution, add a column showing the ratio of b_i/a_{ik}. Find the minimum *positive* value of this ratio and designate the variable in the corresponding row as the outgoing variable.

c. Set up a new matrix with the incoming variable, as determined under (a), substituted for the outgoing variable, as determined under (b). The modification of the table is accomplished by matrix operations so that the entering variable will have a 1 in the row of the departing variable and zeros in the rest of that column. The matrix operations involve row manipulations of multiplying rows by constants and subtracting from or adding to other rows until the necessary 1 and 0 values are reached. This new matrix should have added to it the additional rows and column as explained under parts 3, 4a, and 4b.

d. Apply the optimality test to the new matrix.

e. Continue the iterations until the optimality test indicates that the optimum objective function has been attained.

5. Special cases:

a. If the initial solution obtained by use of the method given in the preceding is not feasible, a feasible solution can be obtained by adding more artificial variables which must then be forced out of the final solution.

b. Degeneracy may occur in the simplex method when the outgoing variable is selected. If there are two or more minimal values of the same size, the problem is degenerate, and a poor choice of the outgoing variable may result in cycling, although cycling almost never occurs in real problems. This can be eliminated by a method of ratioing each element in the rows in question by the positive coefficients of the kth column and choosing the row for the outgoing variable as the one first containing the smallest algebraic ratio.

6. The preceding method for obtaining a maximum as the objective function can be applied for the case when the objective function is a minimum by recognizing that maximizing the negative of a function is equivalent to minimizing the function.

THE SIMPLEX ALGORITHM APPLIED TO THE EXAMPLE SHOWN IN FIGURE 10-10

In the example used previously, and whose graphical solution is shown in Fig. 10-10, the problem in standard linear-programming form is: Find the values

Table 7 Tableau form of matrix for initial feasible solution ($x_1 = x_2 = 0$)

c_j c_i	Solution	b	0 x_3	0 x_4	3 x_1	4 x_2	b_i/a_{ik}
0	x_3	10	1	0	2	⑤	$\frac{10}{5} = 2$
0	x_4	12	0	1	4	3	$\frac{12}{3} = 4$
	z_j	0	0	0	0	0	
	$c_j - z_j$	0	0	0	3	4	
						(k)	

of the variables which represent a solution to

$$2x_1 + 5x_2 + x_3 \quad = 10 \tag{85}$$

$$4x_1 + 3x_2 + \quad x_4 = 12 \tag{86}$$

which maximizes

$$3x_1 + 4x_2 \quad \text{(express as } 3x_1 + 4x_2 + 0x_3 + 0x_4 = z) \tag{87}$$

where $x_1 \geq 0$, $x_2 \geq 0$, $x_3 \geq 0$, and $x_4 \geq 0$.

The next step after the appropriate statement of the linear-programming problem is to establish an initial feasible solution from which further iterations can proceed. For this case, let $x_1 = x_2 = 0$, and $x_3 = 10$, $x_4 = 12$, $z = 0$. (Solution A in Fig. 10-10 or Table 6.) The corresponding matrix in a standard tableau form is shown in Table 7.

The top row, c_j, in the tableau permits a convenient recording of the coefficients on the variables in the objective function, with these values listed at the head of the appropriate columns.

The first column, c_i, gives the coefficients of the variables in the objective function for this first solution. In this case, they are both zero because x_3 and x_4 do not appear in the objective function.

The second column, *Solution*, gives the variables involved in the current solution and shows the row for which the variables involved apply.

The third column, b, gives the list of condition constants for the limiting equations.

The columns following b have x headings and represent variables. The *slack* variables are x_3 and x_4, designated as unity for the appropriate row, while the *structural* variables are x_1 and x_2 with normal matrix form based on the coefficients for x_1 and x_2 in the limiting equations.

The final column on the right, b_i/a_{ik}, is used to record the indicated ratios for each row during the iteration process.

The bottom two rows are included to give a convenient method for recording the objective-function row component z_j and the values of $c_j - z_j$ for each column. By definition of z_j as $\sum_{i=1}^{m} c_i a_{ij}$ for $j = 1, 2, \ldots, n$, the value of z_j is 0 for all columns because both c_i's are 0.

Because row $c_j - z_j$ in Table 7 has at least one positive value in it, a better optimal program is available. The variable at the head of the column (k) with the

Table 8 Tableau form of matrix for first iteration ($x_1 = x_3 = 0$)

c_j			0	0	3	4	
c_i	Solution	b	x_3	x_4	x_1	x_2	b_i/a_{ik}
0	x_3	2	$\frac{1}{5}$	0	$\frac{2}{5}$	1	$\frac{2}{1} = 2$
0	x_4	12	0	1	4	3	$\frac{12}{3} = 4$
	z_j	0	0	0	0	0	
	$c_j - z_j$	0	0	0	3	4	
						(k)	
c_j			0	0	3	4	
c_i	Solution	b	x_3	x_4	x_1	x_2	b_i/a_{ik}
4	x_2	2	$\frac{1}{5}$	0	$\frac{2}{5}$	1	$2/\frac{2}{5} = 5$
0	x_4	6	$-\frac{3}{5}$	1	$\textcircled{\frac{14}{5}}$	0	$6/\frac{14}{5} = \frac{30}{14}$
	z_j	8	$\frac{4}{5}$	0	$\frac{8}{5}$	4	
	$c_j - z_j$	-8	$-\frac{4}{5}$	0	$\frac{7}{5}$	0	
					(k)		

maximum value of $c_j - z_j$ is x_2. Therefore, x_2 will be the incoming variable. The minimum value of b_i/a_{ik} occurs for the x_3 row; so x_3 will be the outgoing variable and the encircled 5 becomes the so-called *pivotal point*.

To eliminate x_3 from the basis, the use of the indicated pivotal point gives, as a first step, the matrix tableau shown in the top part of Table 8 where the corresponding element for the pivotal point has been reduced to 1 by dividing the x_3 row by 5. The bottom portion of Table 8 is the matrix tableau for the next iteration with $x_3 = x_1 = 0$. This is established by a matrix row operation to reduce the other elements in the (k) column to zero (i.e., for this case, the multiplying factor for the x_3 row is -3, and the x_3 result is added to the x_4 row). The values of z_j are $(4)(2) + (0)(6) = 8$, $(4)(\frac{1}{5}) + (0)(-\frac{3}{5}) = \frac{4}{5}$, $(4)(0) + (0)(1) = 0$, $(4)(\frac{2}{5}) + (0)(\frac{14}{5}) = \frac{8}{5}$, and $(4)(1) + (0)(0) = 4$ for the five columns from left to right, respectively.

Therefore, from Table 8, another extreme-point solution is $x_2 = 2$, $x_4 = 6$, $x_1 = x_3 = 0$. (Point B in Fig. 10-10). This is still not the optimal solution because row $c_j - z_j$ has a positive value in it. The encircled $\textcircled{\frac{14}{5}}$ is the pivotal point for the next iteration which will have x_4 as the outgoing variable and x_1 as the incoming variable. The same procedure is followed for this second iteration as for the first iteration. The steps are shown in Table 9, where the pivotal-point element is first reduced to 1 by dividing the x_4 row by $\frac{14}{5}$, and the other elements in the (k) column are then reduced to zero by a matrix row operation involving a multiplication factor of $-\frac{2}{5}$ for the x_4 row and adding the x_4 row to the x_2 row.

The results shown in Table 9 give another extreme point of $x_1 = \frac{15}{7}$, $x_2 = \frac{8}{7}$, $x_3 = x_4 = 0$. The z_j value for the b column is $(4)(\frac{8}{7}) + (3)(\frac{15}{7}) = 11$, and the values for the other four columns from left to right are obtained by a similar addition as $(4)(\frac{2}{7}) + (3)(-\frac{3}{14}) = \frac{1}{2}$, $(4)(-\frac{1}{7}) + (3)(\frac{5}{14}) = \frac{1}{2}$, $(4)(0) + (3)(1) = 3$, and $(4)(1) + (3)(0) = 4$.

Table 9 Tableau form of matrix for second iteration ($x_3 = x_4 = 0$)

c_j c_i	Solution	b	0 x_3	0 x_4	3 x_1	4 x_2	b_i/a_{ik}
4	x_2	2	$\frac{1}{5}$	0	$\frac{2}{5}$	1	$2/\frac{2}{5} = 5$
0	x_4	$\frac{30}{14}$	$-\frac{3}{14}$	$\frac{5}{14}$	1	0	$\frac{30}{14}/1 = \frac{30}{14}$
	z_j	8	$\frac{4}{5}$	0	$\frac{8}{5}$	4	
	$c_j - z_j$	-8	$-\frac{4}{5}$	0	$\frac{7}{5}$	0 (k)	

c_j c_i	Solution	b	0 x_3	0 x_4	3 x_1	4 x_2	b_i/a_{ik}
4	x_2	$\frac{8}{7}$	$\frac{2}{7}$	$-\frac{1}{7}$	0	1	
3	x_1	$\frac{15}{7}$	$-\frac{3}{14}$	$\frac{5}{14}$	1	0	
	z_j	11	$\frac{1}{2}$	$\frac{1}{2}$	3	4	
	$c_j - z_j$	-11	$-\frac{1}{2}$	$-\frac{1}{2}$	0	0	

Because row $c_j - z_j$ has only negative or zero values in it, this is the optimal solution, and the objective function is a maximum of $z = (3)(\frac{15}{7}) + (4)(\frac{8}{7}) = 11$ at $x_1 = \frac{15}{7}$ and $x_2 = \frac{8}{7}$, which is the same solution (point D in Fig. 10-10) that was obtained by the graphical analysis in Fig. 10-10. Note that, in each basic, initial, table matrix where the column for that variable has all zeros except for the variable row which is 1, the b column gives the values of the variables and the objective function for that solution.

The preceding information can serve as an introduction to the methods of linear programming including the step-by-step rule approach used for a simplex algorithm. The reader is referred to any of the many standard texts on linear programming for proof of the theorems and rules used in this treatment and further extensions of the methods of linear programming.[†]

THE STRATEGY OF DYNAMIC PROGRAMMING FOR OPTIMIZATION ANALYSIS

The concept of dynamic programming is based on converting an overall decision situation involving many variables into a series of simpler individual problems with each of these involving a small number of total variables. In its extreme, an

† For example, see W. J. Adams, A. Gewirtz, and L. V. Quintas, "Elements of Linear Programming," Van Nostrand Reinhold Company, New York, 1969; G. B. Dantzig, "Linear Programming and Extensions," 5th Pr., Princeton Univ. Press, Princeton, N.J., 1968; S. I. Gass, "Linear Programming: Methods and Applications," 3d ed., McGraw-Hill Book Company, New York, 1969; R. W. Llewellyn, "Linear Programming," Holt, Rinehart, and Winston, Inc., New York, 1964; G. E. Thompson, "Linear Programming: An Elementary Introduction," Macmillan Book Company, New York, 1971.

optimization problem involving a large number of variables, all of which may be subject to constraints, is broken down into a sequence of problems with each of these involving only one variable. A characteristic of the process is that the determination of one variable leaves a problem remaining with one less variable. The computational approach is based on the principle of optimality, which states that *an optimal policy has the property that, no matter what the initial state and initial decision, the remaining decisions must constitute an optimal policy with regard to the state resulting from the first decision.*

The use of dynamic programming is pertinent for design in the chemical industry where the objective function for a complicated system can often be obtained by dividing the overall system into a series of stages. Optimizing the resulting simple stages can lead to the optimal solution for the original complex problem.

The general formulation for a dynamic-programming problem, presented in a simplified form, is shown in Fig. 10-11. On the basis of the definitions of terms given in Fig. 10-11a, each of the variables, x_{i+1}, x_i, and d_i, may be replaced by vectors because there may be several components or streams involved in the input and output, and several decision variables may be involved. The profit or return P_i is a scalar which gives a measure of contribution of stage i to the objective function.

For the operation of a single stage, the output is a function of the input and the decisions, or

$$x_i = h_i(x_{i+1}, d_i) \tag{88}$$

Similarly, for the individual-stage objective function P_i

$$P_i = g_i(x_{i+1}, x_i, d_i) \tag{89}$$

or, on the basis of the relation shown as Eq. (88),

$$P_i = g_i(x_{i+1}, d_i) \tag{90}$$

For the simple multistage process shown in Fig. 10-11b, the process design strategy to optimize the overall operation can be expressed as

$$\bar{f}_i(x_{i+1}) = \max_{d_i} \left[g_i(x_{i+1}, d_i) + \bar{f}_{i-1}(x_i) \right] = \max_{d_i} \left[Q_i(x_{i+1}, d_i) \right] \tag{91}$$

for

$$x_i = h_i(x_{i+1}, d_i) \qquad i = 1, 2, \ldots, n \text{—subject to } \bar{f}_0 = 0$$

The symbolism $\bar{f}_i(x_{i+1})$ indicates that the maximum (or optimum) return or profit from a process depends on the input to that process, and the terms in the square brackets of Eq. (91) refer to the function that is being optimized. Thus, the expression $Q_i(x_{i+1}, d_i)$ represents the combined return from all stages and must equal the return from stage i, or $g_i(x_{i+1}, d_i)$, plus the maximum return from the preceding stages 1 through $i - 1$, or $\bar{f}_{i-1}(x_i)$.

In carrying out the procedure for applying dynamic programming for the

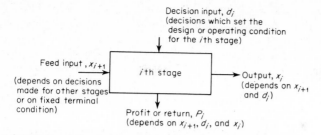

a. General formulation for one stage in a dynamic programming model

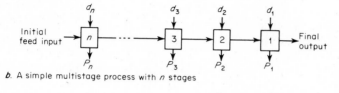

b. A simple multistage process with n stages

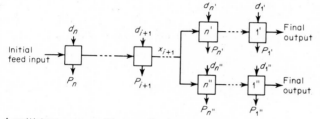

c. A multistage process with separating branches

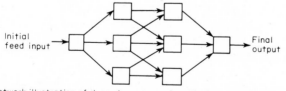

d. A network illustration of dynamic programming showing various paths

Figure 10-11 Illustration of stages involved in dynamic programming.

solution of appropriate plant-design problems, each input x_{i+1} is considered as a parameter. Thus, at each stage, the problem is to find the optimum value of the decision variable d_i for all feasible values of the input variable. By using the dynamic-programming approach involving n stages, a total of n optimizations must be carried out. This approach can be compared to the conventional approach in which optimum values of all the stages and decisions would be made by a basic probability combination analysis. Thus, the conventional method would have a computational effort that would increase approximately exponentially with the number of stages, while the dynamic-programming approach can give a great reduction in necessary computational effort because this effort would only increase about linearly with the number of stages. However, this advantage of dynamic programming is based on a low number of components in the input vector x_{i+1}, and dynamic programming rapidly loses its effectiveness for practical computational feasibility if the number of these components increases above two.

A SIMPLIFIED EXAMPLE OF DYNAMIC PROGRAMMING†

As an illustration of the general procedure and analytical methods used in dynamic programming, consider the example design problem presented in Table 10. The general procedure consists of the following steps:

1. Establish the sequence of single stages into which the process will be divided. These are shown as stages 1 to 5 in Table 10.
2. Decide on the units to be used for expressing the profit function for each individual stage and the overall process. For this case, the problem statement makes it clear that an appropriate unit for this purpose is the profit over a five-year period of operation, and this unit will be used in the solution of the problem.
3. For each stage, determine the possible inputs, decisions, and outputs. These are shown in Table 11.
4. For each stage and for each combination of input decisions, establish the stage output.
5. Establish the optimal return from the overall process and from each stage by the application of the principle shown in Eq. (91).

Steps (1), (2), and (3) are completed, for the indicated example, in Tables 10 and 11. To carry out steps (4) and (5), it is necessary to assume a number of discrete levels for each of the decision variables. The size of the subdivisions for each decision variable, of course, represents an imposed constraint on the system solution, but these constraints are very useful for narrowing down the region which must receive the most careful attention for optimization.

The stage outputs are established in sequence for the subprocesses of stage 1, stage 1–stage 2, stage 1–stage 2–stage 3, stage 1–stage 2–stage 3–stage 4, and finally stage 1–stage 2–stage 3–stage 4–stage 5. The optimum is determined for each subprocess employing all the discrete levels chosen for the variables involved in that subprocess.

The subdivisions for the possible decisions in this example are shown by the data given in Table 10 and are summarized in Table 11. Thus, in stage 5, there are three possible decisions on the choice of mixer, and each of these has four

---

† Based on L. G. Mitten and G. L. Nemhauser, *Chem. Eng. Progr.*, **59**(1):52 (1963).

Table 10 A dynamic-programming model for production of a new chemical with no recycle including specific data for an example

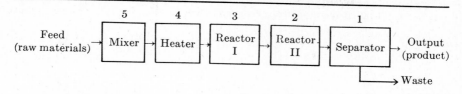

Feed: 50,000 lb/yr of raw material are fed to stage 5 of the above model of the process at a cost of $1 per pound.

The output from the 5-stage process must be at least 15,000 lb of product per year.

The overall objective function of the entire process is to optimize for a maximum profit over a five-year period. Assume equipment life period is 5 years.

a. Anticipated selling price of the product vs. annual production

Production, 1000 lb/yr	47.5	45.0	42.5	40.0	37.5	30.0	25.0	22.5	20.0	15.0
Expected selling price, $/lb	3.2	3.3	3.4	3.6	3.8	4.6	5.0	5.2	5.3	5.5

b. Operating costs for the mixing operation, $1000/yr

Mixing efficiency	1.0	0.8	0.6	0.5
Mixer A	12.0	6.0	3.0	2.0
B	8.0	4.0	2.5	1.5
C	5.0	3.0	2.0	1.0

c. Heater operating costs, $1000/yr

Mixing efficiency	Temperature, °F			
	650	700	750	800
1.0	0.5	1.0	6.0	10.0
0.8	1.0	1.5	8.0	12.0
0.6	1.5	2.5	10.0	16.0
0.5	2.0	3.0	12.0	20.0

d. Reactor I and catalyst costs

Reactor	Initial cost, $1000	Operating cost, $1000/yr
I_A	40.0	4.0
I_B	20.0	2.0
I_C	5.0	1.0
Catalyst		
1	....	10.0
2	....	4.0

e. Percent conversion in reactor I

Temp., °F	650		700		750		800	
Catalyst	1	2	1	2	1	2	1	2
Reactor I_A	30	25	40	30	50	45	60	50
I_B	25	20	30	25	45	40	50	45
I_C	20	15	25	20	40	30	45	40

(continued)

Table 10 A dynamic-programming model for production of a new chemical with no recycle including specific data for an example (Continued)

f. Total conversion from reactor I plus reactor II

Conversion in reactor	I	15	20	25	30	40	45	50	60
Second reactor	II_A	30	40	50	60	80	85	90	95
	II_B	45	60	75	85	90	95	95	95

g. Reactor II costs

Reactor	Initial cost, $1000	Operating cost, $1000/yr
II_A	60.0	10.0
II_B	80.0	20.0

i. Initial investment ($)

Mixer A.	10,000
Mixer B.	15,000
Mixer C.	25,000
Heater:	
700°F or less.	5,000
More than 700°F.	20,000

h. Costs for the separation unit

	One large separator		Two small separators (cost per separator)	
% Conversion	Initial cost, $1000	Operating cost, $1000/yr	Initial cost, $1000	Operating cost, $1000/yr
30	12	2.5	7.5	1.5
40	12	3.0	7.5	1.5
45	12	4.0	7.5	1.5
50	15	4.0	9.0	1.5
60	15	5.0	9.0	2.0
75	20	6.0	12.0	2.0
80	20	6.5	12.0	2.0
85	20	7.0	12.0	2.0
90	20	7.5	12.0	2.5
95	20	8.0	12.0	2.5

possible efficiency decisions. In stage 4, there are four possible decisions on temperature level for the heater. In stage 3, there are three possible reactors and two possible catalysts. Stage 2 has three possible decisions of reactor II_A, II_B, or no reactor. In stage 1, there are two possible decisions of one large separator or two small separators. On an overall basis, therefore, the total possible modes of operation by a completely random approach would be

$$3 \times 4 \times 4 \times 3 \times 2 \times 3 \times 2 = 1728$$

By applying the technique of dynamic programming, the final optimum condition can be established by a stage-by-stage operation so that only about 15 modes of operation must be considered.

SUBPROCESS OF STAGE 1

For the dynamic-programming procedure involving only stage 1, it is desirable to base the analysis on final product sales with consideration of only the first

Table 11 Possible decisions, inputs, and outputs for the example presented in Table 10

For stage number	Decisions	Output from stage = input to next stage (expressed as relevant variable)
5	Type of mixer (that is, A, B, or C) Mixing efficiency (that is, $\eta = 1.0$, 0.8, 0.6, or 0.5)	Mixing efficiency
4	Temperature level at which heater operates (that is, 650, 700, 750, or 800°F)	Temperature
3	Type of reactor (that is, I_A, I_B, or I_C) Type of catalyst (that is, 1 or 2)	Percent conversion
2	Type of reactor (that is, II_A, II_B, or none)	Percent conversion
1	Choice of separators (that is, one large or two small)	

stage. With this basis, all possible conversions of the entering stream must be considered. The data given in Table 10 show that at least 30 percent of the feed must be converted. This immediately indicates that the possibility of not including reactor II at stage 2 can only be considered if the conversion leaving reactor I is 30 percent or higher. Therefore, only those conversions of 30 percent or higher, as shown in Table 10f, need to be considered.

For each conversion (for example, for 50 percent conversion), the five-year profit can be evaluated for the cases of one large separator and two small separators. Therefore, using the data given in Table 10 and neglecting the cost of feed which is a constant,

Five-year profit using one large separator

$$= (5)(50,000)(0.5)(\$5.0) - \$15,000 - (5)(\$4,000) = \$590,000$$

Five-year profit using two small separators

$$= (5)(50,000)(0.5)(\$5.0) - (2)(\$9,000) - (2)(5)(\$1,500) = \$592,000$$

This indicates that the optimal operation of stage 1 with a 50 percent conversion requires the use of two small separators. These calculations are repeated for all feasible conversions, and the results [i.e., the one-stage profits $Q_1(x_2, d_1)$] are presented in Table 12 with the optimum condition for each conversion indicated by an asterisk.

Table 12 One stage profits, $Q_1(x_2, d_1)$, $1000

	Stage 1 decision, d_1	
Stage 1 input, x_1, % conversion	One separator	Two separators
95	700.0	711.0*
90	685.0	693.5*
85	667.5	678.5*
80	667.5	676.0*
75	662.5	668.5*
60	650.0	652.0*
50	590.0	592.0*
45	553.0	555.0*
40	503.0*	500.0
30	388.0*	382.5

Table 13 Two-stage profits, $Q_2(x_3, d_2)$, $1000

	Stage 2 decisions, d_2		
Stage 2 input, x_3, % conversion	Reactor IIA	Reactor IIB	No reactor II
60	601.0	531.0	652.0*
50	583.5	531.0	592.0*
45	568.5*	531.0	555.0
40	566.0*	513.5	503.0
30	542.0*	498.5	388.0
25	482.0	488.5*	
20	393.0	472.0*	
15	278.0	375.0*	

Table 14 Three-stage profits, $Q_3(x_4, d_3)$, $1000

	Stage 3 decisions, d_3			
Stage 3 input, x_4, temperature	Reactor I_A	Reactor I_B	Reactor I_C	Catalyst
800	542.0*	512.0	508.5	1
800	512.0	518.5	536.0	2
750	482.0	488.5	506.0	1
750	488.5	516.0*	512.0	2
700	456.0	462.0*	428.5	1
700	462.0*	438.5	442.0	2
650	432.0*	408.5	412.0	1
650	408.5	422.0	345.0	2

Table 15 Four-stage profits, $Q_4(x_5, d_4)$, $1000

	Stage 4 decisions, d_4, temperature			
Stage 4 input, x_5, mixing efficiency	800	750	700	650
1.0	472.0*	466.0	452.0	424.5
0.8	462.0*	456.0	449.5	422.0
0.6	442.0	446.0*	444.5	419.5
0.5	422.0	436.0	442.0*	417.0

Table 16 Five-stage profits, $Q_5(x_F, d_5)$, $1000

	Stage 5 decisions, d_5			
	Mixing efficiency			
Mixer	1.0	0.8	0.6	0.5
C	422.0	422.0	411.0	412.0
B	417.0	427.0*	418.5	419.5
A	402.0	422.0	421.0	422.0

SUBPROCESS OF STAGE 1–STAGE 2

This subprocess involves making a decision on the type of reactor II (II_A, II_B, or none). Possible conversions for the feed entering stage 2 can be established from Table 10f or Table 10e as 15, 20, 25, 30, 40, 45, 50, or 60 percent. All of these possibilities, including the decisions on conversion and reactor type, must be evaluated. Each result will give an exit conversion which represents the feed to stage 1, but the optimum condition for stage 1 has already been generated for the various feeds. Therefore, the sum of the optimum cost for stage 1 and the developed cost for stage 2 can be tabulated so that the optimum system for stage 1–stage 2 can be chosen for any appropriate feed to stage 2.

For example, if the stage-2 input conversion is chosen as 40 percent, the following data and calculations apply (neglecting cost of feed which remains constant):

Five-year profit using reactor II_A

$$= \$676{,}000 - \$60{,}000 - (5)(\$10{,}000) = \$566{,}000$$

Optimum for
stage 1 with
80% conversion
(Table 12)

Five-year profit using reactor II_B

$$= \$693{,}500 - \$80{,}000 - (5)(\$20{,}000) = \$513{,}500$$

Optimum for
stage 1 with
90% conversion
(Table 12)

Five-year profit using no reactor II = \$503,000

The preceding procedure can be repeated for all feasible combinations for the stage 1–stage 2 process, and the results [i.e., the one-stage–two-stage profits $Q_2(x_3, d_2)$] are tabulated in Table 13.

REMAINING SUBPROCESSES AND FINAL SOLUTION

The same type of optimizing procedure can now be followed for each of the remaining three subsystems, and the results are presented in Tables 14, 15, and 16 with asterisks being used to indicate the optimum sets. The final optimum for the full process can now be established directly from Table 16 as giving a five-year profit of \$427,000. The stage-wise operations should be as follows:

Stage 5: From Table 16, a type B mixer with an efficiency of 80 percent should be used.

Stage 4: From Table 15, the heater should be operated at 800°F.

Stage 3: From Tables 14 and 10, reactor I_A with catalyst 1 should be used giving a 60 percent conversion.

Stage 2: From Table 13, no reactor II should be used.

Stage 1: From Table 12, two small separators should be used.

The preceding example illustrates the technique used in dynamic programming. This technique permits a great saving in the amount of computational effort involved as is illustrated by the fact that the stage-by-stage optimization approach used in the example involved consideration of only about 15 possible modes of operation. This can be compared to the total possible modes of operation of 1728 which would have had to be considered by a totally random approach.†

OTHER MATHEMATICAL TECHNIQUES AND STRATEGIES FOR ESTABLISHING OPTIMUM CONDITIONS

Many mathematical techniques, in addition to the basic approaches already discussed, have been developed for application in various situations that require determination of optimum conditions. A summary of some of the other common and more advanced mathematical techniques, along with selected references for additional information, is presented in the following:

APPLICATION OF LAGRANGE MULTIPLIERS‡

When equality constraints or restrictions on certain variables exist in an optimization situation, a powerful analytical technique is the use of Lagrange multipliers. In many cases, the normal optimization procedure of setting the partial of the objective function with respect to each variable equal to zero and solving the resulting equations simultaneously becomes difficult or impossible mathematically. It may be much simpler to optimize by developing a *Lagrange expression*, which is then optimized in place of the real objective function.

In applying this technique, the Lagrange expression is defined as the real function to be optimized (i.e., the objective function) plus the product of the Lagrangian multiplier (λ) and the constraint. The number of Lagrangian multipliers must equal the number of constraints, and the constraint is in the form of an equation set equal to zero. To illustrate the application, consider the situation in which the aim is to find the positive value of variables x and y which make the product xy a maximum under the constraint that $x^2 + y^2 = 10$. For this simple

---

† For additional basic information on dynamic programming, see R. E. Bellman and S. E. Dreyfus, "Applied Dynamic Programming," Princeton University Press, Princeton, N.J., 1962; R. Aris, "Discrete Dynamic Programming," Blaisdell Press, New York, 1964; L. T. Fan, "The Continuous Maximum Principle," John Wiley & Sons, Inc., New York, 1966; and S. E. Dreyfus and A. M. Law, "The Art and Theory of Dynamic Programming," Academic Press, New York, 1977.

‡ A. H. Boas, How to Use Lagrange Multipliers, *Chem. Eng.*, **70**(1):95 (1963); C. Lanczos, "The Variational Principles of Mechanics," 2d ed., University of Toronto Press, Toronto, Canada, 1964; I. S. Sokolnikoff and R. M. Redheffer, "Mathematics of Physics and Modern Engineering," 2d ed., McGraw-Hill Book Company, New York, 1966.

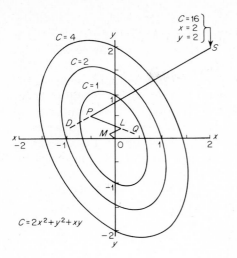

Figure 10-12 Method of steepest descent applied to unimodal surface.

case, the objective function is xy and the constraining equation, set equal to zero, is $x^2 + y^2 - 10 = 0$. Thus, the Lagrange expression is

$$\text{L.E. } (x, y) = xy + \lambda(x^2 + y^2 - 10) \tag{92}$$

Taking the partial of Eq. (92) with respect to x, y, and λ, and setting each result equal to zero gives

$$y + 2\lambda x = 0 \tag{93}$$

$$x + 2\lambda y = 0 \tag{94}$$

$$x^2 + y^2 - 10 = 0 \tag{95}$$

Simultaneous solution of the preceding three equations for x, y, and λ gives, for the case where both x and y are positive, the optimum values of x equal to 2.24 and y equal to 2.24.

METHOD OF STEEPEST ASCENT OR DESCENT†

For the optimization situation in which two or more independent variables are involved, response surfaces can often be prepared to show the relationship among the variables. Figure 10-12 is an example of a unimodal response surface with a single minimum point. Many methods have been proposed for exploring such response surfaces to determine optimum conditions.

† G. E. P. Box and K. B. Wilson, *J. Royal Stat. Soc.,* **B13**:1 (1951); A. H. Boas, *Chem. Eng.,* **70**(5):97 (1963); W. D. Baasel, *Chem. Eng.,* **72**(22):147 (1965); D. J. Wilde and C. S. Beightler, "Foundations of Optimization," Prentice-Hall, Inc., Englewood Cliffs, N.J., 1967.

One of the early methods proposed for establishing optimum conditions from response surfaces is known as the method of *steepest ascent or descent*. The basis of this method is the establishment of a straight line or a two-dimensional plane which represents a restricted region of the curved surface. The gradient at the restricted region is then determined from the linearized approximation, and the desired direction of the gradient is established as that linear direction giving the greatest change in the function being optimized relative to the change in one or more of the independent variables. If the objective function is to be maximized, the line of steepest ascent toward the maximum is sought. For the case of a minimum as the desired objective, the approach would be by means of the steepest descent.

To illustrate the basic ideas involved, consider the case where the objective function to be minimized (C) is represented by

$$C = 2x^2 + y^2 + xy \tag{96}$$

where x and y are the independent variables. Equation (96) is plotted as a contour surface in Fig. 10-12, and the objective is to determine, by the method of steepest descent, the values of x and y which make C a minimum. Arbitrarily, a starting point of $x = 2$, $y = 2$, and $C = 16$ is chosen and designated as point S in Fig. 10-12. The gradient at point S is determined by taking the partial of C with respect to each of the independent variables to give

$$\frac{\partial C}{\partial x} = 4x + y = (4)(2) + 2 = 10 \tag{97}$$

$$\frac{\partial C}{\partial y} = 2y + x = (2)(2) + 2 = 6 \tag{98}$$

Both of these partials are positive which means that both x and y must change in the negative direction to head toward a minimum for C. The direction to be taken is established by recognizing that C must change more rapidly in the x direction than in the y direction in direct ratio to the partial derivatives. Thus, x should decrease faster than y in the ratio of (decrease in x)/(decrease in y) = $\frac{10}{6}$. Assume, arbitrarily, to decrease x linearly from point S in increments of 0.5. Then y must decrease in increments of $(0.5) \frac{6}{10} = 0.3$. Under these conditions, the first line of steepest descent is found as follows and is shown as line SD in Fig. 10-12.

$x_0 =$ 2.00	$y_0 = 2.00$	$C_0 = 16.00$
$x_1 =$ 1.50	$y_1 = 1.70$	$C_1 = 9.94$
$x_2 =$ 1.00	$y_2 = 1.40$	$C_2 = 5.36$
$x_3 =$ 0.50	$y_3 = 1.10$	$C_3 = 2.26$
$x_4 =$ 0.00	$y_4 = 0.80$	$C_4 = 0.64$
$x_5 = -0.50$	$y_5 = 0.50$	$C_5 = 0.50$
$x_6 = -1.00$	$y_6 = 0.20$	$C_6 = 1.84$

The minimum for line SD occurs at x_5, y_5; so a new line is now established using point x_5, y_5 as the starting point. Using the same procedure as was followed for finding line SD, the line PQ is found with a minimum at L. Thus, point L now becomes the new starting point. This same linearization procedure is repeated with each line getting closer to the true minimum of $C = 0$, $x = 0$, $y = 0$.

The method outlined in the preceding obviously can become very tedious mathematically, and computer solution is normally necessary. The method also has limitations based on choice of scale and incremental steps for the variables, extrapolation past the region where the straight line approximates the surface, and inability to handle surfaces that are not unimodal.

EXPLORATION OF RESPONSE SURFACES BY GROUP EXPERIMENTS[†]

In addition to the method of steepest ascent and descent, many other strategies for exploring response surfaces which represent objective functions have been proposed. Many of these are based on making group experiments or calculations in such a way that the results allow a planned search of the surface to approach quickly a unimodal optimum point.

A typical example of an efficient search technique by group experiments is known as the *Five-Point Method* and is explained in the following. The basis of this method is first to select the overall range of the surface to be examined and then to determine the values of the objective function at both extremes of the surface and at three other points at equally spaced intervals across the surface. Figure 10-13 shows a typical result for these initial five points for a simplified two-dimensional case in which only one maximum or minimum is involved.

From these first five calculations, it can be seen that, by keeping the optimum point and the point on each side of it, the search area can be cut in half with assurance that the remaining area still contains the optimum value. In Fig. 10-13, the optimum is represented by the maximum profit, so the middle half of the search area is retained.

Two more calculations or experiments are then made in the remaining search area with these points again being equally spaced so that the remaining search area is again divided into four equal portions. As before, the optimum (highest profit) point is kept along with the points on each side of it, so the search area is again cut in half.

This procedure can be repeated to reduce the search area by a large amount with a relatively few calculations. For example, as shown in the following, 99.9

† W. D. Baasel, Exploring Response Surfaces to Establish Optimum Conditions, *Chem. Eng.*, **72**(22):147 (1965); A. H. Boas, Optimization Techniques, *AACE Bull.*, **8**(2):59 (1966); D. J. Wilde and C. S. Beightler, "Foundations of Optimization," Prentice-Hall, Inc., Englewood Cliffs, N.J., 1967.

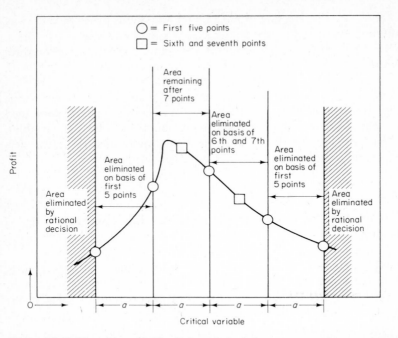

Figure 10-13 Illustration of Five-Point Method for group-experiment exploration of response surface.

percent of the search area can be eliminated by a total of only 23 calculations or experiments:

For step number	Number of new calculations	Total calculations	Fraction of region isolated $= \Delta$
1	5	5	$\frac{1}{2}$
2	2	7	$(\frac{1}{2})^2$
3	2	9	$(\frac{1}{2})^3$
⋮	⋮	⋮	⋮
m_s	2	$n_c = 3 + 2m_s$	$(\frac{1}{2})^{m_s} = \Delta$

$$\text{If} \quad (\tfrac{1}{2})^{m_s} = \Delta, \quad m_s = \frac{\log \Delta}{-\log 2} = \frac{n_c - 3}{2}, \quad \text{and} \quad n_c = 3 - \frac{2 \log \Delta}{\log 2} \quad (99)$$

For the case where Δ is 0.001, or 99.9 percent of the surface has been eliminated, Eq. (99) gives the number of calculations needed (n_c) as 23.

A similar approach is used in the *Golden Section Search Technique* which uses as its basis a symmetrical placement of search points located at an arbitrary distance from each side of the search area.† This method can eliminate 99.9

---

† See D. F. Rudd and C. C. Watson, "Strategy of Process Engineering," John Wiley & Sons, Inc., New York, 1968.

percent of the search area by a total of 17 search points as compared to 23 search points for the simple Five-Point Method.

A so-called *dichotomous search* for the optimum on a surface representing an objective function is conducted by performing the experiments or calculations in pairs. By locating the pairs at appropriate intervals over the surface, inappropriate regions can be eliminated quickly, and a sequential technique can be developed to permit rapid elimination of major portions of the surface. Similarly, the *simplex method*, based on a triangulation of experimental or calculated points, can be used to indicate the desired direction of a search. A highly effective sequential search technique, known as the *Fibonacci search* because the search sequence is based on Fibonacci numbers, can be employed when the objective function has only one optimum and is based on a single independent variable. Experimental errors involved in analyzing response surfaces can be eliminated partially by a so-called *evolutionary operations* (*EVOP*) technique based on measuring the response to the operating conditions a sufficient number of times so that the mean of the sample response approaches the true mean.

GEOMETRIC PROGRAMMING†

A technique for optimization, based on the inequality relating the arithmetic mean to the geometric mean for a set of numbers, has been called *geometric programming*. With this method, the basic idea is to start by finding the optimum way to distribute the total cost among the various factors of the objective function. This is then followed by an analysis of the optimal distribution to establish the final optimum for the objective function. Although this approach can become very involved mathematically and may involve nonlinear equations, it can handle equality and inequality constraints and can often be simpler than a direct nonlinear-programming approach.

OPTIMUM CONDITIONS FOR PRODUCTION, PLANNING, SCHEDULING, AND CONTROL‡

In recent years, refined numerical techniques have been developed for effective planning, scheduling, and control of projects. Two of these methods, critical path method (CPM) and program evaluation and review technique (PERT) have received particular attention and have shown the desirability of applying math-

---

† R. J. Duffin, *J. Soc. Ind. Appl. Math.*, **10**:119 (1962); R. J. Duffin, *J. Oper. Res.*, **10**:668 (1962); D. J. Wilde, *Ind. Eng. Chem.*, **57**(8):31 (1965); R. L. Zahradnik, "Theory and Techniques of Optimization for Practicing Engineers," Barnes & Noble, Inc., New York, 1971.

‡ J. J. Moder and C. R. Phillips, "Project Management with CPM and PERT," 2d ed., Reinhold Publishing Corporation, New York, 1970; L. R. Shaffer, L. B. Ritter, and W. L. Meyer, "The Critical Path Method," McGraw-Hill Book Company, New York, 1965; Federal Electric Corporation, "A Programmed Introduction to PERT," John Wiley & Sons, Inc., New York, 1965.

ematical and graphical analyses to the planning and control of production processes.

The basis of both the critical path method and program evaluation and review technique is a graphical portrait, or network, showing the interdependencies of the various activities in the program leading from the initial input, or startup, to the end objective. PERT is of primary use for organizing and planning projects that involve research and development wherein the activities are usually being attempted for the first time. As a result, estimates of time, cost, and results cannot be made with accuracy, and probability and statistical concepts must be used to develop the predictions. In comparison, CPM is usually applied to projects for which relatively accurate estimations of time, cost, and results can be made, such as for construction projects.

For both CPM and PERT, the overall project is viewed as a series of activities or operations performed in an optimum sequence to reach a desired objective. Each activity is considered as having a beginning and an end so that the overall project consists of a series of these "events." The general technique, then, is to develop a mathematical model to give the best program or interrelated series of events to achieve a desired goal. The major difference in concept between CPM and PERT is that involved in estimating the time duration of activities. Thus, CPM may be relatively specific on time items, while PERT includes measures of the uncertainties involved.

When the series of activities is diagrammed, it can be seen that many possible paths exist between the "start" and the "end." The "critical path" is defined as that path involving the desired (usually shortest) duration for completion of the project. The mathematical concepts of both PERT and CPM are normally of sufficient complexity that a digital computer must be used for the solution. By the appropriate network computations, a final sequential procedure is developed which gives the "critical path" that must be followed from the "start" to the "end" to complete the job in the most efficient manner in a given duration of time.

THE STRATEGY OF ACCOUNTING FOR INFLATION IN DESIGN ESTIMATES

The method of correcting for price changes that have occurred in the past when estimating costs for design purposes has been discussed in Chap. 5 (Cost Estimation). As this discussion showed, the history of cost changes in the United States in the recent past has been strongly inflationary. For example, the Marshall and Swift All-Industry Installed-Equipment Cost Index has doubled from 256 in January of 1967 to 512 in the third quarter of 1977. Similarly, the All Commodities Price Index, as published regularly in the *Monthly Labor Review* of the U.S. Department of Labor Statistics, has approximately doubled in the ten-year period from the base year of 1967 (Index = 100) to 1977 (Index = 194.2 for 1977 and 199.9 for January, 1978).

An effective interest rate of 7.18 percent will cause a doubling of value when compounded for 10 years. Consequently, past history of price changes in the

United States would indicate that a rate of inflation of 7 or more percent can be assumed for at least the near future, and this factor should be taken into account in presenting design estimates of costs.

The critical element of the strategy for accounting for inflation in design estimates is to present the results in the form of *present worth* (*present value, profitability index, discounted cash flow*) with all future dollars discounted to the value of the present dollar at zero time. The discount factor must include both the interest required by the company as minimum return and the estimated interest rate of inflation. If profits on which income taxes are charged are involved, then the present worth based on the after-tax situation should be used.

In order to understand which form of discount factor to use with inflation (or with deflation), the two specific cases for constant annual income in the future and constant annual productivity in the future will be considered. In all cases, effective interest and instantaneous end-of-year cash flow will be assumed.

CASE OF CONSTANT ANNUAL INCOME IN THE FUTURE

Assume that a firm wishes to make an investment now to provide $100,000 in cash at the end of each year for the next ten years. The firm expects to receive a 10 percent return ($i = 0.10$) on its investment irrespective of inflation effects. However, the firm also wishes to account for an assumed annual inflation of 7 percent ($i_{inflation} = 0.07$) so that the dollars it invests now are corrected for the fact that these dollars will be worth less in the future. Under these conditions, the question is how to establish the correct discount factor to determine the investment the firm needs to make at this time. In other words, what is the total present value of the future annual incomes of $100,000 for ten years discounted for both return on investment and inflation?

Consider the case of the first $100,000 coming in at the end of the first year. The present value at zero time of this $100,000 based only on the need to keep the purchasing power of the dollar constant by correcting for inflation is $(\$100,000)(1 + 0.07)^{-1}$ or, in general, $(\$100,000)(1 + i_{inflation})^{-n'}$ where n' is the year referred to. In addition, the firm demands a 10 percent direct return on the investment; so an additional discount factor of $(1 + 0.10)^{-1}$ or, in general, $(1 + i)^{-n'}$ must be applied to the annual income value to give its present value at zero time. Thus, the zero-time present value of the first $100,000 is $(\$100,000) \times (1 + 0.07)^{-1}(1 + 0.10)^{-1}$. The total present value at zero time of all the annual incomes is merely the following sum:

For first year	$(\$100,000)(1 + 0.07)^{-1}(1 + 0.10)^{-1}$
+	
For second year	$(\$100,000)(1 + 0.07)^{-2}(1 + 0.10)^{-2}$
+	
For third year	$(\$100,000)(1 + 0.07)^{-3}(1 + 0.10)^{-3}$
+	+
$\vdots$	$\vdots$
+	+
For tenth year	$(\$100,000)(1 + 0.07)^{-10}(1 + 0.10)^{-10}$

or, in general

$$\text{The total present value} = \sum_{1}^{10} (\$100{,}000)(1 + i_{\text{inflation}})^{-n'}(1 + i)^{-n'}$$

The effective discount factor including both inflation and required return on investment is $[(1 + i_{\text{inflation}})(1 + i)]^{-n'}$ or $[1 + i + i_{\text{inflation}} + (i_{\text{inflation}})(i)]^{-n'}$. Consequently, the effective combined interest (i_{comb}) including both inflation interest and required return on investment is

$$i_{\text{comb}} = i + i_{\text{inflation}} + (i_{\text{inflation}})(i) \tag{100}$$

The preceding situation, of course, is merely a case of an ordinary annuity ($R = \$100{,}000$ each year) at an interest rate of i_{comb} so that Eq. (24) of Chap. 6 (Interest and Investment Costs) applies as follows:

$$\text{Present value} = R\,\frac{(1 + i_{\text{comb}})^{n} - 1}{i_{\text{comb}}(1 + i_{\text{comb}})^{n}} \tag{101}$$

For the example under consideration, R = annual periodic payment = $100,000, n = total life period = 10 years, $i_{\text{comb}} = 0.10 + 0.07 + (0.07)(0.10) = 0.177$, and present value (or necessary investment now) = $\$100{,}000[(1 + 0.177)^{10} - 1]/$ $0.177(1 + 0.177)^{10} = \$452{,}240$.

CASE OF CONSTANT ANNUAL PRODUCTIVITY IN THE FUTURE

For the typical situation of an industrial operation which has been designed to produce a set number of units per year which will be sold at the prevailing price, there would be no special problem with handling inflation except for the influence of income taxes. If the inflationary costs are considered as having the same effects on the selling price of the product as on the costs for the operation, then return on investment before taxes is the same whether or not inflation is taken into account. However, as illustrated by the following example, when income taxes are included in the analysis, the return on investment changes if inflation is taken into account. This is due to the fact that depreciation costs are not changed by inflation in normal accounting procedures.

Example 8. Return on investment before and after taxes with and without inflation An investment of $1,000,000 will give annual returns as shown in the following over a life of five years. Assume straight-line depreciation, negligible salvage value, and 48 percent income taxes. What is the discounted-cash-flow rate of return on the investment (Profitability Index) before and after taxes with

(a) No inflation and annual returns of $300,000 each year (i.e.; cash flow to the company of $300,000) before taxes?

(b) Inflation rate of 7 percent ($i_{\text{inflation}} = 0.07$) and a situation where the increase in profits due to inflation is also at an annual rate of 7 percent so the annual returns remain at the equivalent of $300,000 in zero-time dollars before taxes?

SOLUTION:

(a) For the case of no inflation, Eq. (24) of Chap. 6 [or Eq. (101) of this chapter with $i_{comb} = i$] applies as follows:

$$\text{Present value} = R\frac{(1 + i)^n - 1}{i(1 + i)^n}$$

For return on investment (i) before taxes,

$$\$1,000,000 = \$300,000\frac{(1 + i)^5 - 1}{i(1 + i)^5}$$

By trial and error, or by use of tables of $[(1 + i)^n - 1]/i(1 + i)^n$,

$$i = 0.152 \text{ or } 15.2\% \text{ return}$$

For return on investment (i) after taxes,

$$\text{Depreciation} = \frac{\$1,000,000 - 0}{5} = \$200,000 \text{ per year}$$

Taxable income = $300,000 − $200,000 = $100,000 per year

Taxes at 48% rate = $48,000 per year

Annual cash flow = $300,000 − $48,000 = $252,000

$$\$1,000,000 = \$252,000\frac{(1 + i)^5 - 1}{i(1 + i)^5}$$

$$i = 0.0824 \text{ or } 8.24\% \text{ return}$$

(b) For the case of 7% inflation

Year (n')	Before-tax annual return based on zero-time dollars ($)	Actual dollars received at 7% inflation = 300,000 $(1 + 0.07)^{n'}$ ($)	Depreciation ($)	Taxable income ($)	Income tax at 48% ($)	After-tax annual cash flow ($)
1	300,000	321,000	200,000	121,000	58,080	262,920
2	300,000	343,470	200,000	143,470	68,866	274,604
3	300,000	367,513	200,000	167,513	80,406	287,107
4	300,000	393,239	200,000	193,239	92,755	300,484
5	300,000	420,766	200,000	220,766	105,968	314,798

For return on investment (i) before taxes, the actual annual return based on zero-time dollars is $300,000; so the return on the investment is exactly the same as for the case of no inflation, and $i = 0.152$ or 15.2% return.

For return on investment (i) after taxes, the annual cash flows based on zero-time dollars must have a total present value of $1,000,000. As is shown in the following tabulation, this occurs for a value of $i = 0.0568$:

Year (n')	After-tax annual cash flow (A in $)	Inflation adjustment. After-tax annual cash flow based on zero-time dollars $= A (1 + 0.07)^{-n'}$ (B in $)	Inflation plus return adjustment. Present value at $i = 0.0568$ (5.68%) $= B (1 + 0.0568)^{-n'}$ ($)
1	262,920	245,720	232,521
2	274,604	239,850	214,775
3	287,107	234,365	198,590
4	300,484	229,238	183,811
5	314,798	224,447	170,303
		Total present value =	$1,000,000

Under these conditions, the return on investment after taxes with a 7% inflation rate is $i = 0.0568$ or 5.68% return.

Thus, as would be expected because profits for the inflation case increased at the same rate as the inflation, the before-tax return on the investment was the same for the cases with or without inflation at 15.2%. However, due to the depreciation costs remaining constant in the case of inflation, the after-tax return on the investment was different for the no-inflation case (8.24%) and the inflation case (5.68%).

The preceding example clearly shows that inflation effects can be important in determining returns on investment. The best strategy for handling such effects is to use the discounted-cash-flow or present-worth method for reporting returns on investment with the results based on the after-tax situation. This method of reporting can be handled easily and effectively by use of an appropriately arranged table for the presentation such as Table 2 in Chap. 9 (Profitability, Alternative Investments, and Replacements).†

NOMENCLATURE FOR CHAPTER 10

a = constant, or depreciation factor for installed piping system [See Eq. (82) for definition of a_i]
a' = depreciation factor for pumping installation
A = heat-transfer area, ft^2

---

† For further discussion on this topic, see W. H. Griest, Jr., Making Decisions in an Inflationary Environment, *Chem. Eng. Progr.*, **75**(6):13 (1979).

b = constant, or maintenance factor for installed piping system [See Eq. (82) for definition of b_i]

b' = maintenance factor for pumping installation

B = constant

B' = constant

c = constant

c_i = objective-function coefficient for solution variable

c_j = objective-function coefficient for row in simplex algorithm matrix

c_p = heat capacity, Btu/(lb)(°F)

c_{pipe} = purchase cost of new pipe per foot of pipe length, $/ft

c_T = total cost per unit of production, $/unit of production

C = cost, or objective function

C_A = installed cost of heat exchanger per square foot of heat-transfer area, $/ft^2

C_c = cost for one cleaning, dollars

C_F = fixed costs, $/year

C_{pipe} = installed cost for piping system expressed as dollars per year per foot of pipe length, $/(year)(ft)

$C_{pumping}$ = pumping costs as dollars per year per foot of pipe length when flow is turbulent, $/(year)(ft)

$C'_{pumping}$ = pumping cost as dollars per year per foot of pipe length when flow is viscous, $/(year)(ft)

C_T = total cost for a given unit of time, dollars

C_w = cost of cooling water, $/lb

d = constant, or derivative, or design decision for dynamic programming

D = inside diameter of pipe, ft, or determinant

D_i = inside diameter of pipe, in.

E = efficiency of motor and pump expressed as a fraction

$\bar{f}$ = Fanning friction factor, dimensionless, or function for dynamic programming in Eq. (91) indicating optimum return depends on that input

F = ratio of total cost for fittings and installation to purchase cost for new pipe

g = function

g_c = conversion factor in Newton's law of motion, 32.17 ft · lbm/(s)(s)lbf

h = operating costs which remain constant per unit of production, $/unit of production, or function

i = annual effective interest rate of return, percent/100

i_{comb} = annual effective interest rate of change combining regular return and inflation estimate, percent/100

$i_{inflation}$ = annual effective interest rate of change based on inflation estimate, percent/100

H = total time used for actual operation, emptying, cleaning, and recharging, h

H' = total time available for operation, emptying, cleaning, and recharging, h

H_y = total time of operation per year, h/year

I = row

j = constant

J = frictional loss due to fittings and bends, expressed as equivalent fractional loss in a straight pipe, or column

k = designation for column in simplex algorithm matrix with maximum value of $c_j - z_j$

K = cost of electrical energy, \$/kWh

K_F = annual fixed charges including maintenance, expressed as a fraction of the initial cost for the completely installed equipment

L = length of pipe, ft

L'_e = frictional loss due to fittings and bends, expressed as equivalent pipe length in pipe diameters per unit length of pipe

m = constant, or number of independent equations

m_s = number of steps

M = ratio of total cost for pumping installation to yearly cost of pumping power required, \$/\$

M_B = bottoms flow rate, mol/h

M_D = distillate flow rate, mol/h

n = constant, estimated service life, or number of unknowns or stages

n' = year of project life to which cash flow applies

n_c = number of calculations

N_{Re} = Reynolds number = $DV\rho/\mu$, dimensionless

O_c = organization costs per unit of time, \$/day

P = rate of production, units of production/day, or return as objective function in dynamic programming

P_b = amount of production per batch, lb/batch

P_f = filtrate delivered in filtering time θ_f h, ft^3

P_o = optimum rate of production, units of production/day

q = rate of heat transfer, Btu/h

q_f = rate of fluid flow, ft^3/s

q_r = rate of heat transfer in reboiler, Btu/h

Q = total amount of heat transferred in a given time, Btu

Q_H = total amount of heat transferred in H h, Btu

Q_i = function for dynamic programming indicating combined return from all stages

r = profit per unit of production, \$/unit of production

R = annual periodic payment in ordinary annuity, \$/year

R' = profit per unit of time, \$/day

s = selling price per unit of production, \$/unit of production

S = slack variable

S_b = direct labor cost per hour during operation, \$/h

t = temperature, °F

t_1 = temperature of cooling water entering condenser, °F

t_2 = temperature of cooling water leaving condenser, °F

t' = condensation temperature, °F

U = overall coefficient of heat transfer, Btu/(h)(ft^2)(°F)

V = average linear velocity, ft/s

V_A = volume of A, gal

V_B = volume of B, gal

V_W = volume of water, gal

w = flow rate, lb/h

w_m = thousands of pounds mass flowing per hour, 1000 lb/h

w_s = pounds mass flowing per second, lb/s

x = a variable

X = purchase cost for new pipe per foot of pipe length if pipe diameter is 1 in., \$/ft

X' = purchase cost for new pipe per foot of pipe length if pipe diameter is 1 ft, \$/ft

y = a variable

Y = days of operation per year, days/yr

z = a variable

z_j = objective-function row coefficient component for simplex algorithm matrix defined as $\sum_{j=1}^{m} c_i a_{ij}$ $(j = 1, 2, \ldots, n)$

Z = fractional rate of return on incremental investment

Greek symbols

α = symbol meaning go to next starting point

Δ = fraction of search area eliminated

Δt = temperature-difference driving force (subscript lm designates log mean), °F

θ_b = time in operation, h or h/cycle

θ_c = time for emptying, cleaning, and recharging per cycle, h/cycle

θ_f = filtering time, h or h/cycle

θ_t = total time per complete cycle, h/cycle

λ = Lagrangian multiplier

μ = absolute viscosity, lbm/(s)(ft)

μ_c = absolute viscosity, cP

ρ = density, lbm/ft^3

$\phi, \phi^i, \phi^{ii}, \phi^{iii}$ = function of the indicated variables, or fractional factor for rate of taxation

SUGGESTED ADDITIONAL REFERENCES FOR OPTIMUM DESIGN AND DESIGN STRATEGY

Critical path method (CPM)

Jenett, E., Experience with and Evaluation of Critical Path Methods, *Chem. Eng.*, **76**(3):96 (1969).
Kerridge, A. E., Improve CPM Milestone Networks, *Hydrocarbon Process.*, **57**(9):277 (1978).
Klimpel, R. R., Operations Research: Decision-Making Tool—II, *Chem. Eng.*, **80**(10):87 (1973).
Manganelli, L., Direct Charting of PERT-Time Via Computer, *Petro/Chem. Eng.*, **39**(6):21 (1967).
Moder, J. J., and C. R. Phillips, "Project Management with CPM and PERT," Van Nostrand Reinhold Company, New York, 1970.
Poulton, C., and P. Zakaib, Construction Administration in a Medium-Sized Industrial Company, *1973 Trans. Am. Assoc. Cost Eng.*, (1973), p. 93.
Reynaud, C. B., Network Analysis Works, *Petro/Chem. Eng.*, **40**(11):PM7 (1968).
Robertson, D. C., "Project Planning and Control," CRC Press, Cleveland, Ohio, 1967.
Waligura, C. L., and R. L. Motard, Data Management on Engineering and Construction Projects, *Chem. Eng. Progr.*, **73**(12):62 (1977).

Design strategy

Fair, R. R., Planning for the 1970's, *ASTME Vectors*, **4**(6):30 (1969).
Murphy, T. D., Jr., Design and Analysis of Industrial Experiments, *Chem. Eng.*, **84**(12):168 (1977).
Nelles, M., Accelerated Manufacturing Planning, *ASTME Vectors*, **4**(6):35 (1969).
Rudd, D. F., G. J. Powers, and J. J. Siirola, "Process Synthesis," Prentice-Hall, Englewood Cliffs, New Jersey, 1973.
——— and C. C. Watson, "Strategy of Process Engineering," J. Wiley and Sons, New York, 1968.
Schweitzer, O. R., Reverse Synthesis Simplifies Problem Solving, *Chem. Eng.*, **85**(14):111 (1978).
Ware, C. H., Jr., Improving R & D Effectiveness, *Chem. Eng.*, **85**(5):99 (1978).

Dynamic programming

Beightler, C. S., and W. L. Meier, Jr., Design of an Optimum Branched Allocation System, *I&EC*, **60**(2):44 (1968).
Chen, K., Optimization of Time-Dependent Systems by Dynamic Programming, *ISA Trans.*, **6**(2):157 (1967).
Gluss, B., "An Elementary Introduction to Dynamic Programming: A State Equation Approach," Allyn and Bacon, Boston, Massachusetts, 1972.
Ithara, S., and L. I. Stiel, Optimal Design of Multiple-Effect Evaporators with Vapor Bleed Streams, *I&EC Proc. Design & Develop.*, **7**:6 (1968).
Klimpel, R., Operations Research: Decision-Making Tool—II, *Chem. Eng.*, **80**(10):87 (1973).
Kuester, J. L., and J. H. Mize, "Optimization Techniques with Fortran," McGraw-Hill Book Company, New York, 1973.
Larson, R. E., and J. Casti, in "Basic Analytic and Computational Methods," Marcel Dekker, Inc., New York (1978), p. 344.
Lele, P. T., Optimizing Long-Range Multiplant Capacity Expansion, *The Engineer*, **16**(2):18 (1972).
McDermott, C., E. S. Lee, and L. E. Erickson, Iterative Techniques in Optimization: II. The Linearity Difficulty in Dynamic Programming and Quasilinearization, *AIChE J.*, **16**(4):639 (1970).
Ray, W. H., and J. Szekely, "Process Optimization," J. Wiley and Sons, New York, 1973.
Siddall, J. N., "Analytical Decision Making in Engineering Design," Prentice-Hall, Englewood Cliffs, New Jersey, 1972.
Thiriet, L., and A. Deledicq, Applications of Suboptimality in Dynamic Programming, *I&EC*, **60**(2):23 (1968).
White, D. J., "Dynamic Programming," Holden-Day, San Francisco, California, 1969.

Linear programming

Aronofsky, J. S., Linear Programming Models for Business Systems, *Chem. Eng. Progr.*, **64**(4):87 (1968).

Avriel, M., and D. J. Wilde, Optimal Condenser Design by Geometric Programming, *I&EC Proc. Design & Develop.*, **6**(2):256 (1967).

Boulloud, Ph., Compute Steam Balance by LP, *Hydrocarbon Process.*, **48**(8):127 (1969).

Dano, S., "Linear Programming in Industry: Theory and Applications," 4th ed., Springer-Verlag, New York, 1974.

Flanigan, O., W. W. Wilson, and D. R. Sule, Process-Cost Reduction Through Linear Programming, *Chem. Eng.*, **79**(3):68 (1972).

Fox, R. L., "Optimization Methods for Engineering Design," Addison-Wesley, Reading, Massachusetts, 1971.

Goldstein, R. P., and R. B. Stanfield, Flexible Method for the Solution of Distillation Design Problems Using the Newton-Raphson Technique, *I&EC Proc. Design & Develop.*, **9**(1):78 (1970).

Himmelblau, D. M., "Applied Linear Programming," McGraw-Hill Book Company, New York, 1972.

Klimpel, R. R., Operations Research: Decision-Making Tool, *Chem. Eng.*, **80**(9):103 (1973); **80**(10):87 (1973).

Komatsu, S., Application of Linearization to Design of a Hydrodealkylation Plant, *I&EC*, **60**(2):36 (1968).

Kuester, J. L., and J. H. Mize, "Optimization Techniques with Fortran," McGraw-Hill Book Company, New York, 1973.

Norris, W. E., Linear Programming: Advantages and Shortcomings, *Chem. Eng.*, **76**(26):95 (1969).

Ray, W. H., and J. Szekely, "Process Optimization," J. Wiley and Sons, New York, 1973.

Robers, P. D., and A. Ben-Israel, Interval Programming, *I&EC Proc. Design & Develop.*, **8**(4):496 (1969).

Rothman, S. N., Ethylene Plant Optimization, *Chen. Eng. Progr.*, **66**(6):37 (1970).

Schader, M., Solving Quotation Analysis Problems by Linear Programming, *The Engineer*, **15**(2):10 (1971).

Seinfeld, J. H., and W. L. McBride, Optimization with Multiple Performance Criteria, *I&EC Proc. Design & Develop.*, **9**(1):53 (1970).

Siddall, J. N., "Analytical Decision Making in Engineering Design," Prentice-Hall, Englewood Cliffs, New Jersey, 1972.

Mathematical modelling

April, G. C., and R. W. Pike, Modeling Complex Chemical Reaction Systems, *I&EC Proc. Design & Develop.*, **13**(1):1 (1974).

Bobis, A. H., and A. C. Atkinson, Analyzing R&D Investments Via Dynamic Modeling, *Chem. Eng.*, **77**(5):133 (1970).

Chen, J. W., J. A. Buege, F. L. Cunningham, and J. I. Northam, Scale-up of a Column Adsorption Process by Computer Simulation, *I&EC Proc. Design & Develop.*, **7**:26 (1968).

Chung, S. F., Mathematical Model and Optimization of Drying Process for a Through-Circulation Dryer, *Can. J. Chem. Eng.*, **50**(5):657 (1972).

Fuller, O. M., Differential Scaling: A Method for Chemical Reactor Modelling, *Can. J. Chem. Eng.*, **48**(1):119 (1970).

Goldman, M. R., Simulating Multicomponent Batch Distillation, *British Chem. Eng.*, **15**(11):1450 (1970).

Hanson, D. T., Matrix Reduction by Change of Basis, *Chem. Eng.*, **78**(6):117 (1971).

Hyman, M. H., Fundamentals of Engineering/HPI Plants with the Digital Computer, *Petro/Chem. Eng.*, **40**(6):43 (1968).

———, Simulate Methane Reformer Reactions, *Hydrocarbon Process.*, **47**(7):131 (1968).

———, Fundamentals of Engineering/HPI Plants with the Digital Computer, *Petro/Chem. Eng.*, **76**(14):46 (1969).

Kim, C., and J. C. Friedly, Approximate Dynamic Modeling of Large Staged Systems, *I&EC Proc. Design and Develop.*, **13**(2):177 (1974).

Korchinsky, W. J., Modelling of Liquid-Liquid Extraction Columns: Use of Published Model Correlations in Design, *Can. J. Chem. Eng.*, **52**(4):468 (1974).

Korelitz, T. H., Integrating Plant Design with Process Simulation, *Chem. Eng.*, **78**(13):98 (1971).

Lee, E. S., and E. H. Grey, Optimizing Complex Chemical Plants by Mathematical Modeling Techniques, *Chem. Eng.*, **74**(18):131 (1967).

Lunde, P. J., Modeling, Simulation, and Operation of a Sabatier Reactor, *I&EC Proc. Design & Develop.*, **13**(3):226 (1974).

Mezaki, R., J. R. Kittrell, and W. J. Hill, An Analysis of Kinetic Function Models, *I&EC*, **59**(1):93 (1967).

Murrill, P. W., R. W. Pike, and C. L. Smith, Development of Dynamic Mathematical Models, *Chem. Eng.*, **75**(19):117 (1968).

—— —— and ——, The Basis for Bode Plots, *Chem. Eng.*, **75**(21):177 (1968).

—— —— and ——, Frequency-Response Data Yield Analytic Equations, *Chem. Eng.*, **75**(24):165 (1968).

—— —— and ——, Transient Response in Dynamic-Systems Analyses, *Chem. Eng.*, **75**(27):103 (1968).

—— —— and ——, Transient-Response Data Yield Frequency-Response Models, *Chem. Eng.*, **76**(2):167 (1969).

—— —— and ——, Pulse-Testing Methods, *Chem. Eng.*, **76**(4):105 (1969).

—— —— and ——, Nonperiodic Inputs Provide Data for Frequency-Response Models, *Chem. Eng.*, **76**(5):111 (1969).

—— —— and ——, Random Inputs Yield System Transfer Functions, *Chem. Eng.*, **76**(7):151 (1969).

—— —— and ——, Frequency-Response Data from Statistical Correlations, *Chem. Eng.*, **76**(11):195 (1969).

—— —— and ——, Models for a Process Plant, *Chem. Eng.*, **76**(13):97 (1969).

—— —— and ——, Models for Process Equipment, *Chem. Eng.*, **76**(16):139 (1969).

—— —— and ——, Algorithm for Computing Fourier Integrals Rapidly, *Chem. Eng.*, **76**(18):125 (1969).

Newell, R. B., and D. G. Fisher, Model Development, Reduction, and Experimental Evaluation for an Evaporator, *I&EC Proc. Design & Develop.*, **11**(2):213 (1972).

Paris, J. R., and W. F. Stevens, Mathematical Models for a Packed-bed Chemical Reactor, *Can. J. Chem. Eng.*, **48**(1):100 (1970).

Price, J. Computerized Models—The Scientific Way to Pick a Winner, *Chem. Week*, **106**(15):37 (1970).

Reilly, P. M., and G. E. Blau, The Use of Statistical Methods to Build Mathematical Models of Chemical Reacting Systems, *Can. J. Chem. Eng.*, **52**(3):289 (1974).

Sinibaldi, F. J., Jr., T. L. Koehler, and A. H. Bobis, Transformed Data Simplify and Confirm Math Models, *Chem. Eng.*, **78**(11):139 (1971).

Umeda, T., M. Nishio, and S. Komatsu, A Method for Plant Data Analysis and Parameters Estimation, *I&EC Proc. Design & Develop.*, **10**(2):236 (1971).

Waggoner, R. C., and S. J. Calvin, Rapid Distillation Column Simulation, *Chem. Eng. Progr.*, **72**(9):70 (1976).

HPI Plants with the Digital Computer, *Petro/Chem. Eng.*, **40**(10):37 (1968).

Mathematical techniques

Ekstrom, R. E., Numerical Methods—1: Power Series, *Machine Design*, **39**(25):61 (1967).

——, Numerical Methods—2: Newton's Method, *Machine Design*, **39**(26):197 (1967).

Goldstein, R. P., and R. B. Stanfield, Flexible Method for the Solution of Distillation Design Problems Using the Newton-Raphson Technique, *I&EC Proc. Design & Develop.*, **9**(1):78 (1970).

Grekel, H., and L. E. Childers, Starting a 48-Plant EVOP Program, *Chem. Eng. Progr.*, **69**(8):93 (1973).

Hanson, D. T., Principles of Linear Algebra, *Chem. Eng.*, **77**(7):119 (1970).

———, Gauss-Jordan Reduction Solves Linear Equations, *Chem. Eng.*, **77**(11):153 (1970).

———, Basic Properties of Matrices, *Chem. Eng.*, **77**(13):169 (1970).

———, Rank of a Matrix, *Chem. Eng.*, **77**(16):154 (1970).

———, Properties of Determinants, *Chem. Eng.*, **77**(18):83 (1970).

———, Operations and Concepts for Linear Equations, *Chem. Eng.*, **77**(20):180 (1970).

———, Matrix Reduction by Gauss-Jordan Techniques, *Chem. Eng.*, **77**(25):154 (1970).

———, Basic Theorems of Simultaneous Linear Equations, *Chem. Eng.*, **77**(26):81 (1970).

———, Computational Methods Using Linear Algebra, *Chem. Eng.*, **78**(1):133 (1971).

———, Stoichiometry by Matrix Techniques, *Chem. Eng.*, **78**(8):123 (1971).

Hershey, H. C., J. L. Zakin, and R. Simha, Numerical Differentiation of Equally Spaced and Not Equally Spaced Experimental Data, *I&EC Chem. Fund.*, **6**(3):413 (1967).

Hyman, M. H., Fundamentals of Engineering—HPI Plants with the Digital Computer, *Petro/Chem. Eng.*, **40**(11):33 (1968).

———, HPI Plants with the Digital Computer—Numerical Analysis, *Petro/Chem. Eng.*, **40**(12):36 (1968).

Jackisch, P. F., Shortcuts to Small-Sample Statistics Problems, *Chem. Eng.*, **80**(13):107 (1973).

Jelen, F. C., Series Summation for Economic Analysis, *Hydrocarbon Process.*, **46**(12):153 (1967).

Klimpel, R. R., Operations Research: Decision-Making Tool, *Chem. Eng.*, **80**(9):103 (1973); **80**(10):87 (1973).

Koenig, D. M., Invariant Imbedding: New Design Method in Unit Operations, *Chem. Eng.*, **74**(19):181 (1967).

Lapidus, L., The Control of Nonlinear Systems via Second-Order Approximations, *Chem. Eng. Progr.*, **63**(12):64 (1967).

———, Applied Mathematics, *I&EC*, **60**(12):42 (1968).

Lee, K. F., A. H. Masso, and D. F. Rudd, Branch and Bound Synthesis of Integrated Process Designs, *I&EC Fund.*, **9**(1):48 (1970).

Lynch, E. P., Logic Diagrams for Batch Processes, *Chem. Eng.*, **81**(21):101 (1974).

Murrill, P. W., R. W. Pike, and C. L. Smith, The Basis for Bode Plots, *Chem. Eng.*, **75**(21):177 (1968).

Woolsey, R. E. D., G. A. Kochenburger, K. R. Linck, Can Engineers Find Happiness Without Calculus? *Hydrocarbon Process.*, **50**(8):133 (1971).

Zanker, A., Developing Empirical Equations, *Chem. Eng.*, **83**(2):101 (1976).

Zienkiewicz, O. C., "The Finite Element Method," McGraw-Hill Book Company, New York, 1977.

Optimization

Abramovitz, J. L., and R. Cordero, How to Select Insulation Thickness for Hot Pipes, *Chem. Eng.*, **82**(15):88 (1975).

Avriel, M., and D. J. Wilde, Optimal Condenser Design by Geometric Programming, *I&EC Proc. Design & Develop.*, **6**(2):256 (1967).

——— and ———, Engineering Design under Uncertainty, *I&EC Proc. Design and Develop.*, **8**(1):124 (1969).

Barneson, R. A., N. F. Brannock, J. G. Moore, and C. Morris, Picking Optimization Methods, *Chem. Eng.*, **77**(16):132 (1970).

Barnhart, J. M., Economic Thickness of Thermal Insulation, *Chem. Eng. Progr.*, **70**(8):50 (1974).

Barona, N., and H. W. Prengle, Jr., Design Reactors This Way for Liquid-Phase Processes, *Hydrocarbon Process.*, **52**(3):63 (1973).

Beamer, J. H., and D. J. Wilde, Approximate Design of a Multi-Stage Plant by Differential Optimization, *I&EC Proc. Design & Develop.*, **12**(4):481 (1973).

Beveridge, G. S. G., and R. S. Schechter, "Optimization: Theory and Practice," McGraw-Hill Book Company, New York, 1970.

Biles, W. E., A Response Surface Method for Experimental Optimization of Multi-Response Processes, *I&EC Proc. Design & Develop.*, **14**(2):152 (1975).

Billet, R., and L. Raichle, Optimizing Method for Vacuum Rectification, *Chem. Eng.*, **74**(4):145 (1967); **74**(5):149 (1967).

Billet, R., Cost Optimization of Towers, *Chem. Eng. Progr.*, **66**(1):41 (1970).

Birta, L. G., and P. J. Trushel, An Optimal Control Algorithm Using the Davidon-Fletcher-Power Method with the Fibonacci Search, *AIChE J.*, **16**(3):363 (1970).

Blau, G. E., and D. J. Wilde, A Lagrangian Algorithm for Equality Constrained Generalized Polynomial Optimization, *AIChE J.*, **17**(1):235 (1971).

Brock, J. E., Optimising the Design of Multishell Cylindrical Pressure Vessels, *Brit. Chem. Eng.*, **16**(6):494 (1971).

Brooks, J. R., and D. E. Waite, An Example of Reactor Optimization Using a Parallel-Logic Analog Computer, *Brit. Chem. Eng.*, **15**(5):647 (1970).

Brosilow, C., and E. Nunez, Multi-Level Optimization Applied to a Catalytic Cracking Plant, *Can. J. Chem. Eng.*, **46**(3):205 (1968).

Burley, D. M., "Studies in Optimization," J. Wiley and Sons, New York, 1974.

Bush, M. J., and G. L. Wells, Unit Plot Plans for Plant Layout, *Brit. Chem. Eng.*, **16**(6):514 (1971).

Chen, K., Optimization of Time-Dependent Systems by Dynamic Programming, *ISA Trans.*, **6**(2):157 (1967).

Chen, M. S. K., L. E. Erickson, and L.-T. Fan, Consideration of Sensitivity and Parameter Uncertainty in Optimal Process Design, *I&EC Proc. Design & Develop.*, **9**(4):514 (1970).

Chen, N. H., Optimization by Geometric Programming, *Chem. Proc. Eng.*, **49**(10):109 (1968).

———, Liquid-Solid Filtration: Generalized Design and Optimization Equations, *Chem. Eng.*, **85**(17):97 (1978).

Christensen, J. H., The Structuring of Process Optimization, *AIChE J.*, **16**(2):177 (1970).

Chung, S. F., Mathematical Model and Optimization of Drying Process for a Through-Circulation Dryer, *Can. J. Chem. Eng.*, **50**(5):657 (1972).

Cooney, D. A., Optimization of Linear Fixed-Bed Separations, *I&EC Proc. Design & Develop.*, **9**(4):578 (1970).

Cormack, D. E., and R. Luus, Suboptimal Control of Chemical Engineering Systems, *Can. J. Chem. Eng.*, **50**(3):390 (1972).

Corrigan, T. E., and M. J. Dean, Determining Optimum Plant Size, *Chem. Eng.*, **74**(17):152 (1967).

Curt, R. P., A New Approach to Economic Insulation Thickness, *Hydrocarbon Process.*, **55**(3):137 (1976).

Dickson, A. N., The Optimization of Multi-Stage Flash Distillation Plant, *Chem. Eng. (London)*, 79 (February 1975).

Doshi, K. J., C. H. Katira, and H. A. Stewart, Optimization of a Pressure Swing Cycle, *Chem. Eng. Progr. Symp. Ser.*, **67**(117):90 (1971).

Donovan, J. R., J. S. Palermo, and R. M. Smith, Sulfuric Acid Converter Optimization, *Chem. Eng. Progr.*, **74**(9):51 (1978).

Dyson, D. C., F. J. M. Horn, R. Jackson, and C. B. Schlesinger, Reactor Optimization Problems for Reversible Exothermic Reactions, *Can. J. Chem. Eng.*, **45**(5):310 (1967).

Edler, J., P. N. Nikiforuk, and E. B. Tinker, A Comparison of the Performance of Techniques for Direct On-Line Optimization, *Can. J. Chem. Eng.*, **48**(4):432 (1970).

Fanaritis, J. P., and J. W. Bevevino, How To Select the Optimum Shell-and-Tube Heat Exchanger, *Chem. Eng.*, **83**(14):62 (1976).

Fauth, G. F., and F. G. Shinskey, Advanced Control of Distillation Columns, *Chem. Eng. Progr.*, **71**(6):49 (1975).

Fiero, W. J., and P. E. Kelly, To Optimize the FCC Unit, *Hydrocarbon Process.*, **56**(9):117 (1977).

Flanigan, O., W. W. Wilson, and D. R. Sule, Process-Cost Reduction Through Linear Programming, *Chem. Eng.*, **79**(3):68 (1972).

Fox, R. L., "Optimization Methods for Engineering Design," Addison-Wesley, Reading, Massachusetts, 1971.

Freshwater, D. C., and B. D. Henry, The Optimal Configuration of Multicomponent Distillation Trains, *Chem. Eng. (London)*, 533 (September 1975).

Friedman, P., and K. L. Pinder, Optimization of a Simulation Model of a Chemical Plant, *I&EC Proc. Design & Develop.*, **11**(4):512 (1972).

Frith, J. F., B. M. Bergen, and M. M. Shreehan, Optimize Heat Train Design, *Hydrocarbon Process.*, **52**(7):89 (1973).

Furman, T. T., and R. F. Cheers, *Brit. Chem. Eng.*, **16**(6):478 (1971).

Gaines, L. D., and J. L. Gaddy, Process Optimization by Flow Sheet Simulation, *I&EC Proc. Design & Develop.*, **15**(1):206 (1976).

Gambro, A. J., K. Muenz, and M. Abrahams, Optimize Ethylene Complex, *Hydrocarbon Process.*, **51**(3):73 (1972).

Generoso, E., Jr., and L. B. Hitchcock, Optimizing Plant Expansion—Two Cases, *I&EC*, **60**(11):12 (1968).

Gentry, J. W., A New One-Dimensional Search Technique, *Chem. Eng. Sci.*, **25**(3):425 (1970).

Glass, R. W., and D. F. Bruley, REFLEX Method for Empirical Optimization, *I&EC Proc. Design & Develop.*, **12**(1):6 (1973).

Gottfried, B. S., P. R. Bruggink, and E. R. Harwood, Chemical Process Optimization Using Penalty Functions, *I&EC Proc. Design & Develop.*, **9**(4):581 (1970).

Grossman, I. E., and R. W. H. Sargent, Optimum Design of Chemical Plants with Uncertain Parameters, *AIChE J.*, **24**(6):1021 (1978).

Harter, M. D., Find Optimum Cat Regeneration Cycle, *Hydrocarbon Process.*, **46**(5):181 (1967).

Horn, B. C., On-Line Optimization of Plant Utilities, *Chem. Eng. Progr.*, **74**(6):76 (1978).

Itahara, S., and L. I. Stiel, Optimal Design of Multiple-Effect Evaporators with Vapor Bleed Streams, *I&EC Proc. Design & Develop.*, **7**(1):6 (1968).

Jelen, F. C., "Cost and Optimization Engineering," McGraw-Hill Book Company, New York, 1970.

Jelen, F. C., and M. S. Cole, Can New Equipment Cost Less? *Hydrocarbon Process.*, **50**(7):93 (1971).

Jenssen, S. K., Heat Exchanger Optimization, *Chem. Eng. Progr.*, **65**(7):59 (1969).

Jung, B. S., W. Mirosh, and W. H. Ray, Large-Scale Process Optimization Techniques Applied to Chemical and Petroleum Processes, *Can. J. Chem. Eng.*, **49**(6):844 (1971).

Katta, S., and W. H. Gauvin, Behavior of a Spray in the Near Vicinity of a Pneumatic Atomizer, *Can. J. Chem. Eng.*, **53**(5):556 (1975).

Keefer, D. L., Simpat: Self-Bounding Direct Search Method for Optimization, *I&EC Proc. Design & Develop.*, **12**(1):92 (1973).

Kerr, D., To Find Optimum for Deisohexanizer, *Hydrocarbon Process.*, **48**(10):93 (1969).

King, R. P., Optimal Replacement and Maintenance Policies, *I&EC*, **60**(2):29 (1968).

Klooster, H. J., G. A. Vogt, and G. F. Braun, Optimizing the Design of Relief and Flare Systems, *Chem. Eng. Progr.*, **71**(1):39 (1975).

Kohli, J. P., Design Best Cooling Water System, *Hydrocarbon Process.*, **47**(12):108 (1968).

Kuester, J. L., and J. H. Mize, "Optimization Techniques with Fortran," McGraw-Hill Book Company, New York, 1973.

Larson, R. E., and J. Casti, "Principles of Dynamic Programming," Marcel Dekker, Inc., New York, (1978).

Lee, E. S., Quasilinearization in Optimization: A Numerical Study, *AIChE J.*, **13**(6):1043 (1967).

Lee, E. S., and E. H. Gray, Optimizing Complex Chemical Plants by Mathematical Modeling Techniques, *Chem. Eng.*, **74**(18):131 (1967).

Lee, W., and M. T. Tayyabkhan, Optimize Process Runs . . . Get More Profit, *Hydrocarbon Process.*, **49**(9):286 (1970).

Liddle, C. J., Optimization of Simple Cyclic Processes, *Chem. Proc. Eng.*, **51**(4):103 (1970).

——, Optimum Design of Recycle Reaction Systems, *Chem. Proc. Eng.*, **53**(5):50 (1972).

Llovet, J. E., H. J. Klooster, and D. G. Chapel, Refinery Design (circa 1984), *Chem. Eng. Progr.*, **71**(6):85 (1975).

Lohrisch, F. W., What is Optimum Batch Heat Transfer? *Hydrocarbon Process.*, **46**(7):169 (1967).

Loonkar, Y. R., and J. D. Robinson, Minimization of Capital Investment for Batch Processes, *I&EC Proc. Design & Develop.*, **9**(4):625 (1970).

Luus, R., Optimization of Multistage Recycle Systems by Direct Search, *Can. J. Chem. Eng.*, **53**(2):217 (1975).

Mapstone, G. E., Most Economic Expansion Rate, *Hydrocarbon Process.*, **54**(1):91 (1975).

McDermott, C., E. S. Lee, and L. E. Erickson, Iterative Techniques in Optimization: II. The Linearity Difficulty in Dynamic Programming and Quasilinearization, *AIChE J.*, **16**(4):639 (1970).

Melin, G. A., J. L. Niedzwiecki, and A. M. Goldstein, Optimum Design of Sour Water Strippers, *Chem. Eng. Progr.*, **71**(6):78 (1975).

Menicatti, S., Check Tank Insulation Economics, *Hydrocarbon Process.*, **48**(4):133 (1969).

Menicatti, S., and L. Cappiello, Find Optimum Furnace Efficiency, *Hydrocarbon Process.*, **51**(9):226 (1972).

Mitchell, W. T., Distillation Practices: Distillation Optimization, *Chem. Eng. Progr.*, **68**(10):62 (1972).

Neretnicks, I., The Optimization of Sieve Tray Columns, *Brit. Chem. Eng.*, **15**(2):193 (1970).

Nguyen, H. X., Optimize Water Outlet Temperature, *Hydrocarbon Process.*, **57**(5):245 (1978).

Nishida, N., and A. Ichikawa, Synthesis of Optimal Dynamic Process Systems by a Gradient Method, *I&EC Proc. Design & Develop.*, **14**(3):236 (1975).

Nishida, N., and L. Lapidus, Studies in Chemical Process Design and Synthesis, III. A Simple and Practical Approach to the Optimal Synthesis of Heat Exchanger Networks, *AIChE J.*, **23**(1):77 (1977).

Ogunye, A. F., and W. H. Ray, Optimization of Cyclic Tubular Reactors with Catalyst Decay, *I&EC Proc. Design & Develop.*, **10**(3):410 (1971).

—— and ——, Optimization of Recycle Reactors Having Catalyst Decay, *I&EC Proc. Design & Develop.*, **10**(3):416 (1971).

Pacey, W. C., and A. Rustin, Optimization of a Reaction and Heat Exchange System with Recycle, *Can. J. Chem. Eng.*, **45**(5):305 (1967).

Paynter, J. D., and S. G. Bankoff, Computational Methods in Process Optimization, *Can. J. Chem. Eng.*, **45**(4):226 (1967).

Perry, R. H., and E. Singer, Practical Guidelines for Process Optimization, *Chem. Eng.*, **75**(5):163 (1968).

Pfeiffer, R. R., Planned Equipment Replacement, *AACE Bull.*, **12**(2):35 (1970).

Picciotti, M., Optimize Quench Water Systems, *Hydrocarbon Process.*, **56**(9):179 (1977).

——, Optimize Caustic Scrubbing Systems, *Hydrocarbon Process.*, **57**(5):201 (1978).

Pierru, A., and C. Alexandre, Optimize PVC Reactor, *Hydrocarbon Process.*, **52**(6):97 (1973).

Pinto, A., and P. L. Rogerson, Optimizing the ICI Low-Pressure Methanol Process, *Chem. Eng.*, **84**(14):102 (1977).

Powers, G. J., Heuristic Synthesis in Process Development, *Chem. Eng. Progr.*, **68**(8):88 (1972).

Ray, W. H., and J. Szekely, "Process Optimization; With Applications in Metallurgy and Chemical Engineering," J. Wiley and Sons, New York, 1973.

Raymond, T. C., Heuristic Algorithm for the Traveling-Salesman Problem, *IBM J. Res. & Develop.*, **13**(4):400 (1969).

Reisser, A., Improving the Efficiency of Batch Operations, *Chem. Eng.*, **75**(4):117 (1968).

Rijnsdorp, J. E., Chemical Process Systems and Automatic Control, *Chem. Eng. Progr.*, **63**(7):97 (1967).

Robertson, J. M., Plan Small for Expansion, *Chem. Eng. Progr.*, **63**(9):87 (1967).

——, Design for Expansion, Part 2: Engineering Optimization, *Chem. Eng.*, **75**(10):187 (1968).

Robinson, E. R., Optimum Reflux Policies for Batch Distillation, *Chem. Proc. Eng.*, **52**(5):47 (1971).

Royston, D., Heat Transfer in the Flow of Solids in Gas Suspensions Through a Packed Bed, *I&EC Proc. Design & Develop.*, **10**(2):145 (1971).

Sargent, R. W. H., Integrated Design and Optimization of Processes, *Chem. Eng. Progr.*, **63**(9):71 (1967).

Seinfeld, J. H., and W. L. McBride, Optimization With Multiple Performance Criteria, *I&EC Proc. Design & Develop.*, **9**(1):53 (1970).

Shaner, R. L., Energy Scarcity: A Process Design Incentive, *Chem. Eng. Progr.*, **74**(5):47 (1978).

Shalbenderian, A. P., Plants, Plans, and Probabilities, *Chem. Proc. Eng.*, **48**(9):94 (1967).

Smith, G. M., Energy Evaluation in Sulfuric Acid Plants, *Chem. Eng. Progr.*, **73**(3):77 (1977).

Siddall, J. N., "Analytical Decision Making in Engineering Design," Prentice-Hall, Englewood Cliffs, New Jersey, 1972.

Stuhlbarg, D., How to Find Optimum Exchanger Size for Forced Circulation, *Hydrocarbon Process.*, **49**(1):149 (1970).

Takamatsu, T., I. Hashimoto, and H. Ohno, Optimal Design of a Large Complex System from the Viewpoint of Sensitivity Analysis, *I&EC Proc. Design & Develop.*, **9**(3):368 (1970).

Tarrer, A. R., H. C. Lim, L. B. Koppel, Finding the Economically Optimum Heat Exchanger, *Chem. Eng.*, **78**(22):79 (1971).

Tedder, D. W., and D. F. Rudd, Parametric Studies in Industrial Distillation: Part II. Heuristic Optimization, *AIChE J.*, **24**(2):316 (1978).

Thygeson, J. R., Jr., and E. D. Grossman, Optimization of a Continuous Through-Circulation Dryer, *AIChE J.*, **16**(5):749 (1970).

Uchiyama, T., Best Size for Refinery and Tankers, *Hydrocarbon Process.*, **47**(12):85 (1968).

Umeda, T., Optimal Design of an Absorber-Stripper System, *I&EC Proc. Design & Develop.*, **8**(3):308 (1969).

———, M. Nishio, and S. Komatsu, A Method for Plant Data Analysis and Parameters Estimation, *I&EC Proc. Design & Develop.*, **10**(2):236 (1971).

———, A. Shindo, and E. Tazaki, Optimal Design of Chemical Process by Feasible Decomposition Method, *I&EC Proc. Design & Develop.*, **11**(1):1 (1972).

Wang, B.-C., and R. Luus, Reliability of Optimization Procedures for Obtaining Global Optimum, *AIChE J.*, **24**(4):619 (1978).

Webb, P. U., B. E. Lutter, and R. L. Hair, Dynamic Optimization of Fluid Cat Crackers, *Chem. Eng. Progr.*, **74**(6):72 (1978).

Weisman, J., H. Pulida, and A. Khanna, Optimal Process System Design in Accordance with Demand and Price Forecasts, *I&EC Proc. Design & Develop.*, **14**(1):51 (1975).

Wen, C. Y., and T. M. Chang, Optimal Design of Systems Involving Parameter Uncertainty, *I&EC Proc. Design & Develop.*, **7**(1):49 (1968).

White, C. H., Optimizing Production Rates for Parallel Equipment, *Chem. Eng.*, **78**(13):86 (1971).

Wigham, I., Designing Optimum Cooling Systems, *Chem. Eng.*, **78**(18):95 (1971).

Wilde, D. J., and C. S. Beightler, "Foundations of Optimization," Prentice-Hall, Inc., Englewood Cliffs, New Jersey, 1967.

Yen, Y.-C., Bigger Reactors or More Recycle, *Hydrocarbon Process.*, **49**(1):157 (1970).

Zdonik, S. B., Techniques for Saving Energy in Processes and Equipment, *Chem. Eng.*, **84**(14):99 (1977).

Simulation Programs Speed Optimization, *Chem. Eng. News*, **46**(15):46 (1968).

Program evaluation and review technology (PERT)

Currie, R., PERT—Versatile Planning Tool, *Mfg. Eng. & Mgt.*, **64**(6):47 (1970).

Gans, M., The A to Z of Plant Startup, *Chem. Eng.*, **83**(6):72 (1976).

Klimpel, R. R., Operations Research: Decision-Making Tool—II, *Chem. Eng.*, **80**(10):87 (1973).

Moder, J. J., and C. R. Phillips, "Project Management with CPM and PERT," Van Nostrand, Reinhold Company, New York, 1970.

Smith, C. A., Simplified PERT for Developmental Projects, presented at American Association of Cost Engineers 12th National Meeting, Houston, Texas, June 17–19, 1968.

Planning under uncertainty

Barish, N. N., and S. Kaplan, "Economic Analysis for Engineering and Managerial Decision Making," 2d ed., McGraw-Hill Book Company, New York, 1978.

Cadman, M. H., Dealing with Uncertainty, *Chem. Eng. (London)*, **234**:424 (December 1969).

Cooper, G. T., E. Solomon, and R. E. Templeton, Engineering Planning Under Conditions of Uncertainty—A Simplified Approach to Forecasting Engineering Backlog, *1970 Trans. Am. Assoc. Cost Eng.* (1970), p. 147.

Debreczeni, E. J., Future Supply and Demand for Basic Petrochemicals, *Chem. Eng.*, **84**(12):135 (1977).

Filippello, A. N., Forecasting and Planning, *Chem. Eng.*, **84**(12):114 (1977).

Foord, A., Investment Appraisal for Chemical Engineers, *Chem. Proc. Eng.*, **49**(2):87 (1968).

Holland, F. A., F. A. Watson, and J. K. Wilkinson, Estimating Profitability When Uncertainties Exist, *Chem. Eng.*, **81**(3):73 (1974).

——— ——— and ———, Numerical Measures of Risk, *Chem. Eng.*, **81**(5):119 (1974).

Mai, K. L., Energy and Petrochemical Raw Materials Through 1990, *Chem. Eng.*, **84**(12):122 (1977).

Martino, J. O., Technological Forecasting for the Chemical Process Industries, *Chem. Eng.*, **78**(29):54 (1971).

———, "An Introduction to Technological Forecasting," Gordon and Breach, New York, 1972.

Neely, C., and J. E. Browning, Sources of Capital for Growth of Process Plants, *Chem. Eng.*, **84**(12):142 (1977).

Rose, L. M., "Engineering Investment Decisions—Planning Under Uncertainty," Elsevier Scientific Publishing Company, New York, 1976.

Shahbenderian, A. P., Plants, Plans and Probabilities, *Chem. Proc. Eng.*, **48**(9):94 (1967).

Young, D., and L. E. Contreras, Expected Present Worths of Cash Flows Under Uncertain Timing, *Eng. Econ.*, **20**(4):257 (1975).

Zapp, G. M., and W. T. Hewitt, Industry Outlook for Inorganic Chemicals, *Chem. Eng.*, **84**(12):128 (1977).

Risk analysis

Barish, N. N., and S. Kaplan, "Economic Analysis for Engineering and Managerial Decision Making," 2d ed., McGraw-Hill Book Company, New York, 1978.

Bowen, J. H., Individual Risk vs. Public Risk Criteria, *Chem. Eng. Progr.*, **72**(2):63 (1976).

Browning, R. L., Analyzing Industrial Risks, *Chem. Eng.*, **76**(23):109 (1969).

———, Finding the Critical Path to Loss, *Chem. Eng.*, **77**(2):119 (1970).

Burgoyne, J. H., Risk Management and Chemical Manufacture, *Chem. Eng.* (*London*), **151** (March 1975).

Cooper, D. O., and L. B. Davidson, The Parameter Method for Risk Analysis, *Chem. Eng. Progr.*, **72**(11):73 (1976).

Deshmukh, S. S., Risk Analysis, *Chem. Eng.*, **81**(13):141 (1974).

Gibson, S. B., Risk Criteria in Hazard Analysis, *Chem. Eng. Progr.*, **72**(2):59 (1976).

Herbert, V. D., Jr., and A. Bisio, The Risk and the Benefit: Part 1, Price, *Chemtech*, **6**(3):174 (1976).

Holland, F. A., and F. A. Watson, Project Risk, Inflation, and Profitability, *Chem. Eng.*, **84**(6):133 (1977).

Kletz, T., Evaluate Risk in Plant Design, *Hydrocarbon Process.*, **56**(5):297 (1977).

Malloy, J. B., Risk Analysis of Chemical Plants, *Chem. Eng. Progr.*, **67**(10):68 (1971).

Newendorp, P. D., Decision Analysis: Risk Evaluation Helps Make Better Decisions, *Oil & Gas J.*, **74**(42):55 (1976).

Reilly, P. M., and H. P. Johri, Decision-Making Through Opinion Analysis, *Chem. Eng.*, **76**(7):122 (1969).

Riggs, J. L., "Engineering Economics," McGraw-Hill Book Company, New York, 1977.

Ross, R. C., Uncertainty Analysis Helps in Making Business Decisions, *Chem. Eng.*, **78**(21):149 (1971).

Spooner, J. E., A Mathematical Model for Contingency, *1973 Trans. Am. Assoc. Cost Eng.* (1973), p. 81.

Stuhlbarg, D., Calculating the Calculated Risk, *Chem. Eng.*, **75**(2):152 (1968).

Tarquin, A. J., and L. Blank, "Engineering Economy: A Behavioral Approach," McGraw-Hill Book Company, New York, 1976.

Ward, E. G., Evaluation of Risk in New Capital Projects, *1969 Trans. Am. Assoc. Cost Eng.*, (1969).

Weisman, J., and A. G. Holzman, Optimal Process System Design Under Conditions of Risk, *I&EC Proc. Design & Develop.*, **11**(3):386 (1972).

Zeanah, P. H., Advanced Techniques for Contingency Evaluation, *1973 Trans. Am. Assoc. Cost Eng.* (1973), p. 68.

PROBLEMS

1 A multiple-effect evaporator is to be used for evaporating 400,000 lb of water per day from a salt solution. The total intial cost for the first effect is $18,000, and each additional effect costs $15,000. The life period is estimated to be 10 years, and the salvage or scrap value at the end of the life period may be assumed to be zero. The straight-line depreciation method is used. Fixed charges minus depreciation are 15 percent yearly based on the first cost of the equipment. Steam costs $0.50 per 1000 lb. Annual maintenance charges are 5 percent of the initial equipment cost. All other costs are independent of the number of effects. The unit will operate 300 days per year. If the pounds of water evaporated per pound of steam equals 0.85 × number of effects, determine the optimum number of effects for minimum annual cost.

2 Determine the optimum economic thickness of insulation that should be used under the following conditions: Saturated steam is being passed continuously through a steel pipe with an outside diameter of 10.75 in. The temperature of the steam is 400°F, and the steam is valued at $0.60 per 1000 lb. The pipe is to be insulated with a material that has a thermal conductivity of 0.03 Btu/$(h)(ft^2)(°F/ft)$. The cost of the installed insulation per foot of pipe length is $4.5 × I_t, where I_t is the thickness of the insulation in inches. Annual fixed charges including maintenance amount to 20 percent of the initial installed cost. The total length of the pipe is 1000 ft, and the average temperature of the surroundings may be taken as 70°F. Heat-transfer resistances due to the steam film, scale, and pipe wall are negligible. The air-film coefficient at the outside of the insulation may be assumed constant at 2.0 Btu/$(h)(ft^2)(°F)$ for all insulation thicknesses.

3 An absorption tower containing wooden grids is to be used for absorbing SO_2 in a sodium sulfite solution. A mixture of air and SO_2 will enter the tower at a rate of 70,000 ft^3/min, temperature of 250°F, and pressure of 1.1 atm. The concentration of SO_2 in the entering gas is specified, and a given fraction of the entering SO_2 must be removed in the absorption tower. The molecular weight of the entering gas mixture may be assumed to be 29.1. Under the specified design conditions, the number of transfer units necessary varies with the superficial gas velocity as follows:

$$\text{Number of transfer units} = 0.32 G_s^{0.18}$$

where G_s is the entering gas velocity as lb/$(h)(ft^2)$ based on the cross-sectional area of the empty tower. The height of a transfer unit is constant at 15 ft. The cost for the installed tower is $1 per cubic foot of inside volume, and annual fixed charges amount to 20 percent of the initial cost. Variable operating charges for the absorbent, blower power, and pumping power are represented by the following equation:

$$\text{Total variable operating costs as \$/h} = 1.8 G_s^2 \times 10^{-8} + \frac{81}{G_s} + \frac{4.8}{G_s^{0.8}}$$

The unit is to operate 8000 h/year. Determine the height and diameter of the absorption tower at conditions of minimum annual cost.

4 Derive an expression for the optimum economic thickness of insulation to put on a flat surface if the annual fixed charges per square foot of insulation are directly proportional to the thickness, (a) neglecting the air film, (b) including the air film. The air-film coefficient of heat transfer may be assumed as constant for all insulation thicknesses.

5 A continuous evaporator is operated with a given feed material under conditions in which the concentration of the product remains constant. The feed rate at the start of a cycle after the tubes have been cleaned has been found to be 5000 kg/h. After 48 h of continuous operation, tests have shown that the feed rate decreases to 2500 kg/h. The reduction in capacity is due to true scale formation. If the down time per cycle for emptying, cleaning, and recharging is 6 h, how long should the evaporator be operated between cleanings in order to obtain the maximum amount of product per 30 days?

6 A solvent-extraction operation is carried out continuously in a plate column with gravity flow. The unit is operated 24 h/day. A feed rate of 1500 ft^3/day must be handled 300 days per year. The allowable velocity per square foot of cross-sectional tower area is 40 ft^3 of combined solvent and charge per hour. The annual fixed costs for the installation can be predicted from the following equation:

$$C_F = 8800F_{sf}^2 - 51,000F_{sf} + 110,000 \text{ \$/year}$$

where F_{sf} = cubic feet of solvent per cubic foot of feed. Operating and other variable costs depend on the amount of solvent that must be recovered, and these costs are \$0.04 for each cubic foot of solvent passing through the tower. What tower diameter should be used for optimum conditions of minimum total cost per year?

7 Prepare a plot of optimum economic pipe diameter versus the flow rate of fluid in the pipe under the following conditions:

Costs and operating conditions ordinarily applicable in industry may be used.

The flow of the fluid may be considered as in the turbulent range.

The viscosity of the fluid may range from 0.1 to 20 centipoises.

The plot is to apply for steel pipe.

Express the diameters in inches and use inside diameters.

The plot should cover a diameter range of 0.1 to 100 in.

Express the flow rate in 1000 lb/h.

The plot should cover a flow-rate range of 10 to 100,000 lb/h.

The plot should be presented on log-log coordinates.

One line on the plot should be presented for each of the following fluid densities: 100, 50, 10, 1, 0.1, 0.01, and 0.001 lb/ft^3.

8 For the conditions indicated in Prob. 7, prepare a log-log plot of fluid velocity in feet per second versus optimum economic pipe diameter in inches. The plot should cover a fluid-velocity range of 1 to 100 ft/s and a pipe-diameter range of 1 to 10 in.

9 A continuous evaporator is being used to concentrate a scale-forming solution of sodium sulfate in water. The overall coefficient of heat transfer decreases according to the following expression:

$$\frac{1}{U^2} = 8 \times 10^{-6}\theta_b + 6 \times 10^{-6}$$

where U = overall coefficient of heat transfer, Btu/(h)(ft^2)(°F), and θ_b = time in operation, h.

The only factor which affects the overall coefficient is the scale formation. The liquid enters the evaporator at the boiling point, and the temperature and heat of vaporization are constant. At the operating conditions, 990 Btu are required to vaporize 1 lb of water, the heat-transfer area is 400 ft^2, and the temperature-difference driving force is 70°F. The time required to shut down, clean, and get back on stream is 4 h for each shutdown, and the total cost for this cleaning operation is \$100 per cycle. The labor costs during operation of the evaporator are \$10 per hour. Determine the total time per cycle for minimum total cost under the following conditions:

(*a*) An overall average of 65,000 lb of water per 24-h day must be evaporated during each 30-day period.

(*b*) An overall average of 81,000 lb of water per 24-h day must be evaporated during each 30-day period.

10 An organic chemical is produced by a batch process. In this process, chemicals X and Y react to form chemical Z. Since the reaction rate is very high, the total time required per batch has been found to be independent of the amounts of the materials, and each batch requires 2 h, including time for charging, heating, and dumping. The following equation shows the relation between the pounds of Z produced (lb$_Z$) and the pounds of X (lb$_X$) and Y (lb$_Y$) supplied:

$$\text{lb}_Z = 1.5(1.1 \text{ lb}_X \text{lb}_Z + 1.3 \text{ lb}_Y \text{lb}_Z - \text{lb}_X \text{lb}_Y)^{0.5}$$

Chemical X costs \$0.09 per pound. Chemical Y costs \$0.04 per pound. Chemical Z sells for \$0.80 per pound. If one-half of the selling price for chemical Z is due to costs other than for raw materials, what is the maximum profit obtainable per pound of chemical Z?

11 Derive an expression similar to Eq. (56) for finding the optimum exit temperature of cooling water from a heat exchanger when the temperature of the material being cooled is not constant. Designate the true temperature-difference driving force by $F_G \, \Delta t_{lm}$, where F_G is a correction factor with value dependent on the geometrical arrangement of the passes in the exchanger. Use primes to designate the temperature of the material that is being cooled.

12 Under the following conditions, determine the optimum economic thickness of insulation for a $1\frac{1}{2}$-in. standard pipe carrying saturated steam at 100 psig. The line is in use continuously. The covering specified is light carbonate magnesia, which is marketed in whole-number thicknesses only (i.e., 1 in., 2 in., 3 in., etc.). The cost of the installed insulation may be approximated as $10 per cubic foot of insulation. Annual fixed charges are 20 percent of the initial investment, and the heat of the steam is valued at $0.75 per 1 million Btu. The temperature of the surroundings may be assumed to be 80°F.

L. B. McMillan, *Trans. ASME*, **48**:1269 (1926), has presented approximate values of optimum economic insulation thickness versus the group $(kb_c \, H_y \, \Delta t/a_c)^{0.5}$, with pipe size as a parameter.

k = thermal conductivity of insulation, Btu/(h)(ft^2)(°F/ft)
b_c = cost of heat, $/Btu
H_y = hours of operation per year, h/year
Δt = overall temperature-difference driving force, °F
a_c = cost of insulation, $/(ft^3)(year)

The following data are based on the results of McMillan, and these data are applicable to the conditions of this problem:

$\left(\dfrac{kb_c H_y \, \Delta t}{a_c}\right)^{0.5}$	Optimum economic thickness of insulation, in., for nominal pipe diameter of			
	$\frac{1}{2}$ in.	1 in.	2 in.	4 in.
0.1	....	0.40	0.5	0.6
0.2	0.80	0.95	1.1	1.3
0.3	1.20	1.4	1.6	1.9
0.5	1.85	2.1	2.45	2.9
0.8	2.75	3.1	3.6	4.3
1.2	3.80	4.3	4.9	

13 A catalytic process uses a catalyst which must be regenerated periodically because of reduction in conversion efficiency. The cost for one regeneration is constant at $800. This figure includes all shutdown and startup cost, as well as the cost for the actual regeneration. The feed rate to the reactor is maintained constant at 150 lb/day, and the cost for the feed material is $2.50 per pound. The daily costs for operation are $300, and fixed charges plus general overhead costs are $100,000 per year. Tests on the catalyst show that the yield of product as pounds of product per pound of feed during the first day of operation with the regenerated catalyst is 0.87, and the yield decreases as $0.87/(\theta_D)^{0.25}$, where θ_D is the time in operation expressed in days. The time necessary to shut down the unit, replace the catalyst, and start up the unit is negligible. The value of the product is $14.00 per pound, and the plant operates 300 days per year. Assuming no costs are involved other than those mentioned, what is the maximum annual profit that can be obtained under these conditions?

14 Derive the following equation for the optimum outside diameter of insulation on a wire for maximum heat loss:

$$D_{opt} = \frac{2k_m}{(h_c + h_r)_c}$$

where k_m is the mean thermal conductivity of the insulation and $(h_c + h_r)_c$ is the combined and constant surface heat-transfer coefficient. The values of k_m and $(h_c + h_r)_c$ can be considered as constants independent of temperature level and insulation thickness.†

15 Derive Eq. (49) for the optimum economic pipe diameter and compare this to the equivalent expression presented as Eq. (5-90) in J. H. Perry and C. H. Chilton, ed., "Chemical Engineers' Handbook," 5th ed., p. 5-32, McGraw-Hill Book Company, New York, 1973.

16 Using a direct partial derivative approach for the objective function, instead of the Lagrangian multiplier as was used in Eqs. (92) to (95), determine the optimum values of x and y involved in Eqs. (92) to (95).

17 Find the values of x, y, and z that minimize the function $x + 2y^2 + z^2$ subject to the constraint that $x + y + z = 1$, making use of the Lagrangian multiplier.

18 For the mixing problem referred to in Tables 3 and 4 of this chapter, present the computer solution as

    (*a*) The computer diagram (similar to Table 4) based on the logic given in Table 3.

    (*b*) The computer program (Fortran language preferred).

    (*c*) The printout of the computer solution giving the minimum value of the objective function and the corresponding values of the variables.

    (*d*) The interpretation of the computer solution.

19 For the linear-programming example problem presented in this chapter where the simultaneous-equation solution is presented in Table 5, solve the problem using the simplex algorithm as was done in the text for the example solved in Fig. 10-10. Use as the initial feasible starting solution the case of solution 2 in Table 5 where $x_2 = S_4 = 0$. Note that this starting point should send the solution directly to the optimum point (solution 6) for the second trial.

20 From the data given for the dynamic-programming problem in Table 10 and the appropriate data from Table 13, show how the value of 462 was obtained in Table 14 for 700°F, Reactor I_B, and Catalyst 1.

21 Using the method outlined for steepest descent in Eqs. (96) to (98) and presented in Fig. 10-12, what would be the minimum value of C along the first line of steepest descent if the initial point had been chosen arbitrarily as $x = 2$ and $y = 3$ with x decreasing in increments of 0.5?

22 In order to continue the operation of a small chemical plant at the same capacity, it will be necessary to make some changes on one of the reactors in the system. The decision has been made by management that the unit must continue in service for the next 12 years and the company policy is that no unnecessary investments are made unless at least an 8 percent rate of return (end-of-year compounding) can be obtained. Two possible ways for making a satisfactory change in the reactor are as follows:

    (1) Make all the critical changes now at a cost of $5800 so the reactor will be satisfactory to use for 12 years.

    (2) Make some of the changes now at a cost of $5000 which will permit operation for 8 years and then make changes costing $2500 to permit operation for the last 4 years.

    (*a*) Which alternative should be selected if no inflation is anticipated over the next 12 years?

    (*b*) Which alternative should be selected if inflation at a rate of 7 percent (end-of-year compounding) is assumed for all future costs?

    † See W. H. McAdams, "Heat Transmission," 3d ed., p. 415, McGraw-Hill Book Company, New York, 1954.

ELEVEN

MATERIALS AND FABRICATION SELECTION

Any engineering design, particularly for a chemical process plant, is only useful when it can be translated into reality by using available materials and fabrication methods. Thus, selection of materials of construction combined with the appropriate techniques of fabrication can play a vital role in the success or failure of a new chemical plant or in the improvement of an existing facility.

MATERIALS OF CONSTRUCTION

As chemical process plants turn to higher temperatures and flow rates to boost yields and throughputs, selection of construction materials takes on added importance. This trend to more severe operating conditions forces the chemical engineer to search for more dependable, more corrosion-resistant materials of construction for these process plants, because all these severe conditions intensify corrosive action. Fortunately, a broad range of materials is now available for corrosive service. However, this apparent abundance of materials also complicates the task of choosing the "best" material because, in many cases, a number of alloys and plastics will have sufficient corrosion resistance for a particular application. Final choice cannot be based simply on choosing a suitable material from a corrosion table but must be based on a sound economic analysis of competing materials.

The chemical engineer would hardly expect a metallurgist to handle the design and operation of a complex chemical plant. Similarly, the chemical engineer cannot become a materials specialist overnight. But a good metallurgist must have a working knowledge of the chemical plant environment in which the recommendations will be applied. In the same manner, the chemical engineer should also understand something of the materials that make the equipment and processes possible.

The purpose of this chapter is to provide the process designer with a working knowledge of some of the major forms and types of materials available, what they offer, and how they are specified. With this background, the engineer can consult a materials specialist at the beginning of the design, not when the mistakes already have been made.

METALS

Materials of construction may be divided into the two general classifications of *metals* and *nonmetals*. Pure metals and metallic alloys are included under the first classification. Table 1 presents data showing the comparison of purchased cost for various types of metals in plate form.

Iron and Steel

Although many materials have greater corrosion resistance than iron and steel, cost aspects favor the use of iron and steel. As a result, they are often used as

Table 1 Comparison of purchased cost for metal plate

Material	Ratio = $\dfrac{\text{cost per pound for metal}}{\text{cost per pound for steel}}$
Flange quality steel†	1
304 stainless-steel-clad steel	5
316 stainless-steel-clad steel	6
Aluminum (99 plus)	6
304 stainless steel	7
Copper (99.9 plus)	7
Nickel-clad steel	8
Monel-clad steel	8
Inconel-clad steel	9
316 stainless steel	10
Monel	10
Nickel	12
Inconel	13
Hastelloy C	40

† Purchased cost for steel plate (January, 1979) can be approximated as 26 to 48 cents per pound, depending on the amount purchased.

materials of construction when it is known that some corrosion will occur. If this is done, the presence of iron salts and discoloration in the product can be expected, and periodic replacement of the equipment should be anticipated.

In general, cast iron and carbon steel exhibit about the same corrosion resistance. They are not suitable for use with dilute acids, but can be used with many strong acids, since a protective coating composed of corrosion products forms on the metal surface.

Because of the many types of rolled and forged steel products used in industry, basic specifications are needed to designate the various types. The American Iron and Steel Institute (AISI) has set up a series of standards for steel products.† However, even the relatively simple product descriptions provided by AISI and shown in Table 2 must be used carefully. For instance, the AISI 1020 carbon steel does not refer to all 0.20 percent carbon steels. AISI 1020 is part of the numerical designation system defining the chemical composition of certain "standard steels" used primarily in bar, wire, and some tubular steel products. The system almost never applies to sheets, strip, plates, or structural material. One reason is that the chemical composition ranges of standard steels are unnecessarily restrictive for many applications.

Carbon steel plates for reactor vessels are a good example. This application generally requires a minimum level of mechanical properties, weldability, formability, and toughness as well as some assurance that these properties will

† Specifications and codes on materials have also been established by the Society of Automotive Engineers (SAE), the American Society of Mechanical Engineers (ASME), and the American Society for Testing Materials (ASTM).

Table 2 AISI standard steels†

(XX's indicate nominal carbon content within range)

Carbon steels AISI series designations	Nominal composition or range‡
10XX	Non-resulfurized carbon steels with 44 compositions ranging from 1008 to 1095. Manganese ranges from 0.30 to 1.65%; if specified, silicon is 0.10 max. to 0.30 max., each depending on grade. Phosphorus is 0.040 max., sulfur is 0.050 max.
11XX	Resulfurized carbon steels with 15 standard compositions. Sulfur may range up to 0.33%, depending on grade.
B11XX	Acid Bessemer resulfurized carbon steels with 3 compositions. Phosphorus generally is higher than 11XX series.
12XX	Rephosphorized and resulfurized carbon steels with 5 standard compositions. Phosphorus may range up to 0.12% and sulfur up to 0.35%, depending on grade.

(continued)

Table 2 AISI standard steels† (*continued*)

Alloy steels	Nominal composition or range†
13XX	Manganese, 1.75%. Four compositions from 1330 to 1345.
40XX	Molybdenum, 0.20 or 0.25%. Seven compositions from 4012 to 4047.
41XX	Chromium, to 0.95%, molybdenum to 0.30%. Nine compositions from 4118 to 4161.
43XX	Nickel, 1.83%, chromium to 0.80%, molybdenum, 0.25%. Three compositions from 4320 to E4340.
44XX	Molybdenum, 0.53%. One composition, 4419.
46XX	Nickel to 1.83%, molybdenum to 0.25%. Four compositions from 4615 to 4626.
47XX	Nickel, 1.05%, chromium, 0.45%, molybdenum to 0.35%. Two compositions, 4718 and 4720.
48XX	Nickel, 3.50%, molybdenum, 0.25%. Three compositions from 4815 to 4820.
50XX	Chromium, 0.40%. One composition, 5015.
51XX	Chromium to 1.00%. Ten compositions from 5120 to 5160.
5XXXX	Carbon, 1.04%, chromium to 1.45%. Two compositions, 51100 and 52100.
61XX	Chromium to 0.95%, vanadium to 0.15% min. Two compositions, 6118 and 6150.
86XX	Nickel, 0.55%, chromium, 0.50%, molybdenum, 0.20%. Twelve compositions from 8615 to 8655.
87XX	Nickel, 0.55%, chromium, 0.50%, molybdenum, 0.25%. Two compositions, 8720 and 8740.
88XX	Nickel, 0.55%, chromium, 0.50%, molybdenum, 0.35%. One composition, 8822.
92XX	Silicon, 2.00%. Two compositions, 9255 and 9260.
50BXX	Chromium to 0.50%, also containing boron. Four compositions from 50B44 to 50B60.
51BXX	Chromium to 0.80%, also containing boron. One composition, 51B60.
81BXX	Nickel, 0.30%, chromium, 0.45%, molybdenum, 0.12%, also containing boron. One composition, 81B45.
94BXX	Nickel, 0.45%, chromium, 0.40%, molybdenum, 0.12%, also containing boron. Two compositions, 94B17 and 94B30.

† When a carbon or alloy steel also contains the letter L in the code, it contains from 0.15 to 0.35 percent lead as a free-machining additive, i.e., 12L14 or 41L40. The prefix E before an alloy steel, such as E4340, indicates the steel is made only by electric furnace. The suffix H indicates an alloy steel made to more restrictive chemical composition than that of standard steels and produced to a measured and known hardenability requirement, e.g., 8630H or 94B30H.

‡ For a detailed listing of nominal composition or range, see "Chemical Engineers' Handbook," 5th ed., McGraw-Hill Book Company, New York, 1973.

be uniform throughout. A knowledge of the detailed composition of the steel alone will not assure that these requirements are met. Even welding requirements for plate can be met with far less restrictive chemical compositions than would be needed for the same type of steel used in bar stock suitable for heat treating to a minimum hardness or tensile strength.

Stainless Steels

There are more than 100 different types of stainless steels. These materials are high chromium or high nickel-chromium alloys of iron containing small amounts of other essential constituents. They have excellent corrosion-resistance and heat-resistance properties. The most common stainless steels, such as type 302 or type 304, contain approximately 18 percent chromium and 8 percent nickel, and are designated as 18-8 stainless steels.

The addition of molybdenum to the alloy, as in type 316, increases the corrosion resistance and high-temperature strength. If nickel is not included, the low-temperature brittleness of the material is increased and the ductility and pit-type corrosion resistance are reduced. The presence of chromium in the alloy gives resistance to oxidizing agents. Thus, type 430, which contains chromium but no nickel or molybdenum, exhibits excellent corrosion resistance to nitric acid and other oxidizing agents.

Although fabricating operations on stainless steels are more difficult than on standard carbon steels, all types of stainless steel can be fabricated successfully.†
The properties of types 430F, 416, 410, 310, 309, and 303 make these materials particularly well suited for machining or other fabricating operations. In general, machinability is improved if small quantities of phosphorus, selenium, or sulfur are present in the alloy.

The types of stainless steel included in the 300 series are hardenable only by cold-working; those included in the 400 series are either nonhardenable or hardenable by heat-treating. As an example, type 410, containing 12 percent chromium and no nickel, can be heat-treated for hardening and has good mechanical properties when heat-treated. It is often used as a material of construction for bubble caps, turbine blades, or other items that require special fabrication.

Stainless steels exhibit the best resistance to corrosion when the surface is oxidized to a passive state. This condition can be obtained, at least temporarily, by a so-called "passivation" operation in which the surface is treated with nitric acid and then rinsed with water. Localized corrosion can occur at places where foreign material collects, such as in scratches, crevices, or corners. Consequently, mars or scratches should be avoided, and the equipment design should specify a minimum of sharp corners, seams, and joints. Stainless steels show great susceptibility to stress corrosion cracking. As one example, stress plus contact with small concentrations of halides can result in failure of the metal wall.

The high temperatures involved in welding stainless steel may cause precipitation of chromium carbide at the grain boundary, resulting in decreased corrosion resistance along the weld. The chances of this occurring can be minimized by using low-carbon stainless steels or by controlled annealing.

---

† For a detailed discussion of machining and fabrication of stainless steels, see Selection of Stainless Steels, *Bulletin OLE* 11366, Armco Steel Corporation, Middletown, Ohio 45042; and Fabrication of Stainless Steel, *Bulletin* 031478, United States Steel Corporation, Pittsburgh, Pa. 15230.

Table 3 Classification of stainless steels by alloy content and microstructure

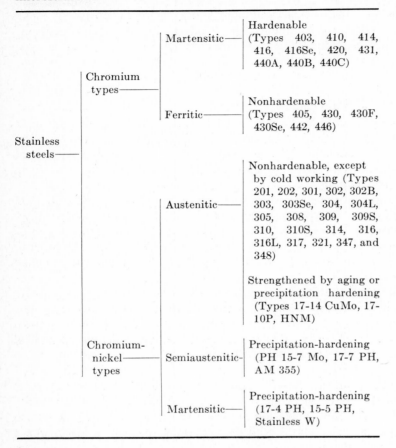

Stainless steels

Chromium types

Martensitic — Hardenable (Types 403, 410, 414, 416, 416Se, 420, 431, 440A, 440B, 440C)

Ferritic — Nonhardenable (Types 405, 430, 430F, 430Se, 442, 446)

Chromium-nickel types

Austenitic — Nonhardenable, except by cold working (Types 201, 202, 301, 302, 302B, 303, 303Se, 304, 304L, 305, 308, 309, 309S, 310, 310S, 314, 316, 316L, 317, 321, 347, and 348)

Strengthened by aging or precipitation hardening (Types 17-14 CuMo, 17-10P, HNM)

Semiaustenitic — Precipitation-hardening (PH 15-7 Mo, 17-7 PH, AM 355)

Martensitic — Precipitation-hardening (17-4 PH, 15-5 PH, Stainless W)

A preliminary approach to the selection of the stainless steel for a specific application is to classify the various types according to the alloy content, microstructure, and major characteristic. Table 3 outlines the information according to the classes of stainless steels—austenitic, martensitic, and ferritic. Table 4 presents characteristics and typical applications of various types of stainless steel while Table 5 indicates resistance of stainless steels to oxidation in air.

Hastelloy

The beneficial effects of nickel, chromium, and molybdenum are combined in Hastelloy C to give an expensive but highly corrosion-resistant material. A typical analysis of this alloy shows 56 percent nickel, 17 percent molybdenum, 16 percent chromium, 5 percent iron, and 4 percent tungsten, with manganese, silicon, carbon, phosphorus, and sulfur making up the balance. Hastelloy C is used where structural strength and good corrosion resistance are necessary under conditions of high temperatures. The material can be machined and is readily fabricated. It is used in the form of valves, piping, heat exchangers, and various types of vessels. Other types of Hastelloys are also available for use under special corrosive conditions.

Copper and its Alloys

Copper is relatively inexpensive, possesses fair mechanical strength, and can be fabricated easily into a wide variety of shapes. Although it shows little tendency to dissolve in nonoxidizing acids, it is readily susceptible to oxidation. Copper is resistant to atmospheric moisture or oxygen because a protective coating composed primarily of copper oxide is formed on the surface. The oxide, however, is soluble in most acids, and thus copper is not a suitable material of construction when it must contact any acid in the presence of oxygen or oxidizing agents. Copper exhibits good corrosion resistance to strong alkalies, with the exception of ammonium hydroxide. At room temperature it can handle sodium and potassium hydroxide of all concentrations. It resists most organic solvents and aqueous solutions of organic acids.

Copper alloys, such as brass, bronze, admiralty, and Muntz metals, can exhibit better corrosion resistance and better mechanical properties than pure copper. In general, high-zinc alloys should not be used with acids or alkalies owing to the possibility of dezincification. Most of the low-zinc alloys are resistant to hot dilute alkalies.

Nickel and its Alloys

Nickel exhibits high corrosion resistance to most alkalies. Nickel-clad steel is used extensively for equipment in the production of caustic soda and alkalies. The strength and hardness of nickel is almost as great as that of carbon steel, and the metal can be fabricated easily. In general, oxidizing conditions promote the corrosion of nickel, and reducing conditions retard it.

Monel, an alloy of nickel containing 67 percent nickel and 30 percent copper, is often used in the food industries. This alloy is stronger than nickel and has better corrosion-resistance properties than either copper or nickel. Another important nickel alloy is Inconel (77 percent nickel and 15 percent chromium). The presence of chromium in this alloy increases its resistance to oxidizing conditions.

Aluminum

The lightness and relative ease of fabrication of aluminum and its alloys are factors favoring the use of these materials. Aluminum resists attack by acids

Table 4 Stainless steels most commonly used in the chemical process industries†

Type§	Composition, %			Other significant elements‡	Major characteristics	Properties	Applications
	Cr	Ni	C max				
301	16.00–18.00	6.00–8.00	0.15		High work-hardening rate combines cold-worked high strength with good ductility.	Good structural qualities.	Structural applications, bins and containers
302	17.00–19.00	8.00–10.00	0.15		Basic, general purpose austenitic type with good corrosion resistance and mechanical properties.	General purpose.	Heat exchangers, towers, tanks, pipes, heaters, general chemical equipment
303	17.00–19.00	8.00–10.00	0.15	S 0.15 min	Free machining modification of type 302; contains extra sulfur.	Type 303Se is also available for parts involving extensive machining.	Pumps, valves, instruments, fittings
304	18.00–20.00	8.00–12.00	0.08		Low carbon variation of type 302, minimizes carbide precipitation during welding.	General purpose. Also available as 304L with 0.03% carbon to minimize carbide precipitation during welding.	Perforated blow-pit screens, heat-exchanger tubing, preheater tubes
305	17.00–19.00	10.00–13.00	0.12		Higher heat and corrosion resistance than type 304.	Good corrosion resistance.	Funnels, utensils, hoods
308	19.00–21.00	10.00–12.00	0.08		High Cr and Ni produce good heat and corrosion resistance. Used widely for welding rod.	In order of their numbers, these alloys show increased resistance to high temperature corrosion. Types 308S, 309S and 310S are also available for welded construction.	Welding rod, more ductile welds for type 430
309	22.00–24.00	12.00–15.00	0.20		High strength and resistance to scaling at high temperatures.		Welding rod for type 304, heat exchangers, pump parts
310	24.00–26.00	19.00–22.00	0.25		Higher alloy content improves basic characteristics of type 309.		Jacketed high-temperature, high-pressure reactors, oil-refining-still tubes

Type	Cr	Ni	C	Other	Characteristics	Properties	Applications
314	23.00–26.00	19.00–22.00	0.25	Si 1.5–3.0	High silicon content.	Resistant to oxidation in air to 2000°F.	Radiant tubes, carburizing boxes, annealing boxes
316	16.00–18.00	10.00–14.00	0.08	Mo 2.00–3.00	Mo improves general corrosion and pitting resistance and high temperature strength over that of type 302.	Resistant to high pitting corrosion. Also available as 316L for welded construction.	Distillation equipment for producing fatty acids, sulfite paper processing equipment
317	18.00–20.00	11.00–15.00	0.08	Mo 3.00–4.00	Higher alloy content improves basic advantages of type 316.	Type 317 has the highest aqueous corrosion resistance of all AISI stainless steels.	Process equipment involving strong acids or chlorinated solvents
321	17.00–19.00	9.00–12.00	0.08	Ti 5 × C, min	Stabilized to permit use in 420°–870°C range without harmful carbide precipitation.	Stabilized with titanium and columbium-tantalum, respectively, to permit their use for large welded structures which cannot be annealed after welding.	Furnace parts in presence of corrosive fumes
347	17.00–19.00	9.00–13.00	0.08	Cb-Ta 10 × C, min.	Characteristics similar to type 321. Stabilized by Cb and Ta.		Like 302 but used where carbide precipitation during fabrication or service may be harmful, welding rod for type 321
403	11.50–13.50		0.15	Si 0.50 max.	Version of type 410 with limited hardenability but improved fabricability.	Not highly resistant to high temperature oxidation in air.	Steam turbine blades
405	11.50–14.50		0.08	Al 0.10–0.30	Version of type 410 with limited hardenability but improved weldability.	Good weldability and cladding properties.	Tower linings, baffles, separator towers, heat exchanger tubing
410	11.50–13.50		0.15		Lowest cost general purpose stainless steel.	Wide use where corrosion is not severe.	Bubble-tower parts for petroleum refining, pump rods and valves, machine parts, turbine blades

(continued)

453

Table 4 Stainless steels most commonly used in the chemical process industries† (continued)

Type§	Composition, %				Major characteristics	Properties	Applications
	Cr	Ni	C max	Other significant elements‡			
416	12.00–14.00		0.15	S 0.15 min	Sulfur added for free machining version of type 410. Type 416Se also available.	The freest machining type of martensitic stainless.	Valve stems, plugs, gates, useful for screws, bolts, nuts, and other parts requiring considerable machining during fabrication
420	12.00–14.00		0.15 min		Similar to type 410 but higher carbon produces higher strength and hardness.	High-spring temper.	Utensils, bushings, valve stems and wear-resisting parts
430	14.00–18.00		0.12		Most popular of nonhardening chromium types. Combines good corrosion resistance (to nitric acid and other oxidizing media).	Good heat resistance and good mechanical properties. Also available in type 430F.	Chemical and processing towers, condensers. Furnace parts such as retorts and low stressed parts subject to temperatures up to 800 °C. Type 430 nitric-acid storage tanks, furnace parts, fan scrolls. Type 430F-pump shafts, instrument parts, valve parts
431	15.00–17.00	1.25–2.50	0.20		High yield point	Very resistant to shock.	Products requiring high yield point and resistance to shock
442	18.00–23.00		0.25		High chromium nonhardenable type	High temperature uses where high sulfur atmospheres make presence of nickel undesirable.	Fume furnaces, flare stacks, materials in contact with high sulfur atmospheres

454

| 446 | 23.00–27.00 | 0.20 | Similar to type 442 but Cr increased to provide maximum resistance to scaling. Especially suited to intermittent high temperatures. | Excellent corrosion resistance to many liquid solutions; fabrication difficulties limit its use primarily to high temperature applications. Useful in high sulfur atmospheres. | Burner nozzles, stack dampers, boiler baffles, furnace linings, glass molds |

† Adapted from Twenty-Second Biennial Materials of Construction Report, *Chem. Eng.*, **73** (23): 187 (1966).
‡ For a detailed listing of nominal composition or range, see the latest issue of "Data on Physical and Mechanical Properties of Stainless and Heat-Resisting Steels," Carpenter Steel Company, Reading, Pa. 19603.

§ In general, stainless steels in the 300 series contain large amounts of chromium and nickel; those in the 400 series contain large amounts of chromium and little or no nickel; those in the 500 series contain low amounts of chromium and little or no nickel; in the 300 series, except for type 309, the nickel content can be 10 percent or less if the second number is zero and greater than 10 percent if the second number is one; in the 400 series, an increase in the number represented by the last two digits indicates an increase in the chromium content. (NOTE: *This is not in agreement with the SAE and general AISI steel numbering systems which use a first digit of four for molybdenum steels and a first digit of five for chromium steels, while the third and fourth digits represent the carbon content.*)

Table 5 Resistance of stainless steels to oxidation in air†

Maximum temperature, °C	Stainless steel type
650	416
700	403, 405, 410, 414
800	430F
850	430, 431
900	302, 303, 304, 316, 317, 321, 347, 348, 17–14 CuMo
1000	302B, 308, 442
1100	309, 310, 314, 329, 446

† Adapted from Twenty-Second Biennial Materials of Construction Report, *Chem. Eng.*, 73(23):187 (1966).

because a surface film of inert hydrated aluminum oxide is formed. This film adheres to the surface and offers good protection unless materials which can remove the oxide, such as halogen acids or alkalies, are present.

Lead

Pure lead has low creep and fatigue resistance, but its physical properties can be improved by the addition of small amounts of silver, copper, antimony, or tellurium. Lead-clad equipment is in common use in many chemical plants. The excellent corrosion-resistance properties of lead are caused by the formation of protective surface coatings. If the coating is one of the highly insoluble lead salts, such as sulfate, carbonate, or phosphate, good corrosion resistance is obtained. Little protection is offered, however, if the coating is a soluble salt, such as nitrate, acetate, or chloride. As a result, lead shows good resistance to sulfuric acid and phosphoric acid, but it is susceptible to attack by acetic acid and nitric acid.

Tantalum

The physical properties of tantalum are similar to those of mild steel, with the exception that its melting point (2996°C) is much higher. It is ordinarily used in the pure form, and it is readily fabricated into many different shapes. The corrosion-resistance properties of tantalum resemble those of glass. The metal is attacked by hydrofluoric acid, by hot concentrated alkalies, and by materials containing free sulfur trioxide. It is resistant to all other acids and is often used for equipment involving contact with hydrochloric acid.

Silver

Because of its low mechanical strength and high cost, silver is generally used only in the form of linings. Silver is resistant to alkalies and many hot organic acids. It also shows fair resistance to aqueous solutions of the halogen acids.

Galvanic Action between Two Dissimilar Metals

When two dissimilar metals are used in the construction of equipment containing a conducting fluid in contact with both metals, an electric potential may be set up between the two metals. The resulting galvanic action can cause one of the metals to dissolve into the conducting fluid and deposit on the other metal. As an example, if a piece of copper equipment containing a solution of sodium chloride in water is connected to an iron pipe, electrolysis can occur between the iron and copper, causing high rates of corrosion. As indicated in Table 6, iron is higher in the electromotive series than copper, and the iron pipe will gradually dissolve and deposit on the copper. The farther apart the two metals are in the electromotive series, the greater is the possible extent of corrosion due to electrolysis.

Table 6 Electromotive series of metals

List of metals arranged in decreasing order of their
tendencies to pass into ionic form by losing electrons

Metal	Ion	Standard electrode potential at 25°C
Lithium	Li^+	$+2.96$
Potassium	K^+	2.92
Calcium	Ca^{++}	2.87
Sodium	Na^+	2.71
Magnesium	Mg^{++}	2.40
Aluminum	Al^{3+}	1.70
Manganese	Mn^{++}	1.10
Zinc	Zn^{++}	0.76
Chromium	Cr^{++}	0.56
Gallium	Ga^{3+}	0.50
Iron	Fe^{++}	0.44
Cadmium	Cd^{++}	0.40
Cobalt	Co^{++}	0.28
Nickel	Ni^{++}	0.23
Tin	Sn^{++}	0.14
Lead	Pb^{++}	0.12
Iron	Fe^{3+}	0.045
Hydrogen	H^+	0.0000
Antimony	Sb^{3+}	-0.10
Bismuth	Bi^{3+}	-0.23
Arsenic	As^{3+}	-0.30
Copper	Cu^{++}	-0.34
Copper	Cu^+	-0.47
Silver	Ag^+	-0.80
Lead	Pb^{4+}	-0.80
Platinum	Pt^{4+}	-0.86
Gold	Au^{3+}	-1.36
Gold	Au^+	-1.50

NONMETALS

Glass, carbon, stoneware, brick, rubber, plastics, and wood are common
examples of nonmetals used as materials of construction. Many of the nonmetals
have low structural strength. Consequently, they are often used in the form of
linings or coatings bonded to metal supports. For example, glass-lined or
rubber-lined equipment has many applications in the chemical industries.

Glass and Glassed Steel

Glass has excellent resistance and is subject to attack only by hydrofluoric acid
and hot alkaline solutions. It is particularly suitable for processes which have

critical contamination levels. Chief drawback is its brittleness and damage by thermal shock. On the other hand, glassed steel combines the corrosion resistance of glass with the working strength of steel. Nucerite is a new ceramic-metal composite made in a similar manner to glassed steel and resists corrosive hydrogen-chloride gas, chlorine, or sulfur dioxide at 650°C. Its impact strength is 18 times that of safety glass and the abrasion resistance is superior to porcelain enamel.

Carbon and Graphite

Generally, impervious graphite is completely inert to all but the most severe oxidizing conditions. This property, combined with excellent heat transfer, has made impervious carbon and graphite very popular in heat exchangers, as brick lining, and in pipe and pumps. One limitation of these materials is low tensile strength. Threshold oxidation temperatures are 350°C for carbon and 400°C for graphite.

Stoneware and Porcelain

Materials of stoneware and porcelain are about as resistant to acids and chemicals as glass, but with the advantage of greater strength. This is offset somewhat by poor thermal conductivity and susceptibility to damage by thermal shock. Porcelain enamels are used to coat steel, but the enamel has slightly inferior chemical resistance.

Brick and Cement Materials

Brick-lined construction can be used for many severely corrosive conditions, where high alloys would fail. Acidproof refractories can be used up to 900°C.

A number of cement materials are used with brick. Standard are phenolic and furane resins, polyesters, sulfur, silicate, and epoxy-based materials. Carbon-filled polyesters and furanes are good against nonoxidizing acids, salts, and solvents. Silica-filled resins should not be used against hydrofluoric or fluorsilicic acids. Sulfur-based cements are limited to 95°C, while resins can be used to about 175°C. The sodium silicate based cements are good against acids to 400°C.

Rubber and Elastomers

Natural and synthetic rubbers are used as linings or as structural components for equipment in the chemical industries. By adding the proper ingredients, natural rubbers with varying degrees of hardness and chemical resistance can be produced. Hard rubbers are chemically saturated with sulfur. The vulcanized products are rigid and exhibit excellent resistance to chemical attack by dilute sulfuric acid and dilute hydrochloric acid.

Natural rubber is resistant to dilute mineral acids, alkalies, and salts, but oxidizing media, oils, benzene, and ketones will attack it. Chloroprene or neoprene rubber is resistant to attack by ozone, sunlight, oils, gasoline, and aromatic or halogenated solvents. Styrene rubber has chemical resistance similar to natural. Nitrile rubber is known for resistance to oils and solvents. Butyl rubber's resistance to dilute mineral acids and alkalies is exceptional; resistance to concentrated acids, except nitric and sulfuric, is good. Silicone rubbers, also known as polysiloxanes, have outstanding resistance to high and low temperatures as well as against aliphatic solvents, oils, and greases. Chlorosulfonated polyethylene, known as hypalon, has outstanding resistance to ozone and oxidizing agents except fuming nitric and sulfuric acids. Oil resistance is good. Fluoroelastomers (Viton A, Kel-F) combine excellent chemical and high-temperature resistance. Polyvinyl chloride elastomer (Koroseal) was developed to overcome some of the limitations of natural and synthetic rubbers. It has excellent resistance to mineral acids and petroleum oils.

Plastics

In comparison with metallic materials, the use of plastics is limited to relatively moderate temperatures and pressures (230°C is considered high for plastics). Plastics are also less resistant to mechanical abuse and have high expansion rates, low strengths (thermoplastics), and only fair resistance to solvents. However, they are lightweight, are good thermal and electrical insulators, are easy to fabricate and install, and have low friction factors.

Generally, plastics have excellent resistance to weak mineral acids and are unaffected by inorganic salt solutions—areas where metals are not entirely suitable. Since plastics do not corrode in the electrochemical sense, they offer another advantage over metals: most metals are affected by slight changes in pH, or minor impurities, or oxygen content, while plastics will remain resistant to these same changes.

The most chemical-resistant plastic commercially available today is tetrafluoroethylene or TFE (Teflon). This thermoplastic is practically unaffected by all alkalies and acids except fluorine and chlorine gas at elevated temperatures and molten metals. It retains its properties up to 260°C. Chlorotrifluoroethylene or CFE (Kel-F) also possesses excellent corrosion resistance to almost all acids and alkalies up to 175°C. FEP, a copolymer of tetrafluoroethylene and hexafluoropropylene, has similar properties to TFE except that it is not recommended for continuous exposures at temperatures above 200°C. Also, FEP can be extruded on conventional extrusion equipment, while TFE parts must be made by complicated "powdered-metallurgy" techniques.

Polyethylene is the lowest-cost plastic commercially available. Mechanical properties are generally poor, particularly above 50°C, and pipe must be fully supported. Carbon-filled grades are resistant to sunlight and weathering.

Unplasticized polyvinyl chlorides (type I) have excellent resistance to oxidizing acids other than concentrated, and to most nonoxidizing acids. Resistance is good to weak and strong alkaline materials. Resistance to chlorinated hydrocarbons is not good.

Acrylonitrile butadiene styrene polymers (ABS) have good resistance to nonoxidizing and weak acids but are not satisfactory with oxidizing acids. Upper temperature limit is about 65°C. Resistance to weak alkaline solutions is excellent. They are not satisfactory with aromatic or chlorinated hydrocarbons but have good resistance to aliphatic hydrocarbons.

Chlorinated polyether can be used continuously up to 125°C, intermittently up to 150°C. Chemical resistance is between polyvinyl chloride and the fluorocarbons. Dilute acids, alkalies, and salts have no effect. Hydrochloric, hydrofluoric, and phosphoric acids can be handled at all concentrations up to 105°C. Sulfuric acid over 60 percent and nitric over 25 percent cause degradation, as do aromatics and ketones.

Acetals have excellent resistance to most organic solvents but are not satisfactory for use with strong acids and alkalies.

Cellulose acetate butyrate is not affected by dilute acids and alkalies or gasoline but chlorinated solvents cause some swelling. Nylons resist many organic solvents but are attacked by phenols, strong oxidizing agents, and mineral acids.

Polypropylene's chemical resistance is about the same as polyethylene, but it can be used at 120°C. Polycarbonate is a relatively high-temperature plastic. It can be used up to 150°C. Resistance to mineral acids is good. Strong alkalies slowly decompose it, but mild alkalies do not. It is partially soluble in aromatic solvents and soluble in chlorinated hydrocarbons.

Among the thermosetting materials are phenolic plastics filled with asbestos, carbon or graphite, and silica. Relatively low cost, good mechanical properties, and chemical resistance (except against strong alkalies) make phenolics popular for chemical equipment. Furane plastics, filled with asbestos, have much better alkali resistance than phenolic asbestos. They are more expensive than the phenolics but also offer somewhat higher strengths.

General-purpose polyester resins, reinforced with fiberglass, have good strength and good chemical resistance, except to alkalies. Some special materials in this class, based on bisphenol, are more alkali resistant. Temperature limit for polyesters is about 95°C. The general area of fiberglass reinforced plastics (FRP) represents a rapidly expanding application of plastics for processing equipment, and it is necessary to solve the problem of development of fabrication standards.

Epoxies reinforced with fiberglass have very high strengths and resistance to heat. Chemical resistance of the epoxy resin is excellent in nonoxidizing and weak acids but not good against strong acids. Alkaline resistance is excellent in weak solutions. Chemical resistance of epoxy-glass laminates may be affected by any exposed glass in the laminate.

Phenolic asbestos, general-purpose polyester glass, saran, and CAB (cellulose acetate butyrate) are adversely affected by alkalies, while thermoplastics generally show poor resistance to organics.

Wood

This material of construction, while fairly inert chemically, is readily dehydrated by concentrated solutions and consequently shrinks badly when subjected to the action of such solutions. It also has a tendency to slowly hydrolyze when in contact with hot acids and alkalies.

LOW- AND HIGH-TEMPERATURE MATERIALS

The extremes of low and high temperatures used in many recent chemical processes has created some unusual problems in fabrication of equipment. For example, some metals lose their ductility and impact strength at low temperatures, although in many cases yield and tensile strengths increase as the temperature is decreased. It is important in low temperature applications to choose materials resistant to shock. Usually a minimum Charpy value of 15 ft·lb (keyhole notch) is specified at the operating temperature. For severe loading, a value of 20 ft·lb is recommended. Ductility tests are performed on notched specimens since smooth specimens usually show amazing ductility. Table 7 provides a brief summary of metals and alloys recommended for low-temperature use.

Among the most important properties of materials at the other end of the temperature spectrum are creep, rupture, and short-time strengths. Stress rupture is another important consideration at high temperatures since it relates stress and time to produce rupture. Ferritic alloys are weaker than austenitic compositions, and in both groups molybdenum increases strength. Higher

Table 7 Metals and alloys for low-temperature process use†

ASTM specification and grade	Recommended minimum service temp, °C
Carbon and alloy steels:	
T-1	−45
A 201, A 212, flange or firebox quality	
A 203, grades A and B ($2\frac{1}{4}$ Ni)	−60
A 203, grades D and E ($3\frac{1}{2}$ Ni)	−100
A 353 (9% Ni)	−195
Copper alloys, silicon bronze, 70–30 brass, copper	
Stainless steels type 302, 304L, 304, 310, 347	−255
Aluminum alloys 5052, 5083, 5086, 5154, 5356, 5454, 5456	

† Based on a series of articles on low-temperature metals appearing in *Chem. Eng.* from May 20 to Oct. 31, 1960. See also R. M. McClintock and H. P. Gibbons, "Mechanical Properties of Structural Materials at Low Temperatures," National Bureau of Standards, June 1960; Vol. 1–24, K. D. Timmerhaus (ed.), "Advances in Cryogenic Engineering," Plenum Press, New York.

strengths are available in Inconel, cobalt-based Stellite 25, and iron-base A286. Other properties which become important at high temperatures include thermal conductivity, thermal expansion, ductility, alloy composition, and stability.

Actually, in many cases strength and mechanical properties become of secondary importance in process applications, compared with resistance to the corrosive surroundings. All common heat-resistant alloys form oxides when exposed to hot oxidizing environments. Whether the alloy is resistant depends upon whether the oxide is stable and forms a protective film. Thus, mild steel is seldom used above 500°C because of excessive scaling rates. Higher temperatures require chromium. This is evident, not only from Table 5, but also from Table 8 which lists the important commercial alloys for high-temperature use.

GASKET MATERIALS

Metallic and nonmetallic gaskets of many different forms and compositions are used in industrial equipment. The choice of a gasket material depends on the corrosive action of the chemicals that may contact the gasket, the location of the gasket, and the type of gasket construction. Other factors of importance are the cost of the materials, pressure and temperature involved, and frequency of opening the joint.

TABULATED DATA FOR SELECTING MATERIALS OF CONSTRUCTION

Table 9 presents information on the corrosion resistance of some common metals, nonmetals, and gasket materials. Table 10 presents similar information for various types of plastics. These tables can be used as an aid in choosing materials of construction, but no single table can take into account all the factors that can affect corrosion. Temperature level, concentration of the corrosive agent, presence of impurities, physical methods of operation, and slight alterations in the composition of the constructional material can affect the degree of corrosion resistance. The final selection of a material of construction, therefore, may require reference to manufacturers' bulletins and consultation with persons who are experts in the particular field of application.†

SELECTION OF MATERIALS

The chemical engineer responsible for the selection of materials of construction must have a thorough understanding of all the basic process information available. This knowledge of the process can then be used to select materials of con-

---

† References to additional information on materials of construction in the chemical process industries are given at the end of the chapter.

Table 8 Alloys for high-temperature process use†

	Nominal composition, %			
	Cr	Ni	Fe	Other
Ferritic steels:				
Carbon steel	....	....	bal.	
2¼ chrome	2¼	....	bal.	Mo
Type 502	5	....	bal.	Mo
Type 410	12	....	bal.	
Type 430	16	....	bal.	
Type 446	27	....	bal.	
Austenitic steels:				
Type 304	18	8	bal.	
Type 321	18	10	bal.	Ti
Type 347	18	11	bal.	Cb
Type 316	18	12	bal.	Mo
Type 309	24	12	bal.	
Type 310	25	20	bal.	
Type 330	15	35	bal.	
Nickel-base alloys:				
Nickel	....	bal.		
Incoloy	21	32	bal.	
Hastelloy B	....	bal.	6	Mo
Hastelloy C	16	bal.	6	W, Mo
60/15	15	bal.	25	
Inconel	15	bal.	7	
80/20	20	bal.		
Hastelloy X	22	bal.	19	Co, Mo
Multimet	21	20	bal.	Co
Rene 41	19	bal.	5	Co, Mo, Ti
Cast irons:				
Ductile iron	....	....	bal.	C, Si, Mg
Ni-Resist, D-2	2	20	bal.	Si, C
Ni-Resist, D-4	5	30	bal.	Si, C
Cast stainless (ACI types):				
HC	28	4	bal.	
HF	21	11	bal.	
HH	26	12	bal.	
HK	26	20	bal.	
HT	15	35	bal.	
HW	12	bal.	28	
Super alloys:				
Inconel X	15	bal.	7	Ti, Al, Cb
A 286	15	25	bal.	Mo, Ti
Stellite 25	20	10	Co-base	W
Stellite 21 (cast)	27.3	2.8	Co-base	Mo
Stellite 31 (cast)	25.2	10.5	Co-base	W

† "Chemical Engineers' Handbook," 5th ed., McGraw-Hill Book Company, New York, 1973.

Table 9 Corrosion resistance of constructional materials†

Code designation for corrosion resistance

A = acceptable, can be used successfully
C = caution, resistance varies widely depending on conditions; used when some corrosion is permissible
X = unsuitable
Blank = information lacking

Code designation for gasket materials

a = asbestos, white (compressed or woven)
b = asbestos, blue (compressed or woven)
c = asbestos (compressed and rubber-bonded)
d = asbestos (woven and rubber-frictioned)
e = GR-S or natural rubber
f = Teflon

Chemical	Iron and steel	Cast iron (Ni-resist)	Stainless steel 18-8	Stainless steel 18-8 Mo	Nickel	Monel	Red brass	Aluminum	Industrial glass	Carbon (Karbate)	Phenolic resins (Haveg)	Acrylic resins (Lucite)	Vinylidene chloride (Saran)	Acceptable nonmetallic gasket materials
					Metals						Nonmetals			
Acetic acid, crude	C	C	C	C	C	C	C	A	A	A	A	A	C	b, c, d, f
Acetic acid, pure	X	X	C	A	C	A	X	A	A	A	A	X	X	b, c, d, f
Acetic anhydride	C	C	A	A	A	A	X	A	A	A	C	X	C	b, c, d, f
Acetone	A	A	A	A	A	A	A	A	A	A	A	X	C	a, e, f
Aluminum chloride	X	C	X	X	C	C	A	A	A	A	A		A	a, c, d, e, f
Aluminum sulfate	X	C	C	A	C	C	X	A	A	A	A	A	A	a, c, d, e, f
Alums	X	C	C	A	C	A	X	C	A		A	A	A	a, f
Ammonia (gas)	A	A	C	A	A	A	X	C	A	A	A		A	b, c, d, e, f
Ammonium chloride	C	A	C	C	A	A	C	C	A		A	A	C	a, c, d, f
Ammonium hydroxide	A	A	A	A	C	C	X	X	A	A	A		C	
Ammonium phosphate (monobasic)	X	C	A	A			X	C	A	A	A			b, c, d, e, f
Ammonium phosphate (dibasic)	C	A	A	A		A	C		A	A	A			a, c, d, e, f
Ammonium phosphate (tribasic)	A	A	A	A	A	A	X	C	A	A	A	A	A	a, c, d, e, f
Ammonium sulfate	C	A	C	C	A	A	C	A	A	A	A	A	A	b, c, d, e, f
Aniline	A	A	A	A		A	X		A	A	C		C	a, f

Chemical													Notes
Benzene, benzol	A	A	A	A	A	A	A	A	—	A	—	C	a, f
Boric acid	X	C	C	A	C	C	C	C	C	X	—	A	a, c, d, e, f
Bromine	C	C	C	O	C	C	C	A	A	X	—	X	b, f
Calcium chloride	A	A	A	A	A	A	A	A	A	A	A	A	b, c, d, e, f
Calcium hydroxide	X	A	A	A	A	A	A	A	A	A	—	—	a, c, d, e, f
Calcium hypochlorite	C	C	C	C	C	C	O	O	C	—	—	C	b, c, d, f
Carbon tetrachloride	C	C	C	A	A	A	A	A	A	A	—	C	a, f
Carbonic acid	C	C	A	A	A	A	A	A	A	A	A	A	a, e, f
Chloracetic acid	X	X	X	X	X	X	X	X	X	A	—	—	b, f
Chlorine, dry	A	A	A	A	A	A	A	A	A	A	—	—	b, e, f
Chlorine, wet	X	X	X	X	X	X	C	X	X	X	X	X	b, e, f
Chromic acid	C	C	C	C	C	O	C	O	C	C	X	X	b, f
Citric acid	X	C	A	A	A	A	A	A	A	A	A	A	b, c, d, e, f
Copper sulfate	X	C	A	A	A	A	A	A	A	A	X	A	b, c, d, e, f
Ethanol	A	A	A	A	A	A	A	A	A	A	—	A	a, c, e, f
Ethylene glycol	A	A	A	A	A	A	A	A	A	A	A	A	a, c, e, f
Fatty acids	C	A	A	A	A	A	C	C	A	A	—	A	a, e, f
Ferric chloride	X	C	X	X	X	X	X	A	A	A	—	A	b, e, f
Ferric sulfate	X	X	C	C	C	C	C	A	A	A	—	A	b, c, e, f
Ferrous sulfate	C	C	A	A	A	A	C	A	A	A	—	A	a, c, e, f
Formaldehyde	C	A	A	A	A	A	A	A	A	A	—	A	b, c, e, f
Formic acid	X	C	C	C	C	C	X	X	X	A	—	C	a, c, e, f
Glycerol	A	—	A	A	A	A	A	A	A	A	A	A	b, c, e, f
Hydrocarbons (aliphatic)	A	A	A	A	A	A	A	A	A	A	A	A	a, c, e, f
Hydrochloric acid	X	X	X	X	X	X	X	X	X	A	—	C	b, c, e, f
Hydrofluoric acid	C	X	X	X	C	C	C	X	C	A	—	C	a, c, d, f
Hydrogen peroxide	C	—	C	C	C	C	C	C	C	A	—	C	b, c, d, f
Lactic acid	X	C	C	C	C	C	A	A	A	A	A	—	b, f
Magnesium chloride	C	A	A	A	A	A	C	A	A	A	—	A	a, e, f
Magnesium sulfate	A	A	A	A	A	A	A	A	A	A	—	A	a, b, c, d, e, f
Methanol	A	A	A	A	A	A	A	A	A	A	—	A	b, c, e, f
Nitric acid	X	X	X	X	X	X	X	X	X	A	A	A	b, c, e, f
Oleic acid	C	C	C	C	C	C	C	A	A	A	A	C	a, c, e, f
Oxalic acid	C	A	A	A	A	A	C	C	C	A	—	A	a, c, e, f
Phenol (carbolic acid)	C	A	A	A	A	A	X	X	X	A	A	—	b, f
Phosphoric acid	C	C	C	C	C	C	X	X	X	O	C	C	a, f
Potassium hydroxide	C	C	C	A	A	A	C	C	C	A	—	A	b, c, f
Sodium bisulfate	O	A	A	A	A	A	C	C	C	—	—	A	a, e, f
													b, c, d, e, f

(continued)

Table 9 Corrosion resistance of constructional materials† (continued)

Columns are grouped under **Metals** (Iron and steel; Cast iron (Ni-resist); Stainless steel 18-8; Stainless steel 18-8 Mo; Nickel; Monel; Red brass; Aluminum) and **Nonmetals** (Industrial glass; Carbon (Karbate); Phenolic resins (Haveg); Acrylic resins (Lucite); Vinylidene chloride (Saran)).

Chemical	Iron and steel	Cast iron (Ni-resist)	Stainless steel 18-8	Stainless steel 18-8 Mo	Nickel	Monel	Red brass	Aluminum	Industrial glass	Carbon (Karbate)	Phenolic resins (Haveg)	Acrylic resins (Lucite)	Vinylidene chloride (Saran)	Acceptable nonmetallic gasket materials
Sodium carbonate	A	A	A	A	A	A	C	C	C	A	A	X	…	a, c, d, e, f
Sodium chloride	A	A	C	C	A	A	C	C	A	A	A	…	C	a, c, d, e, f
Sodium hydroxide	A	A	A	A	A	A	C	X	C	A	X	A	A	a, c, d, f
Sodium hypochlorite	X	C	C	A	C	C	C	X	A	C	X	…	A	b, c, d, f
Sodium nitrate	A	A	A	A	A	A	C	A	A	A	A	…	A	b, c, d, e, f
Sodium sulfate	A	A	A	C	A	A	A	X	A	A	A	…	A	a, c, d, e, f
Sodium sulfide	A	A	A	A	A	A	X	C	A	A	A	…	…	a, e, f
Sodium sulfite	A	A	A	A	A	A	C	C	A	A	A	…	A	a, c, d, e, f
Sodium thiosulfate	C	…	A	A	A	A	C	A	A	A	A	…	…	a, e, f
Stearic acid	A	A	A	A	C	C	C	C	A	A	A	…	…	a, e, f
Sulfur	A	C	C	C	C	C	C	A	A	A	A	…	A	a, f
Sulfur dioxide	C	C	C	C	C	C	C	C	A	X	X	X	C	b, f
Sulfuric acid (98 % to fuming)	A	C	X	C	X	X	X	C	A	X	X	X	C	b, f
Sulfuric acid (75–95 %)	A	C	X	X	X	C	X	X	A	C	C	X	C	b, f
Sulfuric acid (10–75 %)	X	C	X	X	C	C	X	X	A	A	C	C	A	a, b, c, e, f
Sulfuric acid (<10 %)	X	C	X	C	C	C	C	C	A	A	A	A	C	b, c, d, e, f
Sulfurous acid	X	…	C	A	X	X	C	C	A	A	A	…	C	a, f
Trichloroethylene	C	A	C	A	A	A	C	C	…	…	A	…	…	b, c, d, e, f
Zinc chloride	C	C	C	X	A	A	X	C	…	…	A	A	…	b, c, d, e, f
Zinc sulfate	C	A	A	A	A	A	C	C	…	…	…	…	…	

† From miscellaneous sources (see references at end of chapter).

Table 10 Chemical resistance of plastics in various solvents†

Code designation for chemical resistance
S = good to 25°C
S_1 = good to 60°C
S_2 = good above 60°C
F = fair
U = unsatisfactory

	PVC rigid	PVC plasti-cized	Poly-ethylene	Poly-propyl-ene	Meth-acry-lates	Poly-esters	Epox-ies	Fluoro-carbons	Poly-sty-rene	ABS Poly-mers	Acetal poly-mers	Phenol-formal-dehyde	Poly-car-bonate	Cl-poly-ether	Furan	Saran
Acetone	U	U	S	S	U	U	F	S_2	U	U	S_1	U	S	S	F	S
Alcohols, methyl	S	S_1	S_1	S_1	S	S_1	S	S_2	S	S_1	S_1	F	S_1	S_2	S	S_1
ethyl	S	S_1	S_1	S_1	S	S_1	S	S_2	S	S_1	S_1	F	S_1	S_2	S	S_1
butyl	S	S_1	S_1	S_1	S	S_1	S	S_2	S	S_1	S_1	F	S_1	S_2	S	S_1
Aniline	U	U	U	S			S	S_2	U	U	S_1	U	S_1	F	U	U
Benzene	U	U	U	U	U	S_1	S	S_2	U	U	S_1	S	U	S	S	F
Carbon tetrachloride	U	U		U	U	S_1	S	S_2		U		S	U	S_2	S	S
Cyclohexanone	U	U	U				F	S_2	U	U		U				U
Ethyl acetate	U	U	U	U	U	U	F	S_2	U	U	S_1	U	U	S	S	F
Ethylene dichloride	U	U	U	U	U	U	F	S_2	U	U	S	S		S_2		S
Ethyl ether	U	U	S	S	U		S	S_2	U	F	S	S	U	S	S	U
Hexane	S	U	F	F	S	S	S	S_2	U	F	S	S	S_1	S	S	S_1
Kerosene	S	S_1	S	S	S	S	S	S_2	U	F	S	S	S_1	S_2	S	S_1
Lubricating oils	S_1	S	S	S	S	S	S_2	S_2	F	F	S	S	S_1	S_2	S	S_1
Naphthalene	U	U	S	S	U	S	S	S_2	U	U	S	S	S_1	S_2	S	S_1
Triethanolamine	S_1	S	S	S				S_2		U			F	S_2	S	
Xylene	U	U	U	U	U	S_2	S	S_2	U	U	S_1	S	U	S	S	F

† Twentieth Biennial Materials of Construction Report, *Chem. Eng.,* **69**(23):185 (1962).

struction in a logical manner. A brief plan for studying materials of construction is as follows:

1. Preliminary selection
   Experience, manufacturer's data, special literature, general literature, availability, safety aspects, preliminary laboratory tests
2. Laboratory testing
   Reevaluation of apparently suitable materials under process conditions
3. Interpretation of laboratory results and other data
   Effect of possible impurities, excess temperature, excess pressure, agitation, and presence of air in equipment
   Avoidance of electrolysis
   Fabrication method
4. Economic comparison of apparently suitable materials
   Material and maintenance cost, probable life, cost of product degradation, liability to special hazards
5. Final selection

In making an economic comparison, the engineer is often faced with the question of where to use high-cost claddings or coatings over relatively cheap base materials such as steel or wood. For example, a column requiring an expensive alloy-steel surface in contact with the process fluid may be constructed of the alloy itself or with a cladding of the alloy on the inside of carbon-steel structural material. Other examples of commercial coatings for chemical process equipment include baked ceramic or glass coatings, flame-sprayed metal, hard rubber, and many organic plastics. The durability of coatings is sometimes questionable, particularly where abrasion and mechanical-wear conditions exist. As a general rule, if there is little economic incentive between a coated type versus a completely homogeneous material, a selection should favor the latter material, mainly on the basis of better mechanical stability.

ECONOMICS IN SELECTION OF MATERIALS

First cost of equipment or material often is not a good economic criterion when comparing alternate materials of construction for chemical process equipment. Any cost estimation should include the following items:

1. Total equipment or materials costs
2. Installation costs
3. Maintenance costs
4. Estimated life
5. Replacement costs

When these factors are considered, cost comparisons bear little resemblance to first costs. Table 11 presents a typical analysis of comparative costs for alterna-

Table 11 Alternative investment comparison

	Material A	Material B	Material C
Purchased cost	$25,000	$30,000	$35,000
Installation cost	15,000	20,000	25,000
Total installed cost	40,000	50,000	60,000
Additional cost over A		10,000	20,000
Estimated life, years	4	10	10
Estimated maintenance cost/year	5,000	4,500	3,000
Annual replacement cost (installed cost/estimated life)	10,000	5,000	6,000
Total annual cost	15,000	9,500	9,000
Annual savings vs. cost for A		5,500	6,000
Tax on savings, 50%		2,750	3,000
Net annual savings		2,750	3,000
Return on investment over A (net savings/additional cost over A) 100		27.5%	15.0%

tive materials when based on return on investment. One difficulty with such a comparison is the uncertainty associated with "estimated life." Well-designed laboratory and plant tests can at least give order-of-magnitude estimates. Another difficulty arises in estimating the annual maintenance cost. This can only be predicted from previous experience with the specific materials.

Table 11 could be extended by the use of continuous compounding interest methods as outlined in Chaps. 6 and 9 to show the value of money to a company above which (or below which) material A would be selected over B, B over C, etc. Table 11 indicates that material B is always better than material A (this, of course, is inherent in the yearly return on investment method used). However, depending on the value of money to a company, this may not always be true.

FABRICATION OF EQUIPMENT

Fabrication expenses account for a large fraction of the purchased cost for equipment. A chemical engineer, therefore, should be acquainted with the methods for fabricating equipment, and the problems involved in the fabrication should be considered when equipment specifications are prepared.

Many of the design and fabrication details for equipment are governed by various codes, such as the ASME Codes. These codes can be used to indicate definite specifications or tolerance limits without including a large amount of descriptive restrictions. For example, fastening requirements can often be indicated satisfactorily by merely stating that all welding should be in accordance with the ASME Code.

The exact methods used for fabrication depend on the complexity and type of equipment being prepared. In general, however, the following steps are involved in the complete fabrication of major pieces of chemical equipment, such as tanks, autoclaves, reactors, towers, and heat exchangers:

1. Layout of materials
2. Cutting to correct dimensions
3. Forming into desired shape
4. Fastening
5. Testing
6. Heat-treating
7. Finishing

Layout

The first step in the fabrication is to establish the layout of the various components on the basis of detailed instructions prepared by the fabricator. Flat pieces of the metal or other constructional material involved are marked to indicate where cutting and forming are required. Allowances must be made for losses caused by cutting, shrinkage due to welding, or deformation caused by the various forming operations.

After the equipment starts to take shape, the location of various outlets and attachments will become necessary. Thus, the layout operation can continue throughout the entire fabrication. If tolerances are critical, an exact layout, with adequate allowances for deformation, shrinkage, and losses, is absolutely essential.

Cutting

Several methods can be used for cutting the laid-out materials to the correct size. *Shearing* is the cheapest method and is satisfactory for relatively thin sheets. The edge resulting from a shearing operation may not be usable for welding, and the sheared edges may require an additional grinding or machining treatment.

Burning is often used for cutting metals. This method can be employed to cut and, simultaneously, prepare a beveled edge suitable for welding. Carbon steel is easily cut by an oxyacetylene flame. The heat effects on the metal are less than those involved in welding. Stainless steels and nonferrous metals that do not oxidize readily can be cut by a method known as *powder* or *flux burning*. An oxyacetylene flame is used, and powdered iron is introduced into the cut to increase the amount of heat and improve the cutting characteristics. The high temperatures involved may affect the materials, resulting in the need for a final heat-treatment to restore corrosion resistance or removal of the heat-affected edges.

Sawing can be used to cut metals that are in the form of flat sheets. However, sawing is expensive, and it is used only when the heat effects from burning would be detrimental.

Forming

After the constructional materials have been cut, the next step is to form them into the desired shape. This can be accomplished by various methods, such as by rolling, bending, pressing, bumping (i.e., pounding), or spinning on a die. In some cases, heating may be necessary in order to carry out the forming operation. Because of work hardening of the material, annealing may be required before forming and between stages during the forming.

When the shaping operations are finished, the different parts are assembled and fitted for fastening. The fitting is accomplished by use of jacks, hoists, wedges, and other means. When the fitting is complete and all edges are correctly aligned, the main seams can be tack-welded in preparation for the final fastening.

Fastening

Riveting can be used for fastening operations, but electric welding is far more common and gives superior results. The quality of a weld is very important, because the ability of equipment to withstand pressure or corrosive conditions is often limited by the conditions along the welds. Although good welds may be stronger than the metal that is fastened together, design engineers usually assume a weld is not perfect and employ weld efficiencies of 80 to 95 percent in the design of pressure vessels.

The most common type of welding is the *manual shielded-arc* process in which an electrode approximately 14 to 16 in. long is used and an electric arc is maintained manually between the electrode and the material being welded. The electrode melts and forms a filler metal, while, at the same time, the work material fuses together. A special coating is provided on the electrode. This coating supplies a flux to float out impurities from the molten metal and also serves to protect the metal from surrounding air until the metal has solidified and cooled below red heat. The type of electrode and coating is determined by the particular materials and conditions that are involved in the welding operation.

A *submerged-arc* process is commonly used for welding stainless steels and carbon steels when an automatic operation is acceptable. The electrode is a continuous roll of wire fed at an automatically controlled rate. The arc is submerged in a granulated flux which serves the same purpose as the coating on the rods in the shielded-arc process. The appearance and quality of the submerged-arc weld is better than that obtained by an ordinary shielded-arc manual process; however, the automatic process is limited in its applications to main seams or similar long-run operations.

Heliarc welding is used for stainless steels and most of the nonferrous materials. This process can be carried out manually, automatically, or semiautomatically. A stream of helium or argon gas is passed from a nozzle in the electrode holder onto the weld, where the inert gas acts as a shielding blanket to protect the molten metal. As in the shielded-arc and submerged-arc processes, a filler rod is fed into the weld, but the arc in the heliarc process is formed between a tungsten electrode and the base metal.

In some cases, fastening can be accomplished by use of various solders, such as brazing solder (mp, 840 to 905°C) containing about 50 percent each of copper and zinc; silver solders (mp, 650 to 870°C) containing silver, copper, and zinc; or ordinary solder (mp, 220°C) containing 50 percent each of tin and lead. Screw threads, packings, gaskets, and other mechanical methods are also used for fastening various parts of equipment.

Testing

All welded joints can be tested for concealed imperfections by X rays, and code specifications usually require X-ray examination of main seams. Hydrostatic tests can be conducted to locate leaks. Sometimes, delicate tests, such as a helium probe test, are used to check for very small leaks.

Heat-treating

After the preliminary testing and necessary repairs are completed, it may be necessary to heat-treat the equipment to remove forming and welding stresses, restore corrosion-resistance properties to heat-affected materials, and prevent stress-corrosion conditions. A low-temperature treatment may be adequate, or the particular conditions may require a full anneal followed by a rapid quench.

Finishing

The finishing operation involves preparing the equipment for final shipment. Sandblasting, polishing, and painting may be necessary. Final pressure tests at $1\frac{1}{2}$ to 2 or more times the design pressure are conducted together with other tests as demanded by the specified code or requested by the inspector.

SUGGESTED ADDITIONAL REFERENCES FOR MATERIALS AND FABRICATION SELECTION

Corrosion

Arnold, C. G., Process Corrosion Control: Using Real Time Corrosion Monitors in Chemical Plants, *Chem. Eng. Progr.*, **74**(3):43 (1978).

Ashbaugh, W. G., Stress Corrosion Cracking of Process Equipment, *Chem. Eng. Progr.*, **66**(10):44 (1970).

Askew, T., Selecting Economic Boiler-Water Pretreatment Equipment, *Chem. Eng.*, **80**(9):114 (1973).

Banfield, R. H., Corrosion Problems in Chemical Plant, *Anti-Corrosion Methods and Materials*, **14**(9):8 (1967).

Bauman, T. C., and L. T. Overstreet, Corrosion and Piping Materials in the CPI, *Chem. Eng.*, **85**(8):59 (1978).

Berger, D. M., How to Test Linings for Corrosion-Resistant Tanks, *Chem. Eng.*, **82**(2):100 (1975).

——, How Corrosion Theory Relates to Protective Coatings, *Chem. Eng.*, **84**(16):77 (1977); **84**(18):89 (1977).

Boche, A. D., Corrosion-Preventive Plastic Tape, *Chem. Eng. Progr.*, **66**(10):63 (1970).

Bonar, J. A., Fuel Ash Corrosion, *Hydrocarbon Process.*, **51**(8):76 (1972).

Brooke, J. M., Inhibitors: New Demands for Corrosion Control, *Hydrocarbon Process.*, **49**(1):121 (1970); **49**(3):138 (1970).

———, Corrosion Inhibitor Checklist, *Hydrocarbon Process.*, **49**(8):107 (1970).

———, Corrosion Inhibitor Economics, *Hydrocarbon Process.*, **49**(9):299 (1970).

———, How to Field Test Inhibitors, *Hydrocarbon Process.*, **49**(10):117 (1970).

Brooks, W. B., Practical Notes on Plant Corrosion, *Hydrocarbon Process.*, **46**(7):147 (1967).

Browning, J. E., Troubleshooters Swap Data, *Chem. Eng.*, **81**(8):52 (1974).

Burkhalter, L. C., M. F. Shelton, and E. H. Tomlinson, Cathodic Cure for Corrosion, *Chem. Eng.*, **75**(22):164 (1968).

Butwell, K. F., E. N. Hawkes, and B. F. Mago, Corrosion Control in CO_2 Removal Systems, *Chem. Eng. Progr.*, **69**(2):57 (1973).

Cangi, J. W., Process Corrosion Control: Characteristics & Corrosion Properties of Cast Alloys, *Chem. Eng. Progr.*, **74**(3):61 (1978).

Cantwell, J. E., and R. E. Bryant, Alloy Failures in Piping and Flare Tips, *Hydrocarbon Process.*, **52**(5):114 (1973).

Capel, L., and J. M. A. Van der Horst, Stainless Steel Corrosion in Urea Synthesis, *Chem. Eng. Progr.*, **66**(10):38 (1970).

Catlett, R. E., Specifications and the Corrosion Engineer, *Chem. Eng.*, **79**(17):90 (1972).

Cooper, C. M., Naphthenic Acid Corrosion, *Hydrocarbon Process.*, **51**(8):75 (1972).

———, What to Specify for Corrosion Allowance, *Hydrocarbon Process.*, **51**(5):123 (1972).

Cotton, J. B., Using Titanium in the Chemical Plant, *Chem. Eng. Progr.*, **66**(10):57 (1970).

Covington, L. C., R. W. Schutz, and I. A. Franson, Process Corrosion Control: Titanium Alloys for Corrosion Resistance, *Chem. Eng. Progr.*, **74**(3):67 (1978).

Curtis, S. D., and R. M. Silverstein, Corrosion and Fouling Control of Cooling Waters, *Chem. Eng. Progr.*, **67**(7):39 (1971).

Degerbeck, J., Corrosion of Stainless Steel in Seawater, *Chem. Proc. Eng.*, **52**(11):47 (1971).

Desensy, M. G. J., Materials for Seawater Cooling, *Chem. Eng.*, **77**(13):182 (1970).

Dunlop, A. K., Materials Engineering Forum, Using Corrosion Inhibitors, *Chem. Eng.*, **77**(21):108 (1970).

Ehmke, E. F., Air Cooler, Corrosion, *Hydrocarbon Process.*, **54**(7):172 (1975).

Enrico, R. J., Corrosion of Nuclear Materials, *Materials*, **7**(10):464 (1967).

Estefan, S. L., Design Guide to Metallurgy and Corrosion in Hydrogen Processes, *Hydrocarbon Process.*, **49**(12):85 (1970).

Feige, N. G., and R. L. Kane, CPI Applications of Titanium, *Chem. Eng. Progr.*, **66**(10):53 (1970).

Fontana, M. G., Selecting Construction Materials for Pumps, *Chem. Eng. Progr.*, **66**(5):65 (1970).

——— and N. D. Greene, "Corrosion Engineering," McGraw-Hill Book Company, New York, 1968.

French, E. C., Control Corrosion in Sour Water System, *Hydrocarbon Process.*, **55**(8):147 (1976).

Fyfe, D., R. Vanderland, and J. Rodda, Corrosion in Sulfuric Acid Storage Tanks, *Chem. Eng. Progr.*, **73**(3):65 (1977).

Gasper, K. E., Process Corrosion Control: Non-Chromate Methods of Cooling Water Treatment, *Chem. Eng. Progr.*, **74**(3):52 (1978).

Gilbu, A., and J. A. Rolston, Large Diameter Reinforced Plastic Pipe, *Chem. Eng. Progr.*, **70**(1):50 (1974).

Gleason, T. G., How to Avoid Scrubber Corrosion, *Chem. Eng. Progr.*, **71**(3):43 (1975).

Gordon, H. B., How Metals Resist Sea Water, *Materials Eng.*, **65**(5):82 (1967).

Harrell, J. B., Process Corrosion Control: Corrosion Monitoring in the CPI, *Chem. Eng. Progr.*, **74**(3):57 (1978).

Hawk, C. W., Jr., Do You Understand Galvanic Corrosion? *Chem. Eng.*, **84**(12):193 (1977).

Hawkes, E. N., and B. F. Mago, Stop MEA CO_2 Unit Corrosion, *Hydrocarbon Process.*, **50**(8):109 (1971).

Henthorne, M., Cathodic and Anodic Protection for Corrosion Control, *Chem. Eng.*, **78**(29):73 (1971).

————, Control the Process and Control Corrosion, *Chem. Eng.*, **78**(24):139 (1971).

————, Fundamentals of Corrosion, *Chem. Eng.*, **78**(11):127 (1971).

————, Electrochemical Corrosion, *Chem. Eng.*, **78**(13):102 (1971).

————, Polarization Data Yield Corrosion Rates, *Chem. Eng.*, **78**(17):99 (1971).

————, Measuring Corrosion in the Process Plant, *Chem. Eng.*, **78**(19):89 (1971).

————, Stress Corrosion, *Chem. Eng.*, **78**(21):159 (1971).

————, Good Engineering Design Minimizes Corrosion, *Chem. Eng.*, **78**(26):163 (1971).

————, Corrosion Protection via Coatings, *Chem. Eng.*, **79**(1):103 (1972).

————, Paints Prevent Corrosion, *Chem. Eng.*, **79**(3):82 (1972).

————, Materials Selection for Corrosion Control, *Chem. Eng.*, **79**(5):113 (1972).

————, Finding Answers to Corrosion Problems, *Chem. Eng.*, **79**(7):81 (1972).

————, Understanding Corrosion, *Chem. Eng.*, **79**(27):19 (1972).

Holden, H. A., Corrosion Prevention in Chemical Plants, *Chem. Proc. Eng.*, **49**(6):75 (1968).

Husen, C., and C. H. Samans, Metals, *Chem. Eng.*, **77**(22):15 (1970).

Kirchner, R. W., Equipment Testing: Are You Getting What You Pay For? *Chem. Eng.*, **74**(18):107 (1967).

————, Corrosion of Pollution Control Equipment, *Chem. Eng. Progr.*, **71**(3):58 (1975).

Kobrin, G., Microbiology for Oil People . . . Corrosion in Natural Water, *Hydrocarbon Process.*, **55**(8):148 (1976).

Krisher, A. S., Cooling Water Corrosion Control, *Hydrocarbon Process.*, **47**(7):127 (1968).

Landrum, R. J., Designing for Corrosion Resistance, *Chem. Eng.*, **76**(4):118 (1969); **76**(6):172 (1969).

Leonard, R. B., New Corrosion-Resistant Alloys, *Chem. Eng. Progr.*, **65**(7):84 (1969).

Lochmann, W. J., Corrosion Specs. Increase Costs, *Hydrocarbon Process.*, **50**(7):101 (1971).

Lythe, R. G., Alloys for Nitric Acid at High Temperatures, *Chem. Proc. Eng.*, **53**(6):571 (1972).

Marshall, S. P., and J. L. Brandt, Installed Cost of Corrosion-Resistant Piping, *Chem. Eng.*, **78**(19):68 (1971); **78**(22):111 (1971).

———— and ————, 1974 Installed Cost of Corrosion-Resistant Piping, *Chem. Eng.*, **81**(23):94 (1974).

McCoy, J. D., and F. B. Hamel, New Corrosion Rate Data for Hydrodesulfurizing Units, *Hydrocarbon Process.*, **49**(6):116 (1970).

McDowell, D. W., Corrosion: Back to Basics, *Chem. Eng.*, **80**(22):92 (1973).

————, Corrosion in Urea-Synthesis Reactors, *Chem. Eng.*, **81**(10):118 (1974).

————, Handling Mineral Acids, *Chem. Eng.*, **81**(24):118 (1974).

————, Handling Phosphoric Acid and Phosphate Fertilizers, *Chem. Eng.*, **82**(16):119 (1975).

————, Sulfuric Acid Plants—Materials of Construction, *Chem. Eng. Progr.*, **71**(3):69 (1975).

McGill, W. A., and M. J. Weinbaum, Will Aluminum Reduce Pipe and Tubing Corrosion? *Hydrocarbon Process.*, **51**(6):127 (1972).

McMillin, F. A., Crude Unit Corrosion, *Hydrocarbon Process.*, **57**(4):141 (1978).

Miksic, B. A., Volatile Corrosion-Inhibitors Find a New Home, *Chem. Eng.*, **84**(20):115 (1977).

Mills, J. F., and B. D. Oakes, Bromine Chloride: Less Corrosive Than Bromine, *Chem. Eng.*, **80**(18):102 (1973).

Montrone, E. D., and W. P. Long, Choosing Materials for CO_2 Absorption Systems, *Chem. Eng.*, **78**(2):94 (1971).

Moreland, P. J., and J. G. Hines, Corrosion Monitoring . . . Select the Right System, *Hydrocarbon Process.*, **57**(11):251 (1978).

Morris, P. E., and R. M. Kain, An Electrochemical Approach to Alloy Protection, *Chem. Eng. Progr.*, **73**(6):103 (1977).

Morton, T. R., Fiber-Glass-Reinforced Plastics for Corrosion Resistance, *Chem. Eng.*, **80**(29):70 (1973); **81**(2):140 (1974).

Most, C. R., Comparing Coatings for Wear and Corrosion-Resistance, *Chem. Eng.*, **77**(2):140 (1970).

Nester, D. M., Salt Water Cooling Tower, *Chem. Eng. Progr.*, **67**(7):49 (1971).

Norden, R. B., Maintenance Painting, *Chem. Eng.*, **80**(5):85 (1973).

Pierce, R. R., Protecting Metal From Corrosive Atmospheres, *Chem. Eng.*, **78**(1):106 (1971).

Puckorius, P. R., Controlling Corrosive Microorganisms in Cooling-Water Systems, *Chem. Eng.*, **85**(23):171 (1978).

Rak, G. R., Process Corrosion Control: Corrosion Monitoring Equipment, *Chem. Eng. Progr.*, **74**(3):46 (1978).

Reiser, K., The Control of Corrosion in Refinery Distillation Units, *J. Inst. Petrol.*, **53**(527):352 (1967).

Rinckhoff, J. B., Controlling Corrosion in Wet-Gas Sulfuric Acid Plants, *Chem. Eng.*, **74**(24):158 (1967).

Roddy, C. P., Sulfur and Air Heater Corrosion, *Power Eng.*, **72**(1):40 (1968).

Schmeal, W. R., A. J. MacNab, and R. P. Rhodes, Process Corrosion Control: Corrosion in Amine/Sour Gas Treating Contactors, *Chem. Eng. Progr.*, **74**(3):37 (1978).

Silverstein, R. M., and S. D. Curtis, Cooling Water, *Chem. Eng.*, **78**(18):85 (1971).

Skabo, R. R., Internal Maintenance of Vessels, *Chem. Eng. Progr.*, **66**(10):33 (1970).

Szymanski, W. A., and D. W. Kloda, Polyester and Furfuryl Alcohol Resins for Corrosion Control, *Chem. Eng. Progr.*, **70**(1):51 (1974).

Tarlas, H. D., The Selection and Use of Protective Coatings, *Chem. Eng. Progr.*, **66**(8):38 (1970).

Tesmen, A. B., Materials of Construction for Process Plants, *Chem. Eng.*, **80**(7):126 (1973).

Verink, E. D., Jr., Aluminum Alloys for Saline Waters, *Chem. Eng.*, **81**(8):104 (1974).

Webster, H. A., The Costs of Corrosion: A Perspective, *J. APCA*, **26**(4):302 (1976).

Williams, C. P., Explosion Clad Titanium/Steel Plate, *Chem. Eng. Progr.*, **66**(10):48 (1970).

Wismer, M., and J. F. Bosso, Make the Part the Cathode: Key to Resistant Coatings, *Chem. Eng.*, **78**(13):114 (1971).

Woods, G. A., Bacteria: Friends or Foes? *Chem. Eng.*, **80**(6):81 (1973).

———, Selected Bibliography, *Chem. Eng.*, **77**(22):26 (1970).

———, Corrosion Literature, Selected Bibliography, *Chem. Eng.*, **79**(27):31 (1972).

———, Internal Corrosion in Floating Roof Gasoline Tanks, *Hydrocarbon Process.*, **55**(8):145 (1976).

Materials of construction

Adams, L., Supporting Cryogenic Equipment with Wood, *Chem. Eng.*, **78**(11):156 (1971).

Alliger, G., and F. C. Weissert, Elastomer Technology, *I&EC*, **60**(8):51 (1968).

Anderson, C. C., Adhesives, *I&EC*, **60**(8):80 (1968).

Arnold, J. L., Heat Treatment Protects Steel Alloys from Hot Gases, *Chem. Eng.*, **85**(11):205 (1978).

Artus, C. H., All About Chemical Hose, *Chem. Eng.*, **83**(9):112 (1976).

Atkinson, H. E., Fluoroplastic Linings for Corrosive Service, *Chem. Eng.*, **79**(29):76 (1972).

Avery, H. S., Abrasive Wear, *Hydrocarbon Process.*, **51**(8):79 (1972).

———, Thermal Fatigue, *Hydrocarbon Process.*, **51**(8):78 (1972).

Bailey, E. E., and R. W. Heinz, SO_2 Recovery Plants—Materials of Construction, *Chem. Eng. Progr.*, **71**(3):64 (1975).

Bauman, T. C., and L. T. Overstreet, Corrosion and Piping Materials in the CPI, *Chem. Eng.*, **85**(8):59 (1978).

Beerbower, A., L. A. Kaye, and D. A. Pattison, Picking the Right Elastomer to Fit Your Fluids, *Chem. Eng.*, **72**(26):118 (1967).

Berger, D. M., Liquid-Applied Linings for Steel Tanks, *Chem. Eng.*, **82**(27):65 (1975); **83**(2):123 (1976).

———, Applicator's Guide to Zinc-rich Primers, *Chem. Eng.*, **84**(6):147 (1977).

Bhat, V. K., and W. W. Carpenter, Thermo-mechanical Properties of Glass-Lined Equipment, *Chem. Eng.*, **80**(20):118 (1973).

Bland, W. F., New Materials, Fabrication Problems, *Petro/Chem. Eng.*, **39**(12):25 (1967).

Bonar, J. A., Fuel Ash Corrosion, *Hydrocarbon Process.*, **51**(8):76 (1972).

Brautigam, F. C., Welding Practices that Minimize Corrosion, *Chem. Eng.*, **84**(4):97 (1977).

Bravenec, E. V., Why Steels Fracture, *Chem. Eng.*, **79**(15):100 (1972).

Brink, J. A., Jr., W. F. Burggrabe, and L. E. Greenwell, Mist Eliminators for Sulfuric Acid Plants, *Chem. Eng. Progr.*, **64**(11):82 (1968).

Britton, O. J., D. H. Declerck, and F. H. Vorhis, Linings, *Chem. Eng.*, **77**(22):127 (1970).

Brown, R. W., and K. H. Sandmeyer, Applications of Fused-Cast Refractories, *Chem. Eng.*, **76**(13):106 (1969).

Browning, J. E., Troubleshooters Swap Data, *Chem. Eng.*, **81**(8):52 (1974).

Budris, A. R., Try Filled-TFE Bearings for Problem Services, *Chem. Eng.*, **81**(14):102 (1974).

Burkhalter, L. C., M. F. Shelton, and E. H. Tomlinson, Cathodic Cure for Corrosion, *Chem. Eng.*, **75**(22):164 (1968).

Burst, J. F., and J. A. Spieckerman, A Guide to Selecting Modern Refractories, *Chem. Eng.*, **74**(16):85 (1967).

Buttignol, V., and H. L. Gerhart, Polymer Coatings, *I&EC*, **60**(8):68 (1968).

Cameron, J. A., and F. M. Danowski, Jr., How to Select Materials for Centrifugal Compressors, *Hydrocarbon Process.*, **53**(6):115 (1974).

Campbell, R. W., Materials of Construction for Cryogenic Processes, *Chem. Eng.*, **74**(22):188 (1967).

Carlen, J. C., and C. Helmer, Refiners Report Longer Life Using ASTM A 669 Exchanger Tubes, *Hydrocarbon Process.*, **52**(5):157 (1973).

Chambard, J. L., Selecting Materials for High Pressure Fittings, *Chem. Eng. Progr.*, **73**(12):33 (1977).

Clark, C. C., High-Temperature Hydrogen: Threat to Alloy Steels, *Chem. Eng.*, **85**(15):87 (1978).

Clarke, J. S., Why Not Just Build Reactors to the Code? *Hydrocarbon Process.*, **49**(6):79 (1970).

Clayton, C. H., Jr., and T. E. Johnson, Engineering Materials for Pumps and Valves, *Chem. Eng. Progr.*, **74**(9):51 (1978).

Cooper, C. M., Hydrogen Attack, *Hydrocarbon Process.*, **51**(8):69 (1972).

———, Naphthenic Acid Corrosion, *Hydrocarbon Process.*, **51**(8):75 (1972).

Costello, T. M., and J. P. Bressanelli, For High Temperature Service . . . Tests Show Cr-Ni Alloy Superior, *Hydrocarbon Process.*, **55**(6):84 (1976).

Cowan, C. T., Choosing Materials of Construction for Plate Heat Exchangers, *Chem. Eng.*, **82**(12):100 (1975); **82**(14):102 (1975).

Creamer, E. L., Stress Corrosion Cracking of Reformer Tubes, *Chem. Eng. Progr.*, **70**(8):63 (1974).

Criss, G. H., and K. A. Kisko, Strength of Refractory Castables, *Chem. Eng. Progr.*, **66**(8):43 (1970).

Crowley, M. S., and J. F. Wygant, Acid-Resistant Concrete, *Chem. Eng. Progr.*, **64**(2):44 (1968).

Dahlstrom, D. A., and S. S. Davis, Plastics in Continuous Filtration Equipment, *Chem. Eng. Progr.*, **65**(10):80 (1969).

DeFalco, J. J., High-Speed Fans of Reinforced Polyester, *Chem. Eng.*, **79**(19):88 (1972).

Dell, G. J., Construction Materials for Phos-Acid Manufacture, *Chem. Eng.*, **74**(8):203 (1967).

Desensy, M. G. J., Materials for Seawater Cooling, *Chem. Eng.*, **77**(13):182 (1970).

Daly, J. J., Controlled Shotpeening Prevents Stress-Corrosion Cracking, *Chem. Eng.*, **83**(4):113 (1976).

Dorsey, J. S., Using Reinforced Plastics and Process Equipment, *Chem. Eng.*, **82**(19):104 (1975).

Dukes, R. R., and C. H. Schwarting, Choosing Materials for Making Chlorine and Caustic, *Chem. Eng.*, **75**(6):206 (1968); **75**(8):172 (1968).

Elgee, H., Using Wood Tanks, *Chem. Eng.*, **83**(14):95 (1976).

Evans, L. S., "Selecting Engineering Materials for Chemical and Process Plant," J. Wiley and Sons, New York, 1974.

Fairhurst, W., Hydrogen Attack on Pressure Vessels, *Chem. Proc. Eng.*, **51**(10):66 (1970).

Falck, S. B., Process Tank Linings, *Chem. Eng.*, **79**(27):69 (1972).

Fenner, O. H., 23d Biennial Materials of Construction Report—Plastic Materials of Construction, *Chem. Eng.*, **75**(23):126 (1968).

———, Evaluating Plastics and Resins, *Chem. Eng.*, **75**(24):182 (1968).

———, Plastics (Why Materials Fail), *Chem. Eng.*, **77**(22):23 (1970).

———, Plastics Testing, *Chem. Eng.*, **77**(22):53 (1970).

Ferreira, L. E., B. E. Steinkuhler, D. E. Werschky, and C. E. Williams, Alumina Ceramic Solve Problems in the CPI, *Chem. Eng. Progr.*, **64**(10):74 (1968).

Fischer, H. B., Stainless Steels and Other Ferrous Alloys, *I&EC*, **60**(8):45 (1968).

Fontana, M. G., Selecting Construction Materials for Pumps, *Chem. Eng. Progr.*, **66**(5):65 (1970).

Foxall, D. H., and P. T. Gilbert, Selecting Tubes for CPI Heat Exchangers, *Chem. Eng.*, **83**(6):99 (1976); **83**(8):147 (1976).

Friedrich, E. R., Jr., Floors for Process Areas, *Chem. Eng.*, **81**(13):157 (1974).

Furman, H. N., The Synergism of Teflon-Lined FRP, *Chem. Eng. Progr.*, **73**(11):92 (1977).

Fyfe, D., R. Vanderland, and J. Rodda, Corrosion in Sulfuric Acid Storage Tanks, *Chem. Eng. Progr.*, **73**(3):65 (1977).

Gackenbach, R. E., Glassed Steel for Process Equipment, *Chem. Eng.*, **85**(26):132 (1978).

Gall, R. L., and E. J. Piasecki, SO_2 Processing: The Double Alkali Wet Scrubbing System, *Chem. Eng. Progr.*, **71**(5):72 (1975).

Gaugh, R. R., and D. C. Perry, A New Stainless Steel for the CPI, *Chem. Eng.*, **79**(22):84 (1972).

———, New Stainless for High Temperatures, *Hydrocarbon Process.*, **55**(6):87 (1976).

Gilbu, A., and J. A. Rolston, Large Diameter Reinforced Plastic Pipe, *Chem. Eng. Progr.*, **70**(1):50 (1974).

Gleason, T. G., Halt Corrosion in Particulate Scrubbers, *Chem. Eng.*, **84**(23):145 (1977).

Gleekman, L. W., Nonferrous Metals, *Chem. Eng.*, **77**(22):111 (1970); **79**(27):47 (1972).

———, Pipe and Equipment Lined With Elastomers and Non-fluoropolymers, *Chem. Eng.*, **85**(26):118 (1978).

Glickman, M., and A. H. Hehn, Valves, *Chem. Eng.*, **76**(8):131 (1969).

Gulya, J. A., and E. G. Marshall, New Property Data for Hydrocracker Steels, *Hydrocarbon Process.*, **49**(6):91 (1970).

Hauck, J. E., Glass Parts: How to Select and Specify, *Materials Eng.*, **66**(2):85 (1967).

Heckler, N. B., Finding a Profit in Failures, *Chem. Eng.*, **76**(24):100 (1969).

Hegedus, A. K., Guide to Refractory Metals, *Machine Design*, **39**(26):169 (1967).

Hilado, C. J., How to Predict If Materials Will Burn, *Chem. Eng.*, **77**(27):174 (1970).

Hoffman, C. H., Consider Wood for Process-Plant Uses, *Chem. Eng.*, **79**(6):126 (1972).

———, Wood-Tank Engineering, *Chem. Eng.*, **79**(8):120 (1972).

Hresko, G. P., and W. J. McGuire, Hydrogenation Reactors for Higher Temperatures/Pressures, *Petro/Chem. Eng.*, **40**(12):17 (1968).

Hughson, R. V., Predicting and Preventing Brittle Failures, *Chem. Eng.*, **81**(16):114 (1974).

———, High-nickel Alloys for Corrosion Resistance, *Chem. Eng.*, **83**(24):125 (1976).

———, Fluorocarbon-lined Pipe and Equipment, *Chem. Eng.*, **85**(26):124 (1978).

Husen, C., and C. H. Samans, Metals, *Chem. Eng.*, **77**(22):15 (1970).

——— and ———, Avoiding the Problems of Stainless Steels, *Chem. Eng.*, **76**(2):178 (1969).

Hyland, J., Teflon Tank-Linings, *Chem. Eng.*, **80**(13):124 (1973).

Irving, G. M., Construction Materials for Breweries, *Chem. Eng.*, **75**(14):100 (1968).

Ishii, J., K. Maeda, and M. Izumiyama, Try Improved Calorized Steel, *Hydrocarbon Process.*, **55**(6):75 (1976).

Iverson, W. P., Microbiological Corrosion, *Chem. Eng.*, **75**(20):242 (1968).

Kies, F. K., I. A. Franson, and B. Coad, Finding New Uses for Ferritic Stainless Steel, *Chem. Eng.*, **77**(6):150 (1970).

Knoth, R. J., G. E. Lasko, and W. A. Matejka, New Ni-Free Stainless Bids to Oust Austenitic, *Chem. Eng.*, **77**(11):170 (1970).

Kobrin, G., and E. S. Kopecki, Choosing Alloys for Ammonia Service, *Chem. Eng.*, **85**(28):115 (1978).

Kopecki, E. S., Stainless Steel for Saline-Water Service, *Chem. Eng.*, **80**(2):124 (1973).

Krisher, A. S., Metals Testing, *Chem. Eng.*, **77**(22):47 (1970).

Lambertin, W. J., and F. H. Vaughan, Will Existing Equipment Fail by Catastrophic Brittle Fracture? *Hydrocarbon Process.*, **53**(9):217 (1974).

Lancaster, J. F., U.S. vs. European Reactor Steels, *Hydrocarbon Process.*, **49**(6):84 (1970).

Lederman, P. B., Materials: Key to Exploiting the Oceans, *Chem. Eng.*, **75**(12):105 (1968).

Lee, R. P., How Poor Design Causes Equipment Failures, *Chem. Eng.*, **84**(3):129 (1977).

Leonard, R. B., New Corrosion-Resistant Alloys, *Chem. Eng. Progr.*, **65**(7):84 (1969).

Lochmann, W. J., Creep and Creep-Rupture Life, *Hydrocarbon Process.*, **51**(8):71 (1972).

Loginow, A. W., and K. G. Brickner, Designing With a New Stainless Steel, *Chem. Eng.*, **76**(19):152 (1969).

Long, W. P., and E. D. Montrone, Economics, *Chem. Eng.*, **77**(22):41 (1970).

Lula, R. A., From Refining To Coal Gasification . . . New Steels for Fuel Processing, *Hydrocarbon Process.*, **55**(6):89 (1976).

Lyman, W. S., and A. Cohen, Engineering With Copper Alloys, *Chem. Eng.*, **85**(6):99 (1978); **85**(9):147 (1978).

Lythe, R. G., Alloys for Nitric Acid at High Temperatures, *Chem. Proc. Eng.*, **53**(6):571 (1972).

Mack, W. C., Selecting Steel Tubing for the CPI, *Chem. Eng.*, **80**(27):94 (1973).

———, Selecting Steel Tubing for High-temperature Service, *Chem. Eng.*, **83**(12):145 (1976).

MacNab, A. J., Design and Materials Requirements for Coal Gasification, *Chem. Eng. Progr.*, **71**(11):51 (1975).

Mallinson, J. H., Plastics, *Chem. Eng.*, **79**(27):63 (1972).

Margus, E., Polyvinylidene Fluoride for Corrosion-Resistant Pumps, *Chem. Eng.*, **82**(23):133 (1975).

———, Plastic Centrifugal Pumps for Corrosive Service, *Chem. Eng.*, **84**(5):213 (1977).

Masek, J. A., Metallic Piping, *Chem. Eng.*, **75**(13):215 (1968).

McCoy, J. D., and F. B. Hamel, New Corrosion Rate Data for Hydrodesulfurizing Units, *Hydrocarbon Process.*, **49**(6):116 (1970).

McDowell, D. W., Jr., Specifications for Acidproof Brick, *Chem. Eng.*, **81**(12):100 (1974).

———, Handling Mineral Acids, *Chem. Eng.*, **81**(24):118 (1974).

———, Sulfuric Acid Plants—Materials of Construction, *Chem. Eng. Progr.*, **71**(3):69 (1975).

———, Handling Phosphoric Acids and Phosphate Fertilizers, *Chem. Eng.*, **82**(18):121 (1975).

———, Choosing Materials for Ethylbenzene Services, *Chem. Eng.*, **85**(2):159 (1978).

Michels, H. T., and E. C. Hoxie, How to Rate Alloys for SO_2 Scrubbers, *Chem. Eng.*, **85**(13):161 (1978).

Miller, R., Jr., Matching Materials to Temperatures, *Chem. Eng.*, **75**(10):210 (1968).

———, Materials for Making Formaldehyde, *Chem. Eng.*, **75**(2):182 (1968).

Mills, J. F., and B. D. Oakes, Bromine Chloride: Less Corrosive Than Bromine, *Chem. Eng.*, **80**(18):102 (1973).

Mistrot, D. J., Materials—Construction Problems Challenge HPI, *Petro/Chem. Eng.*, **39**(12):19 (1967).

Morgan, J. D., Jr., Supply and Demand of Corrosion-Resistant Materials, *Chem. Eng. Progr.*, **74**(3):25 (1978).

Morton, T. R., Fiber-Glass-Reinforced Plastics for Corrosion Resistance, *Chem. Eng.*, **80**(29):70 (1973).

———, Fiber-Glass-Reinforced Plastics for Corrosion Control, *Chem. Eng.*, **81**(2):140 (1974).

Osborn, O., C. F. Schrieber, and H. G. Smith, Metals Performance in Desalination Plants, *Chem. Eng. Progr.*, **66**(7):74 (1970).

O'Hara, J. B., W. J. Lochmann, and N. E. Jentz, Materials Challenges of Coal Liquefaction, *Chem. Eng.*, **84**(8):147 (1977).

Payne, B. S., Jr., Nucerite—a New Composite, *Chem. Eng. Progr.*, **64**(2):40 (1968).

Pfoutz, B. D., and L. L. Stewart, Electrostatic Precipitators—Materials of Construction, *Chem. Eng. Progr.*, **71**(3):53 (1975).

Pierce, R. R., Protecting Concrete Floors from Chemicals, *Chem. Eng.*, **75**(27):118 (1968).

——— and V. Bressi, Linings for the Process Industry, *Chem. Eng. Progr.*, **69**(6):104 (1973).

Pike, J. J., Choosing Materials for Phos-Acid Concentration, *Chem. Eng.*, **74**(6):192 (1967).

Pitcher, J. H., Stainless Steels: CPI Workhorses, *Chem. Eng.*, **83**(24):119 (1976).

———, Trends in CPI Materials—Steels, *Chem. Eng.*, **79**(27):39 (1972).

Puckett, D. B., What You Should Know about Fiberglass-reinforced Plastic Tanks, *Chem. Eng.*, **83**(20):129 (1976).

Richards, J. F., New Data on Carbon-$\frac{1}{2}$ Mo Heavy Forgings, *Hydrocarbon Process.*, **47**(11):203 (1968).

Rinckhoff, J. B., Making Acid in Iron and Steel Equipment, *Chem. Eng.*, **74**(18):160 (1967).

Roberts, R., Designing for Fluoroplastic Linings, *Chem. Eng.*, **78**(5):138 (1971).

Robertson, A. B., and W. A. Miller, Fluorinated Polymers for Coating CPI Equipment, *Chem. Eng.*, **81**(20):138 (1974).

Robertson, C. A., Materials Engineering of Hydrotreating Equipment, *Chem. Eng. Progr.*, **66**(9):33 (1970).

———, Materials Selection for Pumps, *Hydrocarbon Process.*, **51**(6):93 (1972).

Ross, A. P., Rediscovering Lead, *Chem. Eng.*, **82**(25):79 (1975); **83**(16):107 (1976); **83**(24):175 (1976).

Rothman, R. A., Gaskets and Packings, *Chem. Eng.*, **80**(5):69 (1973).

Rozic, E. J., Jr., Which Type 304 is Right for You? *Petro/Chem. Eng.*, **41**(4):20 (1969).

Sakol, S. L., and R. A. Schwartz, Construction Materials for Wet Scrubbers, *Chem. Eng. Progr.*, **70**(8):63 (1974).

Samans, C. H., Making the Most of Contemporary Steels, *Chem. Eng.*, **75**(4):150 (1968).

Sayers, J. A., Brittle Materials, *Chem. Eng.*, **79**(27):51 (1972).

Schley, J. R., Impervious Graphite for Process Equipment, *Chem. Eng.*, **81**(4):144 (1974); **81**(6):102 (1974).

Schueler, R. C., Metal Dusting, *Hydrocarbon Process.*, **51**(8):73 (1972).

Schumacher, W. J., Wear and Galling Can Knock Out Equipment, *Chem. Eng.*, **84**(10):155 (1977).

Schwartz, C. D., I. A. Franson, and R. J. Hodges, Inventing a New Ferritic Stainless Steel, *Chem. Eng.*, **77**(8):164 (1970).

Seymour, R. B., Plastics Technology, *I&EC*, **60**(8):30 (1968).

Shadduck, A. K., Designing for Reinforced Plastics, *Chem. Eng.*, **78**(18):116 (1971).

Shannon, J., Polybutylene—A New Thermoplastic for Industrial Piping, *Chem. Eng.*, **83**(18):121 (1976).

Sheets, H. D., Jr., M. J. O'Hara, and M. J. Snyder, Refractory Ceramics for the CPI, *Chem. Eng. Progr.*, **64**(2):35 (1968).

Sheppard, W. L., Jr., Membranes Behind Brick, *Chem. Eng.*, **79**(11):122 (1972); **79**(13):110 (1972).

Skavdahl, R. E., and E. L. Zebroski, Finding Materials for Fast Reactors of the Future, *Chem. Eng.*, **76**(17):114 (1969).

Sorell, G., Controlling Corrosion by Process Design, *Chem. Eng.*, **75**(16):162 (1968).

Stindt, W. H., Pump Selection, *Chem. Eng.*, **78**(23):43 (1971).

Szymanski, W. A., and D. W. Kloda, Polyester and Furfuryl Alcohol Resins for Corrosion Control, *Chem. Eng. Progr.*, **70**(1):51 (1974).

Tesmen, A. B., Materials of Construction for Process Plants, *Chem. Eng.*, **80**(4):140 (1973); **80**(7):126 (1973); **80**(11):158 (1973).

Thomas, B., Designing Brick Linings to Resist Hot Chemicals, *Chem. Eng.*, **76**(26):110 (1969).

Tator, K. B., Protective Coatings, *Chem. Eng.*, **79**(27):75 (1972).

Thomas, P. D., Embrittlement, *Hydrocarbon Process.*, **51**(8):65 (1972).

Thornton, D. P., Jr., Corrosion-Free HF Alkylation, *Chem. Eng.*, **77**(15):108 (1970).

Traeger, R. K., Compression Testing of Rigid Polyurethane Foams, *Chem. Eng. Progr.*, **64**(2):56 (1968).

Vandelinder, L. S., Plastics, *Chem. Eng.*, **77**(22):119 (1970).

Verink, E. D., Jr., Aluminum Alloys for Saline Waters, *Chem. Eng.*, **81**(8):104 (1974).

Von Wiesenthal, P., and H. W. Cooper, Guide to Economics of Fired Heater Design, *Chem. Eng.*, **77**(7):104 (1970).

Wagner, J., Jr., Sea Water for Industry Creates Special Problems, *Chem. Eng.*, **64**(10):59 (1968).

Walko, J. F., Controlling Biological Fouling in Cooling Systems, *Chem. Eng.*, **79**(26):104 (1972).

Wanderer, E. T., Aluminum, *I&EC*, **60**(8):63 (1968).

Ward, J. R., Lined-Pipe Systems, *Chem. Eng.*, **75**(13):238 (1968).

Wenschhof, D. E., Developments in Wrought Alloys, *Hydrocarbon Process.*, **55**(6):81 (1976).

Wismer, M., and J. G. Bosso, Make the Part the Cathode: Key to Resistant Coatings, *Chem. Eng.*, **78**(13):114 (1971).

Wright, C. E., Nonmetallic Pipe: Promise and Problems, *Chem. Eng.*, **78**(13):230 (1968).

Wright, L. E., Piping Codes and Standards, *Chem. Eng.*, **75**(13):247 (1968).

Yamartino, J., Installed Cost of Corrosion-resistant Piping—1978, *Chem. Eng.*, **85**(26):138 (1978).

Zeis, L. A., and A. Heinz, Catalyst Tubes in Primary Reformer Furnaces, *Chem. Eng.*, **66**(7):68 (1970).

Graphite Bids Against Stainless, *Chem. Eng.*, **77**(4):126 (1970).

New Stainless Needs No Nickel, *Chem. Eng.*, **77**(26):92 (1970).

Making Sense from Alloy Compositions, *Chem. Eng.*, **78**(25):92 (1971).

What You Should Know About Glass Reinforced Epoxy Tubing, *Hydraulics & Pneumatics*, **21**(10):142 (1968).

Inventory of Current Literature, *Chem. Eng.*, **81**(24):136 (1974).

Materials Engineering Forum, Setting a Value on Creep Strength, *Chem. Eng.*, **76**(11):208 (1969).
New Materials Data for High-Pressure-Design, *Chem. Eng.*, **75**(12):122 (1968).
Process Piping, *Chem. Eng.*, **76**(8):95 (1969).
Using Large-Diameter Polyethylene Pipe, *Chem. Eng.*, **80**(25):132 (1973).
Which Steel for Hydrodealkylation Units? *Hydrocarbon Process.*, **49**(6):131 (1969).

Painting and coating

Berger, D. M., Choosing and Applying Paint, *Chem. Eng.*, **81**(25):112 (1974).
———, Detecting Film Flaws in Coatings, *Chem. Eng.*, **82**(6):79 (1975).
Byrd, J. D., Well-Planned Refinery Painting Program Can Help Cut Cost, *Oil & Gas J.*, **75**(19):85 (1977).
Falck, S. B., Process Tank Linings, *Chem. Eng.*, **79**(27):69 (1972).
Guthrie, K., Costs, *Chem. Eng.*, **76**(8):201 (1969).
Long, G. E., Spraying Theory and Practice, *Chem. Eng.*, **85**(6):73 (1978).
Norden, R. B., Maintenance Painting, *Chem. Eng.*, **80**(5):85 (1973).
Shaw, F. G., Paint and Protective Coatings, *Chem. Eng.* (*London*), 99 (February 1976).
Sisson, B., Calculating the Paint Needed for Tanks, *Chem. Eng.*, **83**(4):122 (1976).
Sommerfield, E. M., Power Plant Painting, *Power Eng.*, **71**(10):38 (1967).
Stark, L., and B. Verzello, Industrial Coatings in the Capital Cost Budget, *1970 Trans. Am. Assoc. Cost. Eng.* (1970), p. 28.
Tator, K. B., Protective Coatings, *Chem. Eng.*, **77**(22):137 (1970).
———, Engineered Painting Pays Off, *Chem. Eng.*, **78**(29):84 (1971).
———, Protective Coatings, *Chem. Eng.*, **79**(27):75 (1972).
Thorsen, T., and R. Oen, How to Measure Industrial Wastewater Flow, *Chem. Eng.*, **82**(4):95 (1975).
Weaver, P. E., How to Specify Coatings, *Hydrocarbon Process.*, **49**(2):127 (1970).
Wood, W., Painting Costs, *AACE Bull.*, **12**(5):138 (1970).
Protecting Maintenance Coating, *I&EC*, **59**(9):70 (1967).

PROBLEMS

1 A new plant requires a large rotary vacuum filter for the filtration of zinc sulfite from a slurry containing 1 kg of zinc sulfite solid per 20 kg of liquid. The liquid contains water, sodium sulfite, and sodium bisulfite. The filter must handle 8000 kg of slurry per hour. What additional information is necessary to design the rotary vacuum filter? How much of this information could be obtained from laboratory or pilot-plant tests? Outline the method for converting the test results to the conditions applicable in the final design.

2 For each of the following materials of construction, prepare an approximate plot of temperature versus concentration in water for sulfuric acid and for nitric acid, showing conditions of generally acceptable corrosion resistance:

   (*a*) Stainless steel type 302.
   (*b*) Stainless steel type 316.
   (*c*) Karbate.
   (*d*) Haveg.
[See *Chem. Eng.*, **69**(23):185 (1962).]

3 A process for sulfonation of phenol requires the use of a 3000-gal kettle. It is desired to determine the most suitable material of construction for this vessel. The time value of money is to be taken into account by use of an interest rate of 10 percent.

   The life of the kettle is calculated by dividing the corrosion allowance of $\frac{1}{8}$ in. by the estimated corrosion rate. The equipment is assumed to have a salvage value of 10 percent of its original cost at the end of its useful life.

For the case in question, corrosion data indicate that only a few corrosion-resistant alloys will be suitable:

Vessel Type	Installed cost	Average corrosion rate, in./yr
Nickel clad	$80,000	0.020
Monel clad	$95,000	0.010
Hastelloy B	$180,000	0.0045

Determine which material of construction should be used with appropriate justification for the selection.

4 Synthesis gas may be prepared by a continuous, noncatalytic conversion of any hydrocarbon by means of controlled partial combustion in a fire-brick lined reactor. Such a process has been developed by Shell Development Company and is described in *Hydrocarbon Process. Petrol Refiner*, **39**(3):151 (1960). A flow sheet, material and energy balance, and equipment size for this process have been considered earlier in Probs. 13 through 18 of Chap. 2. With this information, specify the materials of construction which should be considered for each piece of equipment.

5 What materials of construction should be specified for the thiophane process described in Prob. 20 of Chap. 2? Note the extremes of temperatures and corrosion which are encountered in this process because of the regeneration step and the presence of H_2S and caustic.

6 What materials of construction would be considered for the various water desalination plants now being operated in the United States? [See *Chem. Eng.*, **70**(20):124 (1963) and **70**(21):224 (1963).]

7 A manhole plate for a reactor is to be 2 in. thick and 18 in. in diameter. It has been proposed that the entire plate be made of stainless steel type 316. The plate will have 18 bolt holes, and part of the face will need to be machined for close gasket contact. If the base price for stainless steel type 316 in the form of industrial plates is $2.00 per pound, estimate the purchased cost for the manhole plate.

8 Six tanks of different constructional materials and six different materials to be stored in these tanks are listed in the following columns:

Tanks	*Materials*
Brass-lined	20% hydrochloric acid
Carbon steel	10% caustic soda
Concrete	75% phosphoric acid for
Nickel-lined	food products
Stainless steel type 316	98% sulfuric acid
Wood	Vinegar
	Water

All tanks must be used, and all materials must be stored without using more than one tank for any one material. Indicate the material that should be stored in each tank.

9 For the design of internal-pressure cylindrical vessels, the API-ASME Code for Unified Pressure Vessels recommends the following equations for determining the minimum wall thickness when extreme operating pressures are not involved:

$$t = \frac{PD_m}{2SE} + C \qquad \text{applies when} \qquad \frac{D_o}{D_i} < 1.2$$

or

$$t = \frac{D_i}{2}\left(\sqrt{\frac{SE + P}{SE - P}} - 1\right) + C \qquad \text{applies when} \qquad \frac{D_o}{D_i} > 1.2$$

where t = wall thickness, in.

    P = internal pressure, psig (this assumes atmospheric pressure surrounding the vessel)

    D_m = mean diameter, in.

    D_i = ID, in.

    D_o = OD, in.

    E = fractional efficiency of welded or other joints

    C = allowances, in., for corrosion, threading, and machining

    S = design stress, lb/in.^{2} (for the purpose of this problem, S may be taken as one-fourth of the ultimate tensile strength)

A cylindrical storage tank is to have an ID of 12 ft and a length of 36 ft. The seams will be welded, and the material of construction will be plain carbon steel (0.15 percent C). The maximum working pressure in the tank will be 100 psig, and the maximum temperature will be 25°C. No corrosion problems are anticipated. On the basis of the preceding equations, estimate the necessary wall thickness.

10 A proposal has been made to use stainless-steel tubing as part of the heat-transfer system in a nuclear reactor. High temperatures and extremely high rates of heat transfer will be involved. Under these conditions, temperature stresses across the tube walls will be high, and the design engineer must choose a safe wall thickness and tube diameter for the proposed unit. List in detail all information and data necessary to determine if a proposed tube diameter and gauge number would be satisfactory.

TWELVE

THE DESIGN REPORT

A successful engineer must be able to apply theoretical and practical principles in the development of ideas and methods and also have the ability to express the results clearly and convincingly. During the course of a design project, the engineer must prepare many written reports which explain what has been done and present conclusions and recommendations. The decision on the advisability of continuing the project may be made on the basis of the material presented in the reports. The value of the engineer's work is measured to a large extent by the results given in the written reports covering the study and the manner in which these results are presented.

The essential purpose of any report is to pass on information to others. A good report writer never forgets the words "to others." The abilities, the functions, and the needs of the reader should be kept in mind constantly during the preparation of any type of report. Here are some questions the writer should ask before starting, while writing, and after finishing a report:

What is the purpose of this report?
Who will read it?
Why will they read it?
What is their function?
What technical level will they understand?
What background information do they have now?

The answers to these questions indicate the type of information that should be presented, the amount of detail required, and the most satisfactory method of presentation.

Types of Reports

Reports can be designated as *formal* and *informal*. Formal reports are often encountered as research, development, or design reports. They present the results in considerable detail, and the writer is allowed much leeway in choosing the type of presentation. Informal reports include memorandums, letters, progress notes, survey-type results, and similar items in which the major purpose is to present a result without including detailed information. Stereotyped forms are often used for informal reports, such as those for sales, production, calculations, progress, analyses, or summarizing economic evaluations. Figures 12-1 through 12-3 present examples of stereotyped forms that can be used for presenting the summarized results of economic evaluations.

Although many general rules can be applied to the preparation of reports, it should be realized that each industrial concern has its own specifications and regulations. A stereotyped form shows exactly what information is wanted, and detailed instructions are often given for preparing other types of informal reports. Many companies have standard outlines that must be followed for formal

---

ESTIMATED MANUFACTURING-COST STATEMENT

Product_____

Basis: Capacity_____   Operating rate_____

      M & S index_____   Labor rate_____

Raw materials .... _____

Operating labor .... _____

Operating supervision .... _____

Maintenance and repairs .... _____

Operating supplies .... _____

Power and utilities .... _____

Royalties .... _____

    Direct-production cost .... _____

Depreciation .... _____

Rent .... _____

Taxes (property).. _____

Insurance .... _____

    Fixed charges .... _____

Safety and protection .... _____

Payroll overhead .... _____

Packaging.. _____

Salvage.. _____

Control laboratories .... _____

Plant superintendence .... _____

General plant overhead .... _____

    Plant-overhead cost .... _____

    Factory-manufacturing costs .... _____

  By_____   Date_____

Figure 12-1 Example of form for an informal summarizing report on factory manufacturing cost.

ESTIMATED CAPITAL-INVESTMENT STATEMENT

Product_____

Basis: Capacity_____ M & S index_____

Purchased equipment (delivered) .... _____
Installation of equipment .... _____
   Insulation .... _____
   Instrumentation... _____
Piping .... _____
Electrical installations .... _____
Buildings including services .... _____
Yard improvements .... _____
Service facilities .... _____
Land .... _____
    Total physical cost .... _____
Engineering and construction .... _____
    Direct plant cost .... _____
Contractor's fee ..... _____
Contingency .... _____
    Fixed-capital investment .... _____
Raw-materials inventory...................................... _____
Product and in-process inventory .... _____
Accounts receivable .... _____
Cash.. _____
    Working capital .... _____
    Total capital investment....................... _____

  By_____ Date_____

Figure 12-2 Example of form for an informal summarizing report on capital investment.

ESTIMATED INCOME AND RETURN STATEMENT

Product_____ Sales price_____

Basis: Capacity_____ Operating rate_____

    M & S index_____ Labor rate_____

Direct production costs.. _____
Fixed charges... _____
Plant overhead costs .... _____
   Factory-manufacturing costs .... _____
Administrative costs .... _____
Distribution and selling costs .... _____
Research and development .... _____
Financing .... _____
    General expenses .... _____
    Total product cost .... _____
    Total income... _____

Fixed-capital investment_____ Working capital_____

Total-capital investment_____ Probable accuracy of estimate_____

Gross earnings before taxes_____ Net profit after __% taxes _____

Annual return on capital before taxes_____%

Annual return on capital after __% taxes _____%

  By_____ Date_____

Figure 12-3 Example of form for an informal summarizing report on income and return.

reports. For convenience, certain arbitrary rules of rhetoric and form may be established by a particular concern. For example, periods may be required after all abbreviations, titles of articles may be required for all references, or the use of a set system of units or nomenclature may be specified.

ORGANIZATION OF REPORTS

The organization of a formal report requires careful sectioning and the use of subheadings in order to maintain a clear and effective presentation.† To a lesser degree, the same type of sectioning is valuable for informal reports. The following discussion applies to formal reports, but, by deleting or combining appropriate sections, the same principles can be applied to the organization of any type of report.

A complete design report consists of several essentially independent parts, with each succeeding part giving greater detail on the design and its development. A covering *Letter of Transmittal* is usually the first item in any report. After this come the *Title Page*, the *Table of Contents*, and an *Abstract* or *Summary* of the report. The *Body* of the report is next, and it includes the essential information, presented in the form of discussion, graphs, tables, and figures. The *Appendix*, at the end of the report, gives detailed information which permits complete verification of the results shown in the body. Tables of data, sample calculations, and other supplementary material are included in the Appendix. A typical outline for a design report is as follows:

ORGANIZATION OF A DESIGN REPORT

   I. Letter of transmittal
     A. Indicates why report has been prepared
     B. Gives essential results that have been *specifically requested*
  II. Title page
     A. Includes title of report, name of person to whom report is submitted, writer's name and organization, and date
 III. Table of contents
     A. Indicates location and title of figures, tables, and all major sections
 IV. Summary
     A. Briefly presents essential results and conclusions in a clear and precise manner

---

† T. A. Sherman, "Modern Technical Writing," 2d ed., Prentice-Hall, Inc., Englewood Cliffs, N.J., 1966; R. P. Turner, "Technical Report Writing," Rinehart Press, San Francisco, 1971; "Effective Communication for Engineers" (Collection of articles from *Chem. Eng.*), McGraw-Hill Book Company, New York, 1974; J. W. Souther and M. L. White, "Technical Report Writing," 2d ed., John Wiley & Sons, New York, 1977.

V. Body of report
   A. Introduction
      1. Presents a brief discussion to explain what the report is about and the reason for the report; no results are included
   B. Previous work
      1. Discusses important results obtained from literature surveys and other previous work
   C. Discussion
      1. Outlines method of attack on project and gives design basis
      2. Includes graphs, tables, and figures that are essential for understanding the discussion
      3. Discusses technical matters of importance
      4. Indicates assumptions made and the reasons
      5. Indicates possible sources of error
      6. Gives a general discussion of results and proposed design
   D. Final recommended design with appropriate data
      1. Drawings of proposed design
         a. Qualitative flow sheets
         b. Quantitative flow sheets
         c. Combined-detail flow sheets
      2. Tables listing equipment and specifications
      3. Tables giving material and energy balances
      4. Process economics including costs, profits, and return on investment
   E. Conclusions and recommendations
      1. Presented in more detail than in Summary
   E. Acknowledgment
      1. Acknowledges important assistance of others who are not listed as preparing the report
   G. Table of nomenclature
      1. Sample units should be shown
   H. References to literature (bibliography)
      1. Gives complete identification of literature sources referred to in the report
VI. Appendix
   A. Sample calculations
      1. One example should be presented and explained clearly for each type of calculation
   B. Derivation of equations essential to understanding the report but not presented in detail in the main body of the report
   C. Tables of data employed with reference to source
   D. Results of laboratory tests
      1. If laboratory tests were used to obtain design data, the experimental data, apparatus and procedure description, and interpretation of the results may be included as a special appendix to the design report.

Letter of Transmittal

The purpose of a letter of transmittal is to refer to the original instructions or developments that have made the report necessary. The letter should be brief, but it can call the reader's attention to certain pertinent sections of the report or give definite results which are particularly important. The writer should express any personal opinions in the letter of transmittal rather than in the report itself. Personal pronouns and an informal business style of writing may be used.

Title Page and Table of Contents

In addition to the title of the report, a title page usually indicates other basic information, such as the name and organization of the person (or persons) submitting the report and the date of submittal. A table of contents may not be necessary for a short report of only six or eight pages, but, for longer reports, it is a convenient guide for the reader and indicates the scope of the report. The titles and subheadings in the written text should be shown, as well as the appropriate page numbers. Indentations can be used to indicate the relationships of the various subheadings. A list of tables, figures, and graphs should be presented separately at the end of the table of contents.

Summary

The summary is probably the most important part of a report, since it is referred to most frequently and is often the only part of the report that is read. Its purpose is to give the reader the entire contents of the report in one or two pages. It covers all phases of the design project, but it does not go into detail on any particular phase. All statements must be concise and give a minimum of general qualitative information. The aim of the summary is to present precise quantitative information and final conclusions with no unnecessary details.

The following outline shows what should be included in a summary:

1. A statement introducing the reader to the subject matter
2. What was done and what the report covers
3. How the final results were obtained
4. The important results including quantitative information, major conclusions, and recommendations

An ideal summary can be completed on one typewritten page. If the summary must be longer than two pages, it may be advisable to precede the summary by an *abstract*, which merely indicates the subject matter, what was done, and a brief statement of the major results.

Body of the Report

The first section in the body of the report is the *introduction*. It states the purpose and scope of the report and indicates why the design project originally appeared to be feasible or necessary. The relationship of the information presented in the report to other phases of the company's operations can be covered, and the effects of future developments may be worthy of mention. References to *previous work* can be discussed in the introduction, or a separate section can be presented dealing with literature-survey results and other previous work.

A description of the methods used for developing the proposed design is presented in the next section under the heading of *discussion*. Here the writer shows the reader the methods used in reaching the final conclusions. The validity of the methods must be made apparent, but the writer should not present an annoying or distracting amount of detail. Any assumptions or limitations on the results should be discussed in this section.

The next section presents the *recommended design*, complete with figures and tables giving all necessary qualitative and quantitative data. An analysis of the cost and profit potential of the proposed process should accompany the description of the recommended design.

The body of a design report often includes a section giving a detailed discussion of all *conclusions and recommendations*. When applicable, sections covering *acknowledgment, table of nomenclature,* and *literature references* may be added.

Appendix

In order to make the written part of a report more readable, the details of calculation methods, experimental data, reference data, certain types of derivations, and similar items are often included as separate appendixes to the report. This information is thus available to anyone who wishes to make a complete check on the work; yet the descriptive part of the report is not made ineffective because of excess information.

PREPARING THE REPORT

The physical process of preparing a report can be divided into the following steps:

1. Define the subject matter, scope, and intended audience
2. Prepare a skeleton outline and then a detailed outline
3. Write the first draft
4. Polish and improve the first draft and prepare the final form
5. Check the written draft carefully, have the report typed, and proofread the final report

In order to accomplish each of these steps successfully, the writer must make certain the initial work on the report is started soon enough to allow a thorough job and still meet any predetermined deadline date. Many of the figures, graphs, and tables, as well as some sections of the report, can be prepared while the design work is in progress.

PRESENTING THE RESULTS

Accuracy and logic must be maintained throughout any report. The writer has a moral responsibility to present the facts accurately and not mislead the reader with incorrect or dubious statements. If approximations or assumptions are made, their effect on the accuracy of the results should be indicated. For example, a preliminary plant design might show that the total investment for a proposed plant is $5,500,000. This is not necessarily misleading as to the accuracy of the result, since only two significant figures are indicated. On the other hand, a proposed investment of $5,554,328 is ridiculous, and the reader knows at once that the writer did not use any type of logical reasoning in determining the accuracy of the results.

The style of writing in technical reports should be simple and straightforward. Although short sentences are preferred, variation in the sentence length is necessary in order to avoid a disjointed staccato effect. The presentation must be convincing, but it must also be devoid of distracting and unnecessary details. Flowery expressions and technical jargon are often misused by technical writers in an attempt to make their writing more interesting. Certainly, an elegant or forceful style is sometimes desirable, but the technical writer must never forget that the major purpose is to present information clearly and understandably.

Subheadings and Paragraphs

The use of effective and well-placed subheadings can improve the readability of a report. The sections and subheadings follow the logical sequence of the report outline and permit the reader to become oriented and prepared for a new subject.

Paragraphs are used to cover one general thought. A paragraph break, however, is not nearly as definite as a subheading. The length of paragraphs can vary over a wide range, but any thought worthy of a separate paragraph should require at least two sentences. Long paragraphs are a strain on the reader, and the writer who consistently uses paragraphs longer than 10 to 12 typed lines will have difficulty in holding the readers' attention.

Tables

The effective use of tables can save many words, especially if quantitative results are involved. Tables are included in the body of the report only if they are essential to the understanding of the written text. Any type of tabulated data that is not directly related to the discussion should be located in the appendix.

Every table requires a title, and the headings for each column should be self-explanatory. If numbers are used, the correct units must be shown in the column heading or with the first number in the column. A table should never be presented on two pages unless the amount of data makes a break absolutely necessary.

Graphs

In comparison with tables, which present definite numerical values, graphs serve to show trends or comparisons. The interpretation of results is often simplified for the reader if the tabulated information is presented in graphical form.

If possible, the experimental or calculated points on which a curve is based should be shown on the plot. These points can be represented by large dots, small circles, squares, triangles, or some other identifying symbol. The most probable smooth curve can be drawn on the basis of the plotted points, or a broken line connecting each point may be more appropriate. In any case, the curve should not extend through the open symbols representing the data points. If extrapolation or interpolation of the curve is doubtful, the uncertain region can be designated by a dotted or dashed line.

The ordinate and the abscissa must be labeled clearly, and any nomenclature used should be defined on the graph or in the body of the report. If numerical values are presented, the appropriate units are shown immediately after the labels on the ordinate and abscissa. Restrictions on the plotted information should be indicated on the graph itself or with the title.

The title of the graph must be explicit but not obvious. For example, a log-log plot of temperature versus the vapor pressure of pure glycerol should not be entitled "Log-Log Plot of Temperature versus Vapor Pressure for Pure Glycerol." A much better title, although still somewhat obvious, would be "Effect of Temperature on Vapor Pressure of Pure Glycerol."

Some additional suggestions for the preparation of graphs follow:

1. The independent or controlled variable should be plotted as the abscissa, and the variable that is being determined should be plotted as the ordinate.
2. Permit sufficient space between grid elements to prevent a cluttered appearance (ordinarily, two to four grid lines per inch are adequate).
3. Use coordinate scales that give good proportionment of the curve over the entire plot, but do not distort the apparent accuracy of the results.
4. The values assigned to the grids should permit easy and convenient interpolation.
5. If possible, the label on the vertical axis should be placed in a horizontal position to permit easier reading.
6. Unless families of curves are involved, it is advisable to limit the number of curves on any one plot to three or less.
7. The curve should be drawn as the heaviest line on the plot, and the coordinate axes should be heavier than the grid lines.

Illustrations

Flow diagrams, photographs, line drawings of equipment, and other types of illustrations may be a necessary part of a report. They can be inserted in the body of the text or included in the appendix. Complete flow diagrams, drawn on oversize paper, and other large drawings are often folded and inserted in an envelope at the end of the report.

References to Literature

The original sources of any literature referred to in the report should be listed at the end of the body of the report. References are usually tabulated and numbered in alphabetical order on the basis of the first author's surname, although the listing is occasionally based on the order of appearance in the report.

When a literature reference is cited in the written text, the last name of the author is mentioned and the bibliographical identification is shown by a raised number after the author's name or at the end of the sentence. An underlined number in parentheses may be used in place of the raised number, if desired.

The bibliography should give the following information:

1. For journal articles: (*a*) authors' names, followed by initials, (*b*) journal, abbreviated to conform to the "List of Periodicals" as established by *Chemical Abstracts*, (*c*) volume number, (*d*) issue number, if necessary, (*e*) page numbers, and (*f*) year (in parentheses). The title of the article is usually omitted. Issue number is omitted if paging is on a yearly basis. The date is sometimes included with the year in place of the issue number.

   ```
   Kelly, W. J., Chem. Eng., 85(27):133-137 (1978).
   Kelly, W. J., Chem. Eng., 85:133-137 (Dec. 4, 1978).
   Gregg, D. W., and T. F. Edgar, AIChE J., 24:753-781
       (1978).
   ```

2. For single publications, as books, theses, or pamphlets: (*a*) author's names, followed by initials, (*b*) title (in quotation marks), (*c*) edition (if more than one has appeared), (*d*) volume (if there is more than one), (*e*) publisher, (*f*) place of publication, and (*g*) year of publication. The chapter or page number is often listed just before the publisher's name. Titles of theses are often omitted.

   ```
   Mandel, S., "Proposal and Inquiry Writing," p. 161,
       The Macmillan Company, New York, 1962.
   Fiszbin, M., M.S. Thesis in Chem. Eng., Univ. of
       Colorado, Boulder, Colo., 1978.
   ```

3. For unknown or unnamed authors: (*a*) alphabetize by the journal or organization publishing the information.

   ```
   Chem. Eng., 85(27):81 (1978).
   ```

4. For patents: (*a*) patentees' names, followed by initials, and assignee (if any) in parentheses, (*b*) country granting patent and number, and (*c*) date issued (in parentheses).

   ```
   Fenske, E. R. (to Universal Oil Products Co.),
       U.S. Patent 3,249,650 (May 3, 1966).
   ```

5. For unpublished information: (*a*) "in press" means formally accepted for publication by the indicated journal or publisher; (*b*) the use of "private communication" and "unpublished data" is not recommended unless absolutely necessary, because the reader may find it impossible to locate the original material.

```
Smith, R. E., Chem. Eng. Progr., in press (1978).
```

Sample Calculations

The general method used in developing the proposed design is discussed in the body of the report, but detailed calculation methods are not presented in this section. Instead, sample calculations are given in the appendix. One example should be shown for each type of calculation, and sufficient detail must be included to permit the reader to follow each step. The particular conditions chosen for the sample calculations must be designated. The data on which the calculations are based should be listed in detail at the beginning of the section, even though these same data may be available through reference to one of the tables presented with the report.

Mechanical Details

The final report should be submitted in a neat and businesslike form. Formal reports are usually bound with a heavy cover, and the information shown in the title page is repeated on the cover. If paper fasteners are used for binding in a folder, the pages should be attached only to the back cover.

The report should be typed on a good grade of paper with a margin of at least 1 in. on all sides. Normally, only one side of the page is used and all material, except the letter of transmittal, footnotes, and long quotations, is double-spaced. Starting with the summary, all pages including graphs, illustrations, and tables should be numbered in sequence.

Written material on graphs and illustrations may be typed or lettered neatly in ink. If hand lettering is required, best results are obtained with an instrument such as a LeRoy or Wrico guide.

Short equations can sometimes be included directly in the written text if the equation is not numbered. In general, however, equations are centered on the page and given a separate line, with the equation number appearing at the right-hand margin of the page. Explanation of the symbols used can be presented immediately following the equation.

Proofreading and Checking

Before final submittal, the completed report should be read carefully and checked for typographical errors, consistency of data quoted in the text with those presented in tables and graphs, grammatical errors, spelling errors, and similar obvious mistakes. If excessive corrections or changes are necessary, the appearance of the report must be considered and some sections may need to be retyped.

Nomenclature

If many different symbols are used repeatedly throughout a report, a table of nomenclature, showing the symbols, meanings, and sample units, should be included in the report. Each symbol can be defined when it first appears in the written text. If this is not done, a reference to the table of nomenclature should be given with the first equation.

Ordinarily, the same symbol is used for a given physical quantity regardless of its units. Subscripts, superscripts, and lower- and upper-case letters can be employed to give special meanings. The nomenclature should be consistent with common usage (a list of recommended symbols for chemical engineering quantities is presented in Table 1).†

Abbreviations

Time and space can be saved by the use of abbreviations, but the writer must be certain the reader knows what is meant. Unless the abbreviation is standard, the meaning should be explained the first time it is used. The following rules are generally applicable for the use of abbreviations:

1. Abbreviations are acceptable in tables, graphs, and illustrations when space limitations make them desirable
2. Abbreviations are normally acceptable in the text only when preceded by a number [3 cm/s (three centimeters per second)]
3. Periods may be omitted after abbreviations for common scientific and engineering terms, except when the abbreviation forms another word (e.g., in. for inch)
4. The plural of an abbreviation is the same as the singular (pounds—lb) (kilograms—kg)
5. The abbreviation for a noun derived from a verb is formed by adding n (concentration—concn)
6. The abbreviation for the past tense is formed by adding d (concentrated—concd)
7. The abbreviation for the participle is formed by adding g (concentrating—concg)

---

† The nomenclature presented in Table 1 is consistent with the recommendations of the American Standards Association presented as American Standard ASA Y10.12-1955. Note the use of periods after abbreviations in Table 1 in contrast to the common practice followed in this text of eliminating most periods after abbreviations. See also Appendix A for rules and recommendations relative to the use of the SI system of units.

Table 1 Letter symbols for chemical engineering

Listing is alphabetical by concept within each category. Illustrative units or definitions are supplied where appropriate†

1. General concepts

Concept	Symbol	Unit of definition (U.S. customary system)	Concept	Symbol	Unit of definition (U.S. customary system)
Acceleration	a	(ft/s)/s	Moment of inertia	I	(ft)^4
Of gravity	g	(ft/s)/s	Newton law of motion, conversion factor in	g_c	$g_c = ma/F$, (lb) (ft)/(s)^2 (lbf)
Base of natural logarithms	e		Number		
Coefficient	C		In general	N	
Difference, finite	Δ		Of moles	n	
Differential operator	d		Pressure	p	lbf/ft^2; atm
Partial	∂		Quantity, in general	Q	
Efficiency	η		Ratio, in general	R	
Energy, dimension of	E	Btu (ft) (lbf)	Resistance	R	
Enthalpy	H	Btu	Shear stress	τ	lbf/ft^2
Entropy	S	Btu/°R	Temperature		
Force	F	lbf	Dimension of	θ	
Function	ϕ, ψ, χ		Absolute	T	K (Kelvin); °R (Rankine)
Gas constant, universal	R	To distinguish, use R_0	In general	T, t	°C; °F
Gibbs free energy	G, F	$G = H - TS$, Btu	Temperature difference, logarithmic mean	$\bar{\theta}$	°F
Heat	Q	Btu	Time		
Helmholtz free energy	A	$A = U - TS$, Btu	Dimension of	T	
Internal energy	U	Btu	In general	t, r	s; h
Mass, dimension of	m	lb	Work	W	Btu
Mechanical equivalent of heat	J	(ft) (lbf)/Btu			

(continued)

Table 1 Letter symbols for chemical engineering (*Continued*)

2. Geometrical concepts

Concept	Symbol	Unit of definition (U.S. customary system)		Concept	Symbol	Unit of definition (U.S. customary system)
Linear dimension				Wavelength	λ	cm; ft
Breadth	b	ft		Area		
Diameter	D	ft		In general	A	ft²
Distance along path	s, x	ft		Cross section	S	ft²
Height above datum plane	Z	ft		Fraction free cross-section	σ	
Height equivalent	H	ft (Use subscript p for equilibrium stage and t for transfer unit)		Projected	A_p	ft²
Hydraulic radius	r_H	ft; ft²/ft		Surface		
Lateral distance from datum plane	Y	ft		Per unit mass	A_w, S	ft²/lb
Length, distance or dimension of	L	ft		Per unit volume	A_v, a	ft²/ft³
Longitudinal distance from datum plane	X	ft		Volume		
Mean free path	λ	cm; ft		In general	V	ft³
Radius	r	ft		Fraction voids	ϵ	
Thickness				Humid volume	v_H	ft³/lb dry air
In general	B	ft		Angle	α, θ, ϕ	
Of film	B_f	ft		In x, y plane	α	
				In y, z plane	ϕ	
				in z, x plane	θ	
				Solid angle	ω	
				Other		
				Particle-shape factor	ϕ_s	

3. Intensive properties

Concept	Symbol	Unit of definition (U.S. customary system)		Concept	Symbol	Unit of definition (U.S. customary system)
Absorptivity for radiation	α			Diffusivity		
Activity	a			Molecular, volumetric	D_v, δ	ft³/(h)(ft); ft²/h
Activity coefficient, molal basis	γ			Thermal	α	$\alpha = k/c\rho$, ft²/h
				Emissivity ratio for radiation	ϵ	

Coefficient of expansion		
Linear	α	(ft/ft)/°F
Volumetric	β	(ft³/ft³)/°F
Compressibility factor	z	$z = pV/RT$
Density	ρ	lb/ft³
Helmholtz free energy, per mole	A	Btu/lb mol
Humid heat	c_s	Btu(lb dry air)(°F)
Internal energy, per mole	U	Btu/lb mol
Latent heat, phase change	λ	Btu/lb
Molecular weight	M	lb
Reflectivity for radiation	ρ	
Specific heat	c	Btu/(lb)(°F)
At constant pressure	c_p	Btu/(lb)(°F)
At constant volume	c_v	Btu/(lb)(°F)
Enthalpy, per mole	H	Btu/lb mol
Entropy, per mole	S	Btu/(lb mol)(°R)
Fugacity	f	lbf/ft²; atm
Gibbs free energy, per mole	G, F	Btu/lb mol
Specific heats, ratio of	γ	
Surface tension	σ	lbf/ft
Thermal conductivity	k	Btu/(h)(ft²)(°F/ft)
Transmissivity of radiation	τ	
Vapor pressure	p^*	lbf/ft²; atm
Viscosity		
Absolute or coefficient of	μ	lb/(s)(ft)
Kinematic	ν	ft²/s
Volume, per mole	V	ft³/lb mol

4. Symbols for concentrations

Absorption factor	A	$A = L/K^*V$
Concentration, mass or moles per unit volume	c	lb/ft³; lb mol/ft³
Fraction		
Cumulative beyond a given size	ϕ	
By volume	x_v	
By weight	x_w	
Humidity	H, Y_H	lb/lb dry air
At saturation	H_s, Y^*	lb/lb dry air
At wet-bulb temperature	H_w, Y_w	lb/lb dry air
At adiabatic saturation temperature	H_a, Y_a	lb/lb dry air
Mole or mass fraction		
In heavy or extract phase	x	
In light or raffinate phase	y	
Mole or mass ratio		
In heavy or extract phase	X	
In light or raffinate phase	Y	
Number concentration of particles	n_p	number/ft³
Phase equilibrium ratio	K^*	$K^* = y^*/x$
Relative distribution of two components		

(continued)

Table 1 Letter symbols for chemical engineering (Continued)

Concept	Symbol	Unit of definition (U.S. customary system)	Concept	Symbol	Unit of definition (U.S. customary system)
Mass concentration of particles	c_p	lb/ft³	Between two phases in equilibrium	α	$\alpha = K_i^*/K_j^*$
Moisture content			Between successive stages	β	$\beta_n = (y_i/y_j)_n/(x_j/x_i)_{n+1}$
Total water to bone-dry stock	X_T	lb/lb dry stock	Relative humidity	H_R, R_H	
Equilibrium water to bone-dry stock	X^*	lb/lb dry stock	Slope of equilibrium curve	m	$m = dy^*/dx$
Free water to bone-dry stock	X	lb/lb dry stock	Stripping factor	S	$S = K^*V/L$

5. Symbols for rate concepts

Concept	Symbol	Unit of definition (U.S. customary system)	Concept	Symbol	Unit of definition (U.S. customary system)
Quantity per unit time, in general	q		In general	u	ft/s
Angular velocity	ω		Instantaneous local		
Feed rate	F	lb/h; lb mol/h	Longitudinal (x) component of	u	ft/s
Frequency	f, N_f		Lateral (y) component of	v	ft/s
Friction velocity	u^*	$u^* = (g_c \tau_w \rho)^{1/2}$, ft/s	Normal (z) component of	w	ft/s
Heat transfer rate	q	Btu/h	Volumetric rate of flow	q	ft³/s; ft³/h
Heavy or extract phase rate	L	lb/h; lb mol/h	Quantity per unit time, unit area		
Heavy or extract product rate	B	lb/h; lb mol/h	Emissive power, total	W	Btu/(h) (ft²)
Light or raffinate phase rate	V	lb/h; lb mol/h	Mass velocity, average	G	$G = w/S$, lb (s) (ft²)
Light or raffinate product rate	D	lb/h; lb mol/h	Vapor or light phase	$G, \overline{G}$	lb/(h) (ft²)
Mass rate of flow	w	lb/s; lb/h	Liquid or heavy phase	L, L	lb/(h) (ft²)
			Radiation, intensity of	I	Btu/(h) (ft²)

Quantity	Symbol	Units
Molal rate of transfer	N	lb mol/h
Power	P	(ft)(lbf)/(s)
Revolutions per unit time	n	
Velocity		
Nominal, basis total cross section of packed vessel	vS	ft/s
Volumetric average	$V, \bar{V}$	$(\text{ft}^2/\text{s})/\text{ft}^2$; ft/s
Heat transfer coefficient		
Individual	h	Btu/(h)(ft²)(°F)
Overall	U	Btu/(h)(ft²)(°F)
Mass transfer coefficient		
Individual	k	lb mol/(h)(ft²) (driving force)
Gas film	k_G	
Liquid film	k_L	To define driving force, use subscript:
Overall	K	c for lb mol/ft³
Gas film basis	K_G	p for atm
Liquid film basis	K_L	x for mole fraction
Stefan-Boltzmann constant	σ	0.173×10^{-8} Btu/(h)(ft²)(°R)⁴
Space velocity, volumetric	N_R	(mol/s)/ft³
Quantity per unit time, unit volume	Λ	(ft³/s)/ft³
Quantity reacted per unit time, reactor volume		
Quantity per unit time, unit area, unit driving force, in general	k	
Eddy diffusivity	δ_E	ft²/h
Eddy viscosity	ν_E	ft²/h
Eddy thermal diffusivity	α_E	ft²/h

6. Dimensionless numbers used in chemical engineering

Name	Symbol	Formula	Name	Symbol	Formula
Condensation number	N_{Co}	$\dfrac{h}{k}\left(\dfrac{\nu^2}{a}\right)^{1/3}$; $\dfrac{h}{k}\left(\dfrac{\nu^2}{g}\right)^{1/3}$	Nusselt number	N_{Nu}	$\dfrac{hL}{k}$; $\dfrac{hD}{k}$
Euler number	N_{Eu}	$\dfrac{g_c P}{\rho u^2}$; $\dfrac{g_c \rho P}{G^2}$	Peclet number	N_{Pe}	$\dfrac{Luc\rho}{k}$ or $\dfrac{Lu}{\alpha}$; $\dfrac{DV}{\alpha}$
Fanning friction factor	f	$\dfrac{g_c \rho D(\Delta p_f)}{2G^2(\Delta L)}$	Prandtl number	N_{Pr}	$\dfrac{c\mu}{k}$ or $\dfrac{\nu}{\alpha}$
Fourier number	N_{Fo}	$\dfrac{kt}{c\rho L^2}$ or $\dfrac{\alpha t}{L^2}$	Prandtl velocity ratio	u^+	$\dfrac{\bar{u}}{u^*}$

(continued)

Table 1 Letter symbols for chemical engineering (*continued*)

6. Dimensionless numbers used in chemical engineering (*continued*)

Froude number	N_{Fr}	$\dfrac{u^2}{aL}$; $\dfrac{u^2}{gL}$	Reynolds number	N_{Re}	$\dfrac{Lu\rho}{\mu}$; $\dfrac{DG}{\mu}$
Graetz number	N_{Gz}	$\dfrac{cLG}{k}$ or $\dfrac{LV}{\alpha}$	Reynolds number, local	y^+	$\dfrac{ru^*\rho}{\mu}$
Grashof number	N_{Gr}	$\dfrac{L^3\rho^2\beta g\Delta t}{\mu^2}$ or $\dfrac{L^3\beta g\Delta t}{\nu^2}$	Schmidt number	N_{Sc}	$\dfrac{\mu}{\rho D_v}$
Heat transfer factor	j_H	$\dfrac{h}{cG}\left(\dfrac{c\mu}{k}\right)^{2/3}$ or $(N_{St})(N_{Pr})^{2/3}$	Sherwood number	N_{Sh}	$\dfrac{k_c L}{D_v}$ or $j_M(N_{Re})(N_{Sc})^{1/3}$
Lewis number	N_{Le}	$\dfrac{k}{c\rho D_v}$ or $\dfrac{\alpha}{D_v}$	Stanton number	N_{St}	$\dfrac{h}{c\rho u}$; $\dfrac{h}{cG}$
Mass transfer factor	j_M	$\dfrac{k_c}{u}\left(\dfrac{\mu}{\rho D_v}\right)^{2/3}$	Vapor condensation number	N_{Cv}	$\dfrac{L^3\rho^2 g\lambda}{k\mu\Delta t}$
			Weber number	N_{We}	$\dfrac{Lu^2\rho}{g_c\sigma}$; $\dfrac{DG^2}{g_c\rho\sigma}$

7. Modifying signs for principal symbols

Concept	Remarks	Superscript	Subscript	Concept	Remarks	Superscript	Subscript
Average value	Written over symbol	$-$ (Bar)		Partial molal quantity	Written over small capitals	$-$ (Bar)	
Dimensionless form	Follows symbol	$+$ (Plus)		Sequence in time or space	Follows symbol	$'$ (Prime) $''$ (Double prime)	1, 2, 3, etc.
Equilibrium value	Follows symbol	$*$ (Asterisk)		Standard state	Follows symbol	$°$ (Degree)	
Fluctuating component	Usually applied to local velocity	$'$ (Prime)		First derivative with respect to time	Written over symbol	$\cdot$ (Dot)	
Initial or reference value	Follows symbol		0 (zero)	Second derivative with respect to time	Written over symbol	$\cdot\cdot$ (Double dot)	
Modified form	Follows symbol	$'$ (Prime) $''$ (Double prime)					

† Units shown are for the U.S. Customary system. See Appendix A for units in the SI system. For a revised set of recommendations for symbols for use in chemical engineering based on the SI system, see E. Buck, Letter Symbols for Chemical Engineering, *Chem. Eng. Progr.*, **74**(10): 73 (1978).

Examples of accepted abbreviations are shown in Table 2.

Table 2 Accepted abbreviations

American Chemical Society	ACS	Efficiency	eff
American Institute of Chemical Engineers	AIChE	Electromotive force	emf
		Equivalent	equiv
American Iron and Steel Institute	AISI	Ethyl	Et
American Petroleum Institute	API	Evaporate	evap
American Society of Mechanical Engineers	ASME	Experiment	expt
		Experimental	exptl
American Society for Testing Materials	ASTM	Extract	ext
American wire gauge	AWG	Feet per minute	fpm or ft/min
Ampere	A	Figure	fig.
Angstrom	Å.	Foot	ft
Atmosphere	atm	Foot pound	ft.lb
Average	avg		
Barrel	bbl	Gallon	gal
Baumé	Bé	Gallons per minute	gpm or gal/min
Biochemical oxygen demand	BOD	Grain	spell out
Boiling point	bp	Gram	gm (sometimes g) or spell out
Bottoms	btms		
British thermal unit	Btu	Height equivalent to a theoretical plate	HETP
Brown and Sharpe gauge number	B&S	Height of a transfer unit	HTU
Calorie	cal	Horsepower	hp
Capacity	cap.	Hour	h
Catalytic	cat.	Hundredweight (100 lb)	cwt
Centigrade	C		
Centigram	cg	Inch	in.
Centimeter	cm	Inside diameter	ID or i.d.
Centipoise	cP	Insoluble	insol
Centistoke	cS		
Chemically pure	CP	Kilogram	kg
Concentrate	conc	Kilometer	km
Critical	crit	Kilovolt	kV
Cubic	cu	Kilowatt	kW
Cubic centimeter	cc	Kilowatt-hour	kWh
Cubic foot	cu ft or ft^3		
Cubic foot per minute	cfm or ft^3/min	Liquid	liq
Cubic foot per second	cfs or ft^3/s	Liter	l or spell out
Cubic inch	cu in. or in.^{3}	Logarithm (base 10)	log
		Logarithm (base e)	ln
Degree	deg or °		
Diameter	diam	Maximum	max
Dilute	dil	Melting point	mp
Distill or distillate	dist		

Table 2 Accepted abbreviations (*continued*)

Meter	m	Saybolt Universal seconds	SUS
Methyl	Me	Second	s
Micron	μ or mu	Society of Automotive Engineers	SAE
Mile	mi	Soluble	sol
Miles per hour	mph	Solution	soln
Milliampere	mA	Specific gravity	sp gr
Million electron volts	meV	Specific heat	sp ht
Millivolt	mV	Square	sq
Minute	min	Square foot	sq ft or ft^2
Molecular	mol	Standard	std
		Standard temperature and	
Ounce	oz	pressure	STP
Outside diameter	OD or o.d.		
Overhead	ovhd	Tank	tk
		Technical	tech
Page	p.	Temperature	temp
Pages	pp.	Tetraethyl lead	TEL
Parts per million	ppm	Thousand	M
Pint	pt	Ton	spell out
Pound	lb	Tubular Exchangers Manu-	
Pound centigrade unit	Pcu	facturers Association	TEMA
Pounds per cubic foot	lb/cu ft or lb/ft^3		
Pounds per square foot	lb/sq ft or lb/ft^2	Volt	V
Pounds per square inch	psi or lb/in.^{2}	Volt-ampere	VA
Pounds per square inch absolute	psia	Volume	vol
Pounds per square inch gauge	psig		
		Watt	W
Quart	qt	Watthour	Wh
		Weight	wt
Refractive index	RI or n		
Revolutions per minute	rpm or r/min	Yard	yd

RHETORIC

Correct grammar, punctuation, and style of writing are obvious requirements for any report. Many engineers, however, submit unimpressive reports because they do not concern themselves with the formal style of writing required in technical reports. This section deals with some of the restrictions placed on formal writing and presents a discussion of common errors.

Personal Pronouns

The use of personal pronouns should be avoided in technical writing. Many writers eliminate the use of personal pronouns by resorting to the passive voice. This is certainly acceptable, but, when applicable, the active voice gives the writing a less stilted style. For example, instead of saying "We designed the absorption tower on the basis of . . . ," a more acceptable form would be "The absorption tower was designed on the basis of . . ." or "The basis for the absorption-tower design was . . . ."

The pronoun "one" is sometimes used in technical writing. In formal writing, however, it should be avoided or, at most, employed only occasionally.

Tenses

Both past and present tenses are commonly used in report writing; however, tenses should not be switched in one paragraph or in one section unless the meaning of the written material requires the change. General truths that are not limited by time are stated in the present tense, while references to a particular event in the past are reported in the past tense (e.g., "The specific gravity of mercury is 13.6." "The experiment was performed . . .").

Diction

Contractions such as "don't" and "can't" are seldom used in technical writing, and informal or colloquial words should be avoided. Humorous or witty statements are out of place in a technical report, even though the writer may feel they are justified because they can stimulate interest. Too often, the reader will be devoid of a sense of humor, particularly when engrossed in the serious business of digesting the contents of a technical report. A good report is made interesting by clarity of expression, skillful organization, and the significance of its contents.

Singular and Plural

Many writers have difficulty in determining if a verb should be singular or plural. This is especially true when a qualifying phrase separates the subject and its verb. For example, "A complete list of the results *is* (not *are*) given in the appendix."

Certain nouns, such as "number," "none," and "series," can be either singular or plural. As a general rule, the verb should be singular if the subject is viewed as a unit ("The number of engineers in the United States *is* increasing") and plural if the things involved are considered separately ("A number of the workers *are* dissatisfied"). Similarly, the following sentences are correct: "*Thirty thousand gallons was* produced in *two* hours." "The tests show that *18* batches *were* run at the wrong temperature."

Dangling Modifiers

The technical writer should avoid dangling modifiers that cannot be associated directly with the words they modify. For example, the sentence "Finding the results were inconclusive, the project was abandoned" could be rewritten correctly as "Finding the results were inconclusive, the investigators abandoned the project."

Poor construction caused by dangling modifiers often arises from retention of the personal viewpoint, even though personal pronouns are eliminated. The writer should analyze the work carefully and make certain the association between a modifying phrase and the words referred to is clear.

Compound Adjectives

Nouns are often used as adjectives in scientific writing. This practice is acceptable; however, the writer must use it in moderation. A sentence including "a centrally located natural gas production plant site is ..." should certainly be revised. Prepositional clauses are often used to eliminate a series of compound adjectives.

Hyphens are employed to connect words that are compounded into adjectives—for example, "a hot-wire anemometer," "a high-pressure line"—but no hyphen appears in "a highly sensitive element."

Split Infinitives

Split infinitives are acceptable in some types of writing, but they should be avoided in technical reports. A split infinitive bothers many readers, and it frequently results in misplaced emphasis. Instead of "The supervisor intended to carefully check the data," the sentence should be "The supervisor intended to check the data carefully."

That—Which

Many technical writers tend to overwork the word *that*. Substitution of the word *which* for *that* is often acceptable, even though a strict grammatical interpretation would require repetition of *that*. The general distinction between the pronouns "that" and "which" can be stated as follows: *That* is used when the clause it introduces is necessary to define the meaning of its antecedent; *which* introduces some additional or incidental information.

COMMENTS ON COMMON ERRORS

1. The word *data* is usually plural. Say "data are," not "data is."
2. "Balance" should not be used when "remainder" is meant.
3. Use "different from" instead of "different than."
4. The word "farther" refers to distance, and "further" indicates "in addition to."
5. "Affect," as a verb, means "to influence." It should never be confused with the noun "effect," which means "result."
6. "Due to" should be avoided when "because of" can be used.
7. Use "fewer" when referring to numbers and "less" when referring to quantity or degree.

CHECK LIST FOR THE FINAL REPORT

Before submitting the final draft, the writer should make a critical analysis of the report. Following is a list of questions the writer should ask when evaluating the report. These questions cover the important considerations in report writing and can serve as a guide for both experienced and inexperienced writers.

1. Does the report fulfill its purpose?
2. Will it be understandable to the principal readers?
3. Does the report attempt to cover too broad a subject?
4. Is sufficient information presented?
5. Is too much detail included in the body of the report?
6. Are the objectives stated clearly?
7. Is the reason for the report indicated?
8. Is the summary concise? Is it clear? Does it give the important results, conclusions, and recommendations? Is it a true summary of the entire report?
9. Is there an adequate description of the work done?
10. Are the important assumptions and the degree of accuracy indicated?
11. Are the conclusions and recommendations valid?
12. Are sufficient data included to support the conclusions and recommendations?
13. Have previous data and earlier studies in the field been considered?
14. Is the report well organized?
15. Is the style of writing readable and interesting?
16. Has the manuscript been rewritten and edited ruthlessly?
17. Is the appendix complete?
18. Are tables, graphs, and illustrations presented in a neat, readable, and organized form? Is all necessary information shown?
19. Has the report been proofread? Are pages, tables, and figures numbered correctly?
20. Is the report ready for submittal on time?

SUGGESTED ADDITIONAL REFERENCES FOR REPORT WRITING

Report writing

Buck, E., Letter Symbols for Chemical Engineering, *Chem. Eng. Progr.*, **74**(10):73 (1978).
Fair, J. R., Dictation and the Engineer, *Chem. Eng.*, **76**(14):114 (1969).
Geiger, G. H., "Supplementary Readings in Engineering Design," McGraw-Hill Book Company, New York, 1975.
Hine, E. A., Write in Style: Be Clear and Concise, *Chem. Eng.*, **82**(27):41 (1975).
Hissong, D. W., Write and Present Persuasive Reports, *Chem. Eng.*, **84**(14):131 (1977).
Holden, G. F., Dress up Your Technical Reports, *Chem. Eng.*, **78**(29):80 (1971).
Hughson, R. V., Writing for Publication: Main Road to Recognition, *Chem. Eng.*, **82**(27):49 (1975).
Johnson, T. P., Fast, Functional Writing, *Chem. Eng.*, **76**(14):105 (1969).

Marbach, M. G., Better Visuals Make Speeches Better, *Chem. Eng.*, **84**(6):141 (1977).

Mintz, H. K., How to Write Better Memos, *Chem. Eng.*, **77**(2):136 (1970).

Popper, H., Six Guidelines for Fast, Functional Writing, *Chem. Eng.*, **76**(14):118 (1969).

Power, R. M., Engineer's Can't Write? Wrong! *Hydrocarbon Process.*, **52**(3):105 (1973).

Quackenbos, H. M., Creative Report Writing, *Chem. Eng.*, **79**(15):94 (1972); **79**(16):146 (1972).

Smook, G. A., Illustration Techniques for Technical Reports, *Chem. Eng.*, **79**(4):62 (1972).

Sherman, T. A., and S. S. Johnson, "Modern Technical Writing," 3d ed., Prentice-Hall, Englewood Cliffs, New Jersey, 1975.

Sousa, A. J., Poetry? For Engineers? Words in Their Best Order, *Chem. Eng.*, **77**(13):176 (1970).

Thorne, H. C., Jr., How to do Post-Installation Appraisals, *Hydrocarbon Process.*, **52**(4):203 (1973).

Vervalin, C. H., Steps to More Effective Writing, presented at the 12th National Meeting American Association of Cost Engineers, Houston, Texas, June 17–19, 1968.

Vinci, V., Watch Your Words, *Chem. Eng.*, **78**(18):112 (1971).

———, Ten Report Writing Pitfalls: How to Avoid Them, *Chem. Eng.*, **82**(27):45 (1975).

Woodson, T. T., "Effective Communication for Engineers," McGraw-Hill Book Company, New York, 1975.

PROBLEMS

1 Prepare a skeleton outline and a detailed outline for a final report on the detailed-estimate design of a distillation unit. The unit is to be used for recovering methanol from a by-product containing water and methanol. In the past, this by-product has been sold to another concern, but the head of the engineering-development group feels that the recovery should be accomplished by your company. The report will be examined by the head of the engineering-development group and will then be submitted to the plant management for final approval.

2 List ten words you often misspell and five grammatical errors you occasionally make in formal writing.

Correct the following sentences:

3 "Using the mass transfer coefficients and other physical data, Schmidt and Nusselt numbers were calculated for each experiment."

4 "The excellent agreement between the experimental and theoretical values substantiate the validity of the assumptions."

5 "This property makes the packing more efficient than any packing."

6 "He is an engineer with good theoretical training and acquainted with industrial problems."

7 "Wrought iron is equally as good as stainless steel because no temperatures will be used at above 25°C."

8 "The pressure has got to be maintained constant or the tank will not empty out at a constant rate."

Interpret, rewrite, and improve the following:

9 "The purchasing division has contracted with the *X* Chemical Company to supply 20,000 kg of chemical *A* which corresponds closely to the specifications presented and 10,000 kg of chemical *E*."

10 "A more rigorous derivation would be extremely complicated and would hardly be justified in view of the uncertainties existing with respect to basic information necessary for practical applications of the results."

11 "An important factor in relation to safety precautions is first and foremost giving to workmen some kind of a clear and definite instruction along the line of not coming into the radioactive areas in connection with their work."

CHAPTER THIRTEEN

MATERIALS TRANSFER, HANDLING, AND TREATMENT EQUIPMENT—DESIGN AND COSTS

The design and cost estimation for equipment items and systems used for the transfer, handling, or treatment of materials is vitally involved in almost every type of plant design. The most common means for transferring materials is by pumps and pipes. Conveyors, chutes, gates, hoists, fans, and blowers are examples of other kinds of equipment used extensively to handle and transfer various materials. Many forms of special equipment are used for the treatment of materials, as, for example, filters, blenders, mixers, kneaders, centrifugal separators, crystallizers, crushers, grinders, dust collectors, kettles, reactors, and screens. The design engineer must decide which type of equipment is best suited for the purposes and be able to prepare equipment specifications that will satisfy the operational demands of the process under reasonable cost conditions. Consequently, theoretical design principles, practical problems of operation, and cost considerations are all involved in the final choice of materials transfer, handling, and treatment equipment.

PUMPS AND PIPING

POWER REQUIREMENTS

A major factor involved in the design of pumping and piping systems is the amount of power that is required for the particular operation. Mechanical power must be supplied by the pump to overcome frictional resistance, changes in elevation, changes in internal energy, and other resistances set up in the flow system.

The various forms of energy can be related by the total energy balance or the total mechanical-energy balance. On the basis of 1 lbm† of fluid flowing under steady conditions, the total energy balance may be written in differential form as

$$\frac{g}{g_c} dZ + d(pv) + \frac{V_i \, dV_i}{g_c} + du = \delta Q + \delta W \tag{1}$$

The total mechanical-energy balance in differential form is

$$\frac{g}{g_c} dZ + v \, dp + \frac{V_i \, dV_i}{g_c} = \delta W_o - \delta F \tag{2}$$

where g = local gravitational acceleration, usually taken as 32.17 ft/(s)(s)
g_c = conversion factor in Newton's law of motion, 32.17 ft·lb mass/(s)(s)(lbf)
Z = vertical distance above an arbitrarily chosen datum plane, ft
v = specific volume of the fluid ft³/lbm
p = absolute pressure, lbf/ft²
V_i = instantaneous or point linear velocity, ft/s
u = internal energy, ft·lbf/lbm
Q = heat energy imparted as such to the fluid system from an outside source, ft·lbf/lbm
W = shaft work, gross work input to the fluid system from an outside source, ft·lbf/lbm
W_o = mechanical work imparted to the fluid system from an outside source,‡ ft·lbf/lbm
F = mechanical-energy loss due to friction, ft·lbf/lbm

Integration of these energy balances between "point 1" where the fluid enters the system and "point 2" where the fluid leaves the system gives

Total energy balance:

$$Z_1 \frac{g}{g_c} + p_1 v_1 + \frac{V_1^2}{2\alpha g_c} + u_1 + Q + W = Z_2 \frac{g}{g_c} + p_2 v_2 + \frac{V_2^2}{2\alpha g_c} + u_2 \tag{3}$$

Total mechanical-energy balance:

$$Z_1 \frac{g}{g_c} - \int_1^2 v \, dp + \frac{V_1^2}{2\alpha g_c} + W_o = Z_2 \frac{g}{g_c} + \frac{V_2^2}{2\alpha g_c} + \Sigma F \tag{4}$$

where V is the average linear velocity, ft/s, and α is the correction coefficient to account for use of average velocity, usually taken as 1.0 if flow is turbulent and

---

† The U.S. customary system of units is used here to clarify the role of g and g_c and to show the common usage by design engineers in the U.S.
‡ The mechanical work W_o is equal to the total shaft work W minus the amount of energy transmitted to the fluid as a result of pump friction or pump inefficiency. When W_o is used in the total mechanical-energy balance, pump friction is not included in the term for the mechanical-energy loss due to friction F.

0.5 if flow is viscous. Equations (1) through (4) are sufficiently general for treatment of almost any flow problem and are the basis for many design equations that apply for particular simplified conditions.

Evaluation of the term $\int_1^2 v \, dp$ in Eq. (4) may be difficult if a compressible fluid is flowing through the system, because the exact path of the compression or expansion is often unknown. For noncompressible fluids, however, the specific volume v remains essentially constant and the integral term reduces simply to $v(p_2 - p_1)$. Consequently, the total mechanical-energy balance is especially useful and easy to apply when the flowing fluid can be considered as noncompressible.

Friction

Frictional effects are extremely important in flow processes. In many cases, friction may be the main cause for resistance to the flow of a fluid through a given system. Consider the common example of water passing through a pipe. If no frictional effects were present, pipes of very small diameters could be used for all flow rates. Under these conditions, the pumping-power costs for forcing 100,000 gal of water per hour through a $\frac{1}{8}$-in.-diameter pipe would be the same as the power costs for forcing water at the same mass rate through a pipe of equal length having a diameter of 2 ft. In any real flow process, however, frictional effects *are* present, and they must be taken into consideration.

When a fluid flows through a conduit, the amount of energy lost due to friction depends on the properties of the flowing fluid and the extent of the conduit system. For the case of steady flow through long straight pipes of uniform diameter, the variables that affect the amount of frictional losses are the velocity at which the fluid is flowing (V), the density of the fluid (ρ), the viscosity of the fluid (μ), the diameter of the pipe (D), the length of the pipe (L), and the equivalent roughness of the pipe (ε). By applying the method of dimensional analysis to these variables, the following expression, known as the *Fanning equation*, can be obtained for the frictional effects in the system:

$$dF = \frac{-dp_f}{\rho} = \frac{2 \mathfrak{f} V^2 \, dL}{g_c D} \tag{5}$$

The friction factor $\mathfrak{f}$ is based on experimental data and has been found to be a function of the Reynolds number and the relative roughness of the pipe (ε/D). Figure 13-1 presents a plot of the friction factor versus the Reynolds number in straight pipes. In the viscous-flow region, the friction factor is not affected by the relative roughness of the pipe; therefore, only one line is shown in Fig. 13-1 for Reynolds numbers up to about 2100. In the turbulent-flow region, the relative roughness of the pipe has a large effect on the friction factor. Curves with different parameters of the dimensionless ratio ε/D are presented in Fig. 13-1 for values of Reynolds numbers greater than 2100. A table on the plot indicates values for ε for various pipe-construction materials. As the methods for determining $\mathfrak{f}$ do not permit high accuracy (± 10 percent), the value of the friction

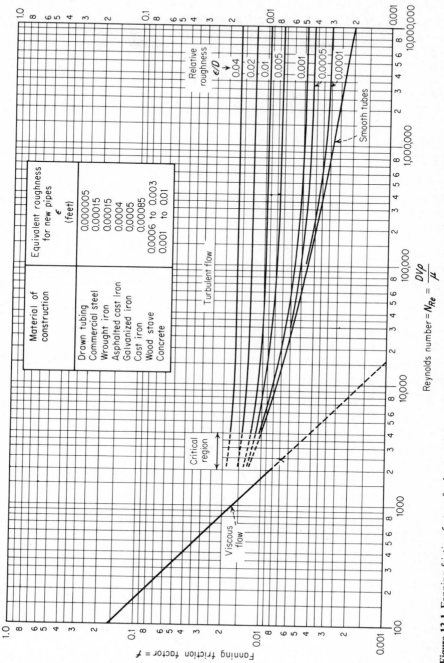

Figure 13-1 Fanning friction factors for long straight pipes. [Based on L. F. Moody, *Trans. ASME*, **66**:671–684 (1944).]

factor should not be read to more than two significant figures, and Fig. 13-1 gives adequate accuracy for determining the numerical size of the friction factor.

The values for equivalent pipe roughness given in Fig. 13-1 are only approximations, even for new pipe, and the values may increase because of surface pitting and corrosion after the pipe is in service. The design engineer, therefore, should recognize the inherent inaccuracies in estimating the effects of pipe roughness, and this matter should be taken into consideration when the final design is prepared.

Curves similar to Fig. 13-1 are sometimes presented in the literature with a different defining value of $\bar{f}$. For example, mechanical engineers usually define the friction factor so that it is exactly four times the friction factor given in Eq. (5).

The Reynolds number range between 2100 and 4000 is commonly designated as the *critical region*. In this range, there is considerable doubt as to whether the flow is viscous or turbulent. For design purposes, the safest practice requires the assumption that turbulent flow exists at all Reynolds numbers greater than 2100.

A mathematical expression for the friction factor can be obtained from the equation for the straight line in the viscous-flow region of Fig. 13-1. Thus, at Reynolds numbers below 2100

$$\bar{f} = \frac{16}{N_{Re}} = \frac{16\mu}{DV\rho} \tag{6}\dagger$$

Approximate equations showing the relationship between the friction factor and the Reynolds number in the turbulent-flow region have been developed. Two of these equations follow:

For smooth pipe or tubes,

$$\bar{f} = \frac{0.046}{(N_{Re})^{0.2}} \tag{7}$$

For new iron or steel pipe,

$$\bar{f} = \frac{0.04}{(N_{Re})^{0.16}} \tag{8}$$

Integrated form of the Fanning equation If the linear velocity, density, and viscosity of the flowing fluid remain constant and the pipe diameter is uniform over a total pipe length L, Eq. (5) can be integrated to give the following result:

$$F = \frac{-\Delta p_f}{\rho} = \frac{2\bar{f}V^2 L}{g_c D} \tag{9}$$

$\dagger$ Substitution of this expression for $\bar{f}$ into Eq. (5) results in the well-known Hagen-Poiseuille law for viscous flow.

In a strict sense, Eq. (9) is limited to conditions in which the flowing fluid is noncompressible and the temperature of the fluid is constant. When dealing with compressible fluids, such as air, steam, or any gas, it is good engineering practice to use Eq. (9) only if the pressure drop over the system is less than 10 percent of the initial pressure. If a change in the fluid temperature occurs, Eq. (9) should not be used in the form indicated unless the total change in the fluid viscosity is less than approximately 50 percent based on the maximum viscosity.†‡ If Eq. (9) is used when pressure changes or temperature changes are involved, the best accuracy is obtained by using the linear velocity, density, and viscosity of the fluid as determined at the average temperature and pressure. Exact results for compressible fluids or nonisothermal flow can be obtained from the Fanning equation by integrating the differential expression, taking all changes into consideration.

For turbulent flow in a conduit of noncircular cross section, an equivalent diameter can be substituted for the circular-section diameter, and the equations for circular pipes can then be applied without introducing a large error. This equivalent diameter is defined as four times the hydraulic radius R_H, where the hydraulic radius is the ratio of the cross-sectional flow area to the wetted perimeter. When the flow is viscous, substitution of $4R_H$ for D does not give accurate results, and exact expressions relating frictional pressure drop and velocity can be obtained only for certain conduit shapes.

Frictional effects due to end losses, fittings, orifices, and other installations If the cross-sectional area of a pipe changes gradually to a new cross-sectional area, the disturbances to the flow pattern can be so small that the amount of mechanical energy lost as friction due to the change in cross section is negligible. If the change is sudden, however, an appreciable amount of mechanical energy can be lost as friction. Similarly, the presence of bends, fittings, valves, orifices, or other installations that disturb the flow pattern can cause frictional losses. All of these effects must be included in the friction term appearing in the total mechanical-energy balance. Recommended expressions for evaluating the important types of frictional losses are presented in Table 1.

---

† Overall effects of temperature on the friction factor are more important in the streamline-flow range where f is directly proportional to the viscosity than in the turbulent-flow range where f is approximately proportional to $\mu^{0.16}$.

‡ For heating or cooling of fluids, a temperature gradient must exist from the pipe wall across the flowing fluid. A simplified design procedure for this case is as follows: When temperature and viscosity changes must be taken into consideration, the friction factor for use in Eq. (9) should be taken as the isothermal friction factor (Fig. 13-1) based on the arithmetic-average temperature of the fluid divided by a correction factor ϕ, where $\phi = 1.1(\mu_a/\mu_w)^{0.25}$ when DG/μ_a is less than 2100 and $\phi = 1.02(\mu_a/\mu_w)^{0.14}$ when DG/μ_a is greater than 2100. [G = mass velocity, lb/(h)(ft² of cross-sectional area); μ_a = viscosity of fluid at average bulk temperature, lb/(s)(ft); μ_w = viscosity of fluid at temperature of wall, lb/(s)(ft).]

Table 1 Expressions for evaluating frictional losses in the flow of fluids through conduits

For noncircular, cross-sectional area and turbulent flow, replace D by $4R_H = 4$ (cross-sectional flow area/wetted perimeter).

Friction caused by	General expression for frictional loss	Limited expression and remarks
Flow through long straight pipe of constant cross-sectional area	$dF = \dfrac{2\mathfrak{f}V^2\,dL}{g_cD}$	For case in which fluid is essentially noncompressible and temperature is constant $$F = \frac{2\mathfrak{f}V^2L}{g_cD}$$
Sudden enlargement	$F_e = \dfrac{(V_1 - V_2)^2}{2\alpha g_c}$	The following values for α may be used in design calculations: turbulent flow, $\alpha = 1$; streamline flow, $\alpha = 0.5$ $V_1 \longrightarrow V_2$
Sudden contraction	$F_c = \dfrac{K_cV_2{}^2}{2\alpha g_c}$	The following values for α may be used in design calculations: turbulent flow, $\alpha = 1$; streamline flow, $\alpha = 0.5$ $A_1 \quad A_2$ $V_2 \longrightarrow$ For $\dfrac{A_2}{A_1} < 0.715$, $K_c = 0.4\left(1.25 - \dfrac{A_2}{A_1}\right)$ For $\dfrac{A_2}{A_1} > 0.715$, $K_c = 0.75\left(1 - \dfrac{A_2}{A_1}\right)$ For conical or rounded shape, $K_c = 0.05$
Fittings, valves, etc.	$F = \dfrac{2\mathfrak{f}V^2L_e}{g_cD}$	L_e/D per fitting (dimensionless) 45° elbows — 15 90° elbows, std. radius — 32 90° elbows, medium radius — 26 90° elbows, long sweep — 20 90° square elbows — 60 180° close-return bends — 75

Table 1 Expressions for evaluating frictional losses in the flow of fluids through conduits (*Continued*)

Friction caused by	General expression for frictional loss	Limited expression and remarks
Fittings, valves, etc.	$F = \dfrac{2fV^2L_e}{g_cD}$	L_e/D per fitting (dimensionless) 180° medium-radius return bends — 50 Tee (used as elbow, entering run) — 60 Tee (used as elbow, entering branch) — 90 Couplings — Negligible Unions — Negligible Gate valves, open — 7 Globe valves, open — 300 Angle valves, open — 170 Water meters, disk — 400 Water meters, piston — 600 Water meters, impulse wheel — 300
Sharp-edged orifice	$-\Delta p_f = F\rho$	$\dfrac{D_0}{D}$ $\quad$ $\dfrac{\Delta p_f(100)}{\Delta p \text{ across orifice}} = \%$ 0.8 — 40 0.7 — 52 0.6 — 63 0.5 — 73 0.4 — 81 0.3 — 89 0.2 — 95 Measured Δp across orifice
Rounded orifice	$F = \dfrac{(V_0 - V_2)^2}{2\alpha g_c}$	The following values for α may be used in design calculations: turbulent flow, $\alpha = 1$; streamline flow, $\alpha = 0.5$ $V_0 \rightarrow \quad V_2 \rightarrow$
Venturi	$-\Delta p_f = F\rho$	$-\Delta p_f = \frac{1}{8}$ to $\frac{1}{10}$ of total pressure drop from upstream section to venturi throat

Design Calculations of Power Requirements

Liquids For noncompressible fluids, the integrated form of the total mechanical-energy balance reduces to

$$W_o = \Delta Z + \Delta\left(\frac{V^2}{2\alpha g_c}\right) + \Delta(pv) + \Sigma F \tag{10}$$

where g is assumed to be numerically equal to g_c. Because the individual terms in Eq. (10) can be evaluated directly from the physical properties of the system and the flow conditions, the design engineer can apply this equation to many liquid-flow systems without making any major assumptions. The following example illustrates the application of Eq. (10) for a design calculation of the size of motor necessary to carry out a given pumping operation.

Example 1. Application of the total mechanical-energy balance to noncompressible-flow systems Water at 61°F is pumped from a large reservoir into the top of an overhead tank using standard 2-in.-diameter steel pipe (ID = 2.067 in.). The reservoir and the overhead tank are open to the atmosphere, and the difference in vertical elevation between the water surface in the reservoir and the discharge point at the top of the overhead tank is 70 ft. The length of the pipeline is 1000 ft. Two gate valves and three standard 90° elbows are included in the system. The efficiency of the pump is 40 percent. This includes losses at the entrance and exit of the pump housing. If the flow rate of water is to be maintained at 50 gpm and the water temperature remains constant at 61°F, estimate the horsepower of the motor required to drive the pump.

SOLUTION
Basis: 1 lb of flowing water
 Total mechanical-energy balance between point 1 (surface of water in reservoir) and point 2 (just outside of pipe at discharge point):

$$W_o = Z_2 - Z_1 + \frac{V_2^2}{2\alpha g_c} - \frac{V_1^2}{2\alpha g_c} + p_2 v_2 - p_1 v_1 + \Sigma F$$

Points 1 and 2 are taken where the linear velocity of the fluid is negligible; therefore

$$\frac{V_2^2}{2\alpha g_c} = 0 \quad \text{and} \quad \frac{V_1^2}{2\alpha g_c} = 0$$

$p_1 = p_2 =$ atmospheric pressure. $v_1 = v_2$, since liquid water can be considered as a noncompressible fluid. $p_2 v_2 - p_1 v_1 = 0$. $Z_2 - Z_1 = 70$ ft·lbf/lbm (assuming $g = g_c$).
 Determination of friction:

$$\text{Average velocity in 2-in. pipe} = \frac{(50)(144)}{(60)(7.48)(2.067)^2(0.785)} = 4.78 \text{ ft/s}$$

Viscosity of water at 61°F = 1.12 cp = (1.12)(0.000672) lb/(ft)(s)

Density of water at 61°F = 62.3 lb/ft^3

$$\text{Reynolds number in 2-in. pipe} = \frac{(2.067)(4.78)(62.3)}{(12)(1.12)(0.000672)} = 68{,}000$$

$$\frac{\varepsilon}{D} = \frac{(0.00015)(12)}{2.067} = 0.00087$$

Friction factor = $\mathfrak{f}$ = 0.0057   (from Fig. 13-1)

$$\text{Total } L_e \text{ for fittings and valves} = \frac{(2)(7)(2.067)}{12} + \frac{(3)(32)(2.067)}{12}$$

$$= 19 \text{ ft}$$

Friction due to flow through pipe and all fittings

$$= \frac{2\mathfrak{f}V^2(L + L_e)}{g_c D}$$

$$= \frac{(2)(0.0057)(4.78)^2(1000 + 19)(12)}{(32.17)(2.067)}$$

$$= 47.9 \text{ ft·lbf/lbm}$$

Friction due to contraction and enlargement (from Table 1)

$$= \frac{(0.5)(4.78)^2}{(2)(1)(32.17)} + \frac{(4.78 - 0)^2}{(2)(1)(32.17)}$$

$$= 0.53 \text{ ft·lbf/lbm}$$

$$\Sigma F = 47.9 + 0.53 = 48.4 \text{ ft·lbf/lbm}$$

From the total mechanical-energy balance, W_o = theoretical mechanical energy necessary from pump = 70 + 48.4 = 118.4 ft·lbf/lbm.

$$\text{hp of motor required to drive pump} = \frac{(118.4)(50)(62.3)}{(0.40)(60)(7.48)(550)} = 3.74 \text{ hp}$$

$$\frac{\text{ft·lbf}}{\text{lbm}} \bigg| \frac{\text{gal}}{\text{min}} \bigg| \frac{\text{min}}{\text{s}} \bigg| \frac{\text{ft}^3}{\text{gal}} \bigg| \frac{\text{lbm}}{\text{ft}^3} \bigg| \frac{\text{s (hp)}}{\text{ft·lbf}} = \text{hp}$$

A 4.0-hp motor would be adequate for a design estimate.

Gases Because of the difficulty that may be encountered in evaluating the exact integral of $v\,dp$ and dF for compressible fluids, use of the total mechanical-energy balance is not recommended for compressible fluids when large pressure drops are involved. Instead, the total energy balance should be used if the necessary data are available.

If g is assumed to be numerically equal to g_c, the integrated form of the total energy balance [Eq. (3)] can be written as

$$W = \Delta Z + \Delta h + \Delta\left(\frac{V^2}{2\alpha g_c}\right) - Q \tag{11}$$

where h = enthalpy = $u + pv$, ft·lbf/lbm.

When Eq. (11) is applied in design calculations, information must be available for determining the change in enthalpy over the range of temperature and pressure involved. The following illustrative examples show how Eq. (11) can be used to calculate pumping power when compressible fluids are involved.

Example 2. Application of total energy balance for the flow of an ideal gas
Nitrogen is flowing under turbulent conditions at a constant mass rate through a long, straight, horizontal pipe. The pipe has a constant inside diameter of 2.067 in. At an upstream point (point 1), the temperature of the nitrogen is 70°F, the pressure is 15 psia, and the average linear velocity of the gas is 50 ft/s. At a given downstream point (point 2), the temperature of the gas is 140°F and the pressure is 50 psia. An external heater is located between points 1 and 2, and 10 Btu is transferred from the heater to each pound of the flowing gas. Except at the heater, no heat is transferred as such between the gas and the surroundings. Under these conditions, nitrogen may be considered to be an ideal gas, and the mean heat capacity C_p of the gas is 7.0 Btu/(lb mol)(°F). Estimate the total amount of energy (as foot-pounds force per pound of the flowing gas) supplied by the compressor located between points 1 and 2.

SOLUTION
Basis: 1 lb of flowing nitrogen
        Total energy balance between points 1 and 2 for horizontal system and turbulent flow:

$$W = h_2 - h_1 + \frac{V_2^2}{2g_c} - \frac{V_1^2}{2g_c} - Q$$

where V_1 = 50 ft/s and V_2 = (50)(600)(15)/(530)(50) = 17 ft/s.
Since nitrogen is to be considered an ideal gas,

$$h_2 - h_1 = \frac{C_p}{M}(T_2 - T_1) = \frac{(7)(140 - 70)(778)}{28} = 13{,}600 \text{ ft·lbf/lbm}$$

$$Q = (10)(778)\text{ft·lbf/lbm}$$

W = total energy supplied by compressor

$$= 13{,}600 + \frac{(17)^2}{(2)(32.17)} - \frac{(50)^2}{(2)(32.17)} - 7780$$

$$= 5790 \text{ ft·lbf/lbm}$$

Example 3. Application of total energy balance for the flow of a nonideal gas (steam turbine) Superheated steam enters a turbine under such conditions that the enthalpy of the entering steam is 1340 Btu/lb. On the same basis, the enthalpy of the steam leaving the turbine is 990 Btu/lb. If the turbine operates under adiabatic conditions and changes in kinetic energy and elevation potential energy are negligible, estimate the maximum amount of energy obtainable from the turbine per pound of entering steam.

SOLUTION

Basis: 1 lb of entering steam

For an adiabatic system and negligible change in potential and kinetic energies, the total energy balance becomes

$$W = h_2 - h_1 = 990 - 1340 = -350 \text{ Btu/lb of steam}$$

Maximum energy obtainable from the turbine = 350 Btu/lb of steam.

When the data necessary for application of the total energy balance are not available, the engineer may be forced to use the total mechanical-energy balance for design calculations, even though compressible fluids and large pressure drops are involved. The following example illustrates the general method for applying the total mechanical-energy balance under these conditions.†

Example 4. Application of the total mechanical-energy balance for the flow of a compressible fluid with high pressure drop Air is forced at a rate of 15 lb/min through a straight, horizontal, steel pipe having an inside diameter of 2.067 in. The pipe is 3000 ft long, and the pump is located at the upstream end of the pipe. The air enters the pump through a 2.067-in.-ID pipe. The pressure in the pipe at the downstream end of the system is 5 psig, and the temperature is 70°F. If the air pressure in the pipe at the entrance to the pump is 10 psig and the temperature is 80°F, determine the following:

(*a*) The pressure in the pipe at the exit from the pump.

(*b*) The mechanical energy as foot-pounds force per minute added to the air by the pump, assuming the pump operation is isothermal.

SOLUTION

Basis: 1 lb mass of flowing air

Because the amount of heat exchanged between the surroundings and the system is unknown, the total energy balance cannot be used to solve this problem. However, an approximate result can be obtained from the total mechanical-energy balance.

Designate point 1 as the entrance to the pump, point 2 as the exit from the pump, and point 3 as the downstream end of the pipe. Under these

---

† For additional discussion and methods for integrating the total mechanical-energy balance when the flow of a compressible fluid and high pressure drop are involved, see J. H. Perry and C. H. Chilton, "Chemical Engineers' Handbook," 5th ed., McGraw-Hill Book Company, New York, 1973.

conditions, the total mechanical-energy balance for the system between points 2 and 3 may be written as follows:

$$\int_2^3 v \, dp + \int_2^3 \frac{V \, dV}{\alpha g_c} = -\int_2^3 \delta F = -\int_2^3 \frac{2 \mathfrak{f} V^2 \, dL}{g_c D} \tag{A}$$

The mass velocity G [as $\text{lbm}/(\text{s})(\text{ft}^2)$] is constant, and

$$V = Gv$$

$$dV = G \, dv$$

Eliminating V and dV from Eq. (A) and dividing by v^2 gives

$$-\int_2^3 \frac{dp}{v} = \int_2^3 \frac{G^2 \, dv}{\alpha g_c v} + \int_2^3 \frac{2 \mathfrak{f} G^2 \, dL}{g_c D} \tag{B}$$

Assume that air acts as an ideal gas at the pressures involved, or

$$v = \frac{RT}{Mp} \tag{C}$$

where M = molecular weight of air = 29 lb/lb mol
    R = ideal-gas-law constant = 1545 $(\text{lbf}/\text{ft}^2)(\text{ft}^3)/(\text{lb mol})(°R)$
    T = temperature, °R

Substituting Eq. (C) into Eq. (B) and integrating gives

$$\frac{M}{2RT_{\text{avg}}} (p_2^2 - p_3^2) = \frac{G^2}{\alpha g_c} \ln \frac{v_3}{v_2} + \frac{2 \mathfrak{f}_{\text{avg}} G^2 L}{g_c D} \tag{D}$$

T_{avg} represents the average absolute temperature between points 2 and 3, and temperature variations up to 20 percent from the average absolute value will introduce only a small error in the final result. The error introduced by using a constant $\mathfrak{f}_{\text{avg}}$ (based on average temperature and pressure) instead of the exact integrated value is not important unless pressure variations are considerably greater than those involved in this problem.

If the pump operation is isothermal, $T_2 = 80°F$ and $T_{\text{avg}} = 75 + 460 = 535°R$. NOTE: If the pump operation were assumed to be adiabatic, a different value for T_{avg} would be obtained.

At 535°R,

$$\mu_{\text{air}} = 0.018 \text{ cp} = (0.018)(0.000672) \text{ lb}/(\text{s})(\text{ft})$$

$$G = \frac{(15)(144)(4)}{(60)(2.067)^2(\pi)} = 10.77 \text{ lb}/(\text{s})(\text{ft}^2)$$

$$N_{\text{Re}} = \frac{DG}{\mu} = \frac{(2.067)(10.77)}{(12)(0.018)(0.000672)} = 153{,}000$$

$$\alpha = 1$$

$$\frac{\varepsilon}{D} = \frac{(0.00015)(12)}{2.067} = 0.00087$$

From Fig. 13-1,

$$\bar{f}_{avg} = 0.0052$$

$$p_3 = (5 + 14.7)(144) = 2840 \text{ lbf/ft}^2$$

Since air is assumed to act as an ideal gas,

$$\frac{v_3}{v_2} = \frac{T_3 p_2}{T_2 p_3} = \frac{(530)p_2}{(540)(2840)} = \frac{p_2}{2890}$$

Substituting into Eq. (D),

$$\frac{29}{(2)(1545)(535)}[p_2^2 - (2840)^2] = \frac{(10.77)^2}{32.17} \ln \frac{p_2}{2890}$$

$$+ \frac{(2)(0.0052)(10.77)^2(3000)(12)}{(32.17)(2.067)}$$

By trial-and-error solution,

(a) $p_2 = $ pressure in pipe at exit from pump

$$= 6750 \text{ psf} = \tfrac{6750}{144} \text{ psia} = 47 \text{ psia}$$

(b) The mechanical energy added by the pump can be determined by making a total mechanical-energy balance between points 1 and 3:

$$\int_1^3 v \, dp + \int_1^3 \frac{V \, dV}{\alpha g_c} = \int_1^3 \delta W_o - \int_1^3 \delta F \tag{E}$$

The friction term, by definition, includes all friction except that occurring at the pump. Therefore,

$$\int_1^3 \delta F = \int_2^3 \delta F$$

and

$$\int_1^2 v \, dp + \int_2^3 v \, dp + \int_1^2 \frac{V \, dV}{\alpha g_c} + \int_2^3 \frac{V \, dV}{\alpha g_c} = W_o - \int_2^3 \delta F \tag{F}$$

Subtracting Eq. (A) from Eq. (F) gives

$$\int_1^2 v \, dp + \int_1^2 \frac{V \, dV}{\alpha g_c} = W_o \tag{G}$$

The value of $\int_1^2 v \, dp$ depends on the conditions or path followed in the pump, and the integral can be evaluated if the necessary p-v relationships are known. Although many pumps and compressors operate near adiabatic conditions, the pump operation will be assumed as isothermal in this example.

For an ideal gas and isothermal compression,

$$\int_1^2 v\, dp = \frac{RT}{M} \ln \frac{p_2}{p_1}$$

$$V_1 = G v_1 = \frac{(10.77)(359)(540)(14.7)}{(29)(492)(24.7)} = 87 \text{ ft/s}$$

$$V_2 = V_1 \frac{p_1}{p_2} = \frac{(87)(24.7)}{47} = 46 \text{ ft/s}$$

$$W_o = \frac{RT}{M} \ln \frac{p_2}{p_1} + \frac{V_2^2}{2g_c} - \frac{V_1^2}{2g_c}$$

$$W_o = \frac{(1545)(540)}{29} \ln \frac{47}{24.7} + \frac{(46)^2}{(2)(32.17)} - \frac{(87)^2}{(2)(32.17)}$$

$$= 18,400 \text{ ft·lbf/lbm}$$

Mechanical energy added to the air by the pump, assuming the pump operation is isothermal $= (18,400)(15) = 276,000$ ft·lbf/min.

The total power supplied to the pump could be determined if the isothermal efficiency of the pump (including any end effects caused by the pump housing) were known.

PIPING STANDARDS

Pipe Strength

Iron and steel pipes were originally classified on the basis of wall thickness as standard, extra-strong, and double-extra-strong. Modern industrial demands for more exact specifications have made these three classifications obsolete. Pipes are now specified according to wall thickness by a standard formula for *schedule number* as designated by the American Standards Association.

The bursting pressure of a thin-walled cylinder may be estimated from the following equation:

$$P_b = \frac{2 S_T t_m}{D_m} \tag{12}$$

where P_b = bursting pressure (difference between internal and external pressures), psi

S_T = tensile strength, psi

t_m = minimum wall thickness, in.

D_m = mean diameter, in.

A safe working pressure P_s can be evaluated from Eq. (12) if the tensile strength is replaced by a safe working fiber stress S_s.

$$P_s = \frac{2 S_s t_m}{D_m} \tag{13}$$

Schedule number is defined by the American Standards Association as the approximate value of

$$1000 \frac{P_s}{S_s} = \text{schedule number} \qquad (14)$$

For temperatures up to 250°F, the recommended safe working stress is 9000 psi for lap-welded steel pipe and 6500 psi for butt-welded steel pipe.† If the schedule number is known, the safe working pressure can be estimated directly from Eq. (14).

Ten schedule numbers are in use at the present time. These are 10, 20, 30, 40, 60, 80, 100, 120, 140, and 160. For pipe diameters up to 10 in., schedule 40 corresponds to the former "standard" pipe and schedule 80 corresponds to the former "extra-strong" pipe. The original "double-extra-strong" pipe is not represented by a definite schedule number.

Nominal Pipe Diameter

Pipe sizes are based on the approximate diameter and are reported as nominal pipe sizes. Although the wall thickness varies depending on the schedule number, the outside diameter of any pipe having a given nominal size is constant and independent of the schedule number. This permits the use of standard fittings and threading tools on pipes of different schedule numbers. A table showing outside diameters, inside diameters, and other dimensions for pipes of different diameters and schedule numbers is presented in the Appendix.

Tubing

Copper tubing and brass tubing are used extensively in industrial operations. Other metals, such as nickel and stainless steel, are also available in the form of tubing. Although pipe specifications are based on standard nominal sizes, tubing specifications are based on the actual outside diameter with a designated wall thickness. Conventional systems, such as the Birmingham wire gauge (BWG), are used to indicate the wall thickness. Common designations of tubing dimensions are given in the Appendix.

Fittings and Other Piping Auxiliaries

Threaded fittings, flanges, valves, flow meters, steam traps, and many other auxiliaries are used in piping systems to connect pieces of pipe together, change the direction of flow, regulate the flow, or obtain desired conditions in a flow system. Flanges are usually employed for piping connections when the pipe diameter is

---

† For allowable stresses at other temperatures and for other materials of construction, see J. H. Perry and C. H. Chilton, "Chemical Engineers' Handbook," 5th ed., p. 6–38, McGraw-Hill Book Company, New York, 1973.

3 in. or larger, while screwed fittings are commonly used for smaller sizes. In the case of cast-iron pipe used as underground water lines, bell-and-spigot joints are ordinarily employed rather than flanges.

The auxiliaries in piping systems must have sufficient structural strength to resist the pressure or other strains encountered in the operation, and the design engineer should provide a wide safety margin when specifying the ratings of these auxiliaries. Fittings, valves, steam traps, and similar items are often rated on the basis of the safe operating pressure as (a) low pressure (25 psi), (b) standard (125 psi), (c) extra-heavy (250 psi), or (d) hydraulic (300 to 10,000 psi). Figures D-5 and D-6 in the Appendix show examples of standard designations used to indicate various types of fittings and auxiliaries in sketches of piping systems.

DESIGN OF PIPING SYSTEMS†

The following items should be considered by the engineer when developing the design for a piping system:

1. Choice of materials and sizes
2. Effects of temperature level and temperature changes
   a. Insulation
   b. Thermal expansion
   c. Freezing
3. Flexibility of the system for physical or thermal shocks
4. Adequate support and anchorage
5. Alterations in the system and the service
6. Maintenance and inspection
7. Ease of installation
8. Auxiliary or stand-by pumps and lines
9. Safety
   a. Design factors
   b. Relief valves and flare systems

In the early years of industrial development in the United States, many plants buried their outside pipelines. The initial cost for this type of installation is low because no supports are required and the earth provides insulation. However, location and repair of leaks are difficult, and other pipes buried in the same trench may make repairs impossible. Above-ground piping systems in industrial plants have proven to be more economical than buried systems, and, except for major water and gas lines, most in-plant piping systems in new plants are now located above ground or in crawl-space tunnels.

† For a detailed treatment of this subject see H. F. Rase, "Piping Design for Process Plants," John Wiley and Sons, Inc., New York, 1963.

Thermal expansion and the resultant pipe stresses must be considered in any piping-system design. For example, if the temperature changes from 50 to 600°F, the increase in length would be 4.9 in. per 100 ft for steel pipe and 7.3 in. per 100 ft for brass pipe. This amount of thermal expansion could easily cause a pipe or wall to buckle if the pipe were fastened firmly at each end with no allowances for expansion. The necessary flexibility for the piping system can be provided by the use of expansion loops, changes in direction, bellows joints, slip joints, and other devices.

The possibility of solidification of the fluid should not be overlooked in the design of a piping system. Insulation, steam tracing, and sloping the line to drain valves are methods for handling this type of problem.

Water hammer may cause extreme stresses at bends in pipelines. Consequently, liquid pockets should be avoided in steam lines through the use of steam traps and sloping of the line in the direction of flow. Quick-opening or quick-closing valves may cause damaging water hammer, and valves of this type may require protection by use of expansion or surge chambers.

A piping system should be designed so that maintenance and inspection can be accomplished easily, and the possibility of future changes in the system should not be overlooked. Personal-safety considerations in the design depend to a large extent on the fluids, pressures, and temperatures involved. For example, an overhead line containing a corrosive acid should be shielded from open walkways, and under no conditions should an unprotected flange in this type of piping system be located immediately over a walkway.

Pipe Sizing

The design engineer must specify the diameter of pipe that will be used in a given piping system, and economic factors must be considered in determining the optimum pipe diameter. Theoretically, the optimum pipe diameter is the one that gives the least total cost for annual pumping power and fixed charges with the particular piping system. Many short-cut methods have been proposed for estimating optimum pipe diameters, and some general "rules of thumb" for use in design estimates of pipe diameters are presented in Table 2.

The derivation of equations for determining optimum economic pipe diameters is presented in Chap. 10 (Optimum Design and Design Strategy). The following simplified equations [Eqs. (45) and (47) from Chap. 10] can be used for making design estimates:

For turbulent flow ($N_{Re} > 2100$) in steel pipes

$$D_{i,opt} = 3.9 q_f^{0.45} \rho^{0.13} \tag{15}$$

For viscous flow ($N_{Re} < 2100$) in steel pipes

$$D_{i,opt} = 3.0 q_f^{0.36} \mu_c^{0.18} \tag{16}$$

Table 2 "Rule-of thumb" economic velocities for sizing steel pipelines

Turbulent flow	
Type of fluid	Reasonable velocity, ft/s
Water or fluid similar to water	3–10
Low-pressure steam (25 psig)	50–100
High-pressure steam (100 psig and up)	100–200
Air at ordinary pressures (25–50 psig)	50–100

The preceding values apply for motor drives. Multiply indicated velocities by 0.6 to give reasonable velocities when steam turbine drives are used.

Viscous flow (liquids)			
Nominal pipe diameter, in.	Reasonable velocity, ft/s		
	μ_c† = 50	μ_c = 100	μ_c = 1000
1	1.5–3	1–2	0.3–0.6
2	2.5–3.5	1.5–2.5	0.5–0.8
4	3.5–5.0	2.5–3.5	0.8–1.2
8		4.0–5.0	1.3–1.8

† μ_c = viscosity, centipoises.

where $D_{i,\text{opt}}$ = optimum inside pipe diameter, in.
q_f = fluid flow rate, ft³/s
ρ = fluid density, lb/ft³
μ_c = fluid viscosity, centipoises

The preceding equations are the basis for the nomograph presented in Fig. 13-2, and this figure can be used for estimating the optimum diameter of steel pipe under ordinary plant conditions. Equations (15) and (16) should not be applied when the flowing fluid is steam, because the derivation makes no allowance for the effects of pressure drop on the value of the flowing material. Equation (15) is limited to conditions in which the viscosity of the fluid is between 0.02 and 20 centipoises.

As discussed in Chap. 10, the constants in Eqs. (15) and (16) are based on average cost and operating conditions. When unusual conditions are involved or when a more accurate determination of the optimum diameter is desired, other equations given in Chap. 10 can be used.

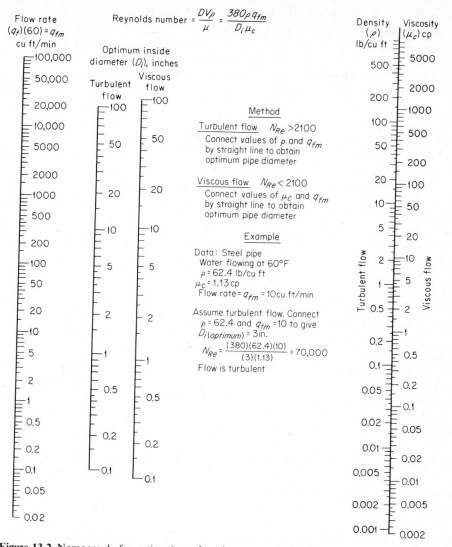

Figure 13-2 Nomograph for estimation of optimum economic pipe diameters with turbulent or viscous flow based on Eqs. (15) and (16).

COSTS FOR PIPING AND PIPING-SYSTEM AUXILIARIES

Piping is a major item in the cost of chemical process plants. These costs in a fluid-process plant can run as high as 80 percent of the purchased equipment cost or 20 percent of the fixed-capital investment. There are essentially two basic methods for preparing piping-cost estimates—the percentage of installed equipment method and the material and labor take-off method. Several variations of each method have appeared in the literature.

The percentage of installed equipment method as described in Chap. 5 is a quick procedure for preliminary or order-of-magnitude type of cost estimates. In the hands of experienced estimators it can be a reasonably accurate method, particularly on repetitive type units. It is not recommended on alteration jobs or on projects where the total installed equipment is less than $100,000.

The material and labor take-off method is the recommended method for definitive estimates where accuracy within 10 percent is required. To prepare a cost estimate by this method usually requires piping drawings and specifications, material costs, fabrication and erection labor costs, testing costs, auxiliaries, supports, and painting requirements. The take off from the drawings must be made with the greatest possible accuracy because it is the basis for determining material and labor costs. In the case of revisions to existing facilities, thorough field study is necessary to determine job conditions and their possible effects.

Although accurate costs for pipes, valves, and piping system auxiliaries can be obtained only by direct quotations from manufacturers, the design engineer can often make satisfactory estimates from data such as those presented in Figs. 13-3 through 13-34. The cost of materials and installation time presented in these figures covers the types of equipment most commonly encountered in industrial operations.

Piping labor consists of cutting, fitting, welding and/or threading, and field assembly. It frequently may be as high as 200 percent of materials cost. This labor is generally calculated on either the "diameter-inch" method or the "lineal-foot" method. In the diameter-inch method, all connections (threaded or welded) are counted and multiplied by the nominal pipe diameter. This diameter-inch factor is then multiplied by labor factors of employee-hour/diameter-inch to yield employee-hours to fabricate and erect a piping system. Such a technique requires less data for each line size and for varying conditions of complexity; however, it also has certain limitations. Many estimators use 1.0 employee-hour/diameter-inch for welding carbon-steel pipe. This factor can give labor cost estimates that are up to 25 to 30 percent too high, particularly if an efficiently operated field-fabrication shop is available. The lineal-foot method, on the other hand, estimates the piping-installation costs by applying employee-hour units to the erection of the pipe (considering the length of the piping system), the installation of valves, fittings, and auxiliaries, and the welding or threading of piping components. Accurate estimation by this method requires that the engineering and design of the system be well along so that piping flowsheets, elevations, isometrics, etc., are available for a material take-off.

PUMPS

Pumps are used to transfer fluids from one location to another. The pump accomplishes this transfer by increasing the pressure of the fluid and, thereby, supplying the driving force necessary for flow. Power must be delivered to the

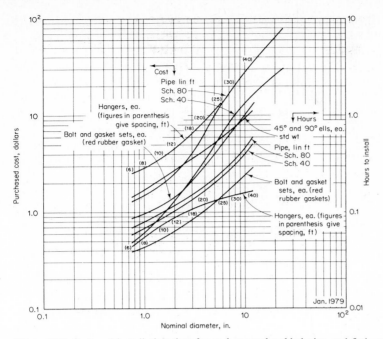

Figure 13-3 Cost and installation time for carbon-steel welded pipe and fittings.

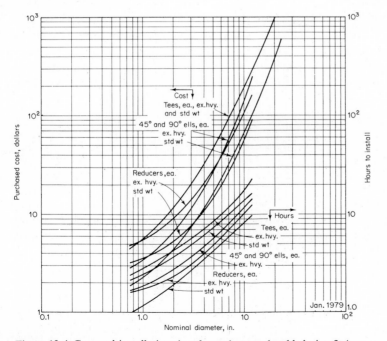

Figure 13-4 Cost and installation time for carbon-steel welded pipe fittings.

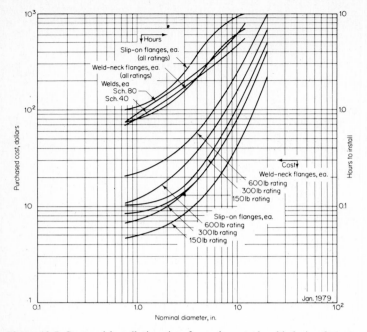

Figure 13-5 Cost and installation time for carbon-steel welded pipe fittings.

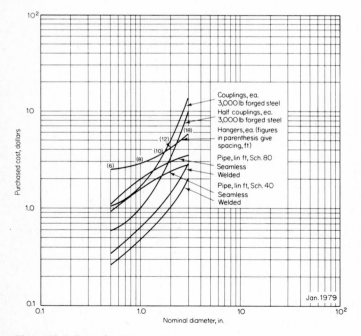

Figure 13-6 Cost of carbon-steel screwed pipe and fittings.

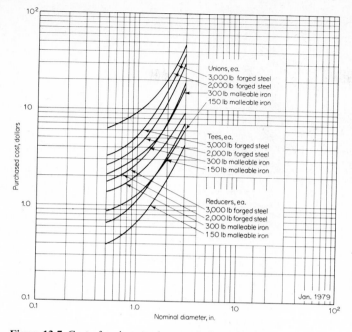

Figure 13-7 Cost of carbon-steel screwed pipe fittings.

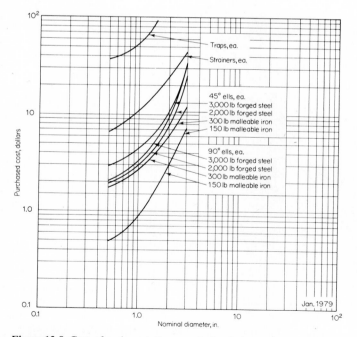

Figure 13-8 Cost of carbon-steel screwed pipe fittings.

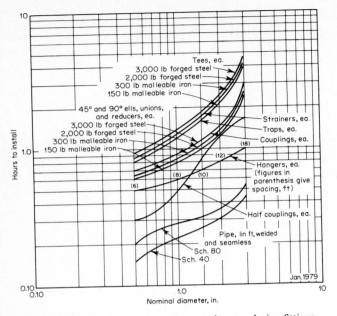

Figure 13-9 Installation time for carbon-steel screwed pipe fittings.

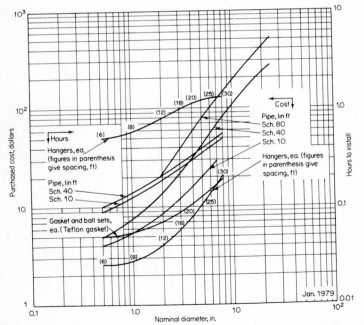

Figure 13-10 Cost and installation time for stainless-steel welded pipe and fittings. Prices are for types 304 and 304L. For types 316 and 347, multiply by 1.25; for type 316L, multiply by 1.45.

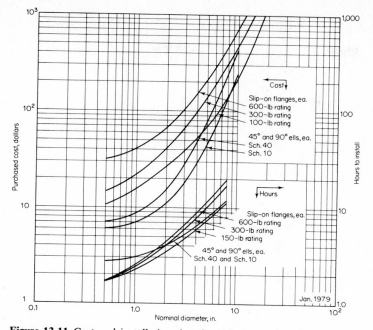

Figure 13-11 Cost and installation time for stainless-steel pipe fittings. Prices are for types 304 and 304 L. For types 316 and 347, multiply by 1.25; for type 316L, multiply by 1.45.

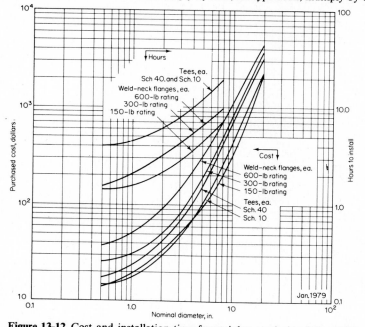

Figure 13-12 Cost and installation time for stainless-steel pipe fittings. Prices are for types 304 and 304L. For types 316 and 347, multiply by 1.25; for type 316L, multiply by 1.45.

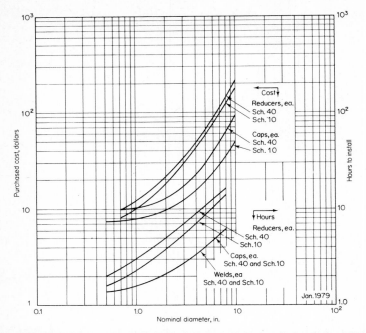

Figure 13-13 Cost and installation time for stainless-steel pipe fittings. Prices are for type 304 and 304L. For types 316 and 347, multiply by 1.25; for type 316L, multiply by 1.45.

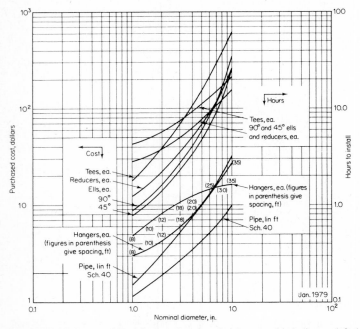

Figure 13-14 Cost and installation time for aluminum welded pipe and fittings.

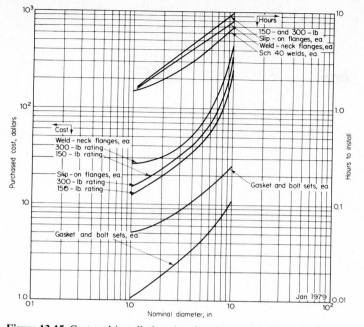

Figure 13-15 Cost and installation time for aluminum welding and pipe fittings.

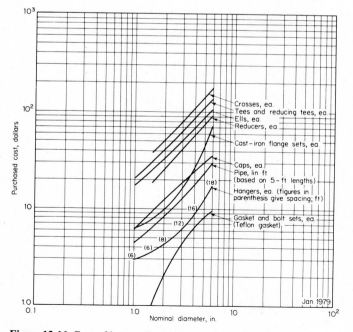

Figure 13-16 Cost of heat-resistant glass pipe and fittings.

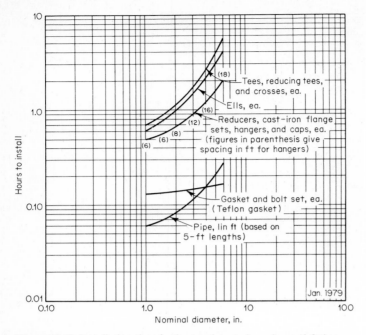

Figure 13-17 Installation time for heat-resistant glass pipe and fittings.

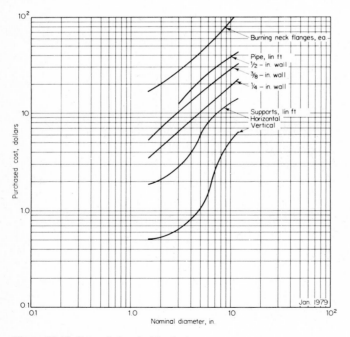

Figure 13-18 Cost of chemical-lead pipe.

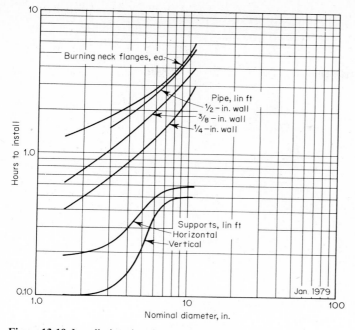

Figure 13-19 Installation time for chemical-lead pipe.

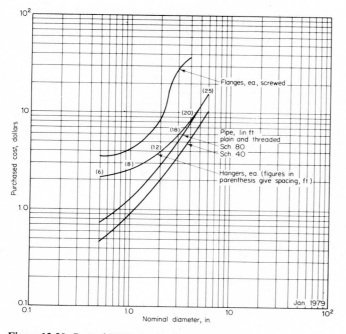

Figure 13-20 Cost of PVC plastic pipe and fittings.

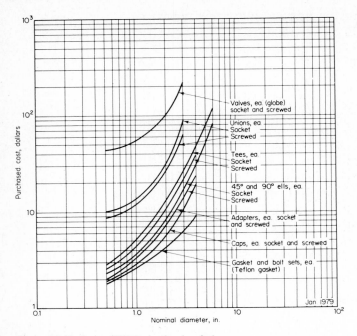

Figure 13-21 Cost of PVC plastic pipe fittings.

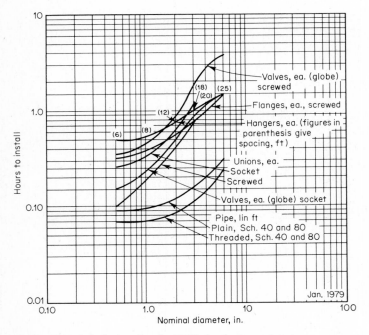

Figure 13-22 Installation time for PVC plastic pipe and fittings.

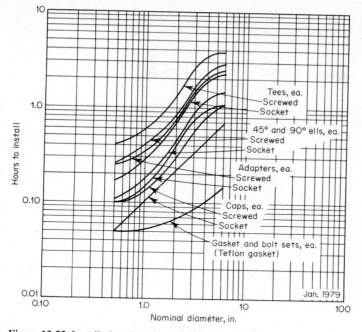

Figure 13-23 Installation time for PVC plastic pipe fittings.

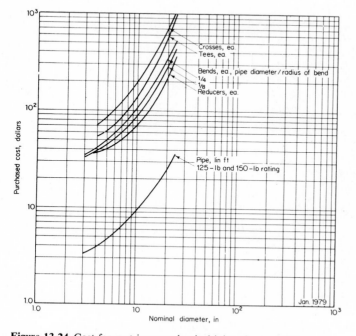

Figure 13-24 Cost for cast-iron mechanical-joint pipe and fittings.

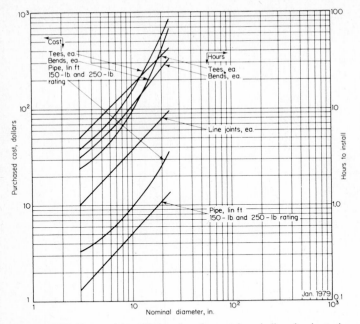

Figure 13-25 Cost and installation time for cast-iron bell-and-spigot pipe and fittings.

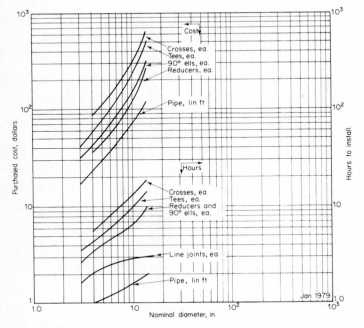

Figure 13-26 Cost and installation time for cast-iron flanged pipe and fittings. Cost of the pipe includes jointing material.

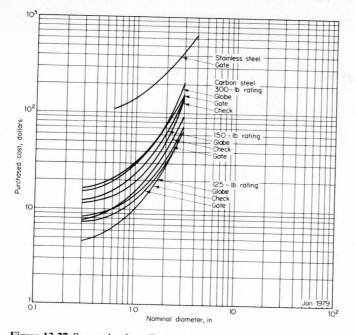

Figure 13-27 Screwed valves. For water, oil, and gas.

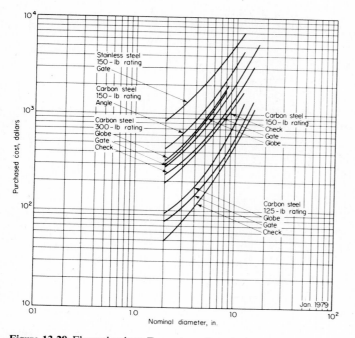

Figure 13-28 Flanged valves. For water, oil, and gas.

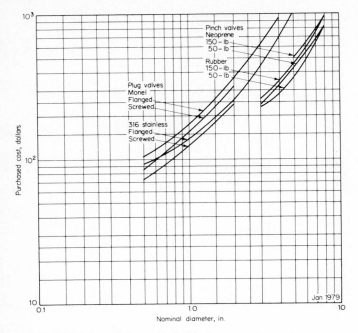

Figure 13-29 Cost of plug and pinch valves.

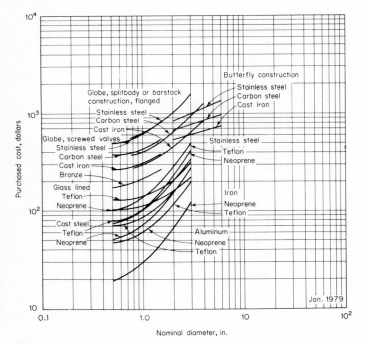

Figure 13-30 Cost of diaphragm valves.

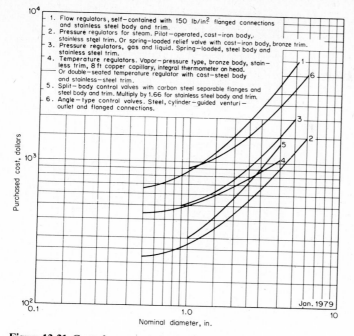

Figure 13-31 Costs for control and relief valves.

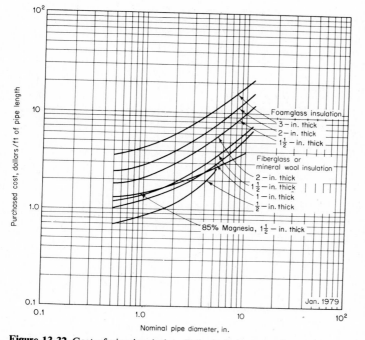

Figure 13-32 Cost of pipe insulation. Price includes cost of standard covering.

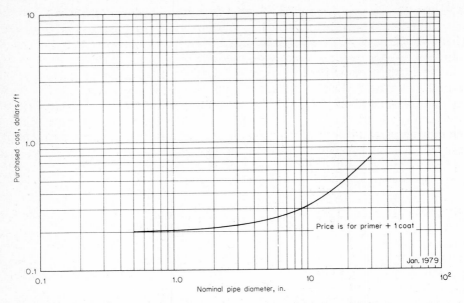

Figure 13-33 Cost of pipe painting. Cost includes material and labor, no overhead.

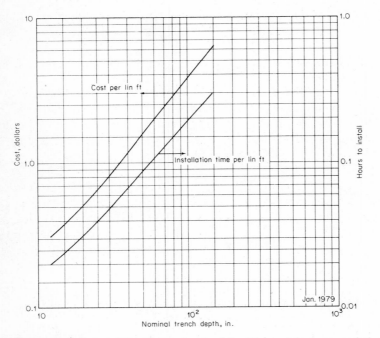

Figure 13-34 Cost and installation time for 2-ft-wide trench in damp sandy loam sloped $\frac{1}{2}$ to 1.

pump from some outside source. Thus, electrical or steam energy may be transformed into mechanical energy which is used to drive the pump. Part of this mechanical energy is added to the fluid as work energy, and the rest is lost as friction due to inefficiency of the pump and drive.

Although the basic operating principles of gas pumps and liquid pumps are similar, the mechanical details are different because of the dissimilarity in physical properties of gases and liquids. In general, pumps used for transferring gases operate at higher speeds than those used for liquids, and smaller clearances between moving parts are necessary for gas pumps because of the lower viscosity of gases and the greater tendency for the occurrence of leaks.

The different types of pumps commonly employed in industrial operations can be classified as follows:†

1. Reciprocating or positive-displacement pumps with valve action: piston pumps, diaphragm pumps, plunger pumps
2. Rotary positive-displacement pumps with no valve action: gear pumps, lobe pumps, screw pumps, eccentric-cam pumps, metering pumps
3. Rotary centrifugal pumps with no valve action: open impeller, closed impeller, volute pumps, turbine pumps
4. Air-displacement systems: air lifts, acid eggs or blow cases, jet pumps, barometric legs

Many different factors can influence the final choice of a pump for a particular operation. The following list indicates the major factors that govern pump selection:

1. The amount of fluid that must be pumped. This factor determines the size of pump (or pumps) necessary.
2. The properties of the fluid. The density and the viscosity of the fluid influence the power requirements for a given set of operating conditions; corrosive properties of the fluid determine the acceptable materials of construction. If solid particles are suspended in the fluid, this factor dictates the amount of clearance necessary and may eliminate the possibility of using certain types of pumps.
3. The increase in pressure of the fluid due to the work input of the pumps. The head change across the pump is influenced by the inlet and downstream-reservoir pressures, the change in vertical height of the delivery line, and frictional effects. This factor is a major item in determining the power requirements.

---

† For a detailed discussion on different types of pumps, see J. H. Perry and C. H. Chilton, "Chemical Engineers' Handbook," 5th ed., McGraw-Hill Book Company, New York, 1973, and R. F. Neerken, Pump Selection for the Chemical Process Industries, *Chem. Eng.*, **81**(4): 104 (Feb. 18, 1974).

4. Type of flow distribution. If nonpulsating flow is required, certain types of pumps, such as simplex reciprocating pumps, may be unsatisfactory. Similarly, if operation is intermittent, a self-priming pump may be desirable, and corrosion difficulties may be increased.
5. Type of power supply. Rotary positive-displacement pumps and centrifugal pumps are readily adaptable for use with electric-motor or internal-combustion-engine drives; reciprocating pumps can be used with steam or gas drives.
6. Cost and mechanical efficiency of the pump.

Reciprocating Pumps

A *reciprocating piston pump* delivers energy to a flowing fluid by means of a piston acting through a cylinder. Although steam is often employed as the source of power for this type of pump, the piston can be activated by other means, such as a rotating crankshaft operated by an electric motor. Thus, reciprocating piston pumps can be classified as *steam pumps* or *power pumps*. They can also be classified as *single-acting* or *double-acting* depending on whether energy is delivered to the fluid on both the forward and backward strokes of the piston.

Specifications for reciprocating steam pumps are expressed in abbreviated form as diameter of the steam cylinder, diameter of the water cylinder, and length of the piston stroke in inches. For example, a $7 \times 4\frac{1}{2} \times 12$ pump has a steam-cylinder diameter of 7 in., a water-cylinder diameter of $4\frac{1}{2}$ in., and a stroke length of 12 in. In general, reciprocating steam pumps having stroke lengths less than 10 in. should not operate at more than 100 strokes per minute because of excessive wear. For longer stroke lengths, reasonable piston speeds are in the range of 50 to 90 ft/min.

In reciprocating steam pumps, the steam is not used expansively as in the common types of steam engines. Instead, essentially the full pressure of the steam is maintained throughout the entire stroke by keeping the steam-inlet ports fully open during the stroke. As the piston moves forward through the cylinder on the delivery side of the pump, the fluid is compressed and forced out of the cylinder. By a system of opening and closing valves, the piston can deliver energy to the fluid with every stroke.

The rate of fluid discharge from the cylinder is zero at the beginning of a piston stroke and increases to a maximum value when the piston reaches full speed. If only one discharge cylinder is used, the flow rate will pulsate. These pulsations can be reduced by placing an air chamber on the discharge line or by using a number of delivery cylinders suitably compounded. *Simplex* pumps have only one delivery cylinder, *duplex* pumps have two cylinders, and *triplex* pumps have three cylinders.

The theoretical fluid displacement of a piston pump equals the total volume swept by the piston on each delivery stroke. Because of leakage past the piston and the valves and failure of the valves to close instantly, this theoretical displacement is not attained in actual practice. The *volumetric efficiency*, defined as the ratio of the actual displacement to the theoretical displacement, is usually in the range of 70 to 95 percent.

When a steam pump is used, the pressure of the steam in pounds per unit area times the area of the piston would be the maximum force that could be exerted on the work-delivery piston if the machine were perfect and no friction were involved. However, friction is involved and work must be done on the liquid (or work-receiving fluid) under conditions in which the steam pressure is a finite amount greater than the liquid pressure. The ratio of the pressure theoretically required on the steam piston to the pressure actually exerted by the steam is known as the *pressure efficiency* or *steam-end efficiency*. It includes the effects of piston and rod friction, momentum changes in acceleration of the piston and fluid, and leakage of fluid past the piston. The pressure efficiency varies from about 50 percent for small pumps up to 80 percent for large pumps.

Another so-called efficiency, known as *hydraulic efficiency*, is sometimes given for reciprocating pumps. This efficiency indicates losses due to velocity changes in the inlet and outlet of the pump, friction, and valves. It is defined as the ratio of the actual head across the pump to the sum of the actual head pumped and the losses in the suction and discharge lines.

For pumps driven by electric motors, overall efficiency is usually defined as the work done on the fluid divided by the electrical energy supplied to the motor. Attempts to apply this type of definition to the overall efficiency of steam pumps can give misleading results because the term "energy supplied by the steam" can have many meanings. For design estimates, an overall efficiency for steam pumps is sometimes defined as the work done on the fluid divided by the ideal work that could have been obtained by the isentropic expansion of the steam from its initial temperature and pressure to its exhaust pressure. The numerical value of this type of overall efficiency is significant only when the operating conditions are specified for the particular pump.

Reciprocating pumps, in general, have the advantage of being able to deliver fluids against high pressures and operate with good efficiency over a wide range of operating conditions. A major disadvantage of piston and plunger pumps is that they cannot be used with fluids which contain abrasive solids. Reciprocating diaphragm pumps, however, are satisfactory for handling fluids with large amounts of suspended solids at low heads.

Rotary Positive-Displacement Pumps

Pumps of this type combine a rotary motion with a positive displacement of the fluid. A common type of rotary gear pump is illustrated in Fig. 13-35. Two intermeshing gears are fitted into a casing with a sufficiently close spacing to seal off effectively each separate tooth space. As the gears rotate in opposite directions, fluid is trapped in each tooth space and is delivered to the exit side of the pump. Similar results can be obtained by using a rotating eccentric cam or separately driven impellers having several lobes.

No priming is required with rotary positive pumps, and they are well

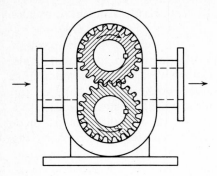

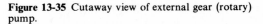

Figure 13-35 Cutaway view of external gear (rotary) pump.

adapted for pumping highly viscous fluids. A constant rate of delivery is obtained, and the fluid may be delivered at high pressures. Because of the small clearance that must be maintained, this type of pump should not be used with nonlubricating fluids or with fluids containing solid particles.

Rotary Centrifugal Pumps

In a centrifugal pump the fluid is fed into the pump at the center of a rotating impeller and is thrown outward by centrifugal force. The fluid at the outer periphery of the impeller attains a high velocity and, consequently, high kinetic energy. The conversion of this kinetic energy into pressure energy supplies the pressure difference between the suction side and the delivery side of the pump.

Different forms of impellers are used in centrifugal pumps. One common type, known as a *closed impeller*, consists of a series of curved vanes attached to a central hub and extending outward between two enclosing plates. An open impeller is similar, except that there are no enclosing plates. Impellers of this type are used in *volute pumps*, which are the simplest form of centrifugal pumps.

Energy losses caused by turbulence at the point where the liquid path changes from radial flow to tangential flow in the pump casing can be decreased by using so-called *turbine pumps*. With this type of centrifugal pump, the liquid flows from the impeller through a series of fixed vanes forming a diffusion ring. The change in direction of the fluid is more gradual than in a volute pump, and a more efficient conversion of kinetic energy into pressure energy is obtained.

For an ideal centrifugal pump, the speed of the impeller (N r/min) should be directly proportional to the rate of fluid discharge (q gpm), or

$$\frac{N_1}{N_2} = \frac{q_1}{q_2} \tag{17}$$

The head (or pressure difference) produced by the pump is a function of the kinetic energy developed at the point of release from the impeller. The head developed by an ideal pump, therefore, should be directly proportional to the square of the impeller speed:

$$\frac{\text{Head}_1}{\text{Head}_2} = \frac{q_1^2}{q_2^2} = \frac{N_1^2}{N_2^2} \tag{18}$$

The power required for a perfect pump is directly proportional to the product of the head and the flow rate; therefore,

$$\frac{\text{Power}_1}{\text{Power}_2} = \frac{q_1^3}{q_2^3} = \frac{N_1^3}{N_2^3} \tag{19}$$

The preceding equations apply for the ideal case in which there are no frictional, leakage, or recirculation losses. In any real pump, however, these losses do occur, and their magnitudes can be determined only by actual tests. As a result, characteristic curves are usually supplied by pump manufacturers to indicate the performance of any particular centrifugal pump. Figure 13-36 shows a typical set of characteristic curves for a centrifugal pump.

The ratings of centrifugal pumps are ordinarily based on the head and capacity of the point of maximum efficiency. The size of the pump is usually specified as the diameter of the discharge opening. The rating for the pump referred to in Fig. 13-36 would be 140 gpm and a head of 40 ft if water (viscosity of 1 centistoke) is the fluid involved. From the data shown in Fig. 13-36, if the head is increased to 50 ft, the capacity will decrease to 80 gpm and the efficiency of the pump will decrease. The capacity could be decreased to 80 gpm at a head of 40 ft by throttling the discharge so that a head of 50 ft is actually generated within the pump, but this would result in a reduction in the pump efficiency. Consequently, the design engineer should always attempt to give the necessary pump specifications as accurately as possible in order to obtain the correct pump which will operate at maximum efficiency.

Figure 13-37 gives values that are suitable for design estimates of centrifugal-pump efficiencies. Because pump and driver efficiencies must both be considered when total power costs are determined, necessary design data on the efficiency of electric motors are presented in Fig. 13-38.

The following list gives the major advantages and disadvantages of centrifugal pumps:

Advantages

1. They are simple in construction and cheap.
2. Fluid is delivered at uniform pressure without shocks or pulsations.
3. They can be coupled directly to motor drives. In general, the higher the speed, the smaller the pump and motor required for a given duty.
4. The discharge line may be partly shut off or completely closed off without damaging the pump.
5. They can handle liquid with large amounts of solids.
6. There are no close metal-to-metal fits.

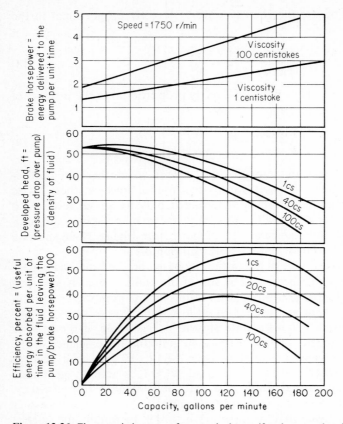

Figure 13-36 Characteristic curves for a typical centrifugal pump showing effect of viscosity.

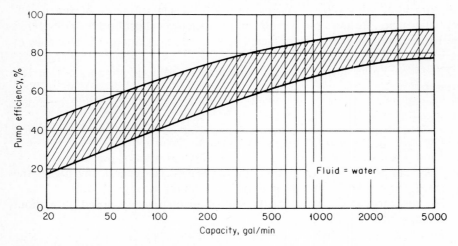

Figure 13-37 Efficiencies of centrifugal pumps.

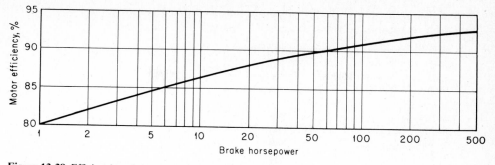

Figure 13-38 Efficiencies of three-phase motors.

7. There are no valves involved in the pump operation.
8. Maintenance costs are lower than for other types of pumps.

Disadvantages

1. They cannot be operated at high heads.
2. They are subject to air binding and usually must be primed.
3. The maximum efficiency for a given pump holds over a fairly narrow range of conditions.
4. They cannot handle highly viscous fluids efficiently.

Air-Displacement Systems

The pumps discussed in the preceding sections depend on the mechanical action of pistons, impellers, plungers, or other devices to move the fluid. Movement of fluids can also be accomplished by the use of air pressure, and many types of air-displacement systems have been developed for this purpose. The most common of these systems are air lifts, acid eggs, and jet pumps.

In the simple *air lift*, compressed air is introduced into the submerged end of the discharge pipe at a distance of H_s ft below the liquid surface. Because the air-and-liquid mixture is lighter than the liquid alone, the mixture rises through the discharge pipe and is expelled into an overhead receiver at a distance of H_t ft above the liquid surface. Although equations for the operation of an air lift can be developed theoretically, the frictional effects are so complex that the following empirical equation is usually assumed for the design basis:[†]

$$V_{air} = 0.8 \frac{H_t}{C_a \log [(H_s + 34)/34]} \tag{20}$$

[†] F. W. O'Neil, "Compressed Air Data," 5th ed., pp. 188–191, Ingersoll-Rand Company, New York, 1954.

where V_{air} is the cubic feet of free air (i.e., at normal atmospheric conditions) required to raise 1 gal of water, and C_a has the following values:

H_t, ft	Recommended value of $H_s/(H_s + H_t)$	C_a (outside air line)
20–125	0.65	348
126–250	0.60	335
251–400	0.50	296
401–650	0.40	246
651–700	0.35	216

Acid eggs or *blow cases* are simply closed vessels with inlet and outlet lines and an air connection. Air is admitted to the vessel and forces the liquid out through the discharge line. Operation of acid eggs is intermittent, and the elevation attained depends on the air pressure. Although these systems are inexpensive and easy to operate, they are inefficient. Their use is limited primarily to batchwise operations with corrosive fluids.

Jet pumps, employing water, steam, or gas as the operating medium, are often used for transferring fluids. The operating medium flows rapidly through an expanding nozzle and discharges into the throat of a venturi. As the operating medium issues from the nozzle, the high velocity head causes a decrease in the pressure head. If the resulting pressure is less than that of the second fluid at that point, the fluid will be sucked into the venturi throat along with the operating medium and discharged from the venturi outlet.

Jet pumps are used to remove air, gases, or vapors from condensers and vacuum equipment, and the steam jets can be connected in series or parallel to handle larger amounts of gas or to develop a greater vacuum. The capacity of steam-jet ejectors is usually reported as pounds per hour instead of on a volume basis. For design purposes, it is often necessary to make a rough estimate of the steam requirements for various ejector capacities and conditions. The data given in Table 3 can be used for this purpose.

Barometric-leg pumps are used for assisting in maintaining a vacuum when condensable vapors are involved. Auxiliary pumps are necessary to remove any fixed gases that may accumulate in the leg. For water condensation, the leg usually empties into an open well, and the vertical length of the leg must be longer than 34 ft.

Gas Compressors[†]

Movement of gases can be accomplished by use of fans, blowers, vacuum pumps, and compressors. Fans are useful for moving gases when pressure differences less

[†] For a detailed discussion on equipment characteristics for preliminary selection of type and size of compressors to handle gases in process plants with ranges from very high pressures to vacuum for various flow conditions, see R. F. Neerken, Compressor Selection for the Chemical Process Industries, *Chem. Eng.*, **82**(2):78 (Jan. 20, 1975).

Table 3 Steam-jet average consumption of steam at 100 psig in pounds per hour†

For larger or smaller capacities, steam consumption is approximately proportional to the capacity.

Capacity		Suction pressure, in. Hg absolute									
Wt of gas-vapor mixture handled, lb/h	Wt % net dry air (non-condensables) in gas–vapor mixture	0.5		1.0		1.5	2.0	3.0	4.0		6.0
		3-stage	2-stage	3-stage	2-stage	2-stage	2-stage	2-stage	2-stage	1-stage	1-stage
10	100	73	99	59	70	58	50	42	38	58	36
10	70	59	84	47	60	49	42	35	31	63	39
10	40	45	68	33	47	38	32	26	23	68	41
10	10	24	45	16	28	21	17	14	12	74	42

† J. H. Perry, "Chemical Engineers' Handbook," 4th ed., McGraw-Hill Book Company, New York, 1963.

than about 0.5 psi are involved. Centrifugal blowers can handle large volumes of gases, but the delivery pressure is limited to approximately 50 psig. Reciprocating compressors can be employed over a wide range of capacities and pressures, and they are used extensively in industrial operations. Sizes of reciprocating compressors ranging from less than 1 to 3000 hp are available, and some types can give delivery pressures as high as 4000 atm.

Compressor efficiencies are usually expressed as isentropic efficiencies, i.e., on the basis of an adiabatic and reversible process. Isothermal efficiencies are sometimes quoted, and design calculations are simplified when isothermal efficiencies are used. In either case, the efficiency is defined as the ratio of the power required for the ideal process to the power actually consumed.

Because the energy necessary for an isentropic compression is greater than that required for an equivalent isothermal compression, the isentropic efficiency is always greater than the isothermal efficiency. For reciprocating compressors, isentropic efficiencies are generally in the range of 70 to 90 percent and isothermal efficiencies are about 50 to 70 percent. Multistage compression is necessary for high efficiency in most large compressors if the ratio of the delivery pressure to the intake pressure exceeds approximately 5 : 1.

Expressions for the theoretical power requirements of gas compressors can be obtained from the basic equations of thermodynamics. For an ideal gas undergoing an isothermal compression (pv = constant), the theoretical power requirement for any number of stages can be expressed as follows:

$$\text{Power} = p_1 v_1 \ln \frac{p_2}{p_1} \tag{21}$$

or

$$\text{hp} = 3.03 \times 10^{-5} p_1 q_{fm_1} \ln \frac{p_2}{p_1} \tag{22}$$

where power = power requirement, ft·lbf/lbm

$\quad p_1$ = intake pressure, lbf/ft^2

$\quad v_1$ = specific volume of gas at intake conditions, ft^3/lbm

$\quad p_2$ = final delivery pressure, lbf/ft^2

$\quad q_{fm_1}$ = cubic feet of gas per minute at intake conditions

Similarly, for an ideal gas undergoing an isentropic compression ($pv^k =$ constant), the following equations apply:

For single-stage compressor

$$\text{Power} = \frac{k}{k-1} p_1 v_1 \left[\left(\frac{p_2}{p_1} \right)^{(k-1)/k} - 1 \right] \tag{23}$$

$$\text{hp} = \frac{3.03 \times 10^{-5} k}{k-1} p_1 q_{fm_1} \left[\left(\frac{p_2}{p_1} \right)^{(k-1)/k} - 1 \right] \tag{24}$$

$$p_2 = p_1 \left(\frac{v_1}{v_2} \right)^k = p_1 \left(\frac{T_2}{T_1} \right)^{k/(k-1)} \tag{25}$$

$$T_2 = T_1 \left(\frac{v_1}{v_2} \right)^{k-1} = T_1 \left(\frac{p_2}{p_1} \right)^{(k-1)/k} \tag{26}$$

For multistage compressor, assuming equal division of work between cylinders and intercooling of gas to original intake temperature,

$$\text{hp} = \frac{3.03 \times 10^{-5} k N_s}{k-1} p_1 q_{fm_1} \left[\left(\frac{p_2}{p_1} \right)^{(k-1)/kN_s} - 1 \right] \tag{27}$$

$$T_2 = T_1 \left(\frac{p_2}{p_1} \right)^{(k-1)/kN_s} \tag{28}$$

where k = ratio of specific heat of gas at constant pressure to specific heat of gas at constant volume

$\quad v_2$ = specific volume of gas at final delivery conditions, ft^3/lbm

$\quad T_1$ = absolute temperature of gas at intake conditions, °R

$\quad T_2$ = absolute temperature of gas at final delivery conditions, °R

$\quad N_s$ = number of stages of compression

Cost of Pumping Machinery

Figures 13-39 through 13-56 give approximate costs for different types of pumps, compressors, blowers, fans, and motors. Although the data from these figures can be used for preliminary design estimates, firm estimates should be based on manufacturers' quotations.

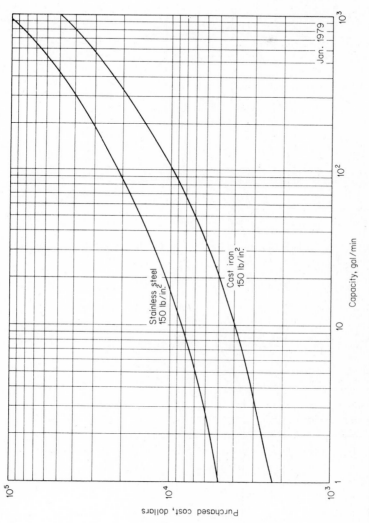

Figure 13-39 Cost of reciprocating pumps. Price includes pump and motor.

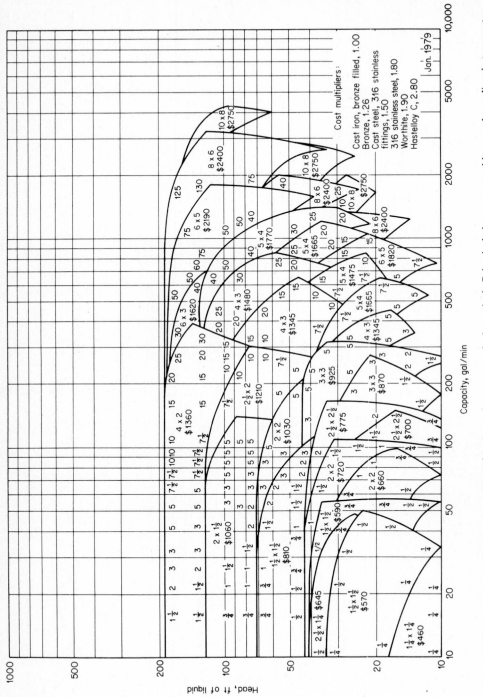

Figure 13-40 Pumps: general-purpose centrifugal (single- and two-stage, single-suction). Price includes pump, steel base, and coupling, but no motor. Small numbers within selection blocks indicate approximate horsepower.

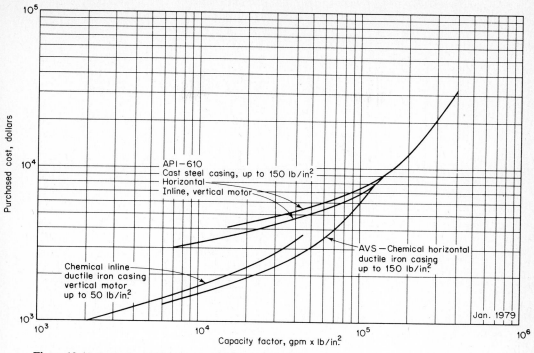

Figure 13-41 Cost of centrifugal pumps. Price includes motor.

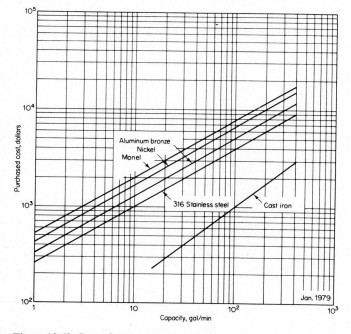

Figure 13-42 Cost of gear pumps, 100 psi discharge pressure. Cost includes pump, base plate, and V-belt drive, but no motor.

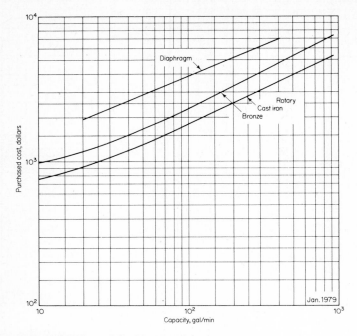

Figure 13-43 Cost of diaphragm and rotary pumps.

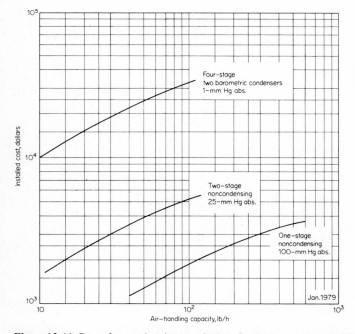

Figure 13-44 Cost of steam-jet ejectors. Carbon-steel construction, 1000 lb/h steam consumption.

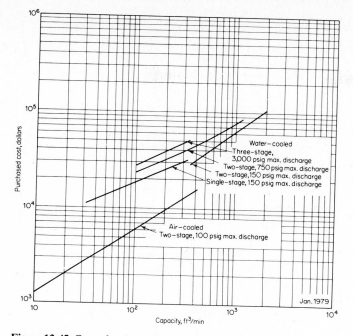

Figure 13-45 Cost of reciprocating compressors.

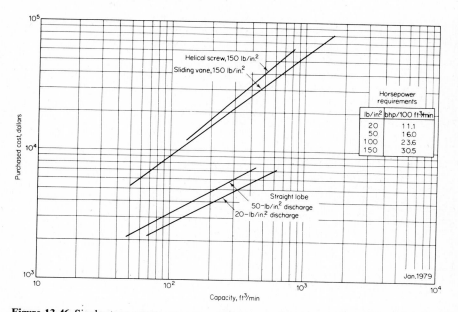

Figure 13-46 Single-stage rotary compressors. Prices are for completely packaged compressor units (freight and installation costs excluded). The straight lobe prices also exclude aftercooler, trap, and controls.

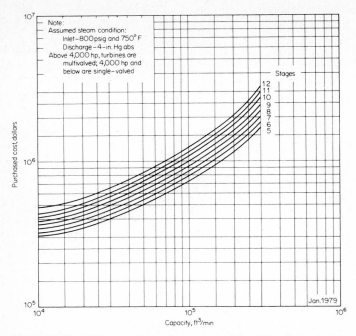

Figure 13-47 Axial compressor with steam turbine drive. Price includes multivalve steam turbine, base plate, and lubrication system.

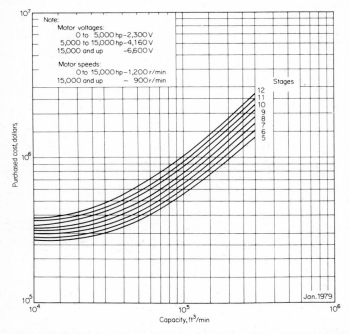

Figure 13-48 Axial compressor with motor-gear drive. Price includes synchronous motor with 0.8 P.F., suitable speed-increasing gear, lubrication system, and base plate for the compressor and drive.

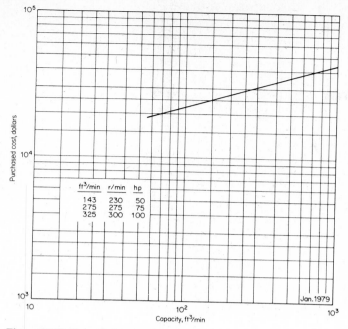

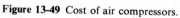

ft³/min	r/min	hp
143	230	50
275	275	75
325	300	100

Jan.1979

Figure 13-49 Cost of air compressors.

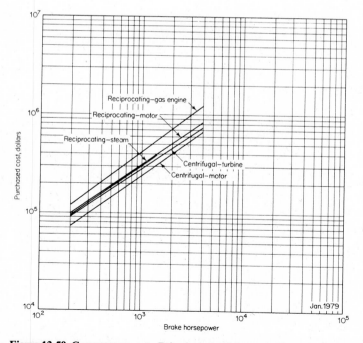

Jan.1979

Figure 13-50 Compressor costs. Price includes drive, gear mounting, base plate, and normal auxiliary equipment; operating pressure to 1000 psig.

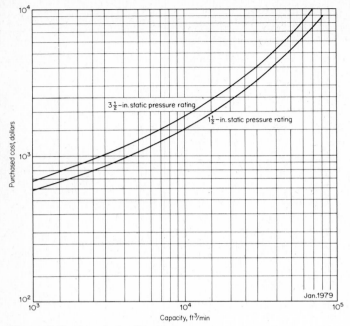

Figure 13-51 Cost of centrifugal fans.

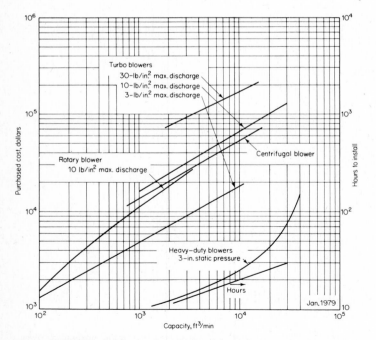

Figure 13-52 Blowers (heavy-duty, industrial type).

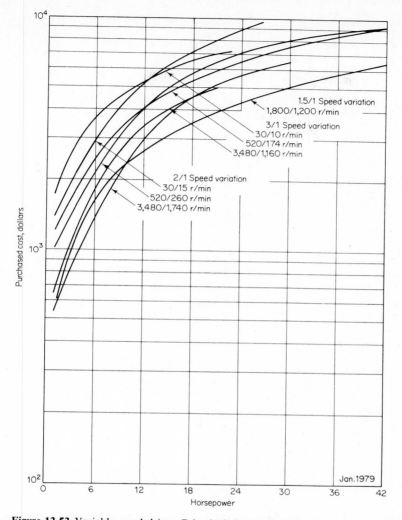

Figure 13-53 Variable-speed drives. Price includes handwheel control with a built-in indicator and TEFC motors as an integral part of the unit.

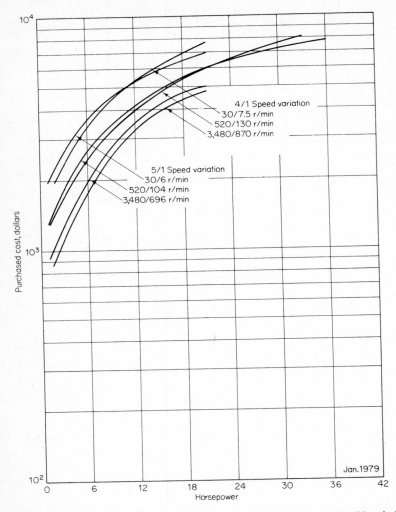

Figure 13-54 Variable-speed drives. Price includes handwheel control with a built-in indicator and TEFC motors as an integral part of the unit.

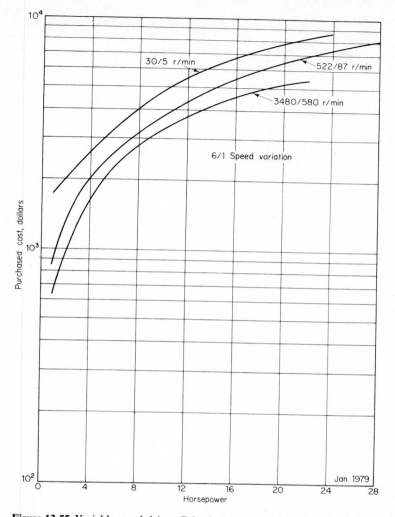

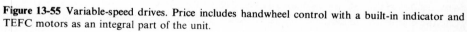

Figure 13-55 Variable-speed drives. Price includes handwheel control with a built-in indicator and TEFC motors as an integral part of the unit.

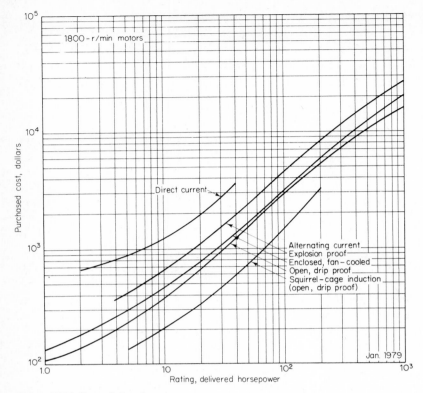

Figure 13-56 Cost of electric motors.

FLOW-MEASURING EQUIPMENT

Orifice meters, venturi meters, rotameters, and displacement meters are used extensively in industrial operations for measuring the rate of fluid flow. Other flow-measuring devices, such as weirs, pitot tubes, anemometers, and wet-test meters, are also useful in industrial operations.† In general, orifice meters are the cheapest and most flexible of the various types of equipment for measuring flow rates. Despite the inherent disadvantage of large permanent pressure drops with orifice installations, they are one of the most common types of flow-measuring equipment in industrial operations. Venturi meters are expensive and must be carefully proportioned and fabricated. However, they do not cause a large permanent pressure drop and are, therefore, very useful when power cost is an important factor.

---

† Detailed descriptions of various types of flow-measuring equipment and derivations of related equations are presented in essentially all books dealing with chemical engineering principles. See J. H. Perry and C. H. Chilton, "Chemical Engineers' Handbook," 5th ed., McGraw-Hill Book Company, New York, 1973.

Basic equations for the design and operation of orifice meters, venturi meters, and rotameters can be derived from the total energy balances presented at the beginning of this chapter. The following equations apply when the flowing fluid is a liquid, and they also give accurate results for the flow of gases if the pressure drop caused by the constriction is less than 5 percent of the upstream pressure:

For orifice meters and venturi meters

$$q_f = C_d S_c \sqrt{\frac{2g_c v(p_1 - p_c)}{1 - (S_c/S_1)^2}} \tag{29}$$

For rotameters

$$q_f = C_d S_c \sqrt{\frac{V_p 2g(\rho_p - 1/v)v}{S_p[1 - (S_c/S_1)^2]}} \tag{30}$$

where q_f = flow rate, ft^3/s

C_d = coefficient of discharge

S_c = cross-sectional flow area at point of minimum cross-sectional flow area, ft^2

S_1 = cross-sectional flow area in upstream section of duct before constriction, ft^2

g_c = gravitational constant in Newton's law of motion, 32.17 ft·lbm/(s)(s)(lbf)

q = local gravitational acceleration, ft/(s)(s)

v = average specific volume of fluid, ft^3/lb

p_1 = static pressure in upstream section of duct before constriction, psf

p_c = static pressure at point of minimum cross-sectional flow area, psf

V_p = volume of plummet, ft^3

S_p = maximum cross-sectional area of plummet, ft^2

ρ_p = density of plummet, lb/ft^3

The value of the coefficient of discharge C_d for orifice meters depends on the properties of the flow system, the ratio of the orifice diameter to the upstream diameter, and the location of the pressure-measuring taps. Values of C_d for sharp-edged orifice meters are presented in Fig. 13-57. These values apply strictly for pipe orifices with *throat taps*, in which the downstream pressure tap is located one-third of one pipe diameter from the downstream side of the orifice plate and the upstream tap is located one pipe diameter from the upstream side. However, within an error of about 5 percent, the values of C_d indicated in Fig. 13-57 may be used for manometer taps located anywhere between the orifice plate and the hypothetical throat taps.

Venturi meters usually have a tapered entrance with an interior total angle of 25 to 30° and a tapered exit with an interior angle of 7°. Under these conditions, the value of the coefficient of discharge may be assumed to be 0.98 if the Reynolds number based on conditions in the upstream section is greater than 5000.

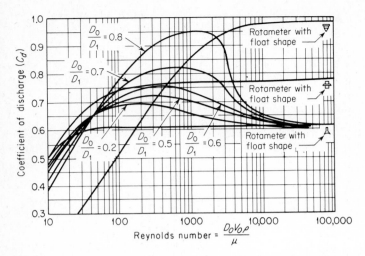

Figure 13-57 Coefficients of discharge for square-edged orifices with centered circular openings and for rotameters. (Subscript o indicates "at orifice or at constriction," and subscript 1 indicates "at upstream section.")

Values of C_d for various plummet shapes in rotameters are presented in Fig. 13-57. The Reynolds number applicable to the rotameter coefficient of discharge is based on the flow conditions through the annular opening between the plummet and the containing tube. The equivalent diameter for use in the Reynolds number consists of the difference between the diameter of the rotameter tube at the plummet location and the maximum diameter of the plummet.

TANKS, PRESSURE VESSELS, AND STORAGE EQUIPMENT

Storage of liquid materials is commonly accomplished in industrial plants by use of cylindrical, spherical, or rectangular tanks. These tanks may be constructed of wood, concrete, fiber reinforced plastic (FRP), or metal. Metal is the most common material of construction, although use of FRP is becoming increasingly important despite lack of rational design procedures. The design of storage vessels involves consideration of details such as wall thickness, size and number of openings, shape of heads, necessary temperature and pressure controls, and corrosive action of the contents.

The same principles of design apply for other types of tanks, including pressure vessels such as those used for chemical reactors, mixers, and distillation columns. For these cases, the shell is often designed and its cost estimated separately with the other components, such as tray assemblies, agitators, linings, and packing units, being handled separately. Process pressure vessels are normally designed in accordance with the *ASME Boiler and Pressure Vessel*

Code.† They are usually cylindrical shells capped with an elliptical or hemispherical head at each end with installation in either a vertical or horizontal position. A major concern in the design is to make certain the walls of the vessel are sufficiently thick to permit safe usage under all operating conditions.

The necessary wall thickness for metal vessels is a function of (1) the ultimate tensile strength or the yield point of the metal at the operating temperature, (2) the operating pressure, (3) the diameter of the tank, and (4) the joint or welding efficiencies.‡ Table 4 presents a summary of design equations and data for use in the design of tanks and pressure vessels based on the ASME Boiler and Pressure Vessel Code as specified in the 1977 Edition, Section VIII, Division 1.

COSTS FOR TANKS, PRESSURE VESSELS, AND STORAGE EQUIPMENT

Cost data for mixing tanks including agitators,§ storage tanks, and pressure tanks are presented in Figs. 13-58 to 13-60 while Table 5 gives costs for selected containers. In determining the total cost for the vessel, allowances must be made for nozzles on the unit, supports and foundations, platforms, labor, and indirect costs as well as for all of the internals in the vessel.¶

Numerous articles have been published which give methods for obtaining vessel costs based on estimates of costs for the individual components, such as for materials, labor, nozzles, manholes, and overhead related to fabrication, to arrive at an estimated cost (f.o.b) at the fabricator's shop. Final installed cost can be obtained by applying factors to account for freight, labor, materials, engineering, and overhead related to getting the unit to the plant and installing it ready for use. These methods take into account the materials of construction to be used as well as operating temperature and pressure.††

---

† The ASME Boiler and Pressure Vessel Code is published by the ASME Boiler and Pressure Vessel Committee, American Society of Mechanical Engineers, New York City, with a new edition coming out every three years. Section VIII of the Code deals specifically with pressure vessels with the basic rules being given in Division 1 and alternative rules being presented in Division 2.

‡ In the design of vacuum vessels, the ratio of length to diameter must also be taken into consideration.

§ For methods of calculating power requirements for agitators, see R. H. Perry and C. H. Chilton, "Chemical Engineers' Handbook," 5th ed., McGraw-Hill Book Company, New York, 1973.

¶ See the information presented in Chap. 15 on costs for plate and packed towers where cost data are given for tower shells including manholes and nozzles, internals such as trays and packing, and auxiliaries such as ladders and insulation. Cost data for reactor vessels are also presented in Chap. 15 under the section on costs for reactor equipment components.

†† For example, see A. Pikulik and H. E. Diaz, Cost Estimation for Major Process Equipment, *Chem. Eng.*, **84**(21):107 (Oct. 10, 1977) for method based on materials cost plus fabrication labor and engineering costs to give an f.o.b. cost. Examples of methods for estimating installed costs are given by J. S. Miller and W. A. Kapella, Installed Cost of a Distillation Column, *Chem. Eng.*, **84**(8):129 (April 11, 1977); K. M. Guthrie, Data and Techniques for Preliminary Cost Estimating, *Chem. Eng.*, **76**(6):114 (March 24, 1969); and K. M. Guthrie, "Process Plant Estimating, Evaluation, and Control," Craftsman Book Company of America, Solana Beach, California, 1974.

Table 4 Design equations and data for pressure vessels

a	= 2 for thicknesses < 1 in. and 3 for thicknesses ⩾ 1 in.
C_c	= allowance for corrosion, in.
D_a	= the major axis of an ellipsoidal head, before corrosion allowance is added, in.
E_J	= efficiency of joints expressed as a fraction
IDD	= inside depth of dish, in.
L_a	= inside radius of hemispherical head or inside crown radius of torispherical head, before corrosion allowance is added, in.
n	= 1.2 for D ⩽ 60 in., 1.21 for D = 61–79 in., 1.22 for D = 80–106 in., and 1.23 for D > 106 in.
OD	= outside diameter, in.
P	= maximum allowable internal pressure, psig
r	= knuckle radius, in.
r_i	= inside radius of the shell, before corrosion allowance is added, in.
S	= maximum allowable working stress, psi
t	= minimum wall thickness, in.
ρ_m	= density of metal, lbm/in.^{3}

Recommended design equations for vessels under internal pressure	Limiting conditions
For cylindrical shells	
$t = \dfrac{Pr_i}{SE_J - 0.6P} + C_c$	$\left\{ \begin{array}{l} t \leqslant r_i/2 \\ \text{or} \quad P \leqslant 0.385SE_J \end{array} \right.$
$t = r_i \left(\dfrac{SE_J + P}{SE_J - P} \right)^{1/2} - r_i + C_c$	$\left\{ \begin{array}{l} t > r_i/2 \\ \text{or} \quad P > 0.385SE_J \end{array} \right.$
For spherical shells	
$t = \dfrac{Pr_i}{2SE_j - 0.2P} + C_c$	$\left\{ \begin{array}{l} t \leqslant 0.356r_i \\ \text{or} \quad P \leqslant 0.665SE_j \end{array} \right.$
$t = r_i \left(\dfrac{2SE_j + 2P}{2SE_j - P} \right)^{1/3} - r_i + C_c$	$\left\{ \begin{array}{l} t > 0.356r_i \\ \text{or} \quad P > 0.665SE_j \end{array} \right.$
For ellipsoidal head	
$t = \dfrac{PD_a}{2SE_j - 0.2P} + C_c$	0.5 (minor axis) = $0.25D_a$
For torispherical (spherically dished) head	
$t = \dfrac{0.885PL_a}{SE_j - 0.1P} + C_c$	r = knuckle radius = 6% of inside crown radius and is not less than $3t$
For hemispherical head Same as for spherical shells with $r_i = L_a$	

Table 4 Design equations and data for pressure vessels (*Continued*)

Properties of vessel heads (Include corrosion allowance in variables)	2:1 Ellipsoidal	Hemi-spherical	Standard ASME torispherical
Capacity as volume in head, in.^{3}	$\dfrac{\pi D_a^3}{24}$	$\tfrac{2}{3}\pi L_a^3$	$0.9\left[\dfrac{2\pi L_a^2}{3}(IDD)\right]$
IDD = inside depth of dish, in.	$\dfrac{D_a}{4}$	L_a	$L_a - [(L_a - r)^2 - (L_a - t - r)^2]^{1/2}$
Approximate weight of dished portion of head, lbm	$\rho_m\left[\dfrac{\pi(nD_a + t)^2 t}{4}\right]$	$\rho_m[2\pi L_a^2 t]$	$\rho_m\left[\dfrac{\pi(OD + OD/24 + at)^2 t}{4}\right]$

Joint efficiencies	Recommended stress values		
	Metal	Temp., °F	S, psi
For double-welded butt joints if fully radiographed = 1.0 if spot examined = 0.85 if not radiographed = 0.70	Carbon steel (SA-285, Gr. C)	−20 to 650 750 850	13,700 12,000 8,300
In general, for spot examined if electric resistance weld = 0.85 if lap welded = 0.80 if single-butt welded = 0.60	Low-alloy steel for resistance to H_2 and $H_2 S$ (SA-387, Gr.12C1.1)	−20 to 800 950 1050 1200	13,700 11,000 5,000 1,000
	High-tensile steel for heavy-wall vessels (SA-302, Gr.B)	−20 to 750 850 950 1000	20,000 16,800 10,000 6,200
	High-alloy steel for cladding and corrosion resistance Stainless 304 (SA-240)	−20 650 800 1000	18,700 11,200 10,500 9,700
	Stainless 316 (SA-240)	−20 650 800 1000	18,700 11,500 11,000 10,600
	Nonferrous metals Copper (SB-11) Aluminum (SB-209, 1100-0)	100 400 100 400	6,700 3,000 2,300 1,000

See the latest ASME Boiler and Pressure Vessel Code for further details.

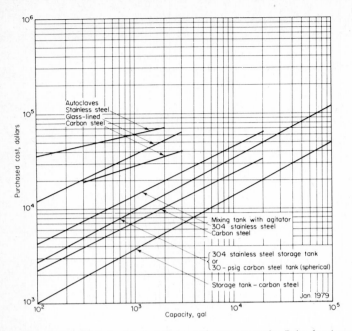

Figure 13-58 Cost of mixing, storage, and pressure tanks. Price for the mixing tank includes the cost of the driving unit.

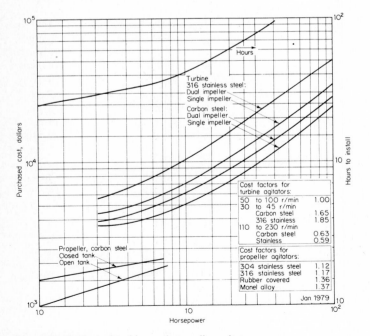

Figure 13-59 Cost of turbine and propeller agitators.

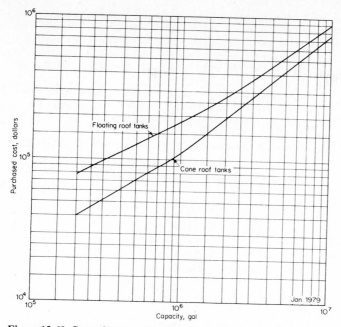

Figure 13-60 Cost of large-volume carbon-steel storage tanks.

Table 5 Approximate costs of small containers for chemical products (Jan., 1979)

Container size, description	Unit cost	Usable volume, ft³
55-gal steel drum, new	$15.00	7.35
55-gal steel drum, used, cleaned	8.40	7.35
55-gal aluminum drum	68.90	7.35
55-gal type 304 stainless steel drum	183.00	7.35
30-gal steel drum	10.00	4.00
16-gal steel drum	5.10	2.14
61-gal fiber drum, dry products only	6.70	8.15
55-gal fiber drum, dry products only	6.10	7.35
47-gal fiber drum, dry products only	6.00	6.28
41-gal fiber drum, dry products only	5.70	5.48
30-gal fiber drum, dry products only	4.80	4.00
15-gal fiber drum, dry products only	2.10	2.00
Multiwall paper bags, polyethylene film	0.27–0.30	1.33
Corrugated cartons, 24 × 16 × 6 in	0.55	1.33
1-gal glass jug, plastic cap	0.67	0.1335
1-gal polyethylene jar or bottle	0.32	0.1335
1-qt glass jar, plastic cap	0.28	0.034
1-qt polyethylene bottle	0.15	0.034
Pallets, expendable, 40 × 48 in to 44 × 50 in	4.60–8.40	
Pallets, warehouse type, 40 × 48 in to 44 × 50 in.	9.30–12.90	

Table 6 Rules of thumb for use in preliminary estimates of costs for pressure vessels

Costs for vessel (January, 1979—Including nozzles, manholes, and saddle or skirt but no special internals such as trays or agitators) as dollars per pound of weight of fabricated unit f.o.b. with carbon steel as the cost basis = $50(W_v)^{-0.34}$ where W_v is the total weight in pounds (applicable in W_v range of 800 lb to 100,000 lb).

To account for extra weight due to nozzles, manholes, and skirts or saddles, increase the weight calculated for the smooth vessel including top and bottom by 15% for vessels to be installed in a horizontal position and by 20% for vessels to be installed in a vertical position.

Steel density can be taken as 489 lb/ft^3 or 0.283 lb/in^3.

Cost factors to convert from carbon steel as the material of construction for the fabricated unit follow:

Shell-material cost factors

Carbon steel	1.0	(basis)
Stainless steel 304	2.0 to 3.5	
Stainless steel 316	2.3 to 4.3	
Monel	4.5 to 9.8	
Titanium	4.9 to 10.6	

Cost factors to convert from an internal pressure of up to 50 psig
for carbon steel at temperatures below 800° F†

Pressure	Pressure factor	Pressure	Pressure factor
Up to 50 psig	1.0 (basis)	800 psig	3.8
100	1.3	900	4.0
200	1.6	1000	4.2
300	2.0	1500	5.4
400	2.4	2000	6.5
500	2.8	3000	8.8
600	3.0	4000	11.3
700	3.3	5000	13.8

In general, the minimum wall thickness, not including allowances for corrosion, for any plate subject to pressure should not be less than $\frac{3}{32}$ in. for welded or brazed construction and not be less than $\frac{3}{16}$ in. for riveted construction except that the thickness of walls for unfired steam boilers should not be less than $\frac{1}{4}$ in.

A corrosion allowance of 0.010 to 0.015 in./yr, or about $\frac{1}{8}$ in. for a 10-year life is a reasonable value.

For high-pressure vessels, hemispherical heads are usually the most economical.

Lang factors to convert from the base cost of the delivered vessel (costed as if it were of carbon-steel material of construction so that weight becomes the primary measure of installation cost) to the cost of the vessel installed with all necessary auxiliaries except special internals such as trays or agitators are 3.0 for vessels installed in a horizontal position and 4.0 for vessels installed in a vertical position.

---

† If the data are available, it is much better to use the design equations presented in Table 4 of this chapter to obtain necessary wall thickness based on the stress value at the operating temperature in place of using the given pressure factors since there is a critical interrelationship among material of construction, operating pressure, and operating temperature in establishing the design and cost of pressure vessels.

The most reliable method for estimating the costs for tanks and pressure vessels is to obtain the assistance of a representative of a vessel fabricator. In many cases, these representatives can give an estimate based on a cost per unit weight for the particular vessel called for, or an actual delivered or installed price can be quoted. In addition, expert help with experience is needed to make good estimates of allowances to use for corrosion and to advise on the most appropriate materials of construction. Nevertheless, rough preliminary estimates of costs for tanks and pressure vessels can be made on the basis of gross methods such as are illustrated in Figs. 13-58 to 13-60. Some general rules of thumb for making cost estimates for pressure vessels are given in Table 6.†‡

FILTERS

The primary factor in the design of filters is the cake resistance or cake permeability. Because the value of the cake resistance can be determined only on the basis of experimental data, laboratory or pilot-plant tests are almost always necessary to supply the information needed for a filter design. After the basic constants for the filter cake have been determined experimentally, the theoretical concepts of filtration can be used to establish the effects of changes in operating variables such as filtering area, slurry concentration, or pressure-difference driving force.

In recent years, there has been considerable advance in the development of filtration theory, but the development has not reached the stage where an engineer can design a filter directly from basic equations as with a fractionation tower or a heat exchanger. Instead, the final design should be carried out by the technical personnel in filtration-equipment manufacturing concerns or by someone who has access to the necessary testing equipment and has an extensive understanding of the limitations of the design equations.

Choice of a filter for a particular operation depends on many factors. Some of the more important of these are:

1. Fixed and operating costs
2. Quantities and value of materials to be handled
3. Properties of the fluid, such as viscosity, density, and corrosive nature

---

† Adapted from R. H. Perry and C. H. Chilton, "Chemical Engineers' Handbook," 5th ed., McGraw-Hill Book Company, New York, 1973; ASME Boiler and Pressure Vessel Code, Section VIII, Div. 1, ASME, New York, 1977; H. Rase, "Chemical Reactor Design for Process Plants—Vol. II—Case Studies and Design Data," John Wiley and Sons, New York, 1977; and other references such as those listed for the preceding paragraph as examples of methods for estimating vessel costs.

‡ An example illustrating the use of Table 4 and Table 6 in the design and costing of a reactor vessel is given in Chap. 15 with the section on *Reactors*.

4. Whether the valuable product is to be the solid, the fluid, or both
5. Concentration, temperature, and pressure of slurry
6. Particle size and shape, surface characteristics of the particles, and compressibility of the solid material
7. Extent of washing necessary for the filter cake

Although a wide variety of filters are available on the market, the types can generally be classified as *batch* or *continuous*.

There are many times when the engineer wishes to make a preliminary design without asking for immediate assistance from a specialist in the field. The theoretical equations presented in the following sections are adequate for this purpose.†

DESIGN EQUATIONS

The rate at which filtrate is obtained in a filtering operation is governed by the materials making up the slurry and the physical conditions of the operation. The major variables that affect the filtration rate are:

1. Pressure drop across the cake and the filter medium
2. Area of the filtering surface
3. Viscosity of the filtrate
4. Resistance of the cake
5. Resistance of the filter medium

The rate of filtrate delivery is inversely proportional to the combined resistance of the cake and filtering medium, inversely proportional to the viscosity of the filtrate, and directly proportional to the available filtering area and the pressure-difference driving force. This statement can be expressed in equation form as

$$\frac{dV}{d\theta} = \frac{A \, \Delta P}{(R_K + R_F)\mu} \tag{31}$$

where V = volume of filtrate delivered in time θ
A = area of the filtering surface
ΔP = pressure drop across filter
R_K = resistance of the cake
R_F = resistance of the filter medium
μ = viscosity of the filtrate

---

† For application of these equations to give optimum values of maximum production and minimum cost, see N. H. Chen, Liquid-Solid Filtration: Generalized Design and Optimization Equations, *Chem. Eng.*, **85**(17):97 (July 31, 1978).

Cake resistance R_K varies directly with the thickness of the cake, and the proportionality can be expressed as

$$R_K = Cl \tag{32}$$

where C is a proportionality constant and l is the cake thickness at time θ.

It is convenient to express R_F in terms of a fictitious cake thickness l_F with resistance equal to that of the filter medium. Thus,

$$R_F = Cl_F \tag{33}$$

Designating w as the pounds of dry-cake solids per unit volume of filtrate, ρ_c as the cake density expressed as pounds of dry-cake solids per unit volume of wet filter cake, and V_F as the fictitious volume of filtrate per unit of filtering area necessary to lay down a cake of thickness l_F, the actual cake thickness plus the fictitious cake thickness is

$$l + l_F = \frac{w(V + AV_F)}{\rho_c A} \tag{34}$$

Equations (31) to (34) can be combined to give

$$\frac{dV}{d\theta} = \frac{A^2 \, \Delta P}{\alpha w(V + AV_F)\mu} \tag{35}$$

where α equals C/ρ_c and is known as the *specific cake resistance*. In the usual range of operating conditions, the value of the specific cake resistance can be related to the pressure difference by the empirical equation

$$\alpha = \alpha'(\Delta P)^s \tag{36}$$

where α' is a constant with value dependent on the properties of the solid material and s is a constant known as the *compressibility exponent of the cake*. The value of s would be zero for a perfectly noncompressible cake and unity for a completely compressible cake. For commercial slurries, the value of s is usually between 0.1 and 0.8.

The following general equation for rate of filtrate delivery is obtained by combining Eqs. (35) and (36):

$$\frac{dV}{d\theta} = \frac{A^2(\Delta P)^{1-s}}{\alpha'w(V + AV_F)\mu} \tag{37}$$

This equation applies to the case of constant-rate filtration. For the more common case of constant-pressure-drop filtration, A, ΔP, s, α', w, V_F, and μ can all be assumed to be constant with change in V, and Eq. (37) can be integrated between the limits of zero and V to give

$$V^2 + 2AV_F V = \frac{2A^2(\Delta P)^{1-s}}{\alpha'w\mu}\theta \tag{38}$$

Batch Filters

Equations (37) and (38) are directly applicable for use in the design of batch filters. The constants α', s, and V_F must be evaluated experimentally, and the general equations can then be applied to conditions of varying A, ΔP, V, θ, w, and μ. One point of caution is necessary, however. In the usual situations, the constants are evaluated experimentally in a laboratory or pilot-plant filter. These constants may be used to scale up to a similar filter with perhaps 100 times the area of the experimental unit. To reduce scale-up errors, the constants should be obtained experimentally with the same slurry mixture, same filter aid, and approximately the same pressure drop as are to be used in the final designed filter. Under these conditions, the values of α' and s will apply adequately to the larger unit. Fortunately, V_F is usually small enough for changes in its value due to scale-up to have little effect on the final results.

The following example illustrates the methods for determining the constants and applying them in the design of a large plate-and-frame filter.

Example 5. Estimation of filtering area required for a plate-and-frame filtration operation A plate-and-frame filter press is to be used for removing the solid material from a slurry containing 5 lb of dry solids per cubic foot of solid-free liquid. The viscosity of the liquid is 1 centipoise, and the filter must deliver at least 400 ft^3 of solid-free filtrate over a continuous operating time of 2 h when the pressure-difference driving force over the filter unit is constant at 25 psi. On the basis of the following data obtained in a small plate-and-frame filter press, estimate the total area of filtering surface required.

Experimental data The following data were obtained in a plate-and-frame filter press with a total filtering area of 8 ft^2:

Total volume of filtrate (V), ft^3	Time from start of filtration (θ), h, at constant pressure difference of		
	$\Delta P = 20$ psi	$\Delta P = 30$ psi	$\Delta P = 40$ psi
5	0.34	0.25	0.21
8	0.85	0.64	0.52
10	1.32	1.00	0.81
12	1.90	1.43	1.17

The slurry (with filter aid) was identical to that which is to be used in the large filter. The filtrate obtained was free of solid, and a negligible amount of liquid was retained in the cake.

SOLUTION An approximate solution could be obtained by interpolating for values of V at $\Delta P = 25$ psi and then using two of these values to set up

Eq. (38) in the form of two equations involving only the two unknowns V_F and $(\Delta P)^{1-s}/\alpha'w\mu$. By simultaneous solution, the values of V_F and $(\Delta P)^{1-s}/\alpha'w\mu$ could be obtained. The final required area could then be determined directly from Eq. (38). Because this method puts too much reliance on the precision of individual experimental measurements, a more involved procedure using all the experimental data is recommended.

The following method can be used to evaluate the constants V_F, s, and α' in Eq. (38):

Rearrange Eq. (38) to give

$$\frac{\theta \, \Delta P}{V/A} = \frac{\alpha'w\mu(\Delta P)^s}{2} \frac{V}{A} + \alpha'w\mu V_F(\Delta P)^s$$

At constant ΔP a plot of $\theta \, \Delta P/(V/A)$ vs. V/A should give a straight line with a slope equal to $\alpha'w\mu(\Delta P)^s/2$ and an intercept at $V/A = 0$ of $\alpha'w\mu V_F(\Delta P)^s$. Figure 13-61 presents a plot of this type based on the experimental data for this problem. Any time the same variable appears in both the ordinate and abscissa of a straight-line plot, an analysis for possible misinterpretation should be made. In this case, the values of θ and ΔP change sufficiently to make a plot of this type acceptable.

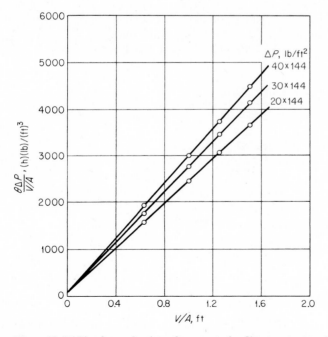

Figure 13-61 Plot for evaluation of constants for filtrate-rate equation in Example 5.

The following slopes and intercepts are obtained from Fig. 13-61:

ΔP, psf	Slope = $\dfrac{\alpha' w\mu(\Delta P)^{s}}{2}$, (h)(lb)/(ft^4)	Intercept = $\alpha' w\mu V_F(\Delta P)^{s}$, (h)(lb)/ft^3
20×144	2380	70
30×144	2680	80
40×144	2920	90

Values of α', s and V_F could now be obtained by simultaneous solution with any three of the appropriate values presented in the preceding list. However, a better idea as to the reliability of the design constants is obtained by using the following procedure:

Take the logarithm of the expressions for the slope and the intercept in Fig. 13-61. This gives

$$\log (\text{slope}) = s \log \Delta P + \log \frac{\alpha' w\mu}{2}$$

$$\log (\text{intercept}) = s \log \Delta P + \log \alpha' w\mu V_F$$

A log-log plot of the Fig. 13-61 slopes versus ΔP should give a straight line with a slope of s and an intercept at $\log (\Delta P) = 0$ of $\log (\alpha' \omega \mu/2)$. In this way, s and α' can be evaluated, and the consistency of the data can be checked. This plot is presented in Fig. 13-62. From the slope and intercept,

$$s = 0.3$$

$$\frac{\alpha' w\mu}{2} = 220$$

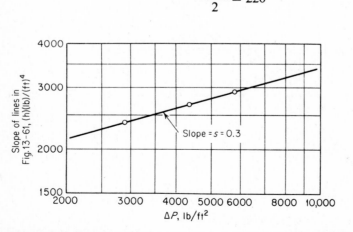

Figure 13-62 Secondary plot for evaluation of constants for filtrate-rate equation in Example 5.

Similarly, a log-log plot of the Fig. 13-61 intercepts versus ΔP should give a straight line with a slope of s and an intercept at $\log (\Delta P) = 0$ of $\log (\alpha' w \mu V_F)$, from which V_F could be evaluated. Because the value of V_F is relatively small, the intercepts read from Fig. 13-61 are not precise. The value of V_F, therefore, will be estimated from the combined results of Figs. 13-61 and 13-62.

$$w = 5 \text{ lb/ft}^3$$

$$\mu = 2.42 \text{ lb/(h)(ft)}$$

$$\alpha' = \frac{(220)(2)}{(5)(2.42)} = 36 \qquad \text{with units equivalent to } \alpha \text{ units of h}^2/\text{lb}$$

On the basis of the Fig. 13-61 intercept for the 30-psi line,

$$V_F = \frac{80}{\alpha' w \mu (\Delta P)^s} = \frac{80}{(36)(5)(2.42)(30 \times 144)^{0.3}}$$
$$= 0.015 \text{ ft}^3/\text{sq ft}$$

Substitution of the constants into Eq. (38) gives the final equation for use in evaluating the total filtering area needed for the large filter:

$$V^2 + 0.03AV = \frac{2A^2(\Delta P)^{1-0.3}}{36w\mu} \theta$$

For the conditions of this problem,

$$V = 400 \text{ ft}^3$$

$$\Delta P = 25 \times 144 \text{ psf}$$

$$w = 5 \text{ lb/ft}^3$$

$$\mu = 2.42 \text{ lb/(h)(ft)}$$

$$\theta = 2 \text{ h}$$

Substituting the indicated values gives

$$(400)^2 + (0.03)(400)A = \frac{2A^2(25 \times 144)^{0.7}(2)}{(36)(5)(2.42)}$$

Solving for A,

$$A = 240 \text{ ft}^2$$

The total area of filtering surface required is approximately 240 ft².

Continuous Filters

Many types of continuous filters, such as rotary-drum or rotary-disk filters, are employed in industrial operations. Development of the general design equations

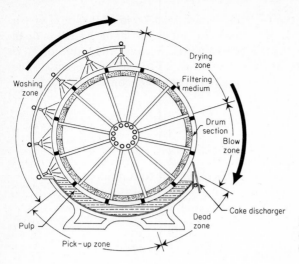

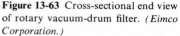

Figure 13-63 Cross-sectional end view of rotary vacuum-drum filter. (*Eimco Corporation.*)

for these units follows the same line of reasoning as that presented in the development of Eq. (38). The following analysis is based on the design variables for a typical rotary vacuum filter of the type shown in Fig. 13-63.

It is convenient to develop the design equations in terms of the total area available for filtering service, even though only a fraction of this area is in direct use at any instant. Designate the total available area as A_D and the fraction of this area immersed in the slurry as ψ_f. The effective area of the filtering surface then becomes $A_D\psi_f$, and Eq. (31) can be expressed in the following form:

$$\frac{dV}{d\theta} = \frac{A_D\psi_f \, \Delta P}{(R_K + R_F)\mu} \tag{39}$$

According to Eqs. (32) and (33),

$$R_K + R_F = C(l + l_F) \tag{40}$$

With a continuous filter, the cake thickness at any given location on the submerged filtering surface does not vary with time. The thickness, however, does vary with location as the cake builds up on the filtering surface during passage through the slurry. The thickness of the cake leaving the filtering zone is a function of the slurry concentration, cake density, and volume of filtrate delivered per revolution. This thickness can be expressed by the following equation:

$$l_{\text{leaving filtering zone}} = \frac{wV_R}{\rho_c A_D} \tag{41}$$

where V_R is the volume of filtrate delivered per revolution and ρ_c is the cake density as pounds of dry-cake solids per unit volume of wet filter cake leaving filter zone.

An average cake thickness during the cake-deposition period can be assumed to be one-half the sum of the thicknesses at the entrance and exit of the filtering zone. Since no appreciable amount of cake should be present on the filter when it enters the filtering zone,

$$l_{\text{avg}} = \frac{wV_R}{2\rho_c A_D} \tag{42}$$

By using the same procedure as was followed in the development of Eq. (34),

$$l + l_F = l_{\text{avg}} + l_F = \frac{w(V_R/2 + A_D\psi_f V_F)}{\rho_c A_D} \tag{43}$$

Combination of Eqs. (39), (40), and (43), with $\alpha = C/\rho_c$, gives

$$\frac{dV}{d\theta} = \frac{2A_D^2 \psi_f \, \Delta P}{\alpha w(V_R + 2A_D\psi_f V_F)\mu} \tag{44}$$

Integration of Eq. (44) between the limits of $V = 0$ and $V = V_R$, and $\theta = 0$ and $\theta = 1/N_R$, where N_R is the number of revolutions per unit time, gives

$$V_R^2 + 2A_D\psi_f V_F V_R = \frac{2A_D^2 \psi_f \, \Delta P}{\alpha w\mu N_R} \tag{45}$$

or, by including Eq. (36),

$$V_R^2 + 2A_D\psi_f V_F V_R = \frac{2A_D^2 \psi_f (\Delta P)^{1-s}}{\alpha' w\mu N_R} \tag{46}$$

The constants in the preceding equations can be evaluated by a procedure similar to that described in Example 5. Equation (46) is often used in the following simplified forms, which are based on the assumptions that the resistance of the filter medium is negligible and the filter cake is noncompressible:

$$\text{Volume of filtrate per revolution} = V_R = A_D\sqrt{\frac{2\psi_f \, \Delta P}{\alpha w\mu N_R}} \tag{47}$$

$$\text{Volume of filtrate per unit time} = V_R N_R = A_D\sqrt{\frac{2\psi_f N_R \, \Delta P}{\alpha w\mu}} \tag{48}$$

$$\text{Weight of dry cake per unit time} = V_R N_R w = A_D\sqrt{\frac{2\psi_f N_R w \, \Delta P}{\alpha\mu}} \tag{49}$$

Example 6. Effect of pressure difference on capacity of a rotary vacuum filter A rotary vacuum filter with negligible filter-medium resistance delivers 100 ft^3 of filtrate per hour when a given $CaCO_3$-H_2O mixture is filtered under known conditions. How many cubic feet of filtrate will be delivered per hour if the pressure drop over the cake is doubled, all other conditions remaining constant? Assume the $CaCO_3$ filter cake is noncompressible.

SOLUTION Equation (48) applies for this case:

$$100 = A_D \sqrt{\frac{2\psi_f N_R}{\alpha w \mu}} \sqrt{\Delta P_1} \tag{A}$$

$$\text{Unknown filtrate rate} = A_D \sqrt{\frac{2\psi_f N_R}{\alpha w \mu}} \sqrt{2\,\Delta P_1} \tag{B}$$

Dividing Eq. (*B*) by Eq. (*A*),

$$\text{Unknown filtrate rate} = 100\sqrt{2} = 141 \text{ ft}^3/\text{h}$$

Air Suction Rate in Rotary Vacuum Filters

A vacuum pump must be supplied for the operation of a rotary vacuum filter, and the design engineer may need to estimate the size of pump and power requirement for a given filtration unit. Because air leakage into the vacuum system may supply a major amount of the air that passes through the pump, design methods for predicting air suction rates must be considered as approximate since they do not account for air leakage.

The rate at which air is drawn through the dewatering section of a rotary vacuum filter can be expressed in a form similar to Eq. (39) as

$$\frac{dV_a}{d\theta} = \frac{A_D \psi_a \, \Delta P}{(R'_F + R'_K)\mu_a} \tag{50}$$

where V_a = volume of air at temperature and pressure of surroundings drawn through cake in time θ

ψ_a = fraction of total surface available for air suction

μ_a = viscosity of air at temperature and pressure of surroundings

The cake resistance R'_K is directly proportional to the cake thickness l, and the filter-medium resistance R'_F can be assumed to be directly proportional to a fictitious cake thickness l'_F. Designating C' as the proportionality constant,

$$R'_K + R'_F = C'(l + l'_F) \tag{51}$$

If the cake is noncompressible, l must be equal to the thickness of the cake leaving the filtering zone. Therefore, by Eq. (41) and using the same procedure as was followed in the development of Eq. (34),

$$l + l'_F = \frac{w(V_R + A_D \psi_a V'_F)}{\rho_c A_D} \tag{52}$$

where V'_F is the fictitious volume of filtrate per unit of air-suction area necessary to lay down a cake of thickness l'_F.

Combination of Eqs. (50) to (52) gives

$$\frac{dV_a}{d\theta} = \frac{A_D^2 \psi_a \, \Delta P}{\beta w(V_R + A_D \psi_a V'_F)\mu_a} \tag{53}$$

where β equals C'/ρ_c and is known as the *specific air-suction cake resistance*.

Integration of Eq. (53) between the limits corresponding to $V_a = 0$ and $V_a = V_{aR}$, where V_{aR} designates the volume of air per revolution, gives

$$V_{aR} = \frac{A_D^2 \psi_a \, \Delta P}{\beta w (V_R + A_D \psi_a V_F') \mu_a N_R} \tag{54}$$

If the cake is compressible, a rough correction for variation in β with change in ΔP can be made by use of the following empirical equation:

$$\beta = \beta' (\Delta P)^{s'} \tag{54a}$$

where β' and s' are constants.

By neglecting the resistance of the filter medium, Eq. (54) can be simplified to

$$\text{Volume of air per revolution} = V_{aR} = \frac{A_D^2 \psi_a \, \Delta P}{\beta w V_R \mu_a N_R} \tag{55}$$

$$\text{Volume of air per unit time} = V_{aR} N_R = \frac{A_D^2 \psi_a \, \Delta P}{\beta w V_R \mu_a} \tag{56}$$

Equations (47), (49), and (56) can be combined to give

$$\text{Volume of air per unit time} = \frac{A_D \psi_a}{\beta \mu_a} \sqrt{\frac{\alpha \mu N_R \, \Delta P}{2 w \psi_f}} \tag{57}$$

$$\frac{\text{Volume of air per unit time}}{\text{Weight of dry cake per unit time}} = \frac{\psi_a}{\psi_f} \frac{\mu}{\mu_a} \frac{\alpha}{2 \beta w} \tag{58}$$

If the constants in the preceding equations are known for a given filter system and the assumption of no air leakage is adequate, the total amount of suction air can be estimated. This value, combined with a knowledge of the air temperature and the pressures at the intake and delivery sides of the vacuum pump, can be used to estimate the power requirements of the vacuum pump by the methods described elsewhere in this chapter.

Example 7. Estimation of horsepower motor required for vacuum pump on a rotary vacuum filter A rotary vacuum-drum filter is to handle a slurry containing 20 lb of water per 1 lb of solid material. Tests on the unit at the conditions to be used for the filtration have shown that the dimensionless ratio of α/β is 0.6 and 19 lb of filtrate (not including wash water) is obtained for each 21 lb of slurry. The temperature of the surroundings and of the slurry is 70°F, and the pressure of the surroundings is 1 atm. The pressure drop to be maintained by the vacuum pump is 5 psi. The fraction of the drum area submerged in the slurry is 0.3, and the fraction of the drum area available for air suction is 0.1. On the basis of the following assumptions, estimate the horsepower of the motor necessary for the vacuum pump if the unit handles 50,000 lb of slurry per hour.

Assumptions:

Resistance of filter medium is negligible.

Any effects caused by air leakage are taken into account in the value given for α/β.

For air at the temperature involved, heat capacity at constant pressure divided by heat capacity at constant volume is 1.4.

The vacuum pump and motor have an overall efficiency of 50 percent based on an isentropic compression.

The value of β is based on the temperature and pressure of the air surrounding the filter.

The filter removes all of the solid from the slurry.

SOLUTION Because the value given for α/β applies at the operating conditions for the filtration and the resistance of the filter medium is negligible, Eq. (58) can be used:

$$\frac{\text{Volume of air per unit time}}{\text{Weight of dry cake per unit time}} = \frac{\psi_a}{\psi_f} \frac{\mu}{\mu_a} \frac{\alpha}{2\beta w}$$

$\psi_a = 0.1$
$\psi_f = 0.3$
μ = viscosity of water at 70°F = 0.982 centipoise = 0.982×2.42 lb/(h)(ft)
μ_a = viscosity of air at 70°F = 0.018 centipoise = 0.018×2.42 lb/(h)(ft)

$$\frac{\alpha}{\beta} = 0.6$$

Density of water at 70°F = 62.3 lb/ft^3
Pounds filtrate per pound of dry-cake solids = 19

$$w = \frac{1}{19/62.3} = 3.28 \text{ lb dry-cake solids/ft}^3 \text{ filtrate}$$

Weight of dry cake/h = $(50{,}000)(\frac{1}{21})$ = 2380 lb/h

$$\text{Volume of air/h} = \frac{(2380)(0.1)(0.982 \times 2.42)(0.6)}{(0.3)(0.018 \times 2.42)(2 \times 3.28)}$$

$$= 3960 \text{ ft}^3/\text{h at 70°F and 1 atm}$$

By Eq. (24)
Theoretical horsepower for isentropic single-stage compression

$$= \frac{3.03 \times 10^{-5}k}{k - 1} P_1 q_{fm_1} \left[\left(\frac{P_2}{P_1} \right)^{(k-1)/k} - 1 \right]$$

k = ratio of heat capacity of gas at constant pressure to heat capacity of gas at constant volume = 1.4
P_1 = vacuum-pump intake pressure = $(14.7 - 5)(144)$ psf

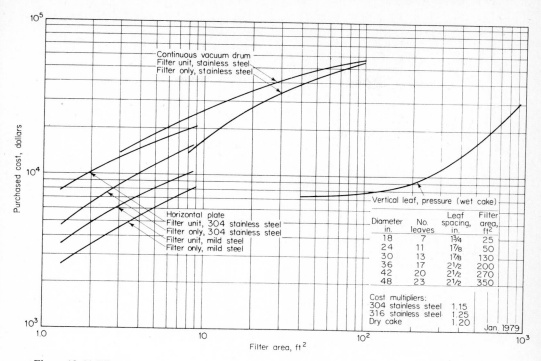

Figure 13-64 Filters.

P_2 = vacuum-pump delivery pressure = $(14.7)(144)$ psf
q_{fm_1} = cubic feet of gas per minute at vacuum-pump intake conditions

$$= \frac{(3960)(14.7)(144)}{(60)(14.7-5)(144)} = 100 \text{ cfm at } 70°F \text{ and } 9.7 \text{ psia}$$

Horsepower of motor required for vacuum pump

$$= \frac{(3.03 \times 10^{-5})(1.4)(14.7-5)(144)(100)}{(0.5)(1.4-1)} \left[\left(\frac{14.7}{9.7}\right)^{(1.4-1)/1.4} - 1 \right]$$

$$= 3.7 \text{ hp}$$

A 4-hp motor would be satisfactory.

Costs for Filters

Information to permit estimation of the cost for various types of filters is presented in Figs. 13-64 through 13-67.

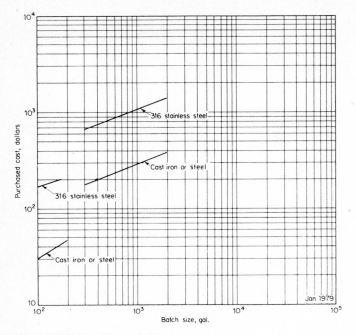

Figure 13-65 Cartridge-type filters.

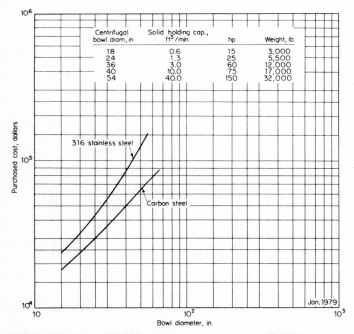

Figure 13-66 Centrifugal filters. Continuous solid bowl. Price does not include motor and drive.

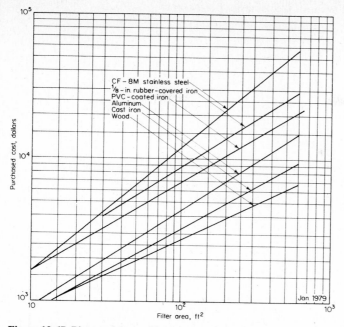

Figure 13-67 Plate-and-frame filters. Order-of-magnitude capital-cost estimating data.

MISCELLANEOUS PROCESSING EQUIPMENT COSTS

Cost data for blenders and mixers, kneaders, centrifugal separators, crystallizers, crushing and grinding equipment, dust collectors, electrostatic precipitators, and screens are presented in Figs. 13-68 through 13-89.

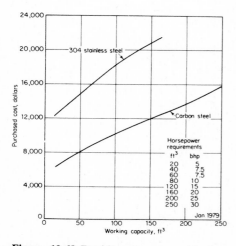

Figure 13-68 Double-cone rotary blenders. Price does not include motor.

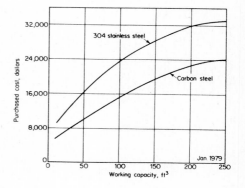

Figure 13-69 Ribbon blenders. Price includes standard floor-mounted support, baffled shell cover, antifriction pillow blocks mounted on outboard bearing shelves, stuffing box, flanged inlet opening, and lever-operated discharge gate.

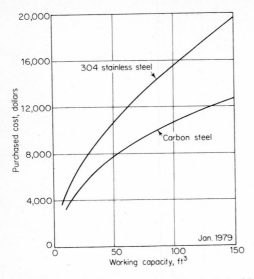

Figure 13-70 Twin-shell blenders. Price includes blender only.

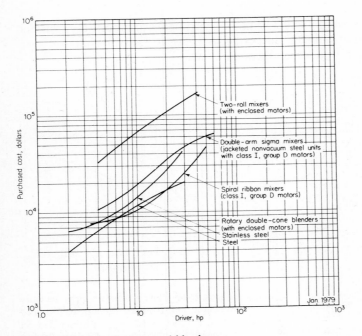

Figure 13-71 Cost of mixers and blenders.

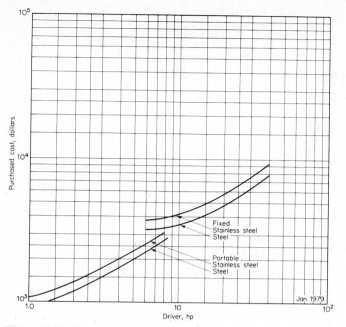

Figure 13-72 Cost of propeller mixers.

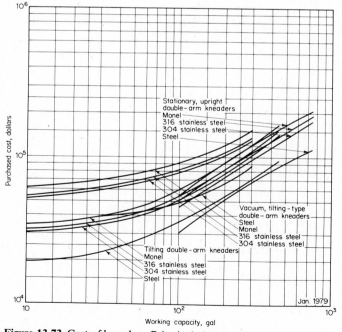

Figure 13-73 Cost of kneaders. Price includes machine, jacket, gear reducer and drive, cover, nozzles, and agitator. Motor and starter are not included in purchase price.

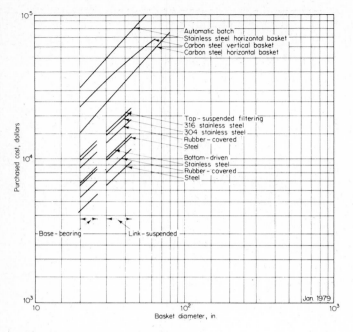

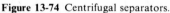

Figure 13-74 Centrifugal separators.

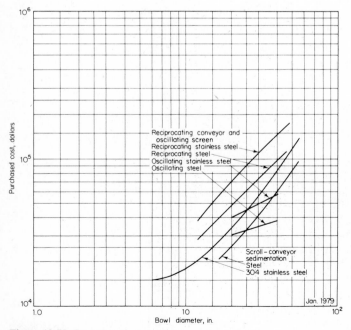

Figure 13-75 Centrifugal separators.

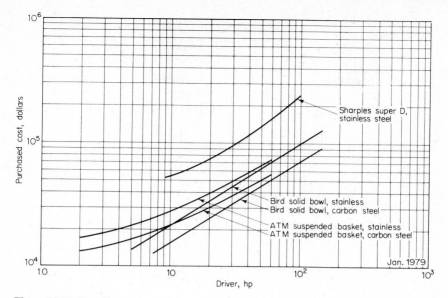

Figure 13-76 Centrifuges.

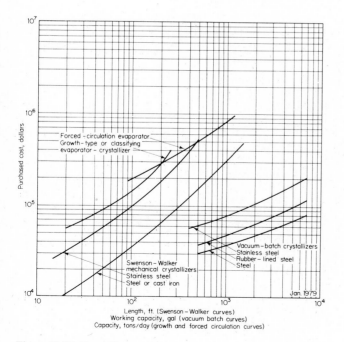

Figure 13-77 Crystallizers.

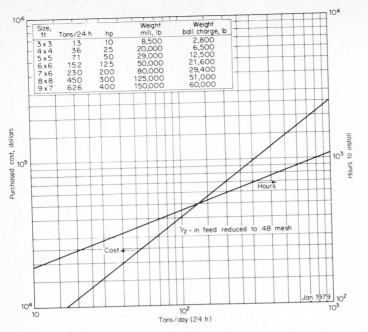

Size, ft	Tons/24 h	hp	Weight mill, lb	Weight ball charge, lb
3 x 3	13	10	8,500	2,800
4 x 4	36	25	20,000	6,500
5 x 5	71	50	29,000	12,500
6 x 6	152	125	50,000	21,600
7 x 6	230	200	80,000	29,400
8 x 8	450	300	125,000	51,000
9 x 7	626	400	150,000	60,000

Figure 13-78 Ball mills. Ball charge is $195/ton. Price includes liner, motor, drive, and guard.

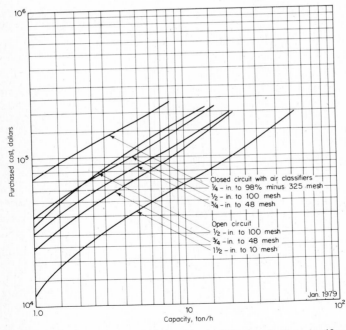

Figure 13-79 Ball mill dry grinding. Price includes installation, classifier, motors, drives, and an average allowance for foundations and erection. Does not include freight, auxiliary equipment, or equipment for handling the material.

594

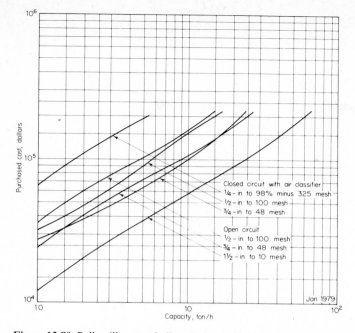

Figure 13-80 Ball mill wet grinding. Price includes installation, classifier, motors, drives, and an average allowance for foundations and erection. Does not include freight, auxiliary equipment, or equipment for handling the material.

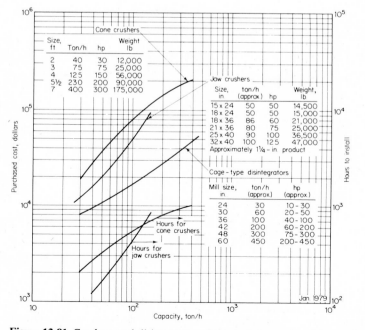

Figure 13-81 Crushers and disintegrators. Price includes motor, drive, and guard.

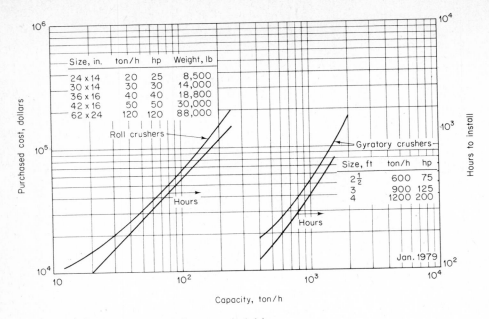

Figure 13-82 Crushers. Price includes motor and drive.

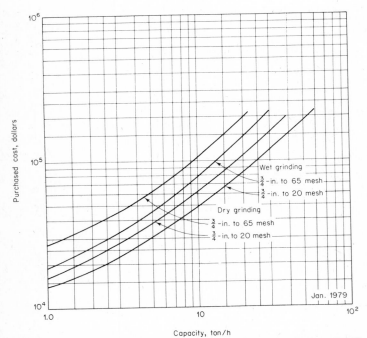

Figure 13-83 Rod mill in open circuit. Price includes installation, classifier, motors, drives, and an average allowance for foundations and erection. Does not include freight, auxiliary equipment, or equipment for handling material.

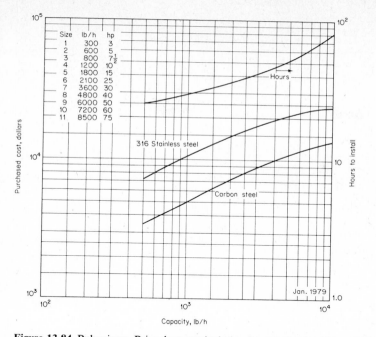

Figure 13-84 Pulverizers. Price does not include motor and drive. Cost of legs for units 1 through 5 is \$110 and of stands for units 6 through 11 is \$325. Add 15% for explosion-proof construction and \$430 for flexible coupling.

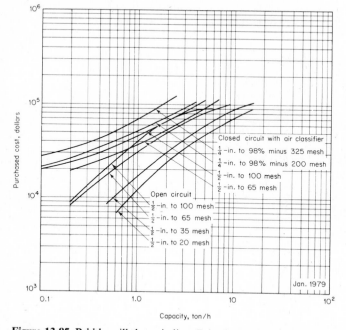

Figure 13-85 Pebble mill dry grinding. Price includes installation, classifier, motors, drives, and an average allowance for foundations and erection. Does not include freight, auxiliary equipment, or equipment for handling the material.

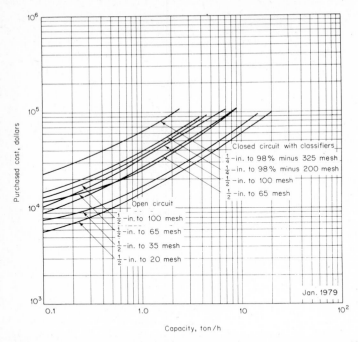

Figure 13-86 Pebble mill wet grinding. Price includes installation, classifier, motors, drives, and an average allowance for foundations and erection. Does not include freight, auxiliary equipment, or equipment for handling the material.

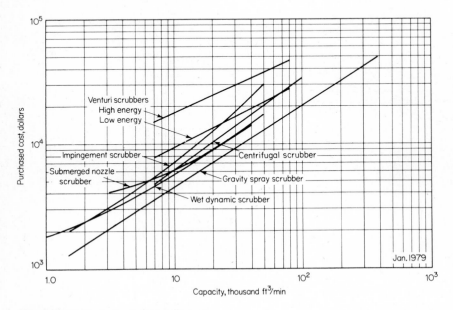

Figure 13-87 Cost of wet dust collectors.

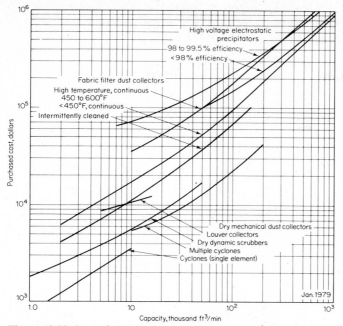

Figure 13-88 Cost of dry, mechanical dust collectors, high-voltage electrostatic precipitators, and fabric-filter dust collectors.

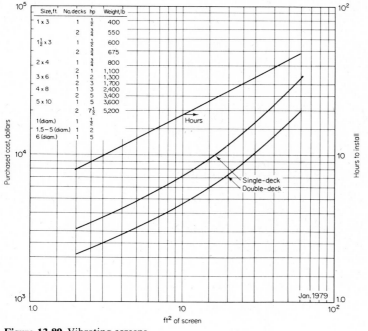

Figure 13-89 Vibrating screens.

MATERIALS-HANDLING EQUIPMENT COSTS

Materials-handling cost for chutes, conveyors, gates, and hoists are given in Figs. 13-90 through 13-96. In addition, Table 7 lists costs for materials-handling equipment by automotive means.

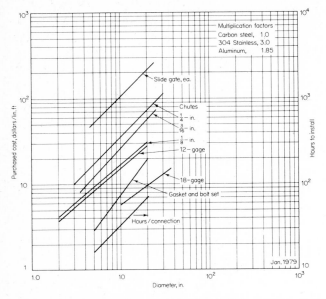

Figure 13-90 Chutes and gates. Price for flexible connections, $43 to $76 each. Transition piece, round to square, five times the per-foot price.

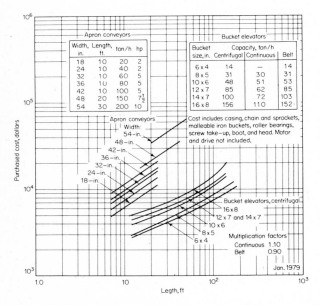

Figure 13-91 Conveying equipment. Cost of apron conveyors and bucket elevators.

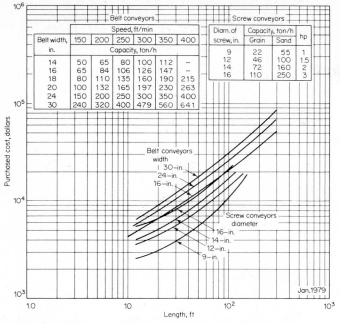

Belt conveyors							Screw conveyors			
	Speed, ft/min						Diam. of screw, in.	Capacity, ton/h		hp
Belt width, in.	150	200	250	300	350	400		Grain	Sand	
	Capacity, ton/h						9	22	55	1
14	50	65	80	100	112	–	12	46	100	1.5
16	65	84	106	126	147	–	14	72	160	2
18	80	110	135	160	190	215	16	110	250	3
20	100	132	165	197	230	263				
24	150	200	250	300	350	400				
30	240	320	400	479	560	641				

Figure 13-92 Conveying equipment. Cost of belt and screw conveyors.

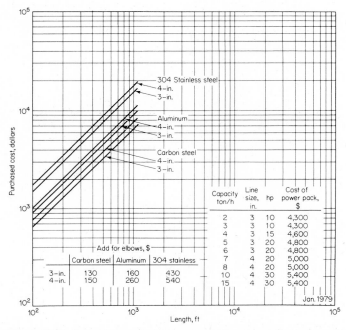

Capacity ton/h	Line size, in.	hp	Cost of power pack $
2	3	10	4,300
3	3	10	4,300
4	3	15	4,600
5	3	20	4,800
6	3	20	4,800
7	4	20	5,000
8	4	20	5,000
10	4	30	5,400
15	4	30	5,400

Add for elbows, $			
	Carbon steel	Aluminum	304 stainless
3-in.	130	160	430
4-in.	150	260	540

Figure 13-93 Conveying equipment: pneumatic systems. Cost of pipeline includes couplings, #11 gauge. Cost of powerpack includes blower, TEFC motor, base, coupling, check valve, pressure gauges, and filter. For free-flowing materials, 20 to 60 lb/ft^3.

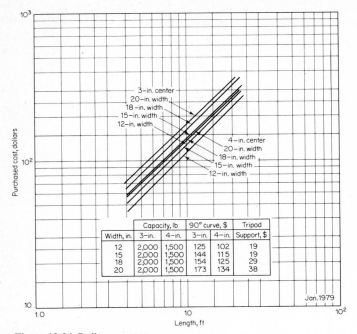

| Width, in. | Capacity, lb | | 90° curve, $ | | Tripod |
	3-in.	4-in.	3-in.	4-in.	Support, $
12	2,000	1,500	125	102	19
15	2,000	1,500	144	115	19
18	2,000	1,500	154	125	29
20	2,000	1,500	173	134	38

Jan.1979

Figure 13-94 Roller conveyors.

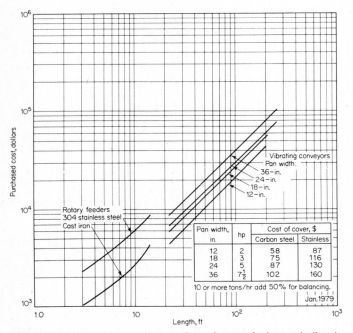

| Pan width, in. | hp | Cost of cover, $ | |
		Carbon steel	Stainless
12	2	58	87
18	3	75	116
24	5	87	130
36	$7\frac{1}{2}$	102	160

10 or more tons/hr add 50% for balancing.

Jan.1979

Figure 13-95 Conveying equipment. Cost of rotary feeders and vibrating conveyors.

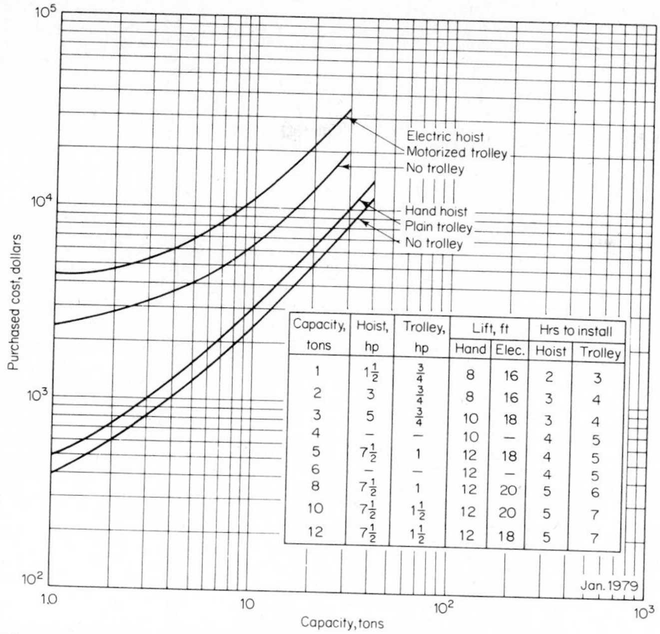
Purchased cost, dollars
10^5
10^4
10^3
10^2
Electric hoist
Motorized trolley
No trolley
Hand hoist
Plain trolley
No trolley
Jan. 1979
1.0
10
10^2
10^3
Capacity, tons

Capacity, tons	Hoist, hp	Trolley, hp	Lift, ft: Hand	Lift, ft: Elec.	Hrs to install: Hoist	Hrs to install: Trolley
1	$1\frac{1}{2}$	$\frac{3}{4}$	8	16	2	3
2	3	$\frac{3}{4}$	8	16	3	4
3	5	$\frac{3}{4}$	10	18	3	4
4	—	—	10	—	4	5
5	$7\frac{1}{2}$	1	12	18	4	5
6	—	—	12	—	4	5
8	$7\frac{1}{2}$	1	12	20	5	6
10	$7\frac{1}{2}$	$1\frac{1}{2}$	12	20	5	7
12	$7\frac{1}{2}$	$1\frac{1}{2}$	12	18	5	7

Figure 13-96 Hoisting equipment. Extra costs include: acid-resistant construction, \$160; dust-tight construction, \$70; power reels, \$350; and chain container, \$65.

Table 7 Costs for automotive materials-handling equipment (Jan. 1979)†

Equipment			Cost, \$
Fork lift trucks			
3,000 lb.	15-in. load center	72-in. lift	14,700
4,000 lb.	15-in. load center	72-in. lift	16,100
5,000 lb.	24-in. load center	72-in. lift	17,900
6,000 lb.	24-in. load center	84-in. lift	20,400
8,000 lb.	24-in. load center	84-in. lift	22,700
10,000 lb.	24-in. load center	92-in. lift	26,300
Hand truck, heavy duty			125
Hydraulic pallet truck, 4000 lb			1,810
Jack lift electric pallet ruck, 4000 lb			6,460
Payloaders			
18 ft^3, gas			12,000
20 ft^3, gas			17.200
1 yd^3, gas			19,700
1.75 yd^3, gas			35,700
1.75 yd^3, diesel			39,900
4.0 yd^3, gas			59,800
4.0 yd^3, diesel			64,600

† Adapted from H. E. Mills, *Chem. Eng.* **71**(6): 133 (1964).

(*Continued*)

Table 7 (*Continued*)

Equipment	Cost, $
Railway tank cars (8000 gal)	
Steel	26,300
Aluminum	38,300
Stainless steel	64,600
Tank trailers (4300 gal unlined)	
Carbon steel	23,800
Aluminum	31,000
Stainless steel	38,300
Tractors	
Gasoline	23,800
Diesel	47,600
Tractor shovel	
$2\frac{1}{4}$ yd^3 bucket 105 hp	49,800
$2\frac{1}{4}$ yd^3 bucket 125 hp	54,400
$3\frac{1}{4}$ yd^3 bucket 150 hp	68,000
Walkie pallet truck, 4000 lb	4,500
Battery	1,200
Charger	950

NOMENCLATURE FOR CHAPTER 13

a = special correction constant for pressure-vessel design, see Table 4

A = area of filtering surface, ft^2

A_D = total available surface for filtration on continuous filter, ft^2

C = a proportionality constant relating cake thickness to resistance to liquid passage

C' = a proportionality constant relating cake thickness to resistance to air passage

C_a = coefficient in air-lift design equation

C_c = allowance for corrosion, in.

C_d = coefficient of discharge for orifice or rotameter

C_p = heat capacity at constant pressure, Btu/(lb mol)(°F)

D = diameter, ft

D_a = inside length of the major axis of an ellipsoidal head, in., see Table 4

D_i = inside diameter, in.

D_m = mean diameter, in.

D_o = orifice diameter, ft

E_J = efficiency of joints, expressed as a fraction

$\mathfrak{f}$ = Fanning friction factor

F = mechanical-energy loss due to friction, ft·lbf/lbm

F_c = frictional loss due to sudden contraction, ft·lbf/lbm

F_e = frictional loss due to sudden expansion, ft·lbf/lbm

g = local gravitational acceleration, ft/(s)(s)

g_c = conversion factor in Newton's law of motion, 32.17 ft·lbm/(s)(s)(lbf)

G = mass velocity, lbm/(h)(ft^2 of cross-sectional area)

h = enthalpy, ft·lbf/lbm

H_s = submergence, ft

H_t = lift, ft

IDD = inside depth of dish, in., see Table 4

k = ratio of specific heat at constant pressure to specific heat at constant volume

K_c = coefficient in expression for frictional loss due to sudden contraction

l = cake thickness at time θ, ft

l_F = fictitious cake thickness for liquid flow with resistance equal to that of filter medium, ft

l'_F = fictitious cake thickness for air flow with resistance equal to that of filter medium, ft

L = length, ft

L_a = inside radius of hemispherical head or inside crown radius of torispherical head, in., see Table 4

L_e = fictitious length of straight pipe with the equivalent resistance of a pipe fitting of same nominal diameter as pipe, ft

M = molecular weight, lbm/lbmol

n = special correction constant for pressure-vessel design, see Table 4

N = speed of impeller, r/min

N_R = number of revolutions per unit time, revolutions/h

N_{Re} = Reynolds number, equals $DV\rho/\mu$, dimensionless

N_s = number of stages of compression

OD = outside diameter, in., see Table 4

p = absolute pressure, lbf/ft^2

p_c = static pressure at point of minimum cross-sectional flow area, lbf/ft^2

P = maximum allowable internal pressure, psig, or total pressure; ΔP means pressure-difference driving force across filter; P_1 refers to vacuum-pump intake pressure; P_2 refers to vacuum-pump delivery pressure, lbf/ft^2

P_b = bursting pressure, psig

P_s = safe working pressure, psig

q = rate of fluid discharge, gpm

q_f = fluid-flow rate, ft^3/s

q_{f_m} = fluid-flow rate, ft^3/min

$q_{f_{m_1}}$ = cubic feet of gas flowing per minute at vacuum-pump intake conditions, ft^3/min

Q = heat energy provided as such to the fluid system from an outside source, ft·lbf/lbm

r = knuckle radius, in., see Table 4

r_i = inside radius of shell, in., see Table 4

R = ideal-gas-law constant, 1545 (lbf/ft^2)(ft^3)/(lb mol)(°R)

R_F = resistance of filter medium to passage of liquid

R'_F = resistance of filter medium to passage of air

R_H = hydraulic radius, ft

R_K = resistance of filter cake to passage of liquid

R_K' = resistance of filter cake to passage of air

s = compressibility exponent of filter cake, defined by Eq. (36)

s' = a constant, defined by Eq. (54a)

S = maximum allowable working stress, lbf/in.^{2}

S_1 = cross-sectional flow area in upstream section of duct before constriction, ft^2

S_c = cross-sectional flow area at point of minimum cross-sectional flow area, ft^2

S_p = maximum cross-sectional area of plummet, ft^2

S_s = safe working fiber stress, lbf/in.^{2}

S_T = tensile strength, lbf/in.^{2}

t = minimum wall thickness, in., see Table 4

T = temperature, °R

u = internal energy, ft·lbf/lbm

v = specific volume of fluid, ft^3/lbm

V = average linear velocity, ft/s, or volume of filtrate delivered in time θ, ft^3

V_a = volume of air at temperature and pressure of surroundings drawn through filter cake in time θ, ft^3

V_{air} = volume of free air required in air lift to raise 1 gal of water, ft^3/gal

V_{aR} = volume of air drawn through filter cake per revolution, ft^3/revolution

V_F = fictitious volume of filtrate per unit of filtering area necessary to lay down a cake of thickness l_F, ft^3/ft^2

V_F' = fictitious volume of filtrate per unit of air-suction area necessary to lay down a cake of thickness l_F', ft^3/ft^2

V_i = instantaneous or point linear velocity, ft/s

V_o = average linear velocity through orifice, ft/s

V_p = volume of plummet, ft^3

V_R = volume of filtrate delivered per revolution, ft^3/revolution

w = weight of dry-cake solids per unit volume of filtrate, lb/ft^3

W = shaft work, gross work input to the fluid system from an outside source, ft·lbf/lbm

W_o = mechanical work imparted to the fluid system from an outside source, ft·lbf/lbm

W_v = weight of vessel, lbm, see Table 6

Z = vertical distance above an arbitrarily chosen datum plane, ft

Greek symbols

α = correction coefficient to account for use of average velocity or specific cake resistance in units of h^2/lb

α' = a constant, defined by Eq. (36)

β = specific air-suction cake resistance, h^2/lb

β' = a constant, defined by Eq. (54a)

Δ = difference

Δp_f = change in pressure due to friction, lbf/ft^2

ε = equivalent roughness, ft

θ = time, h

μ = absolute viscosity of fluid or absolute viscosity of filtrate, lb/(s)(ft) or lb/(h)(ft)

μ_a = absolute viscosity of fluid at average bulk temperature or absolute viscosity of air at temperature and pressure of surroundings, lb/(s)(ft) or lb/(h)(ft)

μ_c = absolute viscosity of fluid, centipoises

μ_w = absolute viscosity of fluid at average temperature of wall, lbm/(s)(ft)

ρ = density of fluid, lbm/ft^3

ρ_c = cake density as weight of dry-cake solids per unit volume of wet filter cake, lbm/ft^3

ρ_m = density of metal for pressure-vessel design, lbm/in.^{3}, see Table 4

ρ_p = density of plummet, lbm/ft^3

ϕ = correction factor for nonisothermal flow = $1.1(\mu_a/\mu_w)^{0.25}$ when DG/μ_a is less than 2100 and $1.02(\mu_a/\mu_w)^{0.14}$ when DG/μ_a is greater than 2100

ψ_a = fraction of total surface area available for air suction

ψ_f = fraction of total available surface for filtration immersed in slurry

SUGGESTED ADDITIONAL REFERENCES FOR MATERIALS-TRANSFER-HANDLING, AND TREATMENT—EQUIPMENT DESIGN AND COSTS

Blender, mixer, and agitator design and costs

Abraham, R. W., Reliability of Rotating Equipment, *Chem. Eng.*, **80**(24):96 (1973).

Bowers, R. H., Some Aspects of Agitator Design, *Chem. Eng. (London)*, **50** (March 1970).

Casto, L. V., Practical Tips on Designing Turbine-Mixer Systems, *Chem. Eng.*, **79**(1):97 (1972).

Chen, S. J., A New Low-Cost Saline Mixing and Processing Device, *Cost Eng.*, **17**(4):8 (1972).

———— and A. R. Macdonald, Motionless Mixers for Viscous Polymers, *Chem. Eng.*, **80**(7):105 (1973).

————, Static Mixing of Polymers, *Chem. Eng. Progr.*, **71**(8):80 (1975).

Connolly, J. R., and R. L. Winter, Approaches to Mixing Operation Scale-Up, *Chem. Eng. Progr.*, **65**(8):70 (1969).

————, How to Make Mixers Meet Process Needs, *Chem. Eng.*, **80**(17):128 (1973).

————, Energy Conservation in Fluid Mixing, *Chem. Eng. Progr.*, **72**(5):52 (1976).

Coyle, C. K., H. E. Hirschland, B. J. Michel, and J. Y. Oldshue, Mixing in Viscous Liquids, *AIChE J.*, **16**(6):903 (1970).

Dickey, D. S., and R. W. Hicks, Liquid Agitation: Fundamentals of Agitation, *Chem. Eng.*, **83**(3):93 (1976).

———— and J. C. Fenic, Dimensional Analysis for Fluid Agitation Systems, *Chem. Eng.*, **83**(1):139 (1976).

Dykman, M., and B. J. Michel, Comparing Mechanical Aerator Designs, *Chem. Eng.*, **76**(5):117 (1969).

Gates, L. E., J. R. Morton, and P. L. Fondy, Selecting Agitator Systems to Suspend Solids in Liquids, *Chem. Eng.*, **83**(11):144 (1976).

———— T. L. Henley, and J. G. Fenic, How to Select the Optimum Turbine Agitator, *Chem. Eng.*, **82**(26):110 (1975).

Guthrie, K. M., Capital Cost Estimating, *Chem. Eng.*, **76**(6):114 (1969).

Harnby, N., Unit Cost of Mixing Dry Solids, *Chem. Proc. Eng.*, **49**(12):53 (1968).

Herman, E. A., The Airmix Mixer, *Cost Eng.*, **15**(3):4 (1970).

Hicks, R. W., and D. S. Dickey, Applications Analysis for Turbine Agitators, *Chem. Eng.*, **83**(24):127 (1976).

——— and L. E. Gates, Fluid Agitation in Polymer Reactors, *Chem. Eng. Progr.*, **71**(8):74 (1975).

——— J. R. Morton, and J. G. Fenic, How to Design Agitators for Desired Process Response, *Chem. Eng.*, **83**(9):102 (1976).

Gates, L. E., T. L. Henley, and J. G. Fenic, How to Select the Optimum Turbine Agitator, *Chem. Eng.*, **82**(26):110 (1975).

Hicks, R. W., and L. E. Gates, How to Select Turbine Agitators for Dispersing Gas into Liquids, *Chem. Eng.*, **83**(15):141 (1976).

——— and ———, How to Select Turbine Agitators for Dispersing Gas into Liquids, *Chem. Eng.*, **88**(15):141 (1976).

Hill, R. S., and D. L. Kime, How to Specify Drive Trains for Turbine Agitators, *Chem. Eng.*, **83**(16):89 (1976).

Ho, F. C., and A. Kwong, A Guide to Designing Special Agitators, *Chem. Eng.*, **80**(17):94 (1973).

Hoyle, D. L., Designing for pH Control, *Chem. Eng.*, **83**(24):121 (1976).

Johnson, R. T., Batch Mixing of Viscous Liquids, *I&EC Process Des. & Dev.*, **6**(3):340 (1967).

Jones, T. E. M., Continuous Mixing, *Chem. Proc. Eng.*, **48**(10):66 (1967).

Khang, S. J., and O. Levenspiel, The Mixing-rate Number Agitator-stirred Tanks, *Chem. Eng.*, **83**(21):141 (1976).

King, G. T., Mixing With Reaction, *Chem. Proc. Eng.*, **48**(10):72 (1967).

Leggett, E., Multistage Mixing Solved by Graphs, *Hydrocarbon Process.*, **48**(5):141 (1969).

Lloyd, P. J., and P. C. M. Yeung, Mixing of Powders, *Chem. Proc. Eng.*, **48**(10):57 (1967).

Meyer, W. S., and D. L. Lime, Cost Estimation for Turbine Agitators, *Chem. Eng.*, **83**(20):109 (1976).

Norris, B. E., Mass Transfer Without Mixing, *Hydrocarbon Process.*, **54**(9):127 (1975).

Oldshue, J. Y., Mixing, *I&EC*, **59**(11):58 (1967).

———, Suspending Solids and Dispersing Gases in Mixing Vessel, *I&EC*, **61**(19):79 (1969).

———, How to Specify Mixers, *Hydrocarbon Process.*, **48**(10):73 (1969).

——— and F. L. Connelly, Gas-Liquid Contacting With Impeller Mixers, *Chem. Eng. Progr.*, **73**(3):85 (1977).

Penney, W. R., Guide to Trouble-Free Mixers, *Chem. Eng.*, **77**(12):171 (1970).

———, Recent Trends in Mixing Equipment, *Chem. Eng.*, **78**(7):86 (1971).

Ramsey, W. D., and G. C. Zoller, How the Design of Shafts, Seals, and Impellers Affects Agitator Performance, *Chem. Eng.*, **83**(18):101 (1976).

Rautzen, R. R., R. R. Corpstein, and D. S. Dickey, How to Use Scale-up Methods for Turbine Agitators, *Chem. Eng.*, **83**(23):119 (1976).

Rees, L. H., Evaluating Homogenizers for Chemical Processes, *Chem. Eng.*, **81**(10):86 (1974).

Root, W. R., Porcupine Processor, *Cost Eng.*, **12**(4):4 (1967).

Root, W. L., and R. A. Nichols, Heat Transfer in Mechanically Agitated Units, *Chem. Eng.*, **80**(7):98 (1973).

Schott, N. R., B. Weinstein, and D. LaBombard, Motionless Mixers in Plastic Processing, *Chem. Eng. Progr.*, **71**(1):54 (1975).

Schweitzer, G. R., Gemco's Versatile Blending System, *Cost Eng.*, **15**(2):7 (1970).

Sheridan, L. A., The Practical Side of Mixing in Extruders, *Chem. Eng. Progr.*, **71**(2):83 (1975).

Simpson, L. L., Turbulence and Industrial Mixing, *Chem. Eng. Progr.*, **70**(10):77 (1974).

Stockdale, C., Influence of Plant Capacity on Mixer Design, *Chem. Proc. Eng.*, **48**(10):62 (1967).

Todd, D. B., Mixing in Starved Twin Screw Extruders, *Chem. Eng. Progr.*, **71**(2):81 (1975).

——— and H. F. Irving, Axial Mixing in a Self-Wiping Reactor, *Chem. Eng. Progr.*, **65**(9):84 (1969).

Valentin, F. H. H., Mixing of Powders, Pastes and non-Newtonian Fluids, *Chem. Proc. Eng.*, **48**(10):69 (1967).

van Klaveran, N., Nomograms for Scale-up of Agitated Vessels, *Chem. Proc. Eng.*, **50**(9):127 (1969).

Wang, R. H., and L. T. Fan, Methods for Scaling-up Tumbling Mixers, *Chem. Eng.*, **81**(11):88 (1974).

Wolh, M. H., Mixing of non-Newtonian Fluids, *Chem. Eng.*, **75**(18):113 (1968).

Zimmerman, O. T., The Kady Mill, *Cost Eng.*, **12**(1):4 (1967).

————, Strong Scott Turbilizer, *Cost Eng.*, **12**(3):7 (1967).

———— and I. Lavine, The Cornell Versator, *Cost Eng.*, **13**(2):4 (1968).

————, Clearfield Muller-Type Mixers, *Cost Eng.*, **13**(3):8 (1968).

————, Prater Blue Streak "Super Twin" Spiral Mixer, *Cost Eng.*, **14**(4):4 (1969).

————, Prater Horizontal Ribbon Mixer-Blender, *Cost Eng.*, **15**(1):4 (1970).

————, The Munson Glass-Batcher, *Cost Eng.*, **15**(4):16 (1970).

————, Ribbon Blenders, *Cost Eng.*, **16**(2):13 (1971).

————, Munson Rotary Mixers, *Cost Eng.*, **16**(3):6 (1971).

Other Motionless Mixers, *Chem. Eng.*, **80**(7):111 (1973).

Centrifugal separator design and costs

Albertson, O. E., and D. R. Vaughn, Handling of Solid Wastes, *Chem. Eng. Progr.*, **67**(9):49 (1971).

Ambler, C. M., Centrifuge Selection, *Chem. Eng.*, **78**(4):55 (1971).

————, How to Select the Optimum Centrifuge, *Chem. Eng.*, **76**(23):96 (1969).

Baumann, D. K., and D. B. Todd, When to Use a Pusher Centrifuge, *Chem. Eng. Progr.*, **69**(9):62 (1973).

Calvert, S., How to Choose a Particulate Scrubber, *Chem. Eng.*, **84**(18):54 (1977).

Day, R. W., The Hydrocyclone in Process and Pollution Control, *Chem. Eng. Progr.*, **69**(9):67 (1973).

————, Techniques for Selecting Centrifuges, *Chem. Eng.*, **81**(10):98 (1974).

Erickson, R. A., Design Factors and Applications of Centrifuges, *Chem. Eng. Progr.*, **67**(9):66 (1971).

Fitch, B., Choosing a Separation Technique, *Chem. Eng. Progr.*, **70**(12):33 (1974).

Guthrie, K. M., Capital Cost Estimating, *Chem. Eng.*, **76**(6):114 (1969).

Hanf, E. B., Entrainment Separator Design, *Chem. Eng. Progr.*, **67**(11):54 (1971).

Keith, F. W., Jr., and R. T. Moll, Matching a Dewatering Centrifuge to Waste Sludge, *Chem. Eng. Progr.*, **67**(9):55 (1971).

Kouloheris, A. P., and R. L. Meek, Centrifugal Washing of Solids, *Chem. Eng.*, **75**(19):121 (1968).

Landis, D. M., Process Control of Centrifuge Operations, *Chem. Eng. Progr.*, **66**(1):51 (1970).

Rao, K. N., and T. C. Rao, Estimating Hydrocyclone Efficiency, *Chem. Eng.*, **82**(11):121 (1975).

Shah, Y. M., and R. T. Price, Calculator Program Solves Cyclone Efficiency Equations, *Chem. Eng.*, **85**(19):99 (1978).

Schnittger, J. R., Integrated Theory of Separation for Bulk Centrifuges, *I&EC Process Des. & Dev.*, **9**(3):407 (1970).

Stevens, M. H., Centrifuges for Power Plants, *Power Eng.*, **72**(12):34 (1968).

Thrush, R. E., and R. W. Honeychurch, How to Specify Centrifuges, *Hydrocarbon Process.*, **48**(10):81 (1969).

Todd, D. B., and C. A. Hopper, Centrifugal Extraction of Waste Dyes, *Chem. Eng. Progr.*, **67**(9):60 (1971).

————, Improving Performance of Centrifugal Extractors, *Chem. Eng.*, **79**(16):152 (1972).

Vesilind, P. A., Estimating Centrifuge Capacities, *Chem. Eng.*, **81**(7):54 (1974).

Weiss, E. G., and H. Allison, Western States "Quadramitic Centrifugal," *Cost Eng.*, **15**(2):4 (1970).

White, W. F., Water Clarification via Centrifuge, *Chem. Eng. Progr.*, **67**(9):45 (1971).

Willus, C. A., and B. Fitch, Flow Patterns in a Disc Centrifuge, *Chem. Eng. Progr.*, **69**(9):73 (1973).

Zanker, A., Hydrocyclones: Dimensions and Performance, *Chem. Eng.*, **84**(10):122 (1977).

Zimmerman, O. T., and I. Lavine, De Laval Peeler Centrifuge, *Cost Eng.*, **12**(1):9 (1967).

———— and ————, Wemco-Conturbex Screening Centrifuge, *Cost Eng.*, **12**(2):12 (1967).

———— and ————, Teknika/Sontrifuge, *Cost Eng.*, **12**(4):10 (1967).

Compressor design and costs

Abidi, F., Pipeline-System Costs Estimated, *Oil & Gas J.*, **73**(41):99 (1975).

Abraham, R. W., Reliability of Rotating Equipment, *Chem. Eng.*, **80**(24):96 (1973).

Bauermeister, K. J., Turbocompressors for Air and Oxygen Service, *Hydrocarbon Process.*, **54**(6):63 (1975).

Birdsall, J. C., Graph Finds Compression Rate Quickly, *Hydrocarbon Process.*, **47**(1):153 (1968).

Blumenfeld, H. J., et al., Compressors, *AACE Cost Eng. Notebook*, A-4.160, August 1972.

Boyce, M. P., How to Achieve Online Availability of Centrifugal Compressors, *Chem. Eng.*, **85**(13):115 (1978).

Branch, S. G., How to Estimate the Size and Cost of Gas Turbines and Gas Engines, *Hydrocarbon Process.*, **46**(10):131 (1967).

Bresler, S. A., Guide to Trouble-Free Compressors, *Chem. Eng.*, **77**(12):161 (1970).

Brown, P. J., Compressor and Engine Maintenance, *Chem. Eng. Progr.*, **68**(6):77 (1972).

Brown, R. N., Can Specifications Improve Compressor Reliability? *Hydrocarbon Process.*, **51**(7):89 (1972).

Bultzo, C., Problems With Rotating and Reciprocating Equipment, *Chem. Eng. Progr.*, **68**(3):66 (1972).

Cameron, J. A., and F. M. Danowski, Jr., How to Select Materials for Centrifugal Compressors, *Hydrocarbon Process.*, **53**(6):115 (1974).

Canova, F., Matching Turbomachinery to a Process, *Chem. Eng.*, **76**(12):178 (1969).

Cavaliere, G. F., and R. A. Gyepes, Cost Evaluation of Intercooler Systems for Air Compressors, *Hydrocarbon Process.*, **52**(10):107 (1973).

Chodnowsky, N. M., Centrifugal Compressors for High-Pressure Service, *Chem. Eng.*, **75**(25):110 (1968).

Cordes, R. J., Use Compressors Safely, *Hydrocarbon Process.*, **56**(10):227 (1977).

Davis, H. M., How to Improve Compressor Operation and Maintenance, *Hydrocarbon Process.*, **53**(1):93 (1974).

Diehl, G. M., How to Control Compressor Noise, *Hydrocarbon Process.*, **54**(7):157 (1975).

Dimoplon, W., What Process Engineers Need to Know About Compressors, *Hydrocarbon Process.*, **57**(5):221 (1978).

Dwyer, J. J., Air Filters and Intercoolers, *Hydrocarbon Process.*, **50**(10):80 (1971).

Guthrie, K. M., Estimating the Cost of High-Pressure Equipment, *Chem. Eng.*, **75**(25):144 (1968).

———, Capital Cost Estimating, *Chem. Eng.*, **76**(6):114 (1969).

Haden, R. C., Centrifugal Compressors in Chlorine Service, *Chem. Eng. Progr.*, **70**(3):69 (1974).

Hallock, D. C., R. M. Farber, and C. C. Davis, Compressors and Drivers for LNG Plants, *Chem. Eng. Progr.*, **68**(9):77 (1972).

Jackson, C., How to Align Barrel-Type Centrifugal Compressors, *Hydrocarbon Process.*, **50**(9):189 (1971).

———, How to Prevent Turbomachinery Thrust Failures, *Hydrocarbon Process.*, **54**(6):73 (1975).

Jacobs, D. L. E., Reliability, Performance Specifications, *Hydrocarbon Process.*, **50**(10):69 (1971).

Klumpar, I. V., Process Economics by Computer, *Chem. Eng.*, **77**(1):107 (1970).

LaCerda, D. J., Better Instrument and Plant Air Systems, *Hydrocarbon Process.*, **47**(6):107 (1968).

Lady, E. R., Compressor Efficiency: Definition Makes a Difference, *Chem. Eng.*, **77**(17):112 (1970).

Lapina, R. P., Can You Rerate Your Centrifugal Compressor? *Chem. Eng.*, **82**(2):95 (1975).

Lewis, P., Increase Reliability by Better Compressor Component Design, *Hydrocarbon Process.*, **53**(5):117 (1974).

Magliozzi, T. L., Control System Prevents Surging in Centrifugal-Flow Compressors, *Chem. Eng.*, **74**(10):139 (1967).

Mehta, D. D., For Process Designers . . . Compressors and Converters, *Hydrocarbon Process.*, **51**(6):129 (1972).

Moens, J. P. C., Adapt Process to Compressor, *Hydrocarbon Process.*, **50**(12):96 (1971).

Moody, J. F., Which Type of Air Compressor? Cost Comparison Helps Selection, *Plant Eng.*, **24**(14):66 (1970).

Morrow, D. R., Performance Curves in Plant Operations, *Chem. Eng. Progr.*, **64**(12):56 (1968).

Neale, D. F., Inspection, Shipping, and Erection, *Hydrocarbon Process.*, **50**(10):81 (1971).

Neerken, R. F., Compressor Selection for the Chemical Process Industries, *Chem. Eng.*, **82**(2):78 (1975).

Nelson, W. E., Compressor Seal Fundamentals, *Hydrocarbon Process.*, **56**(12):91 (1977).

Nisenfeld, A. E., R. Miyasaki, T. Liem, and J. M. Eskes, For Easier Compressor Control, *Hydrocarbon Process.*, **54**(4):153 (1975).

Nisenfeld, A. E., and C. H. Cho, Parallel Compressor Control . . . What Should be Considered, *Hydrocarbon Process.*, **57**(2):147 (1978).

O'Donnell, J. P., Small Turbine-Compressor Use is Growing, *Oil & Gas J.*, **65**(33):105 (1967).

Pikulik, A., and H. E. Diaz, Cost Estimating for Major Process Equipment, *Chem. Eng.*, **84**(21):107 (1977).

Scheel, L. F., New Ideas on Centrifugal Compressors, *Hydrocarbon Process.*, **47**(9):253 (1968).

———, New Ideas on Centrifugal Compressors, *Hydrocarbon Process.*, **47**(10):161 (1968).

———, New Piston Compressor Rating Method, *Hydrocarbon Process.*, **46**(12):133 (1967).

———, New Ideas on Centrifugal Compressors, *Hydrocarbon Process.*, **47**(11):238 (1968).

———, New Ideas on Rotary Compressors, *Hydrocarbon Process.*, **48**(11):271 (1969).

———, Air Compressor Rotor and Components, *Hydrocarbon Process.*, **50**(10):74 (1971).

Schirm, A. C., Bearings, Seals, Testing, *Hydrocarbon Process.*, **50**(10):76 (1971).

Sohre, J. S., Stop Vibration of Centrifugals, *Petro/Chem Eng.*, **40**(12):22 (1968).

Sommerfield, E. M., Compressed Air for Power Plants, *Power Eng.*, **72**(5):51 (1968), **73**(2):48 (1969).

———, Selecting Compressors, *Power Eng.*, **73**(4):45 (1969).

Strub, R. A., and C. Matile, Centrifugal Hypercompressors for Ethylene Service, *Hydrocarbon Process.*, **54**(6):67 (1975).

Tanner, A. L., C. R. Cooper, and E. F. Drucker, Costs Key to Turbine Choice for LNG Plants, *Oil & Gas J.*, **73**(50):60 (1975).

Wagoner, H. E., Jr., Small Gas Turbines Click in Gas-Transmission Service, *Oil & Gas J.*, **66**(11):114 (1968).

White, M. H., Surge Control for Centrifugal Compressors, *Chem. Eng.*, **79**(29):54 (1972).

Wilson, W. E., Where to Use Stream Turbines, *Petro/Chem Eng.*, **40**(13):27 (1968).

Zimmerman, O. T., Air Compressor Packages, *Cost Eng.*, **18**(3):7 (1973).

Compressor Calculation Charts, *Petro/Chem Eng.*, **40**(5):56 (1968).

A Guide to Compressor Selection, *Compressed Air*, **83**(9):15 (1978).

Crushing and grinding design and costs

Austin, L. G., Understanding Ball Mill Sizing, *I&EC Process Des. & Dev.*, **12**(2):121 (1973).

Cowper, N. T., T. L. Thompson, T. C. Aude, and E. J. Wasp, Processing Steps: Keys to Successful Slurry-Pipeline Systems, *Chem. Eng.*, **79**(3):58 (1972).

Furuya, M., Y. Nakajima, and T. Tanaka, Design of Closed-Circuit Grinding System With Tube Mill and Nonideal Classifier, *I&EC Process Des. & Dev.*, **12**(1):18 (1973).

Guthrie, K. M., Capital Cost Estimating, *Chem. Eng.*, **76**(6):114 (1969).

Jenness, R. C., Determining Drive Motor Requirements for Rotary Grinding Mills, *Allis-Chalmers Eng. Rev.*, **38**(2):8 (1973).

Lapple, C. E., Particle-Size Analysis and Analyzers, *Chem. Eng.*, **75**(11):149 (1968).

Nakajima, Y., and T. Tanaka, Solution of Batch Grinding Equation, *I&EC Process Des. & Dev.*, **12**(1):23 (1973).

Narashimhan, K. S., and S. R. S. Sastri, How to Estimate Size Distribution of Crushed Products, *Chem. Eng.*, **82**(12):77 (1975).

Pebworth, J. T., Selecting Mills for Heat-Sensitive Materials, *Chem. Eng.*, **79**(17):81 (1972).

Ratcliffe, A., Trends in Size Reduction of Solids . . . Crushing and Grinding, *Chem. Eng.*, **79**(15):62 (1972).

Sastri, S. R. S., and K. S. Narashimhan, Predicting Grinding-Mill Energy Use, *Chem. Eng.*, **82**(18):103 (1975).

Somasundaran, P., and I. J. Lin, Effect of the Nature of Environment on Comminution Processes, *I&EC Process Des. & Dev.*, **11**(3):321 (1972).

Tanaka, T., Scale-Up Theory of Jet Mills on Basis of Comminution Kinetics, *I&EC Process Des. & Dev.*, **11**(2):238 (1972).

Zanker, A., Shortcut Technique Gives Size Distribution of Comminuted Solids, *Chem. Eng.*, **85**(10):101 (1978).

Zimmerman, O. T., and I. Lavine, The Pulva-Sizer, *Cost Eng.*, **13**(1):6 (1968).

———, Schuty-O'Neil Air-Swept Pulverizer, *Cost Eng.*, **13**(2):8 (1968).

———, Stedman Cage-Type Disintegrators, *Cost Eng.*, **13**(3):12 (1968).

Crystallizer design and costs

Ajinkya, M. B., and W. H. Ray, On the Optimal Operation of Crystallization Processes, *Chem. Eng. Comm.*, **1**(4):181 (1974).

Bennett, R. C., H. Fiedelman, and A. D. Randolph, Crystallizer Influenced Nucleation, *Chem. Eng. Progr.*, **69**(7):86 (1973).

Canning, T. F., Interpreting Population Density Data from Crystallizers, *Chem. Eng. Progr.*, **66**(7):80 (1970).

Chien, H. H. Y., and A. H. Larsen, Calculation of Multistage Multiphase Reacting Systems in Crystallization, *I&EC Process Des. & Dev.*, **13**(3):299 (1974).

Fitch, B., How to Design Fractional Crystallization Processes, *I&EC*, **62**(12):6 (1970).

Guthrie, K. M., Capital Cost Estimating, *Chem. Eng.*, **76**(6):114 (1969).

Larson, M. A., Guidelines for Selecting a Crystallizer, *Chem. Eng.*, **85**(4):90 (1978).

———, and J. Garside, Crystallizer Design Techniques Using the Population Balance, *Chem. Eng. (London)*, 318 (June 1973).

Palermo, J. A., Crystallization, *I&EC*, **60**(4):65 (1968).

Randolph, A. D., Crystallization, *Chem. Eng.*, **77**(10):80 (1970).

Sun, Y-C., Purify Compounds by . . . Evaporative Crystallization, *Hydrocarbon Process.*, **50**(4):159 (1971).

Wey, J.-S., and J. P. Terwilliger, Design Considerations for a Multistage Cascade Crystallizer, *I&EC Fund.*, **15**(3):467 (1976).

Zimmerman, O. T., and I. Lavine, Classifying Evaporator Crystallizer, *Cost Eng.*, **14**(3):4 (1969).

Dust and mist collection equipment design and costs

Bailey, E. E., and R. W. Heinz, SO_2 Recovery Plants—Materials of Construction, *Chem. Eng. Progr.*, **71**(3):64 (1975).

Benson, G. E., S. Plater-Zyberk, and A. A. Metry, Methods and Costs of Controlling Sulfur Emissions from a Petroleum Refinery, *Chem. Eng. Progr. Symp. Ser.*, **73**(165):241 (1977).

Brandt, D. L., and A. N. Mann, Preliminary Estimation of Costs for Air and Water Pollution Control, *1973 Trans. Am. Assoc. Cost Eng.* (1973), p. 157.

Bump, R. L., Electrostatic Precipitators in Industry, *Chem. Eng.*, **84**(2):129 (1977).

Calvert, S., How To Choose a Particulate Scrubber, *Chem. Eng.*, **84**(18):54 (1977).

———, Get Better Performance from Particulate Scrubbers, *Chem. Eng.*, **84**(23):133 (1977).

———, Guidelines for Selecting Mist Eliminators, *Chem. Eng.*, **85**(5):109 (1978).

Caplan, F., Plant Notebook, Sizing Electrostatic Precipitators, *Chem. Eng.*, **85**(9):153 (1978).

Carlton-Jones, D., and H. B. Schneider, Tall Chimneys, *Chem. Eng.*, **75**(22):166 (1968).

Chalker, W. R., Determine Optimum Stack Size, *Petro/Chem Eng.*, **39**(5):35 (1967).

Cheney, W. A., Modular Electrostatic Precipitators, *Plant Eng.*, **26**(15):66 (1972).

Cines, M. R., D. M. Haskell, and C. G. Houser, Molecular Sieves for Removing H_2S from Natural Gas, *Chem. Eng. Progr.*, **72**(8):89 (1976).

Constance, J. D., Estimating Exhaust-Air Requirements for Processes, *Chem. Eng.*, **77**(17):116 (1970).

———, Calculate Effective Stack Height Quickly, *Chem. Eng.*, **79**(19):81 (1972).

Culhane, F. R., Production Baghouses, *Chem. Eng. Progr.*, **64**(1):65 (1968).

Devitt, J. W., L. V. Yerino, T. C. Ponder, Jr., and C. J. Chatlynne, Estimating Costs of Flue Gas Desulfurization Systems for Utility Boilers, *J. Air Poll. Cont. Assoc.*, **26**(3):204 (1976).

Doerschlag, C., and G. Miczek, How to Choose a Cyclone Dust Collector, *Chem. Eng.*, **84**(4):64 (1977).

Edmister, N. G., and F. L. Bunyard, A Systematic Procedure for Determining the Cost of Controlling Particulate Emissions from Industrial Sources, *J. Air Poll. Cont. Assoc.*, **20**(7):446 (1970).

Finney, J. A., Jr., Selecting Precipitators, *Power Eng.*, **72**(12):26 (1968).

Frenkel, D. I., Improving Electrostatic Precipitator Performance, *Chem. Eng.*, **85**(14):105 (1978).

Gilbert, W., Troubleshooting Wet Scrubbers, *Chem. Eng.*, **84**(23):140 (1977).

Gleason, T. G., How to Avoid Scrubber Corrosion, *Chem. Eng. Progr.*, **71**(3):43 (1975).

Hanf, E. W., and J. W. MacDonald, Economic Evaluation of Wet Scrubbers, *Chem. Eng. Progr.*, **71**(3):48 (1975).

Hardison, L. C., Air Pollution Control Equipment, *Petro/Chem. Eng.*, **40**(3):30 (1968).

Harwood, C. F., P. C. Siebert, and D. K. Oestreich, Optimizing Baghouse Performance . . . To Control Asbestos Emissions, *Chem. Eng. Progr.*, **73**(1):54 (1977).

Holland, W. D., and R. E. Conway, Three Multi-Stage Stack Samplers, *Chem. Eng. Progr.*, **69**(6):93 (1973).

Honea, F. I., Energy Requirements for Environmental Control Equipment, *Chem. Eng.*, **81**(22):55 (1974).

Horzella, T. I., Selecting, Installing, and Maintaining Cyclone Dust Collectors, *Chem. Eng.*, **85**(3):85 (1978).

Imperato, N. F., Gas Scrubbers, *Chem. Eng.*, **75**(22):152 (1968).

Jones, H. R., "Air and Gas Cleanup Equipment," Noyes Data Corporation, Park Ridge, New Jersey, 1972.

Katell, S., and K. D. Plants, Here's What SO$_2$ Removal Costs, *Hydrocarbon Process.*, **46**(7):161 (1967).

Kirchner, R. W., Corrosion of Pollution Control Equipment, *Chem. Eng. Progr.*, **71**(3):58 (1975).

Kittleman, T. A., and R. B. Akell, The Cost of Controlling Organic Emissions, *Chem. Eng. Progr.*, **74**(4):87 (1978).

Klooster, H. J., G. A. Vogt, and G. F. Braun, Optimizing the Design of Relief and Flare Systems, *Chem. Eng., Progr.*, **71**(1):39 (1975).

Koch, W. H., and W. Licht, New Design Approach Boosts Cyclone Efficiency, *Chem. Eng.*, **84**(24):80 (1977).

Kopita, R., and T. G. Gleason, Wet Scrubbing of Boiler Flue Gas, *Chem. Eng. Progr.*, **64**(1):74 (1968).

Lasater, R. C., and J. H. Hopkins, Removing Particulates from Stack Gases, *Chem. Eng.*, **84**(22):111 (1977).

Lewandowski, G. A., Specifying Mechanical Design of Electrostatic Precipitators, *Chem. Eng.*, **84**(14):108 (1978).

Lucas, R. L., Gas-Solids Separations: State-of-the Art, *Chem. Eng. Progr.*, **70**(12):52 (1974).

Marchello, J. M., "Control of Air Pollution Sources," Marcel Dekker, Inc., New York, 1976.

Moody, G. B., Mechanical Design of a Tall Stack, *Hydrocarbon Process.*, **48**(9):173 (1969).

Munson, J. S., Dry Mechanical Collectors, *Chem. Eng.*, **75**(22):147 (1968).

Nelson, W. C., How Sulfur and Metals Affect Desulfurization Costs, *Oil & Gas J.*, **75**(18):247 (1977).

Nichols, R. A., Hydrocarbon-Vapor Recovery, *Chem. Eng.*, **80**(6):85 (1973).

O'Connor, J. P., and J. F. Citarella, An Air Pollution Control Cost Study of the Steam-Electric Power Generating Industry, *J. Air. Poll. Cont. Assoc.*, **20**(5):283 (1970).

Parekh, R., Equipment for Controlling Gaseous Pollutants, *Chem. Eng.*, **82**(21):129 (1975).

Peters, J. M., Predicting Efficiency of Fine-Particle Collectors, *Chem. Eng.*, **80**(9):99 (1973).

Pfoutz, B. D., and L. L. Stewart, Electrostatic Precipitators—Materials of Construction, *Chem. Eng. Progr.*, **71**(3):53 (1975).

Rashmi, P., Equipment for Controlling Gaseous Pollutants, *Chem. Eng.*, **82**(21):129 (1975).

Rymarz, T. M., How to Specify Pulse-Jet Filters, *Chem. Eng.*, **82**(7):97 (1975).

——— and D. H. Klipstein, Removing Particulates from Gases, *Chem. Eng.*, **82**(21):113 (1975).

Sakol, S. L., and R. A. Schwartz, Construction Materials for Wet Scrubbers, *Chem. Eng. Progr.*, **70**(8):63 (1974).

Sargent, G. D., Dust Collection Equipment, *Chem. Eng.*, **76**(2):130 (1969).

———, Gas/Solid Separations, *Chem. Eng.*, **78**(4):11 (1971).

Schneider, G. G., T. I. Horzella, J. Cooper, and P. J. Striegl, Selecting and Specifying Electrostatic Percipitators, *Chem. Eng.*, **82**(11):94 (1975).

———— ———— ———— and ————, Electrostatic Precipitators: How They are Used in the CPI, *Chem. Eng.*, **82**(17):97 (1975).

Semrau, K. T., Practical Process Design of Particulate Scrubbers, *Chem. Eng.*, **84**(20):87 (1977).

Shah, I. S., Removing SO_2 and Acid Mist With Venturi Scrubbers, *Chem. Eng. Progr.*, **67**(5):51 (1971).

Shah, Y. M., and R. T. Price, Calculator Program Solves Cyclone Efficiency Equations, *Chem. Eng.*, **85**(19):99 (1978).

Sickles, R. W., Electrostatic Precipitators, *Chem. Eng.*, **75**(22):156 (1968).

Smith, J. L., and H. A. Snell, Selecting Dust Collectors, *Chem. Eng. Progr.*, **64**(1):60 (1968).

Stites, J. G., Jr., W. R. Horlacher, Jr., J. L. Bachofer, Jr., and J. S. Bartman, Removing SO_2 From Flue Gas, *Chem. Eng. Progr.*, **65**(10):74 (1969).

Stockton, E. L., Air Pollution Control Costs, *1969 Trans. Am. Assoc. Cost Eng.*, (1969).

Thomson, S. J., Data Improves Separator Design, *Hydrocarbon Process.*, **52**(10):81 (1973).

Willett, H. P., Cutting Air Pollution Control Costs, *Chem. Eng. Progr.*, **63**(3):80 (1967).

Zanker, A., Hydrocyclones: Dimensions and Performance, *Chem. Eng.*, **84**(10):122 (1977).

Zimmerman, O. T., and I. Lavine, Balanced Air XQ Series Cyclone Dust Collectors, *Cost Eng.*, **14**(2):13 (1969).

————, Desulfurization, *Cost Eng.*, **14**(3):6 (1969).

————, The Pantecon Dust Filter, *Cost Eng.*, **15**(2):14 (1970).

————, Dust Collection in an Industrial Plant, *Cost Eng.*, **15**(3):16 (1970).

————, Fibre-Dyne Fume Scrubber, *Cost Eng.*, **16**(1):4 (1971).

————, Dust Collectors, *Cost Eng.*, **17**(1):4 (1972).

Scrubber System Knocks Out Fly Ash and Sulfur Dioxide, *Chem. Eng.*, **78**(15):56 (1971).

Stack-Gas Cleanup Process Ready for Use, *Oil & Gas J.*, **75**(1):77 (1977).

Ejector design and costs

Davies, G. S., A. K. Mitra, and A. N. Roy, Momentum Transfer Studies in Ejectors, *I&EC Process Des. & Dev.*, **6**(3):293 (1967); **6**(3):299 (1967).

Guthrie, K. M., Capital Cost Estimating, *Chem. Eng.*, **76**(6):114 (1969).

Huff, G. A., Jr., Selecting a Vacuum Producer, *Chem. Eng.*, **83**(6):83 (1976).

Jackson, D. H., Steam Jet Ejectors: Their Uses and Disadvantages. *Chem. Eng. Progr.*, **72**(7):77 (1976).

Jeelani, S. A. K., A. Rajkumar, and K. V. Kasipathi Rao, Designing Air-Jet Ejectors, *Chem. Eng.*, **85**(21):135 (1978).

Khoury, F., M. Heyman, and W. Resnick, Performance Characteristics of Self-Entrainment Ejectors, *I&EC Process Des. & Dev.*, **6**(3):331 (1967).

Newman, E. F., How to Specify Steam-Jet Ejectors, *Chem. Eng.*, **74**(8):203 (1967).

Pikulik, A., and H. E. Diaz, Cost Estimating for Major Process Equipment, *Chem. Eng.*, **84**(21):107 (1977).

Expander and turbine design

Franzke, A., Benefits of Energy-Recovery Turbines, *Chem. Eng.*, **77**(4):109 (1970).

Scheel, L. F., What You Need to Know About Gas Expanders, *Hydrocarbon Process.*, **49**(2):105 (1970).

Swearingen, J. S., Engineer's Guide to Turboexpanders, *Hydrocarbon Process.*, **49**(4):97 (1970).

Wickl, R., and R. Sparmann, Can Steam Turbines Save Energy? *Hydrocarbon Processes.*, **53**(7):109 (1974).

Willoughby, W. W., Steam Rate: Key to Turbine Selection, *Chem. Eng.*, **85**(20):146 (1978).

Electric motors and driver selection and costs

Cates, J. H., Jr., Electric Motors: Principles and Applications in the Chemical Process Plant, *Chem. Eng.*, **79**(28):82 (1972).

Center, C. E., Selection and Costing Electric Motors via the COME Program, *AACE Bull.*, **16**(5):141 (1974).

Constance, J. D., How to Pressure-Ventilate Large Motors for Corrosion, Explosion, and Moisture Protection, *Chem. Eng.*, **85**(5):113 (1978).

Cooke, E. F., A. Guide for Selecting Adjustable-Speed Drives, *Chem. Eng.*, **78**(20):70 (1971).

Culver, R. E., and E. H. Marsh, Estimates for Electrical Process Plant, *Chem. Proc. Eng.*, **48**(5):89 (1967).

Flappert, J. F., Selecting Large Electric Drivers for the HPI, *Hydrocarbon Process.*, **54**(12):103 (1975).

Guthrie, K., Costs, *Chem. Eng.*, **76**(8):201 (1969).

Hale, W. O., Choosing Between Synchronous and Induction Motors, *Power Eng.*, **74**(9):56 (1970).

Hill, R. S., and D. L. Kime, How to Specify Drive Trains for Turbine Agitators, *Chem. Eng.*, **83**(16):89 (1976).

Jenett, E., Hydraulic Power Recovery Systems, *Chem. Eng.*, **75**(13):257 (1968).

———, Major Liquid Power Recovery Turbine/Pump Systems, *Petro/Chem Eng.*, **40**(10):25 (1968).

Mapes, W. H., How to Keep Motors Running, *Chem. Eng.*, **82**(15):107 (1975).

Moore, J. C., Electric Motor Drivers for Centrifugal Compressors, *Hydrocarbon Process.*, **54**(5):133 (1975).

Nailen, R. L., Specifying Motors for a Quiet Plant, *Chem. Eng.*, **77**(11):157 (1970).

———, Guide to Trouble-Free Electric-Motor Drives, *Chem. Eng.*, **77**(12):181, (1970).

———, Pick Motors by Their " Start ", *Hydrocarbon Process.*, **52**(1):117 (1973).

———, How to Improve Your Motor Specs, *Hydrocarbon Process.*, **53**(11):211 (1974).

———, Protect Motors by Heat Sensing, *Hydrocarbon Process.*, **57**(4):175 (1978).

Olson, C. R., and E. S. McKelvy, How to Specify and Cost Estimate Electric Motors, *Hydrocarbon Process.*, **46**(10):118 (1967).

Parkinson, E. A., and A. L. Mular, Mineral Processing Equipment Costs and Preliminary Capital Cost Estimations, *Can. Inst. Mining & Metal.*, Special Vol. 13, (1972).

Platt, H., Drivers for Process Pumps, *Hydrocarbon Process.*, **51**(6):95 (1972).

Pritchett, D. H., Electric Motors: A Guide to Standards, *Chem. Eng.*, **79**(28):88 (1972).

Steen-Johnsen, H., How to Estimate the Size and Cost of Mechanical Drive Steam Turbines, *Hydrocarbon Process.*, **46**(10):126 (1967).

Tyler, D. A., How Noisy is a Refinery? *Hydrocarbon Process.*, **48**(7):173 (1969).

Wilson, W. B., How to Select Major Motor Drives, *Petro/Chem. Eng.*, **40**(10):20 (1968).

———, Where to Use Steam Turbines, *Petro/Chem. Eng.*, **40**(13):27 (1968).

Wolfe, G., Worm-Gear Speed Reducers, *Plant Eng.*, **31**(25):102 (1977).

Woods, D. R., and S. J. Anderson, Evaluation of Capital Cost Data: Drives, *Can. J. Chem. Eng.*, **53**(4):357 (1975).

Zimmerman, O. T., and I. Lavine, Tri-Flo Stainless Steel Centrifugal Pumps, *Cost Eng.*, **14**(2):4 (1969).

———, Ribbon Blenders, *Cost Eng.*, **16**(2):13 (1971).

———, Munson Rotary Mixers, *Cost Eng.*, **16**(3):6 (1971).

Fan selection and costs

Gerchow, F. J., How to Select a Pneumatic-Conveying System, *Chem. Eng.*, **82**(4):72 (1975).

Glass, J., Specifying and Rating Fans, *Chem. Eng.*, **85**(7):120 (1978).

Graham, J. B., Fan Selection and Installation, *Chem. Eng. Progr.*, **71**(10):74 (1975).

Guthrie, K., Costs, *Chem. Eng.*, **76**(8):201 (1969).

Martz, J. W., and R. R. Pfahler, How to Troubleshoot Large Industrial Fans, *Hydrocarbon Process.*, **54**(6):57 (1975).

Patton, P. W., and C. F. Joyce, How to Find the Lowest-Cost Vacuum System, *Chem. Eng.*, **83**(3):84 (1976).

Pollak, R., Selecting Fans and Blowers, *Chem. Eng.*, **80**(2):86 (1973).

Scheel, L. F., New Ideas on Fan Selection, *Hydrocarbon Process.*, **48**(6):125 (1969).

Filter design and costs

Bolto, B. A., K. W. V. Cross, R. J. Eldridge, E. A. Swinton, and D. E. Weiss, Magnetic Filter Aids, *Chem. Eng. Progr.*, **71**(12):47 (1975).

Bonem, J. M., Plant Test Point the Way to Better Filter Performance, *Chem. Eng.*, **85**(4):107 (1978).

Breslau, B. R., E. A. Agranat, A. J. Testa, S. Messinger, and R. A. Cross, Hollow Fiber Ultrafiltration, *Chem. Eng. Progr.*, **71**(12):74 (1975).

Brody, M. A., and R. J. Lumpkins, Performance of Dual-Media Filters, *Chem. Eng. Progr.*, **73**(4):83 (1977).

Brooks, N. S., and E. R. Chow, Cartridge Filtration in Pressurized Water Reactors, *Chem. Eng. Progr.*, **71**(12):62 (1975).

Chen, N. S., Liquid-Solid Filtration: Generalized Design and Optimization Equations, *Chem. Eng.*, **85**(17):97 (1978).

Culver, R. H., Diatomaceous Earth Filtration, *Chem. Eng. Progr.*, **71**(12):51 (1975).

Dahlstrom, D. A., Process Corrosion Control: Predicting Performance of Continuous Filters, *Chem. Eng. Progr.*, **74**(3):69 (1978).

—— and S. S. Davis, Plastics in Continuous Filtration Equipment, *Chem. Eng. Progr.*, **65**(10):80 (1969).

Dollinger, L. L., Jr., How to Specify Filters, *Hydrocarbon Process.*, **48**(10):88 (1969).

Emmett, R. C., and C. E. Silverblatt, When to Use Continuous Filtration, *Chem. Eng. Progr.*, **70**(12):38 (1974).

Fitch, B., Choosing a Separation Technique, *Chem. Eng. Progr.*, **70**(12):33 (1974).

Gerchow, F. J., How to Select a Pneumatic-Conveying System, *Chem. Eng.*, **82**(4):72 (1975).

Grutsch, J. F., and R. C. Mallatt, Filtration & Separation: Optimizing Granular Media Filtration, *Chem. Eng. Progr.*, **73**(4):57 (1977).

Guthrie, K. M., Capital Cost Estimating, *Chem. Eng.*, **76**(6):114 (1969).

Hauslein, R. H., Ultra Fine Filtration of Bulk Fluids, *Chem. Eng. Progr.*, **67**(5):82 (1971).

Hawley, C. W., Filtration: Advances and Guidelines; Economic Filtration Guidelines for Pressurized Gas Systems, *Chem. Eng.*, **83**(4):91 (1976).

Henry, J. D., Jr., L. F. Lawler, and C. H. A. Kuo, A Solid/Liquid Separation Process Based on Cross Flow and Electrofiltration, *AIChE J.*, **23**(6):851 (1977).

Jahreis, C. A., Filtration: Advances and Guidelines; Recent Developments in Pressure Filtration, *Chem. Eng.*, **83**(4):80 (1976).

Johnston, P. R., Submicron Filtration, *Chem. Eng. Progr.*, **71**(12):70 (1975).

Kempling, J. C., and J. Eng, Performance of Dual-Media Filters, *Chem. Eng. Progr.*, **73**(4):87 (1977).

Kinsman, S., Instrumentation for Filtration Tests, *Chem. Eng. Progr.*, **70**(12):48 (1974).

Klinkowski, P. R., Ultrafiltration: An Emerging Unit-Operation, *Chem. Eng.*, **85**(11):165 (1978).

Kos, P., Fundamentals of Gravity Thickening, *Chem. Eng. Progr.*, **73**(11):99 (1977).

Lash, L., Filtration & Separation: Scale-up of Granular Media Filters, *Chem. Eng. Progr.*, **73**(4):67 (1977).

Lloyd, P. J., Particle Characterization, *Chem. Eng.*, **81**(9):120 (1974).

Mais, L. G., Filter Media, *Chem. Eng.*, **78**(4):49 (1971).

Maloney, G. F., Selecting and Using Pressure Leaf Filters, *Chem. Eng.*, **79**(11):88 (1972).

Mixon, F. O., Moving Bed Filter Performance Studies, *Chem. Eng. Progr.*, **71**(12):40 (1975).

Mohler, E. F., Jr., and L. T. Clere, Filtration & Separation: Removing Colloidal Solids via Upflow Filtration, *Chem. Eng. Progr.*, **73**(4):74 (1977).

Monfort, R. H., Straight Line Filters, *Cost Eng.*, **18**(3):4 (1973).

Nickolaus, N., Evaluating Filter Cartridges, *Chem. Eng. Progr.*, **70**(12):46 (1974).

Orr, C., ed., "Filtration: Principles and Practices," Marcel Dekker, Inc., New York, 1977.

Porter, H. F., J. E. Flood, and F. W. Rennie, Filter Selection, *Chem. Eng.*, **78**(4):39 (1971).

Porter, M. C., Selecting the Right Membrane, *Chem. Eng. Progr.*, **71**(12):55 (1975).

Purchas, D. B., Cake Filtration: A Standard Test Method for Any Filter, *Chem. Eng.*, **79**(18):86 (1972).

——, Guide to Trouble-Free Plant Operation . . . Filtration, *Chem. Eng.*, **79**(14):88 (1972).

Rothwell, E., Fabric Dust Filtration, *Chem. Eng. (London)*, 138 (March 1975).

Rymarz, T. M., How to Specify Pulse-Jet Filters, *Chem. Eng.*, **82**(7):97 (1975).

Sargent, G. D., Gas/Solid Separations, *Chem. Eng.*, **78**(4):11 (1971).

Silverblatt, C. E., H. Risbud, and F. M. Tiller, Batch, Continuous Processes for Cake Filtration, *Chem. Eng.*, **81**(9):127 (1974).

Smith, G. R. S., Filter Aid Regeneration and Recovery, *Chem. Eng. Progr.*, **71**(12):37 (1975).

——, Filtration: Advances and Guidelines; How to Use Rotary, Vacuum, Precoat Filters, *Chem. Eng.*, **83**(4):84 (1976).

Soules, W. J., Filter Cartridge Standards, *Chem. Eng. Progr.*, **70**(12):43 (1974).

Tiller, F. M., Bench-Scale Design of SLS Systems, *Chem. Eng.*, **81**(9):117 (1974).

——, Pretreatment of Slurries, *Chem. Eng.*, **81**(9):123 (1974).

—— and J. R. Crump, Solid-Liquid Separation: An Overview, *Chem. Eng. Prog.*, **73**(10):65 (1977).

—— and ——, How to Increase Filtration Rates in Continuous Filters, *Chem. Eng.*, **84**(12):183 (1977).

Walling, J. C., Ins and Outs of Gas Filter Bags, *Chem. Eng.*, **77**(25):162 (1970).

Wrotnowski, A. C., Final Filtration With Felt Bag Strainers, *Chem. Eng. Progr.*, **74**(10):89 (1978).

Yoshida, T., Y. Kousaka, S. Inake, and S. Nakai, Pressure Drop and Collection Efficiency of an Irrigated Bag Filter, *I&EC Process Des. & Dev.*, **14**(2):101 (1975).

Zahka, J., and L. Mir, Ultrafiltration of Latex Emulsions, *Chem. Eng. Progr.*, **73**(12):53 (1977).

Mist Filters, *I&EC*, **59**(10):89 (1967).

Fluid transfer design

Arant, J. B., Special Control Valves Reduce Noise and Vibration, *Chem. Eng.*, **79**(5):92 (1972).

Aude, T. C., N. T. Cowper, T. L. Thompson, and E. J. Wasp, Slurry Piping Systems: Trends, Design Methods, Guidelines, *Chem. Eng.*, **78**(15):74 (1971).

Austin, P. P., How to Simplify Fluid Flow Calculations, *Hydrocarbon Process.*, **54**(9):197 (1975).

Benz, G., Cost Effectiveness in Pumping System Design and Operation, *J. Eng. Power*, **89**(4):600 (1967).

Canon, J., Guidelines for Selecting Process-Control Valves, *Chem. Eng.*, **76**(8):109 (1969).

Churchill, S. W., Friction-Factor Equation Spans all Fluid-Flow Regimes, *Chem. Eng.*, **84**(24):91 (1977).

Constance, J. D., Plant Notebook, Estimating Fluid Friction in Ducts on Nonstandard Shapes, *Chem. Eng.*, **78**(5):146 (1971).

Cowper, N. T., T. L. Thompson, T. C. Aude, E. J. Wasp, Processing Steps: Keys to Successful Slurry-Pipeline Systems, *Chem. Eng.*, **79**(3):58 (1972).

De Gance, A. E., and R. W. Atherton, Chemical Engineering Aspects of Two-Phase Flow, *Chem. Eng.*, **77**(6):135 (1970); **77**(8):151 (1970); **77**(10):113 (1970); **77**(15):95 (1970); **77**(17):119 (1970); **77**(24):101 (1970); **77**(29):87 (1970); **78**(5):125 (1971).

Dickey, D. S., and J. C. Fenic, Dimensional Analysis for Fluid Agitation Systems, *Chem. Eng.*, **83**(1):139 (1976).

Dodge, L., How to Compute and Combine Fluid Flow Resistances in Components, *Hydraulics & Pneumatics*, **21**(9):118 (1968); **21**(11):98 (1968).

Driskell, L. R., New Approach to Control Valve Sizing, *Hydrocarbon Process.*, **48**(7):131 (1969).

Elshout, R., Graphs Determine Time Required to Drain Vessels, *Chem. Eng.*, **75**(20):246 (1968).

Gerchow, F. J., How to Select a Pneumatic-Conveying System, *Chem. Eng.*, **82**(4):72 (1975).

Gill, W. N., R. Cole, J. Estrin, R. J. Nunge, and H. Littman, Fluid Dynamics, *I&EC*, **59**(12):69 (1967).

Giraldo, J., Generalized Method for ΔP Calculations, *Hydrocarbon Process.*, **47**(11):219 (1968).

Glickman, M., and A. H. Hehn, Valves, *Chem. Eng.*, **76**(8):131 (1969).

Gloyer, W., A New Look at Two-Phase Flow, *Chem. Eng.*, **75**(1):93 (1968).

Golan, L. P., and G. Borushko, Multi-Phase Flow in the Annulus of a Double-Pipe Exchanger, *Chem. Eng. Progr.*, **73**(2):79 (1977).

Grote, S. H., Calculating Pressure-Release Times, *Chem. Eng.*, **74**(15):203 (1967).

Holland, F. A., "Fluid Flow for Chemical Engineers," Edward Arnold, London, 1973.

Hooper, W. B., Predicting Flow Patterns in Plant Equipment, *Chem. Eng.*, **82**(16):103 (1975).
Kern, R., How to Size Process Piping for Two-Phase Flow, *Hydrocarbon Process.*, **48**(10):105 (1969).
———, How to Size Flowmeters, *Chem. Eng.*, **82**(5):161 (1975).
———, Piping Design for Two-Phase Flow, *Chem. Eng.*, **82**(13):145 (1975).
Kirkpatrick, D. M., Piping Systems for Difficult-to-Pump Fluids, *Petro/Chem. Eng.*, **39**(13):32 (1967).
Kopalinsky, E. M., and R. A. A. Bryant, Friction Coefficients for Bubbly Two-Phase Flow in Horizontal Pipes, *AIChE J.*, **22**(1):82 (1976).
Kuong, J. F., Nomograph Finds ΔP in Gas Piping, *Hydrocarbon Process.*, **49**(2):156 (1970).
Larson, A. M., Jr., Streamlining Head-Loss Calculations, *Chem. Eng.*, **82**(23):115 (1975).
Leonard, R. E., Disassembly of Tube Fittings, *Chem. Eng.*, **85**(15):93 (1978).
Loeb, M. B., New Graphs for Solving Compressible Flow Problems, *Chem. Eng.*, **76**(11):179 (1969).
Lyons, J. L., and C. Askland, Jr., "Lyons' Encyclopedia of Valves," Van Nostrand Reinhold Company, New York (1978).
McCoy, R. E., Comparing Flow-Ratio Control Systems, *Chem. Eng.*, **76**(13):92 (1969).
Meador, L., and A. Shah, Steam Lines Designed for Two-Phase, *Hydrocarbon Process.*, **48**(1):143 (1969).
Miller, J. E., Miller Number, *Chem. Eng.*, **81**(15):103 (1974).
Nguyen, H. X., Simplify Calculation of Economic Pipe Size, *Hydrocarbon Process.*, **57**(2):143 (1978).
Nguyen, H. H., Estimating Venturi or Orifice Diameter, *Chem. Eng.*, **85**(15):92 (1978).
Paige, P. M., How to Estimate the Pressure Drop of Flashing Mixtures, *Chem. Eng.*, **74**(17):159 (1967).
Pillai, B. C., and M. Raja Rao, Pressure Drop and Flow Characteristics of Packed Fluidized Systems, *I&EC Fund.*, **15**(2):250 (1976).
Roberts, R. N., Pipelines for Process Slurries, *Chem. Eng.*, **74**(16):125 (1967).
Seyer, F. A., and A. B. Metzner, Turbulent Flow Properties of Viscoelastic Fluids, *Can. J. Chem. Eng.*, **45**(3):121 (1967).
Shah, G. C., Inert Gas Injection Reduces Pump Noise, *Chem. Eng.*, **85**(15):93 (1978).
Shotkin, L. M., Stability Considerations in Two-Phase Flow, *Nuclear Sci. & Eng.*, **28**(3):317 (1967).
Sisson, W., How to Size Condensate Piping, *Hydrocarbon Process.*, **50**(4):152 (1971).
Sommerfeld, J. T., Equation for Fluid Friction Factor, *Hydrocarbon Process.*, **46**(7):135 (1967).
Santoleri, J. J., Spray Nozzle Selection, *Chem. Eng. Progr.*, **70**(9):84 (1974).
Tan, S. H., Simplified Flare System Sizing, *Hydrocarbon Process.*, **46**(10):149 (1967).
Thorsen, T., and R. Oen, How to Measure Industrial Wastewater Flow, *Chem. Eng.*, **82**(4):95 (1975).
Turpin, J. L., and R. L. Huntington, Prediction of Pressure Drop for Two-Phase, Two-Component Concurrent Flow in Packed Beds, *AIChE J.*, **13**(6):1196 (1967).
Weismantel, G., Solids Pipelines Advance, *Chem. Eng.*, **80**(25):68 (1973).
Wohl, M. H., Designing for Non-Newtonian Fluids, *Chem. Eng.*, **75**(2):148 (1968).
———, Instruments for Viscometry, *Chem. Eng.*, **75**(7):99 (1968).
———, Isothermal Laminar Flow of Non-Newtonian Fluids in Pipes, *Chem. Eng.*, **75**(8):143 (1968).
———, Dynamics of Flow Between Parallel Plates and in Noncircular Ducts, *Chem. Eng.*, **75**(10):183 (1968).
———, Isothermal Turbulent Flow in Pipes, *Chem. Eng.*, **75**(12):95 (1968).
———, Mixing of Non-Newtonian Fluids, *Chem. Eng.*, **75**(18):113 (1968).
———, Properties of Liquids, *Chem. Eng.*, **76**(8):11 (1969).
Zenz, F. A., How Flow Phenomena Affect Design of Fluidized Beds, *Chem. Eng.*, **84**(27):81 (1977).
Designing for Non-Newtonian Fluids: Rheology of Non-Newtonian Materials, *Chem. Eng.*, **75**(4):130 (1968).
Rational or General Flow Formula and Charts, *Petro/Chem. Eng.*, **40**(5):18 (1968).

Materials handling and costs

Alonso, J. R. F., Estimating the Costs of Gas-Cleaning Plants, *Chem. Eng.*, **78**(28):86 (1971).
Bates, L., Using Helical Screws for Solids Handling, *Chem. Eng.*, **84**(5):183 (1977).
——— and D. Kershaw, Flow Profiles in Hoppers, *Chem. Eng. Progr.*, **71**(2):66 (1975).

Bevan, R. R., From the Receiving Hopper to the Furnace: Coal Handling at the Plant Site, *Chem. Eng.*, **85**(2):120 (1978).

Binning, J. E., Fluidized Solids & Inclined Conveyors, *Chem. Eng. Progr.*, **66**(6):35 (1970).

Brand, R. R., and F. L. Burgess, Jr., Reducing Hazards of Handling Reactive Catalysts, *Chem. Eng.*, **80**(26):248 (1973).

Browning, J. E., Agglomeration: Growing Larger in Applications and Technology, *Chem. Eng.*, **74**(25):147 (1967).

Buffington, M. A., Mechanical Conveyors and Elevators, *Chem. Eng.*, **76**(22):33 (1969).

Burke, A. J., Weighing Bulk Materials in the Process Industries, *Chem. Eng.*, **80**(6):66 (1973).

Caldwell, L. G., Pneumatic Conveying Materials: A Pneumatic Conveying Primer, *Chem. Eng. Progr.*, **72**(3):63 (1976).

Canon, R. M., and G. W. Medlin, Calculating Solids Inventory, *Chem. Eng.*, **85**(15):91 (1978).

Carleton, A. J., and C.-H. Cheng, Pipeline Design for Industrial Slurries, *Chem. Eng.*, **84**(9):95 (1977).

Carr, R. L., Jr., Properties of Solids, *Chem. Eng.*, **78**(22):7 (1969).

Carroll, P. J., Hopper Designs With Vibratory Feeders, *Chem. Eng. Progr.*, **66**(6):44 (1970).

——— and H. Colijn, Vibrations in Solids Flow, *Chem. Eng. Progr.*, **71**(2):53 (1975).

Colijn, H., Mechanical Conveyors and Elevators—CPI Workhorses, *Chem. Eng.*, **85**(24):43 (1978).

Condolios, E., E. E. Chapus, and J. A. Constans, New Trends in Solids Pipelines, *Chem. Eng.*, **74**(10):131 (1967).

Dalstad, J. I., Slurry Pump Selection and Application, *Chem. Eng.*, **84**(9):101 (1977).

Dressel, J. H., and J. Absil, Prepare Coal for Gasification, *Hydrocarbon Process.*, **53**(3):91 (1974).

Eisenhart-Rothe, M. V., and I. A. S. Z. Peschl, In-Plant Bulk Materials Handling, *Chem. Eng.*, **84**(7):97 (1977).

Garabette, L. G., The Handling of Urea and Melamine, *Hydrocarbon Process.*, **47**(10):96 (1968).

Gerchow, F. J., Specifying Components of Pneumatic-Conveying Systems, *Chem. Eng.*, **82**(7):88 (1975).

———, Pneumatic Conveying Materials: Selecting Rotary Feeder Vane Valves, *Chem. Eng. Progr.*, **72**(3):78 (1976).

Gluck, S. E., Design Tips for Pneumatic Conveyers, *Hydrocarbon Process.*, **47**(10):88 (1968).

Guthrie, K. M., Capital Cost Estimating, *Chem. Eng.*, **76**(6):114 (1969).

Harvey, D. J., and J. R. Fowler, Putting Evaporators to Work: Dynamic Process Modeling of a Quadruple Effect Evaporation System, *Chem. Eng. Progr.*, **72**(4):47 (1976).

Haynes, H. D., Large Wheel Loaders Cut Materials Handling Costs, *Chem. Eng.*, **76**(20):150 (1969).

Hyman, H., Working With the Volume Inorganics, *Chem. Eng.*, **80**(23):15 (1973).

Jenike, A. W., and J. R. Johanson, Solids Flow in Bins and Moving Beds, *Chem. Eng. Progr.*, **66**(6):31 (1970).

——— and J. W. Carson, Feeding Solids with Mass-Flow Bins, *Chem. Eng. Progr.*, **71**(2):69 (1975).

Johanson, J. R., Feeding, *Chem. Eng.*, **76**(22):75 (1969).

———, In-Bin Blending, *Chem. Eng. Progr.*, **66**(6):50 (1970).

———, Particle Segregation . . . and What To Do About It, *Chem. Eng.*, **85**(11):183 (1978).

———, Design for Flexibility in Storage and Reclaim, *Chem. Eng.*, **85**(24):19 (1978).

———, Know Your Material—How to Predict and Use the Properties of Bulk Solids, *Chem. Eng.*, **85**(24):9 (1978).

Joklik, O. F., A User's Guide to Catalysts, *Chem. Eng.*, **80**(23):49 (1973).

Jones, V., and E. Fenska, Systems Approach to Pellet Handling, *Hydrocarbon Process.*, **47**(10):104 (1968).

Kiley, L. R., and H. Scheffer, Liquid-Transportation Technology, *Chem. Eng.*, **85**(8):17 (1978).

Kirk, M. M., Cranes, Hoists and Trolleys, *Chem. Eng.*, **74**(5):168 (1967).

Kouloheris, A. P., and R. L. Meek, Centrifugal Washing of Solids, *Chem. Eng.*, **75**(19):121 (1968).

Kraus, M. N., Pneumatic Conveyors, *Chem. Eng.*, **76**(22):59 (1969).

———, Guide to Pneumatic Conveyors, *Chem. Eng.*, **85**(24):63 (1978).

Kravitz, S., How to Estimate Weight of Natural Piles, *Chem. Eng.*, **76**(19): 166 (1969).

Ku, J. C. Z., and D. Bevan, Outdoor Bulk Storage for Hydrophilic Materials, *Chem. Eng.*, **84**(18):69 (1977).

Kurylchek, A. L., Bin Discharging Problems and Solutions, *Chem. Eng. Progr.*, **74**(7):84 (1978).

Leung, L. S., and R. J. Wiles, A Quantitative Design Procedure for Vertical Pneumatic Conveying Systems, *I&EC Fund.*, **15**(4):552 (1976).

Ludwig, E. E., Designing Process Plants to Meet OSHA Standards, *Chem. Eng.*, **80**(20):88 (1973).

Marchello, J. M., and A. Gomezplata, eds., "Gas-Solids Handling in the Process Industries," Marcel Dekker, Inc., New York, 1976.

McDowell, D. W., Jr., Handling Phosphoric Acid and Phosphate Fertilizers, *Chem. Eng.*, **82**(16):119 (1975).

——, Choosing Materials for Sulfuric-Acid Services, *Chem. Eng.*, **84**(14):137 (1977).

Merville, R. L., and E. A. Albrecht, Heated Conveyor Handles Hot Asphalt, *Hydrocarbon Process.*, **47**(10):101 (1968).

Mitchell, J. R., Methods for Conveying and Weighing Solids—Designing for Batch and Continuous Weighers, *Chem. Eng.*, **84**(5):177 (1977).

——, How to Weigh Bulk Solid Materials, *Chem. Eng.*, **85**(24):93 (1978).

Oringer, K., Current Practice in Polymer-Recovery Operations, *Chem. Eng.*, **79**(6):96 (1972).

Paulis, N. J., and D. Silvermetz, Instrumentation for Slurry Systems, *Chem. Eng.*, **84**(9):107 (1977).

Perkins, D. E., and J. E. Wood, Design and Select . . . Pneumatic Conveying Systems, *Hydrocarbon Process.*, **53**(3):75 (1974).

Pietsch, W., Improving Powders by Agglomeration, *Chem. Eng. Progr.*, **66**(1):31 (1970).

La Pushin, G., Transportation and Storage, *Chem. Eng.*, **76**(22):19 (1969).

Roberts, A. W., and G. J. Montagner, Flow in a Hopper Discharge Chute System, *Chem. Eng. Progr.*, **71**(2):71 (1975).

Roberts, E. J., P. Stavenger, J. P. Bowersox, A. K. Walton, and M. Mehta, Solids Concentration, *Chem. Eng.*, **77**(14):52 (1970).

—— —— —— —— and ——, Solid/Solid Separation, *Chem. Eng.*, **78**(4):89 (1971).

Roe, L. A., Mineral Processing Methods: Review and Forecast, *Chem. Eng.*, **83**(13):102 (1977).

Sargent, G. D., Gas/Solid Separations, *Chem. Eng.*, **78**(4):11 (1971).

Scher, J. A., Industrial Wastewater Pumps, *Chem. Eng.*, **82**(21):95 (1975).

Scholtz, G. A., In-Plant Handling of Bulk Materials in Packages and Containers, *Chem. Eng.*, **85**(24):29 (1978).

Shinohara, K., K. Shoji, and T. Tanaka, Mechanism of Size Segregation of Particles in Filling a Hopper, *I&EC Process Des. & Dev.*, **11**(3):369 (1972).

Schofield, C., and H. M. Sutton, Systems Approach for In-Plant Bulk Materials Handling, *Chem. Eng.*, **84**(7):103 (1977).

Schultz, G. A., Specifying Screw Conveyors, *Plant Eng.*, **28**(5):173 (1974).

——, Belt Conveyors for Bulk Materials, *Plant Eng.*, **29**(18):95 (1975).

——, Overhead Trolley Conveyors, *Plant Eng.*, **30**(4):117 (1976).

——, Selecting Vibrating Conveyors, *Plant Eng.*, **30**(17):105 (1976).

Sisson, B., Use Nomograph to Find Slope and Length of Belt Conveyors, *Hydrocarbon Process.*, **47**(10):108 (1968).

Smego, R. A., Materials Management, *Chem. Eng.*, **76**(22):100 (1969).

Smith, B. A., Charts Tell Which Conveyor Horsepower to Use, *Plant Eng.*, **21**(12):122 (1967).

Snow, K., Shipping Hazardous Material, *Chem. Eng.*, **85**(25):102 (1978).

Snow, R. H., and L. T. Work, Size Reduction, *I&EC*, **59**(11):80 (1967).

Solt, P. E., Pneumatic Conveying Materials: Trouble-shooting Pneumatic Conveying Systems, *Chem. Eng. Progr.*, **72**(3):70 (1976).

Strassburger, F., Polymer-Plant Engineering: Materials Handling and Compounding of Plastics, *Chem. Eng.*, **79**(7):81 (1972).

Strop, H. R., and G. O. Briggs, Rubber Handling: Crumb to Bale, *Hydrocarbon Process.*, **53**(3):79 (1974).

Taylor, A. H., Industrial Gases: Industry's Workhorses, *Chem. Eng.*, **80**(23):27 (1973).

Thomson, F. M., Smoothing the Flow of Materials Through the Plant: Feeders, *Chem. Eng.*, **85**(24):77 (1978).

Tiller, F. M., Bench-Scale Design of SLS Systems, *Chem. Eng.*, **81**(9):117 (1974).

Tyson, S. E., Sure Shortcut to Stainless-Steel Specification, *Chem. Eng.*, **76**(21):188 (1969).
Uncles, R. F., Containers and Packaging, *Chem. Eng.*, **76**(22):87 (1969).
VanCleve, W. A., Jr., Proven Catalyst Handling Methods, *Hydrocarbon Process.*, **47**(10):106 (1968).
Vedrilla, S., Typical Phthalic Anhydride Handling, *Hydrocarbon Process.*, **47**(10):102 (1968).
Wahl, E. A., Bin Activators: Key to Practical Storage and Flow of Solids, *Chem. Eng. Progr.*, **69**(1):62 (1973).
Weisselberg, E., Troubleshooting Chemical Solids Handling, *Chem. Eng. Progr.*, **63**(11):72 (1967).
———, Pneumatic Conveying Materials: Rotary Airlocks, *Chem. Eng. Progr.*, **72**(3):76 (1976).
Weston, F. C., Improve Solids Handling Instrumentation Systems, *Hydrocarbon Process.*, **53**(3):87 (1974).
Winters, R. J., Improving Material Flow from Bins, *Chem. Eng. Progr.*, **73**(9):82 (1977).
Witt, W. A., The Super-Flo Conveyor, *Cost Eng.*, **17**(2):12 (1972).
Zenz, F. A., Size Cyclone Diplegs Better, *Hydrocarbon Process.*, **54**(5):125 (1975).
Zimmerman, O. T., and I. Lavine, The Fluidflo Automatic Pump, *Cost Eng.*, **13**(3):4 (1968).
——— and ———, Jones Double-Disk Refiners, *Cost Eng.*, **14**(2):11 (1969).
———, Pipeline Delumper, *Cost Eng.*, **15**(2):18 (1970).
———, Dual-Roll Conveyors, *Cost Eng.*, **15**(3):8 (1970).
———, Wyssmount Multiple Screw Feeder, *Cost Eng.*, **15**(4):6 (1970).
———, The Carman Densifier, *Cost Eng.*, **15**(4):13 (1970).
———, The Tubetransit, *Cost Eng.*, **16**(2):9 (1971).
———, The Tip Trak Bucket Conveyor System, *Cost Eng.*, **17**(3):8 (1972).
———, Miscellaneous Cost Data, *Cost Eng.*, **17**(3):17 (1972).
Processing Agents for In-Process Applications, *Chem. Eng.*, **80**(23):61 (1973).
How to Handle Delayed Coke, *Hydrocarbon Process.*, **53**(3):85 (1974).

Piping design and costs

Abidi, F., Pipeline-System Costs Estimated, *Oil & Gas J.*, **73**(41):99 (1975).
Allen, E. E., Valves Can be Quiet, *Hydrocarbon Process.*, **51**(10):138 (1972).
Aslam, M., Minimum Length of Process Piping, *Hydrocarbon Process.*, **49**(11):203 (1970).
Assini, J., Choosing Welding Fittings and Flanges, *Chem. Eng.*, **81**(18):90 (1974).
Austin, P. P., How to Simplify Fluid Flow Calculations, *Hydrocarbon Process.*, **54**(9):197 (1975).
Aude, T. C., N. T. Cowper, T. L. Thompson, and E. J. Wasp, Slurry Piping Systems: Trends, Design Methods, Guidelines, *Chem. Eng.*, **78**(15):74 (1971).
Badhwar, R. K., Shortcut Design Methods for Piping, Exchangers, Towers, *Chem. Eng.*, **78**(24):113 (1971).
Baird, D., Plastic Piping: What Can it Handle? What Does it Cost? *Plant Eng.*, **23**(18):94 (1969).
Baumann, H. D., How to Estimate Pressure Drop Across Liquid-Control Valves, *Chem. Eng.*, **81**(9):137 (1974).
Bean, D. W., How to Select High-Performance Valves, *Chem. Eng.*, **80**(3):62 (1973).
Benson, R. E., Analyzing Piping Flexibility, *Chem. Eng.*, **80**(25):102 (1973).
Bergomi, J. G., R. J. Carpenter, and J. P. Ford, Sizing of Pumps and Pipes for Coating Colors, *Tappi*, **53**(1):104 (1970).
Bertrem, B. E., Butterfly Valves for Flow of Process Fluids, *Chem. Eng.*, **83**(27):63 (1976).
Birdwell, J. R., and W. W. Shull, Computers Write Our Piping Specs., *Hydrocarbon Process.*, **49**(8):103 (1970).
Bosworth, C. A., Installed Costs of Outside Piping, *Chem. Eng.*, **75**(7):132 (1968).
Brodgesell, A., Valve Selection, *Chem. Eng.*, **78**(23):119 (1971).
Brown, G. W., Valve Problems: Causes and Cures, *Hydrocarbon Process.*, **53**(6):97 (1974).
Burton, B., Butterfly Valves, *Plant Eng.*, **29**(2):133 (1975).
Bush, M. J., and G. L. Wells, Unit Plot Plans for Plant Layout, *Brit. Chem. Eng.*, **16**(4/5):325 (1971).
——— and ———, Unit Plot Plans for Plant Layout, *Brit. Chem. Eng.*, **16**(6):514 (1971).
Canon, J., Guidelines for Selecting Process-Control Valves, *Chem. Eng.*, **76**(10):105 (1969).
Carey, J. A., and D. Hammitt, How to Select Liquid-Flow Control Valves, *Chem. Eng.*, **85**(8):137 (1978).

Carleton, A. J., and D. C.-H. Cheng, Pipeline Design for Industrial Slurries, *Chem. Eng.*, **84**(9):95 (1977).

Chalfin, S., Specifying Control Valves, *Chem. Eng.*, **81**(21):105 (1974).

Chimes, A. R., Confidence Lines for Sizing Plant Piping Systems, *Chem. Eng.*, **81**(16):118 (1974).

Constance, J. D., Sizing Steam Control Valves, *Chem. Eng.*, **81**(18):94 (1974).

Cran, J. E., How to Estimate Piping Costs, *Chem. Eng. (London)*, **218**:110 (May 1968).

Crocker, S., and R. C. King, "Piping Handbook," 5th ed., McGraw-Hill Book Company, New York (1968).

Daniel, P. T., Computer System for Integrated Pipework Design, *Brit. Chem. Eng.*, **14**:7 (1969).

—————— and M. Hall, An Integrated System of Pipework Estimating, Detailing and Control, *Chem. Eng. (London)*, **228**:169 (May 1969).

deLesdernier, D. L., and J. T. Sommerfeld, Computer Program Sizes Pipe, *Hydrocarbon Process.*, **51**(3):112 (1972).

Denton, W. M., Economic Factors in Transporting Petroleum Liquids by Pipe Line, presented at Am. Assoc. of Cost Engrs. 12th National Meeting, Houston, Texas, June 17–19, 1968.

Easton, D. J., How to Specify Valves for Extreme Temperature and Pressure Service, *Petro/Chem. Eng.*, **39**(7):60 (1967).

Edmunds, K. B., Cast Iron Ring Joint Water Pipe, *ACHV*, **65**(1):71 (1968).

——————, Piping Charts, *ACHV*, **65**(2):66 (1968).

Einhauser, L. F., Data Helps Estimate Piping Costs, *Plant Eng.*, **24**(10):102 (1970).

——————, Figuring the Tab for Nonmetallic Piping, *Plant Eng.*, **25**(8):70 (1971).

Evans, F. L., Valves for the HPI, *Hydrocarbon Process.*, **53**(6):87 (1974).

Gallant, R. W., Sizing Pipe for Liquids and Vapors, *Chem. Eng.*, **76**(4):96 (1969).

Gilbu, A., and J. A. Rolston, Large Diameter Reinforced Plastic Pipe, *Chem. Eng. Progr.*, **70**(1):50 (1974).

Glickman, M., and A. H. Hehn, Valves, *Chem. Eng.*, **76**(8):131 (1969).

Gober, H. W., Methods for Sizing Control Valves, *Chem. Eng.*, **74**(23):247 (1967).

Guthrie, K. M., Capital Cost Estimating, *Chem. Eng.*, **76**(6):114 (1969).

——————, Costs, *Chem. Eng.*, **76**(8):201 (1969).

——————, Pump and Valve Costs, *Chem. Eng.*, **78**(23):151 (1971).

Harris, J., and A. K. M.A. Quader, Design Procedures for Pipelines Transporting Non-Newtonian Fluids and Solid-Liquid Systems, *Brit. Chem. Eng.*, **16**(4/5):307 (1971).

Howie, J. A., The Design and Construction of Plastics Pipework Systems, *Brit. Chem. Eng.*, **16**(4/5):317 (1971).

Kannappan, S., and V. H. Helguero, How to Determine Allowable Steam Turbine Piping Loads, *Hydrocarbon Process.*, **53**(8):75 (1974).

Kenney, D. E., Estimating Conceptual Costs for Construction Projects, *Plant Eng.*, **27**(16):72 (1973); **27**(19):140 (1973).

Kent, G. R., Selecting Gaskets for Flanged Joints, *Chem. Eng.*, **85**(7):125 (1978).

——————, Preliminary Pipeline Sizing, *Chem. Eng.*, **85**(21):119 (1978).

Kern, R., How to Size Process Piping for Two-Phase Flow, *Hydrocarbon Process.*, **48**(10):105 (1969).

——————, Pump Piping Design, *Chem. Eng.*, **78**(23):85 (1971).

——————, How Discharge Piping Affects Pump Performance, *Hydrocarbon Process.*, **51**(3):89 (1972).

——————, How to Size Pump Suction Piping, *Hydrocarbon Process.*, **51**(4):119 (1972).

——————, Size Pump Piping and Components, *Hydrocarbon Process.*, **52**(3):81 (1973).

——————, Useful Properties of Fluids for Piping Design, *Chem. Eng.*, **81**(27):58 (1974).

——————, How to Design Piping for Pump-Suction Conditions, *Chem. Eng.*, **82**(9):119 (1975).

——————, How to Size Piping for Pump-Discharge Conditions, *Chem. Eng.*, **82**(11):113 (1975).

——————, Piping Design for Two-Phase Flow, *Chem. Eng.*, **82**(13):145 (1975).

——————, How to Design Piping for Reboiler Systems, *Chem. Eng.*, **82**(16):107 (1975).

——————, How to Design Overhead Condensing Systems, *Chem. Eng.*, **82**(19):129 (1975).

——————, How to Size Piping and Components as Gas Expands at Flow Conditions, *Chem. Eng.*, **82**(22):125 (1975).

————, Pipe Systems for Process Plants, *Chem. Eng.*, **82**(24):209 (1975).

————, How to Compute Pipe Size, *Chem. Eng.*, **82**(1):115 (1975).

————, Measuring Flow in Pipes With Orifices and Nozzles, *Chem. Eng.*, **82**(3):72 (1975).

————, How to Size Flowmeters, *Chem. Eng.*, **82**(5):161 (1975).

————, Control Valves in Process Plants, *Chem. Eng.*, **82**(7):85 (1975).

Kirkpatrick, D. M., Electrical Analogy Simplifies Piping Network Design, *Petro/Chem. Eng.*, **39**(12):40 (1967).

————, Piping Systems for Difficult-to-Pump Fluids, *Petro/Chem. Eng.*, **39**(13):32 (1967).

————, Simpler Sizing of Gas Piping, *Hydrocarbon Process.*, **48**(12):135 (1969).

Knowlton, T. M., and I. Hirsan, L-Valves Characterized for Solids Flow, *Hydrocarbon Process.*, **57**(3):149 (1978).

Krause, A. A., Sizing Valves for Compressible Flow, *Instr. Control Sys.*, **44**(1):71 (1971).

Litz, W. J., Design of Gas Distributors, *Chem. Eng.*, **79**(25):162 (1972).

Loeb, M. B., New Graphs for Solving Compressible Flow Problems, *Chem. Eng.*, **76**(11):179 (1969).

Lovett, O. P., Jr., Control Valves, *Chem. Eng.*, **78**(23):129 (1971).

Mack, W. C., How to Predict Remaining Service Life of Tubing, *Chem. Eng.*, **78**(18):124 (1971).

Marshall, S. P., and K. L. Brandt, Installed Cost of Corrosion-Resistant Piping, *Chem. Eng.*, **78**(19):68 (1971).

———— and ————, 1974 Installed Cost of Corrosion-Resistant Piping, *Chem. Eng.*, **81**(23):94 (1974).

Masek, J. A., Metallic Piping, *Chem. Eng.*, **75**(13):215 (1968).

Mather, L. R., Pipe Network Analysis and Design, *Chem. Eng.* (*London*), **211**:185 (September 1967).

McNulty, F. G., Designing Orifices for Flow Regulation, *Chem. Eng.*, **80**(26):239 (1973).

————, Designing Critical Flow Orifices, *Chem. Eng.*, **81**(16):122 (1974).

Mendel, O., Cost Comparisons for Processing Piping, *Chem. Eng.*, **75**(13):255 (1968).

Moody, G. B., Tubing: An Ideal Pipe Support, *Hydrocarbon Process.*, **49**(4):117 (1970).

Nelson, W. L., What Is Cost of Deeper Pipeline Ditching? *Oil & Gas J.*, **65**(3):97 (1967).

Nicholson, G. D., Why the Swing to Wafer Type Check Valves: *Hydrocarbon Process.*, **53**(6):93 (1974).

Paige, P. M., Shortcuts to Optimum-Size Compressor Piping, *Chem. Eng.*, **74**(6):153 (1967).

Pengelly, D., Select Expansion Joints Properly, *Hydrocarbon Process.*, **57**(3):141 (1978).

Phelps, D. R., Flange Loads In A Pumping System, *Chem. Eng. Progr.*, **66**(5):43 (1970).

Pikulik, A., Selecting and Specifying Valves for New Plants, *Chem. Eng.*, **83**(19):168 (1976).

————, Manually Operated Valves, *Chem. Eng.*, **85**(8):119 (1978).

Pothanikat, J. J., New Approach to Piping System Weight Analysis, *Hydrocarbon Process.*, **55**(4):195 (1976).

Richter, S. H., Control Valve Sizing by Computer, *Petro/Chem. Eng.*, **40**(13):23 (1968).

Roberts, J. A., Computer-Aided Pipe Sketching, *Innovation*, **3**(1):9 (1972).

Rogers, R. L., Piping Stresses and Forces Due to Thermal Expansion, *Petro/Chem. Eng.*, **40**(5):23 (1968).

Ruskin, R. P., Calculating Line Sizes for Flashing Steam-Condensate, *Chem. Eng.*, **82**(17):101 (1975).

Schweitzer, W. J., Cage Trim Valve Promises Savings in Overall Cost, *Plant Eng.*, **80**(11):52 (1976).

Smith, T. J., Reinforced-Plastic Pipe vs. Stainless: Cost Comparison, *Chem. Eng.*, **74**(1):110 (1967).

Simon, H., and S. J. Thomson, Relief System Optimization, *Chem. Eng. Progr.*, **68**(5):52 (1972).

Simpson, L. L., Sizing Piping for Process Plants, *Chem. Eng.*, **75**(13):192 (1968).

————, Process Piping: Functional Design, *Chem. Eng.*, **76**(8):167 (1969).

————, and M. L. Weirick, Designing Plant Piping, *Chem. Eng.*, **85**(8):35 (1978).

Sisson, B., Easy Way to Estimate Steam Line Pipe and Flow, *Plant Eng.*, **21**(5):146 (1967).

Smith, B., Charts Used for Easier Pipe Sizing, *Hydrocarbon Process.*, **48**(5):173 (1969).

Sokullu, E. S., Estimating Piping Costs from Process Flowsheets, *Chem. Eng.*, **76**(3):148 (1969).

Spitzer, H., Pipework Detailing by Computer, *Chem. Proc. Eng.*, **50**(11):86 (1969).

————, Computer Approach to Pipe Detailing, *Chem. Eng.* (*London*), **252**:305 (August 1971).

Styer, R. F., and J. T. Wier, Revised Piping Code Highlights, *Hydrocarbon Process.*, **53**(3):101 (1974).

Surtees, L. S., and P. Rooney, Tips on Specifying FRP Piping, *Chem. Eng.*, **84**(25):215 (1977).

Valkenhoff, B. H., Piping Estimation, *Cost Eng.*, **20**(4):137 (1978).

van Eenennaam, H., and A. C. A. v. Kesteren, Isometric Pipework Drawing by Computer, *Chem. Proc. Eng.*, **52**(9):53 (1971).

Volkin, R. A., Economic Piping of Parallel Equipment, *Chem. Eng.*, **74**(7):148 (1967).

Weaver, R., "Process Piping Design," Vol. 1 and Vol. 2, Gulf Publishing Company, Houston, Texas, 1973.

Wier, J. T., Selecting Valves for the CPI, *Chem. Eng.*, **82**(25):62 (1975).

Wills, J. S., Size Vapor Piping by Computer, *Hydrocarbon Process.*, **49**(5):149 (1970).

Wrasman, T. J., Plastic-Piping Maintenance, *Chem. Eng.*, **78**(3):88 (1971).

Yamartino, J., Installed Cost of Corrosion-Resistant Piping—1978, *Chem. Eng.*, **85**(26):138 (1978).

Yoder, D., Mechanical Aspects of Piping Design, *Chem. Eng.*, **75**(13):251 (1968).

Zimmerman, O. T., Sahara "HP" Deliquescent Air Dryers, *Cost. Eng.*, **17**(3):11 (1972).

Process Piping, *Chem. Eng.*, **76**(8):95 (1969).

Optimum Pipeline Diameters by Nomogram, *Brit. Chem. Eng.*, **16**(4/5):313 (1971).

Completed Pipeline Costs, *Oil & Gas J.*, **74**(34):81 (1976).

Pipeline Equipment Cost Index, *Oil & Gas J.*, **75**(34):74 (1977).

Pump design and costs

Abraham, R. W., Reliability of Rotating Equipment, *Chem. Eng.*, **80**(24):96 (1973).

Aude, T. C., N. T. Cowper, T. L. Thompson, and E. J. Wasp, Slurry Piping Systems: Trends, Design Methods, Guidelines, *Chem. Eng.*, **78**(15):74 (1971).

Benedict, R. M., NPSH and Centrifugal Pumps, *Chem. Eng. Progr.*, **66**(5):58 (1970).

Birk, J. R., and J. H. Peacock, Pump Requirements for the Chemical Process Industries, *Chem. Eng.*, **81**(4):116 (1974).

Booker, J. R., and E. P. Riehl, Designing Large Circulating Water Pumps, *Allis-Chalmers Eng. Rev.*, **39**(3):12 (1974).

Brown, R., and H. Palmour, Lifting Costs Shaved in Sadler West Unit, *Oil & Gas J.*, **66**(42):103 (1968).

Buse, F., The Effects of Dimensional Variations on Centrifugal Pumps, *Chem. Eng.*, **84**(20):93 (1977).

Chambers, A. A., and F. R. Dube, Vacuum Pumps and Systems, *Plant Eng.*, **31**(17):129 (1977).

Condray, K. E., How to Select and Estimate Costs of High-Pressure Vertical-Turbine Pumps, *Oil & Gas J.*, **66**(46):88 (1968).

D'Ambra, F. K., and Z. C. Dobrowolski, Pollution Control for Vacuum Systems, *Chem. Eng.*, **80**(15):95 (1973).

Dalstad, J. I., Slurry Pump Selection and Application, *Chem. Eng.*, **84**(9):101 (1977).

De Santis, G. J., How to Select a Centrifugal Pump, *Chem. Eng.*, **83**(24):163 (1976).

D'Innocenzio, M., Pump Reliability Specifications, *Hydrocarbon Process.*, **51**(6):86 (1972).

Doolin, J. H., Updating Standards for Chemical Pumps, *Chem. Eng.*, **80**(13):117 (1973).

———, Select Pumps to Cut Energy Cost, *Chem. Eng.*, **84**(2):137 (1977).

Duckham, C. B., Design of Centrifugal Pump Installations for Viscous and Non-Newtonian Fluids, *Chem. Proc. Eng.*, **52**(7):66 (1971).

Eng, D., and R. Vach, Computer Program Aids Pump Selection, *Allis-Chalmers Eng. Rev.*, **39**(3):18 (1974).

Glickman, M., Positive Displacement Pumps, *Chem. Eng.*, **78**(23):37 (1971).

Guthrie, K. M., Capital Cost Estimating, *Chem. Eng.*, **76**(6):114 (1969).

———, Costs, *Chem. Eng.*, **76**(8):201 (1969).

———, Estimating the Cost of High-Pressure Equipment, *Chem. Eng.*, **75**(25):144 (1968).

———, Pump and Valve Costs, *Chem. Eng.*, **78**(23):151 (1971).

Hancock, W. P., How to Control Pump Vibration, *Hydrocarbon Process.*, **53**(3):107 (1974).

Hattiangadi, U. S., Specifying Centrifugal and Reciprocating Pumps, *Chem. Eng.*, **77**(4):101 (1970).

Hernandez, L. A., Jr., Controlled-Volume Pumps, *Chem. Eng.*, **75**(22):124 (1968).

Holland, F. A., and F. S. Chapman, "Pumping of Liquids," Reinhold Publishing Corporation, New York, 1967.

Huff, G. A., Jr., Selecting a Vacuum Producer, *Chem. Eng.*, **83**(6):83 (1976).

Jackson, C., Specifying Pump Couplings, Bearings, Seals and Packing, *Hydrocarbon Process.*, **51**(6):90 (1972).

———, How to Prevent Pump Cavitation, *Hydrocarbon Process.*, **52**(5):157 (1973).

James, R., Jr., Maintenance of Rotating Equipment; Pump Maintenance, *Chem. Eng. Progr.*, **72**(2):35 (1976).

Karrasik, I. J., Pump Performance Characteristics, *Hydrocarbon Process.*, **51**(6):101 (1972).

———, Tomorrow's Centrifugal Pump, *Hydrocarbon Process.*, **56**(9):247 (1977).

Kempf, V., Centrifugal Pumps, *Plant Eng.*, **26**(20):77 (1972); **26**(21):82 (1972).

Kern, R., How Discharge Piping Affects Pump Performance, *Hydrocarbon Process.*, **51**(3):89 (1972).

———, Size Pump Piping and Components, *Hydrocarbon Process.*, **52**(3):81 (1973).

Kirkpatrick, D. M., How to Modify Centrifugal Pumps for New Conditions, *Petro/Chem. Eng.*, **49**(1):50 (1968).

Knoll, H., and S. Tinney, Why Use Vertical Inline Pumps? *Hydrocarbon Process.*, **50**(5):131 (1971).

Langitan, F. B., Selecting Submersible Pumps for Specific Production Rates, *Petrol. Eng.*, **47**(7):68 (1975).

Makay, E., How to Avoid Field Problems With . . . Boiler Feed Pumps, *Hydrocarbon Process.*, **55**(12):79 (1976).

Morlok, W. J., Inline or Horizontal . . . Which is Best? *Hydrocarbon Process.*, **57**(6):99 (1978).

Murray, M. G., Jr., How to Specify Better Pump Baseplates, *Hydrocarbon Process.*, **52**(9):157 (1973).

Neerken, R. F., Pump Selection for the Chemical Process Industries, *Chem. Eng.*, **81**(4):104 (1974).

———, Selecting the Right Pump, *Chem. Eng.*, **85**(8):87 (1978).

Panesar, K. S., Consider Suction Specific Speed, *Hydrocarbon Process.*, **57**(6):107 (1978).

———, Select Pumps to Save Energy, *Hydrocarbon Process.*, **57**(10):217 (1978).

Patton, P. W., and C. F. Joyce, How to Find the Lowest-Cost Vacuum System, *Chem. Eng.*, **83**(3):84 (1976).

Penney, W. R., Inert Gas in Liquid Mars Pump Performance, *Chem. Eng.*, **85**(15):63 (1978).

Pietrucha, W. J., Cut Costs With Geared Pumps, *Plant Eng.*, **73**(3):50 (1969).

Pikulik, A., and H. E. Diaz, Cost Estimating for Major Process Equipment, *Chem. Eng.*, **84**(21):107 (1977).

Platt, H., Drivers for Process Pumps, *Hydrocarbon Process.*, **51**(6):95 (1972).

Reynolds, J. A., Saving Energy and Costs in Pumping Systems, *Chem. Eng.*, **83**(1):135 (1976).

Robertson, C. A., Materials Selection for Pumps, *Hydrocarbon Process.*, **51**(6):93 (1972).

Rost, M., and E. T. Visich, Pumps, *Chem. Eng.*, **76**(8):45 (1969).

Rudd, D. F., G. J. Powers, and J. J. Siirola, "Process Synthesis," Prentice-Hall, Englewood Cliffs, New Jersey, 1973.

Scher, J. A., Industrial Wastewater Pumps, *Chem. Eng.*, **82**(21):95 (1975).

Sence, L. H., Reducing Pump Noise, *Chem. Eng. Progr.*, **66**(5):55 (1970).

Simo, F. E., Which Flow Control: Valve or Pump? *Hydrocarbon Process.*, **52**(7):103 (1973).

Stindt, W. H., Pump Selection, *Chem. Eng.*, **78**(23):43 (1971).

Streeter, V. L., Transient Pressures in Centrifugal Pump Systems, *Chem. Eng. Progr.*, **66**(5):60 (1970).

Sturgis, R. P., For Big Savings—Control Costs While Defining Scope, *Chem. Eng.*, **74**(17):188 (1967).

Siirola, J. J., and D. F. Rudd, Computer-Aided Synthesis of Chemical Process Designs, *I&EC Fund.*, **10**:353 (1971).

Sutton, G. P., Save Energy With Pumps, *Hydrocarbon Process.*, **57**(6):103 (1978).

Thurlow, C., III, Centrifugal Pumps, *Chem. Eng.*, **78**(23):29 (1971).

Tinney, W. S., How to Obtain Trouble-Free Performance from Centrifugal Pumps, *Chem. Eng.*, **85**(13):139 (1978).

Van Blarcom, P. P., Bypass Systems for Centrifugal Pumps, *Chem. Eng.*, **81**(3):94 (1974).

Walters, J. K., and A. Wint, Process Design and the Environment, *Chem. Eng. (London)*, **751** (November-December 1976).

Weisman, J., and A. G. Holzman, Optimal Process System Design Under Conditions of Risk, *I&EC Process Des. & Dev.*, **11**(3):386 (1972).

——— H. Pulida, and A. Khanna, Optimal Process System Design in Accordance With Demand and Price Forecasts, *I&EC Process Des. & Dev.*, **14**(1):1 (1975).

Whillier, A., Pump Efficiency Determination in Chemical Plant From Temperature Measurements, *I&EC Process Des. & Dev.*, **7**(2):194 (1968).

Wolfe, G., Fixed Displacement Hydraulic Pumps, *Plant Eng.*, **29**(25):94 (1975).

Yedidiah, S., How to Improve Pump Performance, *Hydrocarbon Process.*, **53**(4):165 (1974).

———, Diagnosing Problems of Centrifugal Pumps, *Chem. Eng.*, **84**(23):125 (1977); **84**(25):193 (1977); **84**(26):141 (1977).

Zimmerman, O. T., and I. Lavine, Tri-Flo Stainless Steel Centrifugal Pumps, *Cost Eng.*, **14**(2):4 (1969).

———, The Grinder Pump Unit, *Cost Eng.*, **15**(3):10 (1970).

Useful Data for Pump Calculations, *Petro/Chem. Eng.*, **40**(5):64 (1968).

Screen design and costs

Fitch, B., Choosing a Separation Technique, *Chem. Eng. Progr.*, **70**(12):33 (1974).

Guthrie, K. M., Capital Cost Estimating, *Chem. Eng.*, **76**(6):114 (1969).

———, Costs, *Chem. Eng.*, **76**(8):201 (1969).

King, E. H., How to Determine Plant Screening Requirements, *Chem. Eng. Progr.*, **73**(5):74 (1977).

Matthews, C. W., Screening, Solid/Solid Separation, *Chem. Eng.*, **78**(4):99 (1971).

———, Trends in Size Reduction of Solids . . . Screening, *Chem. Eng.*, **79**(15):76 (1972).

Tank design and costs

Billet, R., Recent Investigations of Metal Pall Rings, *Chem. Eng. Progr.*, **63**(9):53 (1967).

Boberg, I. E., Choosing Tank Designs, *Chem. Eng.*, **77**(17):134 (1970).

Bolles, W. L., The Solution of a Foam Problem, *Chem. Eng. Progr.*, **63**(9):48 (1967).

Brock, J. E., Optimising the Design of Multishell Cylindrical Pressure Vessels, *Brit. Chem. Eng.*, **16**(6):494 (1971).

Caplan, F., ASME Requirements for Vessels, *Chem. Eng.*, **83**(10):139 (1976).

Clark, F. D., and S. P. Terni, Jr., Thick-Wall Pressure Vessels, *Chem. Eng.*, **79**(7):112 (1972).

Clarke, J. S., New Rules Prevent Tank Failures, *Hydrocarbon Process.*, **50**(5):92 (1971).

Corrigan, T. E., W. E. Lewis, and K. N. McKelvey, What Do Chemical Reactors Cost in Terms of Volume? *Chem. Eng.*, **74**(11):214 (1967).

Derrick, G. C., Estimating The Cost of Jacketed, Agitated and Baffled Reactors, *Chem. Eng.*, **74**(21):272 (1967).

Dimoplon, W., Jr., How to Determine the Geometry of Pressure Vessel Heads, *Hydrocarbon Process.*, **53**(8):71 (1974).

Dorsey, J. S., Using Reinforced Plastics for Process Equipment, *Chem. Eng.*, **82**(19):104 (1975).

Epstein, L. D., Cost of Standard-Sized Reactors and Storage Tanks, *Chem. Eng.*, **78**(24):160 (1971).

Furman, T. T., and R. F. Cheers, *Brit. Chem. Eng.*, **16**(6):478 (1971).

Guthrie, K. M., Estimating the Cost of High-Pressure Equipment, *Chem. Eng.*, **75**(25):144 (1968).

———, Capital Cost Estimating, *Chem. Eng.*, **76**(6):114 (1969).

Karcher, G. G., New Design Calculations for High Temperature Storage Tanks, *Hydrocarbon Process.*, **57**(10):137 (1978).

Kemp, J. B., The Thermal Stress Problem in Pressure Vessels, *Hydrocarbon Process.*, **49**(11):195 (1970).

Kern, R., How to Design Overhead Condensing Systems, *Chem. Eng.*, **82**(19):129 (1975).

Kirby, G. N., How to Specify Fiberglass-Reinforced-Plastics Equipment, *Chem. Eng.*, **85**(13):133 (1978).

Lessi, A., How Weeping Affects Distillation, *Hydrocarbon Process.*, **51**(3):109 (1972).

Loeb, M. B., Optimum Vessel Design for Gas Storage, *Chem. Eng.*, **81**(13):170 (1974).

Ludwig, T. B., Patterson Dimple Jacketed Vessels, *Cost Eng.*, **12**(2):8 (1967).

Mahajan, K. K., Size Vessel Stiffeners Quickly, *Hydrocarbon Process.*, **56**(4):207 (1977).

Markovitz, R. E., Choosing the Most Economical Vessel Head, *Chem. Eng.*, **78**(16):102 (1971).

———, Picking the Best Vessel Jacket, *Chem. Eng.*, **78**(26):156 (1971).

———, Process and Project Data Pertinent to Vessel Design, *Chem. Eng.*, **84**(21):123 (1977).

Martin, J. P., A Proposed Approach to Estimating Vessel Costs by Computer, presented at Am. Assoc. of Cost Engrs. 12th National Meeting, Houston, Texas, June 17–19, 1968.

Nelson, W. L., Cost of Refineries: Storage, Environment, Land, *Oil & Gas J.*, **72**(30):161 (1974).

Newton, P., W. R. Von Tress, and J. S. Bridges, Liquid Storage in the CPI, *Chem. Eng.*, **85**(8):9 (1978).

Manning, W. R. D., Ultra-High Pressure Vessel Design, *Chem. Proc. Eng.*, **48**(3):51 (1967); **48**(4):74 (1967).

McFarland, I., Safety in Pressure Vessel Design, *Chem. Eng. Progr.*, **66**(6):56 (1970).

Phadke, P. S., and P. D. Kilkarni, Estimating the Costs and Weights of Process Vessels, *Chem. Eng.*, **84**(8):157 (1977).

Pikulik, A., and H. E. Diaz, Cost Estimating for Major Process Equipment, *Chem. Eng.*, **84**(21):107 (1977).

Raghavan, K. V., Pressure Vessel Design, *Chem. Proc. Eng.*, **51**(10):81 (1970).

Ross, S., Mechanisms of Foam Stabilization and Antifoaming Action, *Chem. Eng. Progr.*, **63**(9):41 (1967).

Sigalés, B., How to Design Reflux Drums, *Chem. Eng.*, **82**(5):157 (1975).

———, How to Design Settling Drums, *Chem. Eng.*, **82**(13):141 (1975).

———, More on Design of Reflux Drums, *Chem. Eng.*, **82**(20):87 (1975).

Snyder, A. I., Safe Design, Construction and Installation of Pressure Vessels, *Brit. Chem. Eng.*, **16**(6):499 (1971).

Strelzoff, S., and L. C. Pan, Designing Pressure Vessels, *Chem. Eng.*, **75**(23):191 (1968).

——— and E. J. Miller, Trends in Pressure Vessels and Closures, *Chem. Eng.*, **75**(22):143 (1968).

Watkins, R. N., Sizing Separators and Accumulators, *Hydrocarbon Process.*, **46**(11):253 (1967).

Witkin, D. E., A New Code Worth Its Weight in Metal, *Chem. Eng.*, **75**(18):124 (1968).

Zimmerman, O. T., Devine Autoclave Systems, *Cost Eng.*, **12**(3):11 (1967).

——— and I. Lavine, Kabe-O-Rap Tanks, *Cost Eng.*, **13**(1):4 (1968).

Tanks-Storage, Carbon Steel, *AACE Bull.*, **11**(3):108 (1969).

Thickener and clarifier design and costs

Abernathy, M. W., Design Horizontal Gravity Settlers, *Hydrocarbon Process.*, **56**(9):199 (1977).

Barnea, E., New Plot Enhances Value of Batch-Thickening Tests, *Chem. Eng.*, **84**(18):75 (1977).

Dahlstrom, D. A., and C. F. Cornell, Thickening and Clarification, *Chem. Eng.*, **78**(4):63 (1971).

Davies, G. A., G. V. Jeffreys, and F. Ali, Design and Scale-up Gravity Settlers, *Chem. Eng. (London)*, 378 (November 1970).

Fitch, B., Batch Tests Predict Thickener Performance, *Chem. Eng.*, **78**(19):83 (1971).

———, Choosing a Separation Technique, *Chem. Eng. Progr.*, **70**(12):33 (1974).

Guthrie, K., Costs, *Chem. Eng.*, **76**(8):201 (1969).

Mace, G. R., and R. Laks, Heat Transfer—Developments in Gravity Sedimentation, *Chem. Eng. Progr.*, **74**(7):77 (1978).

Manchanda, K. D., and D. R. Woods, Significant Design Variables in Continuous Gravity Decantation, *I&EC Process Des. & Dev.*, **7**(2):183 (1968).

Oldshue, J. Y., and O. B. Mady, Flocculation Performance of Mixing Impellers, *Chem. Eng. Progr.*, **74**(8):103 (1978).

Tarrer, A. R., H. C. Lim, L. B. Koppel, and C. P. L. Grady, Jr., A Model for Continuous Thickening, *I&EC Process Des. & Dev.*, **13**(14):341 (1974).

Waste disposal equipment design and costs

Antoine, R. L., A Low-Energy Approach to Industrial Wastewater Treatment, *Plant Eng.*, **31**(17):99 (1977).

Becker, K. P., and C. J. Wall, Incinerate Refinery Waste in a Fluid Bed, *Hydrocarbon Process.*, **54**(10):88 (1975).

Blecker, H. G., H. S. Epstein, and T. M. Nichols, How to Estimate and Escalate Costs of Wastewater Equipment, *Chem. Eng.*, **81**(22):115 (1974).

Brewer, G. L., Fume Incineration, *Chem. Eng.*, **75**(22):160 (1968).

Bush, K. E., Refinery Wastewater Treatment and Reuse, *Chem. Eng.*, **83**(8):113 (1976).

Ciaccio, L. L., Instrumentation in Water Reuse and Pollution Control, *Am. Lab.*, **3**(12):21 (1971).

Connelly, E. J., Cleaning Water by Ultrafiltration, *Plant Eng.*, **31**(23):145 (1977).

Cummings-Saxton, J., Chemical-industry Costs of Water Pollution Abatement, *Chem. Eng.*, **83**(24):106 (1976).

Dunn, K. S., Incineration's Role in Ultimate Disposal of Process Wastes, *Chem. Eng.*, **82**(21):141 (1975).

Dupre, E. E., Jr., Survey of Wastewater Rates and Charges, *Water Poll. Cont. Fed. J.*, **41**(1):33 (1970).

Eckenfelder, W. W., Jr., Economics of Wastewater Treatment, *Chem. Eng.*, **76**(18):109 (1969).

———— and J. L. Barnard, Treatment-Cost Relationship for Industrial Wastes, *Chem. Eng. Progr.*, **67**(9):76 (1971).

Evans, R. R., and R. S. Millward, Equipment for Dewatering Waste Streams, *Chem. Eng.*, **82**(21):83 (1975).

Ford, D. L., and R. L. Elton, Removal of Oil and Grease from Industrial Wastewaters, *Chem. Eng.*, **84**(22):49 (1977).

Gardiner, W. C., and F. Munoz, Mercury Removed From Waste Effluent via Ion Exchange, *Chem. Eng.*, **78**(19):57 (1971).

Himmelstein, K. J., R. D. Fox, and T. H. Winter, In-Place Regeneration of Activated Carbon, *Chem. Eng. Progr.*, **69**(11):65 (1973).

Jones, H. R., "Wastewater Cleanup Equipment 1973," 2d ed., Noyes Data Corporation, Park Ridge, New Jersey, 1973.

Klett, R. J., Treat Sour Water for Profit, *Hydrocarbon Process.*, **51**(10):97 (1972).

Knowlton, H. E., Why Not Use a Rotating Disk? *Hydrocarbon Process.*, **56**(9):227 (1977).

Koches, C. F., and S. B. Smith, Reactivate Powdered Carbon, *Chem. Eng.*, **79**(9):46 (1972).

Krisher, A. S., Raw Water Treatment in the CPI, *Chem. Eng.*, **85**(19):79 (1978).

Kunin, R., and D. G. Downing, Ion-Exchange System Boasts More Pulling Power, *Chem. Eng.*, **78**(15):67 (1971).

Lamp, G. E., Jr., Package Treatment Plant Prices, *Water Poll. Cont. Fed. J.*, **46**(11):2604 (1974).

Lanouette, K. H., Heavy Metals Removal, *Chem. Eng.*, **84**(22):73 (1977).

————, Treatment of Phenolic Wastes, *Chem. Eng.*, **84**(22):99 (1977).

Larkman, D., Physical/Chemical Treatment, *Chem. Eng.*, **80**(14):87 (1973).

McGovern, J. G., Inplant Wastewater Control, *Chem. Eng.*, **80**(11):137 (1973).

McHarg, W. H., Designing the Optimum System for Biological-Waste-Treatment, *Chem. Eng.*, **80**(29):46 (1973).

Monti, R. P., and P. T. Silverman, Wastewater System Alternates: What Are They . . . and What Cost? *Water and Wastes Eng.*, **11**(5):40 (1974).

Nathan, M. F., Choosing a Process for Chloride Removal, *Chem. Eng.*, **85**(3):93 (1978).

Novak, R. G., J. J. Cudahy, M. B. Denove, R. L. Standifer, and W. E. Wass, How Sludge Characteristics Affect Incinerator Design, *Chem. Eng.*, **84**(10):131 (1977).

Oldshue, J. Y., The Case for Deep Aeration Tanks, *Chem. Eng. Progr.*, **66**(11):73 (1970).

Parlante, R., Comparing Water Treatment Costs, *Plant Eng.*, **23**(10):108 (1969).

Paulson, E. G., Reducing Fluoride in Industrial Wastewater, *Chem. Eng.*, **84**(22):89 (1977).

Prengle, H. W., Jr., J. R. Crump, S. Curtice, T. P. John, A. P. Gutierrez, and T. T. Tseng, Recycle Waste Water by Ion Exchange, *Hydrocarbon Process.*, **54**(4):173 (1975).

———— C. E. Mauk, R. W. Legan, and C. G. Hewes, III, Ozone/UV Process Effective Wastewater Treatment, *Hydrocarbon Process.*, **54**(10):82 (1975).

Quasim, S. R., and A. K. Shah, Cost Analysis of Package Wastewater Treatment Plants, *Water & Sewage Works*, **122**(2):67 (1975).

Rizzo, J. L., and A. R. Shepherd, Treating Industrial Wastewater with Activated Carbon, *Chem. Eng.*, **84**(1):95 (1977).

Robinson, P. E., and F. P. Coughlan, Jr., Municipal-Industrial Waste Treatment Costs, *Tappi*, **54**(12):2005 (1971).

Ross, R. D., "Industrial Waste Disposal," Van Nostrand Reinhold, Cincinnati, Ohio, 1968.

Sawyer, G. A., New Trends in Wastewater Treatment and Recycle, *Chem. Eng.*, **79**(16):120 (1972).

Shell, G. L., J. L. Boyd, and D. A. Dahlstrom, Upgrading Waste Treatments Plants, *Chem. Eng.*, **78**(14):97 (1971).

Snyder, N. W., Pyrolysis of Municipal Wastes to Fuels and Chemicals, *Chem. Eng. Progr. Symp. Ser.*, **73**(162):150 (1977).

Strohl, K. P., Comparing the Costs for Thermal and Catalytic Incinerators, *Chem. Eng.*, **83**(12):153 (1976).

Sweeny, F. J., On-Site Disposal of Sludges from Effluent Treatment Plant Cuts Costs, *Oil & Gas J.*, **73**(27):52 (1975).

Thompson, R. G., Water-Pollution Instrumentation, *Chem. Eng.*, **83**(13):151 (1976).

Truby, R. L., and J. H. Sleigh, Jr., Cleaning Water by Reverse Osmosis, *Plant Eng.*, **28**(25):83 (1974).

Tyer, B. R., and C. H. Hayes, Oxygenation Equipment—Aerators and Pure Oxygen Systems, *Chem. Eng.*, **82**(21):75 (1975).

Vadovic, J. P., Industrial Water Treatment Systems, *Plant Eng.*, **31**(26):77 (1977).

Watkins, J. P., Controlling Sulfur Compounds in Wastewaters, *Chem. Eng.*, **84**(22):61 (1977).

Weinstein, N. J., and R. F. Toro, Costs for Thermal Processing of Solid Wastes, *Public Works*, **107**(5):61 (1976).

Witt, P. A., Jr., Disposal of Solid Wastes, *Chem. Eng.*, **78**(22):62 (1971).

————, Solid Waste Disposal, *Chem. Eng.*, **79**(10):109 (1972).

Zimmerman, O. T., Wastewater Treatment, *Cost Eng.*, **16**(4):11 (1971).

————, Waste Compactors and Shredders, *Cost Eng.*, **18**(3):9 (1973).

————, Aerodyne Oil Skimmer, *Cost Eng.*, **18**(3):17 (1973).

————, Waste Incinerators, *Cost Eng.*, **18**(4):4 (1973).

————, C. L. Parker, and C. V. Fong, Estimation of Operating Costs for Industrial Wastewater Treatment Facilities, *AACE Bull.*, **18**(6):207 (1976).

Miscellaneous Cost Data, *Cost Eng.*, **14**(3):16 (1969).

PROBLEMS

1 A lean oil is to be used as the absorbing medium for removing a component of a gas. As part of the design for the absorption unit, it is necessary to estimate the size of motor necessary to pump the oil to the top of the absorption tower. The oil must be pumped from an open tank with a liquid level 10 ft above the floor and forced through 150 ft of schedule number 40 pipe of 3-in. nominal diameter. There are five 90° elbows in the line, and the top of the tower is 30 ft above the floor level. The operating pressure in the tower is to be 50 psig, and the oil requirement is estimated to be 50 gpm. The viscosity of the oil is 15 centipoises, and its density is 53.5 lb/ft³. If the efficiency of the pumping assembly including the drive is 40 percent, what horsepower motor will be required?

2 Hydrogen at a temperature of 20°C and an absolute pressure of 1380 kPa enters a compressor where the absolute pressure is increased to 4140 kPa. If the mechanical efficiency of the compressor is 55 percent on the basis of an isothermal and reversible operation, calculate the pounds of hydrogen that can be handled per minute when the power supplied to the pump is 224 kW. Kinetic-energy effects can be neglected.

3 For the conditions indicated in Prob. 2, determine the mechical efficiency of the pump on the basis of adiabatic and reversible operation. A single-stage compressor is used, and the ratio of heat capacity at constant pressure to the heat capacity at constant volume for the hydrogen may be assumed to be 1.4.

4 A steel pipe of 4-in. nominal diameter is to be used as a high-pressure steam line. The pipe is butt-welded, and its schedule number is 40. Estimate the maximum steam pressure that can be used safely in this pipe.

5 A preliminary estimate is to be made of the total cost for a completely installed pumping system. A pipeline is to be used for a steady delivery of 250 gpm of water at 60°F. The total length of the line will be 1000 ft, and it is estimated that the theoretical horsepower requirement (100 percent efficiency) of the pump is 10.0 hp. Using the following additional data, estimate the total installed cost for the pumping system.

> Materials of construction—black steel (standard weights are satisfactory)
> Number of fittings (equivalent to tees)—40
> Number of valves (gate)—4
> Insulation—85 percent magnesia, $1\frac{1}{2}$ in. thick
> Pump—centrifugal (no standby pump is needed)
> Motor—ac, enclosed, 3-phase, 1800 r/min

6 A centrifugal pump delivers 100 gpm of water at 60°F when the impeller speed is 1750 r/min and the pressure drop across the pump is 20 psi. If the speed is reduced to 1150 r/min, estimate the rate of water delivery and the developed head in feet if the pump operation is ideal.

7 A two-stage steam jet is to be used on a large vacuum system. It is estimated that 9 kg of air must be removed from the system per hour. The leaving vapors will contain water vapor at a pressure equivalent to the equilibrium vapor pressure of water at 15°C. If a suction pressure of 2.0 in. Hg absolute is to be maintained by the jet, estimate the pounds of steam required per hour to operate the jet.

8 The rate of flow of a liquid mixture is to be measured continuously. The flow rate will be approximately 40 gpm, and rates as low as 30 gpm or as high as 50 gpm can be expected. An orifice meter, a rotameter, and a venturi meter are available. On the basis of the following additional information, would you recommend installation of the orifice meter, the venturi meter, or the rotameter? Give reasons for your choice.

> Density of liquid = 58 lb/ft³
> Viscosity of liquid = 1.2 centipoises
> Diameter of venturi throat = 1 in.
> Upstream diameter of venturi opening = 2 in.
> Manometers connected across the venturi and the orifice contain a nonmiscible liquid (sp gr = 1.56) in contact with the liquid mixture
> The maximum possible reading on these manometers is 15 in.
> Orifice is square-edged with throat taps
> Diameter of orifice opening = 1 in.
> Diameter of upstream chamber for orifice meter = 3 in.

Calibration curve for rotameter is for water at 60°F and gives the following values:

Rotameter reading	Flow rate, ft³/min
2.0	2.0
4.0	4.0
6.0	6.0
8.0	8.0

The density of the plummet in the rotameter is 497 lb/ft³.

9 A spherical carbon-steel storage tank has an inside diameter of 30 ft. All joints are butt-welded with backing strip. If the tank is to be used at a working pressure of 30 psig and a temperature of 80°F, estimate the necessary wall thickness. No corrosion allowance is necessary.

10 Estimate the cost of the steel for the spherical storage tank in the preceding problem if the steel sheet costs $0.15 per pound. On the basis of the data presented in Fig. 13-58, estimate the fraction of the purchased cost of the tank that is due to the cost for the steel.

11 Filtration tests with a given slurry have indicated that the specific cake resistance α is 157 h²/lb. The fluid viscosity is 2.5 lb/(h)(ft), and 3 lb of dry cake is formed per cubic foot of filtrate. The cake may be assumed to be noncompressible, and the resistance of the filter medium may be neglected. If the unit is operated at a constant pressure drop of 5 psi, what is the total filtering area required to deliver 30 ft^3 of filtrate in $\frac{1}{2}$ h?

12 A rotary filter with a total filtering area of 8 ft^2 has been found to deliver 10 ft^3 of filtrate per minute when operating at the following conditions:

Fraction of filtering area submerged in the slurry = 0.2
r/min = 2
Pressure drop = 20 psi

Another rotary filter is to be designed to handle the same slurry mixture. This unit will deliver 100 ft^3 of filtrate per minute and will operate at a pressure drop of 15 psi and a revolving speed of 1.5 r/min. If the fraction of filtering area submerged in the slurry is 0.2, estimate the total filtering area required for the new unit. In both cases, it may be assumed that no solids pass through the filter cloth, the cake is noncompressible, and the resistance of the filtering medium is negligible.

13 A plate-and-frame filter press is used to filter a known slurry mixture. At a constant pressure drop of 10 psi, 50 ft^3 of filtrate is delivered in 10 min, starting with a clean filter. In a second run with the same slurry and filter press, 40 ft^3 of filtrate is obtained in 9 min when the pressure drop is 6 psi, starting with a clean filter. What is the compressibility exponent for the cake if the resistance of the filter medium is negligible?

14 A slurry containing 1 lb of filterable solids per 10 lb of liquid is being filtered with a plate-and-frame filter press having a total filtering area of 250 ft^2. This unit delivers 10,000 lb of filtrate during the first 2 h of filtration, starting with a clean unit and maintaining a constant pressure drop of 10 psi. The resistance of the filter medium is negligible. The time required for washing and dumping is 3 h/cycle. The pressure drop cannot exceed 10 psi, and the unit is always operated at constant pressure drop.

The filter press is to be replaced by a rotary vacuum-drum filter with negligible filter-medium resistance. This rotary filter can deliver the filtrate at a rate of 1000 lb/h when the drum speed is 0.3 r/min. Assuming the fraction submerged and the pressure drop are unchanged, what drum speed in r/min is necessary to make the amount of filtrate delivered in 24 h from the rotary filter exactly equal to the maximum amount of filtrate obtainable per 24 h from the plate-and-frame filter?

15 A plate-and-frame filter press with negligible filter-medium resistance is being used to filter a water slurry of constant composition. Experimental tests show that, during 3 h of continuous operation, 300 ft^3 of filtrate is delivered when the pressure drop is 20 psi and 150 ft^3 of filtrate is delivered when the pressure drop is 5 psi. The unit is to be operated at a constant pressure drop of 15 psi during filtration and washing. The cake will be washed with 10 ft^3 of water at the end of 2 h of continuous filtration. If reverse thorough washing (i.e., wash rate = $\frac{1}{4}$ of final filtrate delivery rate) is used, estimate the time required for washing.

16 A slurry is filtered, and the filter cake is washed by use of a plate-and-frame filter press operated at a constant pressure drop of 40 psi throughout the entire run. Experimental tests have been carried out on this equipment, and the results for the slurry mixture used can be expressed as follows for any one pressure drop:

$$\theta \, \Delta P = k'\left(\frac{V}{A}\right)^2 + k''\frac{V}{A}$$

where k' and k'' are constants. At a pressure drop of 40 psi, 0.02 lb of filtrate is collected in 1.8 min for each square inch of cloth area, and 0.08 lb of filtrate per square inch of cloth area is collected in 22.2 min. Calculate the time required to filter and wash the cake formed when 0.11 lb of filtrate has been collected per square inch of cloth area if an amount of wash water equal to half the filtrate is used. The specific gravities of the filtrate and wash water are 1.0, and both are at the same temperature. Simple forward washing is used so that the washing rate is equal to the filtrate delivery rate at the end of the filtration.

FOURTEEN

HEAT-TRANSFER EQUIPMENT— DESIGN AND COSTS

Equipment for transferring heat is used in essentially all the process industries, and the design engineer should be acquainted with the many different types of equipment employed for this operation. Although relatively few engineers are involved in the manufacture of heat exchangers, many engineers are directly concerned with specifying and purchasing heat-transfer equipment. Process-design considerations, therefore, are of particular importance to those persons who must decide which piece of equipment is suitable for a given process.

Modern heat exchangers range from simple concentric-pipe exchangers to complex surface condensers with thousands of square feet of heating area. Between these two extremes are found the conventional shell-and-tube exchangers, coil heaters, bayonet heaters, extended-surface finned exchangers, plate exchangers, furnaces, and many varieties of other equipment. Exchangers of the shell-and-tube type are used extensively in industry, and they are often named specifically for distinguishing design features. For example, U-tube, fin-tube, fixed-tubesheet, and floating-head exchangers are common types of shell-and-tube exchangers. Figure 14-1 shows design details of a conventional two-pass exchanger of the shell-and-tube type.

Intelligent selection of heat-transfer equipment requires an understanding of the basic theories of heat transfer and the methods for design calculation. In addition, the problems connected with mechanical design, fabrication, and operation must not be overlooked. An outline of heat-transfer theory and design-calculation methods is presented in this chapter, together with an analysis of the general factors that must be considered in the selection of heat-transfer equipment.

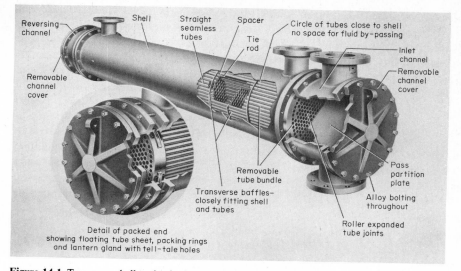

Figure 14-1 Two-pass shell-and-tube heat exchanger showing construction details. (*Ross Heat Exchanger Division of American-Standard.*)

Determination of appropriate coefficients of heat transfer is required for design calculations on heat-transfer operations. These coefficients can sometimes be estimated on the basis of past experience, or they can be calculated from empirical or theoretical equations developed by other workers in the field. Many semiempirical equations for the evaluation of heat-transfer coefficients have been published. Each of these equations has its limitations, and the engineer must recognize the fact that these limitations exist. A summary of useful and reliable design equations for estimating heat-transfer coefficients under various conditions is presented in this chapter. Additional relations and discussion of special types of heat-transfer equipment and calculation methods are presented in the numerous books and articles that have been published on the general subject of heat transfer.†

BASIC THEORY OF HEAT TRANSFER

Heat can be transferred from a source to a receiver by *conduction, convection,* or *radiation.* In many cases, the exchange occurs by a combination of two or three of these mechanisms. When the rate of heat transfer remains constant and is

† For example, J. P. Holman, "Heat Transfer," 4th ed., McGraw-Hill Book Company, New York, 1976; J. R. Weltz, "Engineering Heat Transfer," John Wiley and Sons, Inc., New York, 1974; D. A. Kaminski, "Heat Transfer Data Book," General Electric Company, Schenectady, New York, 1970–1977; W. M. Rohsenow and J. P. Hartnett, "Handbook of Heat Transfer," McGraw-Hill Book Company, New York, 1973; and D. Q. Kern and A. D. Krause, "Extended Surface Heat Exchangers," McGraw-Hill Book Company, New York, 1972.

unaffected by time, the flow of heat is designated as being in a *steady state;* an *unsteady state* exists when the rate of heat transfer at any point varies with time. Most industrial operations in which heat transfer is involved are carried out under steady-state conditions. However, unsteady-state conditions are encountered in batch processes, cooling and heating of materials such as metals or glass, and certain types of regeneration, curing, or activation processes.

Conduction

The transfer of heat through a fixed material is accomplished by the mechanism known as *conduction.* The rate of heat flow by conduction is proportional to the area available for the heat transfer and the temperature gradient in the direction of the heat-flow path. The rate of heat flow in a given direction, therefore, can be expressed as†

$$\frac{dQ}{d\theta} = -kA\frac{dt}{dx} \tag{1}$$

where Q = amount of heat transferred in time θ h, Btu

k = the proportionality constant, designated as thermal conductivity and defined by Eq. (1), Btu/(h)(ft^2)(°F/ft)

A = area of heat transfer perpendicular to direction of heat flow, ft^2

t = temperature, °F

x = length of conduction path in direction of heat flow, ft

The thermal conductivity is a property of any given material, and its value must be determined experimentally. For solids, the effect of temperature on thermal conductivity is relatively small at normal temperatures. Because the conductivity varies approximately linearly with temperature, adequate design accuracy can be obtained by employing an average value of thermal conductivity based on the arithmetic-average temperature of the given material. Values of thermal conductivities for common materials at various temperatures are listed in the Appendix.

For the common case of steady-state flow of heat, Eq. (1) can be expressed as

$$\frac{Q}{\theta} = q = -kA_m\frac{\Delta t}{x} \tag{2}$$

where A_m = mean area of heat transfer perpendicular to direction of heat flow, ft^2

q = rate of heat transfer, Btu/h

Δt = temperature-difference driving force, °F

---

† In accord with common design practice in the United States, the U.S. customary system of units is used in this chapter. See Appendix A for conversion to SI units.

Convection

Transfer of heat by physical mixing of the hot and cold portions of a fluid is known as *heat transfer by convection*. The mixing can occur as a result of density differences alone, as in *natural convection*, or as a result of mechanically induced agitation, as in *forced convection*.

The following equation is used as a basis for evaluating rates of heat transfer by convection:

$$\frac{dQ}{d\theta} = hA\,\Delta t \tag{3}$$

The proportionality constant h is designated as the heat-transfer coefficient, and it is a function of the type of agitation and the nature of the fluid. The heat-transfer coefficient, like the thermal conductivity k, is often determined on the basis of experimental data. For steady-state conditions, Eq. (3) becomes

$$q = hA\,\Delta t \tag{4}$$

Radiation

When radiant heat energy is transferred from a source to a receiver, the method of heat transfer is designated as *radiation*. The rate at which radiant heat energy is emitted from a source is

$$\frac{dQ}{d\theta} = \sigma \varepsilon A T^4 \tag{5}$$

where σ = Stefan-Boltzmann constant
$\quad = 0.1713 \times 10^{-8}$ Btu/(h)(ft^2)($°$R)^4
$\quad \varepsilon$ = emissivity of surface
$\quad A$ = exposed-surface area of heat transfer, ft^2
$\quad T$ = absolute temperature, $°$R

The emissivity depends on the characteristics of the emitting surface and, like the thermal conductivity and the heat-transfer coefficient, must be determined experimentally. Part of the radiant energy intercepted by a receiver is absorbed, and part may be reflected. In addition, the receiver, as well as the source, can emit radiant energy.

The engineer is usually interested in the net rate of heat interchange between two bodies. Some of the radiated energy indicated by Eq. (5) may be returned to the source by reflection from the receiver, and the receiver, of course, emits radiant energy which can be partly or completely absorbed by the source. Equation (5), therefore, must be modified to obtain the net rate of radiant heat exchange between two bodies. The general steady-state equation is

$$q_{\text{from body 1 to body 2}} = 0.171A\left[\left(\frac{T_1}{100}\right)^4 F_{A_1}F_{E_1} - \left(\frac{T_2}{100}\right)^4 F_{A_2}F_{E_2}\right] \tag{6}$$

Table 1 Values of F_A and F_E for use in Eq. (7)

$$\epsilon_1 = \text{emissivity of surface 1}$$
$$\epsilon_2 = \text{emissivity of surface 2}$$

Assumptions: Base area is smooth and no direct self-radiation is intercepted by the base surface; separating medium between surfaces is nonabsorbing; emissivity is equal to absorptivity.

Orientation of surfaces	Area, A	F_A	F_E
Surface A_1 small compared with the totally enclosing surface A_2 (for example, heat loss from equipment to surroundings)	A_1	1	ϵ_1
Two parallel planes of equal area when length and width are large compared with distance between planes	A_1 or A_2	1	$\dfrac{1}{\dfrac{1}{\epsilon_1} + \dfrac{1}{\epsilon_2} - 1}$
Surface A_1 is a sphere with radius r_1 inside a concentric sphere with radius r_2	A_1	1	$\dfrac{1}{\dfrac{1}{\epsilon_1} + \left(\dfrac{r_1}{r_2}\right)^2 \left(\dfrac{1}{\epsilon_2} - 1\right)}$
Surface A_1 is a cylinder with radius r_1 inside a concentric cylinder with radius r_2 when length is large compared with diameter	A_1	1	$\dfrac{1}{\dfrac{1}{\epsilon_1} + \dfrac{r_1}{r_2} \left(\dfrac{1}{\epsilon_2} - 1\right)}$

or, in an alternative form,

$$q_{\text{from body 1 to body 2}} = 0.171 A \left[\left(\frac{T_1}{100}\right)^4 - \left(\frac{T_2}{100}\right)^4\right] F_A F_E \qquad (7)$$

where A represents the area of one of the surfaces, F_A is a correction factor based on the relative orientation of the two surfaces and the surface chosen for the evaluation of A, and F_E is a correction factor based on the emissivities and absorptivities of the surfaces.

Table 1 indicates methods for evaluating F_A and F_E for several simple cases. Table 2 lists the emissivities of surfaces commonly encountered in industrial operations. Additional tables, plots, and methods for evaluating the correction factors are available in the various texts and handbooks on heat transfer.

The design engineer often encounters the situation in which a nonblack body is surrounded completely by a large enclosure containing a nonabsorbing gas. An example would be a steam line exposed to the atmosphere. Under these conditions, little error is introduced by assuming that none of the heat radiated from the source is reflected to it, and Eq. (6) or (7) can be simplified to

$$q_{\text{radiated from pipe}} = 0.171 A_1 \varepsilon_1 \left[\left(\frac{T_1}{100}\right)^4 - \left(\frac{T_2}{100}\right)^4\right] \qquad (8)$$

where subscript 1 refers to the pipe and subscript 2 refers to the surroundings. It

Table 2 Normal emissivities of various surfaces

Surface	Temperature, °F	Emissivity
Aluminum, polished plate	73	0.040
Aluminum, rough plate	78	0.055
Aluminum paint	. . . .	0.3–0.6
Aluminum-surfaced roofing	100	0.22
Asbestos paper	100	0.93
Brass, dull	120–660	0.22
Brass, polished	100–600	0.096
Brick, building	. . . .	0.80–0.95
Brick, refractory	. . . .	0.75–0.90
Copper, oxidized	77	0.78
Iron, cast plate, smooth	73	0.80
Iron, oxidized	212	0.74
Iron, polished	800	0.14
Lead, gray, oxidized	75	0.28
Nickel, pure, polished	440	0.07
	710	0.087
Oil paint	212	0.92–0.96
Roofing paper	69	0.91
Silver, polished	100	0.022
	700	0.031
Steel, polished	212	0.066
Steel sheet, oxidized	75	0.80
Water	32	0.95
	212	0.963
Zinc, galvanized sheet iron, gray oxidized	75	0.276

is often convenient to use Eq. (8) in the following form, analogous to the equation for heat transfer by convection:

$$q_{\text{radiated from pipe}} = h_r A_1 \, \Delta t = h_r A_1 (T_1 - T_2) \tag{9}$$

where h_r is a fictitious heat-transfer coefficient based on the rate at which radiant heat leaves the surface of the pipe. Combination of Eqs. (8) and (9) gives

$$h_r = \frac{0.171\varepsilon_1[(T_1/100)^4 - (T_2/100)^4]}{T_1 - T_2} \tag{10}$$

Equations (8), (9), and (10) are based on the assumption that none of the heat radiated from the source is reflected to it. This, of course, implies that the air surrounding the pipe has no effect. In other words, air is assumed to be nonabsorbing and nonreflecting. This assumption is essentially correct for gases such as oxygen, nitrogen, hydrogen, and chlorine. Other gases, however, such as carbon monoxide, carbon dioxide, sulfur dioxide, ammonia, organic gases, and water vapor, exhibit considerable ability to absorb radiant energy in certain regions of the infrared spectrum. As a result, the design engineer dealing with heat transfer in furnaces or other equipment in which absorbing gases are present may find it necessary to consider radiation from gases, as well as ordinary surface radiation.

Example 1. Combined heat transfer by convection and radiation The OD of an uninsulated steam pipe is 4.5 in. The outside-surface temperature of the pipe is constant at 300°F, and the pipe is located in a large room where the surrounding temperature is constant at 70°F. The heat content of the steam is valued at \$0.80 per 10^6 Btu. The emissivity of the pipe surface is 0.7, and the heat-transfer coefficient for heat loss from the surface by convection is 1.4 Btu/(h)(ft^2)(°F). Under these conditions, determine the cost per year for heat losses from the uninsulated pipe if the length of the pipe is 100 ft.

SOLUTION:

From Eq. (4), heat loss by convection = $hA\,\Delta t_f$

$$A = \frac{(3.14)(4.5)(100)}{12} = 118 \text{ ft}^2$$

$$\Delta t_f = 300 - 70 = 230°F$$

$$q_{\text{convection}} = (1.4)(118)(230) = 38,000 \text{ Btu/h}$$

From Eq. (8),

$$\text{Heat lost by radiation} = 0.171 A_1 \varepsilon_1 \left[\left(\frac{T_1}{100}\right)^4 - \left(\frac{T_2}{100}\right)^4\right]$$

$$= (0.171)(118)(0.7)\left[\left(\frac{300 + 460}{100}\right)^4 - \left(\frac{70 + 460}{100}\right)^4\right]$$

$$= 36,000 \text{ Btu/h}$$

Yearly cost for heat losses per 100 ft of pipe

$$= \frac{(38,000 + 36,000)(24)(365)(0.8)}{10^6} = \$518$$

OVERALL COEFFICIENTS OF HEAT TRANSFER

Many of the important cases of heat transfer involve the flow of heat from one fluid through a solid retaining wall into another fluid. This heat must flow through several resistances in series. The net rate of heat transfer can be related to the total temperature-difference driving force by employing an overall coefficient of heat transfer U; thus, for steady-state conditions,

$$q = UA\,\Delta t_{oa} \tag{11}$$

where Δt_{oa} represents the total temperature-difference driving force. Because there are usually several possible areas on which the numerical value of the overall coefficient can be based, the units of U should include a designation of the base area A. For example, if an overall coefficient is based on the inside area of a pipe, the units of this U should be expressed as Btu/(h)(ft^2 of inside area)(°F).

The overall coefficient (including a dirt or fouling resistance) can be related to the individual coefficients or resistances by the following equation:

$$\frac{1}{U_d} = \frac{A}{h'A'_f} + \frac{A}{h''A''_f} + \frac{Ax_w}{kA_{m_w}} + \frac{A}{h'_d A'_f} + \frac{A}{h''_d A''_f} \tag{12}$$

where A represents the base area chosen for the evaluation of U_d, and the primes refer to the different film resistances involved.

Fouling Factors†

After heat-transfer equipment has been in service for some time, dirt or scale may form on the heat-transfer surfaces, causing additional resistance to the flow of heat. To compensate for this possibility, the design engineer can include a resistance, called a *dirt*, *scale*, or *fouling factor*, when determining an overall coefficient of heat transfer. Equation (12) illustrates the method for handling the fouling factor. In this case, the fouling coefficients h'_d and h''_d are used to account for scale or dirt formation on the heat-transfer surfaces.

Fouling factors, equivalent to $1/h_d$, are often presented in the literature for various materials and conditions. Values of h_d for common process services are listed in Table 3. Because the scale or dirt resistance increases with the time the equipment is in service, some basis must be chosen for numerical values of fouling factors. The common basis is a time period of 1 year, and this condition applies to the values of h_d presented in Table 3. When the correct fouling factors are used, the equipment should be capable of transferring more than the required amount of heat when the equipment is clean. At the end of approximately 1 year of service, the capacity will have decreased to the design value, and a shut-down for cleaning will be necessary. With this approach, numerous shut-downs for cleaning individual units are not necessary. Instead, annual or periodic shutdowns of the entire plant can be scheduled, at which time all heat-transfer equipment can be cleaned and brought up to full capacity.

MEAN AREA OF HEAT TRANSFER

The cross-sectional area of heat transfer A in Eq. (1) can vary appreciably along the length of the heat-transfer path x. Therefore, the shape of the solid through which heat is flowing must be known before Eq. (1) can be integrated to give

---

† For a detailed review on fouling factors related to heat exchangers, see J. W. Suitor, W. J. Marner, and R. B. Ritter, The History and Status of Research in Fouling of Heat Exchangers in Cooling Water Service, *Can. J. Chem. Eng.*, **55**(4):374 (1977).

Table 3 Individual heat-transfer coefficients to account for fouling

	h_d for water, Btu/(h)(ft^2)(°F)			
Temperature of heating medium:	Up to 240°F		240–400°F	
Temperature of water:	125°F or less		Above 125°F	
Water velocity, ft/s:	3 and less	Over 3	3 and less	Over 3
Distilled	2000	2000	2000	2000
Sea water	2000	2000	1000	1000
Treated boiler feedwater	1000	2000	500	1000
Treated make-up for cooling tower	1000	1000	500	500
City, well, Great Lakes	1000	1000	500	500
Brackish, clean river water	500	1000	330	500
River water: muddy, silty†	330	500	250	330
Hard (over 15 grains/gal)	330	330	200	200
Chicago Sanitary Canal	130	170	100	130

	h_d for miscellaneous process services, Btu/(h)(ft^2)(°F)
Organic vapors, liquid gasoline	2000
Refined petroleum fractions (liquid), organic liquids, refrigerating liquids, brine, oil-bearing steam	1000
Distillate bottoms (above 25°API), gas oil or liquid naphtha below 500°F, scrubbing oil, refrigerant vapors, air (dust)	500
Gas oil above 500°F, vegetable oil	330
Liquid naphtha above 500°F, quenching oils	250
Topped crude (below 25°API), fuel oil	200
Cracked residuum, coke-oven gas, illuminating gas	100

† Mississippi, Schuylkill, Delaware, and East rivers and New York Bay.

Eq. (2). The exact value for A_m, based on the limiting areas A_1 and A_2, follows for three cases commonly encountered in heat-transfer calculations:

1. Conduction of heat through a solid of constant cross section (example, a large flat plate)

$$A_m = A_{\text{arith. avg}} = \frac{A_1 + A_2}{2} \tag{13}$$

2. Conduction of heat through a solid when the cross-sectional area of heat transfer is proportional to the radius (example, a long hollow cylinder)

$$A_m = A_{\text{log mean}} = \frac{A_1 - A_2}{\ln (A_1/A_2)} \tag{14}$$

3. Conduction of heat through a solid when the cross-sectional area of heat transfer is proportional to the square of the radius (example, a hollow sphere)

$$A_m = A_{\text{geom. mean}} = \sqrt{A_1 A_2} \tag{15}$$

When the value of A_1/A_2 (or A_2/A_1 if A_2 is larger than A_1) does not exceed 2.0, the mean area based on an arithmetic-average value is within 4 percent of the logarithmic-mean area and within 6 percent of the geometric-mean area. This accuracy is considered adequate for most design calculations. It should be noted that the arithmetic-average value is always greater than the logarithmic mean or the geometric mean.

MEAN TEMPERATURE-DIFFERENCE DRIVING FORCE

When a heat exchanger is operated continuously, the temperature difference between the hot and cold fluids usually varies throughout the length of the exchanger. To account for this condition, Eqs. (4) and (11) can be expressed in a differential form as

$$dq = h \, dA \, \Delta t_f \tag{16}$$

$$dq = U \, dA \, \Delta t_{oa} \tag{17}$$

The variations in Δt and the heat-transfer coefficients must be taken into account when Eqs. (16) and (17) are integrated. Under some conditions, a graphical or stepwise integration may be necessary, but algebraic solutions are possible for many of the situations commonly encountered with heat-transfer equipment.

Constant heat-transfer coefficient The integrated forms of Eqs. (16) and (17) are often expressed in the following simplified forms:

$$q = hA \, \Delta t_{f_m} \tag{18}$$

$$q = UA \, \Delta t_{oa_m} \tag{19}$$

where the subscript m refers to a mean value. Under the following conditions, the correct mean temperature difference is the logarithmic-mean value:

1. U (or h) is constant.
2. Mass flow rate is constant.
3. There is no partial phase change.
4. Specific heats of the fluids remain constant.
5. Heat losses are negligible.
6. Equipment is for counterflow, parallel flow, or any type of flow if the temperature of one of the fluids remains constant (phase change can occur) over the entire heat-transfer area.

$$\Delta t_{oa_m} = \Delta t_{oa_{\log \text{ mean}}} = \frac{\Delta t_{oa_1} - \Delta t_{oa_2}}{\ln \left(\Delta t_{oa_1} / \Delta t_{oa_2} \right)} \tag{20}$$

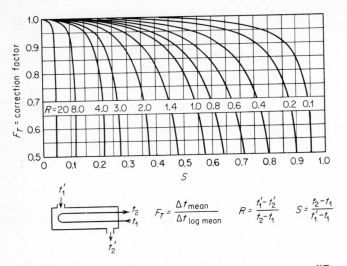

Figure 14-2 Chart for determining correct, mean temperature-difference driving force for an exchanger with one shell pass and two or more even-numbered tube passes. (Correction factor F_T is based on the $\Delta t_{\text{log mean}}$ for counterflow. If F_T is below 0.7, operation of the exchanger may not be practical.)

where Δt_{oa_1} and Δt_{oa_2} represent the two terminal values of the overall temperature-difference driving force. For design calculations, an arithmetic-average temperature difference can be used in place of the logarithmic-mean value if the ratio $\Delta t_{oa_1}/\Delta t_{oa_2}$ (or $\Delta t_{oa_2}/\Delta t_{oa_1}$ if Δt_{oa_2} is greater than Δt_{oa_1}) does not exceed 2.0.

When multipass exchangers are involved and the first five conditions listed in the preceding paragraph apply, values for Δt_{oa_m} can be obtained by integrating Eq. (17) or, more simply, from plots that give the correct Δt_{oa_m} for various types of multipass exchangers as a function of the logarithmic-mean temperature difference for counterflow.†‡ Figure 14-2 is a plot of this type for the common case of an exchanger with one shell pass and two or more even-numbered tube passes.

Variable heat-transfer coefficient If the heat-transfer coefficient varies with temperature, one can assume that the complete exchanger consists of a number of smaller exchangers in series and that the coefficient varies linearly with temperature in each of these sections. When the last five conditions listed in the

---

† J. H. Perry and C. H. Chilton, "Chemical Engineers' Handbook," 5th ed., McGraw-Hill Book Company, New York, 1973.

‡ These plots present graphically the results of integrating Eq. (17).

preceding section hold and the overall coefficient varies linearly with temperature, integration of Eq. (17) gives

$$q = A \frac{U_1 \, \Delta t_{oa_2} - U_2 \, \Delta t_{oa_1}}{\ln \left(U_1 \, \Delta t_{oa_2} / U_2 \, \Delta t_{oa_1} \right)} \tag{21}$$

The values of q and A in Eq. (21) apply to the section of the equipment between the limits indicated by the subscripts 1 and 2. Consequently, the total value of q or A for the entire exchanger can be obtained by summing the quantities for each of the individual sections.

UNSTEADY-STATE HEAT TRANSFER

When heat is conducted through a solid under unsteady-state conditions, the following general equation applies:

$$\frac{\partial t}{\partial \theta} = \frac{1}{\rho c_p} \left[\frac{\partial}{\partial x} \left(k_x \frac{\partial t}{\partial x} \right) + \frac{\partial}{\partial y} \left(k_y \frac{\partial t}{\partial y} \right) + \frac{\partial}{\partial z} \left(k_z \frac{\partial t}{\partial z} \right) \right] \tag{22}$$

where c_p is the heat capacity of the material through which heat is being conducted and x, y, and z represent the cartesian coordinates. The solution of any problem involving unsteady-state conduction consists essentially of integrating Eq. (22) with the proper boundary conditions.

For a homogeneous and isotropic material, Eq. (22) becomes

$$\frac{\partial t}{\partial \theta} = \alpha \left(\frac{\partial^2 t}{\partial x^2} + \frac{\partial^2 t}{\partial y^2} + \frac{\partial^2 t}{\partial z^2} \right) \tag{23}$$

where α = thermal diffusivity = $k/\rho c_p$, ft^2/h.

Many cases of practical interest in unsteady-state heat transfer involve one-dimensional conduction. For one-dimensional conduction in the x direction, Eq. (23) reduces to

$$\frac{\partial t}{\partial \theta} = \alpha \frac{\partial^2 t}{\partial x^2} \tag{24}$$

Solutions of Eqs. (23) and (24) for various shapes and boundary conditions are available in the literature. The simplest types of problems are those in which the surface of a solid suddenly attains a new temperature and this temperature remains constant. Such a condition can exist only if the temperature of the surroundings remains constant and there is no resistance to heat transfer between the surface and the surroundings (i.e., surface film coefficient is infinite). Although there are few practical cases when these conditions occur, the solutions of such problems are of interest to the design engineer because they indicate the results obtainable for the limiting condition of the maximum rate of unsteady-state heat transfer.

Figure 14-3 presents graphically the results of integrating the unsteady-state equations for a sudden change from a uniform surface and bulk temperature to a

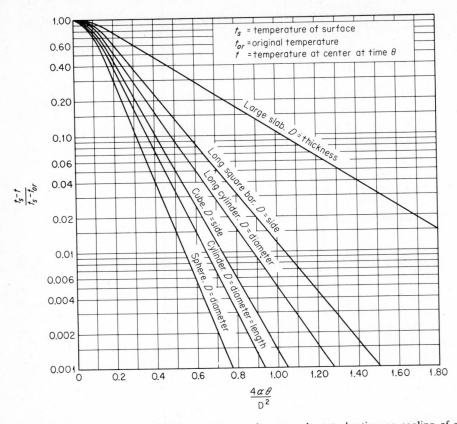

Figure 14-3 Midpoint or midplane temperatures for unsteady-state heating or cooling of solids having negligible surface resistance.

new constant surface temperature when the surface film coefficient is infinite. The reference temperature in this plot is at the center point, center line, or center plane, and results are presented for a slab, a square bar, a cube, a cylinder, and a sphere.

Example 2. Estimation of minimum time required for unsteady-state cooling A solid steel cylinder has a diameter of 3 ft and a length of 3 ft. The temperature of the cylinder is 1000°F. It is suddenly placed in a well-ventilated room where the temperature of the room remains constant at 90°F. Estimate the minimum time (i.e., surface film coefficient is infinite) the cylinder must remain in the room before the temperature at the center can drop to 150°F. Assume the following values for the steel: $k = 24.0$ Btu/(h)(ft^2)(°F/ft); $c_p = 0.12$ Btu/(lb)(°F); $\rho = 488$ lb/ft^3.

SOLUTION

$$\alpha = k/\rho c_p = 24/(488)(0.12) = 0.41 \text{ ft}^2/\text{h}$$

$$t_s = 90°\text{F}$$

$$t = 150°\text{F}$$

$$t_{or} = 1000°\text{F}$$

$$\frac{t_s - t}{t_s - t_{or}} = \frac{-60}{-910} = 0.066$$

From Fig. 14-3,

$$\frac{4\alpha\theta}{D^2} = 0.42$$

$$\theta = \frac{(0.42)(3)^2}{(4)(0.41)} = 2.3 \text{ h}$$

The minimum time before temperature at center of cylinder can drop to 150°F = 2.3 h.

DETERMINATION OF HEAT-TRANSFER COEFFICIENTS

Exact values of convection heat-transfer coefficients for a given situation can be obtained only by experimental measurements under the particular operating conditions. Approximate values, however, can be obtained for use in design by employing correlations based on general experimental data. A number of correlations that are particularly useful in design work are presented in the following sections. In general, the relationships applicable to turbulent conditions are more accurate than those for viscous conditions. Film coefficients obtained from the correct use of equations in the turbulent-flow range will ordinarily be within ±20 percent of the true experimental value, but values determined for viscous-flow conditions or for condensation, boiling, natural convection, and shell sides of heat exchangers may be in error by more than 100 percent. Because of the inherent inaccuracies in the methods for estimating film coefficients, some design engineers prefer to use overall coefficients based on past experience, while others include a large safety factor in the form of fouling factors or fouling coefficients.

Film Coefficients for Fluids in Pipes and Tubes (No Change in Phase)

The following equations are based on the correlations presented by Sieder and Tate:†

For viscous flow $(DG/\mu < 2100)$,

$$\frac{h_i D}{k} = 1.86 \left(\frac{DG}{\mu} \frac{c_p \mu}{k} \frac{D}{L} \right)^{1/3} \left(\frac{\mu}{\mu_w} \right)^{0.14} = 1.86 \left(\frac{4wc_p}{\pi k L} \right)^{1/3} \left(\frac{\mu}{\mu_w} \right)^{0.14} \tag{25}$$

† E. N. Sieder and G. E. Tate, *Ind. Eng. Chem.*, **28**:1429 (1936).

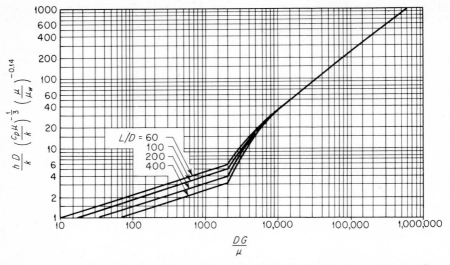

Figure 14-4 Plot for estimating film coefficients for fluids flowing in pipes and tubes. [Based on Eqs. (25) and (26).]

For turbulent flow above the transition region $(DG/\mu > 10{,}000)$,

$$\frac{h_i D}{k} = 0.023 \left(\frac{DG}{\mu}\right)^{0.8} \left(\frac{c_p \mu}{k}\right)^{1/3} \left(\frac{\mu}{\mu_w}\right)^{0.14} \tag{26}†$$

For transition region $(DG/\mu = 2100$ *to* $10{,}000)$, *see Fig. 14-4,*
where D = diameter of pipe or tube (inside), ft
G = mass velocity inside tube, lb/(h)(ft^2 of cross-sectional area)
c_p = heat capacity of fluid at constant pressure, Btu/(lb)(°F)
μ = viscosity of fluid (subscript w indicates evaluation at wall temperature), lb/(h)(ft)
L = heated length of straight tube, ft
w = weight rate of flow per tube, lb/h
k, μ, and c_p are evaluated at the average bulk temperature of the fluid.

Equations (25) and (26) are applicable for organic fluids, aqueous solutions, water, and gases. The two equations are plotted in Fig. 14-4 to facilitate their solution and to indicate the values for use in the transition region.

---

† In some references, the constant 0.027 is used in place of 0.023. The constant 0.023 is recommended in order to make Eq. (26) generally applicable for water, organic fluids, and gases at moderate Δt.

Simplified equations For common gases, Eq. (26) can be simplified to give the following approximate equation:†

$$h_i = \frac{0.014 c_p G^{0.8}}{D^{0.2}} \tag{27}$$

Similarly, for water at ordinary temperatures and pressures,

$$h_i = \frac{150(1 + 0.011 t_b)(V')^{0.8}}{(D')^{0.2}} \tag{28}$$

Equations (27) and (28) are dimensional, and a value of h_i as Btu/(h)(ft^2)(°F) is obtained only if the following units are employed for the indicated variables:

c_p = heat capacity of fluid, Btu/(lb)(°F)
D = diameter, ft
D' = diameter, in.
G = mass velocity inside tube, lb/(h)(ft^2)
t_b = average (i.e., bulk) temperature of water, °F
V' = velocity of water, ft/s

Noncircular Cross Section—Equivalent Diameter

The situation is often encountered in which a fluid flows through a conduit having a noncircular cross section, such as an annulus. The heat-transfer coefficients for turbulent flow can be determined by using the same equations that apply to pipes and tubes if the pipe diameter D appearing in these equations is replaced by an equivalent diameter D_e. Best results are obtained if this equivalent diameter is taken as four times the hydraulic radius, where the hydraulic radius is defined as the cross-sectional flow area divided by the *heated* perimeter. For example, if heat is being transferred from a fluid in a center pipe to a fluid flowing through an annulus, the film coefficient around the inner pipe would be based on the following equivalent diameter:

$$D_e = 4 \times \text{hydraulic radius} = 4 \times \frac{\pi D_2^2/4 - \pi D_1^2/4}{\pi D_1} = \frac{D_2^2 - D_1^2}{D_1}$$

where D_1 and D_2 represent, respectively, the inner and outer diameters of the annulus.

The difference between the hydraulic radii for heat transfer and for fluid flow should be noted. In the preceding example, the correct equivalent diameter for evaluating friction due to the fluid flow in the annulus would be four times the cross-sectional flow area divided by the *wetted* perimeter, or $4 \times (\pi D_2^2/4 - \pi D_1^2/4)/(\pi D_2 + \pi D_1) = D_2 - D_1$.

---

† Equations (27) and (28) are derived by neglecting the viscosity correction factor in Eq. (26) and substituting average values for the physical properties.

Film Coefficients for Fluids Flowing Outside Pipes and Tubes (No Change in Phase)

In the common types of baffled shell-and-tube exchangers, the shell-side fluid flows across the tubes. The equations for predicting heat-transfer coefficients under these conditions are not the same as those for flow of fluids inside pipes and tubes. An approximate value for shell-side coefficients in a cross-flow exchanger with segmental baffles and reasonable clearance between baffles, between tubes, and between baffles and shell can be obtained by using the following correlation:[†][‡]

$$\frac{h_o D_o}{k_f} = \frac{a_o}{F_s}\left(\frac{D_o G_s}{\mu_f}\right)^{0.6}\left(\frac{c_p \mu}{k}\right)^{1/3}_f \tag{29}$$

where $a_o = 0.33$ if tubes in tube bank are staggered and 0.26 if tubes are in line

F_s = safety factor to account for bypassing effects[§]

G_s = shell-side mass velocity across tubes, based on minimum free area between baffles at shell axis,[¶] lb/(h)(ft^2)

Subscript f refers to properties at average film temperature

Equation (29) can be used to obtain approximate film coefficients for hydrocarbons, aqueous solutions, water, and gases when the Reynolds number $(D_o G_s/\mu_f)$ is in the ordinary range of 2000 to 32,000.

Film Coefficients and Overall Coefficients for Miscellaneous Cases

Table 4 presents equations for use in the estimation of heat-transfer film coefficients for condensation, boiling, and natural convection. Values showing the general range of film coefficients for various situations are indicated in Table 5.

The design engineer often prefers to use overall coefficients directly without

---

[†] A. P. Colburn, *Trans. AIChE*, **29**:174 (1933).

[‡] For an alternate approach which takes pressure drop into consideration, see D. A. Donohue, *Ind. Eng. Chem.*, **41**(11):2499 (1949).

[§] Amount of shell-side bypassing between the cross baffles and the shell, between tubes and tube holes in baffles, and between outermost tubes and shell depends on the manufacturing methods and tolerances for the exchanger. The amount of bypassing can have a large influence on the shell-side heat-transfer coefficient. The value of F_s is usually between 1.0 and 1.8, and a value of 1.6 is often recommended.

[¶] The free area for use in Eqs. (29) and (31) occurs where the total cross-sectional area of the shell (normal to direction of flow) is a maximum, and the free area at this axis plane is based on the transverse or diagonal openings that give the smallest free area. The free area S_o for use in evaluating G_s for the common case of a full-packed shell and transverse openings giving the smallest free area can be estimated to be

$$\frac{\text{(ID of shell)(clearance between adjacent tubes)(baffle spacing)}}{\text{Center-to-center distance between adjacent tubes}}$$

attempting to evaluate individual film coefficients. When this is the case, the engineer must predict an overall coefficient on the basis of past experience with equipment and materials similar to those involved in the present problem. Design values of overall heat-transfer coefficients for many of the situations commonly encountered by design engineers are listed in Table 6.

Table 4 Equations and methods for estimating film coefficients of heat transfer for common cases

Type of heat transfer	Limitations
Film-type condensation of vapors: Outside horizontal tubes $$h = 0.725 \left(\frac{k_f{}^3 \rho_f{}^2 g \lambda_c}{N_V D_o \mu_f \, \Delta t_f} \right)^{1/4} = 0.95 \left(\frac{k_f{}^3 \rho_f{}^2 g L}{w \mu_f} \right)^{1/3}$$ For steam, average h at 1 atm pressure $= \dfrac{3100}{(N_V D_o)^{1/4} (\Delta t_f)^{1/3}}$ Vertical tubes $$h = 1.47 \left(\frac{\pi D_o k_f{}^3 \rho_f{}^2 g}{4 w \mu_f} \right)^{1/3}$$ For steam, average h at 1 atm pressure $= \dfrac{4000}{L^{1/4} (\Delta t_f)^{1/3}}$	McAdams† Pure saturated vapors $\dfrac{2w}{L \mu_f} < 2000$ Physical properties are for condensate. McAdams† Pure saturated vapors $4w/\pi D_o \mu_f < 2000$ Physical properties are for condensate.
Boiling liquids outside horizontal tubes: Film boiling (above critical Δt_f) $$h_{co} = 0.62 \left[\frac{k_v{}^3 \rho_v (\rho_L - \rho_v) g \lambda_c}{D_o \mu_v \, \Delta t_f} \right]^{1/4}$$ Nucleate boiling (below critical Δt_f) Value of h depends on Δt_f, type of surface, and materials involved (critical Δt_f for water as temperature drop from heating surface to liquid is approximately 45°F).	Bromley‡ For saturated liquids on submerged surfaces; film coefficient h_{co} is for conduction through the vapor; no radiation effect is included.
Natural convection: General equation $$\frac{h_c L}{k_f} = c_c [(N_{Gr})_f (N_{Pr})_f]^n$$	Perry § Values of c_c and n depend on $N_{Gr} \times N_{Pr}$ and the geometry of the surfaces.

(Continued)

Table 4 Equations and methods for estimating film coefficients of heat transfer for common cases (*Continued*)

Type of heat transfer		Limitations
Simplified equations for air $h_c = K(\Delta t)^{0.25}$		Kern¶ Ordinary air temperature, atmospheric pressure, normal conditions; for more exact equations, see Perry. §
Physical arrangement of equipment	K	
Horizontal plates facing upward	0.38	
Horizontal plates facing downward	0.20	
Vertical plates more than 2 ft high	0.30	
Vertical plates less than 2 ft high	$\dfrac{0.28}{(\text{vertical height, ft})^{0.25}}$	
Vertical pipes more than 1 ft high	$\dfrac{0.22}{(D_o, \text{ft})^{0.25}}$	
Horizontal pipes	$\dfrac{0.27}{(D_o, \text{ft})^{0.25}}$	

λ_c = latent heat of condensation, Btu/lb
g = local gravitational acceleration, ft/(h) (h)
N_V = number of rows of tubes in a vertical tier
D_o = outside diameter of tube, ft
Δt_f = temperature-difference driving force across film, °F
L = heated length of straight tube or length of heat-transfer surface, ft
w = weight rate of flow of condensate per tube from lowest point on condensing surface, lb/(h) (tube)

$(N_{Gr})_f$ = Grashof number = $\dfrac{L^3 \rho_f{}^2 g \beta_f \Delta t_f}{\mu_f{}^2}$

$(N_{Pr})_f$ = Prandtl number = $\left(\dfrac{c_p \mu}{k}\right)_f$

β = coefficient of volumetric expansion, 1/°R
c_c, K, n = constants
Subscript f designates "at average film temperature"
Subscript v designates vapor at average vapor temperature
Subscript L designates liquid at average liquid temperature

---

† W. H. McAdams, "Heat Transmission," 3d ed., McGraw-Hill Book Company, New York, 1954.
‡ L. A. Bromley, *Chem. Eng. Progr.,* 46: 221 (1950).
§ R. H. Perry and C. H. Chilton, "Chemical Engineers' Handbook," 5th ed., McGraw-Hill Book Company, New York, 1973.
¶ D. Q. Kern, "Process Heat Transfer," McGraw-Hill Book Company, New York, 1950.

Table 5 Order of magnitude of individual film coefficients

Condition	h, Btu/(h) (ft^2) ($^\circ$F)
Dropwise condensation of steam	10,000–20,000
Film-type condensation of steam	1,000–3,000
Boiling water	300–9,000
Film-type condensation of organic vapors	200–400
Heating or cooling of water	50–3,000
Heating or cooling of organic solvents	30–500
Heating or cooling of oils	10–120
Superheated steam	5–30
Heating or cooling of air	0.2–20
(Low value for free convection—	
High value for forced convection)	

Table 6 Approximate design values of overall heat-transfer coefficients

The following values of overall heat-transfer coefficients are based primarily on results obtained in ordinary engineering practice. The values are approximate because variation in fluid velocities, amount of noncondensable gases, viscosities, cleanliness of heat-transfer surfaces, type of baffles, operating pressure, and similar factors can have a significant effect on the overall heat-transfer coefficients. The values are useful for preliminary design estimates or for rough checks on heat-transfer calculations.

Upper range of overall coefficients given for coolers may also be used for condensers. Upper range of overall coefficients given for heaters may also be used for evaporators.

Units of coefficients are Btu/(h) (ft^2) ($^\circ$F).

Hot fluid	Cold fluid	Fouling coefficient, h_d	Overall coefficient, U_d
		Coolers	
Water	Water	1000	250–500
Methanol	Water	1000	250–500
Ammonia	Water	1000	250–500
Aqueous solutions	Water	1000	250–500
Light organics—viscosities less than 0.5 cP (benzene, toluene, acetone, ethanol, gasoline, light kerosene, and naphtha)	Water	300	75–150
Medium organics—viscosities between 0.5 and 1.0 cP (kerosene, straw oil, hot gas oil, hot absorber oil, some crude oils)	Water	300	50–125
Heavy organics—viscosities greater than 1.0 cP (cold gas oil, lube oils, fuel oils, reduced crude oils, tars, and asphalts)	Water	300	5–75
Gases	Water	300	2–50
Water	Brine	300	100–200
Light organics	Brine	300	40–100

(*Continued*)

Table 6 Approximate design values of overall heat-transfer coefficients (*Continued*)

Hot fluid	Cold fluid	Fouling coefficient, h_d	Overall coefficient, U_d
		Heaters	
Steam	Water	1000	200–700
Steam	Methanol	1000	200–700
Steam	Ammonia	1000	200–700
Steam	Aqueous solutions:		
	Less than 2.0 cP	1000	200–700
	More than 2.0 cP	1000	100–500
Steam	Light organics	300	100–200
Steam	Medium organics	300	50–100
Steam	Heavy organics	300	6–60
Steam	Gases	300	5–50
Dowtherm	Gases	300	4–40
Dowtherm	Heavy organics	300	6–60
		Exchangers (no phase change)	
Water	Water	1000	250–500
Aqueous solutions	Aqueous solutions	1000	250–500
Light organics	Light organics	300	40–75
Medium organics	Medium organics	300	20–60
Heavy organics	Heavy organics	300	10–40
Heavy organics	Light organics	300	30–60
Light organics	Heavy organics	300	10–40

PRESSURE DROP IN HEAT EXCHANGERS

The major cause of pressure drop in heat exchangers is friction resulting from flow of fluids through the exchanger tubes and shell. Friction due to sudden expansion, sudden contraction, or reversal in direction of flow also causes a pressure drop. Changes in vertical head and kinetic energy can influence the pressure drop, but these two effects are ordinarily relatively small and can be neglected in many design calculations.

Tube-side Pressure Drop

It is convenient to express the pressure drop for heat exchangers in a form similar to the Fanning equation as presented in Chap. 13. Because the transfer of heat is involved, a factor must be included for the effect of temperature change on the friction factor. Under these conditions, the pressure drop through the tube passes (i.e., tube side) of a heat exchanger may be expressed as follows (subscript i refers to inside of tubes at bulk temperature):

$$-\Delta P_i = \frac{B_i 2\mathfrak{f}_i G^2 L n_p}{g_c \rho_i D_i \phi_i} \tag{30}$$

where $\mathfrak{f}_i$ = Fanning friction factor for isothermal flow based on conditions at the arithmetic-average temperature of the fluid; $\mathfrak{f}_i$ as function of $D_i G/\mu_i$ is shown in Fig. 13-1

n_p = number of tube passes

g_c = conversion factor in Newton's law of motion, $(32.17) \times (3600)^2$ ft · lbm/(h)(h)(lbf)

ϕ_i = correction factor for nonisothermal flow is equal to $1.1(\mu_i/\mu_w)^{0.25}$ when $D_i G/\mu_i$ is less than 2100 and $1.02(\mu_i/\mu_w)^{0.14}$ when $D_i G/\mu_i$ is greater than 2100; μ_i is viscosity at arithmetic-average (bulk) temperature of fluid, and μ_w is viscosity of fluid at average temperature of the inside-tube wall surface

B_i = correction factor to account for friction due to sudden contraction, sudden expansion, and reversal of flow direction

$= 1 + (F_e + F_c + F_r)/(2\mathfrak{f}_i G^2 L/g_c \rho_i^2 D_i \phi_i)$†

Shell-side Pressure Drop

The pressure drop due to friction when a fluid is flowing parallel to and outside of tubes can be calculated in the normal manner described in Chap. 13 by using a mean diameter equal to four times the hydraulic radius of the system and by including all frictional effects due to contraction and expansion. In heat exchangers, however, the fluid flow on the shell side is usually across the tubes, and many types and arrangements of baffles may be used. As a result, no single

---

† F_e (friction due to sudden enlargement) and F_c (friction due to sudden contraction) can be estimated by the methods indicated in Chap. 13. $F_e = (V_i - V_{header})^2/2g_c$. $F_c = K_c V_i^2/2g_c$. F_r (friction due to reversal of flow direction) depends on the details of the exchanger construction, but a good estimate for design work is $(0.5V_i^2/2g_c) \times (n_p - 1)/n_p$. Designating S_i/S_H as the ratio of total inside-tube cross-sectional area per pass to header cross-sectional area per pass and K_1 as $[1 - (S_i/S_H)]^2 + K_c + 0.5(n_p - 1)/n_p$,

$$B_i = 1 + \frac{K_1 D_i \phi_i}{4\mathfrak{f}_i L}$$

If the flow is highly turbulent and in smooth tubes, $\mathfrak{f}_i$ may be taken as $0.046(D_i G/\mu_i)^{-0.2}$. Combining the preceding relationships with Eqs. (4), (11), (26), and the rate equation for no phase change [Eq. (36)] gives, for smooth tubes and turbulent flow,

$$B_i = 1 + \frac{0.51 K_1 n_p \Delta t_{fi} (\mu_i/\mu_w)^{0.28}}{(t_2 - t_1)_i (c_p \mu/k)_i^{2/3}}$$

where $t_2 - t_1$ represents the total change in temperature of the fluid as it flows through the exchanger tubes.

explicit equation can be given for evaluating pressure drop on the shell side of all heat exchangers.

For the case of flow across tubes, the following equation can be used to approximate the pressure drop due to friction (subscript o refers to outside of tubes at bulk temperature):

$$-\Delta P_o = \frac{B_o\, 2\bar{f}' N_r G_s^2}{g_c \rho_o} \tag{31}$$

where $\bar{f}'$ = special friction factor for shell-side flow

N_r = number of rows of tubes across which shell fluid flows

B_o = correction factor to account for friction due to reversal in direction of flow, recrossing of tubes, and variation in cross section; when the flow is across unbaffled tubes, B_o can be taken as 1.0; as a rough approximation, B_o can be estimated as equal to the number of tube crosses

The friction factor $\bar{f}'$ is a function of the Reynolds number of the flowing fluid and the arrangement of the tubes. For the common case of $D_o G_s / \mu_f$ in the range of 2000 to 40,000, the friction factor in Eq. (31) may be represented as[†]

$$\bar{f}' = b_o \left(\frac{D_o G_s}{\mu_f} \right)^{-0.15} \tag{32}$$

The following approximate equations for b_o are based on the data of Grimson:[‡][§]
    For staggered tubes,

$$b_o = 0.23 + \frac{0.11}{(x_T - 1)^{1.08}} \tag{33}$$

For tubes in line,

$$b_o = 0.044 + \frac{0.08 x_L}{(x_T - 1)^{0.43 + 1.13/x_L}} \tag{34}$$

where x_T = ratio of pitch (i.e., tube center-to-center distance) transverse to flow to tube diameter, dimensionless; and x_L = ratio of pitch parallel to flow to tube diameter, dimensionless.

Example 3. Estimation of film coefficient and pressure drop inside tubes in a shell-and-tube exchanger A horizontal shell-and-tube heat exchanger with

[†] For general cases with higher Reynolds numbers, see R. H. Perry and C. H. Chilton, "Chemical Engineers' Handbook," 5th ed., McGraw-Hill Book Company, New York, 1973 or D. Q. Kern, "Process Heat Transfer," McGraw-Hill Book Company, New York, 1950. For added pressure drop due to flow through baffle-pass area, see page 5-38 of "Chemical Engineers' Handbook," 5th ed.

[‡] E. D. Grimson, *Trans. ASME*, **59**:583 (1937); **60**:381 (1938).

[§] In Eqs. (33) and (34), best results are obtained if x_T is between 1.5 and 4.0. For design purposes, the range can be extended down to 1.25. When the Reynolds number is between 2000 and 10,000 with tubes in line, $\bar{f}'$ for a given tube spacing can be assumed to be constant at the value obtained when the Reynolds number is 10,000.

two tube passes and one shell pass is being used to heat 70,000 lb/h of 100 percent ethanol from 60 to 140°F at atmospheric pressure. The ethanol passes through the inside of the tubes, and saturated steam at 230°F condenses on the shell side of the tubes. The tubes are steel with an OD of $\frac{3}{4}$ in. and a BWG of 14. The exchanger contains a total of 100 tubes (50 tubes per pass), and the average number of tubes in a vertical tier can be taken as 6. The ratio of total inside-tube cross-sectional area per pass to header cross-sectional area per pass can be assumed to be 0.5. Estimate the film coefficient for the ethanol and the pressure drop due to friction through the tube side of the exchanger.

SOLUTION From the Appendix,

Tube wall thickness = 0.083 in.

Tube ID = 0.584 in.

Tube OD = 0.75 in.

Flow area per tube = 0.268 in.^{2}

Inside surface area per lin ft = 0.1529 ft^2

k for steel = 26 Btu/(h)(ft^2)(°F/ft)

At $(60 + 140)/2 = 100$°F,

μ for ethanol = 0.9 centipoise

k for ethanol = 0.094 Btu/(h)(ft^2)(°F/ft)

c_p for ethanol = 0.62 Btu/(lb)(°F)

ρ for ethanol = 49 lb/ft^3

Mass velocity $= G = \dfrac{(70,000)(144)}{(0.268)(50)} = 752,000$ lb/(h)(ft^2)

$N_{Re} = \dfrac{DG}{\mu} = \dfrac{(0.584)(752,000)}{(12)(0.9)(2.42)} = 16,800$

$\dfrac{c_p \mu}{k} = \dfrac{(0.62)(0.9)(2.42)}{0.094} = 14.4$

Equation (26), for the evaluation of the average coefficient, applies under these conditions. The only remaining unknown in the equation is μ_w.

Evaluation of μ_w. As a first approximation, assume the following:

Δt over ethanol film = 70% of average total $\Delta t = 0.70(230 - 100) = 91$°F

Δt over steam film = 8% of average total $\Delta t = 0.08(230 - 100) = 10$°F

Wall temperature for evaluation of $\mu_w = 100 + 91 = 191$°F

$\mu_w = \mu$ of ethanol at $191°F = 0.4$ centipoise

$\mu/\mu_w = 0.9/0.4 = 2.25$

From Eq. (26),

$$h_i = \frac{k}{D}(0.023)\left(\frac{DG}{\mu}\right)^{0.8}\left(\frac{c_p\mu}{k}\right)^{1/3}\left(\frac{\mu}{\mu_w}\right)^{0.14}$$

$$= \frac{(0.094)(12)(0.023)(16{,}800)^{0.8}(14.4)^{1/3}(2.25)^{0.14}}{0.584}$$

$$= 290 \text{ Btu/(h)(ft}^2)(°F)$$

Check on Δt assumptions. From Table 3, a fouling coefficient of $1000 \text{ Btu/(h)(ft}^2)(°F)$ is adequate for the ethanol. No fouling coefficient will be used for the steam.

From Table 4, the steam film coefficient can be approximated as

$$\frac{3100}{[(6)(0.75/12)]^{1/4}(10)^{1/3}} = 1800 \text{ Btu/(h)(ft}^2)(°F)$$

By Eq. (12) (basing U_d on the inside tube area),

$$\frac{1}{U_d} = \frac{A}{h'A'_f} + \frac{A}{h''A''_f} + \frac{Ax_w}{kA_w} + \frac{A}{h'_d A'_f}$$

$$= \frac{1}{290} + \frac{0.584}{(1800)(0.75)} + \frac{(0.584)(0.083)}{(26)(0.667)(12)} + \frac{1}{1000}$$

$$= 0.00345 + 0.00043 + 0.00023 + 0.001 = 0.00511$$

Percent Δt over ethanol film $= \dfrac{0.00345}{0.00511}(100) = 68\%$ of total Δt

Percent Δt over steam film $= \dfrac{0.00043}{0.00511}(100) = 8.4\%$ of total Δt

The Δt assumptions are adequate.

Determination of pressure drop due to friction. Assume U_d is constant over length of exchanger:

$$U_d = \frac{1}{0.00511} = 196 \text{ Btu/(h)(ft}^2 \text{ of inside area)(°F)}$$

$$\Delta t_{oa,\,m} = \frac{(230-60)+(230-140)}{2} = 130°F = \Delta t_{oa,\,\text{arith avg}}$$

The arithmetic-average value is satisfactory because the temperature of the steam remains constant and the ratio $\Delta t_1/\Delta t_2$ is less than 2.0.

$$q = (70{,}000)(0.62)(140-60) = 3{,}470{,}000 \text{ Btu/h}$$

By Eq. (19),

$$A = \frac{q}{U_d \, \Delta t_{oa, \, m}} = \frac{3,470,000}{(196)(130)} = 136 \text{ ft}^2 \text{ of inside tube area}$$

$$\text{Length per tube} = L = \frac{136}{(0.1529)(100)} = 8.9 \text{ ft}$$

By Eq. (30),

$$-\Delta P_i = \frac{B_i \, 2 \mathfrak{f}_i G_i^2 L n_p}{g_c \rho_i D_i \phi_i}$$

$$B_i = 1 + \frac{F_e + F_c + F_r}{2 \mathfrak{f}_i G_i^2 L / g_c \rho_i^2 D_i \phi_i} = 1 + \frac{0.51 K_1 n_p \, \Delta t_{fi} (\mu_i / \mu_w)^{0.28}}{(t_2 - t_1)_i (c_p \mu / k)_i^{2/3}}$$

$$K_1 = \left(1 - \frac{S_i}{S_H}\right)^2 + K_c + \frac{0.5(n_p - 1)}{n_p}$$

$$= (1 - 0.5)^2 + 0.3 + (0.5)(1)(\tfrac{1}{2}) = 0.8$$

$$K_c \text{ (Chap. 13, Table 1)} = (0.4)(1.25 - 0.5) = 0.3$$

$$B_i = 1 + \frac{(0.51)(0.8)(2)(0.68)(230 - 100)(2.25)^{0.28}}{(140 - 60)(14.4)^{2/3}} = 1.19$$

$$\mathfrak{f}_i \text{ (from Fig. 13-1)} = 0.0066$$

$$n_p = 2$$

$$\phi_i = 1.02 \left(\frac{\mu_i}{\mu_w}\right)^{0.14} = 1.02(2.25)^{0.14} = 1.14$$

$$-\Delta P_i = \frac{(1.19)(2)(0.0066)(752,000)^2 (8.9)(2)}{(3600)^2 (32.17)(49)(0.584/12)(1.14)} = 140 \text{ psf}$$

The pressure drop due to friction through the tube side of the exchanger is approximately 140 psf or 1 psi.

If the cost for power is \$0.05 per kilowatthour, the pumping cost due to frictional pressure drop on the tube side of the exchanger is $(140)(70,000)(0.05)/(49)(2.655 \times 10^6) = \$0.0038$ per hour.

The low power cost and low pressure drop indicate that a higher liquid velocity would be beneficial, since it would give a better heat-transfer coefficient and have little effect on the total cost.

Example 4. Estimation of film coefficient and pressure drop on shell side in a shell-and-tube exchanger A shell-and-tube heat exchanger with one shell pass and one tube pass is being used as a cooler. The cooling medium is water on the shell side of the exchanger. Five segmental baffles with a 25 percent cut are used on the shell side, and the baffles are spaced equally 2 ft apart. The safety factor F_s for use in evaluating the shell-side film coefficient is 1.6. The inside diameter of the shell is 23 in. The OD of the tubes is 0.75 in., and the tubes are staggered. Clearance between tubes is 0.25 in., and the flow of water across the tubes is normal to this clearance. There is a total of 384 tubes in the exchanger, and the shell can be considered to be full-packed. Water flows through the shell side at a rate of 90,000 lb/h. The average temperature of the water is 90°F, and the average temperature of the tube walls on the water side is 100°F. Under these conditions, estimate the heat-transfer coefficient for the water and the pressure drop due to friction on the shell side of the exchanger. Neglect pressure drop due to flow through baffle-pass area.

SOLUTION Average water-film temperature = 95°F. From the Appendix,

μ for water at 95°F = 0.724 centipoise

k for water at 95°F = 0.361 Btu/(h)(ft^2)(°F/ft)

c_p for water at 95°F = 1 Btu/(lb)(°F)

ρ for water at 90°F = 62.1 lb/ft^3

μ for water at 100°F = 0.684 centipoise

Calculation of Reynolds number = $D_o G_s / \mu_f$

$D_o = 0.75/12$ ft

$\mu_f = (0.724)(2.42)$ lb/(h)(ft)

Free area for evaluation of G_s occurs at center plane of shell where number of tube rows normal to flow direction = $23/(0.75 + 0.25) = 23$. Free area is based on the smallest flow area at the center plane. In this case, the transverse openings give the smallest free area.

$$\text{Free area between baffles} = \frac{(23)(0.25)(2)}{12} = 0.96 \text{ ft}^2$$

$$G_s = \frac{90,000}{0.96} = 94,000 \text{ lb/(h)(ft}^2)$$

$$\frac{D_o G_s}{\mu_f} = \frac{(0.75)(94,000)}{(12)(0.724)(2.42)} = 3360$$

$$\left(\frac{c_p \mu}{k}\right)_f = \frac{(1)(0.724)(2.42)}{0.361} = 4.85$$

Equation (29) may be used under these conditions.

$$h_o = \frac{k_f a_o}{D_o F_s}\left(\frac{D_o G_s}{\mu_f}\right)^{0.6}\left(\frac{c_p \mu}{k}\right)_f^{1/3}$$

$a_o = 0.33 \qquad F_s = 1.6$

$$h_o = \frac{(0.361)(12)(0.33)(3360)^{0.6}(4.85)^{1/3}}{(0.75)(1.6)} = 260 \text{ Btu/(h)(ft}^2)(°F)$$

Determination of pressure drop due to friction. Equations (31) to (33) apply for this case:

$$-\Delta P_o = \frac{B_o\, 2\mathfrak{f}' N_r G_s^2}{g_c \rho_o}$$

$$\mathfrak{f}' = \left[0.23 + \frac{0.11}{(x_T - 1)^{1.08}}\right]\left(\frac{D_o G_s}{\mu_f}\right)^{-0.15}$$

$$\mathfrak{f}' = \left[0.23 + \frac{0.11}{(1.0/0.75 - 1)^{1.08}}\right](3360)^{-0.15} = 0.175$$

B_o = roughly the number of tube crosses = number of baffles + 1 = 6. N_r = number of tube rows across which shell fluid flows = 23 minus the tube rows that pass through the cut portions of the baffles. With 25 percent cut baffles, the fluid will flow across approximately one-half of the tubes. In this case, N_r will be taken as 14. This gives some allowance for neglecting friction due to reversal of flow direction and friction due to flow parallel to the tubes:

$$-\Delta P_o = \frac{(6)(2)(0.175)(14)(94,000)^2}{(32.17)(3600)^2(62.1)} = 10 \text{ psf}$$

Pressure drop due to friction on the shell side of the exchanger is approximately 10 psf.

CONSIDERATIONS IN SELECTION OF HEAT-TRANSFER EQUIPMENT

When the design engineer selects heat-transfer equipment, it is necessary to consider the basic process-design variables and also many other factors, such as temperature strains, thickness of tubes and shell, types of baffles, tube pitch, and standard tube lengths. Under ordinary conditions, the mechanical design of an exchanger should meet the requirements of the ASME or API-ASME Safety Codes. The Tubular Exchanger Manufacturers Association (TEMA) publishes standards on general design methods and fabrication materials for tubular heat exchangers.

Tube Size and Pitch

The standard length of tubes in a shell-and-tube heat exchanger is 8, 12, or 16 ft, and these standard-length tubes are available in a variety of different diameters and wall thicknesses. Exchangers with small-diameter tubes are less expensive per square foot of heat-transfer surface than those with large-diameter tubes, because a given surface can be fitted into a smaller shell diameter; however, the small-diameter tubes are more difficult to clean. A tube diameter of $\frac{3}{4}$ or 1 in.

OD is the most common size, but outside diameters ranging from $\frac{5}{8}$ to $1\frac{1}{2}$ in. are found in many industrial installations.

Tube-wall thickness is usually specified by the Birmingham wire gauge, and variations from the nominal thickness may be ± 10 percent for "average-wall" tubes and $+22$ percent for "minimum-wall" tubes. Pressure, temperature, corrosion, and allowances for expanding the individual tubes into the tube sheets must be taken into consideration when the thickness is determined.

Tube pitch is defined as the shortest center-to-center distance between adjacent tubes, while the shortest distance between two tubes is designated as the *clearance*. In most shell-and-tube exchangers, the pitch is in the range of 1.25 to 1.50 times the tube diameter. The clearance should not be less than one-fourth of the tube diameter, and $\frac{3}{16}$ in. is usually considered to be a minimum clearance.

Tubes are commonly laid out on a square pattern or on a triangular pattern, as shown in Fig. 14-5. Although a square pitch has the advantage of easier external cleaning, the triangular pitch is sometimes preferred because it permits the use of more tubes in a given shell diameter. Table 7 indicates the number of tubes that can be placed in an exchanger with conventional tube sizes and pitches.

Shell Size

For shell diameters up to 24 in., nominal pipe sizes apply to the shell. Inside diameters are usually indicated, and schedule number or wall thickness should also be designated. In general, a shell thickness of $\frac{3}{8}$ in. is used for shell diameters between 12 and 24 in. unless the fluids are extremely corrosive or the operating pressure on the shell side exceeds 300 psig.

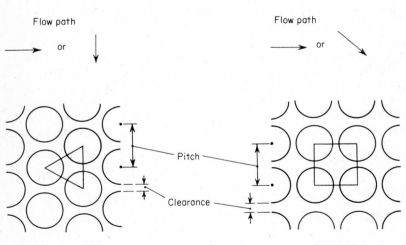

Figure 14-5 Conventional tube-plate layouts.

Table 7 Number of tubes in conventional tubesheet layouts

Shell ID, in.	One-pass		Two-pass		Four-pass	
	Square pitch	Triangular pitch	Square pitch	Triangular pitch	Square pitch	Triangular pitch
$\frac{3}{4}$-in.-OD tubes on 1-in. pitch						
8	32	37	26	30	20	24
12	81	92	76	82	68	76
15¼	137	151	124	138	116	122
21¼	277	316	270	302	246	278
25	413	470	394	452	370	422
31	657	745	640	728	600	678
37	934	1074	914	1044	886	1012
1-in.-OD tubes on 1¼-in. pitch						
8	21	21	16	16	14	16
12	48	55	45	52	40	48
15¼	81	91	76	86	68	80
21¼	177	199	166	·188	158	170
25	260	294	252	282	238	256
31	406	472	398	454	380	430
37	596	674	574	664	562	632

Thermal Strains

Thermal expansion can occur when materials, such as the metal components of a heat exchanger, are heated. For example, in a shell-and-tube heat exchanger, thermal expansion can cause an elongation of both the tube bundle and the shell as the temperature of the unit is increased. Because the tube bundle and the shell may expand by different amounts, some arrangement may be necessary to reduce thermal strains. Figure 14-6 shows a fixed-tube-sheet exchanger with no

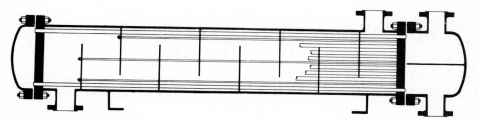

Figure 14-6 Heat exchanger with fixed tubesheets, two tube passes, and one shell pass. (*Struthers-Wells Corporation.*)

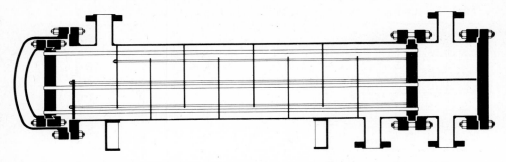

Figure 14-7 Heat exchanger with internal floating head, two tube passes, and one shell pass. (*Struthers-Wells Corporation.*)

allowance for expansion, while Figs. 14-7 and 14-8 illustrate two conventional types of exchangers with a floating head or a slip joint to relieve stresses caused by thermal expansion. Temperature stresses due to tube elongation can also be avoided by using U-shaped tubes, and some exchangers have a U-type bellows loop or ring in the shell to handle thermal elongation of the shell.

Use of the fixed-head type of exchanger should be limited to exchangers with short tubes or to cases in which the maximum temperature difference between shell and tubes is less than 50°F. In general, floating-head exchangers with removable bundles are recommended for most services.

Cleaning and Maintenance

Heat exchangers require periodic cleaning, tube replacements, or other maintenance work. The inside of straight tubes can be cleaned easily by forcing a wire brush or worm through the tubes, but cleaning of the outside of the tubes usually requires removal of the entire tube bundle from the exchanger. Consequently, many exchangers are provided with removable tube bundles, and the pitch and arrangement of the tubes are often dictated by the amount and type of cleaning that are required.

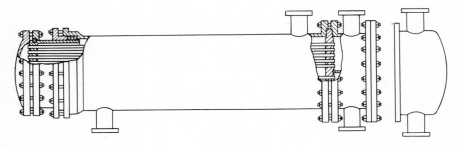

Figure 14-8 Heat exchanger with external floating head, two tube passes, and one shell pass. (*Struthers-Wells Corporation.*)

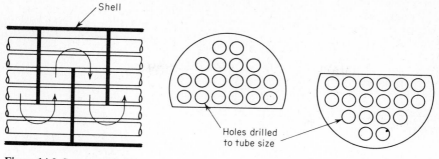

Figure 14-9 Segmental baffles.

Baffles

Although the presence of baffles in the shell side of a shell-and-tube exchanger increases the pressure drop on the shell side, the advantage of better mixing of the fluid and increased turbulence more than offsets the pressure-drop disadvantage. The distance between baffles is known as the *baffle spacing*. In general, baffle spacing is not greater than a distance equal to the diameter of the shell or less than one-fifth of the shell diameter.

The most common type of baffle used in heat exchangers is the *segmental baffle*, illustrated in Fig. 14-9. Many segmental baffles have a baffle height that is 75 percent of the inside diameter of the shell. This arrangement is designated as *25 percent cut segmental baffles*. Other types of baffles include the *disk-and-doughnut baffle* and the *orifice baffle*, shown in Figs. 14-10 and 14-11. Longitudinal baffles of the permanent type (i.e., welded to the shell) or removable type are also found in some exchangers.

Segmental and disk-and-doughnut baffles contain tube-pass holes of size close to that of the diameter of the tubes. With these two types of baffles, the

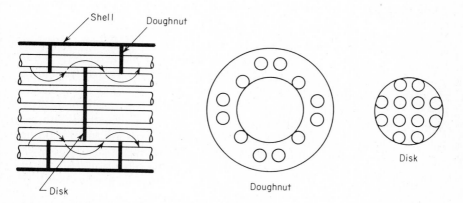

Figure 14-10 Disk-and-doughnut baffles.

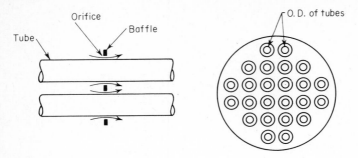

Figure 14-11 Orifice baffle.

clearance between the tubes and the edge of the holes may range from 1 percent to as high as 3 percent of the tube diameter. As a result, some fluid passes through the clearance spaces, but the major portion of the fluid should flow between the baffles in a direction perpendicular to the tubes.

Orifice baffles should be spaced reasonably close together to produce frequent increases in fluid velocity through the orifice openings between the tubes and the baffles. This type of baffle should not be used for fluids with high fouling characteristics.

Straight tie rods are used to hold baffles in place. Usually, four to six rods of $\frac{1}{8}$- to $\frac{1}{2}$-in. diameter are necessary. These are fitted to the fixed tube sheet, and short lengths of pipe sleeves are placed around the tie rods to form shoulders between adjacent baffles. The thickness of baffle plates should be at least twice the thickness of the tube walls and is ordinarily in the range of $\frac{1}{8}$ to $\frac{1}{4}$ in. The tube sheets should have a thickness at least as great as the outside diameter of the tubes. In industrial exchangers, $\frac{7}{8}$ in. is ordinarily considered as a minimum thickness for tube sheets.

Fluid Velocities and Location of Fluids

The major factors involved in determining the best location for fluids in a heat exchanger are the fouling and corrosion characteristics of the fluids, pressure drop across the unit, materials costs, and general physical characteristics of the fluids and the exchanger. When one of the fluids is highly corrosive, it should flow inside the tubes to avoid the expense of corrosion-resistant materials of construction on the shell side. Because cleaning inside tubes is easier than external cleaning, consideration should always be given to locating the fluid with the greatest fouling tendencies inside the tubes. If the other factors are equal and one fluid is under high pressure, the expense of a high-pressure shell construction can be avoided by passing the high-pressure fluid through the tubes.

The velocities of the fluids passing through the shell side and the tube side of an exchanger can have a large influence on the heat-transfer coefficients and the pressure drop. At high velocities, the beneficial effects of large film coefficients can be counterbalanced by the detrimental effects of high pressure drop. If one of the fluids is much more viscous than the other, pressure drop on the tube side may be excessive when the viscous fluid is passed through the tubes at the velocity necessary for adequate rates of heat transfer. The effects of fluid velocities and viscosities, therefore, must be considered carefully before a final decision is made concerning the best routing of the fluids.

Ineffective Surface

In the course of heating a fluid, noncondensable gases, such as absorbed air, may be evolved. If these gases are not removed, they can collect in the exchanger and form an effective blanket around some of the heat-transfer surface. Adequate provision, therefore, should be made for venting noncondensables. The heat-transfer surface can also become ineffective because of build-up of condensate when condensing vapors are involved. Consequently, drains, steam traps with bypasses, and sight glasses to indicate condensate level are often necessary auxiliaries on heat exchangers. When high pressures are used, relief valves or rupture disks may be essential for protection. Vapor blanketing can occur in boilers if the critical temperature-difference driving force is exceeded, and the design engineer must take this factor into consideration. Inadequate baffling on the shell side of an exchanger can result in poor distribution of the shell-side fluid, with a resulting ineffective use of the available surface area.

Use of Water in Heat Exchangers

Because of the abundance of water and its high heat capacity, this material is used extensively as a heat-exchange medium. At high temperatures, water exerts considerable corrosive action on steel, particularly if the water contains dissolved oxygen. Nonferrous metals, therefore, are often employed in heat exchangers when water is one of the fluids. To reduce costs, the water may be passed through the more expensive tubes, and the shell side of the exchanger can be constructed of steel. When water is used in contact with steel, a large corrosion allowance should be included in the mechanical design.

Liquid mixtures that contain solids tend to foul heat exchangers very rapidly, because the solids can settle out and bake into a cake on the hot walls. This difficulty can be reduced by passing the fluid through the exchanger at a velocity which is sufficient to keep the solids in suspension. As a standard practice, a fluid velocity of at least 3 fps should be maintained in an exchanger when the fluid is water. This minimum velocity is particularly important if the water is known to contain any suspended solids. Since it is very difficult to eliminate low-velocity pockets on the shell side of an exchanger, water that contains suspended solids should be passed through the tube side of the exchanger.

As far as possible, suspended solids should be removed from water before it enters a heat exchanger. This can be accomplished by use of settling tanks or filters. Screen filters are commonly used on water lines to remove debris, such as sticks, pebbles, or pieces of algae, but a screen filter will not remove finely dispersed solids.

Another difficulty encountered when water is employed as a heat-transfer medium is the formation of mineral scale. At temperatures higher than 120°F, the formation of scale from water of average mineral and air content tends to become excessive. Consequently, an outlet water temperature above 120°F should be avoided. In many cases, it is advisable to soften the water by a chemical treatment before using it in a heat exchanger. Hot water leaving a heat exchanger may be reused by employing a cooling tower to lower the temperature of the water. This reduces both the softening costs and the amount of water that must be purchased.

Use of Steam in Heat Exchangers

Steam has a high latent heat of condensation per unit weight and is, therefore, very effective as a heating medium. Exhaust steam is often available as a by-product from power plants. Because exhaust steam usually has a saturation temperature in the range of 215 to 230°F, the material is useful for heating only in the delivery-temperature range below about 200°F. If a material must be heated to higher temperatures, process steam is necessary. At a pressure of 200 psig, the condensing temperature of steam is 382°F. Consequently, when delivery temperatures much higher than 300°F are required, the high steam pressure necessary may eliminate steam as a useful heating medium. Under these conditions, another heating medium, such as Dowtherm, which does not require high pressures, might be chosen.

The film coefficients of heat transfer for pure condensing steam are ordinarily in the range of 1000 to 3000 Btu/(h)(ft²)(°F). The condensing-steam film, therefore, is seldom the controlling resistance in a heat exchanger, and approximate values of steam-film coefficients are usually adequate for design purposes.† If the steam is superheated, it is standard design practice to neglect the superheat and assume all the heat is delivered at the condensing temperature corresponding to the steam pressure. Similarly, subcooling of the condensate and pressure drop due to friction on the steam side of the exchanger are usually neglected. Under these conditions of assumed isothermal condensation, the true temperature-difference driving force is the same as the logarithmic-mean value.

Specifications for Heat Exchangers

The design engineer should consider both process design and mechanical design when preparing the specifications for a heat exchanger. The fabricator ordinarily has a plant that is tooled to produce standard parts at the least cost and in the

---

† For design estimates under ordinary conditions, a conservative value of 1500 Btu/(h)(ft²)(°F) is often assumed for the condensing-steam film coefficient.

shortest time. Consequently, attempts to specify every detail on an exchanger should not be made unless one is thoroughly familiar with the fabricator's equipment and methods. On the other hand, the specifications must be sufficiently complete to permit an adequate evaluation by the fabricator.

The following list presents the basic information that should be supplied to a fabricator in order to obtain a price estimate or firm quotation on a proposed heat exchanger:

<table>
<tr><td>

Process information

1. Fluids to be used
   a. Include fluid properties if they are not readily available to the fabricator
2. Flow rates or amounts of fluids
3. Entrance and exit temperatures
4. Amount of vaporization or condensation
5. Operating pressures and allowable pressure drops
6. Fouling factors
7. Rate of heat transfer

</td><td>

Mechanical information

1. Size of tubes
   a. Diameter
   b. Length
   c. Wall thickness
2. Tube layout and pitch
   a. Horizontal tubes
   b. Vertical tubes
3. Maximum and minimum temperatures and pressures
4. Necessary corrosion allowances
5. Special codes involved
6. Recommended materials of construction

</td></tr>
</table>

Some of the preceding information can be presented in the form of suggestions, with an indication of the reasons for the particular choice. This would apply, in particular, to such items as fouling factors, tube layout, codes, and materials of construction. An example of a typical specification table for heat exchangers is presented in Fig. 2-6.

HEAT-TRANSFER EQUIPMENT COSTS

The major factors that can influence the cost for heat-transfer equipment are indicated in the following list:

1. Heat-transfer area
2. Tube diameter and gauge
3. Tube length
4. Pressure
5. Materials of construction for tubes and shell
6. Degree and type of baffling
7. Supports, auxiliaries, and installation
8. Special features, such as floating heads; removable bundles; multipass, finned surfaces; and U bends

Certain manufacturers specialize in particular types of exchangers and can, therefore, give lower quotations on their specialities than other manufacturers. As a result, price quotations should be obtained from several manufacturers before a firm cost is listed.

Figure 14-12 presents average cost data for heat exchanger tubing constructed of steel. The average purchased cost of complete heat exchangers with materials of construction is given in Figs. 14-13 through 14-18, while the cost for coolers and barometric condensers is given in Figs. 14-19 through 14-21. Because of the many possible variations in heat exchangers, these data must be regarded as approximate and should be used only for preliminary rough estimates.

Relative Costs

The relative effect of tube diameter, tube length, and operating pressure on the purchased cost of shell-and-tube heat exchangers is shown in Figs. 14-22 through 14-24. If two floating heads are used in a single shell in place of one floating head, the exchanger cost will be increased by approximately 30 percent,

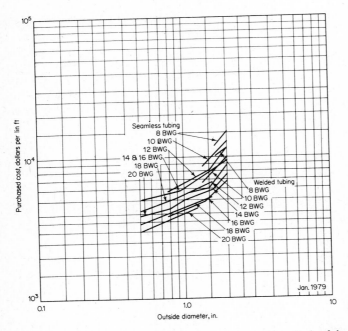

Figure 14-12 Welded and seamless heat-exchanger tubing. Basis of the cost data includes: quantity, 40,000 lb or ft; lengths, cut lengths within range of 10–24 ft; specifications, seamless, ASTM-A-179 and welded ASTM-A-214; wall thickness, minimum wall. Material of construction is low-carbon (less than 0.18 percent) steel.

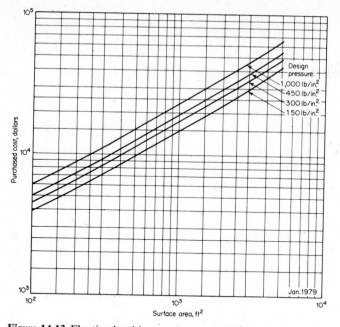

Figure 14-13 Floating-head heat exchangers with $\frac{3}{4}$-in. OD × 1-in. square pitch and 16-ft bundles of carbon-steel construction.

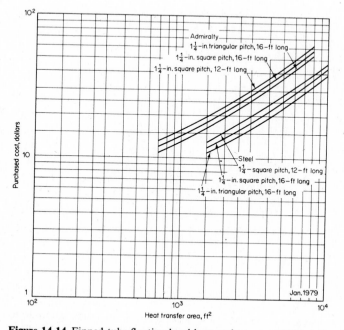

Figure 14-14 Finned-tube floating-head heat exchangers at 150 psi. Cost is for 1-in. OD Turfin tubes.

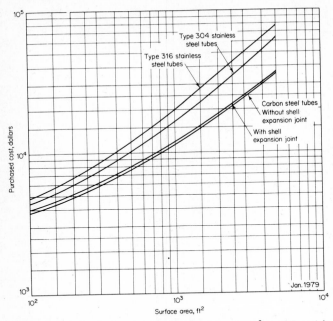

Figure 14-15 Fixed-tube-sheet heat exchangers with $\frac{3}{4}$-in. OD × 1-in. square pitch and 16- or 20-ft bundles and carbon-steel shell operating at 150 psi.

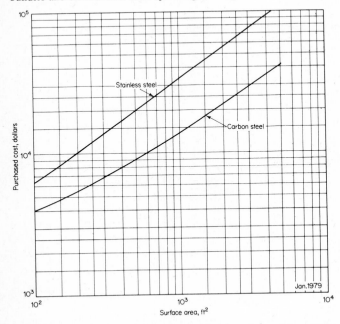

Figure 14-16 U-tube heat exchangers with 1-in. tubes × 1-in. square pitch and 16-ft bundles operating at 150 psi.

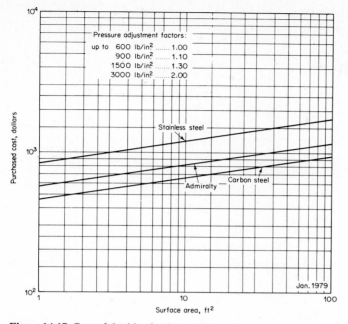

Figure 14-17 Cost of double-pipe heat exchangers.

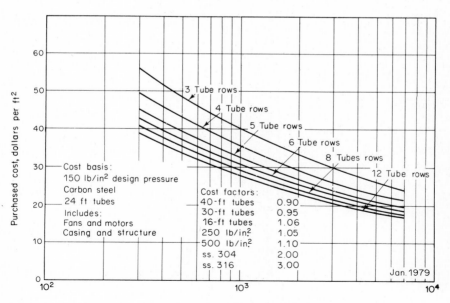

Bare tube surface area, ft.²

Figure 14-18 Cost of air-cooled heat exchangers.

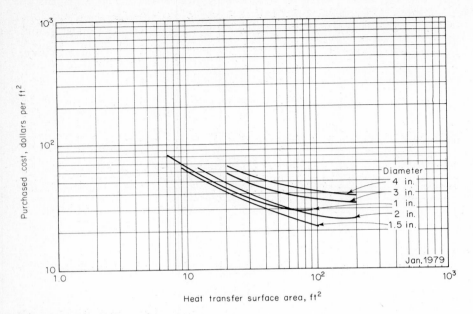

Figure 14-19 Cost of cascade coolers.

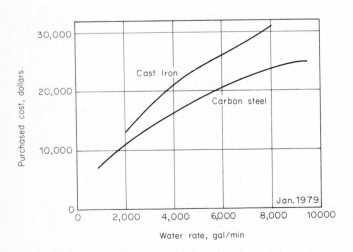

Figure 14-20 Cost of multijet spray-type barometric condensers. Basic construction consists of a shell, water nozzle, case and plate, and spray-type nozzles.

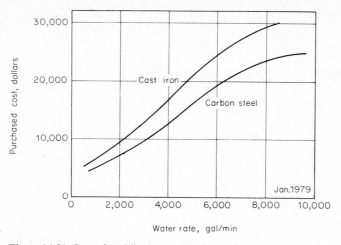

Figure 14-21 Cost of multijet barometric condensers.

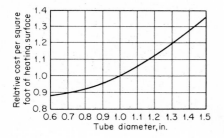

Figure 14-22 Effect of tube diameter on cost of conventional shell-and-tube heat exchangers.

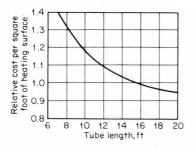

Figure 14-23 Effect of tube length on cost of conventional shell-and-tube heat exchangers.

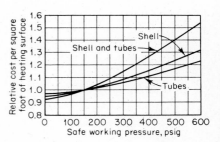

Figure 14-24 Effect of operating pressure on cost of conventional shell-and-tube heat exchangers.

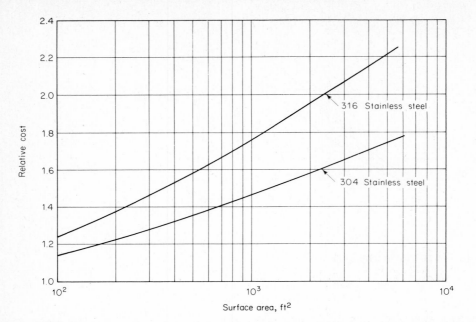

Figure 14-25 Cost of heat exchangers with stainless-steel tubes relative to all-carbon-steel construction.

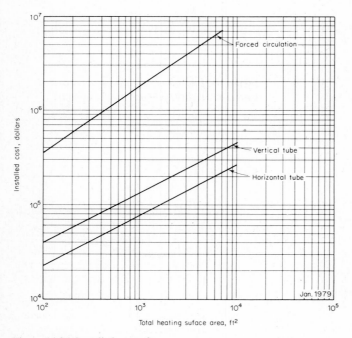

Figure 14-26 Installed cost of evaporators.

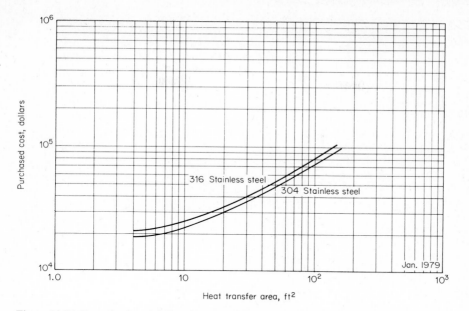

Figure 14-27 Cost of agitated falling-film evaporators, complete with motor and drive.

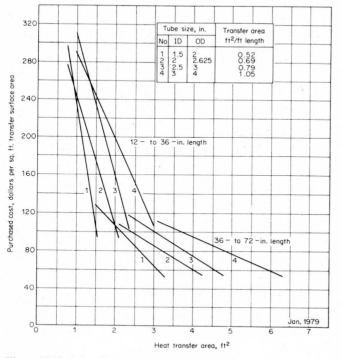

Figure 14-28 Cost of bayonet heaters.

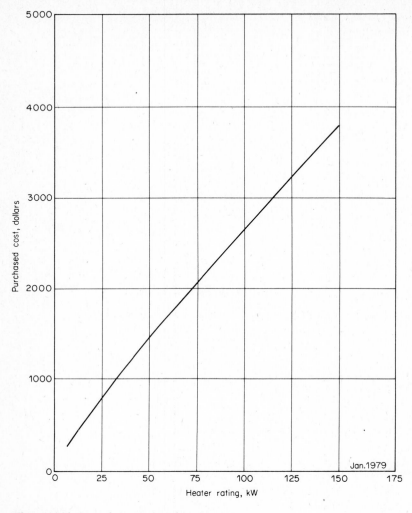

Figure 14-29 Cost of electric immersion heaters.

and a decrease in cost of about 10 to 15 percent occurs if fixed tube sheets or U tubes are used instead of one floating head. Table 8 lists relative costs for tube-and-shell heat exchangers on the basis of the material of construction used for the tubes and the exchanger, and Fig. 14-25 shows the costs of heat exchangers with two types of stainless-steel tubes relative to all-carbon-steel construction.

Costs for forced circulation evaporators, long tube evaporators, and agitated falling-film evaporators are presented in Figs. 14-26 and 14-27. Information on purchased cost of heaters is given in Figs. 14-28 and 14-29, while Figs. 14-30 and 14-31 present costs for process furnaces and direct-fired heaters.

Table 8 Relative costs of heat-exchanger tubing and heat exchangers with 1500 ft² of surface†

Material	Tubing	Heat exchanger
Zirconium—seamless	25.1	7.7
Hastelloy—C-276—welded	18.2	5.9
Zirconium—20 BWG—seamless	15.8	5.2
Inconel—625 welded	15.1	5.0
Carpenter—20-CB3—welded	8.6	3.3
Incoloy—825—welded	7.6	3.0
Monel—400—seamless	7.5	3.0
Titanium—welded	6.8	2.8
E-Brite—26-1—welded	5.2	2.4
Titanium—20 BWG—welded	3.6	1.9
316L stainless steel—welded	3.2	1.8
Cu/Ni—70/30—seamless	2.9	1.8
Cu/Ni—90/30—welded	2.4	1.6
304L stainless steel—welded	2.2	1.6
Carbon steel	1.0	1.0

Basis: Tubing is 1 in. OD × 16 BWG, except as noted. Heat exchangers are TEMA-type BEM with $\frac{3}{4}$ in. OD × 16 BWG × 20 ft welded tubing and mild-carbon steel shells.

† Estimates will be conservative for smaller exchangers and too low for larger exchangers.

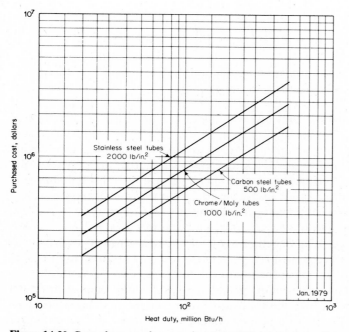

Figure 14-30 Cost of process furnaces, box type with horizontal radiant tubes.

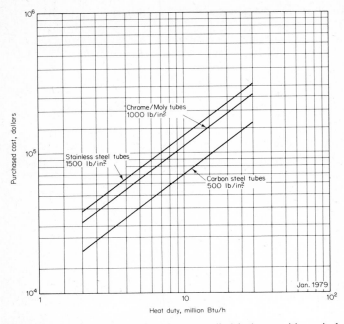

Figure 14-31 Cost of direct-fired heaters, cylindrical type with vertical tubes.

OPTIMUM DESIGN OF HEAT EXCHANGERS

Two types of quantitative problems are commonly encountered by the design engineer when dealing with heat-transfer calculations. The first type is illustrated by Examples 3 and 4 in this chapter. In these examples, all of the design variables are set, and the calculations involve only a determination of the indicated nonvariant quantities. The conditions specified for Examples 3 and 4 give low pressure drops through the exchangers. Alternative conditions could be specified which would give higher pressure drops and less heat-transfer area. By choosing various conditions, the engineer could ultimately arrive at a final design that would give the least total cost for fixed charges and operation. Thus, the second type of quantitative problem involves conditions in which at least one variable is not fixed, and the goal is to obtain an optimum economic design.

In general, increased fluid velocities result in larger heat-transfer coefficients and, consequently, less heat-transfer area and exchanger cost for a given rate of heat transfer. On the other hand, the increased fluid velocities cause an increase in pressure drop and greater pumping costs. The optimum economic design occurs at the conditions where the total cost is a minimum. The basic problem, therefore, is to minimize the sum of the variable annual costs for the exchanger and its operation.

GENERAL CASE†

The design of most heat exchangers involves initial conditions in which the following variables are known:

1. Process-fluid rate of flow
2. Change in temperature of process fluid
3. Inlet temperature of utility fluid (for cooling or heating)

With this information, the engineer must prepare a design for the optimum exchanger that will meet the required process conditions. Ordinarily, the following results must be determined:

1. Heat transfer area
2. Exit temperature and flow rate of utility fluid
3. Number, length, diameter, and arrangement of tubes
4. Tube-side and shell-side pressure drops

The variable annual costs of importance are the fixed charges on the equipment, the cost for the utility fluid, and the power cost for pumping the fluids through the exchanger. The total annual cost for optimization, therefore, can be represented by the following equation:

$$C_T = A_o K_F C_{A_o} + w_u H_y C_u + A_o E_i H_y C_i + A_o E_o H_y C_o \qquad (35)$$

where C_T = total annual variable cost for heat exchanger and its operation, $/year

C_{A_o} = installed cost of heat exchanger per unit of outside-tube heat-transfer area, $/ft^2

C_u = cost of utility fluid, $/lb

C_i = cost for supplying 1 ft·lbf to pump fluid flowing through inside of tubes, $/ft·lbf

C_o = cost for supplying 1 ft·lbf to pump fluid flowing through shell side of unit, $/ft·lbf

A_o = area of heat transfer, ft^2; subscript o designates outside of tubes

K_F = annual fixed charges including maintenance, expressed as a fraction of initial cost for completely installed unit, dimensionless

w_u = flow rate of utility fluid, lb/h

H_y = hours of operation per year, h/year

E_i = power loss inside tubes per unit of outside tube area, ft·lbf/(h)(ft^2)

E_o = power loss outside tubes per unit of outside tube area, ft·lbf/(h)(ft^2)

---

† The analysis presented in this section is similar to that of M. Cichelli and M. Brinn, *Chem. Eng.*, **63**(5):196 (1956). See also Chap. 10 and H. T. Bates, S. S. Patel, and K. G. Shaw, Design Optimization of 1-2 Pass and 2-4 Pass Shell and Tube Heat Exchangers, *Paper 47a*, 61st AIChE Nat. Mtg., Houston, Tex., Feb. 22, 1967.

An optimum design could be developed from Eq. (35) by the laborious procedure of direct trial and error, taking all possible variables into consideration; however, the procedure can be simplified considerably by using the method of partial derivatives.

Under ordinary circumstances, the effect of tube diameter on total cost at the optimum operating conditions is not great, and a reasonable choice of tube diameter, wall thickness, and tube spacing can be specified at the start of the design. Similarly, the number of tube passes is usually specified. If a change in phase of one of the fluids occurs (for example, if the utility fluid is condensing steam), solution of Eq. (35) for optimum conditions can often be simplified. For the case of no change in phase, the solution can become complex, because the velocities and resulting power costs and heat-transfer coefficients can be varied independently over a wide range of values. In the following analysis, the general case of steady-state heat transfer in shell-and-tube exchangers with no change in fluid phase is considered, and a specified tube diameter, wall thickness, number of passes, and arrangement of baffles and tubes are assumed. Simplifications are indicated for the common conditions of turbulent flow.

Choice of Primary Independent Variables

The heat-transfer area A_o can be related to the flow rates and the temperature changes by an overall heat balance and the rate equation. If heat losses are assumed as negligible and q is designated as the rate of heat transfer to the utility fluid,

$$q = w_u c_{p_u}(t_2 - t_1) = w'c'_p(t'_1 - t'_2) = U_o A_o \Delta t_m \tag{36}†$$

where the primes refer to the process fluid. Subscript 1 refers to the entering temperature, and subscript 2 refers to the leaving temperature.

From Eq. (36),

$$w_u = \frac{q}{c_{p_u}(\Delta t_1 - \Delta t_2 + t'_1 - t'_2)} \tag{37}$$

where $\Delta t_1 = t'_2 - t_1$ and $\Delta t_2 = t'_1 - t_2$. Since q, c_{p_u}, Δt_1, t'_1, and t'_2 are constant, w_u is a function only of the independent variable Δt_2 .‡

The area A_o is known if U_o and Δt_m are fixed. The overall coefficient U_o is known if the inside and outside film coefficients, h_i and h_o, are fixed, and, for a

---

† Under these conditions, q is negative if the exchanger is used for heating the process fluid, and q is positive if the process fluid is cooled.

‡ The heat capacity, viscosity, and density of the fluids are taken as constant at an assumed average temperature. This assumption is adequate for most cases. If greater precision is necessary, the results of the initial calculation can be used to obtain new values for the physical constants, and the corresponding optimum conditions can be determined.

given number of tube passes, Δt_m varies only with changes in Δt_2. Therefore, A_o is a function of h_i, h_o, and Δt_2, as shown by the following equations:

$$\frac{\Delta t_m}{q} = \frac{1}{U_o A_o} = \frac{1}{A_o}\left(\frac{D_o}{D_i h_i} + \frac{1}{h_o} + R_{dw}\right) \tag{38}$$

$$\frac{F_T(\Delta t_2 - \Delta t_1)}{q \ln (\Delta t_2/\Delta t_1)} = \frac{1}{U_o A_o} = \frac{1}{A_o}\left(\frac{D_o}{D_i h_i} + \frac{1}{h_o} + R_{dw}\right) \tag{39}$$

where U_o = overall coefficient of heat transfer based on outside tube area A_o

F_T = correction factor on logarithmic-mean temperature difference for counterflow to account for number of passes; $F_T = 1$ if unit is counterflow and single-pass on shell and tube sides (see Fig. 14-2)

R_{dw} = combined resistance of tube wall and scaling or dirt factors,

$$\frac{D_o x_w}{k_w D_{w\log mean}} + \frac{D_o}{D_i h_{d_i}} + \frac{1}{h_{d_o}}$$

For a set diameter and tube arrangement, Eqs. (26) and (29) show that h_i is fixed by the mass velocity G inside the tubes and h_o is fixed by the mass velocity G_s outside the tubes. Similarly, since the heat-transfer area A_o, mass velocities, and flow rates determine the length of the tubes L, Eqs. (30) and (31) show that E_i and E_o are functions of A_o, flow rates, and, respectively, G and G_s,[†] Thus, E_i and E_o are functions of h_i, h_o, and Δt_2.

The variables in Eq. (35) are A_o, w_u, E_i, and E_o, and their values are set if h_i, h_o, and Δt_2 are known. Partial differentiation therefore, of Eq. (35) with respect to the three independent variables h_i, h_o, and Δt_2 would lead to a solution for the optimum conditions. However, the resulting equations are cumbersome, and the procedure is simplified by retaining the following four variables:

1. Tube-inside coefficient of heat transfer, h_i
2. Tube-outside coefficient of heat transfer, h_o
3. Temperature-difference driving force for counterflow based on temperature of utility fluid at exit from exchanger, $\Delta t_2 = t_1' - t_2$
4. Outside tube area of heat transfer, A_o

Optimization Procedure

The first step in the optimization procedure is to express Eq. (35) in terms of the fundamental variables. The following relationships for power loss inside tubes and power loss outside tubes are developed in Table 9 for conditions of turbulent flow and shell-side fluid flowing in a direction normal to the tubes:

$$E_i = \psi_i h_i^{3.5} \tag{40}$$

$$E_o = \psi_o h_o^{4.75} \tag{41}$$

† See Eqs. (B) and (H) in Table 9.

Table 9 Development of equations for optimization of heat-exchanger design

Conditions: Turbulent flow in shell-and-tube heat exchangers with cross flow on the shell side.

Power Loss inside Tubes

(A) $$-\Delta p_i = \frac{B_i 2 \mathfrak{f}_i G^2 L n_p}{g_c \rho_i D_i \phi_i} = \frac{2 \mathfrak{f}_i G_i^2 L n_p}{g_c \rho_i D_i \phi_i} + (F_e + F_c + F_r) n_p \rho_i \qquad (30)$$

$(A1)$ $$\phi_i = 1.02 \left(\frac{\mu_i}{\mu_{w_i}} \right)^{0.14}$$

$(A2)$ $$S_i = \frac{\pi D_i^2 N_t}{4 n_p}$$

$(A3)$ $$A_o = N_t \pi D_o L$$

where S_i = cross-sectional flow area inside tubes per pass

N_t = total number of tubes in exchanger = (number of tubes per pass) n_p

(B) $$E_i = \frac{-\Delta p_i w_i}{\rho_i A_o} = \frac{-\Delta p_i G S_i}{\rho_i A_o} = \frac{-\Delta p_i G D_i^2}{4 \rho_i D_o L n_p}$$

For turbulent flow in tubes,

(C) $$\mathfrak{f}_i = \frac{0.046}{(N_{Re})^{0.2}} = \frac{0.046}{(D_i G / \mu_i)^{0.2}} \qquad \text{[Chap. 13, Eq. (7)]}$$

Combining Eqs. (A), (B), and (C),

(D) $$E_i = \frac{0.023 B_i \mu_i^{0.2} D_i^{0.8} G^{2.8}}{g_c D_o \rho_i^2 \phi_i}$$

(E) $$\frac{h_i D_i}{k_i} = 0.023 \left(\frac{D_i G}{\mu_i} \right)^{0.8} \left(\frac{c_{p_i} \mu_i}{k_i} \right)^{1/3} \left(\frac{\mu_i}{\mu_{w_i}} \right)^{0.14} \qquad (26)$$

$(E1)$ $$G = \left[\frac{h_i D_i^{0.2} \mu_i^{0.8}}{0.023 k_i} \left(\frac{k_i}{c_{p_i} \mu_i} \right)^{1/3} \left(\frac{\mu_{w_i}}{\mu_i} \right)^{0.14} \right]^{1.25}$$

Combining Eqs. $(A1)$, (D), and $(E1)$,

(F) $$E_i = h_i^{3.5} B_i \frac{D_i^{1.5} \mu_i^{1.83} (\mu_{w_i} / \mu_i)^{0.63}}{(1.02)(0.023)^{2.5} g_c D_o \rho_i^2 k_i^{2.33} c_{p_i}^{1.17}}$$

$$\frac{1}{(1.02)(0.023)^{2.5}} = 12,200$$

Power Loss outside Tubes

(G) $$-\Delta p_o = \frac{B_o 2 \mathfrak{f}' N_r G_s^2}{g_c \rho_o} \qquad (31)$$

(H) $$E_o = \frac{-\Delta p_o w_o}{\rho_o A_o} = \frac{-\Delta p_o G_s S_o}{\rho_o A_o} = \frac{-\Delta p_o G_s S_o}{\rho_o N_t \pi D_o L}$$

(I) $$S_o = \frac{N_c D_c L}{n_b}$$

Table 9 Development of equations for optimization of heat-exchanger design (*Continued*)

where S_o = shell-side free-flow area across shell axis

N_c = number of clearances between tubes for flow of shell-side fluid across shell axis

D_c = clearance between tubes to give smallest free area across shell axis

n_b = number of baffle spaces = number of baffles + 1

For turbulent flow across tubes,

(*J*) $$f' = b_o \left(\frac{D_o G_s}{\mu_{f_o}}\right)^{-0.15} \tag{32}$$

See Eqs. (33) and (34) for values of b_o in terms of tube size and arrangement. Combining Eqs. (*G*), (*H*), (*I*), and (*J*),

(*K*) $$E_o = \frac{2 B_o b_o \mu_{f_o}^{0.15} D_c G_s^{2.85} N_r N_c}{\pi g_c D_o^{1.15} \rho_o{}^2 n_b N_t}$$

(*L*) $$\frac{h_o D_o}{k_{f_o}} = \frac{a_o}{F_s} \left(\frac{D_o G_s}{\mu_{f_o}}\right)^{0.6} \left(\frac{c_{p_o} \mu_o}{k_o}\right)_f^{1/3} \tag{29}$$

$a_o = 0.33$ for staggered tubes and 0.26 for in-line tubes

(*L1*) $$G_s = \left[\frac{h_o D_o^{0.4} \mu_{f_o}^{0.6} F_s}{k_{f_o} a_o} \left(\frac{k_o}{c_{p_o} \mu_o}\right)_f^{1/3}\right]^{1.67}$$

Combining Eqs. (*K*) and (*L1*),

(*M*) $$E_o = h_o^{4.75} \frac{B_o}{n_b} \frac{N_r N_c}{N_t} \frac{2 b_o D_c D_o^{0.75} F_s^{4.75} \mu_{f_o}^{1.42}}{\pi a_o^{4.75} g_c \rho_o{}^2 k_{f_o}^{3.17} c_{p_{f_o}}^{1.58}}$$

where

$$\psi_i = B_i \left[\frac{12,200 D_i^{1.5} \mu_i^{1.83} (\mu_{w_i}/\mu_i)^{0.63}}{g_c D_o \rho_i^2 k_i^{2.33} c_{p_i}^{1.17}}\right] \tag{40a}$$

$$\psi_o = \frac{B_o}{n_b} \frac{N_r N_c}{N_t} \left(\frac{2 b_o D_c D_o^{0.75} F_s^{4.75} \mu_{f_o}^{1.42}}{\pi a_o^{4.75} g_c \rho_o^2 k_{f_o}^{3.17} c_{p_{f_o}}^{1.58}}\right) \tag{41a}$$

All the terms in the brackets are set by the design conditions or can be approximated with good accuracy on the first trial. The values of B_i and B_o/n_b are not completely independent of the film coefficients, but they do not vary enough to be critical. As a first approximation, B_i is usually close to 1, and B_o is often taken to be equal to or slightly greater than the number of baffle passes n_b. The value of the safety factor F_s depends on the amount of bypassing and is often taken as 1.6 for design estimates.

The ratio $N_r N_c/N_t$ depends on the tube layout and baffle arrangement. For rectangular tube bundles and no baffles, this ratio is equal to 1.0. For other tube layouts and segmental baffles, the ratio is usually in the range of 0.6 to 1.2.

Equation (35) can now be expressed in terms of the primary variables Δt_2, h_i, h_o, and A_o:

$$C_T = A_o K_F C_{A_o} + \frac{q H_y C_u}{c_{p_u}(\Delta t_1 - \Delta t_2 + t'_1 - t'_2)}$$
$$+ A_o \psi_i h_i^{3.5} H_y C_i + A_o \psi_o h_o^{4.75} H_y C_o \tag{42}$$

Only three of the four variables in Eq. (42) are independent. Under these conditions, optimization can be accomplished by use of the Lagrange multiplier method.† The necessary relationship for applying the constant lagrangian multiplier λ is given by Eq. (43):

$$\lambda \left[\frac{F_T(\Delta t_2 - \Delta t_1)}{q \ln (\Delta t_2 / \Delta t_1)} - \frac{1}{A_o}\left(\frac{D_o}{D_i h_i} + \frac{1}{h_o} + R_{dw} \right) \right] = 0 \tag{43}$$

Equations (42) and (43) can now be added to give the following equation for optimization by partial differentiation with respect to each of the four primary variables:

$$C_T = A_o K_F C_{A_o} + \frac{q H_y C_u}{c_{p_u}(\Delta t_1 - \Delta t_2 + t'_1 - t'_2)}$$
$$+ A_o \psi_i h_i^{3.5} H_y C_i + A_o \psi_o h_o^{4.75} H_y C_o$$
$$+ \lambda \left[\frac{F_T(\Delta t_2 - \Delta t_1)}{q \ln (\Delta t_2 / \Delta t_1)} - \frac{1}{A_o}\left(\frac{D_o}{D_i h_i} + \frac{1}{h_o} + R_{dw} \right) \right] \tag{44}$$

Optimum value of h_o. The following relationship between the optimum values of h_i and h_o is obtained by taking the partial derivative of Eq. (44) with respect to h_i and then with respect to h_o, setting the results equal to zero, and eliminating A_o and λ:

$$\frac{\partial C_T}{\partial h_i} = 3.5 A_{o,\text{opt}} \psi_i h_{i,\text{opt}}^{2.5} H_y C_i + \frac{\lambda D_o}{A_{o,\text{opt}} D_i h_{i,\text{opt}}^2} = 0 \tag{45a}$$

$$\frac{\partial C_T}{\partial h_o} = 4.75 A_{o,\text{opt}} \psi_o h_{o,\text{opt}}^{3.75} H_y C_o + \frac{\lambda}{A_{o,\text{opt}} h_{o,\text{opt}}^2} = 0 \tag{45b}$$

$$h_{o,\text{opt}} = \left(\frac{0.74 \psi_i C_i D_i}{\psi_o C_o D_o} \right)^{0.17} h_{i,\text{opt}}^{0.78} \tag{45c}$$

Optimum value of h_i. The optimum value of h_i can be determined by setting the partial derivatives of Eq. (44) with respect to A_o and with respect to h_i equal to zero and eliminating A_o and λ. This gives a result with $h_{i,\text{opt}}$ and $h_{o,\text{opt}}$ as the only

† See Chap. 10.

unknowns, and simultaneous solution with Eq. (45c) yields Eq. (46b), where $h_{i,\text{opt}}$ is the only unknown.

$$\frac{\partial C_T}{\partial A_o} = K_F C_{A_o} + \psi_i h_{i,\text{opt}}^{3.5} H_y C_i + \psi_o h_{o,\text{opt}}^{4.75} H_y C_o$$

$$+ \frac{\lambda}{A_{o,\text{opt}}^2} \left(\frac{D_o}{D_i h_{i,\text{opt}}} + \frac{1}{h_{o,\text{opt}}} + R_{dw} \right) = 0 \quad (46a)$$

$$h_{i,\text{opt}}^{3.5} \left[2.5 \psi_i H_y C_i + \frac{3.5 \psi_i H_y C_i D_i R_{dw} h_{i,\text{opt}}}{D_o} \right.$$

$$\left. + 2.9 \left(\frac{\psi_i C_i D_i}{D_o} \right)^{0.83} (\psi_o C_o)^{0.17} H_y h_{i,\text{opt}}^{0.22} \right] = K_F C_{A_o} \quad (46b)$$

Optimum value of U_o. A trial-and-error or graphical method can be used to obtain $h_{i,\text{opt}}$ from Eq. (46b). Then, by Eqs. (38) and (45c), the value of $U_{o,\text{opt}}$ can be determined as

$$U_{o,\text{opt}} = \left(\frac{D_o}{D_i h_{i,\text{opt}}} + \frac{1}{h_{o,\text{opt}}} + R_{dw} \right)^{-1} \quad (47)$$

Optimum value of Δt_2. The value of $U_{o,\text{opt}}$ is now known, and $\Delta t_{2,\text{opt}}$ can be determimed by setting the partial derivatives of Eq. (44) with respect to Δt_2 and with respect to A_o equal to zero and eliminating λ. The result can be combined with Eqs. (39) to (41) to give

$$\frac{F_T U_{o,\text{opt}} H_y C_u}{c_{p_u}(K_F C_{A_o} + E_{i,\text{opt}} H_y C_i + E_{o,\text{opt}} H_y C_o)}$$

$$= \left(1 + \frac{t_1' - t_2'}{\Delta t_1 - \Delta t_{2,\text{opt}}} \right)^2 \left(\ln \frac{\Delta t_{2,\text{opt}}}{\Delta t_1} - 1 + \frac{\Delta t_1}{\Delta t_{2,\text{opt}}} \right) \quad (48)$$

Equation (48) can be solved for $\Delta t_{2,\text{opt}}$ by trial and error or by Fig. 14-32 which is a plot of Eq. (48).

Optimum value of A_o. Since $\Delta t_{2,\text{opt}}$ and, therefore, $\Delta t_{m,\text{opt}}$ are now known, $A_{o,\text{opt}}$ can be determined directly from Eq. (38).

Optimum value of G and G_s. Equations $(E1)$ and $(L1)$ in Table 9 give G_{opt} and $G_{s,\text{opt}}$, respectively, in terms of $h_{i,\text{opt}}$ and $h_{o,\text{opt}}$.

Optimum value of w_u. The flow rate of the utility fluid (w_u) is set by the value of Δt_2. Therefore, when $\Delta t_{2,\text{opt}}$ is known, $w_{u,\text{opt}}$ can be calculated from Eq. (37).

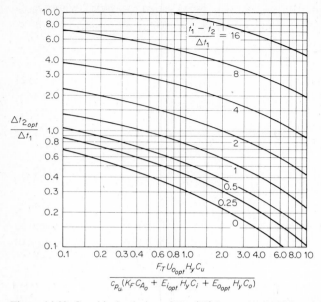

Figure 14-32 Graphical solution of Eq. (48) for evaluation of optimum Δt_2 in heat exchangers.

Optimum values of S_i and N_t. The optimum flow area inside the tubes per pass can be calculated from the following equation:

$$S_{i,\text{opt}} = \frac{w_i}{G_{\text{opt}}} \tag{49}$$

The optimum total number of tubes in the exchanger is

$$N_{t,\text{opt}} = \frac{4 n_p S_{i,\text{opt}}}{\pi D_i^2} \tag{50}$$

Optimum value of L. The optimum length per tube is set by the optimum heat-transfer area and the total number of tubes. Thus, for a given tube diameter

$$L_{\text{opt}} = \frac{A_{o,\text{opt}}}{\pi D_o N_{t,\text{opt}}} \tag{51}$$

Optimum values of S_o, N_c, and n_b. The following equation gives the optimum shell-side free-flow area across the shell axis:

$$S_{o,\text{opt}} = \frac{w_o}{G_{s,\text{opt}}} \tag{52}$$

The number of clearances N_c for flow between tubes across the shell axis is determined by the number of tubes in the shell, the pitch of the tubes, and the arrangement of the tubes. For the common case of a cylindrical shell and transverse clearances giving the minimum free area, the following equations can be used to obtain an approximation of $N_{c,opt}$:[†][‡]

With square pitch and N_t greater than 25

$$N_{c,opt} = 1.37(N_{t,opt})^{0.475} \tag{53}$$

With equilateral triangular pitch and N_t greater than 20

$$N_{c,opt} = 0.94 + \left(\frac{N_{t,opt} - 3.7}{0.907}\right)^{1/2} \tag{54}$$

The optimum number of baffle spaces can be estimated by Eq. (I) in Table 9 as follows:

$$n_{b,opt} = \frac{N_{c,opt} D_c L_{opt}}{S_{o,opt}} \tag{55}$$

SUMMARY OF PROCEDURE FOR GENERAL CASE OF OPTIMUM DESIGN

In the preceding analysis, consideration has been given to the general case in heat-exchanger design in which the following conditions apply:

1. The flow rate and necessary temperature change of the process fluid are known.
2. The inlet temperature of the utility fluid is known.
3. The exchanger is a shell-and-tube type with crossflow baffling, and flow is in the turbulent range on both the tube side and the shell side.
4. No partial phase changes occur.
5. Necessary safety factors are known.

The following information may be specified for the design or can be assumed as a reasonable approximation:

1. Tube diameter, wall thickness, pitch, and arrangement
2. Number of tube passes
3. Heat-transfer resistance caused by tube walls, dirt, and scale

The following information must be assumed for the first trial and then checked when the optimum conditions are obtained (usually, no more than two

---

[†] W. H. McAdams, "Heat Transmission," 3d ed., p. 434, McGraw-Hill Book Company, New York, 1954.

[‡] Equations (53) and (54) are based on the assumption that shell inside diameter/tube pitch = number of tubes in a row across the shell axis = number of clearances across the shell axis. No allowance is made for decrease in available tube space when more than one tube pass is used. The equations are general and are not necessarily limited to optimum conditions.

trials are necessary, and an experienced engineer can often make adequate assumptions on the first trial):

1. Average bulk and film temperatures
2. Values for B_i (usually 1.0), B_o/n_b (usually 1.0), and $N_r N_c/N_t$ (usually 1.0)

The calculation procedure is as follows:

1. Determine $h_{i,\text{opt}}$ from Eq. (46b)
2. Determine $h_{o,\text{opt}}$ from Eq. (45c)
3. Determine $U_{o,\text{opt}}$ from Eq. (47)
4. Determine $\Delta t_{2,\text{opt}}$ from Eq. (48) or Fig. 14-32
5. Determine $A_{o,\text{opt}}$ from Eq. (39)
6. Determine G_{opt} and $G_{s,\text{opt}}$ from Eqs. (E1) and (L1) in Table 9
7. Determine $w_{u,\text{opt}}$ from Eq. (37)
8. Determine $S_{i,\text{opt}}$ and $N_{t,\text{opt}}$ from Eqs. (49) and (50)
9. Determine L_{opt} from Eq. (51)
10. Determine $S_{o,\text{opt}}$, $N_{c,\text{opt}}$, and $n_{b,\text{opt}}$ from Eqs. (52) through (55)
11. Check assumptions; if any are invalid, make new reasonable assumptions and repeat procedure

Example 5. Development of the optimum design for a shell-and-tube heat exchanger A gas under pressure with properties equivalent to air must be cooled from 150 to 100°F. Cooling water is available at a temperature of 70°F. Use of a shell-and-tube floating-head heat exchanger with cooling water as the utility fluid has been proposed. On the basis of the following data and specifications, determine the tube length, number of tubes, and installed cost for the optimum exchanger which will handle 20,000 lb of the gas per hour.

1. Exchanger specifications
    a. Steel shell-and-tube exchanger with cross-flow baffling.
    b. Cooling water passes through shell side of unit.
    c. One tube pass and countercurrent flow.
    d. Tube OD = 1.0 in. Tube ID = 0.782 in.
    e. $\frac{15}{16}$-in. triangular pitch. Tubes are staggered.
2. Costs
    a. Data presented in Fig. 14-13 are applicable.
    b. Installation cost equals 15 percent of purchased cost.
    c. Annual fixed charges including maintenance equal 20 percent of installed cost.
    d. Cost for cooling water (not including pumping cost) is $0.009 per 1000 lb.

e. Cost for energy supplied to force the cooling water and the gas through the exchanger (including effect of pump-and-motor efficiency and cost) is $0.04 per kWh.
3. General
   a. Average absolute pressure of gas in exchanger is 10 atm.
   b. Unit operates 7000 h/year.
   c. In the friction relations, B_i can be taken as 1.2, and B_o can be taken as equal to the number of tube crosses, assuming the optimum number of baffles can be installed in the unit.
   d. The safety factor F_s for the outside film coefficient is 1.3.
   e. Fouling coefficient for cooling water is 1500 Btu/(h)(ft^2)(°F). Fouling coefficient for gas is 2000 Btu/(h)(ft^2)(°F).
   f. At the optimum conditions, flow on tube side and shell side is turbulent.
   g. The factor $N_r N_c / N_t$ can be taken as 1.0.

SOLUTION

Assumptions:

Exit temperature of cooling water $= t_2 = 110°F$
Average Δt over cooling-water film $= 10$ percent of total Δt
Average Δt over air film $= 80$ percent of total Δt
Purchased cost per square foot of outside heat-transfer area $= \$34.00$
Temperatures and physical properties:
$t'_1 = 150°F$, $t'_2 = 100°F$, $t_1 = 70°F$, $t_2 = 110°F$
$\Delta t_1 = t'_2 - t_1 = 30°F$
Average bulk water temperature $= 90°F$
Average bulk gas temperature $= 125°F$
Average water-film temperature $= 90 + (0.1/2)(35) = 91.8°F$
Inside wall temperature $= 125 - (0.8)(35) = 97°F$
From Appendix: For water at 91.8°F, $k = 0.359$ Btu/(h)(ft^2)(°F/ft), $\mu = (0.748)(2.42) = 1.81$ lb/(h)(ft), $c_p = 1$ Btu/(lb)(°F). For water at 90°F, $\rho = 62.1$ lb/ft^3. For air at 125°F, $k = 0.0162$ Btu/(h)(ft^2)(°F/ft), $\mu = (0.019) \times (2.42) = 0.046$ lb/(h)(ft), $c_p = 0.25$ Btu/(lb)(°F), $\rho = (29)(492)(10)/(359) \times (460 + 125) = 0.68$ lb/ft^3. For air at 97°F, $\mu = (0.0185)(2.42) = 0.045$ lb/(h)(ft).
Determination of optimum U_o. From Eq. (40a)

$$\psi_i = B_i \frac{12{,}200 D_i^{1.5} \mu_i^{1.83} (\mu_{w_i}/\mu_i)^{0.63}}{g_c D_o \rho_i^2 k_i^{2.33} c_{pi}^{1.17}}$$

$$\psi_i = \frac{(1.2)(12{,}200)(0.782/12)^{1.5}(0.046)^{1.83}(0.045/0.046)^{0.63}}{(32.17)(3600)^2 (\frac{1}{12})(0.68)^2 (0.0162)^{2.33}(0.25)^{1.17}}$$

$$\psi_i = 4.1 \times 10^{-3}$$

From Eqs. (33) and (41a),

$$b_o = 0.23 + \frac{0.11}{(1.25 - 1)^{1.08}} = 0.72$$

$$\psi_o = \frac{B_o}{n_b} \frac{N_r N_c}{N_t} \frac{2b_o D_c D_o^{0.75} F_s^{4.75} \mu_{f_o}^{1.42}}{\pi a_o^{4.75} g_c \rho_o^2 k_{f_o}^{3.17} c_{p f_o}^{1.58}}$$

$$\psi_o = \frac{(1)(1)(2)(0.72)(0.25/12)(\frac{1}{12})^{0.75}(1.3)^{4.75}(1.81)^{1.42}}{(3.14)(0.33)^{4.75}(32.17)(3600)^2(62.1)^2(0.359)^{3.17}(1)^{1.58}}$$

$$\psi_o = 3.8 \times 10^{-11}$$

$$R_{dw} = \frac{x_w D_o}{k_w D_{w\log \text{mean}}} + \frac{D_o}{D_i h_{d_i}} + \frac{1}{h_{d_o}} = \frac{(0.109/12)(\frac{1}{12})}{(26)(0.891/12)} + \frac{\frac{1}{12}}{(0.782/12)(2000)} + \frac{1}{1500}$$

$$= 1.7 \times 10^{-3}$$

$$C_i = C_o = \frac{\$0.04}{2.655 \times 10^6} \text{ per ft·lbf}$$

From Eq. (46b),

$$(h_{i,\text{opt}})^{3.5}(7000)\left(\frac{0.04}{2.655 \times 10^6}\right)$$

$$\times \left[(2.5)(4.1 \times 10^{-3}) + \frac{(3.5)(4.1 \times 10^{-3})(0.782/12)(1.7 \times 10^{-3})h_{i,\text{opt}}}{\frac{1}{12}} \right.$$

$$\left. + 2.9\left(\frac{4.1 \times 10^{-3} \times 0.782/12}{\frac{1}{12}}\right)^{0.83}(3.8 \times 10^{-11})^{0.17}(h_{i,\text{opt}})^{0.22} \right]$$

$$= (0.2)(34)(1.15)$$

By trial and error,

$$h_{i,\text{opt}} = 85 \text{ Btu/(h)(ft}^2)(°F)$$

From Eq. (45c),

$$h_{o,\text{opt}} = \left[\frac{(0.74)(4.1 \times 10^{-3})(0.782/12)}{(3.8 \times 10^{-11})(\frac{1}{12})}\right]^{0.17}(85)^{0.78} = 670 \text{ Btu/(h)(ft}^2)(°F)$$

From Eq. (47),

$$U_{o,\text{opt}} = \left[\frac{1}{(0.782)(85)} + \frac{1}{670} + 0.0017\right]^{-1}$$

$$= (0.0150 + 0.00149 + 0.0017)^{-1} = (0.0182)^{-1}$$

$$= 55 \text{ Btu/(h)(ft}^2)(°F)$$

$$\text{Percent total } \Delta t \text{ over air film} = \frac{0.0150}{0.0182}(100) = 82 \text{ percent}$$

$$\text{Percent total } \Delta t \text{ over water film} = \frac{0.00149}{0.0182}(100) = 8.2 \text{ percent}$$

Therefore, the percent Δt assumptions are adequate.
Determination of optimum Δt_2:

$$C_u = \$9 \times 10^{-6} \text{ per lb of water}$$

$$F_T = 1$$

$$\frac{F_T U_{o,\text{opt}} H_y C_u}{c_{p_u}(K_F C_{A_o} + E_{i,\text{opt}} H_y C_i + E_{o,\text{opt}} H_y C_o)}$$

$$= \frac{(1)(55)(7000)(9 \times 10^{-6})}{(1)\left[\begin{array}{c}(0.2)(34.00)(1.15) + (4.1 \times 10^{-3})(85)^{3.5}(7000) \\ \times \left(\dfrac{0.04}{2.655 \times 10^6}\right) + (3.8 \times 10^{-11})(670)^{4.75}(7000)\left(\dfrac{0.04}{2.655 \times 10^6}\right)\end{array}\right]}$$

$$= 0.31$$

$$\frac{t_1' - t_2'}{\Delta t_1} = \frac{150 - 100}{30} = 1.67$$

From Fig. 14-32

$$\frac{\Delta t_{2,\text{opt}}}{\Delta t_1} = 1.44$$

$$\Delta t_{2,\text{opt}} = t_1' - t_{2,\text{opt}} = (30)(1.44) = 43°F$$

$$t_{2,\text{opt}} = 150 - 43 = 107°F$$

The assumption of $t_2 = 110°F$ is adequate.
Determination of optimum A_o:

$$\Delta t_{m,\text{opt}} = \frac{30 + 42}{2} = 36°F$$

$$A_{o,\text{opt}} = \frac{q}{U_{o,\text{opt}} \Delta t_{m,\text{opt}}} = \frac{(20,000)(0.25)(50)}{(55)(36)} = 126 \text{ ft}^2$$

From Fig. 14-13, the assumed purchased cost of $34.00 per square foot of outside area is adequate.

Cost of installed exchanger $= (126)(34.00)(1.15) = \$4930$

Determination of optimum number of tubes and tube length: From Eq. ($E1$) in Table 9,

$$G_{i,\text{opt}} = \left[\frac{(85)(0.782/12)^{0.2}(0.046)^{0.8}}{(0.023)(0.0162)}\left(\frac{0.0162}{0.25 \times 0.046}\right)^{1/3}\left(\frac{0.045}{0.046}\right)^{0.14}\right]^{1.25}$$

$$= 133,000 \text{ lb/(h)(ft}^2)$$

$$S_{i,\text{opt}} = \frac{20,000}{133,000} = 0.150 \text{ ft}^2$$

$$\text{Number of tubes} = \frac{(4)(0.150)}{(3.14)(0.782/12)^2} = 45$$

$$\text{Tube length} = \frac{126}{(3.14)(\frac{1}{12})(45)} = 10.7 \text{ ft}$$

A check on the Reynolds numbers indicates that flow on tube side and shell side is turbulent.

For the optimum exchanger:

Installed cost = \$4930

Number of tubes = 45

Tube length = 10.7 ft

NOTE: An optimum design of this type can be used as a guide for the final selection of an exchanger. However, practical factors of operation must also be considered. In this case, a large number of baffles are required to give the optimum conditions. Because the water-film coefficient is large relative to the air-film coefficient, a reduction in the number of baffles would have little effect on the optimum design.

SIMPLIFIED CASES OF OPTIMUM DESIGN

Immaterial Power Costs on Shell Side or Tube Side

In the operation of heat exchangers, the power cost for one of the fluids is often immaterial. For example, process cooling water is ordinarily supplied under sufficient pressure so that no pumping-power costs are necessary for the water. The highest useful water velocity should be employed in order to minimize the heat-transfer resistance caused by the water film, and a reasonable water-film coefficient can be assumed. Thus, if power costs on one side of the exchanger are immaterial, the problem reduces to one in which the film coefficient and, therefore, the fluid velocity on one side of the exchanger are fixed. Under these conditions, the number of independent variables for the development of the type of optimum design discussed in the preceding sections is two. Possible choices of the basic independent variables could be h_i, Δt_2; h_i, A_o; or A_o, Δt_2.†

Shell-side power cost immaterial In this case, the value of C_o in Eqs. (35) and (44) is zero and h_o is constant. The optimization can be accomplished by differentiating Eq. (44) under the given conditions. The following expression for the

---

† An alternative approach for this type of situation is described in Chap. 10 for the case of optimum flow rate of cooling water in a condenser. Equation (56) in Chap. 10 is merely a special case of Eq. (48) where $t_1' - t_2'$ is zero.

evaluation of $h_{i,\text{opt}}$ is obtained by setting the partial derivatives of Eq. (44) with respect to h_i and with respect to A_o equal to zero:

$$h_{i,\text{opt}}^{3.5}\left[2.5\psi_i H_y C_i + \frac{3.5\psi_i H_y C_i D_i}{D_o}\left(\frac{1}{h_o} + R_{dw}\right)h_{i,\text{opt}}\right] = K_F C_{A_o} \qquad (56)$$

The value of $h_{i,\text{opt}}$ can be determined from Eq. (56), and $U_{o,\text{opt}}$ can then be calculated from Eq. (38). When $U_{o,\text{opt}}$ is known, the optimum value of Δt_2 can be obtained from Eq. (48) or Fig. 14-32, and the optimum design is fixed.

Tube-side power cost immaterial Optimization of the heat-exchanger design for this situation is based on the assumption that C_i is zero and h_i is constant. The procedure is similar to that for the case of shell-side power immaterial as described in the preceding paragraph. The optimum value of h_o can be determined from the following equation, which is obtained by setting the partial derivatives of Eq. (44) with respect to h_o and with respect to A_o equal to zero:

$$h_{o,\text{opt}}^{4.75}\left[3.75\psi_o H_y C_o + 4.75\psi_o H_y C_o\left(\frac{D_o}{D_i h_i} + R_{dw}\right)h_{o,\text{opt}}\right] = K_F C_{A_o} \qquad (57)$$

If power costs on both the tube side and the shell side are to be considered, the ratio of these two costs can be evaluated from the following equation:

$$\frac{\text{Power cost on tube side}}{\text{Power cost on shell side}} = \frac{E_i C_i}{E_o C_o} = \frac{\psi_i h_i^{3.5} C_i}{\psi_o h_o^{4.75} C_o} \qquad (58)$$

As in the previous cases, Eq. (58) applies to a crossflow shell-and-tube exchanger with flow in the turbulent range and no phase change. The equation is not limited to optimum conditions.

Flow Rate of Utility-Fluid Fixed

A situation sometimes occurs in which the maximum temperature of a utility coolant fluid is fixed, thereby setting the coolant flow rate. In other cases, the operating circumstances may arbitrarily set the utility-fluid flow rate as well as the process-fluid flow rate. Under these conditions, Δt_2 is constant, and only two independent variables are involved in the optimization procedure. Since Δt_2 was held constant in the development of Eqs. (45c), (46b), and (47), these equations can be used for determining the optimum overall heat-transfer coefficient. This coefficient, combined with the known values of Δt_2 and Δt_1, is sufficient to permit calculation of the optimum heat-transfer area by Eq. (39), and the rest of the design variables are then established.

Change in Phase of One Fluid (Temperature of the Fluid Remains Constant)

The procedure for developing an optimum heat-exchanger design is simplified if the temperature of one of the fluids remains constant. This condition is often encountered when one of the fluids changes phase, as in a condenser or a steam heater.

When the process fluid is the one that changes phase, the three independent variables h_i, h_o, and Δt_2 are involved, and the general procedure for developing the optimum-design equations is similar to that outlined for the case of no phase change.

Under many conditions, however, determination of the optimum overall coefficient is simplified, because the pressure drop and the resulting power costs for a condensing fluid can be taken as zero and the heat-transfer resistance of the condensing-fluid film is relatively small, permitting the assumption of a reasonable constant value for the film coefficient. Thus, if the fluid that changes phase is the process fluid, the number of independent variables in the design is reduced to two, as in the case of immaterial power costs on the shell side or tube side. If the change in phase occurs on the shell side, Eq. (56) can be used to evaluate $h_{i,\text{opt}}$, and $U_{o,\text{opt}}$ can then be obtained from Eq. (38). Similarly, if the phase change is on the tube side, Eq. (57) gives the optimum value of h_o. When $U_{o,\text{opt}}$ is known, Eq. (48) with $t_1' - t_2'$ equal to zero, or the bottom line in Fig. 14-32, can be used to evaluate the optimum value of Δt_2. The procedure for this situation, as applied to a condenser with immaterial power costs for the utility fluid, is presented in detail in Chap. 10.

If the condensing fluid happens to be the utility fluid, as in a steam heater, the value of Δt_2 is set by the condensing temperature, and the independent variables are h_i and h_o. For the common case of negligible power cost and constant value of the film coefficient on the phase-change side of the exchanger, only one independent variable remains. In this situation, the optimum design can be established by setting the partial derivatives of Eq. (44) with respect to A_o and with respect to the independent film coefficient equal to zero. The resulting optimum film coefficient for change in shell-side phase can be calculated from Eq. (56), and that for change in tube-side phase is given by Eq. (57).

Velocity of One Fluid Fixed (h_i or h_o fixed)

Because of fouling effects, there may be a limit on the velocity of one of the fluids in a heat exchanger. For example, the velocity of cooling water in the tubes of a shell-and-tube exchanger is often specified as 3 ft/s. If the velocity of one fluid is specified, the coefficient for that fluid is set, and the independent variables become Δt_2 and the film coefficient of the other fluid.

Inside-tube velocity fixed If the inside-tube velocity is fixed, h_i is constant, and the optimum value of h_o can be established by eliminating λ/A_o from Eqs. (45b) and (46a) to give

$$h_{o,\text{opt}}^{4.75}\left[3.75\psi_o H_y C_o + 4.75\psi_o H_y C_o\left(\frac{D_o}{D_i h_i} + R_{dw}\right)h_{o,\text{opt}}\right] = K_F C_{A_o} + \psi_i h_i^{3.5} H_y C_i$$

$$(59)$$

Trial-and-error or graphical solution of Eq. (59) yields the value of $h_{o,\text{opt}}$, and $U_{o,\text{opt}}$ can be calculated from Eq. (39). Figure 14-32 or Eq. (48) can then be used to obtain $\Delta t_{2,\text{opt}}$, and the optimum design conditions are fixed.

Outside-tube velocity fixed When h_o is constant, the procedure for evaluation of $h_{i,\text{opt}}$ and the other optimum conditions is similar to that described in the preceding paragraph. By eliminating λ/A_o from Eqs. (45a) and (46a), the following expression for evaluation of $h_{i,\text{opt}}$ is obtained:

$$h_{i,\text{opt}}^{3.5}\left[2.5\psi_i H_y C_i + \frac{3.5\psi_i H_y C_i D_i}{D_o}\left(\frac{1}{h_o} + R_{dw}\right)h_{i,\text{opt}}\right] = K_F C_{A_o} + \psi_o h_o^{4.75} H_y C_o \quad (60)$$

GENERAL METHODS FOR PROCESS DESIGN OF HEAT EXCHANGERS

The procedures used for developing the design of heat exchangers vary with the type of problem and the preference of the worker. Some engineers prefer to develop the design for a heat exchanger by a method known as *rating an exchanger*. In this method, the engineer assumes the existence of an exchanger and makes calculations to determine if the exchanger would handle the process requirements under reasonable conditions. If not, a different exchanger is assumed, and the calculations are repeated until a suitable design is developed. For example, with a given set of process requirements, the engineer could assume the existence of an exchanger with a designated tube size, tube spacing, baffle type, baffle spacing, and number of tubes and passes. The engineer might then proceed through the process-design calculations by computing an overall heat-transfer coefficient and evaluating all flow rates, areas, lengths, and pressure drops. Repeated trials may be necessary to obtain an accurate overall coefficient. If the results of the final design indicate that the assumed exchanger has reasonable dimensions, reasonable cost, and acceptable pressure drops, the unit is considered as adequate and the design is complete.

An alternative approach, of course, is to base the design on optimum economic conditions, using the methods described in the preceding sections. No matter which approach is used, the general method of attack for a given set of process conditions consists of the following steps:

1. Determine the rates of flow and rate of heat transfer necessary to meet the given conditions.
2. Decide on the type of heat exchanger to be used, and indicate the basic equipment specifications.
3. Evaluate the overall heat-transfer coefficient and also the film coefficients, if necessary. In many cases, fluid velocities must be determined in order to obtain accurate heat-transfer coefficients.
4. Evaluate the mean temperature-difference driving force.

5. Determine the necessary area of heat transfer and the exchanger dimensions.
6. Analyze the results to see if all dimensions, costs, pressure drops, and other design details are satisfactory.
7. If the results of (6) show that the exchanger is not satisfactory, the specifications given in (2) are inadequate. Choose new specifications and repeat steps 3 through 7 until a satisfactory design is obtained.

NOMENCLATURE FOR CHAPTER 14

a_o = constant in Eq. (29) for evaluating outside film coefficient of heat transfer, dimensionless

A = area of heat transfer, ft^2; subscript m designates mean area; subscript o designates outside area; subscript f designates film area

b_o = constant in Eq. (32) for evaluating shell-side friction factor, dimensionless

B_i = correction factor in Eq. (30) to account for friction due to sudden contraction, sudden expansion, and reversal of flow direction, dimensionless

B_o = correction factor in Eq. (31) to account for friction due to reversal of flow direction, recrossing of tubes, and variation in cross section, dimensionless

c_c = constant, dimensionless; defined in Table 4

c_p = heat capacity, Btu/(lb)(°F); prime refers to process fluid

C_{A_o} = installed cost of heat exchanger per unit of outside-tube heat-transfer area, \$/ft^2

C_i = cost for supplying 1 ft·lbf to pump the fluid through the inside of the tubes, \$/ft·lbf

C_o = cost for supplying 1 ft·lbf to pump the fluid through the shell side of the exchanger, \$/ft·lbf

C_T = total annual variable cost for heat exchanger and its operation, \$/year

C_u = cost of utility fluid, \$/lb

D = diameter or distance, ft

D' = diameter, in.

D_c = clearance between tubes to give smallest free area across shell axis, ft

D_e = equivalent diameter = 4 × hydraulic radius, ft

E = power loss per unit of outside-tube heat-transfer area, ft·lbf/(h)(ft^2); subscript i designates inside tubes, and subscript o designates outside tubes

$\tilde{f}_i$ = Fanning friction factor for isothermal flow, dimensionless

$\tilde{f}'$ = special friction factor for shell-side flow, dimensionless

F_A = correction factor for radiant heat transfer based on relative orientation of surfaces, dimensionless; defined by Eqs. (6) and (7)

F_c = friction due to sudden contraction, ft·lbf/lbm

F_e = friction due to sudden enlargement, ft·lbf/lbm

F_E = correction factor for radiant heat transfer based on emissivities and absorptivities of surfaces, dimensionless; defined by Eqs. (6) and (7)

F_r = friction due to reversal of flow direction, ft·lbf/lbm

F_s = safety factor in Eq. (29) to account for bypassing on shell side of exchanger, dimensionless

F_T = correction factor on logarithmic-mean Δt for counterflow to give mean Δt, dimensionless; defined in Fig. 14-2

g = local gravitational acceleration, ft/(s)(s)

g_c = conversion factor in Newton's law of motion, 32.17 ft·lbm/(s)(s)(lbf)

G = mass velocity inside tubes, lb/(h)(ft^2)

G_s = shell-side mass velocity across tubes based on the minimum free area between baffles across the shell axis, lb/(h)(ft^2)

h = film coefficient of heat transfer, Btu/(h)(ft^2)(°F); subscript c indicates convection; subscript d represents dirt or fouling; subscript co indicates conduction

h_r = film coefficient for heat transfer by radiation, Btu/(h)(ft^2)(°F)

H_y = hours of operation per year, h/year

k = thermal conductivity, Btu/(h)(ft^2)(°F/ft); subscripts x, y, and z refer to direction of heat-flow path

K = dimensional constant; defined in Table 4

K_c = constant in expression for evaluating friction due to sudden contraction, dimensionless

K_F = annual fixed charges including maintenance, expressed as a fraction of the initial cost for the completely installed unit, dimensionless

K_1 = constant for evaluation of B_i, dimensionless; defined with Eq. (30)

L = heated length of straight tube or length of heat-transfer surface, ft; if tubes in parallel are involved, L is the length of one tube

n = constant, dimensionless

n_b = number of baffle spaces = number of baffles plus 1, dimensionless

n_p = number of tube passes, dimensionless

N_c = number of clearances between tubes for flow of shell-side fluid across shell axis, dimensionless

N_{Gr} = Grashof number = $L^3 \rho^2 g \beta \, \Delta t / \mu^2$, dimensionless

N_{Pr} = Prandtl number = $c_p \mu / k$, dimensionless

N_r = number of rows of tubes across which shell fluid flows, dimensionless

N_{Re} = Reynolds number = DG/μ, dimensionless

N_t = total number of tubes in exchanger = number of tubes per pass $\times \, n_p$, dimensionless

N_V = number of rows of tubes in a vertical tier, dimensionless

P, p = pressure, lbf/ft^2

q = rate of heat transfer, Btu/h

Q = amount of heat transferred in time θ, Btu

r = radius, ft

R = temperature ratio for evaluating F_T, dimensionless; defined in Fig. 14-2

R_{dw} = combined resistance of tube wall and scaling or dirt factors, $[\text{Btu}/(\text{h})(\text{ft}^2)(°\text{F})]^{-1}$; defined with Eq. (39).

S = temperature ratio for evaluating F_T, dimensionless; defined in Fig. 14-2

S_H = cross-sectional flow area of header per pass, ft^2

S_i = cross-sectional flow area inside tubes per pass, ft^2

S_o = shell-side free-flow area across the shell axis, ft^2

t = temperature, °F; subscript b refers to average bulk temperature; subscript or refers to original temperature; subscript s refers to surface; in general, primes refer to the process fluid, subscript 1 refers to the entering temperature, and subscript 2 refers to the leaving temperature

t' = temperature of second fluid in a heat exchanger, °F; refers, in general, to process fluid

T = absolute temperature, °R

U = overall coefficient of heat transfer, $\text{Btu}/(\text{h})(\text{ft}^2)(°\text{F})$; subscript d indicates that a dirt or fouling factor is included; subscript o indicates based on outside area and fouling factor included

V = velocity, ft/h; subscript i indicates in tubes

V' = velocity, ft/s

w = weight rate of flow, lb/h; no subscript indicates per tube; subscript u indicates total flow rate of utility fluid; subscript i indicates total flow rate of inside-tube fluid; subscript o indicates total flow rate of outside-tube fluid

w' = total weight rate of flow of process fluid, lb/h

x = length of conduction path, ft

x_L = ratio of pitch parallel to flow to tube diameter, dimensionless

x_T = ratio of pitch transverse to flow to tube diameter, dimensionless

y = length of conduction path, ft

z = length of conduction path, ft

Greek symbols

α = thermal diffusivity = $k/\rho c_p$, ft^2/h

β = coefficient of volumetric expansion, 1/°R

Δ = Δt designates temperature-difference driving force, °F; subscript f designates across film; subscript m designates mean Δt; subscript oa or no subscript designates overall Δt; $\Delta t_1 = t'_2 - t_1$; $\Delta t_2 = t'_1 - t_2$; ΔP and Δp designate pressure drop; $\Delta p = -\Delta P$

ε = emissivity, dimensionless

θ = time, h

λ_c = latent heat of condensation, Btu/lb

λ = Lagrangian multiplier, dimensionless; defined by Eq. (43)

μ = absolute viscosity, lb/(h)(ft)

π = 3.1416 $\cdots$

ρ = density, lb/ft^3

σ = Stefan-Boltzmann dimensional constant for radiant heat transfer; defined by Eq. (5)

ϕ = correction factor for nonisothermal flow, dimensionless; defined with Eq. (30)

ψ_i, ψ_o = dimensional factors for evaluation of E_i and E_o; defined with Eqs. (40) and (41)

Subscripts

f = across film or at average film temperature

i = inside pipe or tube, based on average bulk temperature

L = liquid at average liquid temperature

m = mean

o = outside pipe or tube, based on average bulk temperature

oa = overall

opt = optimum conditions

u = utility fluid

v = vapor at average vapor temperature

w = tube or pipe wall, based on temperature at wall surface

SUGGESTED ADDITIONAL REFERENCES FOR HEAT-TRANSFER EQUIPMENT DESIGN AND COSTS

Air conditioning and refrigeration design and costs

Ballou, D. F., T. A. Lyons, and J. R. Tacquard, Jr., Design and Cost Estimating of Mechanical Refrigeration Systems, *Hydrocarbon Process.*, **46**(6):119 (1967).

Briley, G. C., Conserve Energy . . . Refrigerate with Waste Heat, *Hydrocarbon Process.*, **55**(5):173 (1976).

Goethe, S. P., Central Heating and Refrigeration Systems, *ACHV*, **65**(10):46 (1968).

Guthrie, K. M., Capital Cost Estimating, *Chem. Eng.*, **76**(6):114 (1969).

Kaiser, V., O. Salhi, and C. Pocini, Analyze Mixed Refrigerant Cycles, *Hydrocarbon Process.*, **57**(7):163 (1978).

McCarthy, A. J., and M. E. Hopkins, Simplify Refrigeration Estimating, *Hydrocarbon Process.*, **50**(7):105 (1971).

Mehra, Y. R., How to Estimate Power and Condenser Duty for Ethylene Refrigeration Systems, *Chem. Eng.*, **85**(28):97 (1978).

Mesko, J. E., Economic Advantages of Central Heating and Cooling Systems, *Bldg. Sys. Des.*, **71**(1):15 (1974).

Scheel, L. F., Refrigeration: Centrifugal or Recip? *Hydrocarbon Process.*, **48**(3):123 (1969).

Spencer, E., Estimating the Size and Cost of Steam Vacuum Refrigeration, *Hydrocarbon Process.*, **46**(6):136 (1967).

Vinet, J., C. A. Miller, N. A. Ishkow, and R. J. Tonelli, Refrigeration, Cost Engineers' Notebook, *AACE Bull.*, **14**(5):A-4.181 (1972).

Zafft, R. W., How to Size and Find the Cost of Absorption Refrigeration, *Hydrocarbon Process.*, **46**(6):131 (1967).

Zimmerman, O. T., The Hydro-Miser, *Cost Eng.*, **16**(3):4 (1971).

Air cooler design and costs

Black, G. M., and W. Schoonman, Save Water: Air Condense Steam, *Hydrocarbon Process.*, **49**(10):101 (1970).

Brown, J. W., and G. J. Benkly, Air Coolers in Cold Climates, *Hydrocarbon Process.*, **53**(5):149 (1974).

Dehne, M. F., Air-Cooled Overhead Condensers, *Chem. Eng. Progr.*, **65**(7):51 (1969).

Doyle, P. T., and G. J. Benkly, Use Fanless Air Coolers, *Hydrocarbon Process.*, **52**(7):81 (1973).

Gunter, A. Y., and K. V. Shipes, Hot Air Recirculation by Air Cooler, *Chem. Eng. Progr.*, **68**(2):49 (1972).

Kals, W., Wet-surface Aircoolers, *Chem. Eng.*, **78**(17):90 (1971).

Larinoff, M. W., W. E. Moles, and R. Reichhelm, Design and Specifications of Air-Cooled Stream Condensers, *Chem. Eng.*, **85**(12):86 (1978).

Lerner, J. E., Simplified Air Cooler Estimating, *Hydrocarbon Process.*, **51**(2):93 (1972).

Rossie, J. P., Dry-Type Cooling Systems, *Chem. Eng. Progr.*, **67**(7):58 (1971).

Russell, J. J., and T. C. Carnavos, Air Cooling of Internally Finned Tubes, *Chem. Eng. Progr.*, **73**(2):84 (1977).

Shipes, K. V., Air Coolers in Cold Climates, *Hydrocarbon Process.*, **53**(5):147 (1974).

Rapid Sizing of Air Cooling Units, *Petro/Chem. Eng.*, **40**(5):32 (1968).

Boiler and heater design and costs

Buckham, J. A., S. J. Horn, B. M. Legler, T. K. Thompson, and B. R. Wheeler, Process Heating via Combustion, *Chem. Eng. Progr.*, **64**(7):52 (1968).

Buffington, M. A., How to Select Package Boilers, *Chem. Eng.*, **82**(23):98 (1975).

Bailey, T., and F. M. Wall, Heat Transfer—Ethylene Furnace Design, *Chem. Eng. Progr.*, **74**(7):45 (1978).

Berman, H. L., Fired Heaters: Finding the Basic Design for Your Application, *Chem. Eng.*, **85**(14):99 (1978).

———, Fired Heaters: Construction Materials, Mechanical Features, Performance Monitoring, *Chem. Eng.*, **85**(17):87 (1978).

———, Fired Heaters: How Combustion Conditions Influence Design and Operation, *Chem. Eng.*, **85**(17):129 (1978).

———, Fired Heaters: How to Reduce Your Fuel Bill, *Chem. Eng.*, **85**(20):165 (1978).

Chambers, L. E., and W. S. Potter, Design Ethylene Furnaces: Maximum Ethylene, *Hydrocarbon Process.*, **53**(1):121 (1974).

——— and ———, Design Ethylene Furnaces: Maximum Olefins, *Hydrocarbon Process.*, **53**(3):95 (1974).

——— and ———, Design Ethylene Furnaces: Furnace Costs, *Hydrocarbon Process.*, **53**(8): 99 (1974).

Clayton, W. H., and J. G. Singer, Steam Generator Designs, *Chem. Eng. Progr.*, **69**(7):81 (1973).

Collins, G. K., Horizontal-Thermosiphon-Reboiler Design, *Chem. Eng.*, **88**(15):149 (1976).

Frank, O., and R. D. Prickett, Designing Vertical Thermosyphon Reboilers, *Chem. Eng.*, **80**(20):107 (1973).

Gallagher, J. T., Cost of Direct-Fired Heaters, *Chem. Eng.*, **74**(15):232 (1967).

Ganapathy, V., Basic Data for Steam Generators—At a Glance, *Chem. Eng.*, **84**(12):197 (1977).

———, Short-cut Calculation for Steam Heaters and Boilers, *Chem. Eng.*, **85**(6):105 (1978).

Griffin, J. J., B. R. Kersey, and S. J. Eaton, Heat Transfer—Forced-Draft Firing in Refinery Heaters, *Chem. Eng. Progr.*, **74**(7):57 (1978).

Gunder, P. F., How to Specify Process Heaters and Evaluate Bids, *Hydrocarbon Process.*, **48**(10):116 (1969).

Gupton, P. S., and A. S. Krisher, Waste Heat Boiler Failures, *Chem. Eng. Progr.*, **69**(1):47 (1973).

Guthrie, K. M., Capital Cost Estimating, *Chem. Eng.*, **76**(6):114 (1969).

Helzner, A. E., Operating Performance of Steam-Heated Reboilers, *Chem. Eng.*, **84**(4):73 (1977).

Hinchley, P., How to Avoid Problems of Waste-Heat Boilers, *Chem. Eng.*, **82**(18):94 (1975).

———, Waste Heat Boilers: Problems & Solutions, *Chem. Eng. Progr.*, **73**(3):90 (1977).

Hughmark, G. A., Designing Thermosiphon Reboilers, *Chem. Eng. Progr.*, **65**(7):67 (1969).

Kern, R., Thermosiphon Reboiler Piping Simplified, *Hydrocarbon Process.*, **47**(12):118 (1968).

————, Space Requirements and Layout for Process Furnaces, *Chem. Eng.*, **85**(5):117 (1978).

Lenoir, J. M., Furnace Tubes: How Hot? *Hydrocarbon Process.*, **48**(10):97 (1969).

Lobo, W. E., Design of Furnaces with Flue Gas Temperature Gradients, *Chem. Eng. Progr.*, **70**(1):65 (1974).

Loftus, J., H. C. Schutt, and A. F. Sarofim, Design of Furnaces for Tubular Reactors, *Chem. Eng. Progr.*, **63**(7):47 (1967).

Lowry, J. A., Evaluate Reboiler Fouling, *Chem. Eng.*, **85**(4):103 (1978).

Miller, R., Jr., Process Energy Systems, *Chem. Eng.*, **75**(11):130 (1968).

Monroe, L. R., Process Plant Utilities, Steam, *Chem. Eng.*, **77**(27):130 (1970).

Olds, F. C., On-Site Generation Today, *Power Eng.*, **73**(3):42 (1969).

Orrell, W. H., Physical Considerations in Designing Vertical Thermosiphon Reboilers, *Chem. Eng.*, **80**(21):120 (1973).

O'Sullivan, T. F., H. R. McChesney, and W. H. Pollock, Coal-Fired Process Heaters? *Hydrocarbon Process.*, **57**(7):95 (1978).

Papamarcos, J., Package Boilers: Applications and Trends, *Power Eng.*, **73**(12):34 (1969).

Pikulik, A., and H. E. Diaz, Cost Estimating for Major Process Equipment, *Chem. Eng.*, **84**(21):107 (1977).

Reeling, N. E., Fuels and Boiler Design, *Power Eng.*, **73**(2):52 (1969).

Sarma, N. V. L. S., P. J. Reddy, and P. S. Murti, A Computer Design Method for Vertical Thermosiphon Reboilers, *I&EC Process Des. & Dev.*, **12**(3):278 (1973).

Smith, J. V., Improving the Performance of Vertical Thermosiphon Reboilers, *Chem. Eng. Progr.*, **70**(7):68 (1974).

Streich, H. J., and F. G. Feeley, Jr., Design and Operation of Process Waste Heat Boilers, *Chem. Eng. Progr.*, **68**(7):57 (1972).

Thorngren, J. T., Reboiler Computer Evaluation, *I&EC Process Des. & Dev.*, **11**(1):39 (1972).

van Loosen, J., What You Need to Know About Boilers, *Hydrocarbon Process.*, **47**(6):95 (1968).

Von Wiesenthal, P., and H. W. Cooper, Guide to Economics of Fired Heater Design, *Chem. Eng.*, **77**(7):104 (1970).

Wimpress, N., Generalized Method Predicts Fired-Heater Performance, *Chem. Eng.*, **85**(12):95 (1978).

Womack, J. W., Improved Waste Heat Boiler Operation, *Chem. Eng. Progr.*, **72**(7):56 (1976).

Woodard, A. M., Reduce Process Heater Fuel, *Hydrocarbon Process.*, **53**(7):106 (1974).

————, Upgrading Process Heater Efficiency, *Chem. Eng. Progr.*, **71**(10):53 (1975).

Zimmerman, O. T., Fulton Electric Boilers, *Cost Eng.*, **17**(3):4 (1972).

Cooling tower design and costs

Burger, R., Cooling Tower Drift Elimination, *Chem. Eng. Progr.*, **71**(7):73 (1975).

DeMonbrun, J. R., Factors to Consider in Selecting a Cooling Tower, *Chem. Eng.*, **75**(19):106 (1968).

Donohue, J. M., and C. C. Nathan, Unusual Problems in Cooling Water Treatment, *Chem. Eng. Progr.*, **71**(7):88 (1975).

Elgawhary, A. W., Spray Cooling System Design, *Chem. Eng. Progr.*, **71**(7):83 (1975).

Friar, F., Cooling-Tower Basin Design, *Chem. Eng.*, **81**(15):122 (1974).

Guthrie, K. M., Capital Cost Estimating, *Chem. Eng.*, **76**(6):114 (1969).

Hall, W. A., Cooling Tower Plume Abatement, *Chem. Eng. Progr.*, **67**(7):52 (1971).

Hansen, E. P., and J. J. Parker, Status of Big Cooling Towers, *Power Eng.*, **71**(5):38 (1967).

Holzhauer, R., Industrial Cooling Towers, *Plant Eng.*, **29**(15):60 (1975).

Jordan, D. R., M. D. Bearden, and W. F. McIlhenny, Blowdown Concentration by Electrodialysis, *Chem. Eng. Progr.*, **71**(7):77 (1975).

Juong, J. F., How to Estimate Cooling Tower Costs, *Hydrocarbon Process.*, **48**(7):200 (1969).

Kelly, G. M., Cooling Tower Design and Evaluation Parameters, *ASME Paper*, 75-IPWR-9, 1975.

Kolflat, T. D., Cooling Tower Practices, *Power Eng.*, **78**(1):32 (1974).

Maze, R. W., Practical Tips on Cooling Tower Sizing, *Hydrocarbon Process.*, **46**(2):123 (1967).

————, Air Cooler or Water Tower—Which for Heat Disposal? *Chem. Eng.*, **83**(1):106 (1975).

Meytsar, J., Estimate Cooling Tower Requirements Easily, *Hydrocarbon Process.*, **57**(11):238 (1978).
Nelson, W. L., What Is Cost of Cooling Towers, *Oil & Gas J.*, **65**(47):182 (1967).
Olds, F. C., Cooling Towers, *Power Eng.*, **76**(12):30 (1972).
Paige, P. M. Costlier Cooling Towers Require a New Approach to Water-Systems Design, *Chem. Eng.*, **74**(14):93 (1967).
Park, J. E., and J. M. Vance, Computer Model of Crossflow Towers, *Chem. Eng. Progr.*, **67**(7):55 (1971).
Picciotti, M., Design Quench Water Towers, *Hydrocarbon Process.*, **56**(6):163 (1977).
———, Optimize Quench Water Systems, *Hydrocarbon Process.*, **56**(9):179 (1977).
Rabb, A., Are Dry Cooling Towers Economical? *Hydrocarbon Process.*, **47**(2):122 (1968).
Uchiyama, T., Cooling Tower Estimates Made Easy, *Hydrocarbon Process.*, **55**(12):93 (1976).
Wrinkle, R. B., Performance of Counterflow Cooling Tower Cells, *Chem. Eng. Progr.*, **67**(7):45 (1971).

Evaporator design and costs

Bennett, R. C., Heat Transfer—Recompression Evaporation, *Chem. Eng. Progr.*, **74**(7):67 (1978).
Casten, J. W., Heat Transfer—Mechanical Recompression Evaporators, *Chem. Eng. Progr.*, **74**(7):61 (1978).
Farin, W. G., Low-Cost Evaporation Method Saves Energy By Reusing Heat, *Chem. Eng.*, **83**(5):100 (1976).
Dockendorff, J. D., and P. J. Cheng, Energy-Conscious Evaporators, *Chem. Eng. Progr.*, **72**(5):56 (1976).
Guthrie, K. M., Capital Cost Estimating, *Chem. Eng.*, **76**(6):114 (1969).
Itahara, S., and L. I. Stiel, Optimal Design of Multiple-Effect Evaporators with Vapor Bleed Streams, *I&EC Process Des. & Dev.*, **7**(1):6 (1968).
Levachev, A. G., L. A. Zykov, B. V. Khetrov, and A. Y. Nemtinov, Use of an Analog Computer for Modelling of an Evaporator, *Khim Prom.*, **46**:147 (1970).
Moore, J. G., and E. B. Pinkel, When to Use Single Pass Evaporators, *Chem. Eng. Progr.*, **64**(7):39 (1968).
Newell, R. B., and D. G. Fisher, Model Development, Reduction, and Experimental Evaluation for an Evaporator, *I&EC Process Des. & Dev.*, **11**(2):213 (1972).
Newman, H. H., How to Test Evaporators, *Chem. Eng. Progr.*, **64**(7):33 (1968).
Oden, E. C., Sr., Charts Speed Computations for Multiple-Effect Evaporator Problems, *Chem. Eng.*, **74**(9):159 (1967).
Ritter, R. A., and H. Andre, Evaporator Control System Design, *Can. J. Chem. Eng.*, **48**(6):696 (1970).
Rozycki, J., Energy Conservation via Recompression Evaporation, *Chem. Eng. Progr.*, **72**(5):69 (1976).
Stickney, W. W., and T. M. Vosberg, Putting Evaporators to Work; Treating Chemical Wastes by Evaporation, *Chem. Eng. Progr.*, **72**(4):41 (1976).
Swartz, A., A Guide for Troubleshooting Multiple-Effect Evaporators, *Chem. Eng.*, **85**(11):175 (1978).
Waltrich, P. F., Sizing Vacuum Equipment for Evaporative Coolers, *Chem. Eng.*, **80**(11):164 (1973).
Wetherhorn, D., Evaporators, *Chem. Eng.*, **77**(12):187 (1970).
———, Guide to Trouble-Free Evaporators, *Chem. Eng.*, **77**(12):187 (1970).
Zimmerman, O. T., and I. Lavine, Pfaudler Wiped-Film Evaporator, *Cost Eng.*, **12**(3):4 (1967).
——— and ———, The Votator Turba-Film Processor, *Cost Eng.*, **13**(3):14 (1968).

Heat exchanger design and costs

Bachmann, T. H., and D. D. Lineberry, Refrigerant Systems in Scraped Surface Exchangers, *Chem. Eng. Progr.*, **63**(7):68 (1967).
Badhwar, R. K., Shortcut Design Methods for Piping, Exchangers, and Towers, *Chem. Eng.*, **78**(24):113 (1971).
Barrington, E. A., Acoustic Vibrations in Tubular Exchangers, *Chem. Eng. Progr.*, **69**(7):62 (1973).

Berggren, J. R., and W. J. Sindelar, New Approach to Exchanger Train Design, *Hydrocarbon Process.*, **47**(10):138 (1968).

Blackburn, W. R., How to Write Effective Heat Exchanger Specifications, *Petro/Chem. Eng.*, **39**(9):19 (1967).

Brown, J. W., and G. J. Benkly, Heat Exchangers in Cold Service—A Contractor's View, *Chem. Eng. Progr.*, **70**(7):59 (1974).

Brown, P. M. M., and D. W. France, How to Protect Air-Cooled Heat Exchangers Against Overpressure, *Hydrocarbon Process.*, **54**(8):103 (1975).

Brown, R., A Procedure for Preliminary Estimates, *Chem. Eng.*, **85**(7):108 (1978).

Butterworth, D., and L. B. Cousins, Use of Computer Programs in Heat Exchanger Design, *Chem. Eng.*, **83**(14):72 (1976).

Cariati, V., Costs: Bare or Finned Exchanger Tubes? *Hydrocarbon Process.*, **47**(2):106 (1968).

Clark, D. F., Plate Heat Exchanger Design and Recent Development, *Chem. Eng. (London)*, 275 (May 1974).

Cocks, A. M., Plate Heat Exchanger Design by Computer, *Chem. Eng. (London)*, **228**:193 (May 1969).

Cowan, C. T., Choosing Materials of Construction for Plate Heat Exchangers, *Chem. Eng.*, **82**(12):100 (1975); **82**(14):102 (1975).

Dimoplon, W., A New Approach to Compressible Flow in Exchangers, *Hydrocarbon Process.*, **50**(9):195 (1971).

Fanaritis, J. P., and J. W. Bevevino, How to Select the Optimum Shell-and-Tube Heat Exchanger, *Chem. Eng.*, **83**(14):62 (1976).

Fisher, J., and R. O. Parker, New Ideas on Heat Exchanger Design, *Hydrocarbon Process.*, **48**(7):147 (1969).

Foxall, D. H., and P. T. Gilbert, Selecting Tubes for CPI Heat Exchangers, *Chem. Eng.*, **83**(10):133 (1976).

Franklin, G. M., and W. B. Munn, Problems with Heat Exchangers in Low Temperature Environments, *Chem. Eng. Progr.*, **70**(7):63 (1974).

Gilmour, C. H., Trouble-Shooting Heat-Exchanger Design, *Chem. Eng.*, **74**(13):221 (1967).

Gloyer, W., Thermal Design of Mixed Vapor Condensers, *Hydrocarbon Process.*, **49**(7):107 (1970).

Gottzmann, C. F., P. S. O'Neill, and P. E. Minton, High Efficiency Heat Exchangers, *Chem. Eng. Progr.*, **69**(7):69 (1973).

Guthrie, K. M., Estimating the Cost of High-Pressure Equipment, *Chem. Eng.*, **75**(25):144 (1968).

———, Capital Cost Estimating, *Chem. Eng.*, **76**(6):114 (1969).

Hanger, R. C., How to Design Double Pipe Finned Tube Heat Exchangers, *Petro/Chem. Eng.*, **40**(9):27 (1968).

Hargis, A. M., A. T. Beckmann, and J. J. Loiacono, Applications of Spiral Plate Heat Exchangers, *Chem. Eng. Progr.*, **63**(7):62 (1967).

Harvey, D. C., and G. E. Glass, Uses of Inflated-Plate Heat Exchangers, *Chem. Eng.*, **81**(24):170 (1974).

Hills, D. E. G., Graphite Heat Exchangers, *Chem. Eng.*, **81**(27):80 (1974); **82**(2):116 (1975).

Hood, R. R. Designing Heat Exchangers in Teflon, *Chem. Eng.*, **74**(11):181 (1967).

Huckaba, C. E., N. Master, and J. J. Santoleri, Performance of a Novel Sub-X Heat Exchanger, *Chem. Eng. Progr.*, **63**(7):74 (1967).

Jamin, B., Exchanger Stages Solved by Graph, *Hydrocarbon Process.*, **49**(7):137 (1970).

Jenssen, S. K., Heat Exchanger Optimization, *Chem. Eng. Progr.*, **65**(7):59 (1969).

Jones, P. R., and S. Katell, Computer Usage for Evaluation of Design Parameters and Cost of Heat Exchangers, presented at Am. Assoc. Cost Eng. 12th National Meeting, Houston, Texas, June 17–19, 1968.

Katell, S., and P. R. Jones, Programs for the Price: Optimum Heat Exchangers, *Brit. Chem. Eng.*, **15**(4):491 (1970).

Kern, D. Q., and A. D. Kraus, "Extended Surface Heat Transfer," McGraw-Hill Book Company, New York, 1972.

Kern, R., How to Find the Optimum Layout for Heat Exchangers, *Chem. Eng.*, **84**(19):169 (1977).

Knight, W. P., Plant Operating Data Improve Heat Exchanger Design, *Hydrocarbon Process.*, **54**(5):151 (1975).

Knülle, H. R., Problems With Exchangers in Ethylene Plants, *Chem. Eng. Progr.*, **68**(7):53 (1972).

Kroehle, T. P., Modular Units of Heat Exchangers, *Petro/Chem. Eng.*, **39**(9):28 (1967).

———, Rules for Reducing Exchanger Costs, *Chem. Eng.*, **74**(17):156 (1967).

Lohrisch, F. W., Short Cut Exchanger Tube Side Rating, *Hydrocarbon Process.*, **48**(4):125 (1969).

Lord, R. C., P. E. Minton, and R. P. Slusser, Design of Heat Exchangers, *Chem. Eng.*, **77**(2):96 (1970).

——— ——— and ———, Guide to Trouble-Free Heat Exchangers, *Chem. Eng.*, **77**(12):153 (1970).

Madsen, N., Design Heat Exchangers for Liquids in Laminar Flow, *Chem. Eng.*, **81**(17):92 (1974).

Malek, R. G., A New Approach to Exchanger Tubesheet Design, *Hydrocarbon Process.*, **56**(1):164 (1977).

———, Improved Exchanger Design, *Hydrocarbon Process.*, **52**(5):128 (1973).

Marriott, J., Where and How to Use Plate Heat Exchangers, *Chem. Eng.*, **78**(8):127 (1971).

———, Performance of an Alfaflex Plate Heat Exchanger, *Chem. Eng. Progr.*, **73**(2):73 (1977).

Mathur, J., Performance of Steam Heat-Exchangers, *Chem. Eng.*, **80**(20):101 (1973); **81**(6):86 (1974).

Mikus, H. J., Plate-Fin Heat Exchanger Use, *Petro/Chem. Eng.*, **39**(9):25 (1967).

Miller, R., Jr., Process Energy Systems, *Chem. Eng.*, **75**(11):130 (1968).

Minton, P. E., Designing Spiral-Plate Heat Exchangers, *Chem. Eng.*, **77**(10):103 (1970); **77**(11):145 (1970).

Moore, J. A., Development of a Low Maintenance Heat Recovery Exchanger, *Chem. Eng. Progr.*, **69**(1):43 (1973).

Nelson, W. L., How the Panic of 1974 Hit Heat-Exchanger Prices, *Oil & Gas J.*, **75**(16):73 (1977).

Newell, R. G., Air-Cooled Heat Exchangers in Low Temperature Environments: A Critique, *Chem. Eng. Progr.*, **70**(10):86 (1974).

Nishida, N., and L. Lapidus, Studies in Chemical Process Design and Synthesis: III. A Simple and Practical Approach to the Optimal Synthesis of Heat Exchanger Networks, *AIChE J.*, **23**(1):77 (1977).

O'Neill, P. S., C. F. Gottzmann, and J. W. Terbot, Heat Exchanger for NGL, *Chem. Eng. Progr.*, **67**(7):80 (1971).

Peters, D. L., and F. J. L. Nicole, Efficient Programming for Cost-Optimised Heat Exchanger Design, *Chem. Eng. (London)*, 98 (March 1972).

Petrie, J. C., W. A. Freeby, and J. A. Buckham, In-Bed Heat Exchangers, *Chem. Eng. Progr.*, **64**(7):45 (1968).

Pikulik, A., and H. E. Diaz, Cost Estimating for Major Process Equipment, *Chem. Eng.*, **84**(21):107 (1977).

Pudlock, J. R., Corrosion Resistant Shell and Tube Exchanger Costs Compared, *Oil & Gas J.*, **75**(37):101 (1977).

Ryan, J. E., Have You Looked at Finned Tubing Applications, *Petro/Chem. Eng.*, **8**(41):23 (1969).

Ramalho, R. S., and E. G. Alabastro, Rigorous Design of Multipass Exchangers when Overall Heat Transfer Coefficient is a Parabolic Function of Temperature, *Can. J. Chem. Eng.*, **45**(1):31 (1967).

Rissler, K., Heat Exchanger Design by Computer, *Chem. Proc. Eng.*, **52**(10):61 (1971).

Rozenman, T., and J. Pundyk, Reducing Solidification in Air-Cooled Heat Exchangers, *Chem. Eng. Progr.*, **70**(10):80 (1974).

———, and J. Taborek, The Effect of Leakage Through the Longitudinal Baffle on the Performance of Two-Pass Shell Exchangers, *AIChE Symp. Ser.*, **68**(118):12 (1972).

Rubin, F. L., How to Specify Heat Exchangers, *Chem. Eng.*, **75**(8):130 (1968).

———, Practical Heat Exchanger Design, *Chem. Eng. Progr.*, **64**(12):44 (1968).

Sack, M., Falling Film Shell-and-Tube Heat Exchangers, *Chem. Eng. Progr.*, **63**(7):55 (1967).

Schmidt, J. R., and D. R. Clark, Analog Simulation Techniques for Modelling Parallel Flow Heat Exchangers, *Simulation*, **12**(1):15 (1969).

Schwarz, G. W., Jr., Preventing Vibration in Shell-and-Tube Heat Exchangers, *Chem. Eng.*, **88**(15):134 (1976).

Shipes, K. V., Air-Cooled Exchangers in Cold Climates, *Chem. Eng. Progr.*, **70**(7):53 (1974).

Singh, K. P., How to Locate Impingement Plates in Tubular Heat Exchangers, *Hydrocarbon Process.*, **53**(10): 147 (1974).

Smith, E. C., Application of Air-Cooled Heat Exchangers, in "AIChE Workshop on Ind. Pwr. Des. for Water Pollution Control Proceedings," Vol. 2 (1969), p. 82.

Spencer, R. A., Jr., Predicting Heat-Exchanger Performance by Successive Summation, *Chem. Eng.*, **85**(27): 121 (1978).

Starczewski, J., Find Tube Side Heat Transfer Coefficient by Nomograph, *Hydrocarbon Process.*, **48**(11): 298 (1969).

————. Short-Cut Method to Exchanger Tube-Side Pressure Drop, *Hydrocarbon Process.*, **50**(5): 122 (1971).

————, Short-Cut Method to Exchanger Shell-Side Pressure Drop, *Hydrocarbon Process.*, **50**(6): 147 (1971).

————, Short Cut to Shell-Side Heat-Transfer Coefficient, *Hydrocarbon Process.*, **52**(4): 155 (1973).

Story, G., and I. MacFarland, Sealing Defective Heat Exchanger Tubes, *Chem. Eng. Progr.*, **71**(7): 94 (1975).

Strickland, J. R., Titanium Heat Exchangers, *Oil & Gas J.*, **74**(36): 100 (1976).

Stuhlbarg, D., How to Find Optimum Exchanger Size for Forced Circulation, *Hydrocarbon Process.*, **49**(1): 149 (1970).

Tarrer, A. R., H. C. Lim, and L. B. Koppel, Finding the Economically Optimum Heat Exchanger, *Chem. Eng.*, **78**(22): 79 (1971).

Thompson, J. W., How Not To Buy Heat Exchangers, *Hydrocarbon Process.*, **51**(12): 83 (1972).

Thorngren, J. T., Predict Exchanger Tube Damage, *Hydrocarbon Process.*, **49**(4): 129 (1970).

Usher, J. D., Evaluating Plate Heat-Exchangers, *Chem. Eng.*, **77**(4): 90 (1970).

Verdoglia, L., Graphical Method Shows Relationships in Solving Heat Exchanger Calculations, *Hydrocarbon Process.*, **47**(9): 300 (1968).

Willmott, A. J., The Regenerative Heat Exchanger Computer Representation, *Intern. J. Heat & Mass Transfer*, **12**: 997 (1969).

Woods, D. R., S. J. Anderson, and S. L. Norman, Evaluation of Capital Cost Data: Heat Exchangers, *Can. J. Chem. Eng.*, **54**(6): 469 (1976).

Wood, R. K., and V. A. Sastry, Simulation Studies of a Heat Exchanger, *Simulation*, **18**(3): 105 (1972).

Wright, F. W., Advantages of Plate-in-Frame Heat Exchangers, *Petro/Chem. Eng.*, **39**(10): 31 (1967).

Wohl, M. H., Heat Transfer in Laminar Flow, *Chem. Eng.*, **75**(14): 81 (1968).

Yokell, S., Double-Tubesheet Heat-Exchanger Design Stops Shell-Tube Leakage, *Chem. Eng.*, **80**(11): 133 (1973).

Zimmerman, O. T., Adams Economat, *Cost Eng.*, **5**(4): 4 (1970).

Data Survey—U-Tube Heat Exchanger, *Chem. & Proc. Eng.*, **48**(3): 98 (1967).

New TEMA Standards What's New, What's Changed, *Hydrocarbon Process.*, **47**(9): 277 (1968).

Graphite Heat Exchangers, *Chem. Eng. Progr.*, **66**(7): 113 (1970).

Heat transfer design and costs

Adams, J. A., and D. F. Rogers, "Computer-Aided Heat Transfer Analysis," McGraw-Hill Book Company, New York, 1973.

Alves, G. E., Cocurrent Liquid-Gas Pipeline Contactors, *Chem. Eng. Progr.*, **66**(7): 60 (1970).

Amir, S. J., Calculating Heat Transfer from a Buried Pipeline, *Chem. Eng.*, **82**(16): 123 (1975).

Austin, G. T., Quick Charts for Integrating Heat-Capacity Equations, *Chem. Eng.*, **75**(12): 128 (1968).

Avriel, M., and D. J. Wilde, Engineering Design under Uncertainty, *I&EC Process Des. & Dev.*, **8**(1): 124 (1969).

Balakrishnan, A. R., and D. C. T. Pei, Heat Transfer in Fixed Beds, *I&EC Process Des. & Dev.*, **13**(4): 441 (1974).

Bannon, R. P., and S. Marple, Jr., Heat Transfer-Heat Recovery in Hydrocarbon Distillation, *Chem. Eng. Progr.*, **74**(7): 41 (1978).

Barrington, E. A., Cure Exchanger Acoustic Vibration, *Hydrocarbon Process.*, **57**(7): 193 (1978).

Bell, K. J., Temperature Profiles in Condensers, *Chem. Eng. Progr.*, **68**(7): 81 (1972).

Benkly, G. J., Fit Cooling Systems to the Job, *Hydrocarbon Process.*, **49**(11):219 (1970).

Bertram, C. G., V. J. Desai, and E. Interess, Designing Steam Tracing, *Chem. Eng.*, **79**(7):74 (1972).

Bisi, F., and S. Menicatti, How to Calculate Tank Heat Losses, *Hydrocarbon Process.*, **46**(2):145 (1967).

Bregman, J. I., Useful Energy From Unwanted Heat, *Chem. Eng.*, **78**(2):83 (1971).

Brooke, J. M., Corrosion Inhibitor Economics, *Hydrocarbon Process.*, **49**(9):299 (1970).

Brown, C. L., and D. Figenscher, Preheat Process Combustion Air, *Hydrocarbon Process.*, **52**(7):115 (1973).

Brown, C. W., Electric Pipe Tracing, *Chem. Eng.*, **82**(13):172 (1975).

Brown, R., A Procedure for Preliminary Estimates, *Chem. Eng.*, **85**(7):108 (1978).

Brown, T. R., Heating and Cooling in Batch Processes, *Chem. Eng.*, **80**(12):99 (1973).

Cahill, W. R., Pick Heaters Meet Variety of Water Needs, *Cost Eng.*, **15**(4):7 (1970).

Caplan, F., How to Put Finger on Heat Loss Data Fast, *Plant Eng.*, **21**(12):107 (1967).

Cavaliere, G. F., and R. A. Gyepes, Cost Evaluation of Intercooler Systems for Air Compressors, *Hydrocarbon Process.*, **52**(10):107 (1973).

Chapman, A. J., "Heat Transfer," The Macmillan Company, New York, 1967.

Cheng, C. Y., New Directions in Heat Transfer, *Chem. Eng.*, **81**(17):82 (1974).

Crawford, D. B., and G. P. Eschenbrenner, Heat Transfer Equipment for LNG Projects, *Chem. Eng. Progr.*, **68**(9):62 (1972).

Debrand, S., Heat Transfer During a Flash Drying Process, *I&EC Process Des. & Dev.*, **13**(4):396 (1974).

DeGance, A. E., R. W. Atherton, Transferring Heat In Two-Phase Systems, *Chem. Eng.*, **77**(10):113 (1970).

Dickey, B. R., E. S. Grimmett, and D. C. Kilian, Waste Heat Disposal via Fluidized Beds, *Chem. Eng. Progr.*, **70**(1):60 (1974).

Dickey, D. S., and R. W. Hicks, Liquid Agitation: Fundamentals of Agitation, *Chem. Eng.*, **83**(3):93 (1976).

Dorn, R. K., and M. J. Maddock, Design Pyrolysis Heater for Max Profits, *Hydrocarbon Process.*, **51**(11):79 (1972).

Elshout, R. V. L., Estimate Heat Loss From Insulated Pipe, *Hydrocarbon Process.*, **46**(1):216 (1967).

Emerson, W. H., The Application of a Digital Computer to the Design of Surface Condensers, *Chem. Eng. (London)*, **228**:178 (May 1969).

Epstein, L. D., What Do Jacketed Reactors Cost? *Hydrocarbon Process.*, **51**(12):102 (1972).

Fair, J. R., Designing Direct-Contact Coolers/Condensers, *Chem. Eng.*, **79**(13):91 (1972).

———, Process Heat Transfer by Direct Fluid-Phase Contact, *AIChE Symp. Ser.*, **68**(118):1 (1972).

Fanaritis, J. P., and H. J. Streich, Heat Recovery in Process Plants, *Chem. Eng.*, **80**(12):80 (1973).

Fischer, P., J. W. Suitor, and R. B. Ritter, Fouling Measurement Techniques, *Chem. Eng. Progr.*, **71**(7):66 (1975).

Fogg, R. M., and V. W. Uhl, Heat Transfer Resistance in Half-Tube and Dimpled Jackets, *Chem. Eng. Progr.*, **69**(7):76 (1973).

Frank, O., Estimating Overall Heat Transfer Coefficients, *Chem. Eng.*, **81**(10):126 (1974).

Fried, J. R., Heat-Transfer Agents for High-Temperature Systems, *Chem. Eng.*, **80**(12):89 (1973).

Frikken, D. R., K. S. Rosenberg, and D. E. Steinmeyer, Understanding Vapor-Phase Heat-Transfer Media, *Chem. Eng.*, **82**(12):86 (1975).

Frith, J. F., B. M. Bergen, and M. M. Shreehan, Optimize Heat Train Design, *Hydrocarbon Process.*, **52**(7):89 (1973).

Frost, W., and G. S. Dzakowic, Graphical Estimation of Nucleate Boiling Heat Transfer, *I&EC Process Des. & Dev.*, **6**(3):346 (1967).

Gambill, W. R., An Evaluation of Recent Correlations for High-flux Heat Transfer, *Chem. Eng.*, **74**(18):147 (1967).

Ganapathy, V., Quick Estimation of Gas Heat-Transfer Coefficients, *Chem. Eng.*, **83**(19):199 (1976).

———, Charts Simplify Spiral Finned-Tube Calculations, *Chem. Eng.*, **84**(9):117 (1977).

————, To Get Heat Transfer Coefficients, *Hydrocarbon Process.*, **56**(10):139 (1977); **56**(11):303 (1977); **56**(12):105 (1977).

————, Process-Design Criteria, *Chem. Eng.*, **85**(7):113 (1978).

Ghajar, A. J., and W. G. Tiederman, Prediction of Heat Transfer Coefficients in Drag Reducing Turbulent Pipe Flows, *AIChE J.*, **23**(1):128 (1977).

Gloyer, W., Thermal Design of Mixed Vapor Condensers, *Hydrocarbon Process.*, **49**(6):103 (1970).

————, Wall Viscosity Correction—New Look, *Hydrocarbon Process.*, **46**(10):158 (1967).

Gordon, E., M. H. Hashemi, R. D. Dodge, and J. LaRosa, Heat Transfer—A Versatile Steam Balance Program, *Chem. Eng. Progr.*, **74**(7):51 (1978).

Hinkle, R. E., and J. Friedman, Controlling Heat-Transfer Systems for Glass-Lined Reactors, *Chem. Eng.*, **85**(3):101 (1978).

Holt, A. D., Heating and Cooling of Solids, *Chem. Eng.*, **74**(22):145 (1967).

Hughes, R., and V. Deumaga, Insulation Saves Energy, *Chem. Eng.*, **81**(11):95 (1974).

Hughmark, G. A., Heat and Mass Transfer in the Wall Region of Turbulent Pipe Flow, *AIChE J.*, **17**(1):51 (1971).

————, Heat and Mass Transfer for Turbulent Pipe Flow, *AIChE J.*, **17**(4):902 (1971).

Jimeson, R. M., and G. G. Adkins, Waste Heat Disposal in Power Plants, *Chem. Eng. Progr.*, **67**(7):64 (1971).

Kasper, S., Selecting Heat-Transfer Media By Cost Comparison, *Chem. Eng.*, **75**(25):117 (1968).

Kern, D. Q., Converting Research to Design Use, *Chem. Eng. Progr.*, **65**(7):77 (1969).

———— and A. D. Kraus, "Extended Surface Heat Transfer," McGraw-Hill Book Company, New York, 1972.

Kern, W. I., Continuous Tube Cleaning Improves Performance of Condensers and Heat Exchangers, *Chem. Eng.*, **82**(22):139 (1975).

Khan, R. A., Effect of Noncondensables in Sea Water Evaporators, *Chem. Eng. Progr.*, **68**(7):79 (1972).

Koskinen, E., Indirect Cooling System Guards Environment, *Hydrocarbon Process.*, **52**(6):116 (1973).

Kugelman, A. M., Calculate Reforming Heat Needs, *Hydrocarbon Process.*, **52**(12):67 (1973).

Linnhoff, B., and J. R. Flower, Synthesis of Heat Exchanger Networks: Systematic Generation of Energy Optimal Networks, *AIChE J.*, **24**(4):633 (1978).

———— and ————, Synthesis of Heat Exchanger Networks: Evolutionary Generation of Networks with Various Criteria of Optimality, *AIChE J.*, **24**(4):642 (1978).

Lohrisch, F. W., What Is Optimum Batch Heat Transfer? *Hydrocarbon Process.*, **46**(7):169 (1967).

Lord, R. C., P. E. Minton, and R. P. Slusser, Design Parameters for Condensers and Reboilers, *Chem. Eng.*, **77**(6):127 (1970).

Madsen, N., Design Heat Exchangers for Liquids in Laminar Flow, *Chem. Eng.*, **81**(17):92 (1974).

Malek, R. G., Predict Nucleate Boiling Transfer Rates, *Hydrocarbon Process.*, **52**(2):89 (1973).

Margetts, R. J., Flow Problems in Feed Water Coils, *Chem. Eng. Progr.*, **69**(1):51 (1973).

Markovitz, R. E., Picking the Best Vessel Jacket, *Chem. Eng.*, **78**(26):156 (1971).

————, Improve Heat Transfer with Electropolished Clad Reactors, *Hydrocarbon Process.*, **52**(8):117 (1973).

Marshall, V. C., and N. Yazdani, Design of Agitated Coil-in-tank-Coolers, *Chem. Proc. Eng.*, **51**(4):89 (1970).

Martignon, D. R., Heat More Efficiently—With Electric Immersion Heaters, *Chem. Eng.*, **84**(11):141 (1977).

Milton, R. M., and C. F. Gottzman, High Efficiency Reboilers and Condensers, *Chem. Eng. Progr.*, **68**(9):56 (1972).

Mol, A., Which Heat Recovery System? *Hydrocarbon Process.*, **52**(7):109 (1973).

———— and J. J. Westenbrink, Steam Cracker Quench Coolers: Their Design and Location, *Hydrocarbon Process.*, **53**(2):83 (1974).

Moore, J. A., Development of a Low Maintenance Heat Recovery Exchanger, *Chem. Eng. Progr.*, **69**(1):43 (1973).

Neeld, R. K., and J. T. O'Bara, Jet Trays in Heat Transfer Service, *Chem. Eng. Progr.*, **66**(7): 53 (1970).

Nemunaitis, R. R., and J. S. Eckert, Heat Transfer in Packed Towers, *Chem. Eng. Progr.*, **71**(8): 60 (1975).

Nguyen, H. X., Optimize Water Outlet Temperature, *Hydrocarbon Process.*, **57**(5): 245 (1978).

Niccoli, L. G., R. T. Jaske, and P. A. Witt, System Costs Say Optimize Cooling, *Hydrocarbon Process.*, **49**(10): 97 (1970).

Noren, D. W., How Heat Pipes Work, *Chem. Eng.*, **81**(17): 89 (1974).

Null, H. R., Heat Pumps in Distillation, *Chem. Eng. Progr.*, **72**(7): 58 (1976).

Oleson, K. A., and R. R. Boyle, How to Cool Steam-Electric Power Plants, *Chem. Eng. Progr.*, **67**(7): 70 (1971).

Osburn, J. O., Simplified Calculation of Thermal Radiation Shields, *Chem. Eng.*, **76**(20): 139 (1969).

Rodriguez, F., Approximate LMTD, *Hydrocarbon Process.*, **55**(2)1: 125 (1976).

Root, W. L., III, and R. A. Nichols, Heat Transfer in Mechanically Agitated Units, *Chem. Eng.*, **80**(7): 98 (1973).

Saaski, E. W., and J. L. Franklin, Performance of an Evaporative Heat Transfer Wick, *Chem. Eng. Progr.*, **73**(7): 74 (1977).

Sadek, S. E., Heat Transfer to Air-Solids Suspensions in Turbulent Flow, *I&EC Process Des. & Dev.*, **11**(1): 133 (1972).

Sarkies, E., Waste Heat Recovery, *Power Eng.*, **71**(8): 62 (1967).

Sass, A., Simulation of the Heat-Transfer Phenomena in a Rotary Kiln, *I&EC Process Des. & Dev.*, **6**(4): 532 (1967).

Seifert, W. F., L. L. Jackson, and C. E. Sech, Organic Fluids for High-Temperature Heat-Transfer Systems, *Chem. Eng.*, **79**(24): 96 (1972).

Skelland, A. H. P., Non-Newtonian Flow and Heat Transfer, John Wiley & Sons, New York, 1967.

Stahel, E. P., and J. K. Ferrell, Heat Transfer, *I&EC*, **60**(1): 75 (1968).

Starczewski, J., Short Cut to Tubeside Heat Transfer Coefficient, *Hydrocarbon Process.*, **49**(2): 129 (1970).

Steinmeyer, D. E., Fog Formation in Partial Condensers, *Chem. Eng. Progr.*, **68**(7): 64 (1972).

Strek, F., and S. Masiuk, Heat Transfer in Liquid Mixers, *Intern. Chem. Eng.*, **7**(4): 693 (1968).

Taborek, J., T. Aoki, R. B. Ritter, and J. W. Palen, Heat Transfer: Fouling: The Major Unresolved Problem in Heat Transfer, *Chem. Eng. Progr.*, **68**(2): 59 (1972).

———— ———— ———— ———— and J. G. Knudsen, Predictive Methods for Fouling Behavior, Heat Transfer, *Chem. Eng. Progr.*, **68**(7): 69 (1972).

Uhl, V. W., and W. L. Root, III, Heat Transfer to Granular Solids in Agitated Units, *Chem. Eng. Progr.*, **63**(7): 81 (1967).

Umeda, T., J. Itoh, and K. Shiroko, Heat Transfer—Heat Exchange System Synthesis, *Chem. Eng. Progr.*, **74**(7): 71 (1978).

Walko, J. F., Controlling Biological Fouling in Cooling Systems, *Chem. Eng.*, **79**(24): 128 (1972); **79**(26): 104 (1972).

Waltrich, P. F., Sizing Vacuum Equipment for Evaporative Coolers, *Chem. Eng.*, **80**(11): 164 (1973).

Wasmund, B., and J. W. Smith, Wall to Fluid Heat Transfer in Liquid Fluidized Beds, *Can. J. Chem. Eng.*, **45**(3): 156 (1967).

Weierman, C., Pressure Drop Data for Heavy-Duty Finned Tubes, *Chem. Eng. Progr.*, **73**(2): 69 (1977).

Wigham, I., Designing Optimum Cooling Systems, *Chem. Eng.*, **78**(18): 95 (1971).

Wohl, M. H., Heat Transfer to Non-Newtonian Fluids, *Chem. Eng.*, **75**(15): 127 (1968).

Wolf, W., High Flux Tubing Conserves Energy, *Chem. Eng. Progr.*, **72**(7): 53 (1976).

Yurkanin, R. M., HPI Applications of Electric Process Heating, *Petro/Chem. Eng.*, **39**(9): 32 (1967).

Zanker, A., Predict Fouling by Nomograph, *Hydrocarbon Process.*, **57**(3): 145 (1978).

Miscellaneous Cost Data, *Cost Eng.*, **14**(3): 16 (1969).

Insulation costs

See references in Chap. 5.

PROBLEMS

1 A single-pass shell-and-tube heat exchanger contains 60 steel tubes. The ID of the tubes is 0.732 in., and the OD is 1.0 in. The shell side of the exchanger contains saturated steam at 290°F, and water passes through the tubes. The unit is designed with sufficient tube area to permit 15,000 gph of water to be heated from 70 to 150°F. In the course of this design, an h_d of 1500 Btu/(h)(ft^2)(°F) was assumed to allow for scaling on the water side of the tube. The film coefficient for the steam is 2000 Btu/(h)(ft^2)(°F). No safety factor other than the one scale value was used in carrying out the exchanger design. Estimate the temperature of saturated steam which must be used when the exchanger is new (i.e., no scale present) if the water enters the unit at a rate of 15,000 gph and is heated from 70 to 150°F.

2 A horizontal heat exchanger has seven steel tubes enclosed in a shell having an ID of 5.0 in. The OD of the tubes is 1.0 in., and the tube wall thickness is 0.10 in. Pure ethyl alcohol flows through the 1.0-in.-OD tubes. The ethyl alcohol enters the unit at 150°F and leaves at 100°F. Water at 70°F enters the shell side of the unit and flows countercurrent to the ethyl alcohol. It is necessary to cool 50,000 lb of ethyl alcohol per hour, and it has been decided to use 100,000 lb of water per hour.

Under the following conditions, determine the total pumping cost for the two fluids in the exchanger as dollars per year:

(a) There are no baffles, and flow on shell side can be considered to be parallel to the tubes.
(b) The outside of the shell is insulated, and there is no heat loss from the shell.
(c) The unit operates three hundred 24-h days per year.
(d) The efficiency of both pumps is 60 percent.
(e) Contraction, expansion, and fitting losses can be accounted for by increasing the straight-section frictional pressure drop by 20 percent.
(f) The specific heat of the ethyl alcohol may be assumed to be constant at 0.60. For water, the value may be assumed to be 1.0.
(g) The specific gravity of the ethyl alcohol may be assumed constant at 0.77. For water, the value may be assumed to be 1.0.
(h) No scale is present, and no safety factor is to be applied to the heat-transfer coefficients.
(i) Cost of power is $0.04 per kilowatthour.

3 A heat exchanger is to be constructed by forming copper tubing into a coil and placing it inside an insulated steel shell. If the following data apply, what should be the length of the coil?

(a) Water will flow inside the tubing, and a hydrocarbon vapor will condense on the outside of the tubing.
(b) ID of tubing = 0.5 in.
(c) OD of tubing = 0.6 in.
(d) Condensate rate = 1000 lb/h.
(e) Temperature of condensation = 190°F.
(f) Heat of vaporization of hydrocarbon at 190°F = 144 Btu/lb.
(g) Heat-transfer coefficient for condensing vapor = 250 Btu/(h)(ft^2)(°F).
(h) Inlet water temperature = 50°F.
(i) Outlet water temperature = 90°F.
(j) Heat losses from the shell may be neglected.

4 A solid surface at 1100 K is radiating to a second surface at 330 K. What temperature of the hot surface would be required if it were desired to double the number of joules transmitted per hour, the sink temperature and both emissivities remaining constant? What will be the percentage increase of the radiation coefficient h_r under the changed conditions?

5 A plywood-manufacturing concern is using a binder that requires a temperature of 180°F in order to obtain adequate holding strength. Large slabs of the plywood at an initial uniform temperature of 70°F are placed in a heater, and the heating unit is maintained at a constant temperature. The manufacturer wishes to pass the slabs through the heater continuously at such a rate that each slab remains in the heater for 15 min. If the slabs are 1 in. thick, determine the minimum constant temperature (i.e., with negligible surface resistance) required for the heater if the temperature at the center of each slab is to reach 180°F before leaving the heater. The following average data apply to the plywood: density = 35 lb/ft³ ; thermal conductivity = 0.10 Btu/(h)(ft²)(°F/ft); heat capacity = 0.50 Btu/(lb)(°F). Consider the plywood as homogeneous and isotropic.

6 A heat exchanger with two tube passes has been proposed for cooling distilled water from 93 to 85°F. The proposed unit contains 160 copper tubes, each $\frac{3}{4}$ in. OD, 18 BWG, and 16 ft long. The tubes are laid out on a $\frac{15}{16}$-in. triangular pitch, and the shell ID is 15$\frac{1}{4}$ in. Twenty-five percent cut segmental baffles spaced 1 ft apart are located in the shell. The correction factor F_s, for use in evaluating the outside-tube film coefficient can be assumed to be 1.3. Cooling water at 75°F will be used to remove the heat, and this fluid will flow through the tubes at a velocity of 6.7 fps. Under these conditions, the fouling coefficient is 2000 Btu/(h)(ft²)(°F) for the distilled water and 1000 Btu/(h)(ft²)(°F) for the cooling water. The pressure drop on the tube side and on the shell side cannot exceed 10 psi. Would the proposed unit be satisfactory for cooling 175,000 lb of distilled water per hour?

7 Determine the installed cost, tube length, and number of tubes for the optimum exchanger that will meet the following operating conditions and specifications:

    (a) Twenty thousand pounds of air per hour is cooled from 150 to 100°F inside the tubes of a shell-and-tube exchanger.

    (b) Water is used as the cooling medium. The water enters the unit at 70°F and leaves at 100°F.

    (c) All other conditions are the same as those specified in Example 5 of this chapter, except that B_i is not given.

8 Present a detailed derivation of Eq. (48) in this chapter, using, as a starting point, any of the other equations that are given.

9 Air, for use in a catalytic oxidation process, is to be heated from 200 to 520°F before entering the oxidation chamber. The heating is accomplished by the product gases, which cool from 720 to 400°F. A steel one-pass shell-and-tube exchanger with crossflow on the shell side will be used. The average absolute pressure on both the shell side and tube side can be assumed to be 10 atm, and the hot gases will pass through the tubes. The exchanger must handle 15,000 lb of the colder gas per hour, and it operates continuously for 8000 h/year. The properties of the product gases can be considered as identical to air. The cost for power delivered to either gas is $0.04 per kilowatthour. The OD of the tubes is 1.0 in.; the ID is 0.782 in. The tubes will be arranged in line with a square pitch of 1.5 in. All thermal resistances except those of the gas films may be neglected. The safety factor F_s for the outside film coefficient is 1.4. The terms B_o/n_b and $N_r N_c/N_t$ can both be assumed to be equal to 1.0. The cost data presented in Fig. 14-13 apply. Installation costs are 15 percent of the purchased cost, and annual fixed charges including maintenance are 20 percent of the installed cost. Under these conditions, estimate the tube length and purchased cost for the optimum exchanger.

CHAPTER FIFTEEN

MASS-TRANSFER AND REACTOR EQUIPMENT —DESIGN AND COSTS

The transfer of mass from one phase to another is involved in the operations of distillation, absorption, extraction, humidification, adsorption, drying, and crystallization. The principal function of the equipment used for these operations is to permit efficient contact between the phases. Many special types of equipment have been developed that are particularly applicable for use with a given operation, but finite-stage contactors and continuous contactors are the types most commonly encountered. A major part of this chapter, therefore, is devoted to the design aspects and costs of stagewise plate contactors and continuous packed contactors.

Chemical reactions and the equipment in which such reactions are carried out play an important role in chemical process analysis. This involves mass transfer as well as chemical kinetics, and a portion of this chapter deals with the major aspects of chemical reactor design and the costs of related equipment.

The initial cost for the operating equipment includes expenses for foundations, supports, installation, shell and shell internals, insulation, pumps, blowers, piping, heaters, coolers, and other auxiliaries, such as instruments, controls, heat exchangers, or special accessory equipment. Operating costs include power for circulating the fluids, maintenance, labor, cooling water, steam, and unrecovered materials. As illustrated in Chap. 10, a balance can be made among these various costs to yield an optimum economic design.

FINITE-STAGE AND CONTINUOUS CONTACTORS

Because the equipment for a finite-stage contactor consists of a series of inter-connected individual units or stages, a study of the overall assembly is best made on the basis of the flow and mass-transfer characteristics in each individual stage. Thus, for a sieve-, valve-, or bubble-cap-tray contactor, each tray can be considered as a separate entity, and the total design requires an analysis of the stepwise operation from one tray to another. In a differential-stage contactor, such as a packed column, the contacting operation can be considered as occurring continuously throughout the unit, since there are no fixed locations where the equipment is divided physically into finite sections. The overall analysis for a differential-stage contactor, therefore, can be based on a differential length or height.

Net mass transfer between two phases can occur only when there is a driving force, such as a concentration difference, between the phases. When equilibrium conditions are attained, the driving force and, consequently, the net rate of mass transfer becomes zero. A state of equilibrium, therefore, represents a theoretical limit for mass-transfer operations. This theoretical limit is used extensively in mass-transfer calculations.

A theoretical stage is defined as a contacting stage in which equilibrium is attained between the various phases involved. Thus, in a sieve-tray column, a theoretical or perfect plate is one in which the liquid leaving the tray is in equilibrium with the gas leaving the tray. The same approach is often used for packed columns, where the HETP is defined as the height of the packed column necessary to give a separation equivalent to one theoretical plate. A more rigorous method for evaluating the performance of a continuous contactor requires a differential treatment of the separation process and gives results that can be expressed in terms of the mass-transfer coefficient or as the number of transfer units. The transfer unit is similar to the theoretical plate, but the transfer unit is based on a differential change in equilibrium conditions and actual concentrations, and the theoretical plate is based on finite changes.

The design of most mass-transfer equipment requires evaluation of the number of theoretical stages or transfer units. Methods for carrying out these calculations for various types of mass-transfer operations are presented in many general chemical engineering books, such as those indicated in the Chemical Engineering Series list of books given at the front of this text.

Because methods for determining theoretical stages and transfer units are covered extensively in the types of references mentioned in the preceding paragraph, the details will not be repeated here. Other important decisions, however, must be made in the design of mass-transfer equipment, and an error in these decisions can be just as detrimental as an error in evaluating the number of theoretical stages. The following sections deal with design factors for finite-stage and differential-stage contactors as related to the direct operational characteristics of the equipment.

FINITE-STAGE CONTACTORS

The most common types of finite-stage contactors are bubble-cap-tray, sieve-tray, or valve-tray units, although turbogrid trays and other speciality types of units are also used in industrial operations. Many of the units constructed in the past have used bubble caps for the contactor, but sieve and valve types of contactors are less expensive than bubble caps and are just as efficient; so bubble caps are now seldom used for new equipment. Nevertheless, it is still worthwhile to consider the aspects of bubble-cap-tray design in some detail, since many of the same principles apply to valve-tray, sieve-tray, and bubble-cap-tray design and large numbers of bubble-cap-tray columns remain in service for which the engineer needs to know design principles in order to understand the operation.

Plate or tray towers are particularly useful as compared to packed towers when fluctuations in the vapor or liquid rate may occur or where major changes in the overall capacity of the column are anticipated. The liquid held on each tray makes the finite-stage contactor useful for cases where time must be allowed for a chemical reaction, as in an absorption tower for producing nitric acid by absorption and aqueous reaction of nitrogen dioxide. The problems of poor liquid or gas distribution encountered at various loading capacities with packed towers can be avoided with plate towers, and plate towers are usually easier to clean when solid deposits are involved. On the other hand, packed towers often can operate with lower pressure drop than plate units which can be a major advantage for high-vacuum service, and the cost for a packed tower capable of doing the equivalent job is often considerably cheaper than that for a plate tower.

Critical factors in the design of finite-stage contactors, other than the determination of the number of stages theoretically necessary for the required operation, are (1) diameter of column so that flooding or excessive entrainment will not occur, (2) the operating efficiency of the trays expressing how close the operation comes to the theoretically perfect tray, and (3) pressure drop generated across each tray. Other factors of importance in the design are appropriate dimensions and form of the contactor assemblies, liquid flow patterns on the trays, entrainment, tray spacing, downcomers for carrying fluid from one tray to another, and tray stability. Detailed empirical equations and other relationships have been developed to aid in designing the finite-stage contacting equipment taking the various factors and types of contactors into account. Rules of thumb are also useful for application in developing preliminary designs. These methods are discussed in the following sections.†

---

† For further treatment on design methods for finite-stage contactors used for fractionation including a tabulation of recommended limits on design variables for bubble-cap trays, perforated trays, and valve trays, see M. VanWinkle, "Distillation," Chap. 14, McGraw-Hill Book Company, New York, 1967. See also A. P. Economopoulos, Computer Design of Sieve Trays and Tray Columns, *Chem. Eng.*, **85**(27):109 (Dec. 4, 1978).

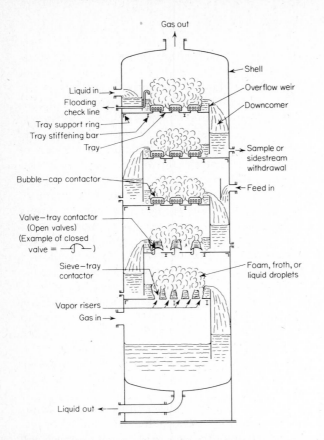

Figure 15-1 Cross-sectional view of finite-stage contactor tower in operation showing an example of a sieve tray, a valve tray, and a bubble-cap tray.

BUBBLE-CAP-TRAY, SIEVE-TRAY, AND VALVE-TRAY UNITS

Examples of the three common forms of finite-stage contactors (Bubble cap, Sieve, and Valve) are shown in Fig. 15-1 which represents a tray tower in operation and illustrates the basic form of each of the three contactors. The tower consists of a series of individual trays, each equipped with a series of contacting units to achieve close contact and resultant mass transfer between a gas phase and a liquid phase. With the bubble-cap contactor, the gas passes upward through the risers into the bubble caps, where the liquid is depressed, permitting the gas to bubble through the slots or notches in the cap into the liquid. As the bubbles are dispersed into and rise through the liquid on the tray, a large amount of interfacial area exists between the gas and liquid phases, thereby permitting effective mass transfer. Liquid flows downward from tray to tray through downcomers, and the necessary gas-liquid contact is made as the liquid passes

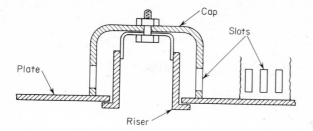

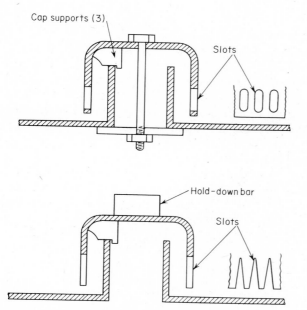

Figure 15-2 Methods for holding bubble caps in place.

across each tray. Figures 15-2 to 15-4 show typical forms of bubble caps and operations including effects of excessive liquid gradient while Fig. 15-5 shows examples of liquid-flow patterns for finite-stage trays.

The sieve-tray contactor shown in Fig. 15-1 is presented in a form known as a *crossflow plate contactor*. The tray consists of a flat plate perforated with many small holes that are drilled or punched in a size range of $\frac{1}{8}$ to $\frac{1}{2}$ in. diameter. Liquid flows across the plate as shown in Fig. 15-1, through the froth or spray which develops, and passes over a weir into the downcomer leading to the tray below. The upward flow of the vapor keeps the liquid from flowing through the holes, and the operation of the tray is basically the same as that of a bubble-cap tray. If the flow of gas is low, some or all of the liquid may drain down through the perforations so that some of the contacting areas may be bypassed. If the entire transfer of the liquid from one tray to another is by this so-called "weep-

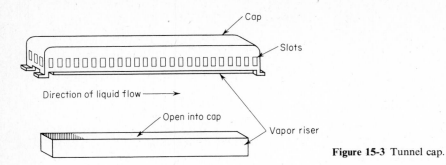

Figure 15-3 Tunnel cap.

ing" action with no downcomer being used, the type of unit is designated as a *counterflow plate contactor*.

Because best results are normally obtained with full crossflow plate operation for a sieve tray, units may be designed with a lift valve over the hole in the plate or over a riser from the plate so that the rising gas lifts this valve to allow the vapors to be passed horizontally into the liquid as is illustrated in Fig. 15-1. The liquid cannot easily flow back down the holes in the plate when the gas flow is low because the valve tends to close with the reduced gas flow.

Many modifications of the three types of contactors just discussed have been developed in an effort to reduce costs, reduce pressure drop, equalize vapor flow through each contactor, increase plate efficiencies, or, in general, improve the operating performance of the tower. An example of this for modification of bubble-cap towers is the old *Uniflux tray* originally developed by Socony-Vacuum, which consisted of a series of interlocking S-shaped sections which were assembled in the form of tunnel caps with slot outlets on one side only. Segmental downcomers, similar to the downcomers in conventional bubble-cap columns, were provided. The vapors issued from the Uniflux caps in only one

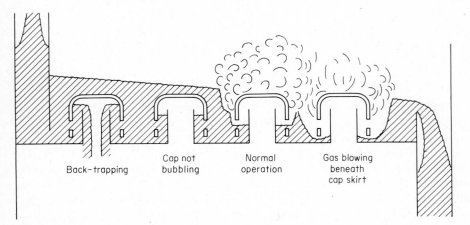

Figure 15-4 Cross-sectional view of bubble-cap tower showing effect of excessive liquid gradient.

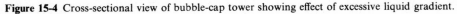

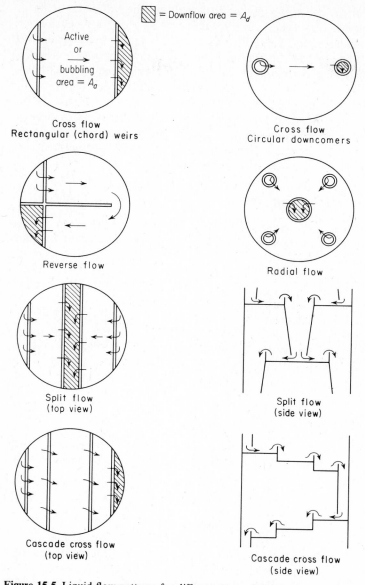

Figure 15-5 Liquid flow patterns for different types of finite-stage trays.

direction in such a manner as to give equal vapor distribution among the caps. Even though the simplified construction resulted in costs 30 to 50 percent less than for equivalent bubble-cap trays of conventional design while providing large capacities and good plate efficiencies, the Uniflux tray has not been able to compete effectively with sieve trays and is no longer being fabricated.

Various types of valve arrangements are used for the valve-tray towers. Some of these are liftable disk-type valves which come as a cap with each riser. As the vapor rate increases, the valve lifts to provide a larger opening until it reaches a limit point which still directs the vapors horizontally into the liquid. *Float-valve trays* are designed to operate on the same principle as the liftable disk-type valves except that a floating pivotal valve is used in each cap.

A variation of the normal sieve tray is the *Turbogrid* tray developed by the Shell Development Company. The Turbogrid tray uses slots instead of circular holes and operates as a counterflow plate unit with no downcomers. A typical tray consists of a flat grid of parallel slots. The slots can be stamped perforations in a flat metal sheet or the open spaces between parallel bars.

MAXIMUM ALLOWABLE VAPOR VELOCITIES

The vapor velocity in a finite-stage contactor column can be limited by the liquid handling capacity of the downcomers or by entrainment of liquid droplets in the rising gases. In most cases, however, downcomer limitations do not set the allowable vapor velocity; instead, the common design basis for choosing allowable vapor velocities is a function of the amount of gas entrainment which can result in improper operation or flooding of the column.

A tower must have sufficient cross-sectional area to handle the rising gases without excessive carry-over of liquid from one tray to another. By assuming that the frictional drag of the vapor on suspended liquid droplets should not exceed the average weight of a droplet, Souders and Brown† derived the following equation applicable for any specific location in the column:

$$V_m = K_v \sqrt{\frac{\rho_L - \rho_G}{\rho_G}} \qquad (1)‡$$

where V_m = maximum allowable superficial vapor velocity (based on cross-sectional area of empty tower), ft/s, and K_v = an empirical constant, ft/s. An alternative form of Eq. (1) in terms of mass velocity follows:

$$G_m = V_m \rho_G = K_v \sqrt{\rho_G(\rho_L - \rho_G)} \qquad (2)$$

where G_m = maximum allowable mass velocity of vapor, lb/(s)(ft^2).

Equation (1) or (2) can be used as an empirical guide for estimating the maximum vapor velocities in plate columns. The constant K_v is a major function of the plate spacing and also varies to a lesser extent with depth of liquid on the tray, ratio of liquid flow rate to gas flow rate, surface tension of the liquid, density of the gas and liquid, and physical arrangement of the tray components.

While it is always best to obtain values for the constant K_v based on data obtained with fluids, equipment, and operating pressures and temperatures simi-

---

† M. Souders and G. G. Brown, *Ind. Eng. Chem.*, **26**:98 (1934).

‡ As a general rule of thumb, V_m should be near 4 ft/s for finite-stage fractionation towers operating at atmospheric pressure.

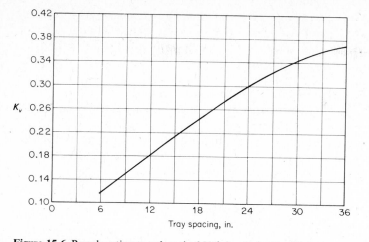

Figure 15-6 Rough estimate values ($\pm 25\%$) for K_v in Eq. (1) for maximum allowable velocities in finite-stage towers *(weir height less than 15% of plate spacing)*.

lar to those involved in the particular design, a rough approximation of K_v can be obtained from Fig. 15-6.† This figure gives values of K_v as a function of tray spacing only and should not be relied on for better than ± 25 percent for maximum allowable velocities.

An alternate approach for estimating maximum allowable velocities has been presented by Fair (see reference given in footnote for preceding paragraph) which is based on data obtained with sieve-tray and other types of finite-stage columns and takes into account the effect of surface tension of the liquid in the column, the ratio of the liquid flow rate to the gas flow rate, gas and liquid densities, and dimensions and arrangement of the contactor.‡ In this method, the basic equation for the maximum allowable vapor velocity, equivalent to Eq. (1), is

$$V'_m = K'_v \left(\frac{\sigma}{20} \right)^{0.2} \sqrt{\frac{\rho_L - \rho_G}{\rho_G}} \tag{3}$$

where V'_m = maximum allowable vapor velocity based on net area for vapor flow which is usually the active or bubbling cross-sectional area of the tower (A_a in Fig. 15-5) plus the area of one downcomer (A_d in Fig. 15-5), ft/s

K'_v = an empirical constant, ft/s

σ = surface tension of the liquid in the tower, dyn/cm.

† Figure 15-6 is adapted from values of constants presented by J. R. Fair, *Petro/Chem. Eng.,* **33**(10):45 (1961) and gives K_v values that are much less conservative than the original values recommended by Souders and Brown.

‡ See also J. H. Perry and C. H. Chilton, "Chemical Engineers Handbook," 5th ed., Sec. 18, McGraw-Hill Book Company, New York, 1973.

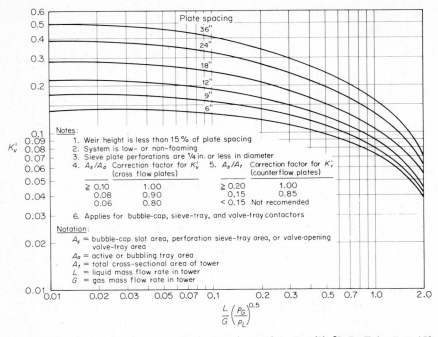

The chart axes and labels include:

K_v' on the left vertical axis, with values 0.6, 0.5, 0.4, 0.3, 0.2, 0.1, 0.09, 0.08, 0.07, 0.06, 0.05, 0.04, 0.03, 0.02, 0.01

Horizontal axis: $\dfrac{L}{G}\left(\dfrac{\rho_G}{\rho_L}\right)^{0.5}$ with values 0.01, 0.02, 0.03, 0.05, 0.07, 0.1, 0.2, 0.3, 0.5, 0.7, 1.0, 2.0

Plate spacing curves: 36", 24", 18", 12", 9", 6"

Notes:
1. Weir height is less than 15% of plate spacing
2. System is low- or non-foaming
3. Sieve plate perforations are ¼ in. or less in diameter
4. A_s/A_a Correction factor for K_v' (cross flow plates) 5. A_s/A_t Correction factor for K_v' (counterflow plates)

A_s/A_a	factor	A_s/A_t	factor
≥ 0.10	1.00	≥ 0.20	1.00
0.08	0.90	0.15	0.85
0.06	0.80	< 0.15	Not recomended

6. Applies for bubble-cap, sieve-tray, and valve-tray contactors

Notation:
A_s = bubble-cap slot area, perforation sieve-tray area, or valve-opening valve-tray area
A_a = active or bubbling tray area
A_t = total cross-sectional area of tower
L = liquid mass flow rate in tower
G = gas mass flow rate in tower

Figure 15-7 Chart for estimating values of K_v' ($\pm 10\%$) in Eq. (3). [J. R. Fair, *Petro/Chem. Eng.*, **33**(10):45 (1961). *With permission.*]

Figure 15-7 presents values of K_v' as a function of plate spacing and also presents information on limitation of the applicability of the results.

Although the allowable velocity given by Eq. (1) may be conservative for many types of operations while Eq. (3) is somewhat less conservative due to the use of the active or bubbling cross-sectional area of the tower along with the other corrections, many engineers employ vapor velocities in the range of 65 to 80 percent of V_m or V_m' to make certain their column will be operable. The actual velocity to be used is applied to set the diameter of the column. The additional dimensions of the trays are then established, and an investigation of other limiting factors, such as downcomer capacity, should be made.

In many cases, the vapor rate changes over the length of the tower, and the theoretical diameter based on the allowable vapor velocity varies. Occasionally, two different diameters are used for different sections of one tower. Cost considerations, however, usually make it impractical to vary the diameter, and the constant diameter should be based on the tower location where allowable velocity and throughput rates require the largest diameter.

Example 1. Determination of distillation-column diameter on basis of allowable vapor velocity A sieve-tray distillation tower is to be operated under the following conditions:

	At top of tower	At bottom of tower
Liquid rate	245 lb mol/h	273 lb mol/h
Vapor rate	270 lb mol/h	310 lb mol/h
Vapor molecular weight	70	110
Temperature	220°F	260°F
Pressure	1.0 atm	1.1 atm
Liquid density	44 lb/ft³	42 lb/ft³
Liquid surface tension	20 dyn/cm	20 dyn/cm

The tray spacing is 24 in. with a weir height of 3 in. or 12.5 percent of the tray spacing. The ideal-gas law can be assumed as applicable to the vapors, and the system is nonfoaming. Total area of sieve holes (A_s) is 10 percent of the active or bubbling tray area (A_a). A rough rule of thumb is that downcomer area (A_d in Fig. 15-5) is 5 percent of total column cross-sectional area ($2A_d + A_a$ in Fig. 15-5). The molecular weights of the liquid and the gas may be assumed as identical at any given point in the column. If the tower diameter is to remain constant over the entire length, compare the minimum diameter estimated using Fig. 15-6 with that obtained by using Fig. 15-7.

SOLUTION From Fig. 15-6, $K_v = 0.30$. At top of tower,

$$\rho_G = \frac{(70)(492)}{(359)(680)} = 0.141 \text{ lb/ft}^3$$

$$\rho_L = 44 \text{ lb/ft}^3$$

$$V_m = 0.30 \sqrt{\frac{44 - 0.141}{0.141}} = 5.29 \text{ ft/s based on total cross-sectional area}$$

$$\text{Minimum diameter} = \left[\frac{(70)(270)(4)}{(0.141)(\pi)(5.29)(3600)} \right]^{1/2} = 3.0 \text{ ft}$$

At bottom of tower,

$$\rho_G = \frac{(110)(492)(1.1)}{(359)(720)(1.0)} = 0.23 \text{ lb/ft}^3$$

$$\rho_L = 42 \text{ lb/ft}^3$$

$$V_m = 0.30 \sqrt{\frac{42 - 0.23}{0.23}} = 4.04 \text{ ft/s}$$

$$\text{Minimum diameter} = \left[\frac{(110)(310)(4)}{(0.23)(\pi)(4.04)(3600)} \right]^{1/2} = 3.6 \text{ ft}$$

The limiting diameter occurs at the bottom of the tower; therefore, the minimum diameter based on the maximum allowable vapor velocity is 3.6 ft obtained with use of Fig. 15-6.

Figure 15-7 can be used to obtain K_v' directly since $A_s/A_a = 0.10$, weir height is 12.5 percent of tray spacing, and system is nonfoaming.

At top of tower,

$$\frac{L}{G}\left(\frac{\rho_G}{\rho_L}\right)^{0.5} = \frac{245}{270}\left(\frac{0.141}{44}\right)^{0.5} = 0.051$$

From Fig. 15-7, $K_v' = 0.36$

$$V_m' = (0.36)\left(\frac{20}{20}\right)^{0.2}\sqrt{\frac{44 - 0.141}{0.141}}$$

$$= 6.35 \text{ ft/s based on active area plus area of one downcomer}$$

With rule of thumb that downcomer area $(A_d) = 5$ percent of total cross-sectional area,

$$V_m = (6.35)\left(\frac{100 - 5}{100}\right) = 6.0 \text{ ft/s based on total cross-sectional area}$$

$$\text{Minimum diameter} = \left[\frac{(70)(270)(4)}{(0.141)(\pi)(6.0)(3600)}\right]^{1/2} = 2.8 \text{ ft}$$

At bottom of tower,

$$\frac{L}{G}\left(\frac{\rho_G}{\rho_L}\right)^{0.5} = \frac{273}{310}\left(\frac{0.23}{42}\right)^{0.5} = 0.065$$

From Fig. 15-7, $K_v' = 0.35$

$$V_m' = (0.35)\left(\frac{20}{20}\right)^{0.2}\sqrt{\frac{42 - 0.23}{0.23}}$$

$$= 4.72 \text{ ft/s based on active area plus area of one downcomer}$$

$$V_m = (4.72)\left(\frac{100 - 5}{100}\right)$$

$$= 4.5 \text{ ft/s based on total cross-sectional area}$$

$$\text{Minimum diameter} = \left[\frac{(110)(310)(4)}{(0.23)(\pi)(4.5)(3600)}\right]^{1/2} = 3.5 \text{ ft}$$

The limiting diameter occurs at the bottom of the tower; therefore, the minimum diameter based on the maximum allowable vapor velocity obtained with use of Fig. 15-7 is 3.5 ft compared to 3.6 ft obtained with use of Fig. 15-6.

PLATE AND COLUMN EFFICIENCIES

As discussed in the first part of this chapter, the design of mass-transfer equipment often requires evaluation of the number of theoretical stages necessary to accomplish a desired separation. To complete the design, information must be available that shows the relationship between these ideal values and the actual performance of the equipment. The translation of ideal stages into actual finite stages can be accomplished by the use of plate efficiencies.

Types of Plate Efficiency

Three kinds of plate efficiencies may be used for expressing the relationship between the performance of theoretical and actual stages. They are (1) overall column efficiency or overall plate efficiency, (2) Murphree plate efficiency, and (3) point efficiency or local efficiency.

The *overall column efficiency* applies to the total number of stages and is defined as the number of theoretical stages required to produce a given separation divided by the number of stages actually necessary to produce the same separation. The *Murphree plate efficiency* applies to a single plate. It is defined as the ratio of the actual change in average vapor composition accomplished by a given plate to the change in average vapor composition if the vapors leaving the plate were in equilibrium with the liquid leaving the plate. The *point efficiency* is similar to the Murphree plate efficiency, except that the point efficiency applies to a single location on a given tray.

Although the overall column efficiency has no fundamental mass-transfer basis, it is widely used because of its simplicity. The number of actual stages required for a given separation is simply equal to the number of theoretical stages divided by the overall column efficiency. The Murphree plate efficiency is more fundamental than the overall value but is less convenient to use because it must be applied to each individual plate. The point efficiency is of considerable theoretical interest but is seldom used in design practice because it requires knowledge of the variations in liquid composition across the tray and integration of the point efficiencies over the entire tray. Point efficiencies are always less than 100 percent, but Murphree plate efficiencies may be greater than 100 percent and are usually greater than point efficiencies on the same tray because of variation in liquid composition across the tray.

Factors Influencing Plate and Column Efficiencies

A comparison of overall column efficiencies for bubble-cap, sieve, and valve finite-stage contactors is presented in Fig. 15-8 which also shows the effects of the superficial vapor velocity (based on cross-sectional area of empty tower) and the gas density.† In general, the three types of contactors give plate efficiencies in the range of 80 to 90 percent when the column is operated at appropriate conditions with sieve and valve contactors generally giving slightly higher efficiencies

† Kastenek et al., *Proc. Intern. Symp. Distillation 1969*, Institute of Chemical Engineers, London (1969).

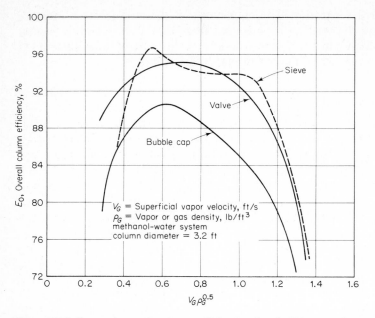

Figure 15-8 Comparison of column efficiencies for bubble-cap, sieve, and valve finite-stage contactors.

than bubble-cap contactors. The major factors that influence plate and column efficiencies are discussed in the following:

Vapor velocity As shown in Fig. 15-8, efficiencies increase with increase in superficial vapor velocity when the velocity is low, reach a fairly constant value over the range of velocities corresponding to normal acceptable operation, and then decrease at higher velocities. Vapor slot velocities appear to have no appreciable effect on efficiency if they are kept in the range indicated by Eqs. (20) and (21) as presented later in this chapter.

Liquid depth above vapor openings Increase in depth of the liquid above the vapor openings tends to increase the efficiency, especially if this liquid depth is less than 1 in.

Plate spacing Because of entrainment carry-over, the effect of plate spacing is related to the superficial vapor velocity. Too small a plate spacing can cause a low efficiency if the vapor velocity is greater than the allowable value.

Length of liquid path The length of the liquid path across a tray is an important factor in determining the degree of liquid concentration gradient across the tray. Thus, if the length of the liquid path is long enough that an appreciable liquid concentration gradient is established, the Murphree plate efficiency is greater than the point efficiency. In general, as the length of the liquid path is increased, the overall column efficiency increases. The effect of liquid-path length is usually negligible if the length is less than 5 ft, but increasing the length to 10 to 15 ft may increase the overall column efficiency by 20 to 40 percent.

Liquid resistance to interphase mass transfer Liquid viscosity, gas solubility in absorbers, and relative volatility in rectification columns are important factors in determining the liquid resistance to interphase mass transfer. Increase in liquid viscosity, decrease in gas solubility for absorbers, and increase in relative volatility for rectification columns cause an increase in the liquid resistance to interphase mass transfer and a resultant reduction in plate efficiency. The ratio of the liquid rate to the gas rate influences the relative importance of the liquid resistance to interphase mass transfer. An increase in the ratio of liquid rate to gas rate reduces the importance of the liquid resistance and can cause an increase in the plate efficiency.

Other factors Design details of the column, such as vapor-opening dimensions, plate layout, or the total number of trays can have an effect on the efficiencies. The exact influence of these factors is best determined by experimental tests.

In multicomponent mass-transfer operations, the assumption is usually made that the same plate efficiency applies to all components being separated. The overall column efficiency is then considered in terms of the key components, and the same efficiency is assumed for the lighter and heavier components. This assumption is not necessarily correct because of the different properties of the components. More exact results can be obtained by using Murphree plate efficiencies and accounting for the difference in efficiencies.

Correlations for Estimation of Plate Efficiencies

Overall column efficiencies are based on performance data, and no exact correlation of the results obtained with various mixtures and types of columns can be presented. Many generalized correlations have been developed, however, and these are useful for making design estimates when no other data are available.

For standard types of finite-stage contactor columns operated in the range of allowable velocities where the overall column efficiencies are essentially constant, O'Connell has correlated efficiency data on the basis of liquid viscosity and relative volatility (or gas solubility).† The results for fractionators and absorbers are presented in Fig. 15-9. This correlation is based, primarily, on experimental data obtained with bubble-cap columns having a liquid path of less than 5 ft and operated at a reflux ratio near the minimum value. Figure 15-9 is adequate for design estimates with most types of commercial equipment and mixtures, although efficiencies determined directly from equipment operating near the conditions involved in the design are always to be preferred.

† H. E. O'Connell, *Trans. AIChE*, **42**:751 (1946).

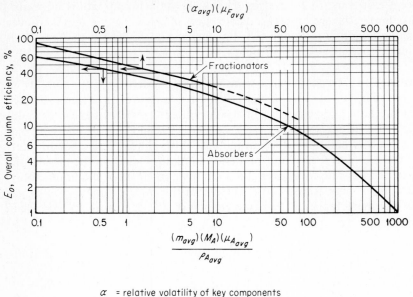

$$\alpha \quad = \text{relative volatility of key components}$$

μ_F = molal avg. viscosity of feed, cp
μ_A = molal avg. viscosity of liquid, cp
M_A = average molecular weight of liquid
ρ_A = liquid density, lb/cu ft
m = mole fraction of solute in gas in equilibrium
with liquid/mole fraction of solute in liquid

avg = at average column temperature and pressure

Figure 15-9 Overall column efficiencies for finite-stage contactor fractionators and absorbers.

The correlation shown by Fig. 15-9 can be extended to include the effects of liquid submergence and the ratio of liquid flow rate to vapor flow rate by use of the following simplified equations for estimating overall column efficiencies:†‡
    For fractionators,

$$\log E_o = 1.67 - 0.25 \log \left(\mu_{F,\,\text{avg}}\alpha_{\text{avg}}\right) + 0.30 \log \frac{L'_M}{V'_M} + 0.09\left(S_m + \frac{c}{2}\right) \quad (4)$$

For absorbers,

$$\log E_o = 1.60 - 0.38 \log \frac{m_{\text{avg}} M_A \mu_{A,\,\text{avg}}}{\rho_{A,\,\text{avg}}} + 0.25 \log \frac{L'_M}{V'_M} + 0.09\left(S_m + \frac{c}{2}\right) \quad (5)$$

† J. C. Chu et al., *J. Appl. Chem.*, **1**:529 (1951). For sieve-tray units, replace $(S_m + c/2)$ by the weir height. For valve-tray units, replace $(S_m + c/2)$ by liquid height above the base of the valve opening.
‡ For other empirical methods for estimating overall efficiencies, see S. Bakowski, *Brit. Chem. Eng.*, **8**:384 and 472 (1963) and S. Bakowski, *Brit. Chem. Eng.*, **14**:945 (1969).

where E_o = overall column efficiency, percent

S_m = static submergence, ft

c = slot height, ft

L'_M = molal liquid flow rate, lb mol/h

V'_M = molal vapor flow rate, lb mol/h

$\mu_{F,\text{avg}}$, α_{avg}, m_{avg}, M_A, $\mu_{A,\text{avg}}$, $\rho_{A,\text{avg}}$ are defined in Fig. 15-9.

Equations (4) and (5) are based on the relationships shown by Fig. 15-9 plus additional experimental results on the effects of liquid submergence and the ratio L'_M/V'_M. Use of these equations should be limited to conditions in which the ratio L'_M/V'_M is in the range of 0.4 to 8 and the static submergence is less than 1.5 in.

A simplified empirical equation for finite-stage contactor columns operating on petroleum and similar hydrocarbons has been presented by Drickamer and Bradford.† Their results are based on plant tests with 54 refinery columns used for distillation or absorption of hydrocarbons. The columns were of the standard bubble-cap or perforated-plate type operated under typical refinery conditions. The results were correlated on the basis of the single variable, liquid viscosity, to give

$$E_o = 17 - 61.1 \log \mu_{F,\text{avg}} \qquad (6)$$

Use of Eq. (6) should be limited to commercial towers for which no other data are available. It gives adequate results for the fractionation of petroleum and similar hydrocarbons, but it is not recommended if the relative volatility of the key components is greater than 4.0 or if the value of $\mu_{F,\text{avg}}$ is outside the range of 0.07 to 1.4 centipoises.

Example 2. Estimation of overall column efficiency A continuous fractionation unit has been designed to operate on a liquid feed containing components, A, B, C, and D. Calculations have shown that 20 theoretical stages are necessary in the column, not including the reboiler. On the basis of the following data, estimate the overall column efficiency and the number of actual trays needed in the column by (a) Fig. 15-9, (b) Eq. (4), and (c) Eq. (6):

Com-ponent	Mole fraction in			Viscosity of liquid at 260°F, centipoises
	Feed	Overhead	Bottoms	
A	0.10	0.25		0.040
B	0.30	0.70	0.03	0.110
C	0.40	0.05	0.64	0.138
D	0.20		0.33	0.175

† H. G. Drickamer and J. R. Bradford, *Trans. AIChE*, **39**:319 (1943); *Petrol. Refiner*, **22**(10):105 (1943).

Materials B and C are considered as the key components.

Relative volatility of the key components is independent of concentration and equals 1.94 at 260°F and the average column pressure. Feed temperature = 85°F. Overhead temperature = 240°F. Bottoms temperature = 280°F.

The fractionating unit is a sieve-tray column with a standard type of tray design. The vapor velocity through the tower is about 90 percent of the maximum allowable value.

Tower diameter = 4.8 ft. Weir height = 2 in. The ratio L'_M/V'_M is 0.7 in the rectifying section of the column and 1.2 in the stripping section.

SOLUTION Average column temperature = $(280 + 240)/2 = 260°F$.

Molal average viscosity of feed at average column temperature = $\mu_{F,\,avg} = (0.040)(0.10) + (0.110)(0.30) + (0.138)(0.40) + (0.175)(0.20) = 0.127$ centipoise.

Average relative volatility of key components at average column temperature = $\alpha_{avg} = 1.94$.

Since L'_M/V'_M is different in the rectifying and stripping sections, an average value must be used, or else a separate efficiency for each section can be obtained. In this case an arithmetic-average value is adequate:

$$\frac{L'_M}{V'_M} = \frac{0.7 + 1.2}{2} = 0.95$$

$$\text{Weir height} = \frac{2}{12} = 0.167 \text{ ft.}$$

(a) Efficiency and total number of trays by Fig. 15-9:

$$(\alpha_{avg})(\mu_{F,\,avg}) = (1.94)(0.127) = 0.246$$

From Fig. 15-9,

$$E_o = 70\%$$

$$\text{Number of actual trays needed in column} = \frac{20}{0.7} = 29$$

(b) Efficiency and total number of trays by Eq. (4):

$$\log E_o = 1.67 - 0.25 \log 0.246 + 0.3 \log 0.95 + 0.09(0.167)$$

$$E_o = 68\%$$

$$\text{Number of actual trays needed in column} = \frac{20}{0.68} = 29$$

(c) Efficiency and total number of trays by Eq. (6):

$$E_o = 17 - 61.1 \log 0.127 = 72\%$$

$$\text{Number of actual trays needed in column} = \frac{20}{0.72} = 28$$

PRESSURE DROP OVER FINITE-STAGE CONTACTORS

As the gas passes through a finite-stage contactor, the pressure of the gas decreases because of the following causes:

1. Pressure drop through the contactor assembly
   a. Contraction as the gas enters the riser or sieve opening
   b. Friction in the riser or sieve opening and in the annular space if a bubble-cap unit is used
   c. Friction due to change in direction of gas flow for bubble-cap and valve units
   d. Passage of the gas through the slots for a bubble-cap unit
2. Pressure drop due to liquid head above slots, sieve openings, or valve openings.

Because the pressure drop is uniform throughout the gas space above or below a tray, the gas pressure drop across a given tray must be the same irrespective of the location on the tray. A design estimate of the pressure drop, therefore, can be obtained on the basis of an average contactor assembly, such as a bubble-cap, a sieve tray, or a valve unit. Because of the liquid gradient which may exist with bubble-cap or valve-tray units, the rate of gas flow through the individual caps or valves may vary from the liquid-inlet to the liquid-outlet location on the tray. For a stable plate, an average contactor assembly for use in pressure-drop calculations is one located at the point of average liquid gradient. The flow rate of gas through this assembly can then be assumed to be the flow rate per assembly assuming each assembly delivers the same amount of gas.

In making pressure-drop calculations for bubble-cap, sieve-tray, or valve contactor units, the same general principles are applied with the only differences being based on the geometrical arrangement of the individual contacting units. A rule of thumb for correctly designed bubble caps or valve trays is that the total pressure drop per tray will be about two times the pressure drop equivalent to the average head of liquid above the top of the bubble-cap slots or the top portion of the valve opening. The equivalent rule of thumb for total pressure drop over a correctly designed sieve tray is to multiply the total head of liquid on the tray by 2.0 to give the liquid head equivalent to the total pressure drop for that plate.

For a typical bubble-cap column, the following pressure drops per tray are considered reasonable, and they also would be order-of-magnitude values for sieve trays or valve trays:

Total pressure	Pressure drop per tray
30 mm Hg	3 mm Hg or less
1 atm	0.07 − 0.12 psi
300 psi	0.15 psi

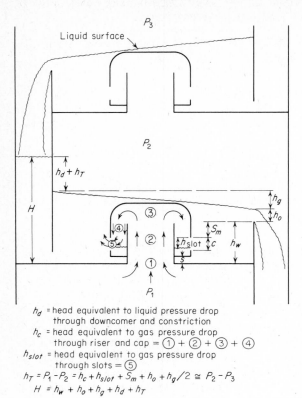

h_d = head equivalent to liquid pressure drop
    through downcomer and constriction

h_c = head equivalent to gas pressure drop
    through riser and cap = ① + ② + ③ + ④

h_{slot} = head equivalent to gas pressure drop
    through slots = ⑤

$h_T = P_1 - P_2 = h_c + h_{slot} + S_m + h_o + h_g/2 \cong P_2 - P_3$

$H = h_w + h_o + h_g + h_d + h_T$

Figure 15-10 Cross-sectional view of bubble-cap tower showing flow and nomenclature for pressure-drop calculations. (Units of all symbols are feet of liquid).

Figures 15-10 and 15-11 present a cross-sectional view of an average bubble cap and a typical sieve tray showing the various causes of pressure drop expressed as liquid head equivalent to the pressure drop. The total pressure drop across the tray in U.S. conventional pressure units is related to the total liquid head h_T (expressed in feet) by the following equation:

$$\Delta p_T = \frac{h_T \rho_L g}{144 g_c} \tag{7}$$

where Δp_T = total pressure drop of gas across tray, psi

    ρ_L = density of the liquid, lb/ft³

    g = local acceleration due to gravity, ft/(s)(s)

    g_c = conversion factor in Newton's law of motion, 32.17 ft · lbm/(s)(s)(lbf)

In most cases, g and g_c are considered to be numerically equal.

Pressure drop through the contactor assembly Causes for pressure drop through the contactor assembly are shown in Fig. 15-10 for bubble caps as (1) contrac-

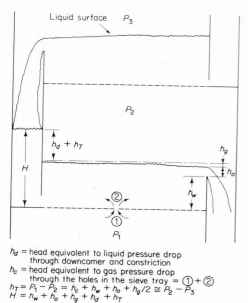

h_d = head equivalent to liquid pressure drop
through downcomer and constriction
h_c = head equivalent to gas pressure drop
through the holes in the sieve tray = ①+②
$h_T = P_1 - P_2 = h_c + h_w + h_o + h_g/2 \cong P_2 - P_3$
$H = h_w + h_o + h_g + h_d + h_T$

Figure 15-11 Cross-sectional view of sieve-tray tower showing flow and nomenclature for pressure-drop calculations. (Units of all symbols are feet of liquid.)

tion, (2) friction in riser, (3) reversal of flow direction, and (4) friction in annular space. Similarly, Fig. 15-11 shows for sieve trays that this cause for pressure drop is (1) contraction and (2) friction in the sieve hole. The total pressure drop due to the preceding causes is primarily a function of the kinetic head. The pressure drop as feet of liquid equivalent to one kinetic head is

$$h_H = \frac{V_c^2}{2g} \frac{\rho_G}{\rho_L} \tag{8}$$

where V_c = maximum linear velocity of gas in riser, reversal area, annulus, or sieve hole, ft/s
ρ_G = gas density, lb/ft³
ρ_L = density of liquid, lb/ft³
g = local acceleration due to gravity, ft/(s)(s)

The total pressure drop through the riser and bubble cap is usually in the range of four to eight kinetic heads, depending on the cap design. A factor of six kinetic heads is a reasonable average figure. With this factor, the pressure drop due to gas flow through the riser and bubble cap expressed as liquid head is

$$h_c = \frac{3V_c^2}{g} \frac{\rho_G}{\rho_L} \tag{9}$$

For sieve trays, the number of kinetic heads equivalent to the total pressure drop through the plate itself is a function of the ratio of the sieve-hole diameter to the tray thickness and the ratio of the hole area per tray to the active area per

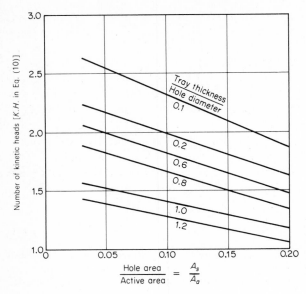

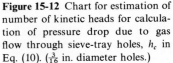

Figure 15-12 Chart for estimation of number of kinetic heads for calculation of pressure drop due to gas flow through sieve-tray holes, h_c in Eq. (10). ($\frac{3}{16}$ in. diameter holes.)

tray as shown in Fig. 15-5. This pressure drop for a reasonable sieve-tray design is generally in the range of 1 to 3 kinetic heads, and Fig. 15-12 can be used to choose the most reasonable number to use in preliminary designs.† Designating the number of kinetic heads obtained from Fig. 15-12 as $K.H.$, the pressure drop due to gas flow through the holes for a sieve tray expressed as liquid head is

$$h_c = (K.H.)\frac{V_c^2 \rho_G}{2g\rho_L} \tag{10}$$

It should be noted that this sieve-tray value for pressure drop through the contactor assembly is considerably less than for the equivalent case of bubble caps with the same gas velocity in the cap as in the sieve hole as shown by the difference in Eq. (9) and Eq. (10) where the $K.H.$ range is from 1 to 3. Equivalent values for valve trays would normally be closer to those for bubble-cap trays, but the actual result depends so much on the design of the valve that it is recommended that vendors be contacted for appropriate values even in preliminary designs.‡

Pressure drop through bubble-cap slots is a direct function of the vertical distance the liquid is depressed below the top of the slots. In Fig. 15-10, this distance is represented by h_{slot}. The difference in pressure between the inside of

---

† Adapted from I. Liebson, R. Kelley, and L. Bullington, How to Design Perforated Trays, *Petrol. Refiner*, **36**(2):127 (1957); **36**(3):288 (1957).

‡ For further information on valve-tray design, see W. L. Bolles, Estimating Valve Tray Performance, *Chem. Eng. Progr.*, **72**(9):43 (1976).

the cap and the liquid outside the cap at the top of the slots is defined as the pressure drop through the slots. Consequently, h_{slot} is the liquid head equivalent to the pressure drop through the slots.

At low gas velocities, intermittent bubbling through the slots is obtained because of liquid surface-tension effects, and the pressure drop is a function of the surface tension and the slot dimensions. When the velocity increases sufficiently, the gas issues from the slots in a steady stream, and the effect of surface tension becomes unimportant.

Under normal operating conditions, the slots remain partly open continuously and the gas is delivered in a steady stream. For this situation, the following equations, developed by Cross and Ryder,† can be used to estimate the average slot opening and the pressure drop through the slots:

For rectangular-shaped slots

$$h_{\text{slot}} = 1.5\left(\frac{Q_s}{b}\right)^{2/3}\left[\frac{\rho_G}{(\rho_L - \rho_G)g}\right]^{1/3} \tag{11}$$

For triangular-shaped slots

$$h_{\text{slot}} = 1.85\left(\frac{cQ_s}{b}\right)^{2/5}\left[\frac{\rho_G}{(\rho_L - \rho_G)g}\right]^{1/5} \tag{12}$$

where Q_s = volumetric flow rate of gas per slot, ft³/s
b = width of slot at base, ft
c = height of slot, ft

Equations (11) and (12) apply when h_{slot} is less than c, and a design value for h_{slot} of $0.5c$ is often recommended.

Pressure drop due to liquid head above slots, sieve holes, or valve openings Reference to Fig. 15-10 shows that the total head above bubble-cap slots for an average cap is the sum of static submergence S_m, height of liquid crest above weir h_o, and average liquid gradient $0.5h_g$. The same would hold for valve trays with static submergence being the distance to the top of the open valve. For sieve trays, liquid gradient is usually ignored and liquid head above the sieve holes is merely the weir height h_w plus the height of liquid crest above the weir h_o. The static submergence is determined directly by the construction details of the plate, and design methods are available for estimating the liquid crest over the weir and the liquid gradient.

The head of liquid over the weir, based on a straight segmental downcomer, can be estimated by the following modification of the Francis weir formula:

$$h_o = \left(\frac{1.7Q_L}{l_w\sqrt{g}}\right)^{2/3} \tag{13}$$

where Q_L = volumetric flow rate of liquid, ft³/s, and l_w = weir length, ft.

† C. A. Cross and H. Ryder, *J. Appl. Chem.*, **2**:51 (1952).

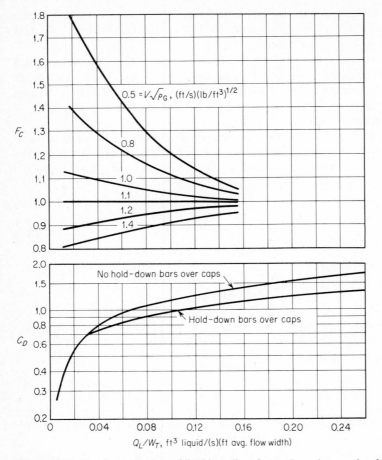

Figure 15-13 Plot for evaluation of liquid-gradient factor C_D and correction factor F_C in Eq. (14).

The liquid gradient across the tray (h_g) can be approximated by use of the following equation developed by Davies.†

For bubble caps arranged on equilateral-triangle centers and caps covered with liquid to a depth not greater than 1 in.,

$$(h_g F_C)^{1.5}(1.5r - 1.4) + (h_g F_C)^{0.5}(3r)\left[h_w + h_o + \frac{s(l_r - l_c)}{l_c}\right] = \frac{0.42 Q_L r^{1.5}}{C_D l_c} \quad (14)$$

where F_C = a correction factor to account for variations in gas rate and gas density; values of F_C are given in Fig. 15-13 as a function of $V\sqrt{\rho_G}$ and Q_L/W_T

† J. A. Davies, *Petrol. Refiner*, **29**(9):121 (1950); *Ind. Eng. Chem.*, **39**:774 (1947).

r = number of rows of caps perpendicular to direction of liquid flow

s = skirt clearance, ft

l_r = total free space between risers perpendicular to direction of liquid flow, average of various rows, ft

l_c = total free space between caps perpendicular to direction of liquid flow, average of various rows, ft

C_D = liquid-gradient factor; value of C_D can be obtained from Fig. 15-13

V = superficial linear gas velocity (based on cross-sectional area of empty tower), ft/s

W_T = average width of tray normal to direction of liquid flow, ft, computed as average of total tray width at each rows of caps

For caps arranged on square centers and for tunnel caps, replace the constants 0.42 and 1.4 in Eq. (14) by 0.37 and 1.0, respectively.

All the variables in Eq. (14) except the liquid gradient are fixed by the tray design and the operating conditions, and the value of h_g can be determined by a trial-and-error or graphical procedure. Equation (14) is based on the assumption that the liquid level is below the top of the caps. In most bubble-cap towers, however, the liquid level is above the caps. Equation (14), therefore, is conservative for design estimates, since it tends to give high values for h_g.

Evaluation of total pressure drop per tray The total pressure drop of the gas across a bubble-cap tray as indicated by Fig. 15-10 equals the sum of the pressure drop through the cap assembly and the pressure drop due to liquid head above the slots, or

$$h_T = h_c + h_{slot} + S_m + h_o + 0.5h_g \qquad (15)$$

where h_T = head of liquid equivalent to total pressure drop of gas across tray, ft.

Similarly, the total pressure drop of the gas across a sieve tray, as indicated in Fig. 15-11, becomes

$$h_T = h_c + h_w + h_o + 0.5h_g \qquad (16)$$

Because liquid gradient is very small in most sieve tray towers, the last term in Eq. (16) is often dropped.

Because the liquid in most finite-stage contactors is usually aerated, the clear-liquid density is greater than the density of the aerated liquid on the tray. Therefore, if ρ_L is taken as the clear-liquid density, the values of Δp_T and h_T given by Eqs. (7) and (15) may be high, and these values are often multiplied by 0.7 to give an approximate correction for the liquid aeration.[†]

Liquid head in downcomer If the head of liquid in the downcomer is greater than the tray spacing plus the weir height, flooding will occur and liquid will build up

---

[†] For further treatment on aeration and additional methods for estimating pressure drop with finite-stage contactors, see J. H. Perry and C. H. Chilton, "Chemical Engineers' Handbook," 5th ed., Sect. 18, McGraw-Hill Book Company, New York, 1973.

on the trays. In design practice, the height of liquid in the downcomer (based on clear-liquid density) should be less than 50 percent of the tray spacing.

Liquid head in the downcomer is composed of five individual heads as follows:

1. Weir height h_w, which is established by the tray design.
2. Height of crest over weir h_o, which can be calculated by Eq. (13).
3. Liquid gradient h_g, which can be calculated by Eq. (14).
4. Head of liquid equivalent to the frictional flow resistance in the downcomer and in the passage of the fluid from the downcomer onto the plate. This head is usually relatively small, but it can be estimated as three kinetic heads for the liquid, based on the linear liquid velocity at the minimum cross-sectional area for downcoming liquid flow, or

$$h_d = \frac{3}{2g} \left(\frac{Q_L}{A_d} \right)^2 \tag{17}$$

where h_d = head of liquid equivalent to liquid pressure drop due to flow through downcomer and constriction, ft, and A_d = minimum cross-sectional area for downcoming liquid flow, ft^2.

5. Head of liquid which the gas must overcome as the gas passes through the next tray above. Assuming the same gas pressure drop over adjacent trays, this liquid head is equal to h_T.

The total head in the downcomer, H, as shown in Figs. 15-10 and 15-11, is equal to the sum of the preceding five heads, or

$$H = h_w + h_o + h_g + h_d + h_T \tag{18}$$

Combination of Eqs. (18) and (15) gives for bubble caps

$$H = 2h_o + 1.5h_g + h_w + h_d + h_c + h_{\text{slot}} + S_m \tag{19}$$

Examples 3 and 4 presented in the following illustrate methods for estimating pressure drop with bubble-cap contactors and with sieve-tray contactors. The examples also give information as to typical design conditions for the two types of contactors.

Example 3. Determination of pressure drop and liquid height in downcomer for bubble-cap plate The following specifications apply to a bubble-cap plate:

    Diameter = 10.0 ft
    Tray spacing = 26 in.
    Liquid crossflow
    Weir length = 6.2 ft
    Weir height = 3.0 in.

Skirt clearance = 0.5 in.
Static submergence = 0.5 in.
Rectangular slots
Height of slots = 1.5 in.
Width of slots = 0.3 in.
Total riser cross-sectional area = 9 ft^2
Caps are bolted to tray (no hold-down bars)
Clearance between bottom of downcomer and plate = 2.5 in.
Number of rows of caps perpendicular to direction of liquid flow = 11

For an average row of caps perpendicular to direction of liquid flow, total free space between risers = 4.4 ft, total free space between caps = 2.7 ft, width of tray = 9 ft

Cross-sectional areas for vapor flow through riser, direction-reversal space, annular cap space, and slots are equal.

Caps are arranged on equilateral-triangle centers staggered perpendicular to direction of liquid flow, and liquid depth above caps is less than 1 in.

This bubble-cap plate is to be used under the following conditions:

Vapor density = 0.15 lb/ft^3
Liquid density = 50 lb/ft^3
Superficial vapor velocity based on cross-sectional area of empty tower = 1.8 ft/s
Liquid flow rate = 1 ft^3/s
Surface tension of liquid is such that Eq. (11) is applicable.

Estimate the gas pressure drop across the tray, the percent of this pressure drop due to liquid head above the top of the bubble-cap slots, and the liquid head in the downcomer.

SOLUTION Liquid head equivalent to pressure drop through riser and cap (h_c): By Eq. (9),

$$h_c = \frac{3V_c^2}{g}\frac{\rho_G}{\rho_L}$$

Cross-sectional area of empty tower = $(10)^2\frac{\pi}{4} = 78.5$ ft^2

V_c = linear velocity of gas in riser, reversal area, annular cap space, and slots

$$= \frac{(1.8)(78.5)}{9} = 15.7 \text{ ft/s}$$

$$h_c = \frac{(3)(15.7)^2(0.15)}{(32.17)(50)} = 0.069 \text{ ft}$$

Liquid head equivalent to pressure drop through slots (h_{slot}): By Eq. (11),

$$h_{slot} = 1.5 \left(\frac{Q_s}{b}\right)^{2/3} \left[\frac{\rho_G}{(\rho_L - \rho_G)g}\right]^{1/3}$$

Q_s = volumetric flow rate of gas per slot

$$= \frac{(15.7)(1.5)(0.3)}{144} = 0.049 \text{ ft}^3/\text{s}$$

$$b = \text{slot width} = \frac{0.3}{12} \text{ ft}$$

$$h_{slot} = 1.5 \left(\frac{0.049}{0.3/12}\right)^{2/3} \left[\frac{0.15}{(50 - 0.15)(32.17)}\right]^{1/3} = 0.107 \text{ ft}$$

Height of liquid crest over weir (h_o): By Eq. (13),

$$h_o = \left(\frac{1.7Q_L}{l_w\sqrt{g}}\right)^{2/3}$$

Q_L = liquid flow rate = 1.0 ft^3/s

l_w = weir length = 6.2 ft

$$h_o = \left[\frac{(1.7)(1.0)}{(6.2)(32.17)^{1/2}}\right]^{2/3} = 0.132 \text{ ft}$$

Liquid gradient (h_g): By Eq. (14),

$$(h_g F_C)^{1.5}(1.5r - 1.4) + (h_g F_C)^{0.5}(3r)\left[h_w + h_o + \frac{s(l_r - l_c)}{l_c}\right] = \frac{0.42Q_L r^{1.5}}{C_D l_c}$$

where r = rows of caps perpendicular to direction of fluid flow = 11

h_w = weir height = 3.0/12 = 0.25 ft

h_o = 0.132 ft

s = skirt clearance = 0.5/12 = 0.0416 ft

l_r = average total free space between risers perpendicular to direction of liquid flow = 4.4 ft

l_c = average total free space between bubble caps perpendicular to direction of liquid flow = 2.7 ft

V = superficial linear gas velocity based on empty tower = 1.8 ft/s

W_T = average tray width perpendicular to direction of fluid flow = 9 ft

$Q_L/W_T = \frac{1}{9} = 0.111$ ft^3/(s)(ft)

$V\sqrt{\rho_G} = (1.8)(0.15)^{1/2} = 0.7$ (ft/s)(lb/ft^3)$^{1/2}$

From Fig. 15-13,

$$F_C = 1.12$$

$$C_D = 1.2$$

$$(h_g)^{1.5}(1.12)^{1.5}[(1.5)(11) - 1.4] + (h_g)^{0.5}(1.12)^{0.5}(3)(11)$$

$$\times \left[0.25 + 0.132 + \frac{(0.0416)(4.4 - 2.7)}{2.7} \right] = \frac{(0.42)(1.0)(11)^{1.5}}{(1.2)(2.7)}$$

By trial and error,

$$h_g = 0.088 \text{ ft}$$

Pressure drop across tray based on clear-liquid density (Δp_T):

By Eq. (15), liquid head equivalent to the pressure drop across the tray is

$$h_T = h_c + h_{\text{slot}} + S_m + h_o + 0.5h_g$$

$$S_m = \text{static submergence} = \frac{0.5}{12} = 0.0416 \text{ ft}$$

$$h_T = 0.069 + 0.107 + 0.0416 + 0.132 + (0.5)(0.088) = 0.3936 \text{ ft}$$

By Eq. (7),

$$\Delta p_T = \frac{h_T \rho_L}{144} \frac{g}{g_c} = \frac{(0.3936)(50)(32.17)}{(144)(32.17)} = 0.137 \text{ psi}$$

Percent of Δp_T due to liquid head (average) above top of plate

$$= \frac{S_m + h_o + 0.5h_g}{h_T}(100) = \frac{0.0416 + 0.132 + (0.5)(0.088)}{0.3936}(100) = 55\%$$

Liquid head in downcomer (H): By Eq. (18),

$$H = h_w + h_o + h_g + h_d + h_T$$

By Eq. (17), head due to liquid flow through downcomer constriction of area A_d is

$$h_d = \frac{3}{2g}\left(\frac{Q_L}{A_d}\right)^2$$

$$A_d = (\text{weir length})(\text{clearance between downcomer and plate})$$

$$= (6.2)\left(\frac{2.5}{12}\right) = 1.29 \text{ ft}^2$$

$$h_d = \frac{(3)}{(2)(32.17)}\left(\frac{1.0}{1.29}\right)^2 = 0.028 \text{ ft}$$

$$H = 0.25 + 0.132 + 0.088 + 0.028 + 0.3936 = 0.8916 \text{ ft}$$

Liquid head in downcomer based on clear-liquid density $= (0.8916) \times (12) = 10.7$ in.

Example 4. Determination of pressure drop and liquid height in downcomer for sieve tray The conditions of operation and design as given in Example 3 apply for a sieve-tray column with the exception that the design conditions for the bubble caps and weir height are replaced by the following design conditions for the sieve tray:

Sieve-tray holes are drilled on equilateral-triangle pattern with hole diameter = $\frac{3}{16}$ in.

Plate thickness = $\frac{3}{16}$ in.

Active area of plate is 88 percent of the total column cross-sectional area

The total area of sieve holes is 5 percent of the active area of the plate

Weir height = 2.0 in.

Liquid gradient (h_g) is negligible.

Estimate the gas pressure drop across the tray, the percent of the pressure drop due to liquid head above the sieve holes, and the liquid head in the downcomer.

SOLUTION Liquid head equivalent to pressure through the holes (h_c): By Eq. (10),

$$h_c = (K.H.)\frac{V_c^2 \rho_G}{2g\rho_L}$$

$$V_c = \frac{1.8}{(0.88)(0.05)} = 40.9 \text{ ft/s}$$

K.H. from Fig. 15-12 (with area of sieve hole/active area of plate = 0.05 and tray thickness/hole diameter = 1.0) = 1.5

$$h_c = \frac{(1.5)(40.9)^2(0.15)}{(2)(32.17)(50)} = 0.117 \text{ ft}$$

Height of liquid crest over weir (h_o): Same as calculated in Example 3 by Eq. (13),

$$h_o = 0.132 \text{ ft}$$

$$\text{Height of weir} = h_w = \frac{2.0}{12} = 0.167 \text{ ft}$$

Pressure drop across tray based on clear-liquid density (Δp_T):

By Eq. (16), liquid head equivalent to the pressure drop across the tray [assuming liquid gradient (h_g) is negligible] is

$$h_T = h_c + h_w + h_o = 0.117 + 0.167 + 0.132 = 0.416 \text{ ft}$$

By Eq. (7),

$$\Delta p_T = \frac{h_T \rho_L g}{144 g_c} = \frac{(0.416)(50)(32.17)}{(144)(32.17)} = 0.144 \text{ psi}$$

Percent of Δp_T due to liquid head above the sieve holes

$$= \frac{h_o + h_w}{h_T}(100) = \frac{0.132 + 0.167}{0.416}(100) = 72\%$$

Liquid head in downcomer (H): By Eq. (18), neglecting liquid gradient,

$$H = h_w + h_o + h_d + h_T$$

From Example 3, head due to liquid flow through downcomer constriction (h_d) by Eq. (17) is

$$h_d = 0.028 \text{ ft}$$

$$H = 0.167 + 0.132 + 0.028 + 0.416 = 0.743 \text{ ft}$$

Liquid head in downcomer based on clear-liquid density $= (0.743) \times (12) = 8.9$ in.

OTHER DESIGN FACTORS FOR FINITE-STAGE CONTACTORS

In addition to the critical design factors for finite-stage contactors of number of theoretical trays, maximum allowable vapor velocity, column efficiency, and pressure drop as discussed earlier, a number of other factors are of importance in the development of the design. These factors are discussed in the following sections.

Bubble Caps and Risers

Bubble-cap assemblies in the form of round bell caps are commonly used with diameters ranging from 4 to 7 in. A 6-in.-diameter cap with a 4-in.-diameter riser is a standard size used in many industrial operations. Cap diameters as large as 8 in. have been employed successfully in some operations, and 3-in. caps with 2-in. risers are used in many vacuum towers. Tunnel caps are ordinarily 3 to 6 in. wide and 12 or more in. long. Comparison of the two types shows that tunnel caps have the advantage of a smaller number of units for installation for a given slot and riser area, but the round caps are more adaptable to changes in tray layout and can be purchased and stocked in one standard size.

The caps are slotted around the lower periphery and can be anchored to the plate with teeth touching the plate or suspended to permit a so-called "skirt clearance" between the plate and the cap. Slots can be of a saw-tooth type or in the form of punched holes, usually rectangular or triangular. In common practice, a skirt clearance in the range of 0.5 to 1.5 in. is recommended to prevent plugging of the slots by residue build-up. Since the purpose of the slots is to disperse the gas into the liquid in the form of small bubbles, sufficient slot area should be provided in order that no gas may pass through the skirt clearance.

As illustrated in Fig. 15-2, the caps can be held in place by bolts or by a hold-down bar; however, the hold-down bar is seldom used because it interferes with the flow distribution. Since periodic removal or maintenance of the caps may be necessary, the preferred method for holding caps and trays in place permits removal of the caps by one person working in the crawl space above the plate.

Slot velocities and relative dimensions After the design of the cap has been established, the next step is to determine the number of bubble caps to be used per tray. This number is set by the allowable gas velocity through the slots. If the velocity is too high, pressure drop may be excessive and the liquid may be blown away from the cap, thus resulting in poor efficiency as shown by Fig. 5-8. On the other hand, if the velocity is too low, the gas bubbles will have little opportunity to disperse through the liquid, and the efficiency of the tray will be low. Davies has recommended the following equations for use in preliminary estimates of slot velocities in rectification columns:[†]

$$\text{Maximum linear slot velocity, ft/s} = \frac{12}{\rho_G^{0.5}} \tag{20}$$

$$\text{Minimum linear slot velocity, ft/s} = \frac{3.4}{\rho_G^{0.5}} \tag{21}$$

where ρ_G = gas density, lb/ft^3.

In common design practice, the riser area, total slot area, and passage area in the annular space under the cap are approximately equal in order to reduce pressure losses caused by expansion and contraction. Gas bubbles issuing from the slots are seldom projected more than about 1 in. from the cap, and a clearance between caps in the range of 1 to 3 in. is usually sufficient to eliminate large amounts of undesirable bubble coalescence. The caps should be spaced evenly over the entire tray. A clearance between the shell and adjacent caps of less than 2 in. is recommended in order to reduce the possibility of liquid bypassing the bubble-contact regions. For towers 3 or more feet in diameter, a plate layout with the total riser area in the range of 10 to 20 percent of the tower cross-sectional area is common, with the greater riser area being more easily obtainable in larger-diameter towers.

Sieve Trays

The common sieve trays are constructed of flat metal sheets with holes drilled or punched to form the sieve plate. Hole sizes in the range of $\frac{1}{8}$ to $\frac{1}{2}$ in. diameter are used with $\frac{3}{16}$ in. being a common size. The design range for ratio of tray thickness to sieve hole diameter is usually 0.1 to 0.7. The liquid head on a sieve tray should be in the range of 2 in. to 4 in. The sieve holes should be spaced as equilateral triangles with a pitch-to-diameter ratio greater than 2.0 but less than 5.0 with an optimum of about 3.8. The downcomer inlet velocity should not exceed 0.4 feet per second to allow for adequate area for vapor disengaging. At the time of installation of new trays, they should be inspected to make certain that smooth and clean holes are provided with a minimum of surface irregularities on the side toward the flowing gas.

[†] J. A. Davies, *Petrol. Refiner*, **29**(9):121 (1950).

Valve Trays

The major advantage of valve trays over sieve trays is that high efficiencies can be maintained over a wider range of operating liquid and gas throughputs than with sieve trays. *Turnover ratios*, defined as the ratio of the maximum allowable throughput rate to the minimum allowable throughput rate, as high as 10 can be obtained with valve trays, while the ratio is much less with sieve trays. In general, good design information on the various devices used as the contactor for valve trays is proprietary in that they are manufactured by individual concerns which hold patent rights and related private knowledge on the design and operation of the units.†

Valve-tray perforations are larger than those for sieve trays being in the order of 1.5 in. for circular holes and 6 in. long for the case of rectangular-slot holes. There are many methods to control the movement of the valve lids, but the lids generally are allowed a vertical movement in the range of $\frac{1}{4}$ to $\frac{1}{2}$ in., and the weight of the lid varies depending on the intended use of the valves.

Basically, the overall model for valve trays is similar to that for sieve trays with the exceptions that the vapor flow area in the contactor can vary for valve units depending on the extent of the valve opening and the number of kinetic heads for calculation of the gas flow pressure drop through the contactor (h_c) is usually greater for valve trays than for sieve trays. Thus, many of the basic principles of fluid mechanics as used for sieve-tray design also apply for valve-tray design so that rough preliminary designs for valve-tray columns can often be made by engineers for use in considering alternatives before contacting a specific manufacturer. Because valve units are more complex mechanically than sieve trays, valve trays are more expensive to fabricate, and this extra cost must be recognized in pricing the units. On the other hand, proprietary knowhow will become available for the valve units if they are purchased, and this advantage to the user plus the advantage of wider range of turnover ratios may swing the balance in favor of the more expensive valve trays.

Shell and Trays

Cylindrical shells are commonly used for finite-stage contactor columns. The shell may be constructed of short sections that are bolted together or in the form of one long cylinder. Manholes should be provided to give access to the individual trays for cleaning, maintenance, and installation. Adequate foundations and tower supports must be provided. In general, the size of the foundations is set by the size necessary to resist the overturning stresses on the column, since this load usually exceeds the direct load caused by the weight of the column assembly.

---

† Examples of special information which is available on valve-tray design are Glitch, Inc., "Ballast Tray Design Manual," Bulletin 4900, Dallas, Tex.; Koch Engineering Co., "Flexitray Design Manual," Bulletin 960, Wichita, Kans.; and Nutter Engineering Co., "Float Valve Tray Design Manual," Tulsa, Okla.

Both the corrosion characteristics of the fluids involved and costs dictate the acceptable materials of construction for the shell and trays. Bubble caps, valves, and trays are usually made of a suitable metal to facilitate fabrication, but the shell material can be glass, plastic, impervious carbon, wood, glass-lined or resin-lined steel, or metal. Despite the additional weight involved, trays for bubble-cap units are often made of cast iron, usually 0.5 in. thick. With cast-iron trays, the risers may be fabricated as a permanent part of the tray. Lighter-gauge alloy metals may be cheaper than cast iron, and the final decision must be made on the basis of the situation for each individual case.

Supporting beams are used to stiffen the trays, and the trays must be fastened securely to prevent movement caused by gas surges. To allow for thermal expansion and to facilitate installation, slotted boltholes should be provided in the supporting rings, and there should be adequate clearance between the tray and shell wall.

Tray level The calculations for finite-stage contactor operations are based on the assumption that the trays are perfectly level. A tolerance of $\pm \frac{1}{8}$ in. change in vertical tray level is usually specified for the location of tray supports and fastening of the trays to these supports. Variations from the horizontal level are also caused by foundation settling, tray deflections caused by operating conditions, tower deflection due to wind loads, and uneven corrosion of the tray or tray supports. For design purposes, a maximum change in tray level of $\frac{1}{2}$ in. is a safe assumption, and this figure can be used as a safety factor in setting the seal dimensions of the tray.

Leakage and weep holes To ensure a minimum of liquid leakage from one tray to another, bubble-cap risers should be fitted firmly or sealed into the trays, and an effective seal is necessary around the tray supports. In large columns, the trays are often made in sections for ease of installation. Each of these sections must be installed carefully to minimize leakage.

Provision for draining the liquid from a bubble-cap tray when the unit is not in operation is made by the use of *weep holes*. These holes are usually located near the overflow weir so that any delivery during operation follows approximately the same path as the overflow fluid. The weep holes must be large enough to prevent plugging, but they should not deliver an excessive amount of fluid during operation. A size in the range of $\frac{1}{4}$- to $\frac{5}{8}$-in. diameter is usually adequate.

Liquid Flow

High tray efficiency is achieved when all the contactor units are delivering gas uniformly into liquid that is evenly distributed over the entire tray surface. The beneficial effects of a liquid concentration gradient are obtained if liquid crossflow is used, where the liquid enters on one side of the tray and makes one pass across the tray. For tower diameters larger than 4 ft, better liquid distribution and less change in liquid head can often be achieved by using split flow, radial flow, or cascade flow, as illustrated in Fig. 15-5.

Because of the flow resistance of the caps and risers in bubble-cap columns, there is a decrease in liquid depth as the liquid passes across the tray. Figure 15-4 shows an extreme example in which this liquid gradient is so great that only one out of the four bubble caps is operating normally. In general, the dimensionless ratio of total liquid gradient to pressure-drop head caused by the bubble-cap assembly should be less than 0.4 in order to ensure adequate vapor distribution, and, for single-pass crossflow trays, the rate of liquid flow across the tray should be less than 0.22 ft^3/(s)(ft) diameter.

A total liquid gradient of 0.5 in. over one tray is usually acceptable, but a different flow pattern should be considered if the value approaches 1 in. The liquid gradient can be reduced by decreasing the number of rows of caps through which the liquid flows or by decreasing the velocity of liquid flow past the caps. A higher skirt clearance or a higher weir will sometimes be sufficient to reduce excessive liquid gradients to acceptable values.

Entrainment

As a gas passes through the contactor unit into the liquid, a large amount of turbulence is set up, and liquid particles can become entrained with the gas. Carry-over of these liquid particles from one tray to the tray above is known as *entrainment*. It is often defined as the weight of liquid entrained per unit weight of gas. Liquid can be entrained by the gas as a result of violent splashing of the liquid or because of extensive foaming or frothing.

Entrainment has an adverse effect on the column operation in that it reduces the concentration change per tray and, consequently, decreases the efficiency. Nonvolatile impurities can be carried up a tower by entrainment, resulting in off-color or impure overhead products.

The major factors that determine the amount of entrainment are plate spacing, depth of liquid on the tray, and vapor velocity in the space between the plates. Slot or sieve-hole vapor velocity and liquid flow rate have some effect on the entrainment, but they are not of major importance.

Tray Spacing

Tray spacing in large columns is usually determined by the need for easy access for maintenance and inspection. With columns less than 3 ft in diameter, the minimum value of tray spacings is about 6 in., and greater values are used in most cases. In the petroleum industry, an 18-in. spacing is considered to be the minimum value below which entrainment or tendency toward flooding becomes excessive, and a tray spacing of 20 to 24 in. is often set as a minimum for reasons of accessibility. Table 1 presents recommended tray spacings for petroleum rectification columns of various diameters.

Downcomers and Weirs

Downcomers for conducting the liquid from one tray to the next tray below may be in the form of circular pipes or segments of the tower isolated from the rising

Table 1. Recommended tray spacings for towers.

Tower diameter, ft	Tray spacing, in.
	6 minimum
4 or less	18–20 (no manways in trays)
6–10	24
12–24	36

gas by means of vertical or angled plates. Some vapor is entrained in the liquid as the liquid enters the downcomer, and sufficient residence time should be provided in the downcomer to permit escape of the entrained vapor. A residence time of 5 s, evaluated as the volume of the downcomer divided by the volumetric flow rate of the downcoming fluid, is enough to permit release of most of the entrained vapor. Vapor release is also accomplished by providing a calming section before the liquid flows into the downcomer. This is supplied by locating the bubble caps, valves, or sieve holes 3 to 5 in. from the downcomer weir or by blanking off the cap or valve slots that face toward the downcomer. The liquid head in the downcomer should not be greater than one-half the plate spacing.

The discharge end of the downcomer must project far enough into the tray liquid so that no gas bubbles can enter the open end and bypass the bubble caps. When the liquid contains no sediment, a seal pot or discharge weir is often placed around the discharge end of the downcomer to make certain that no free vapor can enter the open end. The distance between the liquid level on the loaded discharge plate and the bottom of the downcomer when no liquid is flowing is known as the *downcomer liquid seal*. A downcomer liquid seal, based on a perfectly level tray, in the range of $\frac{1}{2}$ to $1\frac{1}{2}$ in. is usually satisfactory.

An extension of the downcomer plate can be used as an overflow entrance weir, or a separate overflow weir may be provided. Since adequate vapor-disengaging space is necessary, extension of circular-pipe downcomers to form an overflow weir is not recommended if the column diameter is greater than 3 ft. Straight rectangular weirs are often used, and, on crossflow trays, these generally have a length in the range of 0.6 to 0.8 times the column diameter.

The height of the overflow weir is a major factor in determining the head of liquid on the tray. Although plate efficiency is increased slightly as the liquid head is increased, the beneficial effect is seldom enough to counterbalance the detrimental pressure-drop effects caused by high heads. This is particularly true when pressure drop is an important factor, as with towers operating under vacuum. The distance between the top of the bubble-cap slots and the liquid surface when the static liquid is just ready to flow over the overflow weir is known as the *static submergence*. Table 2 shows recommended values of static submergence for various operating pressures.

Example 5. Determination of holdup time in downcomer A valve-tray tower with 24-in. plate spacing and liquid crossflow contains straight segmental

Table 2. Typical values of static submergence for bubble-cap plate columns

Operating pressure	Static submergence, in.
Vacuum (30 mm Hg abs.)	0
Atmospheric	$\frac{1}{2}$
100 psig	1
300 psig	$1\frac{1}{2}$
500 psig	$1\frac{1}{2}$

downcomers. The overflow weir at the downcomer entrance is formed by an extension of the downcomer plate. The height of this weir is 3 in. The inside diameter D of the tower is 5 ft, and the weir length is 0.6D. If liquid with a density of 55 lb/ft^3 flows across the plate at a rate of 30,000 lb/h, estimate the residence or holdup time in the downcomer from this plate.

SOLUTION:
Volumetric flow rate of downcoming fluid $= 30,000/(55)(3600) = 0.152$ ft^3/s
Weir length $= l_w = (0.6)(5) = 3$ ft
Perpendicular distance from weir to center of tower $= d_{wc} = [(D/2)^2 - (l_w/2)^2]^{1/2} = [(\frac{5}{2})^2 - (\frac{3}{2})^2]^{1/2} = 2$ ft
If θ designates the angle in degrees subtended from the center of the tower by the weir,

$$l_w = D \sin \frac{\theta}{2}$$

$$\sin \frac{\theta}{2} = 0.6$$

$$\theta = 73.7°$$

Downcomer cross-sectional area is

$$\frac{\pi D^2 \theta}{1440} - \frac{l_w d_{wc}}{2} = \frac{(3.14)(5)^2(73.7)}{1440} - \frac{(3)(2)}{2} = 1.02 \text{ ft}^2$$

Alternatively, downcomer cross-sectional area is

$$\frac{\pi D^2}{8} - \frac{l_w d_{wc}}{2} - \frac{D^2}{4} \left(\sin^{-1} \frac{2d_{wc}}{D} \right)_{\text{radians}} = 1.02 \text{ ft}^2$$

$$\text{Effective height of downcomer} = \frac{24 + 3}{12} = 2.25 \text{ ft}$$

$$\text{Residence time} = \frac{\text{volume of downcomer}}{\text{volumetric flow rate of downcoming fluid}}$$

$$= \frac{(2.25)(1.02)}{0.152} = 15 \text{ s}$$

NOTE: The preceding answer is based on the assumption that the descending liquid occupies all the available volume in the downcomer.

Plate Stability

The term *plate stability* refers to the ability of the plate to maintain satisfactory operating characteristics when flow rates change or when unsteady-state conditions exist. Vapor and liquid distribution on the plate are the primary factors that determine plate stability. A stable plate is one in which all the contacting units are handling appropriate amounts of vapor, and the best efficiencies are obtained when the vapor flow is evenly distributed among the contacting units. Thus, an ideal sieve plate would have all the vapor holes discharging uniformly even when fluctuations in flow rates occur. In an actual bubble-cap column, the vapor is not distributed uniformly among the caps primarily because of the gradient in liquid height across the tray.

CONTINUOUS CONTACTORS—PACKED TOWERS

As illustrated in Fig. 15-14, the common type of packed tower consists of a cylindrical shell containing an inert packing material. These towers usually operate with countercurrent fluid flow. For the case in which liquid and vapor phases are involved, the liquid descends through the column in the form of films distributed over the packing surface, and the vapors rise through the spaces between the packing particles. Consequently, a large amount of vapor-liquid contact area becomes available, thus resulting in efficient mass-transfer operations.

The design of a packed tower requires consideration of mechanical factors, such as pressure drop, flow capacities, and foundation load. In addition, consideration must be given to the factors that influence the effectiveness of contact between the fluid phases. A satisfactory packing should have the following properties:

1. *Low pressure drop.* Since the pressure drop through the packing is a direct function of the fluid velocities, a large free cross-sectional area should be available between the packing particles in order to give a low pressure drop.
2. *High capacity.* The packing should permit high fluid rates without excessive pressure drop or build-up of liquid in the tower. Because flooding or carryover of liquid out of the tower can occur above certain limiting fluid velocities, a large free cross-sectional area is desirable for high capacities.

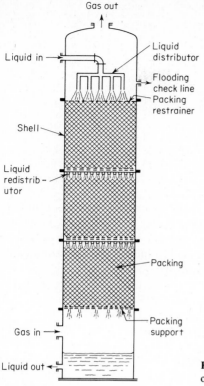

Gas out

Liquid in →

Shell —

Liquid redistrib– utor

Gas in →

Liquid out ◄—

Liquid distributor

Flooding check line

Packing restrainer

Packing

Packing support

Figure 15-14 Cross-sectional view of packed tower in operation.

3. *Low weight and low liquid holdup.* The total weight of the column and the resultant foundation load is low if the weight of the packing and the liquid holdup in the tower are low. The amount of liquid holdup, however, must be sufficiently great to retain an effective driving force for mass transfer.

4. *Large active surface area per unit volume.* To give high efficiencies, the packing must provide a large amount of contact area between the two fluid phases. This can be accomplished by using irregularly shaped packings that permit extensive distribution of liquid over surface area which can be contacted directly by the second fluid.

5. *Large free volume per unit total volume.* This property is particularly important if time must be available for a gas-phase chemical reaction, such as the oxidation of nitric oxide in the aqueous absorption of nitrogen dioxide.

6. *Miscellaneous.* High durability. High corrosion resistance. Low side thrust on tower walls. Low cost.

TYPES OF PACKING

Although many different types of packing are available for obtaining efficient contact between two fluid phases, the types can generally be classified as *random*

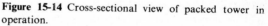

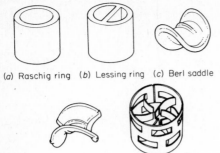

(a) Raschig ring   (b) Lessing ring   (c) Berl saddle

(d) Intalox saddle   (e) Pall ring

Figure 15-15 Single pieces of typical random packings.

or *stacked*. A random packing is one that is merely dumped into a containing shell, and the individual pieces are not arranged in any particular pattern. Pall rings, Intalox saddles, Raschig rings, and Berl saddles, as shown in Fig. 15-15, are the most common of the random packings used in industrial operations. Pall rings and Intalox saddles are generally replacing the older Raschig rings and Berl saddles because, in most cases, the Pall rings and Intalox saddles permit a more economical tower design than the other packings. Pall rings, made of metal or plastic, have the same general form as Raschig rings being open cylinders with the height equal to the diameter. However, Pall rings are stamped during the forming operation so that part of the original cylinder wall is cut with the projections bent inward leaving holes in the wall. The projections nearly touch at the center and the result is an opening of the ring and utilization of the interior of the ring to give improved vapor-liquid contact. They are available in sizes from $\frac{5}{8}$ to 3 in. or more.

Saddle-shaped packings, such as Intalox saddles and Berl saddles, are available in sizes from $\frac{1}{4}$ to 2 in. These packings are formed from chemical stoneware, plastics, or any other material that can be shaped by punch dies. They form an interlocking structure which gives less side thrust and more active surface area per unit volume than Raschig rings.

Raschig rings, as illustrated in Fig. 15-15, are simply hollow cylinders with the outside diameter equal to the height. They are usually made of inert materials that are cheap and light, such as porcelain, chemical stoneware, or carbon. Other materials of construction, such as clay, plastic, steel, and metal alloys, are also used. Raschig rings are available in sizes ranging from $\frac{1}{4}$ to 3 in. or more. Because breakage of fragile packing may occur when the pieces are dropped into an open shell, the initial packing charge is sometimes made by filling the empty tower with water and then dumping the packing slowly into the water.

Additional active surface can be provided by adding a single web or cross web on the inside of a Raschig ring. When a single web is present, the packing is known as Lessing rings. With a solid cross web, the packing is known as cross-partition rings which are normally available in sizes ranging from 3 to 6 in. and are almost always used as a stacked packing.

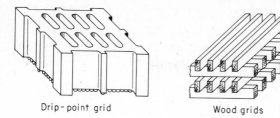

Drip-point grid   Wood grids   **Figure 15-16** Stacked grid packings.

Stacked packings, in general, give lower pressure drops for equivalent fluid capacities than random packings. However, this advantage is gained at the expense of higher initial costs due to the extra installation labor. The ring packings of nominal sizes 3 in. or larger are often used as a stacked packing. Other examples of stacked packings are shown in Fig. 15-16. A list of the important physical characteristics for common packings used in industrial operations is presented in Table 3.

LIQUID DISTRIBUTION

In absorption and distillation towers, the packing should be wetted completely by the descending liquid in order to provide the maximum amount of active surface area. For this reason, the entering liquid should be distributed uniformly over the top of the packing at a flow rate sufficient to permit wetting of all the packing surface. The packing support plate should provide ample space for passage of both liquid and gas with good distribution of the gas. A cone-shaped packing support is often used in small columns to distribute the entering gas uniformly throughout the packing and reduce the possibility of liquid flooding at the support plate.

Even with distributors at the top and bottom of the tower, the descending liquid has a tendency to channel or flow toward the column wall, thus resulting in packing regions where the surface is dry. Stacked packings do not give as good liquid distribution as random packings, and, in general, distribution difficulties tend to become excessive if the ratio of tower diameter to packing diameter is less than 7.

Loss of active surface area due to improper liquid distribution can be reduced by using intermediate redistributor plates, as shown in Fig. 15-14. Complete wetting of the packing by preflooding the column at the start of the operation or use of a self-wetting packing, such as protruded packing, can decrease the detrimental effects of poor liquid distribution.

PRESSURE DROP

The primary factors that affect pressure drop in packed towers are (1) fluid-flow rates, (2) density and viscosity of the fluids, and (3) size, shape, orientation, and surface of the packing particles. Figure 15-17 illustrates the effects of fluid rates

Table 3 Physical characteristics of commercial tower packings

For use when direct experimental data are not available

Packing	Nominal size, in.	Approximate average weight per ft³ of tower volume, lb	Approximate average total surface area of packing, ft²/ft³ of tower volume	Percent free-gas space $= \epsilon$ $\times 100$	Packing factor, a_p/ϵ^3, effective surface area/(void fraction)³, ft² per ft³ of tower volume (dry-packed values for use with Fig. 15–20)
Random packings:					
Stoneware	$\frac{1}{4}$	46	240	73	768
Raschig rings	$\frac{3}{8}$	51	134	68	494
	$\frac{1}{2}$	50	122	64	517
	$\frac{3}{4}$	44	80	73	199
	1	40	58	73	150
	$1\frac{1}{2}$	42	35	68	108
	2	37	28	74	46
	3	40	19	74	
Carbon Raschig	$\frac{1}{2}$	27	114	74	373
rings	1	27	57	74	170
	$1\frac{1}{2}$	34	38	67	92
	2	27	29	74	56
Steel Raschig rings	$\frac{1}{2}$	132	118	73	
(wall thickness	1	73	57	85	
$= \frac{1}{1.6}$ in.)	2	38	31	92	
Lessing rings	1	50	69	66	
(porcelain)	$\frac{1}{2}$	58	40	60	
	2	49	32	68	
Intalox saddles	$\frac{1}{2}$	34	190	78	
(porcelain)	1	34	78	78	100
	$1\frac{1}{2}$	30	60	81	52
Berl saddles	$\frac{1}{4}$	56	274	58	4225
(porcelain)	$\frac{1}{2}$	54	155	60	574
	1	45	79	69	229
	$1\frac{1}{2}$	38	52	70	79
Pall rings	$\frac{5}{8}$	37	104	93	70
(steel)	1	30	63	94	48
	$1\frac{1}{2}$	26	39	95	28
	2	24	31	96	20
Pall rings	$\frac{5}{8}$	$7\frac{1}{4}$	104	87	97
(polypropylene)	1	$5\frac{1}{2}$	63	90	52
	$1\frac{1}{2}$	$4\frac{3}{4}$	39	91	32
	2	$4\frac{1}{2}$	31	92	25
Stacked packings (stacked for maximum surface area per unit of tower volume):					
Raschig rings	2		32	80	
Cross partition	$4 \times 4 \times \frac{3}{8}$	81	32	53	
rings (porcelain)	$6 \times 6 \times \frac{5}{8}$	70	21	48	

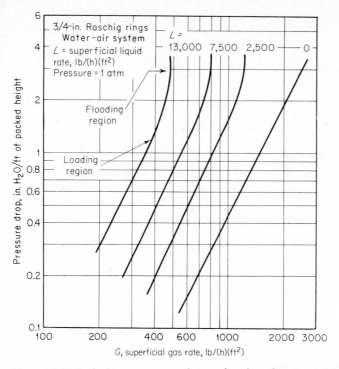

Figure 15-17 Packed-tower pressure drop as function of gas rate and liquid rate.

at constant operating pressure, and Fig. 15-19 shows how increased gas velocity due to reduction in operating pressure can affect the pressure drop.

As indicated by Figs. 15-17 through 15-19, a log-log plot of pressure drop per foot of packed height versus gas rate gives a straight-line relationship over the lower range of pressure drops. The point where the line first starts to curve upward is often designated as the *loading point*, to indicate that liquid is starting to build up in the column and is reducing the effective free space for gas flow. At the *flooding point* the pressure-drop-versus-gas-rate curve becomes almost vertical, and a liquid layer starts to build up on top of the packing. The flooding point represents the upper limiting conditions of pressure drop and fluid rates for practical tower operation.

Estimation of Pressure Drop in Packed Towers

Many methods have been proposed for estimating pressure drop in packed towers. Most of these methods are based primarily on experimental data obtained with countercurrent flow of water and air through various types of packed towers. Because of the empirical nature of these correlations and the fact that the effects of some of the variables are not included, it is always best to

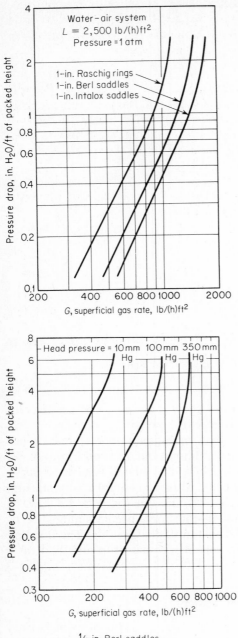

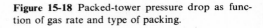

Figure 15-18 Packed-tower pressure drop as function of gas rate and type of packing.

Figure 15-19 Pressure drop in packed distillation tower as function of gas rate and operating pressure.

predict pressure drops in the design of packed towers on the basis of experimental data obtained with equipment operating under conditions equivalent to those involved in the design.† If such data are not available, approximations can be made by using the methods outlined in the following discussion.

Irrigated packings Because the operation of packed towers in the range between the loading region and the flooding point may give erratic fluctuations, many packed towers are operated in the preloading range. In this range, the slope of the straight lines in Figs. 15-17 through 15-19 is approximately 2.0, and the pressure drop can therefore be taken as directly proportional to the square of the gas mass velocity.

Leva‡ has correlated experimental data to obtain the following empirical equation for estimating pressure drop in packed beds under preloading conditions with simultaneous counterflow of liquid and gas:

$$\frac{\Delta p}{h} = \gamma (10)^{\Phi L/\rho_L} \frac{G^2}{\rho_G} \tag{22}$$

where Δp = pressure drop, psf
  h = packed height, ft
  γ, Φ = constants with value dependent on packing size and type (see Table 4)
  L = superficial liquid mass velocity (based on cross-sectional area of empty tower), $lb/(h)(ft^2)$
  ρ_L = density of liquid, lb/ft^3
  G = superficial gas mass velocity (based on cross-sectional area of empty tower), $lb/(h)(ft^2)$
  ρ_G = density of gas, lb/ft^3

The effect of liquid viscosity is not included in Eq. (22). Although an increase in liquid viscosity tends to cause an increase in pressure drop if flow rates are held constant, the results of Eq. (22) can be accepted as a reasonable approximation if the liquid involved has a viscosity less than about 2 centipoises.

At the flooding point for a given packing and set of fluids, the pressure drop per foot of packed height remains approximately constant with variations in fluid rates and operating pressure. Table 5 presents typical pressure drops for various fluids and packings at the flooding point.§ The table also indicates a method for estimating pressure drop when a packed column is operated in the range between the loading point and the flooding point.

† For a compilation of various experimental data, see J. H. Perry and C. H. Chilton, "Chemical Engineers' Handbook," 5th ed., Sect. 18, McGraw-Hill Book Company, New York, 1973.
  ‡ M. Leva, *Chem. Eng. Progr. Symp. Ser.*, **50**(10):51 (1954).
  § F. A. Zenz, *Chem. Eng.*, **60**(8):176 (1953).

Table 4. Constants for estimating pressure drop in packed towers by Eq. (22)

Apply only to preloading range. Above this range, pressure drop predicted by Eq. (22) will be too low.

Packing	Nominal size, in.	$\gamma \times 10^8$	$\Phi \times 10^3$	Valid for range of L, lb/(h)(ft²)
Raschig rings	$\frac{1}{2}$	139	7.2	300–9000
	$\frac{3}{4}$	33	4.5	2000–11,000
	1	32	4.3	400–27,000
	$1\frac{1}{2}$	12	4.0	700–18,000
	2	11	2.3	700–22,000
Berl saddles	$\frac{1}{2}$	60	3.4	300–14,000
	$\frac{3}{4}$	24	3.0	400–14,000
	1	16	3.0	700–29,000
	$1\frac{1}{2}$	8	2.3	700–22,000
Intalox saddles	1	12	2.8	2500–14,000
	$1\frac{1}{2}$	6	2.3	2500–14,000

Table 5. Pressure drop at flooding point in packed towers

Values are based on countercurrent flow of liquid and air at atmospheric pressure. In the range between the loading region and the flooding point, an approximation of pressure drop Δp at gas mass velocity G can be obtained from $\Delta p = \Delta p_F (G/G_m)^{3.2}$. The pressure drop at the loading point is usually in the range of one-fourth to one-sixth of the pressure drop at the flooding point.

Packing	Nominal size, in.	Pressure drop at flooding point, $\Delta p_F/h$, in. water/ft of packed height						
		Kinematic viscosity of liquid, centistokes = centipoises/specific gravity						
		0.6	1.0	3.0	6.0	10	20	40
Raschig rings	$\frac{1}{4}$	4.4	4.0	3.3	3.0	2.8	2.6	2.4
	$\frac{1}{2}$	3.9	3.5	2.9	2.7	2.5	2.3	2.1
	$\frac{3}{4}$	3.3	3.0	2.5	2.3	2.1	1.9	1.8
	1	3.3	3.0	2.5	2.3	2.1	1.9	1.8
	$1\frac{1}{2}$	2.8	2.5	2.1	1.9	1.8	1.6	1.5
	2	2.8	2.5	2.1	1.9	1.8	1.6	1.5
Berl saddles	$\frac{1}{4}$	1.4	1.3	1.2	1.1	1.0	1.0	0.9
	$\frac{1}{2}$	2.1	2.0	1.8	1.7	1.6	1.5	1.4
	$\frac{3}{4}$	2.7	2.5	2.3	2.1	2.0	1.9	1.8
	1	2.7	2.5	2.3	2.1	2.0	1.9	1.8
	$1\frac{1}{2}$	2.4	2.2	2.0	1.8	1.7	1.6	1.5

Dry Packings The following equation, by Ergun,† can be used to estimate pressure drop caused by the flow of a gas through dry packings:

$$\frac{\Delta p}{h} = \frac{1 - \varepsilon}{\varepsilon^3} \frac{G^2}{d_p g_c \rho_G} \left[\frac{150(1 - \varepsilon)\mu_G}{d_p G} + 1.75 \right] \tag{23}$$

where ε = fractional void volume in bed, ft^3 void/ft^3 of packed-tower volume

μ_G = absolute viscosity of gas, lb/(ft)(h)

d_p = effective diameter of packing particle, ft, or diameter of a sphere with same surface-to-volume ratio as packing particle

= $6(1 - \varepsilon)/a_p$

a_p = surface area of packing per unit of packed-tower volume, ft^2/ft^3

Equation (23) accounts for simultaneous kinetic and viscous energy losses and is applicable to the single-phase flow of liquids as well as gases.‡

Example 6. Estimation of pressure drop in packed tower A column 2 ft in diameter is packed with $\frac{3}{4}$-in. stoneware Raschig rings. Air, at an average pressure of 1 atm and an average temperature of 70°F, flows through the tower at a superficial mass velocity of 600 lb/(h)(ft^2). Estimate the pressure drop through the dry packing by Eq. (23). If water at 70°F flows countercurrent to the air at a rate of 8000 lb/h, estimate the pressure drop by Eq. (22). Express the answers as inches of water per foot of packed height, and compare the calculated results with the values shown in Fig. 15-17.

SOLUTION For dry packing,

$$\frac{\Delta p}{h} = \left(\frac{1 - \varepsilon}{\varepsilon^3} \right) \frac{G^2}{d_p g_c \rho_G} \left[\frac{150(1 - \varepsilon)\mu_G}{d_p G} + 1.75 \right]$$

From Table 3,

$$\varepsilon = 0.73$$

$$a_p \text{ (based on total surface area)} = 80 \text{ ft}^2/\text{ft}^3$$

$$d_p = \frac{6(1 - \varepsilon)}{a_p}$$

$$= \frac{6(1 - 0.73)}{80} = 0.0202 \text{ ft}$$

---

† S. Ergun, *Chem. Eng. Progr.*, **48**:89 (1952).

‡ For additional information and methods for estimating pressure drop and flooding velocities in packed towers, see the articles by Eckert and coauthors as follows: J. S. Eckert, Selecting the Proper Distillation Column Packing, *Chem. Eng. Progr.*, **66**(3):39 (1970); J. S. Eckert, E. Foote, and L. Walter, What Affects Packing Performance? *Chem. Eng. Progr.*, **62**(1):59 (1966); J. S. Eckert, A New Look at Distillation—Four Tower Packings—Comparative Performance, *Chem. Eng. Progr.*, **59**(5):76 (1963); J. S. Eckert, Design Techniques for Sizing Packed Towers, *Chem. Eng. Progr.*, **57**(9):54 (1961); J. S. Eckert, E. Foote, and L. Huntington, Pall Rings—New Type of Tower Packing, *Chem. Eng. Progr.*, **54**(1):70 (1958).

$$\mu_G \text{ (at } 70°F) = 0.018 \text{ centipoise}$$

$$= 0.018 \times 2.42 = 0.044 \text{ lb/(ft)(h)}$$

$$\rho_G = \frac{(29)(492)}{(359)(530)}$$

$$= 0.075 \text{ lb/ft}^3$$

$$\frac{\Delta p}{h} = \frac{1 - 0.73}{(0.73)^3} \frac{(600)^2}{(0.0202)(32.17)(3600)^2(0.075)} \left[\frac{150(1 - 0.73)(0.044)}{(0.0202)(600)} + 1.75\right]$$

$$= 0.67 \text{ lb/(ft}^2)(\text{ft}) = \frac{(0.67)(33.93)(12)}{(14.7)(144)} = 0.13 \text{ in. water/ft}$$

According to Fig. 15-17, $\Delta p/h$ = about 0.15 in. water/ft for $L = 0$ (dry packing). For irrigated packing, Eq. (22) applies.

$$\frac{\Delta p}{h} = \gamma(10)^{\Phi L/\rho_L} \frac{G^2}{\rho_G}$$

From Table 4,

$$\gamma = 33 \times 10^{-8} \qquad \Phi = 4.5 \times 10^{-3}$$

$$\rho_L = 62.3 \text{ lb/ft}^3$$

$$L = \frac{8000}{(\pi/4)(2)^2} = 2550 \text{ lb/(h)(ft}^2)$$

$$\frac{\Delta p}{h} = 33 \times 10^{-8}(10)^{(0.0045)(2550)/62.3} \frac{(600)^2}{(0.075)} = 2.42 \text{ lb/(ft}^2)(\text{ft})$$

$$= \frac{(2.42)(33.93)(12)}{(14.7)(144)} = 0.47 \text{ in. water/ft}$$

According to Fig. 15-17, $\Delta p/h$ = about 0.48 in. water/ft.

ALLOWABLE VAPOR VELOCITY

The limiting vapor velocity for practical operation of a packed tower is set by the flooding point. As in the case of towers utilizing finite-stage contactors, the design engineer can determine the necessary diameter of a packed tower on the basis of the limiting vapor velocity. The maximum allowable vapor velocity is commonly designated as the superficial velocity at the flooding conditions. Because the tower operation may become unstable as the flooding point is approached, the design value for allowable vapor velocity is usually estimated to be 50 to 70 percent of the maximum allowable velocity, and this allowable velocity is used to establish the column diameter.

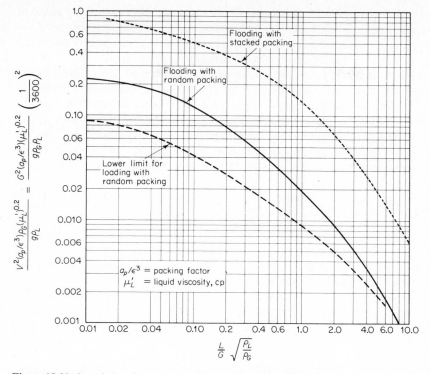

Figure 15-20 Correlation for estimating flooding rate in packed towers.

General Correlation for Flooding Rate

Figure 15-20 presents a correlation for flooding conditions in packed towers.†‡ The flooding line for random packings is based on experimental data obtained with 15 different liquids, 3 gases, and various random packings, such as Raschig

---

† The correlation presented in Fig. 15-20 is based on the method proposed by T. K. Sherwood, G. H. Shipley, and F. A. L. Holloway, *Ind. Eng. Chem.*, **30**:765 (1938), with the modification of experimental determination of appropriate a_p/ϵ^3 values as given by Lobo, et al., *Trans. AIChE*, **41**:693 (1945).

‡ Many other correlations for flooding rates in packed towers have been proposed. Some examples follow: The results of B. C. Sakiadis and A. I. Johnson, *Ind. Eng. Chem.*, **46**:1229 (1954), apply to spray columns and packed towers. E. H. Hoffing and F. J. Lockhart, *Chem. Eng. Progr.*, **50**:94 (1954); F. R. Dell and H. R. C. Pratt, *Trans. Inst. Chem. Engrs. (London)*, **29**:89 (1951), and *J. Appl. Chem.*, **2**:429 (1952), have presented correlations for liquid-liquid and gas-liquid systems. J. H. Perry and C. H. Chilton, "Chemical Engineers' Handbook," 5th ed., Sect. 18, McGraw-Hill Book Company, New York, 1973, gives packing characteristics for flooding. F. G. Eichel, *Chem. Eng.*, **73**(19):197 (1966), discusses the effects of packing size on column capacity. H. X. Nguyen, *Chem. Eng.*, **85**(26):181 (Nov. 20, 1978) presents methods showing how computer programs can expedite packed-tower design.

rings, Berl saddles, spheres, and helices. Best results are obtained if the value of the packing factor (a_p/ε^3) is based on direct experimental measurements with the particular packed tower; however, in the absence of experimental values, the data given in Table 3 can be used to approximate a_p/ε^3.

The dashed line in Fig. 15-20 represents an approximate lower limit of the loading range for random packings and can be used as a rough check to determine what percent of the maximum allowable velocity will give operating conditions in the preloading range.[†] As indicated by the upper line in Fig. 15-20, stacked packings have higher flow capacities than random packings. This increased capacity can be attributed partly to the decreased holdup and lower pressure drop due to the regular arrangement of the packing.

Simplified Approximate Method for Estimating Flooding Velocities in Packed Distillation Towers[‡]

At the flooding velocity in a packed tower, liquid starts to build up over the packing. It can be assumed, therefore, that flooding occurs when the downward pressure of the descending films or droplets equal the upward velocity pressure of the rising vapor. With this assumption and considering the gas density as negligible in comparison to the liquid density, the pressure-balance equation can be written as

$$\left(\frac{V_m^2}{2g}\right)\rho_G = H_p\rho_L \tag{24}$$

where H_p is a constant for each packing and can be considered as the mean diameter or film length of the liquid droplets. If K_p is defined as $(2gH_p)^{0.5}$, Eq. (24) can be expressed in a form similar to Eq. (1) as

$$V_m = K_p\sqrt{\frac{\rho_L}{\rho_G}} \tag{25}$$

Since the superficial gas mass velocity equals $V\rho_G$, Eq. (25) can be written in the following alternative form:

$$G_m = V_m\rho_G = K_p\sqrt{\rho_L\rho_G} \tag{26}$$

Equations (24) through (26) do not include any effects due to liquid viscosity or liquid flow rate. Consequently, the equations are most useful when experimental data for the packing are available at known L/G ratios and liquid viscosities. For design purposes, liquid-viscosity effects can be assumed negligible if the viscosity is under 2 centipoises. Because of errors that may be introduced by neglecting the influence of the liquid rate, use of Eq. (25) or (26) should be limited to cases in which the liquid rate is less than 1.5 times the gas rate.

---

† M. Leva, *Chem. Eng. Progr. Symp. Ser.*, **50**(10):51 (1954). Leva states that the correlation presented in Fig. 15-20 is improved if the ordinate group is multiplied by $(\rho_{\text{water}}/\rho_L)^2$.

‡ M. S. Peters and M. R. Cannon, *Ind. Eng. Chem.*, **44**:1452 (1952).

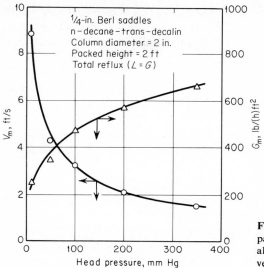

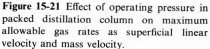

Figure 15-21 Effect of operating pressure in packed distillation column on maximum allowable gas rates as superficial linear velocity and mass velocity.

Effect of Operating Pressure on Flooding Velocities

Because many packed towers are operated at reduced pressures, the effect of total operating pressure on flooding rates is of particular interest to both the design engineer and the operations engineer. As a rough rule of thumb, the maximum allowable vapor velocity V_m in packed distillation columns is from 1 to 3 ft/s at atmospheric pressure and increases with decreasing pressure in direct proportion to the square root of the pressure ratios. Similarly, the maximum allowable mass velocity G_m decreases with decreasing pressure in direct proportion to the square root of the pressure ratios. A typical example of the effect of operating pressure on maximum allowable vapor velocity and maximum allowable gas mass velocity is presented in Fig. 15-21.

The relationship between the maximum allowable rates in a packed distillation tower at two different pressures for a given packing, reflux ratio, and set of fluids can be estimated by writing Eq. (25) or Eq. (26) for each pressure and eliminating the constant K_p. This procedure gives

$$V_{m_1} = V_{m_2} \sqrt{\frac{\rho_{G_2}\rho_{L_1}}{\rho_{G_1}\rho_{L_2}}} \tag{27}$$

$$G_{m_1} = G_{m_2} \sqrt{\frac{\rho_{G_1}\rho_{L_1}}{\rho_{G_2}\rho_{L_2}}} \tag{28}$$

where subscript 1 designates average conditions at one operating pressure and subscript 2 designates average conditions at the second operating pressure.

If the vapor acts as a perfect gas, $\rho_G = M_G P/RT$, and Eqs. (27) and (28) can be expressed as

$$V_{m_1} = V_{m_2}\left(\frac{P_2}{P_1}\right)^{0.5}\left(\frac{\rho_{L_1}}{\rho_{L_2}}\right)^{0.5}\left(\frac{T_1}{T_2}\right)^{0.5}\left(\frac{M_{G_2}}{M_{G_1}}\right)^{0.5} \tag{29}$$

$$G_{m_1} = G_{m_2}\left(\frac{P_1}{P_2}\right)^{0.5}\left(\frac{\rho_{L_1}}{\rho_{L_2}}\right)^{0.5}\left(\frac{T_2}{T_1}\right)^{0.5}\left(\frac{M_{G_1}}{M_{G_2}}\right)^{0.5} \tag{30}$$

where R = ideal-gas-law constant
P = total pressure
T = absolute temperature
M_G = molecular weight of gas

Example 7. Estimation of maximum allowable gas rate in packed tower A random-packed distillation tower is being operated at an average pressure equal to 1 atm. The reflux ratio is such that an average value of 1.0 may be assumed for L/G. The column is being operated at 60 percent of the maximum allowable gas rate. Under these conditions, the average gas rate through the tower is 100 lb/h and the average superficial gas velocity is 1.5 ft/s. If the operating pressure is reduced until the average pressure in the tower is 100 mm Hg and the ratio L/G is unchanged, estimate the maximum allowable gas rate as pounds per hour by (*a*) use of the simplified approximate Eq. (26), and (*b*) use of Fig. 15-20. The following data apply:

	At average temperature and atmospheric pressure	At average temperature and 100 mm Hg
ρ_G, lb/ft³	0.20	0.031
ρ_L, lb/ft³	50	52
μ'_L, centipoises	0.5	0.7

SOLUTION (*a*) By Eq. (26)

$$G_m = K_p\sqrt{\rho_L\rho_G}$$

At a pressure of 1 atm

$$G_m = \frac{100}{(0.60)(\text{cross-sectional area})}\ \text{lb}/(\text{h})(\text{ft}^2)$$

$$K_p = \frac{100}{(0.60)(\text{cross-sectional area})[(50)(0.20)]^{1/2}}$$

At pressure = 100 mm Hg

$$G_m = K_p\sqrt{\rho_L\rho_G} = \frac{(100)[(52)(0.031)]^{1/2}}{(0.60)(\text{cross-sectional area})[(50)(0.20)]^{1/2}}$$

Maximum allowable gas rate at 100 mm Hg pressure by Eq. (26) is

$$G_m(\text{cross-sectional area}) = \frac{100}{0.60}\sqrt{\frac{(52)(0.031)}{(50)(0.20)}} = 67 \text{ lb/h}$$

The same result could be obtained directly from Eq. (28).

(b) At pressure = 1 atm

$$\frac{L}{G}\sqrt{\frac{\rho_G}{\rho_L}} = (1)\sqrt{\frac{0.20}{50}} = 0.063$$

From Fig. 15-20,

$$\frac{V_m^2(a_p/\varepsilon^3)\rho_G(\mu_L')^{0.2}}{g\rho_L} = 0.15$$

$$V_m = \frac{1.5}{0.60} = 2.5 \text{ ft/s}$$

$$\frac{a_p}{\varepsilon^3} = \frac{(0.15)(32.17)(50)}{(2.5)^2(0.20)(0.5)^{0.2}} = 222$$

At a pressure of 100 mm Hg

$$\frac{L}{G}\sqrt{\frac{\rho_G}{\rho_L}} = (1)\sqrt{\frac{0.031}{52}} = 0.024$$

From Fig. 15-20,

$$\frac{V_m^2(a_p/\varepsilon^3)\rho_G(\mu_L')^{0.2}}{g\rho_L} = 0.20$$

$$V_m = \left[\frac{(0.20)(32.17)(52)}{(222)(0.031)(0.7)^{0.2}}\right]^{1/2} = 7.2 \text{ ft/s}$$

$$\text{Cross-sectional area of empty tower} = \frac{100}{(1.5)(0.20)(3600)}$$

$$= 0.0927 \text{ ft}^2$$

Maximum allowable gas rate at 100 mm Hg by Fig. 15-20 is

$$V_m\rho_G(\text{cross-sectional area}) = (7.2)(0.031)(0.0927)(3600) = 74 \text{ lb/h}$$

PACKING EFFICIENCIES

The ability of a given packing to achieve effective mass transfer between a gas phase and a liquid phase is commonly expressed in an empirical form as the height of packing equivalent to one transfer unit (HTU) or the height of packing equivalent to one theoretical plate (HETP). HTU can be expressed on the basis of the number of transfer units calculated from gas-phase driving-force data

(NTU_G) or liquid-phase driving-force data (NTU_L) as shown in the following.†

$$NTU_G = \int_{y_1}^{y_2} \frac{dy}{y - y_L^*} \tag{31}$$

$$NTU_L = \int_{x_1}^{x_2} \frac{dx}{x_G^* - x} \tag{32}$$

$$HTU_G = \frac{Z}{NTU_G} \; ; \quad HTU_L = \frac{Z}{NTU_L} \tag{33}$$

where y, x = gas- and liquid-phase concentrations, respectively

y_L^*, x_G^* = gas- and liquid-phase concentrations, respectively, corresponding to values in equilibrium with the liquid- and gas-phase concentrations

Z = height of packing corresponding to distance from the lower to upper limits indicated by the subscripts on y and x

Numerous empirical methods and results giving data for HTU and HETP values for various packings have been presented in the literature.‡

Empirical Prediction of HTU

The following equations giving empirical correlations of data for HTU_G based on a large amount of published data for Raschig rings and Berl saddles have been developed by Cornell et al.:§

For Raschig rings:

$$HTU_G = \frac{\psi Sc_G^{0.5}}{(Lf_1 f_2 f_3)^{0.6}} \left(\frac{D'}{12}\right)^{1.24} \left(\frac{Z}{10}\right)^{1/3} \tag{34}$$

For Berl saddles:

$$HTU_G = \frac{\psi Sc_G^{0.5}}{(Lf_1 f_2 f_3)^{0.5}} \left(\frac{D'}{12}\right)^{1.11} \left(\frac{Z}{10}\right)^{1/3} \tag{35}$$

where HTU_G = height of a gas-phase transfer unit, ft

ψ = parameter for a given packing material, see Figs. 15-22 and 15-23 for values

Sc_G = gas-phase Schmidt number = $\mu_G / \rho_G D_G$

L = superficial liquid mass flow rate, lb/(h)(ft^2)

$f_1 = (\mu_L/2.42)^{0.16}$

$f_2 = (62.4/\rho_L)^{1.25}$

$f_3 = (72.8/\sigma)^{0.8}$

---

† HTU can also be based on overall gas or overall liquid values which combine HTU_G and HTU_L. See any standard chemical engineering text on mass transfer for details.

‡ See J. H. Perry and C. H. Chilton, "Chemical Engineers' Handbook," 5th ed., Sect. 18, McGraw-Hill Book Company, New York, 1973, for a detailed summary.

§ D. Cornell, W. G. Knapp, and J. R. Fair, Mass Transfer Efficiency—Packed Columns—Part 1, *Chem. Eng. Progr.*, **56**(7):68 (1960).

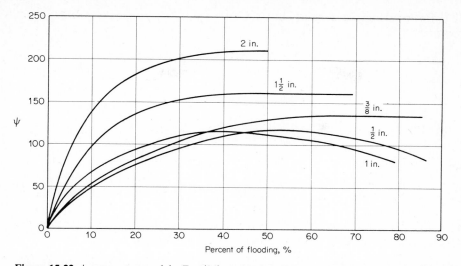

Figure 15-22 ψ parameter used in Eq. (34) to develop HTU_G values for several sizes of Raschig ring tower packing.

$$D' = \text{column diameter, in.}$$
$$Z = \text{packed height, ft}$$
$$\mu_G = \text{gas viscosity, lb/(ft)(h)}$$
$$\mu_L = \text{liquid viscosity, lb/(ft)(h)}$$
$$\rho_G = \text{gas density, lb/ft}^3$$
$$\rho_L = \text{liquid density, lb/ft}^3$$
$$\sigma = \text{surface tension, dyn/cm}$$
$$D_G = \text{gaseous-diffusion coefficient, ft}^2/\text{h}$$

To apply Eqs. (34) and (35), values of ψ are first obtained from Fig. 15-22 or Fig. 15-23 depending on the size of Raschig rings or Berl saddles being used.

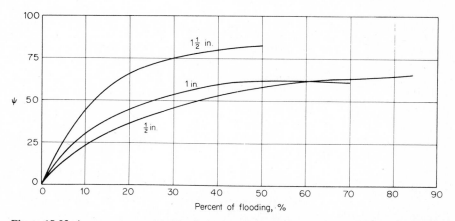

Figure 15-23 ψ parameter used in Eq. (35) to develop HTU_G values for several sizes of Berl saddle tower packing.

Table 6 Packing constants for use in Eq. (36)

	Size, in.	K_1	K_2	K_3
		Summary of packing constant		
Rings	$\frac{1}{4}$			
	$\frac{3}{8}$	2.10	−0.37	1.24
	$\frac{1}{2}$	8.53	−0.34	1.24
	1.0	0.57	−0.10	1.24
	2.0	0.42	0	1.24
Saddles	$\frac{1}{2}$	5.62	−0.45	1.11
	1.0	0.76	−0.14	1.11
McMahon	$\frac{1}{4}$	0.017	+0.50	1.00
	$\frac{3}{8}$	0.20	+0.25	1.00
	$\frac{1}{2}$	0.33	+0.20	1.00
Protruded packing	0.16	0.39	+0.25	0.30
	0.24	0.076	+0.50	0.30
	0.48	0.45	+0.30	0.30
	1.0	3.06	+0.12	0.30
Stedman	2	0.077	+0.48	0.24
	3	0.363	+0.26	0.24
	6	0.218	+0.32	0.24

Empirical Prediction of HETP

On the basis of analysis of published data on distillation operations with packed towers, Murch has presented the following empirical equation as a correlation to be used for estimating HETP:†

$$HETP = K_1 G^{K_2} D'^{K_3} Z^{1/3} \frac{\alpha \mu'_L}{\rho'_L} \tag{36}$$

where $HETP$ = height equivalent to a theoretical plate, in.
K_1, K_2, and K_3 = empirical constants as a function of the packing, see Table 6 for values of these constants
G = superficial gas mass flow rate, lb/(h)(ft²)
D' = column diameter, in.
Z = height of packing, ft
α = relative volatility
μ'_L = liquid viscosity, centipoise
ρ'_L = liquid density, g/cc.

† D. Murch, Height of Equivalent Theoretical Plate in Packed Fractionation Columns—An Empirical Correlation, *Ind. Eng. Chem.*, **45**:2616 (1953). See also M. S. Kuk, Key Design Variables in Packed Towers, *Chem. Eng. Progr.*, **75**(5):68 (1979).

RELATIVE MERITS OF PLATE AND PACKED TOWERS

The choice between use of a plate tower or a packed tower for a given mass-transfer operation should, theoretically, be based on a detailed cost analysis for the two types of contactors. Thus, the optimum economic design for each type would be developed in detail, and the final choice would be based on a consideration of costs and profits at the optimum conditions. In many cases, however, the decision can be made on the basis of a qualitative analysis of the relative advantages and disadvantages, and the need for a detailed cost comparison is eliminated. The following general advantages and disadvantages of plate and packed towers should be considered when a choice must be made between the two types of contactors:

1. Stage efficiencies for packed towers must be based on experimental tests with each type of packing. The efficiency varies, not only with the type and size of packing, but also with the fluid rates, the fluid properties, the column diameter, the operating pressure, and, in general, the extent of liquid dispersion over the available packing surface.
2. Because of liquid-dispersion difficulties in packed towers, the design of plate towers is considerably more reliable and requires less safety factor when the ratio of liquid mass velocity to gas mass velocity is low.
3. Plate towers can be designed to handle wide ranges of liquid rates without flooding.
4. If the operation involves liquids that contain dispersed solids, use of a plate tower is preferred because the plates are more accessible for cleaning.
5. Plate towers are preferred if interstage cooling is required to remove heats of reaction or solution, because cooling coils can be installed on the plates or the liquid-delivery line from plate to plate can be passed through an external cooler.
6. The total weight of a dry plate tower is usually less than that of a packed tower designed for the same duty. However, if liquid holdup during operation is taken into account, both types of towers have about the same weight.
7. When large temperature changes are involved, as in distillation operations, plate towers are often preferred because thermal expansion or contraction of the equipment components may crush the packing.
8. Design information for plate towers is generally more readily available and more reliable than that for packed towers.
9. Random-packed towers are seldom designed with diameters larger than 4 ft, and diameters of commercial plate towers are seldom less than 2 ft.
10. Packed towers prove to be cheaper and easier to construct than plate towers if highly corrosive fluids must be handled.
11. Packed towers are usually preferred if the liquids have a large tendency to foam.
12. The amount of liquid holdup is considerably less in packed towers.
13. The pressure drop through packed towers may be less than the pressure

drop through plate towers designed for the same duty. This advantage, plus the fact that the packing serves to lessen the possibility of tower-wall collapse, makes packed towers particularly desirable for vacuum operation.

MASS-TRANSFER EQUIPMENT COSTS

PLATE AND PACKED TOWERS

The purchased cost for plate and packed towers can be divided into the following components: (1) cost for shell, including heads, skirts, manholes, and nozzles; (2) cost for internals, including trays and accessories, packing, supports, and distributor plates; (3) cost for auxiliaries, such as platforms, ladders, handrails, and insulation.

The cost for fabricated tower shells is quite often estimated on the basis of weight. Figure 15-24 indicates the approximate cost for fabricated steel towers

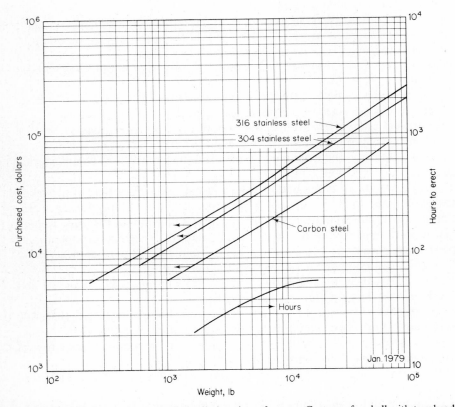

Figure 15-24 Fabricated costs and installation time of towers. Costs are for shell with two heads and skirt, but without trays, packing, or connections.

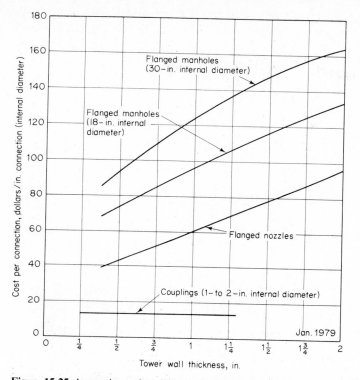

Figure 15-25 Approximate installed cost of steel-tower connections. Values apply to 300-lb connections. Multiply costs by 0.9 for 150-lb connections and by 1.2 for 600-lb connections.

based on weight. This cost is for the tower shell only, without trays, packing, or connections.† The cost for installed connections, such as manholes and nozzles, may be approximated from the data in Fig. 15-25.

Tray costs are shown in Fig. 15-26 for conventional installations. The approximate prices for various types and sizes of the most common industrial packings are indicated in Table 7. Costs for other packing materials are listed in Table 8. For rough estimates, the cost for a distributor plate in a packed tower can be assumed to be the same as that for one bubble-cap tray.

Table 9 presents data that can be used for estimating the cost of such auxiliaries as ladders, platforms, and handrails. The installed cost of several industrial-type insulations, all with aluminum jackets, for towers and tanks may be approximated from Fig. 15-27. Relative fabricated costs for various materials used in tray-tower construction are listed in Table 10.

---

† See also the information presented in Chap. 13 on costs for tanks, pressure vessels, and storage equipment along with information on design methods for tanks and pressure vessels.

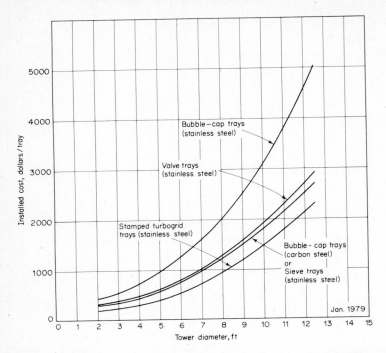

Figure 15-26 Cost of trays in plate towers. Price includes tray deck, bubble caps, risers, downcomers, and structural-steel parts.

Table 7 Costs of tower packings, uninstalled, January, 1979

Prices in dollars per ft³, 100 ft³ orders, f.o.b. manufacturing plant

	Size, in.			
	1	$1\frac{1}{2}$	2	3
Raschig rings:				
Chemical porcelain	10.3	8.3	7.6	6.3
Carbon steel	23.7	15.5	13.3	10.9
Stainless steel	85.5	64.6	57.0	——
Carbon	32.3	28.5	20.9	19.0
Intalox saddles:				
Chemical stoneware	14.2	10.4	9.5	8.6
Chemical porcelain	15.2	11.4	10.4	9.5
Polypropylene	17.1	——	10.6	5.5
Berl saddles:				
Chemical stoneware	22.0	17.0	——	——
Chemical porcelain	27.0	20.0	——	——
Pall rings:				
Carbon steel	19.0	12.9	11.8	——
Stainless steel	72.0	55.0	47.5	——
Polypropylene	17.1	11.6	10.6	——

Table 8 Costs of miscellaneous packing materials, January, 1979

Materials	Cost, $/ft^3
Activated carbon	28
Alumina	26
Calcium chloride	7
Coke	3
Crushed limestone	7
Resin	158
Silica gel	52

Table 9 Cost of tower auxiliaries

Item	Cost, January, 1979	Amount for typical tower	
Ladder	$0.43/lb	30 lb/ft of height	
Platforms and handrails	0.43/lb	Tower diameter, ft	Weight, lb
		4	1700
		6	2300
		8	2800
		10	3300

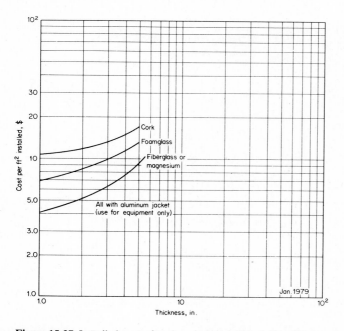

Figure 15-27 Installed cost of various industrial insulations for towers and tanks.

Table 10 Relative fabricated cost for metals used in tray-tower construction

Materials of construction	Relative cost per ft² of tray area (based on carbon steel = 1)
Sheet-metal trays:	
Steel	1
4–6 % chrome—$\frac{1}{2}$ moly alloy steel	2
11–13 % chrome type 410 alloy steel	2.5
Red brass	3
Stainless steel type 304	4
Stainless steel type 347	4.8
Monel	7.0
Stainless steel type 316	5.2
Inconel	8.2
Cast-iron trays	2.8

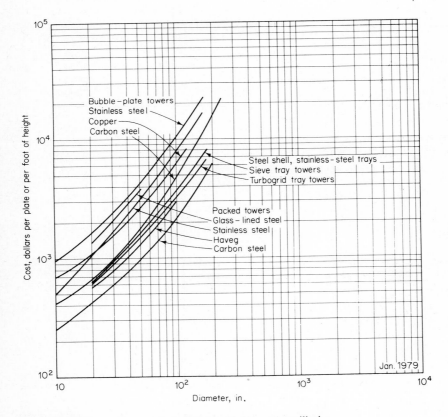

Figure 15-28 Cost of towers including installation and auxiliaries.

The complete installed cost of various types of towers is given in Fig. 15-28. This cost includes all the components normally associated with a tower as outlined above.

Example 8. Estimation of cost for bubble-cap tower A distillation tower contains 18 steel bubble-cap trays. An 18-in. manhole is located above each tray, and one manhole is located below the bottom tray. The ID of the tower is 6 ft, and the total height including the skirt is 50 ft. The shell is steel (density = 490 lb/ft^3) with a $\frac{5}{8}$-in. wall thickness. Six 1-in. couplings and the following flanged nozzles are attached to the tower: one 10-in. vapor-line nozzle; three 4-in. nozzles; and six 2-in. nozzles. On the basis of the data presented in Fig. 15-24, Fig. 15-25, and Fig. 15-26, estimate the cost of the tower with trays installed, but not including cost for auxiliaries or tower installation. The total weight of the shell, including heads and skirt, may be assumed to be 1.12 times the weight of the cylindrical shell. Material of construction is carbon steel.

SOLUTION Total weight of shell = $(6)(\frac{5}{8})(\frac{1}{12})(3.14)(50)(490)(1.12) = 27,000$ lb:

Purchased cost of steel shell = (27,000 lb—From Fig. 15-24)	$42,000
Cost for 19 installed 18-in. manholes = (19)(80)(18)	27,360
Cost for 1 installed 10-in. nozzle = (1)(47)(10)	470
Cost for 3 installed 4-in. nozzles = (3)(47)(4)	565
Cost for 6 installed 2-in. nozzles = (6)(47)(2)	565
Cost for 6 installed 1-in. couplings = (6)(13)(1)	80
Cost for 18 installed steel bubble-cap trays = (18)(760)	13,680
Total	$84,720

Estimated total purchased cost for tower = $85,000.

DRYERS

Cost data for drum dryers, rotary dryers, tray dryers, and tumble dryers are presented in Figs. 15-29 through 15-33. The data available cover both atmospheric and pressure systems.

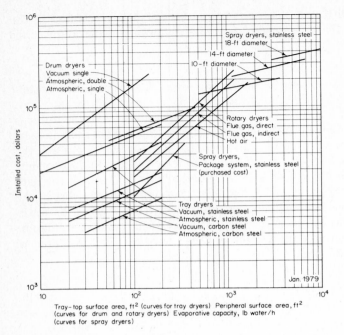

Figure 15-29 Installed cost of dryers.

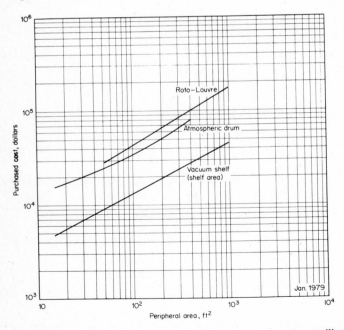

Figure 15-30 Costs for dryers, steel construction. Price includes auxiliaries (motor, drive, fan, vacuum pump, condenser, and receiver).

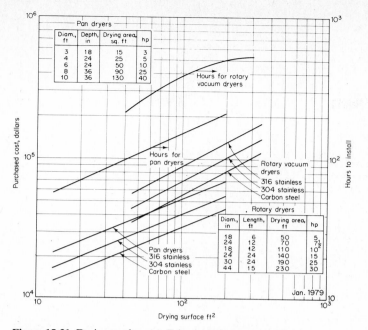

Figure 15-31 Drying equipment. Price includes motor and drive.

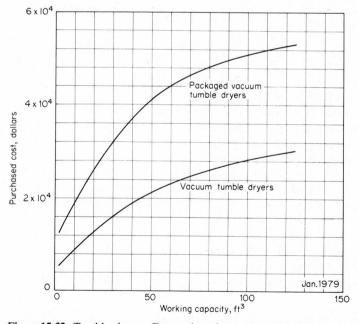

Figure 15-32. Tumble dryers. For packaged vacuum tumble dryers, price includes the following equipment: jacketed shell for steam or electric water heater, water pump, vacuum pump, condenser, receiver, piping, and temperature-pressure- and motor controls. For the vacuum tumble dryers, price includes frame and neoprene-lined butterfly outlet valve.

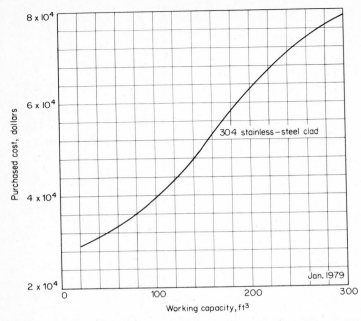

Figure 15-33 Cost of Dyna-Cone vacuum dryers. Price includes power supply, butterfly discharge valve, fan-cooled motor, internal brake, gear, coupling, belt or chain guards, replaceable cartridge filter, and 304 stainless-steel thermocouple well.

REACTORS

Application of the principles of chemical reaction kinetics for the design of chemical reactors is an area of activity that is particularly appropriate for chemical engineers.† For reactor design, the fundamental ideas of process analysis, equipment design, economic evaluation, and optimization are all combined to give a mathematical model of the reactor system. Prior to the ready availability of large computers, mathematical models for application to overall reactor systems were generally restrained to a relatively simple form so that solution of the resulting equations could be accomplished by standard computational techniques.

Now that sophisticated computer equipment is widely available, the chemical engineer can concentrate much more heavily on analysis of the system and development of the best mathematical model. Solution of complex mathematical models for reactor-design systems, which would have been impossible before the

† For an overall treatment of principles and techniques as applied to reactor design, see H. F. Rase, "Chemical Reactor Design for Process Plants, Vol. I—Principles and Techniques; Vol. II—Case Studies and Design Data," John Wiley & Sons, New York, 1977.

availability of computers, can now be accomplished with ease so that the actual solution techniques are no longer nearly so important as the basic analysis involved in developing the mathematical model.†

The basic mathematical model for a reactor system is developed from (1) material balances including inflow, outflow, reaction rates, mixing effects, and diffusional effects; (2) energy balances including heats of reaction, heat transfer, and latent and sensible heat effects; (3) reaction-rate expressions incorporating mechanism definition and temperature functionality; (4) economic evaluations; and (5) special constraints on the design system. The material balances and the energy balances referred to in (1) and (2) are developed from the fundamental conservation equations based on a differential volume of the reacting system, or,

Material balance:

$$
\begin{array}{l}
\text{Rate of accumulation} \\
\text{of the } i\text{th species in} \\
\text{the volume element}
\end{array}
=
\begin{array}{l}
\text{net rate of inward} \\
\text{flow of the } i\text{th} \\
\text{species to the} \\
\text{volume element}
\end{array}
+
\begin{array}{l}
\text{rate of genera-} \\
\text{tion of the } i\text{th} \\
\text{species in the} \\
\text{volume element}
\end{array}
\quad (37)
$$

Energy balance:

$$
\begin{array}{l}
\text{Rate of energy} \\
\text{accumulation in} \\
\text{the volume element}
\end{array}
=
\begin{array}{l}
\text{net rate of inward} \\
\text{energy flow to the} \\
\text{volume element}
\end{array}
+
\begin{array}{l}
\text{rate of generation} \\
\text{of energy in the} \\
\text{volume element}
\end{array}
\quad (38)
$$

There are many types of chemical reactors which operate under various conditions, such as batch, flow, homogeneous, heterogeneous, steady state, etc. Thus, one general mathematical description which would apply to all types of reactors would be extremely complex. The general approach for reactor design, therefore, is to develop the appropriate mathematical model which will describe the specific reaction system for that particular form of reactor under consideration. For example, if the reaction system is to be evaluated for steady-state operating conditions, the derivative terms on the left-hand sides of Eqs. (37) and (38) are zero which results in a simplified mathematical model.

It is convenient to use different forms for expressing the rate of generation of the ith species and the accompanying rate of energy generation due to the heat of reaction which are appropriate to the particular reaction system under consideration. For example, if a batch reactor of volume V_b involving a homogeneous reaction is under consideration, the convenient form for expressing the rate of reaction, based on material i as a reactant, is

$$
r_i = -\frac{1}{V_b}\frac{dN_i}{dt} = k_i(C_a, C_b, \ldots, C_i, \ldots, C_{n-1}, C_n) \quad (39)
$$

† For examples of computer-aided-design programs which are available for reactor-design applications, see J. N. Peterson, C.-C. Chen, and L. B. Evans, Computer Programs for Chemical Engineers: 1978, Part 2, *Chem. Eng.*, **85**(15):69 (July 3, 1978).

where N_i = amount of base-reactant material i present in the volume V_b at time t
   t = time
   k_i = function for species i related to the specific reaction-rate constant
$C_a, \ldots, C_n$ = concentration of species $a, \ldots, n$ in the reactor at time t

On the other hand, for a flow reactor where concentration varies throughout the reactor, it is usually more convenient to base the rate on a differential volume of the reactor, and a commonly used rate form is

$$r_i = -\mathrm{div}\ (\mathbf{U}C_i) \tag{40}$$

where $\mathbf{U}$ is the velocity vector with normal x, y, and z directed velocity elements using Cartesian coordinates. For the case of a plug-flow reactor and z representing the length of the reactor in the direction of flow, Eq. (40) reduces to the familar form of

$$r_i = -\frac{d(U_z C_i)}{dz} \tag{41}$$

Similar forms of simplified expressions can be developed for back-mix tank reactors and other types of reactor systems.†‡ A typical example of the analysis used in developing the mathematical model for a reactor design is presented in the following section.

AN EXAMPLE OF MATHEMATICAL MODELING FOR REACTOR DESIGN

Consider the case of a reactor design dealing with a unit that is to be used to chlorinate benzene with ferric chloride as the catalyst.§ A semibatch reactor is to be used containing an initial charge of pure benzene into which dry chlorine gas is fed at a steady state. The reaction vessel is operated at a constant pressure of two atmospheres and is equipped with coils to maintain a constant temperature of 55°C. The liquid is agitated to give a homogeneous mixture. The unit contains a reflux condenser which returns all vaporized benzene and chlorobenzenes to the reactor while allowing the generated hydrogen chloride gas and excess chlorine gas to leave the system. Monochlorobenzene, dichlorobenzene, and trichlorobenzene

---

† Design equations for use with various idealized types of reactor systems are presented later in this chapter.
‡ For further development of standard mathematical forms of rate expressions and basic developments for design expressions, see any standard text on chemical engineering reaction kinetics, such as J. M. Smith, "Chemical Engineering Kinetics," 2d ed., McGraw-Hill Book Company, New York, 1970; O. Levenspiel, "Chemical Reaction Engineering," 2d ed., John Wiley and Sons, Inc., New York, 1972, or J. J. Carberry, "Chemical and Catalytic Reaction Engineering," McGraw-Hill Book Company, New York, 1976.
§ Adapted from information and methods presented by R. B. MacMullin, *Chem. Eng. Progr.,* **44**:183 (1948) and A. Carlson, *Instr. and Control Systems,* **38**(4):147 (1965).

are produced through successive irreversible reactions as follows, where the k's represent the forward rate constants for the equations as written:

$$C_6H_6 + Cl_2 \xrightarrow{k_1} C_6H_5Cl + HCl$$

$$C_6H_5Cl + Cl_2 \xrightarrow{k_2} C_6H_4Cl_2 + HCl$$

$$C_6H_4Cl_2 + Cl_2 \xrightarrow{k_3} C_6H_3Cl_3 + HCl$$

For this semibatch system, it is desired to know at what time the reaction should be stopped in order to obtain a maximum yield from the original benzene of any one of the three products, based on the data and assumptions given in Table 11.

Table 11 Basic data and assumptions for reactor-design example dealing with optimization of yields for benzene chlorinator

Data for chlorobenzene reactor-design example

The feed rate of the dry chlorine is 1.4 lb mole of chlorine/(h) (lb mole of initial benzene charge).

The following rate constants are estimated values for the catalyst used at 55°C (assumed for this problem):

$$k_1 = 510 \text{ (lb mol/ft}^3)^{-1}(\text{h})^{-1}$$
$$k_2 = 64 \text{ (lb mol/ft}^3)^{-1}(\text{h})^{-1}$$
$$k_3 = 2.1 \text{ (lb mol/ft}^3)^{-1}(\text{h})^{-1}$$

There is negligible liquid or vapor holdup in the reflux condenser.

Volume changes in the reacting mixture are negligible, and the volume of liquid in the reactor remains constant at 1.46 ft^3/lb mole of initial benzene charge.

Hydrogen chloride has a negligible solubility in the liquid mixture.

The chlorine gas fed to the system goes into the liquid solution immediately up to its solubility limit of 0.12 lb mole of chlorine/lb mole of original benzene, and this value then remains constant.

Each reaction is second order as written.

Nomenclature

F = feed rate of chlorine as (lb mol/h)/(lb mole of initial benzene charge) = 1.4
V_b = volume of reactor liquid contents = $1.46N_0$; ($V_b/N_0 = 1.46$)
N_0 = lb moles of benzene present at time 0
N_B = lb moles of benzene present at time t
N_{Cl} = lb moles of chlorine present in liquid at time t
N_D = lb moles of dichlorobenzene present at time t
N_M = lb moles of monochlorobenzene present at time t
N_T = lb moles of trichlorobenzene present at time t
t = time in hours

$$B = \frac{N_B}{N_0} \qquad D = \frac{N_D}{N_0} \qquad T = \frac{N_T}{N_0}$$

$$C = \frac{N_{Cl}}{N_0} \qquad M = \frac{N_M}{N_0}$$

Equation (37) forms the basis for the initial design analysis of this system. Equation (38) is not used for the kinetic analysis because the system is isothermal. Because there is no input or output flow of benzene or chlorobenzenes, the material balances, by Eq. (37), become

For benzene,

$$-\frac{1}{V_b}\frac{dN_B}{dt} = k_1\frac{N_B}{V_b}\frac{N_{Cl}}{V_b} \tag{42}$$

For monochlorobenzene,

$$\frac{1}{V_b}\frac{dN_M}{dt} = k_1\frac{N_B}{V_b}\frac{N_{Cl}}{V_b} - k_2\frac{N_M}{V_b}\frac{N_{Cl}}{V_b} \tag{43}$$

For dichlorobenzene,

$$\frac{1}{V_b}\frac{dN_D}{dt} = k_2\frac{N_M}{V_b}\frac{N_{Cl}}{V_b} - k_3\frac{N_D}{V_b}\frac{N_{Cl}}{V_b} \tag{44}$$

For trichlorobenzene,

$$\frac{1}{V_b}\frac{dN_T}{dt} = k_3\frac{N_D}{V_b}\frac{N_{Cl}}{V_b} \tag{45}$$

For chlorine, subject to the solubility limit of $N_{Cl}/N_0 = 0.12$, there is an input flow term in Eq. (37). Therefore, the rate expression for the consumption of chlorine can be written as

$$\frac{1}{V_b}\frac{dN_{Cl}}{dt} = \frac{N_0 F}{V_b} - k_1\frac{N_B}{V_b}\frac{N_{Cl}}{V_b} - k_2\frac{N_M}{V_b}\frac{N_{Cl}}{V_b} - k_3\frac{N_D}{V_b}\frac{N_{Cl}}{V_b} \tag{46}$$

subject to a maximum value of

$$\frac{N_{Cl}}{N_0} = 0.12 \tag{47}$$

Equations (42) to (47) can be converted to a more convenient form for solution by rearranging with use of the nomenclature and data shown in Table 11 to bring in the primary variables being sought (namely, B, M, D, T, and C). The result is

$$\frac{dB}{dt} = -\frac{k_1 N_0}{V_b} BC \tag{48}$$

$$\frac{dM}{dt} = \frac{k_1 N_0}{V_b} BC - \frac{k_2 N_0}{V_b} MC \tag{49}$$

$$\frac{dD}{dt} = \frac{k_2 N_0}{V_b} MC - \frac{k_3 N_0}{V_b} DC \tag{50}$$

$$\frac{dT}{dt} = \frac{k_3 N_0}{V_b} DC \tag{51}$$

$$\frac{dC}{dt} = F - \frac{k_1 N_0}{V_b} BC - \frac{k_2 N_0}{V_b} MC - \frac{k_3 N_0}{V_b} DC \tag{52}$$

subject to a maximum of $C = 0.12$ \hfill (53)

Table 12 Information flow diagram of mathematical model for chlorobenzene reactor-design example

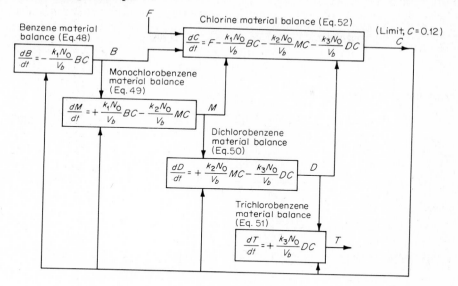

The solution can be checked at any point by an overall material balance which indicates that

$$1 = B + M + D + T \tag{54}$$

Thus, the result of the mathematical analysis is five differential equations [Eqs. (48) through (52)] with five unknowns (B, M, D, T, C). The solution to the problem is now completed by simultaneous solution of the above equations for the five variables as a function of time to give results from which the optimum times can be chosen.

Without the availability of a computer, the calculations for this solution would be very tedious and time consuming.† With computers available, the design engineer's task is essentially finished with the correct generation of the mathematical model as given in Eqs. (48) to (53) and summarized as an information-flow diagram in Table 12.

For this example, the final result can be obtained by use of either an analog or digital computer. The details for obtaining the solution with either type of computer are presented in the following.

---

† For example, see the amount of labor involved in the hand solution presented by J. M. Smith, "Chemical Engineering Kinetics," 2d ed., pp. 79–82, McGraw-Hill Book Company, New York, 1970, in which the same problem is solved with the simplification that no excess chlorine passes through the system.

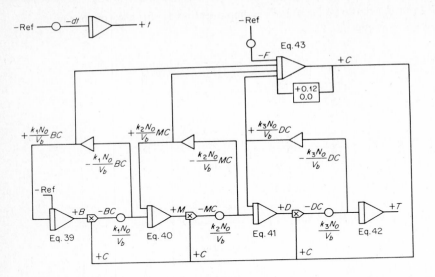

Figure 15-34 Analog-computer diagram for chlorobenzene reaction-kinetics problem (scaling not shown).

ANALOG COMPUTER SOLUTION†

The analog computer diagram for the solution of the example chlorobenzene problem is presented in Fig. 15-34. Figure 15-35 shows the resultant plot from the analog computer and indicates that the maximum monochlorobenzene concentration occurs after 0.78 hours of operating time, the maximum dichlorobenzene concentration occurs after 1.75 hours of operating time, and the trichlorobenzene concentration would become a maximum as time approaches infinity.

† Analog computers perform mathematical operations on continuous variables using dc voltage as the analog for the continuous variables. A major value of the analog computer is the conceptual simplicity of the operations, but this is counterbalanced by the mechanical problems of scaling the results from the analog voltage to real values of the variables, patching the plugboard, and debugging the program and operation. With the major development of digital computers and availability of their programs, the use of analog computers in industry has almost disappeared. However, for teaching concepts, the analog computer and its results are very useful, and this is the reason for presenting this solution here. For information on analog computers and applications, see A. Carlson, G. Hannauer, and P. J. Holsberg, "Handbook of Analog Computation," 2d ed., Electronic Associates, Inc., Princeton, N.J., 1965 or M. L. James, G. M. Smith, and J. C. Wolford, "Analog Computer Simulation of Engineering Systems," 2d ed., Intext Educational Publishers, Scranton, Pa., 1971.

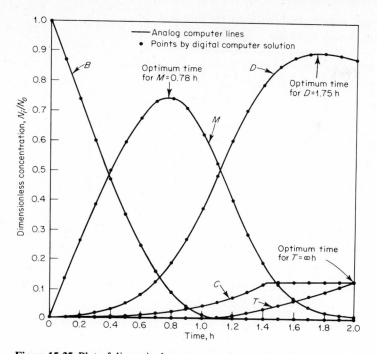

Figure 15-35 Plot of dimensionless concentration vs. time for chlorobenzene reaction-kinetics problem showing scaled analog-computer plot and digital-computer result.

DIGITAL COMPUTER SOLUTION

The digital simulation language MIMIC was chosen for use in solving this problem.† Table 13 gives the source program and Table 14 is a printout of the function language generated from the source language. The printed-out results are shown in Table 15 and are plotted as the solid points in Fig. 15-35.

EQUATIONS FOR REACTOR-DESIGN APPLICATION

The common, idealized designations for types of reactors are *batch*, *plug-flow*, and *back-mix*. In an idealized *batch reactor*, the reactants are mixed at the same time and no reaction mixture is removed during the reaction. Complete mixing

---

† MIMIC is a digital simulation program for solving a set of ordinary nonlinear differential equations. It uses the same basic techniques of automatic iteration computing as are used on an analog computer. For additional information on MIMIC, see R. G. Franks, "Mathematical Modeling in Chemical Engineering," John Wiley and Sons, Inc., New York, 1967. For information on the use of FORTRAN in place of MIMIC, see R. G. Franks, "Modeling and Simulation in Chemical Engineering," Wiley-Interscience, New York, 1972.

Table 13 MIMIC source-language program for solution to chlorobenzene reactor-design example

---

```
INTRODUCE CONSTANTS AND PARAMETERS
            CON(K1,K2,K3,V,BIC,CLIM)
            CON(DT,STOP)
            PAR(F)
REACTION RATES
      R1        K1*B*C/V
      R2        K2*M*C/V
      R3        K3*D*C/V
MATERIAL BALANCES
BENZENE
      B         INT(-R1,BIC)
MONOCHLOROBENZENE
      M         INT(R1-R2,O.)
DICHLOROBENZENE
      D         INT(R2-R3,O.)
TRICHLOROBENZENE
      TB        INT(R3,O.)
CHLORINE
      DCL       LIN(F-R1-R2-R3,C,O.,CLIM)
      C         INT(DCL,O.)
PRINTOUT STATEMENTS
            HDR(TIME,BENZ,MONO,DICHLO,TRI,CL2)
            HDR(HR.)
            HDR
            OUT (T,B,M,D,TB,C)
            OUT
            FIN(T,STOP)
            END
END COMPILE BEGIN SORT
END SORT BEGIN ASSEMBLY
```

---

is assumed during the entire reaction so that the entire reactor contents are at the same temperature and concentration at any one instant. The composition (and possibly the temperature) changes with time. The idealized *plug-flow* reactor is a tubular reactor with the reacting fluid moving through it with no back mixing or radial concentration gradients. Conditions are in a steady state so the concentrations and temperature profile along the length of the reactor do not change with time. An idealized *back-mix flow reactor* is equivalent to a *continuous stirred-tank reactor* (*CSTR*) where the contents of the reactor are completely mixed so that the complete contents of the reactor are at the same concentration and temperature as the product stream. The unit is operated at steady state so the flow rates of the inlet and outlet streams as well as the reactor conditions remain unchanged with time.

Table 14 Print-out of the function-language program generated from the source-language program shown in Table 13 for the chlorobenzene reactor-design problem. (Note that, for this table, $V_b/N_o = 1.46$.)

IFN	LCV	RESULT	FTN	A	B	C	D	E	F
1			CON	K1	K2	K3	V	BIC	CLIM
2			CON	DT	STOP				
3			PAR	F					
4	(005)		MPY	C	K1	B			
5	R1		DIV	(005)	V				
6	(007)		MPY	C	K2	M			
7	R2		DIV	(007)	V				
8	(009)		MPY	C	K3	D			
9	R3		DIV	(009)	V				
10	(011)		NEG	R1					
11	B		INT	(011)	BIC				
12	(013)		SUB	R1	R2				
13	M		INT	(013)	O.				
14	(015)		SUB	R2	R3				
15	D		INT	(015)	O.				
16	TB		INT	R3	O.				
17	(018)		SUB	F	R1				
18	(019)		SUB	(018)	R2				
19	(020)		SUB	(019)	R3				
20	DCL		LIN	(020)	C	O.	CLIM		
21	C		INT	DCL	O.				
22			FIN	T	STOP				
23			HDR	TIME	BENZ	MONO	DICHLO	TRI	CL2
24			HDR	HR.					
25			HDR						
26			OUT	T	B	M	D	TB	C
27			OUT						
28			END						

K1	K2	K3	V	BIC	CLIM
5.10000E 02	6.40000E 01	2.10000E 00	1.46000E 00	1.00000E 00	1.20000E−01

DT	STOP
1.00000E−01	2.0000E 00

END ASSEMBLY

Table 15 Printed-out results from computer program of Table 14 for chlorobenzene reactor-design example (Note: $8.65758E\text{-}01 = 8.65758 \times 10^{-1}$)

BEGIN RUN   1

F

1.40000E 00

TIME HR.	BENZ	MONO	DICHLO	TRI	CL2
0.	1.00000E 00	0.	0.	0.	0.
1.00000E−01	8.65758E−01	1.33006E−01	1.23554E−03	2.47758E−07	4.52194E−03
2.00000E−01	7.30722E−01	2.63774E−01	5.50053E−03	2.43779E−06	5.21724E−03
3.00000E−01	5.99766E−01	3.86611E−01	1.36120E−02	1.00033E−05	6.13489E−03
4.00000E−01	4.74175E−01	4.99064E−01	2.67307E−02	2.92531E−05	7.38617E−03
5.00000E−01	3.55853E−01	5.97523E−01	4.65514E−02	7.22168E−05	9.15741E−03
6.00000E−01	2.47682E−01	6.76562E−01	7.55913E−02	1.62902E−04	1.17658E−02
7.00000E−01	1.53973E−01	7.28138E−01	1.17537E−01	3.51431E−04	1.57346E−02
8.00000E−01	8.05234E−02	7.41485E−01	1.77245E−01	7.44649E−04	2.17855E−02
9.00000E−01	3.25819E−02	7.06837E−01	2.59021E−01	1.55867E−03	3.04442E−02
1.00000E 00	9.36515E−03	6.25617E−01	3.61858E−01	3.15808E−03	4.11915E−02
1.10000E 00	1.80804E−03	5.15588E−01	4.76591E−01	6.01058E−03	5.31961E−02
1.20000E 00	2.22085E−04	3.97632E−01	5.91500E−01	1.06448E−02	6.74335E−02
1.30000E 00	1.53360E−05	2.84486E−01	6.97721E−01	1.77752E−02	8.67440E−02
1.40000E 00	4.66958E−07	1.83564E−01	7.87924E−01	2.85097E−02	1.15057E−01
1.50000E 00	7.14556E−09	1.08640E−01	8.48706E−01	4.26513E−02	1.20007E−01
1.60000E 00	1.08018E−10	6.41986E−02	8.78209E−01	5.75899E−02	1.20007E−01
1.70000E 00	1.63290E−12	3.79369E−02	8.89197E−01	7.28638E−02	1.20007E−01
1.80000E 00	2.46843E−14	2.24180E−02	8.89354E−01	8.82255E−02	1.20007E−01
1.90000E 00	3.73149E−16	1.32475E−02	8.83219E−01	1.03531E−01	1.20007E−01
2.00000E 00	5.64085E−18	7.82833E−03	8.73473E−01	1.18696E−01	1.20007E−01

BATCH REACTORS

For batch reactors, Eq. (39) can be rearranged to give a basic design equation as[†]

$$\int_o^t dt = t = -\int_{N_{io}}^{N_{ie}} \frac{dN_i}{V_b r_i} \tag{55}$$

where N_{ie} = moles of material i left at the end of reaction time t
N_{io} = moles of material i at the start of the reaction

In terms of conversions (X_i), where X_{ie} represents the total fractional conversion of material i during the reaction time t with N_{io} being a constant,

$$N_i = N_{io} - N_{io}X_i = N_{io}(1 - X_i) \quad \text{and} \quad dN_i = -N_{io}\, dX_i \tag{56}$$

Substituting Eq. (56) into Eq. (55) gives

$$\int_o^t dt = t = N_{io} \int_o^{X_{ie}} \frac{dX_i}{V_b r_i} \tag{57}$$

Equation (57) represents the basic design equation for use with batch reactors. In case the reactor volume (V_b) remains constant during the entire reaction, integration of the equation can be simplified by recognizing that N_{io}/V_b is merely the concentration of the reactant at the start of the reaction and removing the V_b term from under the integral sign.

TUBULAR PLUG-FLOW REACTORS

Tubular reactors are often designed for steady-state operation on the basis of idealized plug flow so that Eq. (41) applies, or

$$\int_o^z dz = z = -\int_{(U_z C_i)_o}^{(U_z C_i)_e} \frac{d(U_z C_i)}{r_i} \tag{58}$$

To simplify this expression, designate the constant cross-sectional area of the flow reactor as A_R, the total volume of the reactor to point z as V_R so that $V_R = zA_R$, and the constant feed rate of material i as F_i. Then multiply both sides of Eq. (58) by A_R and note that $A_R U_z C_i$ equals the moles of material i flowing at length z in the reactor per unit of time which is also $F_i - F_i X_{iz}$ where X_{iz} represents the fractional conversion of material i in the entering feed at any distance z along the reactor length. Because $d(F_i - F_i X_{iz}) = -F_i\, dX_{iz}$, the result of the preceding definitions and operations can be written as

$$\int_o^{zA_R} d(zA_R) = \int_o^{V_R} dV_R = V_R = F_i \int_o^{X_{iz}} \frac{dX_{iz}}{r_i} \quad \text{or} \quad \frac{V_R}{F_i} = \int_o^{X_{iz}} \frac{dX_{iz}}{r_i} \tag{59}$$

[†] Note that r_i as used in this chapter represents the reaction rate expressed as the rate of *disappearance* of reactant i. The form $-r_i$ represents the rate of *generation* of the ith species.

As an alternate expression for plug-flow reactors, the more general case can be considered in which the reactant i enters the reactor already partly converted with this initial conversion, designated as X_{io}, taking the place of the zero conversion assumed initially in Eq. (59). The resulting design equation equivalent to Eq. (59) is

$$\frac{V_R}{F_i} = \int_{X_{io}}^{X_{iz}} \frac{dX_{iz}}{r_i} \tag{60}$$

Equations (59) and (60) are in a form often used for tabular, idealized, plug-flow reactors when the reaction rate is based on the volume of the reactor. When the reaction is heterogeneous, such as one occurring on the surface of a catalyst, it is common practice to base the reaction rate on the mass of the catalyst rather than on the volume of the reactor and substitute r_{ic} for r_i. The resulting design equation equivalent to Eq. (60) is

$$\frac{W_c}{F_i} = \int_{X_{io}}^{X_{iz}} \frac{dX_{iz}}{r_{ic}} \tag{61}$$

where W_c = mass of the catalyst in the reactor
$\quad r_{ic}$ = rate of the reaction as moles of material i converted per unit of time per unit of catalyst mass

Space velocity In design treatment for plug-flow reactors, the concept of *space velocity* is often used where space velocity is defined as the ratio of the volumetric feed rate to the volume of the reactor, or

$$\text{space velocity} = \frac{F_i v_F}{V_R} = \frac{F_i}{V_R C_{iF}} \tag{62}$$

where v_F = the volume of the feed per mole of material i in the feed
$\quad C_{iF}$ = the concentration of material i in the feed as moles of i per unit volume = $1/v_F$

BACK-MIX REACTORS

For the steady-state operation of idealized back-mix reactors, Eq. (37) becomes

$$\begin{matrix} \text{(Rate of accumulation} \\ \text{of } i) \end{matrix} = \begin{matrix} \text{(inflow rate of } i \\ \text{− outflow rate of } i) \end{matrix} + \begin{matrix} \text{(rate of generation} \\ \text{of } i) \end{matrix}$$

$$0 \quad = \quad F_i - F_i(1 - X_{ie}) \quad + \quad (-r_i)V_B \tag{63}$$

Therefore,

$$F_i X_{ie} = r_i V_B \quad \text{or} \quad \frac{V_B}{F_i} = \frac{X_{ie}}{r_i} \tag{64}$$

where X_{ie} = final conversion based on fraction of entering material i converted
$\quad V_B$ = total volume of the back-mix reactor

In case the feed reactant material (i in F_i) on which the conversion is based enters the reactor with part of the material already converted, the initial or entering conversion becomes X_{io} instead of zero as was the case for Eq. (64). Under these conditions, Eq. (64) assumes the more general form of

$$\frac{V_B}{F_i} = \frac{X_{ie} - X_{io}}{r_i} \tag{65}$$

EXPRESSIONS FOR r_i

It is normally necessary to use a simplified or empirical expression for the reaction rate r_i in terms of constants and reactant and product concentrations which can be assumed from the stoichiometry of a proposed reaction mechanism or developed purely empirically on the basis of experimental data. One of the key components of the rate expression is the specific rate constant k_i which must almost always be determined directly from laboratory or other rate data.

The most common way of presenting a rate constant is in the form of an Arrhenius equation as

$$k = Ae^{-E/RT} \tag{66}$$

where k = specific rate constant with appropriate units to fit the rate equation
A = frequency factor, same units as k
E = activation energy, units so that E/RT is dimensionless
R = perfect-gas-law constant, units so that E/RT is dimensionless
T = absolute temperature, units so that E/RT is dimensionless

It is important to be certain that the rate equation, as a function of concentrations, and the rate constant used in the design equations apply for the reaction conditions of temperature, pressure, and concentration to be used in the reactor.

MECHANICAL FEATURES OF DESIGN

Chemical reactors basically come in the form of tanks, such as for batch reactors or back-mix flow reactors, large cylinders, such as for fluidized-bed or plug-flow reactors, or multiple tubes inside a cylindrical container, such as for plug-flow reactors when special needs exist for temperature control. High pressure and extremes of temperature as well as corrosive action of the materials involved can introduce complications in the design which must be handled by the design engineer.

Tank reactors must be designed with closed ends which can withstand the operating conditions. Cooling or heating can be provided by internal coils or pipes or by jacketing. Vents, pressure releases, sampling inlets, and seals and auxiliaries for agitation must be provided, as well as adequate foundations, supports, and facilities for personnel access to handle operations, inspection, and maintenance.

Some flow reactors are designed as large cylinders closed at both ends through which the reactant flows while others are designed in the form of multiple tubes located in parallel with the reactant flowing through the inside of the tubes. This latter arrangement is the form usually used for thermal cracking units in the petroleum industry since it is well adapted to allow the necessary heating to attain the appropriate temperatures for reaction. In both of the forms discussed in the preceding, the simple assumption of ideal plug-flow operation is often assumed for the design even though it is clear that some back mixing will occur, especially in the large-cylinder flow units.

Many reactors fall in the classification of fluid-solid catalytic units where the catalyst may be retained in a *fixed-bed* position in the reactor with the reactant flowing through the catalyst bed, or the unit may be operated as a *fluidized-bed* reactor with the catalyst particles being suspended in the flowing fluid due to motion of the fluid. A third type of reactor is one in which the catalyst particles fall slowly through the fluid by gravity in the form of a so-called *moving bed*.

Fixed-bed reactors can be designed as units in large cylinders with jacketed or internal-coil heating or cooling as needed. They can also be designed as multiple units with heating or cooling between the separate beds to maintain the necessary temperature levels. The design for units of this type is often based on the assumption of ideal plug-flow although a series of back-mix reactors can also be used for the design basis.

The fluidized-bed reactor involves a rapid movement of the solid catalytic particles throughout the bed so that the operation can come close to one of uniform temperature throughout the reactor. The actual flow pattern for the operation of a fluidized bed is very complex and is between that for the ideal back-mix reactor and the ideal plug-flow reactor so that special methods for design may be required to approximate the real situation.†

The fluidized-bed reactor allows catalyst to be regenerated while the unit is in operation by continuously removing a portion of the catalyst from the reactor for regeneration treatment and subsequent flow-back into the reactor. Because there is a tendency for the catalyst particles to be carried over in the product stream, auxiliary units, such as cyclone separators or dust collectors, must be provided for separating out the solid particles or catalyst fines.

ESTIMATION OF COSTS FOR REACTOR EQUIPMENT COMPONENTS

The design and costing of reactor vessels are handled in a manner similar to that for regular mixing and pressure vessels as described in Chap. 13 or for heat exchangers as described in Chap. 14. The most reliable results are obtained by direct fabricator quotations or by the assistance of a fabricator representative who is an expert in the design of reactor vessels.

---

† For a detailed treatment of the principles of fluidized-bed operation and design, see D. Kunii and O. Levenspiel, "Fluidization Engineering," John Wiley and Sons, Inc., New York, 1969.

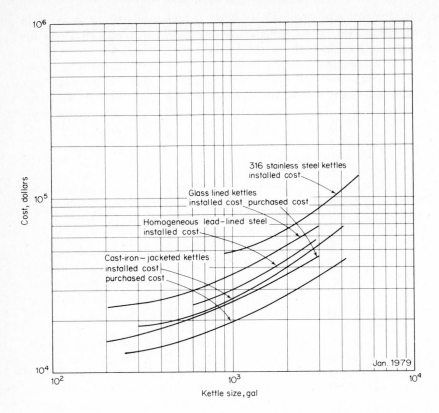

Figure 15-36 Cost of kettles. Purchased cost includes kettle, jacket, agitator, thermometer well, drive and support, manhole cover, stuffing box, and clamps where necessary. Installed cost includes above plus installation costs (labor, material, and overhead), design, and expense. No piping, structural changes, instrumentation, or electrical charges are included.

One critical item in the design of reactor vessels, in addition to establishing the type and size of the reactor by methods mentioned previously, is to select the correct materials of construction and wall thicknesses to handle the given operating conditions. Table 4 in Chap. 13 gives design equations and stress values for various materials of construction which can be applied to the design of reactor vessels, while Table 6 of Chap. 13 presents some rules of thumb that can be helpful in the design and costing of reactor vessels.† Figures 13-58 and 13-59 give data which can be used for estimating costs of reactor vessels. Costs for tubular flow reactors with internal tubes can often be estimated by considering the reactor unit as equivalent to a heat exchanger so that the data presented in Chap. 14

† See also the information presented earlier in this chapter on costs for plate and packed vessels and the cost information given in Chap. 13 for tanks, pressure vessels, and storage equipment.

can be used for the cost estimate. Examples of costs for kettle reactors are given in Fig. 15-36, while Fig. 15-37 gives data on cost and installation time for jacketed and agitated reactors.

Example of Design and Costing for Reactor Vessel

The following example illustrates how a pressure vessel used as a reactor can be designed and its cost estimated using the design equations and rules-of-thumb presented in Chap. 13: Consider the case where a reactor is to be designed and a preliminary cost estimate (January, 1979) is to be made for the installed reactor for the following conditions:

The reactor will be cylindrical with a 9 ft inside diameter and a length of 30 ft. It will be constructed of carbon steel (SA-285, Gr. C) and will operate at a temperature of 750°F and an internal pressure of 100 psig. It will require double-welded butt joints and will be spot-examined by the radiograph technique. It will be operated as a flow reactor in a horizontal position with the reacting materials

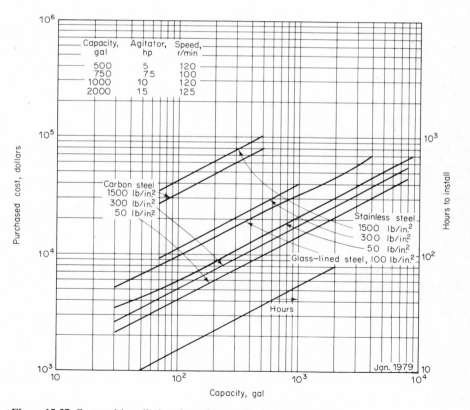

Figure 15-37 Cost and installation time of jacketed and agitated reactors.

flowing through it. There are no major heating or cooling needs, and, therefore, no special internal accessories are needed. Heads for the reactor will be hemispherical and the same thickness as the shell. The corrosion allowance for the wall thickness is $\frac{1}{8}$ in. Neglect freight and other delivery charges.

The solution for this case will be made by the use of Tables 4 and 6 in Chap. 13. For carbon steel (SA-285, Gr. C) at 750°F, the stress value is 12,000 psi and the joint efficiency is 0.85 for the given fabrication conditions as shown in Table 4 of Chap. 13.

The minimum wall thickness is (see Table 4 of Chap. 13 for equations and nomenclature)

$$\text{Thickness} = \frac{P\ (\text{inside radius})}{SE_J - 0.6P} + C_c = \frac{\overbrace{(100)}^{P}[\overbrace{(9)(12)}^{\text{inside radius}}(0.5) + 0.125]}{\underbrace{(12,000)}_{S}\underbrace{(0.85)}_{E_J} - (0.6)\underbrace{(100)}_{P}} + \overbrace{0.125}^{C_c}$$

$$= 0.66 \text{ in.}$$

The vessel weight, with density of steel = 489 lb/ft^3, is (see Tables 4 and 6 of Chap. 13)

$$\text{Weight of shell} = \pi(9)(30)\left(\frac{0.66}{12}\right)(489) = 22,800 \text{ lb}$$

$$\text{Weight of two heads} = (2\pi)\left[\frac{(9)(12)}{2}\right]^2 (0.66)\left[\frac{489}{(12)^3}\right](2) = 6,840 \text{ lb}$$

Total weight including 15 percent increase (see Table 6 of Chap. 13) for nozzles, manholes, and saddles = (22,800 + 6,840)(1.15) = 34,100 lb.

F.o.b. cost for reactor based on Table 6 of Chap. 13 = (cost per lb)× (34,100) = (50)(34,100)$^{-0.34}$ (34,100) = (1.44)(34,100) = $49,000. (See Fig. 15-24 for an approximate check on this number.)

The cost for installation (Lang factor of 3.0 by Table 6 of Chap. 13) will be twice the cost of the f.o.b. unit if it is constructed of carbon steel. In this case, the material of construction is carbon steel, so it is not necessary to use a materials cost-conversion factor to obtain the f.o.b. cost; thus, the estimated total installed cost of the reactor is $49,000 + (2)($49,000) = $147,000.

NOMENCLATURE FOR CHAPTER 15

a_p = surface area of packing per unit of packed-tower volume, ft^2/ft^3

A = frequency factor, same units as k

A_a = active or bubbling tray area, ft^2

A_d = minimum cross-sectional area for downcoming liquid flow in finite-stage tower, ft^2

A_R = cross-sectional area of flow reactor, ft^2

A_s = bubble-cap slot area, perforation sieve-tray area, or valve-opening valve-tray area, ft^2

A_t = total cross-sectional area of tower, ft^2

b = width of slot at base, ft

B = N_B/N_O in Eqs. (48), (49), (52), and (54)

c = height of slot, ft

C = N_{Cl}/N_O in Eqs. (48) to (53)

$C_a, C_b, C_i, C_{n-1}, C_n$ = concentrations of species $a, \ldots, C_n$ at time t, mol/ft^3

C_D = liquid gradient factor; see Fig. 15-13

C_{iF} = concentration of material i in feed, mol/ft^3

d_p = effective diameter of packing particle, ft; defined with Eq. (23)

d_{wc} = perpendicular distance from weir to center of tower, ft

D' = inside diameter of tower, in.

D = inside diameter of tower, ft, or N_D/N_O in Table 11 and Eqs. (50), (51), (52), and (54)

D_G = gaseous-diffusion coefficient, ft^2/h

E = activation energy, units so that E/RT is dimensionless

E_o = overall column efficiency, percent

$f_1 = (\mu_L/2.42)^{0.16}$

$f_2 = (62.4/\rho_L)^{1.25}$

$f_3 = (72.8/\sigma)^{0.8}$

F = feed rate of chlorine as (lb mol/h)/(lb mol of initial benzene charge)

F_C = correction factor for evaluation of liquid gradient; see Fig. 15-13

F_i = feed rate of material i, mol/h

g = local acceleration due to gravity; usually taken as 32.17 ft/(s)(s)

g_c = conversion factor in Newton's law of motion, 32.17 ft·lbm/(s)(s)(lbf)

G = superficial mass velocity of gas (based on cross-sectional area of empty tower), lb/(h)(ft^2)

G_m = maximum allowable superficial mass velocity of gas (based on cross-sectional area of empty tower) lb/(s)(ft^2) or lb/(h)(ft^2)

h = packed height, ft

h_c = head of liquid equivalent to pressure drop due to gas flow through riser and cap or through sieve holes, ft of liquid

h_d = head of liquid equivalent to pressure drop due to liquid flow through downcomer and constriction, ft of liquid

h_g = liquid gradient across tray, ft of liquid

h_H = head of liquid equivalent to pressure drop due to one kinetic head, ft of liquid

h_o = head of liquid over weir, ft

h_{slot} = distance liquid is depressed below top of slot = head of liquid equivalent to pressure drop due to gas flow through slots, ft of liquid

h_T = head of liquid equivalent to total pressure drop of gas across tray, ft of liquid

h_w = height of weir, ft

H = total head of liquid in downcomer, ft

H_p = a constant for evaluating maximum allowable gas rates in packed towers, ft

$HETP$ = height of packing equivalent to one theoretical plate, in.

HTU_G = height of a gas-phase transfer unit, ft

k = specific reaction rate constant with appropriate units to fit the rate equation

k_i = specific reaction rate constant for species i

k_1, k_2, k_3 = specific reaction rate constants for species 1, 2, and 3

K_1, K_2, K_3 = empirical constants as a function of the packing, see Table 6 and Eq. (36)

K_p = a constant for evaluating maximum allowable gas rates in packed towers, ft/s

K'_v = a constant for evaluating maximum allowable gas rates in finite-stage towers, ft/s, see Eq. (3)

K_v = a constant for evaluating maximum allowable gas rates in finite-stage towers, ft/s, see Eq. (1)

$K.H.$ = number of kinetic heads equivalent to pressure drop through sieve-tray plate, see Fig. 15-12

l_c = total free space between bubble caps perpendicular to direction of liquid flow, average of various rows, ft

l_r = total free space between bubble-cap risers perpendicular to direction of liquid flow, average of various rows, ft

l_w = weir length, ft

L = superficial mass velocity of liquid (based on cross-sectional area of empty tower), lb/(h)(ft^2)

L'_M = molal liquid flow rate, lb mol/h

m_{avg} = mole fraction of solute in gas in equilibrium with liquid/mole fraction of solute in liquid at average column temperature and pressure

$M = N_M/N_O$ in Eqs. (49), (50), (52), and (54)

M_A = average molecular weight of liquid absorbent, lb/lb mol

M_G = molecular weight of gas, lb/lb mol

N_B = lb mol of benzene present at time t

N_{Cl} = lb mol of chlorine present at time t

N_D = lb mol of dichlorobenzene present at time t

N_i = amount of base material i present in the volume V_b at time t

N_{ie} = moles of material i left at end of reaction time t

N_{io} = moles of material i at the start of the reaction

N_M = lb mol of monochlorobenzene present at time t

N_O = lb mol of benzene present at start

N_T = lb mol of trichlorobenzene present at time t

NTU = number of transfer units, see Eqs. (31) and (32)

p = pressure; Δp is pressure drop in packed tower, psf or in. water

p_F = pressure at flooding; Δp_F is pressure drop in packed column at flooding point, psf or in. water

p_T = pressure; Δp_T is total pressure drop of gas across finite-stage tray, psi

P = total pressure, expressed as equivalent ft of liquid or psf

Q_L = volumetric flow rate of liquid, ft^3/s

Q_s = volumetric flow rate of gas, ft^3/s

r = number of rows of caps perpendicular to direction of liquid flow

r_i = rate of reaction of material i expressed as rate of disappearance of reactant i, mol/(ft^3)(h)

r_{ic} = rate of reaction as moles of material i converted per unit of time per unit of catalyst mass

R = ideal-gas-law constant, 1545 psf $\cdot$ ft^3/(lb mol)($^\circ$R)

s = skirt clearance, ft

Sc_G = gas-phase Schmidt number = $\mu_G/\rho_G D_G$

S_m = static submergence, distance from top of bubble-cap slot to liquid surface when liquid is just ready to flow over the weir, ft

t = time, h

T = absolute temperature, $^\circ$R, or N_T/N_O in Table 11 and Eqs. (51) and (54)

U = velocity vector, ft/s

U_z = velocity vector in z direction, ft/s

v_F = volume of feed per mole of material i in the feed, ft^3/mol

V = superficial linear velocity of gas (based on cross-sectional area of empty tower), ft/s

V_b = volume of batch reactor, ft^3

V_B = volume of back-mix reactor, ft^3

V_c = linear velocity of gas in riser, reversal area, or annulus of bubble cap (maximum value) or in sieve hole

V'_m = maximum allowable superficial linear velocity of gas (based on "net" cross-sectional area of tower for vapor flow), ft/s, see Eq. (3)

V_m = maximum allowable superficial linear velocity of gas (based on cross-sectional area of empty tower), ft/s, see Eq. (1)

V'_M = molal gas flow rate, lb mol/h

V_R = volume of tubular reactor to point z, ft^3

W_c = mass of catalyst in reactor, lb

W_T = average width of tray normal to direction of liquid flow (computed as the average of the total tray width at each row of caps), ft

X_i = fractional conversion of material i

X_{ie} = fractional conversion of material i in reaction time t

X_{io} = fractional conversion of material i at start of reaction or at $t = 0$ or $z = 0$

X_{iz} = fractional conversion of material i at distance z along reactor length

y, x = gas- and liquid-phase concentrations, respectively

y_L^*, x_G^* = gas- and liquid-phase concentrations, respectively, corresponding to values in equilibrium with the liquid- and gas-phase concentrations

z = length of reactor in direction of flow, ft

Z = packed height, ft

Greek symbols

α = relative volatility

α_{avg} = relative volatility of key components at average column temperature and pressure

γ = constant, value dependent on packing size and type; see Table 4

ε = fractional void volume in packed bed, ft^3 void/ft^3 of packed-tower volume

θ = angle subtended from center of tower by chord weir, deg

$\mu_{A,\text{avg}}$ = molal average viscosity of liquid absorbent at average column temperature and pressure, centipoises

$\mu_{F,\text{avg}}$ = molal average viscosity of feed at average column temperature and pressure, centipoises

μ_G = absolute viscosity of gas, lb/(ft)(h)

μ_L = liquid viscosity, lb/(ft)(h)

μ'_L = liquid viscosity, centipoises

π = constant = 3.1416

$\rho_{A,\text{avg}}$ = density of liquid absorbent at average column temperature and pressure, lb/ft^3

ρ_G = gas density, lb/ft^3

ρ_L = liquid density, lb/ft^3

ρ'_L = liquid density, g/cc

σ = surface tension of liquid, dyn/cm

ψ = parameter for a given packing material, see Figs. 15-22 and 15-23

Φ = constant, value dependent on packing size and type; see Table 4

SUGGESTED ADDITIONAL REFERENCES FOR MASS-TRANSFER EQUIPMENT—DESIGN AND COSTS

Dryer design and costs

Alexis, R. W., Gas Dryer Designed for Multi-Uses, *Hydrocarbon Process.*, **50**(5):145 (1971).

Belcher, D. W., D. A. Smith, and E. M. Cook, Analyzing Suspended-Particle Dryers With Psychrometric Charts, *Chem. Eng.*, **84**(2):112 (1977).

Calus, W. F., Drying—Unit Operations Review, *Chem. Proc. Eng.*, **48**(5):50 (1967).

Chung, S. F., Mathematical Model and Optimization of Drying Process for a Through-Circulation Dryer, *Can. J. Chem. Eng.*, **50**(5):657 (1972).

Clark, W. E., Fluid-Bed Drying, *Chem. Eng.*, **74**(6):177 (1967).

Cook, E. M., Influences of Solvent Properties on Dryer Design, *Chem. Eng. Progr.*, **74**(4):75 (1978).

Davis, K. G., and K. D. Manchanda, Unit Operations for Drying Fluids, *Chem. Eng.*, **81**(19):102 (1974).

Dittman, F. W., How to Classify a Drying Process, *Chem. Eng.*, **84**(2):106 (1977).

———, Establishing the Parameters for a Spray Dryer, *Chem. Eng.*, **84**(2):108 (1977).

———, Drying Slabs, Sheets, and Beds, *Chem. Eng.*, **84**(2):118 (1977).

Emery, D. L., Drying Natural Gas With Alumina, *Hydrocarbon Process.*, **48**(3):130 (1969).

Gauvin, W. H., and S. Katta, Basic Concepts of Spray Dryer Design, *AIChE J.*, **22**(4):713 (1976).

Guthrie, K. M., Capital Cost Estimating, *Chem. Eng.*, **76**(6):114 (1969).

Hales, G. E., Tips on Gas Dehydration, *Hydrocarbon Process.*, **50**(6):151 (1971).

———, Drying Reactive Fluids With Molecular Sieves, *Chem. Eng. Progr.*, **67**(11):49 (1971).

Keey, R. B., "Drying-Principles and Practice," Pergamon Press, Oxford, England, 1972.

———, "Introduction to Industrial Drying Operations," Pergamon Press, Oxford, England, 1978.

LaCerda, D. J., Better Instrument and Plant Air Systems, *Hydrocarbon Process.*, **47**(6):107 (1968).

Lazaridis, A., Moisture Content of Air, *Chem. Eng.*, **80**(20):124 (1973).

Long, G. E., Spraying Theory and Practice, *Chem. Eng.*, **85**(6):73 (1978).

Martin, C. G., Some Comments on Dryer Design, *Chem. Eng.* (*London*), 367 (July–August 1973).

Masters, K, "Spray-Drying: An Introduction to Principles, Operational Practice and Applications," CRC Press, Cleveland, Ohio, 1972.

McAllister, D. G., Jr., How to Select a System for . . . Drying Instrument Air, *Chem. Eng.*, **80**(5):39 (1973).

Papee, D., C. Meniere, A. Bellier, and F. Jouanneault, Optimize Alumina Gas Drying Systems, *Hydrocarbon Process.*, **46**(10):142 (1967).

Noden, D., Industrial Dryers—Selection, Sizing and Costs, *Chem. Proc. Eng.*, **50**(10):67 (1969).

Pippitt, R. R., Maintenance of Rotating Equipment: Maintenance of Kilns, Calciners and Dryers, *Chem. Eng. Progr.*, **72**(2):41 (1976).

Sloan, C. E., Drying Systems and Equipment, *Chem. Eng.*, **74**(13):169 (1967).

Spotts, M. R., and P. F. Waltrich, Vacuum Dryers, *Chem. Eng.*, **84**(2):120 (1977).

Tsao, G. T., and T. D. Wheelock, Drying Theory and Calculations, *Chem. Eng.*, **74**(13):201 (1967).

Updegrove, L. B., Batch Drying in a Nauta Mixer, *Chem. Eng. Progr.*, **73**(4):107 (1977).

Weiner, A. L., Dynamic Fluid Drying, *Chem. Eng.*, **81**(19):92 (1974).

Wunder, J. W. J., How to Design a Natural-Gas Drier, *Oil and Gas J.*, **60**(32):137 (1967).

Zimmerman, O. T., Sahara "HP" Deliquescent Air Dryers, *Cost Eng.*, **17**(3):11 (1972).

Mass-transfer design

Abernathy, M. W., Design Horizontal Gravity Settlers, *Hydrocarbon Process.*, **56**(9):199 (1977).

Adler, P., Right Absorber Lean Oil Cuts Operating Costs, *Chem. Eng.*, **84**(26):125 (1977).

Agarwal, J. C., and I. V. Klumpar, Multistage-Leaching Simulation, *Chem. Eng.*, **83**(11):135 (1976).

Alonso, J. R. F., Calculate Particle and Droplet Size Simply, *Hydrocarbon Process.*, **56**(1):141 (1977).

Anderson, R. H., G. Garrett, and M. Van Winkle, Efficiency Comparison of Valve and Sieve Trays in Distillation Columns, *I&EC Fund.*, **15**(1):96 (1976).

Apelblat, A., Calculating Stagewise Unit Operation, *Brit. Chem. Eng.*, **12**(9):1378 (1967).

Armstrong, M., and A. E. Schofield, The Design of Air Separation Distillation Columns Using a Computer, *Chem. Eng. (London)*, **229**:184 (May 1969).

Badhwar, R. K., Quick Sizing of Distillation Columns, *Chem. Eng. Progr.*, **66**(3):56 (1970).

Bailes, P. J., C. Hanson, and M. A. Hughes, Liquid-Liquid Extraction: The Process, The Equipment, *Chem. Engr.*, **83**(2):86 (1976).

———— and ———, Liquid-Liquid Extraction: Nonmetalic Materials, *Chem. Eng.*, **83**(10):115 (1976).

———— and ———, Liquid-Liquid Extraction: Metals, *Chem. Eng.*, **83**(18):86 (1976).

Barnes, F. J., D. N. Hanson, and C. J. King, Calculation of Minimum Reflux for Distillation Columns With Multiple Feeds, *I&EC Process Des. & Dev.*, **11**(1):136 (1972).

Barrere, C. A., Jr., Natural Gas Adsorption Plant Design, *Hydrocarbon Process.*, **47**(10):141 (1968).

Barrett, E. C., and S. G. Dunn, Design of Direct Contact Humidifiers and Dehumidifiers Using Tray Columns, *I&EC Process Des. & Dev.*, **13**(4):353 (1974).

Benzer, W. C., Steels, *Chem. Eng.*, **77**(22):101 (1970).

Berg, L., Selecting the Agent for Distillation, *Chem. Eng. Progr.*, **65**(9):52 (1969).

Biddulph, M. W., When Distillation Can be Unstable, *Hydrocarbon Process.*, **54**(9):123 (1975).

———, Predicted Comparisons of the Efficiency of Large Valve Trays and Large Sieve Trays, *AIChE J.*, **23**(5):770 (1977).

———, Tray Efficiency is not Constant, *Hydrocarbon Process.*, **56**(10):145 (1977).

Billet, R., Cost Optimization of Towers, *Chem. Eng. Progr.*, **66**(1):41 (1970).

———, Gauze-Packed Columns for Vacuum Distillation, *Chem. Eng.*, **79**(4):68 (1972).

Billingsley, D. S., On the Numerical Solution of Problems in Multicomponent Distillation at the Steady State, *AIChE J.*, **16**(3):441 (1970).

———, and G. W. Boynton, Iterative Methods for Solving Problems in Multicomponent Distillation at the Steady State, *AIChE J.*, **17**(1):65 (1971).

——— and ———, Iterative Methods for Solving Problems in Multicomponent Distillation at the Steady State, *AIChE J.*, **17**(1):68 (1971).

Bischoff, K. B., and D. M. Himmelblau, Mass Transfer, *I&EC*, **60**(1):66 (1968).

Bojnowski, J. H., J. W. Crandall, and R. M. Hoffman, Modernized Separation System Saves More Than Energy, *Chem. Eng. Progr.*, **71**(10):50 (1975).

Bolles, W. L., and J. R. Fair, Distillation, *I&EC*, **59**(11):86 (1967).

——— and ———, Distillation, *I&EC*, **60**(12):29 (1968).

———, Multipass Flow Distribution and Mass Transfer Efficiency for Distillation Plates, *AIChE J.*, **22**(1):153 (1976).

———, Estimating Valve Tray Performance, *Chem. Eng. Progr.*, **72**(9):43 (1976).

Boston, J. F., and S. L. Sullivan, Jr., A New Class of Solution Methods for Multicomponent, Multistage Separation Processes, *Can. J. Chem. Eng.*, **52**(1):52 (1974).

Bourne, J. R., U. von Stockar, and G. C. Coggan, Gas Absorption With Heat Effects. 1. A New Computational Method, *I&EC Process Des. & Dev.*, **13**(2):115 (1974).

Boyd, D. M., Fractionation Column Control, *Chem. Eng. Progr.*, **71**(6):55 (1975).

Boynton, G. W., Iteration Solves Distillation, *Hydrocarbon Process.*, **49**(1):153 (1970).

Brannock, N. F., B. S. Verneuil, and Y. L. Wang, Rigorous Distillation Simulation, *Chem. Eng. Progr.*, **73**(10):83 (1977).

Brosens, J. R., For Quicker Distillation Estimates, *Hydrocarbon Process.*, **48**(10):102 (1969).

Broughton, D. B., Bulk Separations via Adsorption, *Chem. Eng. Progr.*, **73**(10):49 (1977).

Brown, B. T., H. A. Clay, and J. M. Miles, Drying Liquid Hydrocarbons via Fractional Distillation, *Chem. Eng. Progr.*, **66**(8):54 (1970).

Buckley, P. S., R. K. Cox, and W. L. Luyben, How to Use a Small Calculator in Distillation Column Design, *Chem. Eng. Progr.*, **74**(6):49 (1978).

Cadman, T. W., R. R. Rothfus, and R. I. Kermode, Design and Effectiveness of Feedforward Control Systems for Multicomponent Distillation Columns, *I&EC Fund.*, **6**(3):421 (1967).

Chase, J. D., Sieve-Tray Design, *Chem. Eng.*, **74**(16):105 (1967).

Chen, N. H., Optimum Theoretical Stages in Counter-current Leaching, *Chem. Eng.*, **77**(18):71 (1970).

Chien, H. H. Y., A Rigorous Method for Calculating Minimum Reflux Rates in Distillation, *AIChE J.*, **24**(4):606 (1978).

Coggan, G. C., and J. R. Bourne, The Design of Gas Absorbers With Heat Effects, *Trans. Inst. Chem. Eng.*, **47**:96 (1969).

Cummings, W. P., Save Energy in Adsorption, *Hydrocarbon Process.*, **54**(2):97 (1975).

Davis, K. G., and K. D. Manchanda, Unit Operations for Drying Fluids, *Chem. Eng.*, **81**(19):102 (1974).

Davison, R. R., H. R. James, and G. L. Tromblee, A Method for Multicomponent Equilibrium Calculations by Using Dew Point Pressure-Vapor Composition Data, *AIChE J.*, **16**(4):696 (1970).

Deam, J. R., and R. N. Maddox, How to Figure Three-Phase Flash, *Hydrocarbon Process.*, **48**(7):163 (1969).

Delnicki, W. V., and J. L. Wagner, Performance of Multiple Downcomer Trays, *Chem. Eng. Progr.*, **66**(3):50 (1970).

Dickey, D. S., and R. W. Hicks, Liquid Agitation: Fundamentals of Agitation, *Chem. Eng.*, **83**(3):93 (1976).

Doshi, K. J., C. H. Katira, and H. A. Stewart, Optimization of a Pressure Swing Cycle, *AIChE Symp. Ser.*, **67**(117):90 (1971).

Douglas, J. M., Rule of Thumb for Minimum Trays, *Hydrocarbon Process.*, **56**(11):291 (1977).

———, Criteria for Relative Volatility, *Hydrocarbon Process.*, **57**(2):155 (1978).

Eckhart, R. A., and A. Rose, New Method for Distillation Prediction, *Hydrocarbon Process.*, **47**(5):165 (1968).

Eckert, J. S., Selecting the Proper Distillation Column Packing, *Chem. Eng. Progr.*, **66**(3):39 (1970).

———, Extraction Variables Defined, *Hydrocarbon Process.*, **55**(3):117 (1976).

Economopoulos, A. P., A Fast Computer Method for Distillation Calculations, *Chem. Eng.*, **85**(10):91 (1978).

———, Computer Design of Sieve Trays and Tray Columns, *Chem. Eng.*, **85**(27):109 (1978).

Eduljee, H. E., Equations Replace Gilliland Plot, *Hydrocarbon Process.*, **54**(9):120 (1975).

Ellerbe, R. W., Batch Distillation Basics, *Chem. Eng.*, **80**(12):110 (1973).

———, Steam-Distillation Basics, *Chem. Eng.*, **81**(5):105 (1974).

Ewiler, D. W., F. W. Bonnet, and F. W. Leavitt, Slotted Sieve Trays, *Chem. Eng. Progr.*, **67**(9):86 (1971).

Eyre, D. V., Some Aspects of Distillation Near the Critical Region, *Chem. Eng. Progr.*, **63**(7):93 (1967).

Fair, J. R., and W. L. Bolles, Modern Design of Distillation Columns, *Chem. Eng.*, **75**(10):156 (1968).

———, Sorption Processes for Gas Separation, *Chem. Eng.*, **76**(15):90 (1969).

———, Comparing Trays and Packings, *Chem. Eng. Progr.*, **66**(3):45 (1970).

———, B. B. Crocker, and H. R. Null, Trace-Quantity Engineering, *Chem. Eng.*, **79**(17):60 (1972).

———, Advances in Distillation System Design, *Chem. Eng. Progr.*, **73**(11):78 (1977).

Fauth, G. F., and F. G. Shinskey, Advanced Control of Distillation Columns, *Chem. Eng. Progr.*, **71**(6):49 (1975).

Feintuch, H. M., Distillation Distributor Design, *Hydrocarbon Process.*, **56**(10):150 (1977).

Frank, J. C., G. R. Geyer, and H. Kehde, Styrene-Ethyl-Benzene Separation With Sieve Trays, *Chem. Eng. Progr.*, **65**(2):79 (1969).

Frank, O., Shortcuts for Distillation Design, *Chem. Eng.*, **84**(6):111 (1977).

Fredenslund, A., M. L. Bichelsen, and J. M. Prausnitz, Multicomponent Distillation Column Design, *Chem. Eng. Progr.*, **72**(9):67 (1976).

Freshwater, D. C., and E. Ziogon, Reducing Energy Requirements in Unit Operations, *Chem. Eng. J.*, **11**(3):215 (1976).

Fulham, M. J., and V. G. Hulbert, Gamma Scanning of Large Towers, *Chem. Eng. Progr.*, **71**(6):73 (1975).

Furzer, I. A., and G. E. Ho, Distillation in Packed Columns: The Relationship Between HTU and Packed Height, *AIChE J.*, **13**(3):614 (1967).

———, Natural Draft Cooling Tower, *I&EC Process Des. & Dev.*, **7**(4):555 (1968).

———, Natural Draft Cooling Tower, *I&EC Process Des. & Dev.*, **7**(4):561 (1968).

Gallun, S. E., and C. D. Holland, Solve More Distillation Problems for Highly Nonideal Mixtures, *Hydrocarbon Process.*, **55**(1):137 (1976).

Gardner, R. J., R. A. Crane, and J. F. Hannan, Hollow Fiber Permeator for Separating Gases, *Chem. Eng. Progr.*, **73**(10):76 (1977).

Gaumer, L. S., Flash Column Performance in a Helium Recovery Plant, *Chem. Eng. Progr.*, **63**(5):72 (1967).

Gentry, J. W., An Improved Method for Numerical Solution of Distillation Processes, *Can. J. Chem. Eng.*, **48**(4):451 (1970).

Gerster, J. A., Azeotropic and Extractive Distillation, *Chem. Eng. Progr.*, **65**(9):43 (1969).

Geyer, G. R., and P. E. Kline, Energy Conservation Schemes for Distillation Processes, *Chem. Eng. Progr.*, **72**(5):49 (1976).

Goldman, M. R., and E. R. Robinson, The Computer Simulation of Batch Distillation Processes, *Brit. Chem. Eng.*, **13**(12):602 (1968).

———, Simulating Multi-Component Batch Distillation, *Brit. Chem. Eng.*, **15**(11):1450 (1970).

Gomezplata, A., and T. M. Regan, Mass Transfer, *I&EC*, **60**(12):53 (1968).

Guerreri, G., Easier Way to Figure Turboexpanders, *Hydrocarbon Process.*, **47**(2):101 (1968).

———, Short-Cut Distillation Design: Beware! *Hydrocarbon Process.*, **48**(8):137 (1969).

———, B. Peri, and F. Seneci, Comparing Distillation Designs, *Hydrocarbon Process.*, **51**(12):77 (1972).

——— and C. J. King, Design Falling Film Absorbers, *Hydrocarbon Process.*, **53**(1):131 (1974).

Hafslund, E. R., Propylene-Propane Extractive Distillation, *Chem. Eng. Progr.*, **65**(9):58 (1969).

Hanf, E. B., Entrainment Separator Design, *Chem. Eng. Progr.*, **67**(11):54 (1971).

Hanson, C., Solvent Extraction, *Chem Eng.*, **75**(18):76 (1968).

———, Recent Research in Solvent Extraction, *Chem. Eng.*, **75**(19):135 (1968).

Hariu, O. H., and R. C. Sage, Crude Split Figured by Computer, *Hydrocarbon Process.*, **48**(4):143 (1969).

Harris, R. E., Distillation Designs Using FLOWTRAN, *Chem. Eng. Progr.*, **68**(10):56 (1972).

Hendry, J. E., and R. R. Hughes, Generating Separation Process Flowsheets, *Chem. Eng. Progr.*, **68**(6):71 (1972).

Hengstebeck, R. J., Finding Feedplates From Plots, *Chem. Eng.*, **75**(16):143 (1968).

———, An Improved Shortcut for Calculating Difficult Multicomponent Distillations, *Chem. Eng.*, **76**(1):115 (1969).

Hess, F. E., and C. D. Holland, Solve More Distillation Problems. Evaluate Complex or Interlinked Columns, *Hydrocarbon Process.*, **55**(6):125 (1976).

——— ——— R. McDaniel, and N. J. Tetlow, Solve More Distillation Problems, *Hydrocarbon Process.*, **56**(5):241 (1977).

——— ——— ——— and S. E. Gallun, Solve More Distillation Problems: Which Method to Use? *Hydrocarbon Process.*, **56**(6):181 (1977).

Holland, C. D., and G. P. Pendon, Solve More Distillation Problems, *Hydrocarbon Process.*, **53**(7):148 (1974).

——— and P. T. Eubank, Solve More Distillation Problems, *Hydrocarbon Process.*, **53**(11):176 (1974).

———, G. P. Pendon, and S. E. Gallun, Solve More Distillation Problems, *Hydrocarbon Process.*, **54**(1):101 (1975).

——— and M. S. Kuk, Solve More Distillation Problems, *Hydrocarbon Process.*, **54**(7):121 (1975).

———, "Fundamentals and Modeling of Separation Processes," Prentice-Hall, Englewood Cliffs, New Jersey, 1975.

Hopke, S. W., and C. J. Lin, Improve Absorber Predictions, *Hydrocarbon Process.*, **53**(6):136 (1974).

Huber, W. F., Jr., Figure Staged Process by Matrix, *Hydrocarbon Process.*, **56**(8):121 (1977).

Huckabay, H. K., and R. L. Garrison, Packed Tower Transfer Rate by Graphs, *Hydrocarbon Process.*, **48**(6):153 (1969).

Hughmark, G. A., Liquid-Liquid Spray Column Drop Size, Holdup, and Continuous Phase Mass Transfer, *I&EC Fund.*, **6**(3):408 (1967).

———, Heat and Mass Transfer in the Wall Region of Turbulent Pipe Flow, *AIChE J.*, **17**(1):51 (1971).

————, Heat and Mass Transfer for Turbulent Pipe Flow, *AIChE J.*, **17** (4):902 (1971).

Hutchins, R. A., Activated-Carbon Systems, *Chem. Eng.*, **80**(19):133 (1973).

Jamison, R. H., Internal Design Techniques, *Chem. Eng. Progr.*, **65**(3):46 (1969).

Jashnani, I. L., and R. Lemlich, Transfer Units in Foam Fractionation, *I&EC Process Des. & Dev.*, **12**(3):312 (1973).

Jelinek, J., and V. Hlavacek, Calculation of Multistage Multicomponent Liquid-Liquid Extraction by Relaxation Method, *I&EC Fund.*, **15**(4):481 (1976).

Johnston, W. A., Designing Fixed-Bed Adsorption Columns, *Chem. Eng.*, **79**(26):87 (1972).

Jones, D. W., and J. B. Jones, Tray Performance Evaluation, *Chem. Eng. Progr.*, **71**(6):65 (1975).

Kahre, L. C., and R. W. Hankinson, Predict K's for Butadiene Systems, *Hydrocarbon Process.*, **51**(3):94 (1972).

Kalika, P. W., How Water Recirculation and Steam Plumes Influence Scrubber Design, *Chem. Eng.*, **76**(16):133 (1969).

Katzen, R., V. B. Diebold, and G. D. Moon, Jr., A Self-descaling Distillation Tower, *Chem. Eng. Progr.*, **64**(1):79 (1968).

Kaup, E., Design Factors in Reverse Osmosis, *Chem. Eng.*, **80**(8):46 (1973).

Kern, R., Plant Layout: Layout Arrangements for Distillation Columns, *Chem Eng.*, **84**(17):153 (1977).

Kesler, M. G., B. I. Lee, M. J. Fish, and S. T. Hadden, Correlation Improves K-value Predictions, *Hydrocarbon Process.*, **56**(5):257 (1977).

Kimura, S., S. Sourirajan, and H. Ohya, Stagewise Reverse Osmosis Process Design, *I&EC Process Des. & Dev.*, **8**(1):79 (1969).

King, C. J., "Separation Processes," McGraw-Hill Book Company, New York, 1970.

King, R. W., Distillation of Heat Sensitive Materials, *Brit. Chem. Eng.*, **12**(4):568 (1967); **12**(5):722 (1967).

Kister, H. Z., and I. D. Doig, Distillation Pressure Ups Thruput, *Hydrocarbon Process.*, **56**(7):132 (1977).

———— and ————, Entrainment Flooding Prediction, *Hydrocarbon Process.*, **56**(9):149 (1977).

Koppel, P. M., Fast Way to Solve Problems for Batch Distillations, *Chem. Eng.*, **79**(23):109 (1972).

Kunin, R., and R. L. Gustafson, Ion Exchange, *I&EC*, **59**(11):95 (1967).

Lashmet, P. K., and S. Z. Szczepanski, Efficiency Uncertainty and Distillation Column Overdesign Factors, *I&EC Process Des. & Dev.*, **13**(2):103 (1974).

Lemieux, E. J., and L. J. Scotti, Perforated Tray Performance, *Chem. Eng. Progr.*, **65**(3):52 (1969).

Lemlich, R., Foam Fractionation, Questions and Answers on . . ., *Chem. Eng.*, **75**(27):95 (1968).

Liddle, C. J., Improved Shortcut Method for Distillation Calculations, *Chem. Eng.*, **75**(22):137 (1968).

————, How to Design Desorption Systems Based on Pressure Reduction, *Chem. Eng.*, **77**(15):87 (1970).

Lockhart, F. J., Stagewise Still Calculations by Temperature Profiles, *Chem. Eng.*, **76**(19):131 (1969).

Lövland, J., A Graphical Solution of Cyclic Extraction, *I&EC Process Des. & Dev.*, **7**(1):65 (1968).

Lukchis, G. M., Adsorption Systems: Design by Mass-Transfer-Zone Concept, *Chem. Eng.*, **80**(13):111 (1973).

————, Adsorption Systems: Equipment Design, *Chem. Eng.*, **80**(16):83 (1973).

————, Adsorbent Regeneration, *Chem. Eng.*, **80**(18):83 (1973).

Luyben, W. L., Some Practical Aspects of Optimal Batch Distillation Design, *I&EC Process Des. & Dev.*, **10**(1):54 (1971).

MacFarland, S. A., P. M. Sigmund, and M. Van Winkle, Predict Distillation Efficiency, *Hydrocarbon Process.*, **51**(7):111 (1972).

McLaren, D. B., and J. C. Upchurch, Distillation, *Chem. Eng.*, **77**(12):139 (1970).

Manley, D. B., A Better Method for Calculating Relative Volatility, *Hydrocarbon Process.*, **51**(1):113 (1972).

Mapstone, G. E., Reflux Versus Trays by Chart, *Hydrocarbon Process.*, **47**(5):169 (1968).

Markin, A., and J. T. Sommerfeld, Packing Pressure Drop by Computer, *Hydrocarbon Process.*, **48**(9):206 (1969).

Mass, J. H., Optimum-Feed-Stage Location in Multicomponent Distillations, *Chem. Eng.*, **80**(9):96 (1973).

May, R. A., and F. J. M. Horn, Stage Efficiency of a Periodically Operated Distillation Column, *I&EC Process Des. & Dev.*, **7**(1):61 (1968).

Meier, W., W. D. Stoecker, and B. Weinstein, Performance of a New, High Efficiency Packing, *Chem. Eng. Progr.*, **73**(11):71 (1977).

Metcalfe, R. S., P. Barton, and R. H. McCormick, Low-Pressure Vapor-Liquid Trays, *Hydrocarbon Process.*, **55**(4):97 (1976).

Michaels, A. S., New Separation Technique for the CPI, *Chem. Eng. Progr.*, **64**(12):31 (1968).

Mikolaj, P. G., Computer Aid for Design of Crude Oil Distillation Units, *Brit. Chem. Eng.*, **15**(5):638 (1970).

Miller, D. N., Scale-Up of Agitated Vessels, *I&EC Process Des. & Dev.*, **10**(3):365 (1971).

———, Scale-Up of Agitated Vessels Gas-Liquid Mass Transfer, *AIChE J.*, **20**(3):445 (1974).

Miranda, J. G., Designing Parallel-Plate Separators, *Chem. Eng.*, **84**(3):105 (1977).

Mitchell, W. T., Distillation Practices: Distillation Optimization, *Chem. Eng. Progr.*, **68**(10):62 (1972).

Mottola, A. C., Diffusivities Streamline Wet Scrubber Design, *Chem. Eng.*, **84**(27):77 (1977).

Naphtali, L. M., and D. P. Sandholm, Multicomponent Separation Calculations by Linearization, *AIChE J.*, **17**(1):148 (1971).

Nemunaitis, R. R., J. S. Eckert, E. H. Foote, and L. R. Rollison, Packed Liquid-Liquid Extractors, *Chem. Eng. Progr.*, **67**(11):60 (1971).

———, Sieve Trays? Consider Viscosity, *Hydrocarbon Process.*, **50**(11):235 (1971).

Neretnicks, I., The Optimisation of Sieve Tray Columns, *Brit. Chem. Eng.*, **15**(2):193 (1970).

Nguyen, H.-X., Calculating Actual Stages in Countercurrent Leaching, *Chem. Eng.*, **85**(25):121 (1978).

———, Computer Program Expedites Packed-tower Design, *Chem. Eng.*, **85**(26):181 (1978).

Nisenfeld, A. E., Reflux or Distillate: Which to Control? *Chem. Eng.*, **76**(21):169 (1969).

——— and J. Harbison, Benefits of Better Distillation Control, *Chem. Eng. Progr.*, **74**(7):88 (1978).

Nishida, N., A. Ichikawa, and E. Tazaki, Synthesis of Optimal Process Systems with Uncertainty, *I&EC Process Des. & Dev.*, **13**(3):209 (1974).

Null, H. R., Heat Pumps in Distillation, *Chem. Eng. Progr.*, **72**(7):58 (1976).

Oldshue, J. Y., and F. L. Connelly, Gas-Liquid Contacting With Impeller Mixers, *Chem. Eng. Progr.*, **73**(3):85 (1977).

Osborne, A., The Calculation of Unsteady State Multicomponent Distillation Using Partial Differential Equations, *AIChE J.*, **17**(3):696 (1971).

Owens, W. R., and R. N. Maddox, Short-cut Absorber Calculations, *I&EC*, **60**(12):14 (1968).

Petterson, W. C., and T. A. Wells, Energy-Saving Schemes in Distillation, *Chem. Eng.*, **84**(20):78 (1977).

Prahl, W. H., Pressure Drop in Packed Columns, *Chem. Eng.*, **76**(17):89 (1969).

———, Liquid Density Distorts Packed Column Correlation, *Chem. Eng.*, **77**(24):109 (1970).

Prasher, B. D., and G. B. Wills, Mass Transfer in an Agitated Vessel, *I&EC Process Des. & Dev.*, **12**(3):351 (1973).

Pratt, H. R. C., A Simplified Analytical Design Method for Differential Extractors with Back-mixing. Linear Equilibrium Relationship, *I&EC Process Des. & Dev.*, **14**(1):74 (1975).

———, A Simplified Analytical Design Method for Differential Extractors with Backmixing Curved Equilibrium Line, *I&EC Fund.*, **15**(1):34 (1976).

Prengle, H. W., Jr., and N. Barona, Make Petrochemicals by Liquid Phase Oxidation: Kinetics, Mass Transfer, and Reactor Design, *Hydrocarbon Process.*, **49**(11):159 (1970).

Reissinger, R.-H., and J. Schröter, Selection Criteria for Liquid-Liquid Extractors, *Chem. Eng.*, **85**(25):109 (1978).

Ripps, D. L., Minimum Reflux Figured an Easier Way, *Hydrocarbon Process.*, **47**(12):84 (1968).

Robinson, E. R., and M. R. Goldman, The Simulation of Multi-Component Batch Distillation Processes on a Small Digital Computer, *Brit. Chem. Eng.*, **14**(6):318 (1969).

Rosen, A. M., and V. S. Krylov, Theory of Scaling-Up and Hydrodynamic Modelling of Industrial Mass Transfer Equipment, *Chem. Eng. J.*, **7**(2):85 (1974).

Saletan, D. I., Specifying Distillation Safety Factors, *Chem. Eng. Progr.*, **65**(5):81 (1969).

Sargent, R. W. H., and B. A. Murtagh, The Design of Plate Distillation Columns for Multi-Component Mixtures, *Trans. Inst. Chem. Eng.*, **47**:85 (1969).

Scheiman, A. D., Find Minimum Reflux by Heat Balance, *Hydrocarbon Process.*, **48**(9):187 (1969).

Scott, B. D., and H. S. Myers, Bubble Tray Efficiency Studies, *Chem. Eng. Progr.*, **69**(10):73 (1973).

Sewell, A., Practical Aspects of Distillation Column Design, *Chem. Eng. (London)*, 442 (August 1975).

Shah, Y. T., Computer Simulation of Transport Processes, *I&EC Prod. Res. & Develop.*, **11**(3):269 (1972).

Sherwood, T. K., R. L. Pigford, and C. R. Wilke, "Mass Transfer," McGraw-Hill Book Company, New York, 1975.

Skelland, A. H. P., and W. L. Conger, A Rate Approach to Design of Perforated-Plate Extraction Columns, *I&EC Process Des. & Dev.*, **12**(4):448 (1973).

Smith, S., Use of Reflux in Solvent Extraction, *Brit. Chem. Eng.*, **12**(9):1361 (1967).

Smith, V. C., and W. V. Delnicki, Optimum Sieve Tray Design, *Chem. Eng. Progr.*, **71**(8):68 (1975).

Sofer, S. S., and R. G. Anthony, Distillation: Real vs. Theory, *Hydrocarbon Process.*, **52**(3):93 (1973).

Stewart, R. R., E. Weisman, B. M. Goodwin, and C. E. Speight, Effect of Design Parameters in Multicomponent Batch Distillation, *I&EC Process Des. & Dev.*, **12**(2):131 (1973).

Stragio, V. A., and R. E. Treybal, Reflux—Stages Relations for Distillation, *I&EC Process Des. & Dev.*, **13**(3):279 (1974).

Standart, G., Distillation Practices: Distillation and Absorption Research in Europe, *Chem. Eng. Progr.*, **68**(10):66 (1972).

Stupin, W. J., and F. J. Lockhart, Distillation Practices: Thermally Coupled Distillation—A Case History, *Chem. Eng. Progr.*, **68**(10):71 (1972).

Surowiec, A. J., Estimate Reboiled Absorbers, *Hydrocarbon Process.*, **48**(9):211 (1969).

Symoniak, M. F., A Correlation for Sizing Adsorption Systems, *Chem. Eng.*, **76**(21):172 (1969).

Tanagi, T., and B. D. Scott, Effect of Liquid Mixing on Sieve Trays, *Chem. Eng. Progr.*, **69**(10):75 (1973).

Tassios, D., Choosing Solvents for Extractive Distillation, *Chem. Eng.*, **76**(3):118 (1969).

Taylor, D. L., and W. C. Edmister, Solutions for Distillation Processes Treating Petroleum Fractions, *AIChE J.*, **17**(6):1324 (1971).

Taylor, J. H., Systems Design for Centrifugal Molecular Distillation, *Chem. Eng.*, **75**(18):109 (1968).

Tedder, D. W., and D. F. Rudd, Parametric Studies in Industrial Distillation: Design Comparisons, *AIChE J.*, **24**(2):303 (1978).

—— and ——, Parametric Studies in Industrial Distillation: Heuristic Optimization, *AIChE J.*, **24**(2):316 (1978).

—— and ——, Parametric Studies in Industrial Distillation: Design Methods and Their Evaluation, *AIChE J.*, **24**(2):323 (1978).

Thibodeaux, L. J., Continuous Crosscurrent Mass Transfer in Towers, *Chem. Eng.*, **76**(12):165 (1969).

Thorngren, J. T., Valve Tray Flooding Generalized, *Hydrocarbon Process.*, **57**(8):111 (1978).

Tierney, J. W., and J. A. Bruno, Equilibrium Stage Calculations, *AIChE J.*, **13**(3):556 (1967).

Timmins, R. S., L. Mir, and J. M. Ryan, Large-Scale Chromatography: New Separation Tool, *Chem. Eng.*, **76**(11):170 (1969).

Todd, W. G., and M. Van Winkle, Fractionation Efficiency, *I&EC Process Des. & Dev.*, **11**(4):578 (1972).

Tondeur, D., and G. Klein, Multicomponent Ion Exchange in Fixed Beds, *I&EC Fund.*, **6**(3):351 (1967).

Treybal, R. E., A Simple Method for Batch Distillation, *Chem. Eng.*, **77**(21):95 (1970).

Truman, A. H., Batch Distillation of Phthalic Anhydride, *Chem. Eng. Progr.*, **66**(3):62 (1970).

Van Winkle, M., and W. G. Todd, Optimum Fractionation Design by Simple Graphical Methods, *Chem. Eng.*, **78**(21):136 (1971).

—— and ——, Minimizing Distillation Costs via Graphical Techniques, *Chem. Eng.*, **79**(5):105 (1972).

Vashist, P. N., and R. B. Beckmann, Liquid-liquid Extraction, *I&EC*, **59**(11):71 (1967).

Vermuelen, T., Process Arrangements for Ion Exchange and Adsorption, *Chem. Eng. Progr.*, **73**(10):57 (1977).

Wade, H. L., and C. J. Ryskamp, Tray Flooding Sets Crude Thruput, *Hydrocarbon Process.*, **56**(11):281 (1977).

Walker, G. J., Design Sour Water Strippers Quickly, *Hydrocarbon Process.*, **48**(6):121 (1969).

Waterman, W. W., J. P. Frazier, and G. M. Brown, Compute Best Distillation Feed Point, *Hydrocarbon Process.*, **47**(6):155 (1968).

Watkins, R. N., How to Design Crude Distillation, *Hydrocarbon Process.*, **48**(12):93 (1969).

——, "Petroleum Refinery Distillation," Gulf Publishing Company, Houston, Texas, 1973.

Weiler, D. W., W. V. Delnicki, and B. L. England, Flow Hydraulics of Large Diameter Trays, *Chem. Eng. Progr.*, **69**(10):67 (1973).

Weiner, A. L., Dynamic Fluid Drying, *Chem. Eng.*, **81**(19):92 (1974).

Wheeler, D. E., Design Criteria for Chimney Trays, *Hydrocarbon Process.*, **47**(7):119 (1968).

Wichterle, I., R. Kobayashi, and P. S. Chappelear, Caution! Pinch Point in Y-X Curve! *Hydrocarbon Process.*, **50**(11):233 (1971).

Wilckens, K. L., and J. P. Perez, Easy Way to Estimate TBP Data, *Hydrocarbon Process.*, **50**(4):150 (1971).

Winter, G. R., and K. D. Uitti, Froth Initiators Can Improve Tray Performance, *Chem. Eng. Progr.*, **72**(9):50 (1976).

Wnek, W. J., and R. H. Snow, Design of Cross-Flow Cooling Towers and Ammonia Stripping Towers, *I&EC Process Des. & Dev.*, **11**(3):343 (1972).

Wood, C. E., Tray Selection for Column Temperature Control, *Chem. Eng. Progr.*, **64**(1):85 (1968).

Woosley, R. D., G. K. Baker, and D. J. Stubblefield, Use of the Computer to Select Design Parameters for Solid Adsorbent Dehydration of Gas Streams, *AIChE Symp. Ser.*, **67**(117):98 (1971).

Yon, C. M., and P. H. Turnock, Multicomponent Adsorption Equilibria on Molecular Sieves, *AIChE Symp. Ser.*, **67**(117):75 (1971).

Young, G. C., and J. H. Weber, Murphree Point Efficiencies in Multicomponent Systems, *I&EC Process Des. & Dev.*, **11**(3):440 (1972).

Zanker, A., Quick Calculation for Minimum Theoretical Plates, *Chem. Eng.*, **81**(18):96 (1974).

——, Minimum Reflux From a Nomograph, *Chem. Eng.*, **83**(8):156 (1976).

——, Nomograph Replaces Gilliland Plot, *Hydrocarbon Process.*, **56**(5):263 (1977).

——, Designing Spouted Beds, *Chem. Eng.*, **84**(25):207 (1977).

Zenz, F. A., Designing Gas-Absorption Towers, *Chem. Eng.*, **79**(25):120 (1972).

Reactor design and costs

Albright, L. F., and C. G. Bild, Designing Reaction Vessels for Polymerization, *Chem. Eng.*, **82**(19):121 (1975).

Andersen, T. S., Evaluation of Models for Tubular, Laminar Flow Reactors, *AIChE J.*, **16**(4):543 (1970).

April, G. C., and R. W. Pike, Modeling Complex Chemical Reaction Systems, *I&EC Process Des. & Dev.*, **13**(1):1 (1974).

Aris, R., "Elementary Chemical Reactor Analysis," Prentice-Hall, Englewood Cliffs, New Jersey, 1969.

Barona, N., and H. W. Prengle, Jr., Design Reactors This Way for Liquid-Phase Processes, *Hydrocarbon Process.*, **52**(3):63 (1973); **52**(12):73 (1973).

Brooks, J. R., and D. E. Waite, An Example of Reactor Optimization Using a Parallel-Logic Analog Computer, *Brit. Chem. Eng.*, **15**(5):647 (1970).

Buckley, P. S., Protective Controls for Chemical Reactor, *Chem. Eng.*, **77**(8):145 (1970).

Chalfant, J. R., and C. H. Barron, A New Approach to Nonlinear Chemical Reaction Rate Equations, *Chem. Eng. Symp. Ser.*, **72**(63):79 (1967).

Cooper, A. R., and G. V. Jeffreys, "Chemical Kinetics and Reactor Design," Prentice-Hall, Englewood Cliffs, New Jersey, 1971.

——, "Chemical Kinetics and Reactor Design," Prentice-Hall, Englewood Cliffs, New Jersey, 1973.

Corrigan, T. E., W. E. Lewis, and K. N. McKelvey, What do Chemical Reactors Cost in Terms of Volume? *Chem. Eng.*, **74**(11):214 (1967).

———— G. A. Lessells, and M. J. Dean, Application of Chemical Reactor Technology to a Commercial Reactor, *I&EC*, **60**(4):63 (1968).

———— and M. J. Dean, Estimate Reactor Backmixing Effect, *Hydrocarbon Process.*, **47**(7):149 (1968).

Devia, N., and W. L. Luyben, Reactors: Size Versus Stability, *Hydrocarbon Process.*, **57**(6):119 (1978).

Dyson, D. C., F. J. M. Horn, R. Jackson, and C. B. Schlesinger, Reactor Optimization Problems for Reversible Exothermic Reactions, *Can. J. Chem. Eng.*, **45**(5):310 (1967).

Elhalwagi, M. M., An Engineering Concept of Reaction Rate, *Chem. Eng.*, **78**(12):75 (1971).

Epstein, L. D., Cost of Standard-Sized Reactors and Storage Tanks, *Chem. Eng.*, **78**(24):160 (1971).

Eschenbrenner, G. P., and G. A. Wagner, III, A New High Capacity Ammonia Converter, *Chem. Eng. Progr.*, **68**(1):63 (1972).

Fair, J. R., Designing Gas-Sparged Reactors, *Chem. Eng.*, **74**(15):207 (1967).

Fournier, C. D., and F. R. Groves, Jr., Isothermal Temperatures for Reversible Reactions, *Chem. Eng.*, **77**(3):121 (1970).

———— and ————, Rapid Method for Calculating Reactor Temperature Profiles, *Chem. Eng.*, **77**(13):157 (1970).

Frank, A., Homogeneous Second-Order Chemical Reaction in Countercurrent Flow Systems, *Chem. Eng. Progr. Symp. Ser.*, **72**(63):54 (1967).

Fuller, O. M., Differential Scaling: A Method for Chemical Reactor Modelling, *Can. J. Chem. Eng.*, **48**(1):119 (1970).

Hazbun, E. A., and J. W. White, Simulation, Design, and Analysis of Reactors, *Chem. Eng. Progr.*, **70**(7):83 (1974).

Hill, C. G., Jr., "An Introduction to Chemical Engineering Kinetics and Reactor Design," J. Wiley and Sons, New York, 1977.

Hlavacek, V., Aspects in Design of Catalytic Reactors, *I&EC*, **62**(7):8 (1970).

Holland, C. D., and R. F. Anthony, "Fundamentals of Chemical Reaction Engineering," Prentice-Hall, Englewood Cliffs, New Jersey, 1978.

Huang, I-Der, Two-Point Plot Proves Key to Kinetic Relations, *Chem. Eng.*, **74**(4):135 (1967).

Hyman, M. H., Fundamentals of Engineering HPI Plants with the Digital Computer, *Petro/Chem. Eng.*, **9**(40):31 (1969).

Joklik, O. F., A User's Guide to Catalysts, *Chem. Eng.*, **80**(23):49 (1973).

Kantyka, T. A., Reactor Development and Design—Application of Mathematical Models, *Chem. Eng. (London)*, 141 (March 1974).

Kao, Y. K., and S. G. Bankoff, Optimal Design of a Chemical Reactor with Fast and Slow Reactions, *Chem. Eng. Comm.*, **1**(3):141 (1974).

Koo, L., and E. N. Ziegler, Reactor Design Made Simple, *Chem. Eng.*, **76**(4):91 (1969).

Levenspiel, O., "Chemical Reaction Engineering: An Introduction to the Design of Chemical Reactors," 2d ed., J. Wiley and Sons, New York, 1972.

Levine, R., Stages Needed by Backmixed Reactors, *Hydrocarbon Process.*, **46**(7):159 (1967).

————, A New Design Approach for Backmixed Reactors, *Chem. Eng.*, **75**(14):62 (1968); **75**(16):145 (1968); **75**(17):167 (1968).

Lunde, P. J., Modeling, Simulation, and Operation of a Sabatier Reactor, *I&EC Process Des. & Dev.*, **13**(3):226 (1974).

Nishida, N., Y. A. Liu, and A. Ichikawa, Studies in Chemical Process Design and Synthesis. Optimal Synthesis of Dynamic Process Systems with Uncertainty, *AIChE J.*, **22**(3):539 (1976).

Paris, J. R., and W. F. Stevens, Mathematical Models for a Packed-bed Chemical Reactor, *Can. J. Chem. Eng.*, **48**(1):100 (1970).

Pickert, P. E., A. P. Bolton, and M. A. Lanewala, Process Design with Molecular Sieve Catalysts, *Chem. Eng.*, **75**(16):139 (1968).

Peterson, R. P., D. E. Smith, and D. R. May, Reactor Design From Reaction Kinetics, *Chem. Eng.*, **76**(17):101 (1969).

Powers, G. J., and J. F. Mayer, Design of a Catalytic Reactor-Separator System with Uncertainty in Catalyst Activity and Selectivity, *I&EC Process Des. & Dev.*, **14**(1):41 (1975).

Prengle, H. W., Jr., and N. Barona, Make Petrochemicals by Liquid Phase Oxidation: Kinetics, Mass Transfer and Reactor Design, *Hydrocarbon Process.*, **49**(11):159 (1970).

Priestley, A. J., and J. B. Agnew, Sensitivity Analysis in the Design of a Packed Bed Reactor, *I&EC Process Des. & Dev.*, **14**(2):171 (1975).

Raines, G. E., and T. E. Corrigan, The Use of the Axial Dispersion Model to Predict Conversions of First-and Second-Order Reactions, *Chem. Eng. Progr. Symp. Ser.*, **72**(63):90 (1967).

Rase, H. F., "Chemical Reactor Design for Process Plants" Vols. 1 and 2, J. Wiley and Sons, New York, 1977.

Richardson, J. T., SNG Catalyst Technology, *Hydrocarbon Process.*, **52**(12):91 (1973).

Sanderson, R. V., and H. H. Y. Chien, Simultaneous Chemical and Phase Equilibrium Calculation, *I&EC Process Des. & Dev.*, **12**(1):81 (1973).

Satterfield, C. N., Trickle-Bed Reactors, *AIChE J.*, **21**(2):209 (1975).

Schlegel, W. F., Polymer-Plant Engineering: Reaction, Polymer Recovery, *Chem. Eng.*, **79**(6):88 (1972).

Silverstein, J., and R. Shinnar, Design of Fixed Bed Catalytic Microreactors, *I&EC Process Des. & Dev.*, **14**(2):127 (1975).

Small, W. M., Scaleup Problems in Reactor Design, *Chem. Eng. Progr.*, **65**(7):81 (1969).

Storey, S. H., and F. Van Zeggeren, Solving Complex Chemical Equilibria by a Method of Nested Iterations, *Can. J. Chem. Eng.*, **45**(5):323 (1967).

Townsend, D. I., Hazard Evaluation of Self-Accelerating Reactions, *Chem. Eng. Progr.*, **73**(9):80 (1977).

Weber, A. P., Residence-Time Spectrum in Continuous-Flow Reactors, *Chem. Eng.*, **76**(24):79 (1969).

Wen, C.-Y., Noncatalytic Heterogeneous Solid Fluid Reaction Models, *I&EC*, **60**(9):34 (1968).

——— and L. T. Fan, "Models for Flow Systems and Chemical Reactors," Marcel Dekker, Inc., New York, 1975.

Yen, Y. C., Bigger Reactors or More Recycle, *Hydrocarbon Process.*, **49**(1):157 (1970).

Wirges, H.-P., and S. R. Shah, For a Given Kinetic Duty . . . Select Optimum Reactor Quickly, *Hydrocarbon Process.*, **55**(4):135 (1976).

Tower design and costs

Amundson, N. R., A. J. Pontinen, and J. W. Tierney, Multicomponent Distillation on a Large Digital Computer: Generalization with Side-Stream Stripping, *AIChE J.*, **5**(3):295 (1959).

Badhwar, R. K., Shortcut Design Methods for Piping, Exchangers, Towers, *Chem. Eng.*, **78**(24):113 (1971).

Billet, R., Gauze-Packed Columns for Vacuum Distillation, *Chem. Eng.*, **79**(4):68 (1972).

Boyd, D. M., Fractionation Column Control, *Chem. Eng. Progr.*, **71**(6):55 (1975).

Broughton, D. B., and K. D. Uitti, Estimate Tower for Naphtha Cuts, *Hydrocarbon Process.*, **50**(10):109 (1971).

Buchanan, J. E., Holdup in Irrigated Ring-Packed Towers Below the Loading Point, *I&EC Fund.*, **6**(3):400 (1967).

Chase, J. D., Sieve-tray Design, *Chem. Eng.*, **74**(18):139 (1967).

Chimes, A. R., Demonstrated Data for Tower Design, *Chem. Eng.*, **82**(23):137 (1975).

Eckert, J. S., How Tower Packings Behave, *Chem. Eng.*, **82**(2):70 (1975).

Eduljee, H. E., Equations Replace Gilliland Plot, *Hydrocarbon Process.*, **54**(9):120 (1975).

Ellerbe, R. W., Batch Distillation Basics, *Chem. Eng.*, **80**(12):110 (1973).

Fair, J. R., Advances in Distillation System Design, *Chem. Eng. Progr.*, **73**(11):78 (1977).

Fauth, G. F., and F. G. Shinskey, Advanced Control of Distillation Columns, *Chem. Eng. Progr.*, **71**(6):49 (1975).

Garvin, R. G., and E. R. Norton, Sieve Tray Performance Under GS Process Conditions, *Chem. Eng. Progr.*, **64**(3):99 (1968).

Guthrie, K. M., Capital Cost Estimating, *Chem. Eng.*, **76**(6):114 (1969).

Holland, C. D., and G. P. Pendon, Solve More Distillation Problems, *Hydrocarbon Process.*, **53**(7):148 (1974).

Hulswitt, C. E., Adiabatic and Falling Film Absorption of Hydrogen Chloride, *Chem. Eng. Progr.*, **69**(2):50 (1973).

Hutchins, R. A. Activated Carbon Systems, *Chem. Eng.*, **80**(19):133 (1973).

Interess, E., Practical Limitations on Tray Design, *Chem. Eng.*, **78**(26):167 (1971).

Jakob, R. R., Estimate Number of Crude Trays, *Hydrocarbon Process.*, **50**(5):149 (1971).

Kern, R., How to Design Piping for Reboiler Systems, *Chem. Eng.*, **82**(16):107 (1975).

Kitterman, L., and M. Ross, Tray Guides to Avoid Tower Problems, *Hydrocarbon Process.*, **46**(5):216 (1967).

Leva, M., Film Tray Equipment for Vacuum Distillation, *Chem. Eng. Progr.*, **67**(3):65 (1971).

Love, F. S., Troubleshooting Distillation Problems, *Chem. Eng. Progr.*, **71**(6):61 (1975).

Lubowicz, R. E., and P. Reich, High Vacuum Distillation Design, *Chem. Eng. Progr.*, **67**(3):59 (1971).

Lukchis, G. M., Adsorbent Regeneration, *Chem. Eng.*, **80**(18):83 (1973).

Luyben, W. L., Azeotropic Tower Design by Graph, *Hydrocarbon Process.*, **52**(1):109 (1973).

MacFarland, S. A., P. M. Sigmund, and M. Van Winkle, Predict Distillation Efficiency, *Hydrocarbon Process.*, **51**(7):111 (1972).

Madsen, N., Finding the Right Reflux Ratio, *Chem. Eng.*, **78**(25):73 (1971).

McLaren, D. B., and J. C. Upchurch, Guide to Trouble-Free Distillation, *Chem. Eng.*, **77**(12):139 (1970).

Mihm, J. C., Optimize Absorption Type Gas Processing Plants, *Pet. Eng.*, **45**(2):66 (1973).

Miller, J. S., and W. A. Kapella, Installed Cost of a Distillation Column, *Chem. Eng.*, **84**(8):129 (1977).

Nemunaitis, R. R., J. S. Eckert, E. H. Foote, and L. R. Rollison, Packed Liquid-Liquid Extractors, *Chem. Eng. Progr.*, **67**(11):60 (1971).

——— and J. S. Eckert, Heat Transfer in Packed Towers, *Chem. Eng. Progr.*, **71**(8):60 (1975).

Nygren, P. G., and G. K. S. Connolly, Vacuum Distillation—Selecting Vacuum Fractionation Equipment, *Chem. Eng. Progr.*, **67**(3):49 (1971).

Onda, K., E. Sada, K. Takahashi, and S. A. Mukhtar, Plate and Column Efficiencies of Continuous Rectifying Columns for Binary Mixtures, *AIChE J.*, **17**(5):1141 (1971).

Pikulik, A., and H. E. Diaz, Cost Estimating for Major Process Equipment, *Chem. Eng.*, **84**(21):107 (1977).

Rowland, C. H., and E. A. Grens, II, Design Absorbers: Use Real Stages, *Hydrocarbon Process.*, **50**(9):201 (1971).

Ruziska, P. A., Packings for Hot Carbonate Systems, *Chem. Eng. Progr.*, **69**(2):67 (1973).

Shih, Y.-P., T.-C. Chou, and C.-T. Chen, Best Distillation Feed Points, *Hydrocarbon Process.*, **50**(7):93 (1971).

Smith, V. C., and W. V. Delnicki, Optimum Sieve Tray Design, *Chem. Eng. Progr.*, **71**(8):68 (1975).

Sommerville, R. F., New Method Gives Quick, Accurate Estimate of Distillation Costs, *Chem. Eng.*, **79**(9):71 (1972).

Tang, S. S., Shortcut Method for Calculating Tower Deflections, *Hydrocarbon Process.*, **47**(11):230 (1968).

Taylor, D. L., and W. C. Edmister, Solutions for Distillation Processes Treating Petroleum Fractions, *AIChE J.*, **17**(6):1324 (1971).

Van Winkle, M., and W. G. Todd, Optimum Fractionation Design by Simple Graphical Methods, *Chem. Eng.*, **78**(21):136 (1971).

——— and ———, Minimizing Distillation Costs via Graphical Techniques, *Chem. Eng.*, **79**(5):105 (1972).

Vermeulen, T., and G. Klein, Recent Background Developments for Adsorption Column Design, *AIChE Symp. Ser.*, **67**(117):65 (1971).

Wall, K. J., Design of Packed Distillation Columns, *Chem. Proc. Eng.*, **48**(7):56 (1967).

Yen, I.-K., Predicting Packed-Bed Pressure Drop, *Chem. Eng.*, **74**(6):173 (1967).

Youness, A., New Approach to Tower Deflection, *Hydrocarbon Process.*, **49**(6):121 (1970).

Zanker, A., Estimating Cooling Tower Costs from Operating Data, *Chem. Eng.*, **79**(13):118 (1972).

———, Simplified Calculations for Packed Towers, *Chem. Eng.*, **82**(20):100 (1975).

———, Designing Spouted Beds, *Chem. Eng.*, **84**(25):207 (1977).

Zenz, F. A., L. Stone, and M. Crane, Find Sieve Tray Weepage Rates, *Hydrocarbon Process.*, **46**(12):138 (1967).
————, Designing Gas-Adsorption Towers, *Chem. Eng.*, **79**(25):120 (1972).
Zimmerman, O. T., and I. Lavine, Peabody Gas Scrubbers, *Cost Eng.*, **13**(2):16 (1968).

PROBLEMS†

1 A sieve-tray tower has an ID of 5 ft, and the combined cross-sectional area of the holes on one tray is 10 percent of the total cross-sectional area of the tower. The height of the weir is 1.5 in. The head of liquid over the top of the weir is 1 in. Liquid gradient is negligible. The diameter of the perforations is $\frac{3}{16}$ in., and the superficial vapor velocity (based on the cross-sectional area of the empty tower) is 3.4 ft/s. The pressure drop due to passage of gas through the holes may be assumed to be equivalent to 1.4 kinetic heads (based on gas velocity through holes). (Tray thickness = hole diameter and active area = 90 percent of total area—see Fig. 15-12). If the liquid density is 50 lb/ft^3 and the gas density is 0.10 lb/ft^3, estimate the pressure drop per tray as pounds force per square inch.

2 An existing bubble-cap distillation tower has an ID of 8 ft and a tray spacing of 30 in. The tower is operating under a pressure of 70 psig, and a slot liquid seal of 2 in. is maintained. At the point of maximum volumetric vapor flow, the molecular weight of the vapor is 100, the rate of vapor flow is 1500 lb mol/h, the liquid density is 55 lb/ft^3, and the temperature is 175°F. The pressure drop through the tower is negligible, and the ideal gas law is applicable to the rising vapors. Approximately what percent of the maximum allowable flow rate is being used in the tower?

3 A mixture of benzene and toluene containing 60 mole percent benzene is to be separated into an overhead product containing 96 mole percent benzene and a bottoms containing 25 mole percent benzene. A valve-tray tower has been designed which is intended to accomplish this separation at atmospheric pressure. At the operating conditions chosen, calculations have shown that 6.1 theoretical stages are necessary. The temperature is 181°F at the top of the column and 213°F in the reboiler. Assuming the reboiler acts as one theoretical stage, estimate the number of actual trays required. Benzene and toluene mixtures may be considered as ideal. At 197°F, the vapor pressure of pure benzene is 1070 mm Hg and the vapor pressure of pure toluene is 429 mm Hg.

4 For the conditions given in Example 3, what weir height would be necessary to reduce the dimensionless ratio of liquid gradient/pressure-drop head caused by bubble-cap assemblies to the recommended maximum value of 0.4?

5 Air is passing through a bubble-cap plate at a superficial vapor velocity of 2 ft/s. Water is flowing across the tray at a rate of 0.1 ft^3/(s)(ft diameter). The water and air are at a temperature of 70°F, and the pressure is 1 atm. Assuming stable operation of the plate and a liquid gradient of 0.4 in., estimate the percent of the total pressure drop across the tray due to the head of liquid above the slots. The following specifications apply to the unit:

Inside diameter = 6 ft
Minimum total cross-sectional area of riser, direction-reversal space in cap, or annular cap space = 15 percent of tower cross-sectional area
Total slot area = 10 percent of tower cross-sectional area
Length of chord weir = 0.7 × tower diameter
Slots in caps are triangular, and slot width at base = 0.4 in.
Slot height = 1.5 in.
Static submergence = 1.0 in.

6 A random-packed absorption tower is to be used for removing a hydrocarbon from a gas mixture by countercurrent scrubbing with an oil. Stoneware Raschig rings of 1-in. nominal size will be used

† See Appendix C for reactor-design problems.

as the packing. The tower must handle a gas rate of 2000 lb/h and a liquid rate of 6000 lb/h. A gas velocity equal to 70 percent of the maximum allowable velocity at the given liquid and gas rates will be used. The gas density is 0.075 lb/ft³. The liquid density is 55 lb/ft³. The viscosity of the oil is 20 centipoises. Under these conditions, estimate the necessary tower diameter and the pressure drop through the tower as inches of water per foot of packed height.

7 A blower is being used to force 200 lb of air per hour through a 12-in. inside-diameter tower countercurrent to water flowing at a rate of 6000 lb/h. The tower is packed to a height of 10 ft with 1-in. Berl saddles. The average pressure in the tower is 1 atm, and the temperature is 75°F. The blower-motor combination used at the gas inlet has an overall efficiency of 50 percent. Estimate the power cost per 8000 operating hours for forcing the air through the packing if the cost of power is $0.04 per kilowatthour.

8 A random-packed distillation tower with an inside diameter of 6 in. is being operated at a condenser pressure of 100 mm Hg. The following data are obtained during operation:

Packed height = 3 ft
Gas rate = 100 lb/h
Liquid rate = 90 lb/h
Average molecular weight of gas = 100
Average temperature = 110°F
Total pressure drop = 12 mm Hg
Liquid viscosity = 1.2 centipoises
Liquid density = 55 lb/ft³

A larger distillation tower is needed for the same mixture, and it has been decided to use a 12-in. inside-diameter tower with the same type and size of packing. The packed height will be 3 ft. The condenser pressure will be 50 mm Hg, and the average temperature can be assumed to be 95°F. At this temperature, the liquid viscosity is 1.4 centipoises, the liquid density 57 lb/ft³. The 12-in. tower will be operated at the same percent of the maximum allowable gas mass velocity as the 6-in. tower, and the same ratio of L/G will be used. The average molecular weight of the gas is unchanged. With the given packing and fluids, the pressure drop at the flooding point can be assumed to be 16 mm Hg/ft of packed height. Neglecting the effect of temperature on fluid densities at the flooding point, estimate the gas rate as pounds per hour at the indicated operating conditions in the 12-in. tower.

9 A special test on the 6-in. distillation tower described in Prob. 8 indicates that the total pressure drop is 2.9 mm Hg when the gas rate is 50 lb/h. For this test, the condenser pressure was maintained at 100 mm Hg, the L/G ratio was the same as shown in Prob. 8, and temperature change was negligible. Operation can be considered to be under preloading conditions. Estimate the total pressure drop at the operating conditions indicated in Prob. 8 for the 12-in. column.

10 A random-packed tower is to be used for contacting 3000 lb/h of air with 4000 lb/h of water. The tower will be packed with 1-in. chemical porcelain Raschig rings. The operating pressure is 1 atm, and the temperature is 70°F. Assuming column operation is in the preloading range, estimate the optimum diameter per foot of packed height to give minimum annual cost for fixed charges and blower operating charges. The following additional data apply:

Annual fixed charges = 0.2 times cost of installed unit
Cost of installed unit, complete with distributor plates, supports, and all auxiliaries = 2.0 times purchased cost of shell and packing.
Cost of delivered power = $0.04/kWh
Operating time = 8000 h/year
Purchased cost of shell = $20/ft³ of packed volume
Purchased cost of packing = $11.00/ft³

SIXTEEN

STATISTICAL ANALYSIS IN DESIGN

Application of statistical analysis in process design has become so extensive that every chemical engineer must have a basic knowledge of this branch of mathematics. Statistical design and analysis help the engineer to make decisions by getting the most out of the data. In many cases, where too little is known about a chemical process to permit a mathematical model, a statistical approach may indicate the manner of the direction in which to proceed with the design.

Besides analyzing and correlating data by statistical means, the chemical engineer also uses statistics in the development of quality control to establish acceptable limits of process variables and in the design of laboratory, pilot plant, and process plant (evolutionary operation) experiments. In the latter application, statistical strategy in the design of experiments enables the engineer to set experimental variables at levels that will yield maximum information with a minimum amount of data.

The basis for any statistical analysis is the fact that all data are to some extent, one way or another, subject to chance errors. These chance errors may arise whether the problem involves an estimation, the development of a reliable model, or the testing of a hypothesis. For example, since no experimentally determined value in the laboratory is absolute, it is generally necessary to determine by statistical analysis the reliability of the newly obtained data. The eventual use of these data by the engineer in design work may often determine the necessary confidence level that will be required concerning the true value of the data.

The specific techniques of statistical analysis in design require extensive discussions and can only be mentioned here for lack of space. The objective of this presentation is primarily to describe and briefly illustrate the concepts of statistical analysis with applications in the chemical process industry.

BASIC CONCEPTS

Statistical methods are based on the single concept of variability. It is through this fundamental concept that a basis is determined for design of experiments and analysis of data. Full utilization of this concept makes it possible to derive maximum information from a given set of data and to minimize the amount of data necessary to derive specific information.

AVERAGES

One of the most useful quantities for providing a measure of the location of data, making inferences about the true mean, μ, and comparing sets of data by statistical methods is the sample mean, $\bar{x}$, defined as the sum of all the values of a set of data divided by the number of observations, n, making up the data sample. Mathematically, the mean can be symbolized by

$$\bar{x} = \frac{1}{n} \sum x_i \tag{1}$$

The value of the mean obtained in this manner is not necessarily the true mean μ, unless all the values being averaged represent the total population. Another measure of location that is frequently used is the sample *median*. The median is defined as that value of the data that has the same number of values below it as above it once the data are arranged in order of value. A median is not affected by extremely large or small values and is frequently used when little is known about the underlying probability laws. A third measure of location of data is the *mode*. This term is defined as the most frequently occurring value in a set of data and finds application when there is a large frequency of one value to be averaged with one or a few numerically large values.

Example 1. Evaluation of mean and median Five analyses have been run on a natural gas sample to determine the methane content with the following results: 92.4, 92.8, 92.3, 93.0, and 92.5 percent. What are the sample mean and the sample median?

SOLUTION

$$\bar{x} = \frac{1}{n} \sum x_i$$

$$\bar{x} = \frac{1}{5}(92.4 + 92.8 + 92.3 + 93.0 + 92.5)$$

$$\bar{x} = 92.6$$

Retabulating the data in order of magnitude results in:

$$92.3, 92.4, \underline{92.5}, 92.8, 93.0$$

The median value is underlined.

In addition to the sample mean, there is also a harmonic mean and a geometric mean. The *harmonic mean* $\bar{H}$ is the reciprocal of the arithmetic mean of the reciprocals, and is used whenever reciprocals are involved, e.g., rates. For example, to obtain the average flow rate in a pipe delivering A gal at R_1 gpm and B gal at R_2 gpm, it is necessary to use the harmonic mean.

$$\bar{H} = \frac{n}{\sum \frac{1}{x_i}} \qquad (2)$$

Example 2. Evaluation of harmonic mean What is the average rate of flow for a pump delivering 100 gal at 50 gpm and 100 gal at 10 gpm?

SOLUTION

$$\bar{H} = \frac{n}{\sum \frac{1}{x_i}}$$

$$n = 2$$

$$\bar{H} = \frac{2}{\frac{1}{50} + \frac{1}{10}}$$

$$\bar{H} = 16.7 \text{ gpm}$$

This value could also have been obtained by determining the total time necessary to pump 100 gal at the two rates and dividing this into the total number of gallons pumped. Thus,

$$\tau_1 = \frac{100}{50} = 2 \text{ min}$$

$$\tau_2 = \frac{100}{10} = 10 \text{ min}$$

$$\text{Average rate} = \frac{100 + 100}{12} = 16.7 \text{ gpm}$$

Note that the mean would be 30 gpm, which is nearly twice as much as the harmonic mean.

The *geometric mean* $\bar{G}$ is the nth root of the product of n terms, and is used when averaging numbers involving growth data, e.g., population growth.

VARIABILITY OF DATA

There are a variety of ways in which to characterize the variability of data. One of the more useful quantities, although not the simplest, is the *true standard deviation* σ, which is defined as the square root of the sum of the squares of the deviations of the data points from the true mean divided by the number of observations n.

$$\sigma = \sqrt{\frac{\sum (x_i - \bar{x})^2}{n}} \tag{3}$$

Units for the standard deviation are the same as for the individual observation. Because the true mean is practically never known, the true standard deviation is generally a theoretical quantity. However, σ may be approximated by the estimated standard deviation, $s(x)$, where

$$s(x) = \sqrt{\frac{\sum (x_i - \bar{x})^2}{n - 1}} \tag{4}$$

In this relationship $(n - 1)$ appears in the denominator rather than n in order to make $s(x)$ an unbiased estimator of σ.

In statistical analysis it is necessary to use a quantity called *degrees of freedom*, designated henceforth as *d.f.* This quantity allows for a mathematical correction of the data for constraints placed upon the data. In this case, in the calculation of the estimated standard deviation, the number of observations n is fixed and the estimated standard deviation is calculated from the mean. Only $(n - 1)$ of the observations or sample terms can be varied and the last term is fixed by $\bar{x}$ and n. Thus, there are only $(n - 1)$ degrees of freedom in estimating the standard deviation from a sample of the population data.

Squaring the true standard deviation gives a term called the *true variance*, σ^2. It can be shown that the standard deviation of means $\bar{\sigma}$, calculated for samples taken from the total population of data, will have a true standard deviation equal to $\sigma/\sqrt{n}$, where n is the sample size. In other words, the spread of these means is less than the spread of the overall data around the group mean.

In the same manner, squaring the estimated standard deviation results in an *estimated variance* $s^2(x)$. An estimated standard deviation of the mean has the same relation to the estimated standard deviation of the population as the true standard deviation of the mean, that is, $s(\bar{x})$ is given by $s(x)/\sqrt{n}$.

Example 3. Determining the estimated standard deviation What is the estimated standard deviation for five weighings of a sample with the following values: 18.5, 18.4, 18.6, 18.3, and 18.7? What is the estimated standard deviation of the mean for this sample?

SOLUTION

$$\bar{x} = \frac{18.5 + 18.4 + 18.6 + 18.3 + 18.7}{5} = 18.5$$

$$s(x) = \sqrt{\frac{\Sigma(x_i - \bar{x})^2}{n-1}} = \sqrt{\frac{[\Sigma x^2 - (\Sigma x)^2/n]}{n-1}}$$

$$= \left[\frac{(18.5)^2 + (18.4)^2 + (18.6)^2 + (18.3)^2 + (18.7)^2 - (92.5)^2/5}{5-1}\right]^{1/2}$$

$$= 0.158$$

$$s(\bar{x}) = \frac{s(x)}{\sqrt{n}}$$

$$= \frac{0.158}{\sqrt{5}}$$

$$= 0.071$$

Another measure of the spread of the data is the *mean deviation, md*. This is simply the arithmetic mean of the deviation from the mean without regard to sign.

$$md = \frac{\Sigma|\bar{x} - x_i|}{n} \tag{5}$$

For a normal distribution, the true standard deviation is approximately 1.25 times the mean deviation.

Example 4. Evaluation of mean deviation Find the mean deviation of the following results found for five different samples: 0.57, 0.73, 0.69, 0.63, and 0.70.

SOLUTION

$$\bar{x} = 0.66$$

$$md = \frac{\Sigma|\bar{x} - x_i|}{n}$$

$$= \frac{\begin{array}{c}|0.57 - 0.66| + |0.73 - 0.66| + |0.69 - 0.66| \\ + |0.63 - 0.66| + |0.70 - 0.66|\end{array}}{5}$$

$$= 0.052$$

The spread of the data would be reported as ± 0.05 since the third decimal point would have very little meaning.

The last measure of dispersion to be noted here is the *range*, which is the difference between the largest and smallest value in any data that are reported. For small quantities of data (less than 10) the range is a useful number for comparison and is used in sampling and quality control. However, since this measure of variance is only affected by two of the data points, it obviously loses power as the size of the data sample or the number of observations increases.

THEORETICAL AND EMPIRICAL FREQUENCY DISTRIBUTION

When dealing with large amounts of data, it is convenient to form an array of the data in such a way that the frequencies of occurrence of given values or ranges of values can be tabulated and graphed. This grouping is accomplished by the designation of ranges which are called *class intervals*. The relative frequency of the class intervals is called an empirical distribution or an empirical frequency distribution and is used to estimate the theoretical frequency distribution.

Statistical formulas are based on various mathematical distribution functions representing these frequency distributions. The most widely used of all continuous frequency distributions is the *normal distribution*, the common bell-shaped curve. It has been found that the normal curve is the model of experimental errors for repeated measurements of the same thing. Assumption of a normal distribution is frequently and often indiscriminately made in experimental work because it is a convenient distribution on which many statistical procedures are based. However, some experimental situations subject to random error can yield data that are not adequately described by the normal distribution curve.

Usefulness of the normal distribution curve lies in the fact that from two parameters, the true mean μ and the true standard deviation σ, the distribution of all of the data can be established. The true mean determines the value on which the bell-shaped curve is centered, and σ determines the "spread" of the curve. A large σ gives a broad flat curve, while a small σ yields a tall narrow curve with most probability concentrated on values near the mean. It is impossible to find the exact value of the true mean from information provided by a sample. But an interval within which the true mean most likely lies can be found with a definite probability, for example, 0.95 or 0.99. The 95 *percent confidence level* indicates that while the true mean may or may not lie within the specified interval, the odds are 19 to 1 that it does.† Assuming a normal distribution, the 95 percent limits are $\bar{x} \pm 1.96\bar{\sigma}$ where $\bar{\sigma}$ is the true standard deviation of the sample mean. Thus if a process gave results that were known to fit a normal distribution curve having a mean of 11.0 and a standard deviation of 0.1, it would be clear from Fig. 16-1 that there is only a 5 percent chance of a result falling outside the range of 10.804 and 11.196.

---

† Note that the value of the true mean is always fixed, but that the interval specified by the confidence level because of its dependence upon the mean does not necessarily always include the true mean.

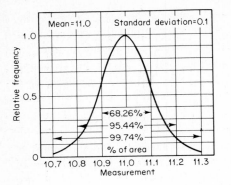

Figure 16-1 Probability curve.

As suggested above it is customary to work at the 95 percent or sometimes at the 99 percent probability level. The 95 percent probability level, which gives a 5 percent chance of a Type I error, represents the usual optimum for minimizing the two types of statistical error. A Type I error is a false rejection by a statistical test of the null hypothesis when it is true. Conversely, a Type II error is a false acceptance of the null hypothesis by a statistical test. The probability level at which statistical decisions are made will obviously depend on which type of error is more important.

STATISTICAL TESTS

There are a large number of statistical distributions which have for the most part been derived from the normal frequency distribution. The principles underlying only three of these distributions will be presented here. These distributions form the basis for the most frequently used statistical tests of significance.

There are two important statistical tests which can be used to determine whether the difference between two sets of data are real and significant or are just due to chance errors. Both assume that the experimental results are independently and normally distributed. One is the t-test and the other is the chi-squared test. The t-test applies only to continuous data—usually a measurement. The chi-squared test in many cases can approximate frequency or count type of data.

Statistics also provides a function for testing whether the scatter of two sets of data—as measured by the variance—is such as would be expected from two samples from the same population. This test is known as the F-test.

Tabular values for the above statistical tests are available in most statistical texts or engineering handbooks.†

---

† See, for example, R. E. Walpole and R. H. Myers, "Probability and Statistics for Engineers and Scientists," 2d ed., Macmillan Publishing Company, Inc., New York, 1978; A. H. Bowker and G. J. Lieberman, "Engineering Statistics," 2d ed., Prentice-Hall, Inc., Englewood Cliffs, New Jersey, 1972; and B. W. Lindgren, G. W. McElrath, and D. A. Berry, "Introduction to Probability and Statistics," 4th ed., Macmillan Publishing Company, Inc., New York, 1978.

THE t-TEST

The t-test compares the observed difference between averages with the inherent variability within the data—as measured by the standard deviation—to tell whether the difference is significant. Success of this test depends on the fact that the means of groups of samples from a distribution will tend to locate themselves around a normal distribution.†

The t-test consists of setting up a hypothesis for the value of the mean, calculating t from the observed mean and the estimated standard deviation, and comparing the calculated t with tabulated values of t. If the calculated t exceeds the tabulated value at the proper degree of freedom, then the original hypothesis may be rejected with the chance that the rejection is wrong corresponding to the probability level. Three common hypotheses and their corresponding t-equations are given below:

Hypothesis A: $\mu = \mu_0$ (one sample test)

$$\text{d.f.} = n - 1$$

$$t = \frac{\bar{x} - \mu_0}{s(\bar{x})}$$

Hypothesis B: $\mu_1 - \mu_2 = 0$ (two sample test-matched pairs)

$$\text{d.f.} = n - 1$$

$$t = \frac{\bar{x} - 0}{s(\bar{x})}$$

Hypothesis C: $\mu_1 = \mu_2$ (two sample test)

$$\text{d.f.} = n_1 + n_2 - 2$$

$$t = \frac{\bar{x}_1 - \bar{x}_2}{\bar{s}(\bar{x})}$$

where

$$s(\bar{x}) = \frac{s(x)}{\sqrt{n}} \tag{6}$$

$$\bar{s}(\bar{x}) = \bar{s}(x)\sqrt{\frac{1}{n_1} + \frac{1}{n_2}} \tag{7}$$

$$\bar{s}(x) = \sqrt{\frac{\Sigma(\bar{x}_1 - x_1)^2 + \Sigma(\bar{x}_2 - x_2)^2}{n_1 + n_2 - 2}} \tag{8}$$

---

† Alternative tests which are sometimes used in place of the t-tests are the Wilcoxon's tests. These tests are described by J. L. Hodges and E. L. Lehman, "Basic Concepts of Probability and Statistics," 2d ed., Holden-Day, Inc., San Francisco, 1970.

Tabular values for the t-test are normally presented for various degrees of freedom. The probability level is the chance of obtaining a value larger than the tabular values by chance alone. As the sample size (degrees of freedom) increases, the t-values approach the probability values for the normal distribution curve.

The following examples illustrate the use of the t-test.

Example 5. Example of t-test using Hypothesis A Consider the analysis of a gas sample in Example 1. Is the mean significantly different from an assumed true mean of 92.5?

SOLUTION Before a statistical test is applied, the level of significance is generally selected. It is customary to prescribe a significance of 95 percent or more before rejecting a specific hypothesis. The 95 percent level will be chosen in this example. This means that, if the calculated absolute value of t is larger than that which would occur by chance alone 1 time in 19, the hypothesis would be assumed to be false. From Example 1

$$\bar{x} = 92.6$$

$$s(x) = 0.292$$

$$s(\bar{x}) = 0.130$$

Assuming Hypothesis A

$$\mu = \mu_0 = 92.5$$

$$\text{d.f.} = n - 1 = 5 - 1 = 4$$

$$t = \frac{\bar{x} - \mu_0}{s(\bar{x})}$$

$$t = \frac{92.6 - 92.5}{0.130} = 0.77$$

The tabulated value of t (from a t-table for a "double-sided" or "two-tailed" test which effectively treats the observed difference between $\bar{x}$ and μ of 0.1 as ± 0.1) for a 95 percent significance level and four degrees of freedom is 2.78. Since the calculated value of t is less than 2.78, the hypothesis that $\mu = 92.5$ is not rejected.

Looking at it another way, the calculated value of t for four degrees of freedom corresponds to a probability level of approximately 0.49 (using the same t-table). This means that rejection of the hypothesis that $\mu = \mu_0$ can be expected to be in error approximately 49 percent of the time, or conversely, there is approximately a 51 percent chance that μ is not significantly different from μ_0.

To determine if there is any difference in the measurements for two sets of data made under similar conditions, it is useful to try to pair the measurements and use the t-test on the average differences between the measurements. The hypothesis to be tested under such conditions is whether $\mu_1 - \mu_2$ is different from zero.

Example 6. Example of t-test using Hypothesis B Seven samples of catalyst are analyzed for carbon content by two technicians with the following results. Is there any difference between the two analyses?

Sample	Analysis 1	Analysis 2	Difference
1	15.1	14.6	−0.5
2	14.7	14.2	−0.5
3	15.2	15.0	−0.2
4	15.3	15.3	0.0
5	14.9	14.0	−0.9
6	14.7	14.6	−0.1
7	15.1	14.5	−0.6

SOLUTION

$$\bar{x}_{\text{difference}} = -0.4$$

$$s(x) = 0.316$$

$$s(\bar{x}) = 0.119$$

Assuming Hypothesis B

$$\mu_1 - \mu_2 = 0$$

$$\text{d.f.} = 6$$

$$t = \frac{0 - \bar{x}}{s(\bar{x})}$$

$$t = \frac{0 - (-0.4)}{0.119} = 3.36$$

The tabulated t-value ("double-sided" test) for a 95 percent significance level and six degrees of freedom is 2.45. Thus, there is greater than a 95 percent probability that the two analyses are not the same, i.e., the hypothesis may be rejected with a likelihood of less than 5 percent error. Note that this test has not indicated whether one analysis was better than the other, but rather whether or not there was any significant difference between the two.

Hypothesis C, useful for determining if differences exist between two materials, assumes that $\bar{x}_1 = \bar{x}_2$. In this case, rather than using the estimated standard deviation of the mean, $s(\bar{x})$, an overall better estimate is often obtained if a pooled estimate of the standard deviation of the mean, $\bar{s}(\bar{x})$, is used.† A handy rule of thumb for samples of 5 to 10 items is not to pool the standard deviations if the ratio of the individual standard deviations is greater than 2.2.

Example 7. Example of t-test using Hypothesis C Hydrogen gas from two sources of supply gives the following hydrogen content. Is there a significant difference between the hydrogen content of the two streams?

	Source 1	Source 2
	65.0	64.0
	64.5	69.0
	74.5	61.5
	64.0	69.0
	75.0	67.5
	74.0	
	67.0	

SOLUTION

For source 1	For source 2
$\bar{x}_1 = 69.1$	$\bar{x}_2 = 66.2$
$n = 7$	$n = 5$
$s(x_1) = 5.11$	$s(x_2) = 3.32$
$s^2(x_1) = 26.1$	$s^2(x_2) = 11.0$

Assuming Hypothesis C,

$$\mu_1 = \mu_2$$

$$\text{d.f.} = n_1 + n_2 - 2 = (7 + 5 - 2) = 10$$

$$\bar{s}(x) = \sqrt{\frac{\Sigma(\bar{x}_1 - x_1)^2 + \Sigma(\bar{x}_2 - x_2)^2}{n_1 + n_2 - 2}} = \sqrt{\frac{(n_1 - 1)s^2(x_1) + (n_2 - 1)s^2(x_2)}{n_1 + n_2 - 2}}$$

$$= \sqrt{\frac{(6)(26.1) + (4)(11.0)}{10}} = 4.47$$

$$\bar{s}(\bar{x}) = \bar{s}(x)\sqrt{\frac{1}{n_1} + \frac{1}{n_2}}$$

$$= 4.47\sqrt{\frac{1}{7} + \frac{1}{5}} = 2.62$$

$$t = \frac{\bar{x}_1 - \bar{x}_2}{\bar{s}(\bar{x})}$$

$$= \frac{(69.1 - 66.2)}{2.62} = 1.11$$

† To determine if pooling is permissible use the F-test. If it is not, use Welch's approximation. The latter is outlined by K. A. Brownlee, "Statistical Theory and Methodology in Science and Engineering," 2d ed., John Wiley and Sons, New York, 1965.

From the t-table ("double-sided" test) for ten degrees of freedom, the 1.11 corresponds to a probability level of between 0.20 and 0.30. This indicates that it would not be reasonable to reject the hypothesis. In other words, the hydrogen content of the two streams does not show a significant difference. A t-value of 2.23 or 3.17 would have been necessary to have rejected the hypothesis with a confidence level of 95 or 99 percent, respectively.

THE CHI-SQUARED TEST

The chi-squared test provides a test for determining whether the observed count (enumeration data with distinct integers which would not produce a continuous distribution curve) differs significantly from an expected count. Fundamentally, if samples are withdrawn from a distribution of known frequency, the resulting distribution frequency of the samples will be different from the parent distribution. As the samples get larger, the frequency distribution will approach more closely the frequency distribution of the total population. Chi-squared is a mathematical function which gives the probability of getting differences between the two frequencies, for varying sample sizes, when expressed in the following form:

$$\chi^2 = \sum \frac{(f - f')^2}{f'} \qquad (9)$$

where f is the observed frequency and f' is the expected frequency.

The procedure used in making the chi-squared test is quite similar to that described for the t-test. After making a hypothesis that the observed frequencies are equal to the expected frequencies, chi-squared is calculated by Eq. (9), and the result is compared at the proper level of degrees of freedom with the tabulated values. Any tabular value of chi-squared is the maximum value which could occur by chance alone, at a particular probability level. The degrees of freedom have the same meaning as for the t-test and represent the number of independent sums of squares.

There are two restrictions to the chi-squared test. One is that the test should not be used when expected frequencies are less than five. The other restriction is that when only one degree of freedom is involved, the difference $(f - f')$ is reduced by 0.5 before squaring. This is necessary to correct for a bias in the numerical answer when the chi-squared distribution, which is a continuous mathematical function, is applied to discrete numbers. Adjustment is not necessary when the degrees of freedom exceed one.

Example 8. Testing homogeneity of data The feed to a reactor is handled by four pumps which have experienced a total of 40 failures during the past year. The distribution of failures per pump is as follows:

Pump 1: 16
Pump 2: 9
Pump 3: 6
Pump 4: 9

Is the maintenance supervisor justified in saying that pump 1 has had too many failures during the past year compared to the other pumps?

SOLUTION The expected failures, assuming equal failure rates for all four pumps (null hypothesis), from the total of 40 failures divided among the four pumps would be 10

$$\chi^2 = \sum \frac{(f - f')^2}{f'}$$

$$= \frac{(16 - 10)^2 + (9 - 10)^2 + (6 - 10)^2 + (9 - 10)^2}{10} = \frac{54}{10} = 5.4$$

With four possible categories of numbers and with the total fixed, there are three degrees of freedom. This corresponds to a probability level of approximately 0.25 and indicates that, if all the pumps were operating under similar conditions, a chi-squared value as large as 5.4 would be expected one time in four by chance alone. Thus there is a 25 percent chance that the supervisor could be wrong in the analysis of the operation of pump 1.

Example 9. Effect of additional data on the homogeneity test Suppose that the pump failures in Example 8 are analyzed for the past two years and the total pump failures are found to be 80 with the same relative distribution as previously.

Pump 1:32
Pump 2:18
Pump 3:12
Pump 4:18

Are the pump failures distributed in such a manner as to indicate that all pumps are operating essentially the same?

SOLUTION The expected failures would be $\frac{80}{4} = 20$

$$\chi^2 = \frac{(32 - 20)^2 + (18 - 20)^2 + (12 - 20)^2 + (18 - 20)^2}{20} = \frac{216}{20} = 10.8$$

This chi-squared value, at three degrees of freedom, is between the 0.02 and 0.01 level and there is approximately only 1 chance in 75 of obtaining this distribution if the pumps were in fact operating the same.

Example 9 illustrates that the same relative distribution of data produces a different statistical conclusion when there are different amounts of data available. Chi-squared varies directly with the sample size when the relative distribution within the sample is unchanged. Thus, it is possible to determine the sample necessary to give a significant test.

A chi-squared test applied to data which can be arranged in various categories is usually called a *contingency test*, and χ^2 is used to determine whether the data are homogeneous within themselves.

THE *F*-TEST

The analysis of variance as performed by the *F*-test makes it possible to separate the total variance of a process into its component parts. The procedure for this test is similar to that followed in the *t*-test. Assuming a normal distribution, a null hypothesis that $\sigma_1^2 = \sigma_2^2$ is set up and *F* is calculated from the two estimated variances, $F = s_1^2/s_2^2$. If the calculated *F* exceeds the expected value, then with the indicated probability of error, the null hypothesis may be rejected.

With the *F*-test the degrees of freedom corresponding to the two variances need not be identical. Most statistics texts will tabulate *F* values for the 0.05 and 0.01 probability level. The degrees of freedom for the variance in the numerator are normally listed across the top of the table while the degrees of freedom for the variance in the denominator are given in the left-hand column. The *F* tables contain the maximum values that occur by chance alone at a given probability level. For example, if the numerator variance had 10 degrees of freedom and the denominator had 12, the *F* tables would indicate that the hypothesis of equal variances could be rejected with a 5 percent chance of being wrong for *F* values exceeding 2.76. Likewise, for the same degrees of freedom, an *F* value exceeding 4.30 would indicate that there would only be a 1 percent chance of error by rejecting the hypothesis of equal variances.

Example 10. Comparison of two variances A simplified analytical procedure is proposed for a routine laboratory test. It is necessary to determine not only whether the new procedure gives the same results as the old, i.e., whether the means of a duplicate set are the same, but also whether the precision of the new test is as good as the current test. The data for the two tests are as follows:

Current method	Revised method
79.7	79.2
79.5	79.7
79.6	79.5
79.5	79.4
79.7	80.0
	79.6
	79.8

SOLUTION For current method:

$$\bar{x}_1 = 79.6$$

$$\text{d.f.} = n - 1 = 4$$

$$s_1^2(x) = \frac{\Sigma(x - \bar{x})^2}{n - 1} = 0.01$$

For revised method:

$$\bar{x}_2 = 79.6$$

$$\text{d.f.} = n - 1 = 6$$

$$s_2^2(x) = 0.07$$

F-test hypothesis:

$$\frac{s_2^2(x)}{s_1^2(x)} = \frac{0.07}{0.01} = 7.0$$

The sample means of the two procedures are the same, but the difference in variance (or scatter) is significant at the 5 percent probability level. This is evident from the F table which indicates a value of 6.2 at the 5 percent probability level for the degrees of freedom present.

To determine whether the mean obtained in the current procedure is significantly different from the mean of the revised method, it would be necessary to proceed as in Example 7 and use the t-test. However, the t-test would not indicate which method is more precise.

The F-test provides a statistical measure of the variability for two sets of measurements. If there are more than two sets, the chi-squared test is used to determine whether the variability of the separate sets forms a homogeneous group. The relationship for chi-squared under these conditions is given as follows:

$$\chi^2 = 2.303 \frac{\log \bar{s}^2(x)\Sigma(k - 1) - \Sigma(k - 1) \log s^2(x)}{C} \tag{10}$$

where $\bar{s}^2(x)$ is the pooled variance $= \dfrac{\Sigma(k - 1)s^2(x)}{\Sigma(k - 1)}$

$$C = 1 + \frac{1}{3(n - 1)}\left[\Sigma \frac{1}{(k - 1)} - \frac{1}{\Sigma(k - 1)}\right]$$

k = number of datum points to calculate $s^2(x)$
n = number of sets of data
d.f. = $n - 1$

The use of the chi-squared test in one case and the use of the F-test in another case to test variances is not surprising when it is realized that these two tests are interrelated. To find the confidence limits for the variance of a normal population, the following definition of the chi-squared is used:

$$\chi^2 = \frac{\text{d.f. } s^2}{\sigma^2} \tag{11}$$

At infinite degrees of freedom $s^2 = \sigma^2$. Therefore χ^2 corresponds to F times the degrees of freedom for values at infinite degrees of freedom in the denominator.

The *t*-test provides a method for comparing two means while the *F*-test permits a comparison of two or more means. It can be shown that if a group of samples of size n from a total population has a sample variance $s^2(x)$, then the means of the samples have a variance

$$s^2(\bar{x}) = \frac{s^2(x)}{n}$$

Thus, to determine whether the means of several sets of measurements are similar, it is necessary to calculate the pooled variance of the sets of samples, $\bar{s}^2(x)$, and the variance of the means of the sets, $s^2(\bar{x})$. If the means have only a normal amount of variation, then $ns^2(\bar{x})/\bar{s}^2(x)$ is usually 1.0. If the means are different, then $s^2(\bar{x})$ will include not only the variance within the sets of samples, but will include the variance due to the difference of the means, and the ratio of the variances will be different from the tabulated F values. This test is the simplest form of the analysis of variance.

Example 11. Analysis of variance for three sets of data Three different reactors used at different locations but using the same process give the following yields. It is desired to determine whether the means for the three reactors are similar.

	Yields	
1	*2*	*3*
11.9	11.9	11.2
11.8	12.4	9.8
10.9	12.0	10.6
11.4	12.1	10.4

SOLUTION The analysis of variance is made in the following manner:

Source of variance	Degrees of freedom	Sum of squares
Between means	$n - 1$	$\sum \dfrac{(\text{set sums})^2}{k} - \dfrac{(\Sigma x)^2}{\Sigma k}$
Within sets of data	$\sum k - n$	$\sum x^2 - \sum \dfrac{(\text{set sums})^2}{k}$
Total	$\sum k - 1$	$\sum x^2 - \dfrac{(\Sigma x)^2}{\Sigma k}$

where n = number of sets of data
k = number of datum points in each set of data

For the three sets of data

Reactors	1	2	3
Set sums	46.0	48.4	42.0
Mean	11.5	12.1	10.5

$$\sum \frac{(\text{set sums})^2}{k} = \frac{(46.0)^2}{4} + \frac{(48.4)^2}{4} + \frac{(42.0)^2}{4} = 1555.64$$

$$\frac{(\Sigma x)^2}{\Sigma k} = \frac{(46.0 + 48.4 + 42.0)^2}{4 + 4 + 4} = 1550.41$$

$$\Sigma x^2 = (11.9)^2 + (11.8)^2 + \cdots + (10.6)^2 + (10.4)^2 = 1557.40$$

$$s^2(x) = \frac{\text{sum of squares}}{\text{d.f.}}$$

Between means

$$s^2(x) = \frac{1555.64 - 1550.41}{3 - 1} = 2.62$$

Within sets of data

$$s^2(x) = \frac{1557.40 - 1555.64}{12 - 3} = 0.186$$

$$F = \frac{s^2(x) \text{ between means}}{s^2(x) \text{ within sets of data}}$$

$$= \frac{2.62}{0.186} = 41.1$$

F tables tabulated for 0.05 and 0.01 probability levels with d.f. (numerator) of 2 and d.f. (denominator) of 9 indicate values of 4.26 and 8.02, respectively. Since the calculated value is larger than either one of the tabulated values, it may be concluded that the means of the three reactors are significantly different at both the 0.05 and 0.01 probability levels.

The analysis-of-variance test has great utility in statistical analysis because it can be applied in suitable cases to any number of classifications. For example, if experiments are run at several temperatures and in several different reactors, the yield data could be analyzed to see if the reactor mean results are similar and to determine the variance of the individual measurements. If the data are divided into groups according to the classification, the mean square of each category is the corresponding sum of squares divided by the proper degrees of freedom. An F-test applied to each classification, comparing it with the variance within sets, will establish whether the variance due to this classification exists, i.e., whether the means are significantly different. If the variance due to one classification does exist, it may be calculated from the mean squares in the following manner:

Mean squares for	*Estimates variance of*
Rows	$s^2 + $ (no. in row) (s^2 rows)
Columns	$s^2 + $ (no. in column) (s^2 columns)
Within sets of samples	s^2

In a set of experiments with two variables, the effect of the variables might be different at various levels of each other. This effect is called "interaction." The effect of this interaction is included in the analysis of variance in the "within-set" variance.

CONFIDENCE-INTERVAL ESTIMATION

Statistical analysis development involves the theoretical determination of the distributions of certain quantities—such as mean, standard deviation, and variance—which would be expected to occur by chance alone. Thus, in analyzing experimental data, statistical theory serves as a powerful tool for determining with a reasonable degree of certainty whether certain observed differences might have been due to chance.

In terms of the previously mentioned normal distribution, the probability that a randomly selected observation x from a total population of data will be within so many units of the true mean μ can be calculated. However, this leads to an integral which is difficult to evaluate. To overcome this difficulty, tables have been developed in terms of $\mu \pm Z\sigma$.[†] This means that, if the true standard deviation σ of a particular normal distribution under study is known and assuming that the difference between the sample $\bar{x}$ and the true mean μ is only the result of chance and that the individual observations are normally distributed, then a confidence interval in estimating μ can be determined. This measure was referred to previously as the confidence level.

Since, by definition,

$$\pm Z = \frac{\bar{x} - \mu}{\sigma/\sqrt{n}} \tag{12}$$

it follows by rearrangement, that

$$\mu = \bar{x} \pm \frac{Z\sigma}{\sqrt{n}} \tag{13}$$

Therefore, corresponding to a selected probability level which determines the value of Z, it can be stated that the confidence interval of μ will be given by

$$\bar{x} - \frac{Z\sigma}{\sqrt{n}} < \mu < \bar{x} + \frac{Z\sigma}{\sqrt{n}} \tag{14}$$

[†] See for example table 1-19 in J. H. Perry and C. H. Chilton, "Chemical Engineers' Handbook," 5th ed., McGraw-Hill Book Company, New York, 1973.

The *confidence interval* for μ is defined as the random interval in which μ is predicted to lie for a selected probability level. A high probability level will give a larger confidence interval for μ.

If the true standard deviation is not known, a corresponding confidence interval can still be determined. However, this estimate must utilize the *t*-distribution instead of the *Z*-distribution since the *t*-concept includes the additional variation introduced by the estimate of standard deviation. In this case the rearranged *t*-equation is used.

$$\mu = \bar{x} \pm \frac{ts(x)}{\sqrt{n}} \tag{15}$$

This indicates that the true mean lies within some range of the observed mean, the range being set by the degrees of freedom used in getting s and the probability level of t.

$$\bar{x} - \frac{ts}{\sqrt{n}} < \mu < \bar{x} + \frac{ts(x)}{\sqrt{n}} \tag{16}$$

Example 12. Confidence interval for the true mean Set the confidence interval for the true mean in Example 5 for probability levels of 95 and 99 percent.

SOLUTION

From Example 5	From t-table
$\bar{x} = 92.6$	$t_{0.95} = 2.78$
d.f. $= 4$	$t_{0.99} = 4.60$
$s(x) = 0.292$	

$$\mu = \bar{x} \pm \frac{ts(x)}{\sqrt{n}}$$

Probability level	Confidence interval
95 percent	$92.6 \pm \dfrac{(2.78)(0.292)}{\sqrt{5}} = 92.2 \text{ to } 93.0$
99 percent	$92.6 \pm \dfrac{(4.60)(0.292)}{\sqrt{5}} = 92.0 \text{ to } 93.2$

This shows that for a 95 percent confidence level it would be more correct to report the gas analysis in Example 5 as 92.6 ± 0.4 percent rather than simply as 92.6 percent.

Assuming a normal distribution, a confidence interval for σ^2 can also be obtained. In this case the chi-squared distribution is used as the basis for evaluating the chance deviation of the estimated variance from the true variance. Rearranging Eq. (11) results in a confidence interval for the true variance in terms of the degrees of freedom, the estimated variance, and the chi-squared distribution as

$$\text{d.f.} \frac{s^2}{\chi_P^2} \leqslant \sigma^2 \leqslant \text{d.f.} \frac{s^2}{\chi_{1-P}^2} \tag{17}$$

where P is the probability listed in the χ^2 tables. Note that the χ^2 table lists values for low and high probability values. In other words, 90 percent of the chi-squared distribution lies in the range 0.95 to 0.05, and 95 percent of this distribution lies in the range 0.975 and 0.025, etc. Setting the 90 percent probability range of the true variance from the estimated variance means that σ^2 has only a 5 percent chance of being smaller than $\text{d.f.}(s^2)/\chi_{P=0.05}^2$ and a 5 percent chance of being larger than $\text{d.f.}(s^2)/\chi_{P=0.95}^2$.

Often it is of interest in dealing with a normal population to determine what percentage of the observations can be expected to fall within some specified interval when the mean and standard deviation are not known. Estimates of this type are called *tolerance intervals*. Such intervals are in a form of $\bar{x} \pm Ks(x)$ similar to the previously described confidence intervals of $\bar{x} \pm ts(x)/\sqrt{n}$. The value of K (which is a function of n, α, and γ) is selected in such a manner that it can be said with a probability of 100-α (corresponding to an error of α percent) that the interval will cover at least a proportion γ of the population.[†]

REGRESSION ANALYSIS

The statistical techniques which have been discussed to this point were primarily concerned with the testing of hypotheses. A more important and useful area of statistical analysis in engineering design is the development of mathematical models to represent physical situations. This type of analysis, called *regression analysis*, is concerned with the development of a specific mathematical relationship including the mathematical model and its statistical significance and reliability. It can be shown to be closely related to the Analysis of Variance model.

REGRESSION OF TWO VARIABLES

The usual procedure when dealing with one dependent and one independent variable, e.g., reaction rate and pressure, is to plot the data and draw what appears to be the best line through the points. The " best " line in a statistical sense is the line which minimizes the sum of the squares of deviations of observed points from the line. This is similar in principle to the way the mean of a group of data minimizes the scatter of the data when measured as the sum of the squares of deviation.

[†] Values of K can be found in A. H. Bowker and G. J. Lieberman, "Engineering Statistics," 2d ed., page 314, Prentice-Hall, Inc., Englewood Cliffs, New Jersey, 1972.

If the regression, however, is to provide the maximum amount of information on precision, differences between sets of data, and interaction between variables, it will be necessary to use the statistical technique of regression analysis.

The application of regression analysis involves four steps:

1. Selection of a mathematical model to represent the data.
2. Calculation of the coefficients in the mathematical model selected.
3. Statistical test of the mathematical model to represent the physical situation.
4. Evaluation of the mathematical model to determine the direction for further improvements.

The appropriate model to be used for a specific application depends upon what is theoretically known about the process, the number of variables and data points, and the relative reliability of the data. There are no specific rules for selecting a model. Rather, only experience by the engineer will dictate the appropriate course of selection. In general, the selection of nonlinear models should be avoided unless there is a sound theoretical basis for selecting a particular type. ("Nonlinear," in this case, refers to the unknown regression coefficients.)

Assuming, for the moment, that the relationship is best represented by a straight line, the "least-squares" line is the regression giving the minimum squared deviation of the data from the line. The theoretical least-squares line can be represented mathematically as

$$y = \alpha + \beta(x - \bar{x}) + e_i \tag{18}$$

where α, β are unknown constants and e_i is the error term which is normally and independently distributed with zero mean and constant unknown variance. Note that the independent variable x is assumed to be known without error. The least-squares solution yields

$$y = a + b(x - \bar{x}) \tag{19}$$

where $a = \bar{y}$, the mean value of y

$$b = \frac{\Sigma(\bar{x} - x)(\bar{y} - y)}{\Sigma(\bar{x} - x)^2}, \text{ the slope of the least-squares line}$$

If the mathematical model indicates that the regression will go through the origin, e.g., in the case of the yield from a reactor as a function of time, the calculations can obviously be simplified over those shown in Eq. (19). The slope expression is obtained by differentiating the linear equation and setting the sum of squares of the derivative at a minimum. Terms in the slope expression may be simplified in the following manner:

$$\Sigma(\bar{x} - x)(\bar{y} - y) = \Sigma xy - \frac{\Sigma x \Sigma y}{n} \tag{20}$$

$$\Sigma(\bar{x} - x)^2 = \Sigma x^2 - \frac{(\Sigma x)^2}{n} \tag{21}$$

A least-squares line may be drawn in several simple ways once the calculation is completed. The line goes through the $\bar{x}$, $\bar{y}$ point and the y intercept. If the intercept is off the figure, the line may be drawn through the $\bar{x}$, $\bar{y}$ point with a slope of b.

Example 13. Linear regression of two variables Determine the best linear regression to represent the following data.

y	x
3.0	1
4.1	2
5.2	3
6.0	4
6.7	5

SOLUTION

$$\Sigma x = 15 \qquad \bar{x} = \frac{15}{5} = 3$$

$$\Sigma x^2 = 55$$

$$\Sigma(\bar{x} - x)^2 = \Sigma x^2 - \frac{(\Sigma x)^2}{n} = 55 - \frac{(15)^2}{5} = 10$$

$$\Sigma y = 25 \qquad \bar{y} = \frac{25}{5} = 5$$

$$\Sigma y^2 = 133.74$$

$$\Sigma(\bar{y} - y)^2 = \Sigma y^2 - \frac{(\Sigma y)^2}{n} = 133.74 - \frac{(25)^2}{5} = 8.74$$

$$\Sigma xy = 84.3$$

$$\Sigma(\bar{x} - x)(\bar{y} - y) = \Sigma xy - \frac{\Sigma x \Sigma y}{n} = 84.3 - \frac{(15)(25)}{5} = 9.3$$

$$y = a + b(x - \bar{x})$$

$$a = \bar{y} = 5$$

$$b = \frac{\Sigma(\bar{x} - x)(\bar{y} - y)}{\Sigma(\bar{x} - x)^2} = \frac{9.3}{10} = 0.93$$

$$y = 5.0 + 0.93(x - 3)$$

$$y = 5.0 + 0.93x - 2.79$$

$$y = 2.21 + 0.93x$$

The linear regression representing the above data is shown on Fig. 16-2.

Quite often the regression will not be too good and the question arises as to when a regression is not significant, i.e., when is the amount of variance removed by the regression less than could be expected by chance. In such cases, one can

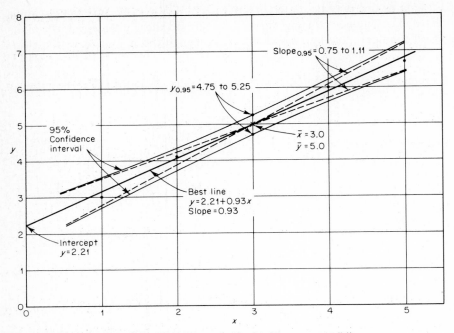

Figure 16-2 Correlation of data from Example 13 with " least-squares " line.

test the hypothesis that the regression coefficient β is zero or alternately find a confidence interval for β and see whether it includes zero.

Returning once again to an analysis of the least-squares line, it is now possible to make confidence statements concerning the slope, the intercept, and any value of y predicted by the equation. The statistical quantities required to make these statements are given in the next four equations and are the residual variance $s^2(Y)$, the variance of the slope $s^2(b)$, the variance of the mean $s^2(y)$, and the variance of a single point, $s^2(y_i)$, respectively.

$$s^2(Y) = \frac{1}{\text{d.f.}\dagger} [\Sigma(\bar{y} - y)^2 - b\Sigma(\bar{x} - x)(\bar{y} - y)] \tag{22}$$

$$s^2(b) = \frac{s^2(Y)}{\Sigma(\bar{x} - x)^2} \tag{23}$$

$$s^2(\bar{y}) = \frac{s^2(Y)}{n} \tag{24}$$

$$s^2(y_i) = s^2(Y)\left[\frac{1}{n} + \frac{(\bar{x} - x_i)^2}{\Sigma(\bar{x} - x)^2}\right] \tag{25}$$

† The degrees of freedom for a linear correlation with one dependent and one independent variable is $n - 2$.

The confidence interval on the slope can be calculated from the variance of the slope in the same manner as was used to determine the confidence range for the mean. Thus for a 95 percent confidence interval on the slope

$$b - t_{0.95} s(b) < b_{0.95} < b + t_{0.95} s(b) \tag{26}$$

Similar expressions for the confidence interval of the mean $\bar{y}$, or with any point included in the data, y_i, can be formulated.

Example 14. Analysis of linear regression What is the confidence range of the slope and the mean of the dependent variable for a 95 percent confidence level in Example 13?

SOLUTION

$$s^2(Y) = \frac{1}{\text{d.f.}} [\Sigma(\bar{y} - y)^2 - b\Sigma(\bar{x} - x)(\bar{y} - y)]$$

$$= \frac{1}{3}[8.74 - (0.93)(9.3)] = 0.03$$

$$s^2(b) = \frac{s^2(Y)}{\Sigma(\bar{x} - x)^2}$$

$$= \frac{0.03}{10} = 0.003$$

$$s^2(\bar{y}) = \frac{s^2(Y)}{n}$$

$$= \frac{0.03}{5} = 0.006$$

The 95 percent confidence interval of the slope is then

$$b_{0.95} = b \pm t_{0.95} s(b)$$

$$= 0.93 \pm (3.18)\sqrt{0.003} = 0.93 \pm 0.175$$

and the 95 percent confidence interval of the mean $\bar{y}$ is

$$\bar{y}_{0.95} = \bar{y} \pm t_{0.95} s(\bar{y})$$

$$= 5.0 \pm (3.18)\sqrt{0.006} = 5.0 \pm 0.25$$

Thus,

$$b_{0.95} = 0.75 \text{ to } 1.11$$

$$\bar{y}_{0.95} = 4.75 \text{ to } 5.25$$

This confidence limit is also shown on Fig. 16-2 to provide a comparison with the best line developed previously in Example 13.

If the calculated regression is to be used to predict what will occur if another set of data points is to be obtained, it will be necessary to add the residual variance $s^2(Y)$ to Eq. (25) to obtain the variance of the predicted point. Thus,

$$s^2(y_p) = s^2(Y)\left[\frac{1}{n} + \frac{(\bar{x} - x_i)^2}{\Sigma(\bar{x} - x)^2}\right] + s^2(Y) \tag{27}$$

The significance of the regression can also be tested by determining whether the slope is significantly different from zero by means of

$$t = \frac{b - 0}{s(b)} \tag{28}$$

in a manner similar to that used in testing the mean $\bar{x}$ difference.

The variance of the slope permits a comparison of two regression lines in the same way that two means were compared by the t-test. Thus,

$$t = \frac{b_1 - b_2}{\bar{s}(b)} \tag{29}$$

where $\bar{s}(b)$ is given by

$$\bar{s}(b) = \sqrt{\left[\frac{(\text{d.f.})_1 s^2(Y_1) + (\text{d.f.})_2 s^2(Y_2)}{(\text{d.f.})_1 + (\text{d.f.})_2}\right]\left[\frac{1}{\Sigma(\bar{x}_1 - x_1)^2} + \frac{1}{\Sigma(\bar{x}_2 - x_2)^2}\right]}$$

REGRESSION OF MULTIPLE VARIABLES

A similar regression procedure may be followed for any number of independent variables. The method is fairly involved for more than two independent variables and under such circumstances it would be best to refer to a good text in statistical analysis. However, since regressions with two independent variables occur quite frequently in chemical engineering design, a brief description of the regression procedure is given in the next few paragraphs.

The expression relating two independent variables x_1 and x_2 to a dependent variable y with a linear model may be expressed as

$$y = \alpha + \beta_1(x_1 - \bar{x}_1) + \beta_2(x_2 - \bar{x}_2) + e_i \tag{30}$$

The least-squares solution yields

$$y = a + b_1(x_1 - \bar{x}_1) + b_2(x_2 - \bar{x}_2) \tag{31}$$

where $a = \bar{y}$, the mean value of y

$$b_1 = C_{11}\Sigma'x_1 y + C_{12}\Sigma'x_2 y$$

$$C_{11} = \frac{\Sigma'x_2^2}{K}$$

$$C_{12} = -\frac{\Sigma'x_1 x_2}{K}$$

$$b_2 = C_{12}\Sigma'x_1 y + C_{22}\Sigma'x_2 y$$

$$C_{22} = \frac{\Sigma'x_1^2}{K}$$

$$K = \Sigma'x_1^2 \Sigma'x_2^2 - (\Sigma'x_1 x_2)^2$$

and the summations are further defined as

$$\Sigma'y^2 = \Sigma y^2 - \bar{y}\Sigma y$$

$$\Sigma'x_1^2 = \Sigma x_1^2 - \bar{x}_1 \Sigma x_1$$

$$\Sigma'x_2^2 = \Sigma x_2^2 - \bar{x}_2 \Sigma x_2$$

$$\Sigma'x_1 y = \Sigma x_1 y - \bar{x}_1 \Sigma y$$

$$\Sigma'x_2 y = \Sigma x_2 y - \bar{x}_2 \Sigma y$$

$$\Sigma'x_1 x_2 = \Sigma x_1 x_2 - \bar{x}_1 \Sigma x_2$$

while the residual variance is given by

$$s^2(Y) = \frac{\Sigma(y_i - \bar{y})^2 - (b_1\Sigma'x_1 y + b_2\Sigma'x_2 y)}{n - 3} \tag{32}$$

and the variance of the two coefficients b_1 and b_2 is

$$s^2(b_1) = C_{11}s^2(Y) \tag{33}$$

$$s^2(b_2) = C_{12}s^2(Y) \tag{34}$$

NONLINEAR REGRESSIONS OF VARIABLES

Regressions obviously do not have to be linear, although the linear ones are most frequently employed and definitely the most easily handled. In fact, it is often possible to modify nonlinear relationships to give linear correlations. The most common example is to transform an exponential equation into a linear logarithmic equation. If such a transformation is not possible, the regression may become quite involved either because of the stepwise solution which is necessary or the tedious iterative method which must be followed.

USE IN EVALUATION OF PROCESS DATA

The use and benefits of regression analysis can be appreciable, particularly in the evaluation of process data. In these applications, processes having as many as fifty variables, which are continuously changing over months of operation, can be evaluated by this technique. For these, the daily log records for say 400 to 500 data points are analyzed through the selected model (usually linear as a first approximation) to determine the relative effects of each variable on the response. This analysis in many cases has led to qualitative and often to quantitative determination of key operating variables whose effect had been masked on individual data point comparisons by the simultaneous changes in other less important, but unknown, variables.

STATISTICAL STRATEGY IN DESIGN OF EXPERIMENTS

In many exploratory experimental programs, the researcher is often confronted with the problem of determining the possible effect of a large number of factors. In such situations, it is necessary to establish an acceptable procedure for choosing the conditions of each experimental run. Statistical strategy in the design of experimental programs involves the systematic and controlled procedure for developing the correct combinations of variable conditions to determine a reliable analysis. Three basic types of statistically designed experiments are most often used in the chemical industry. These are:

1. Factorial design
2. Fractional factorial design
3. Box-Wilson design

Before investigating these methods briefly, it will be necessary to become familiar with the terminology which is used in this particular application of statistical analysis. (The terminology is not always very meaningful in industrial problems since the early work in this area was originally developed in agricultural experimentation.)

Experimental variables are usually called *factors*. The particular value of the variable is called the *level* of the factor. The combination of factors used in a particular experimental run is called a *treatment*. The result of the run is designated as the *effect*. If the material being processed is limited in quantity so that it may be necessary to use several batches of material that are similar but not identical in characteristics, then each batch is called a *block*. Repetitions of the same experiment at the same conditions are known as *replications*.

Applications of these terms to an experimental program can best be illustrated by considering the optimization of operating conditions for a reaction using small bench-scale reactors. If it is desired to determine the effect on the yield of varying temperature, pressure, and residence time by studying this at two temperatures, two pressures, and two contact times, the experimental program will be equivalent to one of three factors, each at two levels. The treatments will be various selected combinations of the levels of the factors. The effects are the yields of each run. If more than one reactor is used in the study and the reactors are suspect of small differences, then data obtained from each reactor may be established as a block. Choice of blocks is based on previous experience or suspected differences. The purpose of the blocks is to eliminate the effect of a variable that often cannot be measured accurately nor controlled by establishing groups that are similar in this variable.

FACTORIAL DESIGN

Factorial design is useful in making a preliminary survey of system factors (variables). It does not yield a quantitative relationship showing the effect of a factor; it merely determines whether there is any effect. The chemical engineer can therefore use factorial design as a first step to determine the influential factors and then use regression analysis on further experimental runs to develop a quantitative relationship among the important factors.

Factorial design can be applied to any number of factors and levels. Frequently, in exploratory work, only two levels of each factor are chosen for the factorial design. Two factors and two levels constitute a 2^2 factorial design and permit setting up a 2×2 analysis-of-variance table with one factor on columns and another on rows. By determining the row effect and the column effect, it is possible to determine, for example in a reactor, whether a variation in temperature or pressure affects the reaction yield. A choice of substantially different values of temperature and pressure would be desirable to ensure conclusions which would be based on a wide range of factors.

In the traditional approach to experimentation, involving a large number of factors, it is normal to vary one factor while holding all the others constant to determine the effect of that factor. The experiment is then repeated for all other factors, each time holding all others constant except the one being studied. In factorial design, rather than varying only one factor at a time, several factors are varied at once in a prearranged pattern to obtain more information with possibly fewer runs. Besides giving an estimate of the size of the error variation in the data, this method also permits evaluation of possible interactions between factors with a minimum of experimentation.

The advantages of factorial design can best be illustrated by considering a simple experiment involving the effect of three factors (A, B, C) each at two levels (1, 2) on the yield of a reaction X. In the traditional design, the effect of varying C is determined by the difference in the yields while holding factors A and B constant. The effects of A and B on the yield are obtained similarly. This requires a minimum of four experimental runs. Four replicate runs must be made to obtain an estimate of the experimental error.

In the 2^3 factorial design, a total of eight experimental runs is also made as represented by the corners of the cube in Fig. 16-3. The effect of varying C is

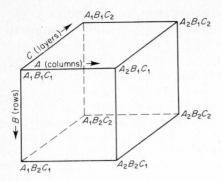

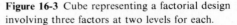

Figure 16-3 Cube representing a factorial design involving three factors at two levels for each.

determined by comparing the average of the C_1 plane to that of the C_2 plane. That is, the average of $X(A_1 B_1 C_1)$, $X(A_2 B_1 C_1)$, $X(A_1 B_2 C_1)$, and $X(A_2 B_2 C_1)$ is compared with the average of $X(A_1 B_1 C_2)$, $X(A_2 B_1 C_2)$, $X(A_1 B_2 C_2)$, and $X(A_2 B_2 C_2)$. Although A and B also vary, their variation is the same in the two sets, C_1 and C_2, so that their effects cancel out. The effects of varying A and B are determined similarly from comparing the other two planes.

One advantage of the factorial design over the traditional design is readily apparent. In the traditional design, the difference between the means of sets of two observations was used to determine a main effect; in the factorial design, the difference in the means of sets of four observations was used. The greater number of observations gives a better estimate of the true difference in the means. Note that, in both cases, the same number of experimental runs was required.

Another advantage of the factorial design is that interaction between factors can be estimated. To determine the interaction between A and B, it is necessary to average the effect of C. That is, in Fig. 16-3, the top and bottom planes are averaged. This results in

$$X(A_1 B_1 \bar{C}) \quad \text{and} \quad X(A_2 B_1 \bar{C}) \quad B_1 \text{ held constant}$$

$$X(A_1 B_2 \bar{C}) \quad \text{and} \quad X(A_2 B_2 \bar{C}) \quad B_2 \text{ held constant}$$

The effect of varying A at B_1 is obtained from a difference of the first two values

$$X(A_1 - A_2)_{B_1} = X(A_1 B_1 \bar{C}) - X(A_2 B_1 \bar{C}) \tag{35}$$

while the effect of varying A at B_2 is evaluated from the difference of the last two values

$$X(A_1 - A_2)_{B_2} = X(A_1 B_2 \bar{C}) - X(A_2 B_2 \bar{C}) \tag{36}$$

If $X(A_1 - A_2)$ at B_1 varies significantly from $X(A_1 - A_2)$ at B_2, there is an interaction between A and B. The interactions between A and C, and B and C, are evaluated in a similar manner.

The analysis of results from a 2^3 factorial experiment requires a three-way analysis of variance. That is, in addition to having row and column effects, there are also layer effects and interactions among the three. The references at the end of the chapter should be consulted for a further analysis of this statistical strategy in the design of experiments and the correct procedures for calculating the analysis of variance.

FRACTIONAL FACTORIAL DESIGN

Procedures are available for 2^n designs involving two levels of n factors where n can, in principle, be any large number. However, the required number of runs for large n may be prohibitive. For a five-factor experiment, 2^5 runs are required in a single block; but block size must be held to a minimum to control known sources of error. For this reason, fractional factorials which utilize some integer fraction (a multiple of the number of levels) of the total factorial experiments are used. The five-factor experiment at two levels would involve a total of 32 experiments in the factorial design whereas a $\frac{1}{2}$ fractional factorial design would involve only 16 experiments. Likewise a 2^{10} factorial design could be analyzed by a $\frac{1}{2}$ fractional factorial design with 32 tests. This simplification reduces the block size at the expense of the accuracy of some of the higher-order interactions.

As the number of factors increases, the number of possible interactions rapidly grows. Higher-order interactions are possible. For example, a 2^4 design requires a four-way analysis of the variance. There are four main effects, six second-order interactions (such as row × column), and four third-order interactions (such as row × column × layer). (Speaking of rows, columns, and layers in a four-way analysis of variance is not too meaningful because a fourth dimension is needed and the geometrical analogy breaks down.)

Various types of fractional factorial designs have been very effectively tabulated by the Department of Commerce.† Unfortunately, designs for mixed levels are not widely documented.

BOX-WILSON DESIGN

The Box-Wilson experimental designs are a general series of experiments that have been developed to efficiently serve as a basis for deriving the mathematical model of a physical process. Their usefulness is enhanced in the study of industrial applications because most physical situations can usually be approximated by a quadratic function over a reasonable range of the factors. For a two-factor system, the generally used form of this model is

$$y = b_0 + b_1 x_1 + b_2 x_2 + b_{11} x_1^2 + b_{12} x_1 x_2 + b_{22} x_2^2 \tag{37}$$

Similar equations can be formulated for models with more than two factors but are obviously more complex.

---

† Fractional Factorial Experiment Designs for Factors at Two(Three) Levels, U.S. Department of Commerce, *Applied Mathematics Series*, *No.* 48 (and *No.* 54), Apr. 15, 1957; National Bureau of Standards Report No. 5199, Statistical Engineering Laboratory, NBS Project 1103-40-5147.

A technique for designing which experimental tests should be carried out to evaluate the coefficients of the model is the Box-Wilson composite rotatable design. For the purpose of the experiment, the independent factors are each specified at five levels. The specific values of these five levels for each factor depend on the number of factors included in the model and the range over which they are to be studied. The design principle includes three types of combinations, the axial, factorial, and center points. Axial points include each factor at its extreme levels with the other factors at their center-point level. The center point is a single test at the average level of each factor. For purposes of estimating experimental error, the center point is usually repeated three to five times during the experiment. Designs for any number of factors can be developed from these principles.

When the experimental tests have been completed, a regression analysis is carried out to determine the coefficients in the assumed model. An analysis of variance by means of the F-test is usually carried out after the model has been developed to determine its significance.

EVOLUTIONARY OPERATION

The technique of evolutionary operation (EVOP) involves systematic small changes in process variables during the operation of the process. The results of previous small changes are used to suggest further changes so as to approach optimum operating conditions by a series of small steps. None of the individual changes which are made, however, are sufficient to upset the process to the point of producing an unacceptable product.

With two independent variables and one response, a response surface may be visualized much like a contour map of a mountain peak. The two independent variables would be represented by the two horizontal dimensions, and the vertical distance would represent the response. (For more than two variables, this geometric analogy breaks down.) Evolutionary operation assumes that an optimum exists. To determine this optimum, experimental runs are made at four points which are removed from the original operating conditions on the response surface. The optimum or highest response of these tests is then used to select four new operating points, and the process is repeated until the optimum is reached. Since the response surface must be visualized, small changes are made to explore it gradually.

The basic assumption made in this simple-minded procedure above is that a smooth response surface exists which gradually rises to a single optimum. The procedure outlined would not necessarily locate the highest of several optimums. Here is where judgment of the experiments and possible past experience must be used to place the first runs on the response surface of the highest " peak." Actually, the optimum may be on a rather flat plateau of the surface, where several sets of near-optimum conditions may be preferable to the true optimum from other considerations.

OTHER STATISTICAL METHODS

Propagation of error, sequential analysis, and quality control are additional statistical techniques with which the chemical engineer in design should be acquainted. The intent of this section will only be to outline the value of these tools and leave the details to the suggested references at the end of the chapter.

PROPAGATION OF ERROR

Cost estimation of a process design generally involves the summation of costs of individual pieces of equipment making up the process. If the precision of the cost estimate for the individual items is known, what is the precision of the total estimate?

There is no rigorous mathematical method of combining the individual precisions, but the variances of the individual measurements may be combined. The precision, at the 95 percent confidence level, is defined as approximately twice the true standard deviation. Since the latter is the square root of the true variance, this permits evaluation of the precision for the combined operation.

The variance of a sum or difference is the sum of the variances assuming independence. Thus, if

$$y_t = f(x_1)$$

the variance of the combined operation y_t is given approximately by

$$\sigma^2(y_t) = \Sigma \left(\frac{\partial f}{\partial x_i}\right)^2 \sigma^2(x_i) \tag{38}$$

The use of this relation is best illustrated in the following example.

Example 15. Propagation of error What is the precision† of a cost estimate including three items if the individual costs and precisions of these items are as follows:

Item	Cost	Precision
A_1	$1000	±$50
A_2	1000	±$100
A_3	2000	±$100

† Precision in this example refers to a confidence interval of 2σ, equivalent to approximately a 95 percent confidence level.

SOLUTION

$$C_t = A_1 + A_2 + A_3 \qquad \text{where } C_t \text{ is total cost}$$

$$\sigma^2(C_t) = \sigma^2(A_1) + \sigma^2(A_2) + \sigma^2(A_3)$$

$$C_t \pm c_t = (A_1 \pm a_1) + (A_2 \pm a_2) + (A_3 \pm a_3)$$

Assume that $a_1 = 2\sigma(A_1)$, $a_2 = 2\sigma(A_2)$, $a_3 = 2\sigma(A_3)$, and $c_t = 2\sigma(C_t)$

$$\sigma^2(C_t) = \frac{a_1^2}{4} + \frac{a_2^2}{4} + \frac{a_3^2}{4}$$

$$\sigma(C_t) = \sqrt{\frac{a_1^2}{4} + \frac{a_2^2}{4} + \frac{a_3^2}{4}}$$

$$c_t = 2\sigma(C_t) = \sqrt{a_1^2 + a_2^2 + a_3^2}$$

$$c_t = 150$$

therefore

$$C_t \pm c_t = (1000 + 1000 + 2000) \pm 150$$

$$= \$4000 \pm \$150$$

In this case, the precision of the total cost estimate is better percent-wise than that of either of the three individual items of equipment.

SEQUENTIAL ANALYSIS

In the usual statistical test, a hypothesis is made about the population from which the sample is taken. A statistical quantity is calculated from the sample and the hypothesis is either accepted or rejected on the basis of the test. When the hypothesis is rejected, it is with a known probability, established by the significance level of the test, of a false rejection. If the hypothesis is accepted, it is normally on the basis of having insufficient evidence for its rejection, and not with an established probability of a false acceptance. In some cases the probability of false acceptance of the hypothesis, the Type II error, can be controlled by the size of the sample.

In sequential testing, a hypothesis is established and limits are set for both Type I and Type II errors. The data are then accumulated sequentially, and, as each new observation is obtained, the information from all the data is tested against the hypothesis with one of three actions possible: the hypothesis may be rejected, the hypothesis may be accepted, or no decision may be made and another datum point is obtained. Data are collected until the hypothesis is either rejected or accepted.

In general, the method involves plotting some cumulative function of the data as ordinate against the number of observations as abscissa. Two parallel guide lines are drawn, depending upon the particular test involved and the probability levels selected. When the plot of the cumulated function of the data crosses either of the parallel guide lines, the indicated decision is made. As long as the plot stays within the parallel lines, no decision is indicated without a greater probability of error than originally established. The slope of the guide lines is a function of the hypothesis tested. The heights of the guide lines or the magnitude of the control values are established by both the hypothesis being tested and the probability of its false acceptance and false rejection.

QUALITY CONTROL

Statistical analysis, as applied to production or other processes in which quantities of materials are continuously being tested or measured, is known as quality control. In this statistical method, some measurable attribute of the processed material is used as a criterion of the quality of the product. Random samples are drawn from the production line in succeeding time intervals, and the means of small groups of these samples are compared with some standard. Statistical methods, particularly the t-test, provide a method of determining when the measured mean differs from the control value by an amount greater than would be expected by chance.

NOMENCLATURE FOR CHAPTER 16

$$a = \text{estimate of } \alpha \text{ in regression equation}$$
$$b = \text{estimate of } \beta \text{ in regression equation}$$
$$b_1, b_2 = \text{estimates of } \beta_1 \text{ and } \beta_2 \text{ in multiple regression equation}$$
$$C = \text{constant defined by Eq. (10)}$$
$$C_{11}, C_{12}, C_{22} = \text{constants defined by Eq. (31)}$$
$$\text{d.f.} = \text{degrees of freedom}$$
$$F = F\text{-test}$$
$$f = \text{observed frequency}$$
$$f' = \text{expected frequency}$$
$$\bar{G} = \text{geometric mean}$$
$$\bar{H} = \text{harmonic mean}$$
$$k = \text{number of datum points in Eq. (10)}$$
$$md = \text{mean deviation}$$
$$n = \text{sample size}$$
$$P = \text{probability level}$$
$$s(b) = \text{estimated standard deviation of slope}$$
$$\bar{s}(b) = \text{pooled estimate of standard deviation of slope}$$

$s^2(b)$ = estimated variance of slope
$s(x)$ = estimated standard deviation
$s(\bar{x})$ = estimated standard deviation of mean
$\bar{s}(x)$ = pooled estimate of standard deviation of sample
$\bar{s}(\bar{x})$ = pooled estimate of standard deviation of mean
$s^2(x)$ = estimated variance
$\bar{s}^2(x)$ = pooled estimate of variance of sample
$s^2(Y)$ = estimated residual variance from correlation
$s^2(\bar{y})$ = estimated variance of mean
$s^2(y_i)$ = estimated variance of a datum point
t = t-test
$\bar{x}$ = mean
$\bar{y}$ = mean of dependent variable
Z = defined by Eq. (12)

Greek symbols

α = percent of Type I error
β_1 = unknown parameter in multiple regression problem
β_2 = unknown parameter in multiple regression problem
γ = proportion of total statistical population
σ = true standard deviation
$\bar{\sigma}$ = true standard deviation of the sample mean
σ^2 = true variance
τ = time
μ = true mean
μ_0 = some fixed number
χ^2 = chi-squared test

ADDITIONAL REFERENCES FOR STATISTICAL ANALYSIS

Probability techniques

Drake, A. W., "Fundamentals of Applied Probability Theory," McGraw-Hill Book Company, New York, 1967.
Holland, F. A., F. A. Watson, and J. K. Wilkinson, Probability Techniques for Estimates of Profitability, *Chem. Eng.*, **81**(1):105 (1974).
———— ———— and ————, Numerical Measures of Risk, *Chem. Eng.*, **81**(5):119 (1974).

Reliability considerations

Abraham, R. W., Reliability of Rotating Equipment, *Chem. Eng.*, **80**(24):96 (1973).
Allen, D. H., Economic Aspects of Plant Reliability, *Chem. Eng.* (*London*), 467 (October 1973).
Anyakora, S. N., G. F. M. Engel, and F. P. Lees, Some Data on the Reliability of Instruments in the Chemical Plant Environment, *Chem. Eng.* (*London*), **255**:396 (November 1971).
Bently, D. E., and C. P. Reid, Unreliability Factors in Chemical Plants, *Chem. Eng. Progr.*, **66**(12):50 (1970).

Boyce, M. P., How to Achieve Online Availability of Centrifugal Compressors, *Chem. Eng.*, **85**(13):115 (1978).

Browning, R. L., Human Factors in the Fault Tree, *Chem. Eng. Progr.*, **72**(6):72 (1976).

Bultzo, C., Problems with Rotating and Reciprocating Equipment, *Chem. Eng. Progr.*, **68**(3):66 (1972).

Cason, R. L., Will PM Help or Hurt? *Hydrocarbon Process.*, **50**(1):95 (1971).

———, Estimate the Downtime Your Improvements Will Save, *Hydrocarbon Process.*, **51**(1):73 (1972).

Cherry, D. H., J. C. Grogan, W. A. Holmes, and F. A. Perris, Availability Analysis for Chemical Plants, *Chem. Eng. Progr.*, **74**(1):55 (1978).

Cornett, C. L., and J. L. Jones, Reliability Revisited, *Chem. Eng. Progr.*, **66**(12):29 (1970).

Coulter, K. E., and V. S. Morello, Improving Onstream Time in Process Plants, *Chem. Eng. Progr.*, **68**(3):56 (1972).

Cox, N. D., Evaluating the Profitability of Standby Components, *Chem. Eng. Progr.*, **72**(6):76 (1976).

Dailey, L. W., Plant/Equipment Reliability: Maximizing Onstream Time for Large Plants, *Chem. Eng. Progr.*, **66**(12):34 (1970).

D'Innocenzio, M., Pump Reliability Specifications, *Hydrocarbon Process.*, **51**(6):86 (1972).

Finley, H. F., Maintenance Management for Today's High Technology Plants, *Hydrocarbon Process.*, **57**(1):101 (1978).

Forman, E. R., How to Improve Online Control-Valve Performance, *Chem. Eng.*, **85**(13):128 (1978).

Garner, N. R., and J. A. Huetinck, Equipment Aging Analysis, *Chem. Eng. Progr.*, **67**(1):45 (1971).

Jackson, C., Vibration Measurement on Turbomachinery, *Chem. Eng. Progr.*, **68**(3):60 (1972).

Kirby, G. N., How to Specify Fiberglass-Reinforced-Plastics Equipment, *Chem. Eng.*, **85**(13):133 (1978).

Kuist, B. B., and H. E. Fife, Process Continuity in Polymer Plants, *Chem. Eng. Progr.*, **67**(1):33 (1971).

Lancaster, J. F., What Causes Equipment to Fail? *Hydrocarbon Process.*, **54**(1):74 (1975).

Lenz, R. E., Plant/Equipment Reliability: Reliability Design in Process Plants, *Chem. Eng. Progr.*, **66**(12):42 (1970).

Loftus, J., Reliability in Ethylene Plants, *Chem. Eng. Progr.*, **66**(12):53 (1970).

McFatter, W. E., Reliability Experiences in a Large Refinery, *Chem. Eng. Progr.*, **68**(3):52 (1972).

McIntire, J. R., Measure Refinery Reliability, *Hydrocarbon Process.*, **56**(5):121 (1977).

Morrison, W. G., Improving the Reliability of Electrical Distribution Systems, *Chem. Eng.*, **85**(21):102 (1978).

Neale, D. F., The Monitoring of Critical Equipment, *Chem. Eng. Progr.*, **70**(10):53 (1974).

Perkins, R. L., Maintenance Costs and Equipment Reliability, *Chem. Eng.*, **82**(7):126 (1975).

Richardson, J. A., and L. R. Templeton, The Impact of Plant Unreliability Upon Cost, *Chem. Eng. (London)*, 625 (December 1973).

Spence, L. A., Methods of Obtaining Stable Control, *Chem. Eng. Progr.*, **68**(3):50 (1972).

Tinney, W. S., How to Obtain Trouble-Free Performance from Centrifugal Pumps, *Chem. Eng.*, **85**(13):139 (1978).

Tucker, W., and W. E. Cline, Large Plant Reliability, *Chem. Eng. Progr.*, **67**(1):37 (1971).

Turner, B., Learn from Equipment Failure, *Hydrocarbon Process.*, **56**(11):317 (1977).

Ufford, P. S., Equipment Reliability Analysis for Large Plants, *Chem. Eng. Progr.*, **68**(3):47 (1972).

Williams, H. W., and B. H. Russell, NASA Reliability Techniques in the Chemical Industry, *Chem. Eng. Progr.*, **66**(12):45 (1970).

Wood, D. R., E. J. Muehl, and A. E. Lyon, Determining Process Plant Reliability, *Chem. Eng. Progr.*, **70**(10):62 (1974).

Statistical techniques

Anderson, V. L., and R. A. McLean, "Design of Experiments," Marcel Dekker, Inc., New York, 1974.

Belz, M. H., "Statistical Methods for the Process Industries," J. Wiley and Sons, New York, 1973.

Bethea, R. M., B. S. Duran, and T. L. Boullion, "Statistical Methods for Engineers and Scientists," Marcel Dekker, Inc., New York, 1975.

Dennee, R. F., Statistics in Cost Engineering, *AACE Bull.*, **10**(2):41 (1968).

Drake, A. W., "Fundamentals of Applied Probability Theory," McGraw-Hill Book Company, New York, 1967.

Gin, G. T., Statistics and Cost Engineering, *1970 Trans. Am. Assoc. Cost Eng.*, (1970), p. 38.

Hemphill, R. B., A Method for Predicting the Accuracy of a Construction Cost Estimate, presented at American Association of Cost Engineers 12th National Meeting, Houston, Texas, June 17–19, 1968.

Holland, F. A., F. A. Watson, and J. K. Wilkinson, Statistical Techniques Improve Decision-Making, *Chem. Eng.*, **80**(29):61 (1973).

—— —— and ——, Estimating Profitability When Uncertainties Exist, *Chem. Eng.*, **81**(3):73 (1974).

Hunter, W. G., and M. E. Hoff, Planning Experiments to Increase Research Efficiency, *I&EC*, **59**(3):43 (1967).

——, W. J. Hill, and T. L. Henson, Designing Experiments for Precise Estimation of All or Some of the Constants in a Mechanistic Model, *Can. J. Chem. Eng.*, **47**(1):76 (1969).

Isaacson, W. B., Statistical Analyses for Multivariable Systems, *Chem. Eng.*, **77**(14):69 (1970).

Jackisch, P. F., Shortcuts to Small-Sample Statistics Problems, *Chem. Eng.*, **80**(13):107 (1973).

Kortlandt, D., and R. L. Zwart, Find Optimum Timing for Sampling Product Quality, *Chem. Eng.*, **78**(25):66 (1971).

Larsen, H. R., Uses for Statistical Methods, *Hydrocarbon Process.*, **47**(2):92 (1968).

Lee, S. M., Analyzing Time-Based Data, *Chem. Eng.*, **76**(23):104 (1969).

Lipson, C., and N. J. Sheth, "Statistical Design and Analysis of Engineering Experiments," McGraw-Hill Book Company, New York, 1973.

Mezaki, R., J. R. Kittrell, and W. J. Hill, An Analysis of Kinetic Function Models, *I&EC*, **59**(1):93 (1967).

Murphy, J. S., Resolving Problems with OR, *Chem. Eng.*, **75**(3):114 (1968).

Murphy, T. D., Jr., Design and Analysis of Industrial Experiments, *Chem. Eng.*, **84**(12):168 (1977).

Pavelic, V., and U. Saxena, Basics of Statistical Experiment-Design, *Chem. Eng.*, **76**(21):175 (1969).

Powers, G. J., and S. A. Lapp, Computer-Aided Fault Tree Synthesis, *Chem. Eng. Progr.*, **72**(4):89 (1976).

Pratt, J. W., "Statistical and Mathematical Aspects of Pollution Problems," Marcel Dekker, Inc., New York, 1974.

Reilly, P. M., and G. E. Blau, The Use of Statistical Methods to Build Mathematical Models of Chemical Reacting Systems, *Can. J. Chem. Eng.*, **52**(3):289 (1974).

Sater, V. E., and F. D. Stevenson, Use of Statistical Experimental Design in a Kinetics Study, *I&EC Process Des. & Dev.*, **11**(3):355 (1972).

Shahbenderian, A. P., Plants, Plans, and Probabilities, *Chem. Proc. Eng.*, **48**(9):94 (1967).

Smith, W. N., Cost of Being Out? Inventory is Key, *Hydrocarbon Process.*, **47**(7):155 (1968).

——, Find Process Variables with Statistics, *Hydrocarbon Process.*, **47**(2):98 (1968).

Spooner, J. E., A Mathematical Model for Contingency, *1973 Trans. Am. Assoc. Cost Eng.* (1973), p. 81.

Thorngren, J. T., Probability Technique Improves Investment Analysis, *Chem. Eng.*, **74**(17):143 (1967).

Vance, F. P., Process Improved? Statistics Tell, *Hydrocarbon Process.*, **46**(7):131 (1967).

Walpole, R. E., and R. H. Meyers, "Probability and Statistics for Engineers and Scientists," 2d ed., Macmillan, New York, 1978.

Wilson, C., "Applied Statistics for Engineers," Applied Science Publishers, Ltd., London, 1972.

Zeanah, P. H., Advanced Techniques for Contingency Evaluation, *1973 Trans. Am. Assoc. Cost Eng.* (1973), p. 68.

PROBLEMS

1 Compute the mean, median, and mode for the following distribution of scores:

Scores	Frequency
110–119	1
100–109	0
90–99	2
80–89	5
70–79	10
60–69	13
50–59	9
40–49	4
30–39	5
20–29	0
10–19	1

Does this distribution fit a normal distribution curve?

2 If a normal distribution has a mean of 118 and an estimated standard deviation of 11, what percentage of the scores will be found between score limits of 100 and 118? What percentage will be expected to lie below a score of 140? What Z values correspond to scores of 100, 118, and 140?

3 If material is received from two sources of supply and separate analysis on samples from the two shipments gives the results listed below, would there be a justification (0.95 probability level) for saying there was a difference between the two materials?

Shipment 1	Shipment 2
84.0	77.5
77.0	74.0
76.0	79.0
84.5	71.5
74.5	77.0
73.0	75.5
85.0	78.5

4 If the scatter in the individual samples of the previous problem is reduced to that listed below, would there be a justification (0.95 probability level) for saying there was a difference between the two materials? (Note that mean values of the two shipments have remained unchanged.)

Shipment 1	Shipment 2
80.0	74.5
77.5	76.5
79.0	77.0
79.5	75.0
78.0	76.5
79.5	77.0
80.5	76.5

5 The analysis of a flue gas gives the following CO_2 volume content: 23.8, 23.4, 24.1, 24.6, 23.9, and 23.7. What is the confidence interval for the true mean for a probability level of 0.95 and 0.99? How much could the confidence range be improved if the number of analyses were doubled assuming the original distribution remained unchanged?

6 The analysis of an acid gas stream from a gas treating plant is known to give an average deviation of 0.5, i.e., hundreds of analyses have verified this fact. In a series of analyses the following results were obtained: 22.5, 24.7, 23.2, 21.9, 22.7, and 22.9. Should the second analysis be discarded based on a 0.95 probability level?

7 The mean difference between duplicate determinations of water in a solvent was found to be 1.0 ppm. A single determination of a sample gave 15 ppm water. What is the confidence interval for probability levels of 0.95 and 0.99?

8 Two shifts give the following C_4 content from samples taken in the overhead of a butane splitter in a refinery. Is there a significant difference (0.99 probability level) between the analyses obtained during the two shifts?

Hour	C_4 content, %	
	Shift 1	Shift 2
1	27.7	27.8
2	28.3	27.0
3	28.0	27.6
4	28.4	27.4
5	27.9	28.0
6	28.1	27.8
7	28.4	27.2
8	27.8	27.7

9 Information on the effect of temperature on a specific reaction has been obtained from four different reactors as follows:

Temperatures, °F	Yields in reactor			
	1	2	3	4
300	10.4	10.8	10.6	10.8
325	10.9	11.6	11.7	12.1
350	12.1	12.9	12.8	13.5

By a two-way analysis of variance determine whether the variance between the reactors and between the temperatures is highly significant (0.01 probability level). Even though no replicates were made at any condition, estimate the precision of a result by the within set variance.

10 The total gas stream to a reactor unit is 5000 scfh of fresh feed, 12,500 scfh of recycle, and 100 scfh for an instrument bleed stream. If the standard deviation of each measurement is 5 percent of the individual flow, what are the 95 percent precision limits of each stream, the 95 percent precision limits for total flow, and the 95 percent confidence interval for total flow?

11 The pressure drop in a process is the summation of the pressure drop in the lines, that contributed by two control valves, the pressure drop in the column, and the pressure drop in the condenser. What is the precision of the calculated total pressure drop in psia calculated at a 95 percent confidence level for the following data?

	Pressure drop	Standard deviation
Line	5.2 psia	0.05 psia
Control valves (each)	10.0 in. H_2O	0.10 in. H_2O
Column	6.8 psia	0.10 psia
Condenser	2.5 psia	0.05 psia

12 Five different batches of a commercial plastic are sampled to determine the variation of inert additives between batches. Since each batch consists of several thousand pounds of material, several samples are taken from each batch to determine the variation in the batch. With the following information, determine whether or not the variability of the separate batches forms a homogeneous test. Use a 95 percent probability level in the statistical analysis.

Batch	Number of samples	Variance in inert material, %
1	7	5.2
2	6	7.6
3	7	6.1
4	5	5.9
5	8	8.0
6	6	4.3
7	7	5.7

13 The tabulated data represent analyses of nitrogen in a crude nitrogen stream made by five different operators with three different types of analyzers. Establish if there is any significant difference (0.95 probability level) between operators and analysis equipment.

	Analyzers		
Operator	1	2	3
A	96.6	97.0	97.6
B	96.4	96.0	97.2
C	97.0	95.0	96.4
D	96.2	95.8	97.4
E	96.8	97.0	97.8

14 Find the best straight line through the following data points.

x	1.00	1.50	2.00	2.50	3.00	3.50	4.00	4.50
y	2.21	3.07	4.19	5.51	6.20	7.13	8.92	9.49

What are the 95 percent confidence limits for the slope of this straight line? Is the predicted value of y at $x = 0$ significantly different from zero at the 95 percent confidence level?

15 Determine whether a linear correlation of y with x accounts for all but the error variation in y from the following replicate values of y at different levels of x.

x	1.00	1.50	2.00	2.50	3.00	3.50	4.00	4.50
y	2.67	4.13	4.62	6.31	6.95	8.27	10.02	10.90
	2.43	3.98	5.43	6.07	7.20	8.41	10.46	10.76
	2.80	4.09	5.07	6.18	7.25	8.25	9.97	11.15

16 Experimental data for the heat capacity in cal/(g·mol)(°C) as a function of temperature are given as follows:

Temperature, K	Heat capacity
300	19.65
400	26.74
500	32.80
600	37.74
700	41.75
800	45.06
900	47.83
1000	50.16

Fit the data to an empirical heat-capacity equation of the form

$$C_p = a + b_1 T + b_2 T^2$$

by simple regression techniques.

THE INTERNATIONAL SYSTEM OF UNITS (SI)

As the International System of Units, or the so-called SI units, become accepted in the U.S., there will be a long transition period when both the U.S. customary system and SI will be in use simultaneously. The design engineer, in particular, will need to be able to think and work in both systems because of the wide variety of persons involved in design considerations. Accordingly, this text has used a mixture of the two systems.

The purpose of this Appendix is to provide a description of SI units along with rules for conversion and rules for usage in the written form. Conversion factors are given with one table presenting a full and detailed list with extensive footnote explanations and another table giving conversion factors in a simplified form for units commonly encountered by chemical engineers.

The name SI is derived from Système International d'Unités and has evolved from an original basis of a given length (meter) and mass (kilogram) established by members of the Paris Academy of Science in the late eighteenth century. The original system was known as the metric system, but there are differences in the modern SI system and the old metric system based primarily on new names being added for derived terms.

The current International System of Units (SI) is a metric system of measurement which has been adopted internationally by the General Conference of Weights and Measures and is described in an International Standard

(ISO 1000)† and in numerous other publications.‡ Usage differences among countries have been resolved by a series of international conferences resulting in a set of seven base units, two supplementary units, and derived units as given in the following:

SI BASE UNITS ON WHICH THE ENTIRE SYSTEM IS FOUNDED

Name, Symbol	Definition
meter,§ m—(length)	The meter is the length equal to 1 650 763.73 wavelengths in vacuum of the radiation corresponding to the transition between the levels $2p_{10}$ and $5d_5$ of the krypton-86 atom.
kilogram, kg—(mass)	The kilogram is a unit of mass (not force). A prototype of the kilogram made of platinum-iridium is kept at the International Bureau of Weights and Measures, Sèvres, France. (The kilogram is the only base unit having a prefix and defined by an artifact.)
second, s—(time)	The second is the duration of 9 192 631 770 periods of the radiation corresponding to the transition between the two hyperfine levels of the ground state of the cesium-133 atom.
ampere, A—(electric current)	The ampere is that constant current which, if maintained in two straight parallel conductors of infinite length and of negligible circular cross section, and placed 1 meter apart in a vacuum, would produce between these conductors a force equal to 2×10^{-7} m·kg/s² (newton) per meter of length.

---

† International Standard, "SI Units and Recommendations for the Use of Their Multiples and of Certain Other Units," ISO 1000-1973(E), American National Standards Institute, 1430 Broadway, New York City, NY 10018.

‡ The basic English document for SI is the National Bureau of Standards Special Publication 330 which can be obtained from the Superintendent of Documents, U.S. Government Printing Office, Washington, D.C. 20402 as document SD Catalog No. C13.10:330/3. This is the authorized English translation of the official document of the international body. For guidance in U.S. usage, the most widely recognized document in use is the ASTM Standard for Metric Practice E 380 available from the American Society for Testing and Materials, 1916 Race St., Philadelphia, PA. 19103.

§ The spelling of *metre* (and *litre*) is commonly accepted internationally and is recommended by the ASTM. However, the spelling as *meter* (and *liter*) is widely used in the United States and is the spelling used in this book.

SI BASE UNITS ON WHICH THE ENTIRE SYSTEM IS FOUNDED (Continued)

Name, Symbol	Definition
kelvin, K—(temperature)	The kelvin is the fraction 1/273.16 of the thermodynamic temperature of the triple point of water.
	The kelvin is a unit of thermodynamic temperature (T). The word (or symbol) "degree" is not used with kelvin.
	The Celsius (formerly centigrade) temperature is also used. Celsius temperature (symbol t) is defined by the equation $t = T - T_0$, where T_0 equals 273.15 K. A degree Celsius (°C) is thus equal to one kelvin.
	The term centigrade should not be used because of possible confusion with the French unit of angular measurement, the grade.
mole, mol—(amount of substance)	The mole is the amount of substance of a system which contains as many elementary entities as there are atoms in 0.012 kilograms of carbon 12.
	When the mole is used, the elementary entities must be specified and may be atoms, molecules, ions, electrons, other particles, or specified groups of such particles.
candela, cd—(luminous intensity)	The candela is the luminous intensity, in the perpendicular direction, of a surface of 1/600 000 square meter of a blackbody at the temperature of freezing platinum (2045 K) under a pressure of $101\,325\ \mathrm{m^{-1} \cdot kg \cdot s^{-2}}$.

SUPPLEMENTARY UNITS

Name, Symbol	Definition
radian, rad—(plane angle)	The radian is the plane angle between two radii of a circle which cuts off, on the circumference, an arc equal in length to the radius.

(Continued)

Name, Symbol	**Definition**
steradian, sr—(solid angle)	The steradian is the solid angle which, having its vertex in the center of a sphere, cuts off an area of the surface of the sphere equal to that of a square with sides of length equal to the radius of the sphere.

DERIVED UNITS

Derived units are algebraic combinations of the seven base units or two supplementary units with some of the combinations being assigned special names and symbols. Examples are shown in Table 1.

For the chemical engineer, the seven base units and two supplementary units are no problem because they have been used regularly in technical work of a chemical nature. However, the SI units for some of the derived terms, such as for pressure, are not familiar or in common usage in the U.S. customary system of

Table 1 Common derived units with special names and symbols acceptable in SI

Name	Symbol	Quantity	Expression in terms of SI base units	Expression in terms of other units
becquerel	Bq	radioactivity	s^{-1}	
coulomb	C	quantity of electricity or electric charge	$A \cdot s$	
farad	F	electric capacitance	$m^{-2} \cdot kg^{-1} \cdot s^4 \cdot A^2$	C/V
gray	Gy	absorbed radiation	$m^2 \cdot s^{-2}$	J/kg
henry	H	electric inductance	$m^2 \cdot kg \cdot s^{-2} \cdot A^{-2}$	Wb/A
hertz	Hz	frequency	s^{-1}	
joule	J	energy, work, or quantity of heat	$m^2 \cdot kg \cdot s^{-2}$	N·m
lumen	lm	luminous flux	$cd \cdot sr$	
lux	lx	illuminance	$m^{-2} \cdot cd \cdot sr$	lm/m²
newton	N	force	$m \cdot kg \cdot s^{-2}$	$J \cdot m^{-1}$
ohm	Ω	electric resistance	$m^2 \cdot kg \cdot s^{-3} \cdot A^{-2}$	V/A
pascal	Pa	pressure or stress	$m^{-1} \cdot kg \cdot s^{-2}$	N/m²
siemens	S	electric conductance	$m^{-2} \cdot kg^{-1} \cdot s^3 \cdot A^2$	A/V
tesla	T	magnetic flux density	$kg \cdot s^{-2} \cdot A^{-1}$	Wb/m²
volt	V	electric potential, potential difference, or electromotive force	$m^2 \cdot kg \cdot s^{-3} \cdot A^{-1}$	W/A
watt	W	power or radiant flux	$m^2 \cdot kg \cdot s^{-3}$	J/s
weber	Wb	magnetic flux	$m^2 \cdot kg \cdot s^{-2} \cdot A^{-1}$	V·s

units. The SI pressure unit is the pascal (Pa) (rhymes with *rascal*) which is a newton (N) per square meter, or $N \cdot m^{-2}$. Since a newton is an SI derived unit of force as mass (kg) times acceleration (m/s^2), the net expression of the pascal in terms of SI base units is

$$Pa = N \cdot m^{-2} = kg \cdot m \cdot s^{-2} \cdot m^{-2} = m^{-1} \cdot kg \cdot s^{-2}.$$

Chemical engineers have commonly used atmospheres as a unit for pressure. Although the unit of atmosphere (1 atm = 101.325 kPa) was internationally authorized as an SI derived unit, this authorization was granted for a limited time only and its use should be minimized.

Another common set of units used by chemical engineers is the calorie (or British thermal unit) for energy. The units of calorie [1 cal = 4.1868 J where J is the symbol for joule (rhymes with *pool*) which is a newton-meter with base units of $m^2 \cdot kg \cdot s^{-2}$] and British thermal unit (1 Btu = $1.055\,056 \times 10^3$ J) are not acceptable with SI units.

In the SI system, the kilogram is restricted to the unit of mass so that it is not acceptable to use a unit of force as kilogram-force which would be analogous to the U.S. customary unit of pound-force. The newton is the unit of force in the SI system and should be used in place of kilogram-force. Confusion can occur because the term *weight* is used to mean either *force* or *mass*. In common everyday use, the term weight normally means mass, but, in physics, weight usually means the force exerted by gravity. Because of the ambiguity involved in the dual use of the term *weight*, the term should be avoided in technical practice unless the conditions are such that the meaning is totally clear.

Table 1 lists common derived SI units with special names. The table also gives the approved SI symbol and the expression for the term in base units and in terms of other units. Table 2 gives examples of other derived units which are commonly used in chemical engineering including a description and SI units. Table 3 shows units which are not officially recognized as usable with SI but which are authorized for use to a certain extent, while Table 4 gives units which are not acceptable for use with SI.

ADVANTAGES AND GUIDELINES FOR THE SI SYSTEM

An advantage of the SI system is its total coherence in that all of the units are related by unity. Thus, as can be seen from Table 1, a force of one newton exerted over a length of one meter gives an energy of one joule, while one joule occurring over a time period of one second results in a power of one watt. Mass is always measured in kilograms and force in newtons when dealing with the SI system so that the confusion often found in the U.S. customary system of using both pounds force and pounds mass is eliminated.

A fundamental characteristic of the SI system is the fact that each defined quantity has only one unit. Thus, the fundamental SI unit of energy is the joule and the fundamental SI unit of power is the watt. While a joule is defined as a newton meter, it refers to a unit force moving through a unit distance. The expression "newton meter" is used in the SI system to refer to torque in which

Table 2 Other derived units commonly used in chemical engineering with description in terms of acceptable SI units

Quantity	Description	Symbol	Expression in terms of SI base units
acceleration	meter per second squared	m/s^2	$m \cdot s^{-2}$
area	square meter	m^2	m^2
coefficient of heat transfer (U.S. symbol of h or U)	watt per square meter kelvin	$W/(m^2 \cdot K)$ $J/(m^2 \cdot K \cdot s)$	$kg \cdot s^{-3} \cdot K^{-1}$
concentration (of amount of substance)	mole per cubic meter	mol/m^3	$mol \cdot m^{-3}$
current density	ampere per square meter	A/m^2	$A \cdot m^{-2}$
density (mass density) (U.S. symbol of ρ)	kilogram per cubic meter	kg/m^3	$kg \cdot m^{-3}$
electric charge density	coulomb per cubic meter	C/m^3	$m^{-3} \cdot s \cdot A$
electric field strength	volt per meter	V/m	$m \cdot kg \cdot s^{-3} \cdot A^{-1}$
electric flux density	coulomb per square meter	C/m^2	$m^{-2} \cdot s \cdot A$
energy density	joule per cubic meter	J/m^3	$m^{-1} \cdot kg \cdot s^{-2}$
force	newton	N or J/m	$m \cdot kg \cdot s^{-2}$
heat capacity or entropy	joule per kelvin	J/K	$m^2 \cdot kg \cdot s^{-2} \cdot K^{-1}$
heat flow rate (U.S. symbol of Q or q)	watt	W or J/s	$m^2 \cdot kg \cdot s^{-3}$
heat flux density or irradiance	watt per square meter	W/m^2	$kg \cdot s^{-3}$
luminance	candela per square meter	cd/m^2	$cd \cdot m^{-2}$
magnetic field strength	ampere per meter	A/m	$A \cdot m^{-1}$
modulus of elasticity or Young's modulus	gigapascal	GPa	$10^{-9} \cdot m^{-1} \cdot kg \cdot s^{-1}$
molar energy	joule per mole	J/mol	$m^{-2} \cdot kg \cdot s^{-2} \cdot mol^{-1}$
molar entropy or molar heat capacity	joule per mole kelvin	$J/(mol \cdot K)$	$m^2 \cdot kg \cdot s^{-2} \cdot K^{-1} \cdot mol^{-1}$
moment of force or torque	newton meter	$N \cdot m$	$m^2 \cdot kg \cdot s^{-2}$
moment of inertia	kilogram meter squared	$kg \cdot m^2$	$kg \cdot m^2$
momentum	kilogram meter per second	$kg \cdot m/s$	$kg \cdot m \cdot s^{-1}$
permeability	henry per meter	H/m	$m \cdot kg \cdot s^{-2} \cdot A^{-2}$
permittivity	farad per meter	F/m	$m^{-3} \cdot kg^{-1} \cdot s^4 \cdot A^2$
power	kilowatt	kW	$10^{-3} \cdot m^2 \cdot kg \cdot s^{-3}$
pressure (U.S. symbol of P or p)	kilopascal	kPa	$10^{-3} \cdot m^{-1} \cdot kg \cdot s^{-2}$
specific energy	joule per kilogram	J/kg	$m^2 \cdot s^{-2}$
specific heat capacity or specific entropy (U.S. symbol of c_p, c_v, or s)	joule per kilogram kelvin	$J/(kg \cdot K)$	$m^2 \cdot s^{-2} \cdot K^{-1}$
specific volume	cubic meter per kilogram	m^3/kg	$m^3 \cdot kg^{-1}$
stress	megapascal	MPa	$10^{-6} \cdot m^{-1} \cdot kg \cdot s^{-2}$
surface tension	newton per meter	N/m	$kg \cdot s^{-2}$
thermal conductivity (U.S. symbol of k)	watt per meter kelvin	$W/(m \cdot K)$	$m \cdot kg \cdot s^{-3} \cdot K^{-1}$
torque	newton meter	$N \cdot m$	$m^2 \cdot kg \cdot s^{-2}$
velocity or speed	meter per second	m/s	$m \cdot s^{-1}$
viscosity—absolute or dynamic (U.S. symbol of μ)	pascal second	$Pa \cdot s$	$m^{-1} \cdot kg \cdot s^{-1}$
viscosity—kinematic (U.S. symbol of ν)	square meter per second	m^2/s	$m^2 \cdot s^{-1}$
volume	cubic meter	m^3	m^3
wave number	1 per meter	$1/m$	m^{-1}
work energy (U.S. symbol of W in foot pounds force)	Joule	J or $N \cdot m$	$m^2 \cdot kg \cdot s^{-2}$

Table 3 Non-SI units which are acceptable for use

The following common units have been authorized for use with SI to a certain extent and continue in use on an unofficially accepted basis.

Name	Symbol	Value in SI units
time−minute, hour, day, year	min, h, d, yr	60s, 3 600s, 86 400s, ≈ 365d
angle−degree, minute, second	°, ′, ″	$(\pi/180)$ rad, $(1/60)°$, $(1/60)′$
liter†	l (or L)	$1\ dm^3$
nautical mile	nautical mile	1 852 m
knot (1 nautical mile per hour)	knot	0.513 9 m/s
hectare	ha	$10^4\ m^2$
ångström	Å	$0.1\ nm = 10^{-10}\ m$
are	a	$10^2\ m^2$
atmosphere pressure	atm	101.325 kPa
bar pressure	bar	$10^5\ Pa$
galileo or gal	Gal	$10^{-2}\ m/s^2$
metric ton	t	$10^3\ kg$

† The SI unit of volume is the cubic meter, and this unit or one of its regular multiples is preferred for all cases. However, the special name *liter* has been approved for the cubic decimeter, but the use of this unit is restricted to the measurements of liquids and gases. No prefix other than *milli* should be used with liter.

Table 4 Common units which are not acceptable with SI

Despite the fact that the following units have been used commonly in the past, they are not acceptable with SI

Name	Symbol	Value in SI units
British thermal unit	Btu	$1.055\ 056 \times 10^3\ J$
calorie	cal	4.186 8 J
dyne	dyn	$10^{-5}\ N$
erg	erg	$10^{-7}\ J$
fermi	Fm	$10^{-15}\ m$
gamma	γ	$10^{-9}\ T$
gauss	Gs, G	$10^{-4}\ T$
kilogram-force	kgf	9.806 65 N
lambda	λ	10^{-6} liter
maxwell	Mx	$10^{-8}\ Wb$
metric carat		200 mg
micron	μ	1 micrometer
oersted	Oe	$(1000/4\pi)A \cdot m^{-1}$
phot	ph	$10^4\ 1x$
poise	P	0.1 Pa·s
stere	st	$1\ m^3$
stilb	sb	$1\ cd/cm^2$
stokes	St	$1\ cm^2/s$
torr		101 325/760 Pa
X unit		1.002×10^{-4} nm (approximately)

there is no indication of motion or movement. Thus, the SI system is very explicit that joule and newton meter are different units.

The SI system has a series of approved prefixes and symbols for decimal multiples as shown in Table 5.

The common usage of "psi" and "atmosphere" for units of pressure will be replaced by the pascal in the SI system. Because a pascal, as a force of one newton against an area of one square meter, is a very small unit, it is convenient to deal with kilopascals (kPa) rather than pascals in many cases.† The following conversion factors are useful for making the transition from the U.S. customary system to the SI system for pressure designations:

To convert to kPa from	multiply by
psi	6.895
atmosphere	101.325
torr	0.133 3
bar	100.000

RULES FOR USE OF SI UNITS

1. **Periods.** A period is never used after a symbol of an SI unit unless it is used to designate the end of a sentence.

2. **Capitalization.** Capitals are not used to start units that are written out except at the beginning of a sentence. However, when the units are expressed as symbols, the first letter of the symbol is capitalized when the name of the unit was derived from the name of a person. For example, it is correct to write

5 pascals or 5 Pa
5 newtons or 5 N
5 meters or 5 m
300 kelvins or 300 K

But note that the following temperature forms are correct:

200 degrees Celsius or 300°C
100 degrees Fahrenheit or 100°F

In the SI system, it is very important to follow the precise, agreed-upon use of uppercase and lowercase letters. This importance is shown by the following examples taken from Tables 3 and 5 and base-unit definitions:

G for giga; g for gram
K for kelvin; k for kilo
M for mega; m for milli
N for newton; n for nano
T for tera; t for metric ton

---

† To give an idea as to the approximate magnitude of a pressure of one pascal, it would be equivalent to the extra pressure exerted on the palm of an open hand when a person blows a sharp breath on the hand.

Table 5 SI unit prefixes

Multiplication factor	Prefix	Symbol	Pronunciation (USA) (1)	Meaning (in USA)	Meaning (in other countries)
$1\,000\,000\,000\,000\,000\,000 = 10^{18}$	exa (2)	E	ex'a (a as in about)	One quintillion times (3)	trillion
$1\,000\,000\,000\,000\,000 = 10^{15}$	peta (2)	P	as in petal	One quadrillion times (3)	thousand billion
$1\,000\,000\,000\,000 = 10^{12}$	tera	T	as in terrace	One trillion times (3)	billion
$1\,000\,000\,000 = 10^{9}$	giga	G	jig'a (a as in about)	One billion times (3)	milliard
$1\,000\,000 = 10^{6}$	mega	M	as in megaphone	One million times	
$1\,000 = 10^{3}$	kilo	k	as in kilowatt	One thousand times	
$100 = 10^{2}$	hecto	h (4)	heck'toe	One hundred times	
$10 = 10$	deka	da (4)	deck'a (a as in about)	Ten times	
$0.1 = 10^{-1}$	deci	d (4)	as in decimal	One tenth of	
$0.01 = 10^{-2}$	centi	c (4)	as in sentiment	One hundredth of	
$0.001 = 10^{-3}$	milli	m	as in military	One thousandth of	
$0.000\,001 = 10^{-6}$	micro	μ (5)	as in microphone	One millionth of	
$0.000\,000\,001 = 10^{-9}$	nano	n	nan'oh (an as in ant)	One billionth of (3)	milliardth
$0.000\,000\,000\,001 = 10^{-12}$	pico	p	peek'oh	One trillionth of (3)	billionth
$0.000\,000\,000\,000\,001 = 10^{-15}$	femto	f	fem'toe (fem as in feminine)	One quadrillionth of (3)	thousand billionth
$0.000\,000\,000\,000\,000\,001 = 10^{-18}$	atto	a	as in anatomy	One quintillionth of (3)	trillionth

1. The first syllable of every prefix is accented to assure that the prefix will retain its identity. Therefore, the preferred pronunciation of kilometer places the accent on the first syllable, not the second.
2. Approved by the 15th General Conference of Weights and Measures (CGPM), May–June 1975.
3. These terms should be avoided in technical writing because the denominations above one million are different in most other countries, as indicated in the last column.
4. While hecto, deka, deci, and centi are SI prefixes, their use should generally be avoided except for the SI unit-multiples for area and volume and non-technical use of centimeter, as for body and clothing measurement. The prefix hecto should be avoided also because the longhand symbol h may be confused with k.
5. Although SI rules prescribe vertical (roman) type, the sloping (*italics*) form is usually acceptable in the USA for the Greek letter μ because of the scarcity of the upright style.

3. **Plurals.** As indicated in some of the preceding examples, the plural is used in the normal grammatical sense when the units are written out as words, but plurals are never used with the unit symbols. For numerical values greater than 1, equal to 0, or less than −1, the names of units are plural. All other values take the singular form for the unit names. For example, the following forms are correct:

200	kilograms	or	200 kg
1.05	meters	or	1.05 m
0	degrees Celsius	or	0°C
−2	degrees Celsius	or	−2°C
3	kelvins	or	3 K
0.9	meter	or	0.9 m
−0.5	degree Celsius	or	−0.5°C
1	kelvin	or	1 K
−1	degree Celsius	or	−1°C

An "s" is added to form the plurals of unit names as illustrated in the preceding except that hertz, lux, and siemens remain unchanged and henry becomes henries.

4. **Groupings of numbers and decimal points.** The common U.S. practice of using commas to separate multiples of 1000 is not followed with SI which uses a space instead of a comma to separate the multiples of 1000. For decimals, the space is filled on both sides of the decimal point. The decimal point is placed on the line as a regular period for U.S. usage rather than at mid-line height or use of a comma as is frequent European practice. When writing numbers with values less than one, a zero should be placed ahead of the decimal.

Numbers with many digits should be set off in groups of three digits away from the decimal point on both the left and the right. For example, the following forms are correct:

$$57\,321\,684.521\,69$$
$$0.431\,684\,2$$

If there are only four digits to the left or right of the decimal point, the use of the space is optional unless there is a column of figures which is aligned on the decimal point with one or more numbers having more than four digits to the left or right of the decimal point. Thus, the following forms are correct:

3200	or	3 200
0.6854	or	0.685 4

$$\begin{array}{r} 13.6 \\ +\,15\,057 \\ +\underline{3\,200} \\ 18\,270.6 \end{array}$$

5. **Spacing, hyphens, and italics.** When a unit symbol is given after a number, a space is always left between the number and the symbol with the exception of cases where the symbol appears in the superscript position, such as degree, minute, and second of plane angles. The symbol for degree Celsius may be written either with or without a space before the degree symbol. For example,

68 kHz

60 mm

10^6 N

plane angle of 20° 45′ 26″

20°C   or   20 °C   (20°C is preferred and 20° C is *not* acceptable)

For both symbols and names of units having prefixes, no space is left between letters making up the symbol or name. For example,

kA, kiloampere; mg, milligram

The symbols when printed are always given as roman (vertical) type. Sloping letters (or *italics*) are reserved for quantity symbols such as m for mass, l for length, or general algebraic quantities such as a, b, or c. When the algebraic quantity is used, there is no space used between the algebraic quantity and the numerical coefficient. For example,

5 m means a distance of 5 meters,

but $5m$ means 5 times the algebraic quantity m

When a quantity is used in an adjectival sense, a hyphen should be used between the number and the symbol except for symbols appearing in the superscript position. For example,

He bought a 35-mm film; but,
     the width of the film is 35 mm.
He bought a 5-kg ham; but,
     the mass of the ham is 5 kg.
However, it is correct to write—
He bought a 100°C thermometer which covers a temperature range of 100°C.

A space should be left on each side of signs for multiplication, division, addition, and subtraction except within a compound symbol. The product dot (as in N · m) is used for the derived unit symbol with no space on either side. The product dot should not be used as a multiplier symbol for calculations. For example,

Write 6 m x 8 m (not 6 mx8 m or 6 m · 8 m)

kg/m^3 or $kg \cdot m^{-3}$

$m^2 \cdot kg \cdot s^{-2}$

6. **Prefixes.** In general, it is desirable to keep numerical values between 0.1 and 1000 by the use of appropriate prefixes shown prior to the unit symbol. Prefixes and symbols along with pronunciations and meanings as acceptable in

the SI system are given in Table 5. Some typical examples are

$$5\,527 \text{ Pa} = 5.527 \times 10^3 \text{ Pa} = 5.527 \text{ kPa}$$
$$0.051 \text{ m} = 51 \times 10^{-3} \text{ m} = 51 \text{ mm}$$
$$0.235 \times 10^{-6} \text{ s} = 0.235 \text{ } \mu\text{s} = 235 \times 10^{-3} \text{ } \mu\text{s} = 235 \text{ ns}$$

Two or more SI prefixes should not be used simultaneously for the designation of a unit. For example,

$$\text{write 1 pF instead of 1 } \mu\mu\text{F}$$

For cases that fall outside the range covered by single prefixes, the situation should be handled by expressing the value with powers of ten as applied to the base unit.

With reference to the spelling with prefixes, there are three cases where the final vowel in a prefix is omitted. These are megohm, kilohm, and hectare. In all other cases, both vowels are retained and both are pronounced. No space or hyphen should be used.

7. **Combination of units.** It is desirable to avoid the use of prefixes in the denominator of compound units with the one exception of the base unit kg. For example,

$$\text{use kN/m instead of N/mm}$$
$$\text{use kg/s instead of g/ms}$$

The single exception is to use J/kg instead of mJ/g.

Use a solidus (/) to indicate a division factor. Avoid the use of a double solidus. For example,

$$\text{write J/(s·m)}^2 \quad \text{or} \quad \text{J·s}^{-2}\text{·m}^{-2} \text{ instead of J/s}^2/\text{m}^2$$

When the denominator of a unit expression is a product, it should normally be shown in parentheses. For example,

$$W/(\text{m}^2 \cdot \text{K})$$

If an expression is given for units raised to a power, such as square millimeters, the power number refers to the entire unit and not just to the last symbol. For example,

$$\text{mm}^2 \text{ means (mm)}^2 \text{ instead of milli(square meters) or m(m}^2)$$

Symbols and unit names should not be used together in the same expression. For example,

$$\text{write joules per kilogram or J/kg instead of}$$
$$\text{joules/kilogram or joules/kg or joules·kg}^{-1}$$

8. **Guidelines for calculations.** It is generally desirable to carry out calculations in base units and then convert the final answers to appropriate-size numbers by use of correct prefixes.

9. **Confusion of meaning of billion in U.S.A. and other countries.** In the

United States, *billion* means a thousand million (prefix *giga*), but, in most other countries, it means a million million (prefix *tera*). Because of possible confusion as to the meaning, the term *billion* should be avoided in technical writing. As shown in Table 5, the same possible confusion exists with *quintillion* (prefix *exa* in U.S.A.), *quadrillion* (prefix *peta* in U.S.A.) and *trillion* (prefix *tera* in U.S.A.).

10. **Round-offs in conversions.** In making a conversion of a number to new units, the number of significant digits should not be increased or decreased. It is, therefore, necessary to use sufficient precision in the conversion factor to preserve the precision of the quantity converted.

11. **Conversions between SI and U.S. customary units.** SI and U.S. customary units can be presented with the U.S. customary units first followed by SI units in parentheses, as 2.45 in (62.2 mm) or as the preferred SI units first with the U.S. customary units in parentheses, such as 170 kPa (24.7 psi).

Table 6 presents a detailed list of conversion factors that can be used to convert between U.S.-British units and SI units, while Table 7 gives a simplified and abbreviated list of equivalences for converting unacceptable units commonly used by chemical engineers into acceptable SI units.

Table 6 Conversion factors for converting from U.S. customary units to SI units—alphabetical listing in detail†

To convert from	To	Multiply by
abampere	ampere (A)	1.000 000*E+01
abcoulomb	coulomb (C)	1.000 000*E+01
abfarad	farad (F)	1.000 000*E+09
abhenry	henry (H)	1.000 000*E−09
abmho	siemens (S)	1.000 000*E+09
abohm	ohm (Ω)	1.000 000*E−09
abvolt	volt (V)	1.000 000*E−08
acre foot (U.S. survey)‡	meter^3 (m^3)	1.233 489 E+03
acre (U.S. survey)‡	meter^2 (m^2)	4.046 873 E+03
ampere hour	coulomb (C)	3.600 000*E+03
are	meter^2 (m^2)	1.000 000*E+02
ångstrom	meter (m)	1.000 000*E−10
astronomical unit	meter (m)	1.495 979 E+11
atmosphere (standard)	pascal (Pa)	1.013 250*E+05
atmosphere (technical = 1kgf/cm^2)	pascal (Pa)	9.806 650*E+04
bar	pascal (Pa)	1.000 000*E+05
barn	meter^2 (m^2)	1.000 000*E−28
barrel (for petroleum, 42 gal)	meter^3 (m^3)	1.589 873 E−01
board foot	meter^3 (m^3)	2.359 737 E−03
British thermal unit (International Table)§	joule (J)	1.055 056 E+03
British thermal unit (mean)§	joule (J)	1.055 87 E+03
British thermal unit (thermochemical)§	joule (J)	1.054 350 E+03

† See end of Table   ‡ See end of Table   § See end of Table   **(*Continued*)**

Table 6 (*Continued*)

To convert from	To	Multiply by
British thermal unit (39°F)	joule (J)	1.059 67 E+03
British thermal unit (59°F)	joule (J)	1.054 80 E+03
British thermal unit (60°F)	joule (J)	1.054 68 E+03
Btu (International Table)·ft/h·ft²·°F		
(k, thermal conductivity)	watt per meter kelvin (W/m·K)	1.730 735 E+00
Btu (thermochemical)·ft/h·ft²·°F		
(k, thermal conductivity	watt per meter kelvin (W/m·K)	1.729 577 E+00
Btu (International Table)· in./h·ft²·°F		
(k, thermal conductivity)	watt per meter kelvin (W/m·K)	1.442 279 E−01
Btu (thermochemical)· in./h·ft²·°F		
(k, thermal conductivity)	watt per meter kelvin (W/m·K)	1.441 314 E−01
Btu (International Table)·in./s·ft²·°F		
(k, thermal conductivity)	watt per meter kelvin (W/m·K)	5.192 204 E+02
Btu (thermochemical)· in./s·ft²·°F		
(k, thermal conductivity)	watt per meter kelvin (W/m·K)	5.188 732 E+02
Btu (International Table)/h	watt (W)	2.930 711 E−01
Btu (thermochemical)/h	watt (W)	2.928 751 E−01
Btu (thermochemical)/min	watt (W)	1.757 250 E+01
Btu (thermochemical)/s	watt (W)	1.054 350 E+03
Btu (International Table)/ft²	joule per meter² (J/m²)	1.135 653 E+04
Btu (thermochemical)/ft²	joule per meter² (J/m²)	1.134 893 E+04
Btu (International Table)/ft²·h	watt per meter² (W/m²)	3.154 591 E+00
Btu (thermochemical)/ft²·h	watt per meter² (W/m²)	3.152 481 E+00
Btu (thermochemical/ft²·min	watt per meter² (W/m²)	1.891 489 E+02
Btu (thermochemical)/ft²·s	watt per meter² (W/m²)	1.134 893 E+04
Btu (thermochemical)/in.²·s	watt per meter² (W/m²)	1.634 246 E+06
Btu (International Table)/h·ft²·°F		
(C, thermal conductance)	watt per meter² kelvin (W/m²·K)	5.678 263 E+00
Btu (thermochemical)/h·ft²·°F		
(C, thermal conductance)	watt per meter² kelvin (W/m²·K)	5.674 466 E+00
Btu (International Table)/s·ft²·°F	watt per meter² kelvin (W/m²·K)	2.044 175 E+04
Btu (thermochemical)/s·ft²·°F	watt per metre² kelvin (W/m²·K)	2.042 808 E+04
Btu (International Table)/lb	joule per kilogram (J/kg)	2.326 000*E+03
Btu (thermochemical)/lb	joule per kilogram (J/kg)	2.324 444 E+03
Btu (International Table)/lb·°F		
(c, heat capacity)	joule per kilogram kelvin (J/kg·K)	4.186 800*E+03
Btu (thermochemical)/lb·°F		
(c, heat capacity)	joule per kilogram kelvin (J/kg·K)	4.184 000 E+03
bushel (U.S.)	meter³ (m³)	3.523 907 E−02
caliber (inch)	meter (m)	2.540 000*E−02
calorie (International Table)	joule (J)	4.186 800*E+00
calorie (mean)	joule (J)	4.190 02 E+00
calorie (thermochemical)	joule (J)	4.184 000*E+00
calorie (15°C)	joule (J)	4.185 80 E+00
calorie (20°C)	joule (J)	4.181 90 E+00
calorie (kilogram, International Table)§	joule (J)	4.186 800*E+03

§ See end of Table

Table 6 (*Continued*)

To convert from	To	Multiply by
calorie (kilogram, mean) §	joule (J)	4.190 02 E+03
calorie (kilogram, thermochemical) §	joule (J)	4.184 000*E+03
cal (thermochemical)/cm²	joule per meter² (J/m²)	4.184 000*E+04
cal (International Table)/g	joule per kilogram (J/kg)	4.186 800*E+03
cal (thermochemical)/g	joule per kilogram (J/kg)	4.184 000*E+03
cal (International Table)/g·°C	joule per kilogram kelvin (J/kg·K)	4.186 800*E+03
cal (thermochemical)/g·°C	joule per kilogram kelvin (J/kg·K)	4.184 000*E+03
cal (thermochemical)/min	watt (W)	6.973 333 E−02
cal (thermochemical)/s	watt (W)	4.184 000*E+00
cal (thermochemical)/cm²·min	watt per meter² (W/m²)	6.973 333 E+02
cal (thermochemical)/cm²·s	watt per meter² (W/m²)	4.184 000*E+04
cal (thermochemical)/cm·s·°C	watt per meter kelvin (W/m·K)	4.184 000*E+02
carat (metric)	kilogram (kg)	2.000 000*E−04
centimeter of mercury (0°C)	pascal (Pa)	1.333 22 E+03
centimeter of water (4°C)	pascal (Pa)	9.806 38 E+01
centipoise	pascal second (Pa·s)	1.000 000*E−03
centistokes	meter² per second (m²/s)	1.000 000*E−06
circular mil	metre² (m²)	5.067 075 E−10
clo	kelvin meter² per watt (K·m²/W)	2.003 712 E−01
cup	meter³ (m³)	2.365 882 E−04
curie	becquerel (Bq)	3.700 000*E+10
day (mean solar)	second (s)	8.640 000 E+04
day (sidereal)	second (s)	8.616 409 E+04
degree (angle)	radian (rad)	1.745 329 E−02
degree Celsius	kelvin (K)	$t_K = t°_C + 273.15$
degree Fahrenheit	degree Celsius	$t°_C = (t°_F − 32)/1.8$
degree Fahrenheit	kelvin (K)	$t_K = (t°_F + 459.67)/1.8$
degree Rankine	kelvin (K)	$t_K = t°_R/1.8$
°F·h·ft²/Btu (International Table) (*R*, thermal resistance)	kelvin meter² per watt (K·m²/W)	1.761 102 E−01
°F·h·ft²/Btu (thermochemical) (*R*, thermal resistance)	kelvin meter² per watt (K·m²/W)	1.762 280 E−01
denier	kilogram per meter (kg/m)	1.111 111 E−07
dyne	newton (N)	1.000 000*E−05
dyne·cm	newton meter (N·m)	1.000 000*E−07
dyne/cm²	pascal (Pa)	1.000 000*E−01
electronvolt	joule (J)	1.602 19 E−19
EMU of capacitance	farad (F)	1.000 000*E+09
EMU of current	ampere (A)	1.000 000*E+01
EMU of electric potential	volt (V)	1.000 000*E−08
EMU of inductance	henry (H)	1.000 000*E−09
EMU of resistance	ohm (Ω)	1.000 000*E−09
ESU of capacitance	farad (F)	1.112 650 E−12
ESU of current	ampere (A)	3.335 6 E−10
ESU of electric potential	volt (V)	2.997 9 E+02
ESU of inductance	henry (H)	8.987 554 E+11
ESU of resistance	ohm (Ω)	8.987 554 E+11

§ See end of Table

(*Continued*)

Table 6 (*Continued*)

To convert from	To	Multiply by
erg	joule (J)	1.000 000*E−07
erg/cm² ·s	watt per meter² (W/m²)	1.000 000*E−03
erg/s	watt (W)	1.000 000*E−07
faraday (based on carbon-12)	coulomb (C)	9.648 70 E+04
faraday (chemical)	coulomb (C)	9.649 57 E+04
faraday (physical)	coulomb (C)	9.652 19 E+04
fathom	meter (m)	1.828 8 E+00
fermi (femtometer)	meter (m)	1.000 000*E−15
fluid ounce (U.S.)	meter³ (m³)	2.957 353 E−05
foot	meter (m)	3.048 000*E−01
foot (U.S. survey)‡	meter (m)	3.048 006 E−01
foot of water (39.2°F)	pascal (Pa)	2.988 98 E+03
ft²	meter² (m²)	9.290 304*E−02
ft²/h (thermal diffusivity)	meter² per second (m²/s)	2.580 640*E−05
ft²/s	meter² per second (m²/s)	9.290 304*E−02
ft³ (volume; section modulus)	meter³ (m³)	2.831 685 E−02
ft³/min	meter³ per second (m³/s)	4.719 474 E−04
ft³/s	meter³ per second (m³/s)	2.831 685 E−02
ft⁴ (moment of section)	meter⁴ (m⁴)	8.630 975 E−03
ft/h	meter per second (m/s)	8.466 667 E−05
ft/min	meter per second (m/s)	5.080 000*E−03
ft/s	meter per second (m/s)	3.048 000*E−01
ft/s²	meter per second² (m/s²)	3.048 000*E−01
footcandle	lux (lx)	1.076 391 E+01
footlambert	candela per meter² (cd/m²)	3.426 259 E+00
ft·lbf	joule (J)	1.355 818 E+00
ft·lbf/h	watt (W)	3.766 161 E−04
ft·lbf/min	watt (W)	2.259 697 E−02
ft·lbf/s	watt (W)	1.355 818 E+00
ft·poundal	joule (J)	4.214 011 E−02
free fall, standard (*g*)	meter per second² (m/s²)	9.806 650*E+00
gal	meter per second² (m/s²)	1.000 000*E−02
gallon (Canadian liquid)	meter³ (m³)	4.546 090 E−03
gallon (U.K. liquid)	meter³ (m³)	4.546 092 E−03
gallon (U.S. dry)	meter³ (m³)	4.404 884 E−03
gallon (U.S. liquid)	meter³ (m³)	3.785 412 E−03
gal (U.S. liquid)/day	meter³ per second (m³/s)	4.381 264 E−08
gal (U.S. liquid)/min	meter³ per second (m³/s)	6.309 020 E−05
gal (U.S. liquid)/hp·h (SFC, specific fuel consumption)	kilogram per joule (kg/J)	1.410 089 E−09
gamma	tesla (T)	1.000 000*E−09
gauss	tesla (T)	1.000 000*E−04
gilbert	ampere	7.957 747 E−01
gill (U.K.)	meter³ (m³)	1.420 654 E−04
gill (U.S.)	meter³ (m³)	1.182 941 E−04
grad	degree (angular)	9.000 000*E−01
grad	radian (rad)	1.570 796 E−02

‡ See end of Table

Table 6 (*Continued*)

To convert from	To	Multiply by
grain (1/7000 lb avoirdupois)	kilogram (kg)	6.479 891*E−05
grain (lb avoirdupois/7000)/gal (U.S. liquid)	kilogram per meter³ (kg/m³)	1.711 806 E−02
gram	kilogram (kg)	1.000 000*E−03
g/cm³	kilogram per meter³ (kg/m³)	1.000 000*E+03
gram-force/cm²	pascal (Pa)	9.806 650*E+01
hectare	meter² (m²)	1.000 000*E+04
horsepower (550 ft·lbf/s)	watt (W)	7.456 999 E+02
horsepower (boiler)	watt (W)	9.809 50 E+03
horsepower (electric)	watt (W)	7.460 000*E+02
horsepower (metric)	watt (W)	7.354 99 E+02
horsepower (water)	watt (W)	7.460 43 E+02
horsepower (U.K.)	watt (W)	7.457 0 E+02
hour (mean solar)	second (s)	3.600 000 E+03
hour (sidereal)	second (s)	3.590 170 E+03
hundredweight (long)	kilogram (kg)	5.080 235 E+01
hundredweight (short)	kilogram (kg)	4.535 924 E+01
inch	meter (m)	2.540 000*E−02
inch of mercury (32°F)	pascal (Pa)	3.386 38 E+03
inch of mercury (60°F)	pascal (Pa)	3.376 85 E+03
inch of water (39.2°F)	pascal (Pa)	2.490 82 E+02
inch of water (60°F)	pascal (Pa)	2.488 4 E+02
in.²	meter² (m²)	6.451 600*E−04
in.³ (volume; section modulus)	meter³ (m³)	1.638 706 E−05
in.³/min	meter³ per second (m³/s)	2.731 177 E−07
in.⁴ (moment of section)	meter⁴ (m⁴)	4.162 314 E−07
in./s	meter per second (m/s)	2.540 000*E−02
in./s²	meter per second² (m/s²)	2.540 000*E−02
kayser	1 per meter (1/m)	1.000 000*E+02
kelvin	degree Celsius	$t_{\circ C} = t_K - 273.15$
kilocalorie (International Table)	joule (J)	4.186 800*E+03
kilocalorie (mean)	joule (J)	4.190 02 E+03
kilocalorie (thermochemical)	joule (J)	4.184 000*E+03
kilocalorie (thermochemical)/min	watt (W)	6.973 333 E+01
kilocalorie (thermochemical)/s	watt (W)	4.184 000*E+03
kilogram-force (kgf)	newton (N)	9.806 650*E+00
kgf·m	newton meter (N·m)	9.806 650*E+00
kgf·s²/m (mass)	kilogram (kg)	9.806 650*E+00
kgf/cm²	pascal (Pa)	9.806 650*E+04
kgf/m²	pascal (Pa)	9.806 650*E+00
kgf/mm²	pascal (Pa)	9.806 650*E+06
km/h	meter per second (m/s)	2.777 778 E−01
kilopond	newton (N)	9.806 650*E+00
kW·h	joule (J)	3.600 000*E+06
kip (1000 lbf)	newton (N)	4.448 222 E+03
kip/in.² (ksi)	pascal (Pa)	6.894 757 E+06
knot (international)	meter per second (m/s)	5.144 444 E−01

(Continued)

Table 6 (*Continued*)

To convert from	To	Multiply by
lambert	candela per meter^2 (cd/m^2)	$1/\pi$ *E+04
lambert	candela per meter^2 (cd/m^2)	3.183 099 E+03
langley	joule per meter^2 (J/m^2)	4.184 000*E+04
league	meter (m)	[see footnote ‡]
light year	meter (m)	9.460 55 E+15
liter	meter^3 (m^3)	1.000 000*E−03
maxwell	weber (Wb)	1.000 000*E−08
mho	siemens (S)	1.000 000*E+00
microinch	meter (m)	2.540 000*E−08
micron	meter (m)	1.000 000*E−06
mil	meter (m)	2.540 000*E−05
mile (international)	meter (m)	1.609 344*E+03
mile (statute)	meter (m)	1.609 3 E+03
mile (U.S. survey)‡	meter (m)	1.609 347 E+03
mile (international nautical)	meter (m)	1.852 000*E+03
mile (U.K. nautical)	meter (m)	1.853 184*E+03
mile (U.S. nautical)	meter (m)	1.852 000*E+03
mi^2 (international)	meter^2 (m^2)	2.589 988 E+06
mi^2 (U.S. survey)‡	meter^2 (m^2)	2.589 998 E+06
mi/h (international)	meter per second (m/s)	4.470 400*E−01
mi/h (international)	kilometer per hour (km/h)	1.609 344*E+00
mi/min (international)	meter per second (m/s)	2.682 240*E+01
mi/s (international)	meter per second (m/s)	1.609 344*E+03
millibar	pascal (Pa)	1.000 000*E+02
millimeter of mercury (0°C)	pascal (Pa)	1.333 22 E+02
minute (angle)	radian (rad)	2.908 882 E−04
minute (mean solar)	second (s)	6.000 000 E+01
minute (sidereal)	second (s)	5.983 617 E+01
month (mean calendar)	second (s)	2.628 000 E+06
oersted	ampere per meter (A/m)	7.957 747 E+01
ohm centimeter	ohm meter (Ω·m)	1.000 000*E−02
ohm circular-mil per foot	ohm millimeter^2 per meter (Ω·mm^2/m)	1.662 426 E−03
ounce (avoirdupois)	kilogram (kg)	2.834 952 E−02
ounce (troy or apothecary)	kilogram (kg)	3.110 348 E−02
ounce (U.K. fluid)	meter^3 (m^3)	2.841 307 E−05
ounce (U.S. fluid)	meter^3 (m^3)	2.957 353 E−05
ounce-force	newton (N)	2.780 139 E−01
ozf·in.	newton meter (N·m)	7.061 552 E−03
oz (avoirdupois)/gal (U.K. liquid)	kilogram per meter^3 (kg/m^3)	6.236 021 E+00
oz (avoirdupois)/gal (U.S. liquid)	kilogram per meter^3 (kg/m^3)	7.489 152 E+00
oz (avoirdupois)/in.^{3}	kilogram per meter^3 (kg/m^3)	1.729 994 E+03
oz (avoirdupois)/ft^2	kilogram per meter^2 (kg/m^2)	3.051 517 E−01
oz (avoirdupois)/yd^2	kilogram per meter^2 (kg/m^2)	3.390 575 E−02
parsec	meter (m)	3.085 678 E+16
peck (U.S.)	meter^3 (m^3)	8.809 768 E−03
pennyweight	kilogram (kg)	1.555 174 E−03

‡ See end of Table

Table 6 (*Continued*)

To convert from	To	Multiply by
perm (0°C)	kilogram per pascal second meter² (kg/Pa·s·m²)	5.721 35 E−11
perm (23°C)	kilogram per pascal second meter² (kg/Pa·s·m²)	5.745 25 E−11
perm·in. (0°C)	kilogram per pascal second meter (kg/Pa·s·m)	1.453 22 E−12
perm·in. (23°C)	kilogram per pascal second meter (kg/Pa·s·m)	1.459 29 E−12
phot	lumen per meter² (lm/m²)	1.000 000*E+04
pica (printer's)	meter (m)	4.217 518 E−03
pint (U.S. dry)	meter³ (m³)	5.506 105 E−04
pint (U.S. liquid)	meter³ (m³)	4.731 765 E−04
point (printer's)	meter (m)	3.514 598*E−04
poise (absolute viscosity)	pascal second (Pa·s)	1.000 000*E−01
pound (lb avoirdupois)	kilogram (kg)	4.535 924 E−01
pound (troy or apothecary)	kilogram (kg)	3.732 417 E−01
lb·ft² (moment of inertia)	kilogram meter² (kg·m²)	4.214 011 E−02
lb·in.² (moment of inertia)	kilogram meter² (kg·m²)	2.926 397 E−04
lb/ft·h	pascal second (Pa·s)	4.133 789 E−04
lb/ft·s	pascal second (Pa·s)	1.488 164 E+00
lb/ft²	kilogram per meter² (kg/m²)	4.882 428 E+00
lb/ft³	kilogram per meter³ (kg/m³)	1.601 846 E+01
lb/gal (U.K. liquid)	kilogram per meter³ (kg/m³)	9.977 633 E+01
lb/gal (U.S. liquid)	kilogram per meter³ (kg/m³)	1.198 264 E+02
lb/h	kilogram per second (kg/s)	1.259 979 E−04
lb/hp·h (SFC, specific fuel consumption)	kilogram per joule (kg/J)	1.689 659 E−07
lb/in.³	kilogram per meter³ (kg/m³)	2.767 990 E+04
lb/min	kilogram per second (kg/s)	7.559 873 E−03
lb/s	kilogram per second (kg/s)	4.535 924 E−01
lb/yd³	kilogram per meter³ (kg/m³)	5.932 764 E−01
poundal	newton (N)	1.382 550 E−01
poundal/ft²	pascal (Pa)	1.488 164 E+00
poundal·s/ft²	pascal second (Pa·s)	1.488 164 E+00
pound-force (lbf)	newton (N)	4.448 222 E+00
lbf·ft	newton meter (N·m)	1.355 818 E+00
lbf·ft/in.	newton meter per meter (N·m/m)	5.337 866 E+01
lbf·in.	newton meter (N·m)	1.129 848 E−01
lbf·in./in.	newton meter per meter (N·m/m)	4.448 222 E+00
lbf·s/ft²	pascal second (Pa·s)	4.788 026 E+01
lbf/ft	newton per meter (N/m)	1.459 390 E+01
lbf/ft²	pascal (Pa)	4.788 026 E+01
lbf/in.	newton per meter (N/m)	1.751 268 E+02
lbf/in.² (psi)	pascal (Pa)	6.894 757 E+03
lbf/lb (thrust/weight [mass] ratio)	newton per kilogram (N/kg)	9.806 650 E+00
quart (U.S. dry)	meter³ (m³)	1.101 221 E−03
quart (U.S. liquid)	meter³ (m³)	9.463 529 E−04

(*Continued*)

(Continued)

To convert from	To	Multiply by
rad (radiation dose absorbed)	gray (Gy)	1.000 000*E−02
rhe	1 per pascal second (1/Pa·s)	1.000 000*E+01
rod	meter (m)	[see footnote ‡]
roentgen	coulomb per kilogram (C/kg)	2.58 E−04
second (angle)	radian (rad)	4.848 137 E−06
second (sidereal)	second (s)	9.972 696 E−01
section	meter² (m²)	[see footnote ‡]
shake	second (s)	1.000 000*E−08
slug	kilogram (kg)	1.459 390 E+01
slug/ft·s	pascal second (Pa·s)	4.788 026 E+01
slug/ft³	kilogram per meter³ (kg/m³)	5.153 788 E+02
statampere	ampere (A)	3.335 640 E−10
statcoulomb	coulomb (C)	3.335 640 E−10
statfarad	farad (F)	1.112 650 E−12
stathenry	henry (H)	8.987 554 E+11
statmho	siemens (S)	1.112 650 E−12
statohm	ohm (Ω)	8.987 554 E+11
statvolt	volt (V)	2.997 925 E+02
stere	meter³ (m³)	1.000 000*E+00
stilb	candela per meter² (cd/m²)	1.000 000*E+04
stokes (kinematic viscosity)	meter² per second (m²/s)	1.000 000*E−04
tablespoon	meter³ (m³)	1.478 676 E−05
teaspoon	meter³ (m³)	4.928 922 E−06
tex	kilogram per meter (kg/m)	1.000 000*E−06
therm	joule (J)	1.055 056 E+08
ton (assay)	kilogram (kg)	2.916 667 E−02
ton (long, 2240 lb)	kilogram (kg)	1.016 047 E+03
ton (metric)	kilogram (kg)	1.000 000*E+03
ton (nuclear equivalent of TNT)	joule (J)	4.184 E+09
ton (refrigeration)	watt (W)	3.516 800 E+03
ton (register)	meter³ (m³)	2.831 685 E+00
ton (short, 2000 lb)	kilogram (kg)	9.071 847 E+02
ton (long)/yd³	kilogram per meter³ (kg/m³)	1.328 939 E+03
ton (short)/h	kilogram per second (kg/s)	2.519 958 E−01
ton-force (2000 lbf)	newton (N)	8.896 444 E+03
tonne	kilogram (kg)	1.000 000*E+03
torr (mm Hg, 0°C)	pascal (Pa)	1.333 22 E+02
township	meter² (m²)	[see footnote ‡]
unit pole	weber (Wb)	1.256 637 E−07
W·h	joule (J)	3.600 000*E+03
W·s	joule (J)	1.000 000*E+00
W/Cm²	watt per meter² (W/m²)	1.000 000*E+04
W/in²	watt per meter² (W/m²)	1.550 003 E+03
yard	meter (m)	9.144 000*E−01
yd²	meter² (m²)	8.361 274 E−01
yd³	meter³ (m³)	7.645 549 E−01
yd³/min	meter³ per second (m³/s)	1.274 258 E−02
year (calendar)	second (s)	3.153 600 E+07
year (sidereal)	second (s)	3.155 815 E+07
year (tropical)	second (s)	3.155 693 E+07

* Exact equivalence.

† Adapted from ASTM Standard for Metric Practice E 380-76. The conversion factors are listed in standard form for computer readout as a number greater than one or less than ten with six or less decimal points. The number is followed by the letter E (for exponent), a plus or minus symbol, and two digits which indicate the power of 10 by which the number must be multiplied. An asterisk (*) after the sixth decimal place indicates that the conversion factor is exact and that all subsequent digits are zero. All other conversion factors have been rounded to the figures given. Where less than six decimal places are shown, more precision is not warranted.

For example, 1.013 250*E + 05 is exactly 1.013 250 $\times$ 10⁵ or 101 325.0.

1.589 873 E − 01 has the last digit rounded off to 3 and is

1.589 873 $\times$ 10^{-1} or 0.158 987 3.

‡ Since 1893, the U.S. basis of length measurement has been derived from metric standards. In 1959, a small refinement was made in the definition of the yard to resolve discrepancies both in this country and abroad which changed its length from 3600/3937 m to 0.9144 m exactly. This resulted in the new value being shorter by two parts in a million. At the same time, it was decided that any data in feet derived from and published as a result of geodetic surveys within the U.S. would remain with the old standard (1 ft = 1200/3937 m) until further decision. This foot is named the U.S. survey foot. As a result, all U.S. land measurements in U.S. customary units will relate to the meter by the old standard. All the conversion factors in this table for units referenced to this footnote are based on the U.S. survey foot rather than on the international foot.

Conversion factors for the land measures given below may be determined from the following relationships:

1 league	= 3 miles (exactly)
1 rod	= $16\frac{1}{2}$ feet (exactly)
1 section	= 1 square mile (exactly)
1 township	= 36 square miles (exactly)

§ By definition, one calorie (International Table) is exactly 4.186 8 absolute joules which converts to 1.055 056 $\times$ 10³ joules for one Btu (International Table). Also, by definition, one calorie (thermochemical) is exactly 4.184 absolute joules which converts to 1.054 350 $\times$ 10³ joules for one Btu (thermochemical). A *mean* calorie is $\frac{1}{100}$ th of the heat required to raise the temperature of one gram of water at one atmosphere pressure from 0°C to 100°C and equals 4.190 02 absolute joules. In all cases, the relationship between calorie and British thermal unit is established by 1 cal/(g·°C) = 1 Btu/(lb·°F). A *mean* Btu, therefore, is $\frac{1}{180}$ th of the heat required to raise the temperature of one pound of water at one atmosphere pressure from 32°F to 212°F and equals 1.055 87 $\times$ 10³ joules. When values are given as Btu or calories, the type of unit (International Table, thermochemical, mean, or temperature of determination) should be given. In all cases for this table, conversions involving joules are based on the absolute joule.

Table 7 Abbreviated list of equivalences for converting units commonly used by chemical engineers to acceptable SI units

Unacceptable unit	Acceptable SI unit with unit conversion factor
ångström	0.1 nm*
atmosphere (standard)	101.325 kPa
Btu†	1.055 056 kJ
Btu/(lb$_m$·°F) (heat capacity)	4.186 8 kJ/(kg·K)*
Btu/h	0.293 971 1 W
Btu/ft^2	11.356 53 kJ/m^2
Btu/(ft^2·h·°F) (heat transfer coefficient)	5.678 263 J/(m^2·s·K)
Btu/(ft^2·h) (heat flux)	3.154 591 J/(m^2·s)
Btu/(ft·h·°F) (thermal conductivity)	1.730 735 J/(m·s·K)
calorie†	4.186 8 J*
cal/(g·°C) (heat capacity)	4.186 8 kJ/(kg·K)*
centipoise (absolute viscosity)	1.0 mPa·s*
centistoke (kinematic viscosity)	1.0 × 10^{-6} m^2/s*
t (°F)	$(t + 459.67)/(1.8)$ K
t (°R)	$t/(1.8)$ K*
dyne	10.0 μN*
erg	100 pJ*
foot‡	0.3048 m*
ft^2	9.290 304 × 10^{-2} m^2*
ft^3	2.831 685 × 10^{-2} m^3
gallon (U.S. liquid)	3.785 412 × 10^{-3} m^3
horsepower (550 ft·lb$_f$/s)	745.699 9 W
inch	2.54 × 10^{-2} m*
in. Hg (60°F) (inches mercury pressure)	3.376 85 kPa
in. H$_2$O (60°F) (inches water pressure)	0.248 84 kPa
kg$_f$ (kilogram force)	9.806 65 N*
mile	1 609.344 m*
mm Hg (0°C) (millimeters mercury pressure)	0.133 322 kPa
poise (absolute viscosity)	0.1 Pa·s*
lb$_f$ (pounds force)	4.448 222 N
lb$_m$ (pounds mass—avoirdupois)	0.453 592 4 kg
psi (pounds per square inch pressure)	6.894 757 kPa
stoke (kinematic viscosity)	1.0 × 10^{-4} m^2/s*
yard	0.9144 m*

* Exact equivalence.

† British thermal unit and calorie are reported as the International Table values as adopted in 1956 for all cases in this table. The exact conversion factor for Btu (International Table) to kJ is 1.055 055 852 62. The Btu (thermochemical) is 1.054 350 kJ and the calorie (thermochemical) is exactly 4.184 J. (See footnote § for Table 6).

‡ The foot is reported as the International Table value and holds for all cases of length in this table.

AUXILIARY, UTILITY,
AND CHEMICAL COST DATA†

CONTENTS

AUXILIARY COST DATA

† Costs reported are Jan. 1, 1979 values except Table 1.

INSTRUMENTATION

UTILITIES

CHEMICALS

Table 1 Hourly wage rates of craft labor in selected U.S. cities*

Employer's burden includes state and federal taxes, insurance, etc. (Hourly wage including fringe benefits + employer's burden, July 1978)

City	Bricklayers	Carpenters	Electricians	General construction laborers	Structural iron workers	Painters	Plasterers	Plumbers or pipefitters	Sheet-metal workers
Atlanta, Ga.	11.26 + 2.40	11.11 + 2.37	12.75 + 2.10	7.28 + 1.39	11.47 + 4.27	11.15 + 2.46	10.90 + 2.64	12.56 + 2.26	10.62 + 1.93
Boston, Mass.	12.83 + 2.73	13.25 + 2.83	15.13 + 2.49	10.09 + 1.93	13.65 + 5.08	12.73 + 2.80	12.28 + 2.97	14.89 + 2.68	13.98 + 2.54
Chicago, Ill.	13.35 + 2.85	13.65 + 2.92	15.41 + 2.54	10.82 + 2.07	14.06 + 5.23	11.76 + 2.59	12.53 + 3.03	14.02 + 2.52	14.26 + 2.59
Dallas, Tex.	11.00 + 2.34	11.11 + 2.37	12.61 + 2.08	7.38 + 1.41	10.65 + 3.96	10.67 + 2.35	10.78 + 2.61	11.97 + 2.16	11.32 + 2.06
Denver, Colo.	12.57 + 2.68	12.31 + 2.66	14.71 + 2.43	8.42 + 1.61	11.60 + 4.32	12.24 + 2.70	11.40 + 2.76	13.43 + 2.42	12.30 + 2.24
Detroit, Mich.	15.18 + 3.24	14.85 + 3.17	15.69 + 2.59	12.06 + 2.31	15.15 + 5.04	14.06 + 3.10	14.28 + 3.45	15.48 + 2.79	13.84 + 2.51
Houston, Tex.	13.09 + 2.79	12.71 + 2.72	13.45 + 2.22	9.36 + 1.79	10.92 + 4.06	11.51 + 2.54	11.90 + 2.88	13.14 + 2.37	11.60 + 2.11
Los Angeles, Calif.	14.40 + 3.07	14.99 + 3.20	17.09 + 2.82	12.58 + 2.41	14.33 + 5.33	14.42 + 3.18	15.66 + 3.79	17.96 + 3.23	15.52 + 2.82
Newark, N.J.	13.74 + 2.93	13.65 + 2.92	15.69 + 2.59	9.98 + 1.91	18.02 + 6.71	11.88 + 2.62	13.66 + 3.30	14.75 + 2.66	
Philadelphia, Pa.	13.74 + 2.93	13.92 + 2.97	14.99 + 2.47	10.71 + 2.05	14.33 + 5.33	11.88 + 2.62	11.90 + 2.88	15.04 + 2.71	14.12 + 2.57
San Francisco, Calif.	16.10 + 3.43	15.92 + 3.40	17.51 + 2.89	13.21 + 2.53	14.47 + 5.38	15.51 + 3.42	16.16 + 3.91	21.46 + 3.87	14.82 + 2.69
Seattle, Wash.	13.09 + 2.79	12.18 + 2.60	15.27 + 2.52	11.75 + 2.25	12.42 + 4.62	12.36 + 2.72	12.66 + 3.06	16.79 + 3.02	13.56 + 2.46

* Adapted from G. Moselle, Ed., "National Construction Estimator," 26th ed., Craftsman Book Company of America, Solana Beach, California, 1978. Units are dollars per hour.

Table 2 Building and construction costs

Item	Unit	Cost, $	Employee—hr to install
Floors			
Asphalt tile	ft²	0.64	†
Concrete, prestressed, 4-in. thick	ft²	2.25	†
Steel grating	ft²	8.00	†
Wood deck, 2-in. thick	ft²	2.30	†
Foundations, includes excavation, backfill, and forming:			
flat slab, 1 yd³ concrete, 5.3 ft² forms, 100 lb reinforcing steel‡	yd³	47	6
Pits and basins: 1 yd³ concrete, 17.5 ft² forms, 115 lb reinforcing steel‡	yd³	60	8
Walls and piers: 1 yd³ concrete, 57.0 ft² forms, 160 lb reinforcing steel‡	yd³	125	16
Lumber			
Structural, plain	MBF	560	30
Structural, creosoted	MBF	700	30
Plywood: $\frac{3}{8}$-in.	ft²	0.20	1
$\frac{1}{2}$-in.	ft²	0.25	1
$\frac{5}{8}$-in.	ft²	0.31	1
$\frac{3}{4}$-in.	ft²	0.38	1
Piling (20–25 ton load, 60 ft long)			
Wood, treated	Ea.	415	†
Wood, untreated	Ea.	350	†
Composite	Ea.	730	†
Concrete	Ea.	900	†
Test pile	Ea.	2750	†
Load test	Ea.	3300	†
Equipment "on and off site"	Per job	4800	†
Roofs			
Aluminum, corrugated, 0.032-in. thick	ft²	1.20	†
Built-up, 5-ply	ft²	0.56	†
Reinforced concrete, 4-in. thick	ft²	2.25	†
Steel, 20 gauge	ft²	0.74	†
Transite, $\frac{3}{8}$-in. thick	ft²	1.52	†
Sprinkler systems, exposed. Add 28% for concealed systems			
Wet system	ft²	1.25	†
Dry system	ft²	1.40	†
4-in. alarm valve, wet system	Ea.	720	40
4-in. alarm valve, dry system	Ea.	1100	50
6-in. alarm valve, wet system	Ea.	860	40
6-in. alarm valve, dry system	Ea.	1200	50
Compressor to operate 500 heads	Ea.	650	16
(Valves and compressor not included in ft² price)			
Structural steel			
Grating, $1\frac{1}{4}$-in. standard	ft²	6.00	0.33

† Labor cost is included in material cost.

‡ To adjust: forms, $2/ft² ; reinforcing steel, 29¢/lb.

Table 2 Building and construction costs (*Continued*)

Item	Unit	Cost, $	Employee— hr to install
Grating, expanded metal	ft²	2.80	0.17
Grating, checker plate, $\frac{5}{16}$-in.	ft²	4.00	0.13
Handrail, standard $1\frac{1}{2}$-in. pipe, 2 rails, welded	Lineal ft	10.60	0.67
Handrail, bar type	Lineal ft	4.50	0.70
Steel ladder, with safety cage (Add 25% for aluminum)	Lineal ft	36.50	0.50
Steel ladder, without safety cage (Add 25% for aluminum)	Lineal ft	21.25	0.35
Operating platforms, including stairs	ft²	25.00	1.8
Stairway	Vert. ft	54.00	
Stair treads, 12-in. wide, galvanized	Ea.	18.00	1
Building steel, shop fab	Lb	0.38	0.01
Platform and support steel, shop fab	Lb	0.47	0.02
Toeplate steel, 4 in. $\times \frac{1}{4}$-in.	Lineal ft	1.90	0.10
Walls			
Siding			
Transite	ft²	1.60	†
Aluminum	ft²	1.40	†
Steel-coated	ft²	1.20	†
Brick			
4-in.	ft²	2.75	†
8-in.	ft²	5.50	†
10-in.	ft²	6.90	†
12-in.	ft²	8.30	†
Concrete block			
6-in.	ft²	1.80	†
8-in.	ft²	2.20	†
12-in.	ft²	3.00	†
Windows, industrial			
Steel	ft²	5.50	†
Aluminum	ft²	6.40	†
Wood	ft²	4.50	†
Total cost of erected buildings (median values)			
Laboratory: steel frame, masonry walls, floor and roof; heating, lighting, and plumbing	ft²	56	†
Office: steel frame, masonry walls, floor, and roof; heating, lighting, and plumbing	ft²	38	†
Process building: multilevel, 12-ft clearance, steel platforms, heating, lighting, and plumbing			
Masonry construction	ft²	25	†
Aluminum on steel	ft²	22	†
Transite on steel	ft²	20	†
Open structure: 3-level, steel, with lighting and plumbing	ft²	18	†
Warehouse: single story, 15-ft clearance. Steel frame, masonry walls, floor, and roof; heating, lighting, and plumbing	ft²	17	†

† Labor cost is included in material cost.

(*Continued*)

Table 2 Building and construction costs (*Continued*)

Item	Unit	Cost, $	Employee–hr to install
Doors			
Metal: Steel frame, 8 × 8 ft, automatic	Ea.	1390	30
Steel rolling, 12 × 12 ft, manual	Ea.	880	24
Swing, with frame, 3 × 7 ft, $1\frac{3}{4}$-in. thick	Ea.	152	†
Wood: Sectional, overhead, 12 × 10 ft	Ea.	430	6
Swing exterior, with frame, 3 × 7 ft	Ea.	120	4
Excavation			
Machine	yd³	2–4	†
Hand	yd³	14–27	†

Table 3 Costs for yard improvements

Item	Unit	Cost, $	Employee–hr to install
Docks and wharfs			
All concrete, 100 lineal ft wide	Lineal ft	7000	†
Timber, with wood deck, 100 lineal ft wide	Lineal ft	6200	†
Dredging	yd³	2.60	†
Pipe bridges (includes structural steel and foundations)			
Heavy duty	Lineal ft	40	2.5
Light duty	Lineal ft	25	1.8
Pipe column, steel, extra heavy			
16 ft–4 in. diameter	Ea.	84	1.2
16 ft–6 in. diameter	Ea.	170	1.4
22 ft–6 in. diameter	Ea.	230	1.9
Plant fence			
7 ft chain link (3-strand B.W.)	Lineal ft	6.60	†
3- to 4–ft wide person gate	Ea.	135	†
8-ft wide equipment gate	Ea.	220	†
30-ft wide double gate (manual)	Ea.	460	†
20-ft wide gate (automatic)	Ea.	3000	†
Relocate plant fence	Lineal ft	3.30	†
Railroads			
Track (90 lb)	Lineal ft	33	†
Track (75 lb)	Lineal ft	26	†
Turnout	Ea.	5760	†
Ties, creosoted (6 in. × 8 in. × 8 ft)	Ea.	34	†
Grade and ballast	Lineal ft	9.60	†
Cars, ore, 24-in. track			
Capacity, ft³ Wt, lb			
12 660	Ea.	690	†
16 700	Ea.	740	†
20 930	Ea.	825	†

† Labor cost is included with material cost

Table 3 Costs for yard improvements (*Continued*)

Item	Unit	Cost, $	Employee—hr to install
Locomotives, mine, battery type			
Size, tons Wt, lb			
9 19,000	Ea.	63,000	
12 26,000	Ea.	72,000	
Locomotive, mine, Diesel type			
1½ 3,000 2 cyl	Ea.	19,500	
3 7,000 4 cyl	Ea.	24,300	
Roads and walkways			
Concrete, slab, mesh reinforced			
Thickness, in. Sub-base, in.			
4 6	yd²	12.00	†
6 6	yd²	12.20	†
8 6	yd²	14.80	†
Paving asphalt			
3 12	yd²	7.20	†
Sewers			
Reinforced concrete, thickness, in.			
12	Lineal ft	11.70	†
15	Lineal ft	13.00	†
18	Lineal ft	14.60	†
24	Lineal ft	20.10	†
30	Lineal ft	26.50	†
Vitrified clay, thickness, in.			
4	Lineal ft	7.80	†
6	Lineal ft	8.90	†
8	Lineal ft	10.90	†
12	Lineal ft	16.90	†
15	Lineal ft	21.70	†
18	Lineal ft	27.20	†
24	Lineal ft	39.90	†
Site development			
Clearing and grubbing	yd²	0.46	†
Grade out	yd³	1.42	†
Cut, fill, and compact	yd³	2.32	†
New fill compacted	yd³	4.60	†
Topsoil	yd³	12.00	†
Gravel fill	yd³	13.00	†
Crushed stone	yd³	13.00	†
Seeding	yd²	0.41	†
Sluiceway			
Open, piled, and sheathed	Lineal ft	325	†
Wells			
200 gpm, 400 ft deep, 15 hp	Ea.	16,000	†
500 gpm, 200 ft deep, 40 hp	Ea.	21,000	†
1200 gpm, 400 ft deep, 75 hp	Ea.	29,000	†

† Labor cost is included with material cost

Table 4 Cost of electrical installations

Main transformer stations, installed cost, $/kVa

Capacity, kVA	Primary voltage, 25 kV		Primary voltage, 46 kV		Primary voltage, 69 kV	
	A	*B*	*A*	*B*	*A*	*B*
3,000	80	65	93	78	103	91
5,000	50	43	56	48	63	54
10,000	33	28	35	30	39	33
20,000	24	22	26	24	30	28

A: Two three-phase transformers, 60 cycles, two main incoming supply lines
B: Two three-phase transformers, 60 cycles, one main incoming supply line

Secondary transformer stations, installed cost, $/kva

Capacity, kVA	2400/240 volts	4200/600 volts	13,200/600 volts
600	21	55	69
1000	17	40	50
1500	15	31	39
2000	13	26	32

Distribution feeders, installed cost, $/kVa main transformer capacity

Capacity, kVA	Primary voltage, 42 kV		Primary voltage, 13.2 kV	
	A	*B*	*A*	*B*
3,000	32–64	10–22		
5,000	22–44	8–32		
10,000			12–26	4–10
20,000			8–16	2–6

A: Underground
B: Overhead

Lighting, installed cost, $/kW

Avg. watts per ft²	Explosion proof	Vapor proof	Standard
1	4600	2920	2400
2	3120	1900	1500
3	2600	1460	950
4	2400	1150	780
5	2300	1050	740

Table 5 Rates for various industrial utilities

Utility	Cost (Jan. 1979)
Steam:	
500 psig	$ 2.00–$2.40/1000 lb
100 psig	1.00–2.00/1000 lb
Exhaust	0.50–0.89/1000 lb
Electricity:	
Purchased	0.020–0.080/kWh†
Self-generated	0.015–0.041/kWh
Cooling water:	
Well	0.06–0.30/1000 gal
River or salt	0.04–0.12/1000 gal
Tower	0.04–0.16/1000 gal
Process water:	
City	0.20–0.90/1000 gal
Filtered and softened	0.30–0.80/1000 gal
Distilled	1.40–2.40/1000 gal
Compressed air:	
Process air	0.04–0.12/1000 ft^3 (at SC)‡
Filtered and dried for instruments	0.08–0.24/1000 ft^3 (at SC)‡
Natural gas:	
Interstate (major pipeline companies)	1.50–1.95/1000 ft^3 (at SC)‡
Intrastate (new contracts)	1.38–2.15/1000 ft^3 (at SC)‡
Intrastate (renegotiated or amended contracts)	1.75–2.30/1000 ft^3 (at SC)‡
Manufactured gas	1.00–3.00/1000 ft^3 (at SC)‡
Fuel oil	11.00–16.00/bbl
Coal	20.00–34.00/ton
Refrigeration (ammonia), to 34°F	1.20/ton-day (288,000 Btu removed)

† Highly dependent upon location. Convert to mils by 1000 mils = one dollar.

‡ For these cases, standard conditions are designated as a pressure of 29.92 in. Hg and a temperature of 60°F.

Table 6 Costs for selected industrial chemicals†

Chemical and conditions	Cost quote unit	Cost, $
Acetaldehyde, 99%, tanks	lb	0.20
Acetic acid, tech., tanks	lb	0.18
Acetic anhydride, tanks	lb	0.27
Acetone, tanks	lb	0.17
Acrylonitrile, tanks	lb	0.25
Allyl alcohol, tanks	lb	0.47
Ammonia, anhyd., fertilizer, tanks	ton	120
Aniline, tanks	lb	0.34
Benzaldehyde, tech., drums	lb	0.67
Benzene, indust., barges	gal	0.80
Benzoic acid, tech., bags	lb	0.36
n-Butyl alcohol, syn., ferment., tanks	lb	0.21
Butyric acid, tanks	lb	0.33
Calcium carbonate, nat., dry-ground, carlots	ton	19
Carbon tetrachloride, tech., drums, carlots	lb	0.14
Chlorine, tanks	ton	135
Chloroform, tech., tanks	lb	0.20
Copper chloride (cupric), anhyd., carlots	lb	0.90
Ethyl alcohol, 190 proof, USP tax free, tanks	gal	1.12
Ethyl ether, refined, tanks	lb	0.24
Ethylene, contract, delivered	lb	0.13
Ethylene oxide, tanks, delivered	lb	0.26
Formaldehyde, 37% by wt., inhibited 7% methanol, tanks	lb	0.06
Glycerine, syn., 99.5%, tanks	lb	0.53
Hexane, indust., tanks	gal	0.44
Hydrochloric acid, CP, USP consumers, carboys	lb	0.23
Isobutyl alcohol, tanks, delivered	lb	0.19
Lime, chemical, pebble (quicklime), bulk	ton	32
Mercury metal, 76-lb flasks	flask	193
Methanol, syn., tanks	gal	0.44
Nitric acid, 36 Be, tanks	100 lb	5.75
Oxalic acid, bags, carlots	lb	0.44
Pentaerythritol, tech., bags, carlots	lb	0.49
Phenol, syn., tanks	lb	0.19
Propylene glycol, indust., tanks	lb	0.26
Soda ash, dense, 58%, paper bags, carlots	ton	57
Sodium hydroxide, tech. (caustic soda), bulk	100 lb	350
Sulfur, crude, 99.5%, 50-lb bags, mines	100 lb	4.70
Sulfuric acid, 100%, tanks, works	ton	54
Toluene, petroleum, indust., tanks	gal	0.64
Urea, 46% N, indust., bulk, delivered	ton	160

† Obtained from "Chemical Marketing Reporter," published by Schnell Publishing Company, Inc., New York, N.Y. Unless otherwise indicated, prices are for large lots, f.o.b., New York City or eastern U.S.A.

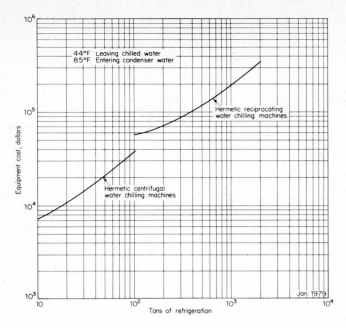

Figure B-1 Cost of air conditioning. Price includes compressor, motor, starter, controls, cooler, condenser, and refrigerant. Does not include cooling tower, pumps, foundations, ductwork, and installation costs.

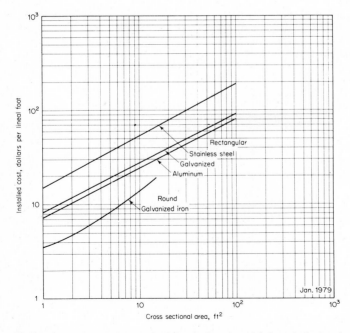

Figure B-2 Cost of ductwork. Price is for shop-fabricated ductwork with hangers and supports, installed.

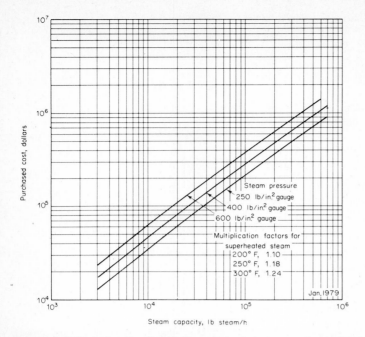

Figure B-3 Cost of package boiler plants. Price includes complete boiler, feed-water deaerator, boiler feed pumps, chemical injection system, stack, and shop assembly labor.

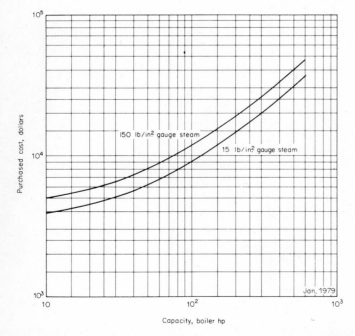

Figure B-4 Cost of steam generators. Price is for packaged unit with steel tubes, gas-fired. Multiply by 1.02 for oil-fired boiler. Multiply horsepower by 33,500 for number of Btu's.

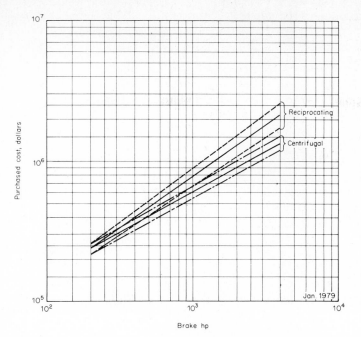

Figure B-5 Cost of compressor plants. (Dotted lines indicate range of costs for the two types of plants.)

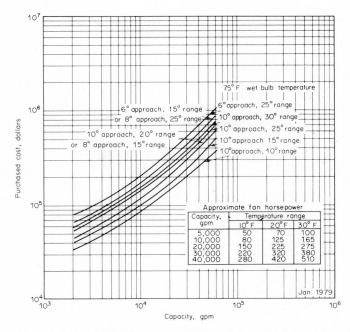

Figure B-6 Cost of cooling towers. Prices are for conventional, wood-frame, induced-draft, cross-flow, cooling towers. Price does not include external piping, power wiring, special foundation work, or field labor.

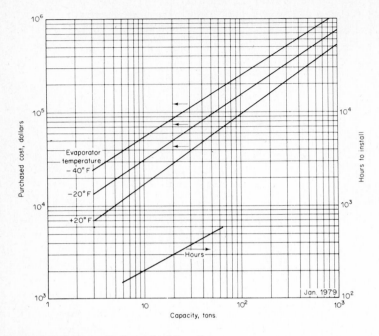

Figure B-7 Cost of industrial refrigeration.

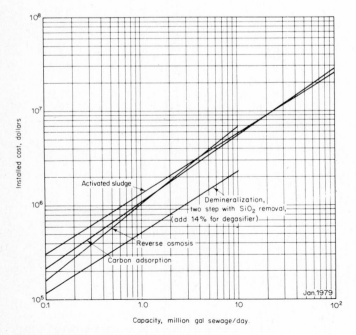

Figure B-8 Cost of wastewater treatment plants.

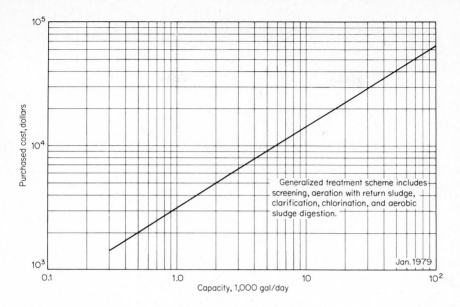

Figure B-9 Cost of small packaged wastewater treatment plants.

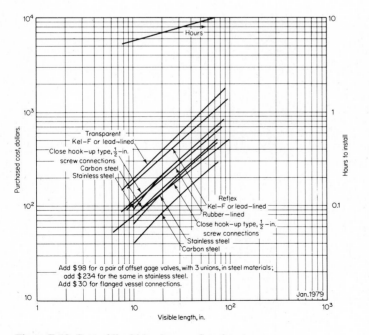

Figure B-10 Cost of liquid level gages, flat-glass type.

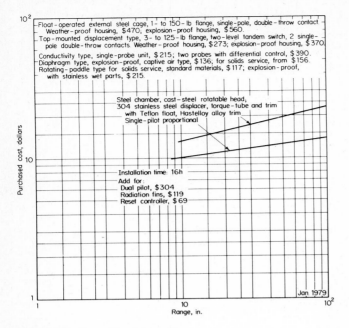

Float-operated external steel cage, 1- to 150-lb flange, single-pole, double-throw contact. Weather-proof housing, $470; explosion-proof housing, $560.
Top-mounted displacement type, 3- to 125-lb flange, two-level tandem switch, 2 single-pole double-throw contacts. Weather-proof housing, $273; explosion-proof housing, $370.
Conductivity type, single-probe unit, $215; two probes with differential control, $390.
Diaphragm type, explosion-proof, captive air type, $136; for solids service, from $156.
Rotating-paddle type for solids service, standard materials, $117; explosion-proof, with stainless wet parts, $215.

Steel chamber, cast-steel rotatable head;
304 stainless steel displacer, torque-tube and trim
with Teflon float, Hastelloy alloy trim
Single-pilot proportional

Installation time: 16h
Add for:
Dual pilot, $304
Radiation fins, $119
Reset controller, $69

Purchased cost, dollars

Jan. 1979

Range, in.

Figure B-11 Cost of level controllers.

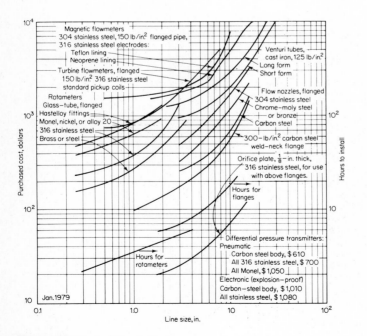

Magnetic flowmeters
304 stainless steel, 150 lb/in² flanged pipe,
316 stainless steel electrodes:
Teflon lining
Neoprene lining
Turbine flowmeters, flanged
150 lb/in² 316 stainless steel
standard pickup coils
Rotameters
Glass-tube, flanged
Hastelloy fittings
Monel, nickel, or alloy 20
316 stainless steel
Brass or steel

Venturi tubes,
cast iron, 125 lb/in²
Long form
Short form

Flow nozzles, flanged
304 stainless steel
Chrome-moly steel
or bronze
Carbon steel

300-lb/in² carbon steel
weld-neck flange

Orifice plate, $\frac{1}{8}$-in. thick,
316 stainless steel, for use
with above flanges.

Hours for
flanges

Hours for
rotameters

Differential pressure transmitters:
Pneumatic
Carbon steel body, $610
All 316 stainless steel, $700
All Monel, $1,050
Electronic (explosion-proof)
Carbon-steel body, $1,010
All stainless steel, $1,080

Purchased cost, dollars

Hours to install

Jan. 1979

Line size, in.

Figure B-12 Cost of flow indicators.

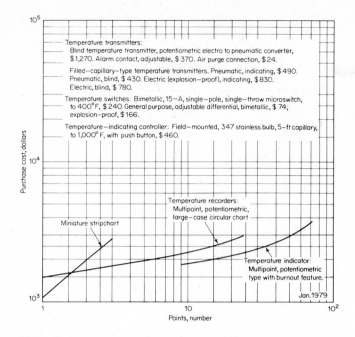

Figure B-13 Cost of temperature recorders and indicators.

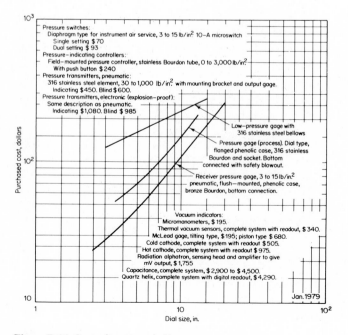

Figure B-14 Cost of pressure indicators.

DESIGN PROBLEMS

CONTENTS

MAJOR PROBLEMS†

Problem 1. Plant for Solvent Rendering of Raw Tissues

MEMORANDUM
TO: Assistant Design Engineer
FROM: A. B. White, Head
Process Development Division

You are to prepare a complete preliminary-estimate design for a new plant for the solvent rendering of raw tissues by an extraction process. Some pieces of equipment are now on hand which we believe can be used in the new plant. Please submit a complete report on the design which you feel will be most favorable for our company. We are particularly interested in total investment, yearly profit before taxes, and the probable percent return on the investment. We shall be interested in the reason for your particular choice of solvent. The design report should also include the number of operators necessary and the approximate operating procedure.

We are considering expansion of our present plant to include an extraction.

† Time period of 30 days recommended for individual student solution.

process for treating 50 tons of raw fish per day. The average analysis of the fish as received by our plant is as follows:

	% by weight
Water	70
Soluble oils	10
Insolubles	20

The general process has been described in some detail in the literature and in patents (see references at end of problem); however, our proposed method is briefly presented in the following:

THE GENERAL PROCESS

The process is to be carried out at a sufficiently low temperature for the biological substances to remain essentially unchanged, except for the removal of oil and water. The top temperature limit for the insoluble materials and the oils is 90°C.

The raw fish as received are ground and delivered to a slurry tank where the pulped material is agitated. The slurry is then sent to cookers where the oil is extracted by a suitable solvent and the water is evaporated along with the solvent. The evolved vapors are condensed, and the solvent and the water are separated by decanting.

The liquid-and-solid mixture from the batch cookers is filtered, and the solid is desolventized to give a fish-meal product.

The filtrate, containing solvent and dissolved fish oil, is sent to a steam-jacketed kettle where the oil is recovered by distilling off the solvent. The oils are finally passed through a steam stripper where the final traces of solvent are removed and the oil is deodorized. The fish-oil product is obtained from the bottom of the steam stripper.

A general flow diagram is included with this problem (Fig. C-1), but this diagram does not show any of the details.

Solvents The following solvents should be considered for possible use: trichloro-ethylene, ethylene dichloride, perchloroethylene, and carbon tetrachloride. The solvents may be purchased in tank-car lots since storage space is available. Note that some of these are on the toxic list.

Special conditions To ensure adequate water removal, the cooker must be operated for $\frac{1}{2}$ hr at the boiling temperature of the pure solvent.

The final dry fish meal may not contain more than 1 percent oil by weight.

The weight ratio of solvent to insolubles at the end of each cooker batch must be 3:1. The fish-meal filter cake must be washed with a weight of solvent at least equal to the weight of the dry fish meal in order to ensure adequate oil removal.

The mixture in the slurry tank contains 50 wt % solvent and 50 wt % fish.

Cooling water is available at 15°C.

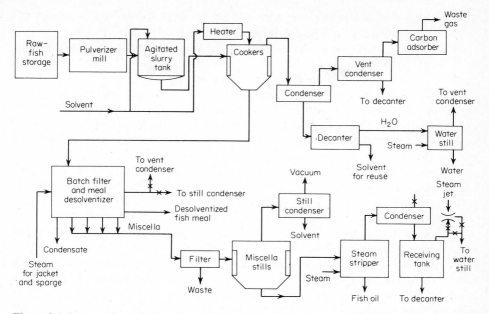

Figure C-1 Suggested rough flow diagram showing extraction process for fish-oil recovery. (Details are not included, and changes may be necessary.)

The solid material remaining after filtration contains 1.0 lb of solvent per pound of dry fish meal.

Tanks larger than 9 ft in diameter or 11 ft high cannot be used under the plant conditions.

The steam pressure in the plant is 30 psig.

Superheated solvent vapor may be used to provide additional heat for the cooker.

Assumptions and data The effective molecular weight of the oil may be assumed to be 150. The vapor pressure of the oil is negligible under the conditions of the process.

Heat capacities and specific volumes for individual components making up the mixtures may be considered as additive. The following values apply between 20 and 100°C.

$$c_p \text{ of insolubles} = 0.2 \text{ Btu/(lb)(°F)}$$
$$c_p \text{ of oil} = 0.3 \text{ Btu/(lb)(°F)}$$
$$\text{sp gr of oil} = 0.7$$
$$\text{Equivalent sp gr of insolubles} = 0.8$$

The only source of solvent loss is from the waste water discarded from the water still.

The liquid leaving all condensers is 5°C below the boiling point.

The temperature of the slurry entering the cookers is 25°C.

The temperature of the liquid entering the Miscella stills is 30°C.

The liquid enters the steam stripper at the same temperature it had leaving the Miscella still.

The desolventized fish meal contains no solvent.

The total fixed-capital investment for the installed equipment may be obtained by multiplying the cost of all the primary and accessory equipment by 4. This factor takes care of construction costs and all other fixed costs such as piping, valves, instrumentation, new buildings, etc. (This, of course, does not include the original solvent cost or any raw materials.)

Overall heat-transfer coefficient in all condensers = 150 Btu/(h)(ft^2)(°F).

Overall heat-transfer coefficient in cookers and in Miscella stills = 100 Btu/(h)(ft^2)(°F) (applies for walls or tubes).

Coils may not be used in the cookers.

The liquid leaving the Miscella still may be assumed to contain 8 percent solvent by weight.

The steam stripper is steam-jacketed, and the oil must leave the bottom at a temperature less than 100°C. In order to ensure complete removal of the solvent from the oil and to deodorize the oil, use excess steam amounting to 30 percent of the weight of the oil.

No agitators are necessary in the cookers or Miscella tanks, because the boiling supplies sufficient agitation.

The total amount of steam theoretically necessary for the process should be increased by 20 percent to take care of unaccounted-for heat losses.

The plant operates 300 days per year.

Equipment The following equipment is now available at the plant:

> Fish-pulverizing mill
> All necessary storage tanks and receiving tanks
> The slurry tank complete with agitator
> Batch filter and meal desolventizer
> Decanter still
> Water still
> Vacuum ejectors (valued at $3000)
> Waste filter
> Steam stripper
> Vent condenser and carbon adsorber

These pieces of equipment are not installed. Preliminary estimates have indicated that all of these pieces of equipment have adequate capacity for the present design. They are valued at a total of $180,000. For the purpose of your cost estimates, you may assume that $180,000 will cover the cost of the listed equipment. *However*, your design should include complete specifications for *all* the equipment necessary for the new plant. In this way, we will be able to have a final check to show us if the present equipment on hand is satisfactory.

It will be necessary to purchase the cookers, the Miscella stills, the condensers and the heaters for these pieces of equipment, and all pumps.

Costs The necessary working-capital investment which must be kept on hand for the new plant is estimated to be $90,000.

Annual fixed charges = 25 percent of total investment.

Annual direct production costs for this process (with the exception of raw material and steam costs) plus costs for plant overhead, administration, office help, distribution, and contingencies are $300,000 per year. This $300,000 includes the cost of operating any vacuum equipment and is unchanged even if vacuum is not used in the process.

Tanks Suitable for cookers or Miscella stills (diameter less than 9 ft):

	Cost for one 1000-gal tank (installed)	
Material	Without steam jacket	With steam jacket
Carbon steel	$ 7,000	$12,000
Stainless steel	18,000	27,000

The tanks are available in standard sizes of (in gallons) 100, 200, 300, 500, 750, 1000, 1500, 2000, 5000, 10,000.

Assume that the cost of each tank varies as $(volume)^{2/3}$.

If heating coils are to be inserted in a tank, increase the cost of the tank by 10 percent and by 10 times the cost of the pipe or tubing making up the coil. (This is an approximation to take care of extra costs for fabrication, special materials, etc. It applies only to a reasonable length of coil.)

Heat exchangers. Shell-and-tube type (for condensers and heaters):

	Dollars per ft^2 of tube surface (installed) (pressures from 25 psig to 0 in. Hg)					
	Square feet					
Material	50	100	200	300	500	1000
Carbon steel	100	80	58	48	40	30
Stainless steel	140	112	81	67	56	42

Steam cost. $0.80/1000 lb.

Materials

    Cost of raw fish (delivered at plant) = $90.00/ton
    Selling price for fish meal of plant-product grade = $0.24/lb
    Selling price for fish oil of plant-product grade = $0.40/lb

References
1. E. Levin and R. K. Finn, A Process for Dehydrating and Defatting Tissues at Low Temperature, *Chem. Eng. Progr.*, **51**:223 (1955).
2. E. Levin and F. Lerman, An Azeotropic Extraction Process for Complete Solvent Rendering of Raw Tissues, *J. Am. Oil Chem. Soc.*, **28**:441 (1951).
3. E. M. Worsham and E. Levin, Simultaneous Defatting and Dehydrating of Fatty Substances, U.S. Patent No. 2,503, 312 (Apr. 11, 1950).
4. E. Levin, Production of Dried, Defatted Enzymatic Material, U.S. Patent No. 2,503,313 (Apr. 11, 1950).
5. E. Levin, Drying and Defatting Tissues, U.S. Patent No. 2,619,425 (Nov. 25, 1952).
6. Defatting Distillery, *Ind. Eng. Chem.*, **42**:14A (June, 1950).
7. E. W. McGovern, Chlorohydrocarbon Solvents, *Ind. Eng. Chem.*, **35**:1230 (1943).

Problem 2. Pentaerythritol Plant

You are a design engineer with the ABC Chemical Company, located at Wilmington, Del. Mr. Charles B. King, Supervisor of the Process Development Division, has asked you to prepare a preliminary design of a new plant that will produce pentaerythritol.

The following memo has been given to you by Mr. King:

Dear Sir:

We are considering construction of a pentaerythritol plant at Louisiana, Missouri. At the present time, we have an anhydrous ammonia plant located at Louisiana, Missouri, and we own sufficient land at this site to permit a large expansion.

As yet, we have not decided on the final plant capacity, but we are now considering construction of a small plant with a capacity of 60 tons of technical grade pentaerythritol per month. Any work you do on the design of this plant should be based on this capacity.

Sufficient water, power, and steam facilities for the new plant are now available at the proposed plant site. Some standby equipment is located at our existing plant in Louisiana, and we may be able to use some of this in the pentaerythritol plant. However, you are to assume that all equipment must be purchased new.

Please prepare a preliminary design for the proposed plant. This design will be surveyed by the Process Development Division and used as a basis for further decisions on the proposed plant. Make your report as complete as possible, including a detailed description of your recommended process, specifications and cost estimates for the different pieces of equipment, total capital investment, and estimated return on the investment assuming we can sell all our product at the prevailing market price. We shall also be interested in receiving an outline of the type and amount of labor required, the operating procedure, and analytical procedures necessary.

Present what you consider to be the best design. Although our legal department will conduct a patent survey to determine if any infringements are involved, it would be helpful if you would indicate it if you know that any part of your design might involve patent infringements.

Enclosed you will find information on our utilities situation, amortization policy, labor standards, and other data.

C. B. King, Supervisor
Process Development Division

Enclosure: Information for Use in Design of Proposed Pentaerythritol Plant

Utilities. Water—available for pumping at 65°F; cost, $0.09 per 1000 gal. Steam—available at 100 psig; cost, $0.80 per 1000 lb. Electricity—cost, $0.04 per kWh.

Storage. No extra storage space is now available.

Transportation facilities. A railroad spur is now available at the plant site.

Labor. Chief operators, $12.00 per hour; all helpers and other labor, $8.00 per hour.

Amortization. Amortize in 10 years.

Return on investment. For design calculations, we require at least a 20 percent return before income taxes on any unnecessary investment.

Income taxes. The income-tax load for our company amounts to 46 percent of all profits.

Raw-materials costs. All raw materials must be purchased at prevailing market prices.

Heats of reaction. With sodium hydroxide as the catalyst, 64,000 Btu is released per pound mole of acetaldehyde reacted. With calcium hydroxide as the catalyst, 60,000 Btu is released per pound mole of acetaldehyde reacted.

Heat-transfer coefficients. Steam to reactor liquid, 200 Btu/(h)(ft^2)(°F). Water to reactor liquid, 150 Btu/(h)(ft^2)(°F).

Problem 3. Formaldehyde Plant

Illinois Chemical Process Corporation
Plant Development Division
Urbana, Illinois

Gentlemen:

We are considering entering the field of production of formaldehyde, and we should like to have you submit a complete preliminary design for a 70-ton-per-day formaldehyde plant (based on 37 wt % formaldehyde) to us.

We have adequate land available for the construction of the plant at Centralia, Illinois, and sufficient water and steam are available for a plant of the desired capacity. Please make your report as complete as possible, including a detailed description of your recommended process, specifications and cost estimates for the different pieces of equipment, total capital investment, and estimated return on the investment assuming we can sell all our production at the prevailing market price. We shall also be interested in receiving an outline of the amount and type of labor required, the operating procedure, and analytical procedures necessary.

Enclosed you will find information concerning our utilities situation, amortization policy, labor standards, product specification, etc.

Very truly yours,

A. B. Blank
Technical Superintendent
ABB:jf   Centralia Chemical Company
Enclosure   Centralia, Illinois

Enclosure

Utilities. Water—available for pumping at 70°F; cost, $0.09 per 1000 gal. Steam—available at 200 psig; cost, $0.80 per 1000 lb. Electricity—cost, $0.04 per kWh.

Storage. No extra storage space is now available.

Transportation facilities. A railroad spur is now available at the plant site.

Land. The plant site is on land we own which is now of no use to us. Therefore, do not include the cost of land in your analysis.

Labor. Chief operators, $12.00 per hour; all helpers and other labor, $8.00 per hour.

Amortization. Amortize in 10 years.

Return on investment. For design comparison, we require at least a 15 percent return per year on any unnecessary investment.

Product specification. All formaldehyde will be sold as N.F. solution containing 37.0 wt % formaldehyde and 8 wt % methanol. All sales may be considered to be by tank car.

MINOR PROBLEMS†

Problem 1. Optimum Temperature for Sulfur Dioxide Reactor

The head of your design group has asked you to prepare a report dealing with the use of a special catalyst for oxidizing SO_2 to SO_3. This catalyst has shown excellent activity at low temperatures, and it is possible that it may permit you to get good SO_3 yields by using only one standard-size reactor instead of the conventional two.

Your report will be circulated among the other members of your design group and will be discussed at the group meetings. This report is to be submitted to the head of your design group.

Some general remarks concerning your report follow:

A. The report should include the following:
   1. Letter of transmittal (the letter to the head of your group telling him you are submitting the report and giving any essential results if applicable)
   2. Title page
   3. Table of Contents
   4. Summary (a concise presentation of the results)
   5. Body of report
      a. Introduction (a brief discussion to explain what the report is about and reason for the report; no results should be included here)
      b. Discussion (outline of method of attack on the problem; do not include any detailed calculations; this should bring out technical matters not important enough to be included in the Summary; indicate assumptions and reasons; include any literature survey results of importance; indicate possible sources of error, etc.)
      c. Final recommended conditions (or design if applicable) with appropriate data (a drawing is not necessary in this case although one could be included if desired)
   6. Appendix
      a. Sample calculations (clearly presented and explained)
      b. Table of nomenclature (if necessary)

---

† Time period of 14 days recommended for individual student solution.

   c. Bibliography (if necessary)
   d. Data employed
B. The outline as presented above can be changed if desired (for example, a section on conclusions and recommendations might be included)
C. The report can be made more effective by appropriate subheadings under the major divisions

The Problem

A new reactor has recently been purchased as a part of a contact sulfuric acid unit. This reactor is used for oxidizing SO_2 to SO_3 employing a vanadium oxide catalyst.

   Using the following information and data, determine the temperature at which the reactor should be operated to give the maximum conversion of SO_2 to SO_3, and indicate the value (as percent) of this maximum conversion. Ten thousand pounds of SO_2 enters the reactor per day.

   Air is used for the SO_2 oxidation, and it has been decided to use 300 percent excess air. Preheaters will permit the air and SO_2 to be heated to any desired temperature, and cooling coils in the reactor will maintain a constant temperature in the reactor. The reactor temperature and the entering air-and-SO_2 temperature will be the same. The operating pressure may be assumed to be 1 atm.

   The inside dimensions of the reactor are 5 by 5 by 8 ft. One-half of the inside reactor volume is occupied by the catalyst.

   Your laboratory has tested a special type of vanadium oxide catalyst, and, on their recommendation, you have decided to use it in the reactor. This catalyst has a void fraction of 60 percent (i.e., free space in catalyst/total volume of catalyst = 0.6).

   The reaction $2SO_2 + O_2 \rightarrow 2SO_3$ proceeds at a negligible rate except in the presence of a catalyst.

   Your laboratory has run careful tests on the catalyst. The results indicate that the reaction is not third-order but is a complex function of the concentrations. Your laboratory reports that the reaction rate is proportional to the SO_2 concentration, inversely proportional to the square root of the SO_3 concentration, and independent of the oxygen concentration.[†]

   This may be expressed as

$$\frac{dx}{dt} = k\frac{a - x}{x^{1/2}}$$

where $a = SO_2$ originally present as pound mol/ft^3
   $x =$ number of lb moles of SO_2 converted in t sec of catalyst contact time per cubic foot of initial gas
   $k =$ specific reaction-rate constant, (lb mol/ft^3)$^{1/2}$/s; this may be assumed to be constant at each temperature up to equilibrium conditions

---

   † Modern tests indicate that this may not be the case with some vanadium oxide reactors. However, the information given above should be used for the solution of this problem.

Table 1

$k \times 10^4$, (lb mol/ft^3)$^{1/2}$/s†	Temperature, °C
14	350
30	400
60	450
100	500
210	550

† Applicable only to the conditions of this problem.

The laboratory has obtained the data given in Table 1 for the reaction-rate constant. These data are applicable to your catalyst and your type of reactor.

The data of standard-state free-energy values at different temperatures given in Table 2 were obtained from the literature. These data apply to the reaction

$$SO_2 + \tfrac{1}{2}O_2 \rightleftharpoons SO_3$$

The fugacities of the gases involved may be assumed to be equal to the partial pressures.

Table 2

ΔF^0, cal/g mol	Temperature, °C
−9120	350
−8000	400
−7000	450
−5900	500
−4900	550

Problem 2. Heat-Exchanger Design

TO: Assistant Process Engineer
FROM: Dr. A. B. Green, Chief Design Engineer
            Mountain View Chemical Company
            Boulder, Colorado

We are in the process of designing a catalytic cracker for our petroleum division. As part of this work, will you please submit a design for a single-pass heat exchanger based on the information given below?

It is estimated that 200,000 gph of oil A must be heated from 100 to 230°F. The heating agent will be saturated steam at 50 psig.

The engineering department has indicated that the exchanger will cost $40.00 per square foot of inside heating area. This cost includes all installation.

You can neglect any resistance due to the tube walls or steam film; so all the heat-flow resistance will be in the oil film.

The cost of power is 5 cents per kilowatthour, and the efficiency of the pump and motor installation is estimated to be 60 percent.

Do not consider the cost of steam or exchanger insulation in your analysis.

The oil will flow inside the tubes in the heat exchanger. Following are data on oil A which have been obtained from the Critical Tables and the Chemical Engineers' Handbook:

Avg viscosity of oil A between 100 and 230°F = 6 centistokes
Avg density of oil A between 100 and 230°F = 0.85 g/cm^3
Avg sp ht of oil A between 100 and 230°F = 0.48
Avg thermal conductivity of oil A between 100 and 230°F = 0.08 Btu/ (h)(ft^2)(°F/ft)

We recommend that the Reynolds number be kept above 5000 in this type of exchanger.

The tubes in the exchanger will be constructed from standard steel tubing. Tube sizes are available in $\frac{1}{2}$-in.-diameter steps. Tubing wall is 16 BWG.

We are particularly interested in the diameter of the tubes we should use, the length of the tubes, and the total cost of the installed unit. In case the exchanger length is unreasonable for one unit, what would you recommend?

Assume the unit will operate continuously for 300 days each year. Fixed charges are 16 percent.

Remember that our company demands a 20 percent return on all extra fixed-capital investments.

You may base your calculations on a total of 100 tubes in the exchanger.

Please submit this information as a complete formal report. Include a short section outlining what further calculations would have been necessary if the specification of 100 tubes in a one-pass exchanger had not been given.

Following are recommended assumptions:

You may assume that the Fanning friction factor can be represented by

$$\hat{f} = \frac{0.04}{(N_{Re})^{0.16}}$$

You may assume that the oil-film heat-transfer coefficient is constant over the entire length of the exchanger.

For simplification, assume that the oil-film heat-transfer coefficient may be represented by (standard heat-transfer nomenclature)

$$h = 0.023 \frac{k}{D} (N_{Re})^{0.8}(N_{Pr})^{0.4}$$

where all variable values are at the average value between 100 and 230°F.

Problem 3. Design of Sulfur Dioxide Absorber

You are a member of a group of design engineers designing equipment for the recovery of SO_2 from stack gases.

The group leader has asked you to determine the optimum size of the SO_2 absorption tower. Specifically, he has asked you to determine the height and cross-sectional area of the optimum absorption tower and to present your recommendations in the form of a formal report.

Your group has held several meetings to discuss the proposed overall design. Following is a list of conditions, assumptions, and data on which the group has decided to base the design:

100,000 ft^3 of gas per minute at 300°F and 1 atm are to be treated.
The entering gas contains 0.3 percent by volume SO_2 and 11.0 percent CO_2, with the balance being N_2, O_2, and H_2O.
The average molecular weight of the entering gas = 29.4.
The mole percent SO_2 in the exit gas is to be 0.01 percent.
The entering and exit pressures of the absorption column may be assumed to be 1 atm for purposes of calculating the SO_2 pressures.

The zinc oxide process will be used for recovering the SO_2. In this process, a solution of H_2O, $NaHSO_3$, and Na_2SO_3 is circulated through the absorption tower to absorb the SO_2. This mixture is then treated with ZnO, and the $ZnSO_3$ formed is filtered off, dried, and calcined to yield practically pure SO_2. The ZnO from the calciner is reused, and the sulfite-bisulfite liquor from the filter is recycled.

The absorption tower will contain nonstaggered wood grids of the following dimensions:

Clearance = 1.5 in.
Height = 4 in.
Thickness = $\frac{1}{4}$ in.
Free cross-sectional area = 85.8%
Active absorption area per cubic foot of volume = a = 13.7 ft^2/ft^3

The average density of the gas at the tower entrance can be assumed to be 0.054 lb/ft^3. The sulfite-bisulfite liquid has a density of 70 lb/ft^3 and can be considered as having a zero equilibrium SO_2 vapor pressure at both the inlet and outlet of the tower.

The sulfite-bisulfite liquid must be supplied at a rate of 675 $lb/(h)(ft^2$ of column cross-sectional area).

The optimum design can be assumed to be that corresponding to a minimum total power cost for circulating the liquid and forcing the gas through the tower. You may assume that this optimum corresponds to the optimum that would be obtained if fixed charges were also considered.

The following simplified equations are applicable for grids of the dimensions to be used:

$$K_g = 0.00222(G_0)^{0.8}$$

$$\frac{\Delta h_w}{L} = 0.23 \times 10^{-7}(G_0)^{1.8}$$

where K_g = molar absorption coefficient, lb mole of component absorbed/ $(h)(ft^2)(atm)_{\text{log mean}}$

G_0 = superficial mass velocity of gas in tower, $lb/(h)(ft^2)$

Δh_w = pressure drop through tower, in. of water

L = height of tower, ft

The liquid is put into the absorption tower by means of a nozzle at the top of the tower. The pressure just before the nozzle is 35 psig. Assume the pump for the liquid must supply power to lift the liquid to the top of the tower and compress the liquid to 35 psig. Use a 10 percent safety factor on the above pumping-power requirements to take care of the friction in the lines and other minor losses.

The gas blower has an overall efficiency of 55 percent.

The pump has an overall efficiency of 65 percent.

Problem 4. Utilization of Liquid Methane Refrigeration for Liquefaction of Nitrogen and Oxygen

Management of a natural-gas transmission company is considering using liquid methane as a heat sink in producing 210 tons/day of liquid nitrogen and 64 tons/day of liquid oxygen for a neighboring customer. Accordingly, one of the company's engineers has outlined a scheme for doing this. The process description is as follows:

Air is compressed from atmospheric conditions to 20 atm and then purified. The clean dry gas is then chilled by counter-current heat exchange with liquid methane boil-off. The partially liquefied air stream serves as the reboiler stream for a dual-pressure air separation column. Before entering the 5-atm high-pressure lower section of the dual-pressure column, the high-pressure air stream is bled through a J-T valve. The bottoms of the lower column is enriched to approximately 40 mole % oxygen and is the feed for the 1-atm low-pressure upper column. The condensing vapors in the high-pressure column serve as a reboiler for the low-pressure column. High-purity N_2 is withdrawn from the top of the high-pressure column, reduced in pressure, and used as reflux for the low-pressure column. High-purity liquid oxygen is withdrawn as bottoms from the low-pressure column while high-purity nitrogen vapor is withdrawn from the top of this same column. The latter stream is warmed to room temperature, compressed to 20 atm, and then cooled in this same heat exchanger down to a low temperature. This high-pressure pure nitrogen stream is then condensed by counter-current exchange with the liquid methane. After liquefaction, it is expanded to atmospheric pressure, and the resulting vapor is recycled and combined with the overhead stream from the low-pressure column to complete the liquefaction cycle all over again.

As a chemical engineer, you have been asked to analyze the process and make recommendations. Start by making as complete a flow sheet and material balance as possible assuming an 85 percent operating factor. Outline the types of equipment necessary for the process. Determine approximate duties of heat exchangers and sizes of major pieces of equipment. Instrument the plant as completely as possible outlining special problems which might be encountered. List all of the additional information which would be needed in order to finish completely the design evaluation.

Problem 5. Production of High-Purity Anhydrous Ammonia

A chemical company is considering the production of 1000 tons/day of high-purity anhydrous ammonia. The method selected is a high-pressure steam methane reforming process. The process description is as follows:

The steam methane reforming process produces ammonia by steam reforming natural or refinery gas under pressure, followed by carbon monoxide shift, purification of raw synthesis gas, and ammonia synthesis. In the process, saturated and unsaturated hydrocarbons are decomposed by steam according to the basic equation:

$$CH_4 + H_2O \rightarrow CO + 3H_2$$

Feed streams high in olefins or sulfur require pretreatment. The primary reformer converts about 70 percent of a natural-gas feed into raw synthesis gas, in the presence of steam using a nickel catalyst. In the secondary reformer, air is introduced to supply the nitrogen required for ammonia. The heat of combustion of the partially reformed gas supplies the energy to reform the remainder of the gas after reacting with the oxygen in the air. High-pressure reforming conserves 30–40 percent in compressor horsepower over usual practices giving low-pressure synthesis gases.

Next, the mixture is quenched and sent to the shift converter. Here CO is converted to CO_2 and H_2. When heat is still available after satisfying the water requirement for the shift reaction, a waste-heat boiler may be installed. Shift reactor effluent, after heat recovery, is cooled and compressed, then goes to the gas-purification section. CO_2 is removed from the synthesis gas in a regenerative MEA (monoethanolamine) or other standard recovery system. After CO_2 removal, CO traces left in the gas stream are removed by methanation. The resulting pure synthesis gas passes to the oil separator, is mixed with a recycle stream, cooled with ammonia refrigeration, and goes to the secondary separator where anhydrous ammonia (contained in the recycle stream) is removed. Synthesis gas is then passed through heat exchange and charged to the catalytic ammonia converter. Product gas from the converter is cooled and exchanged against converter feed gas. Anhydrous liquid ammonia then separates out in the primary separator and, after further cooling, goes to the anhydrous ammonia product flash drum. The feed to the reforming section is normally in excess of 300 psig. The pressure, however, is not fixed and may be varied to provide optimum design for specific local conditions. Temperatures in primary and secondary reformers are 1400 to 1800°F, while shift reaction temperatures are 700 to 850°F in the first stage and 450 to 550°F in the second stage. Ammonia synthesis is normally performed at 3000 psig. Temperature in the quench-type ammonia converter is accurately and flexibly controlled inside the catalyst mass to allow a catalyst basket temperature gradient giving a maximum yield of ammonia per pass, regardless of production rate.

Analyze this process and make as complete a flow sheet and material balance as possible assuming a 90 percent operating factor. Outline the types of equipment necessary for the process. Determine approximate duties of heat exchangers and sizes of the major pieces of equipment. What additional information will be needed in order to finish completely the preliminary design evaluation?

PRACTICE-SESSION PROBLEMS†

Problem 1. Cost for Hydrogen Recovery by Activated-Carbon Adsorption Process

What is the cost, as cents per 1000 ft^3 (at SC) of 95 mol % H_2, for recovering hydrogen of 95 mol % purity from 10 million ft^3 (at SC) of gaseous feed per day if the following conditions apply?

Feed composition by volume = 72.5 percent H_2 and 27.5 percent CH_4.

A hydrogen-recovery method based on selective adsorption by activated carbon will be used.

For the carbon adsorption, three separate beds will be needed so that one bed can be in continuous use while the other two are being desorbed or reactivated. Base the recovery on a single pass and an absorption cycle of 1 hour per bed.

0.00838 mol of material is adsorbed per hour per pound of activated carbon.

The composition of the adsorbed phase is 96.8 mol % CH_4, the balance being hydrogen.

The cost of activated carbon is $2.25 per pound.

The annual amount of additional carbon necessary is 15 percent of the initial charge.

The total plant cost (equipment, piping, instrumentation, etc.) equals $2,930,000.

The capital investment equals the total plant cost plus process materials (process materials are considered as auxiliaries—i.e., process materials are only the initial charge of activated carbon).

The capital investment must be completely paid off (no scrap value) in 5 years.

The operating cost per year, including labor, fuel, water, feed, regeneration,

---

† The following problems are recommended for solution by students working in groups of two or three during a 3-h practice session.

fixed charges minus depreciation, and repairs and maintenance, equals $1,390,000 (operating costs not included in this value are the cost of the additional make-up carbon necessary and depreciation).

The plant operates 350 days/year.

Problem 2. Adsorption-Tower Design for Hydrogen Purification by Activated Carbon

Your plant is producing 10 million ft^3 (measured at SC) per day of a gas containing 72.5 vol % H$_2$ and 27.5 vol % CH$_4$. It is proposed to pass this gas through activated carbon to obtain a product gas containing 95 vol % H$_2$. The activated carbon shows a preferential selective adsorption of the CH$_4$.

An adsorption-desorption-regeneration cycle using three fixed beds will be used. One bed will be regenerated and purged over an 8-h period. During this period, the other two beds will be on alternate 1-h adsorption and 1-h desorption cycles to permit a continuous operation. A single pass of the gas will be used.

Each individual bed may be designed to include a number of carbon-packed towers in parallel. The diameter of the individual towers must all be the same, and the diameter may be 6, 9, 12, or 15 ft.

Determine the following:

1. The number of individual units (or towers) in each bed to give the minimum total cost for the towers.
2. The height and diameter of the towers for the conditions in part 1.

Data and information previously obtained for the chosen conditions of the process (i.e., adsorption at 400 psia, desorption at 20 psia, and an average adsorption temperature of 110°F) *are as follows:*

0.0063 lb mol of material is adsorbed per hour per pound of activated carbon. The composition of the adsorbed phase is 96.8 mol % CH$_4$, the balance being H$_2$. The carbon has a bulk density of 0.30 g/cm^3.

The gas velocity in the adsorbers should not exceed 1 ft/s based on the cross-sectional area of the empty vessel. This applies to all the cycles including the adsorption and the regeneration and purge.

The feed gases enter the adsorbers at 400 psia and 80°F. The product gases leave the adsorbers at almost 400 psia and a temperature of 140°F. Regeneration includes 30 ft^3 (SC) of flue gas per pound of carbon and 80 standard cubic feet of purging air per pound of carbon. The flue gas is at 600°F (its maximum temperature) and 5 psig, and the air is at 90°F and 5 psig. The air may reach a maximum temperature of 600°F at the start of the purging.

For each head (either top or bottom), add equivalent cost of 5 ft additional length per vessel. Cost data are given in Table 3.

Table 3 Cost data

Column diameter, ft	6	9	12	15
Dollars per foot of length	720	1320	2020	3600
Cost per tower for skirt or support, dollars	1200	2000	2800	3200

Problem 3. Design of Rotary Filter for Sulfur Dioxide Recovery System

As a member of a design group working on the design of a recovery system for SO_2, you have been asked to estimate the area necessary for a rotary vacuum filter to handle a zinc sulfite filtration. You are also to determine the horsepower of the motor necessary for the vacuum pump. Do not include any safety factors in your results.

The following conditions have been set by your group:

A slurry containing 20 lb of liquid per pound of dry $ZnSO_3 \cdot 2.5H_2O$ is to be filtered on a continuous rotary filter to give a cake containing 0.20 lb of H_2O per pound of dry hydrate.

One hundred pounds of $ZnSO_3 \cdot 2.5H_2O$ in the slurry mixture will be delivered to the filter per minute.

A drum speed of 0.33 rpm will be used, and a vacuum of 10.2 in. mercury below atmospheric pressure will be used.

The fraction of drum area submerged will be assumed to be 0.25.

The fraction of drum area available for air suction will be assumed to be 0.10.

The temperature of the slurry is 110°F, and the air into the vacuum pump can be considered to be at 110°F. C_p/C_v for air at 110°F = 1.4. The temperature and pressure of the air surrounding the filter are 70°F and 1 atm, respectively.

The specific cake resistance and the specific air-suction cake resistance can be assumed to be independent of pressure drop, drum speed, temperature, fraction of drum submerged, fraction of drum available for suction, and slurry concentration. However, to eliminate possible errors due to this assumption, a lab test should be run at conditions approximating the planned design. The results of these tests can be used as a basis for the design (i.e., cake and filtrate compositions and densities can be assumed to be the same for the design as those found in the lab).

The vacuum pump and motor have an efficiency (isentropic) of 85 percent.

It can be assumed that all of the $ZnSO_3 \cdot 2.5H_2O$ is filtered off and none remains in the filtrate.

The resistance of the filter medium can be assumed to be negligible.

Laboratory data compiled at the request of the SO_2-recovery design group (results of filtration of zinc sulfite slurry on an Oliver rotary vacuum filter) are as follows:

Total area of filter	4.15 ft^2
Fraction of area submerged	0.20
Fraction of area available for air suction	0.10

Vacuum	9 in. Hg below atmospheric pressure
Slurry concentration	12 lb liquid/lb dry $ZnSO_3 \cdot 2.5H_2O$
Temperature	110°F
Drum speed	0.40 r/min
Density of wet cake leaving filtering zone	100 lb/ft^3
Pounds of liquid per pound of dry $ZnSO_3 \cdot 2.5H_2O$ in wet cake leaving filtering zone	0.6
Density of filtrate	68.8 lb/ft^3
Viscosity of filtrate	0.6 centipoise
Pounds of water per pound of dry $ZnSO_3 \cdot 2.5H_2O$ in final cake	0.20
Time interval	5.0 min
Volume of filtrate	0.95 ft^3
Volume of air at SC	8.5 ft^3

Problem 4. Return on Investment for Chlorine Recovery System

The off-gas from a chloral production unit contains 15 vol % Cl_2, 75 vol % HCl, and 10 vol % $EtCl_2$. This gas is produced at a rate of 150 cfm based on 70°F and 2 psig. It has been proposed to recover part of the Cl_2 by absorption and reaction in a partially chlorinated alcohol (PCA). The off-gas is to pass continuously through a packed absorption tower counter-current to the PCA, where Cl_2 is absorbed and partially reacts with the PCA. The gas leaving the top of the tower passes through an alcohol condenser and thence to an existing HCl recovery unit.

The reaction between absorbed Cl_2 and the PCA is slow, and only part of the absorbed Cl_2 reacts in the tower. Part of this PCA from the bottom of the tower is sent to a retention system where the reaction is given time to approach equilibrium. The rest of the PCA is sent to the chloral production unit. Ethyl alcohol is added to the PCA going to the retention system. Then the PCA is recycled from the retention system to the top of the absorption column, and ethyl alcohol is added at a rate sufficient to keep the recycle rate and recycle concentration constant.

Your preliminary calculations have set the optimum conditions of operations, and all costs have been determined except the cost for the absorption tower. Using the following data and information, determine the yearly percent return on the capital expenditure:

The gases leaving the top of the tower contain 2 vol % Cl_2 (on the basis of no PCA vapors present in the gas).

The PCA entering the top of the tower contains 0.01 mol % free Cl_2.

The PCA leaving the bottom of the tower contains 0.21 mol % free Cl_2.

The recycle rate is 200 gpm (entering the top of the absorption tower).

The gas rate at the top of the column is 27.2 lb mol/h.

Table 4

$m_2 G_2 / L_2$ based on conditions at dilute end of column	Height of a transfer unit (log-mean method), ft
0.178	3.6
0.571	4.8
1.04	6.2
1.53	7.6

The PCA entering the top of the tower has a density of 68.5 lb/ft^3 and an average molecular weight of 70.

The column is operated at a temperature of 35°C and a pressure of 1 atm.

The gases enter the bottom of the tower at 70°F and 2 psig.

No Cl$_2$ is absorbed in the alcohol condenser.

The column is packed with 1-in. porcelain Raschig rings, and a porcelain tower is used.

Laboratory tests with a small column packed with Raschig rings have been conducted. These tests were carried out at 35°C using PCA and off-gas having the same concentrations as those in your proposed design. These data have been scaled up to apply to 1-in. Raschig rings and are applicable to your column. The results are given in Table 4.

The absorption of Cl$_2$ in the PCA follows Henry's law, and the following relation may be used for determining the number of transfer units:

$$N_t = \frac{y_1 - y_2}{\Delta y_m}$$

where Δy_m = log-mean driving force = $\dfrac{\Delta y_1 - \Delta y_2}{\ln (\Delta y_1 / \Delta y_2)}$

$\Delta y_1 = y_1 - y_1^*$
$\Delta y_2 = y_2 - y_2^*$

Equilibrium data for Cl$_2$ and the PCA at 35°C and 1 atm are given in Table 5. These data apply over the entire length of the column.

The maximum allowable velocity (based on conditions of the gas at the inlet

Table 5

Free Cl$_2$ in PCA, x, mole %	Free Cl$_2$ in gas (based on PCA vapor-free gas), y, mole %
0.1	2.0
0.2	4.0
0.3	6.0

to the tower) is 1.5 ft/s. The column will operate at 60 percent of the maximum allowable velocity.

Necessary cost data are given in the following:

1. Tower:

Diameter, in	12	18	24	30	36	42
Porcelain, $/ft of length	280	400	540	720	920	1140
Top or bottom heads, porcelain, each, $	240	340	460	600	760	940

2. Tower packing. One-inch Raschig rings, porcelain = $15.00 per cubic foot.
3. Capital expenditure minus cost of absorption tower = $50,000.
4. Net annual savings (taking alcohol loss and all other costs, such as interest, rent, taxes, insurance, depreciation, maintenance, and other overhead expenses into consideration) = $25,000.

NOTE: This $25,000 has been determined by including an accurate estimate of the absorption-tower cost and can be taken as the actual net savings.

Nomenclature for Prob. 4

G = molar gas flow, mol/h
L = molar liquid flow, mol/h
m = slope of equilibrium curve, y/x
N_t = number of transfer units
y = mole fraction of chlorine in gas
y^* = equilibrium mole fraction of chlorine in gas

Subscripts

1 = bottom of tower
2 = top of tower

Problem 5. Economic Analysis of Chlorine Recovery System

The data shown in Table 6 have been obtained for a chlorine recovery system in which all values have been determined at the optimum operating conditions of retention time and recycle rates.

Determine the following:

1. The percent Cl_2 in exit gas at break-even point.
2. The maximum net annual savings and percent Cl_2 in exit gas where it occurs.
3. The maximum percent return on the capital expenditure and percent Cl_2 in exit gas where it occurs.
4. Which investment would you recommend and why?

Table 6

Mole % Cl_2 in exit gas	Total capital expenditure, dollars	Total annual operating costs (fixed costs, production costs, overhead, etc.), dollars	Gross annual savings by using process, dollars
0.2	368,000	336,000	456,000
1.0	304,000	316,000	444,000
2.0	272,000	296,000	420,000
5.0	212,000	268,000	332,000
10.0	136,000	236,000	188,000

Problem 6. Optimum Thickness of Insulation

Insulation is to be purchased for 3 miles of 10-in.-OD pipe carrying saturated steam at 250°F. The average air temperature for the year for the surroundings is 45°F.

It is estimated that the life period of the installation will be 20 years with negligible scrap value. The sum of fixed charges excluding depreciation is 10 percent, and maintenance is estimated to be 2 percent annually.

One company has submitted a bid which includes installation at a cost of $0.15D^{1.3}$ per lineal foot, where D is the outside diameter of the lagging in inches. Using the following data and equations, what thickness of insulation should be used for this job in order to give a 50 percent return on the full investment?

The line will be in continuous service 365 days/year. The steam is valued at $0.80 per 1000 lb. The thermal conductivity of the lagging is 0.04 Btu/(h)(ft^2) × (°F/ft). Steam-film and pipe resistance may be neglected.

Heat losses by conduction and convection from the surface may be calculated by use of the equation

$$h_c = 0.42\left(\frac{\Delta t_s}{D_o'}\right)^{0.25}$$

where Δt_s = temperature difference between surface of lagging and air, °F

D_o' = OD, in.

h_c = Btu/(h)(ft^2)(°F)

An average value of $h_r = 1.2$ Btu/(h)(ft^2)(°F) may be used to determine heat losses by radiation. This is an adjusted value such that total heat loss per hour may be calculated by the equation

$$Q = (h_c + h_r)A \, \Delta t_s$$

A mathematical setup with all necessary numbers and a description of the method for final solution will be satisfactory.

Problem 7. Capacity of Plant for Producing Acetone from Isopropanol

Acetone is produced by the dehydrogenation of isopropanol according to the following reaction:

$$\underset{\substack{CH_3 \\ CH_3}}{\overset{\substack{CH_3 \; H}}{>}}\!\!C\!<\overset{}{\underset{OH}{}} \quad \xrightarrow{\text{catalyst}} \quad CH_3\!-\!\overset{\overset{\displaystyle O}{\|}}{C}\!-\!CH_3 + H_2$$

The reverse of the above reaction can be neglected.

The catalyst used for the process decreases in activity as the amount of isopropanol fed increases. This effect on the reaction rate is expressed in the following:

$$k = \frac{0.000254NT}{V}\left(2.46 \ln \frac{1}{1-\alpha} - \alpha\right)$$

where α = fraction of isopropanol converted to acetone and side products
 = (moles isopropanol converted)/(moles isopropanol supplied)
$\cdot k$ = reaction-rate constant, s^{-1}
N = lb mol of isopropanol fed to converter per hour
T = absolute temperature, $°R$
V = catalyst volume, ft^3

The feed rate of isopropanol (N) is maintained constant throughout the entire process.

The fresh catalyst has an activity such that $k = 0.30 \text{ s}^{-1}$.

After 10,000 lb of isopropanol per cubic foot of catalyst has been fed, $k = 0.15 \text{ s}^{-1}$.

A plot of *log k* versus *total isopropanol fed as pound moles* is a straight line.

The maximum production of acetone is 76.1 lb mol/h. This maximum production can be considered as at zero time (i.e., when $k = 0.30 \text{ s}^{-1}$).

A constant temperature of 572°F and a constant pressure of 1 atm are maintained throughout the entire process.

The catalyst volume $V = 250 \text{ ft}^3$.

$$\text{Efficiency} = \frac{\text{moles acetone produced}}{\text{moles isopropanol consumed}}(100) = 98\% \text{ at } 572°F$$

The catalyst can be restored to its original activity with a 48-hour reactivation.

The catalyst will be thrown away after the last operating period each year.

The unit will operate 350 days/year (this includes reactivation shutdowns). The other 15 days are used for repairs and replacing the old catalyst.

If nine catalyst reactivations are used per year, what will the production of acetone be in pounds per year?

Outline your method of solution. The actual mathematical calculations are not necessary.

Problem 8. Equipment Design for the Production of Acetone from Isopropanol

You are designing a plant for the production of acetone from isopropanol. Acetone is produced according to the following reaction:

$$CH_3\text{-}\underset{\underset{CH_3}{|}}{\overset{\overset{H}{|}}{C}}\text{-}OH \xrightarrow{\text{catalyst}} CH_3\text{-}\overset{\overset{O}{\|}}{C}\text{-}CH_3 + H_2$$

The reaction can be assumed as irreversible.

The catalyst used in this process decreases in activity as the process proceeds. This effect on the reaction rate can be expressed as follows:

$$k = \frac{0.000254}{V} NT \left(2.46 \ln \frac{1}{1-\alpha} - \alpha \right)$$

where α = fraction of isopropanol converted to acetone and side products
$\quad$ = (moles isopropanol converted)/(moles isopropanol supplied)
$\quad k$ = reaction-rate constant, s^{-1}
$\quad N$ = lb mole of isopropanol fed to converter per hour
$\quad T$ = absolute temperature, °R
$\quad V$ = catalyst volume, ft^3

The catalyst must be regenerated periodically throughout the operation.
The fresh catalyst has an activity such that $k = 0.30 \ s^{-1}$.

The products from the reactor (unconverted isopropanol, acetone, side products, and some water from the impure isopropanol feed) are sent to a continuous distillation column where purified acetone is removed.

A distillation column, calandria, and condenser are now available in your plant, and you are to determine if these can be used for the purification step.

Using the following data, determine whether or not the column, calandria, and condenser are usable. If these are not satisfactory, determine the purchase cost of the necessary equipment. Do not buy any new equipment unless you need it. However, the present equipment may be used in another part of the plant at a later date.

Feed rate of isopropanol (N) is kept constant at 97 moles of isopropanol per hour during the entire operation.

A constant temperature of 572°F and a constant pressure of 1 atm are maintained throughout the entire operation.

The catalyst volume $V = 250 \ ft^3$.

$$\text{Efficiency of conversion} = \frac{\text{moles acetone produced}}{\text{moles isopropanol consumed}} (100)$$

$$= 98\% \text{ at } 572°F$$

Table 7

Per 1000 lb/h of acetone distilled over and removed as product

Column cross-sectional area, ft²	6.4 (based on 12-in. plate spacing and 2-in. liquid depth)
Condenser area, ft²	220
Steam, lb/h	1100

The present distillation tower contains 50 actual plates. Your calculations have indicated that a reflux ratio of 1 : 1 will allow 98 percent of the acetone to be removed, assuming a 40 percent plate efficiency. The product material may be assumed as 100 percent acetone. These conditions are satisfactory, and you have made calculations at these conditions, giving the results shown in Table 7.

The overall heat-transfer coefficient in the calandria is 250 Btu/(h)(ft²)(°F).

Saturated steam is available at 50 psig. Neglect any sensible heat transfer from steam condensate to boiling liquid. The temperature-difference driving force Δt in the calandria may be assumed to be 90°F.

Your calculations have indicated that the present condenser on the column is satisfactory and may be used for this process.

The necessary data on the present equipment are given in Table 8. Table 9 shows the installed-cost data for any new equipment which must be purchased.

Steel Distillation Column

Estimate on the basis of $100.00 per square foot of tray area. Column diameters should be even multiples of 6 in.

Table 8

Number of plates in column	50
Column diameter	72 in.
Plate spacing	12 in.
Liquid depth on each plate	2 in.
Calandria heat-transfer area	104 ft²
Calandria shell working pressure	60 psig

Table 9 Steel heat exchangers

Surface, ft²	Cost, $/ft²
100	80.00
200	58.00
300	48.00
500	40.00

Problem 9. Quick-Estimate Design of Debutanizer

You are the chief design engineer at a large petroleum refinery. The head projects engineer has asked you to make a preliminary design estimate for a proposed debutanizer. You are to present the following preliminary information to a group meeting 3 h from now:

1. Number of plates for proposed debutanizer column
2. Diameter of proposed column
3. Outside tube area required for heating coils in reboiler (coils to contain saturated steam at 250 psia, and average heat of vaporization of hydrocarbons in reboiler may be taken as 5000 cal/g mol)

The following information has been supplied you by the projects engineer:

Charge stock from catalytic cracker to debutanizer = 5620 BPSD (barrels per service day). (See Table 10.)
Debutanizer to operate at 165 psia.
Two fractions are to be obtained—OVHD and BTMS.
OVHD is to contain 98.5 percent of the butanes and lighter components with a contamination of 1.5 mol % pentanes and pentenes.

The debutanizer must fractionate between NC_4 and IC_5.
A search through your debutanizer-design card file gives the information shown in Cards A to C.

Table 10 Debutanizer charge stock from catalytic cracker

$^\circ$API (60°F) = 100.3
Viscosity = 42 SSU at 120°F

Component	Mole %	Molecular weight
C_2''	0.1	28
C_2	1.2	30
H_2S	2.1	34
C_3''	16.3	42
C_3	6.9	44
IC_4''	6.5	56
NC_4''	14.3	56
IC_4	10.8	58
NC_4	3.9	58
C_5''	11.9	70
IC_5	9.7	72
NC_5	2.3	72
C_6	11.8	86
C_7	2.1	96
C_8	0.1	112
	100.0	

Card A — Debutanizer: 30 trays

Feed 574 BPSD	Reflux 635 BPSD	OVHD 354 BPSD	BTMS 220 BPSD
5897 lb/h	5630 lb/h	3114 lb/h	2783 lb/h
84.6 mol/h	100.6 mol/h	56.0 mol/h	28.6 mol/h

Mole %		Composition same as OVHD	Mole %		Mole %	
C_3''	12.4	Reflux ratio = 1.8 : 1	C_3''	18.8	C_4's	1.4
C_3	6.1		C_3	9.2	C_5—400°F	98.6
C_4's	47.9		C_4's	71.6		100.0
C_5—400°F	33.6		C_5—400°F	0.4		
	100.0			100.0		

°API = 68.8	°API = 100.7	°API = 100.7	°API = 39.7
Basis: feed	119 mol %	66.2 mol %	33.8 mol %

Card B — Values of K at 165 psia

Listings derived principally from the data of Scheibel and Jenny, *Ind. Eng. Chem.*, **37**:80 (1945).

Hydrocarbon	260°F	280°F	285°F	290°F	300°F
n-Butane	1.75	2.00	2.06	2.13	2.25
n-Pentene	1.15	1.30	1.34	1.39	1.47
Isopentane	1.03	1.19	1.23	1.28	1.37
n-Pentane	0.90	1.06	1.11	1.16	1.25
n-Hexane	0.46	0.58	0.61	0.64	0.70
n-Heptane	0.24	0.31	0.32	0.35	0.39
n-Octane	0.13	0.17	0.18	0.19	0.21

Card C — Recommended overall heat-transfer coefficients (based on outside coil heating area)

	Debutanizer service	Transfer rate, Btu/(h)(ft²)(°F)
Condenser	Butane to water	100–110
Reboiler	Gasoline to steam	120–140
Reboiler	Gasoline to hot oil	50–75
Exchanger	Gasoline to gasoline	80
Preheater	Gasoline to steam	110–120
Cooler	Gasoline to water	75–90

Problem 10. Economic Analysis of Formaldehyde-Pentaerythritol Plant

You are a member of a firm doing consulting work on chemical engineering design. The G. I. Treyz Chemical Company of Cooks Falls, N.Y., has asked your firm to determine the advisability of adding a pentaerythritol production unit to its present formaldehyde plant. You have been sent to Cooks Falls to analyze the situation and obtain the necessary details.

A conference with G. Victor Treyz, owner of the plant, supplies you with the following information:

The present formaldehyde plant cost $700,000 several years ago and is in satisfactory operating condition. It produces formaldehyde by the oxidation of methanol. The yearly fixed charges (interest, rent, taxes, insurance, and depreciation) at the plant amount to 15 percent of the initial investment.

The plant capacity is 100,000 lb of formalin (37.2 percent HCHO, 8 percent CH_3OH as inhibitor, and 54.8 percent H_2O by weight) per day.

Miscellaneous costs (salaries, labor, office expenses, lab supplies, maintenance, repairs, communication, sales, silver catalyst replacement, etc.) amount to $225,000 per year.

The overall efficiency of conversion of methanol into formaldehyde equals 80 percent, i.e., 0.8 lb of CH_3OH converted to HCHO per pound of CH_3OH decomposed.

The total cost of utilities (fuel, electricity, steam, water, etc.) equals 5 percent of the total cost of producing the formaldehyde.

Mr. Treyz has included all the smaller costs such as insurance benefits, etc., in the overhead cost so that the total cost of the present operation is the sum of fixed charges, overhead cost, utilities, and methanol.

The proposed PE plant is to produce 6000 lb of final pentaerythritol per day using the inhibited formalin produced at the plant. The basic reaction involving lime, acetaldehyde, and formaldehyde can be assumed to be going to completion. Only 70 percent of the PE produced in the reaction is obtained as the final product. Any costs due to the presence of excess $Ca(OH)_2$ can be neglected. The calcium formate formed must be discarded.

The initial installed cost of the proposed plant is $600,000. It can be assumed that the yearly cost of the new operation minus raw-materials costs will be 40 percent of the initial installed cost. This 40 percent includes fixed charges, overhead, utilities, and all other expenses except raw materials.

Both plants operate 350 days/year.

Mr. Treyz feels he can sell 6000 lb of PE per day at the present market price. He can also sell 100,000 lb of formalin per day, but he is willing to make the new investment if it will give him better than a 30 percent yearly return on the PE initial plant investment.

Following is a list of prices supplied by Mr. Treyz. All these prices are f.o.b. Cooks Falls, N.Y., and they are to be used for the cost estimate.

Methanol, carload lots	$ 0.45/gal = $0.068/lb
Formaldehyde (or Formalin), 37.2% HCHO, 8% CH_3OH	0.060/lb
Acetaldehyde	0.20/lb
Lime as pure CaO	35.00/ton
Pentaerythritol	0.50/lb

From the preceding information, determine the following:

1. The present profit per year on the formaldehyde unit.
2. The total profit per year if the PE unit were in operation.
3. Yearly percent return on the PE initial plant investment.
4. Should the Treyz Company make the investment?

Problem 11. Operating Time for Catalytic Polymerization Reactor to Reach Minimum Allowable Conversion

A catalytic polymerization plant is to operate continuously with a special catalyst with properties and results as given below. Under the following specified operating conditions, how many days could each reactor remain on stream from the time of fresh (catalyst age zero) catalyst charge until the percent propylene conversion through the catalyst bed drops to 93.75 percent. What will the pressure drop across the catalyst bed be at that time? How often should a new recharged unit come on stream if all units are to operate on identical staggered schedules (i.e., how many days will each unit be down for dumping and recharging)?

Total feed stream is 15,000 barrels per day and is 40 percent by volume propylene and 60 percent by volume propane.

Five reactors are available, each holding 20,000 lb of the catalyst.

All reactors will operate on identical staggered schedules.

The unit will be operated with the same flow rate in four of the reactors while the fifth one is down for dumping and catalyst recharging, with this dumping and recharging requiring at least 5 days but no more than 8 days.

The temperature in the reactors will be maintained constant at 430°F. 0.715 barrel of polymer is obtained for every barrel of propylene converted. One barrel equals 42 gallons.

$$\text{Catalyst age factor} = A = \frac{\text{gallons of polymer produced since catalyst charging}}{\text{pounds of catalyst charged}}$$

$$A = \frac{[\int_0^D F(t)\, dt][\text{conversion factors to give gallons of polymer}]_{\text{arith. ave}}}{\text{pounds of catalyst}}$$

$t = $ time, days

$D = $ time in days at which $\Delta P/F^2$ is to be calculated

$F = F(t) = $ reactor total feed rate as thousands of barrels/day

$\Delta P = $ pressure drop across reactors, psi

At the operating pressure and temperature of 430°F, the following data apply for the special catalyst used:

A plot of $\log \Delta P/F^2$ versus catalyst age, A, is linear with $\Delta P/F^2 = 0.2$ at $A = 0$ and $\Delta P/F^2 = 100$ at $A = 84.90$.

(Resultant equation is $\log \Delta P/F^2 = 0.03179A - 0.699$.)

The conversion of propylene to polymer at the time of fresh catalyst charge (catalyst age zero) is 97.66 percent.

The following data apply for the catalyst and indicated temperature (linear interpolation is satisfactory):

A	$\dfrac{\text{\% conversion of propylene at catalyst age } A}{\text{\% conversion of propylene at catalyst age zero}}$
0	1.00
10	0.995
20	0.99
30	0.98
40	0.97
50	0.96
60	0.935
70	0.91
80	0.87
90	0.82
100	0.76

Reference. AIChE Student Contest Problem, 1974.

Problem 12. Sizing and Costing of Multicomponent Distillation Column for Biphenyl Recovery Unit

The feed to a multicomponent distillation column is as follows (in order of decreasing volatility):

Component	Feed rate as lb mol/h
Toluene	0.488
Naphthalene	1.599
Biphenyl	16.835
O-Methyl biphenyl	0.208
P-Methyl biphenyl	0.333
M-Methyl biphenyl	0.121
Diphenylenemethane (Fluorene)	1.029
Phenanthrene	0.769
M-Terphenyl	1.919
	23.301

A steel, bubble-cap distillation column with ten trays operated at an average pressure of 400 mm Hg, an average temperature of 280°C, and a reflux ratio of 8 to 1 will give an overhead product containing 98 percent of the entering biphenyl and a negligible amount of methyl biphenyl and heavier. Based on the following data and assumptions, estimate the cost of the distillation tower including installation and auxiliaries (*not* including reboiler and condenser) at the present time:

Pressure drop in column can be neglected and average temperature and pressure can be used for calculations.

Ideal gas laws apply to the gas mixture under these conditions.

Liquid feed is at its boiling point.

Column operates adiabatically and constant molal overflow assumption is acceptable.

The average molecular weight of the gas can be taken as that of biphenyl = $C_6H_5 \cdot C_6H_5$ = 154.2.

The average specific gravity of the liquid in the column is 0.72.

The tray spacing is 24 inches with a 2-inch slot liquid seal.

Assume the surface tension of the liquid is 20 dyn/cm.

Size the column by using the K_v from both of the plots given in Chap. 15 for maximum allowable vapor velocity, and use an 85 percent safety factor on the maximum allowable vapor velocity. For comparison purposes, give an answer for each of the two K_v estimates.

Use Fig. 15-28 in Chap. 15 which gives cost data per plate for the final cost estimate.

Reference. AIChE Student Contest Problem, 1975.

Problem 13. Design of Reactor for Coal Conversion to Nonpolluting Fuel Oil (Plus Partial Solution)

A plant is being designed to produce low-sulfur oil from coal under the conditions as outlined in the following. A major factor in the design is to minimize the volume of the reactor, and you are to carry out some preliminary studies for the reactor system. Specifically, you are to determine the total volume of the reactor if it is operated isothermally at 800°F for the case of a single, ideal, plug-flow reactor operation and for the case of a single, back-mix (continuous-stirred-tank reactor) reactor system with the conditions and assumptions as outlined in the following.

Operating Conditions

Plant is to produce 50,000 barrels (based on 60°F)/day of low-sulfur oil (0.4 wt % sulfur) from coal. Following are specifications for the coal feed and the product oil:

Coal Feed		Oil Product	
Bulk density, lb/ft³ = 45.0		4.4° API = Density of 64.97 lb/ft³ at 60°F	
		Density = 44.8 lb/ft³ at 800°F	

†Proximate Analysis, wt %		Boiling distribution: true boiling point cut, wt %	
Moisture	1.5	C_5—400°F	8.1
Ash	10.3	400–650°F	32.1
Volatile matter	35.5	650–975°F	22.1
Fixed Carbon	52.7	975°F+	37.7
Total	100.0	Total	100.0

†Ultimate Analysis, wt %		Ultimate Analysis, wt %	
Carbon	70.2	Carbon	90.2
Hydrogen	4.6	Hydrogen	8.5
Nitrogen	1.0	Nitrogen	0.8
Sulfur	3.6	Sulfur	0.4
Oxygen	10.5	Oxygen	—
Ash	10.1	Ash	0.1
Total	100.0	Total	100.0

Coal in the slurry is 35 percent by weight with the balance being recycled oil of the same composition as the product oil.

A nickel-molybdenum on alumina catalyst in the form of $\frac{1}{8}$-in. spheres is used with a desulfurization activity (A_s) of 1.25 and a bulk density of 42.0 lb/ft³.

The Following Assumptions Apply for the Reactor System

Pressure = 2500 psia and negligible pressure drop across the reactor.

25,000 ft³ of gas at SC (SC = 60°F and 1 atm) flow to reactor per barrel of slurry feed (based on 60°F).

One barrel = 42 gal.

85 percent of the gas to the reactor is hydrogen, and the other 15 percent is methane with negligible H_2S content.

Yield of product is 4.2 barrels of product oil (at 60°F) per ton of coal (as received).

Average molecular weight of fuel oil is 301.

No hydrogen or methane or hydrogen sulfide is dissolved in the slurry.

Necessary heating and cooling units are available so reactors can be assumed to operate isothermally at 800°F.

---

† See Perry's "Chemical Engineers' Handbook" for discussion of these and methods of analysis (5th ed., p. 9-4). Proximate and ultimate analyses in this case were carried out with air-dried coal samples; so the oxygen and hydrogen in the "moisture" reported in the proximate analysis are included in the ultimate analysis.

Partial pressure of hydrogen for plug flow reactor can be assumed as constant at the arithmetic average of entrance and exit pressures.

Assume the term $(1 + K_{HS} P_{HS})$ in the rate equation stays constant for the plug-flow reactor at the arithmetic average of the entering and exit values.

Assume negligible volume change during the reaction so that $C_s = C_{s_o}(1 - X_s)$.

15 wt % of the fuel oil passing through the reactor is vaporized in the reactor section, and this can be doubled to 30 percent on a molal basis considering different volatilities of the components.

The carbon in the coal which is lost to the gas stream is converted to CH_4, C_2H_6, C_3H_8, and C_4H_{10} in equal volume amounts so that the average carbon to hydrogen ratio of the resultant gas is 0.35714 on a mole basis considering hydrogen as being 1.008 lb per mole of hydrogen.

All the nitrogen in the coal that is lost is converted to gaseous NH_3.

All the sulfur in the coal that reacts goes to H_2S.

Reactor sizing will be based on the rate equation for the desulfurization reaction as follows:

$$-r_s = k_s A_s \frac{C_s^2 P_H}{C_{s_o}(1 + K_{HS} P_{HS})}$$

where $-r_s$ = rate of sulfur removal, lb·mol/h·lb catalyst.

k_s = reaction rate constant, ft^3/h·lb cat.·(psia)

k_s = exp $(14.76 - 55,000/RT) = 7.405 \times 10^{-4}$ at 800°F

K_{HS} = adsorption constant for H_2S inhibition, (psia)$^{-1}$

K_{HS} = 0.10 exp $(1200/RT) = 0.162$ at 800°F

R = 1.987 Btu/(lb mol)(°R) = Gas-law constant

T = temperature, °R

A_s = desulfurization activity

C_s = sulfur concentration in slurry, lb mol/ft^3

P_H = hydrogen partial pressure, psia

P_{HS} = hydrogen sulfide partial pressure, psia

The reactor performance equations are as follows:

<table>
<tr><td>For plug-flow</td><td>For back-mix (CSTR)</td></tr>
<tr><td>$$\frac{W}{Q} = C_{s_o} \int_{X_{s_i}}^{X_{s_f}} \frac{dX_s}{-r_s}$$</td><td>$$\frac{W}{Q} = C_{s_o} \frac{(X_{s_f} - X_{s_i})}{-r_s}$$</td></tr>
</table>

where W = catalyst charge, lb.

X_s = fractional conversion of sulfur

C_{s_o} = concentration of sulfur in slurry feed to reactor, lb mol/ft^3

Q = volumetric feed rate of slurry, ft^3/h

i = inlet value

f = final value

Suggestions

Base material balances on one ton (2000 lb) of coal as received.

Integrate rate expression for plug flow analytically (not graphically or by approximations).

See information as provided for initial part of solution presenting necessary material balances for the conditions given for this problem, and understand what was done.

Reference. AIChE Student Contest Problem, 1976.

Problem 13. Partial Solution

Following is the first part of the solution to Problem 13 dealing with the design of a reactor for coal conversion to nonpolluting fuel oil:

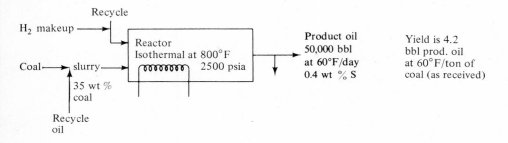

Material Balances

Choose as BASIS: 1 ton of coal as received = 2000 lb = 4.2 bbl of product oil at 60°F

$$\text{Fuel oil produced} = 4.2 \text{ bbl} \times 42 \frac{\text{gal}}{\text{bbl}} \times \frac{1}{7.48} \frac{\text{ft}^3}{\text{gal}} \times 64.97 \frac{\text{lb}}{\text{ft}^3}$$

$$= 1532 \frac{\text{lb prod. oil}}{\text{ton coal}}$$

Material balance, overall—2000 lb coal

Mtl.	Feed		Prod. Oil		Difference	
	wt %	lb	wt %	lb	lb	
C	70.2	1404	90.2	1381.95	22.05	
H	4.6	92	8.5	130.23	−40.25	gain in H
N	1.0	20	0.8	12.26	7.74	
S	3.6	72	0.4	6.13	65.87	
O	10.5	210	–	–	–	
Ash	10.1	202	0.1	1.53	200.47	
Total	100	2000	100	1532.1		

Hydrogen material balance

22.05 lb of C in coal is burned to CH_4, C_2H_6, C_3H_8, or C_4H_{10} as 0.35714 mol C per mol H

$$\text{Hydrogen used to burn C} \quad = \frac{22.05}{12.011} \text{ mol} \times \frac{1}{0.35714} \frac{\text{mol H}}{\text{mol C}} \times 1.008 \frac{\text{lb H}}{\text{mol H}}$$

$$= 5.19 \text{ lb H}$$

$$\text{Hydrogen used to make } NH_3 = \frac{7.74}{14.007} \text{ mol N} \times 3 \frac{\text{mol H}}{\text{mol N}} \times 1.008 \frac{\text{lb H}}{\text{mol H}}$$

$$= 1.67 \text{ lb H}$$

$$\text{Hydrogen used to make } H_2S \ = \frac{65.87}{32.064} \text{ mol S} \times 2 \frac{\text{mol H}}{\text{mol S}} \times 1.008 \frac{\text{lb H}}{\text{mol H}}$$

$$= 4.14 \text{ lb H}$$

$$\text{Hydrogen used to make } H_2O = \frac{210}{16} \text{ mol O} \times 2 \frac{\text{mol H}}{\text{mol O}} \times 1.008 \frac{\text{lb H}}{\text{mol H}}$$

$$= 26.46 \text{ lb H}$$

$$\text{Hydrogen gain} \qquad\qquad = 40.25 \text{ lb}$$

$$\text{Total H used} \qquad\qquad = \frac{40.25}{77.64} \text{ lb H}$$

$$\frac{77.64}{2.016} = 38.537 \qquad 38.537 \text{ lb mol } H_2$$

Material balance, at reactor inlet, for slurry concentration, basis—2000 lb coal

For slurry, 35 wt % is coal and 65 wt % is oil.

$$2000 \text{ lb coal} \times \frac{0.65 \text{ lb oil}}{0.35 \text{ lb coal}} = 3714 \text{ lb of recycle oil}$$

$$= \frac{3714 \text{ lb}}{44.8 \text{ lb/ft}^3 \text{ at } 800°F}$$

$$= 82.9 \text{ ft}^3 \text{ oil at } 800°F/2000 \text{ lb coal}$$

$$\text{Volume of coal} = \frac{2000 \text{ lb}}{45 \text{ lb/ft}^3} = 44.44 \text{ ft}^3$$

Total volume of slurry to reactor

$$= 82.9 + 44.44 = 127.34 \text{ ft}^3/2000 \text{ lb of coal fed}$$

Sulfur content of fuel oil in slurry

$$= 3714 \times 0.004 = 14.856 \text{ lb S}$$

Sulfur content of coal in slurry

$$= 2000 \times 0.036 = \underline{72.00 \text{ lb S}}$$

$$\text{Total} = 86.86$$

$$= \frac{86.86}{32.066} = 2.709 \text{ lb mol S}$$

Concentration of sulfur entering reactor in slurry

$$= C_{S_o} = \frac{2.709}{127.34} = 0.0213 \text{ lb mol/ft}^3$$

At reactor outlet

$$\text{Oil} = \underset{3714}{\text{Recycl.}} + \underset{1532}{\text{Prod.}} = 5246 \text{ lb} = \frac{5246 \text{ lb}}{44.8 \text{ lb/ft}^3} = 117.1 \text{ ft}^3 \text{ at } 800°\text{F}$$

Sulfur in outlet oil $= 5246 \times 0.004 = 20.98 \text{ lb} = \dfrac{20.98}{32.066} = 0.65444 \text{ lb mol}$

Concentration of sulfur in oil leaving reactor

$$= C_{S_f} = \frac{0.65444}{117.1} = 0.00559 \text{ lb mol/ft}^3$$

$$X_{S_f} = 1 - \frac{C_{S_f}}{C_{S_o}} = 1 - \frac{0.00559}{0.0213} = 0.738 \cong 0.74$$

$$X_{S_f} = \frac{C_{S_o} - C_{S_f}}{C_{S_o}} \text{ assuming constant fluid volumetric flow rate}$$

Material balance for gas at entrance and exit of reactor, basis—2000 lb coal
25,000 SCF of gas/bbl of slurry at 60°F is given as condition

$$\text{bbl of slurry at } 60°\text{F}/2000 \text{ lb coal} = \left(\overset{\text{ft}^3 \text{ coal}}{\frac{2000}{45}} + \overset{\text{ft}^3 \text{ oil at } 60°\text{F}}{\frac{3714}{64.97}} \right) \times 7.48 \text{ gal/ft}^3 \times \frac{1 \text{ bbl}}{42 \text{ gal}}$$

$$= 18.1 \text{ bbl/2000 lb coal}$$

Gas to reactor $= 25,000 \times 18.1 \qquad = 452,450 \text{ SCF of gas/2000 lb coal}$

Gas is 85% H_2; So $0.85 \times 452,450 = 384,583 \text{ SCF } H_2/2000 \text{ lb coal}$

$$\frac{359 \times 520}{492} = 380 \text{ ft}^3/\text{mol at SC of } 60°\text{F and 1 atm}$$

$$\frac{384{,}583}{380} = 1014 \text{ lb mol H}_2/2000 \text{ lb coal}$$

15% of the gas is CH_4, so $452{,}450 \times \dfrac{0.15}{380} = 179 \text{ lb mol CH}_4/2000 \text{ lb coal}$

$$\text{Total} = 1193$$

$$P_H \text{ at entrance to reactor} = \frac{1014}{1193}(2500) = 2125 \text{ psia}$$

$$P_{HS} \text{ at entrance} \qquad = 0 \text{ (given)}$$

Material balance, gas at exit of reactor, basis—2000 lb coal

Amount of fuel oil entering reactor is 3714 lb or

$$\frac{3714}{\text{avg mol wt } 301} = 12.34 \text{ lb mol}/2000 \text{ lb coal}$$

Fuel oil is 15 wt % or *30 mol %* vaporized; so $12.34 \times 0.30 = 3.7$ moles of fuel oil are vaporized in reactor leaving 8.64 moles of fuel oil in liquid and 3.7 moles of fuel oil in gas at reactor exit.

Assume no moles of H_2 or CH_4 or NH_3 or H_2S are dissolved in the liquid.

Total moles of H_2 in gas at exit $= 1014 - \underset{\substack{\text{used in} \\ \text{reactor}}}{38.537} = 975.5 \text{ mol}$

Total moles of H_2O in gas at exit $= \dfrac{26.46}{2.016} = 13.1 \text{ mol}$

Total moles of NH_3 in gas at exit $= \dfrac{7.74}{14.007} = 0.55 \text{ mol}$

Total moles of H_2S in gas at exit $= \dfrac{4.14}{2.016} = 2.06 \text{ mol}$

Total moles of CH_4 in gas at exit $= = 179.0 \text{ mol}$

Total moles of fuel oil in gas at exit $= 3.7 \text{ mol}$

Total exit gas moles $ = 1173.91 \text{ mol}/2000 \text{ lb coal}$

Total entering gas moles $ = 1193 \text{mol}/2000 \text{ lb coal}$

P_H at reactor exit $ = \dfrac{975.5}{1173.91}(2500) \cong 2075 \text{ psia}$

P_{HS} at reactor exit $ = \dfrac{2.06}{1173.91}(2500) \cong 4.4 \text{ psia}$

This completes the major work on material balances needed for solving this problem.

Now proceed, using these results and other information given in the problem, to complete the reactor design analysis requested.

Problem 14. Material Balance for Alkylation Plant Evaluation

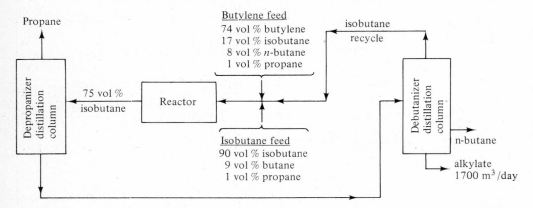

The simplified diagram of a catalytic alkylation unit is shown above.

In the reactor, butylene and isobutane react to form C_8 "alkylate" according to the following reaction:

$$C_4H_8 + C_4H_{10} \rightarrow C_8H_{18}$$

The unit is to produce product alkylate at a rate of 1700 m³/day (10,693 bbl/day).

The yield is 1.72 m³ alkylate per m³ butylene consumed; 1.10 m³ of isobutane are consumed per m³ butylene consumed.

The reactor effluent is to contain 75 vol % isobutane.

It may be assumed that the recycle is pure isobutane and that propane, alkylate, and n-butane are completely recovered as pure products in the columns. Propane and n-butane do not react.

Under these conditions, (a) How much of each of the feed streams is required in m³/day and in bbl/day? (b) How much isobutane must be recycled in m³/day and in bbl/day?

Reference. AIChE Student Contest Problem, 1977.

Problem 15. Paraffin Removal by Extractive Distillation with Dimethylphthalate

Figure C-2 represents a flow sheet prepared by a junior design engineer of part of an aromatics plant which provides paraffin removal by extractive distillation with

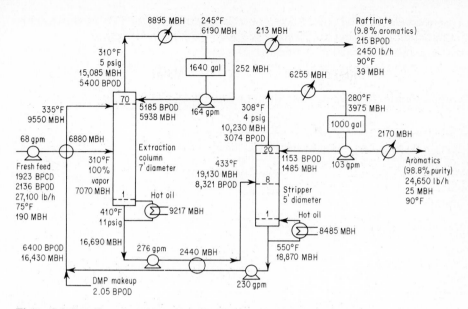

Figure C-2 Paraffin removal by extractive distillation with dimethylphthalate.

dimethylphthalate (DMP). Check this design and make recommendations concerning the design conclusions of the junior design engineer who worked on the project. If there is an inconsistency in the results, indicate where the error is, what must be done to correct the error, and what would be the magnitude of the error if it were not corrected. The basis for the design is as follows:

1. 90 percent operating factor for the process
2. 60 percent tray efficiency of the columns
3. The fresh feed is of the following composition

Component		BPCD	Relative volatility to Ox
C$_8$ paraffin	C$_8$–Pn	11	
C$_8$ napthene	C$_8$–N	24	1.52
C$_9$ paraffin	C$_9$–Pn	68	
C$_9$ napthene	C$_9$–N	9	
ethylbenzene	EB	112	1.22
para-xylene	Px	358	1.155
281°F paraffin	281°F–Pn	49	1.15
meta-xylene	Mx	739	1.135
ortho-xylene	Ox	500	1.0
292°F paraffin	292°F–Pn	34	0.90
Heavies	300°F–C$_9$	19	0.70
		1923	

4. The addition of DMP enhances the relative volatility of the paraffins and naphthenes an average of 25 percent above their normal values
5. $(L/D)_{min}$ of the extraction column is 20
6. $(L/D)_{min}$ of the stripper is 0.5
7. $(L/D)_{act} = 1.2(L/D)_{min}$
8. Dimethylphthalate properties, formula $C_6H_4(CO_2CH_3)_2$

Temp, °F	Vapor press, psia	Liquid enthalpy, Btu/lb	Vapor enthalpy, Btu/lb
60	. . . .	0	
100	. . . .	12	
200	. . . .	42	
300	0.2	73	
400	1.7	110	251
500	8.2	148	281
550	17.0	169	293
600	. . . .	191	306

9. Since the DMP makeup is so small compared to the other streams, it may be neglected in the overall heat and material balance.
10. Pump sizes may be assumed correct even though each one includes a fixed safety factor.

Duties of condensers and reboilers are given in MBH (million Btu per hour). BPCD is the designation for barrels per calendar day and BPOD is that for barrels per operating day. Note that the number of plates in each column is indicated by the number at the top of the column. Significant temperatures are given wherever necessary.

Problem 16. Optimum Operating Range for Commercial Production of Styrene

The following laboratory runs reported in Tables 11, 12, and 13 covering styrene production by catalytic dehydrogenation are to be analyzed with respect to three variables. These are (1) product value, (2) equipment and operating costs, and (3) regeneration costs. With these three variables in mind, determine the optimum operating range which should be prescribed for commercial production of styrene.

Liquid hourly space velocity (L.H.S.V.) = volumes of liquid charged per hour per gross volume of catalyst. Liquid volume has been corrected to 60°F in all cases.

Process period = length of time charging stock is passed through the catalyst bed between successive regenerations.

Laboratory data
Table 11 Effect of pressure at 550°C block temperature, 1.0 L.H.S.V., and 30-min process period

Run No. (arbitrary)	Pressure at tube outlet, mm of Hg, absolute	% conversion	Once-through yields, wt % on ethylbenzene		Styrene ultimate yield, wt %	100 C/ C_8H_8†
			Styrene	Carbon		
5	80	29.0	25.6	0.51	88	2.0
6	110	29.9	27.4	0.59	92	2.2
7	250	29.6	25.6	0.96	86	3.7
8‡	742	21.9	18.3	1.5	84	8.2
9	747	25.8	21.0	1.2	81	5.7
10‡	1530§	16.9	11.4	2.0	67	18
11‡	3320¶	12.7	6.3	2.0	50	41

† Pounds of catalyst carbon per 100 lb of styrene.
‡ The superatmospheric pressure unit was used in these runs (8, 10, 11), the quartz tube unit in the others.
§ Equivalent of 15 psig.
¶ Equivalent to 50 psig.

Table 12 Effect of pressure at 600°C block temperature and 30-min process period

Run No. (arbitrary)	Pressure at tube outlet, mm of Hg, absolute	L.H.S.V. at 60°F	% conversion	Once-through yields, wt % on ethylbenzene		Styrene ultimate yield, wt %	100 C/ C_8H_8†
				Styrene	Carbon		
12	80	0.99	54.6	49.4	0.77	90	1.6
13	80	1.50	45.3	42.0	0.66	92	1.6
14	80	1.98	41.9	39.0	0.36	93	0.92
15	80	4.02	33.2	31.4	0.17	95	0.54
16	250	0.645	60.5	45.7	4.4	76	9.7
17	250	0.98	60.9	49.5	3.0	80	6.1
18	250	1.49	56.9	48.9	1.5	85	3.1
19	250	2.19	53.0	48.1	1.1	91	2.3
20	250	2.51	50.7	44.2	0.92	87	2.1
21	250	3.30	44.9	41.0	0.56	91	1.4
22	250	4.05	41.1	37.4	0.46	92	1.2
23	250	7.98	27.6	25.7	0.14	93	0.55
24	250	9.92	26.4	24.5	0.15	93	0.61
25	742	0.67	44.7	25.8	6.0	58	23
26	748	0.99	43.8	28.6	4.2	65	15
27	741	1.49	43.8	32.3	2.5	73	7.7
28	751	2.21	40.8	33.0	1.8	80	5.5
29	742	3.97	38.4	32.3	0.99	84	3.1
30	751	7.96	26.5	23.8	0.26	90	1.1

† Pounds of catalyst carbon per 100 lb of styrene.

Table 13 Effect of temperature and space velocity at 250-mm outlet pressure and 30-min process period

Run No. (arbitrary)	Block temperature °C	L.H.S.V. at 60°F	% conversion	Once-through yields, wt % on ethylbenzene		Styrene ultimate yield, wt %	100 C/ C₈H₈†
				Styrene	Carbon		
31	550	0.67	40.5	34.9	1.4	86	3.9
7	550	1.02	29.6	25.6	0.96	86	3.7
32	550	2.02	27.9	25.6	0.42	92	1.7
33	550	4.06	19.2	18.0	0.25	94	1.4
34	550	8.02	11.6	10.9	0.16	94	1.5
35	570	0.69	49.9	42.2	2.1	84	5.0
36	570	0.98	46.2	40.6	1.1	87	2.7
37	570	1.50	45.1	40.9	0.74	90	1.8
16–24	600	See Table 12.					
38	630	1.48	66.1	54.4	3.8	81	7.0
39	630	2.17	66.1	54.4	2.9	81	5.3
40	630	2.98	58.8	53.2	1.5	90	2.8

† Pounds of catalyst carbon per 100 lb of styrene.

Conversion = the percentage of ethylbenzene charged which is transformed into other products.

Ultimate yields. Having determined once-through styrene yield, the ultimate yield obtainable under the conditions employed in any run can be calculated providing the percentage of unreacted ethylbenzene is known. The latter could be approximated by considering the vacuum distillates as binary mixtures of styrene and ethylbenzene. Fractional distillation data, however, have shown that small percentages of lower-boiling materials, definitely identified by physical constants and through preparation of their nitro derivatives as benzene and toluene, are also present. Hence, in all ultimate-yield calculations, allowance has been made for the formation of these by-products.

TABLES OF PHYSICAL PROPERTIES
AND CONSTANTS

CONTENTS

Table 1 General engineering conversion factors and constants†

Length		Mass	
1 inch..............	2.54 centimeters	1 pound †.............	16.0 ounces
1 foot..............	30.48 centimeters	1 pound †.............	453.6 grams
1 yard..............	91.44 centimeters	1 pound †.............	7000 grains
1 meter............	100.00 centimeters	1 ton (short)..........	2000 pounds †
1 meter............	39.37 inches	1 kilogram.............	1000 grams
1 micron...........	10^{-6} meter	1 kilogram.............	2.205 pounds †
1 mile..............	5280 ft		
1 kilometer........	0.6214 mile	† Avoirdupois.	

Volume

1 cubic inch..................	16.39 cubic centimeters
1 liter.......................	61.03 cubic inches
1 liter.......................	1.057 quarts
1 cubic foot..................	28.32 liters
1 cubic foot..................	1728 cubic inches
1 cubic foot..................	7.481 U.S. gallons
1 U.S. gallon.................	4.0 quarts
1 U.S. gallon.................	3.785 liters
1 U.S. bushel.................	1.244 cubic feet

Density

1 gram per cubic centimeter....................... 62.43 pounds per cubic foot
1 gram per cubic centimeter....................... 8.345 pounds per U.S. gallon
1 gram mole of an ideal gas at 0°C and 760 mm Hg is equivalent to 22.414 liters
1 pound mole of an ideal gas at 0°C and 760 mm Hg is equivalent to 359.0 cubic feet
Density of dry air at 0°C and 760 mm Hg....................................
           1.293 grams per liter = 0.0807 pound per cubic foot
Density of mercury.................. 13.6 grams per cubic centimeter (at −2°C)

Pressure

1 pound per square inch.................	2.04 inches of mercury
1 pound per square inch.................	51.71 millimeters of mercury
1 pound per square inch.................	2.31 feet of water
1 atmosphere.......................	760 millimeters of mercury
1 atmosphere.......................	2116.2 pounds per square foot
1 atmosphere.......................	33.93 feet of water
1 atmosphere.......................	29.92 inches of mercury
1 atmosphere.......................	14.7 pounds per square inch

Temperature scales

Degrees Fahrenheit (F)...........	1.8 (degrees C) + 32
Degrees Celcius (C)	(degrees F − 32)/1.8
Degrees Kelvin (K).............	degrees C + 273.15
Degrees Rankine (R)...........	degrees F + 459.7

---

† See also Tables 6 and 7 in Appendix A for SI Conversion factors and more exact conversion factors.

Table 1 General engineering conversion factors and constants† (*Continued*)

Power

1 kilowatt	737.56 foot-pounds force per second
1 kilowatt	56.87 Btu per minute
1 kilowatt	1.341 horsepower
1 horsepower	550 foot-pounds force per second
1 horsepower	0.707 Btu per second
1 horsepower	745.7 watts

Heat, energy, and work equivalents

	cal	Btu	ft·lb	kWh
cal	1	3.97×10^{-3}	3.086	1.162×10^{-6}
Btu	252	1	778.16	2.930×10^{-4}
ft·lb	0.3241	1.285×10^{-3}	1	3.766×10^{-7}
kWh	860,565	3412.8	2.655×10^{6}	1
hp-hr	641,615	2545.0	1.980×10^{6}	0.7455
Joules	0.239	9.478×10^{-4}	0.7376	2.773×10^{-7}
liter-atm	24.218	9.604×10^{-2}	74.73	2.815×10^{-5}

	hp-hr	Joules	liter-atm
cal	1.558×10^{-6}	4.1840	4.129×10^{-2}
Btu	3.930×10^{-4}	1055	10.41
ft-lb	5.0505×10^{-7}	1.356	1.338×10^{-2}
kWh	1.341	3.60×10^{6}	35,534.3
hp-hr	1	2.685×10^{6}	26,494
Joules	3.725×10^{-7}	1	9.869×10^{-3}
liter-atm	3.774×10^{-5}	101.33	1

Constants

e 2.7183
π 3.1416

Gas-law constants:

R ..... 1.987 (cal)/(g mol) (K)
R ..... 82.06 (cm^3) (atm)/(g mol) (K)
R ..... 10.73 $(lb/in.^2)$ (ft^3)/(lb mol) (°R)
R ..... 0.730 (atm) (ft^3)/(lb mol) (°R)
R ..... 1545.0 (lb/ft^2) (ft^3)/(lb mol) (°R)
g_C .... 32.17 (ft) (lbm)/(s) (s) (lbf)

Analysis of air

By weight: oxygen, 23.2%; nitrogen, 76.8%
By volume: oxygen, 21.0%; nitrogen, 79.0%
Average molecular weight of air on above basis = 28.84 (usually rounded off to 29)
True molecular weight of dry air (including argon) = 28.96

Viscosity

1 centipoise ..... 0.01 g/(s) (cm)
1 centipoise ..... 0.000672 lb/(s) (ft)
1 centipoise ..... 2.42 lb/(h) (ft)

† See also Tables 6 and 7 in Appendix A for SI Conversion factors and more exact conversion factors.

Table 2 Viscosities of gases
Coordinates for use with Fig. D-1

No.	Gas	X	Y	No.	Gas	X	Y
1	Acetic acid	7.7	14.3	29	Freon 113	11.3	14.0
2	Acetone	8.9	13.0	30	Helium	10.9	20.5
3	Acetylene	9.8	14.9	31	Hexane	8.6	11.8
4	Air	11.0	20.0	32	Hydrogen	11.2	12.4
5	Ammonia	8.4	16.0	33	$3H_2 + 1N_2$	11.2	17.2
6	Argon	10.5	22.4	34	Hydrogen bromide	8.8	20.9
7	Benzene	8.5	13.2	35	Hydrogen chloride	8.8	18.7
8	Bromine	8.9	19.2	36	Hydrogen cyanide	9.8	14.9
9	Butene	9.2	13.7	37	Hydrogen iodide	9.0	21.3
10	Butylene	8.9	13.0	38	Hydrogen sulfide	8.6	18.0
11	Carbon dioxide	9.5	18.7	39	Iodine	9.0	18.4
12	Carbon disulfide	8.0	16.0	40	Mercury	5.3	22.9
13	Carbon monoxide	11.0	20.0	41	Methane	9.9	15.5
14	Chlorine	9.0	18.4	42	Methyl alcohol	8.5	15.6
15	Chloroform	8.9	15.7	43	Nitric oxide	10.9	20.5
16	Cyanogen	9.2	15.2	44	Nitrogen	10.6	20.0
17	Cyclohexane	9.2	12.0	45	Nitrosyl chloride	8.0	17.6
18	Ethane	9.1	14.5	46	Nitrous oxide	8.8	19.0
19	Ethyl acetate	8.5	13.2	47	Oxygen	11.0	21.3
20	Ethyl alcohol	9.2	14.2	48	Pentane	7.0	12.8
21	Ethyl chloride	8.5	15.6	49	Propane	9.7	12.9
22	Ethyl ether	8.9	13.0	50	Propyl alcohol	8.4	13.4
23	Ethylene	9.5	15.1	51	Propylene	9.0	13.8
24	Fluorine	7.3	23.8	52	Sulfur dioxide	9.6	17.0
25	Freon 11	10.6	15.1	53	Toluene	8.6	12.4
26	Freon 12	11.1	16.0	54	2,3,3-Trimethylbutane	9.5	10.5
27	Freon 21	10.8	15.3	55	Water	8.0	16.0
28	Freon 22	10.1	17.0	56	Xenon	9.3	23.0

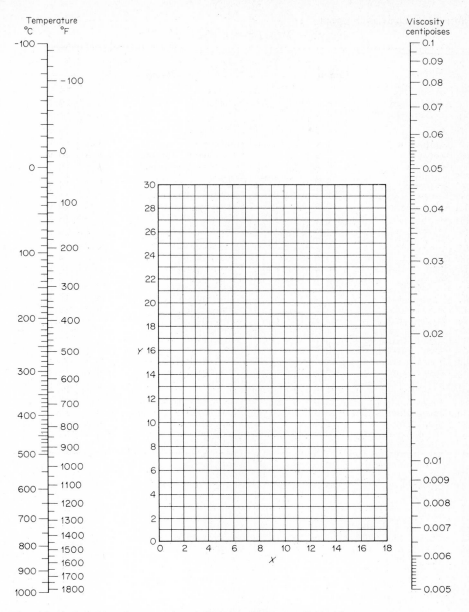

Figure D-1 Viscosities of gases at 1 atm. (For coordinates see Table 2.)

Table 3 Viscosities of liquids

Coordinates for use with Fig. D-2

No.	Liquid	X	Y	No.	Liquid	X	Y
1	Acetaldehyde	15.2	4.8	56	Freon 22	17.2	4.7
2	Acetic acid, 100%	12.1	14.2	57	Freon 113	12.5	11.4
3	Acetic acid, 70%	9.5	17.0	58	Glycerol, 100%	2.0	30.0
4	Acetic anhydride	12.7	12.8	59	Glycerol, 50%	6.9	19.6
5	Acetone, 100%	14.5	7.2	60	Heptene	14.1	8.4
6	Acetone, 35%	7.9	15.0	61	Hexane	14.7	7.0
7	Allyl alcohol	10.2	14.3	62	Hydrochloric acid, 31.5%	13.0	16.6
8	Ammonia, 100%	12.6	2.0	63	Isobutyl alcohol	7.1	18.0
9	Ammonia, 26%	10.1	13.9	64	Isobutyric acid	12.2	14.4
10	Amyl acetate	11.8	12.5	65	Isopropyl alcohol	8.2	16.0
11	Amyl alcohol	7.5	18.4	66	Kerosene	10.2	16.9
12	Aniline	8.1	18.7	67	Linseed oil, raw	7.5	27.2
13	Anisole	12.3	13.5	68	Mercury	18.4	16.4
14	Arsenic trichloride	13.9	14.5	69	Methanol, 100%	12.4	10.5
15	Benzene	12.5	10.9	70	Methanol, 90%	12.3	11.8
16	Brine, CaCl₂, 25%	6.6	15.9	71	Methanol, 40%	7.8	15.5
17	Brine, NaCl, 25%	10.2	16.6	72	Methyl acetate	14.2	8.2
18	Bromine	14.2	13.2	73	Methyl chloride	15.0	3.8
19	Bromotoluene	20.0	15.9	74	Methyl ethyl ketone	13.9	8.6
20	Butyl acetate	12.3	11.0	75	Naphthalene	7.9	18.1
21	Butyl alcohol	8.6	17.2	76	Nitric acid, 95%	12.8	13.8
22	Butyric acid	12.1	15.3	77	Nitric acid, 60%	10.8	17.0
23	Carbon dioxide	11.6	0.3	78	Nitrobenzene	10.6	16.2
24	Carbon disulfide	16.1	7.5	79	Nitrotoluene	11.0	17.0
25	Carbon tetrachloride	12.7	13.1	80	Octane	13.7	10.0
26	Chlorobenzene	12.3	12.4	81	Octyl alcohol	6.6	21.1
27	Chloroform	14.4	10.2	82	Pentachloroethane	10.9	17.3
28	Chlorosulfonic acid	11.2	18.1	83	Pentane	14.9	5.2
29	Chlorotoluene, ortho	13.0	13.3	84	Phenol	6.9	20.8
30	Chlorotoluene, meta	13.3	12.5	85	Phosphorus tribromide	13.8	16.7
31	Chlorotoluene, para	13.3	12.5	86	Phosphorus trichloride	16.2	10.9
32	Cresol, meta	2.5	20.8	87	Propionic acid	12.8	13.8
33	Cyclohexanol	2.9	24.3	88	Propyl alcohol	9.1	16.5
34	Dibromoethane	12.7	15.8	89	Propyl bromide	14.5	9.6
35	Dichloroethane	13.2	12.2	90	Propyl chloride	14.4	7.5
36	Dichloromethane	14.6	8.9	91	Propyl iodide	14.1	11.6
37	Diethyl oxalate	11.0	16.4	92	Sodium	16.4	13.9
38	Dimethyl oxalate	12.3	15.8	93	Sodium hydroxide, 50%	3.2	25.8
39	Diphenyl	12.0	18.3	94	Stannic chloride	13.5	12.8
40	Dipropyl oxalate	10.3	17.7	95	Sulfur dioxide	15.2	7.1
41	Ethyl acetate	13.7	9.1	96	Sulfuric acid, 110%	7.2	27.4
42	Ethyl alcohol, 100%	10.5	13.8	97	Sulfuric acid, 98%	7.0	24.8
43	Ethyl alcohol, 95%	9.8	14.3	98	Sulfuric acid, 60%	10.2	21.3
44	Ethyl alcohol, 40%	6.5	16.6	99	Sulfuryl chloride	15.2	12.4
45	Ethyl benzene	13.2	11.5	100	Tetrachloroethane	11.9	15.7
46	Ethyl bromide	14.5	8.1	101	Tetrachloroethylene	14.2	12.7
47	Ethyl chloride	14.8	6.0	102	Titanium tetrachloride	14.4	12.3
48	Ethyl ether	14.5	5.3	103	Toluene	13.7	10.4
49	Ethyl formate	14.2	8.4	104	Trichloroethylene	14.8	10.5
50	Ethyl iodide	14.7	10.3	105	Turpentine	11.5	14.9
51	Ethylene glycol	6.0	23.6	106	Vinyl acetate	14.0	8.8
52	Formic acid	10.7	15.8	107	Water	10.2	13.0
53	Freon 11	14.4	9.0	108	Xylene, ortho	13.5	12.1
54	Freon 12	16.8	5.6	109	Xylene, meta	13.9	10.6
55	Freon 21	15.7	7.5	110	Xylene, para	13.9	10.9

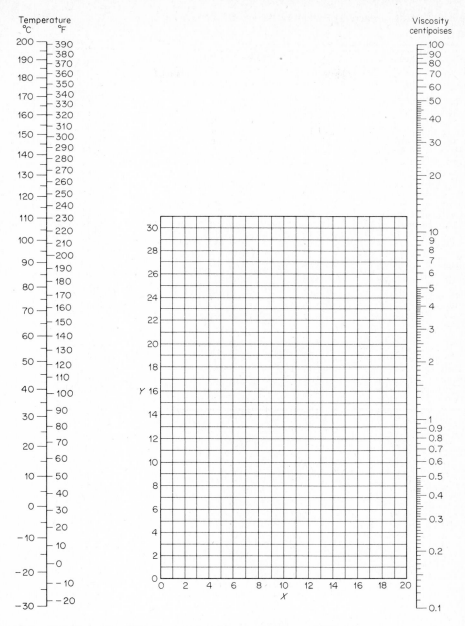

Figure D-2 Viscosities of liquids at 1 atm. (For coordinates see Table 3.)

Table 4 Density, viscosity, and thermal conductivity of water

Temperature, °F	Density of liquid water, lb/ft³	Viscosity of water, centipoises	Thermal conductivity of water, Btu/(h)(ft³)(°F/ft)
32	62.42	1.794	0.320
40	62.43	1.546	0.326
50	62.42	1.310	0.333
60	62.37	1.129	0.340
70	62.30	0.982	0.346
80	62.22	0.862	0.352
90	62.11	0.764	0.358
100	62.00	0.684	0.363
110	61.86	0.616	0.367
120	61.71	0.559	0.371
130	61.55	0.511	0.375
140	61.38	0.470	0.378
150	61.20	0.433	0.381
160	61.00	0.401	0.384
170	60.80	0.372	0.386
180	60.58	0.347	0.388
190	60.36	0.325	0.390
200	60.12	0.305	0.392
210	59.88	0.287	0.393
212	59.83	0.284	0.393

Table 5 Thermal conductivity of metals

Metal	k, Btu/(h)(ft^2)($°$F/ft)		
	At 32°F	At 212°F	At 572°F
Aluminum	117	119	133
Brass (70-30)	56	60	66
Cast iron	32	30	26
Copper	224	218	212
Lead	20	19	18
Nickel	36	34	32
Silver	242	238	. . .
Steel (mild)	. . .	26	25
Tin	36	34	. . .
Wrought iron	. . .	32	28
Zinc	65	64	59

Table 6 Thermal conductivity of nonmetallic solids

Material	Tempera-ture, °F	k, Btu/(h)(ft^2)($°$F/ft)
Asbestos-cement boards	68	0.43
Bricks:		
Building	68	0.40
Fire clay	392	0.58
	1832	0.95
Sil-O-Cel	400	0.042
Calcium carbonate (natural)	86	1.3
Calcium sulfate (building plaster)	77	0.25
Celluloid	86	0.12
Concrete (stone)	. . . .	0.54
Cork board	86	0.025
Felt (wool)	86	0.03
Glass (window)	. . . .	0.3–0.61
Rubber (hard)	32	0.087
Wood (across grain):		
Maple	122	0.11
Oak	59	0.12
Pine	59	0.087

Table 7 Thermal conductivity of liquids

Liquid	Tempera-ture, °F	k, Btu/(h)(ft^2)(°F/ft)
Acetic acid:		
100%	68	0.099
50%	68	0.20
Acetone	86	0.102
	167	0.095
Benzene	86	0.092
	140	0.087
Ethyl alcohol:		
100%	68	0.105
	122	0.087
40%	68	0.224
Ethylene glycol	32	0.153
Glycerol:		
100%	68	0.164
	212	0.164
40%	68	0.259
n-Heptane	86	0.081
Kerosene	68	0.086
Methyl alcohol:		
100%	68	0.124
	122	0.114
40%	68	0.234
n-Octane	86	0.083
Sodium chloride brine, 25%	86	0.330
Sulfuric acid:		
90%	86	0.210
30%	86	0.300
Toluene	86	0.086
Water	32	0.320
	200	0.392

Table 8 Thermal conductivity of gases

Gas	Tempera-ture, °F	k, Btu/(h)(ft^2)(°F/ft)
Air	32	0.0140
	212	0.0183
	392	0.0226
Ammonia	32	0.0128
	122	0.0157
Carbon dioxide	32	0.0085
	212	0.0133
Chlorine	32	0.0043
Hydrogen	32	0.100
	212	0.129
Methane	32	0.0175
	122	0.0215
Nitrogen	32	0.0140
	212	0.0180
Oxygen	32	0.0142
	212	0.0185
Sulfur dioxide	32	0.0050
	212	0.0069
Water vapor	200	0.0159
	600	0.0256

c_p = Heat capacity = Btu/(lb)(°F) = cal/(g)(°C)

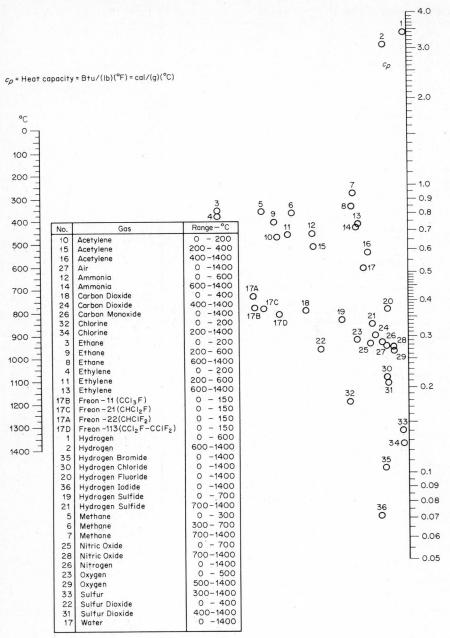

No.	Gas	Range – °C
10	Acetylene	0 – 200
15	Acetylene	200 – 400
16	Acetylene	400 – 1400
27	Air	0 – 1400
12	Ammonia	0 – 600
14	Ammonia	600 – 1400
18	Carbon Dioxide	0 – 400
24	Carbon Dioxide	400 – 1400
26	Carbon Monoxide	0 – 1400
32	Chlorine	0 – 200
34	Chlorine	200 – 1400
3	Ethane	0 – 200
9	Ethane	200 – 600
8	Ethane	600 – 1400
4	Ethylene	0 – 200
11	Ethylene	200 – 600
13	Ethylene	600 – 1400
17B	Freon – 11 (CCl$_3$F)	0 – 150
17C	Freon – 21 (CHCl$_2$F)	0 – 150
17A	Freon – 22 (CHClF$_2$)	0 – 150
17D	Freon – 113 (CCl$_2$F–CClF$_2$)	0 – 150
1	Hydrogen	0 – 600
2	Hydrogen	600 – 1400
35	Hydrogen Bromide	0 – 1400
30	Hydrogen Chloride	0 – 1400
20	Hydrogen Fluoride	0 – 1400
36	Hydrogen Iodide	0 – 1400
19	Hydrogen Sulfide	0 – 700
21	Hydrogen Sulfide	700 – 1400
5	Methane	0 – 300
6	Methane	300 – 700
7	Methane	700 – 1400
25	Nitric Oxide	0 – 700
28	Nitric Oxide	700 – 1400
26	Nitrogen	0 – 1400
23	Oxygen	0 – 500
29	Oxygen	500 – 1400
33	Sulfur	300 – 1400
22	Sulfur Dioxide	0 – 400
31	Sulfur Dioxide	400 – 1400
17	Water	0 – 1400

Figure D-3 Heat capacities c_p of gases at 1 atm pressure.

Heat capacity = Btu/(lb)(°F) = cal/(g)(°C)

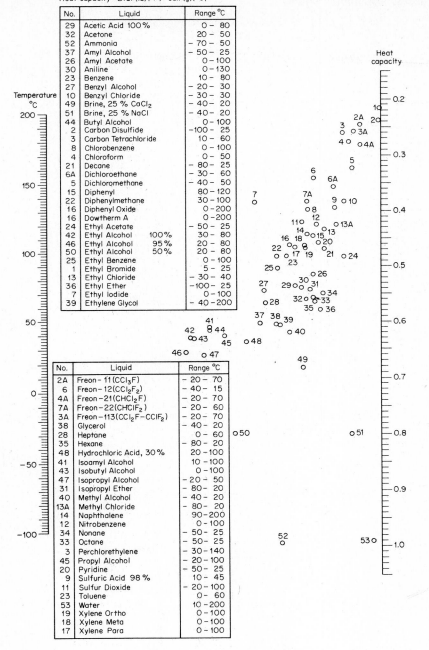

No.	Liquid	Range °C
29	Acetic Acid 100%	0 – 80
32	Acetone	20 – 50
52	Ammonia	– 70 – 50
37	Amyl Alcohol	– 50 – 25
26	Amyl Acetate	0 – 100
30	Aniline	0 – 130
23	Benzene	10 – 80
27	Benzyl Alcohol	– 20 – 30
10	Benzyl Chloride	– 30 – 30
49	Brine, 25 % CaCl$_2$	– 40 – 20
51	Brine, 25 % NaCl	– 40 – 20
44	Butyl Alcohol	0 – 100
2	Carbon Disulfide	–100 – 25
3	Carbon Tetrachloride	10 – 60
8	Chlorobenzene	0 – 100
4	Chloroform	0 – 50
21	Decane	– 80 – 25
6A	Dichloroethane	– 30 – 60
5	Dichloromethane	– 40 – 50
15	Diphenyl	80 – 120
22	Diphenylmethane	30 – 100
16	Diphenyl Oxide	0 – 200
16	Dowtherm A	0 – 200
24	Ethyl Acetate	– 50 – 25
42	Ethyl Alcohol 100%	30 – 80
46	Ethyl Alcohol 95%	20 – 80
50	Ethyl Alcohol 50%	20 – 80
25	Ethyl Benzene	0 – 100
1	Ethyl Bromide	5 – 25
13	Ethyl Chloride	– 30 – 40
36	Ethyl Ether	–100 – 25
7	Ethyl Iodide	0 – 100
39	Ethylene Glycol	– 40 –200

No.	Liquid	Range °C
2A	Freon – 11 (CCl$_3$F)	– 20 – 70
6	Freon – 12 (CCl$_2$F$_2$)	– 40 – 15
4A	Freon – 21 (CHCl$_2$F)	– 20 – 70
7A	Freon – 22 (CHClF$_2$)	– 20 – 60
3A	Freon – 113 (CCl$_2$F – CClF$_2$)	– 20 – 70
38	Glycerol	– 40 – 20
28	Heptane	0 – 60
35	Hexane	– 80 – 20
48	Hydrochloric Acid, 30 %	20 –100
41	Isoamyl Alcohol	10 –100
43	Isobutyl Alcohol	0 –100
47	Isopropyl Alcohol	– 20 – 50
31	Isopropyl Ether	– 80 – 20
40	Methyl Alcohol	– 40 – 20
13A	Methyl Chloride	– 80 – 20
14	Naphthalene	90 –200
12	Nitrobenzene	0 –100
34	Nonane	– 50 – 25
33	Octane	– 50 – 25
3	Perchlorethylene	– 30 –140
45	Propyl Alcohol	– 20 –100
20	Pyridine	– 50 – 25
9	Sulfuric Acid 98 %	10 – 45
11	Sulfur Dioxide	– 20 –100
23	Toluene	0 – 60
53	Water	10 –200
19	Xylene Ortho	0 –100
18	Xylene Meta	0 –100
17	Xylene Para	0 –100

Figure D-4 Heat capacities of liquids.

Table 9 Specific gravities of liquids

The values presented in this table are based on the density of water at 4°C and a total pressure of 1 atm.

$$\text{Specific gravity} = \frac{\text{density of material at indicated temperature}}{\text{density of liquid water at 4°C}}$$

Pure liquid	Formula	Temperature, °C	Specific gravity
Acetaldehyde	CH_3CHO	18	0.783
Acetic acid	CH_3CO_2H	0	1.067
		30	1.038
Acetone	CH_3COCH_3	20	0.792
Benzene	C_6H_6	20	0.879
n-Butyl alcohol	$C_2H_5CH_2CH_2OH$	20	0.810
Carbon tetrachloride	CCl_4	20	1.595
Ethyl alcohol	CH_3CH_2OH	10	0.798
		30	0.791
Ethyl ether	$(CH_3CH_2)_2O$	25	0.708
Ethylene glycol	$CH_2OH·CH_2OH$	19	1.113
Glycerol	$CH_2OH·CHOH·CH_2OH$	15	1.264
		30	1.255
Isobutyl alcohol	$(CH_3)_2CHCH_2OH$	18	0.805
Isopropyl alcohol	$(CH_3)_2CHOH$	0	0.802
		30	0.777
Methyl alcohol	CH_3OH	0	0.810
		20	0.792
Nitric acid	HNO_3	10	1.531
		30	1.495
Phenol	C_6H_5OH	25	1.071
n-Propyl alcohol	$CH_3CH_2CH_2OH$	20	0.804
Sulfuric acid	H_2SO_4	10	1.841
		30	1.821
Water	H_2O	4	1.000
		100	0.958

Table 10 Specific gravities of solids

The specific gravities as indicated in this table apply at ordinary atmospheric temperatures. The values are based on the density of water at 4°C.

$$\text{Specific gravity} = \frac{\text{density of material}}{\text{density of liquid water at 4°C}}$$

Density of liquid water at 4 °C = 1.000 g/cm^3 = 62.43 lb/ft^3

Substance	Specific gravity
Aluminum, hard-drawn	2.55–2.80
Brass, cast-rolled	8.4–8.7
Copper, cast-rolled	8.8–8.95
Glass, common	2.4–2.8
Gold, cast-hammered	19.25–19.35
Iron:	
Gray cast	7.03–7.13
Wrought	7.6–7.9
Lead	11.34
Nickel	8.9
Platinum, cast-hammered	21.5
Silver, cast-hammered	10.4–10.6
Steel, cold-drawn	7.83
Tin, cast-hammered	7.2–7.5
White oak timber, air-dried	0.77
White pine timber, air-dried	0.43
Zinc, cast-rolled	6.9–7.2

Table 11 Properties of saturated steam†

Values in table based on zero enthalpy of liquid water at 32°F

| Tempera-ture, °F | Absolute pressure, psi | Volume of vapor, ft³/lb | Enthalpy | | Latent heat of evaporation, Btu/lb |
			Liquid, Btu/lb	Vapor, Btu/lb	
32	0.0885	3306	0.00	1075.8	1075.8
35	0.0999	2947	3.02	1077.1	1074.1
40	0.1217	2444	8.05	1079.3	1071.3
45	0.1475	2036.4	13.06	1081.5	1068.4
50	0.1781	1703.2	18.07	1083.7	1065.6
55	0.2141	1430.7	23.07	1085.8	1062.7
60	0.2563	1206.7	28.06	1088.0	1059.9
65	0.3056	1021.4	33.05	1090.2	1057.1
70	0.3631	867.9	38.04	1092.3	1054.3
75	0.4298	740.0	43.03	1094.5	1051.5
80	0.5069	633.1	48.02	1096.6	1048.6
85	0.5959	543.5	53.00	1098.8	1045.8
90	0.6982	468.0	57.99	1100.9	1042.9
95	0.8153	404.3	62.98	1103.1	1040.1
100	0.9492	350.4	67.97	1105.2	1037.2
105	1.1016	304.5	72.95	1107.3	1034.3
110	1.2748	265.4	77.94	1109.5	1031.6
115	1.4709	231.9	82.93	1111.6	1028.7
120	1.6924	203.27	87.92	1113.7	1025.8
125	1.9420	178.61	92.91	1115.8	1022.9
130	2.2225	157.34	97.90	1117.9	1020.0
135	2.5370	138.95	102.90	1119.9	1017.0
140	2.8886	123.01	107.89	1122.0	1014.1
145	3.281	109.15	112.89	1124.1	1011.2
150	3.718	97.07	117.89	1126.1	1008.2
155	4.203	86.52	122.89	1128.1	1005.2
160	4.741	77.29	127.89	1130.2	1002.3
165	5.335	69.19	132.89	1132.2	999.3
170	5.992	62.06	137.90	1134.2	996.3
175	6.715	55.78	142.91	1136.2	993.3

Table 11 Properties of saturated steam† (*Continued*)

Tempera-ture, °F	Absolute pressure, psi	Volume of vapor, ft^3/lb	Enthalpy		Latent heat of evaporation, Btu/lb
			Liquid, Btu/lb	Vapor, Btu/lb	
180	7.510	50.23	147.92	1138.1	990.2
185	8.383	45.31	152.93	1140.1	987.2
190	9.339	40.96	157.95	1142.0	984.1
195	10.385	37.09	162.97	1144.0	981.0
200	11.526	33.64	167.99	1145.9	977.9
210	14.123	27.82	178.05	1149.7	971.6
212	14.696	26.80	180.07	1150.4	970.3
220	17.186	23.15	188.13	1153.4	965.2
230	20.780	19.382	198.23	1157.0	958.8
240	24.969	16.323	208.34	1160.5	952.2
250	29.825	13.821	218.48	1164.0	945.5
260	35.429	11.763	228.64	1167.3	938.7
270	41.858	10.061	238.84	1170.6	931.8
280	49.203	8.645	249.06	1173.8	924.7
290	57.556	7.461	259.31	1176.8	917.5
300	67.013	6.466	269.59	1179.7	910.1
310	77.68	5.626	279.92	1182.5	902.6
320	89.66	4.914	290.28	1185.2	894.9
330	103.06	4.307	300.68	1187.7	887.0
340	118.01	3.788	311.13	1190.1	879.0
350	134.63	3.342	321.63	1192.3	870.7
360	153.04	2.957	332.18	1194.4	862.2
370	173.37	2.625	342.79	1196.3	853.5
380	195.77	2.335	353.45	1198.1	844.6
39C	220.37	2.0836	364.17	1199.6	835.4
400	247.31	1.8633	374.97	1201.0	826.0

† Abridged from "Thermodynamic Properties of Steam," by J. H. Keenan and F. G. Keyes, copyright, 1937, by Joseph H. Keenan and Frederick G. Keyes. Published by John Wiley and Sons, Inc., New York. See also 1967 A.S.M.E. Steam Tables.

Table 12 Heat-exchanger and condenser-tube data

Tube OD, in.	BWG	Wall thick-ness, in.	ID, in.	Flow area per tube, in.^{2}	Surface per lin ft, ft^2		Weight per lin ft, lb steel
					Outside	Inside	
½	12	0.109	0.282	0.0625	0.1309	0.0748	0.493
	14	0.083	0.334	0.0876	0.1309	0.0874	0.403
	16	0.065	0.370	0.1076	0.1309	0.0969	0.329
	18	0.049	0.402	0.127	0.1309	0.1052	0.258
	20	0.035	0.430	0.145	0.1309	0.1125	0.190
¾	10	0.134	0.482	0.182	0.1963	0.1263	0.965
	11	0.120	0.510	0.204	0.1963	0.1335	0.884
	12	0.109	0.532	0.223	0.1963	0.1393	0.817
	13	0.095	0.560	0.247	0.1963	0.1466	0.727
	14	0.083	0.584	0.268	0.1963	0.1529	0.647
	15	0.072	0.606	0.289	0.1963	0.1587	0.571
	16	0.065	0.620	0.302	0.1963	0.1623	0.520
	17	0.058	0.634	0.314	0.1963	0.1660	0.469
	18	0.049	0.652	0.334	0.1963	0.1707	0.401
1	8	0.165	0.670	0.335	0.2618	0.1754	1.61
	9	0.148	0.704	0.389	0.2618	0.1843	1.47
	10	0.134	0.732	0.421	0.2618	0.1916	1.36
	11	0.120	0.760	0.455	0.2618	0.1990	1.23
	12	0.109	0.782	0.479	0.2618	0.2048	1.14
	13	0.095	0.810	0.515	0.2618	0.2121	1.00
	14	0.083	0.834	0.546	0.2618	0.2183	0.890
	15	0.072	0.856	0.576	0.2618	0.2241	0.781
	16	0.065	0.870	0.594	0.2618	0.2277	0.710
	17	0.058	0.884	0.613	0.2618	0.2314	0.639
	18	0.049	0.902	0.639	0.2618	0.2361	0.545
1¼	8	0.165	0.920	0.665	0.3271	0.2409	2.09
	9	0.148	0.954	0.714	0.3271	0.2498	1.91
	10	0.134	0.982	0.757	0.3271	0.2572	1.75
	11	0.120	1.01	0.800	0.3271	0.2644	1.58
	12	0.109	1.03	0.836	0.3271	0.2701	1.45
	13	0.095	1.06	0.884	0.3271	0.2775	1.28
	14	0.083	1.08	0.923	0.3271	0.2839	1.13
	15	0.072	1.11	0.960	0.3271	0.2896	0.991
	16	0.065	1.12	0.985	0.3271	0.2932	0.900
	17	0.058	1.13	1.01	0.3271	0.2969	0.808
	18	0.049	1.15	1.04	0.3271	0.3015	0.688

Table 12 Heat-exchanger and condenser-tube data (*Continued*)

Tube OD, in.	BWG	Wall thick-ness, in.	ID, in.	Flow area per tube, in.²	Surface per lin ft, ft²		Weight per lin ft, lb steel
					Outside	Inside	
1½	8	0.165	1.17	1.075	0.3925	0.3063	2.57
	9	0.148	1.20	1.14	0.3925	0.3152	2.34
	10	0.134	1.23	1.19	0.3925	0.3225	2.14
	11	0.120	1.26	1.25	0.3925	0.3299	1.98
	12	0.109	1.28	1.29	0.3925	0.3356	1.77
	13	0.095	1.31	1.35	0.3925	0.3430	1.56
	14	0.083	1.33	1.40	0.3925	0.3492	1.37
	15	0.072	1.36	1.44	0.3925	0.3555	1.20
	16	0.065	1.37	1.47	0.3925	0.3587	1.09
	17	0.058	1.38	1.50	0.3925	0.3623	0.978
	18	0.049	1.40	1.54	0.3925	0.3670	0.831

Table 13 Steel-pipe dimensions

Nominal pipe size, in.	OD, in.	Schedule No.	ID, in.	Flow area per pipe, in.^{2}	Surface per lin ft, ft^2		Weight per lin ft, lb steel
					Outside	Inside	
$\frac{1}{8}$	0.405	40†	0.269	0.058	0.106	0.070	0.25
		80‡	0.215	0.036	0.106	0.056	0.32
$\frac{1}{4}$	0.540	40	0.364	0.104	0.141	0.095	0.43
		80	0.302	0.072	0.141	0.079	0.54
$\frac{3}{8}$	0.675	40	0.493	0.192	0.177	0.129	0.57
		80	0.423	0.141	0.177	0.111	0.74
$\frac{1}{2}$	0.840	40	0.622	0.304	0.220	0.163	0.85
		80	0.546	0.235	0.220	0.143	1.09
$\frac{3}{4}$	1.05	40	0.824	0.534	0.275	0.216	1.13
		80	0.742	0.432	0.275	0.194	1.48
1	1.32	40	1.049	0.864	0.344	0.274	1.68
		80	0.957	0.718	0.344	0.250	2.17
$1\frac{1}{4}$	1.66	40	1.380	1.50	0.435	0.362	2.28
		80	1.278	1.28	0.435	0.335	3.00
$1\frac{1}{2}$	1.90	40	1.610	2.04	0.498	0.422	2.72
		80	1.500	1.76	0.498	0.393	3.64
2	2.38	40	2.067	3.35	0.622	0.542	3.66
		80	1.939	2.95	0.622	0.508	5.03
$2\frac{1}{2}$	2.88	40	2.469	4.79	0.753	0.647	5.80
		80	2.323	4.23	0.753	0.609	7.67
3	3.50	40	3.068	7.38	0.917	0.804	7.58
		80	2.900	6.61	0.917	0.760	10.3
4	4.50	40	4.026	12.7	1.178	1.055	10.8
		80	3.826	11.5	1.178	1.002	15.0
6	6.625	40	6.065	28.9	1.734	1.590	19.0
		80	5.761	26.1	1.734	1.510	28.6
8	8.625	40	7.981	50.0	2.258	2.090	28.6
		80	7.625	45.7	2.258	2.000	43.4
10	10.75	40	10.02	78.8	2.814	2.62	40.5
		60	9.75	74.6	2.814	2.55	54.8
12	12.75	30	12.09	115	3.338	3.17	43.8
16	16.0	30	15.25	183	4.189	4.00	62.6
20	20.0	20	19.25	291	5.236	5.05	78.6
24	24.0	20	23.25	425	6.283	6.09	94.7

† Schedule 40 designates former "standard" pipe.
‡ Schedule 80 designates former "extra-strong" pipe.

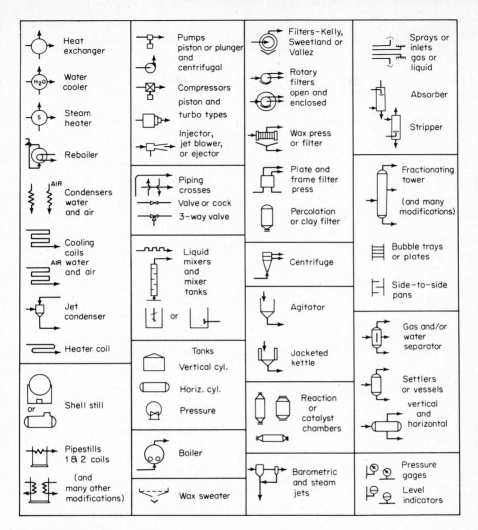

Figure D-5 Equipment symbols.

Symbol	Description	Symbol	Description	Symbol	Description

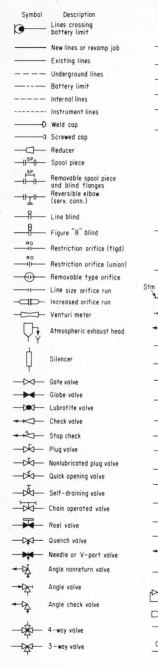
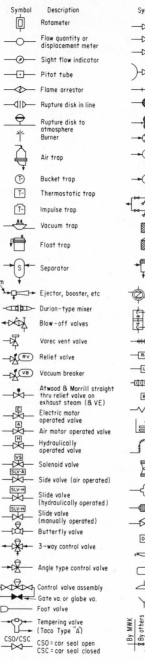
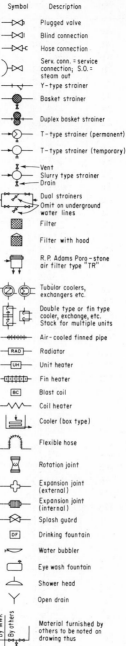

Figure D-6 Flow sheet symbols, particularly for detailed equipment flow sheets. (*Courtesy of the M. W. Kellogg Co.*)

Table 14 International atomic weights

Substance	Symbol	Atomic weight	At. No.	Substance	Symbol	Atomic weight	At. No.
Actinium	Ac	227.03	89	Mercury	Hg	200.59	80
Aluminum	Al	26.98	13	Molybdenum	Mo	95.94	42
Americium	Am	243	95	Neodymium	Nd	144.24	60
Antimony	Sb	121.75	51	Neon	Ne	20.179	10
Argon	A	39.948	18	Neptunium	Np	237	93
Arsenic	As	74.92	33	Nickel	Ni	58.70	28
Astatine	At	210	85	Niobium			
Barium	Ba	137.33	56	(Columbian)	Nb	92.91	41
Berkelium	Bk	247	97	Nitrogen	N	14.007	7
Beryllium	Be	9.012	4	Nobelium	No	259	102
Bismuth	Bi	208.98	83	Osmium	Os	190.2	76
Boron	B	10.81	5	Oxygen	O	16	8
Bromine	Br	79.904	35	Palladium	Pd	106.4	46
Cadmium	Cd	112.41	48	Phosphorus	P	30.975	15
Calcium	Ca	40.08	20	Platinum	Pt	195.09	78
Californium	Cf	251	98	Plutonium	Pu	244	94
Carbon	C	12.011	6	Polonium	Po	209	84
Cerium	Ce	140.12	58	Potassium	K	39.098	19
Cesium	Cs	132.91	55	Praseodymium	Pr	140.91	59
Chlorine	Cl	35.457	17	Promethium	Pm	145	61
Chromium	Cr	52.00	24	Protactinium	Pa	231	91
Cobalt	Co	58.93	27	Radium	Ra	226.03	88
Copper	Cu	63.55	29	Radon	Rn	222	86
Curium	Cm	247	96	Rhenium	Re	186.2	75
Dysprosium	Dy	162.50	66	Rhodium	Rh	102.91	45
Einsteinium	Es	254	99	Rubidium	Rb	85.47	37
Erbium	Er	167.26	68	Ruthenium	Ru	101.1	44
Europium	Eu	152.0	63	Samarium	Sm	150.4	62
Fermium	Fm	257	100	Scandium	Sc	44.96	21
Fluorine	F	19.00	9	Selenium	Se	78.96	34
Francium	Fr	223	87	Silicon	Si	28.09	14
Gadolinium	Gd	157.25	64	Silver	Ag	107.868	47
Gallium	Ga	69.72	31	Sodium	Na	22.990	11
Germanium	Ge	72.59	32	Strontium	Sr	87.62	38
Gold	Au	197.0	79	Sulfur	S	32.06	16
Hafnium	Hf	178.49	72	Tantalum	Ta	180.95	73
Helium	He	4.003	2	Technetium	Tc	97	43
Holmium	Ho	164.93	67	Tellurium	Te	127.60	52
Hydrogen	H	1.008	1	Terbium	Tb	158.93	65
Indium	In	114.82	49	Thallium	Tl	204.37	81
Iodine	I	126.90	53	Thorium	Th	232.04	90
Iridium	Ir	192.2	77	Thulium	Tm	168.93	69
Iron	Fe	55.85	26	Tin	Sn	118.69	50
Krypton	Kr	83.80	36	Titanium	Ti	47.90	22
Lanthanum	La	138.91	57	Tungsten (Wolfram)	W	183.85	74
Lawrencium	Lr	260	103	Uranium	U	238.03	92
Lead	Pb	207.2	82	Vanadium	V	50.94	23
Lithium	Li	6.941	3	Xenon	Xe	131.30	54
Lutetium	Lu	174.97	71	Ytterbium	Yb	173.04	70
Magnesium	Mg	24.31	12	Yttrium	Y	88.91	39
Manganese	Mn	54.94	25	Zinc	Zn	65.38	30
Mendelevium	Md	258	101	Zirconium	Zr	91.22	40

NAME INDEX

Adams, W. J., 409
Aggarwal, D. K., 199
Allen, D. H., 164, 178
Aris, R., 418

Baasel, W. D., 164, 181, 419, 421
Bakowski, S., 726
Bart, E. N., 44
Bates, H. T., 679
Bauman, H. C., 163, 173
Beightler, C. S., 419, 421
Bellman, R. E., 418
Berry, D. A., 816
Boas, A. H., 418, 419, 421
Bolles, W. L., 733
Bowker, A. H., 816, 829
Box, G. E. P., 419
Bradford, J. R., 727
Bridgewater, A. V., 189
Brinn, M., 679
Bromley, L. A., 650
Brown, G. G., 718
Brownlee, K. A., 820
Buck, E., 501
Buehner, F. W., 71
Bullington, L., 732

Cannon, M. R., 760
Carberry, J. J., 778

Carlson, A., 782
Chen, C.-C., 42, 777
Chen, N. H., 576
Chilton, C. H., 164, 382, 519, 545, 566, 569, 575, 642, 650, 654, 719, 735, 755, 759, 764, 827
Chu, J. C., 726
Chichelli, M., 679
Clark, J. P., 56
Colburn, A. P., 381, 648
Cole, M. S., 241
Cornell, D., 764
Cross, C. A., 733

Dantzig, G. B., 405
Davies, J. A., 734, 742
DeGarmo, E. P., 240
Dell, F. R., 759
Diaz, H. E., 569
Dodge, W. J., 189
Donohue, D. A., 648
Drayer, D. E., 185
Dreyfus, S. E., 418
Drickamer, H. G., 727
Duffin, R. J., 423

Eckert, J. S., 757
Economopoulos, A. P., 713
Edmondson, C. H., 186
Eichel, F. G., 759

SUBJECT INDEX